environment
THE SCIENCE BEHIND THE STORIES

6TH EDITION

Jay Withgott

Matthew Laposata

330 Hudson Street, NY NY 10013

Courseware Portfolio Management, Director: Beth Wilbur
Courseware Portfolio Specialist: Alison Rodal
Courseware Editorial Coordinator: Alison Cagle
Courseware Director, Content Development: Ginnie Simione Jutson
Courseware Sr. Analyst, Content Development: Sonia DiVittorio
Development Editors: Debbie Hardin, Jennifer Angel
Managing Producer, Science: Michael Early
Content Producer, Science: Margaret Young
Rich Media Content Producer: Chloe Veylit
Production Management: Norine Strang, Cenveo Publisher Services

Composition: Cenveo Publisher Services
Design Manager: Mark Ong
Interior and Cover Design: Lisa Buckley
Illustrators: ImagineeringArt.com Inc.
Rights & Permissions Manager: Ben Ferrini
Rights & Permissions Project Manager: Eric Schrader
Manufacturing Buyer: Maura Zaldivar-Garcia, LSC Communications
Photo Researcher: Kristin Piljay
VP Product Marketing: Christy Lesko
Executive Product Marketing Manager: Christa Pesek Pelaez

Cover Photo Credit: Chris Cheadle / Alamy Stock Photo

Library of Congress Cataloging-in-Publication Data
Names: Withgott, Jay author. | Laposata, Matthew, author.
Title: Environment : the science behind the stories / Jay Withgott, Matthew
   Laposata.
Description: 6th Edition. | San Francisco : Pearson Education, Inc., [2018] |
   Previous edition: 2014. | Includes bibliographical references and index.
Identifiers: LCCN 2016038664 (print) | LCCN 2016038997 (ebook) | ISBN
   9780134204888 (Student Edition : alk. paper) | ISBN 0134204883 (Student
   Edition : alk. paper) | ISBN 9780134580562 (NASTA) | ISBN 0134580567
   (NASTA) | ISBN 9780134407593
Subjects: LCSH: Environmental sciences.
Classification: LCC GE105 .B74 2016 (print) | LCC GE105 (ebook) | DDC
   363.7--dc23
LC record available at https://lccn.loc.gov/2016038664

5   18

ISBN 10: 0-13-420488-3; ISBN 13: 978-0-13-420488-8 (Student Edition)
ISBN 10: 0-13-448599-8; ISBN 13: 978-0-13-448599-7 (Books a la Carte)
ISBN 10: 0-13-458056-7; ISBN 13: 978-0-13-458056-2 (NASTA)

www.pearsonhighered.com

About the Authors

Jay Withgott has authored *Environment: The Science Behind the Stories* as well as its brief version, *Essential Environment*, since their inception. In dedicating himself to these books, he works to keep abreast of a diverse and rapidly changing field and continually seeks to develop new and better ways to help today's students learn environmental science.

As a researcher, Jay has published scientific papers in ecology, evolution, animal behavior, and conservation biology in journals ranging from *Evolution* to *Proceedings of the National Academy of Sciences*. As an instructor, he has taught university lab courses in ecology and other disciplines. As a science writer, he has authored articles for numerous journals and magazines including *Science, New Scientist, BioScience, Smithsonian,* and *Natural History*. By combining his scientific training with prior experience as a newspaper reporter and editor, he strives to make science accessible and engaging for general audiences. Jay holds degrees from Yale University, the University of Arkansas, and the University of Arizona.

Jay lives with his wife, biologist Susan Masta, in Portland, Oregon.

Matthew Laposata is a professor of environmental science at Kennesaw State University (KSU). He holds a bachelor's degree in biology education from Indiana University of Pennsylvania, a master's degree in biology from Bowling Green State University, and a doctorate in ecology from The Pennsylvania State University.

Matt is the coordinator of KSU's two-semester general education science sequence titled Science, Society, and the Environment, which enrolls roughly 6000 students per year. He focuses exclusively on introductory environmental science courses and has enjoyed teaching and interacting with thousands of students during his nearly two decades in higher education. He is an active scholar in environmental science education and has received grants from state, federal, and private sources to develop innovative curricular materials. His scholarly work has received numerous awards, including the Georgia Board of Regents' highest award for the Scholarship of Teaching and Learning.

Matt resides in suburban Atlanta with his wife, Lisa, and children, Lauren, Cameron, and Saffron.

about our SUSTAINABILITY INITIATIVES

This book is carefully crafted to minimize environmental impact. The materials used to manufacture this book originated from sources committed to responsible forestry practices. The paper is Forest Stewardship Council® (FSC®) certified. The printing, binding, cover, and paper come from facilities that minimize waste, energy consumption, and the use of harmful chemicals.

Pearson closes the loop by recycling every out-of-date text returned to our warehouse. We pulp the books, and the pulp is used to produce items such as paper coffee cups and shopping bags. In addition, Pearson has become the first climate-neutral educational publishing company.

The future holds great promise for reducing our impact on Earth's environment, and Pearson is proud to be leading the way. We strive to publish the best books with the most up-to-date and accurate content, and to do so in ways that minimize our environmental impact.

FSC
www.fsc.org
MIX
Paper from
responsible sources
FSC® C132124

PEARSON

Brief Contents

Contents

PART 1

Foundations of Environmental Science

1 Science and Sustainability: An Introduction to Environmental Science   2

2 Earth's Physical Systems: Matter, Energy, and Geology   20

3 Evolution, Biodiversity, and Population Ecology   46

4 Species Interactions and Community Ecology   72

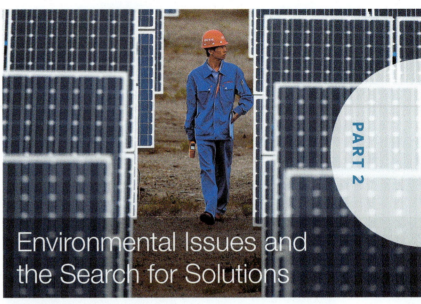

PART 2

Environmental Issues and the Search for Solutions

Dear Student,

You are coming of age at a unique and momentous time in history. Within your lifetime, our global society must chart a promising course for a sustainable future. The stakes could not be higher.

Today we live long lives enriched with astonishing technologies, in societies more free, just, and equal than ever before. We enjoy wealth on a scale our ancestors could hardly have dreamed of. However, we have purchased these wonderful things at a steep price. By exploiting Earth's resources and ecological services, we are depleting our planet's ecological bank account. We are altering our planet's land, air, water, nutrient cycles, biodiversity, and climate at dizzying speeds. More than ever before, the future of our society rests with how we treat the world around us.

Your future is being shaped by the phenomena you will learn about in your environmental science course. Environmental science gives us a big-picture understanding of the world and our place within it. Environmental science also offers hope and solutions, revealing ways to address the problems we create. Environmental science is not simply a subject you learn in college. Rather, it provides you a solid understanding of some of the most important issues of the 21st century, and it relates to everything around you over your entire lifetime.

We have written this book because today's students will shape tomorrow's world. At this unique moment in history, students of your generation are key to achieving a sustainable future for our civilization. The many environmental challenges that face us can seem overwhelming, but you should feel encouraged and motivated. Remember that each dilemma is also an opportunity. For every problem that human carelessness has created, human ingenuity can devise a solution. Now is the time for innovation, creativity, and the fresh perspectives that a new generation can offer. Your own ideas and energy can, and *will,* make a difference.

—*Jay Withgott and Matthew Laposata*

Dear Instructor,

You perform one of our society's most vital functions by educating today's students—the citizens and leaders of tomorrow—on the processes that shape the world around them, the nature of scientific inquiry, and the pressing environmental challenges facing us in our new century. We have written this book to assist you in this endeavor because we feel that the crucial role of environmental science in today's world makes it imperative to engage, educate, and inspire a broad audience of students.

In *Environment: The Science Behind the Stories*, we strive to show students how science informs our efforts to create a sustainable society. We also aim to encourage critical thinking and to maintain a balanced approach as we flesh out the vibrant social debate that accompanies environmental issues. As we assess the challenges facing our civilization and our planet, we focus on providing realistic, forward-looking solutions, for we truly feel there are many reasons for optimism.

In crafting the sixth edition of this text, we have incorporated the most current information from this dynamic discipline and have tailored our presentation to best promote student learning. We have examined every line of text and every figure with great care to make sure all content is accurate, clear, and up-to-date. Moreover, we have introduced a number of changes that are new to this edition.

New to This Edition

This sixth edition includes an array of revisions that enhance our content and presentation while strengthening our commitment to teach science in an engaging and accessible manner.

- central **CASE STUDY** Five *Central Case Studies* are completely new to this edition, complementing the 10 new case studies added in the fifth edition. All other case studies have been updated as needed to reflect recent developments. These updates provide fresh stories and new ways to frame emerging issues in environmental science. Students will compare organic farming with agriculture that uses genetically modified organisms, learn of the approaches California is taking to tackle chronic water shortages, examine how Miami is coping with sea level rise, and visit college campuses to see how students are promoting recycling and sustainable dining.
  - **Chapter 9:** Farm to Table (And Back Again) at Kennesaw State University
  - **Chapter 10:** Can Organic Farming and GMOs Coexist?

- **Chapter 15:** Conserving Every Drop in California
- **Chapter 18:** Rising Seas Threaten South Florida
- **Chapter 22:** A Mania for Recycling on Campus

- closing **THE LOOP** Also new to this edition, each chapter now concludes with a brief section that "closes the loop" by revisiting the *Central Case Study* while reviewing key principles from the chapter. This new *Closing the Loop* section enhances our long-standing and well-received approach of integrating each *Central Case Study* throughout its chapter.

- **THE SCIENCE** behind the story Six *Science Behind the Story* boxes are new to this edition, expanding our library of recently added examples of this feature. These new boxes, along with others that have been updated, provide a current and exciting selection of scientific studies to highlight. Students will follow researchers as they determine whether fracking is inducing earthquakes in Oklahoma; conduct a wide-ranging analysis of genetically modified crops; use DNA fingerprinting to combat poaching; ascertain if endocrine-disrupting chemicals are present in bottled water; predict the future of drought in the American West; and use toxic by-products of mining to reduce water use in hydraulic fracturing.
  - **Chapter 2:** Are the Earthquakes Rattling Oklahoma Caused by Human Activity?
  - **Chapter 10:** What Are the Impacts of GM Crops?
  - **Chapter 11:** Can Forensic DNA Analysis Help Save Elephants?
  - **Chapter 15:** Are We Destined for a Future of "Megadroughts" in the United States?
  - **Chapter 15:** Is Your Bottled Water as Safe as You Think It Is?
  - **Chapter 23:** Can Acid Mine Drainage Reduce Fracking's Environmental Impact?

- **New and revised DATA Q, FAQ, and Weighing the Issues items** Incorporating feedback from instructors across North America, we have examined each example of these three features that boost student engagement, and have revised them and added new examples as appropriate.

- **End-of-chapter elements** Several new approaches are introduced in our redesigned end-of-chapter material. Our *Reviewing Objectives* section is now more streamlined and focused on the learning objectives, and also incorporates visual icons as mileposts to help students connect to the material's location in the chapter. New "Case Study Connection" questions encourage students to craft solutions to issues raised in the chapter's *Central Case Study*.

- **Currency and coverage of topical issues** To live up to our book's hard-won reputation for currency, we've incorporated the most recent data possible throughout,

and we've enhanced coverage of emerging issues. As climate change and energy concerns play ever-larger roles in today's world, our coverage of these topics has kept pace. This edition highlights how renewable energy is growing, yet also how we continue reaching further for fossil fuels with deep offshore drilling, hydraulic fracturing for oil and shale gas, and extraction of oil sands. The text tackles the complex issue of climate change directly, while connections to this issue proliferate among topics throughout our book.

This edition also evolves and improves its coverage of a diversity of topics including the valuation of ecosystem services, invasive species and their ecological impacts, hormone-disrupting substances, fresh water shortages, advanced biofuels, plastic pollution in the oceans, sustainable agriculture, campus sustainability, green-collar jobs, and technologies that help reduce environmental impacts. We continue to use sustainability as an organizing theme throughout the book.

- **Enhanced style and design** We have significantly updated and improved the look and clarity of our visual presentation throughout the text. A more open layout, striking visuals, and an inviting new style all make the book more engaging for students. Over 40% of the photos, graphs, and illustrations in this edition are new or have been revised to reflect current data or for enhanced clarity or pedagogy.

Existing Features

We have also retained the major features that made the first five editions of our book unique and that are proving so successful in classrooms across North America:

- **A focus on science and data analysis** We have maintained and strengthened our commitment to a rigorous presentation of modern scientific research while simultaneously making science clear, accessible, and engaging to students. Explaining and illustrating the *process* of science remains a foundational goal of this endeavor. We also continue to provide an abundance of clearly cited, data-rich graphs, with accompanying tools for data analysis. In our text, our figures, and our online features, we aim to challenge students and to assist them with the vital skills of data analysis and interpretation.

- **An emphasis on solutions** For many students, today's deluge of environmental dilemmas can lead them to believe that there is no hope or that they cannot personally make a difference in tackling these challenges. We have aimed to counter this impression by highlighting innovative solutions being developed around the world. While being careful not to paint too rosy a picture of the challenges that lie ahead, we demonstrate that there is ample reason for optimism, and we encourage action. Our campus sustainability coverage (Chapters 1 and 24, and *Central Case Studies* in Chapters 9 and 22) shows students how their peers are applying principles and lessons from environmental science to forge sustainable solutions on their own campuses.

- **Central Case Studies integrated throughout the text** We integrate each chapter's *Central Case Study* into the main text, weaving information and elaboration throughout the chapter. In this way, compelling stories about real people and real places help to teach foundational concepts by giving students a tangible framework with which to incorporate novel ideas.

- **The Science Behind the Story** Because we strive to engage students in the scientific process of testing and discovery, we feature *The Science Behind the Story* boxes in each chapter. By guiding students through key research efforts, this feature shows not merely *what* scientists discovered, but *how* they discovered it.

- **DATA** These data analysis questions help students to actively engage with graphs and other data-driven figures. This feature accompanies several figures in each chapter, challenging students to practice quantitative skills of interpretation and analysis. To encourage students to test their understanding as they progress through the material, answers are provided in Appendix A. Students can practice data analysis skills further with new *Interpreting Graphs and Data: DataQs* in MasteringEnvironmentalScience.

- **FAQ** The *FAQ* feature highlights questions frequently posed by students in introductory environmental science courses, thereby helping to address widely held misconceptions and to fill in common conceptual gaps in knowledge. By also including questions students sometimes hesitate to ask, the *FAQs* show students that they are not alone in having these questions, thereby fostering a spirit of open inquiry in the classroom.

- **weighing the ISSUES** These questions aim to help develop the critical-thinking skills students need to navigate multifaceted issues at the juncture of science, policy, and ethics. They serve as stopping points for students to reflect on what they have read, wrestle with complex dilemmas, and engage in spirited classroom discussion.

- **Diverse end-of-chapter features** In addition to our new and revised end-of-chapter features detailed above, several hallmark features help students review and apply the concepts in each chapter. *Reviewing Objectives* summarizes each chapter's main points and relates them to the chapter's learning objectives, enabling students to confirm that they have understood the most crucial ideas. *Testing Your Comprehension* provides concise study questions on key topics, while *Seeking Solutions* encourages broader creative thinking that supports our emphasis on finding solutions. "Think It Through" questions in this section personalize the quest for creative solutions by placing students in a scenario and empowering them

to make decisions. *Calculating Ecological Footprints* enables students to quantify the impacts of their own choices and measure how individual impacts scale up to the societal level.

MasteringEnvironmental Science®

With this edition we continue to offer expanded opportunities through *MasteringEnvironmentalScience*, our powerful yet easy-to-use online learning and assessment platform. We have developed new content and activities specifically to support features in the textbook, thus strengthening the connection between these online and print resources. This approach encourages students to practice their science literacy skills in an interactive environment with a diverse set of automatically graded exercises. Students benefit from self-paced activities that feature immediate wrong-answer feedback, while instructors can gauge student performance with informative diagnostics. By enabling assessment of student learning outside the classroom, *MasteringEnvironmentalScience* helps the instructor to maximize the impact of in-classroom time. As a result, both educators and learners benefit from an integrated text and online solution.

NEW TO THIS EDITION *MasteringEnvironmentalScience* for this edition of *Environment: The Science Behind the Stories* offers new resources that are designed to grab student interest and help them develop quantitative reasoning skills.

- *NEW GraphIt* activities help students put data analysis and science reasoning skills into practice through a highly interactive and engaging format. Each of the 10 *GraphIts* prompts students to manipulate a variety of graphs and charts, from bar graphs to line graphs to pie charts, and develop an understanding of how data can be used in decision making about environmental issues. Topics range from agriculture to fresh water to air pollution. These mobile-friendly activities are accompanied by assessment in *MasteringEnvironmentalScience*.

- *NEW Everyday Environmental Science* videos highlight current environmental issues in short (5 minutes or less) video clips and are produced in partnership with BBC News. These videos will pique student interest, and can be used in class or assigned as a high-interest out-of-class activity.

- *NEW Dynamic Study Modules* help students study effectively on their own by continuously assessing their activity and performance in real time. Students complete multiple sets of questions for any given topic, to demonstrate concept mastery with confidence. Each *Dynamic Study Module* question set concludes with an explanation of concepts students may not have mastered. They are available as graded assignments prior to class, and are accessible on smartphones, tablets, and computers.

EXISTING FEATURES *MasteringEnvironmentalScience* retains its popular existing features:

- *Process of Science* activities help students navigate the scientific method, guiding them through in-depth explorations of experimental design using *Science Behind the Story* features from the current and former editions. These activities encourage students to think like a scientist and to practice basic skills in experimental design.

- *Interpreting Graphs and Data: Data Q* activities pair with the in-text *Data Analysis Questions* and coach students to further develop skills related to presenting, interpreting, and thinking critically about environmental science data.

- *"First Impressions" Pre-Quizzes* help instructors determine their students' existing knowledge of environmental issues and core content areas at the outset of the academic term, providing class-specific data that can then be employed for powerful teachable moments throughout the term. Assessment items in the Test Bank connect to each quiz item, so instructors can formally assess student understanding.

- *Video Field Trips* enable students to visit real-life sites that bring environmental issues to life. Students can tour a power plant, a wind farm, a wastewater treatment facility, a site combating invasive species, and more—all without leaving campus.

Environment: The Science Behind the Stories has grown from our collective experiences in teaching, research, and writing. We have been guided in our efforts by input from the hundreds of instructors across North America who have served as reviewers and advisers. The participation of so many learned, thoughtful, and committed experts and educators has improved this volume in countless ways.

We sincerely hope that our efforts are worthy of the immense importance of our subject matter. We invite you to let us know how well we have achieved our goals and where you feel we have fallen short. Please write to us in care of our editor, Alison Rodal (**alison.rodal@pearson.com**), at Pearson Education. We value your feedback and are eager to learn how we can serve you better.

—*Jay Withgott and Matthew Laposata*

Instructor Supplements

A robust set of instructor resources and multimedia accompanies the text and can be accessed through the Pearson Instructor Resource Center or *MasteringEnvironmentalScience*. Organized chapter-by-chapter, everything you need to prepare for your course is offered in one convenient set of files. Resources include the following: Video Field Trips, Everyday Environmental Science Videos, PowerPoint Lecture presentations, Instructor's Guide, Active Lecture "clicker" questions to facilitate class discussions (for use with or without clickers), and an image library that includes all art and tables from the text.

The Test Bank files, offered in both MS Word and Test-Gen, include hundreds of multiple-choice questions plus unique graphing and scenario-based questions to test students' critical-thinking abilities.

***MasteringEnvironmentalScience*® for *Environment: The Science Behind the Stories* (0-134-51016-X)**

The *MasteringEnvironmentalScience* platform is the most effective and widely used online tutorial, homework, and assessment system for the environmental sciences.

Help students connect
current environmental issues ...

Now in its Sixth Edition, *Environment: The Science Behind the Stories,* draws students into the science behind the issues with **updated central case studies** integrated into each chapter, a focus on building **science literacy skills,** and **captivating media** that brings concepts to life.

CHAPTER 18

Global Climate Change

NEW! Closing the Loop sections at the end of each chapter bring the Central Case Studies full circle.

closing
THE LOOP

Many factors influence Earth's climate, and human activities have come to play a major role. Climate change is well underway, and additional greenhouse gas emissions will intensify global warming and cause increasingly severe and varied impacts. Sea level rise and other consequences of global climate change are affecting locations worldwide from Miami to the Maldives, from Alaska to Bangladesh, and from New York to the Netherlands. As scientists and policymakers come to better understand anthropogenic climate change and its consequences, more and more of them are urging immediate action.

Policymakers at the international and national levels have struggled to take meaningful steps to slow greenhouse gas emissions, so increasingly, people at the local and regional levels are the ones making a difference. In South Florida, citizens and local leaders are investing time, thought, money, and creativity into finding solutions to rising sea levels. They are seeking to mitigate climate change by reducing greenhouse gas emissions and to adapt to climate change by building pumping systems, raising streets and foundations, and tailoring financial and insurance incentives to guide development toward upland areas. Like people anywhere who love their homes, residents of South Florida are girding themselves for a long battle to protect their land, communities, and quality of life while our global society inches its way toward emissions reductions. For all of us across the globe, taking steps to mitigate and adapt to climate change represents the foremost challenge for our future.

Bulldozing beach sand off a Fort Lauderdale boulevard after a storm surge

central
CASE STUDY

Rising Seas Threaten South Florida

"Miami, as we know it today, is doomed. It's not a question of if. It's a question of when."
University of Miami geologist Dr. Harold Wanless

"Miami Beach is not going to sit back and go underwater."
Philip Levine, mayor of Miami Beach

It happens now in Miami at least six times a year. Salty water bubbles up from drains, seeps up from the ground, fills the streets, and spills across lawns and sidewalks. Under the dazzling sun of a South Florida sky, floodwaters stall car traffic, creep into doorways, force businesses to close, and keep people from crossing the street. Employees struggle to get to work while tourists stand around, baffled.

The flooding is most severe in Miami Beach, the celebrated strip of glamorous hotels, clubs, shops, and restaurants that rises from a seven-mile barrier island just offshore from Miami. The carefree affluent image of Miami Beach, with its sun and fun, is increasingly jeopardized by the grimy reality of these unwelcome saltwater intrusions. By 2030, flooding is predicted to strike Miami and Miami Beach about 45 times per year—becoming no longer a curious inconvenience, but an existential threat.

These mysterious floods that seem to come out of nowhere are a recent phenomenon, so Miami-area residents are just now coming to realize that their coastal metropolis is slowly being swallowed by the ocean. The cause? Rising sea levels driven by global climate change.

The world's oceans rose 20 cm (8 in.) in the 20th century as warming temperatures expanded the volume of seawater and caused glaciers and ice sheets to melt, discharging water into the oceans. These processes are accelerating today, and scientists predict that sea level will rise another 26–98 cm (10–39 in.) or more in this century as climate change intensifies.

As sea levels rise, coastal cities across the globe—from Venice to Amsterdam to New York to San Francisco—are facing challenges. In the United States, scientists find that the Atlantic Seaboard and the Gulf Coast are especially vulnerable. The hurricane-prone shores of Florida, Louisiana, Texas, and the Carolinas are at risk, as are coastal cities such as Houston and New Orleans. From Cape Cod to Corpus Christi, millions of Americans who live in shoreline communities are beginning to suffer significant expense, disruption to daily life, and property damage as beaches erode, neighborhoods flood, aquifers are fouled, and storms strike with more force.

Perhaps nowhere in America is more vulnerable to sea-level rise than Miami and its surrounding communities in South Florida. Six million people live in this region, and three-quarters of them inhabit low-lying coastal areas that also hold most of the region's wealth and property. Experts calculate that Miami alone has more than $400 billion in assets at risk from sea-level rise—more than

A motorist stranded in Miami floodwaters ▲

Upon completing this chapter, you will be able to:

- Describe Earth's climate system and explain the factors that influence global climate
- Identify greenhouse gases and characterize human influences on the atmosphere and on climate
- Summarize how researchers study climate
- Outline current and expected future trends and impacts of climate change in the United States and around the world
- Suggest and assess ways we may respond to climate change

NEW and UPDATED!
Central Case Studies begin and are woven throughout each chapter, drawing students in and provide a contextual framework to make science memorable and engaging.

NEW!
Case Study Connection questions in the end of chapter material prompt students to think critically.

---

4. **CASE STUDY CONNECTION** You are the city manager for a coastal U.S. city that scientists predict will be hit hard by sea level rise, with risks and impacts trailing those in Miami by just a few years. You have just returned from a professional conference in Florida, where you toured Miami Beach and learned of the efforts being made there to adapt to climate change. What steps would you take to help your own city prepare for rising sea level? How would you explain the risks and impacts of climate change to your fellow city leaders to gain their support? Of the measures being taken in Florida communities, which would you choose to study closely, which would you want to begin right away, and which would be highest priority in the long run? Explain your choices.

Bring real conversations into the classroom ...

NEW and UPDATED! Science Behind the Story features highlight the process of science and profile the current scientific research behind today's most pressing environmental issues.

THE SCIENCE
behind the story

Go to Process of Science on MasteringEnvironmentalScience®

Can Forensic DNA Analysis Help Save Elephants?

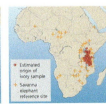

Confiscated tusks being destroyed in Kenya, to discourage poaching

As any television buff knows, forensic science is a crucial tool in solving mysteries and fighting crime. In recent years, conservation biologists have been using forensics to unearth secrets and catch bad guys in the multi-billion-dollar illegal global wildlife trade. One such detective story centers on the poaching of Africa's elephants for ivory.

Each year, tens of thousands of elephants are slaughtered illegally by poachers, simply for their tusks (FIGURE 1). Customs agents and law enforcement authorities manage to discover and confiscate tons of tusks being shipped internationally in the ivory trade—51 tons in 2013 alone. Yet only a small percentage of tusks are found and confiscated, and poachers are rarely apprehended, so the organized international crime syndicates that run these lucrative operations have been largely unhindered thus far.

Enter conservation biologist Samuel Wasser of the University of Washington in Seattle. By bringing the tools of genetic analysis, he and his students and colleagues have been shedding light on where elephants are being killed and where tusks are being shipped, thereby helping law enforcement efforts. In 2015, Wasser's team published a summary of nearly 20 years of work in the journal *Science*, revealing two major "poaching hotspots" in Africa.

The researchers began by accumulating 1350 reference samples of DNA from elephants at 71 locations across 29 African nations. Two subspecies of African elephant exist—savanna elephants, which live in open savannas, and forest elephants, which live in the forests of West and Central Africa. Of the 1350 genetic samples (from tissues or dung), 1001 came from savanna elephants and 349 came from forest elephants.

Wasser's teams sequenced DNA from the reference samples and compiled data on 16 highly variable stretches of DNA. For

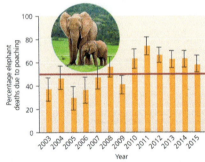

FIGURE 1 Since 2010, more than half of African elephant deaths have been due to poaching, a level scientists believe is unsustainable. Values above the horizontal line in this graph are thought to cause declines in the population. *Data from MIKE (Monitoring the Illegal Killing of Elephants), 2016.* Trends in levels of illegal killing of elephants in Africa to 31 December 2015.

each of the 71 geographic locations, they measured frequencies of alleles (different versions of genes) in these variable DNA stretches. By this process they [...] cies for 71 geographic areas f[...] would act as a kind of referen[...] compare any samples from tu[...] trade.

Working with law enforcem[...] access 20% of all ivory seizure[...] 1996 and 2005, 28% made bet[...] made between 2012 and 201[...] from these tusks and sequenc[...] were then able to compare the [...] tions and look for matches. In t[...] ticated statistical techniques, t[...] geographic origin of each of t[...] estimated distance of about 30[...]

FIGURE 2 Dr. Samuel Wasser worked with law enforcement officials to obtain samples of confiscated ivory.

For instance, ivory seized in the Philippines from 1996 to 2005 all appeared to come from forest elephants in an area of eastern Democratic Republic of Congo—just one small portion of their large geographic range. This region was difficult to patrol, however, due to its remoteness and to warfare occurring at the time, so little could be done with the information.

In contrast, when customs agents seized 6.5 tons of tusks in Singapore in 2002, Wasser's team determined that their DNA matched known samples from savanna elephants in Zambia,

indicating that many more elephants were being killed there than Zambia's government had realized. The Zambian government responded; it replaced its wildlife director and began imposing harsher sentences on poachers and ivory smugglers.

For shipments seized between 2006 and 2014, the genetic research indicated a surprisingly clear pattern of two main poaching hotspots. Most forest elephant tusks seized originated from a small region of West-Central Africa where two protected areas overlap the boundaries of four nations (FIGURE 3a). As for savanna elephant tusks, most came from animals in southern Tanzania and northern Mozambique during the early portion of the period. Later in the period, tusks originated from throughout Tanzania (FIGURE 3b), pointing to a shift northward and an increase in poaching in the parks of central and northern Tanzania.

In most cases, shipments seized in ports such as Hong Kong, Malaysia, Taiwan, and Sri Lanka were labeled with their shipping origin (often a coastal port in Kenya, Tanzania, Togo, or other African countries). With the additional help of Wasser's research, authorities could learn the entire route of the ivory shipments, from where the elephants were killed to where the tusks were exported to where they were imported. Combined with other data on poaching collected by international survey efforts, the genetic information from DNA forensic studies is painting a clearer picture of the crime network threatening elephants, and is giving law enforcement authorities more and better information with which to work.

FIGURE 3 Genetic analysis of confiscated ivory reveals where elephants were killed. For example, analysis of a shipment confiscated in Hong Kong in 2013 (a) shows that it came from forest elephants killed in a small area of West-Central Africa. Likewise, tusks from a shipment confiscated in Uganda in 2013 (b) were found to have come from various areas within Tanzania. *Adapted from Wasser, S., et al., 2015. Genetic assignment of large seizures of elephant ivory reveals Africa's major poaching hotspots. Science 349: 84–87.*

• Estimated origin of ivory sample
+ Forest elephant reference site

(a) Origin of tusks confiscated in Hong Kong

• Estimated origin of ivory sample
+ Savanna elephant reference site

(b) Origin of tusks confiscated in Uganda

... and encourage students to think scientifically

FAQ

How big is a billion?

Human beings have trouble conceptualizing huge numbers. As a result, we often fail to recognize the true magnitude of a number such as 7 billion. Although we know that a billion is bigger than a million, we tend to view both numbers as impossibly large and therefore similar in size. For example, guess (without calculating) how long it would take a banker to count out $1 million if she did it at a rate of a dollar a second for 8 hours a day, 7 days a week. Now guess how long it would take to count $1 billion at the same rate. The difference between your estimate and the answer may surprise you. Counting $1 million would take a mere 35 days, whereas counting $1 billion would take 95 years! Living 1 million seconds takes only 12 days, while living for 1 billion seconds requires more than 31 years. You couldn't live for 7 billion seconds—that would take 221 years. Examples like these can help us appreciate the *b* in *billion*.

FAQs probe common misconceptions students often hold about environmental issues.

First Impression questions, assignable in Mastering, gather data on your students' misconceptions and relate to chapter FAQs.

Weighing the Issues activities encourage students to grapple with environmental problem solving, and apply what they have learned as they go through each chapter.

More species have been identified and classified in this group than in any other.

- Mammals
- Plants
- Fishes
- Insects
- Bacteria

The global human population recently surpassed 7 billion people, and its annual growth is approximately _____ at the present.

- 1%
- 4%
- 7%
- 10%
- 12%

Submit   My Answers   Give Up

weighing the ISSUES

What Are the Consequences of Low Fertility?

In the United States, Canada, and almost every European nation, the total fertility rate is now at or below the replacement fertility rate (although some of these nations are still growing because of immigration). What economic or social consequences do you think might result from below-replacement fertility rates? Would you rather live in a society with a growing population, a shrinking population, or a stable population? Why?

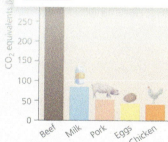

(a) Land required to produce 1 kg of protein

(b) Water required to produce 1 kg of protein

(c) Greenhouse gas emissions released in producing 1 kg of protein

FIGURE 10.9 Producing different types of animal products requires different amounts of (a) land and (b) water—and releases different amounts of (c) greenhouse gas emissions. Raising cattle for beef exerts the greatest impacts in all three ways. *Data (a, b) from Smil, V., 2001. Feeding the world: A challenge for the twenty-first century. Cambridge, MA: MIT Press; and (c) from FAO, 2015. Global Livestock Environmental Assessment Model (GLEAM).*

UPDATED! DataQs are data analysis questions paired with select figures in each chapter and are designed to help students develop their scientific literacy skills.

DATA Q Answer the following in terms of protein, pound for pound.
- How many times more land does it take to produce beef than chicken?
- How many times more water does beef require, compared with chicken?
- How many times more greenhouse gas emissions does beef release, relative to chicken?

Go to **Interpreting Graphs & Data** on **Mastering**EnvironmentalScience®.

Continuous Learning
Before, During, and After Class

BEFORE CLASS

Give students a preview of what's to come with activities that introduce them to key concepts.

Dynamic Study Modules enable students to study more effectively on their own.

With the **Dynamic Study Modules mobile app**, students can quickly access the concepts they need to be more successful in the course.

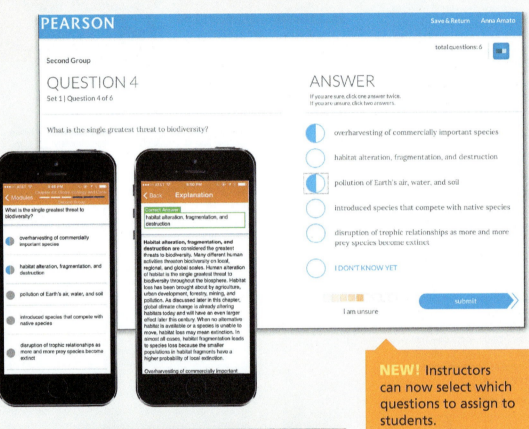

PEARSON

Save & Return   Anna Amato

total questions: 6

Second Group

QUESTION 4
Set 1 | Question 4 of 6

What is the single greatest threat to biodiversity?

ANSWER
If you are sure, click one answer twice.
If you are unsure, click two answers.

- overharvesting of commercially important species
- habitat alteration, fragmentation, and destruction
- pollution of Earth's air, water, and soil
- introduced species that compete with native species
- disruption of trophic relationships as more and more prey species become extinct
- I DON'T KNOW YET

I am unsure

submit

NEW! Instructors can now select which questions to assign to students.

00:04 / 06:49

NEW! Case Study Tour Videos use Google Earth and vibrant images to introduce students to each Central Case Study in the text. These dynamic videos bring each story to life and are assignable in Mastering.

with MasteringEnvironmentalScience

MasteringEnvironmentalScience®

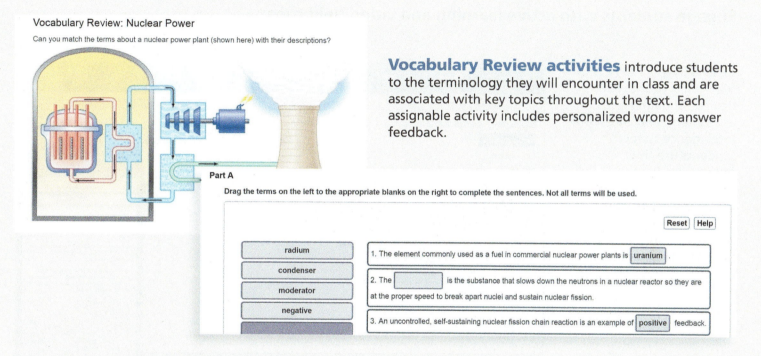

Vocabulary Review: Nuclear Power

Can you match the terms about a nuclear power plant (shown here) with their descriptions?

Part A

Drag the terms on the left to the appropriate blanks on the right to complete the sentences. Not all terms will be used.

Reset | Help

| radium |
| condenser |
| moderator |
| negative |

1. The element commonly used as a fuel in commercial nuclear power plants is [uranium].

2. The [ ] is the substance that slows down the neutrons in a nuclear reactor so they are at the proper speed to break apart nuclei and sustain nuclear fission.

3. An uncontrolled, self-sustaining nuclear fission chain reaction is an example of [positive] feedback.

Vocabulary Review activities introduce students to the terminology they will encounter in class and are associated with key topics throughout the text. Each assignable activity includes personalized wrong answer feedback.

Decline of Pollinators Poses Threat to World Food Supply, Report Says

By JOHN SCHWARTZ   FEB. 26, 2016   307

Current Events activities let instructors assign real news coverage of current environmental topics into their course.

Current Events activities expose students to a variety of current environmental issues. They are assignable and are updated at the start of each semester with new content.

Beekeepers using a smoker to calm colonies befor
Columbia Falls, Me. Plants that depend on pollinat
production volume with a value of as much as $57

The birds and the bees need help. Also, th
and bats. Without an international effort, a new report warns, increasing

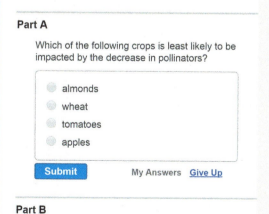

Current Events: Decline of Pollinators Poses Threat to World Food Supply, Report Says (New York Times, 02/26/2016)

Read this *New York Times* article and then answer the questions.

Decline of Pollinators Poses Threat to World Food Supply, Report Says (02/26/2016)

Registration with *The New York Times* provides instant access to breaking news on NYTimes.com. To register, go to http://www.nytimes.com/register. Visit http://www.nytimes.com/content/help/rights/terms/terms-of-service.html to review the current NYT Terms of Service.

Part A

Which of the following crops is least likely to be impacted by the decrease in pollinators?

- almonds
- wheat
- tomatoes
- apples

Submit   My Answers   Give Up

Part B

Continuous Learning
Before, During, and After Class

DURING CLASS

Engage students with active learning and video field trips.

Learning Catalytics is a "bring your own device" engagement, assessment, and classroom intelligence system. Students use their own device (laptop, smartphone, or tablet) to respond to open-ended questions and then discuss answers in groups based on their responses.

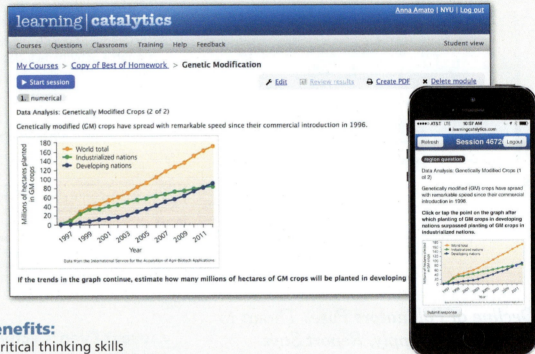

Learning Catalytics benefits:
- Developing higher-level critical thinking skills
- Promoting active learning and student engagement
- A team-based approach to learning
- Peer instruction methods
- Understand student misconceptions and adjust teaching in real time

"*My students are so busy and engaged answering Learning Catalytics questions during lecture that they don't have time for Facebook.*"

Declan De Paor, *Old Dominion University*

Mastering**EnvironmentalScience**®

Video Field Trips developed exclusively for Pearson Environmental Science provide fascinating behind-the-scenes tours of real environmental concerns and the strategies and solutions employed to address them. These popular, short videos engage students as they learn about bee colony collapse, take a tour of a water desalination plant, a wind farm, and more.

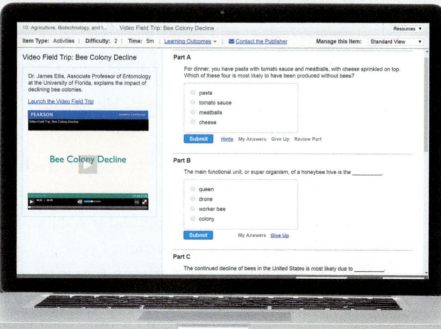

Continuous Learning
Before, During, and After Class

AFTER CLASS

Help students put scientific thinking into practice.

Process of Science coaching activities

help students practice the process of science demonstrated in the *Science Behind the Stories* and encourage students to put scientific inquiry skills into action.

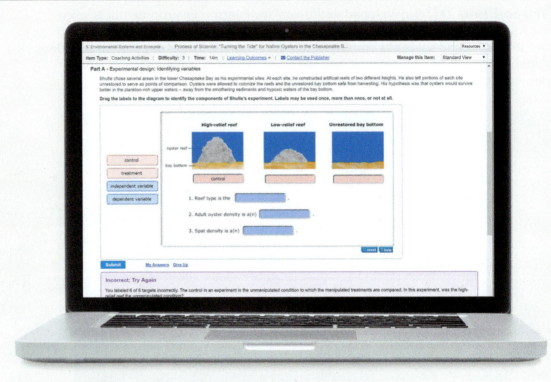

Process of Science coaching activities include:

- "Turning the Tide" for Native Oysters in the Chesapeake Bay
- Determining Zebra Mussels' Impacts on Fish Communities
- Did Soap Operas Reduce Fertility in Brazil?
- Using Forensics to Uncover Illegal Whaling
- And more!

NEW! GraphIt activities are interactive. mobile-friendly and assignable. Each GraphIt will encourage students to graph, interpret and analyze data on topics ranging from agriculture to water availability.

GraphIt topics include:

- Agriculture
- Freshwater Availability
- Nutrient Cycling
- Carrying Capacity
- Renewable Energy
- And more!

with **MasteringEnvironmentalScience**

MasteringEnvironmentalScience®

Concept Review coaching activities

created by coauthor Matt Laposata guide students through understanding tough topics.

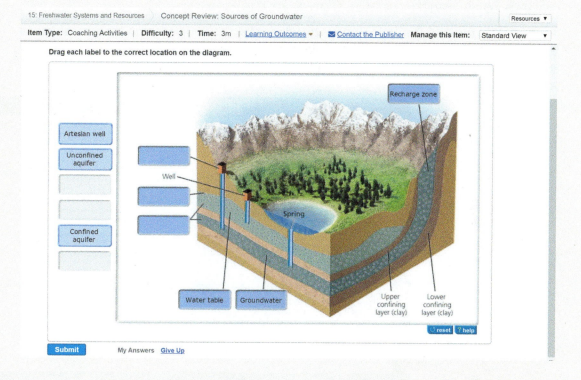

Item Type: Coaching Activities | **Difficulty:** 3 | **Time:** 3m | Learning Outcomes ▼ | ✉ Contact the Publisher | **Manage this Item:** Standard View ▼

Drag each label to the correct location on the diagram.

Recharge zone

Artesian well

Unconfined aquifer

Well

Confined aquifer

Spring

Water table Groundwater

Upper confining layer (clay) Lower confining layer (clay)

↻ reset ? help

Submit   My Answers Give Up

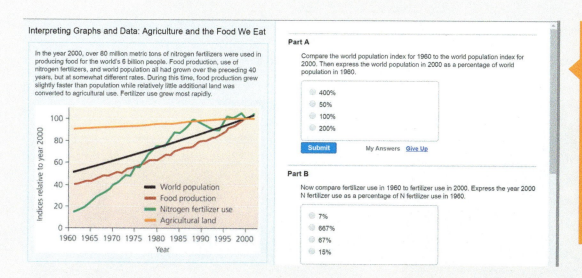

Interpreting Graphs and Data: Agriculture and the Food We Eat

In the year 2000, over 80 million metric tons of nitrogen fertilizers were used in producing food for the world's 6 billion people. Food production, use of nitrogen fertilizers, and world population all had grown over the preceding 40 years, but at somewhat different rates. During this time, food production grew slightly faster than population while relatively little additional land was converted to agricultural use. Fertilizer use grew most rapidly.

Indices relative to year 2000 (y-axis: 0 to 100)
— World population
— Food production
— Nitrogen fertilizer use
— Agricultural land
Year (x-axis: 1960 1965 1970 1975 1980 1985 1990 1995 2000)

Part A

Compare the world population index for 1960 to the world population index for 2000. Then express the world population in 2000 as a percentage of world population in 1960.

○ 400%
○ 50%
○ 100%
○ 200%

Submit   My Answers Give Up

Part B

Now compare fertilizer use in 1960 to fertilizer use in 2000. Express the year 2000 N fertilizer use as a percentage of N fertilizer use in 1960.

○ 7%
○ 667%
○ 67%
○ 15%

NEW and EXPANDED! Interpreting Graphs and Data activities help students practice quantitative literacy and scientific reasoning skills. Each activity includes personalized feedback for wrong answers.

MasteringEnvironmentalScience®

MasteringEnvironmentalScience delivers engaging, dynamic learning opportunities—focusing on course objectives and responsive to each student's progress—that are proven to help students absorb course material and understand challenging environmental processes and concepts.

NEW! Everyday Environmental Science video activities connect course topics with current stories in the news.

Produced by the BBC, these high-quality videos highlight current environmental issues and can be assigned for pre- and post-lecture homework, or can be shown in class to engage students in the topic at hand.

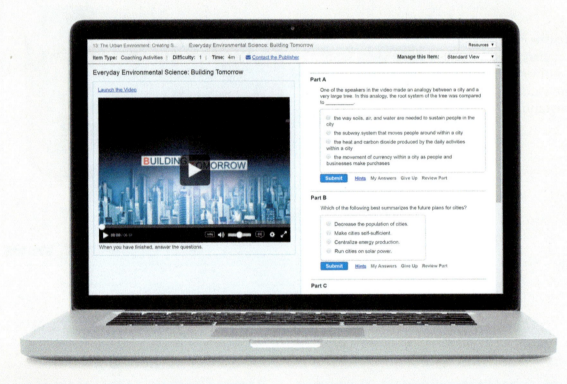

Everyday Environmental Science: Genetically Modified Fruit

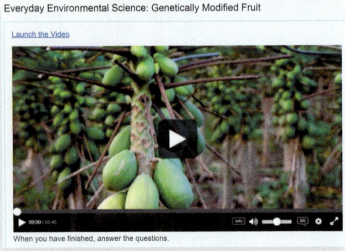

Everyday Environmental Science: Humboldt Current

Access the complete textbook online or offline with eText 2.0

NEW! The **Sixth Edition** is available in Pearson's fully-accessible eText 2.0 platform, live for Fall 2017 classes.

NEW! The **eText 2.0 mobile app** offers offline access and can be downloaded for most iOS and Android phones and tablets from the iTunes or Google Play stores.

Powerful interactive and customization functions include night mode, seamlessly integrated video and other rich media, instructor and student note-taking, highlighting, bookmarking, search, and links to glossary terms.

Accessible (screen-reader ready).

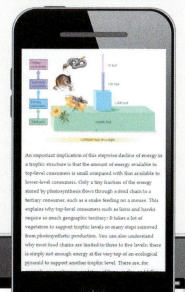

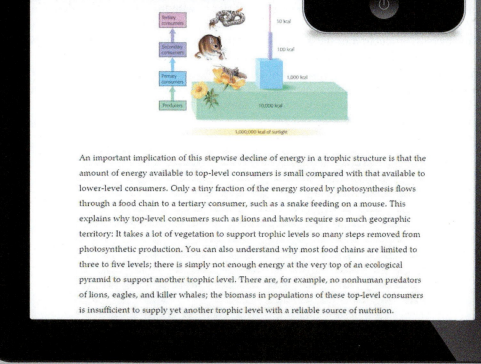

An important implication of this stepwise decline of energy in a trophic structure is that the amount of energy available to top-level consumers is small compared with that available to lower-level consumers. Only a tiny fraction of the energy stored by photosynthesis flows through a food chain to a tertiary consumer, such as a snake feeding on a mouse. This explains why top-level consumers such as lions and hawks require so much geographic territory: It takes a lot of vegetation to support trophic levels so many steps removed from photosynthetic production. You can also understand why most food chains are limited to three to five levels; there is simply not enough energy at the very top of an ecological pyramid to support another trophic level. There are, for example, no nonhuman predators of lions, eagles, and killer whales; the biomass in populations of these top-level consumers is insufficient to supply yet another trophic level with a reliable source of nutrition.

Acknowledgments

This textbook results from the collective labor and dedication of innumerable people. The two of us are fortunate to be supported by a tremendous publishing team.

Development editors Sonia DiVittorio and Debbie Hardin provided structure for our work and beauty and coherence for our chapter layouts. Their careful edits, along with those of Jennifer Angel, ensured clarity and accuracy in content. Project managers Brett Coker and Margaret Young choreographed the delicate scheduling dance required for timely project completion. As program manager, Anna Amato managed the complex logistics inherent to an endeavor of this size. Media producer Chloe Veylit managed the many moving parts of the *MasteringEnvironmentalScience* course for this new edition, including production of our new *GraphIt* activities—with help from instructional designer Sarah Young Dualan and executive media producer Laura Tommasi. Executive editor Alison Rodal facilitated the interactions of the editorial team and effectively collaborated with marketing managers Christa Pesek Pelaez and Mary Salzman, as well as the many sales representatives across the country, to communicate our vision, deliver our text to instructors, and work with instructors to ensure their satisfaction.

Editorial assistant Alison Cagle provided timely and effective assistance, while executive editorial manager Ginnie Simione Jutson oversaw our development needs. Bonnie Boehme provided meticulous copy editing of our text, and photo researcher Kristin Piljay helped secure quality photos. Wynne Au Yeung and Alicia Elliot of Imagineering did an exceptionally good job overseeing production of the art program, and Lisa Buckley designed our engaging new textbook style. Senior production manager Norine Strang worked with our compositor to help guide our book through production. We wish to thank our editor-in-chief Beth Wilbur for her continued support of this book through its six editions and for helping to invest the resources that our books enjoy.

As always, a select number of top instructors from around North America produced the supplementary materials that support the text. Our thanks go to Danielle DuCharme for updating our Instructor's Guide, to Todd Tracy for his work with the Test Bank, to James Dauray for revising the PowerPoint lectures, and to Jennifer Biederman for updating the Active Lecture clicker questions. We also wish to thank Shamili Sandiford for her creativity and effort in developing the *GraphIt* activities, and Stephan Fitzpatrick, Karyn Alme, and Donna Bivans for their work on *MasteringEnvironmental Science*.

In the lists of reviewers that follow, we acknowledge the many instructors and outside experts who have helped us to maximize the quality and accuracy of our content and presentation through their chapter reviews, feature reviews, class tests, focus group participation, and other services.

Finally, we each owe personal debts to the people nearest and dearest to us. Jay thanks his parents and his many teachers and mentors over the years for making his own life and education so enriching. He gives loving thanks to his wife, Susan, who has endured this book's writing and revision over the years with patience and understanding, and who has provided caring support throughout. Matt thanks his family, friends, and colleagues, and is grateful for his children, who give him three reasons to care passionately about the future. Most importantly, he thanks his wife, Lisa, for enriching his existence with love, joy, and wisdom—and for providing him a lifetime of experiences that would have been impossible without her. The talents, input, and advice of Susan and of Lisa have been vital to this project, and without their support our own contributions would not have been possible.

We dedicate this book to today's students, who will shape tomorrow's world.

—*Jay Withgott and Matthew Laposata*

Reviewers

We wish to express special thanks to the dedicated reviewers who shared their time and expertise to help make this sixth edition the best it could be. Their efforts built on those of the roughly 600 instructors and outside experts who have reviewed material for the previous five editions of this book through chapter reviews, pre-revision reviews, feature consultation, student reviews, class testing, and focus groups. Our sincere gratitude goes out to all of them.

Doug Nesmith, *Baylor University*
Richard Orndorff, *Eastern Washington University*
Clayton A. Penniman, *Central Connecticut State University*
Julie Stoughton, *University of Nevada, Reno*
Todd Tracy, *Northwestern College*
Tom Tyning, *Berkshire Community College*
Sharon Walsh, *New Mexico State University, Grants*
Lorne Wolfe, *Georgia Southern University*
Brian Wolff, *Normandale Community College*

Reviewers for the Sixth Edition

Karyn Alme, *Kennesaw State University*
Betsy Bancroft, *Gonzaga University*
Brad Basehore, *Harrisburg Area Community College*
Jill Bessetti, *Columbia College of Missouri Online*
Jennifer Biederman, *Winona State University*
Donna Bivans, *Pitt Community College*
Scott Brame, *Clemson University*
Geoffrey L. Buckley, *Ohio University*
James Dauray, *College of Lake County*
Danielle Ducharme, *Waubonsee Community College*
Eden Effert, *Eastern Illinois University*
Richard Feldman, *Marist College*
Jonathan Fingerut, *Saint Joseph's University*
Eric Fitch, *Marietta College*
Stephan Fitzpatrick, *Georgia Perimeter College*
Markus Flury, *Washington State University*
Steven Frankel, *Northeastern Illinois University*
Michael Freake, *Lee University*
Karen Gaines, *Eastern Illinois University*
Sandi Gardner, *Baker College*
Paul Gier, *Huntingdon College*
Jennifer Hanselman, *Westfield State University*
Keith Hench, *Kirkwood Community College*
Steven Jakobi, *Alfred State College*
Richard Jurin, *University of Northern Colorado*
Karen M. Klein, *Northampton Community College*
George Kraemer, *Purchase College*
Diana Kropf-Gomez, *Richland College, Dallas County Community College District*
James Kubicki, *University of Texas, El Paso*
Andrew Lapinski, *Reading Area Community College*
Hugh Lefcort, *Gonzaga University*
Maureen Leupold, *Genesee Community College*
Kurt Leuschner, *College of the Desert*
Jeffrey Mahr, *Georgia Perimeter College*
Chuck McClaugherty, *University of Mount Union*
Myra Morgan, *Massachusetts' National Math & Science Initiative*
George Myer, *Temple University*

Reviewers for Previous Editions

Matthew Abbott, *Des Moines Area Community College*; David Aborne, *University of Tennessee–Chattanooga;* Charles Acosta, *Northern Kentucky University;* Jeffrey Albert, *Watson Institute of International Studies;* Shamim Ahsan, *Metropolitan State College of Denver;* Isoken T. Aighewi, *University of Maryland–Eastern Shore;* John V. Aliff, *Georgia Perimeter College;* Mary E. Allen, *Hartwick College;* Deniz Z. Altin, *Georgia Perimeter College*; Dula Amarasiriwardena, *Hampshire College;* Gary I. Anderson, *Santa Rosa Junior College;* Mark W. Anderson, *The University of Maine;* Corey Andries, *Albuquerque Technical Vocational Institute;* David M. Armstrong, *University of Colorado–Boulder;* David L. Arnold, *Ball State University;* Joseph Arruda, *Pittsburg State University;* Eric Atkinson, *Northwest College;* Thomas W. H. Backman, *Linfield College;* Timothy J. Bailey, *Pittsburg State University;* Stokes Baker, *University of Detroit;* Kenneth Banks, *University of North Texas;* Narinder Bansal, *Ohlone College;* Jon Barbour, *University of Colorado–Denver;* Reuben Barret, *Prairie State College;* Morgan Barrows, *Saddleback College;* Henry Bart, *LaSalle University;* James Bartalome, *University of California–Berkeley;* Marilynn Bartels, *Black Hawk College;* David Bass, *University of Central Oklahoma;* Christy Bazan, *Illinois State University;* Christopher Beals, *Volunteer State Community College;* Laura Beaton, *York College, City University of New York*; Hans T. Beck, *Northern Illinois University;* Richard Beckwitt, *Framingham State College;* Barbara Bekken, *Virginia Polytechnic Institute and State University;* Elizabeth Bell, *Santa Clara University;* Timothy Bell, *Chicago State University;* David Belt, *Johnson County Community College;* Gary Beluzo, *Holyoke Community College;* Terrence Bensel, *Allegheny College;* Bob Bennett, *University of Arkansas;* William B. N. Berry, *University of California, Berkeley;* Jill Bessetti, *Columbia College;* Kristina Beuning, *University of Wisconsin–Eau Claire;* Peter Biesmeyer, *North Country Community College;* Donna Bivans, *Pitt Community College;* Grady Price Blount, *Texas A&M University–Corpus Christi;* Marsha Bollinger, *Winthrop University;* Lisa K. Bonneau, *Metropolitan Community College–Blue River;* Bruno Borsari, *Winona State University;* Richard D. Bowden, *Allegheny College;* Frederick J. Brenner, *Grove City College;* Nancy Broshot, *Linfield College;* Bonnie L. Brown, *Virginia Commonwealth University;* David Brown, *California State University–Chico;* Evert Brown, *Casper College;* Hugh Brown, *Ball State University;* J. Christopher Brown, *University of Kansas;* Dan Buresh, *Sitting Bull College;* Dale Burnside, *Lenoir-Rhyne University*; Lee Burras, *Iowa State University;* Hauke Busch, *Augusta State University;* Christina Buttington, *University of Wisconsin–Milwaukee;* Charles E. Button, *University of Cincinnati, Clermont College;* John S. Campbell, *Northwest College;* Myra Carmen Hall, *Georgia Perimeter College*; Mike Carney, *Jenks High School;* Kelly S. Cartwright, *College of Lake County;* Jon Cawley, *Roanoke College;* Michelle Cawthorn, *Georgia*

Southern University; Linda Chalker-Scott, University of Washington; Brad S. Chandler, Palo Alto College; Paul Chandler, Ball State University; David A. Charlet, Community College of Southern Nevada; Sudip Chattopadhyay, San Francisco State University; Tait Chirenje, Richard Stockton College; Luanne Clark, Lansing Community College; Richard Clements, Chattanooga State Technical Community College; Philip A. Clifford, Volunteer State Community College; Kenneth E. Clifton, Lewis and Clark College; Reggie Cobb, Nash Community College; John E. Cochran, Columbia Basin College; Donna Cohen, MassBay Community College; Luke W. Cole, Center on Race, Poverty, and the Environment; Mandy L. Comes, Rockingham Community College; Thomas L. Crisman, University of Florida; Jessica Crowe, South Georgia College; Ann Cutter, Randolph Community College; Gregory A. Dahlem, Northern Kentucky University; Randi Darling, Westfield State College; Mary E. Davis, University of Massachusetts, Boston; Thomas A. Davis, Loras College; Lola M. Deets, Pennsylvania State University–Erie; Ed DeGrauw, Portland Community College; Roger del Moral, University of Washington; Bob Dennison, Lead teacher at HISD and Robert E. Lee High School; Michael L. Denniston, Georgia Perimeter College; Doreen Dewell, Whatcom Community College; Craig Diamond, Florida State University; Darren Divine, Community College of Southern Nevada; Stephanie Dockstader, Monroe Community College; Toby Dogwiler, Winona State University; Jeffrey Dorale, University of Iowa; Tracey Dosch, Waubonsee Community College; Michael L. Draney, University of Wisconsin–Green Bay; Iver W. Duedall, Florida Institute of Technology; Jeffrey R. Dunk, Humboldt State University; Jean W. Dupon, Menlo College; Robert M. East, Jr., Washington & Jefferson College; Margaret L. Edwards-Wilson, Ferris State University; Dee Eggers, University of North Carolina–Asheville; Anne H. Ehrlich, Stanford University; Jane Ellis, Presbyterian College; Amy Ellwein, University of New Mexico; Thomas R. Embich, Harrisburg Area Community College; Kenneth Engelbrecht, Metropolitan State College of Denver; James English, Gardner-Webb University; Bill Epperly, Robert Morris College; JodyLee Estrada Duek, Pima Community College; Corey Etchberger, Johnson County Community College; W.F.J. Evans, Trent University; Paul Fader, Freed Hardeman University; Joseph Fail, Johnson C. Smith University; Bonnie Fancher, Switzerland County High School; Jiasong Fang, Iowa State University; Marsha Fanning, Lenoir Rhyne College; Leslie Fay, Rock Valley College; Debra A. Feikert, Antelope Valley College; M. Siobhan Fennessy, Kenyon College; Francette Fey, Macomb Community College; Steven Fields, Winthrop University; Brad Fiero, Pima Community College; Jonathan Fingerut, Saint Joseph's University; Robyn Fischer, Aurora University; Dane Fisher, Pfeiffer University; David G. Fisher, Maharishi University of Management; Eric J. Fitch, Marietta College; Linda M. Fitzhugh, Gulf Coast Community College; Doug Flournoy, Indian Hills Community College–Ottumwa; Johanna Foster, Johnson County Community College; Chris Fox, Catonsville Community College; Nancy Frank, University of Wisconsin–Milwaukee; Steven Frankel, Northeastern Illinois University; Michael Freake, Lee University; Arthur Fredeen, University of Northern British Columbia; Chad Freed, Widener University; Robert Frye, University of Arizona; Laura Furlong, Northwestern College; Karen Gaines, Eastern Illinois University; Navida Gangully, Oak Ridge High School; Sandi B. Gardner, Triton College; Katharine A. Gehl, Asheville Buncombe Technical Community College; Kristen S. Genet, Anoka Ramsey Community College; Stephen Getchell, Mohawk Valley Community College; Marcia Gillette, Indiana University–Kokomo; Scott Gleeson, University of Kentucky; Sue Glenn, Gloucester County College; Thad Godish, Ball State University; Nisse Goldberg, Jacksonville University; Michele Goldsmith, Emerson College; Jeffrey J. Gordon, Bowling Green State University; John G. Graveel, Purdue University; Jack Greene, Millikan High School; Cheryl Greengrove, University of Washington; Amy R. Gregory, University of Cincinnati, Clermont College; Carol Griffin, Grand Valley State University; Carl W. Grobe, Westfield State College; Sherri Gross, Ithaca College; David E. Grunklee, Hawkeye Community College; Judy Guinan, Radford University; Gian Gupta, University of Maryland, Eastern Shore; Mark Gustafson, Texas Lutheran University; Daniel Guthrie, Claremont College; Sue Habeck, Tacoma Community College; David Hacker, New Mexico Highlands University; Greg Haenel, Elon University; Mark Hammer, Wayne State University; Grace Hanners, Huntingtown High School; Jennifer A. Hanselman, Westfield State University; Michael Hanson, Bellevue Community College; Alton Harestad, Simon Fraser University; Barbara Harvey, Kirkwood Community College; David Hassenzahl, University of Nevada Las Vegas; Jill Haukos,

South Plains College; Keith Hench, Kirkwood Community College; George Hinman, Washington State University; Jason Hlebakos, Mount San Jacinto College; Joseph Hobbs, University of Missouri–Columbia; Jason Hoeksema, Cabrillo College; Curtis Hollabaugh, University of West Georgia; Robert D. Hollister, Grand Valley State University; David Hong, Diamond Bar High School; Catherine Hooey, Pittsburgh State University; William Hopper, Florida Memorial University; Kathleen Hornberger, Widener University; Debra Howell, Chabot College; April Huff, North Seattle Community College; Pamela Davey Huggins, Fairmont State University; Barbara Hunnicutt, Seminole Community College; Jonathan E. Hutchins, Buena Vista University; James M. Hutcheon, Asian University for Women; Don Hyder, San Juan College; Daniel Hyke, Alhambra High School; Juana Ibanez, University of New Orleans; Walter Illman, University of Iowa; Neil Ingraham, California State University, Fresno; Daniel Ippolito, Anderson University; Bonnie Jacobs, Southern Methodist University; Jason Janke, Metropolitan State College of Denver; Jack Jeffrey, formerly USFWS, Pepeekeo, Hawaii; Nan Jenks-Jay, Middlebury College; Linda Jensen-Carey, Southwestern Michigan College; Stephen R. Johnson, William Penn University; Gail F. Johnston, Lindenwood University; Gina Johnston, California State University, Chico; Paul Jurena, University of Texas–San Antonio; Richard R. Jurin, University of Northern Colorado; Thomas M. Justice, McLennan Community College; Stanley S. Kabala, Duquesne University; Brian Kaestner, Saint Mary's Hall; Steve Kahl, Plymouth State University; David M. Kargbo, Temple University; Susan Karr, Carson–Newman College; Jerry H. Kavouras, Lewis University; Carol Kearns, University of Colorado–Boulder; Richard R. Keenan, Providence Senior High School; Dawn G. Keller, Hawkeye Community College; Myung-Hoon Kim, Georgia Perimeter College; Kevin King, Clinton Community College; John C. Kinworthy, Concordia University; Karen Klein, Northampton Community College; Cindy Klevickis, James Madison University; Ned J. Knight, Linfield College; David Knowles, East Carolina University; Penelope M. Koines, University of Maryland; Alexander Kolovos, University of North Carolina–Chapel Hill; Erica Kosal, North Carolina Wesleyan College; Steven Kosztya, Baldwin Wallace College; Robert J. Koester, Ball State University; Tom Kozel, Anderson College; George Kraemer, Purchase College; Robert G. Kremer, Metropolitan State College of Denver; Jim Krest, University of South Florida–South Florida; Sushma Krishnamurthy, Texas A&M International University; Diana Kropf-Gomez, Dallas County Community College; Jerome Kruegar, South Dakota State University; James Kubicki, The Pennsylvania State University; Max Kummerow, Curtin University; Frank T. Kuserk, Moravian College; Diane M. LaCole, Georgia Perimeter College; Troy A. Ladine, East Texas Baptist University; William R. Lammela, Nazareth College; Vic Landrum, Washburn University; Tom Langen, Clarkson University; Andrew Lapinski, Reading Area Community College; Matthew Laposata, Kennesaw State University; Michael T. Lares, University of Mary; Kim D. B. Largen, George Mason University; John Latto, University of California–Berkeley; Lissa Leege, Georgia Southern University; Hugh Lefcort, Gonzaga University; James Lehner, Taft School; Kurt Leuschner, College of the Desert; Stephen D. Lewis, California State University, Fresno; Chun Liang, Miami University; John Logue, University of South Carolina–Sumter; John F. Looney, Jr., University of Massachusetts–Boston; Joseph Luczkovich, East Carolina University; Linda Lusby, Acadia University; Richard A. Lutz, Rutgers University; Jennifer Lyman, Rocky Mountain College; Les M. Lynn, Bergen Community College; Timothy F. Lyon, Ball State University; Sue Ellen Lyons, Holy Cross School; Ian R. MacDonald, Texas A&M University; James G. March, Washington and Jefferson College; Blase Maffia, University of Miami; Robert L. Mahler, University of Idaho; Jeffrey Mahr, Georgia Perimeter College, Newton; Keith Malmos, Valencia Community College; Kenneth Mantai, State University of New York–Fredonia; Anthony J.M. Marcattilio, St. Cloud State University; Heidi Marcum, Baylor University; Nancy Markee, University of Nevada–Reno; Patrick S. Market, University of Missouri–Columbia; Michael D. Marlen, Southwestern Illinois College; Kimberly Marsella, Skidmore College; Steven R. Martin, Humboldt State University; John Mathwig, College of Lake County; Allan Matthias, University of Arizona; Robert Mauck, Kenyon College; Brian Maurer, Michigan State University; Bill Mautz, University of New Hampshire; Kathy McCann Evans, Reading Area Community College; Chuck McClaugherty, University of Mount Union; Debbie McClinton, Brevard Community College; Paul McDaniel, University of Idaho; Jake McDonald, University of New Mexico; Gregory McIsaac, Cornell University; Robert M.L. McKay, Bowling Green State University; Annabelle McKie-Voerste, *Dalton State College;* Dan McNally, *Bryant University;* Richard McNeil, *Cornell University;* Julie Meents, *Columbia College;* Alberto Mestas-Nuñez, *Texas A&M University–Corpus Christi;* Mike L. Meyer, *New Mexico Highlands University;* Steven J. Meyer, *University of Wisconsin–Green Bay;* Patrick Michaels, *Cato Institute;* Christopher Migliaccio, *Miami Dade Community College;* William C. Miller, *Temple University;* Matthew R. Milnes, *University of California, Irvine;* Kiran Misra, *Edinboro University of Pennsylvania;* Mark Mitch, *New England College;* Paul Montagna, *University of Texas–Austin;* Lori Moore, *Northwest Iowa Community College;* Brian W. Moores, *Randolph-Macon College;* James T. Morris, *University of South Carolina;* Sherri Morris, *Bradley University;* Mary Murphy, *Penn State Abington;* William M. Murphy, *California State University–Chico;* Carla S. Murray, *Carl Sandburg College;* Rao Mylavarapu, *University of Florida;* Jane Nadel-Klein, *Trinity College;* Muthena Naseri, *Moorpark College;* Michael J. Neilson, *University of Alabama–Birmingham;* Benjamin Neimark, *Temple University;* Michael Nicodemus, *Abilene Christian University;* Richard A. Niesenbaum, *Muhlenberg College;* Moti Nissani, *Wayne State University;* Richard B. Norgaard, *University of California–Berkeley;* John Novak, *Colgate University;* Mark P. Oemke, *Alma College;* Niamh O'Leary, *Wells College;* Bruce Olszewski, *San Jose State University;* Brian O'Neill, *Brown University;* Nancy Ostiguy, *Penn State University;* Anthony Overton, *East Carolina University;* David R. Ownby, *Stephen F. Austin State University;* Gulni Ozbay, *Delaware State University;* Eric Pallant, *Allegheny College;* Philip Parker, *University of Wisconsin–Platteville;* Tommy Parker, *University of Louisville;* Brian Peck, *Simpson College;* Brian D. Peer, *Simpson College;* Clayton Penniman, *Central Connecticut State University;* Christopher Pennuto, *Buffalo State College;* Donald J. Perkey, *University of Alabama–Huntsville;* Barry Perlmutter, *College of Southern Nevada;* Shana Petermann, *Minnesota State Community and Technical College–Moorhead;* Craig D. Phelps, *Rutgers University;* Neal Phillip, *Bronx Community College*; Frank X. Phillips, *McNeese State University;* Raymond Pierotti, *University of Kansas;* Elizabeth Pixley, *Monroe Community College;* John Pleasants, *Iowa State University*; Thomas E. Pliske, *Florida International University;* Gerald Pollack, *Georgia Perimeter College*; Mike Priano, *Westchester Community College;* Daryl Prigmore, *University of Colorado;* Avram G. Primack, *Miami University of Ohio;* Alison Purcell, *Humboldt State University*; Sarah Quast, *Middlesex Community College;* Daniel Ratcliff, *Rose State College;* Patricia Bolin Ratliff, *Eastern Oklahoma State College;* Loren A. Raymond, *Appalachian State University;* Barbara Reynolds, *University of North Carolina–Asheville;* Thomas J. Rice, *California Polytechnic State University;* Samuel K. Riffell, *Mississippi State University;* Kayla Rihani, *Northeast Illinois University;* James Riley, *University of Arizona;* Gary Ritchison, *Eastern Kentucky University;* Virginia Rivers, *Truckee Meadows Community College;* Roger Robbins, *East Carolina University;* Tom Robertson, *Portland Community College, Rock Creek Campus;* Mark Robson, *University of Medicine and Dentistry of New Jersey;* Carlton Lee Rockett, *Bowling Green State University;* Angel M. Rodriguez, *Broward Community College;* Deanne Roquet, *Lake Superior College;* Armin Rosencranz, *Stanford University;* Irene Rossell, *University of North Carolina at Asheville;* Robert E. Roth, *The Ohio State University;* George E. Rough, *South Puget Sound Community College;* Dana Royer, *Wesleyan University;* Steven Rudnick, *University of Massachusetts–Boston;* John Rueter, *Portland State University;* Christopher T. Ruhland, *Minnesota State University;* Dork Sahagian, *Lehigh University;* Shamili A. Sandiford; *College of DuPage;* Robert Sanford, *University of Southern Maine;* Ronald Sass, *Rice University;* Carl Schafer, *University of Connecticut;* Jeffery A. Schneider, *State University of New York–Oswego;* Kimberly Schulte, *Georgia Perimeter College*; Edward G. Schultz, III, *Valencia Community College;* Mark Schwartz, *University of California–Davis;* Jennifer Scrafford, *Loyola College;* Richard Seigel, *Towson University;* Julie Seiter, *University of Nevada–Las Vegas;* Wendy E. Sera, *NDAA's National Ocean Service;* Maureen Sevigny, *Oregon Institute of Technology;* Rebecca Sheesley, *University of Wisconsin–Madison;* Andrew Shella, *Terra State Community College;* Pamela Shlachtman, *Miami Palmetto Senior High School;* Brian Shmaefsky, *Kingwood College;* William Shockner, *Community College of Baltimore County;* Christian V. Shorey, *University of Iowa;* Elizabeth Shrader, *Community College of Baltimore County*; Robert Sidorsky, *Northfield Mt. Hermon High School;* Linda Sigismondi, *University of Rio Grande;* Gary Silverman, *Bowling Green State University;* Jeffrey Simmons, *West Virginia Wesleyan College;* Cynthia Simon, *University of New England;* Jan Simpkin, *College of Southern Idaho;* Michael Singer, *Wesleyan University;* Diane Sklensky, *Le Moyne College;* Jeff Slepski, *Mt. San Jacinto College;* Ben Smith, *Palos Verdes Peninsula High School;* Mark Smith, *Chaffey College;* Patricia L. Smith, *Valencia Community College;* Rik Smith, *Columbia Basin College;* Sherilyn Smith, *Le Moyne College;* Debra Socci, *Seminole Community College;* Roy Sofield, *Chattanooga State Technical Community College;* Annelle Soponis, *Reading Area Community College*; Douglas J. Spieles, *Denison University;* Ravi Srinivas, *University of St. Thomas;* Bruce Stallsmith, *University of Alabama–Huntsville;* Jon G. Stanley, *Metropolitan State College of Denver;* Ninian Stein, *Wheaton College*; Jeff Steinmetz, *Queens University of Charlotte;* Michelle Stevens, *California State University–Sacramento;* Richard Stevens, *Louisiana State University*; Bill Stewart, *Middle Tennessee State University;* Dion C. Stewart, *Georgia Perimeter College;* Julie Stoughton, *University of Nevada–Reno;* Richard J. Strange, *University of Tennessee;* Robert Strikwerda, *Indiana University–Kokomo;* Richard Stringer, *Harrisburg Area Community College;* Norm Strobel, *Bluegrass Community Technical College*; Andrew Suarez, *University of Illinois;* Keith S. Summerville, *Drake University;* Ronald Sundell, *Northern Michigan University;* Bruce Sundrud, *Harrisburg Area Community College;* Jim Swan, *Albuquerque Technical Vocational Institute;* Mark L. Taper, *Montana State University;* Todd Tarrant, *Michigan State University;* Max R. Terman, *Tabor College;* Julienne Thomas, *Robert Morris College;* Patricia Terry, *University of Wisconsin–Green Bay;* Jamey Thompson, *Hudson Valley Community College;* Rudi Thompson, *University of North Texas;* Todd Tracy, *Northwestern College;* Amy Treonis, *Creighton University;* Adrian Treves, *Wildlife Conservation Society;* Frederick R. Troeh, *Iowa State University;* Virginia Turner, *Robert Morris College;* Michael Tveten, *Pima Community College;* Thomas Tyning, *Berkshire Community College;* Charles Umbanhowar, *St. Olaf College;* G. Peter van Walsum, *Baylor University;* Callie A. Vanderbilt, *San Juan College;* Elichia A. Venso, *Salisbury University;* Rob Viens, *Bellevue Community College;* Michael Vorwerk, *Westfield State College;* Caryl Waggett, *Allegheny College;* Maud M. Walsh, *Louisiana State University;* Sharon Walsh, *New Mexico State University;* Daniel W. Ward, *Waubonsee Community College;* Darrell Watson, *The University of Mary Hardin Baylor;* Phillip L. Watson, *Ferris State University;* Lisa Weasel, *Portland State University;* Kathryn Weatherhead, *Hilton Head High School;* John F. Weishampel, *University of Central Florida;* Peter Weishampel, *Northland College;* Barry Welch, *San Antonio College;* Kelly Wessell, *Tompkins Cortland Community College;* James W.C. White, *University of Colorado;* Susan Whitehead, *Becker College;* Jeffrey Wilcox, *University of North Carolina at Asheville;* Richard D. Wilk, *Union College;* Donald L. Williams, *Park University;* Justin Williams, *Sam Houston University;* Ray E. Williams, *Rio Hondo College;* Roberta Williams, *University of Nevada–Las Vegas;* Dwina Willis, *Freed-Hardeman University;* Shaun Wilson, *East Carolina University;* Tom Wilson, *University of Arizona;* James Winebrake, *Rochester Institute of Technology;* Danielle Wirth, *Des Moines Area Community College;* Lorne Wolfe, *Georgia Southern University;* Brian G. Wolff, *Minnesota State Colleges and Universities;* Marjorie Wonham, *University of Alberta;* Wes Wood, *Auburn University;* Jessica Wooten, *Franklin University;* Jeffrey S. Wooters, *Pensacola Junior College;* Joan G. Wright, *Truckee Meadows Community College;* Michael Wright, *Truckee Meadows Community College;* S. Rebecca Yeomans, *South Georgia College;* Karen Zagula, *Waketech Community College*; Lynne Zeman, *Kirkwood Community College;* Zhihong Zhang, *Chatham College.*

Class Testers for Previous Editions

David Aborne, *University of Tennessee–Chattanooga;* Reuben Barret, *Prairie State College;* Morgan Barrows, *Saddleback College;* Henry Bart, *LaSalle University;* James Bartalome, *University of California–Berkeley;* Christy Bazan, *Illinois State University;* Richard Beckwitt, *Framingham State College;* Elizabeth Bell, *Santa Clara University;* Peter Biesmeyer, *North Country Community College;* Donna Bivans, *Pitt Community College;* Evert Brown, *Casper College;* Christina Buttington, *University of Wisconsin–Milwaukee;*

Tait Chirenje, *Richard Stockton College;* Reggie Cobb, *Nash Community College;* Ann Cutter, *Randolph Community College;* Lola Deets, *Pennsylvania State University–Erie;* Ed DeGrauw, *Portland Community College;* Stephanie Dockstader, *Monroe Community College;* Dee Eggers, *University of North Carolina–Asheville;* Jane Ellis, *Presbyterian College;* Paul Fader, *Freed Hardeman University;* Joseph Fail, *Johnson C. Smith University;* Brad Fiero, *Pima Community College, West Campus;* Dane Fisher, *Pfeiffer University;* Chad Freed, *Widener University;* Stephen Getchell, *Mohawk Valley Community College;* Sue Glenn, *Gloucester County College;* Sue Habeck, *Tacoma Community College;* Mark Hammer, *Wayne State University;* Michael Hanson, *Bellevue Community College;* David Hassenzahl, *Oakland Community College;* Kathleen Hornberger, *Widener University;* Paul Jurena, *University of Texas–San Antonio;* Dawn Keller, *Hawkeye Community College;* David Knowles, *East Carolina University;* Erica Kosal, *Wesleyan College;* John Logue, *University of Southern Carolina Sumter;* Keith Malmos, *Valencia Community College;* Nancy Markee, *University of Nevada–Reno;* Bill Mautz, *University of New Hampshire;* Julie Meents, *Columbia College;* Lori Moore, *Northwest Iowa Community College;* John Novak, *Colgate University;* Brian Peck, *Simpson College;* Elizabeth Pixley, *Monroe Community College;* Sarah Quast, *Middlesex Community College;* Roger Robbins, *East Carolina University;* Mark Schwartz, *University of California–Davis;* Julie Seiter, *University of Nevada–Las Vegas;* Brian Shmaefsky, *Kingwood College;* Diane Sklensky, *Le Moyne College;* Mark Smith, *Fullerton College;* Patricia Smith, *Valencia Community College East;* Sherilyn Smith, *Le Moyne College;* Jim Swan, *Albuquerque Technical Vocational Institute;* Amy Treonis, *Creighton University;* Darrell Watson, *The University of Mary Hardin Baylor;* Barry Welch, *San Antonio College;* Susan Whitehead, *Becker College;* Justin Williams, *Sam Houston University;* Roberta Williams, *University of Nevada–Las Vegas;* Tom Wilson, *University of Arizona.*

Suppliers of Student Reviews for Previous Editions

Christine Brady, *California State Polytechnic University, Pomona*
Steven Rudnick, *University of Massachusetts–Boston*
Ninian R. Stein, *Wheaton College*
Todd Tracy, *Northwestern College*
Lorne Wolfe, *Georgia Southern University*

environment
THE SCIENCE BEHIND THE STORIES

6TH EDITION

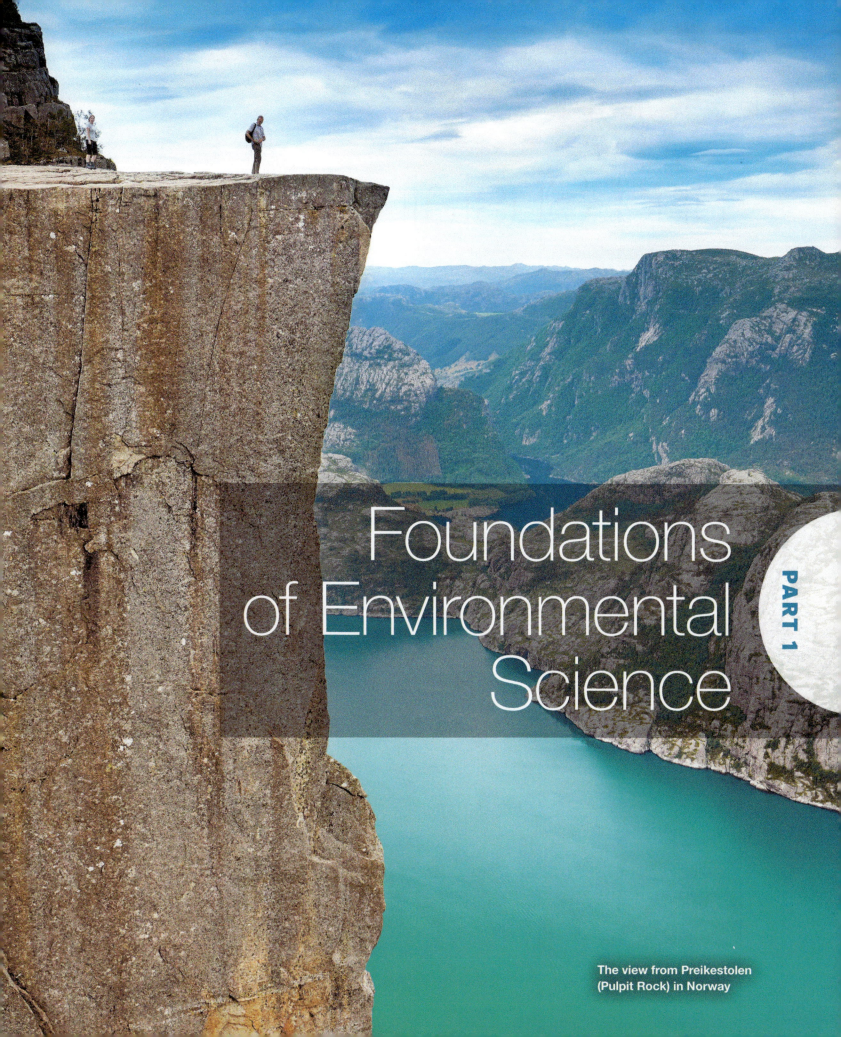

Foundations of Environmental Science

PART 1

The view from Preikestolen (Pulpit Rock) in Norway

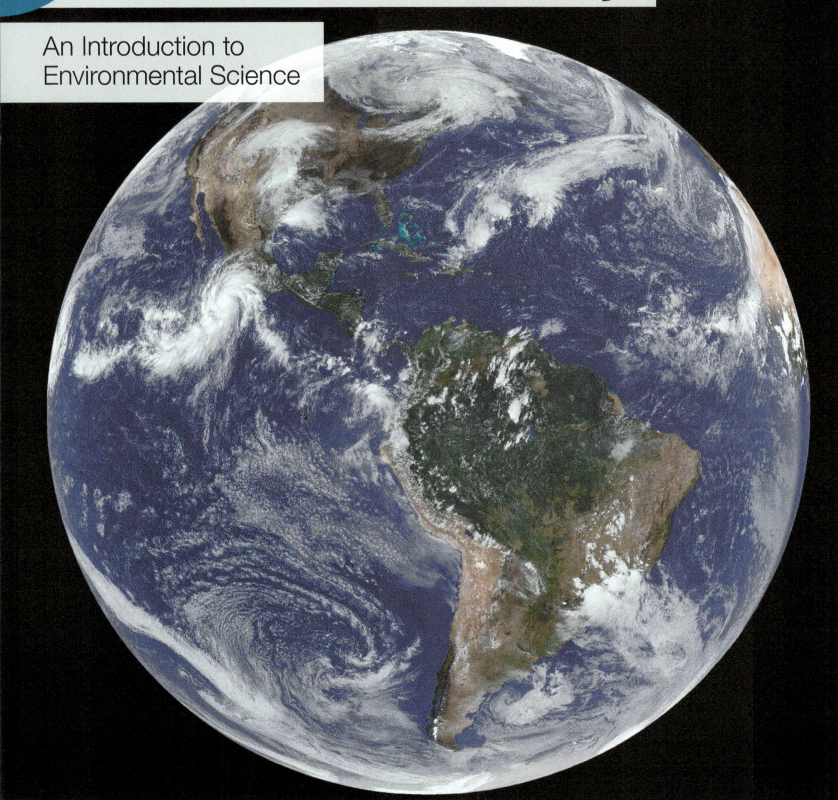

CHAPTER 1

Science
and Sustainability

An Introduction to
Environmental Science

Our Island, Earth

Viewed from space, our home planet resembles a small blue marble suspended in a vast inky-black void. Earth may seem enormous to us as we go about our lives on its surface, but the astronaut's view reveals that our planet is finite and limited. With this perspective, it becomes clear that as our population, technological power, and resource consumption all increase, so does our capacity to alter our surroundings and damage the very systems that keep us alive. Learning how to live peacefully, healthfully, and sustainably on our diverse and complex planet is our society's prime challenge today. The field of environmental science is crucial in this endeavor.

Our environment surrounds us

A photograph of Earth offers a revealing perspective, but it cannot convey the complexity of our environment. Our **environment** consists of all the living and nonliving things around us. It includes the continents, oceans, clouds, and ice caps you can see in a photo of Earth from space, as well as the animals, plants, forests, and farms of the landscapes in which we live. In a more inclusive sense, it also encompasses the structures, urban centers, and living spaces that people have created. In its broadest sense, our environment includes the complex webs of social relationships and institutions that shape our daily lives.

People commonly use the term *environment* in the first, most narrow sense—to mean a nonhuman or "natural" world apart from human society. This is unfortunate, because it masks the vital fact that people exist within the environment and are part of nature. As one of many species on Earth, we share dependence on a healthy, functioning planet. The limitations of language make it all too easy to speak of "people and nature" or "humans and the environment" as though they were separate and did not interact. However, the fundamental insight of environmental science is that we are part of the "natural" world and that our interactions with the rest of it matter a great deal.

Environmental science explores our interactions with the world

Understanding our relationship with the world around us is vital because we depend utterly on our environment for air, water, food, shelter, and everything else essential for life. Throughout human history, we have modified our environment. By doing so, we have enriched our lives; improved our health; lengthened our life spans; and secured greater material wealth, mobility, and leisure time. Yet many of the changes we have made to our surroundings have degraded the natural systems that sustain us. Air and water pollution, soil erosion, species extinction, and other impacts compromise our well-being and jeopardize our ability to survive and thrive in the long term.

Environmental science is the scientific study of how the natural world works, how our environment affects us, and how we affect our environment. Understanding these interactions helps us devise solutions to society's many pressing challenges. It can be daunting to reflect on the sheer magnitude of dilemmas that confront us, but these problems also bring countless opportunities for creative solutions.

Environmental scientists study the issues most centrally important to our world and its future. Right now, global conditions are changing more quickly than ever. Right now, we are gaining scientific knowledge more rapidly than ever. And right now, there is still time to tackle society's biggest challenges. With such bountiful opportunities, this particular moment in history is an exciting time to be alive—and to be studying environmental science.

Upon completing this chapter, you will be able to:

- Describe the field of environmental science

- Compare renewable and nonrenewable resources, and explain the importance of natural resources and ecosystem services to our lives

- Discuss population growth, resource consumption, and their consequences

- Explain what is meant by an ecological footprint

- Describe the scientific method and the process of science

- Identify and illustrate major pressures on the global environment

- Discuss the concept of sustainability, and cite sustainable solutions being pursued on campuses and in the wider world

• Solar energy
• Wind energy
• Wave energy
• Geothermal energy

(a) Inexhaustible renewable natural resources

• Fresh water
• Forest products
• Biodiversity
• Soils

(b) Exhaustible renewable natural resources

• Crude oil
• Natural gas
• Coal
• Minerals

(c) Nonrenewable natural resources

FIGURE 1.1 Natural resources may be renewable or nonrenewable. Perpetually renewable, or inexhaustible, resources such as sunlight and wind energy **(a)** will always be there for us. Renewable resources such as timber, soils, and fresh water **(b)** are replenished on intermediate timescales, if we are careful not to deplete them. Nonrenewable resources such as minerals and fossil fuels **(c)** exist in limited amounts that could one day be gone.

We rely on natural resources

Islands are finite and bounded, and their inhabitants must cope with limitations in the materials they need. On our island—planet Earth—there are limits to many of our **natural resources,** the substances and energy sources that we take from our environment and that we rely on to survive (**FIGURE 1.1**).

Natural resources that are replenished over short periods are known as **renewable natural resources.** Some renewable natural resources, such as sunlight, wind, and wave energy, are perpetually renewed and essentially inexhaustible. Others, such as timber, water, animal populations, and fertile soil, renew themselves over months, years, or decades. These types of renewable resources may be used at sustainable rates, but they may become depleted if we consume them faster than they are replenished. **Nonrenewable natural resources,** such as minerals and fossil fuels, are in finite supply and are formed far more slowly than we use them. Once we deplete a nonrenewable resource, it is no longer available.

We rely on ecosystem services

If we think of natural resources as "goods" produced by nature, then we soon realize that Earth's natural systems also provide "services" on which we depend. Our planet's ecological systems purify air and water, cycle nutrients, regulate climate, pollinate plants, and recycle our waste. Such essential services are commonly called **ecosystem services** (**FIGURE 1.2**). Ecosystem services arise from the normal functioning of natural systems and are not meant for our benefit, yet we could not survive without them. The ways that ecosystem services support our lives and civilization are countless and profound (pp. 116–117, 148–149).

Just as we may deplete natural resources, we may degrade ecosystem services when, for example, we destroy habitat or generate pollution. In recent years, our depletion of nature's goods and our disruption of nature's services have intensified, driven by rising resource consumption and a human population that grows larger every day.

Population growth amplifies our impact

For nearly all of human history, fewer than a million people populated Earth at any one time. Today our population has grown beyond 7 *billion* people. **FIGURE 1.3** shows just how recently and suddenly this monumental change has taken place. For every one person who used to exist more than 10,000 years ago, several thousand people exist today!

FIGURE 1.2 We rely on the ecosystem services that natural systems provide. For example, forested hillsides help people living below by purifying water and air, cycling nutrients, regulating water flow, preventing flooding, and reducing erosion, as well as by providing game, wildlife, timber, recreation, and aesthetic beauty.

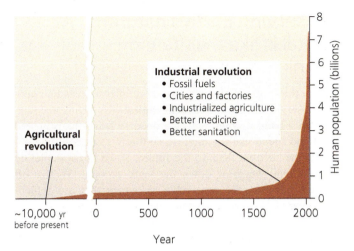

FIGURE 1.3 The global human population increased after the agricultural revolution and then skyrocketed as a result of the industrial revolution. Note that the tear in the graph represents the passage of time and a change in *x*-axis values. *Data compiled from U.S. Census Bureau, U.N. Population Division, and other sources.*

DATA Q For every person alive in the year 1800, about how many people are alive today?

NOTE: Each **DATA Q** in this book asks you to examine the figure carefully so that you understand what it is showing. Once you take the time to understand what it shows, the rest is a breeze!

Because this is the first **DATA Q** of our book, let's walk through it together. You would first note that in the graph, time is shown on the *x* axis and population size on the *y* axis. You would find the year 1800 (three-fifths of the way between 1500 and 2000 on the *x* axis) and trace straight upward to determine the approximate value of the data in that year. You'd then do the same for today's date at the far right end of the graph. To calculate roughly how many people are alive today for every one person alive in 1800, you would simply divide today's number by the number for 1800.

For each **DATA Q**, you can check your answers in **APPENDIX A** in the back of the book.

Go to **Interpreting Graphs & Data** on Mastering EnvironmentalScience®.

Two phenomena triggered our remarkable increase in population size. The first was our transition from a hunter-gatherer lifestyle to an agricultural way of life. This change began about 10,000 years ago and is known as the **agricultural revolution.** As people began to grow crops, domesticate animals, and live sedentary lives on farms and in villages, they produced more food to meet their nutritional needs and began having more children.

The second phenomenon, known as the **industrial revolution,** began in the mid-1700s. It entailed a shift from rural life, animal-powered agriculture, and handcrafted goods toward an urban society provisioned by the mass production of factory-made goods and powered by **fossil fuels** (non-renewable energy sources including oil, coal, and natural gas; pp. 521–523). Industrialization brought dramatic advances in technology, sanitation, and medicine. It also enhanced food

production through the use of fossil-fuel-powered equipment and synthetic pesticides and fertilizers (pp. 212, 238).

The factors driving population growth have brought us better lives in many ways. Yet as our world fills with people, population growth has begun to threaten our well-being. We must ask how well the planet can accommodate 7 billion of us—or the 9 billion forecast by 2050. Already our sheer numbers are putting unprecedented stress on natural systems and the availability of resources.

Resource consumption exerts social and environmental pressures

Besides stimulating population growth, industrialization increased the amount of resources each of us consumes. By mining energy sources and manufacturing more goods, we have enhanced the material affluence of many of the world's people. In the process, however, we have consumed more and more of the planet's limited resources.

One way to quantify resource consumption is to use the concept of the ecological footprint, developed in the 1990s by environmental scientists Mathis Wackernagel and William Rees. An **ecological footprint** expresses the cumulative area of biologically productive land and water required to provide the resources a person or population consumes and to dispose of or recycle the waste the person or population produces (**FIGURE 1.4**). It measures the total area of Earth's

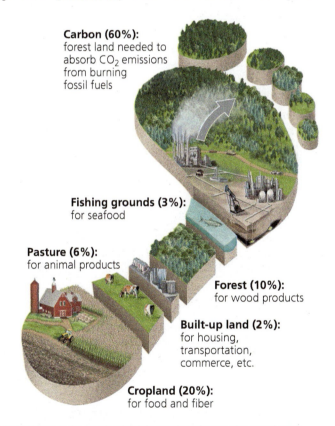

Carbon (60%): forest land needed to absorb CO_2 emissions from burning fossil fuels

Fishing grounds (3%): for seafood

Pasture (6%): for animal products

Forest (10%): for wood products

Built-up land (2%): for housing, transportation, commerce, etc.

Cropland (20%): for food and fiber

FIGURE 1.4 An ecological footprint shows the total area of biologically productive land and water used by a given person or population. Shown is a breakdown of major components of the average person's footprint. *Data from Global Footprint Network, 2016.*

biologically productive surface that a given person or population "uses" once all direct and indirect impacts are summed together.

For humanity as a whole, Wackernagel and his colleagues at the Global Footprint Network calculate that our species is now using 64% more of the planet's renewable resources than are available on a sustainable basis. That is, we are depleting renewable resources by using them 64% faster than they are being replenished. To look at this another way, it would take 1.64 years for the planet to regenerate the renewable resources that people use in just 1 year. The practice of consuming more resources than are being replenished is termed **overshoot** because we are overshooting, or surpassing, Earth's capacity to sustainably support us (**FIGURE 1.5**).

Some scientists have criticized the methods by which the Global Footprint Network calculates footprints, and many question how well its methods measure overshoot. Indeed, any attempt to boil down complicated issues to a single number is fraught with peril, even if the general concept is sound and useful. Yet some things are clear; for instance, people from wealthy nations such as the United States have much larger ecological footprints than do people from poorer nations. Using the Global Footprint Network's calculations, if all the world's people consumed resources at the rate of Americans, humanity would need the equivalent of four planet Earths!

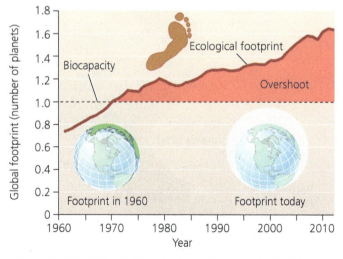

FIGURE 1.5 **Analyses by one research group indicate that we have overshot Earth's biocapacity—its capacity to support us—by 64%.** We are using renewable natural resources 64% faster than they are being replenished. *Data from Global Footprint Network, 2016.*

 DATA How much larger is the global ecological footprint today than it was half a century ago?

Go to **Interpreting Graphs & Data** on Mastering EnvironmentalScience®.

Conserving Earth's natural capital is like maintaining a bank account

We can think of our planet's vast store of resources and ecosystem services—Earth's **natural capital**—as a bank account. To keep a bank account full, we need to leave the principal intact and spend just the interest, so that we can continue living off the account far into the future. If we begin depleting the principal, we draw down the bank account. To live off nature's interest—the renewable resources that are replenished year after year—is sustainable. To draw down resources faster than they are replaced is to eat into nature's capital, the bank account for our planet and our civilization. Currently we are drawing down Earth's natural capital—and we cannot get away with this for long.

Environmental science can help us learn from mistakes

Historical evidence suggests that civilizations can crumble when pressures from population and consumption overwhelm resource availability. Historians have inferred that environmental degradation contributed to the fall of the Greek and Roman empires; the Angkor civilization of Southeast Asia; and the Maya, Anasazi, and other civilizations of the Americas. In Syria, Iraq, and elsewhere in the Middle East, areas that today are barren desert had earlier been lush enough to support the origin of agriculture and thriving ancient societies. Easter Island has long been held up as a society that self-destructed after depleting its resources, although new research paints a more complex picture (see **THE SCIENCE BEHIND THE STORY**, pp. 8–9).

In today's globalized society, the stakes are higher than ever because our environmental impacts are global. If we cannot forge sustainable solutions to our problems, then the resulting societal collapse will be global. Fortunately, environmental science holds keys to building a better world. By studying environmental science, you will learn to evaluate the whirlwind of changes taking place around us and to think critically and creatively about ways to respond.

The Nature of Environmental Science

Environmental scientists aim to comprehend how Earth's natural systems function, how these systems affect people, and how we influence those systems. Many environmental scientists are motivated by a desire to develop solutions to environmental problems. These solutions (such as new technologies, policy decisions, or resource management strategies) are *applications* of environmental science. The study of such applications and their consequences is also part of environmental science.

Environmental science is interdisciplinary

Studying our interactions with our environment is a complex endeavor that requires expertise from many academic disciplines, including ecology, earth science, chemistry, biology, geography, economics, political science, demography, ethics, and others. Environmental science is **interdisciplinary,** bringing techniques, perspectives, and research results from multiple disciplines together into a broad synthesis (**FIGURE 1.6**).

Traditional established disciplines are valuable because their scholars delve deeply into topics, developing expertise in particular areas and uncovering new knowledge. In contrast, interdisciplinary fields are valuable because their practitioners consolidate and synthesize the specialized knowledge from many disciplines and make sense of it in a broad context to better serve the multifaceted interests of society.

Environmental science is especially broad because it encompasses not only the **natural sciences** (disciplines that examine the natural world) but also the **social sciences** (disciplines that address human interactions and institutions). Most environmental science programs focus more on the natural sciences, whereas programs that emphasize the

social sciences often use the term **environmental studies.** Whichever approach one takes, these fields bring together many diverse perspectives and sources of knowledge.

An interdisciplinary approach to addressing environmental problems can produce effective solutions for society. For example, we used to add lead to gasoline to make cars run more smoothly, even though research showed that lead emissions from tailpipes caused health problems, including brain damage and premature death. In 1970 air pollution was severe in many American cities, and motor vehicles accounted for 78% of U.S. lead emissions. In response, environmental scientists, engineers, medical researchers, and policymakers all merged their knowledge and skills into a process that eventually brought about a ban on leaded gasoline. By 1996 all gasoline sold in the United States was unleaded, and the nation's largest source of atmospheric lead pollution had been completely eliminated.

FAQ

Aren't environmental scientists also environmentalists?

Not necessarily. Although environmental scientists search for solutions to environmental problems, they strive to keep their research rigorously objective and free from advocacy. Of course, like all human beings, scientists are motivated by personal values and interests—and like any human endeavor, science can never be entirely free of social influence. However, whereas personal values and social concerns may help shape the questions scientists ask, scientists do their utmost to carry out their work impartially and to interpret their results with wide-open minds. Remaining open to whatever conclusions the data demand is a hallmark of the effective scientist.

Environmental science is not the same as environmentalism

Although many environmental scientists are interested in solving problems, it would be incorrect to confuse environmental science with environmentalism or environmental activism. They are very different. Environmental science involves the scientific study of the environment and our interactions with it. In contrast, **environmentalism** is a social movement dedicated to protecting the natural world—and, by extension, people—from undesirable changes brought about by human actions.

The Nature of Science

Science is a systematic process for learning about the world and testing our understanding of it. The term *science* is also used to refer to the accumulated body of knowledge that arises from this dynamic process of observing, questioning, testing, and discovery.

Knowledge gained from science can be applied to address society's needs—for instance, to develop technology

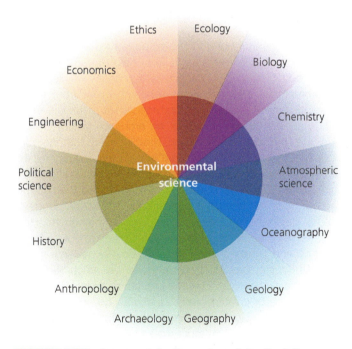

FIGURE 1.6 Environmental science is an interdisciplinary pursuit. It draws from many different established fields of study across the natural sciences and social sciences.

THE SCIENCE
behind the story

What Are the Lessons of Easter Island?

**Terry Hunt and Carl Lipo
on Easter Island**

A mere speck of land in the vast Pacific Ocean, Easter Island is one of the most remote spots on the globe. Yet this far-flung island—called Rapa Nui by its inhabitants—is the focus of an intense debate among scientists seeking to solve its mysteries and decipher the lessons it offers. The debate shows how, in science, new information can challenge existing ideas—and also how interdisciplinary research helps us to tackle complex questions.

Ever since European explorers stumbled upon Rapa Nui on Easter Sunday, 1722, outsiders have been struck by the island's barren landscape. Early European accounts suggested that the 2000–3000 people living on the island at the time seemed impoverished, subsisting on a few meager crops and possessing only stone tools. Yet the forlorn island also featured hundreds of gigantic statues of carved rock. How could people without wheels or ropes, on an island without trees, have moved 90-ton statues 10 m (33 ft) high as far as 10 km (6.2 mi) from the quarry where they were chiseled to the sites where they were erected? Apparently some calamity must have befallen a once-mighty civilization on the island.

Researchers who set out to solve Rapa Nui's mysteries soon discovered that the island had once been lushly forested. Scientist John Flenley and his colleagues drilled cores deep into lake sediments and examined ancient pollen grains preserved there, seeking to reconstruct, layer by layer, the history of vegetation in the region. Finding a great deal of palm pollen, they inferred that when Polynesian people colonized the island (A.D. 300–900, they estimated), it was covered with palm trees similar to the Chilean wine palm—a tree that can live for centuries.

By studying pollen and the remains of wood from charcoal, archaeologist Catherine Orliac found that at least 21 other plant species—now gone—had also been common. Clearly the island had once supported a diverse forest. Forest plants would have provided fuelwood, building material for houses and canoes, fruit to eat, fiber for clothing—and, researchers guessed, logs and fibrous rope to help move statues.

But pollen analysis also showed that trees began declining after human arrival and were replaced by ferns and grasses.

Then between 1400 and 1600, pollen levels plummeted. Charcoal in the soil proved the forest had been burned, likely in slash-and-burn farming. Researchers concluded that the islanders, desperate for forest resources and cropland, had deforested their own island.

With the forest gone, soil eroded away (data from lake bottoms showed a great deal of accumulated sediment). Erosion would have lowered yields of bananas, sugarcane, and sweet potatoes, perhaps leading to starvation and population decline.

Further evidence indicated that wild animals disappeared. Archaeologist David Steadman analyzed 6500 bones and found that at least 31 bird species had provided food for the islanders. Today, only one native bird species is left. Remains from charcoal fires show that early islanders feasted on fish, sharks, porpoises, turtles, octopus, and shellfish—but in later years they consumed little seafood.

As resources declined, researchers concluded, people fell into clan warfare, revealed by unearthed weapons and skulls with head wounds. Rapa Nui appeared to be a tragic case of ecological suicide: A once-flourishing civilization depleted its resources and destroyed itself. In this interpretation—advanced by Flenley and writer Paul Bahn, and popularized by scientist Jared Diamond in his best-selling 2005 book *Collapse*—Rapa Nui seemed to offer a clear lesson: We on our global island, planet Earth, had better learn to use our limited resources sustainably.

When Terry Hunt and Carl Lipo began research on Rapa Nui in 2001, they expected simply to help fill gaps in a well-understood history. But science is a process of discovery, and sometimes evidence leads researchers far from where they anticipated. For Hunt, an anthropologist at the University of Hawai'i at Manoa, and Lipo, an archaeologist at California State University, Long Beach, their work led them to conclude that the traditional "ecocide" interpretation didn't tell the whole story. First, their radiocarbon dating (dating of items using radioisotopes of carbon; p. 24) indicated that people had not colonized the island until about A.D. 1200, suggesting that deforestation occurred rapidly after their arrival. How could so few people have destroyed so much forest so fast?

Hunt and Lipo's answer: rats. When Polynesians settled new islands, they brought crop plants, as well as chickens and other domestic animals. They also brought rats—intentionally as a food source or unintentionally as stowaways. In either case, rats can multiply quickly, and they soon overran Rapa Nui.

Researchers found rat tooth marks on old nut casings, and Hunt and Lipo suggested that rats ate so many palm nuts and shoots that the trees could not regenerate. With no young trees growing, the palm went extinct once mature trees died.

Diamond and others counter that plenty of palm nuts on Easter Island escaped rat damage, that most plants on other islands survived rats introduced by Polynesians, and that more than 20 additional plant species went extinct on Rapa Nui. Moreover, people brought the rats, so even if rats destroyed the forest, human colonization was still to blame.

Despite the forest loss, Hunt and Lipo argue that islanders were able to persist and thrive. Archaeology shows how islanders adapted to Rapa Nui's poor soil and windy weather by developing rock gardens to protect crop plants and nourish the soil. Hunt and Lipo contended that tools that previous researchers viewed as weapons were actually farm implements; lethal injuries were rare; and no evidence of battle or defensive fortresses was uncovered.

Hunt, Lipo, and others also unearthed old roads and inferred how the famous statues were transported. It had been thought that a powerful central authority must have forced armies of laborers to roll them over countless palm logs, but Hunt and Lipo concluded that small numbers of people could have moved them by tilting and rocking them upright—much as we might move a refrigerator. Indeed, the distribution of statues on the island suggested the work of family groups. Islanders had adapted to their resource-poor environment by becoming a peaceful and cooperative society, Hunt and Lipo maintained, with the statues providing a harmless outlet for competition over status and prestige.

Altogether, the evidence led Hunt and Lipo to propose that far from destroying their environment, the islanders had acted as responsible stewards. The collapse of this sustainable civilization, they argue, came with the arrival of Europeans, who unwittingly brought contagious diseases to which the islanders had never been exposed. Indeed, historical journals of sequential European voyages depict a society falling into disarray as if reeling from epidemics.

Peruvian ships then began raiding Rapa Nui and taking islanders away into slavery. Foreigners acquired the land, forced the remaining people into labor, and introduced thousands of sheep, which destroyed the few native plants left on the island. Thus, the new hypothesis holds that the collapse of Rapa Nui's civilization resulted from a barrage of disease, violence, and slave raids following foreign contact. Before that, Hunt and Lipo say, Rapa Nui's people boasted 500 years of a peaceful and resilient society.

Hunt and Lipo's interpretation, put forth in a 2011 book, *The Statues That Walked*, would represent a paradigm shift (p. 14) in how we view Easter Island. Debate between the two camps remains heated, however, and interdisciplinary research continues as scientists look for new ways to test the differing hypotheses. This is an example of how science advances, and in the long term, data from additional studies should lead us closer and closer to the truth.

Like the people of Rapa Nui, we are all stranded together on an island with limited resources. What is the lesson of Easter Island for our global island, Earth? Perhaps there are two: Any island population must learn to live within its means—but with care and ingenuity, there is hope that we can.

Were the haunting statues of Easter Island (Rapa Nui) erected by a civilization that collapsed after devastating its environment or by a sustainable civilization that fell because of outside influence?

(a) Prescribed burning

(b) An electric car

FIGURE 1.7 Scientific knowledge is applied in engineering and technology and in policy and management decisions. Prescribed burning **(a)** is a forest management practice informed by ecological research. Energy-efficient electric automobiles **(b)** are technological advances made possible by materials and energy research.

or to inform policy and management decisions (**FIGURE 1.7**). From the food we eat to the clothing we wear to the health care we depend on, virtually everything in our lives has been improved by the application of science. Many scientists are motivated by the potential for developing useful applications. Others are motivated simply by a desire to understand how the world works.

Scientists test ideas by critically examining evidence

Science is all about asking and answering questions. Scientists examine how the world works by making observations, taking measurements, and testing whether their ideas are supported by evidence. The effective scientist thinks critically and does not simply accept conventional wisdom from others. The scientist becomes excited by novel ideas but is skeptical and judges ideas by the strength of evidence that supports them.

A great deal of scientific work is **descriptive science,** research in which scientists gather basic information about organisms, materials, systems, or processes that are not yet well known. In this approach, researchers explore new frontiers of knowledge by observing and measuring phenomena to gain a better understanding of them.

Once enough basic information is known about a subject, scientists can begin posing questions that seek deeper explanations about how and why things are the way they are. At this point scientists may pursue **hypothesis-driven science,** research that proceeds in a more targeted and structured manner, using experiments to test hypotheses within a framework traditionally known as the scientific method.

The scientific method is a traditional approach to research

The **scientific method** is a technique for testing ideas with observations. There is nothing mysterious or intimidating about the scientific method; it is merely a formalized version of the way any of us might naturally use logic to resolve a question. Because science is an active, creative process, innovative researchers may depart from the traditional scientific method when particular situations demand it. Moreover, scientists in different fields approach their work differently because they deal with dissimilar types of information. Nonetheless, scientists of all persuasions broadly agree on fundamental elements of the process of scientific inquiry. As practiced by individual researchers or research teams, the scientific method (**FIGURE 1.8**) typically follows the steps outlined below.

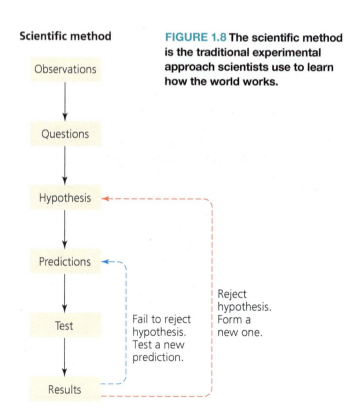

Scientific method

FIGURE 1.8 The scientific method is the traditional experimental approach scientists use to learn how the world works.

Make observations Advances in science generally begin with the observation of some phenomenon that the scientist wishes to explain. Observations set the scientific method in motion and also play a role throughout the process.

Ask questions Curiosity is in our human nature. Just observe young children exploring a new environment—they want to touch, taste, watch, and listen to everything, and as soon as they can speak, they begin asking questions. Scientists, in this respect, are kids at heart. Questions environmental scientists might ask include: Why is the ocean salty? Why are storms becoming more severe? What is causing algae to cover local ponds? When pesticides poison fish or frogs, are people also affected? How can we help restore populations of plants or animals?

Develop a hypothesis Scientists address their questions by devising explanations that they can test. A **hypothesis** is a statement that attempts to explain a phenomenon or answer a scientific question. For example, a scientist investigating why algae are growing excessively in local ponds might observe that chemical fertilizers are being applied on farm fields nearby. The scientist might then propose a hypothesis as follows: "Agricultural fertilizers running into ponds cause the amount of algae in the ponds to increase."

Make predictions The scientist next uses the hypothesis to generate **predictions,** specific statements that can be directly and unequivocally tested. In our algae example, a researcher might predict: "If agricultural fertilizers are added to a pond, the quantity of algae in the pond will increase."

Test the predictions Scientists test predictions by gathering evidence that could potentially refute the predictions and thus disprove the hypothesis. The strongest form of evidence comes from experiments. An **experiment** is an activity designed to test the validity of a prediction or a hypothesis. It involves manipulating **variables,** or conditions that can change.

For example, a scientist could test the prediction linking algal growth to fertilizer by selecting two identical ponds and adding fertilizer to one of them. In this example, fertilizer input is an **independent variable,** a variable the scientist manipulates, whereas the quantity of algae that results is the **dependent variable,** a variable that depends on the fertilizer input. If the two ponds are identical except for a single independent variable (fertilizer input), then any differences that arise between the ponds can be attributed to changes in the independent variable. Such an experiment is known as a **controlled experiment** because the scientist attempts to control for the effects of all variables except the one he or she is testing. In our example, the pond left unfertilized serves as a **control,** an unmanipulated point of comparison for the manipulated **treatment** pond.

Whenever possible, it is best to replicate one's experiment—that is, to stage multiple tests of the same comparison. Our scientist could perform a replicated experiment on, say, 10 pairs of ponds, adding fertilizer to one of each pair.

FIGURE 1.9 Researchers gather data to test predictions in experiments. Here, a scientist samples algae from a pond.

Analyze and interpret results Scientists record **data,** or information, from their studies (**FIGURE 1.9**). Researchers particularly value quantitative data (information expressed using numbers), because numbers provide precision and are easy to compare. The scientist conducting the fertilization experiment, for instance, might quantify the area of water surface covered by algae in each pond or might measure the dry weight of algae in a certain volume of water taken from each. It is vital, however, to collect data that are representative. Because it is impractical to measure a pond's total algal growth, our researcher might instead sample from multiple areas of each pond. These areas must be selected in a random manner; choosing areas with the most growth or the least growth, or areas most convenient to sample, would not provide a representative sample.

To summarize and present the data they obtain, scientists often use graphs. Graphs help to make patterns and trends in the data visually apparent and easy to understand. **FIGURE 1.10** (on page 12) shows just a few examples of how different types of graphs can be used to present data. Each of these types of graphs is explained further in APPENDIX B: HOW TO INTERPRET GRAPHS at the back of this book. The ability to interpret graphs is a skill you will find useful throughout your life. We encourage you to consult Appendix B closely as you begin your environmental science course.

You also will note that many of the graphs in this book are accompanied by "DATA Q" questions to help you build your graph-reading skills. You can check your answers to these questions by referring to APPENDIX A: ANSWERS TO DATA ANALYSIS QUESTIONS at the back of this book.

Even with the precision that numerical data provide, experimental results may not be clear-cut. Data from treatments and controls may vary only slightly, or replicates may yield different results. Researchers must therefore analyze their data using statistical tests. With these mathematical methods, scientists can determine objectively and precisely the strength and reliability of patterns they find.

If experiments disprove a hypothesis, the scientist will reject it and may formulate a new hypothesis to replace it. If experiments fail to disprove a hypothesis, this lends support to the hypothesis but does not *prove* it is correct. The scientist may choose to generate new predictions to test the hypothesis in different ways and further assess its likelihood of being true. In this way, the scientific method loops back on itself, giving rise to repeated rounds of hypothesis revision and experimentation (see Figure 1.8).

If repeated tests fail to reject a hypothesis, evidence in favor of it accumulates, and the researcher may eventually

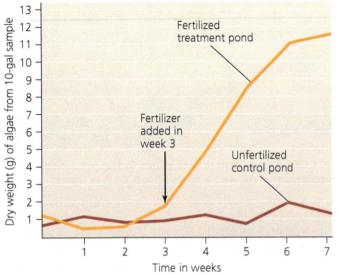

(a) Line graph of algal density through time in a fertilized treatment pond and an unfertilized control pond

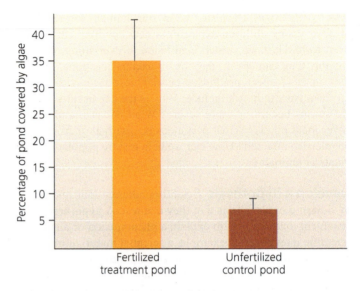

(b) Bar chart of mean algal density in several fertilized treatment ponds and unfertilized control ponds

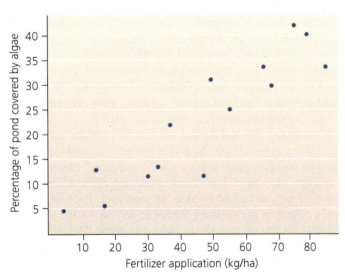

(c) Scatter plot of algal density correlated with fertilizer use on surrounding farmland

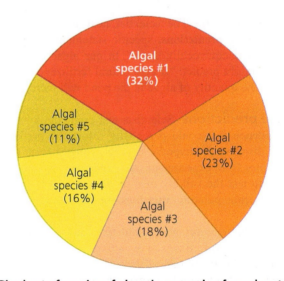

(d) Pie chart of species of algae in a sample of pond water

FIGURE 1.10 Scientists use graphs to present and visualize their data. For example, in **(a)**, a line graph shows how the amount of algae increased when fertilizer was added to a treatment pond in an experiment yet stayed the same in an unfertilized control pond. In **(b)**, a bar chart shows how fertilized ponds, on average, have several times more algae than unfertilized ponds. In **(c)**, a scatter plot shows how ponds with more fertilizer tend to contain more algae. In **(d)**, a pie chart shows the relative abundance of five species of algae in a sample of pond water. See **APPENDIX B** to learn more about how to interpret these types of graphs.

 • In part **(a)**, is time in weeks shown on the *x* axis or the *y* axis?
• In part **(b)**, what is the dependent variable?
• In part **(c)**, do the data show a positive correlation or a negative correlation?
• In part **(d)**, which species is most numerous? Which is least numerous?
• What are the thin black lines atop the colored bars in part **(b)** called? Explain what these lines indicate.

Go to **Interpreting Graphs & Data** on **Mastering**EnvironmentalScience®.

conclude that the hypothesis is well supported. Ideally, the scientist would want to test all possible explanations. For instance, our researcher might formulate an additional hypothesis, proposing that algae increase in fertilized ponds because chemical fertilizers diminish the numbers of fish or invertebrate animals that eat algae. It is possible, of course, that both hypotheses could be correct and that each may explain some portion of the initial observation that local ponds were experiencing algal blooms.

We test hypotheses in different ways

An experiment in which the researcher actively chooses and manipulates the independent variable is known as a *manipulative experiment*. A manipulative experiment provides strong evidence because it can reveal causal relationships, showing that changes in an independent variable cause changes in a dependent variable. In practice, however, we cannot run manipulative experiments for all questions, especially for processes involving large spatial scales or long timescales. For example, to study global climate change (Chapter 18), we cannot run a manipulative experiment adding carbon dioxide to 10 treatment planets and 10 control planets and then compare the results! Thus, it is common for researchers to run *natural experiments*, which compare how dependent variables are expressed in naturally occurring, but different, contexts. In such experiments, the independent variable varies naturally, and researchers test their hypotheses by searching for **correlation,** or statistical association among variables.

For instance, let's suppose our scientist studying algae surveys 50 ponds, 25 of which happen to be fed by fertilizer runoff from nearby farm fields and 25 of which are not. Let's say he or she finds seven times more algal growth in the fertilized ponds. The scientist may conclude that algal growth is correlated with fertilizer input—that is, that one tends to increase along with the other.

This type of evidence is not as strong as the causal demonstration that manipulative experiments can provide, but often a natural experiment is the only feasible approach. Because many questions in environmental science are complex and exist on large scales, they must be addressed with correlative data. As such, environmental scientists cannot always provide black-and-white answers to questions from policymakers and the public. Nonetheless, good correlative studies can make for very strong science, and they preserve the real-world complexity that manipulative experiments often sacrifice. Whenever possible, scientists try to integrate natural experiments and manipulative experiments to gain the advantages of each.

The scientific process continues beyond the scientific method

Scientific research takes place within the context of a community of peers. To have impact, a researcher's work must be published and made accessible to this community (**FIGURE 1.11**).

Peer review When a researcher's work is complete and the results are analyzed, he or she writes up the findings and submits them to a journal (a scholarly publication in which scientists share their work). The journal's editor asks several other scientists who specialize in the subject area to examine the manuscript, provide comments and criticism (generally anonymously), and judge whether the work merits publication in the journal. This procedure, known as **peer review,** is an essential part of the scientific process.

Peer review is a valuable guard against faulty research contaminating the literature (the body of published studies) on which all scientists rely. However, because scientists are human, personal biases and politics can sometimes creep into the review process. Fortunately, just as individual scientists strive to remain objective in conducting their research, the scientific community does its best to ensure fair review of all work.

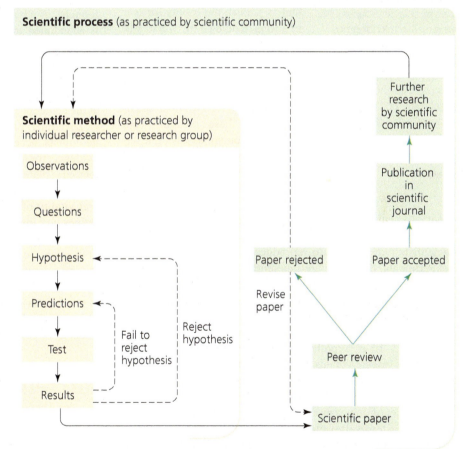

FIGURE 1.11 The scientific method that research teams follow is part of a larger framework—the overall process of science carried out by the scientific community. This process includes peer review and publication of research, acquisition of funding, and the elaboration of theory through the cumulative work of many researchers.

Note that scientists are not paid money for peer review; their services are entirely voluntary. Moreover, researchers generally have to pay the journals that publish their papers.

Conference presentations Scientists frequently present their work at professional conferences, where they interact with colleagues and receive comments on their research. Such interactions can help improve a researcher's work and foster collaboration among researchers, thus enhancing the overall quality and impact of science.

Grants and funding To fund their research, most scientists need to spend a great deal of time requesting money from private foundations or from government agencies such as the National Science Foundation. Grant applications undergo peer review just as scientific papers do, and competition for funding is generally intense.

Scientists' reliance on funding sources can occasionally lead to conflicts of interest. A researcher who obtains data showing his or her funding source in an unfavorable light may be reluctant to publish the results for fear of losing funding—or, worse yet, could be tempted to doctor the results. This situation can arise, for instance, when an industry funds research to test its products for health or safety. Most scientists resist these pressures, but whenever you are assessing a scientific study, it is always a good idea to note where the researchers obtained their funding.

Repeatability The careful scientist may test a hypothesis repeatedly in various ways. Following publication, other scientists may attempt to reproduce the results in their own experiments. Scientists are inherently cautious about accepting a novel hypothesis, so the more a result can be reproduced by different research teams, the more confidence scientists will have that it provides a correct explanation.

Theories If a hypothesis survives repeated testing by numerous research teams and continues to predict experimental outcomes and observations accurately, it may be incorporated into a theory. A **theory** is a widely accepted, well-tested explanation of one or more cause-and-effect relationships that has been extensively validated by a great amount of research. Whereas a hypothesis is a simple explanatory statement that may be disproven by a single experiment, a theory consolidates many related hypotheses that have been supported by a large body of data.

Note that scientific use of the word *theory* differs from popular usage of the word. In everyday language when we say something is "just a theory," we are suggesting it is a speculative idea without much substance. However, scientists mean just the opposite when they use the term. In a scientific context, a theory is a conceptual framework that explains a phenomenon and has undergone extensive and rigorous testing, such that confidence in it is extremely strong.

For example, Darwin's theory of evolution by natural selection (pp. 48–49) has been supported and elaborated by many thousands of studies over 160 years of intensive research. Observations and experiments have shown repeatedly and in great detail how plants and animals change over generations, or evolve, expressing characteristics that best promote survival and reproduction. Because of its strong support and explanatory power, evolutionary theory is the central unifying principle of modern biology. Other prominent scientific theories include atomic theory, cell theory, big bang theory, plate tectonics, and general relativity.

Science goes through paradigm shifts

As the scientific community accumulates data in an area of research, interpretations sometimes may change. Thomas Kuhn's influential 1962 book *The Structure of Scientific Revolutions* argued that science goes through periodic upheavals in thought, in which one scientific **paradigm,** or dominant view, is abandoned for another. For example, before the 16th century, European scientists believed that Earth was at the center of the universe. Their data on the movements of planets fit that concept somewhat well—yet the idea eventually was disproved after Nicolaus Copernicus showed that placing the sun at the center of the solar system explained the data much better.

Another paradigm shift occurred in the 1960s, when geologists accepted plate tectonics (p. 32). By this time, evidence for the movement of continents and the action of tectonic plates had accumulated and become overwhelmingly convincing. Paradigm shifts demonstrate the strength and vitality of science, showing science to be a process that refines and improves itself through time.

Understanding how science works is vital to assessing how scientific interpretations progress through time as information accrues. This is especially relevant in environmental science—a young field that is changing rapidly as we obtain vast amounts of new information, as human impacts on our planet multiply, and as we gather lessons from the consequences of our actions.

Sustainability and Our Future

Throughout this book you will encounter environmental scientists asking questions, testing hypotheses, conducting experiments, analyzing data, and drawing conclusions about the causes and consequences of environmental change. Environmental scientists, who study the condition of our environment and the nature of our impacts, are addressing the most centrally important issues of our time.

weighing the ISSUES

Follow the Money

Let us say you are a research scientist wanting to study the impacts of chemicals released into lakes by pulp-and-paper mills. Obtaining research funding has been difficult. Then a large pulp-and-paper company contacts you and offers to fund your research examining how its chemical effluents affect water bodies. What are the benefits and drawbacks of this offer? Would you accept the offer? Why or why not?

Achieving sustainable solutions is vital

Society's primary challenge today is finding how to live within our planet's means, such that Earth and its resources can sustain us—and all life—for the future. This is the challenge of **sustainability,** a guiding principle of modern environmental science and a concept you will encounter throughout this book. Sustainability means leaving our children and grandchildren a world as rich and full as the world we live in now. It means conserving Earth's resources so that our descendants may enjoy them as we have. It means developing solutions that work in the long term. Sustainability requires maintaining fully functioning ecological systems, because we cannot sustain human civilization without sustaining the natural systems that nourish it.

Population and consumption drive environmental impact

Every day, we add more than 200,000 people to the planet. This is like adding a city the size of Augusta, Georgia, on Monday; Akron, Ohio, on Tuesday; Richmond, Virginia, on Wednesday; Rochester, New York, on Thursday; Amarillo, Texas, on Friday; and on and on, day after day. The rate of population growth is now slowing, but our absolute numbers continue to increase. This ongoing rise in human population (Chapter 8) amplifies nearly all of our impacts.

Our consumption of resources has risen even faster than our population. The modern rise in affluence has been a positive development for humanity, and our conversion of the planet's natural capital has made life better for most of us so far. However, like rising population, rising per capita consumption magnifies the demands we make on our environment.

The world's citizens have not benefited equally from society's overall rise in affluence. Today the 20 wealthiest nations boast more than 55 times the per capita income of the 20 poorest nations—three times the gap that existed just two generations ago. Within the United States, the richest 10% of citizens now claim half of the total income and more than 70% of the total wealth. The ecological footprint of the average citizen of a developed nation such as the United States is considerably larger than that of the average resident of a developing country (FIGURE 1.12).

Our growing population and consumption are intensifying the many environmental impacts we examine in this book, including erosion and other impacts from agriculture (Chapters 9 and 10), deforestation (Chapter 12), toxic substances (Chapter 14), freshwater depletion (Chapter 15), fisheries declines (Chapter 16), air and water pollution (Chapters 15–17), waste generation (Chapter 22), mineral extraction and mining impacts (Chapter 23), and, of course, global climate change (Chapter 18). These impacts degrade our health and quality of life, and they alter the ecosystems and landscapes in which we live. They also are

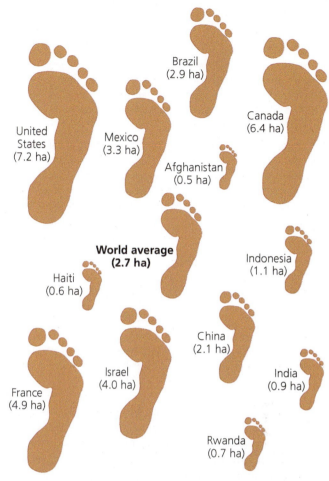

FIGURE 1.12 People of some nations have much larger ecological footprints than people of others. Shown are ecological footprints for average citizens of several nations, along with the world's average per capita footprint of 2.7 hectares. One hectare (ha) = 2.47 acres. *Data from WWF, 2014. Living planet report 2014. Gland, Switzerland: WWF International.*

DATA Q Which nation shown here has the largest footprint? How many times larger is it than that of the nation shown here with the smallest footprint?

Go to **Interpreting Graphs & Data** on Mastering Environmental Science®.

driving the loss of Earth's biodiversity (Chapter 11)—perhaps our greatest problem, because extinction is irreversible. Once a species becomes extinct, it is lost forever.

The most comprehensive scientific assessment of the condition of the world's ecological systems and their capacity to continue supporting our civilization was completed in 2005, when more than 2000 leading environmental scientists from nearly 100 nations completed the Millennium Ecosystem

weighing the **ISSUES**

Ecological Footprints

What do you think accounts for the variation in per capita ecological footprints among societies? Do you feel that people with larger footprints have an ethical obligation to reduce their environmental impact, in order to leave more resources available for people with smaller footprints? Why or why not?

Assessment (**TABLE 1.1**). The Millennium Ecosystem Assessment makes clear that our degradation of environmental systems is causing harm—but that with care and diligence we can still reverse many of these trends and restore health to the natural systems that support us.

Energy choices will shape our future

Our reliance on fossil fuels amplifies virtually every impact we exert on our environment. Yet fossil fuels have also helped to bring us the material affluence our society enjoys today. By exploiting the richly concentrated energy in coal, oil, and natural gas, we have been able to power the machinery of the industrial revolution, produce chemicals that boost crop yields, run vehicles and transportation networks, and manufacture and distribute countless consumer products.

However, in extracting coal, oil, and natural gas, we are splurging on a one-time bonanza, because these fuels are nonrenewable and in finite supply. Attempts to reach further for new fossil fuel sources all seem to threaten more impacts for relatively less fuel. The energy choices we make now will greatly influence the quality of our lives for the rest of the 21st century and beyond.

Sustainable solutions abound

Humanity's challenge is to develop solutions that enhance our quality of life while protecting and restoring the environment that supports us. Many workable solutions are at hand:

- Renewable energy sources (Chapters 20 and 21) are beginning to replace fossil fuels.

- Energy-efficiency efforts continue to gain ground (Chapter 19).

- Scientists and farmers are pursuing soil conservation, high-efficiency irrigation, and organic agriculture (Chapters 9 and 10).

- Laws and new technologies have reduced air and water pollution in wealthier societies (Chapters 15–17).

- Conservation biologists are helping to protect habitat and safeguard endangered species (Chapter 11).

- Better waste management is helping us to conserve resources (Chapter 22).

- Governments, businesses, and individuals are taking steps to reduce emissions of the greenhouse gases that drive climate change (Chapter 18).

These are a few of the many efforts we will examine while exploring sustainable solutions in the course of this book.

Students are promoting solutions on campus

As a college student, you can help to design and implement sustainable solutions on your own campus. Proponents of **campus sustainability** seek ways to help colleges and universities reduce their ecological footprints. Although we tend to think of colleges and universities as enlightened and progressive institutions that benefit society, they also are centers of lavish resource consumption. Classrooms, offices, research labs, residence halls, dining halls, sports arenas, and vehicle fleets all consume resources and generate waste. Together the 4500 campuses in the United States spend $400 billion each year on products and services and generate about 2% of U.S. carbon emissions. Reducing the ecological footprint of a campus can be challenging, but students, faculty, staff, and administrators on thousands of campuses are working together to make the operations of educational institutions more sustainable (**FIGURE 1.13**).

Students are running recycling programs, planting trees and restoring native plants, growing organic gardens, and fostering sustainable dining halls. They are finding ways to improve energy efficiency and water conservation and are pressing for new buildings to meet certification guidelines for sustainable construction. To help address climate change, students are urging their institutions to reduce greenhouse gas emissions, divest from fossil fuel corporations, and use and invest in renewable energy.

In response, nearly 700 university presidents have signed onto the American College and University Presidents' Climate Commitment—a public pledge to inventory emissions, set target dates and milestones for becoming carbon-neutral, take immediate steps to lower emissions, and integrate sustainability into the curriculum.

In our final chapter and throughout this book you will encounter examples of campus sustainability efforts (for example, pp. 209, 605, and 649). Should you wish to pursue such efforts on your own campus, information and links available in the **SELECTED SOURCES AND REFERENCES** at the back of this book point you toward organizations and resources that can help.

(a) Urging divestment from fossil fuels

(b) Recycling at Davidson College

(c) Promoting energy efficiency at Amherst College

FIGURE 1.13 Students are helping to make their campuses more sustainable in all kinds of ways.

Environmental science prepares you for the future

By taking a course in environmental science, you are preparing yourself for a lifetime in a world increasingly dominated by concerns over sustainability.

The course for which you are using this book right now likely did not exist a generation ago. But as society's concerns have evolved, colleges and universities have adapted their academic curricula. As our society comes to appreciate the challenges of creating a sustainable future, colleges and universities are helping students learn how to confront these challenges.

Yet at most schools, fewer than half of students take even a single course on the basic functions of Earth's natural systems, and still fewer take courses on the links between human activity and sustainability. As a result, many educators worry that most students graduate lacking **environmental literacy,** a basic understanding of Earth's physical and living systems and how we interact with them. By taking an environmental science course, you join a privileged minority, benefiting from a valuable education that most of your peers are missing. Your course will equip you with a better understanding of how the world works. You will be better qualified for the green-collar job opportunities of today and tomorrow. And you will be better prepared to navigate the many challenges of creating a sustainable future.

closing
THE LOOP

Finding effective ways of living peacefully, healthfully, and sustainably on our diverse and complex planet will require a thorough scientific understanding of both natural and social systems. Environmental science helps us understand our intricate relationship with our environment and informs our attempts to solve and prevent environmental problems. Although many of today's trends may cause concern, a multitude of inspiring success stories give us reason for optimism. Identifying a problem is the first step toward devising a solution, and addressing environmental problems can move us toward health, longevity, peace, and prosperity. Science in general, and environmental science in particular, can help us develop balanced, workable, sustainable solutions and create a better world now and for the future.

REVIEWING Objectives

You should now be able to:

Describe the field of environmental science

Environmental science is the study of how the natural world works, how our environment affects us, and how we affect our environment. Environmental science uses approaches and insights from many disciplines in the natural sciences and social sciences. (pp. 3, 6–7)

Compare renewable and nonrenewable resources, and explain the importance of natural resources and ecosystem services to our lives

Renewable resources are unlimited or naturally replenished quickly. Nonrenewable resources are limited or replenished by the environment very slowly. Ecosystem services are benefits we receive from the normal functioning of natural systems. Natural resources and ecosystem services are essential to human life and civilization. (p. 4)

Discuss population growth, resource consumption, and their consequences

Today's human population size and per-person level of resource consumption both are higher than ever. Population growth magnifies our environmental impact by adding to the number of people putting demands on resources. Growing per capita consumption amplifies our environmental impact by increasing the demand each person makes on resources. (pp. 4–6)

Explain what is meant by an ecological footprint

An ecological footprint quantifies resource consumption by expressing the total area of biologically productive land and water required to provide the resources and waste disposal for a person or population. (pp. 5–6)

Describe the scientific method and the process of science

Science is a process of using observations to test ideas. The scientific method consists of making observations, formulating questions, stating a hypothesis, generating predictions, testing predictions, and analyzing results. Scientific research occurs within a larger process of science that includes peer review, journal publication, and interaction with colleagues. (pp. 7–14)

Identify and illustrate major pressures on the global environment

Rising population and intensifying consumption magnify human impacts on the environment, which include resource depletion, air and water pollution, climate change, habitat destruction, and biodiversity loss. (pp. 14–16)

Discuss the concept of sustainability, and cite sustainable solutions being pursued on campuses and in the wider world

Sustainability is a term that describes living within our planet's means, such that Earth's resources can sustain us—and all life—for the future. Today we are actively developing sustainable solutions (such as replacing fossil fuels with renewable energy) that promote our quality of life while protecting and restoring our environment. Many college students are taking action to promote sustainable solutions on their campuses, including recycling, energy efficiency, water conservation, transportation alternatives, and. (pp. 14–17)

TESTING Your Comprehension

1. What do renewable resources and nonrenewable resources have in common? How are they different? Identify two renewable and two nonrenewable resources.

2. How and why did the agricultural revolution affect human population size? How and why did the industrial revolution affect human population size? Explain what benefits and what environmental impacts have resulted from these two revolutions.

3. What is an *ecological footprint*? Describe how *natural capital* is similar to a bank account.

4. What is *environmental science*? Name several disciplines that environmental science draws upon.

5. Compare and contrast the two meanings of *science*. Name three applications of science.

6. Describe the scientific method. What is its typical sequence of steps?

7. Explain the difference between correlation and causation, and state how these concepts relate to manipulative and natural experiments.

8. What occurs before a researcher's results are published in a scientific journal? Why is this process important?

9. Give examples of three major environmental problems, along with their causes. How are these problems interrelated? Now name one potential solution for each.

10. Describe in your own words what you think is meant by the term *sustainability*. List three ways that students, faculty, and administrators are seeking to make their campuses more sustainable.

SEEKING Solutions

1. Resources such as soils, timber, fresh water, and biodiversity are renewable if we use them in moderation but can become nonrenewable if we overexploit them (see Figure 1.1). For each of these four resources, describe one way we sometimes overexploit the resource and name one thing we could do to conserve the resource. For each, what might constitute sustainable use? (Feel free to look ahead and peruse coverage of these issues throughout this book.)

2. What do you think is the lesson of Easter Island? What more would you like to learn or understand about this island, its history, or its people? What similarities do you perceive between Easter Island and our own modern society? What differences do you see between their predicament and ours?

3. What environmental problem do *you* feel most acutely yourself? Do you think there are people in the world who do not view your issue as a problem? Who might they be, and why might they take a different view?

4. If the human population were to stabilize tomorrow and never reach 8 billion people, would that solve all our environmental problems? Why or why not? What conditions might get better, and what challenges might remain?

5. Find out what sustainability efforts are being made on your campus. What results have these efforts produced so far? What further efforts would you like to see pursued on your campus? Do you foresee any obstacles to these efforts? How could these obstacles be overcome? How could you become involved?

6. **THINK IT THROUGH** You have become head of a major funding agency that grants money to scientists pursuing research in environmental science. You must give your staff several priorities to determine what types of scientific research to fund. What environmental problems would you most like to see addressed with research? Describe the research you think would need to be completed to develop workable solutions. What else, beyond scientific research, might be needed to develop sustainable solutions?

CALCULATING Ecological Footprints

Mathis Wackernagel and his colleagues at the Global Footprint Network continue to refine the method of calculating ecological footprints—the amount of biologically productive land and water required to produce the energy and natural resources we consume and to absorb the wastes we generate. According to their most recent data, there are 1.8 hectares (4.4 acres) available for each person in the world, yet

we use on average 2.7 ha (6.7 acres) per person, creating a global ecological deficit, or overshoot (p. 6), of 50%.

Compare the ecological footprints of each nation listed in the table. Calculate their proportional relationships to the world population's average ecological footprint and to the area available globally to meet our ecological demands.

NATION	ECOLOGICAL FOOTPRINT (HECTARES PER PERSON)	PROPORTION RELATIVE TO WORLD AVERAGE FOOTPRINT	PROPORTION RELATIVE TO WORLD AREA AVAILABLE
Bangladesh	0.7	0.3 (0.7 ÷ 2.7)	0.4 (0.7 ÷ 1.8)
Tanzania	1.2		
Colombia	1.8		
Thailand	2.4		
Mexico	3.3		
Sweden	5.7		
United States	7.2		
World average	2.7	1.0 (2.7 ÷ 2.7)	1.5 (2.7 ÷ 1.8)
Your personal footprint			

Data from Global Footprint Network, in: WWF, 2014. *Living planet report 2014*. Gland, Switzerland: WWF International.

1. Why do you think the ecological footprint for people in Bangladesh is so small?

2. Why do you think the ecological footprint is so large for people in the United States?

3. Based on the data in the table, how do you think average per capita income is related to ecological footprints? Name some ways in which you believe a wealthy society can decrease its ecological footprint.

4. Go to an online footprint calculator such as the one at http://www.footprintnetwork.org/en/index.php/GFN/

page/personal_footprint and take the test to determine your own personal ecological footprint. Enter the value you obtain in the table and calculate the other values as you did for each nation. How does your footprint compare to that of the average person in the United States? How does it compare to that of people from other nations? Name three actions you could take to reduce your footprint. (*Note*: Save this number—you will calculate your footprint again in Chapter 24 at the end of your course!)

MasteringEnvironmentalScience®

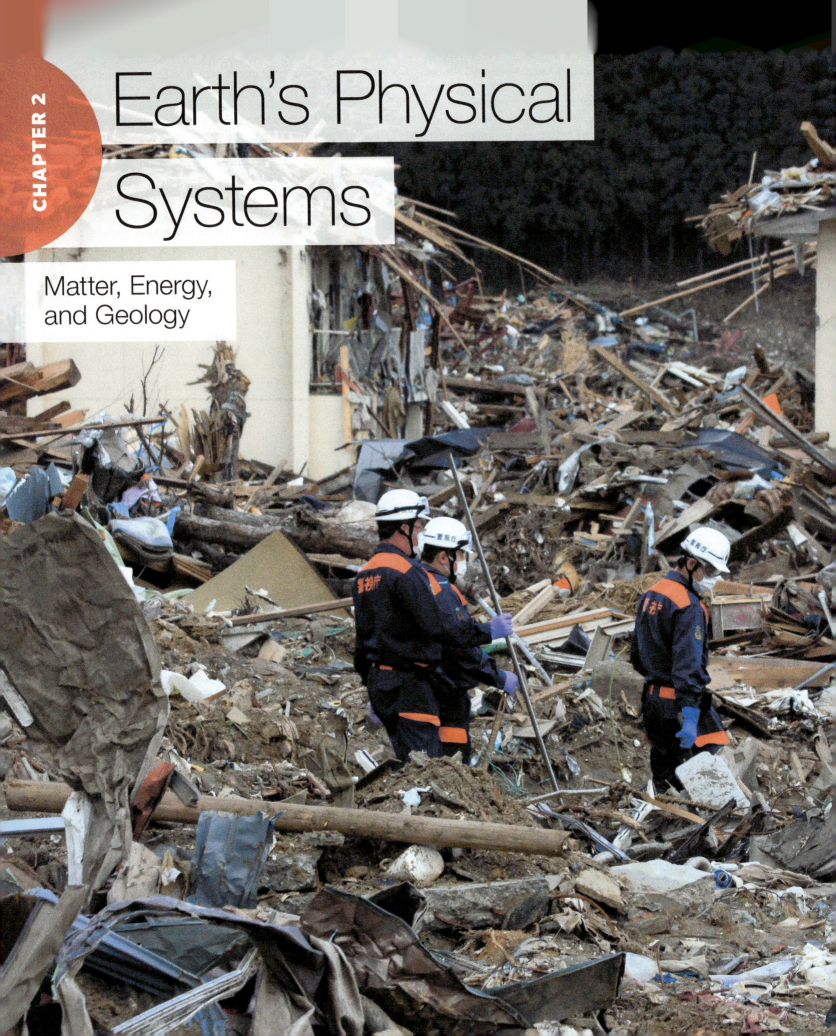

Earth's Physical Systems

Matter, Energy, and Geology

Rescue workers in northeastern Japan

Upon completing this chapter, you will be able to:

- Explain the fundamentals of matter and chemistry and apply them to real-world situations

- Differentiate among forms of energy and explain the first and second laws of thermodynamics

- Distinguish photosynthesis, cellular respiration, and chemosynthesis and summarize their importance to living things

- Explain how plate tectonics and the rock cycle shape the landscape around us

- Identify major types of geologic hazards and describe ways to minimize their impacts

Fukushima
Daiichi

JAPAN

The Tohoku Earthquake:
Has It Shaken the World's Trust in Nuclear Power?

> **This used to be one of the best places for a business. I'm amazed at how little is left.**
>
> Takahiro Chiba, surveying the devastated downtown area of Ishinomaki, Japan, where his family's sushi restaurant was located

> **Fukushima should not just contain lessons for Japan, but for all 31 countries with nuclear power.**
>
> Tatsujiro Suzuki, Vice-chairman, Japan Atomic Energy Commission

At 2:46 p.m. on March 11, 2011, the land along the northeastern coast of the Japanese island of Honshu began to shake violently—and continued to shake for six minutes. These tremors were caused when a large section of the seafloor along a fault line 125 km (77 mi) offshore suddenly lurched, releasing huge amounts of energy through the crust and generating an earthquake of magnitude 9.0 on the Richter scale (a scale used to measure the strength of earthquakes).

The Tohoku earthquake, as it was later named, was not the first major earthquake to strike Japan. The city of Kobe experienced substantial damage from a quake in 1995 that claimed more than 5500 lives. And in 1923, an earthquake devastated the cities of Tokyo and Yokohama, resulting in more than 142,000 deaths. Losses of life and property from the Tohoku quake were far less extensive than the losses from these earlier events, thanks to new stringent building codes that enable buildings to resist crumbling and toppling over during earthquakes. But even when the earth stopped shaking, the residents of northeastern Japan knew that further danger might still await them—from a tsunami.

A **tsunami** ("harbor wave" in English) is a powerful surge of seawater generated when an offshore earthquake displaces large volumes of rocks and sediment on the ocean bottom, suddenly pushing the overlying ocean water upward. This upward movement of water creates waves that speed outward from the earthquake site in all directions. These waves are hardly noticeable at sea, but can rear up to staggering heights when they enter the shallow waters near shore and can sweep inland with great force. The Japanese had built seawalls to protect against tsunamis, but the Tohoku quake caused the island of Honshu to sink perceptibly, thereby lowering the height of the seawalls by up to 2 m (6.5 ft) in some locations. Waves reaching up to 15 m (49 ft) in height then overwhelmed these defenses (**FIGURE 2.1**). The raging water swept up to 9.6 km (6 mi) inland; scoured buildings from their foundations; and inundated towns, villages, and productive agricultural land. As the water's energy faded, the water receded, carrying structural debris, vehicles, livestock, and human bodies out to sea.

When the tsunami overtopped the 5.7-m (19-ft) seawall protecting the Fukushima Daiichi nuclear power plant, it

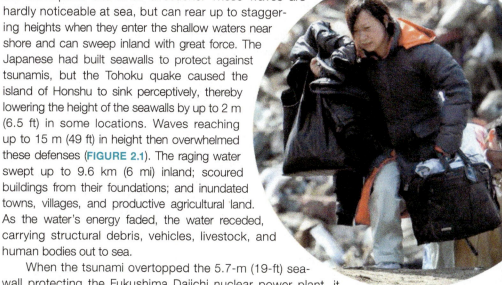

One of the 300,000 people displaced by the Tohoku quake ▲

21

FIGURE 2.1 Tsunami waves overtop a seawall following the Tohoku earthquake in 2011. The tsunami caused a greater loss of life and property than the earthquake that generated it and led to a meltdown at the Fukushima Daiichi nuclear power plant.

flooded the diesel-powered emergency generators responsible for circulating water to cool the plant's nuclear reactors. With the local electrical grid knocked out by the earthquake and the backup generators off-line, the nuclear fuel in the cores of the three active reactors at the plant began to overheat. The water that normally kept the nuclear fuel submerged within the reactor cores boiled off, exposing the nuclear material to the air and further elevating temperatures inside the cores. As the overheated nuclear fuel melted (an event called a nuclear meltdown), chemical reactions within the reactors generated hydrogen gas, which set off explosions in each of the three reactor buildings, releasing radioactive material into the air. To prevent a full-blown catastrophe that could render large portions of their nation uninhabitable, Japanese authorities flooded the reactor cores with seawater pumped in from the ocean.

The 1–2–3 punch of the earthquake–tsunami–nuclear accident left 18,000 people dead and caused hundreds of billions of dollars in material damage. Around 340,000 people were displaced from their homes, and a 20-km (12-mi) area around the Fukushima Daiichi plant has been permanently evacuated due to unsafe levels of radioactive fallout in the soil. Some of the greatest concerns center on contaminated food and water, so crops and seafood from the region will require testing for radiation for many years to come (see **THE SCIENCE BEHIND THE STORY**, pp. 24–25).

After the accident, public opposition to nuclear power in Japan ran high, and the government ordered the immediate shutdown and reinspection of its 48 nuclear reactors. At the time of the earthquake, these reactors supplied 30% of the nation's electricity.

Remediation efforts at the Fukushima Daiichi nuclear power plant are expected to take decades and cost around $17 billion, and fears of radiation poisoning in the region will linger for generations. Given its global implications for the future of energy production, the events of March 11, 2011, will be keenly recalled in all corners of our world.

Matter, Chemistry, and the Environment

The tragic events in northeastern Japan were the result of large-scale forces generated by the powerful geologic processes that shape the surface of our planet. Environmental scientists regularly study these types of processes to understand how our planet works. Because all large-scale processes are made up of small-scale components, however, environmental science— the broadest of scientific fields—must also study small-scale phenomena. At the smallest scale, an understanding of matter itself helps us to fully appreciate all the processes of our world.

All material in the universe that has mass and occupies space—solid, liquid, and gas alike—is called **matter.** In our quick tour of matter in the pages that follow, we examine types of matter and some of the important ways they interact—phenomena that together we term **chemistry.** Once you examine any environmental issue, from acid rain to toxic chemicals to climate change, you will likely discover chemistry playing a central role.

Matter is conserved

To appreciate the chemistry involved in environmental science, we must begin with the fundamentals. Matter may be transformed from one type of substance into others, but it cannot be created or destroyed. This principle is referred to as the **law of conservation of matter.** In environmental science, this principle helps us understand that the amount of matter stays constant as it is recycled in ecosystems and nutrient cycles (p. 117). The law of conservation of matter also makes it clear that we cannot simply wish away "undesirable" matter, such as nuclear waste and toxic pollutants. Because harmful substances can't be destroyed, we must take steps to minimize their impacts on the environment.

Atoms and elements are chemical building blocks

The nuclear reactors at Fukushima Daiichi used the element **uranium** to power its reactors. An **element** is a fundamental type of matter, a chemical substance with a given set of properties that cannot be broken down into substances with other properties. Chemists currently recognize 92 elements occurring in nature, as well as more than 20 others that they have created in the lab. Elements especially abundant on our planet include **oxygen, hydrogen, silicon, nitrogen,** and **carbon** (TABLE 2.1). Each element is assigned an abbreviation, or chemical symbol (for instance, H for hydrogen and O for oxygen). The periodic table of the elements (see **APPENDIX D**) organizes the elements according to their chemical properties and behavior.

TABLE 2.1 Earth's Most Abundant Chemical Elements, by Mass

EARTH'S CRUST		OCEANS		AIR		ORGANISMS	
Oxygen (O)	49.5%	Oxygen (O)	88.3%	Nitrogen (N)	78.1%	Oxygen (O)	65.0%
Silicon (Si)	25.7%	Hydrogen (H)	11.0%	Oxygen (O)	21.0%	Carbon (C)	18.5%
Aluminum (Al)	7.4%	Chlorine (Cl)	1.9%	Argon (Ar)	0.9%	Hydrogen (H)	9.5%
Iron (Fe)	4.7%	Sodium (Na)	1.1%	Other	<0.1%	Nitrogen (N)	3.3%
Calcium (Ca)	3.6%	Magnesium (Mg)	0.1%			Calcium (Ca)	1.5%
Sodium (Na)	2.8%	Sulfur (S)	0.1%			Phosphorus (P)	1.0%
Potassium (K)	2.6%	Calcium (Ca)	<0.1%			Potassium (K)	0.4%
Magnesium (Mg)	2.1%	Potassium (K)	<0.1%			Sulfur (S)	0.3%
Other	1.6%	Bromine (Br)	<0.1%			Other	0.5%

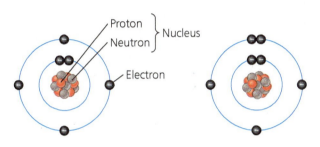
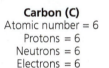
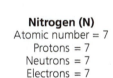
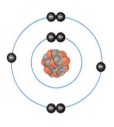

Carbon (C)
Atomic number = 6
Protons = 6
Neutrons = 6
Electrons = 6

Nitrogen (N)
Atomic number = 7
Protons = 7
Neutrons = 7
Electrons = 7

Oxygen (O)
Atomic number = 8
Protons = 8
Neutrons = 8
Electrons = 8

FIGURE 2.2 In an atom, protons and neutrons stay in the nucleus, and electrons move about the nucleus. Each chemical element has its own particular number of protons. Carbon possesses six protons, nitrogen seven, and oxygen eight. These schematic diagrams are meant to clearly show and compare numbers of electrons for these three elements. In reality, however, electrons do *not* orbit the nucleus in rings as shown; they move through space in more complex ways.

Atoms are the smallest units that maintain the chemical properties of the element. Atoms of each element contain a specific number of **protons** (positively charged particles) in the atom's nucleus (its dense center), and this number is called the element's atomic number. (Elemental carbon, for instance, has six protons in its nucleus; thus, its atomic number is 6.) Most atoms also contain **neutrons** (particles lacking electrical charge) in their nuclei, and an element's mass number denotes the combined number of protons and neutrons in the atom. An atom's nucleus is surrounded by negatively charged particles known as **electrons,** which balance the positive charge of the protons (**FIGURE 2.2**).

Isotopes Although all atoms of a given element contain the same number of protons, they do not necessarily contain the same number of neutrons. Atoms of the same element with differing numbers of neutrons are referred to as **isotopes** (**FIGURE 2.3a**). Thus, isotopes of a given element have the same atomic number but different mass numbers. Isotopes are denoted by their elemental symbol preceded by the mass number. For example, ^{14}C (carbon-14) is an isotope of carbon with eight neutrons (and six protons) in the nucleus rather than the six neutrons (and six protons) of ^{12}C (carbon-12), the most abundant carbon isotope.

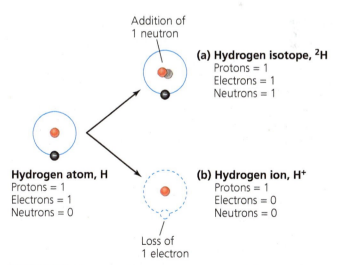

Addition of 1 neutron

(a) Hydrogen isotope, ^2H
Protons = 1
Electrons = 1
Neutrons = 1

Hydrogen atom, H
Protons = 1
Electrons = 1
Neutrons = 0

(b) Hydrogen ion, H⁺
Protons = 1
Electrons = 0
Neutrons = 0

Loss of 1 electron

FIGURE 2.3 Hydrogen has a mass number of 1 because a typical atom of this element contains one proton and no neutrons. Deuterium (hydrogen-2 or ^2H), an isotope of hydrogen **(a)**, contains a neutron as well as a proton and thus has greater mass than a typical hydrogen atom; its mass number is 2. The hydrogen ion, H⁺ **(b)**, occurs when an electron is lost; it therefore has a positive charge.

What Is Fukushima's Nuclear Legacy in the Pacific Ocean?

**Dr. Ken Buesseler,
Woods Hole
Oceanographic Institution**

When Dr. Ken Buesseler, a senior scientist at the Woods Hole Oceanographic Institution in Massachusetts, observed the events unfolding in northeastern Japan in 2011, he saw an opportunity to use his expertise to determine the movements and fate of the radioisotopes released into the Pacific Ocean from the crippled Fukushima Daiichi nuclear power plant. Quick action was needed, so Buesseler scrambled to assemble a team of scientists and secure resources to mount an expedition to sample the waters offshore from Fukushima before the radioisotopes dispersed throughout the Pacific. Dr. Buesseler had done this sort of research before when he used radioisotopes released from the Chernobyl nuclear accident (Chapter 20) in 1986 as tracers to study the poorly understood currents in the nearby Black Sea.

Scientists equate the input of radioisotopes into waters to pouring dye into the ocean. By knowing the half-lives of relevant radioisotopes and their daughter isotopes and by tracking their quantities across bodies of water, scientists can study large-scale circulation patterns in inland seas and the ocean.

Thanks to a $3.7 million award from the Gordon and Betty Moore Foundation, Buesseler and his colleagues were able to charter the research vessel *Ka'imikai-o-Kanaloa* from the University of Hawai'i and zigzag their way through the waters of the north Pacific in June 2011—starting 640 km (400 mi) off the coast of Japan and working their way in to 32 km (20 mi)

offshore. At 30 sites along the cruise route, the scientists tested the air and water for radioactivity, collected water samples to measure levels of several radioisotopes (**FIGURE 1**), and tested plankton and other free-swimming organisms to determine the extent of radioisotope uptake.

Buesseler's group found water radiation levels of more than 100,000 becquerels per cubic meter (Bq/m³; a becquerel is a unit of measurement for radioactivity) in early April, up from a pre-accident level of about 1.5 becquerels per cubic meter. These post-accident radiation levels were about 100 times greater than those found in the Black Sea after the Chernobyl incident (**FIGURE 2**). So, although the release of radioisotopes from Chernobyl was roughly five times larger than that from Fukushima, the impact on the ocean was far greater from Fukushima. Despite this, the team concluded that radioactivity levels

FIGURE 1 An international team of scientists tracked the fate of radioactive material released into the ocean from the Fukushima nuclear power plant.

Some isotopes, called **radioisotopes,** are **radioactive** and "decay" by changing their chemical identity as they shed subatomic particles and emit high-energy radiation. The radiation released by radioisotopes harms organisms because it focuses a great deal of energy in a very small area, which can be damaging to living cells. Intense radiation exposure can kill cells outright or cause changes to the cell's DNA that can increase the probability of the organism developing cancerous tumors later.

The greatest danger from radioisotopes occurs when they enter the bodies of organisms through the lungs, skin, or digestive system. It is therefore important after nuclear

accidents, like that at Fukushima, to regularly test food and water supplies for radioisotopes and to determine the eventual fate of radioactive particles released into the environment (see THE SCIENCE BEHIND THE STORY, above).

Radioisotopes decay into lighter and lighter radioisotopes until they become stable isotopes (isotopes that are not radioactive). Each radioisotope decays at a rate determined by that isotope's **half-life,** the amount of time it takes for one-half the atoms to give off radiation and decay. Different radioisotopes have very different half-lives, ranging from fractions of a second to billions of years. The radioisotope uranium-235 (^{235}U),

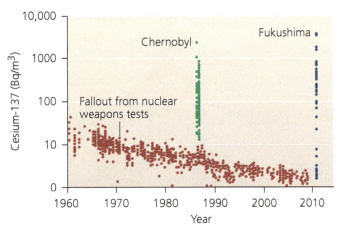

FIGURE 2 Ocean concentrations of radioactive cesium-137 show evidence of the nuclear accidents at Chernobyl and Fukushima. While Chernobyl released five times the radioactive material as Fukushima, the impacts on the ocean were greater from the Fukushima accident due to its coastal location. The EPA standard for safe drinking water is 10,000 Bq/m³. Note the *y* axis is logarithmic, such that each unit is 10 times greater than the previous unit. *Source: Ken Buesseler, Woods Hole Oceanographic Institution.*

DATA Q What was the trend in ocean radioactivity levels from cesium-137 from 1960 to 2010? Did releases from the Chernobyl accident significantly alter this trend? What long-term trends would you therefore expect to see following the releases from Fukushima in 2011?

Go to **Interpreting Graphs & Data** on **Mastering**EnvironmentalScience®

in waters in their sampling areas did not pose an immediate threat to humans or other free-swimming organisms.

The study, published in the *Proceedings of the National Academy of Sciences* in 2012, found that prevailing currents in the area had a major effect on the distribution of radioactive material originating from Fukushima. The Kuroshio current, which flows from south to north along the eastern coast of Japan, helped carry radioactive material quickly away from the coastline and out to sea, but the current also concentrated

radioactivity in some areas near shore. Buoys deployed by the expedition revealed that the current created an eddy—a mass of swirling water—near the coast of Fukushima that mixed minimally with waters in the Kuroshio current. This served to concentrate the water near shore, likely creating a "hotspot" of radioactivity in coastal waters.

Another finding was that radioactivity levels in the water remained high many months after the accident. This observation suggested radioactive material was still leaking from the plant into the ocean and/or radioactively contaminated groundwater might have been feeding into the ocean, a hypothesis verified by Japanese authorities in 2013. Due to the relatively long half-life of cesium radioisotopes (half-life of 30 years), radioactive material could accumulate in offshore sediments and be ingested by species that live on the ocean bottom or that filter water for food. Shellfish, bottom-dwelling fish, and other bottom-dwelling creatures are therefore likely to experience greater exposure to radioactive compounds, and for a longer duration, than species living in the open water. Sediments in areas where radiation was concentrated are likely to remain contaminated for several decades, posing a long-term threat to humans consuming seafood from these areas. Accordingly, as of 2016, a ban on fishing for 32 species remains in place in the waters off Fukushima, including many species of groundfish (p. 415) such as Pacific cod, flounder, and halibut.

Many questions remain about the long-term impacts of this accident, such as how radioactive contaminants will move through aquatic food webs. Scientists are particularly concerned about contamination of coastal waters by the radioisotope strontium-90 leaking from storage tanks at the Fukushima Daiichi site. Although radioactive cesium can be excreted by marine organisms after uptake, strontium mimics the element calcium in biochemical pathways, and so is stored in bone in marine life. This poses risks for human consumption, especially for small fish, such as sardines, that are eaten whole.

Accordingly, continuous monitoring of radioisotope concentrations in aquatic organisms, likely for many decades, will be necessary to determine threats to human health and to ecosystems.

used in commercial nuclear power plants like Fukushima Daiichi, decays into a series of daughter isotopes (in other words, the atoms that are formed as radioisotopes lose protons and neutrons during the process of radioactive decay), eventually forming lead-207 (^{207}Pb), and has a half-life of about 700 million years. Other radioisotopes released into the environment from the Fukushima nuclear power plant accident included iodine-131 (half-life of 8 days), cesium-134 (half-life of 2 years), strontium-90 (half-life of 29 years), and cesium-137 (half-life of 30 years), showing the long-term implications of nuclear accidents.

Ions Atoms may also gain or lose electrons, thereby becoming **ions,** electrically charged atoms or combinations of atoms (**FIGURE 2.3b**). Ions are denoted by their elemental symbol followed by their ionic charge. For instance, a common ion used by mussels and clams to form shells is Ca^{2+}, a calcium atom that has lost two electrons and thus has a charge of positive 2. The damaging radiation emitted by radioisotopes is called **ionizing radiation** (see Figure 2.10, p. 30) because it generates ions when it strikes molecules, and these ions affect the stability and functionality of biological molecules. Ionizing radiation can be put to positive use, though. We can

Thanks to comic books and movies, many people believe that when an organism is exposed to high-energy radiation from a source outside the body, such as the sun or nuclear waste, the organism becomes a *source* of ionizing radiation—that is, it becomes "radioactive." In reality, this does not happen.

An irradiated organism suffers damage from radiation, but it does not absorb the ionizing radiation, store it, and then re-emit it to the environment. The radiation simply enters the organism's cells, causes damage, and passes through the organism. So even after experiencing substantial impacts from radiation poisoning, the organism is no more radioactive than it was before exposure because it was only exposed to radiation (a form of energy) and was not contaminated with radioactive particles (a form of matter) that emit harmful radiation. Hence, when we irradiate raw meat, the process sterilizes it of microbes but the meat does not become radioactive.

irradiate raw meat to kill harmful microbes that may be lurking within it, such as *Salmonella* or disease-causing strains of *Escherichia coli (E. coli)*.

Atoms bond to form molecules and compounds

Atoms can bond together and form **molecules,** combinations of two or more atoms. Common molecules containing only a single element include those of hydrogen (H_2) and oxygen (O_2), each of which exists as a gas at room temperature. A molecule composed of atoms of two or more different elements is called a **compound.** One compound is **water;** it is composed of two hydrogen atoms bonded to one oxygen atom, and it is denoted by the chemical formula H_2O. Another compound is **carbon dioxide,** consisting of one carbon atom bonded to two oxygen atoms; its chemical formula is CO_2.

Atoms bond together because of an attraction for one another's electrons. Because the strength of this attraction varies among elements, atoms may be held together in different ways. When electrons are shared between atoms, a **covalent bond** forms. For example, two atoms of hydrogen share electrons equally as they bind together to form hydrogen gas, H_2. However, in a water molecule, oxygen attracts shared electrons more strongly than does hydrogen. The result is that water has a partial negative charge at its oxygen end and partial positive charges at its hydrogen ends. This arrangement allows water molecules to adhere to one another in a type of weakly attractive interaction called a **hydrogen bond** (**FIGURE 2.4**).

In compounds in which the strength of attraction is sufficiently unequal, an electron may be transferred from one atom to another. This creates oppositely charged ions that form **ionic bonds** due to their differing electrical charges. Such associations are called ionic compounds, or salts. Table salt (NaCl) contains ionic bonds between positively charged sodium ions (Na^+), each of which donates an electron, and negatively charged chloride ions (Cl^-), each of which receives an electron.

Elements, molecules, and compounds can also come together in mixtures without chemically bonding or reacting.

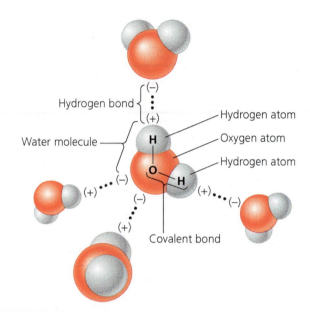

FIGURE 2.4 By enabling water molecules to adhere loosely to one another, hydrogen bonds give water several unique properties crucial for life.

Homogeneous mixtures of substances are called solutions, a term most often applied to liquids, but also applicable to some gases and solids. Air in the atmosphere is a solution formed of constituents such as nitrogen, oxygen, and water vapor. Other solutions include ocean water, plant sap, petroleum, and metal alloys such as brass.

Water's chemistry facilitates life

Water's ability to form loose connections of hydrogen bonds gives it several properties that help to support life and stabilize Earth's climate. Water is called the universal solvent because its partial electrical charges enable it to bond well with ions and other partially charged molecules. This allows water to hold in solution, or dissolve, many other molecules—including biologically important molecules in cells. Water is also an effective buffer to temperature change. Heating weakens hydrogen bonds before it speeds molecular motion, so water can absorb a great deal of heat with only small changes in its temperature. This capacity to resist change helps stabilize water bodies, organisms, and climate systems. In addition, water's properties protect aquatic life in cold climates. Water molecules in ice are farther apart than in liquid water, so ice is less dense. This enables ice to float on water, insulating lakes and ponds from frigid air and preventing them from freezing solid in winter.

Hydrogen ions determine acidity

In any aqueous solution, a small number of water molecules split apart, each forming a hydrogen ion (H^+) and a hydroxide ion (OH^-). The product of hydrogen and hydroxide ion concentrations is always the same; as one increases, the other decreases. Pure water contains equal numbers of these ions. Solutions in which the H^+ concentration is greater than the OH^- concentration are **acidic,** whereas solutions in which the

OH⁻ concentration exceeds the H⁺ concentration are **basic,** or alkaline.

The **pH** scale (**FIGURE 2.5**) quantifies the acidity or alkalinity of solutions. It runs from 0 to 14; pure water has a hydrogen ion concentration of 10^{-7} and a pH of 7. Solutions with a pH less than 7 are acidic, and those with a pH greater than 7 are basic. The pH scale is logarithmic, so each step on the scale represents a 10-fold difference in hydrogen ion concentration. Thus, a substance with a pH of 6 contains 10 times as many hydrogen ions as a substance with a pH of 7 and 100 times as many hydrogen ions as a substance with a pH of 8.

Most biological systems have a pH between 6 and 8, and substances that are strongly acidic (battery acid) or strongly basic (sodium hydroxide) are harmful to living things. Human activities can change the pH of water or soils and make conditions less amenable to life. Examples include the acidification of soils and water from acid rain (pp. 469–472) and from acidic mine drainage (p. 634).

Matter is composed of organic and inorganic compounds

Beyond their need for water, living things also depend on organic compounds. **Organic compounds** consist of carbon atoms (and generally hydrogen atoms) joined by covalent bonds, and they may also include other elements, such as nitrogen, oxygen, sulfur, and phosphorus. Inorganic compounds, in contrast, lack carbon–carbon bonds.

Carbon's unusual ability to bond together in chains, rings, and other structures to build elaborate molecules has resulted in millions of different organic compounds. One class of such compounds that is important in environmental science is

(a) Methane, CH_4

(b) Ethane, C_2H_6

(c) Naphthalene, $C_{10}H_8$

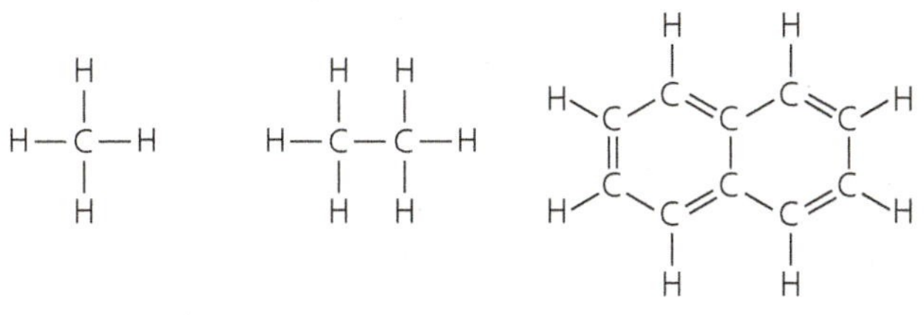

FIGURE 2.6 Hydrocarbons have a diversity of chemical structures. The simplest hydrocarbon is methane **(a)**. Many hydrocarbons consist of linear chains of carbon atoms with hydrogen atoms attached; the shortest of these is ethane **(b)**. The air pollutant naphthalene **(c)** is a ringed hydrocarbon.

hydrocarbons, which consist solely of bonded atoms of carbon and hydrogen (although other elements may enter these compounds as impurities) (**FIGURE 2.6**). Fossil fuels and the many petroleum products we make from them (Chapter 19) consist largely of hydrocarbons.

Some hydrocarbons from petroleum have become ubiquitous in our modern lifestyle because they are moldable into nearly any shape and resist chemical breakdown. Although these **plastics** have many benefits for consumer products, they can be a persistent source of pollution due to their longevity in the environment (pp. 428–429).

Macromolecules are building blocks of life

Just as carbon atoms in hydrocarbons may be strung together in chains, organic compounds sometimes combine to form long chains of repeated molecules. These chains are called **polymers.** There are three types of polymers that are essential to life: proteins, nucleic acids, and carbohydrates. Along with lipids (which are not polymers), these types of molecules are referred to as **macromolecules** because of their large sizes.

Proteins consist of long chains of organic molecules called amino acids. The many types of proteins serve various functions. Some help produce tissues and provide structural support; for example, animals use proteins to generate skin, hair, muscles, and tendons. Some proteins help store energy, whereas others transport substances. Some act in the immune system to defend the organism against foreign attackers. Still others are hormones, molecules that act as chemical messengers within an organism. Proteins can also serve as enzymes, molecules that catalyze, or promote, certain chemical reactions.

Nucleic acids direct the production of proteins. The two nucleic acids—**deoxyribonucleic acid (DNA)** and **ribonucleic acid (RNA)**—carry the hereditary information for organisms and are responsible for passing traits from parents to offspring. Nucleic acids are composed of a series of nucleotides, each of which contains a sugar molecule, a phosphate group, and a nitrogenous base. DNA contains four types of nucleotides and can be pictured as a ladder twisted

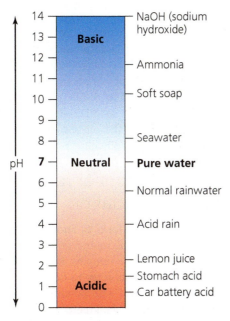

pH		
14	**Basic**	NaOH (sodium hydroxide)
13		
12		Ammonia
11		
10		Soft soap
9		
8		Seawater
7	**Neutral**	**Pure water**
6		Normal rainwater
5		
4		Acid rain
3		
2		Lemon juice
1	**Acidic**	Stomach acid
0		Car battery acid

FIGURE 2.5 The pH scale measures how acidic or basic (alkaline) a solution is. The pH of pure water is 7, the midpoint of the scale. Acidic solutions have a pH less than 7, whereas basic solutions have a pH greater than 7.

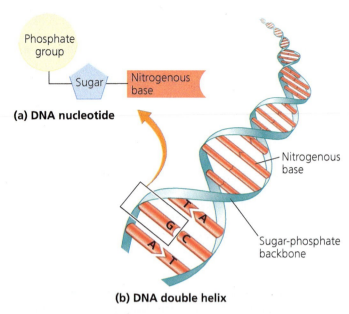

(a) DNA nucleotide

Phosphate group

Sugar

Nitrogenous base

Nitrogenous base

Sugar-phosphate backbone

(b) DNA double helix

FIGURE 2.7 Nucleic acids encode genetic information in the sequence of nucleotides, small molecules that pair together like rungs of a ladder. DNA includes four types of nucleotides **(a)**, each with a different nitrogenous base: adenine (A), guanine (G), cytosine (C), and thymine (T). Adenine (A) pairs with thymine (T), and cytosine (C) pairs with guanine (G). In RNA, thymine is replaced by uracil (U). DNA **(b)** twists into the shape of a double helix.

into a spiral, giving the molecule a shape called a double helix (**FIGURE 2.7**). Regions of DNA coding for particular proteins that perform particular functions are called **genes.**

Carbohydrates include simple sugars that are three to seven carbon atoms long. Glucose ($C_6H_{12}O_6$) fuels living cells and serves as a building block for complex carbohydrates, such as starch. Plants use starch to store energy, and animals eat plants to acquire starch. Plants and animals also use complex carbohydrates to build structure. Insects and crustaceans form hard shells from the carbohydrate chitin. Cellulose, the most abundant organic compound on Earth, is a complex carbohydrate found in the cell walls of leaves, bark, stems, and roots.

Lipids include fats and oils (for energy storage), phospholipids (for cell membranes), waxes (for structure), and steroids (for hormone production). Although chemically diverse, these compounds are grouped together because they do not dissolve in water.

Energy: An Introduction

Creating and maintaining organized complexity—of a cell, an organism, or an ecological system—requires energy. Energy is needed to organize matter into complex forms, to build and maintain cellular structure, to govern species' interactions, and to drive the geologic forces that shape our planet. Energy is involved in nearly every chemical, biological, and physical phenomenon.

But what is energy? **Energy** is the capacity to change the position, physical composition, or temperature of matter—in other words, a force that can accomplish **work** (when a force

acts on an object, causing it to move). As we saw with Japan's 2011 earthquake and tsunami, geologic events involve some of the most dramatic releases of energy in nature. Energy is omnipresent in living things as well. A sparrow in flight expends energy to propel its body through the air (change of position). When the sparrow lays an egg, its body uses energy to create the calcium-based eggshell and color it with pigment (change in composition). The sparrow sitting on its nest transfers energy from its body to heat the developing chicks inside its eggs (change of temperature).

Energy comes in different forms

Energy manifests itself in different ways and can be converted from one form to another. Two major forms of energy that scientists commonly distinguish are **potential energy,** energy of position or composition; and **kinetic energy,** energy of motion. Consider river water held behind a dam. By preventing water from moving downstream, the dam causes the water to accumulate potential energy. When the dam gates are opened, the potential energy is converted to kinetic energy as the water rushes downstream.

Energy conversions take place at the atomic level every time a chemical bond is broken or formed. Chemical energy is essentially potential energy stored in the bonds among atoms. Bonds differ in their amounts of chemical energy, depending on the atoms they hold together. Converting molecules with high-energy bonds (such as the carbon–carbon bonds of fossil fuels) into molecules with lower-energy bonds (such as the bonds in water or carbon dioxide) releases energy and produces motion, action, or heat. Just as automobile engines split the hydrocarbons of gasoline to release chemical energy and generate movement, our bodies split glucose molecules in our food for the same purpose (**FIGURE 2.8**).

Besides occurring as chemical energy, potential energy can occur as nuclear energy, the energy that holds atomic nuclei together. Nuclear power plants use this energy when they break apart the nuclei of large atoms within their reactors. Mechanical energy, such as the energy stored in a compressed spring, is yet another type of potential energy. Kinetic energy can also express itself in different forms, including thermal energy, light energy, sound energy, and electrical energy—all of which involve the movement of atoms, subatomic particles, molecules, or objects.

Energy is always conserved, but it changes in quality

Energy can change from one form to another, but it cannot be created or destroyed. Just as matter is conserved (p. 22), the total energy in the universe remains constant and thus is said to be conserved. Scientists refer to this principle as the **first law of thermodynamics.** The potential energy of the water behind a dam will equal the kinetic energy of its eventual movement downstream. Likewise, we obtain energy from the food we eat and then expend it in exercise, apply it toward maintaining our body and all its functions, or store it in fat. We do not somehow create additional energy or end up with

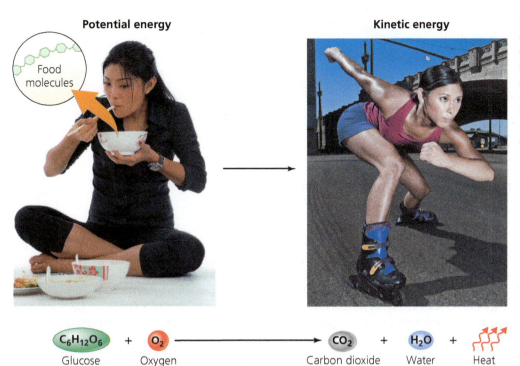

Potential energy

Food molecules

Kinetic energy

FIGURE 2.8 **Energy is released when potential energy is converted to kinetic energy.** Potential energy stored in sugars (such as glucose) in the food we eat, combined with oxygen, becomes kinetic energy when we exercise, releasing carbon dioxide, water, and heat as by-products.

$C_6H_{12}O_6$ + O_2 → CO_2 + H_2O + Heat
Glucose + Oxygen → Carbon dioxide + Water + Heat

less energy than the food gives us. Any particular system in nature can temporarily increase or decrease in energy, but the total amount in the universe always remains constant.

Although the overall amount of energy is conserved in any conversion of energy, the **second law of thermodynamics** states that the nature of energy will change from a more-ordered state to a less-ordered state as long as no force counteracts this tendency. That is, systems tend to move toward increasing disorder, or entropy. For instance, a log of firewood—the highly organized and structurally complex product of many years of tree growth—transforms in the campfire to a residue of carbon ash, smoke, and gases such as carbon dioxide and water vapor, as well as the light and the heat of the flame (**FIGURE 2.9**). With the help of oxygen, the complex biological polymers that make up the wood are converted into a disorganized assortment of rudimentary molecules and heat and light energy. When energy transforms from a more-ordered state to a less-ordered state, it cannot accomplish tasks as efficiently. For example, the potential energy available in ash (a less-ordered state of wood) is far lower than that available in a log of firewood (the more-ordered state of wood).

The second law of thermodynamics specifies that systems tend to move toward disorder, so then how does any system ever hold together? The order of an object or system can be increased by the input of energy from outside the system. Living organisms, for example, maintain their highly ordered structure by consuming energy. When they die and these inputs of energy cease, the organisms undergo decomposition and attain a less-ordered state.

Some energy sources are easier to harness than others

The nature of an energy source helps determine how easily people can harness it. Sources such as fossil fuels and the electricity we produce in power plants contain concentrated energy that we can readily release. It is relatively easy for us to gain large amounts of energy efficiently from such sources. In contrast, sunlight and the heat stored in ocean water are more diffuse. Each day the world's oceans absorb heat energy from the sun equivalent to that of 250 billion barrels of oil—more than 3000 times as much as our global society uses in a

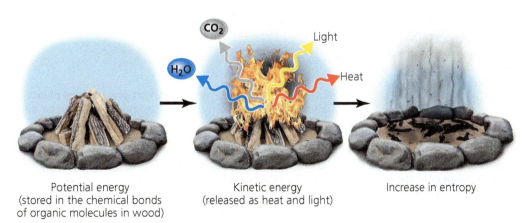

CO_2

Light

H_2O

Heat

Potential energy
(stored in the chemical bonds of organic molecules in wood)

Kinetic energy
(released as heat and light)

Increase in entropy

FIGURE 2.9 **The burning of firewood demonstrates energy conversion from a more-ordered to a less-ordered state.** This increase in entropy reflects the second law of thermodynamics.

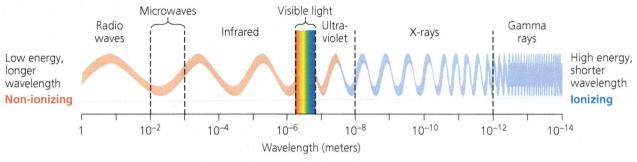

FIGURE 2.10 The sun emits radiation from many portions of the electromagnetic spectrum. Visible light makes up only a small proportion of this energy.

year. However, because this energy is spread across such vast areas, it is difficult to harness.

In each attempt we make to harness energy, some portion escapes. We can express our degree of success in capturing energy in terms of the **energy conversion efficiency,** the ratio of the useful output of energy to the amount we need to input. When we burn gasoline in an automobile engine, only about 16% of the energy released is used to power the automobile, and the rest of the energy is converted to heat and escapes without being used (p. 504).

FAQ

What part of the universe is offsetting all the complexity on Earth?

The second law of thermodynamics tells us that systems, such as our universe, tend to move toward increasing disorder, or entropy. But Earth is a highly ordered planet populated by huge numbers of highly ordered organisms. What part of the universe is offsetting this localized decrease in entropy to satisfy the second law?

It's our local star, the sun. The decrease in entropy in the universe that occurs on Earth is being offset by a larger *increase in entropy* in the sun. The sun provides our planet with energy inputs that resist entropy and maintain order. But in doing so, it consumes its nuclear fuel and becomes increasingly disordered over time. In several billion years, once the sun has exhausted its supplies of nuclear fuel, it will no longer be able to resist entropy. It will then burn out and attain a highly disordered state that will no longer be able to produce substantial amounts of energy.

Light energy from the sun powers most living systems

The energy that powers Earth's biological systems comes primarily from the sun. The sun releases radiation across large portions of the electromagnetic spectrum, although our atmosphere filters much of this out and we see only some of this radiation as visible light (**FIGURE 2.10**). Most of the sun's energy is reflected, or else absorbed and re-emitted, by the atmosphere, land, or water (p. 481). Solar energy drives winds, ocean currents, weather, and climate patterns. A small amount (less than 1% of the total) powers plant growth, and a still smaller amount flows from plants into the organisms that eat them and the organisms that decompose dead organic matter. A minuscule percentage of this energy is eventually deposited belowground in the chemical bonds in fossil fuels (which are derived from ancient plants; pp. 517–518).

Some organisms use the sun's radiation directly to produce their own food. Such organisms, called **autotrophs** or primary producers, include green plants, algae, and cyanobacteria. In the process of **photosynthesis** (**FIGURE 2.11**), autotrophs use sunlight to power a series of chemical reactions that transform molecules with lower-energy bonds—water and carbon dioxide—into sugar molecules with many high-energy bonds. Photosynthesis is an example of a process that moves toward a state of lower entropy, and so it requires a substantial input of outside energy, in this case from sunlight.

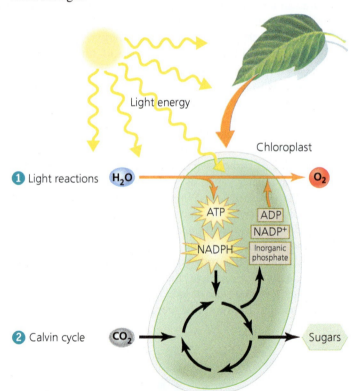

FIGURE 2.11 In photosynthesis, autotrophs such as plants, algae, and cyanobacteria use sunlight to convert water and carbon dioxide into oxygen and sugar. In the light reactions, water is converted to oxygen in the presence of sunlight, creating high-energy molecules (ATP and NADPH). These molecules help drive reactions in the Calvin cycle, in which carbon dioxide is used to produce sugars. Molecules of ADP, NADP+, and inorganic phosphate created in the Calvin cycle, in turn, help power the light reactions, creating an endless loop.

Photosynthesis produces food for plants and animals

Photosynthesis occurs within cell organelles called chloroplasts, where the light-absorbing pigment chlorophyll (which makes plants green) uses solar energy to initiate a series of chemical reactions called **1** light reactions. During these reactions, water molecules split and react to form hydrogen ions (H^+) and molecular oxygen (O_2), thus creating the oxygen that we breathe. The light reactions also produce small, high-energy molecules (ATP and NADPH) that are used to fuel reactions in the **2** Calvin cycle. In these reactions, carbon atoms from carbon dioxide are linked together to manufacture sugars.

Photosynthesis is a complex process, but the overall reaction can be summarized with the following equation:

$$6\,CO_2 + 12\,H_2O + \text{the sun's energy} \longrightarrow \underset{\text{(sugar)}}{C_6H_{12}O_6} + 6\,O_2 + 6\,H_2O$$

The number preceding each molecular formula indicates how many of those molecules are involved in the reaction. Note that the sums of the numbers on each side of the equation for each element are equal; that is, there are 6 C, 24 H, and 24 O atoms on each side. This illustrates how chemical equations are balanced, with each atom recycled and matter conserved. No atoms are lost; they are simply rearranged among molecules. Note also that water appears on both sides of the equation. The reason is that for every 12 water molecules that are input and split in the process, 6 water molecules are newly created. We can streamline the photosynthesis equation by showing only the net loss of 6 water molecules:

$$6\,CO_2 + 6\,H_2O + \text{the sun's energy} \longrightarrow \underset{\text{(sugar)}}{C_6H_{12}O_6} + 6\,O_2$$

Thus, in photosynthesis, water, carbon dioxide, and light energy from the sun are transformed to produce sugar (glucose) and oxygen. To accomplish this, green plants draw up water from the ground through their roots, absorb carbon dioxide from the air through their leaves, and harness the power of sunlight with chlorophyll. With these ingredients, green plants create sugars for their growth and maintenance, and they release oxygen as a by-product.

Animals, in turn, depend on the sugars and oxygen from photosynthesis. Animals survive by eating plants, or by eating animals that have eaten plants, and by taking in oxygen. In fact, animals appeared on Earth's surface only after the planet's atmosphere had been supplied with oxygen by cyanobacteria, the earliest autotrophs.

Cellular respiration releases chemical energy

Organisms make use of the chemical energy created by photosynthesis in a process called **cellular respiration,** which is vital to life. To release the chemical energy of glucose, cells use oxygen to convert glucose back into its original starting materials, water and carbon dioxide. The energy released during this process is used to power all of the biochemical reactions that sustain life. The net equation for cellular respiration is the exact opposite of that for photosynthesis:

$$\underset{\text{(sugar)}}{C_6H_{12}O_6} + 6\,O_2 \longrightarrow 6\,CO_2 + 6\,H_2O + \text{energy}$$

However, the energy released per glucose molecule in respiration is only two-thirds of the energy input per glucose molecule in photosynthesis—a prime example of the second law of thermodynamics. Cellular respiration is a continuous process occurring in all living things and is essential to life. Thus, it occurs in the autotrophs that create glucose and also in **heterotrophs,** organisms that gain their energy by feeding on other organisms. Heterotrophs include most animals, as well as the fungi and microbes that decompose organic matter.

Geothermal energy also powers Earth's systems

Although the sun is life's primary energy source, it is not the only source of energy for our planet. An additional, although minor, energy source is the gravitational pull of the moon, which in conjunction with the sun's gravitational pull causes ocean tides (p. 424). Another significant energy source is geothermal heating emanating from inside Earth, powered primarily by radioactivity (p. 24). Radiation from radioisotopes deep inside our planet heats the inner Earth, and this heat gradually makes its way to the surface. There it heats **magma** (rock heated to a molten, liquid state) that erupts from volcanoes, drives plate tectonics (p. 32), and heats groundwater. This heated groundwater can then be harnessed to produce electricity in some locations (pp. 595–597).

Geothermal energy also powers biological communities. On the ocean floor, jets of geothermally heated water gush into the icy-cold depths. These **hydrothermal vents** can host entire communities of specialized organisms that thrive in the extreme high-temperature, high-pressure conditions. Gigantic clams; immense tubeworms; and odd mussels, shrimps, crabs, and fish all flourish in the seemingly hostile environment of near-scalding water that shoots out of tall chimneys of encrusted minerals.

Hydrothermal vents are so deep underwater that they completely lack sunlight, so the energy flow of these communities cannot be fueled through photosynthesis. Instead, bacteria in deep-sea vents use the chemical-bond energy of hydrogen sulfide (H_2S) to transform inorganic carbon into organic carbon compounds in a process called **chemosynthesis,** defined by the following equation:

$$6\,CO_2 + 6\,H_2O + 3\,H_2S \longrightarrow \underset{\text{(sugar)}}{C_6H_{12}O_6} + 3\,H_2SO_4$$

Chemosynthesis occurs in various ways, but note how this particular reaction for chemosynthesis closely resembles the photosynthesis reaction. Energy from chemosynthesis passes through the deep-sea-vent animal community as heterotrophs such as clams, mussels, and shrimp gain nutrition from chemoautotrophic bacteria. When they were first discovered, hydrothermal vent communities excited scientists because they were novel and unexpected, and they showed just how much we still have to learn about the workings of our planet.

Geology: The Physical Basis for Environmental Science

A good way to understand how our planet functions is to examine the rocks, soil, and sediments beneath our feet. The physical processes that take place at and below Earth's surface shape the landscape and lay the foundation for most environmental systems and for life.

Understanding the physical nature of our planet also benefits our society, because without the study of Earth's rocks and the processes that shape them, we would have no metals for consumer products, no energy from fossil fuels, and no uranium for nuclear power plants. Our planet is dynamic, and this dynamism is what motivates **geology,** the study of Earth's physical features, processes, and history. A human lifetime is just a blink of an eye in the long course of geologic time, and the Earth we experience is merely a snapshot in our changing planet's long history. We can begin to grasp this long-term dynamism as we consider two processes of fundamental importance—plate tectonics and the rock cycle.

Earth consists of layers

Our planet consists of multiple layers (**FIGURE 2.12**). At Earth's center is a dense **core** consisting mostly of iron, solid in the inner core and molten in the outer core. Surrounding the core is a thick layer of less dense, elastic rock called the **mantle.** A portion of the upper mantle called the **asthenosphere** contains especially soft rock, melted in some areas. The harder rock above the asthenosphere is the **lithosphere.** The lithosphere includes both the uppermost mantle and the entirety of Earth's third major layer, the **crust,** the thin, brittle, low-density layer of rock that covers Earth's surface. The intense heat in the inner Earth rises from core to mantle to crust, and it eventually dissipates at the surface.

The heat from the inner layers of Earth also drives convection currents that flow in loops in the mantle, pushing the mantle's soft rock cyclically upward (as it warms) and downward (as it cools), like a gigantic conveyor belt system. As the mantle material moves, it drags large plates of lithosphere along its surface. This movement is known as **plate tectonics,** a process of extraordinary importance to our planet.

Plate tectonics shapes Earth's geography

Our planet's surface consists of about 15 major tectonic plates, which fit together like pieces of a jigsaw puzzle (**FIGURE 2.13**). Imagine peeling an orange and then placing the pieces back onto the fruit; the ragged pieces of peel are like the lithospheric plates riding atop Earth's surface. However, the plates are thinner relative to the planet's size, more like the skin of an apple. These plates move at rates of roughly 2–15 cm (1–6 in.) per year. This slow movement has influenced Earth's climate and life's evolution throughout our planet's history as the continents combined, separated, and recombined in various configurations. By studying ancient rock formations throughout the world, geologists have determined that at least twice, all landmasses were joined together in a "supercontinent." Scientists have dubbed the landmass that resulted about 225 million years ago Pangaea (see inset in Figure 2.13).

There are three types of plate boundaries

The processes that occur at each type of plate boundary all have major consequences.

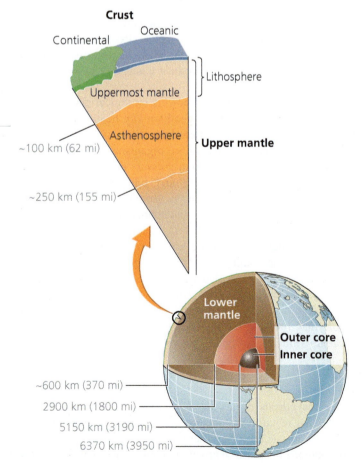

FIGURE 2.12 Earth's three primary layers—core, mantle, and crust—are themselves layered. The inner core of solid iron is surrounded by an outer core of molten iron, and the rocky mantle includes the molten asthenosphere near its upper edge. At Earth's surface, dense and thin oceanic crust abuts lighter, thicker continental crust. The lithosphere consists of the crust and the uppermost mantle above the asthenosphere.

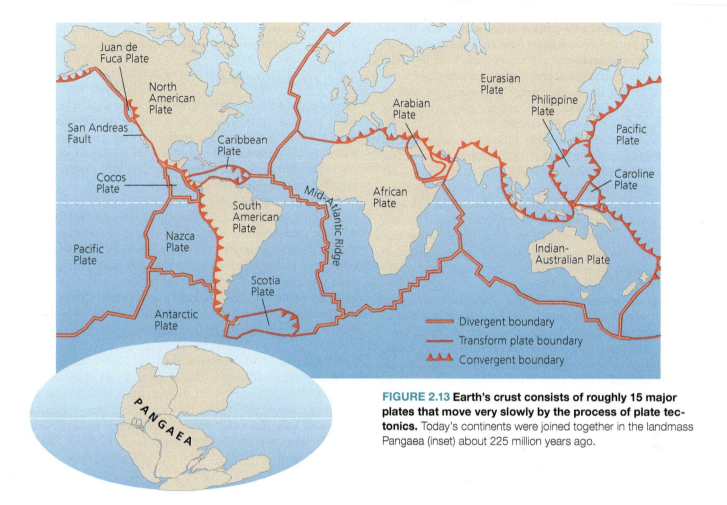

FIGURE 2.13 **Earth's crust consists of roughly 15 major plates that move very slowly by the process of plate tectonics.** Today's continents were joined together in the landmass Pangaea (inset) about 225 million years ago.

At **divergent plate boundaries,** tectonic plates push apart from one another as magma rises upward to the surface, creating new lithosphere as it cools (**FIGURE 2.14a**). An example is the Mid-Atlantic Ridge, part of a 74,000-km (46,000-mi) system of divergent plate boundaries slicing across the floors of the world's oceans.

Where two plates meet, they may slip and grind alongside one another, forming a **transform plate boundary** (**FIGURE 2.14b**). This movement creates friction that generates earthquakes (p. 37) along strike-slip faults. The Tohoku earthquake, for example, occurred at such a fault off the coast of Japan. Faults are fractures in Earth's crust, and at strike-slip faults, each landmass moves horizontally in opposite directions. The Pacific Plate and the North American Plate, for example, slide past one another along California's San Andreas Fault. Southern California is slowly inching its way northward along this fault, and so the site of Los Angeles will eventually reach that of modern-day San Francisco.

Convergent plate boundaries, where two plates come together, can give rise to different outcomes (**FIGURE 2.14c**). As plates of newly formed lithosphere push outward from divergent plate boundaries, this oceanic lithosphere gradually cools, becoming denser. After millions of years, it becomes denser than the asthenosphere beneath it and dives downward

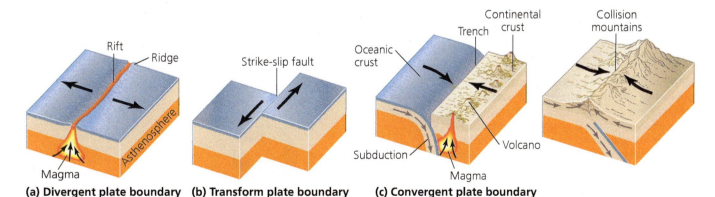

(a) Divergent plate boundary   (b) Transform plate boundary   (c) Convergent plate boundary

FIGURE 2.14 **There are three types of boundaries between tectonic plates, generating different geologic processes.**

into the asthenosphere in a process called **subduction.** As the lithospheric plate descends, it slides beneath a neighboring plate that is less dense, forming a convergent plate boundary. The subducted plate is heated and pressurized as it sinks, and water vapor escapes, helping to melt rock (by lowering its melting temperature). The molten rock rises, and this magma may erupt through the surface via volcanoes (p. 38).

When one plate of oceanic lithosphere is subducted beneath another plate of oceanic lithosphere, the resulting volcanism may form arcs of islands, such as Japan and the Aleutian Islands of Alaska. Subduction zones may also create deep trenches, such as the Mariana Trench, our planet's deepest abyss. When oceanic lithosphere slides beneath continental lithosphere, volcanic mountain ranges form that parallel coastlines (Figure 2.14c, left). An example is South America's Andes Mountains, where the Nazca Plate slides beneath the South American Plate.

When two plates of continental lithosphere meet, the continental crust on both sides resists subduction and instead crushes together, bending, buckling, and deforming layers of rock from both plates in a **continental collision** (Figure 2.14c, right). Portions of the accumulating masses of buckled crust are forced upward as they are pressed together, and mountain ranges result. The Himalayas, the world's highest mountains, result from the Indian-Australian Plate's collision with the Eurasian Plate beginning 40–50 million years ago, and these mountains are still rising today as these plates converge.

Tectonics produces Earth's landforms

Tectonic movements build mountains; shape the geography of oceans, islands, and continents; and give rise to earthquakes and volcanoes. The topography created by tectonic processes, in turn, shapes climate by altering patterns of rainfall, wind, ocean currents, heating, and cooling—all of which affect rates of weathering and erosion and the ability of plants and animals to inhabit different regions. Thus, the locations of biomes (pp. 92–98) are influenced by plate tectonics. Moreover, tectonics has affected the history of life's evolution; the convergence of landmasses into supercontinents such as Pangaea is thought to have contributed to widespread extinctions by reducing the area of

species-rich coastal regions and by creating an arid continental interior with extreme temperature swings.

The rock cycle alters rock

Just as plate tectonics shows geology's dynamism on a large scale, the rock cycle shows it on a smaller one. We tend to think of rock as pretty solid stuff. Yet over geologic time, rocks and the minerals that make them up are heated, melted, cooled, broken down, and reassembled in a very slow process called the **rock cycle** (FIGURE 2.15).

A **rock** is any solid aggregation of minerals. A **mineral** is any naturally occurring solid element or inorganic compound with a crystal structure, a specific chemical composition, and distinct physical properties. The type of rock in a given region affects soil characteristics and thereby influences the region's plant community. Understanding the rock cycle enables us to better appreciate the formation and conservation of soils, mineral resources, fossil fuels, groundwater sources, and other natural resources (all of which we discuss in later chapters).

Igneous rock All rocks can melt. At high enough temperatures, rock will enter the molten, liquid state called magma. If magma is released through the lithosphere (as in a volcanic eruption), it may flow or spatter across Earth's surface as **lava.** Rock that forms when magma or lava cools is called **igneous rock** (from the Latin *ignis*, meaning "fire") (Figure 2.15a).

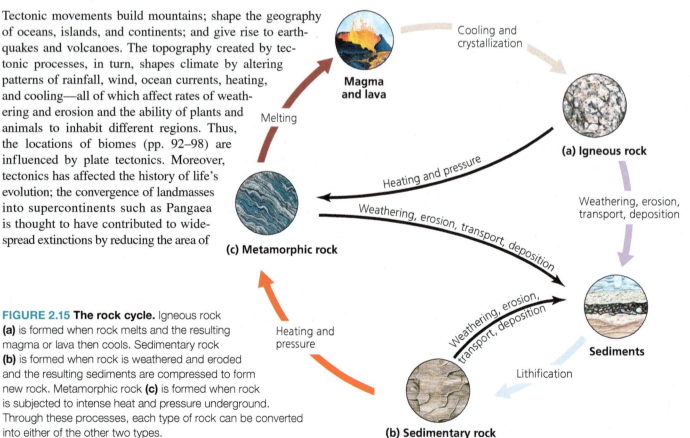

FIGURE 2.15 The rock cycle. Igneous rock **(a)** is formed when rock melts and the resulting magma or lava then cools. Sedimentary rock **(b)** is formed when rock is weathered and eroded and the resulting sediments are compressed to form new rock. Metamorphic rock **(c)** is formed when rock is subjected to intense heat and pressure underground. Through these processes, each type of rock can be converted into either of the other two types.

(a) Intrusive igneous rock: Granite at Yosemite National Park

(b) Extrusive igneous rock: Basalt in the Canary Islands

(c) Sedimentary rock: Sandstone in Arizona

(d) Metamorphic rock: Gneiss in Utah

FIGURE 2.16 Examples of rock types. The towering rock formations of Yosemite National Park are made of granite **(a)**, a type of intrusive igneous rock. The lava flows on the Canary Islands form basalt **(b)**, a type of extrusive igneous rock. The layered formation in Paria Canyon, Arizona, is an example of sandstone **(c)**, a type of sedimentary rock. Gneiss (pronounced "nice") **(d)**, at Antelope Island, Utah, is a type of metamorphic rock.

Igneous rock comes in two main classes because magma can solidify in different ways. When magma cools slowly and solidifies while it is below Earth's surface, it forms intrusive igneous rock. This process created the famous rock formations at Yosemite National Park (**FIGURE 2.16a**). Granite is the best-known type of intrusive rock. A slow cooling process allows minerals of different types to aggregate into large crystals, giving granite its multicolored, coarse-grained appearance. In contrast, when molten rock is ejected from a volcano, it cools quickly, so minerals have little time to grow into coarse crystals. This quickly cooled molten rock is classified as extrusive igneous rock, and its most common representative is basalt, the principal rock type of the Japanese islands (**FIGURE 2.16b**).

Sedimentary rock

All exposed rock weathers away with time. The relentless forces of wind, water, freezing, and thawing strips off one tiny grain (or large chunk) after another. Through weathering (p. 213) and erosion (pp. 221–222), particles of rock blown by wind or washed away by water come to rest downhill, downstream, or downwind from their sources, eventually forming **sediments.** Alternatively, some sediments form chemically from the precipitation of substances out of solution.

Sediment layers accumulate over time, causing the weight and pressure of overlying layers to increase. **Sedimentary rock** (Figure 2.15b) is formed as sediments are physically pressed together (compaction) and as dissolved minerals seep through sediments and act as a kind of glue, binding sediment particles together (cementation). The formation of rock through these processes is termed lithification. Examples of sedimentary rock include sandstone, made of cemented sand particles; shale, comprising still smaller mud particles; and limestone, formed as dissolved calcite precipitates from water or as calcite from marine organisms settles to the bottom.

These processes also create the fossils of organisms (p. 56) we use to learn about the history of life on Earth and the fossil fuels we use for energy. Because sedimentary layers pile up in chronological order (**FIGURE 2.16c**), scientists can assign relative dates to fossils they find in sedimentary rock.

Metamorphic rock Geologic forces may bend, uplift, compress, or stretch rock. When any type of rock is subjected to great heat or pressure, it may alter its form, becoming **metamorphic rock** (from the Greek for "changed form") (Figure 2.15c). The forces that metamorphose rock generally occur deep underground, at temperatures lower than the rock's melting point, but high enough to change its appearance and physical properties. Metamorphic rock (**FIGURE 2.16d**) includes rock such as slate, formed when shale is subjected to heat and pressure, and marble, formed when limestone is heated and pressurized.

Geologic processes occur across "deep time"

Geologic processes occur at timescales that are difficult to conceptualize. But it is only by appreciating the long time periods within which our planet's geologic forces operate that we can realize how exceedingly slow processes such as plate tectonics or the formation of sedimentary rock can reshape our planet. This lengthy timescale is referred to as deep time, or geologic time.

The geologic timescale (**APPENDIX E**) shows the full span of Earth's history—all 4.5 *billion* years of it—and focuses on the most recent 543 million years. Geologists have subdivided Earth's history into 3 eras and 11 periods. The Quaternary period, the most recent, occupies a thin slice of time at the top of the scale because this period began "only" 1.8 million years ago.

Geologists divide this immensely long timescale using evidence from stratigraphy, the study of strata, or layers, of sedimentary rock. Where scientists find fossil evidence for major and sudden changes in the physical, chemical, or biological conditions present on Earth between one set of layers and the next, they assign a boundary between geologic time periods. For instance, fossil evidence for mass extinctions (pp. 58, 278–282) determines several boundaries, such as that between the Permian and Triassic periods.

We live in the Holocene epoch, the most recent slice of the Quaternary period. The Holocene epoch began about 11,500 years ago with a warming trend that melted glaciers and brought Earth out of its most recent ice age. Since then, Earth's climate has been remarkably constant, and this constancy provided our species with the long-term stability we needed to develop agriculture and civilization. However, since the industrial revolution, human activity has had major impacts on Earth's basic processes, including a sharp increase in soil erosion from clearing forests and cultivating land; an alteration in the composition of the atmosphere through emitting greenhouse gases, which elevates Earth's average temperature; and a recent explosion in human population, which has intensified all impacts on Earth. All of these activities have set into motion a new mass extinction event (p. 282). These realizations have led some geologists in 2000 to propose naming a new geologic era, encompassing the past 200 years, after ourselves—the Anthropocene (**FIGURE 2.17**).

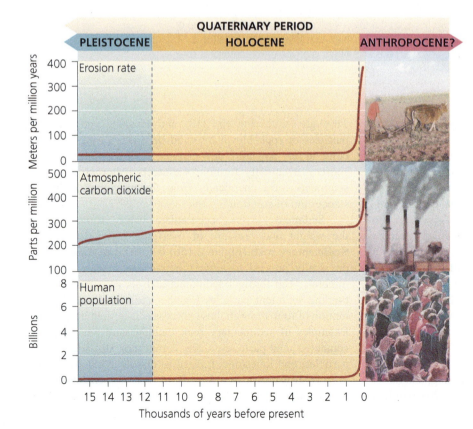

FIGURE 2.17 Global soil erosion rates (top) and atmospheric carbon dioxide concentrations (middle) have increased sharply in just the past few hundred years, along with human population (bottom). These patterns have persuaded some geologists that we should recognize a new epoch in Earth history: the Anthropocene. *Adapted from Zalasiewicz, J., et al., 2008. Are we now living in the Anthropocene? GSA Today 18(2): 4-8, Figure 1.*

Geologic and Natural Hazards

Although plate tectonics have shaped our planet, some tectonic movement can also pose hazards to us. Earthquakes and volcanoes are examples of geologic hazards. We can see how such hazards relate to tectonic processes by examining a map of the circum-Pacific belt, or "ring of fire" (FIGURE 2.18). Nine out of 10 earthquakes and more than half the world's volcanoes occur along this 40,000-km (25,000-mi) arc of subduction zones and fault systems. Like many locations along the circum-Pacific belt, Japan has experienced earthquakes and volcanism frequently in its past.

Earthquakes result from movement at plate boundaries and faults

Along tectonic plate boundaries, Earth may relieve built-up pressure in fits and starts. Each release of energy causes what we know as an **earthquake.** Most earthquakes are barely perceptible, but as shown by the Tohoku quake of 2011, they are occasionally powerful enough to cause significant losses of human life and property (TABLE 2.2). Earthquakes can also occur in the interior tectonic plates, when faults are formed by continental plates being stretched and pulled apart by geologic forces within the earth. Such earthquakes are not only rare but poorly understood. The New Madrid seismic zone, which lies beneath the lower Mississippi River basin in the central United States, is one area where such an "intraplate" earthquake may occur. And human activities

TABLE 2.2 Examples of Large Earthquakes

YEAR	LOCATION	FATALITIES	MAGNITUDE[1]
1556	Shaanxi Province, China	830,000	~8
1755	Lisbon, Portugal	70,000[2]	8.7
1906	San Francisco, California	3,000	7.8
1923	Kwanto, Japan	143,000	7.9
1964	Anchorage, Alaska	128[2]	9.2
1976	Tangshan, China	255,000+	7.5
1985	Michoacan, Mexico	9,500	8.0
1989	Loma Prieta, California	63	6.9
1994	Northridge, California	60	6.7
1995	Kobe, Japan	5,502	6.9
2004	Northern Sumatra	228,000[2]	9.1
2005	Kashmir, Pakistan	86,000	7.6
2008	Sichuan Province, China	50,000+	7.9
2010	Port-au-Prince, Haiti	236,000	7.0
2010	Maule, Chile	500	8.8
2011	Northern Japan	18,000[2]	9.0
2015	Kathmandu, Nepal	8,900	7.8

[1]Measured by moment magnitude; each full unit is roughly 32 times as powerful as the preceding full unit.
[2]Includes deaths from the resulting tsunami.

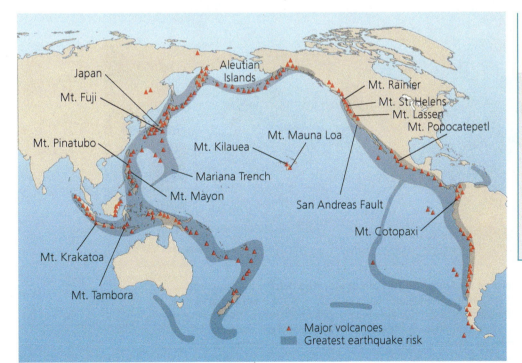

FIGURE 2.18 Most of our planet's volcanoes and earthquakes occur along the circum-Pacific belt, or "ring of fire."

DATA What similarities do you note between the "ring of fire" around the edges of the Pacific Ocean and the boundaries of the tectonic plates shown in Figure 2.13? Which type of plate boundary (see Figure 2.14) is most common along the length of the "ring of fire"?

Go to **Interpreting Graphs & Data** on **Mastering**EnvironmentalScience*

Japan
Mt. Fuji
Aleutian Islands
Mt. Rainier
Mt. St. Helens
Mt. Lassen
Mt. Popocatepetl
Mt. Pinatubo
Mt. Mauna Loa
Mt. Kilauea
Mariana Trench
Mt. Mayon
San Andreas Fault
Mt. Krakatoa
Mt. Cotopaxi
Mt. Tambora

▲ Major volcanoes
Greatest earthquake risk

may also be inducing earthquakes in areas far from the boundaries of tectonic plates (see **THE SCIENCE BEHIND THE STORY**, pp. 40–41).

To minimize damage from earthquakes, engineers have developed ways to protect buildings from shaking. They do this by strengthening structural components while also designing points at which a structure can move and sway harmlessly with ground motion. Just as a flexible tree trunk bends in a storm while a brittle one breaks, buildings with built-in flexibility are more likely to withstand an earthquake's violent shaking. Such designs continue to figure in the building codes used in California, Japan, and other quake-prone regions. Many portions of the United States face danger from earthquakes, so the areas that need to consider such codes are more widespread than most people imagine (**FIGURE 2.19**).

Such quake-resistant designs are more expensive to build than conventional designs, so many buildings in poorer nations do not have such protections. For example, Haiti suffered a 7.0 magnitude earthquake in 2010 that devastated huge portions of the capital city of Port-au-Prince and claimed an estimated 230,000 lives. Although the Tohoku earthquake released more than 950 times the energy than the earthquake that struck Haiti, mortality and property damage from the Tohoku quake (not including the damage and loss of life caused by the subsequent tsunami) were minimized thanks to Japan's earthquake-conscious building.

Volcanoes arise from rifts, subduction zones, or hotspots

Where molten rock, hot gas, or ash erupts through Earth's surface, a **volcano** is formed, often creating a mountain over time as cooled lava accumulates. As we have seen, lava can extrude along mid-ocean ridges or over subduction zones as one tectonic plate dives beneath another. Due to its position along subduction zones, Japan has more than 100 active volcanoes, which is 10% of the world total and more than any other nation. Mount Fuji, one of Japan's most prominent and recognizable natural features, is one such active volcano.

Lava may also be emitted at hotspots, localized areas where plugs of molten rock from the mantle erupt through the crust. As a tectonic plate moves across a hotspot, repeated eruptions from this source may create a linear series of volcanoes. The formation of the Hawaiian Islands over millennia provides an example of this process (**FIGURE 2.20a**). At some volcanoes, lava flows slowly downhill, such as at Mount Kilauea in Hawai'i (**FIGURE 2.20b**), which has been erupting continuously since 1983! At other times, a volcano may let loose large amounts of ash and cinder in a sudden explosion, such as during the 1980 eruption of Mount Saint Helens (Figure 17.11, p. 453). Sometimes a volcano can unleash a pyroclastic flow—a fast-moving cloud of toxic gas, ash, and rock fragments that races down the slopes at speeds up to 725 km/hr (450 mph), enveloping everything in its path. Such a flow

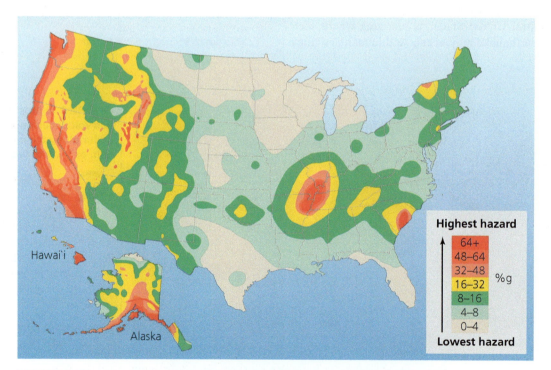

Hawai'i

Alaska

Highest hazard

%g
64+
48–64
32–48
16–32
8–16
4–8
0–4

Lowest hazard

FIGURE 2.19 Many parts of the United States are at elevated risk for earthquakes. The West Coast faces threats from earthquakes due to its position at the boundary of tectonic plates. Portions of the continental interior have elevated risk due to naturally occurring intraplate earthquakes or human-induced earthquakes, typically from wastewater injection or hydraulic fracturing. The units for the figure are %g, a measure of acceleration related to the force of gravity. *Source: U.S. Geological Survey.*

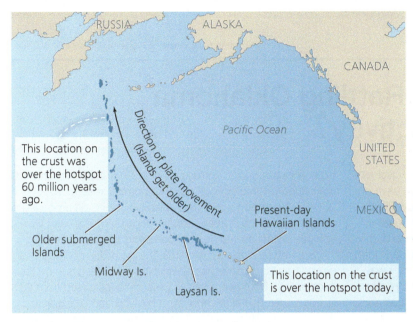

(a) Current and former Hawaiian Islands, formed as crust moves over a volcanic hotspot

(b) Mt. Kilauea erupting

FIGURE 2.20 The Hawaiian Islands are the product of a hotspot on Earth's mantle. The Hawaiian Islands **(a)** have been formed by repeated eruptions from a hotspot of magma in the mantle as the Pacific Plate passes over the hotspot. The Big Island of Hawai'i is most recently formed, and it is still volcanically active. The other islands are older and have already begun eroding away. To their northwest stretches a long series of former islands, now submerged. The active volcano Kilauea **(b)**, on the Big Island's southeast coast, is currently located above the edge of the hotspot.

buried the inhabitants of the ancient Roman cities of Pompeii and Herculaneum in A.D. 79, when Mount Vesuvius erupted.

Volcanic eruptions affect people as well as the environment (**TABLE 2.3**). Ash blocks sunlight, while sulfur emissions lead to a sulfuric acid haze that blocks radiation and cools the atmosphere. Large eruptions can actually depress temperatures throughout the world. When Indonesia's Mount Tambora erupted in 1815, it cooled average global

TABLE 2.3 Examples of Notable Volcanic Eruptions

YEAR	LOCATION	IMPACTS	MAGNITUDE[1]
640,000 B.P.[2]	Yellowstone Caldera, Wyoming, United States	Most recent "mega-eruption" at site of Yellowstone National Park	8
6870 B.P.	Mount Mazama, Oregon, United States	Created Crater Lake	7
A.D. 79	Mount Vesuvius, Italy	Buried Pompeii and Herculaneum	5
1815	Mount Tambora, Indonesia	Created "year without a summer"; killed at least 70,000 people	7
1883	Krakatau, Indonesia	Killed over 36,000 people; heard 5000 km (3000 mi) away; affected weather for 5 years	6
1980	Mount Saint Helens, Washington, United States	Blew top off mountain; sent ash 19 km (12 mi) into sky and into 11 U.S. states; 57 people killed	5
1983–present	Kilauea, Hawai'i, United States	Continuous lava flow	1
1991	Mount Pinatubo, Philippines	Sulfuric aerosols lowered world temperature 0.5°C (0.9°F)	6
2010	Eyjafjallajokull, Iceland	Ash cloud disrupted air travel throughout Europe	1

[1]Measured by the Volcanic Explosivity Index, which ranges from 0 (least powerful) to 8 (most powerful). The scale is logarithmic, such that each unit is 10 times greater than the previous unit.
[2]B.P. = years before the present.

Are the Earthquakes Rattling Oklahoma Caused by Human Activity?

Geophysicist Katie Keranen, Cornell University

In November 2011, a series of earthquakes and aftershocks struck the small town of Prague, Oklahoma (population 2300), roughly 100 km (60 mi) east of Oklahoma City. The shaking damaged 14 homes, buckled the pavement of a local highway, and caused several injuries. One of the tremors measured 5.7 on the Richter scale—the largest earthquake ever recorded in the state.

Earthquakes are uncommon in Oklahoma, especially ones of such magnitude. But scientists were not completely surprised by the 2011 event, because they had been noting an increase in earthquake activity (**FIGURE 1**). For example, between 1978 and 2008, Oklahoma

experienced a yearly average of only ~1.5 earthquakes of 3.0 magnitude or greater. In 2009, that number rose to 20, and by 2014 had jumped to 585. And scientists had even proposed an explanation for the increase—the injection of wastewater from oil and gas extraction into porous rock layers beneath the state.

Spurred by high energy prices, the extraction of crude oil and natural gas from conventional wells (pp. 523–524) and hydraulic fracturing (p. 530) operations in Oklahoma increased from 2010 to 2013, with gas extraction rising by 17% and oil extraction by 65%. And as Oklahoma's oil and gas output increased, so did its disposal of wastewater by injection, increasing by 20% from 2010 to 2013. Oil and gas are the modified remains of ancient marine organisms, and deposits often contain briny water that is separated from the oil and gas after the fuels are extracted. This salty wastewater, which can also contain toxic and radioactive compounds, is then typically disposed of by trucking it to a facility far from the oil and gas wells. Once there, it is injected into porous rock formations thousands of feet underground, well below the more shallow rock layers that contain groundwater aquifers. This approach is designed to dispose of wastewater in

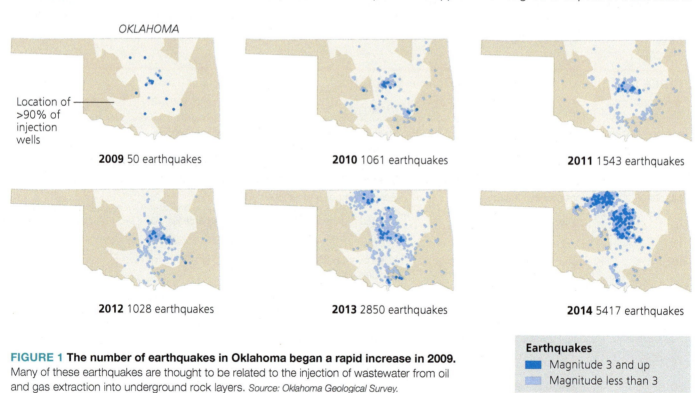

FIGURE 1 The number of earthquakes in Oklahoma began a rapid increase in 2009. Many of these earthquakes are thought to be related to the injection of wastewater from oil and gas extraction into underground rock layers. *Source: Oklahoma Geological Survey.*

a manner that prevents it from contaminating sources of drinking water, both aboveground and belowground.

Wastewater had been injected in this manner beneath Oklahoma for decades without any measurable increase in seismic activity, but scientists were aware that continued pumping of wastewater into rock formations could lead to earthquakes. As the pores within rocks become saturated with water, pressure grows in the underground rock layers, causing the rocks to expand. These expanding rock layers then push against existing faults in the earth, which "lubricates" them and causes them to slip and produce earthquakes. The rock formations beneath Oklahoma facilitate this process, as many of the porous layers of limestone into which wastewaters are injected are located near stressed rocks around faults.

Scientists were increasingly convinced that the increased seismic activity in Oklahoma was due to wastewater injections; however, convincing legislators, regulators, and the public proved challenging. Oil and gas extraction is big business in Oklahoma. By some estimates, one in five jobs in the state is connected to the industry. Landowners benefit from royalties earned by fossil fuel extractions on their land. Tax revenue from sales of oil and gas is the state's third-largest revenue source— behind only sales taxes and personal income taxes.

With such widespread economic benefits, the oil and gas industry enjoys high levels of support in Oklahoma government. When calls arose following the Prague quake to temporarily halt further wastewater injections, the government urged a cautious approach and echoed the industry's position that greater study was needed before decisive action should be taken.

Decisive action was eventually spurred in part by a scientific study published in 2013 that directly linked the Prague earthquake to nearby injections of wastewater from oil and gas extraction. Although the connection between underground fluid injection and earthquakes was well known, studies directly connecting specific events with injection sites were rare. The study, led by geophysicist Katie Keranen of the University of Oklahoma (now at Cornell University) and published in the journal *Geology*, measured the aftershocks produced by the 2011 earthquake to determine the location of the fault that produced the quake.

Reacting quickly to the initial earthquake, the researchers deployed seismic sensors near Prague, and gathered detailed readings on two major tremors and 1183 aftershocks that followed. Analysis of the patterns revealed that the tip of the fault that ruptured was within about 200 m (650 ft) of an active wastewater injection well and occurred at depths consistent with injected rock layers. Keranen's work also showed that it is possible for nearly two decades to pass between the initiation of wastewater injection and a subsequent seismic event, calling into question the safety of many other injection sites. A subsequent study led by Keranen found that earthquakes could

be induced as far as 30 km (19 mi) from injection well fields, and concluded that up to 20% of the induced seismic activity over an area covering 2000 km² (770 mi²) in the central United States could be traced to the activity of four, high-volume wastewater disposal wells in Oklahoma.

As research continued into the connection between wastewater injection and seismic activity, and the issue gained public attention, the state incrementally increased its regulation of wastewater injection wells. The government mandated regular monitoring of well pressures in injection sites, and directed operators to slow injection rates and quantities if underground conditions were deemed conducive to initiating a seismic event. Nevertheless, research has shown that induced earthquakes are likely to occur long after wastewater injection has ceased— even in locations far from injection sites. Therefore, the south central United States, much like the U.S. West Coast, will be a hotbed for seismic study in coming decades (**FIGURE 2**).

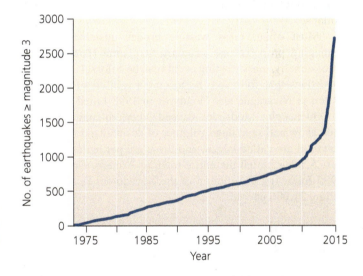

FIGURE 2 Larger earthquakes are becoming more common in the south central United States. Much of the increase is centered in Oklahoma, where the local geology, coupled with the use of seismic-inducing activities such as oil extraction and wastewater injection, has led to more frequent tremors. *Source: Rubinstein, J.L., and A.B. Mahani, 2015. Myths and facts on wastewater injection, hydraulic fracturing, enhanced oil recovery, and induced seismicity. Seismol. Res. Lett. 86: 1–8.*

DATA Q Based on trends from 1973 to 2008, how many earthquakes would be predicted to occur in 2015 in the south central United States? By what percentage was the actual number of earthquakes in 2015 higher than what was expected based on historical data?

Go to **Interpreting Graphs & Data** on Mastering EnvironmentalScience®

temperatures by 0.4–0.7°C (0.7–1.3°F), enough of a rise to cause crop failures worldwide and make 1816 "the year without a summer."

One of the world's largest volcanoes—so large it is called a supervolcano—lies in the United States. The entire basin of Yellowstone National Park is an ancient supervolcano that has at times erupted so massively as to cover large parts of the continent deeply in ash. Although another eruption is not expected imminently, the region is still geothermally active, as evidenced by its numerous hot springs and geysers.

Landslides are a form of mass wasting

Another type of geologic hazard, the **landslide,** occurs when large amounts of rock or soil collapse and flow downhill. Landslides are a severe and often sudden manifestation of the more general phenomenon of **mass wasting,** the downslope movement of soil and rock due to gravity. Mass wasting occurs naturally, and heavy rains may saturate soils and trigger mudslides of soil, rock, and water. However, mass wasting can also be brought about by human land use practices that expose or loosen soil, making slopes more prone to collapse.

Most often, mass wasting erodes unstable hillsides, damaging property one structure at a time (**FIGURE 2.21**). Occasionally, mass wasting events can be colossal and deadly; mudslides that followed the torrential rainfall of Hurricane Mitch in Nicaragua and Honduras in 1998 killed more than 11,000 people. Mudslides caused when volcanic eruptions melt snow and send huge volumes of destabilized mud racing downhill are called lahars, and these are particularly dangerous. A lahar buried the entire town of Armero, Colombia, in 1985 following an eruption of the Nevado del Ruiz volcano, killing 21,000 people.

Tsunamis can follow earthquakes, volcanoes, or landslides

Earthquakes, volcanic eruptions, and large coastal landslides can all displace huge volumes of ocean water instantaneously and trigger a tsunami. The 2011 tsunami that inundated portions of northeastern Japan (**FIGURE 2.22**) was not the only recent major tsunami event. In December 2004, a massive tsunami, triggered by an earthquake off Sumatra, devastated the coastlines of countries all around the Indian Ocean, including Indonesia, Thailand, Sri Lanka, India, and several African nations. Roughly 228,000 people were killed and 1–2 million were displaced. Coral reefs, coastal forests, and wetlands were damaged, and saltwater contaminated soil and aquifers, making it difficult to restore the affected areas.

Those of us who live in the United States and Canada should not consider tsunamis to be something that occurs only in faraway places. Residents of the Pacific Northwest—such as the cities of Seattle, Washington, and Portland, Oregon—could be at risk if there is a slip in the Cascadia subduction

FIGURE 2.21 A landslide near the town of Oso, Washington, in 2014 destroyed around 50 homes and claimed 43 lives. A rain-saturated hillside suddenly gave way, sending tons of rock and soil streaming into homes in the valley below.

FIGURE 2.22 A man surveys the destruction caused by the Tohoku earthquake and tsunami in Japan in 2011. Ocean waters pushed miles inland in some locations, scouring everything in their path, and pulling debris out to sea as the waters receded.

zone that lies 1100 km (700 mi) offshore. The tsunami produced by such a slip would inundate 1.1 million km² (440,000 mi²) of coastal land and cause massive destruction over an area that is currently home to seven million people.

One of the best protections against tsunamis is advance warning. Since the 2004 tsunami, nations and international agencies have stepped up efforts to develop systems to give coastal residents notice of approaching tsunamis so they can move inland or to higher ground. We can also lessen the impacts of tsunamis when they occur if we preserve coastal ecosystems, such as coral reefs and mangrove forests (pp. 425–426), which help protect coastlines by absorbing wave energy.

We can worsen or lessen the impacts of natural hazards

Aside from geologic hazards, people face other types of natural hazards. Heavy rains can lead to flooding that ravages low-lying areas near rivers and streams (p. 396). Coastal erosion can eat away at beaches (p. 495). Wildfire can threaten life and property in fire-prone areas. Tornadoes and hurricanes (p. 452) can cause extensive damage and loss of life.

Although we refer to such phenomena as "natural hazards," the magnitude of their impacts on us often depends on choices we make. We sometimes worsen the impacts of so-called natural hazards in various ways. For example, we live in areas that are prone to hazards, such as the floodplains of rivers or in coastal areas susceptible to flooding. People also use and engineer landscapes around us in ways that can increase the frequency or severity of natural hazards.

Damming and diking rivers to control floods can sometimes lead to catastrophic flooding (p. 397), and the clear-cutting of forests on slopes (p. 311) can induce mass wasting and increase water runoff. Human-induced climate change (Chapter 18) can cause sea levels to rise and promote coastal flooding, and can increase the risks of drought, fire, flooding, and mudslides by altering precipitation patterns.

We can often reduce or lessen the impacts of hazards through the thoughtful use of technology, engineering, and policy, informed by a solid understanding of geology and ecology. Examples include building earthquake-resistant structures; designing early warning systems for earthquakes, tsunamis, and volcanoes; and conserving reefs and shoreline vegetation to protect against tsunamis and coastal erosion. In addition, better forestry, agriculture, and mining practices can help prevent mass wasting. Zoning regulations, building codes, and insurance incentives that discourage development in areas prone to landslides, floods, fires, and storm surges can help keep us out of harm's way. Finally, addressing global climate change may help reduce the frequency of natural hazards in many regions.

closing
THE LOOP

The tragic events at Fukushima following the Tohoku earthquake of 2011 were the product of natural forces—an earthquake and subsequent tsunami—coupled with an accident involving one of humanity's most advanced technologies: nuclear power. These events highlight why knowledge of matter, energy, and geologic forces is vital to understanding environmental impacts in our complex, modern world. Understanding chemistry, for example, helps us predict how radioisotopes released from the Fukushima Daiichi plant will behave in the environment. Understanding energy and how it behaves helps us realize why we need to generate power to fuel our modern lifestyle. And understanding the physical processes that shape our planet helps us prepare for geologic events like those that that afflicted Japan in 2011.

Our understanding of matter and the need for energy can also inform our debate about nuclear energy. After the meltdown at Fukushima Daiichi, for example, Germany shut down half of their nuclear power plants and has plans to decommission all of its reactors by 2022. Switzerland opted to phase out the operation of its five nuclear reactors over the next 20 years. And in Japan, public opposition to restarting its nuclear reactors remains high—although in 2015, Japan restarted the first of its recertified reactors and has plans to restart up to 14 more over the following three years.

As we have seen, the legacy of the Tohoku earthquake and nuclear meltdown at Fukushima Daiichi may be very different from one nation to the next. But one common thread we can take away is that our world contains natural geologic hazards, such as earthquakes and tsunamis, as well as hazards posed by human activities such as nuclear power. And although we cannot ever fully eliminate hazards, we can certainly lessen their potential impacts by thinking carefully about what we build, where we build it, and what we will do should a disaster strike.

REVIEWING Objectives

You should now be able to:

Explain the fundamentals of matter and chemistry and apply them to real-world situations

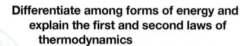

Everything in the universe that has mass and occupies space is matter, which can neither be created nor destroyed. Matter comprises atoms and elements, and changes at the atomic level can result in alternate forms of elements, such as ions and isotopes. Atoms bond with one another to form molecules and compounds conserved in such reactions. Carbon-based organic compounds are particularly important because they are the building blocks of life and provide energy in the form of fossil fuels (pp. 22–28).

Differentiate among forms of energy and explain the first and second laws of thermodynamics

Energy is the capacity to change the position, physical composition, or temperature of matter. Energy can convert from one form to another—for instance, from potential to kinetic energy, and vice versa—and, like matter, energy is conserved during conversions. Systems tend to increase in entropy, or disorder, unless energy is added to build or maintain order and complexity (pp. 28–30).

Distinguish photosynthesis, cellular respiration, and chemosynthesis and summarize their importance to living things

In photosynthesis, autotrophs use carbon dioxide, water, and solar energy to produce oxygen and chemical energy in sugars. In cellular respiration, organisms extract energy from sugars by converting them in the presence of oxygen into carbon dioxide and water, and use this energy to combat entropy and sustain life. In chemosynthesis, specialized autotrophs use carbon dioxide, water, and chemical energy from minerals (instead of energy from sunlight) to produce sugars (pp. 30–32).

Explain how plate tectonics and the rock cycle shape the landscape around us

The crust of our planet is modified by plate tectonics, which shapes Earth's physical geography and produces earthquakes and volcanoes. On a smaller scale, the rock cycle is the mechanism whereby rocks transform from one type to another (pp. 32–36).

Identify major types of geologic hazards and describe ways to minimize their impacts

Volcanoes, earthquakes, mass wasting, and tsunamis are all natural geologic hazards that affect people and the environment. We can minimize their impact by making good decisions about where and how to build (via strict zoning regulations and engineering codes), and by making informed policy decisions regarding global climate change (pp. 37–43).

TESTING Your Comprehension

1. What are the basic building blocks of matter? Provide several examples using chemicals common in Earth's physical or biological systems.

2. How does an ion differ from an isotope? How do atoms, molecules, and compounds differ?

3. Describe two major forms of energy, and give examples of each.

4. Explain the first and second laws of thermodynamics.

5. Describe the three major sources of energy that power Earth's environmental systems.

6. What substances are produced by photosynthesis? By cellular respiration? By chemosynthesis?

7. Name the primary layers that make up our planet. Which portions does the lithosphere include?

8. Describe what occurs at a divergent plate boundary. What happens at a transform plate boundary? Compare and contrast the types of processes that can occur at a convergent plate boundary.

9. Name the three main types of rocks, and describe how each type may be converted to the others via the rock cycle.

10. What causes earthquakes? What are tsunamis, and what causes them? How does a Hawaiian volcano such as Kilauea differ from a volcano in the Andes Mountains?

SEEKING Solutions

1. Think of an example of an environmental problem not mentioned in this chapter. How could chemistry help us address the problem?

2. Think about the ways we harness and use energy sources in our society—both renewable sources such as solar energy and nonrenewable sources such as coal, oil, and natural gas. What implications does the first law of thermodynamics have for our energy usage? How is the second law of thermodynamics relevant to our use of energy?

3. Referring to the chemical reactions for photosynthesis and respiration, provide an argument for why increasing amounts of carbon dioxide in the atmosphere due to global climate change (Chapter 18) might potentially increase amounts of oxygen in the atmosphere. Now give an argument for why increasing amounts of carbon dioxide might potentially decrease amounts of atmospheric oxygen. What would you need to know to determine which of these two outcomes might occur?

4. Describe how plate tectonics accounts for the formation of (a) mountains, (b) volcanoes, and (c) earthquakes.

5. **CASE STUDY CONNECTION** Imagine that you live in a coastal community in a region prone to earthquakes; you are a board member of your regional electrical utility. The utility is considering constructing a nuclear power plant 10 miles up the coast from your town. Some of your fellow citizens support the project because it will bring employment to the region and provide a carbon-free source of electricity. However, some residents fear a repeat of the events of 2011 in northeastern Japan if there should be a significant earthquake along one of the onshore or offshore faults. List three specific pieces of information that you would insist on obtaining from geologists and other scientists before casting your vote on the project.

6. **THINK IT THROUGH** You serve as city manager and have been asked to develop basic emergency plans for your city in case of natural disaster. For each of the following natural hazards, describe one thing that you would recommend the town do to minimize the impact of such a hazard on lives and property: (a) earthquakes, (b) landslides, (c) flooding.

CALCULATING Ecological Footprints

The second law of thermodynamics has profound implications for human impacts on the environment, as it affects the efficiency with which we produce our food. In ecological systems, a rough rule of thumb is that when energy is transferred from plants to plant-eaters or from prey to predator, the efficiency is only about 10% (p. 79). Much of this inefficiency is a consequence of the second law of thermodynamics. Another way to think of this is that eating 1 Calorie of meat from an animal is the ecological equivalent of eating 10 Calories of plant material. So when we raise animals for meat using grain, it is less energetically efficient than if we ate the grain directly.

Humans are considered omnivores because we can eat both plants and animals. The choices we make about what to eat have significant ecological consequences. With this in mind, calculate the ecological energy requirements for four different diets, each of which provides a total of 2000 dietary Calories per day.

DIET	SOURCE OF CALORIES	NUMBER OF CALORIES CONSUMED	ECOLOGICALLY EQUIVALENT CALORIES	TOTAL ECOLOGICALLY EQUIVALENT CALORIES
100% plant 0% animal	Plant Animal			
90% plant 10% animal	Plant Animal	1800 200	1800 2000	3800
50% plant 50% animal	Plant Animal			
0% plant 100% animal	Plant Animal			

1. How many ecologically equivalent Calories would it take to support you for a year for each of the four diets listed?

2. How does the ecological impact from a diet consisting strictly of animal products (e.g., dairy products, eggs, and meat) compare with that of a strictly vegetarian diet? How many additional ecologically equivalent Calories do you consume each day by including as little as 10% of your Calories from animal sources?

3. What percentages of the Calories in your own diet do you think come from plant versus animal sources? Estimate the ecological impact of your diet, relative to a strictly vegetarian one.

4. List the major factors influencing your current diet (e.g., financial considerations, convenience, access to groceries, taste preferences). Do you envision your diet's distribution of plant and animal Calories changing in the near future? Why or why not?

MasteringEnvironmentalScience®

CHAPTER 3

Evolution, Biodiversity, and Population Ecology

Saving Hawaii's Native Forest Birds

HAWAI'I

Pacific Ocean

> **When an entire island avifauna . . . is devastated almost overnight because of human meddling, it is, quite simply, a tragedy.**
> H. Douglas Pratt, ornithologist and expert on Hawaiian birds

> **To keep every cog and wheel is the first precaution of intelligent tinkering.**
> Aldo Leopold

Jack Jeffrey stopped in his tracks. "I hear one!" he said. "Over there in those trees!"

Jeffrey led his group of ecotourists through a lush woodland of ferns, shrubs, and vines toward an emphatic chirping sound. They ducked under twisting gnarled limbs covered with moss and lichens, beneath stately ancient 'ōhi'a-lehua trees offering bright red flowers loaded with nectar and pollen. At last, in the branches of a fast-growing koa tree, they spotted the bird—an 'akiapōlā'au, one of fewer than 1500 of its kind left alive in the world.

The 'akiapōlā'au (or "aki" for short) is a sparrow-sized wonder of nature—one of many exquisite birds that evolved on the Hawaiian Islands and exists only there (see inset photo). For millions of years, this chain of islands in the middle of the Pacific Ocean has acted as a cradle of evolution, generating new and unique species. Yet today many of these species are going from the cradle to the grave. Half of Hawaii's native bird species (70 of 140) have gone extinct in recent times, and many of those that remain—like the aki—teeter on the brink of extinction.

The aki is a type of Hawaiian honeycreeper. The Hawaiian honeycreepers include 18 living species (and at least 38 species recently extinct), all of which originated from a single ancestral species that reached Hawai'i several million years ago. As new volcanic islands emerged from the ocean and then eroded away, and as forests expanded and contracted over millennia, populations were split and new honeycreeper species evolved.

As honeycreeper species diverged from one another, they evolved different colors, sizes, body shapes, feeding behaviors, mating preferences, and bill shapes. Bills in some species became short and straight, allowing birds to glean insects from leaf surfaces. In other species, bills became long and curved, enabling birds to probe into flowers to sip nectar. The bills of still other species became thick and strong for cracking seeds. Some birds evolved highly specialized bills: The aki uses the short, straight lower half of its bill to peck into dead branches to find beetle grubs, then uses the long, curved upper half to reach in and extract them.

Hawaii's honeycreepers thrived for several million years in the island's forests, amid a unique community of plants found nowhere else in the world. Yet today these native Hawaiian forests are under siege. The crisis began 750 or more years ago as Polynesian settlers colonized the islands, cutting down trees and introducing non-native animals. Europeans arrived more than 200 years ago and did more of the same.

The endangered 'akiapōlā'au ▲

Upon completing this chapter, you will be able to:

- Explain natural selection and cite evidence for this process
- Describe how evolution generates and shapes biodiversity
- Discuss major factors behind species extinction and cite Earth's known mass extinction events
- List the levels of ecological organization
- Describe the characteristics of populations that help predict population growth
- Explain how logistic growth, limiting factors, carrying capacity, and other fundamental concepts affect population ecology
- Identify and discuss challenges and current efforts in conserving biodiversity

Native Hawaiian forest at Hakalau Forest National Wildlife Refuge

Pigs, goats, and cattle ate their way through the native plants, transforming luxuriant forests into ragged grasslands. Rats, cats, dogs, and mongooses destroyed the eggs and young of native birds. Foreign plants from Asia, Europe, and America, whose seeds accompanied the people and animals, spread across the altered landscape.

Foreign diseases also arrived, including strains of pox and malaria that target birds. The native fauna were not adapted to resist these novel pathogens. Avian pox and avian malaria, carried by introduced mosquitoes, killed off native birds everywhere except on high mountain slopes, where it was too cold for mosquitoes. Today few native forest birds exist anywhere on the Hawaiian Islands below 1500 m (4500 ft) in elevation.

The aki being watched by Jeffrey's group inhabits the Hakalau Forest National Wildlife Refuge (NWR), high atop the slopes of Mauna Kea, a volcano on the Island of Hawai'i, the largest island in the chain (**FIGURE 3.1**). At Hakalau, native birds find a rare remaining patch of disease-free native forest.

Jeffrey was a biologist at Hakalau for 20 years before his retirement, and led innovative projects to save native plants and birds from extinction. Staff and volunteers at Hakalau fenced out pigs and planted thousands of native plants in areas deforested by cattle grazing. Young restored native forest is now regrowing on thousands of acres. More birds are using this restored forest every year.

However, today global climate change is presenting new challenges. As temperatures climb, mosquitoes move upslope and malaria and pox spread deeper into the remaining forests,

FIGURE 3.1 The Hakalau Forest NWR is located on the slopes of Mauna Kea on the Island of Hawai'i.

so that even protected areas such as Hakalau are not immune. The next generation of managers will need to innovate further strategies to fend off extinction for the island's native species.

Plenty of challenges remain, but the restoration successes at Hakalau Forest so far provide hope that through responsible management we can restore Hawaii's native flora and fauna, prevent further impacts, and preserve the priceless bounty of millions of years of evolution on this extraordinary chain of islands.

Evolution: The Source of Earth's Biodiversity

The animals and plants native to the Hawaiian Islands help reveal how our world became populated with the remarkable diversity of life we see today—a rich cornucopia of millions of species (**FIGURE 3.2**).

A **species** is a particular type of organism. More precisely, it is a population or group of populations whose members share characteristics and can freely breed with one another and produce fertile offspring. A **population** is a group of individuals of a given species that live in a particular region at a particular time. Over vast spans of time, the process of evolution has shaped populations and species, giving us the vibrant abundance of life that enriches Earth today.

In its broad sense, the term *evolution* means change over time. In its biological sense, **evolution** consists of change in populations of organisms from generation to generation. Such change originates in genes (p. 28) and often leads to modifications in appearance or behavior.

Evolution is one of the best-supported and most illuminating concepts in all of science, and it is the very foundation of modern biology. Perceiving how species adapt to their environments and change over time is crucial for comprehending ecology and learning the history of life. Evolutionary processes influence many aspects of environmental science, including agriculture, pesticide resistance, medicine, and environmental health.

Natural selection shapes organisms

Natural selection is a primary mechanism of evolution. In the process of **natural selection,** inherited characteristics that enhance survival and reproduction are passed on more frequently to future generations than characteristics that do not, thereby altering the genetic makeup of populations through time.

Natural selection is a simple concept that offers a powerful explanation for patterns evident in nature. The idea of natural selection follows logically from a few straightforward facts that are readily apparent to anyone who observes the life around us:

- Organisms face a constant struggle to survive and reproduce.
- Organisms tend to produce more offspring than can survive to maturity.
- Individuals of a species vary in their attributes.

Variation is due to differences in genes, the environments in which genes are expressed, and the interactions between genes and environment. As a result of this variation, some individuals of a species will be better suited to their environment than others and so will survive longer and be better able to reproduce.

Attributes are passed from parent to offspring through the genes, and a parent that produces many offspring will pass on more genes to the next generation than a parent that produces few or no offspring. In the next generation, therefore, the genes of better-adapted individuals will outnumber those of individuals that are less well adapted. From one generation to another through time, characteristics, or traits, that lead to better reproductive success in a given environment will evolve in the population. This process is termed **adaptation,** and a trait that promotes reproductive success is also called an *adaptation* or an *adaptive trait*.

The concept of natural selection was first proposed in the 1850s by **Charles Darwin** and, independently, by **Alfred Russel Wallace,** two exceptionally keen British naturalists. By this time, scientists and amateur naturalists were widely discussing the idea that populations evolve, yet no one could say how or why. After spending years studying and cataloging an immense variety of natural phenomena—in his English garden and across the world to the Galápagos Islands—Darwin finally concluded that natural selection helped explain the world's great variety of living things. Once he came to this conclusion, however, Darwin put off publishing his findings, fearing the social disruption that might ensue if people felt their religious convictions were threatened. Darwin was at last driven to go public when Wallace wrote to him from the Asian tropics, independently describing the idea of natural selection. Their shared ideas were presented together at a scientific meeting in 1858, and the next year Darwin published his groundbreaking book, *On the Origin of Species*.

With natural selection, humanity at last uncovered a precise and viable mechanism to explain how and why organisms evolve through time. Once geneticists determined the basis of inheritance, this understanding launched evolutionary biology. In the century and a half since Darwin and Wallace, legions of researchers have refined our understanding of evolution, powering dazzling progress in biology that has helped shape our society.

Selection acts on genetic variation

For an organism to pass a trait along to future generations, genes in its DNA (pp. 27–28) must code for the trait. In an organism's lifetime, its DNA will be copied millions of times by millions of cells. Amid all this copying, sometimes a mistake is made. Accidental changes in DNA, called **mutations,** give rise to genetic variation among individuals. If a mutation

occurs in a sperm or egg cell, it may be passed on to the next generation. Most mutations have little effect, but some can be deadly and others can be beneficial. Those that are not lethal provide the genetic variation on which natural selection acts.

Genetic variation is also generated as organisms mix their genetic material through sexual reproduction. When organisms reproduce sexually, each parent contributes to the genes of the offspring. This process produces novel combinations of genes, generating variation among individuals.

When natural selection acts on genetic variation by favoring certain variants, it can drive a feature in a particular direction (**FIGURE 3.3**). Because such evolutionary change generally requires a great deal of time, a species cannot always adapt to environmental conditions that change quickly. For instance, the warming of our global climate today (Chapter 18) is occurring too rapidly for most species to adapt, and we may lose many species to extinction as a result.

However, genetic variation can sometimes help protect a population against novel challenges. One of the honeycreeper species of the Hakalau Forest, the 'amakihi, has recently been discovered in 'ōhi'a trees at low elevations where avian malaria has killed off other honeycreepers. Researchers determined that some of the 'amakihis living there when malaria arrived had genes that by chance gave them a natural

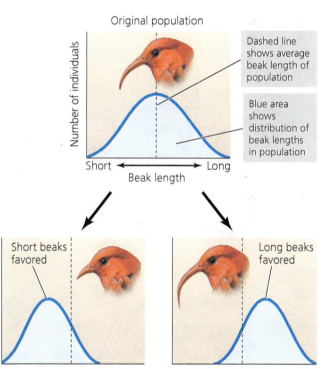

(a) Selection for shorter beaks   (b) Selection for longer beaks

FIGURE 3.3 Natural selection can drive a feature in various directions. Let's consider the 'i'iwi, a Hawaiian honeycreeper, and assume its population possesses genetic variation for the length of its curved bill. In an environment where flowers grow short nectar tubes **(a)**, birds with short bills could feed perfectly well and avoid investing extra energy in growing a long bill, so natural selection would favor a decrease in bill length across the population. In an environment where flowers have long tubes **(b)**, birds with long bills could feed more effectively and pass on more genes, causing the population to shift toward longer average bill length.

FIGURE 3.2 Hawai'i hosts a treasure trove of biodiversity, including ❶ happyface spider, ❷ 'i'iwi, ❸ nēnē, and ❹ Haleakala silversword.

resistance to the disease. These resistant birds survived malaria's onslaught, and their descendants that carried the malaria-resistant genes reestablished a population that is growing today. Scientists hope that perhaps some individuals of the rarer native birds of Hakalau might also harbor resistance genes that could help them persist in the face of malaria.

Selective pressures from the environment influence adaptation

Environmental conditions determine what pressures natural selection will exert, and these selective pressures affect which members of a population will survive and reproduce. Over many generations, this results in the evolution of traits that enable success within the environment in question. Closely related species that live in very different environments tend to diverge in their traits as differing selective pressures drive the evolution of different adaptations (**FIGURE 3.4a**). Conversely, sometimes very unrelated species living in similar environments in separate locations may independently acquire similar traits as they adapt to similar selective pressures; this is called **convergent evolution** (**FIGURE 3.4b**).

Of course, environments change over time, and organisms may move to new locations and encounter new conditions. In either case, a trait that promotes success at one time or place may not do so at another. Hawaiian honeycreepers such as the 'apapane and the 'i'iwi fly long distances in search of flowering trees. This behavior had long helped them to find the best resources across a diverse landscape. However, once malaria arrived, the behavior become counterproductive, as birds from malaria-free mountain forests flew downslope into death zones and were bitten by mosquitoes. As environmental conditions vary in time and space, adaptation becomes a moving target.

Evidence of selection is all around us

The results of natural selection are all around us, visible in every adaptation of every organism. Researchers have documented selection through innumerable observations and experiments at field sites from the Galápagos Islands to the rainforests to the oceans to our cities. Scientists also have demonstrated the rapid evolution of traits by selection in countless lab experiments with fast-reproducing organisms such as bacteria, yeast, and fruit flies.

The evidence for selection that may be most familiar to us is that which Darwin himself cited prominently in his work 160 years ago: our breeding of domesticated animals. In dogs, cats, and livestock, we have conducted selection under our own direction, that is, **artificial selection.** We have chosen animals with traits we like and bred them, while culling out individuals

Cactus in Arizona

Euphorb (spurge) in the Canary Islands

(a) Divergent evolution of Hawaiian honeycreepers

(b) Convergent evolution of cactus and spurge

FIGURE 3.4 Natural selection can cause closely related species to diverge or distantly related species to converge. Hawaiian honeycreepers **(a)** diversified as they adapted to different food resources and habitats, as indicated by their diversity of plumage colors and bill shapes. In contrast, cacti of the Americas and euphorbs of Africa **(b)** became similar to one another as they independently adapted to arid environments. These plants each evolved succulent tissues to hold water, thorns to keep thirsty animals away, and photosynthetic stems without leaves to reduce surface area and water loss.

with traits we do not like. Through such *selective breeding*, we have been able to augment particular traits we prefer.

Consider the great diversity of dog breeds (**FIGURE 3.5a**). People generated every type of dog alive today by starting with a single ancestral species and selecting for particular desired traits as individuals were bred together. From Great Dane to Chihuahua, all dogs are able to interbreed and produce viable offspring, yet breeders maintain striking differences among them by allowing only like individuals to breed.

Artificial selection through selective breeding has also given us the many crop plants and livestock we depend on for food and fiber, all of which people domesticated from wild species and carefully bred over years, centuries, or millennia (**FIGURE 3.5b**). Through selective breeding, we have created corn with bigger, sweeter kernels; wheat and rice with larger and more numerous grains; and apples, pears, and oranges with better taste. We have diversified single types into many—for instance, breeding

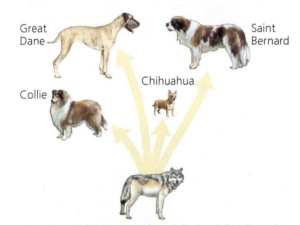

(a) Ancestral wolf (*Canis lupus*) and derived dog breeds

Great Dane
Saint Bernard
Chihuahua
Collie

Cabbage
Broccoli
Brussels sprouts
Cauliflower

(b) Ancestral *Brassica oleracea* and derived crops

FIGURE 3.5 Selective breeding, or artificial selection, has given us our many breeds of dogs and varieties of crops. The ancestral wild species for dogs **(a)** is the gray wolf (*Canis lupus*). By breeding like with like and selecting for traits we preferred, we have produced breeds as different as Great Danes and Chihuahuas. By this same process we have created an immense variety of crop plants **(b)**. Cabbage, brussels sprouts, broccoli, and cauliflower were all generated from a single ancestral species, *Brassica oleracea*.

variants of wild cabbage (*Brassica oleracea*) to create broccoli, cauliflower, cabbage, and brussels sprouts. Our entire agricultural system is based on artificial selection. Thus, we depend on a working understanding of evolution for the very food we eat.

Understanding evolution is vital for modern society

As with selective breeding, evolutionary processes play key roles in today's society and in our everyday lives. Many medical advances have resulted from our knowledge of evolution. Understanding evolution helps us determine how infectious diseases spread and how they gain or lose potency. Deciphering how viruses evolve allows scientists to track the constantly evolving strains of influenza, HIV, and other pathogens. Armed with such information, biomedical experts can predict which flu strains will most likely spread in a given year, and then design effective vaccines targeting them. Comprehending evolution also enables us to detect and respond to the evolution of antibiotic resistance by dangerous bacteria.

In agriculture, our understanding of evolution helps us prevent antibiotic resistance in feedlots and pesticide resistance in crop-eating insects (p. 247). It informs our technology as well; from studying how organisms adapt to challenges and evolve new abilities, we develop ideas on how to design technologies and engineering solutions.

Evolution generates biodiversity

Just as selective breeding helps us create new types of pets, farm animals, and crop plants, natural selection can elaborate and diversify traits in wild organisms, forming new species and whole new types of organisms. Life's complexity can be expressed as **biological diversity**, or **biodiversity.** These terms refer to the variety of life across all levels, including the diversity of species, genes, populations, and communities (p. 271).

Scientists have identified and described about 1.8 million species, but many more remain undiscovered or unnamed. Estimates for the actual number of species in the world vary, but they range from 3 million up to 100 million. Hawaii's

FAQ

Isn't evolution based on just one man's beliefs?

Because Charles Darwin contributed so much to our early understanding of evolution, many people assume the concept itself hinges on his ideas. But scientists and laypeople had been observing nature, puzzling over fossils, and discussing the notion of evolution long before Darwin. Once he and Alfred Russel Wallace independently proposed the concept of natural selection, scientists finally gained a way of explaining how and why organisms change across generations. Later, geneticists discovered Gregor Mendel's research on inheritance and worked out how traits are passed on—and modern evolutionary biology was born. Twentieth-century scientists Ronald Fisher, Sewall Wright, Theodosius Dobzhansky, George Gaylord Simpson, Ernst Mayr, and others ran experiments and developed sophisticated mathematical models, documenting phenomena with extensive evidence and building evolutionary biology into one of science's strongest fields. Since then, evolutionary research by thousands of scientists has driven our understanding of biology and has facilitated spectacular advances in agriculture, medicine, and technology.

insect fauna provides one example of how much we have yet to learn. Scientists studying fruit flies in the Hawaiian Islands have described more than 500 species of them, but they have also identified about 500 others that have not yet been formally named and described. Still more fruit fly species probably exist but have not yet been found.

Subtropical islands such as Hawai'i are by no means the only places rich in biodiversity. Step outside just about anywhere and you will find many species within close reach. Plants poke up from cracks in asphalt in every city in the world. A handful of backyard soil may contain an entire miniature world of life, including insects, mites, millipedes, nematode worms, plant seeds, fungi, and millions of bacteria. (We will examine Earth's biodiversity in detail in Chapter 11.)

FAQ

Are humans still evolving?

As our civilization has developed, it's become easy to feel removed from nature. If we delude ourselves into believing we're above nature, we might be inclined to believe that human beings are no longer subject to natural selection, only to cultural forces. It's true that our day-to-day environment has changed from a wholly "natural" one in our hunter-gatherer days to a largely human-constructed one in our modern urban-techno-industrial society. But when environments change, new selective pressures come into play. For instance, in the not-too-distant past, many people died prematurely from diseases, injuries, and infection. In today's world, medical advances save and extend our lives. Ironically, as a result, illnesses and disabilities that can be inherited get passed on to more children than in the past, and the frequencies with which these maladies occur increase. At the same time, humanity's old foes continue to have an impact: In developing countries, major killers of young people, like malaria, measles, and tuberculosis, still exert strong selection on the human genome. Natural selection never stops, and all organisms—people included—are always evolving.

Speciation produces new types of organisms

How did Earth come to have so many species? The process by which new species are generated is termed **speciation.** Speciation can occur in a number of ways, but the main mode is generally thought to be *allopatric speciation*, whereby species form from populations that become physically separated over some geographic distance. Imagine a population of organisms. Individuals within the population possess many similarities that unify them as a species because they are able to breed with one another and share genetic information. However, if the population is split into two or more isolated areas, individuals from one area cannot reproduce with individuals from the others.

When a mutation arises in the DNA of an organism in one of these newly isolated populations, it cannot spread to the other populations. Over time, each population will independently accumulate its own set of mutations. Eventually, the populations may diverge, growing so different that their members can no longer mate with one another and produce viable offspring. (This can occur because of changes in reproductive organs, hormones, courtship behavior, breeding timing, or other factors.) Populations that no longer exchange genetic information will embark on their own independent evolutionary paths as separate species (**FIGURE 3.6**).

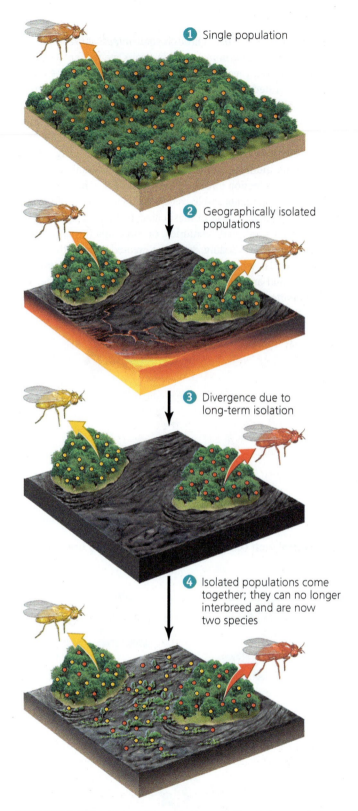

1 Single population

2 Geographically isolated populations

3 Divergence due to long-term isolation

4 Isolated populations come together; they can no longer interbreed and are now two species

FIGURE 3.6 The long, slow process of allopatric speciation begins when a geographic barrier splits a population—as when forest 1 is destroyed by lava flowing from a volcano but isolated patches of forest 2 are left. Hawaiian fruit flies are weak fliers and become isolated in such forested patches, called *kipukas*. Over centuries, each population accumulates its own set of genetic changes 3, until individuals become unable to breed with individuals from the other population. The two populations now represent separate species and will remain so even if the geographic barrier disappears 4 and the new species intermix.

The populations will continue diverging in their characteristics as chance mutations accumulate that cause them to differ more and more. And if environmental conditions happen to differ for the two populations, then natural selection may accelerate the divergence.

For speciation to occur, populations must remain isolated for a very long time, generally thousands of generations. Populations can undergo long-term geographic isolation in various ways. Lava flows can destroy forest, leaving small isolated patches intact (as shown in Figure 3.6). Glacial ice sheets may move across continents during ice ages and split populations in two. Major rivers may change course or mountain ranges may be uplifted, dividing regions and their organisms. Sea level may rise, flooding low-lying regions and isolating areas of higher ground as islands. Drying climate may partially evaporate lakes, subdividing them into smaller bodies of water. Warming or cooling climate may cause plant communities to shift, creating new patterns of plant and animal distribution.

Alternatively, sometimes organisms colonize newly created areas, establishing isolated populations. Hawai'i provides an example. As shown in Figure 2.20 (Chapter 2, p. 39), the Pacific tectonic plate moves over a volcanic "hotspot" that extrudes magma into the ocean, building volcanoes that form islands once they break the water's surface. The plate inches northwest, dragging each island with it, while new islands are formed at the hotspot. The result, over millions of years, is a long string of islands, called an *archipelago*. As each new island is formed, plants and animals that colonize it may undergo allopatric speciation if they are isolated enough from their source population (see **THE SCIENCE BEHIND THE STORY**, pp. 54–55).

We can infer the history of life's diversification by comparing organisms

Innumerable speciation events have generated complex patterns of diversity beyond the species level. Scientists represent this history of divergence by using branching, treelike diagrams called **phylogenetic trees.** Similar to family genealogies, these diagrams illustrate hypotheses proposing how divergence took place (**FIGURE 3.7**). Phylogenetic trees can show relationships among species, groups of species, populations, or genes. Scientists construct these trees by analyzing patterns of similarity among the genes or external traits of present-day organisms and by inferring which groups share similarities because they are related.

Once we have a phylogenetic tree, we can map traits onto the tree according to which organisms possess them, and we can thereby trace how the traits have evolved.

Knowing how organisms are related to one another also helps scientists to classify them and name them. *Taxonomists* use an organism's genetic makeup and physical appearance to determine its species identity. These scientists then group species by their similarity into a hierarchy of categories meant to reflect evolutionary

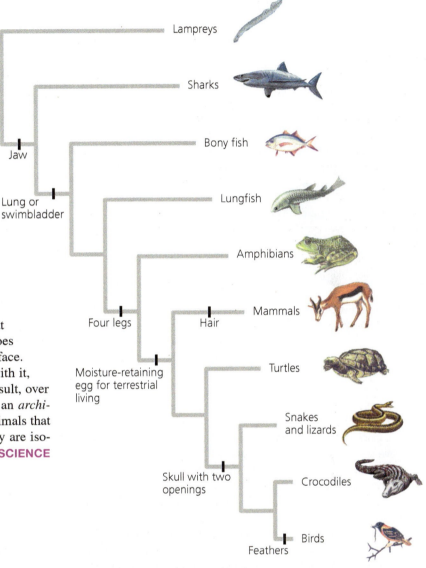

FIGURE 3.7 Phylogenetic trees show the history of life's divergence. The tree here illustrates relationships among groups of vertebrates—just one small portion of the huge and complex "tree of life." Each branch results from a speciation event; as you follow the tree left to right from its trunk to the tips of its branches, you proceed forward in time, tracing life's history. Major traits are "mapped" onto the tree to indicate when they originated.

DATA Find the hash mark indicating the origin of jaws. Which group or groups of vertebrates possess jaws? Which group(s) diverged before jaws originated? Are birds more closely related to amphibians or to crocodiles? Explain how you know this.

Go to **Interpreting Graphs & Data** on MasteringEnvironmentalScience®

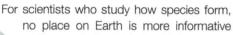

THE SCIENCE
behind the story

How Do Species Form in Hawaii's "Natural Laboratory" of Evolution?

Dr. Heather Lerner of Earlham College

For scientists who study how species form, no place on Earth is more informative than an isolated chain of islands. The Hawaiian Islands—the most remote on the planet—are often called a "natural laboratory of evolution."

The key to this laboratory lies in the process that drives Hawaii's geologic history. Turn back to Figure 2.20 (p. 39), and examine it closely. Deep beneath the Pacific Ocean, a volcanic "hotspot" spurts magma as the Pacific Plate slides across it in tectonic motion like a conveyor belt. Mountains of lava accumulate underwater until eventually a volcano rises above the waves, building an island. As the tectonic plate moves northwest, it carries each newly formed island with it, creating a long chain, or *archipelago*. Over several million years, each island gradually subsides, erodes, and disappears beneath the waves. As old islands disappear on the northwest end of the chain, new islands are formed on the southeast end.

Geologists analyzing radioisotopes (p. 24) in the islands' rocks have determined that this process has been going on for at least 85 million years. They estimate that Kaua'i was formed about 5.1 million years ago (mya), and the Island of Hawai'i just 0.43 mya.

Despite the remoteness of the Hawaiian archipelago, over time a few plants and animals found their way there, establishing populations. As some individuals hopped to neighboring islands, populations that were adequately isolated evolved into new species. Such speciation by "island-hopping" has driven the radiation of Hawaiian honeycreepers and many other organisms.

For instance, the barren and windswept high volcanic slopes of Hawai'i are graced by some of the most striking flowering plants in the world, the silverswords (see Figure 3.2). These spectacular plants have spiky, silvery leaves and tall stalks that explode into bloom with flowers once in the plant's long life before it dies. Researchers have discovered that Hawaii's 28 species of silverswords all evolved from a modest tarweed plant from California that reached Hawai'i and diversified by island-hopping. University of California, Berkeley, botanist Bruce Baldwin and other researchers analyzed genetic relationships and learned that the silverswords' radiation was rapid (on a geologic timescale), taking place in just 5 million years.

The best-understood radiation has occurred with the Hawaiian fruit flies. Some of these insects speciate within islands in kipukas (see Figure 3.6), but most have done so by island-hopping. By combining genetic analysis and geologic dating, researchers determined that the process began 25 mya on islands that today are beneath the ocean. From a single original fruit fly species, an estimated 1000 species have evolved—fully one-sixth of all the world's fruit fly species.

Other groups have undergone adaptive radiations on the Hawaiian Islands as well, including damselflies, crickets, mirid bugs, spiders, and multiple families of plants. Scientists propose that once a species colonizes an island, it may spread and evolve rapidly because competitors are few and there tend to be unoccupied niches (pp. 60, 75).

The Hawaiian honeycreepers are so diverse that researchers have long puzzled over what type of bird gave rise to their radiation—and whether there was just one colonizing ancestor or many. In 2011, to clarify how the honeycreeper radiation took place, one research team combined genetic sequencing technology with resources from museum collections and our knowledge of Hawaiian geology.

Heather Lerner and five colleagues first took tissue samples from bird specimens in museum collections. Working with Robert Fleischer and Helen James at the Smithsonian Institution, Lerner, now at Earlham College in Indiana, sampled 19 species of honeycreepers plus 28 diverse types of finches from around the Pacific Rim that experts had identified as possible ancestors.

Lerner's team obtained data from 13 genes and from mitochondrial genomes by sequencing DNA (pp. 27–28) from each tissue sample. They ran the data through computer programs to analyze how the DNA sequences—and thus the birds—were related to one another, then produced phylogenetic trees (p. 53) showing the relationships (**FIGURE 1**). They published their results in the journal *Current Biology*.

Lerner's team found that the Hawaiian honeycreepers apparently derive from one ancestor and are most related to the Eurasian rosefinches, indicating that honeycreepers evolved after some rosefinch-like bird arrived from Asia. Today's rosefinches are partly nomadic; when food supplies crash, flocks fly long distances to find food. Perhaps a wandering flock of ancestral rosefinches was caught up in a storm long ago and blown to Hawai'i.

Once this common ancestor of today's rosefinches and honeycreepers arrived on an ancient Hawaiian island, its progeny adapted to conditions there by natural selection, resulting in modified bill shape, diet, and coloration. Every once in a great while, wandering birds colonized other islands, founding

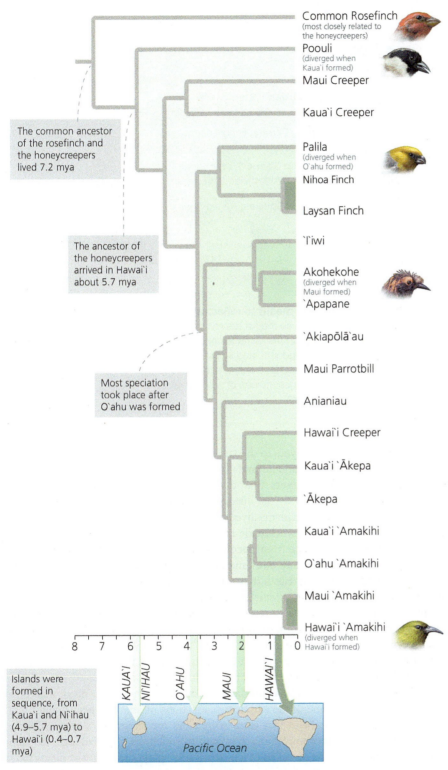

The common ancestor of the rosefinch and the honeycreepers lived 7.2 mya

The ancestor of the honeycreepers arrived in Hawai'i about 5.7 mya

Most speciation took place after O'ahu was formed

Common Rosefinch
(most closely related to the honeycreepers)
Poouli
(diverged when Kaua'i formed)
Maui Creeper
Kaua'i Creeper
Palila
(diverged when O'ahu formed)
Nihoa Finch
Laysan Finch
`I'iwi
Akohekohe
(diverged when Maui formed)
`Apapane
`Akiapōlā`au
Maui Parrotbill
Anianiau
Hawai'i Creeper
Kaua'i `Ākepa
`Ākepa
Kaua'i `Amakihi
O`ahu `Amakihi
Maui `Amakihi
Hawai'i `Amakihi
(diverged when Hawai'i formed)

8 7 6 5 4 3 2 1 0

KAUA'I NI'IHAU O'AHU MAUI HAWAI'I

Islands were formed in sequence, from Kaua'i and Ni'ihau (4.9–5.7 mya) to Hawai'i (0.4–0.7 mya)

Pacific Ocean

FIGURE 1 Using gene sequences, researchers generated this phylogenetic tree showing relationships among the Hawaiian honeycreepers. They then matched the history of the birds' diversification with the known geologic history of the islands' formation. *Adapted from Lerner, H.R.L., et al. 2011. Multilocus resolution of phylogeny and timescale in the extant adaptive radiation of Hawaiian honeycreepers. Curr. Biol. 21: 1838–1844.*

populations that each adapted to local conditions and might eventually evolve into separate species.

Because the age of each island is known, Lerner's team could calibrate rates of evolutionary change in the birds' DNA sequences, and thus measure the age of each divergence. That is, they could tell how "old" each bird species is. They found that the rosefinch-like ancestor arrived by 5.7 mya, about the time that the oldest of today's main islands (Kaua'i and Ni'ihau) were forming. After O'ahu emerged 4.0–3.7 mya, the speciation process went into overdrive, giving rise to many new species with distinctively different colors, bill shapes, and habits. By the time Maui arose 2.4–1.9 mya, most of the major differences in body form and appearance had evolved (see bottom portion of Figure 1).

Thus, most major innovations arose midway through the island-formation process, when O'ahu and Kaua'i were the main islands in the chain. After this burst of innovation, major changes were fewer, perhaps because most evolutionary possibilities had been explored, or perhaps because the newer islands of Maui and Hawai'i were too close together to isolate populations adequately.

The team's data show that the age of each honeycreeper species does not neatly match the age of the island or islands it inhabits today. Instead, the island-hopping process was complex, with some birds hopping "backward" from newer islands to older ones. Moreover, within each island there is a great deal of variation in climate, topography, and vegetation, because windward slopes catch moisture from trade winds over the ocean and become lush and green, whereas leeward slopes in the rainshadow (p. 92) are arid. The varied habitats and rugged topography create barriers that can lead to speciation within islands.

For all these reasons, the "natural laboratory" of Hawai'i still has much to teach us about how the honeycreepers and other groups have evolved, and how new species are formed.

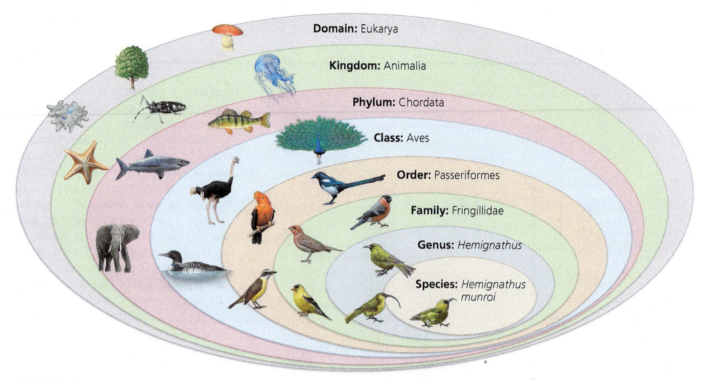

FIGURE 3.8 Taxonomists classify organisms using a hierarchical system meant to reflect evolutionary relationships. Species similar in appearance, behavior, and genetics (because they share recent common ancestry) are placed in the same genus. Organisms of similar genera are placed in the same family. Families are placed within orders, orders within classes, classes within phyla, phyla within kingdoms, and kingdoms within domains. For example, honeycreepers belong to the class Aves, along with peacocks, loons, and ostriches. However, the differences between these types of birds, which have diverged across millions of years of evolution, are great enough that they are placed in different orders, families, and genera.

relationships. Related species are grouped together into *genera* (singular, *genus*), related genera are grouped into families, and so on (**FIGURE 3.8**). Each species is given a two-part Latin or Latinized scientific name denoting its genus and species.

For instance, the ʻakiapōlāʻau, *Hemignathus munroi*, is similar to other Hawaiian honeycreepers in the genus *Hemignathus*. These species are closely related in evolutionary terms, as indicated by the genus name they share. They are more distantly related to honeycreepers in other genera, but all honeycreepers are classified together in the family Fringillidae. This system of naming and classification was devised by Swedish botanist Carl Linnaeus (1707–1778) long before Darwin's work on evolution. Today biologists use evolutionary information from phylogenetic trees to help classify organisms under the Linnaean system's rules.

Fossils reveal life's long history

Scientists also decipher life's history by studying fossils. As organisms die, some are buried by sediment. Under certain conditions, the hard parts of their bodies—such as bones, shells, and teeth—may be preserved, as sediments are compressed into rock (p. 35). Minerals replace the organic material, leaving behind a **fossil,** an imprint in stone of the dead organism (**FIGURE 3.9a**). Over millions of years, geologic processes have buried sediments and later brought sedimentary

rock layers to the surface, revealing assemblages of fossilized plants and animals from different time periods. By dating the rock layers that contain fossils, paleontologists (scientists who study the history of Earth's life) can learn when particular organisms lived (**FIGURE 3.9b**). The cumulative body of fossils worldwide is known as the **fossil record.**

The fossil record shows that the number of species existing at any one time has generally increased, but that the species on Earth today are just a small fraction of all species that have ever existed. During life's 3.5 billion years on Earth, complex structures have evolved from simple ones, and large sizes from small ones. However, simplicity and small size have also evolved when favored by natural selection; it is easy to argue that Earth still belongs to the bacteria and other microbes, some of them little changed over eons.

Even enthusiasts of microbes, however, must marvel at some of the exquisite adaptations of animals, plants, and fungi: the heart that beats so reliably for an animal's entire lifetime; the complex organ system to which the heart belongs; the stunning plumage of a peacock in display; the ability of plants to lift water and nutrients from the soil, gather light from the sun, and produce food; and the human brain and its ability to reason. All these adaptations and more have resulted as evolution has generated new species and whole new branches on the tree of life.

Although speciation generates Earth's biodiversity, it is only part of the equation; the fossil record teaches us that the

(a) Fossil of an extinct trilobite

(b) Paleontologist cleaning teeth of fossilized mastodon skull

FIGURE 3.9 The fossil record helps reveal the history of life on Earth. Trilobites **(a)** were once abundant, but today we know these extinct animals only from their fossils. Finding, excavating, and preparing fossils is hard work; here **(b)** a paleontologist cleans the teeth of a mastodon skull.

vast majority of species that once lived are now gone. The disappearance of a species from Earth is called **extinction.** From studying the fossil record, paleontologists calculate that the average time a species spends on Earth is 1–10 million years. The number of species in existence at any one time is equal to the number added through speciation minus the number removed by extinction.

Extinction occurs naturally, but human impact can profoundly affect the rate at which it occurs. Our planet's biological diversity is being lost at a frightening pace (Chapter 11). This loss affects people directly, because other organisms provide us with life's necessities—food, fiber, medicine, and vital ecosystem services (pp. 4, 116–117, 149). Species extinction brought about by human impact may well be the single biggest problem we face, because the loss of a species is irreversible.

Some species are especially vulnerable to extinction

In general, extinction occurs when environmental conditions change rapidly or drastically enough that a species cannot adapt genetically to the change; the slow process of natural selection simply does not have enough time to work. Small populations are vulnerable to extinction because fluctuations in their size

could, by chance, bring the population size to zero. Small populations also sometimes lack enough genetic variation to buffer them against environmental change. Species narrowly specialized to some particular resource or way of life are also vulnerable, because environmental changes that make that resource or role unavailable can spell doom. Species that are **endemic** to a region, meaning that they occur nowhere else on the planet, also face elevated risks of extinction because when some event influences their region, it may affect all members of the species.

Island-dwelling species are particularly vulnerable. Because islands are smaller than mainland areas and are isolated by water, only some species reach islands, whereas many others never do. As a result, some of the pressures and challenges faced daily by organisms on the mainland simply don't exist on islands. For instance, only one land mammal—a bat—ever reached Hawai'i naturally, so Hawaii's birds evolved for millions of years without needing to defend against the threat of predation by mammals. Likewise, Hawaii's plants did not need to invest in defenses (such as thick bark, spines, or chemical toxins) against mammals that might eat them. Because defenses are costly to invest in (from an energy standpoint), through adaptation most island birds and plants lost any defenses their mainland ancestors may have had.

Eventually, Hawai'i was colonized by people—and Hawaii's native organisms (**FIGURE 3.10a**) were completely unprepared for what people brought to the islands (**FIGURE 3.10b**). Rats, cats, and mongooses preyed on

(a) Hawaiian Petrel, a native species at risk

(b) Mongoose, an introduced species that preys on natives

FIGURE 3.10 Island-dwelling species that have lost defenses are vulnerable to extinction when enemies are introduced. The Hawaiian petrel **(a)**, a seabird that nests in the ground, is endangered as a result of predation by mammals introduced to Hawai'i, such as **(b)** the Indian mongoose.

FIGURE 3.11 Small range sizes can leave species vulnerable to extinction if severe changes occur in their local environment. The Shenandoah salamander (*Plethodon shenandoah*) lives on just three peaks in Shenandoah National Park in Virginia.

ground-nesting seabirds, ducks, geese, and flightless rails, driving many of these birds extinct. Livestock ate through the vegetation, turning luxuriant forests into desolate grasslands. Half of Hawaii's native birds were driven extinct soon after human arrival.

On a mainland, "islands" of habitat (such as forested mountaintops) can host endemic species that are vulnerable to extinction. In the United States, 40 salamander species are restricted to areas the size of a typical county, and some live atop single mountains (**FIGURE 3.11**). Many other amphibians are limited to very small ranges. The Yosemite toad is restricted to a small region of the Sierra Nevada in California, the Houston toad occupies just a few patches of Texas woodland, and the Florida bog frog lives in a tiny area of Florida wetland.

Earth has seen several episodes of mass extinction

Most extinction occurs gradually, one species at a time, at a rate referred to as the **background extinction rate.** However, the fossil record reveals that Earth has seen at least five events of staggering proportions that killed off massive numbers of species at once. These episodes, called **mass extinction events,** have occurred at widely spaced intervals in our planet's history and have wiped out 50–95% of Earth's species each time (see Table 11.5, p. 279).

The best-known mass extinction occurred 66 million years ago and brought an end to dinosaurs and many other types of animals (although today's birds are descendants of a type of dinosaur that survived). Evidence suggests that the impact of a gigantic asteroid caused this event. Still more catastrophic was the mass extinction at the end of the Permian period 250 million years ago (see **APPENDIX E** for Earth's geologic periods). Paleontologists estimate that 75–95% of all species perished during this event, described by one researcher as the "mother of all mass extinctions." Hypotheses as to what caused the end-Permian extinction event include

massive volcanism, an asteroid impact, methane releases and global warming, or some combination of factors.

The sixth mass extinction is upon us

Many biologists have concluded that Earth is currently entering its sixth mass extinction event—and that we are the cause. Indeed, the Millennium Ecosystem Assessment (pp. 15–16) estimated that today's extinction rate is 100–1000 times higher than the background rate, and rising (p. 282). Changes to our planet's natural systems set in motion by human population growth, development, and resource depletion have driven many species extinct and are threatening countless more. As we alter and destroy natural habitats; overhunt and overharvest populations; pollute air, water, and soil; introduce invasive non-native species; and alter climate, we set in motion processes that combine to threaten Earth's biodiversity (pp. 282–288). Because we depend on what nature has to offer, biodiversity loss and extinction ultimately threaten our own survival.

Ecology and the Organism

Extinction, speciation, and other evolutionary forces play key roles in ecology. **Ecology** is the scientific study of the interactions among organisms and of the relationships between organisms and their environments. Ecology allows us to explain and predict the distribution and abundance of organisms in nature. It is often said that ecology provides the stage on which the play of evolution unfolds. The two are intertwined in many ways.

We study ecology at several levels

Life exists in a hierarchy of levels, from atoms, molecules, and cells (pp. 22–28) up through the **biosphere,** the cumulative total of living things on Earth and the areas they inhabit. **Ecologists** are scientists who study relationships at the higher levels of this hierarchy (**FIGURE 3.12**), namely at the levels of the organism, population, community, ecosystem, landscape, and biosphere.

At the level of the organism, ecology describes relationships between an organism and its physical environment. Organismal ecology helps us understand, for example, what aspects of a Hawaiian honeycreeper's environment are important to it, and why. In contrast, **population ecology** examines the dynamics of population change and the factors that affect the distribution and abundance of members of a population. It helps us understand why populations of some species decline while populations of others increase.

In ecology, a **community** consists of an assemblage of populations of interacting species that inhabit the same area. A population of 'akiapōlā'au, a population of koa trees, a population of wood-boring grubs, and a population of ferns, together with all the other interacting plant, animal, fungal, and microbial populations in the Hakalau Forest, would be considered a community. **Community ecology** (Chapter 4) focuses on patterns of species diversity and on interactions

Biosphere

The sum total of living things on Earth and the areas they inhabit

Landscape

A geographic region including an array of ecosystems

Ecosystem

A functional system consisting of a community, its nonliving environment, and the interactions between them

Community

A set of populations of different species living together in a particular area

Population

A group of individuals of a species that live in a particular area

Organism

An individual living thing

FIGURE 3.12 Green sea turtles are part of a coral reef community that inhabits reef ecosystems along Hawaii's coasts.

among species, ranging from one-to-one interactions to complex interrelationships involving the entire community.

Ecosystems (p. 110) encompass communities and the abiotic (nonliving) material and forces with which community members interact. Hakalau's cloud-forest ecosystem consists of its community plus the air, water, soil, nutrients, and energy used by the community's organisms. **Ecosystem ecology** (Chapter 5) addresses the flow of energy and nutrients by studying living and nonliving components of systems in conjunction. Today's warming climate (Chapter 18) is having ecosystem-level consequences as it affects Hakalau and other ecosystems throughout the world.

Today, concerns such as climate change and habitat loss, together with technologies such as satellite imagery, are invigorating the study of how ecosystems are arrayed across the landscape. **Landscape ecology** (p. 114) helps us understand how and why ecosystems, communities, and populations are distributed across geographic regions. Indeed, as new technologies help scientists study the complex dynamics of natural systems at a global scale, ecologists are expanding their horizons to the biosphere as a whole.

Each organism has habitat needs

At the level of the organism, each individual relates to its environment in ways that tend to maximize its survival and reproduction. One key relationship involves the specific environment in which an organism lives, its **habitat.** A species' habitat consists of the living and nonliving elements around it, including rock, soil, leaf litter, humidity, plant life, and more. The ʻakiapōlāʻau (**FIGURE 3.13**) lives in a habitat of cool, moist, montane forest of native koa and ʻōhiʻa trees, where it is high enough in elevation to be safe from avian pox and malaria.

Each organism thrives in certain habitats and not in others, leading to nonrandom patterns of **habitat use.** Mobile organisms actively select habitats in which to live from among the range of options they encounter, a process called **habitat selection.** In the case of plants and of stationary animals (such as sea anemones in the ocean), whose young disperse and settle passively, patterns of habitat use result from success in some habitats and failure in others.

FIGURE 3.13 The ʻakiapōlāʻau fills a unique niche by virtue of its odd bill. The bill's straight bottom half and long, curved top half allow the bird to specialize on digging grubs out from native trees in its montane forest habitat.

Habitats are scale-dependent. A tiny soil mite may use less than a square meter of soil in its lifetime, whereas an eagle, elephant, or whale may traverse many miles of air, land, or water in just a day. Species also may have different habitat needs in different seasons; many migratory birds use distinct breeding, wintering, and migratory habitats.

Likewise, the criteria by which organisms favor some habitats over others can vary greatly. A soil mite may assess habitats in terms of the chemistry, moisture, and texture of the soil. For a whale, water temperature, salinity, and the density of marine microorganisms might be critical characteristics.

Habitat is a vital concept in environmental science. Because habitats provide everything an organism needs, including nutrition, shelter, breeding sites, and mates, the organism's very survival depends on the availability of suitable habitats. Often this need results in conflict with people who want to alter a habitat for their own purposes.

Organisms have roles in communities

Another way in which an organism relates to its environment is through its **niche,** its functional role in a community. A species' niche reflects its use of habitat and resources, its consumption of certain foods, its role in the flow of energy and matter, and its interactions with other organisms. The niche is a multidimensional concept, a kind of summary of everything an organism does. The pioneering ecologist Eugene Odum once wrote that "habitat is the organism's address, and the niche is its profession."

Organisms vary in the breadth of their niches. A species with narrow breadth, and thus very specific requirements, is said to be a **specialist.** One with broad tolerances, a "jack-of-all-trades" able to use a wide array of resources, is a **generalist.** A native Hawaiian honeycreeper like the 'akiapōlā'au (see Figure 3.13) is a specialist, because its unique bill is exquisitely adapted for feeding on grubs that tunnel through the wood of certain native trees. In contrast, the common myna (a bird introduced to Hawai'i from Asia) is a generalist; its unremarkable bill allows it to eat many types of foods in many habitats. As a result, the common myna has spread through virtually all areas of the Hawaiian Islands where human development has altered the landscape.

Generalists like the myna succeed by being able to live in many different places and withstand variable conditions, yet they do not thrive in any single situation as well as a specialist adapted for those specific conditions. (A jack-of-all-trades, as the saying goes, is a master of none.) Specialists succeed over evolutionary time by being extremely good at the things they do, yet they are vulnerable when conditions change and threaten the habitat or resource on which they have specialized. An organism's habitat preferences, niche, and degree of specialization each reflect adaptations of the species and are products of natural selection.

Population Ecology

A population, as we have seen, consists of individuals of a species that inhabit a particular area at a particular time. Population ecologists try to understand and predict how populations change over time. The ability to predict a population's growth or decline is useful in monitoring and managing wildlife, fisheries, and threatened and endangered species (see **THE SCIENCE BEHIND THE STORY**, pp. 62–63). It is also crucial for understanding the dynamics of our human population (Chapter 8)—a central element of environmental science and one of the prime challenges for our society today.

Populations possess features that help predict their dynamics

All populations—from humans to honeycreepers—exhibit attributes that help population ecologists predict their dynamics.

Population size Expressed as the number of individual organisms present at a given time, **population size** may increase, decrease, undergo cyclical change, or remain stable over time. Populations generally grow when resources are abundant and natural enemies are few. Populations can decline in response to loss of resources, natural disasters, or impacts from other species.

The passenger pigeon, now extinct, illustrates extremes in population size (**FIGURE 3.14**). Not long ago it was the most abundant bird in North America; flocks of passenger pigeons literally darkened the skies. In the early 1800s, ornithologist Alexander Wilson watched a flock of 2 billion birds 390 km (240 mi) long that took 5 hours to fly

FIGURE 3.14 Flocks of passenger pigeons literally darkened the skies as billions of birds passed overhead. Still, hunting and deforestation drove North America's most numerous bird to extinction within decades.

over and that he described as sounding like a tornado. Passenger pigeons nested in gigantic colonies in the forests of the upper Midwest and southern Canada. Once settlers began cutting the forests, the birds were easy targets for market hunters, who gunned down thousands at a time. The birds were shipped to market by the wagonload and sold for food. By 1890, the population had declined to such a low number that the birds could not form the large colonies they evidently needed to breed. In 1914, the last passenger pigeon on Earth died in the Cincinnati Zoo, bringing the continent's most numerous bird species to extinction within just a few decades.

Hawai'i offers a story with a happier ending. Hawaii's state bird is the nēnē (pronounced "nay-nay"), also called the Hawaiian goose (see Figure 3.2). Before people reached the Hawaiian Islands, nēnēs were common throughout the island chain, and the nēnē population is thought to have numbered at least 25,000 birds. After human arrival, the nēnē was nearly driven to extinction by hunting; livestock and plants that people introduced (which destroyed and displaced the vegetation it fed on); and rats, cats, dogs, pigs, and mongooses that preyed on its eggs and young. By the 1950s, these impacts had eliminated nēnēs from all islands except the Island of Hawai'i, where the population size was down to just 30 individuals. Fortunately, since then, dedicated conservation efforts have turned this decline around. Biologists and wildlife managers have labored to breed nēnēs in captivity and reintroduce them to protected areas. These efforts are succeeding, and today nēnēs live in at least seven regions on four of the Hawaiian Islands, with a population size of more than 2000 birds.

Population density The flocks and breeding colonies of passenger pigeons showed high population density, another attribute that ecologists assess. **Population density** describes the number of individuals per unit area in a population. High population density makes it easier for organisms to group together and find mates, but it can also lead to competition and conflict if space, food, or mates are in limited supply. Overcrowding among individuals can also increase the transmission of infectious disease. In contrast, at low population densities, individuals benefit from more space and resources but may find it harder to locate mates and companions.

Population distribution **Population distribution** describes the spatial arrangement of organisms in an area. Ecologists define three distribution types: random, uniform, and clumped (**FIGURE 3.15**). In a *random distribution*, individuals are located haphazardly in no particular pattern. This type of distribution can occur when the resources an organism needs are plentiful throughout an area and other organisms do not strongly influence where members of a population settle.

A *uniform distribution*, in which individuals are evenly spaced, can occur when individuals compete for space. Animals may hold and defend territories. Plants need space for their roots to gather moisture, and they may exude

(a) Random: Distribution of organisms displays no pattern

(b) Uniform: Individuals are spaced evenly

(c) Clumped: Individuals concentrate in certain areas

FIGURE 3.15 Individuals in a population can be spatially distributed in three fundamental ways.

chemicals that poison one another's roots as a means of competing for space. As a result, competing individuals may end up distributed at equal distances from one another.

A *clumped distribution* often results when organisms seek habitats or resources that are unevenly spaced. Many desert plants grow in patches near isolated springs, ponds, or streambeds. Hawaiian honeycreepers tend to cluster near flowering trees that offer nectar. People frequently aggregate in villages, towns, or cities.

Sex ratio A population's **sex ratio** is its proportion of males to females. In monogamous species (in which each sex takes

How Are Bird Populations Changing at Hakalau Forest?

Conservationist Jack Jeffrey in the Hakalau Forest

Are populations of honeycreepers increasing or decreasing? It's a high-stakes question. The answer could tell us whether efforts to save them are on the right track, or whether the birds may be headed for extinction.

At Hakalau Forest National Wildlife Refuge, biologists have been working for years to understand the dynamics of bird populations. But monitoring populations is not easy, and there is still debate and uncertainty about trends in these birds' populations.

The Hakalau Refuge is home to nine species of native forest birds, including four federally endangered ones: the Hawai'i 'ākepa; the Hawai'i creeper; the 'akiapōlā'au; and the 'io, or Hawaiian hawk. A number of non-native birds are now found here as well. Much of the region's native forest had been cleared for cattle ranching years earlier, while free-roaming pigs and invasive plants had degraded the rest.

The biologists employed to manage Hakalau after the refuge was established in 1985 built fences to keep pigs and cattle out of the forest. They labored to restore the area by removing invasive weeds and by planting a half-million native plants. Gradually the forest recovered, and stands of young trees took root at higher elevations. But has the restored forest brought higher populations of native forest birds?

In 1987, federal biologists and trained observers began conducting periodic surveys of birds on the refuge. Following standard protocols, they performed "point counts" by walking transects (linear routes), stopping for 8 minutes in predetermined spots, and counting numbers of each species seen or heard. With 343 points along 15 transects across the refuge, this sampling allowed them to estimate population densities for each bird species. Researchers then analyzed changes in these samples through time to make inferences about changes in population densities and population sizes.

After 21 years of point counts, refuge biologists summarized their analyses. At a 2008 workshop and in a 2009 technical report, they concluded that populations of most native birds were either stable or slowly increasing across most of the refuge. Densities of birds varied from year to year, which

Their report included a series of graphs similar to that shown in **FIGURE 1**. Densities of birds varied from year to year, which

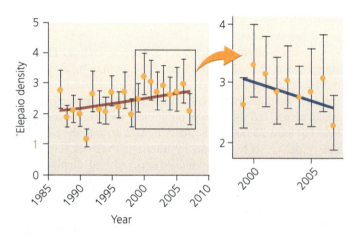

FIGURE 1 Population density data for the Hawai'i 'elepaio, a native forest bird, typify data gathered at Hakalau Forest NWR. Over a 21-year period, regression shows an increase (**red line**). However, over the most recent 9 years, regression shows a decrease (**blue line**). Thin black bars indicate 95% confidence intervals. *Data from Camp, Richard J., et al., 2009.* Passerine bird trends at Hakalau Forest National Wildlife Refuge, Hawai'i. *Hawai'i Cooperative Studies Unit Technical Report HCSU-011.*

 What trend do the data illustrate during the first 12 years of the period shown? Draw a line through the first 12 data points that best represents this trend. How does it differ in slope from the red line? Explain why there is a difference.

Go to **Interpreting Graphs & Data** on **Mastering**EnvironmentalScience®

is a normal and expected result of varying conditions (such as weather). Because of such variation, long-term studies are necessary; scientists expect that over time, year-to-year variation will be overshadowed by long-term trends that reveal the actual growth or decline of the population.

To interpret trends, researchers use a statistical method called *regression*. Linear regression allows scientists to analyze how values change through time and to determine the straight line that, when drawn through data points on a graph, most accurately represents their trend (see Figure 1). With the data from Hakalau, linear regression led managers to conclude that over the 21-year period:

- In high-elevation pastures that managers were restoring to forest, populations of birds that used the young trees were rising sharply.
- In middle-elevation open forest being managed, populations of all native birds were either stable or increasing.

(a) Hawai`i `ākepa

(b) Japanese white-eye

FIGURE 2 Work by biologist Leonard Freed suggests that the `ākepa is suffering competition from the non-native Japanese white-eye.

The long-term picture thus appeared bright. However, during the most recent 9 years of the 21-year period, regression indicated that populations of many species were declining (see Figure 1).

Biologist Leonard Freed of the University of Hawai`i at Manoa argued that federal biologists were overemphasizing the positive long-term trends and ignoring the negative near-term trends. For years, Freed and his colleagues had conducted research at Hakalau, focusing on the breeding biology of the Hawai`i `ākepa (**FIGURE 2a**). Their research suggested that the `ākepa began suffering competition for food once the non-native Japanese white-eye (**FIGURE 2b**) became abundant in the forest. Freed maintained that competition from the Japanese white-eye, along with attacks from parasitic lice in the nests, stunted the growth of the young birds and threatened the `ākepa population. Freed urged that white-eyes be trapped and killed to save the `ākepa. Refuge biologists called Freed's results controversial and said they needed validation before approving of such a response.

The debate came to a head in a pair of papers published back-to-back in the ornithological journal *Condor* in 2010. The team of federal biologists, led by Richard Camp and including Jack Jeffrey, presented their data and acknowledged that many populations on the refuge showed downward trends in the most recent 9 years. Freed and his colleague Rebecca Cann reanalyzed the federal data using alternative methods and questioned whether the earlier apparent increases were reliable.

At stake is how best to manage the forest and its birds. If Camp and his colleagues are right, then management actions taken so far seem to have been effective, boosting populations or holding them stable in the face of dire threats. If Freed and Cann are right, then management strategies may need to be rethought.

Many factors could account for the apparent recent declines in populations of `ākepas and other native honeycreepers, including the simple fact that thicker forest vegetation has made it harder for counters to detect birds. Of concern, though, is the possibility that challenges from outside the refuge—such as avian malaria and pox moving upslope with climate warming—might eventually overwhelm even the best management efforts.

To clarify the situation, researchers today continue to survey Hakalau's birds, adding to their valuable long-term database of population trends. But government budget cuts are threatening their ability to analyze the data—as well as their capacity to safeguard the refuge. Amid shortfalls in funding and staffing several years ago, pigs broke through fences and began degrading the newly restored forest. When budgets are cut, refuge staff must work even harder just to regain the progress previously made.

Despite the apparent recent declines in bird populations at Hakalau, populations there seem to be faring better than elsewhere on the Island of Hawai`i (**FIGURE 3**). Moreover, the reforestation of Hakalau's upper zone is creating a new habitat into which birds are moving. This success is a hopeful sign that research and careful management can help undo past damage and preserve endangered island species.

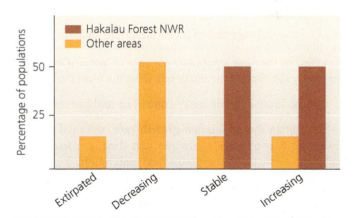

**FIGURE 3 At Hakalau Forest NWR, native forest bird populations were judged to be stable or increasing, whereas at four other protected areas on the Island of Hawai`i, most populations were judged to be decreasing, and some have recently vanished.** *Data from Camp, Richard J., et al., 2009.* Passerine bird trends at Hakalau Forest National Wildlife Refuge, Hawai`i. *Hawai`i Cooperative Studies Unit Technical Report HCSU-011.*

a single mate), a 1:1 sex ratio maximizes population growth, whereas an unbalanced ratio leaves many individuals without mates. Most species are not monogamous, however, so sex ratios may vary from one species to another.

Age structure **Age structure** describes the relative numbers of individuals of different ages within a population. By combining this information with data on the reproductive potential of individuals in each age class, a population ecologist can predict how the population may grow or shrink.

Many plants and animals continue growing in size as they age, and in these species, older individuals often reproduce more. Older, larger trees in a population produce more seeds than smaller, younger trees of the same species. Older, larger fish produce more eggs than smaller, younger fish. Birds use the experience they gain with age to become more successful breeders at older ages.

Human beings are unusual because we often survive past our reproductive years. A human population made up largely of older (post-reproductive) individuals will tend to decline over time, whereas one with many young people (of reproductive or pre-reproductive age) will tend to increase. (We will use diagrams to explore these ideas further in Chapter 8, pp. 193–194, as we study human population growth.)

Populations may grow, shrink, or remain stable

Let's now take a more quantitative approach and examine some simple mathematical concepts used by population ecologists and by **demographers** (scientists who study human populations). Population change is determined by four factors:

- Natality (births within the population)
- Mortality (deaths within the population)
- Immigration (arrival of individuals from outside the population)
- Emigration (departure of individuals from the population)

We can measure a population's **rate of natural increase** simply by subtracting the death rate from the birth rate:

(birth rate) – (death rate) = rate of natural increase

To obtain the **population growth rate,** the total rate of change in a population's size per unit time, we must also include the effects of immigration and emigration:

(birth rate − death rate) + (immigration rate − emigration rate) = population growth rate

The rates in these formulas are often expressed in numbers per 1000 individuals per year. For example, a population with a birth rate of 18 per 1000/yr, a death rate of 10 per 1000/yr, an immigration rate of 5 per 1000/yr, and an emigration rate of 7 per 1000/yr would have a population growth rate of 6 per 1000/yr:

(18/1000 − 10/1000) + (5/1000 − 7/1000) = 6/1000

Given these rates, a population of 1000 in 1 year will grow to 1006 in the next. If the population is 1,000,000, it will reach 1,006,000 the next year. Such population increases are often expressed as percentages, as follows:

population growth rate × 100%

Thus, a growth rate of 6/1000 would be expressed as:

6/1000 × 100% = 0.6%

By measuring population growth in percentages, we can compare changes among populations of far different sizes. We can also project changes into the future. Understanding and predicting such changes helps wildlife and fisheries managers regulate hunting and fishing to ensure sustainable harvests, informs conservation biologists trying to protect rare and declining species, and assists policymakers planning for human population growth in cities, regions, and nations.

Unregulated populations increase by exponential growth

When a population increases by a fixed percentage each year, it is said to undergo **exponential growth.** Imagine you put money in a savings account at a fixed interest rate and leave it untouched for years. As the principal accrues interest and grows larger, you earn still more interest, and the sum grows by escalating amounts each year. The reason is that a fixed percentage of a small number makes for a small increase, but the same percentage of a large number produces a large increase. Thus, as savings accounts (or populations) grow larger, each incremental increase likewise becomes larger in absolute terms. Such acceleration is a characteristic of exponential growth.

We can visualize changes in population size by using population growth curves. The J-shaped curve in **FIGURE 3.16** shows

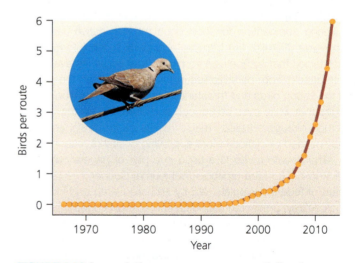

FIGURE 3.16 A population may grow exponentially when colonizing an unoccupied environment or exploiting an unused resource. The Eurasian collared dove is spreading across the United States, propelled by exponential growth. *Data from Sauer, J.R., et al., 2014.* The North American Breeding Bird Survey, results and analysis 1966–2013. v. 01.30.2015. Laurel, MD: USGS Patuxent Wildlife Research Center.

exponential growth. Populations increase exponentially unless they are constrained. Each organism reproduces by a certain amount, and as populations grow, there are more individuals reproducing by that amount. If there are adequate resources and no external limits, ecologists expect exponential growth.

Normally, exponential growth occurs in nature only when a population is small, competition is minimal, and environmental conditions are ideal for the organism in question. Most often, these conditions occur when the organism arrives in a new environment that contains abundant resources. Mold growing on a piece of fruit or bacteria decomposing a dead animal are cases in point. Plants colonizing regions during primary succession (p. 83) after glaciers recede or volcanoes erupt may also show exponential growth. In Hawai'i, many species that colonized the islands underwent exponential growth for a time after their arrival. One current example of exponential growth in mainland North America is the Eurasian collared dove (see Figure 3.16). Unlike its extinct relative the passenger pigeon, this species arrived here from Europe, thrives in human-disturbed areas, and has spread across the continent in a matter of years.

Limiting factors restrain growth

Exponential growth rarely lasts long. If even a single species were to increase exponentially for very many generations, it would blanket the planet's surface! Instead, every population eventually is constrained by **limiting factors**—physical, chemical, and biological attributes of the environment that restrain population growth. Together, these limiting factors determine the **carrying capacity,** the maximum population size of a species that a given environment can sustain.

Ecologists use the S-shaped curve in **FIGURE 3.17** to show how an initial exponential increase is slowed and eventually brought to a standstill by limiting factors. This phenomenon is called **logistic growth.** A logistic growth curve rises sharply at first but then begins to level off as the effects

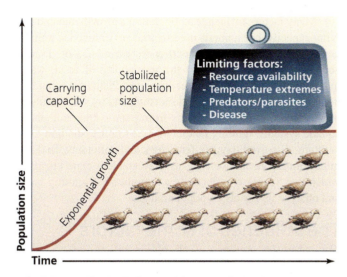

FIGURE 3.17 The logistic growth curve shows how population size may increase rapidly at first, then slow down, and finally stabilize at a carrying capacity.

of limiting factors become stronger. Eventually the collective force of these factors stabilizes the population size at its carrying capacity.

We can witness this process by taking a closer look at the data for the Eurasian collared dove, as gathered by thousands of volunteer birders and analyzed by government biologists in the Breeding Bird Survey, a long-running citizen science project. The dove first appeared in Florida a few decades ago and then spread west and north. Today its numbers are growing fastest in western areas it has recently reached and slower in southeastern areas where it has been present for longer. In Florida, it has apparently reached carrying capacity (**FIGURE 3.18**). Populations of other European birds that spread across North America in the past, such as the house sparrow and European starling, have peaked and are today beginning to decline.

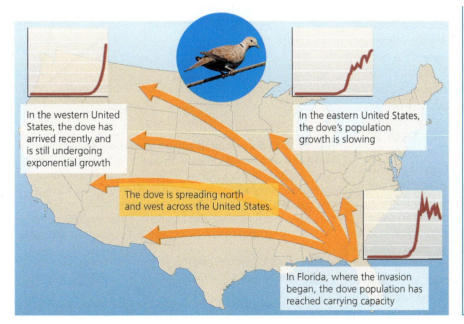

In the western United States, the dove has arrived recently and is still undergoing exponential growth

In the eastern United States, the dove's population growth is slowing

The dove is spreading north and west across the United States.

In Florida, where the invasion began, the dove population has reached carrying capacity

FIGURE 3.18 Exponential growth slows over time and gives way to logistic growth. By breaking down the continent-wide data for the Eurasian collared dove from Figure 3.16, we can track its spread west and north from Florida, where it first arrived. Today its population growth is fastest in the west, slower in the east (where the species has been present longer), and stable in Florida (where it has apparently reached carrying capacity). *Data from Sauer, J.R., et al., 2014.* The North American Breeding Bird Survey, results and analysis 1966–2013.

DATA Looking several decades into the future, what do you predict the population growth graph for the Eurasian collared dove in the western United States will look like? Explain why.

Go to **Interpreting Graphs & Data** on **Mastering**EnvironmentalScience®

Many factors influence a population's growth rate and carrying capacity. For animals in terrestrial environments, limiting factors include temperature extremes; prevalence of disease; abundance of predators; and the availability of food, water, mates, shelter, and suitable breeding sites. Plants are often limited by amounts of sunlight and moisture and by soil chemistry, in addition to disease and attack from plant-eating animals. In aquatic systems, limiting factors include salinity, sunlight, temperature, dissolved oxygen, fertilizers, and pollutants. To determine limiting factors, ecologists may conduct experiments in which they increase or decrease a hypothesized limiting factor and observe its effects on population size.

The influence of some factors depends on population density

A population's density can enhance or diminish the impact of certain limiting factors. Recall that high population density can help organisms find mates but can also increase competition and the risk of predation and disease. Such limiting factors are said to be **density-dependent** because their influence rises and falls with population density. The logistic growth curve in Figure 3.17 represents the effects of density dependence. The larger the population size, the stronger the effects of the limiting factors.

Density-independent factors are those whose influence is independent of population density. Temperature extremes and catastrophic events such as floods, fires, and landslides are examples of density-independent factors, because they can eliminate large numbers of individuals without regard to their density.

The logistic curve is a simplified model, and real populations in nature can behave differently. Some may cycle above and below the carrying capacity. Others may overshoot the carrying capacity and then crash, destined either for extinction or for recovery.

weighing the ISSUES

Carrying Capacity and Human Population Growth

The global human population has surpassed 7 billion, far exceeding our population's size throughout our history on Earth. Name some specific means by which we have apparently raised Earth's carrying capacity for our species. Do you think we can continue to raise our carrying capacity? How might we do so? What limiting factors exist for the human population today? Might Earth's future carrying capacity for us decrease? Why or why not?

Carrying capacities can change

Because environments are complex and ever-changing, carrying capacity can vary. If a fire destroys a forest, for example, the carrying capacities for most forest animals will decline, whereas carrying capacities for species that benefit from fire (such as fire-adapted grasses or trees with specially adapted seeds) will increase. Our own species has proven capable of intentionally altering our environment to raise our carrying capacity. When our ancestors began to build shelters and use fire for heating and cooking, they eased the limiting factors of cold climates and were able to expand into new territory. As human civilization developed, we overcame

limiting factors through the development of new technologies and cultural institutions. People have managed so far to increase the planet's carrying capacity for our species, but we have done so by appropriating immense proportions of the planet's natural resources. In the process, we have reduced carrying capacities for countless other organisms that rely on those same resources.

Life history strategies vary among species

Population ecology, organismal ecology, and evolution all come together in **life history theory,** a scientific approach that explains how natural selection influences patterns in reproduction, survival, and life span. In an environment of limited resources and limiting factors, an organism faces trade-offs in how it can apportion its energy. Over time, each species has evolved its own way of allocating its investment among reproduction, parental care, and survival.

For example, many fish, plants, frogs, and insects mature rapidly and devote their energy to producing many offspring in a short time. Most often such species do not provide parental care to these offspring, but simply leave their survival to chance. The vast majority of offspring die, giving rise to a Type III *survivorship curve* (**FIGURE 3.19**). However, because there are so many offspring, the few survivors are enough to sustain the population. Such species are often called *r-selected,* and they are adapted to do well in variable or

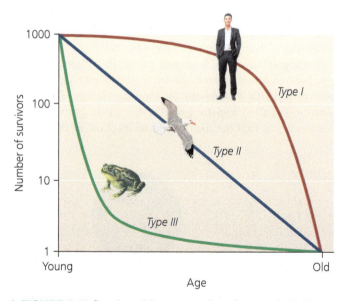

FIGURE 3.19 Survivorship curves show how an individual's likelihood of survival varies with age. In a Type I survivorship curve, survival rates are high when organisms are young and decrease sharply when organisms are old. In a Type II survivorship curve, survival rates are equivalent regardless of an organism's age. In a Type III survivorship curve, most mortality takes place at young ages, and survival rates are greater at older ages.

 DATA Q Which organism has the highest rate of survival at a young age: a toad, a bird, or a human being?

Go to **Interpreting Graphs & Data** on **Mastering**EnvironmentalScience®

unpredictable environments. The abbreviation *r* denotes the per capita rate at which a population increases in the absence of limiting factors. Population sizes of r-selected species fluctuate greatly, such that they are often well below carrying capacity. This is why natural selection in these species favors traits that lead to rapid population growth.

Other species are said to be *K-selected* (because their populations tend to stabilize near carrying capacity, commonly abbreviated *K*). Large animals such as giraffes, elephants, whales, and humans produce relatively few offspring during their lifetimes and require a long time to gestate and raise their young. However, the considerable energy and resources they devote to caring for them helps give these few offspring a high likelihood of survival, giving rise to a Type I survivorship curve (see Figure 3.19)—especially in stable environments. Because their populations stay close to carrying capacity, these slow-maturing, long-lived organisms must compete to hold their own in a crowded world. Thus, natural selection favors investing in high-quality offspring that can be good competitors.

It is important to note, however, that *these are two extremes on a continuum* and that most species fall somewhere between the extremes of r-selected and K-selected species. Moreover, many organisms show combinations of traits that do not correspond to a place on the continuum. A redwood tree, for instance, is large and long-lived, yet it produces many small seeds and offers no parental care.

Conserving Biodiversity

Populations have always been affected by environmental change, but today human development and resource extraction are speeding the rate of change and bringing new types of impacts. Fortunately, committed people are taking action to safeguard biodiversity and to preserve and restore Earth's ecological and evolutionary processes. (We will explore these efforts more fully in Chapter 11.) For now, let's see how Hawaiians are confronting the threats to their biodiversity.

Introduced species pose challenges

By introducing species into areas where they do not occur naturally, human beings often set in motion cascades of impacts on native populations, communities, and ecosystems. Some introduced species thrive in their new surroundings, killing or displacing native species (pp. 86–90, 286–287). Island-dwelling organisms are particularly vulnerable to introduced species. Because island inhabitants have evolved in isolation in small areas amid a limited community of other species, they tend to lack defenses against mainland species that are well adapted to deal with a broad array of enemies.

As we have seen, the Hawaiian Islands have been transformed by introduced species. Cattle, goats, sheep, and pigs eat native vegetation, endangering plant populations. Alien grasses, shrubs, and trees spread across the landscapes that livestock have altered. Birds suffering already from habitat loss and predation by non-native mammals now also struggle against diseases like pox and malaria. Pigs have worsened the malaria problem, because they dig holes in the forest floor, where rainwater forms shallow pools in which mosquitoes breed.

As a result, biologists and land managers have found that trying to help a species in trouble often means trying to eradicate or control populations of another that is doing too well. For instance, in many areas in Hawai'i feral pigs are being hunted and areas are then fenced off once they are free of pigs.

Innovative solutions are working

Amid the challenges of Hawaii's extinction crisis, hard work is resulting in some inspirational success stories, and several species have been saved from imminent extinction. Early work at Hawai'i Volcanoes National Park inspired the conservation work at Hakalau Forest, as well as efforts by managers and volunteers from the Hawai'i Division of Forestry and Wildlife, The Nature Conservancy of Hawai'i, Kamehameha Schools, and local watershed protection groups. Across Hawai'i, many people are protecting land, removing alien mammals and weeds, and restoring native habitats. Offshore, Hawaiians are striving to protect their fabulous coral reefs, sea grass beds, and beaches from pollution and overfishing. The northwesternmost Hawaiian Islands are now part of the largest federally declared marine reserve (p. 440) in the world.

Hawaii's citizens are reaping economic benefits from their conservation efforts. The islands' wildlife and natural areas draw visitors from around the world, a phenomenon called **ecotourism** (FIGURE 3.20). A large percentage of Hawaii's tourism is ecotourism, and altogether tourism draws more than 7 million visitors to Hawai'i each year, creates thousands of jobs, and pumps $12 billion annually into the state's economy.

FIGURE 3.20 Hawai'i protects some of its diverse natural areas, helping to stimulate its economy with ecotourism. Here, a scuba diver observes raccoon butterflyfish at a coral reef along the Kona coast of the Island of Hawai'i.

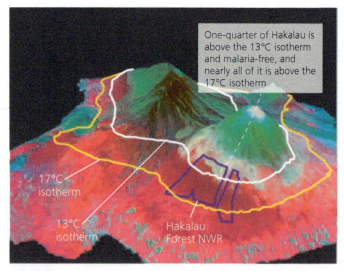

One-quarter of Hakalau is above the 13°C isotherm and malaria-free, and nearly all of it is above the 17°C isotherm

17°C isotherm

13°C isotherm

Hakalau Forest NWR

(a) Today

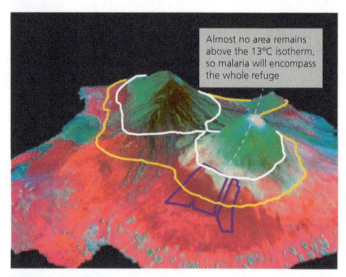

Almost no area remains above the 13°C isotherm, so malaria will encompass the whole refuge

(b) With 2°C of climate warming

(c) Cloud cover viewed from the summit of Mauna Kea

Climate change poses an extra challenge

Traditionally, people sought to conserve populations of threatened species by preserving and managing tracts of land (or areas of ocean) designated as protected areas. However, global climate change (Chapter 18) now threatens this strategy. As temperatures climb and rainfall patterns shift, conditions within protected areas may become unsuitable for the species they were meant to protect.

Hawaii's systems are especially vulnerable. As discussed earlier, at Hakalau Forest on the slopes of Mauna Kea, mosquitoes and malaria are moving upslope toward the refuge as temperatures rise, exposing more and more birds to disease (**FIGURE 3.21**). Some research suggests that climate change will lower the cloud layer atop Mauna Kea, reducing rainfall at high elevations and pushing the upper limit of the forest downward. If this comes to pass, Hakalau's honeycreepers may become trapped within a shrinking band of forest by disease from below and drought from above.

The challenges posed by climate change mean that scientists and managers need to come up with new ways to help save declining populations. (We will learn about the many efforts being made across the world in our exploration of biodiversity and conservation biology in Chapter 11.) In Hawai'i, management and ecotourism can help preserve natural systems, but resources to preserve habitat and protect endangered species will likely need to be stepped up. Restoring altered communities to their former condition—as is being done at Hakalau Forest—will also be necessary. The restoration of ecological communities is just one phenomenon we will examine in our next chapter, as we shift from populations to communities.

FIGURE 3.21 Researchers have modeled how a warming climate will affect the native birds of Hakalau Forest NWR. Avian malaria cannot survive where temperatures dip below 13°C, and it peaks where summer temperatures average 17°C. Today **(a)** 24% of Hakalau lies above (cooler than) the 13°C isotherm and is free of malaria. If climate warms by 2°C **(b)**, however, then the isotherms move upslope, and only 1% of Hakalau will remain cooler than 13°C and malaria-free. *Data from Benning, T.L., et al., 2002. Interactions of climate change with biological invasions and land use in the Hawaiian Islands: Modeling the fate of endemic birds using a geographic information system. Proc Natl. Acad. Sci. 99: 14246–14249.*

closing
THE LOOP

The honeycreepers of Hakalau Forest National Wildlife Refuge, along with many other Hawaiian species, help to illuminate the fundamentals of evolution and population ecology that are integral to environmental science. Island chains like Hawaiʻi can be viewed as "laboratories of evolution," showing us how populations evolve and how new species arise. But just as islands are crucibles of speciation, today they are also hotspots of extinction. In our age of global human travel, non-native species brought to Hawaiʻi have overrun the islands' landscapes and upset their ecological balance. As populations of native species dwindle, biologists monitor the changes and conservationists race to restore habitats, fight invasive species, and save native plants and animals. Scientists and conservationists working with Hawaii's embattled flora and fauna are taking their understanding of how ecological processes function at the population level and are applying it to protect biodiversity threatened by the mass extinction event that is now underway. If we can study, understand, and act to address the challenges facing populations and biodiversity in places like Hawaiʻi, we can do so anywhere in the world—and do so we must if we are to protect Earth's biodiversity.

REVIEWING Objectives

You should now be able to:

Explain natural selection and cite evidence for this process

Natural selection is the process whereby inherited traits that enhance survival and reproduction are passed on more frequently to offspring than traits that do not enhance survival and reproduction. Evidence of natural selection can be found in countless adaptations in the diversity of wild species that exist today. It can also be found in our crop plants, pets, and farm animals, all of which have been bred by artificial selection. (pp. 48–51)

Describe how evolution generates and shapes biodiversity

Species can form in various ways; most commonly, geographic isolation over many generations leads to speciation, producing new types of organisms that enhance Earth's biological diversity. Natural selection can be a diversifying force as populations of organisms adapt to their environments. Phylogenetic trees and the fossil record teach us about the history of life by chronicling how populations of organisms have evolved. (pp. 51–56)

Discuss major factors behind species extinction and cite Earth's known mass extinction events

Although extinction occurs naturally, human impact is profoundly accelerating the rate of extinction. Island species and ecologically specialized species are especially vulnerable. When species that are vulnerable or rare encounter rapid environmental change, this heightens extinction risk. Five episodes of mass extinction are known—caused likely by asteroid impact or geologic factors. Humans may be initiating a sixth mass extinction. (pp. 57–58)

List the levels of ecological organization

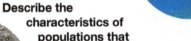

Ecologists study organisms, populations, communities, ecosystems, landscapes, and the biosphere. Habitat, niche, and specialization are vital ecological concepts. (pp. 58–60)

Describe the characteristics of populations that help predict population growth

Populations are characterized by size, density, distribution, sex ratio, and age structure. Rates of birth, death, immigration, and emigration determine how a population will change. (pp. 60–64)

Explain how logistic growth, limiting factors, carrying capacity, and other fundamental concepts affect population ecology

Populations unrestrained by limiting factors undergo exponential growth. Logistic growth results from density dependence; growth slows as population size increases and approaches a carrying capacity. Reproductive strategies differ among species, and carrying capacities can change—all of which affect population ecology. (pp. 64–67)

Identify and discuss challenges and current efforts in conserving biodiversity

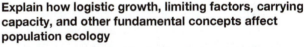

Habitat loss, the introduction of non-native species, and climate change are among the major challenges to biodiversity. People are striving to protect and restore populations, species, and habitats, even as human impacts continue to complicate efforts. (pp. 67–68)

TESTING Your Comprehension

1. Define the concept of natural selection in your own words, and explain how it follows logically from a few common observations of nature.

2. Describe an example of evidence for natural selection and an example of evidence for artificial selection.

3. Describe the steps involved in allopatric speciation.

4. Name three organisms that have become extinct or are threatened with extinction. For each, give a probable reason for its decline.

5. Define the terms *species*, *population*, and *community*. How does a species differ from a population? How does a population differ from a community?

6. Define and contrast the concepts of *habitat* and *niche*.

7. List and describe each of the five major attributes of populations that help ecologists predict population growth or decline. Briefly explain how each attribute shapes population dynamics.

8. Can a species undergo exponential growth indefinitely? Explain your answer.

9. Describe how limiting factors relate to carrying capacity.

10. Explain the difference between K-selected species and r-selected species. For each, provide an example that was not mentioned in the chapter.

SEEKING Solutions

1. In what ways has artificial selection changed people's quality of life? Give examples. How might artificial selection be used to improve our quality of life further? Can you envision a way it could be used to reduce our environmental impact?

2. In your region, what species are threatened with extinction? Why are they vulnerable? Suggest steps that could be taken to increase their populations.

3. Do you think the human species can continue raising its global carrying capacity? How so, or why not? Do you think we *should* try to keep raising our carrying capacity? Why or why not?

4. What are some advantages of ecotourism for a state like Hawai'i? What might be a potential disadvantage? Describe a source of ecotourism that exists—or could exist—in your own region.

5. **CASE STUDY CONNECTION** Describe two of the threats facing native species at Hakalau Forest NWR or elsewhere in Hawai'i, and two actions people have taken to address these threats. What new trend is now jeopardizing some of these efforts? What steps do you think might be required in the future to safeguard native species, populations, and communities in mountainous habitats in places like Hawai'i?

6. **THINK IT THROUGH** You are a population ecologist studying animals in a national park, and park managers are asking for advice on how to focus their limited conservation funds. How would you rate the following three species, from most vulnerable (and thus most in need of conservation attention) to least vulnerable? Give reasons for your choices.

   - A bird that is a generalist in its use of habitats and resources
   - A salamander endemic to the park that lives in high-elevation forest
   - A fish that specializes on a few types of invertebrate prey and has a large population size

CALCULATING Ecological Footprints

Professional demographers delve into the latest statistics from cities, states, and nations to bring us updated estimates on human populations. The table that follows shows population estimates for two consecutive years that take into account births, deaths, immigration, and emigration. To calculate the population growth rate (p. 64) for each region, divide the 2015 data by the 2014 data, subtract 1.00, and multiply by 100:

pop. growth rate = [(2015 pop. / 2014 pop.) − 1] × 100

Texas was one of the fastest-growing U.S. states during this particular 1-year time period, and Pennsylvania was one of the slowest-growing states. You can find data for your own state, city, or metropolitan area by exploring the Web pages of the U.S. Census Bureau.

REGION	2014 POPULATION	2015 POPULATION	POPULATION GROWTH RATE
Hawai'i	1,420,257	1,431,603	
Texas	26,979,078	27,469,114	
Pennsylvania	12,793,767	12,802,503	
Your state or city			
United States	318,907,401	321,418,820	0.79%
World	7,238,184,000	7,336,435,000	

Data: U.S. Census Bureau and Population Reference Bureau.

1. Assuming the population growth rate for the United States remained at the calculated rate, what would the population of the United States have been in 2016?

2. The birth rate of the United States in 2014 and 2015 was 12.5 births per 1000 people, and the death rate was 8.2 deaths per 1000 people. Using the formula on p. 64, what was the rate of natural increase for the United States between 2014 and 2015? Now subtract this rate from the population growth rate to obtain the net migration rate. Was the United States experiencing more immigration or more emigration during this period?

3. At the growth rate you calculated for Texas, the population of that state will double in just 38 years. What impacts would you expect this to have on (1) food supplies, (2) drinking water supplies, (3) forests and other natural areas, and (4) wildlife populations?

4. How does your own state, city, or metropolitan area compare in its growth rate with other regions in the table? What steps could your region take to lessen potential impacts of population growth on (1) food supplies, (2) drinking water supplies, (3) forests and other natural areas, and (4) wildlife populations?

MasteringEnvironmentalScience®

Students Go to **MasteringEnvironmentalScience** for assignments, the etext, and the Study Area with practice tests, videos, current events, and activities.

Instructors Go to **MasteringEnvironmentalScience** for automatically graded activities, current events, videos, and reading questions that you can assign to your students, plus Instructor Resources.

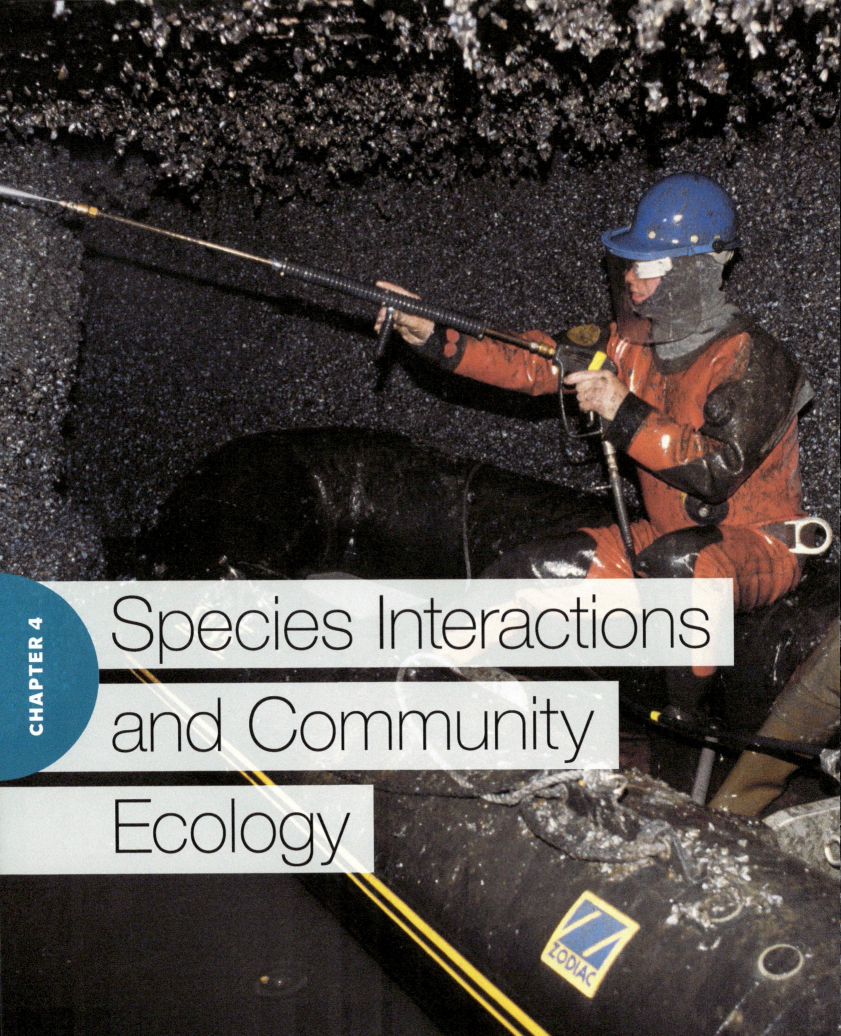

CHAPTER 4

Species Interactions and Community Ecology

CANADA

Great Lakes

UNITED STATES

Black and White, and Spread All Over:
Zebra Mussels Invade the Great Lakes

> We are seeing changes in the Great Lakes that are more rapid and more destructive than any time in [their] history.
>
> Andy Buchsbaum, National Wildlife Federation

> When you tear away the bottom of the food chain, everything that is above it is going to be disrupted.
>
> Tom Nalepa, National Oceanic and Atmospheric Administration

Things were looking up for the Great Lakes. Their waters were becoming cleaner and cleaner as regulation by the U.S. and Canadian governments brought industrial pollution under control starting in the 1970s. People ventured back onto the lakes for recreation, and populations of fish began to rebound.

Then the zebra mussel arrived. Black-and-white-striped shellfish the size of a dime, zebra mussels attach to hard surfaces and feed on algae by filtering water through their gills. This mollusk, given the scientific name *Dreissena polymorpha*, is native to the Caspian and Black seas where Europe meets Asia. In 1988, it was discovered in North American waters at Lake St. Clair, which connects Lake Erie with Lake Huron. People brought it to this continent by accident when ships from Europe discharged ballast water containing the mussels or their larvae. (To maintain stability at sea, ships take water into their hulls as they begin their voyage and then discharge that water at their destination.)

Within just two years of their discovery, zebra mussels had multiplied and reached all five of the Great Lakes. The next year, they invaded New York's Hudson River to the east, and the Illinois River in Chicago to the west. From the Illinois River and its canals, they soon reached the Mississippi River, gaining access to a vast watershed covering 40% of the United States. In just three more years, they spread to 19 U.S. states and 2 Canadian provinces. Today, they have colonized waters in 30 U.S. states.

How could a mussel spread so quickly? The zebra mussel's larval stage is well adapted to disperse long distances. Its tiny larvae drift freely for several weeks, traveling as far as the currents take them. Adults that adhere to boats and ships may be transported from place to place, even overland to isolated lakes and ponds. Once established, the mussels had free rein: In North America they were free of the predators, competitors, and parasites that had evolved with them in the Old World and limited their population growth there.

Why all the fuss? For one thing, zebra mussels clog water intake pipes at factories, power plants, municipal water supplies, and wastewater treatment facilities (**FIGURE 4.1a**). At one Michigan power plant, workers counted 700,000 mussels per square meter of pipe. Clusters of these organisms also

Zebra mussels encrusting a bicycle pulled from a lake

Upon completing this chapter, you will be able to:

- Summarize and compare the major types of species interactions

- Describe feeding relationships and energy flow, and use them to identify trophic levels and navigate food webs

- Discuss characteristics of a keystone species

- Characterize disturbance, succession, and notions of community change

- Predict the potential impacts of invasive species in communities, and suggest responses to biological invasions

- Explain the goals and methods of restoration ecology

- Identify and describe the terrestrial biomes of the world

Workers use a pressure hose to remove zebra mussels from the pump room of a power plant near Detroit.

damage boat engines, degrade docks, foul fishing gear, and sink buoys that ships use for navigation. In these ways, zebra mussels cost Great Lakes economies an estimated $5 billion in the first decade of the invasion, and continue to impose costs of hundreds of millions of dollars each year.

Zebra mussels also exert severe ecological impacts. They eat primarily **phytoplankton:** microscopic photosynthetic algae, protists, and cyanobacteria that drift in open water. Because each mussel filters a liter or more of water every day, zebra mussels can consume enough phytoplankton to deplete populations. Phytoplankton is the foundation of the Great Lakes food web, so its depletion is bad news for **zooplankton,** the tiny aquatic animals that eat phytoplankton—and for the fish that eat both. Researchers are finding that water bodies with zebra mussels contain fewer zooplankton and open-water fish than water bodies without them. And although zebra mussels benefit many bottom-feeding animals, they can suffocate native mollusks by attaching to their shells (**FIGURE 4.1b**).

Zebra mussels are just one of many invasive species affecting ecological communities and human economies today—but the zebra mussel story has a few extra twists. For one, on the heels of its invasion came another: The quagga mussel (*Dreissena bugensis*), a close relative of the zebra mussel from Ukraine, is also spreading through the Great Lakes and beyond. This species, named after an extinct type of zebra, is replacing the zebra mussel in many locations.

Moreover, today there are signs that the zebra mussel invasion may be losing steam. In a number of areas, zebra mussel populations have apparently peaked and begun to decline. In some cases they are being displaced by quagga mussels, but in others predation by fish, crabs, or ducks may be driving down their numbers as native predators develop a taste for the invader.

In areas where zebra mussels are declining, some of the native fish and invertebrates that suffered from their arrival are now recovering. This is welcome news, especially because the Great Lakes today are being bombarded with a variety of other non-native species, including several voracious species of Eurasian carp that are eating their way through the lakes' native flora and fauna.

(a) Zebra mussels clog water intake pipes of power plants and industrial facilities

(b) Zebra mussels foul, starve, and suffocate native clams by adhering to their shells and sealing them shut

FIGURE 4.1 Invasive species can have severe economic and ecological impacts.

Ecologists are monitoring populations closely to see how the situation develops. No one expects zebra mussels to disappear, but there is now hope that some of their impacts may be reversed and that the ecological communities of the Great Lakes can begin to be restored.

Species Interactions

By interacting with many species in a variety of ways, zebra mussels have set in motion an array of changes in the communities they have invaded. Interactions among species are the threads in the fabric of ecological communities. Ecologists organize species interactions into several fundamental categories (**TABLE 4.1**).

Competition can occur when resources are limited

When multiple organisms seek the same limited resource, their relationship is said to be one of **competition.** Competing organisms do not usually fight with one another directly and

TABLE 4.1 Species Interactions: Effects on Their Participants

TYPE OF INTERACTION	EFFECT ON SPECIES 1	EFFECT ON SPECIES 2
Mutualism	+	+
Predation, parasitism, herbivory	+	−
Competition	−	−

"+" denotes a positive effect; "−" denotes a negative effect.

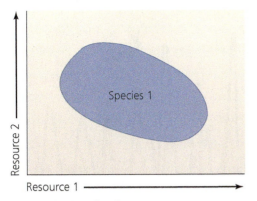

(a) Fundamental niche

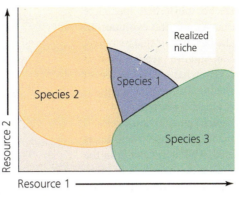

(b) Realized niche

FIGURE 4.2 Competition may force an organism to play a more limited ecological role. With no competitors (a), an organism can exploit its full fundamental niche. When competitors restrict what an organism can do or what resources it can use (b), the organism is limited to a realized niche, which covers only part of its fundamental niche.

physically, as the term might imply. Instead, competition is generally more subtle and indirect, taking place as organisms vie with one another to procure resources. Such resources include food, water, space, shelter, mates, sunlight, and more. Competitive interactions can take place between members of the same species (intraspecific competition) or between members of different species (interspecific competition).

If individuals of the same species are competing for limited resources, then competition becomes more intense when there are more individuals per unit area (denser populations). This is density dependence (p. 66), and it can limit the growth of a population.

Whereas intraspecific competition is a population-level phenomenon, interspecific competition affects communities. Interspecific competition can give rise to different types of outcomes. If one species is a very effective competitor, it may exclude other species from resource use entirely. This outcome, called **competitive exclusion,** occurred in parts of the Great Lakes as zebra mussels displaced native mussels—and it is happening now where quagga mussels are displacing zebra mussels.

Alternatively, if no single competitor fully excludes others, the species may continue to live side by side. This result, called **species coexistence,** may produce a stable point of equilibrium, at which the relative population sizes of each remain fairly constant through time.

Coexisting species that use the same resources tend to adjust to their competitors to minimize competition with them. Individuals may do this by changing their behavior to use only a portion of the total array of resources they are capable of using. In such cases, individuals do not fulfill their entire niche. A species' niche reflects its functional role in a community, including its resource use, habitat use, food consumption, and other attributes (p. 59)—it is a kind of multidimensional summary of everything an organism does.

The full niche of a species is called its **fundamental niche** (FIGURE 4.2a). An individual that plays only part of its role because of competition or another type of species interaction is said to display a **realized niche** (FIGURE 4.2b), the portion of its fundamental niche that is actually "realized," or fulfilled. The quagga mussel can occupy a wider range of water conditions and substrates than the zebra mussel, so it is thought to have a larger fundamental niche. The quagga mussel also appears to be reducing the zebra mussel's realized niche, as it displaces the zebra mussel in many areas.

Species experience similar adjustments over evolutionary time. Over many generations, the process of natural selection (pp. 48–51) may respond to competition by favoring individuals that use slightly different resources or that use shared resources in different ways. For example, if two bird species eat the same type of seeds, natural selection might drive one species to specialize in eating larger seeds and the other to specialize in eating smaller seeds. Or one bird species might become more active in the morning and the other more active in the evening, minimizing interference. This process is called **resource partitioning** because the species partition, or divide, the resources they use in common by specializing in different ways (FIGURE 4.3).

Resource partitioning can lead to **character displacement,** in which competing species come to diverge in their physical characteristics because of the evolution of traits best suited

White-breasted nuthatch climbs down trunk looking for insects

Yellow-bellied sapsucker drills rows of holes and consumes sap and insects stuck in sap

Pileated woodpecker digs deeply into wood to find large insects

Brown creeper climbs up trunk looking for tiny insects

FIGURE 4.3 When species compete, they may partition resources. Birds that forage for insects on tree trunks use different portions of the trunk and seek different foods in different ways.

to the range of resources they use. For birds that specialize on larger seeds, natural selection may favor the evolution of larger bills that enable them to make best use of this resource, whereas for birds specializing on smaller seeds, smaller bills may be favored. This is precisely what extensive recent research has revealed about the finches from the Galápagos Islands that were first described by Charles Darwin (p. 49).

In competitive interactions, each participant exerts a negative effect on other participants, because each takes resources the others could have used. This is reflected in the two minus signs shown for competition in Table 4.1. In other types of interactions, some participants benefit while others are harmed; that is, one species exploits the other (note the +/– interactions in Table 4.1). Such exploitative interactions include predation, parasitism, and herbivory.

Predators kill and consume prey

Every living thing needs to procure food, and for most animals, this means eating other living organisms. **Predation** is the process by which individuals of one species—the predator—hunt, capture, kill, and consume individuals of another species, the prey. Interactions among predators and prey structure the food webs we will examine shortly, and they help shape community composition by influencing the relative numbers of predators and prey.

Predation by zebra mussels on phytoplankton has reduced phytoplankton populations by up to 90%, according to studies in the Great Lakes and Hudson River. Zebra mussels also consume the smaller types of zooplankton, and this predation has diminished zooplankton populations by up to 70% in Lake Erie and the Hudson River. Most predators are also prey, however. Zebra mussels have become a food source for muskrats, crayfish, and a variety of North American fish and ducks.

Predation can sometimes drive population dynamics. An increase in the population size of prey creates more food for predators, which may survive and reproduce more effectively

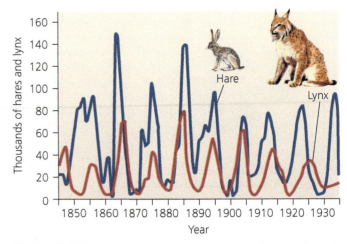

FIGURE 4.4 Predator-prey systems occasionally show paired cycles. A classic case is that of snowshoe hares and the lynx that prey on them in Canada. These data come from fur-trapping records of the Hudson Bay Company and represent numbers of each animal trapped. *Data from MacLulich, D.A., 1937. Fluctuation in the numbers of varying hare* (Lepus americanus). *Univ. Toronto Stud. Biol. Ser. 43. Toronto, Canada: University of Toronto Press.*

as a result. As the predator population rises, intensified predation drives down the population of prey. Diminished numbers of prey in turn cause some predators to starve, and the predator population declines. This allows the prey population to rise again, starting the cycle anew (**FIGURE 4.4**).

Predation also has evolutionary ramifications. Individual predators that are more adept at capturing prey may live longer lives, reproduce more, and be better providers for their offspring. Natural selection (p. 48) will thereby lead to the evolution of adaptations (p. 49) that enhance hunting skills. Prey, however, face an even stronger selective pressure—the risk of immediate death. As a result, predation pressure has driven the evolution of an elaborate array of defenses against being eaten (**FIGURE 4.5**).

A gecko's camouflage hides it from predators.

A yellowjacket's coloring signals that it is dangerous.

This caterpillar's false eyespots startle predators by mimicking a snake's head.

(a) Cryptic coloration (camouflage)   **(b) Warning coloration**   **(c) Mimicry**

FIGURE 4.5 Natural selection to avoid predation has resulted in fabulous adaptations. Some prey species use cryptic coloration **(a)** to blend into their background. Others are brightly colored **(b)** to warn predators they are toxic, distasteful, or dangerous. Others use mimicry **(c)** to fool predators.

Parasites exploit living hosts

Organisms can exploit other organisms without killing them. **Parasitism** is a relationship in which one organism, the parasite, depends on another, the host, for nourishment or some other benefit while doing the host harm. Unlike predation, parasitism usually does not result in an organism's immediate death.

Many types of parasites live inside their hosts. For example, tapeworms live in their hosts' digestive tracts, robbing them of nutrition. The larvae of parasitoid wasps burrow into the tissues of caterpillars and consume them from the inside. Other parasites are free-living. Cuckoos of Eurasia and cowbirds of the Americas lay their eggs in other birds' nests and let the host species raise the parasite's young. Still other parasites live on the exterior of their hosts, such as fleas or ticks that suck blood through the skin. The sea lamprey is a tube-shaped vertebrate that grasps the bodies of fish with a suction-cup mouth and a rasping tongue, sucking their blood for days or weeks (**FIGURE 4.6**). Sea lampreys invaded the Great Lakes from the Atlantic Ocean after people dug canals to connect the lakes for shipping, and the lampreys soon devastated economically important fisheries of chubs, lake herring, whitefish, and lake trout.

Parasites that cause disease in their hosts are called **pathogens.** Common human pathogens include the protists that cause malaria and amoebic dysentery, the bacteria that cause pneumonia and tuberculosis, and the viruses that cause hepatitis and AIDS.

Just as predators and prey evolve in response to one another, so do parasites and hosts, in a reciprocal process of adaptation and counter-adaptation called **coevolution.** Hosts and parasites may become locked in a duel of escalating adaptations, known as an *evolutionary arms race.* Like rival nations racing to stay ahead of one another in military technology, host and parasite repeatedly evolve new responses to the other's latest advance. In the long run, though, it may not

FIGURE 4.7 In herbivory, animals feed on plants. This beetle is feeding on leaves.

be in a parasite's best interest to do its host too much harm. In many cases, a parasite may leave more offspring by allowing its host to live longer.

Herbivores exploit plants

In **herbivory,** animals feed on the tissues of plants. Insects that feed on plants are the most common type of herbivore; nearly every plant in the world is attacked by insects (**FIGURE 4.7**). Herbivory generally does not kill a plant outright but may affect its growth and reproduction.

Like animal prey, plants have evolved an impressive arsenal of defenses against the animals that feed on them. Many plants produce chemicals that are toxic or distasteful to herbivores. Others arm themselves with thorns, spines, or irritating hairs. In response, herbivores evolve ways to overcome these defenses, and the plant and the animal may embark on an evolutionary arms race.

Some plants recruit certain animals as allies to assist in their defense. Such plants may encourage ants to take up residence by providing swelled stems for the ants to nest in, or nectar-bearing structures for them to feed from. In return, the ants protect the plant by attacking insects that land or crawl on it. Other plants respond to herbivory by releasing volatile chemicals when they are bitten or pierced. The airborne chemicals attract predatory insects that may attack the herbivore. Such cooperative strategies—essentially trading food for protection—are examples of mutualism.

Mutualists help one another

Unlike exploitative interactions, **mutualism** is a relationship in which two or more species benefit from interacting with one another. Generally, each partner provides some resource or service that the other needs.

FIGURE 4.6 A parasite benefits at the expense of its host. Parasitic sea lampreys feed by sucking blood from fish in the Great Lakes.

to entice them. The pollinators receive food, and the plants are pollinated and reproduce. Bees pollinate three-fourths of the crops we rely on to survive—from soybeans to potatoes to tomatoes to beans to cabbage to oranges!

FIGURE 4.8 In mutualism, organisms of different species benefit one another. Hummingbirds visit flowers to gather nectar, and in the process they transfer pollen between flowers, helping the plant to reproduce.

Many mutualistic relationships—like many parasitic relationships—occur between organisms that live in close physical contact. Physically close association is called **symbiosis,** and symbiosis can be either mutualistic or parasitic. (Indeed, biologists hypothesize that many mutualistic associations evolved from parasitic ones.) Most terrestrial plant species depend on mutualisms with fungi; plant roots and some fungi together form symbiotic associations called *mycorrhizae*. In these relationships, the plant provides energy and protection to the fungus while the fungus helps the plant absorb nutrients from the soil. In the ocean, coral polyps, the tiny animals that build coral reefs (p. 426), provide housing and nutrients for specialized algae known as zooxanthellae in exchange for food the algae produce through photosynthesis (p. 30).

You, too, are engaged in a symbiotic mutualism. Your digestive tract is filled with microbes that help you digest food and carry out other bodily functions—microbes for which you are providing a place to live. Without these mutualistic microbes, none of us would survive for long.

Not all mutualists live in close proximity. **Pollination** (**FIGURE 4.8**), a mutualistic interaction vital to agriculture and our food supply (pp. 219–220), involves free-living organisms that may encounter each other only once. Bees, birds, bats, and other creatures transfer pollen (containing male sex cells) from flower to flower, fertilizing ovaries (containing female sex cells) that grow into fruits with seeds. Most pollinating animals visit flowers for their nectar, a reward the plant uses

Ecological Communities

A **community** is an assemblage of populations of organisms living in the same area at the same time (as we saw in Figure 3.12, p. 59). Members of a community interact in the ways discussed above, and these species interactions have indirect effects that ripple outward to affect other community members. Species interactions help determine the structure, function, and species composition of communities. Community ecology (p. 58) is the scientific study of species interactions and the dynamics of communities. Community ecologists study which species coexist, how they interact, how communities change through time, and why these patterns occur.

Energy passes among trophic levels

Some of the most important interactions among community members involve who eats whom. As organisms feed on one another, matter and energy move through the community from one **trophic level,** or rank in the feeding hierarchy, to another (**FIGURE 4.9**).

Producers **Producers,** or *autotrophs* ("self-feeders"), make up the first trophic level. Terrestrial green plants, cyanobacteria, and algae capture solar energy and use photosynthesis to produce sugars (pp. 30–31). The chemosynthetic bacteria of hot springs and deep-sea hydrothermal vents use geothermal energy in a similar way to produce food (p. 31).

Consumers Organisms that consume producers are known as **primary consumers** and make up the second trophic level. Herbivorous grazing animals, such as deer and grasshoppers, are primary consumers. The third trophic level consists of **secondary consumers,** which prey on primary consumers. Wolves that prey on deer are considered secondary consumers, as are rodents and birds that prey on grasshoppers. Predators that feed at still higher trophic levels are known as **tertiary consumers.** Examples of tertiary consumers include hawks and owls that eat rodents that have eaten grasshoppers.

Detritivores and decomposers Detritivores and decomposers consume nonliving organic matter. **Detritivores,** such as millipedes and soil insects, scavenge the waste products or dead bodies of other community members. **Decomposers,** such as fungi and bacteria, break down leaf litter and other nonliving matter into simpler constituents that can be taken up and used by plants. These organisms enhance the topmost soil layers (pp. 212, 214) and play essential roles

Aquatic examples

Terrestrial examples

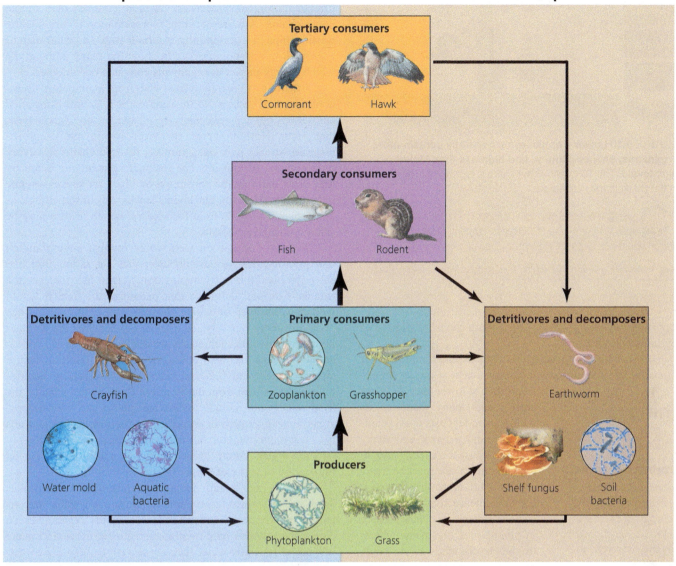

FIGURE 4.9 Ecologists organize species hierarchically by trophic level. The diagram shows aquatic (**left**) and terrestrial (**right**) examples at each level. Arrows indicate the direction of energy flow. Producers generate food by photosynthesis, primary consumers (herbivores) feed on producers, secondary consumers eat primary consumers, and tertiary consumers eat secondary consumers. Detritivores and decomposers feed on nonliving organic matter and return nutrients to the soil or the water column for use by producers.

as the community's recyclers, making nutrients from organic matter available for reuse by living members of the community.

In Great Lakes communities, phytoplankton are the main producers, conducting photosynthesis using sunlight that penetrates the water. Zooplankton are primary consumers, feeding on the phytoplankton. Fish that eat phytoplankton are primary consumers, and fish that eat zooplankton are secondary consumers. Tertiary consumers include birds and larger fish that feed on plankton-eating fish. (The left side of Figure 4.9 shows these relationships in very generalized form.) Zebra mussels and quagga mussels, by eating both phytoplankton and zooplankton, function on multiple trophic levels. When

an organism dies and sinks to the bottom, detritivores scavenge its tissues and decomposers recycle its nutrients.

Energy, numbers, and biomass decrease at higher trophic levels

At each trophic level, organisms use energy in cellular respiration (p. 31). More energy goes toward maintenance than to building new tissues, and most is given off as heat. Only a small portion of the energy is transferred to the next trophic level through predation, herbivory, or parasitism. A general rule of thumb is that each trophic level contains about 10% of the energy of the trophic level below it (although the actual

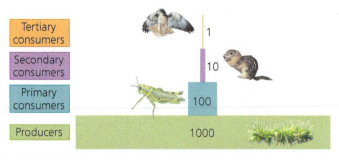

FIGURE 4.10 Lower trophic levels generally contain more organisms, energy content, and biomass than higher trophic levels. The 10:1 ratio shown here is typical, but the shape of the pyramid may vary greatly.

DATA Using the ratios shown in this example, let's suppose that a system has 3000 grasshoppers. How many rodents would you expect? How many hawks would you expect?

Go to **Interpreting Graphs & Data** on **Mastering**EnvironmentalScience®.

proportion can vary greatly). This pattern can be visualized as a pyramid (**FIGURE 4.10**).

This pyramid-like pattern also tends to hold for numbers of organisms; in general, fewer organisms exist at higher trophic levels than at lower ones. A grasshopper eats many plants in its lifetime, a rodent eats many grasshoppers, and a hawk eats many rodents. Thus, for every hawk in a community there must be many rodents, still more grasshoppers, and an immense number of plants. Moreover, because the difference in numbers of organisms among trophic levels tends to be large, the same pyramid-like relationship often holds true for **biomass,** the collective mass of living matter in a given place and time.

This pyramid pattern illustrates why eating at lower trophic levels—being vegan or vegetarian, for instance—decreases a person's ecological footprint. Each amount of meat or other animal product we eat requires the input of a considerably greater amount of plant material (see Figure 10.8, p. 240). Thus, when we eat animal products, we use up far more energy per calorie gained than when we eat plant products.

Food webs show feeding relationships and energy flow

As energy is transferred from lower trophic levels to higher ones, it is said to pass up a **food chain,** a linear series of feeding relationships. Plant, grasshopper, rodent, and hawk make up a food chain—as do phytoplankton, zooplankton, fish, and fish-eating birds.

weighing the
ISSUES

The Footprints of Our Diets

What proportion of your diet would you estimate consists of meat, milk, eggs, or other animal products? Would you choose to decrease this proportion to reduce your ecological footprint? Describe other ways in which you could reduce your footprint through your food choices.

Thinking in terms of food chains is conceptually useful, but ecological systems are far more complex than simple linear chains. A more accurate representation of the feeding relationships in a community is a **food web**—a visual map of energy flow that uses arrows to show the many paths along which energy passes as organisms consume one another.

FIGURE 4.11 portrays a food web from a temperate deciduous forest of eastern North America. It is greatly simplified and leaves out the vast majority of species and interactions that occur. Note, however, that even within this simplified diagram we can pick out a number of food chains involving different sets of species. For instance, grasses are eaten by deer mice that may be consumed by black rat snakes—while in another food chain, blackberry leaves are consumed by caterpillars that provide food for spiders, which in turn may be eaten by American toads.

A Great Lakes food web would involve phytoplankton that photosynthesize, zooplankton that eat them, fish that eat phytoplankton and zooplankton, larger fish that eat the smaller fish, and lampreys that parasitize the fish. It would include a number of native mussels and clams and, since 1988, the zebra mussels and quagga mussels that are displacing them. It would include diving ducks that formerly fed on native bivalves and also now prey on the mussels. It would also show that crayfish and other bottom-dwelling invertebrate animals feed from the refuse of the mussels. Finally, the food web would include underwater plants and macroscopic algae, whose growth is enhanced as the non-native mussels filter out phytoplankton, allowing sunlight to penetrate deeply into the water. (Jump ahead to Figure 4.14a, p. 87, for an illustration of some of these effects.)

Overall, zebra and quagga mussels alter Great Lakes food webs by shifting productivity from open-water regions to benthic (bottom) and littoral (nearshore) regions. In so doing, the mussels help benthic and littoral fishes and make life harder for open-water fishes (see **THE SCIENCE BEHIND THE STORY**, pp. 84–85).

Some organisms play outsized roles

"Some animals are more equal than others," George Orwell wrote in his classic novel *Animal Farm.* Orwell was making wry sociopolitical commentary, but his remark hints at a truth in ecology. In communities, some species exert greater influence than do others. A species that has strong or wide-reaching impact far out of proportion to its abundance is often called a **keystone species.** A keystone is the wedge-shaped piece at the top of a stone arch that holds the structure together. Remove the keystone, and the arch will collapse (**FIGURE 4.12a**, p. 82). In an ecological community, removal of a keystone species will likewise have major consequences.

Often, secondary or tertiary consumers near the tops of food chains are considered keystone species. Top predators control populations of herbivores, which otherwise would multiply and greatly modify the plant community (**FIGURE 4.12b**). Thus, predators at high trophic levels can indirectly promote populations of organisms at low trophic levels by keeping species at intermediate trophic levels in check,

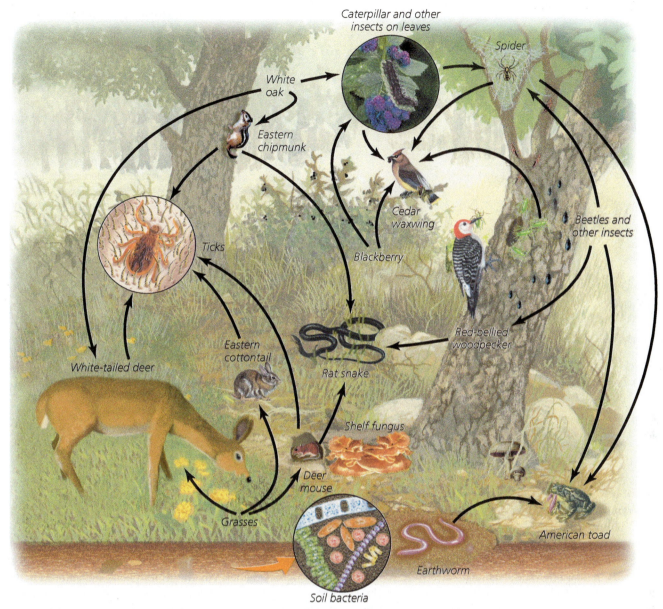

FIGURE 4.11 Food webs represent feeding relationships in a community. This food web shows organisms on several trophic levels in eastern North America's temperate deciduous forest. Arrows indicate the direction of energy flow as a result of predation, parasitism, or herbivory. Like most food web diagrams, this one is simplified. The actual community contains many more species and interactions than can be shown.

a phenomenon ecologists refer to as a **trophic cascade.** For example, government bounties in the United States long promoted the hunting of wolves and mountain lions, which were largely exterminated by the mid-20th century. In the absence of these predators, deer populations grew unnaturally dense and have overgrazed forest-floor vegetation and eliminated tree seedlings, causing major changes in forest structure.

The removal of top predators in the United States was an uncontrolled large-scale experiment with unintended consequences, but ecologists have verified the keystone species concept in controlled scientific experiments. Classic research by marine biologist Robert Paine established that the predatory sea star *Pisaster ochraceus* shapes the community composition of intertidal organisms (p. 424) on North America's Pacific coast. When *Pisaster* is present in this community,

species diversity is high, with various types of barnacles, mussels, and algae. When *Pisaster* is removed, the mussels it preys on become numerous and displace other species, suppressing species diversity. The recent outbreak of sea star wasting disease that has killed most sea stars along the Pacific coast is replicating Paine's experiment on a massive scale, and could soon lead to major changes in coastal communities from California to Alaska.

Animals at high trophic levels—such as wolves, sea stars, sharks, and sea otters (see Figure 4.12)—are often viewed as keystone species that can trigger trophic cascades. However, other types of organisms also exert strong community-wide effects. "Ecosystem engineers" physically modify environments. Beavers build dams across streams, creating ponds and swamps by flooding land. Prairie dogs dig burrows

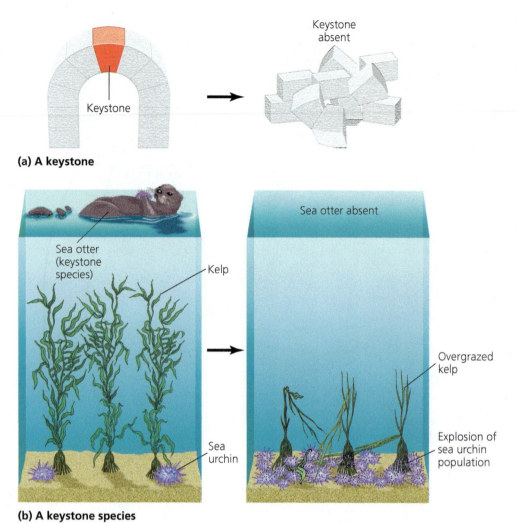

(a) A keystone

Keystone

Keystone absent

(b) A keystone species

Sea otter (keystone species)

Kelp

Sea urchin

Sea otter absent

Overgrazed kelp

Explosion of sea urchin population

FIGURE 4.12 Sea otters are a keystone species. A keystone **(a)** is the wedge-shaped piece atop a stone arch that holds the structure together. A keystone species **(b)** is a species that exerts great influence on a community's composition and structure. Sea otters consume sea urchins that eat kelp in coastal waters of the Pacific. Otters keep urchin numbers down, allowing lush underwater forests of kelp to grow, providing habitat for many species. When otters are absent, urchins increase and devour the kelp, destroying habitat and depressing species diversity.

that aerate the soil and serve as homes for other animals. Ants disperse seeds, redistribute nutrients, and selectively protect or destroy insects and plants near their colonies. And zebra and quagga mussels alter the communities they invade by filtering plankton from the water.

Less conspicuous organisms at low trophic levels can exert still more impact. Remove the fungi that decompose dead matter, or the insects that control plant growth, or the phytoplankton that support the marine food chain, and a community may change rapidly indeed. However, because there are usually more species at lower trophic levels, it is less likely that any single one of them alone has wide influence. Often if one species is removed, other species that remain may be able to perform many of its functions.

Communities respond to disturbance in various ways

The removal of a keystone species is just one type of disturbance that can modify the composition, structure, or function of an ecological community. In ecological terms, a **disturbance** is an event that has rapid and drastic impacts on environmental conditions, resulting in changes to the community and ecosystem. A disturbance can be localized, such as

when a tree falls in a forest, creating a gap in the canopy that lets in light and alters conditions for plants and animals in the gap. Or it can be as large and severe as a hurricane, tornado, or volcanic eruption. Some disturbances are sudden, such as landslides or floods, whereas others are more gradual, such as climate change. Some disturbances recur regularly and are considered normal aspects of a system (such as periodic fire, seasonal storms, or cyclical insect outbreaks). Organisms may, in fact, adapt to regular and predictable types of disturbance. For instance, many plants that grow in fire-prone regions have evolved ways of surviving fire and have seeds that depend on fire to germinate in the nutrient-rich soil that fire leaves behind. Today, human impacts are major sources of disturbance for ecological communities worldwide—from habitat alteration to pollution to the introduction of non-native species such as the zebra mussel.

Communities are dynamic systems and may respond to disturbance in several ways. A community that resists change and remains stable despite disturbance is said to show **resistance** to the disturbance. Alternatively, a community may show **resilience,** meaning that it changes in response to disturbance but later returns to its original state. Or, a community may be modified by disturbance permanently and never return to its original state.

Succession follows severe disturbance

If a disturbance is severe enough to eliminate all or most of the species in a community, the affected site may then undergo a somewhat predictable series of changes that ecologists have traditionally called **succession.** In the conventional view of this process, there are two types of succession (**FIGURE 4.13**). **Primary succession** follows a disturbance so severe that no vegetation or soil life remains from the community that had occupied the site. In primary succession, a community is built essentially from scratch. In contrast, **secondary succession** begins when a disturbance dramatically alters an existing community but does not destroy all life and organic matter in the soil. In secondary succession, vestiges of the previous community remain, and these building blocks help shape the process.

At terrestrial sites, primary succession takes place after a bare expanse of rock, sand, or sediment becomes newly exposed to the atmosphere. This can occur when glaciers retreat, lakes dry up, or volcanic lava or ash spreads across the landscape. Species that arrive first and colonize the new substrate are referred to as **pioneer species.** Pioneer species are well adapted for colonization, having traits such as spores or seeds that can travel long distances.

The pioneers best suited to colonizing bare rock are the mutualistic aggregates of fungi and algae known as lichens. In lichens, the algal component provides food and energy via photosynthesis while the fungal component grips the rock and captures moisture. As lichens grow, they secrete acids that break down the rock surface, beginning the process that forms soil. Small plants and insects arrive, providing more nutrients and habitat. As time passes, larger plants and animals establish themselves, vegetation increases, and species diversity rises.

Secondary succession begins when a fire, a storm, logging, or farming removes much of the biotic community. Consider a farmed field in eastern North America that has

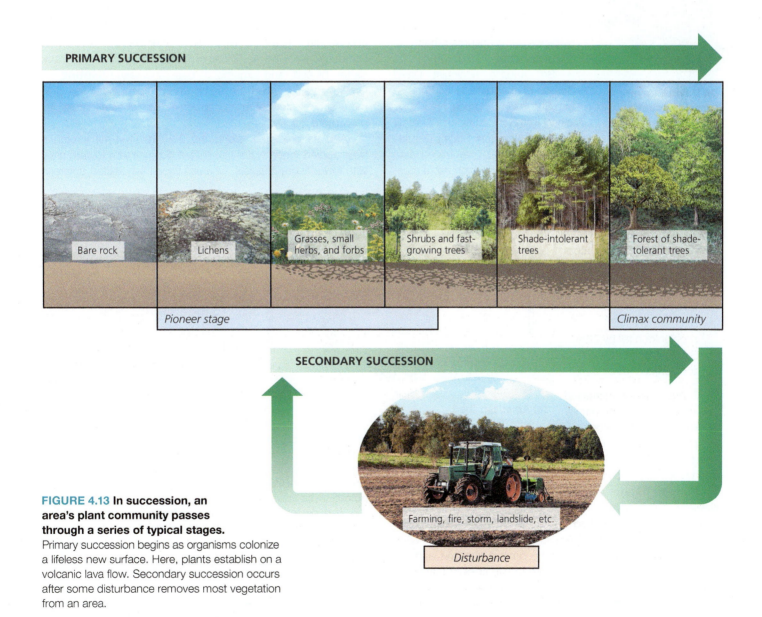

PRIMARY SUCCESSION

Bare rock | Lichens | Grasses, small herbs, and forbs | Shrubs and fast-growing trees | Shade-intolerant trees | Forest of shade-tolerant trees

Pioneer stage | *Climax community*

SECONDARY SUCCESSION

Farming, fire, storm, landslide, etc.

Disturbance

FIGURE 4.13 In succession, an area's plant community passes through a series of typical stages. Primary succession begins as organisms colonize a lifeless new surface. Here, plants establish on a volcanic lava flow. Secondary succession occurs after some disturbance removes most vegetation from an area.

How Do Zebra Mussels Affect Fish Communities?

Dr. David Strayer samples aquatic invertebrates.

When zebra mussels appeared in the Great Lakes, people feared for sport fisheries and predicted that fish population declines could cost billions of dollars. The mussels would deplete the phytoplankton and zooplankton that fish relied on for food, people reasoned.

However, food webs are complicated, and disentangling them to infer the impacts of any one species is difficult. Thus, even 15 years after the arrival of zebra mussels, there was no solid evidence of widespread harm to fish populations.

So, aquatic ecologist David Strayer of the Institute of Ecosystem Studies in Millbrook, New York, joined Kathryn Hattala and Andrew Kahnle of New York State's Department of Environmental Conservation (DEC). They mined data sets on fish populations in the Hudson River, which zebra mussels had invaded in 1991.

Strayer and other scientists had been studying this community for years. Their data showed that after zebra mussels invaded the Hudson:

- Biomass of phytoplankton fell by 80%.
- Biomass of small zooplankton fell by 76%.
- Biomass of large zooplankton fell by 52%.

Zebra mussels increased filter-feeding in the community 30-fold, depleting phytoplankton and small zooplankton and leaving larger zooplankton with less to eat. Overall, zooplankton and invertebrate animals of the open water declined by 70%.

However, Strayer had also found that benthic, or bottom-dwelling, invertebrates in shallow water (especially in the nearshore, or littoral, zone) had increased, because the mussels' shells provide habitat structure and their feces provide nutrients.

These contrasting trends in the benthic shallows and the open deep water led Strayer's team to hypothesize that zebra mussels would harm open-water fish that ate plankton but would help littoral-feeding fish. They predicted that larvae and juveniles of open-water fish species would decline in number, grow more slowly, and shift downriver toward saltier water, where mussels are absent. Conversely, they predicted that larvae and juveniles of littoral fish species would increase in number, grow more quickly, and shift upriver to regions of high zebra mussel density.

To test their predictions, the researchers analyzed data from fish surveys carried out by DEC scientists over 26 years. Strayer's team compared data on abundance, growth, and distribution of young fish before and after the zebra mussel's arrival in 1991.

The results supported their predictions. Larvae and juveniles of open-water fish, such as American shad, blueback herring, and alewife, declined in abundance after zebra mussels were introduced (**FIGURE 1a**). Those of littoral fish, such as tessellated darter, bluegill, and largemouth bass, increased (**FIGURE 1b**). Growth rates showed the same trend: Open-water fish grew more slowly after zebra mussels invaded, whereas littoral fish grew more quickly.

And as predicted, along the 248-km (154-mi) stretch of river studied, open-water fish shifted downstream toward areas with fewer zebra mussels, whereas littoral fish shifted upstream

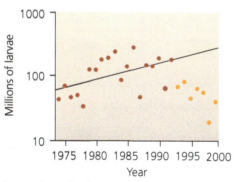

(a) American shad

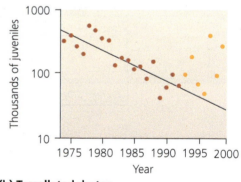

(b) Tessellated darter

FIGURE 1 Zebra mussels harm open-water fish but help littoral fish. Larvae of American shad **(a)**, an open-water fish, were increasing before zebra mussels invaded (**red points and trend line**). After zebra mussels invaded, shad larvae decreased (**orange points**). In contrast, juveniles of the tessellated darter **(b)**, a littoral fish, were decreasing (**red points and trend line**) but increased after zebra mussels invaded (**orange points**). *Source: Strayer, D., et al., 2004. Effects of an invasive bivalve* (Dreissena polymorpha) *on fish in the Hudson River estuary. Canadian Journal of Fisheries and Aquatic Sciences 61: 924–941.* © 2004. Reprinted by permission of NRC Research Press.

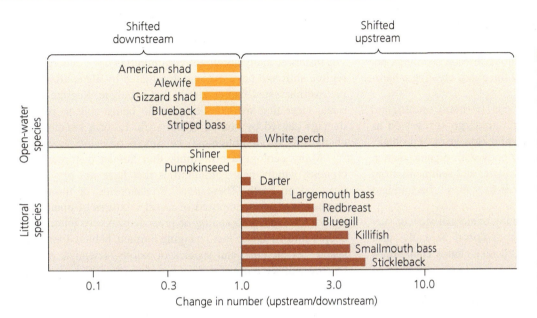

toward areas with more zebra mussels (**FIGURE 2**). Overall, the data supported the hypothesis that the fish community would respond to changes in food resources caused by zebra mussels.

Yet then, surprisingly, some of the zebra mussel's impacts began to reverse. Populations of native mussels and clams in the Hudson that had crashed after the zebra mussel invaded (likely because of competition for food) began to stabilize and persist at about 4–22% of their pre-invasion population sizes. Crustaceans, flatworms, and other invertebrates also rebounded. Several types of zooplankton began to increase (**FIGURE 3**).

To determine why these changes were occurring, Strayer's research teams stepped up monitoring efforts and ran experiments placing cages (to keep out predators) around some areas of mussels. They found that mussels within the cages grew larger than mussels outside, indicating that predators were feeding on mussels and preventing most of them from growing

to full size. Strayer's teams determined that the zebra mussel's survival rate had fallen to less than 1% of what it was during the early years of the invasion! Predators such as the blue crab were eating more and more of the mussels, and large zebra mussels were becoming rare. As the average size of mussels decreased, their filtering capacity fell by more than 80%, and more zooplankton began to survive.

Do these trends suggest that the zebra mussel might just fade away and prove harmless in the long run? Strayer thinks it is too early to answer that question; much remains unknown, and he is continuing his research. He cautions that zebra mussels remain abundant, that phytoplankton levels have not yet bounced back, and that there is no guarantee that zebra mussel impacts will continue to diminish. Nonetheless, the apparent turnaround in the Hudson River is intriguing ecologists and providing hope that the Hudson's native systems may recover.

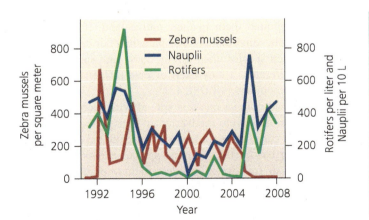

FIGURE 3 Zooplankton bounced back once zebra mussels declined. Large, mature zebra mussels have decreased in density since their initial sudden increase upon colonizing the Hudson River (**red line**). Two types of zooplankton, nauplii and rotifers (**blue and green lines**), declined following the zebra mussel's introduction, but they recovered after 2005, once large mussels disappeared. *Source: Pace, M.L, et al., 2010. Recovery of native zooplankton associated with increased mortality of an invasive mussel. Ecosphere 1(1): Article 3.*

DATA Q What were the densities of nauplii, rotifers, and zebra mussels in 1991? In 2000? In 2008? What pattern do you see in these data? How would you explain this pattern?

Go to **Interpreting Graphs & Data** on **Mastering**EnvironmentalScience®.

been abandoned. After farming ends, the site is colonized by pioneer species of grasses, herbs, and forbs that disperse well or were already in the vicinity. Soon, shrubs and fast-growing trees such as aspens and poplars begin to grow. As time passes, pine trees rise above the pioneer trees and shrubs, forming a pine-dominated forest. This pine forest gains an understory of hardwood trees, because pine seedlings do not grow well under a canopy but some hardwood seedlings do. Eventually the hardwoods outgrow the pines, creating a hardwood forest.

Processes of succession occur in many ecological systems. For instance, a pond may undergo succession as algae, microbes, plants, and zooplankton grow, reproduce, and die, gradually filling the water body with organic matter. The pond acquires further organic matter and sediments from streams and runoff, and eventually it may fill in, becoming a bog (p. 392) or a terrestrial system.

FAQ

If we disturb a community, won't it return to its original state if we just leave the area alone?

Probably not, if the disturbance has been substantial. For example, if soil has become compacted, or if water sources have dried up, then the plant species that grew at the site originally may no longer be able to grow. Different plant species may take their place—and among them, a different suite of animal species may find habitat. Because species interact and because they rely on particular habitat conditions, a change in one aspect of a community can lead to a cascade of other changes. Sometimes a whole new community may arise. For instance, in some grasslands, livestock grazing and fire suppression have led shrubs and trees to invade, forming shrublands. And removing sea otters from marine environments can lead to the loss of kelp forests (p. 426). In the past, people did not realize how permanent such changes could be, because we tended to view natural systems as static, predictable, and liable to return to equilibrium. Today ecologists recognize that systems are highly dynamic and can sometimes undergo rapid, extreme, and long-lasting change.

In the traditional view of succession described here, the process leads to a **climax community,** which remains in place until some disturbance restarts succession. Early ecologists felt that each region had its own characteristic climax community, determined by climate.

Communities may undergo shifts

Today, ecologists recognize that community change is far more variable and less predictable than early models of succession suggested. Conditions at one stage may promote progression to another stage, or organisms may, through competition, inhibit a community's progression to another stage. The trajectory of change can vary greatly according to chance factors, such as which particular species happen to gain an early foothold. And climax communities are not determined solely by climate but vary with soil conditions and other factors from one time or place to another. Ecologists came to modify their views about how communities respond to disturbance after observing changes in long-term field studies at locations such as Mount Saint Helens following its eruption (see **THE SCIENCE BEHIND THE STORY**, pp. 88–89).

Once a community is disturbed and changes are set in motion, there is no guarantee that the community will ever return to its original state. Instead, sometimes communities may undergo a **regime shift,** also known as a *phase shift,* in which the character of the community fundamentally changes. This can occur if some crucial climatic threshold is passed, a keystone species is lost, or a non-native species invades. For instance, many coral reef communities have undergone a regime shift and have become dominated by algae after people overharvested fish or turtles that eat algae. Regime shifts make clear that we cannot count on being able to reverse damage caused by human disturbance. Instead, some of the changes we set in motion may become permanent.

Many ecologists now think that human disturbance is creating wholly new communities that have not previously occurred on Earth. These *novel communities,* or **no-analog communities,** are composed of novel mixtures of plants and animals and have no analog or precedent. As we enter more deeply into an age of fast-changing climate, habitat alteration, species extinctions, and species invasions, scientists predict that we will see more and more novel communities.

Invasive species pose threats to community stability

Traditional concepts of communities involve species native to an area. But what if a species not native to the area (a non-native, alien, or exotic species) arrives from elsewhere? In our age of global mobility and trade, people have moved countless organisms from place to place, intentionally or by accident. As a result, today most non-native arrivals in a community are **introduced species,** species introduced by people.

Most introduced species fail to establish populations, but some turn invasive, spreading widely and coming to dominate communities. Such **invasive species** often thrive in disturbed systems, and in turn disturb them further. By altering communities, invasive species are one of the central ecological forces in today's world.

Introduced species may become invasive when limiting factors (p. 65) that regulate their population growth are absent. Plants and animals brought to a new area may leave behind the predators, parasites, herbivores, and competitors that had exploited them in their native land. If few organisms in the new environment eat, parasitize, or compete with the introduced species, then it may thrive and spread. As the species proliferates, it may exert diverse influences on other community members (**FIGURE 4.14**).

Zebra and quagga mussels spread with global trade; as noted earlier, they were inadvertently transported in the ballast water of cargo ships. Indeed, the unregulated exchange of ballast water has ferried countless species across the oceans. Other types of freshwater invaders include predatory fishes released from aquaria, bait buckets, stocking, or aquaculture; aquatic plants that become ecosystem engineers; crayfish that disrupt the benthic portions of food webs; and disease pathogens.

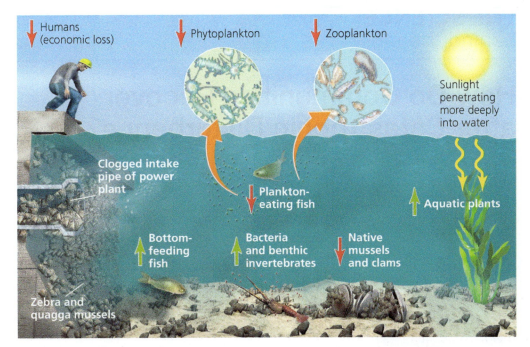

(a) Impacts of zebra and quagga mussels on a Great Lakes nearshore community

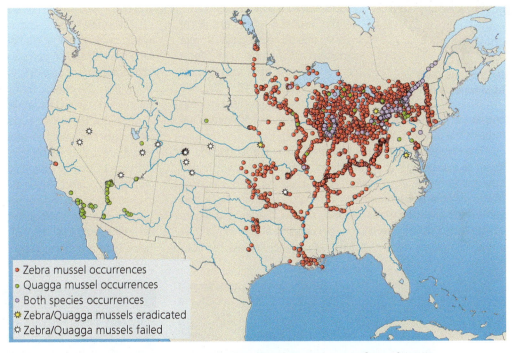

- Zebra mussel occurrences
- Quagga mussel occurrences
- Both species occurrences
- Zebra/Quagga mussels eradicated
- Zebra/Quagga mussels failed

(b) Occurrence of zebra and quagga mussels in North America, as of March 2016

FIGURE 4.14 The zebra mussel and the quagga mussel are modifying ecological communities. By filtering phytoplankton and small zooplankton from open water, they exert impacts **(a)** on other species, both negative (red downward arrows) and positive (green upward arrows). The map **(b)** shows known occurrences of zebra and quagga mussels as of March 2016. In just two decades they spread across North America, assisted by accidental transportation on boats. *Source (b): U.S. Geological Survey.*

DATA The mapped distribution of occurrences of zebra mussels and quagga mussels shows some differences between the two species. Let's focus on two of these differences and try to suggest scientific hypotheses for them.

- Which of the two species occurs in more locations, according to the map's data? Suggest two different hypotheses for why this difference may exist.

- How does the geographic distribution of quagga mussels differ from that of zebra mussels? Describe these spatial patterns shown by the map's data, and then suggest two different hypotheses for why this difference between species may exist.

Go to **Interpreting Graphs & Data** on **Mastering**EnvironmentalScience*.

How Do Communities Recover after Catastrophic Disturbance?

**Dr. Virginia Dale
at Mount Saint Helens**

Step outside, look around, and you'll likely find secondary succession occurring somewhere nearby. The nearest weedy lot or overgrown field will show how plants colonize a disturbed area and begin building a new community from the foundations of the old. But finding primary succession is not as easy. It's unusual to come across a place where all life has been extinguished and a brand new community is being built from scratch as new organisms arrive from far away.

The eruption of Mount Saint Helens offered ecologists a rare opportunity to study how communities recover from catastrophic disturbance. On May 18, 1980, this volcano in the state of Washington erupted in sudden and spectacular violence, with 500 times the force of the atomic blast at Hiroshima.

The massive explosion obliterated an entire landscape of forest as a scalding mix of gas, steam, ash, and rock was hurled outward for miles. A pyroclastic flow (p. 38) sped downslope, along with the largest landslide in recorded history. Rock and ash rained down for miles around, and mudslides and lahars (p. 42) raced down river valleys, devastating everything in their paths. Altogether, 4.1 km³ (1.0 mi³) of material was ejected from the mountain, severely altering 1650 km² (637 mi²), an area larger than the entire city of Houston, Texas.

In the aftermath of the blast, ecologists moved in to take advantage of the natural experiment of a lifetime. For them, the eruption provided an extraordinary chance to study how primary succession unfolds on a fresh volcanic surface. Which organisms would arrive first? What kind of community would emerge? How long it would take? These researchers set up study plots to examine how populations, communities, and ecosystems would respond.

Today, more than 35 years later, the barren gray moonscape that resulted from the blast is a vibrant green (**FIGURE 1**), carpeted with colorful flowers each summer. And what ecologists have learned has modified our view of primary succession and informed the entire study of disturbance ecology.

Given the ferocity and scale of the eruption, most scientists initially presumed that life had been wiped out completely over a large area. Based on traditional views of succession, they expected that pioneer species would colonize the area gradually, spreading slowly from the outside margins inward, and that over many years a community would be rebuilt in a systematic and predictable way.

Instead, researchers discovered that some plants and animals had survived the blast. Some were protected by deep snowbanks. Others were sheltered on steep slopes facing away from the blast. Still others were dormant underground when the eruption occurred. These survivors, it turned out, would play key roles in rebuilding the community.

Many of the ecologists drawn to Mount Saint Helens studied plants. Virginia Dale of Oak Ridge National Laboratory in Tennessee and her colleagues examined the debris avalanche,

(a) 1980

(b) 2012

FIGURE 1 Mount Saint Helens **(a)** after the eruption in 1980 and **(b)** 32 years later.

a landslide of rock and ash as deep as a 15-story building. This region appeared barren, yet small numbers of plants of 20 species had survived, growing from bits of root or stem carried down in the avalanche. However, most plant regrowth occurred from seeds blown in from afar. Dale's team used sticky traps to sample these seeds as they chronicled the area's recovery.

Plant regrowth was very slow at Dale's study plots for several years and then accelerated. After 20 years, 150 species of plants covered 65% of the ground. One important pioneer species was the red alder. This tree germinates on debris, grows quickly, deals well with browsing by animals, and produces many seeds at a young age. As a result, it has become the dominant tree species on the debris avalanche. Because it fixes nitrogen (pp. 121, 124), the red alder enhances soil fertility and thereby helps other plants grow. Researchers predict that red alder will remain dominant for years or decades and that conifers such as Douglas fir (which today are moving in and beginning to seed) will eventually outgrow them and establish a conifer-dominated forest.

Patterns of plant growth have varied in different areas. Roger del Moral of the University of Washington and his colleagues compared ecological responses on a variety of surfaces, including barren pumice, mixed ash and rock, mudflows, and the "blowdown zone" where trees were toppled like matchsticks. Numbers of species and percentage of plant cover increased in different ways on each surface (FIGURE 2), affected by a diversity of factors. Windblown seeds accounted for most regrowth, but plants that happened to survive in sheltered "refugia" within the impact zone helped to repopulate areas nearby.

Chance played a large role in determining which organisms survived and how vegetation recovered, del Moral and others found. Had the eruption occurred in late summer instead of spring, there would have been no snow, and many of the plants that survived would have died. Had the eruption occurred at night instead of in the morning, nocturnal animals would have been hit harder.

Animals played major roles in the recovery right from the beginning. In fact, researchers such as Patrick Sugg and John Edwards of the University of Washington showed that insects and spiders arrived in the impact zone in great numbers before plants did. Insects fly, while spiders disperse by "ballooning" on silken threads, so in summer the atmosphere is filled with an "aerial plankton" of windblown arthropods. Trapping and monitoring at Mount Saint Helens in the months following the eruption showed that insects and spiders landed in the impact zone by the billions. Researchers estimated that more than 1500 species arrived in the first few years, surviving by scavenging or by preying on other arthropods. Most individuals soon died, but the nutrients from their bodies enriched the soil, helping the community to develop.

Once plants took hold, animals began exerting influence through herbivory. Caterpillars fed on plants, occasionally extinguishing small populations as they began to establish. As elk from surrounding forests moved into the region, Dale and her team fenced off "exclosure" plots to compare how plants grew

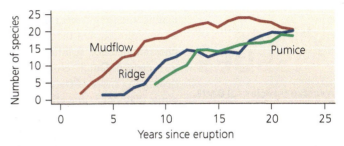

(a) Species richness

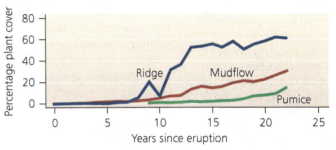

(b) Percentage plant cover

FIGURE 2 **Plants recovered differently at mudflow, ridge, and pumice sites at Mount Saint Helens.** In the 25 years after the eruption, **(a)** species richness of plants and **(b)** percentage of ground covered by plants both increased. *Data from del Moral, R., et al., 2005. Proximity, microsites, and biotic interactions during early succession. In V. Dale et al. (Eds.), Ecological responses to the 1980 eruption of Mount St. Helens (pp. 93–109). New York, NY: Springer.*

DATA Q Which substrate gained the greatest species richness, and what was its highest number of species? On which substrate was the increase in percentage plant cover the slowest?

Go to **Interpreting Graphs & Data** on Mastering**Environmental**Science®.

within the ungrazed exclosures versus in plots grazed by elk outside the exclosures. Both types of plots saw increases in plant cover and similar amounts of species diversity, because elk herbivory can spur plant growth that compensates for what they eat. Non-native species did best in the grazed plots, whereas species important for forest recovery did best in the ungrazed exclosures. Overall, elk herbivory did not stall plant regrowth.

All told, research at Mount Saint Helens has shown that succession is not a simple and predictable process. Instead, communities recover from disturbance in ways that are dynamic, complex, and highly dependent on chance factors. The results also show life's resilience. Even when the vast majority of organisms perish in a natural disaster, a few may survive, and their descendants may eventually build a new community.

Ecological change at Mount Saint Helens is still in its early stages and will continue for many decades more. All along, ecologists will continue to study and learn from this tremendous natural experiment.

Examples abound of invasive species that have had major ecological impacts (pp. 286–287). The chestnut blight, an Asian fungus, killed nearly every mature American chestnut, a dominant tree species of eastern North American forests, between 1900 and 1930. Asian trees had evolved defenses against the fungus over millennia of coevolution, but the American chestnut had not. A different fungus caused Dutch elm disease, destroying most of the American elms that once gracefully lined the streets of many U.S. cities. Fish introduced into streams for sport compete with and exclude native fish. Grasses introduced in the American West for ranching have overrun entire regions, pushing out native vegetation. Hundreds of island-dwelling animals and plants worldwide have been driven extinct by goats, pigs, and rats introduced by human colonists (Chapter 3).

Ecologists tend to view the impacts of invasive species—and introduced species in general—in a negative light. Yet we enjoy the beauty of introduced ornamental plants in our gardens. Some organisms are introduced intentionally to control pests through biocontrol (p. 248). And some introduced species that have turned invasive provide economic benefits, such as the European honeybee, which pollinates many of our crops (pp. 220, 245). Whatever view one takes, the impacts of introduced species on native populations and communities are significant, and they grow year by year with our increasing mobility and the globalization of our society.

We can respond to invasive species with control, eradication, or prevention

Scientific research and media attention to zebra and quagga mussels helped put invasive species on the map as a major environmental and economic issue. In 1990 the U.S. Congress passed legislation that led to the National Invasive Species Act of 1996. Among other things, this law directed the Coast Guard to ensure that ships dump their freshwater ballast at sea and exchange it with saltwater before entering the Great Lakes.

Since this act was passed, funding has become available for the control and eradication of invasive species. Eradication (total elimination of the invasive population) is difficult, so managers usually aim merely to control populations—that is, to limit their growth, spread, and impact. Managers have tried

to control zebra mussels by removing them manually; applying toxic chemicals; drying them out; depriving them of oxygen; introducing predators and diseases; and stressing them with heat, sound, electricity, carbon dioxide, and ultraviolet light. However, most of these are localized and short-term fixes. With one invasive species after another, managers are finding that control and eradication are so difficult and expensive that trying to prevent invasions in the first place (through strategies such as ballast water regulations) represents a better investment.

To prevent invasions, it helps to be able to predict where a given species might spread. By analyzing the biology of the organism, scientists can try to model the environmental conditions in which it will thrive. To predict where zebra and quagga mussels might do best, researchers applied knowledge of how these mussels use calcium from water to create their shells. The researchers mapped low-risk and high-risk regions across North America—and these predictions mostly conformed to the areas of actual spread. One unanswered question is why quagga mussels have leapfrogged zebra mussels by spreading into some western states (see Figure 4.14b).

Altered communities can be restored

Invasive species are adding to the transformations that people have forced on ecological systems through habitat alteration, deforestation, pollution, overharvesting, and climate change. Because ecological systems support our civilization and all of life, when degraded systems cease to function, our health and well-being are threatened.

This realization has given rise to the science of **restoration ecology.** Restoration ecologists research the historical conditions of ecological communities as they existed before our industrialized civilization altered them. They then try to devise ways to restore altered areas to an earlier condition. In some cases, the intent is primarily to restore the functionality of the system—to reestablish a wetland's ability to filter pollutants and recharge groundwater, for example, or a forest's ability to cleanse the air, build soil, and provide habitat for wildlife. In other cases, the aim is to return a community to its natural "presettlement" condition. Either way, the science of restoration ecology informs the practice of **ecological restoration,** the on-the-ground efforts to carry out these visions and restore communities.

In general, ecological restoration involves trying to undo certain impacts of human disturbance and to reestablish species, populations, communities, and natural ecological processes. Specifically, restoration often involves removing invasive species (**FIGURE 4.15a**) and planting native vegetation that had originally grown on a site. Sometimes it means reintroducing natural processes such as fire or flooding. It may also mean modifying the landscape to reduce erosion or influence patterns of water flow.

Many ecological restoration efforts are under way today. For instance, nearly all the tallgrass prairie in the

(a) Removing invasive garlic mustard to restore oak woodland in Nebraska

(b) Demonstrating the restored prairie at Fermilab, outside Chicago, Illinois

Former canal

Restored river

(c) Florida's Kissimmee River now winds freely toward the Everglades, next to a decommissioned canal bed

(d) Removing an invasive Burmese python from the Florida Everglades

FIGURE 4.15 Ecological restoration is being used to restore natural communities. Restoration often involves **(a)** the removal of invasive species and **(b)** the planting of native species. Today's largest-scale restoration projects are in Florida, where canals are being filled in **(c)** to restore natural water flow to the Everglades, and where invasive species **(d)** are being controlled to protect native species while wetland communities are restored.

United States was converted to agriculture in the 19th century. Now, people are restoring patches of prairie by planting native prairie vegetation, weeding out invaders and competitors, and introducing prescribed fire to mimic the fires that historically maintained this community. The region outside Chicago, Illinois, boasts several of the largest prairie restoration projects so far, including more than 400 ha (1000 acres) of restored prairie at the Fermilab nuclear accelerator (**FIGURE 4.15b**).

The world's largest restoration project is the ongoing effort to restore the Florida Everglades, a vast ecosystem of marshes and seasonally flooded grasslands. This wetland system has been drying out for decades because the water that feeds it has been manipulated for flood control and overdrawn for irrigation and development. Economically

important fisheries have suffered greatly as a result, and the region's famed populations of wading birds have dropped by 90–95%. The 30-year, $7.8 billion restoration project intends to restore natural water flow by undoing the damming and diversions of hundreds of miles of canals and levees. Because the Everglades provides drinking water for millions of Florida citizens, as well as considerable tourism revenue, restoring its ecosystem services (pp. 4, 116, 149) should prove economically beneficial as well as ecologically valuable.

Efforts in Florida have been successful so far: Canals have been filled in and stretches of the Kissimmee River now flow freely (**FIGURE 4.15c**), bringing water to Lake Okeechobee and then to the Everglades. But many challenges remain. Invasive species such as the Burmese python (**FIGURE 4.15d**) are spreading faster than they can be controlled, eating their way through the native fauna. Moreover, the ambitious project has struggled against budget shortfalls and political interference, and most of its goals have not yet been met. (We will explore ecological restoration projects further in Chapter 11; p. 296.)

As our population grows and development spreads, ecological restoration is becoming an increasingly vital conservation strategy. However, restoration is difficult, time-consuming, and expensive, and it is not always successful. It is therefore best, whenever possible, to protect natural systems from degradation in the first place.

Earth's Biomes

Across the world, each location is home to different assemblages of species, leading to endless variety in community composition. However, communities in far-flung places often share strong similarities in their structure and function. This allows us to classify communities into broad types. A **biome** is a major regional complex of similar communities—a large-scale ecological unit recognized primarily by its dominant plant type and vegetation structure. The world contains a number of terrestrial biomes, each covering large geographic areas (**FIGURE 4.16**).

Climate helps determine biomes

Which biome covers each portion of the planet depends on a variety of physical factors, including temperature, precipitation, soil conditions, and the circulation patterns of wind in the atmosphere and water in the oceans. Among these factors, temperature and precipitation exert the greatest influence (**FIGURE 4.17**). Global climate patterns cause biomes to occur in large patches in different parts of the world. For instance, temperate deciduous forest occurs in Europe, China, and eastern North America. Note in Figure 4.16 that patches of any given terrestrial biome tend to occur at similar latitudes. This is due to Earth's north–south gradients in temperature and to atmospheric circulation patterns (p. 451).

Mountainous areas host complex mixes of biomes because climate varies with elevation. At higher altitudes, temperature, atmospheric pressure, and oxygen all decline, whereas ultraviolet radiation increases. A hiker scaling one of the great peaks of the Andes in Ecuador, near the equator, might climb from tropical rainforest through cloud forest up to alpine tundra and glaciers. Mountain ranges can also alter climate through the **rainshadow** effect. When moisture-laden air ascends a steep slope, it releases precipitation as it cools. By the time it flows over the top and down the other side, it can be very dry, creating an arid region in the rainshadow.

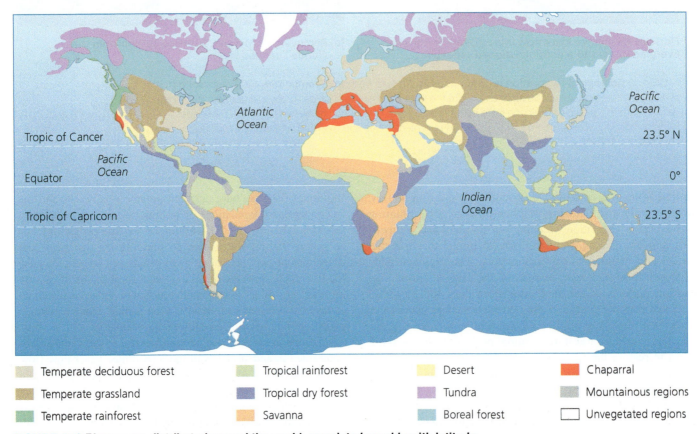

▨ Temperate deciduous forest	▨ Tropical rainforest
▨ Temperate grassland	▨ Tropical dry forest
▨ Temperate rainforest	▨ Savanna

▨ Desert	▨ Chaparral
▨ Tundra	▨ Mountainous regions
▨ Boreal forest	▨ Unvegetated regions

FIGURE 4.16 Biomes are distributed around the world, correlated roughly with latitude.

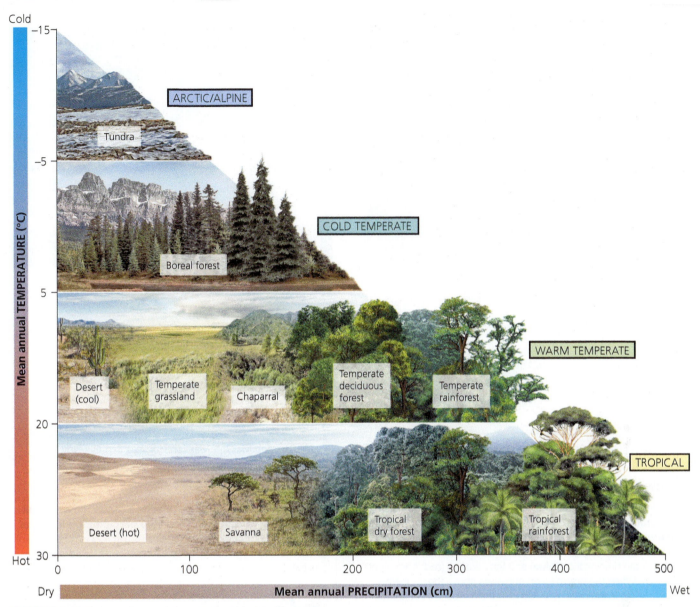

FIGURE 4.17 Temperature and precipitation are the main factors determining where biomes occur. As precipitation increases, vegetation becomes taller and more luxuriant. As temperature increases, types of plant communities change. For instance, deserts occur in dry regions; tropical rainforests occur in warm, wet regions; and tundra occurs in the coldest regions.

Because biomes are largely a function of climate, and because temperature and precipitation are the best indicators of an area's climate, scientists use **climate diagrams,** or *climatographs*, to depict such information. As we tour the world's terrestrial biomes on the following pages, you will see climatographs from specific localities. The data in each graph are typical of the climate for the biome the locality lies within.

Aquatic and coastal systems resemble biomes

In our discussion of biomes, we focus exclusively on terrestrial systems because the biome concept, as traditionally developed and applied, has done so. However, areas equivalent to biomes also exist in the oceans, along coasts, and in freshwater systems. One might consider the shallows along the world's coastlines to represent one aquatic system, the continental shelves another, and the open ocean, the deep sea, coral reefs, and kelp forests as still others. Many coastal systems—such as salt marshes, rocky intertidal communities, mangrove forests, and estuaries—share terrestrial and aquatic components. And freshwater systems such as those of the Great Lakes are widely distributed throughout the world.

Aquatic systems are shaped not by air temperature and precipitation but by water temperature, salinity, dissolved nutrients, wave action, currents, depth, light levels, and type of substrate (e.g., sandy, muddy, or rocky bottom). Marine communities are also more clearly delineated by their animal life than by their plant life. (We will examine freshwater, marine, and coastal systems in the greater detail they deserve in Chapters 15 and 16.)

(a) Temperate deciduous forest

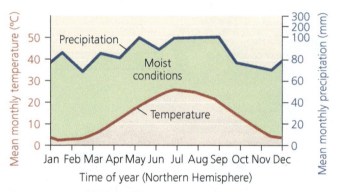

(b) Washington, D.C., USA

FIGURE 4.18 Temperate deciduous forests (a) experience fairly stable precipitation but temperatures that vary with the seasons. Scientists use climate diagrams **(b)** to illustrate average monthly precipitation and temperature. In these diagrams, the lines indicate precipitation (**blue**) and temperature (**red**) from month to month. When the precipitation curve lies above the temperature curve (as is the case year-round in the temperate deciduous forest around Washington, D.C., shown here), the region experiences relatively "moist" conditions, indicated with green coloration. *Climatograph here and in the following figures adapted from Breckle, S.-W. 2002. Walter's vegetation of the Earth: The ecological systems of the geo-biosphere, 4th ed. Berlin, Heidelberg: Springer-Verlag.*

(a) Temperate grassland

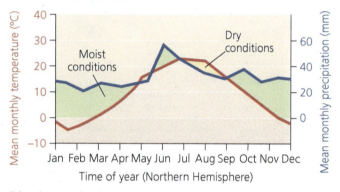

(b) Odessa, Ukraine

FIGURE 4.19 Temperate grasslands experience seasonal temperature variation and too little precipitation for trees to grow. This climatograph indicates "moist" conditions (**green**) as well as "dry" (**yellow**, when the temperature curve is above the precipitation curve). *Climatograph adapted from Breckle, S.-W. 2002.*

DATA Q How would you explain the dry conditions from July to September? Given the temperature and precipitation patterns shown, what role do you think evaporation might play, and why?

Go to **Interpreting Graphs & Data** on Mastering EnvironmentalScience®.

We can divide the world into ten terrestrial biomes

Temperate deciduous forest The **temperate deciduous forest** (**FIGURE 4.18**) that dominates the landscape around the southern Great Lakes is characterized by broad-leafed trees that are *deciduous*, meaning they lose their leaves each fall and remain dormant during winter, when hard freezes would endanger leaves. These midlatitude forests occur in much of Europe and eastern China as well as in eastern North America—all areas where precipitation is spread relatively evenly throughout the year.

Soils of the temperate deciduous forest are relatively fertile, but this biome consists of far fewer tree species than are found in tropical rainforests. Oaks, beeches, and maples are a few of the most common types of trees in these forests. Some typical animals of the temperate deciduous forest of eastern North America are shown in Figure 4.11 (p. 81).

Temperate grassland Traveling westward from the Great Lakes, temperature differences between winter and summer become more extreme, rainfall diminishes, and we find **temperate grasslands** (**FIGURE 4.19**). This is because the limited precipitation in the Great Plains region west of the Mississippi River supports grasses more easily than trees. Also known as steppe or prairie, temperate grasslands were once widespread in much of North and South America and central Asia.

Vertebrate animals of North America's native grasslands include American bison, prairie dogs, pronghorn antelope, and ground-nesting birds such as meadowlarks and prairie chickens.

(a) Temperate rainforest

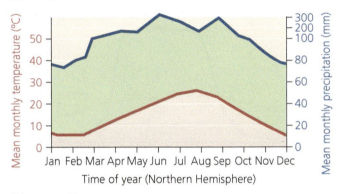

(b) Nagasaki, Japan

FIGURE 4.20 Temperate rainforests receive a great deal of precipitation and have moist, mossy interiors. *Climatograph adapted from Breckle, S.-W. 2002.*

(a) Tropical rainforest

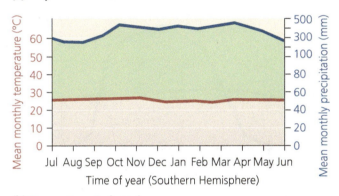

(b) Bogor, Java, Indonesia

FIGURE 4.21 Tropical rainforests, famed for their biodiversity, grow under constant, warm temperatures and a great deal of rain. *Climatograph adapted from Breckle, S.-W. 2002.*

People have converted most of the world's grasslands for farming and ranching, however, so most of these animals exist today at only a small fraction of their historic population sizes.

Temperate rainforest Further west in North America, the topography becomes varied, and biome types intermix. The coastal Pacific Northwest region, with its heavy rainfall, features **temperate rainforest** (**FIGURE 4.20**). Coniferous trees such as cedars, spruces, hemlocks, and Douglas fir grow very tall in the temperate rainforest, and the forest interior is shaded and damp. Moisture-loving animals such as the bright yellow banana slug are common.

The soils of temperate rainforests are fertile but are susceptible to landslides and erosion if forests are cleared. We extract lumber and other commercially valuable products from temperate rainforests. Timber harvesting has eliminated most old-growth trees in these forests, driving species such as the spotted owl and marbled murrelet toward extinction. Local people generally support timber extraction, but they also suffer the consequences of overharvesting.

Tropical rainforest In tropical regions we see the same pattern found in temperate regions: Areas of high rainfall

grow rainforests, areas of intermediate rainfall support dry or deciduous forests, and areas of low rainfall are dominated by grasses. However, tropical biomes differ from their temperate counterparts in other ways because they are closer to the equator and therefore warmer on average year-round. For one thing, they hold far greater biodiversity.

Tropical rainforest (**FIGURE 4.21**)—found in Central America, South America, Southeast Asia, west Africa, and other tropical regions—is characterized by year-round rain and uniformly warm temperatures. Tropical rainforests have dark, damp interiors; lush vegetation; and highly diverse communities, with more species of insects, birds, amphibians, and other animals than any other biome. These forests are not dominated by single species of trees, as are forests closer to the poles, but instead consist of very high numbers of tree species intermixed, each at a low density. A tree may be draped with vines, enveloped by strangler figs, and loaded with epiphytes (orchids and other plants that grow in trees), such that trees occasionally collapse under the weight of all the life they support.

Despite this profusion of life, tropical rainforests have poor, acidic soils that are low in organic matter. Nearly all nutrients present in this biome occur in the plants, not in the

(a) Tropical dry forest

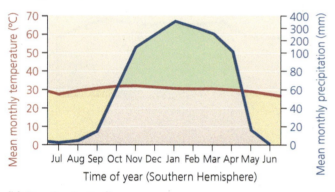

(b) Darwin, Australia

FIGURE 4.22 Tropical dry forests experience significant seasonal variation in precipitation but relatively stable, warm temperatures. *Climatograph adapted from Breckle, S.-W. 2002.*

(a) Savanna

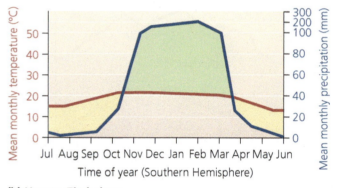

(b) Harare, Zimbabwe

FIGURE 4.23 Savannas are grasslands with clusters of trees. They experience slight seasonal variation in temperature but significant variation in rainfall. *Climatograph adapted from Breckle, S.-W. 2002.*

soil. An unfortunate consequence is that once tropical rainforests are cleared, the nutrient-poor soil can support agriculture for only a short time (p. 215). As a result, farmed areas are abandoned quickly, and farmers move on and clear more forest.

Tropical dry forest Tropical areas that are warm year-round but where rainfall is lower overall and highly seasonal give rise to **tropical dry forest,** or *tropical deciduous forest* (**FIGURE 4.22**), a biome widespread in India, Africa, South America, and northern Australia. Wet and dry seasons each span about half a year in tropical dry forest. Organisms that inhabit tropical dry forest have adapted to seasonal fluctuations in precipitation and temperature. For instance, many plants are deciduous and leaf out and grow profusely with the rains, then drop their leaves during dry times of the year.

Rains during the wet season can be heavy and, coupled with erosion-prone soils, can lead to severe soil loss where people have cleared forest. Across the globe, we have converted a great deal of tropical dry forest to agriculture. Clearing for farming or ranching is straightforward because vegetation is lower and canopies less dense than in tropical rainforest.

Savanna Drier tropical regions give rise to **savanna** (**FIGURE 4.23**), tropical grassland interspersed with clusters of acacias or other trees. The savanna biome is found across stretches of Africa, South America, Australia, India, and other dry tropical regions. Precipitation usually arrives during distinct rainy seasons, whereas in the dry season grazing animals concentrate near widely spaced water holes. Common herbivores on the African savanna include zebras, gazelles, and giraffes. Predators of these grazers include lions, hyenas, and other highly mobile carnivores.

Desert Where rainfall is very sparse, **desert** (**FIGURE 4.24**) forms. The driest biome on Earth, most deserts receive well under 25 cm (9.8 in.) of precipitation per year, much of it during isolated storms months or years apart. Some deserts, such as Africa's Sahara and Namib deserts, are mostly bare sand dunes; others, such as the Sonoran Desert of Arizona and northwest Mexico, receive more rain and are more heavily vegetated.

Deserts are not always hot; the high desert of the western United States is positively cold in winter. Because deserts have low humidity and little vegetation to insulate them from temperature extremes, sunlight readily heats them in the

(a) Desert

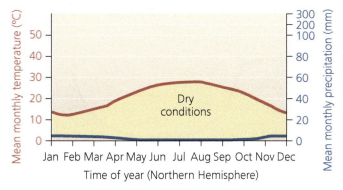

(b) Cairo, Egypt

FIGURE 4.24 Deserts are dry year-round. The precipitation curve is consistently below the temperature curve in this climatograph for Cairo, Egypt, indicating that the region experiences "dry" conditions all year. *Climatograph adapted from Breckle, S.-W. 2002.*

(a) Tundra

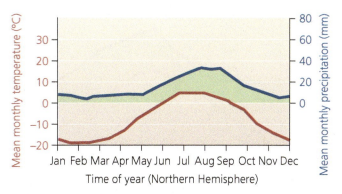

(b) Vaigach, Russia

FIGURE 4.25 Tundra is a cold, dry biome found near the poles. Alpine tundra occurs atop high mountains at lower latitudes. *Climatograph adapted from Breckle, S.-W. 2002.*

daytime, but heat is quickly lost at night. As a result, temperatures vary greatly from day to night and from season to season. Desert soils can be saline and are sometimes known as lithosols, or stone soils, for their high mineral and low organic-matter content.

Desert animals and plants show many adaptations to deal with a harsh climate. Most reptiles and mammals, such as rattlesnakes and kangaroo mice, are active in the cool of night. Many Australian desert birds are nomadic, wandering long distances to find areas of recent rainfall and plant growth. Desert plants tend to have thick, leathery leaves to reduce water loss, or green trunks so that the plant can photosynthesize without leaves, minimizing the surface area prone to water loss. The spines of cacti and other desert plants guard them from being eaten by herbivores desperate for the precious water these plants hold. Such traits have evolved by convergent evolution in deserts across the world (see Figure 3.4b, p. 50).

Tundra Nearly as dry as desert, **tundra** (FIGURE 4.25) occurs at very high latitudes in northern Russia, Canada, and Scandinavia. Extremely cold winters with little daylight and

summers with lengthy days characterize this landscape of lichens and low, scrubby vegetation without trees. The great seasonal variation in temperature and day length results from this biome's high-latitude location, angled toward the sun in summer and away from the sun in winter.

Because of the cold climate, underground soil remains permanently frozen and is called permafrost. During winter, surface soil freezes as well. When the weather warms, the soil melts and produces pools of surface water, forming ideal habitat for mosquitoes and other insects. The swarms of insects benefit bird species that migrate long distances to breed during the brief but productive summer. Caribou also migrate to the tundra to breed, then leave for the winter. Only a few animals, such as polar bears and musk oxen, can survive year-round in tundra.

Most tundra remains intact and relatively unaltered by human occupation and development. However, atmospheric circulation patterns (p. 451) bring our airborne pollutants to this biome, and global climate change is heating high-latitude regions more intensely than other areas (p. 501). Climate change is melting sea ice, altering seasonal cycles to which animals have adapted, and melting

(a) Boreal forest

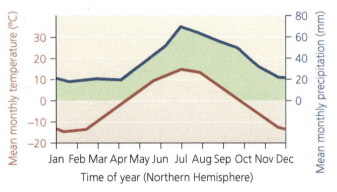

(b) Archangelsk, Russia

FIGURE 4.26 Boreal forest experiences long, cold winters; cool summers; and moderate precipitation. *Climatograph adapted from Breckle, S.-W. 2002.*

(a) Chaparral

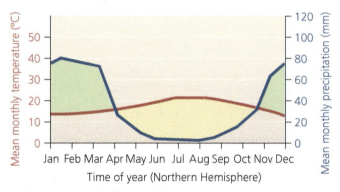

(b) Los Angeles, California, USA

FIGURE 4.27 Chaparral is a seasonally variable biome dominated by shrubs, influenced by marine weather, and dependent on fire. *Climatograph adapted from Breckle, S.-W. 2002.*

permafrost, releasing methane gas that further worsens climate change.

Tundra also occurs as alpine tundra at the tops of tall mountains in temperate and tropical regions. Here, high elevation creates conditions similar to those of high latitude.

Boreal forest The northern coniferous forest, or **boreal forest,** often called *taiga* (**FIGURE 4.26**), extends across much of Canada, Alaska, Russia, and Scandinavia. A few species of evergreen trees, such as black spruce, dominate large stretches of forest, interspersed with bogs and lakes. Boreal forests develop in cooler, drier regions than do temperate forests, and they experience long, cold winters and short, cool summers.

Soils are typically nutrient-poor and somewhat acidic. As a result of strong seasonal variation in day length, temperature, and precipitation, many organisms compress a year's

worth of feeding and breeding into a few warm, wet months. Year-round residents of boreal forest include mammals such as moose, wolves, bears, lynx, and rodents. Many insect-eating birds migrate here from the tropics to breed during the brief, intensely productive, summers.

Chaparral In contrast to the boreal forest's broad, continuous distribution, **chaparral** (**FIGURE 4.27**) is limited to small patches widely flung around the globe. Chaparral consists mostly of evergreen shrubs and is densely thicketed. This biome is highly seasonal, with mild, wet winters and warm, dry summers—a climate induced by oceanic influences and often termed "Mediterranean." Besides ringing the Mediterranean Sea, chaparral occurs along the coasts of California, Chile, and southern Australia. Chaparral communities experience frequent fire, and their plants are adapted to resist fire or even to depend on it for germination of their seeds.

closing
THE LOOP

Ecological communities are shaped by many forces. Within communities, species interact through competition, predation, parasitism, herbivory, and mutualism. Communities are stable only until disturbed—and in today's world, invasive species such as the zebra mussel and quagga mussel are a major and growing form of disturbance. Our economies are organized around established ecological systems and are supported by their services, so when invasive species alter ecological systems, our economies suffer, too.

Scientists, policymakers, and managers are trying to limit the spread of zebra and quagga mussels by preventing their transport to new areas, largely by raising public awareness. Experts most fear them establishing in the Columbia River basin in the Pacific Northwest, where they could do hundreds of millions of dollars of damage to salmon runs and hydroelectric power facilities.

Meanwhile, throughout the Great Lakes and other waterways of the East and Midwest where these invasive mussels are firmly established, people are forced to bear the costs and adapt as best as they can. Fortunately, in recent years some ecological communities are beginning to show resilience to the mussels. Some birds, crabs, and other species are learning to prey on them, the mussels are no longer growing as large, and some of the species they have harmed are rebounding. Indeed, for any invasive species, eventually its rate of spread slows and its population size levels off. Often, native species discover them and become predators, parasites, or competitors. Some long-established invasive species in North America have begun to decline, and a few have even disappeared.

No one knows what the future holds in the case of zebra mussels and quagga mussels, but they and many other species that people have moved from place to place are creating a topsy-turvy world of novel and modified communities. In response, through ecological restoration, we are increasingly attempting to undo some of the changes we have set in motion.

REVIEWING Objectives

You should now be able to:

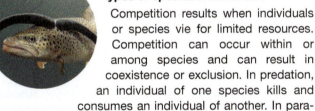

Summarize and compare the major types of species interactions

Competition results when individuals or species vie for limited resources. Competition can occur within or among species and can result in coexistence or exclusion. In predation, an individual of one species kills and consumes an individual of another. In parasitism, an individual of one species derives benefit by harming (but usually not killing) an individual of another. Herbivory is an exploitative interaction in which an animal feeds on a plant. In mutualism, species benefit from one another. (pp. 74–78)

Describe feeding relationships and energy flow, and use them to identify trophic levels and navigate food webs

Energy is transferred among trophic levels in food chains. Lower trophic levels generally contain more energy, biomass, and individuals. Food webs illustrate feeding relationships and energy flow among species in a community. (pp. 78–81)

Discuss characteristics of a keystone species

Keystone species exert impacts on communities that are far out of proportion to their abundance. Top predators are frequently considered keystone species. Other types of organisms (such as ecosystem engineers) can also exert strong effects on communities. (pp. 80–82)

Characterize disturbance, succession, and notions of community change

Disturbances are varied, and communities respond to disturbance in different ways. Succession describes a typical pattern of community change through time. Primary succession begins with an area devoid of life. Secondary succession begins with an area that has been severely disturbed but where remnants of the original community remain. Ecologists today view succession as being less predictable and deterministic than they did in the past. If disturbance is severe enough, communities may undergo regime shifts involving irreversible change—or novel communities may form. (pp. 82–86, 88–89)

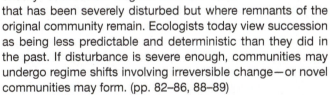

Predict the potential impacts of invasive species in communities, and suggest responses to biological invasions

People have introduced countless species to new areas. Some of these non-native species may become invasive if they do not encounter limiting factors on their population growth. Invasive species such as the zebra mussel have altered the composition, structure, and function of communities. We can respond to invasive species with prevention, control, and eradication measures. (pp. 86–87, 90)

Explain the goals and methods of restoration ecology

Restoration ecology is the science of restoring communities to a previous, more functional or more "natural" condition, variously defined as before human impact or before recent industrial impact. The growing practice of ecological restoration, informed by the science of restoration ecology, helps us restore ecological systems. Common strategies include removing invasive species, reintroducing native species, and reestablishing ecological processes. (pp. 90–92)

Identify and describe the terrestrial biomes of the world

Biomes represent major classes of communities spanning large geographic areas. Their distribution is determined by temperature, precipitation, and other factors. Terrestrial biomes include temperate deciduous forest, temperate grassland, temperate rainforest, tropical rainforest, tropical dry forest, savanna, desert, tundra, boreal forest, and chaparral. (pp. 92–98)

TESTING Your Comprehension

1. Explain how competition leads to a realized niche. How does competition promote resource partitioning?

2. Compare and contrast the three main types of exploitative species interactions (predation, parasitism, and herbivory), explaining how they differ.

3. Give examples of symbiotic and nonsymbiotic mutualisms. Describe at least one way in which a mutualism affects your daily life.

4. Compare and contrast trophic levels, food chains, and food webs. How are these concepts related, and how do they differ?

5. What is meant by the term *keystone species*, and what types of organisms are most often considered keystone species?

6. Describe the process of primary succession. How does it differ from secondary succession? Give an example of each.

7. Why do scientists consider invasive species to be a problem? What makes a species "invasive," and what ecological effects can invasive species have? Give examples.

8. What is restoration ecology? Why is it an important scientific pursuit in today's world?

9. What factors most strongly influence the type of biome that forms in a particular place on land? What factors determine the type of aquatic system that may form in a given location?

10. Draw a typical climate diagram for a tropical rainforest. Label all parts of the diagram and describe all of the types of information an ecologist could glean from such a diagram. Now draw a climate diagram for a desert. How does it differ from your rainforest climatograph, and what does this tell you about how the two biomes differ?

SEEKING Solutions

1. Suppose you spot two species of birds feeding side by side, eating seeds from the same plant. You begin to wonder whether competition is at work. Describe how you might design scientific research to address this question. What observations would you try to make at the outset? Would you try to manipulate the system to test your hypothesis that the two birds are competing? If so, how?

2. Using the concepts of trophic levels and energy flow, explain why the ecological footprint of a vegetarian person is smaller than that of a meat-eater.

3. Spend some time outside on your campus, in your yard, or in the nearest park or natural area. Find at least 10 species of organisms (plants, animals, or others), and observe each one long enough to watch it feed or to make an educated guess about how it derives its nutrition. Now, using Figure 4.11 as a model, draw a simple food web involving all the organisms you observed.

4. Can you think of one organism not mentioned in this chapter as a keystone species that you believe may be

a keystone species? For what reasons do you suspect this? How could you experimentally test whether an organism is a keystone species?

5. **CASE STUDY CONNECTION** Describe five ecological changes to freshwater communities in the Great Lakes or the Hudson River that have occurred since the invasion of the zebra mussel. Describe one economic impact of the invasion. What is one way to prevent the mussel from spreading to new areas?

6. **THINK IT THROUGH** A federal agency has put you in charge of devising responses to the invasions of zebra mussels and quagga mussels. Based on what you know from this chapter, how would you seek to control the spread of these species and reduce their impacts? What strategies would you consider pursuing immediately? For which strategies would you commission further scientific research? For each of your ideas, name one benefit or advantage, and identify one obstacle it might face in being implemented.

CALCULATING Ecological Footprints

Environmental scientists David Pimentel, Rodolfo Zuniga, and Doug Morrison of Cornell University reviewed scientific estimates for the economic and ecological costs inflicted by introduced and invasive species in the United States. They found that approximately 50,000 species had been introduced in the United States and that these accounted for more than $120 billion in economic costs each year. These costs include direct losses and damage, as well as costs required to control the species. (The researchers did not quantify monetary estimates for losses of biodiversity, ecosystem services, and aesthetics, which they said would drive total costs several times higher.) Calculate values missing from the table to determine the number of introduced species of each type of organism and the annual cost that each imposes on our economy.

GROUP OF ORGANISM	PERCENTAGE OF TOTAL INTRODUCED	NUMBER OF SPECIES INTRODUCED	PERCENTAGE OF TOTAL ANNUAL COSTS	ANNUAL ECONOMIC COSTS
Plants	50.0	25,000	27.2	
Microbes	40.0		20.2	
Arthropods	9.0		15.7	
Fish	0.28		4.2	
Birds	0.19		1.5	$1.9 billion
Mollusks	0.18		1.7	
Reptiles and amphibians	0.11		0.009	
Mammals	0.04	20	29.4	$37.5 billion
TOTAL	100	50,000	100	$127.4 billion

Source: Pimentel, D., R. Zuniga and D. Morrison, 2005. Update on the environmental and economic costs associated with alien-invasive species in the United States. *Ecological Economics* 52: 273–288.

1. Of the 50,000 species introduced into the United States, half are plants. Describe two ways in which non-native plants might be brought to a new environment. How might we help prevent non-native plants from establishing in new areas and posing threats to native communities?

2. Organisms that damage crop plants are the most costly of introduced species. Weeds, pathogenic microbes, and arthropods that attack crops together account for half of the costs documented by Pimentel and his colleagues. What steps can we—farmers, governments, and all of us as a society—take to minimize the impacts of invasive species on crops?

3. How might your own behavior influence the influx and ecological impacts of non-native species like those listed above? Name three things you could personally do to help reduce the impacts of invasive species.

MasteringEnvironmentalScience®

Students Go to **MasteringEnvironmentalScience** for assignments, the etext, and the Study Area with practice tests, videos, current events, and activities.

Instructors Go to **MasteringEnvironmentalScience** for automatically graded activities, current events, videos, and reading questions that you can assign to your students, plus Instructor Resources.

CHAPTER 5

Environmental Systems
and Ecosystem Ecology

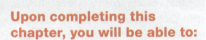

central
CASE STUDY

The Vanishing Oysters of the Chesapeake Bay

> **I'm 60. Danny's 58. We're the young ones.**
> Grant Corbin, Oysterman in Deal Island, Maryland

> **The Bay continues to be in serious trouble. And it's really no question why this is occurring. We simply haven't managed the Chesapeake Bay as a system the way science tells us we must.**
> Will Baker, President, Chesapeake Bay Foundation

A visit to Deal Island, Maryland, on the Chesapeake Bay reveals a situation that is all too common in modern America: The island, which was once bustling with productive industries and growing populations, is falling into decline. Economic opportunities in the community are few, and its populace is increasingly "graying" as more and more young people leave to find work elsewhere. In 1930, Deal Island had a population of 1237 residents. In 2010, it was a mere 471 people—and only 75 of them were under age 18, indicating that the population of Deal Island was aging as young people moved away to work and start families.

Unlike other parts of the country with similar stories of economic decline, the demise of Deal Island and other bayside towns was not caused by the closing of a local factory, steel mill, or corporate headquarters. It was caused by the collapse of the Chesapeake Bay oyster fishery.

The Chesapeake Bay was once a thriving system of interacting plants, animals, and microbes. Blue crabs, scallops, and fish such as giant sturgeon, striped bass, and shad thrived in the bay. Nutrients carried to the bay by streams in its roughly 168,000 km² (64,000 mi²) **watershed**—the land area that funnels water to the bay through rivers—nourished fields of underwater grasses that provided food and refuge to juvenile fish, shellfish, and crabs. Hundreds of millions of oysters kept the bay's water clear by filtering nutrients and phytoplankton (microscopic photosynthetic algae, protists, and cyanobacteria that drift near the surface) from the water column.

Oysters had been eaten locally since the region was populated, but the intensive harvest of bay oysters for export didn't begin until the 1830s. By the 1880s the bay boasted the world's largest oyster fishery. People flocked to the Chesapeake to work on oystering ships or in canneries, dockyards, and shipyards. Bayside towns like Deal Island prospered along with the oyster industry and developed a unique maritime culture that defined the region.

But by 2010 the bay's oyster populations had been reduced to a mere 1% of their abundance prior to the start of commercial harvesting, and the oyster industry in the area was all but wiped out. Perpetual overharvesting, habitat destruction, virulent oyster diseases, and water pollution had nearly eradicated this economically and ecologically important organism from bay waters. The monetary losses associated with the oyster fishery collapse in the Chesapeake Bay have been staggering,

Sorting oysters from the Chesapeake Bay ▲

Upon completing this chapter, you will be able to:

- Describe environmental systems

- Define ecosystems and discuss how living and nonliving entities interact in ecosystem-level ecology

- Discuss the fundamentals of landscape ecology, GIS, and ecological modeling

- Explain ecosystem services and discuss how they benefit our lives

- Compare and contrast how water, carbon, nitrogen, and phosphorus cycle through the environment

- Explain how human activities affect biogeochemical cycles

Chesapeake Bay oystermen haul in their catch.

costing the economies of Maryland and Virginia an estimated $4 billion from 1980 to 2010.

In addition to overharvesting, one of the biggest impacts on oysters in recent decades is the pollution of the bay with high levels of the nutrients nitrogen and **phosphorus** from agricultural fertilizers, animal manure, stormwater runoff, and atmospheric compounds produced by fossil fuel combustion. Oysters naturally filter nutrients from water, but with so few oysters in the bay, elevated nutrient levels have caused phytoplankton populations to increase. When phytoplankton die, settle to the bay bottom, and are decomposed by bacteria, oxygen in the water is depleted (a condition called **hypoxia**), which creates "dead zones" in the bay. Grasses, oysters, and other immobile organisms perish in dead zones when deprived of oxygen. Crabs, fish, and other mobile organisms are forced to flee to habitats where oxygen levels are higher, but they often face smaller food supplies and increased predation pressure in the new locations. Hypoxia, along with other human impacts on the Chesapeake Bay, have landed it on the Environmental Protection Agency's (EPA's) list of dangerously polluted waters in the United States.

However, recent events in the Chesapeake have offered hope for the recovery of the Chesapeake Bay system. The EPA agreed in 2010, for the first time in the region, to hold bay states to strict pollutant "budgets" that aim to substantially reduce inputs of nitrogen and phosphorus into the bay by 2025. Further, oyster restoration efforts are finally showing promise (see **THE SCIENCE BEHIND THE STORY**, pp. 106–107) in the Chesapeake. If these initiatives can begin to restore the bay to health, Deal Island and other oyster-fishing communities may again enjoy the prosperity they once did on the scenic shores of the Chesapeake.

Earth's Environmental Systems

Understanding the rise and fall of the oyster industry in the Chesapeake Bay involves comprehending the complex networks of systems that make up Earth's environment. These include physical systems ranging in scale from tiny matter and molecules to enormous masses of magma and mountains (Chapter 2). The Earth's networks further include biological systems ranging in hierarchy from organisms and populations (Chapter 3) to communities of interacting species (Chapter 4). In ecosystems they involve the interaction of living creatures with the nonliving entities around them. Earth's systems also encompass cycles involving rock, air, and water that shape our landscapes and guide the flow of chemical elements and compounds that support life and regulate climate. We depend on these systems for our very survival.

Assessing questions holistically by taking a systems approach is helpful in environmental science, in which so many issues are multifaceted and complex. Taking a broad and integrative approach poses challenges, because systems often show behavior that is difficult to predict. However, environmental scientists are rising to the challenge of studying systems holistically, helping us to develop comprehensive solutions to complicated problems such as those faced in the Chesapeake Bay.

Systems involve feedback loops

Let's take a step back and define a system. A **system** is a network of relationships among parts, elements, or components that interact with and influence one another through the exchange of energy, matter, or information. Earth's environmental systems receive inputs of energy, matter, or information; process these inputs; and produce outputs. As a system, for example, the Chesapeake Bay receives inputs of freshwater, sediments, nutrients, and pollutants from the rivers that empty into it. Oystermen, crabbers, and fishermen harvest some of the bay system's output: matter and energy in the form of seafood. This output subsequently becomes input to the nation's economic system and to the body systems of people who eat the seafood.

Sometimes a system's output can serve as input to that same system, a circular process described as a **feedback loop,** which can be either negative or positive. In a **negative feedback loop** (**FIGURE 5.1**), output that results from a system moving in one direction acts as input that moves the system in the other direction. Input and output essentially neutralize one another's effects, stabilizing the system. As an example, negative feedback regulates body temperature in humans: If we get too hot, our sweat glands pump out water that then evaporates, which cools us down. Or we just move into the shade. If we get too cold, we shiver, creating heat, or we just move into the sun or put on a sweater. Most systems in nature involve such negative feedback loops. Negative feedback enhances stability, and over time only those systems that are stable will persist.

Rather than stabilizing a system, **positive feedback loops** drive the system further toward an extreme. In positive feedback, increased output from a system leads to increased input, leading to further increased output, and so on. Exponential growth in a population (pp. 64–65) is one such example. The more individuals there are, the more offspring

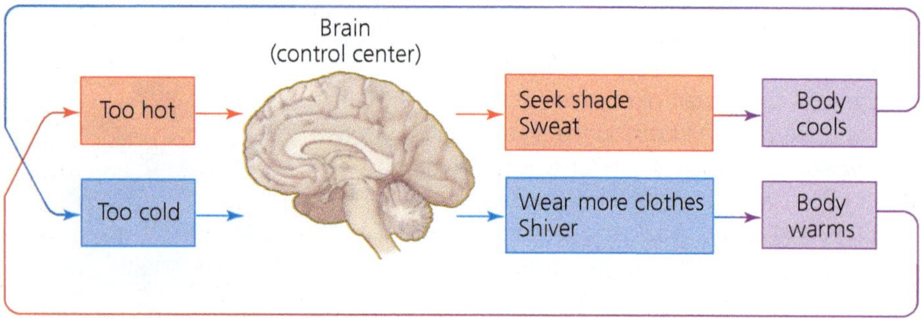

FIGURE 5.1 Negative feedback loops exert a stabilizing influence on systems. The human body's response to heat and cold involves a negative feedback loop that keeps core body temperatures relatively stable.

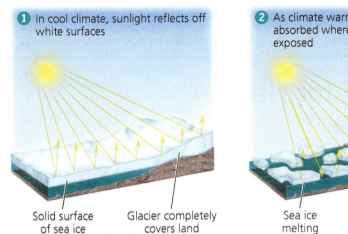

① In cool climate, sunlight reflects off white surfaces

② As climate warms, sunlight is absorbed where dark surfaces are exposed

③ Light absorption speeds warming, melting more ice and in turn exposing more dark surfaces

Solid surface of sea ice   Glacier completely covers land

Sea ice melting   Glacier melting

More water exposed   More land exposed

FIGURE 5.2 Positive feedback loops push systems away from equilibrium. For example, when Arctic glaciers and sea ice melt because of global warming, darker surfaces are exposed, which absorb more sunlight, causing further warming and further melting.

that can be produced. Another positive feedback cycle that is of great concern to environmental scientists today involves the melting of glaciers and sea ice in the Arctic as a result of global warming (p. 494). Because ice and snow are white, they reflect sunlight and keep surfaces cool. But if the climate warms enough to melt the ice and snow, darker surfaces of land and water are exposed, and these darker surfaces absorb sunlight. This absorption warms the surface, causing further melting, which in turn exposes more dark surface area, leading to further warming (**FIGURE 5.2**).

Runaway cycles of positive feedback are rare in untouched nature, but they are common in natural systems altered by human impact, and such feedback loops can destabilize those systems. In the Chesapeake Bay, for example, oysters clean the bay's water by filtering phytoplankton, sediments, and nutrients from the water column. When oyster populations were decimated by overharvesting, however, water filtration by oysters declined. With less filtering, concentrations of nutrients and sediments increased in bay waters. Spurred by elevated nutrient levels, phytoplankton populations in the bay exploded, causing hypoxia on the bay bottom and killing many of the remaining oysters. This additional oyster mortality further reduced water filtration in the bay, leading to worsening water quality and even more oyster deaths—an example of a positive feedback loop.

Systems exhibit several defining properties

In a system stabilized by negative feedback, when processes move in opposing directions at equivalent rates so that their effects balance out, they are said to be in **dynamic equilibrium.** Processes in dynamic equilibrium can contribute to **homeostasis,** the tendency of a system to maintain constant or stable internal conditions. A system (such as an organism) in homeostasis keeps its internal conditions within a narrow range that allows it to function. However, the steady state of a homeostatic

system may itself change slowly over time. For instance, Earth has experienced gradual changes in atmospheric composition and ocean chemistry over its long history, yet life persists and our planet remains, by most definitions, a homeostatic system.

It is difficult to understand systems fully just by focusing on their individual components because systems can have **emergent properties,** characteristics not evident in the components alone. Stating that systems possess emergent properties is a lot like saying, "The whole is more than the sum of its parts." For example, if you were to reduce a tree to its component parts (leaves, branches, trunk, bark, roots, fruit, and so on) you would not be able to predict the whole tree's emergent properties, which include the role the tree plays as habitat for birds, insects, fungi, and other organisms. You could analyze the tree's chloroplasts (photosynthetic cell organelles), diagram its branch structure, and evaluate the nutritional content of its fruit, but you would still be unable to understand the tree as habitat, as part of a forest landscape, or as a reservoir for carbon storage.

FAQ

But isn't positive feedback "good" and negative feedback "bad"?

Sometimes scientific terminology goes against the way we use terms in everyday language. In daily life, positive feedback (such as a compliment of a paper you've written) is something that acts as a stabilizing force ("Keep up the good work, and you'll succeed") whereas negative feedback (such as criticism of a paper) acts as a destabilizing force ("You need to change your approach if you're going to succeed").

In environmental systems, it's the opposite! Negative feedback resists change in systems, and in doing so it enhances stability, typically keeping conditions within ranges beneficial to life. Positive feedback exerts destabilizing effects that push conditions in systems to extremes, threatening organisms adapted to the system's normal conditions. Thus, negative feedback in environmental systems typically aids living things, whereas positive feedback often harms them.

Are We "Turning the Tide" for Native Oysters in Chesapeake Bay?

In 2001, the Eastern oyster (*Crassostrea virginica*) was in dire trouble in the Chesapeake Bay. Populations had dropped by 99%, and the Chesapeake's oyster industry, once the largest in the world, had collapsed. Poor water quality, reef destruction, virulent diseases spread by transplanted oysters, and 200 years of overharvesting all contributed to the collapse.

Restoration efforts had largely failed. Moreover, when scientists or resource managers proposed rebuilding oyster populations by significantly restricting oyster harvests or establishing oyster reef "sanctuaries," these initiatives were typically defeated by the politically powerful oyster industry. All this had occurred in a place whose very name (derived from the Algonquin word *Chesepiook*) means "great shellfish bay."

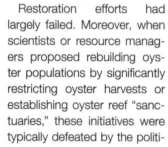

David Schulte of the U.S. Army Corps of Engineers

With the collapse of the native oyster fishery and with political obstacles blocking restoration projects for native oysters, support grew among the oyster industry, state resource managers, and some scientists for the introduction of Suminoe oysters (*Crassostrea ariakensis*) from Asia. This species seemed well suited for conditions in the bay and showed resistance to the parasitic diseases that were ravaging native oysters. Proponents argued that introducing Suminoe oysters would reestablish thriving oyster populations in the bay and revitalize the oyster fishery.

Proponents maintained in addition that introducing oysters would also improve the bay's water quality, because as oysters feed, they filter phytoplankton and sediments from the water column. Filter-feeding by oysters is an important ecological service in the bay because it reduces phytoplankton densities, clarifies waters, and supports the growth of underwater grasses that provide food and refuge for waterfowl and young crabs. Because introductions of invasive species can have profound ecological impacts (pp. 86–90), the Army Corps of Engineers was directed to coordinate an environmental impact statement (EIS, p. 171) on oyster restoration approaches in the Chesapeake.

It was in this politically charged, high-stakes environment that Dave Schulte, a scientist with the Corps and doctoral student at the College of William and Mary, set out to determine whether there was a viable approach to restoring native oyster

populations. The work he and his team began would help turn the tide in favor of native oysters in the bay's restoration efforts.

One of the biggest impacts on native oysters was the destruction of oyster reefs by a century of intensive oyster harvesting. Oysters settle and grow best on the shells of other oysters, and over long periods this process forms reefs (underwater outcrops of living oysters and oyster shells) that solidify and become as hard as stone. Throughout the bay, massive reefs that at one time had jutted out of the water at low tide had been reduced to rubble on the bottom from a century of repeated scouring by metal dredgers used by oyster-harvesting ships. The key, Schulte realized, was to construct artificial reefs like those that once existed, to get oysters off the bottom—away from smothering sediments and hypoxic waters—and up into the plankton-rich upper waters.

Armed with the resources available to the Corps, Schulte opted to take a landscape ecology approach and restore patches of reef habitat on nine complexes of reefs covering a total of 35.3 hectares (87 acres) in an oyster sanctuary near the mouth of the Great Wicomico River (**FIGURE 1**) in the lower Chesapeake Bay. This

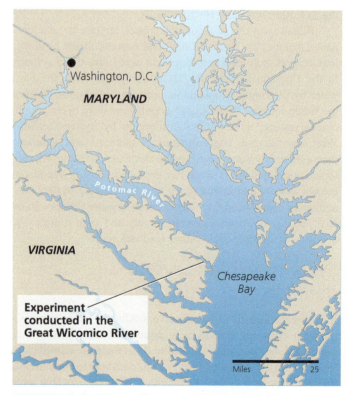

FIGURE 1 Schulte's study was conducted in the Great Wicomico River in Virginia in the lower Chesapeake Bay.

FIGURE 2 A water cannon blows oyster shells off a barge and onto the river bottom to create an artificial oyster reef for the experiment.

approach was very different from the smaller-scale restoration efforts of the past. Artificial reefs of two heights were constructed in 2004 (**FIGURE 2**), and oysters were allowed to colonize the reefs, safe from harvesting. Oyster populations on the constructed reefs were sampled in 2007, and the results were stunning. The reef complex supported an estimated 185 million oysters, a number nearly as large as the wild population of 200 million oysters estimated to live on the remaining degraded habitat in all of Maryland's waters.

Higher constructed reefs supported an average of more than 1000 oysters per square meter—four times more than the lower constructed reefs and 170 times more than unrestored bottom (**FIGURE 3**). Like natural reefs, the constructed reefs began to solidify, providing a firm foundation for the settlement of new oysters. In 2009, Schulte's research made a splash when his team published its findings in the journal *Science*, bringing international attention to their study.

After reviewing eight alternative approaches to oyster restoration that involved one or more oyster species, the Corps advocated an approach that avoided the introduction of non-native oysters. Instead it proposed a combination of native oyster restoration, a temporary moratorium on oyster harvests (accompanied by a compensation program for the oyster industry), and enhanced support for oyster aquaculture in the bay region.

Schulte's restoration project cost roughly $3 million and will require substantial investments if it is to be repeated elsewhere in the bay. This is particularly true in upper portions of the bay, where oyster reproduction levels are lower (requiring restored reefs to be "seeded" with oysters), water conditions are poorer, and oysters are less resistant to disease. Many scientists contend that expanded reef restoration efforts are worth the cost because they enhance oyster populations and provide

a vital service to the bay through water filtering. Some scientists also see value in promoting oyster farming, in which restoration efforts would be supported by businesses instead of taxpayers.

These efforts are encouraged by the success of the project to date. By summer 2015 the majority of high-relief reef acreage was thriving, despite pressures from poachers and hypoxic conditions in several years. Moreover, many of the low-relief reefs that were originally constructed eventually accumulated enough new shell to be as tall as the high-relief reefs in the initial experiment—the reef treatment that showed the highest oyster densities in the original experiment. Further, oyster reproduction rates in 2012 were among the highest Schulte had seen during the project, and a follow-up study in 2013 found that spat (young, newly settled oysters) from the sanctuary reefs were seeding other parts of the Great Wicomico River and increasing oyster populations outside protected areas.

Protected sites for oyster restoration efforts are now being established elsewhere in the bay. Maryland recently designated 3640 hectares (9000 acres) of new oyster sanctuaries—25% of existing oyster reefs in state waters—and seeded these reefs with more than a billion hatchery-raised spat. This movement toward increased protection for oyster populations, coupled with findings of increased disease resistance in bay oysters, has given new hope that native oysters may once again thrive in the waters of the "great shellfish bay."

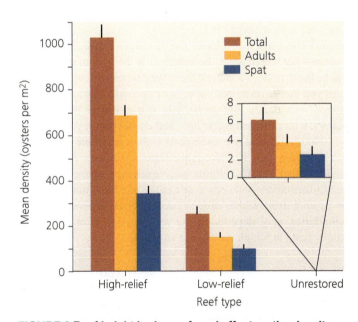

FIGURE 3 Reef height had a profound effect on the density of adult oysters and spat (newly settled oysters). Schulte's work suggested that native oyster populations could rebound in portions of Chesapeake Bay if they were provided elevated reefs and were protected from harvest. *Data from Schulte, D.M., R.P. Burke, and R.N. Lipcius, 2009. Unprecedented restoration of a native oyster metapopulation. Science 325: 1124–1128.*

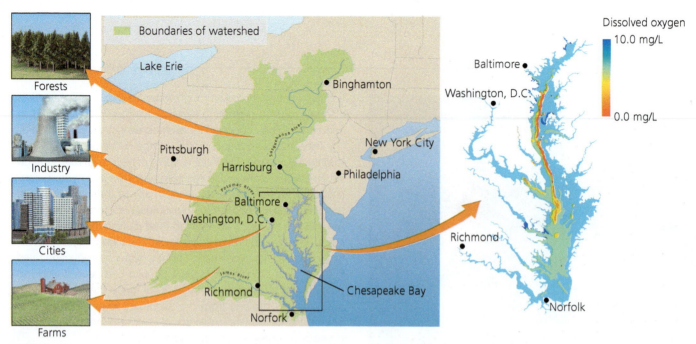

FIGURE 5.3 **The Chesapeake Bay watershed encompasses 168,000 km² (64,000 mi²) of land area in six states and the District of Columbia.** Tens of thousands of streams carry water, sediment, and pollutants from a variety of sources downriver to the Chesapeake, where nutrient pollution has given rise to large areas of hypoxic waters. The zoomed-in map (**at right**) shows dissolved oxygen concentrations in the Chesapeake Bay in 2014. Oysters, crabs, and fish typically require a minimum of 3 mg/L of oxygen and are therefore excluded from large portions of the bay where oxygen levels are too low. *Figure at right adapted from National Oceanic and Atmospheric Administration and U.S. Geological Survey, https://coastalscience.noaa.gov/news/?p=15670*

Systems seldom have well-defined boundaries, so deciding where one system ends and another begins can be difficult. Consider a smartphone. It is certainly a system—a network of circuits and parts that interact and exchange energy and information—but where are its boundaries? Is the system merely the phone itself, or does it include the other phones you call, the websites you access on it, and the cellular networks that keep it connected? What about the energy grid that recharges the phone's battery, with its transmission lines and distant power plants?

No matter how we attempt to isolate or define a system, we soon see that it has connections to systems larger and smaller than itself. Systems may exchange energy, matter, and information with other systems, and they may contain or be contained within other systems. Thus, where we draw boundaries may depend on the spatial (space) or temporal (time) scale at which we choose to focus.

Environmental systems interact

The Chesapeake Bay and the rivers that empty into it are examples of interacting systems. On a map, the rivers that feed into the bay are a branched and braided network of water channels surrounded by farms, cities, and forests (**FIGURE 5.3**). But where are the boundaries of this system? For a scientist interested in **runoff** (precipitation that flows over land and enters waterways) and the flow of water, sediment, or pollutants, it may make the most sense to define the bay's watershed as a system. However, for a scientist interested in hypoxia and

the bay's dead zones, it may be best to define the watershed together with the bay as the system of interest, because their interaction is central to the problem being investigated. Thus, in environmental science, identifying the boundaries of systems depends on the questions being asked.

If the question we are asking about the Chesapeake Bay relates to the dead zones in the bay, which are due to the extremely high levels of nitrogen and phosphorus delivered to its waters from the six states in its watershed and the 15 states in its **airshed**—the geographic area that produces air pollutants likely to end up in a waterway—then we'll want to define the boundaries of the system to include both the watershed and the airshed of the bay. In 2015, the bay received an estimated 121 million kg (267 million lb) of nitrogen and 7.2 million kg (15.8 million lb) of phosphorus. Runoff from agriculture was a major source of these nutrients, contributing 32% of the nitrogen (**FIGURE 5.4a**) and 45% of the phosphorus (**FIGURE 5.4b**) entering the bay. Roughly one-third of nitrogen inputs come from atmospheric sources within this system.

Elevated nitrogen and phosphorus inputs cause phytoplankton in the bay's waters to flourish. High densities lead to elevated mortality in phytoplankton populations. Dead phytoplankton settle to the bottom of the bay. The remains of dead phytoplankton are joined on the bottom by the waste products of zooplankton, tiny creatures that feed on phytoplankton. The increased organic material causes an explosion in populations of bacterial decomposers, which deplete the oxygen in bottom waters while consuming this material.

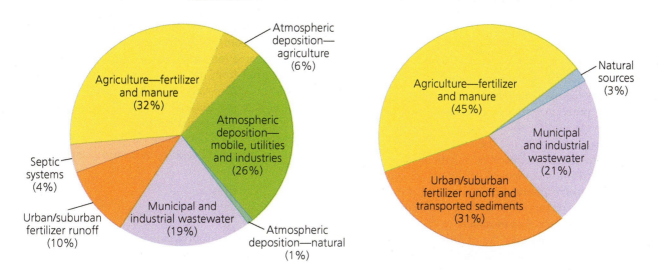

(a) Sources of nitrogen entering the Chesapeake Bay

(b) Sources of phosphorus entering the Chesapeake Bay

FIGURE 5.4 The Chesapeake Bay receives inputs of (a) nitrogen and (b) phosphorus from many sources in its watershed. *Data for (a) are from Chesapeake Bay Program Office, 2012, and (b) are from Chesapeake Bay Program, Watershed Model Phase 4.3 (Chesapeake Bay Program Office, 2009).Totals for nitrogen do not equal 100% due to rounding.*

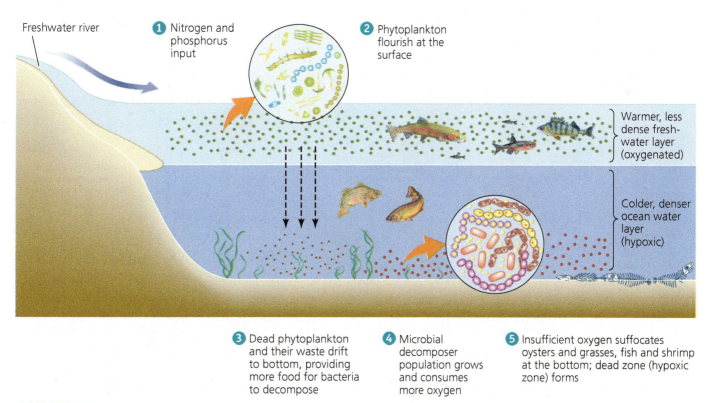

FIGURE 5.5 Excess nitrogen and phosphorus cause eutrophication in aquatic systems such as the Chesapeake Bay. Coupled with stratification (layering) of water, eutrophication can severely deplete dissolved oxygen. ❶ Nutrients from river water ❷ boost growth of phytoplankton, ❸ which die and are decomposed at the bottom by bacteria. Stability of the surface layer prevents deeper water from absorbing oxygen to replace ❹ oxygen consumed by decomposers, and ❺ the oxygen depletion suffocates or drives away bottom-dwelling marine life. This process gives rise to hypoxic zones like those in the bay.

Deprived of oxygen, organisms living in the area will flee if they can or will suffocate if they cannot. Oxygen replenishes slowly at the bottom because fresh water entering the bay from rivers remains naturally stratified in a layer at the surface and is slow to mix with the denser, saltier bay water. This limits the amount of oxygenated surface water that reaches the bottom-dwelling life that needs it. The process of nutrient overenrichment, blooms of algae, increased production of organic matter, and subsequent ecosystem degradation is known as **eutrophication** (**FIGURE 5.5**).

We may perceive Earth's systems in various ways

Categorizing environmental systems can help make Earth's dazzling complexity comprehensible to the human brain and accessible to problem solving. For instance, scientists sometimes divide Earth's components into "structural spheres." The **lithosphere** (p. 32) is the rock and sediment beneath our feet, the planet's uppermost mantle and crust. The **atmosphere** (p. 446) is composed of the air surrounding our planet. The **hydrosphere** encompasses all water—salt or fresh, liquid, ice, or vapor—in surface bodies, underground, and in the atmosphere. The **biosphere** (p. 58) consists of all the planet's organisms and the abiotic (nonliving) portions of the environment with which they interact.

Picture a robin plucking an earthworm from the ground after a rain. You are witnessing an organism (the robin) consuming another organism (the earthworm) by removing it from part of the lithosphere (the soil) that the earthworm had

been modifying, after rain (from the hydrosphere) moistened the ground. The robin might then fly through the air (the atmosphere) to a tree (an organism), in the process respiring (combining oxygen from the atmosphere with glucose from the organism, and adding water to the hydrosphere and carbon dioxide and heat to the atmosphere). Finally, the bird might defecate, adding nutrients from the organism to the lithosphere below. The study of such interactions among living and nonliving things is a key part of ecology at the ecosystem level.

Ecosystems

An **ecosystem** consists of all organisms and nonliving entities that occur and interact in a particular area at the same time. The ecosystem concept builds on the idea of the biological community (Chapter 4), but ecosystems include abiotic components as well as biotic ones. In ecosystems, energy flows and matter cycles among these components.

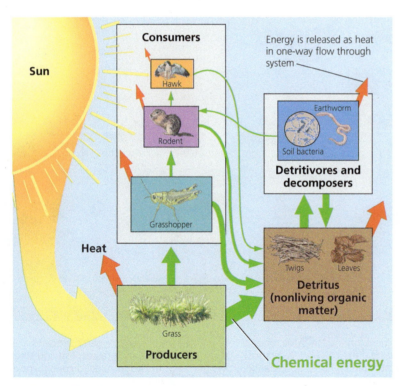

(a) Energy flowing through an ecosystem

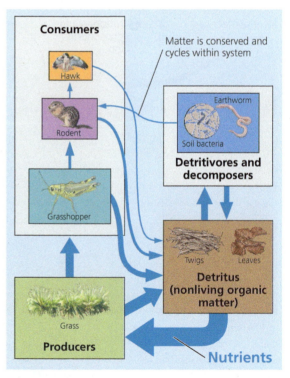

(b) Matter cycling within an ecosystem

FIGURE 5.6 In systems, energy flows in one direction, whereas matter is recycled. (a) Light energy from the sun (**yellow arrow**) drives photosynthesis in producers, which begins the transfer of chemical energy (**green arrows**) among trophic levels (pp. 78–80) and detritus. Energy exits the system through respiration in the form of heat (**orange arrows**). **(b)** Nutrients (**blue arrows**) move among trophic levels and detritus. In both diagrams, box sizes conceptually represent quantities of energy or matter content, and arrow widths represent relative magnitudes of energy or matter transfer. Such values may vary greatly among ecosystems. For simplicity, various abiotic components (such as water, air, and inorganic soil content) of ecosystems have been omitted.

 Based on the figure, which transfer of chemical energy is the largest in ecosystems? Which transfer is the largest for nutrient cycling in ecosystems?

Go to **Interpreting Graphs & Data** on **Mastering**EnvironmentalScience*

Ecosystems are systems of interacting living and nonliving entities

The ecosystem concept originated early in the 20th century from scientists such as British ecologist Arthur Tansley, who recognized that biological entities are tightly intertwined with chemical and physical entities. Tansley argued that there was so much interaction between organisms and their abiotic environments that it made the most sense to view living and nonliving elements together. For instance, in the Chesapeake Bay **estuary**—a water body where rivers flow into the ocean, mixing fresh water with saltwater—aquatic organisms are affected by the flow of water, sediment, and nutrients from the rivers that feed the bay and from the land that feeds those rivers. In turn, the photosynthesis, respiration, and decomposition that these organisms undergo influence the chemical and physical conditions of the Chesapeake's waters.

Ecologists soon began analyzing ecosystems as an engineer might analyze the operation of a machine. In this view, ecosystems are systems that receive inputs of energy, process and transform that energy while cycling matter internally, and produce outputs (such as heat, water flow, and animal waste products) that enter other ecosystems.

Energy flows in one direction through ecosystems. Most arrives as radiation from the sun, powers the system, and exits in the form of heat (**FIGURE 5.6a**). Matter, in contrast, is generally recycled within ecosystems (**FIGURE 5.6b**). Energy and matter pass among producers, consumers, and decomposers through trophic interactions in food-web relationships (p. 80). Matter is recycled because when organisms die and decay, their nutrients remain in the system. In contrast, most energy that organisms take in drives cellular respiration and is released to the environment as heat.

Energy is converted to biomass

As autotrophs, such as green plants and phytoplankton, convert solar energy to the energy of chemical bonds in sugars through photosynthesis (p. 30), they perform **primary production.** Specifically, the total amount of chemical energy produced by autotrophs is termed **gross primary production.** Autotrophs use most of this production to power their own metabolism by cellular respiration (p. 31). The energy that remains after respiration and that is used to generate biomass (such as leaves, stems, and roots) ecologists call **net primary production.** Thus, net primary production equals gross primary production minus the energy used in respiration.

Another way to think of net primary production is that it represents the energy or biomass available for consumption by heterotrophs. Some of this plant biomass is eaten by herbivores. Heterotrophs use the energy they gain from plant biomass for their own metabolism, growth, and reproduction. Some of this energy is used by heterotrophs to generate biomass in their bodies (such as skin, muscle, or bone), which is termed **secondary production.** Plant matter not eaten by herbivores becomes fodder for detritivores and decomposers once the plant dies or drops its leaves.

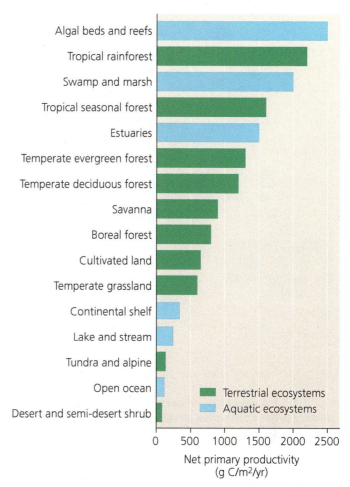

FIGURE 5.7 Net primary productivity varies greatly between ecosystem types. Freshwater wetlands, tropical forests, coral reefs, and algal beds show high values on average, whereas deserts, tundra, and the open ocean show low values. *Data from Whittaker, R.H., 1975. Communities and ecosystems, 2nd ed. New York, NY: Macmillan.*

Ecosystems vary in the rate at which autotrophs convert energy to biomass. The rate at which this conversion occurs is termed **productivity,** and ecosystems whose plants convert solar energy to biomass rapidly are said to have high **net primary productivity.** Freshwater wetlands, tropical forests, coral reefs, and algal beds tend to have the highest net primary productivities, whereas deserts, tundra, and open ocean tend to have the lowest (**FIGURE 5.7**). Variation among ecosystems and among biomes (Chapter 4) in net primary productivity results in geographic patterns across the globe (**FIGURE 5.8**, p. 112). In terrestrial ecosystems, net primary productivity tends to increase with temperature and precipitation. In aquatic ecosystems, net primary productivity tends to rise with light and the availability of nutrients.

Nutrients influence productivity

Nutrients are elements and compounds that organisms require for survival. Organisms need several dozen naturally occurring nutrients to survive. Elements and compounds required in relatively large amounts (such as nitrogen, carbon, and phosphorus)

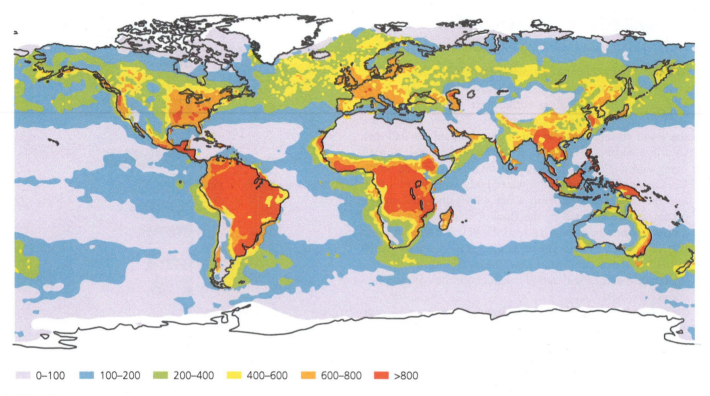

| | 0–100 | | 100–200 | | 200–400 | | 400–600 | | 600–800 | | >800 |

FIGURE 5.8 A world map of net primary production based on satellite data shows that on land, net primary production varies geographically with temperature and precipitation. In the world's oceans, net primary production is highest around the margins of continents, where nutrients (of both natural and human origin) run off from land. *Data from Field, C.B., et al., 1998. Primary production of the biosphere: Integrating terrestrial and oceanic components. Science 281: 237–240. Reprinted with permission from AAAS.*

 Reference the figure to answer the following questions.
- Worldwide, dead zones tend to be associated with areas with bigger human footprints. Provide one example of a nation with a large human footprint that has very few marine dead zones off its coast.
- Provide one explanation for this exception to the general rule of bigger human footprint equals many marine dead zones.

Go to **Interpreting Graphs & Data** on Mastering EnvironmentalScience®

are called **macronutrients.** Nutrients needed in small amounts (such as zinc, copper, and iron) are called **micronutrients.**

Nutrients stimulate production by plants, and lack of nutrients can limit production. The availability of nitrogen or phosphorus frequently is a limiting factor (p. 65) for plant or algal growth. When these nutrients are added to a system, producers show the greatest response to whichever nutrient has been in shortest supply. Nitrogen tends to be limiting in marine systems, and phosphorus in freshwater systems, though both contribute to eutrophication in all waters. Thus eutrophication in the Chesapeake Bay is largely driven by excess nitrogen, whereas eutrophication in the freshwater ponds and lakes in the bay's watershed are spurred by increases in phosphorus.

Canadian ecologist David Schindler and others demonstrated the effects of phosphorus on freshwater systems in the 1970s by experimentally manipulating entire lakes. In one experiment, his team bisected a 16-ha (40-acre) lake in Ontario with a plastic barrier. To one half the researchers added carbon, nitrate, and phosphate; to the other they added only carbon

and nitrate. Soon after the experiment began, they witnessed a dramatic increase in algae in the half of the lake that received phosphate, whereas the other half (the control for the experiment, p. 11) continued to host algal levels typical for lakes in the region (**FIGURE 5.9**). This difference held until shortly after they stopped fertilizing seven years later. At that point, algae decreased to normal levels in the half that had previously received phosphate. Such experiments showed clearly that phosphorus addition can markedly increase primary productivity in freshwater lakes.

Similar experiments in coastal ocean waters show nitrogen to be the more important limiting factor for primary productivity. For open-ocean waters far from shore, research indicates that adding the micronutrient iron is highly effective for stimulating phytoplankton growth.

Increased nutrient pollution from farms, cities, and industries has led to the development of more than 500 documented hypoxic dead zones globally today (**FIGURE 5.10**), including that of the Chesapeake Bay as well as a large dead zone that forms each year in the Gulf of Mexico off the Louisiana coast

After phosphate additions, the water in this portion of the lake turned opaque due to a dramatic and prolonged algal bloom

FIGURE 5.9 The upper portion of this lake in Ontario was experimentally treated with the addition of phosphate. This treated portion experienced an immediate, dramatic, and prolonged algal bloom, identifiable by its opaque waters.

near the mouth of the Mississippi River (pp. 420–421). Some are seasonal (like the Chesapeake Bay's), some occur irregularly, and others are permanent. The increase in the number of dead zones—there were 162 documented in the 1980s and 49 in the 1960s—reflects how the activities of people are changing nutrient concentrations in waters around the world. Scientists calculate that the amount of marine life missing from the oceans as a result of dead zones likely exceeds the total amount of shellfish harvested each year from the entire United States—a harvest worth more than $2 billion.

The good news is that in locations where people have reduced nutrient runoff, dead zones have begun to disappear. In New York City, hypoxic zones at the mouths of the Hudson and East rivers were nearly eliminated once sewage treatment (p. 409) was improved. The Black Sea, which borders Ukraine, Russia, Turkey, and eastern Europe, had long suffered one of the world's worst hypoxic zones. Then in the 1990s, after the Soviet Union collapsed, industrial agriculture in the region declined drastically. With fewer fertilizers draining into it, the Black Sea began to recover, and today fisheries are reviving. However, agricultural collapse is not a strategy anyone would choose to alleviate hypoxia. Rather, scientists are proposing a variety of innovative and economically acceptable ways to reduce nutrient runoff.

Ecosystems interact with one another

Whether we stand at a river's mouth or peruse a satellite image, we can think of ecosystems at different scales. An ecosystem can be as small as a puddle of water or as large as a bay, lake, or forest. For some purposes, scientists even view the entire biosphere as a single all-encompassing ecosystem. The term ecosystem is most often used, however, to refer to systems of moderate geographic extent that are somewhat self-contained. For example, the tidal marshes in the Chesapeake where river water empties into the bay are an ecosystem, as are the sections of the bay dominated by oyster reefs.

Adjacent ecosystems may share components and interact

weighing the
ISSUES

Ecosystems Where You Live

Think about the area where you live, and briefly describe its ecosystems. How do these systems interact? For instance, does any water pass from one to another? Describe the boundaries of watersheds in your region. If one ecosystem were greatly modified (say, if a large apartment complex were built atop a wetland or amid a forest), what impacts on nearby systems might result? (Note: If you live in a city, realize that urban areas can be thought of as ecosystems, too.)

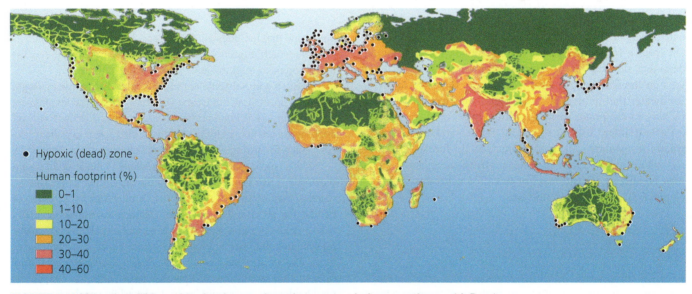

- Hypoxic (dead) zone

Human footprint (%)
- 0–1
- 1–10
- 10–20
- 20–30
- 30–40
- 40–60

FIGURE 5.10 More than 500 marine dead zones have been recorded across the world. Dead zones (shown by dots on the map) occur mostly offshore from areas of land with the greatest human ecological footprints (here, expressed on a scale of 0 to 100, with higher numbers indicating bigger human footprints). *Data from World Resources Institute, 2016, http://www.wri.org/our-work/project/eutrophication-and-hypoxia, and Diaz, R. and R. Rosenberg, 2008. Spreading dead zones and consequences for marine ecosystems. Science 321: 926–929. Reprinted with permission from AAAS.*

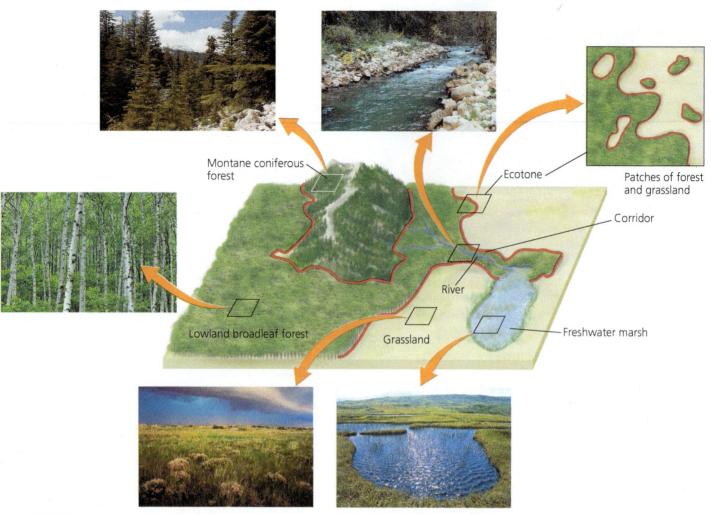

FIGURE 5.11 Landscape ecology deals with spatial patterns above the ecosystem level. This generalized diagram of a landscape shows a mosaic of patches of five ecosystem types (three terrestrial types, a marsh, and a river). Thick red lines indicate ecotones. A stretch of lowland broadleaf forest running along the river serves as a corridor connecting the large region of forest on the left to the smaller patch of forest alongside the marsh. The inset shows a magnified view of the forest-grassland ecotone and how it consists of patches on a smaller scale.

extensively. For instance, a pond ecosystem is very different from a forest ecosystem that surrounds it, but salamanders that develop in the pond live their adult lives under logs on the forest floor until returning to the pond to breed. Rainwater that nourishes forest plants may eventually make its way to the pond, carrying with it nutrients from the forest's leaf litter. Likewise, rivers, tidal marshes, and open waters in estuaries all may interact, as do forests and prairie where they converge. Areas where ecosystems meet may consist of transitional zones called **ecotones,** in which elements of each ecosystem mix.

Landscape ecologists study geographic patterns

Because components of different ecosystems may intermix, ecologists often find it useful to view these systems on a larger geographic scale that encompasses multiple ecosystems. For instance, if you are studying large mammals such as black bears, which move seasonally from mountains to valleys or between mountain ranges, you had better consider the overall

landscape that includes all these areas. If you study fish such as salmon, which migrate between marine and freshwater ecosystems, you need to know how these systems interact.

In such a broad-scale approach, called **landscape ecology,** scientists study how landscape structure affects the abundance, distribution, and interaction of organisms. Landscape-level approaches are also helping scientists, citizens, and policymakers to plan for sustainable regional development (pp. 336–341).

For a landscape ecologist, a landscape is made up of a spatial array of **patches.** Depending on the researcher's perspective, patches may consist of ecosystems or may simply be areas of habitat for a particular organism. Patches are spread spatially over a landscape in a **mosaic.** This metaphor reflects how natural systems often are arrayed across landscapes in complex patterns, like an intricate work of art. Thus, a forest ecologist may refer to a mosaic of forested patches left standing in an agricultural landscape. An amphibian biologist might speak of a mosaic of patches of pond habitat that frogs use for reproduction.

FIGURE 5.11 illustrates a landscape consisting of five ecosystem types, with ecotones along their borders (indicated by thick red lines). At this scale, we perceive a mosaic consisting of four patches and a river. However, we can view a landscape at different scales. The figure's inset shows a magnified view of an ecotone. At this finer resolution, we see that the ecotone consists of patches of forest and grassland in a complex arrangement. The scale at which an ecologist focuses will depend on the questions he or she is interested in and on the organisms he or she is studying.

Populations can be dispersed among patches

Every organism has specific habitat needs, so when its habitat is distributed in patches across a landscape, individuals may need to expend energy and risk predation traveling from one to another. If the patches are far apart, the organism's population may become divided into subpopulations, each occupying a different patch in the mosaic. Such a network of subpopulations, most of whose members stay within their respective patches but some of whom move among patches or mate with members of other patches, is called a **metapopulation.** When patches are still more isolated from one another, individuals may not be able to travel between them at all. In such a case, smaller subpopulations may be at risk of extinction.

Because of this extinction risk, metapopulations and landscape ecology are of great interest to **conservation biologists** (p. 289), scientists who study the loss, protection, and restoration of biodiversity. Of particular concern is the fragmentation of habitat into small and isolated patches (p. 283)—something that often results from human development pressures. Establishing corridors of habitat (see Figure 5.11) to link patches and allow animals to move among them is one approach that conservation biologists pursue as they attempt to maintain biodiversity in the face of human impact (pp. 323–326).

Technology helps us practice landscape ecology

A common tool for research in landscape ecology is the **geographic information system (GIS).** A GIS consists of computer software that takes multiple types of data (for instance, on geology, hydrology, topography, vegetation, animal populations, and human infrastructure from both on-ground studies and satellite imagery) and combines them, layer by layer, on a common set of geographic coordinates (**FIGURE 5.12**). The idea is to create a complete picture of a landscape and to analyze how elements of the different data sets are arrayed spatially and how they may be correlated.

GIS has become a valuable tool used by geographers, landscape ecologists, resource managers, and conservation biologists. GIS is being used to guide restoration efforts in the Chesapeake Bay. The *ChesapeakeStat* website, which was launched in 2010, enables scientists, educators, policymakers, and citizens to create customized composite maps that overlay parameters important to the bay's health. This tool is being used to assess the bay's current status, the effects of restoration efforts, and progress toward long-term goals.

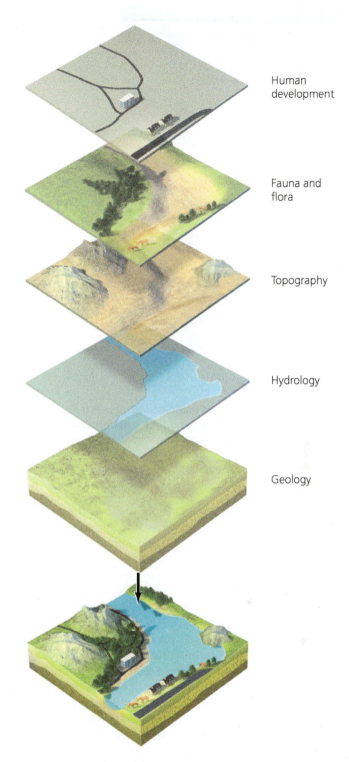

Human development

Fauna and flora

Topography

Hydrology

Geology

FIGURE 5.12 Geographic information systems (GIS) allow us to layer different types of data on natural landscape features and human land uses to produce maps integrating this information. GIS can be used to explore correlations among these data sets and to help in regional planning.

Modeling helps ecologists understand systems

Another way in which ecologists seek to make sense of the complex systems they study is by working with models. In science, a **model** is a simplified representation of a complex

FIGURE 5.13 **Ecological modelers observe relationships among variables in nature and then construct models to explain those relationships and make predictions.** They test and refine the models by gathering new data from nature and seeing how well the models predict those data.

natural process, designed to help us understand how the process occurs and to make predictions. **Ecological modeling** is the practice of constructing and testing models that aim to explain and predict how ecological systems function.

Because ecological processes involve so many factors, ecological models can be mathematically complicated. However, the general approach of ecological modeling is easy to understand (**FIGURE 5.13**). Researchers gather data from nature on relationships that interest them and then form a hypothesis about the nature of those relationships. They construct a model that attempts to explain the relationships in a generalized way so that people can use the model to make predictions about how the system will behave under different conditions. Modelers test their predictions by gathering new data from natural systems, and they use these new data to refine the model, improving its accuracy.

Note that the process illustrated in Figure 5.13 resembles the scientific method; models are essentially hypotheses about how systems function. Accordingly, the use of models is a key part of ecological research and environmental regulation today. As just one example, the National Oceanic and Atmospheric Administration (NOAA) uses the Chesapeake Bay Fisheries Ecosystem model to examine predator–prey relationships among fish species in the bay, the effects of hypoxia on fish populations, and how the distribution of underwater grasses influences blue crab populations. Data from scientific journal articles and direct measurements are used to establish the model's parameters, which are then used to predict the effects of differing fish harvest levels on species and ecosystems in the Chesapeake Bay.

Ecosystem services sustain our world

Human society depends on healthy, functioning ecosystems. When Earth's ecosystems function normally and undisturbed, they provide goods and services that we could not survive without. As we've seen, we rely not just on natural resources (which can be thought of as goods from nature) but also on the *ecosystem services* (p. 4) that our planet's systems provide (**TABLE 5.1**).

Ecological processes form the soil that nourishes our crops, purify the water we drink, pollinate the food plants we eat, and break down (some of) the waste and pollution we generate (**FIGURE 5.14**). The negative feedback cycles that are typical of ecosystems regulate and stabilize the climate and help to dampen the impacts of the disturbances we create in natural systems. Ecosystems also provide services that enhance the quality of our lives, ranging from recreational opportunities to pleasing scenery to inspiration and spiritual renewal.

One of the most important ecosystem services is the cycling of nutrients. Through the processes that take place within and among ecosystems, the chemical elements and compounds that sustain life—carbon, nitrogen, phosphorus, water, and many more—cycle through our environment in complex ways.

TABLE 5.1 Ecosystem Services

Ecological processes do many things that benefit us:

- Regulate oxygen, carbon dioxide, stratospheric ozone, and other atmospheric gases
- Regulate temperature and precipitation by means of ocean currents, cloud formation, and so on
- Protect against storms, floods, and droughts, mainly by means of vegetation
- Store and regulate water supplies in watersheds and aquifers
- Prevent soil erosion
- Form soil by weathering rock and accumulating organic material
- Cycle carbon, nitrogen, phosphorus, sulfur, and other nutrients
- Filter waste, remove toxins, recover nutrients, and control pollution
- Pollinate plant crops and wild plants so they reproduce
- Control crop pests with predators and parasites
- Provide habitat for organisms to breed, feed, rest, migrate, and winter
- Produce fish, game, crops, nuts, and fruits that people eat
- Supply lumber, fuel, metals, fodder, and fiber
- Furnish medicines, pets, ornamental plants, and genes for resistance to pathogens and crop pests
- Provide recreation such as ecotourism, fishing, hiking, birding, hunting, and kayaking
- Provide aesthetic, artistic, educational, spiritual, and scientific amenities

FIGURE 5.14 Ecological processes naturally provide countless services that we call *ecosystem services*. Our society, indeed our very survival, depends on these services.

Biogeochemical Cycles

Just as nitrogen and phosphorus from fertilizer on Pennsylvania corn fields end up in Chesapeake Bay oysters, all nutrients move through the environment in intricate ways. As we have discussed, whereas energy enters an ecosystem from the sun, flows from organism to organism, and dissipates to the atmosphere as heat, the physical matter of an ecosystem is circulated over and over again.

Nutrients circulate through ecosystems in biogeochemical cycles

Nutrients move through ecosystems in **nutrient cycles,** also known as **biogeochemical cycles.** In these pathways, chemical elements or molecules travel through the atmosphere, hydrosphere, lithosphere, and biosphere in dynamic equilibrium. A carbon atom in your fingernail today might have been in the muscle of a cow a year ago, may have resided in a blade of grass a month before that, and may have been part of a dinosaur's tooth 100 million years ago. After we die, the nutrients in our bodies will disperse into the environment, and could be incorporated into other organisms far into the future.

Nutrients and other materials move from one **reservoir,** or *pool*, to another, remaining for varying amounts of time (the **residence time**) in each. You, the cow, the grass, and the dinosaur are each reservoirs for carbon atoms, as are sedimentary rocks and the atmosphere. The rate at which materials move between reservoirs is termed a **flux.** When a reservoir releases more materials than it accepts, it is called a **source,** and when a reservoir accepts more materials than it releases, it is called a **sink. FIGURE 5.15** illustrates these concepts in a simple manner.

Human activity can affect residence times, fluxes, and the ratio of nutrients in reservoirs. For example, by using fossil fuels for energy, we have reduced the residence time for

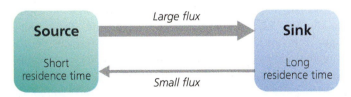

FIGURE 5.15 The main components of a biogeochemical cycle are reservoirs and fluxes. A source releases more materials than it accepts, and a sink accepts more materials than it releases.

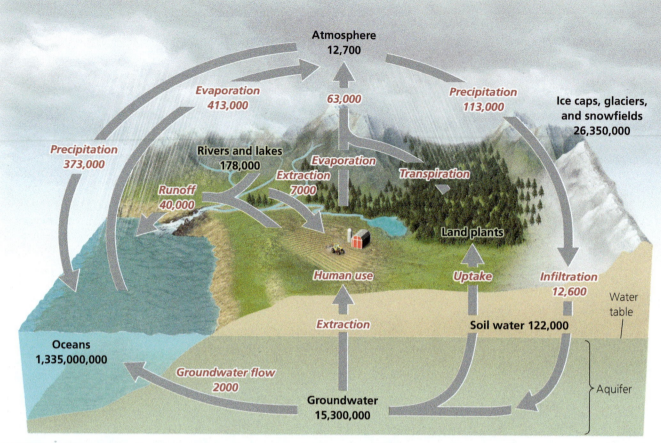

FIGURE 5.16 The water cycle, or hydrologic cycle, summarizes the many routes that water molecules take as they move through the environment. Gray arrows represent fluxes among reservoirs for water. Oceans hold 97% of our planet's water, whereas most fresh water resides in groundwater and ice caps. Water vapor in the atmosphere condenses and falls to the surface as precipitation, then evaporates from land and transpires from plants to return to the atmosphere. Water flows downhill into rivers, eventually reaching the oceans. In the figure, reservoir names are printed in black type, and numbers in black type represent reservoir sizes expressed in units of cubic kilometers (km³). Processes give rise to fluxes, both of which are printed in italic red type and expressed in km³ per year.
Data from Schlesinger, W.H., 2013. Biogeochemistry: An analysis of global change, 3rd ed. London, England: Academic Press.

carbon in underground fossil fuel reserves, sped up the flux of carbon from the ground to the atmosphere, and altered carbon ratios by increasing amounts in the atmosphere.

As we discuss several important biogeochemical cycles, think about how they involve negative feedback loops that promote dynamic equilibrium. Also consider how some human actions can generate destabilizing positive feedback loops that affect these cycles.

The water cycle affects all other cycles

Water is so integral to life and to Earth's fundamental processes that we frequently take it for granted. Water is the essential medium for all manner of biochemical reactions (p. 26), and it plays key roles in nearly every environmental system, including each of the nutrient cycles we are about to discuss. Water carries nutrients, sediments, and pollutants from the continents to the oceans via surface runoff, streams, and rivers. These materials can then be carried thousands of miles on ocean currents. Water also carries atmospheric pollutants to the surface when they dissolve in falling rain or snow. Water is needed for human activities, and many regions of the modern world, such as the western United States, regularly deal with water shortages (pp. 383–385).

The **hydrologic cycle,** or *water cycle* (**FIGURE 5.16**), summarizes how water—in liquid, gaseous, and solid forms—flows through our environment.

The oceans are the main reservoir in the water cycle, holding 97% of all water on Earth. The fresh water we depend on for our survival accounts for less than 3%, and two-thirds of this small amount is tied up in glaciers, snowfields, and ice caps (p. 385). Thus, considerably less than 1% of the planet's water is in forms that we can readily use—groundwater, surface fresh water, and rain from atmospheric water vapor.

Evaporation and transpiration Water moves from oceans, lakes, ponds, rivers, and moist soil into the atmosphere by **evaporation,** the conversion of a liquid to a gaseous form. Warm temperatures and strong winds speed rates of evaporation. A greater degree of exposure has the same effect; an area logged of its forest or converted to agriculture or residential use will lose water more readily than a comparable area that remains vegetated. Water also enters the atmosphere by **transpiration,** the release of water vapor by plants through their leaves, or by evaporation from the surfaces of organisms (such as sweating in humans). Transpiration and evaporation act as natural processes of distillation, because water escaping into the air as a gas leaves behind its dissolved substances.

Precipitation, runoff, and surface water Water returns from the atmosphere to Earth's surface as **precipitation** when water vapor condenses and falls as rain or snow. This moisture may be taken up by plants and used by animals, but much of it flows as runoff into streams, rivers, lakes, ponds, and oceans. Amounts of precipitation vary greatly from region to region, helping give rise to our planet's variety of biomes (p. 92).

Groundwater Some precipitation and surface water soaks down through soil and rock and becomes **groundwater**, water found within the soil. Groundwater recharges **aquifers**, sponge-like regions of rock and soil that are underground reservoirs of water. The upper limit of groundwater held in an aquifer is referred to as the **water table.** (See Figure 15.5, p. 389, for an illustration of these features.) Aquifers can hold groundwater for long periods of time, and can sometimes take hundreds or thousands of years to recharge fully after being depleted. Groundwater becomes surface water when it emerges from springs or flows into streams, rivers, lakes, or the ocean from the soil (p. 389).

Our impacts on the water cycle are extensive

Human activity affects every aspect of the water cycle. By damming rivers, we slow the movement of water from the land to the sea, and we increase evaporation by holding water in reservoirs. We remove natural vegetation by clear-cutting and developing land, which increases surface runoff, decreases infiltration and transpiration, and promotes evaporation. Our withdrawals of surface water and groundwater for agriculture, industry, and domestic uses deplete rivers, lakes, and streams and lower water tables. This can lead to water shortages and conflict over water supplies (p. 404). And by emitting into the atmosphere pollutants that dissolve in water droplets, we change the chemical nature of precipitation, in effect sabotaging the natural distillation process that evaporation and transpiration provide. (We will revisit the water cycle, water resources, and human impacts in more detail in Chapter 15.)

weighing the

ISSUES

Water shortages and you

Has your region faced any water shortages or experienced conflicts over water use with neighboring regions? If not, have you recently heard of such conflicts in other regions? Given your knowledge of the water cycle, what solutions would you propose for addressing water shortages in your region?

The carbon cycle circulates a vital organic nutrient

As the definitive component of organic molecules, carbon is an ingredient in carbohydrates, fats, and proteins and occurs in the bones, cartilage, and shells of all living things. From DNA to fossil fuels, from plastics to pharmaceuticals, carbon (C) atoms are everywhere. The **carbon cycle** describes the routes that carbon atoms take through the environment (**FIGURE 5.17**, p. 120).

Photosynthesis, respiration, and food webs Producers—including plants, algae, and cyanobacteria—pull carbon dioxide out of the atmosphere and out of surface water to use in photosynthesis. Photosynthesis breaks the bonds in carbon dioxide (CO_2) and water (H_2O) to produce oxygen (O_2) and carbohydrates (e.g., glucose, $C_6H_{12}O_6$). Autotrophs use some of the carbohydrates to fuel cellular respiration, thereby releasing some of the carbon back into the atmosphere and oceans as CO_2. When producers are eaten by primary consumers, which in turn are eaten by secondary and tertiary consumers, more carbohydrates are broken down in cellular respiration, producing carbon dioxide and water. The same process occurs as decomposers consume waste and dead organic matter. Cellular respiration from all these organisms releases carbon back into the atmosphere and oceans.

Organisms use carbon for structural growth, so a portion of the carbon an organism takes in becomes incorporated into its tissues (such as net primary production in plants). The abundance of plants and the fact that they take in so much carbon dioxide for photosynthesis makes plants a major reservoir for carbon. Because CO_2 is a greenhouse gas of primary concern (p. 482), much research on global climate change is directed toward measuring the amount of CO_2 that plants store. Scientists are working hard to better understand exactly how this portion of the carbon cycle influences Earth's climate (see **THE SCIENCE BEHIND THE STORY**, pp. 122–123).

Sediment storage of carbon As aquatic organisms die, their remains may settle in sediments in ocean basins or freshwater wetlands. As sediment accumulates, older layers are buried more deeply, experiencing high pressure over long periods of time. These conditions can convert soft tissues into fossil fuels—coal, oil, and natural gas—and can turn shells and skeletons into sedimentary rock, such as limestone. Sedimentary rock makes up the largest reservoir in the carbon cycle. Although any given carbon atom spends a relatively short time in the atmosphere, carbon trapped in sedimentary rock may reside there for hundreds of millions of years.

Carbon trapped in sediments and fossil fuel deposits may eventually be released into the oceans or atmosphere by geologic processes such as uplift, erosion, and volcanic eruptions. It also reenters the atmosphere when we extract and burn fossil fuels.

The oceans The world's oceans are the second-largest reservoir in the carbon cycle. They absorb carbon-containing compounds from the atmosphere, from terrestrial runoff, from undersea volcanoes, and from the waste products and detritus of marine organisms. Some carbon atoms absorbed by the oceans—in the form of carbon dioxide, carbonate ions (CO_3^{2-}), and bicarbonate ions (HCO_3^-)—combine with calcium ions (Ca^{2+}) to form calcium carbonate ($CaCO_3$), an essential ingredient in the skeletons and shells of microscopic marine organisms. As these organisms die, their calcium carbonate shells sink to the ocean floor and begin to form sedimentary rock. The rates at which the oceans absorb and release carbon depend on many factors, including temperature and the numbers of marine organisms converting CO_2 into carbohydrates and carbonates.

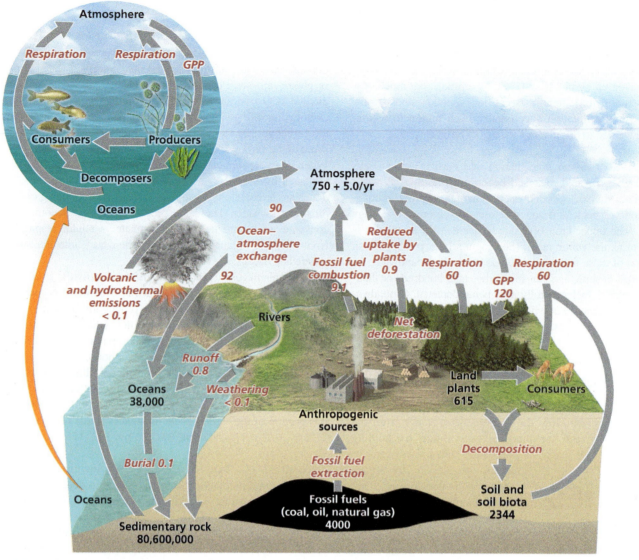

FIGURE 5.17 The carbon cycle summarizes the many routes that carbon atoms take as they move through the environment. Gray arrows represent fluxes among reservoirs for carbon. In the carbon cycle, plants use carbon dioxide from the atmosphere for photosynthesis (gross primary production, or "GPP" in the figure). Carbon dioxide is returned to the atmosphere through cellular respiration by plants, their consumers, and decomposers. The oceans sequester carbon in their water and in deep sediments. The vast majority of the planet's carbon is stored in sedimentary rock. In the figure, reservoir names are printed in black type, and numbers in black type represent reservoir sizes expressed in petagrams (units of 10^{15} g) of carbon. Processes give rise to fluxes, both of which are printed in italic red type and expressed in petagrams of carbon per year. *Data from Schlesinger, W.H., 2013. Biogeochemistry: An analysis of global change, 3rd ed. London, England: Academic Press.*

We are shifting carbon from the lithosphere to the atmosphere

By mining fossil fuel deposits, we are essentially removing carbon from an underground reservoir with a residence time of millions of years. By combusting fossil fuels, we release carbon dioxide and greatly increase the flux of carbon from the ground to the air. In addition, cutting down forests removes carbon from the pool of vegetation and releases it to the air. And if less vegetation is left on the surface, there are fewer plants to draw CO_2 back out of the atmosphere.

As a result, scientists estimate that today's atmospheric carbon dioxide reservoir is the largest that Earth has experienced in the past 800,000 years, and likely in the past 20 million years. The ongoing flux of carbon into the atmosphere is the driving force behind today's anthropogenic global climate change (Chapter 18).

Some of the excess CO_2 in the atmosphere is now being absorbed by ocean water. This is causing ocean water to become more acidic, leading to problems that threaten many marine organisms (pp. 432–433).

Our understanding of the carbon cycle is not yet complete. Scientists remain baffled by the so-called missing carbon sink. Of the carbon dioxide we emit by fossil fuel combustion and deforestation, researchers have measured how much goes into the atmosphere and oceans, but there remain roughly 2.3–2.6 billion metric tons unaccounted for. Many scientists think this CO_2 must be taken up by plants or soils of the

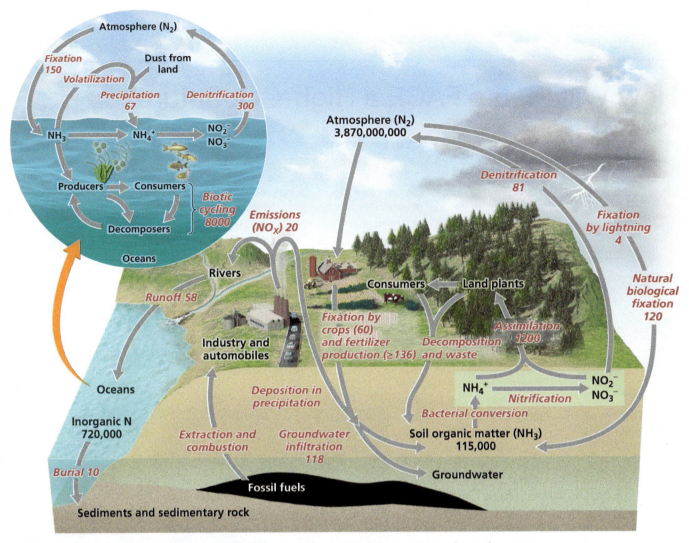

FIGURE 5.18 The nitrogen cycle summarizes the many routes that nitrogen atoms take as they move through the environment. Gray arrows represent fluxes among reservoirs for nitrogen. In the nitrogen cycle, specialized bacteria play key roles in "fixing" atmospheric nitrogen and converting it to chemical forms that plants can use. Other types of bacteria convert nitrogen compounds back to the atmospheric gas, N_2. In the oceans, inorganic nitrogen is buried in sediments, whereas nitrogen compounds are cycled through food webs as they are on land. In the figure, reservoir names are printed in black type, and numbers in black type represent reservoir sizes expressed in teragrams (units of 10^{12} g) of nitrogen. Processes give rise to fluxes, both of which are printed in italic red type and expressed in teragrams of nitrogen per year. *Data from Schlesinger, W.H., 2013.* Biogeochemistry: An analysis of global change, *3rd ed. London, England: Academic Press.*

temperate and boreal forests (pp. 94–98). They'd like to know for sure, though, because if certain forests are acting as a major sink for carbon (and thus restraining global climate change), conserving these ecosystems is particularly vital.

The nitrogen cycle involves specialized bacteria

Nitrogen (N) makes up 78% of our atmosphere by mass and is the sixth most abundant element on Earth. It is an essential ingredient in the proteins, DNA, and RNA that build our bodies. Despite its abundance in the air, nitrogen gas (N_2) is chemically inert and cannot cycle out of the atmosphere and into living organisms without assistance from lightning, highly specialized

bacteria, or human intervention. For this reason, the element is relatively scarce in the lithosphere and hydrosphere and in organisms. However, once nitrogen undergoes the right kind of chemical change, it becomes biologically active and available to organisms, and it can act as a potent fertilizer. Its scarcity makes biologically active nitrogen a limiting factor for plant growth. For all these reasons the **nitrogen cycle** (**FIGURE 5.18**) is of vital importance to us and to all other organisms.

Nitrogen fixation   To become biologically available, inert nitrogen gas (N_2) must be "fixed," or combined with hydrogen in nature to form ammonia (NH_3), whose water-soluble ions of ammonium (NH_4^+) can be taken up by plants. **Nitrogen fixation** can be accomplished in two ways: by the

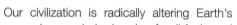

How Will Plants FACE a High-CO$_2$ Future?

Our civilization is radically altering Earth's carbon cycle by burning fossil fuels and deforesting landscapes. Today the atmosphere contains more than 40% more carbon dioxide (CO$_2$) than it did just two centuries ago, and the amount is rapidly increasing. Rising CO$_2$ concentrations are warming our planet, and global climate change brings many unwelcome consequences (Chapter 18).

Plants and other autotrophs remove carbon dioxide from the atmosphere to use in photosynthesis, and all organisms add CO$_2$ to the atmosphere by cellular respiration. Will more CO$_2$ mean more plant growth, and will more plants be able to absorb and store much of the extra CO$_2$? Perhaps, but before we rely on forests and phytoplankton to save us from our own emissions, we'd better be sure they can do so.

Historically, if a researcher wanted to measure how plants respond to increased carbon dioxide, he or she would alter gas levels in a small enclosure such as a lab or a greenhouse. But can we really scale up results from such small indoor experiments and trust that they will show how entire forests will behave? Many scientists thought not, so they pioneered Free-Air CO$_2$ Enrichment (FACE). In FACE experiments, ambient levels of CO$_2$ encompassing areas of forest (or other vegetation) outdoors are precisely controlled. With their large scale and open-air conditions, FACE experiments include most factors that influence a plant community in the wild, such as variation in temperature, sunlight, precipitation, herbivorous insects, disease pathogens, and competition among plants. By measuring how plants respond

Aspen FACE site researcher Dr. Mark Kubiske of the U.S. Forest Service

to changing gas compositions in such real-world conditions, we can better learn how ecosystems may change in the carbon dioxide–soaked world that awaits us.

Dozens of organizations have sponsored FACE facilities—there are 36 sites in 17 nations so far, including U.S. sites in Arizona, California, Illinois, Minnesota, Nevada, North Carolina, Tennessee, Wisconsin, and Wyoming. The sites cover a variety of ecosystems, from forests to grasslands to rice paddies; and the plots range in size from 1 m to 30 m (3–98 ft) in diameter.

To understand how a typical FACE study works, consider the Aspen FACE Experiment at the Harshaw Experimental Forest (where aspen trees are common) near Rhinelander, Wisconsin. Here, tall steel and plastic towers and pipes ring 12 circular plots of forest 30 m (98 ft) in diameter (**FIGURE 1**).

FIGURE 1 At the Aspen FACE facility in Wisconsin, tall towers and pipes control the atmospheric composition around selected patches of trees.

The pipes release CO_2, bathing the plants in an atmosphere 50% richer in CO_2 than today's (equal to what is expected for the year 2050). Sensors monitor wind conditions, and computers control for the influence of wind by adjusting CO_2 releases, keeping ambient concentrations stable within each plot. Because aspen trees can sprout from the root systems of existing trees, researchers were able to study the effects of treatments on *clones*, groups of genetically identical trees.

The pipes at the Aspen plots also release tropospheric ozone (O_3, a major pollutant in urban smog; p. 462), and researchers study how this gas and CO_2 affect plant growth, leaf and root conditions, soil carbon content, and much more. Pipes at some plots release normal air, serving as controls for the treatment plots.

Researchers using the Aspen FACE facility have been asking whether forest trees will sequester more carbon as CO_2 levels rise, whether this will change as trees grow, how CO_2 interacts with ozone, and how these gases affect trees' interactions with insects and diseases.

A number of things have been learned so far from these investigations, including:

- Insects and diseases that attack aspen and birch trees increase as atmospheric levels of ozone and CO_2 rise.
- High CO_2 concentrations delay aspen leaf aging, which makes some aspens vulnerable to frost damage in winter.
- Elevated CO_2 levels increase photosynthesis and tree growth—but moderate levels of ozone offset this increased growth (**FIGURE 2**). Because many modelers have not taken ozone into account when estimating how much carbon trees can sequester, the Aspen FACE data suggest that their models may overestimate the amount of CO_2 that trees will pull out of the air.

Together, such results indicate that rising carbon dioxide levels could have a variety of negative impacts on trees and forests. Thus, the old expectation that more CO_2 makes for happier plants appears to be an oversimplification. Indeed, research from other FACE sites is showing that increased growth from enhanced CO_2 is often temporary and that growth rates later flatten out or decline. This occurred when factors other than CO_2 that affect tree growth, such as levels of nutrients or water in the soil, became limiting and halted the increase in growth rate caused by elevated CO_2 concentrations. Recent work on crop plants has even shown that high CO_2 makes some crops less nutritious.

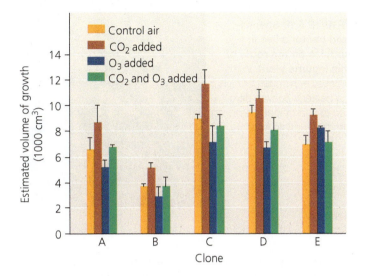

FIGURE 2 **Data from five clones of aspens show that during the study period, trees supplied with carbon dioxide grew more than did the control trees, whereas those supplied with ozone grew less. Trees supplied with both gases did not grow differently from the controls.** *Data from Isebrands, J.G., et al., 2001. Growth responses of* Populus tremuloides *clones to interacting elevated carbon dioxide and tropospheric ozone.* Environmental Pollution *115 (3): 359–371, Fig. 2.*

Obtaining solid answers to questions related to long-term effects such as these often takes years or decades, and FACE experiments gathered data, over several decades at some sites, on tree growth. In 2008, the U.S. Department of Energy (DOE), which funds Aspen and other major sites, announced it would cease funding and advised scientists to cut the trees down and dig up the soil to analyze carbon content. This analysis would provide urgently needed data on carbon sequestration, the DOE said, and then millions of dollars could be shifted toward a new and improved generation of FACE experiments.

The trees at the Aspen site were harvested for analysis in 2009, and researchers are now regrowing forest on the site with funding from the U.S. Forest Service. The knowledge gains from this project, and other FACE sites in which growth is ongoing, have been substantial. Data from the Aspen FACE project alone have generated more than 190 peer-reviewed scientific articles and are helping scientists today better predict how forest ecosystems will respond to the atmosphere of tomorrow.

FIGURE 5.19 Specialized bacteria live in nodules on the roots of this legume plant. In the process of nitrogen fixation, the bacteria convert nitrogen to a form that the plant can take up into its roots.

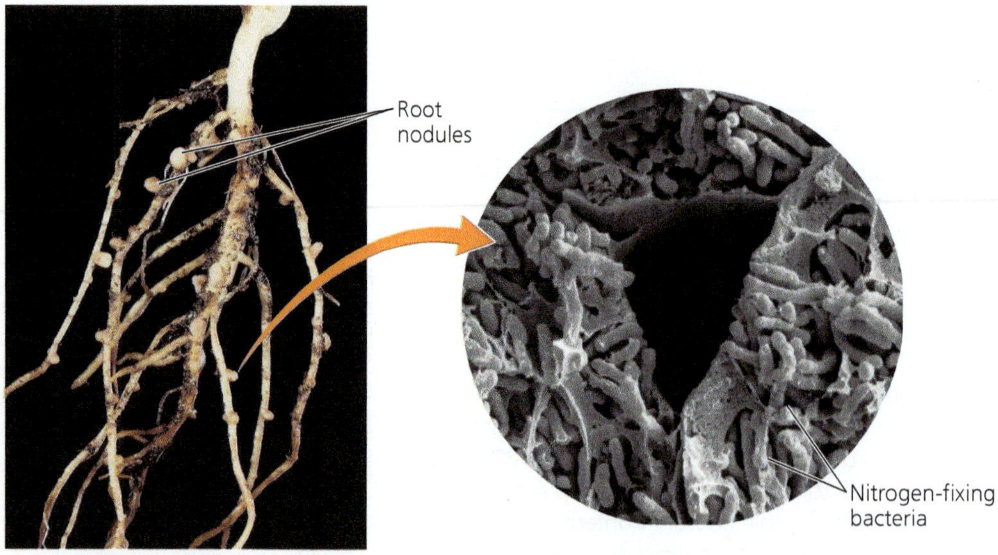

Root nodules

Nitrogen-fixing bacteria

intense energy of lightning strikes, or when air in the top layer of soil comes in contact with particular types of **nitrogen-fixing bacteria.** These bacteria live in a mutualistic relationship (pp. 77–78) with many types of plants, including soybeans and other legumes, providing them nutrients by converting nitrogen to a usable form. Some farmers nourish soils by planting crops that host nitrogen-fixing bacteria among their roots (**FIGURE 5.19**).

Nitrification and denitrification Other types of specialized bacteria then perform a process known as **nitrification.** In this process, ammonium ions are first converted into nitrite ions (NO_2^-), then into nitrate ions (NO_3^-). Plants can take up these ions, which also become available after atmospheric deposition on soils or in water or after application of nitrate-based fertilizer.

Animals obtain the nitrogen they need by consuming plants or other animals. Decomposers obtain nitrogen from dead and decaying plant and animal matter and from animal urine and feces. Once decomposers process nitrogen-rich compounds, they release ammonium ions, making these available to nitrifying bacteria to convert again to nitrates and nitrites.

The next step in the nitrogen cycle occurs when **denitrifying bacteria** convert nitrates in soil or water to gaseous nitrogen via a multistep process. Denitrification thereby completes the cycle by releasing nitrogen back into the atmosphere as a gas.

We have greatly influenced the nitrogen cycle

Historically, nitrogen fixation was a **bottleneck,** a step that limited the flux of nitrogen out of the atmosphere. This changed with the research of two German chemists early in the 20th century. Fritz Haber found a way to combine nitrogen and hydrogen gases to synthesize ammonia, a key ingredient in modern explosives and agricultural fertilizers, and Carl Bosch devised methods to produce ammonia on an industrial scale. The **Haber-Bosch process** enabled people to overcome the limits on productivity long imposed by nitrogen scarcity in nature. By enhancing agriculture, the new fertilizers contributed to the past century's enormous increase in human population. Farmers, homeowners, and landscapers alike all took advantage of fertilizers, dramatically altering the nitrogen cycle. Today, using the Haber-Bosch process, our species is fixing at least as much nitrogen as is being fixed naturally. We have effectively doubled the rate of nitrogen fixation on Earth, overwhelming nature's denitrification abilities.

By fixing atmospheric nitrogen to create fertilizers, we increase nitrogen's flux from the atmosphere to Earth's surface. We also enhance this flux by cultivating legume crops whose roots host nitrogen-fixing bacteria. Moreover, we reduce nitrogen's return to the air when we destroy wetlands whose plants host denitrifying bacteria that convert nitrates to nitrogen gas.

When our farming practices speed runoff and allow soil erosion, nitrogen flows from farms into terrestrial and aquatic ecosystems, leading to nutrient pollution, eutrophication, and hypoxia. These impacts have become painfully evident to oystermen and scientists in the Chesapeake Bay, but hypoxia in waters is by no means the only human impact on the nitrogen cycle. When we burn forests and fields, we force nitrogen out of soils and vegetation and into the atmosphere. When we burn fossil fuels, we release nitric oxide (NO) into the atmosphere, where it reacts to form nitrogen dioxide (NO_2). This compound is a precursor to nitric acid (HNO_3), a key component of acid precipitation (pp. 469–472). We introduce another nitrogen-containing gas, nitrous oxide (N_2O), when anaerobic bacteria break down the tremendous volume of animal waste produced in agricultural feedlots (p. 241). Oddly enough, the overapplication of nitrogen-based fertilizers can strip soil of other essential nutrients, such as calcium and potassium, because fertilizer flushes them out. As these examples show, human activities have affected the nitrogen cycle in diverse and often far-reaching ways.

The phosphorus cycle circulates a limited nutrient

The element phosphorus (P) is a key component of cell membranes and of several molecules vital for life, including DNA, RNA, ATP, and ADP (pp. 27–30). Although phosphorus is indispensable for life, the amount of phosphorus in organisms is dwarfed by the vast amounts in rocks, soil, sediments, and the oceans. Unlike the carbon and nitrogen cycles, the **phosphorus cycle** (FIGURE 5.20) has no appreciable atmospheric component besides the transport of tiny amounts in windblown dust and sea spray.

Geology and phosphorus availability The vast majority of Earth's phosphorus is contained within rocks and is released only by weathering (p. 213), which releases phosphate ions (PO_4^{3-}) into water. Phosphates dissolved in lakes or in the oceans precipitate into solid form, settle to the bottom, and reenter the lithosphere's phosphorus reservoir in sediments. Because most phosphorus is bound up in rock and only slowly released, environmental concentrations of phosphorus available to organisms tend to be very low. This scarcity explains why phosphorus is frequently a limiting factor for plant growth and why an influx of phosphorus can produce immediate and dramatic effects.

Food webs Aquatic producers take up phosphates from surrounding waters, whereas terrestrial producers take up phosphorus from soil water through their roots. Primary consumers acquire phosphorus from plant tissues and pass it on to secondary and tertiary consumers (Chapter 4). Consumers also pass phosphorus to the soil through the excretion of waste. Decomposers break down phosphorus-rich organisms and their wastes and, in so doing, return phosphorus to the soil.

We affect the phosphorus cycle

People increase phosphorus concentrations in surface waters through runoff of the phosphorus-rich fertilizers we apply to

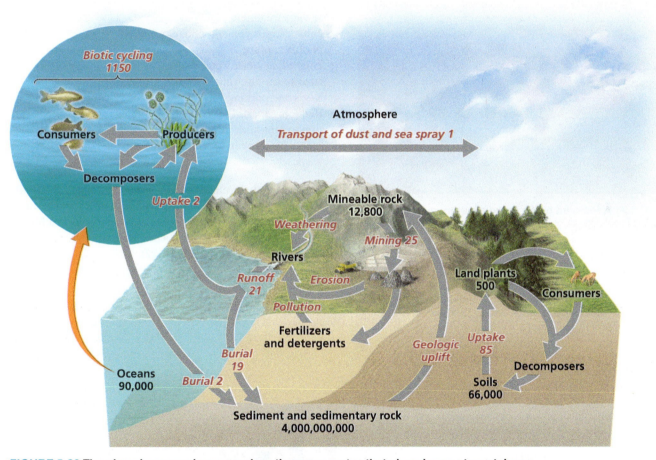

FIGURE 5.20 The phosphorus cycle summarizes the many routes that phosphorus atoms take as they move through the environment. Gray arrows represent fluxes among reservoirs for phosphorus. Most phosphorus resides underground in rock and sediment. Rocks containing phosphorus are uplifted geologically and slowly weathered away. Small amounts of phosphorus cycle through food webs, where this nutrient is often a limiting factor for plant growth. In the figure, reservoir names are printed in black type, and numbers in black type represent reservoir sizes expressed in teragrams (units of 10^{12} g) of phosphorus. Processes give rise to fluxes, both of which are printed in italic red type and expressed in teragrams of phosphorus per year. *Data from Schlesinger, W.H., 2013. Biogeochemistry: An analysis of global change, 3rd ed. London, England: Academic Press.*

lawns and farmlands. A 2008 study determined that an average hectare of land in the Chesapeake Bay region received a net input of 4.52 kg (10 lb) of phosphorus per year, promoting phosphorus accumulation in soils, runoff into waterways, and phytoplankton blooms and hypoxia in the bay. People also add phosphorus to waterways through releases of treated wastewater rich in phosphates from the detergents we use to wash our clothes and dishes.

FIGURE 5.21 A forested buffer lining a waterway in agricultural land in Maryland. Lining waterways with trees or grasses can reduce the amount of nutrients and eroded soil entering waterways from farm fields by trapping runoff.

Managing nutrient enrichment requires diverse approaches

With our reliance on synthetic fertilizers for food production and fossil fuels for energy, managing nutrient enrichment of ecosystems will be a challenge for many years to come. But there are a number of approaches available to control nutrient pollution in the Chesapeake Bay watershed, Mississippi River watershed, and other waterways affected by eutrophication, including:

- Reducing fertilizer use on farms and lawns and timing its application to reduce water runoff
- Planting and maintaining vegetation "buffers" around streams that trap nutrient and sediment runoff (**FIGURE 5.21**)
- Using natural and constructed wetlands (p. 409) to filter stormwater and farm runoff
- Improving technologies in sewage treatment plants to enhance nitrogen and phosphorus capture
- Upgrading stormwater systems to capture runoff from roads and parking lots
- Reducing fossil fuel combustion to minimize atmospheric inputs of nitrogen to waterways

Some of these methods cost more than others for similar results. For example, planting vegetation buffers and restoring wetlands can reduce nutrient inputs into waterways at a fraction of the cost of some other approaches, such as upgrading wastewater treatment plants (**FIGURE 5.22**).

FIGURE 5.22 Costs for reducing nitrogen inputs into the Chesapeake Bay vary widely. Approaches that slow runoff to waterways avoid nitrogen inputs for a few dollars per pound, whereas upgrades to wastewater treatment plants (WWTP), enhanced nutrient management plans (NMP—careful regulation of nutrient applications), and stormwater upgrades can be considerably more expensive. *Data from Jones, C., et al. 2010. How Nutrient Trading Could Help Restore the Chesapeake Bay. WRI Working Paper. Washington D.C.: World Resources Institute.*

DATA Q For what it costs to remove 1 lb of nitrogen by using enhanced nutrient management programs (NMP), how many pounds of nitrogen could be kept out of waterways by planting forested buffers around streams instead?

Go to **Interpreting Graphs & Data** on
Mastering EnvironmentalScience®

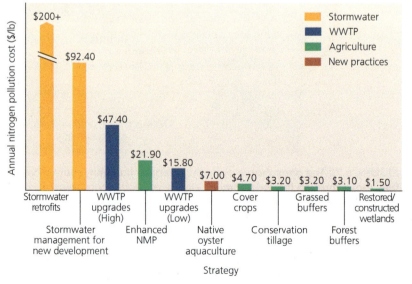

THE LOOP

The Chesapeake Bay finally has prospects for recovery as the federal government and bay states are now managing the bay as a holistic system. Thinking in terms of systems is important in understanding Earth's dynamics, so that we may learn how to avoid disrupting its processes and how to minimize any disruptions we cause through human impacts on ecosystems and alterations of biogeochemical cycles. By studying the environment from a systems perspective and by integrating scientific findings with the policy process, people who care about the Chesapeake Bay and other waterways are working to "bring back" degraded habitats to their former health.

Arriving at this endpoint was not easy, though. After 25 years of failed pollution control agreements and nearly $6 billion spent on cleanup efforts, the Chesapeake Bay Foundation (CBF), a nonprofit organization dedicated to conserving the bay, sued the Environmental Protection Agency in January 2009 for failing to use its available powers under the Clean Water Act to clean up the bay. The CBF's lawsuit spurred action.

In May 2009, via an executive order, President Barack Obama directed the EPA and other federal agencies to establish a comprehensive plan to restore the Chesapeake Bay.

One year later, the EPA and the CBF announced a plan in which the EPA agreed to work with surrounding states to provide aggressive pollution regulation in the bay. In December 2010, a comprehensive "pollution budget" was developed and implemented with the assistance of the District of Columbia, Delaware, Maryland, New York, Pennsylvania, Virginia, and West Virginia that set concrete goals for bay restoration and established timelines for program implementation.

This systemic approach to restoring the Chesapeake Bay is succeeding. The CBF's "2014 State of the Bay" concluded that of 13 parameters assessed annually for the bay—from water quality to wildlife habitat to fisheries—9 were rated higher in 2014 than in 2008, 2 showed no change, and only 2 were rated lower. Overall, the bay's health rating in 2014 was the highest it had been since CBF's founding in 1964 and showed a 14% improvement since 2008.

Although this progress is certainly encouraging, the Chesapeake Bay remains highly degraded and much work is still needed. Nonetheless, the 17 million people living in the Chesapeake Bay watershed have reason to hope that the Chesapeake Bay of tomorrow may be healthier than it is today, thanks to the collaborative efforts of concerned citizens, advocacy organizations, and the federal and bay-state governments.

REVIEWING Objectives

You should now be able to:

Describe environmental systems

Earth's natural systems are complex networks of interacting components that generally involve feedback loops, show dynamic equilibrium, and exhibit emergent properties. (pp. 104–110)

Define ecosystems and discuss how living and nonliving entities interact in ecosystem-level ecology

Ecosystems consist of all organisms and nonliving entities that occur and interact in a particular area at the same time. Matter is recycled in ecosystems, but energy flows in one direction—from producers to higher trophic levels. Input of nutrients can boost productivity—which is the generation of biomass—but an excess of nutrients can alter ecosystems and cause severe ecological and economic consequences. (pp. 110–114)

Discuss the fundamentals of landscape ecology, GIS, and ecological modeling

Landscape ecology studies how the habitat patches that make up landscape structure influence organisms.

Geographic information systems (GIS)—software that overlays multiple types of data on a common set of geographic coordinates—enables landscape ecology to be used increasingly in conservation biology and regional planning. Ecological modeling helps ecologists comprehend the complex systems they study and predict how these systems will react to disturbance. (pp. 114–116)

Explain ecosystem services and discuss how they benefit our lives

Ecosystems provide the "goods" we know of as natural resources. Natural ecological processes provide services that we depend on for everyday living, such as cycling nutrients and purifying drinking water. (pp. 116–117)

Compare and contrast how water, carbon, nitrogen, and phosphorus cycle through the environment

Major reservoirs of the hydrologic cycle include the oceans and ice caps, and large fluxes include evaporation, precipitation, and runoff. Most of the carbon on Earth is

contained in sedimentary rock, and fluxes include photo-synthesis, cellular respiration, and fossil fuel combustion. The major reservoir of nitrogen is the atmosphere, and important fluxes include nitrogen fixation (both by bacteria and by humans) and denitrification. Phosphorus is most abundant in sedimentary rock, and fluxes include the weathering of rocks and human applications of fertilizers. (pp. 117–126)

Explain how human activities affect biogeochemical cycles

People are affecting Earth's biogeochemical cycles by shifting carbon from fossil fuel reservoirs into the atmosphere, shifting nitrogen from the atmosphere to the planet's surface, and depleting groundwater supplies, among other impacts. Policies that seek to minimize human alterations of cycles, such as the remediation efforts embraced in the Chesapeake Bay, can help us address nutrient pollution. (pp. 119–126)

TESTING Your Comprehension

1. Which type of feedback loop is more common in nature, and which more commonly results from human action? How might the emergence of a positive feedback loop affect a system in homeostasis?

2. Describe how hypoxic conditions can develop in ecosystems such as the Chesapeake Bay.

3. What is the difference between an ecosystem and a community?

4. Describe the typical movement of energy through an ecosystem. Now describe the typical movement of matter through an ecosystem.

5. Explain net primary productivity. Name one ecosystem with high net primary productivity and one with low net primary productivity.

6. Why are patches in a landscape mosaic often important to people who are interested in conserving populations of rare animals?

7. What is the difference between evaporation and transpiration? Give examples of how the water cycle interacts with the carbon, phosphorus, and nitrogen cycles.

8. What role does each of the following play in the carbon cycle?

   - Cars
   - Photosynthesis
   - The oceans
   - Earth's crust

9. Distinguish the function performed by nitrogen-fixing bacteria from that performed by denitrifying bacteria.

10. How has human activity altered the carbon cycle? The phosphorus cycle? The nitrogen cycle? What environmental problems have arisen from these changes?

SEEKING Solutions

1. Once vegetation is cleared from a riverbank, water begins to erode away the bank. This erosion may dislodge more vegetation. Would you expect this to result in a feedback process? If so, which type—negative or positive? Explain your answer. How might we halt or reverse this process?

2. Consider the ecosystem(s) that surround(s) your campus. Describe one way in which energy flows through and matter is recycled. Now pick one type of nutrient and briefly describe how it moves through this (these) ecosystem(s). Does the landscape contain patches? Can you describe any ecotones?

3. For a conservation biologist interested in sustaining populations of the organisms below, why would it be helpful to take a landscape ecology perspective? Explain your answer in each case.

   - A forest-breeding warbler that suffers poor nesting success in small, fragmented forest patches
   - A bighorn sheep that must move seasonally between mountains and lowlands

   - A toad that lives in upland areas but travels cross-country to breed in localized pools each spring

4. How do you think we might solve the problem of eutrophication in the Chesapeake Bay? Assess several possible solutions, your reasons for believing they might work, and the likely hurdles we might face. Explain who should be responsible for implementing these solutions, and why.

5. **CASE STUDY CONNECTION** You are an oysterman in the Chesapeake Bay, and your income is decreasing because the dead zone is making it harder to harvest oysters. One day your senator comes to town, and you have a one-minute audience with her. What steps would you urge her to take in Washington, D.C., to try to help alleviate the dead zone and bring back the oyster fishery? Now suppose you are a Pennsylvania farmer who has learned that the government is offering incentives to farmers to help reduce fertilizer runoff into the Chesapeake Bay. What types of approaches described in the text might you be willing to try, and why?

6. THINK IT THROUGH You are a resource manager assigned to devise a plan to protect a rare species of frog that lives in an area slated for development. You are provided with GIS layers of the area as shown in Figure 5.12. The frogs live in the leaf litter in forests, require ponds for breeding, and cannot travel over large open areas when migrating to breeding ponds. What layers (of those shown in the figure) would be most important for your analysis? Why? How would you use principles of landscape ecology when creating your plan?

CALCULATING Ecological Footprints

For many Americans, their desire is to own a suburban home with a weed-free, green lawn. Nationwide, Americans tend about 40.5 million acres of lawn grass. But conventional lawn care involves inputs of fertilizers, pesticides, and irrigation water, not to mention the use of gasoline or electricity for mowing and other care—all of which raise health concerns and affect environmental cycles. Using the figures for a typical lawn in the table, calculate the total amount of fertilizer, water, and gasoline used in lawn care across the nation each year.

	ACREAGE OF LAWN	FERTILIZER USED (LB)	WATER USED (GAL)	GASOLINE USED (GAL)
For the typical quarter-acre lawn	0.25	37	15,700	4.9
For all lawns in your hometown				
For all lawns in the United States	40,500,000			

Data: Chameides, B., 2008. http://www.huffingtonpost.com/bill-chameides/stat-grok-lawns-by-the-nu_b_115079.html

1. How much fertilizer is applied each year on lawns throughout the United States? Where does the nitrogen for this fertilizer come from? What becomes of the nitrogen and phosphorus that are applied to a suburban lawn but not taken up by grass?

2. Leaving grass clippings on a lawn decreases the need for fertilizer by 50%. What else might a homeowner do to decrease fertilizer use in a yard and the environmental impacts of nutrient pollution?

3. How much gasoline could Americans save each year if they did not take care of lawns? At today's gas prices, how much money would this save annually?

MasteringEnvironmentalScience®

CHAPTER 6

Ethics, Economics, and Sustainable Development

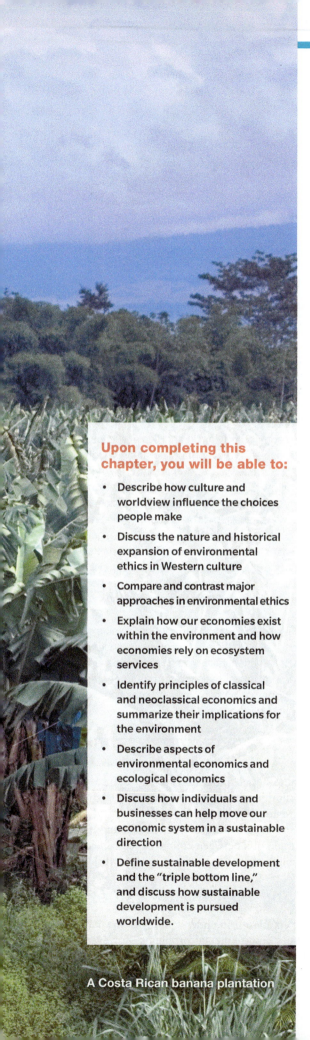

A Costa Rican banana plantation

Costa Rica Values Its Ecosystem Services

> **Costa Rica's PSA program has been one of the conservation success stories of the last decade.**
> Stefano Pagiola, The World Bank

> **In the last 25 years, my home country has tripled its GDP while doubling the size of its forests.**
> Carlos Manuel Rodriguez, former Minister of Energy and the Environment, Costa Rica

Few nations have transformed their path of development in just a few decades—but Costa Rica has. In the 1980s, this small Central American country was losing its forests as fast as any place on Earth. Yet today this nation of 4.9 million people has regained much of its forest cover, boasts a world-class park system, and stands as a global model for sustainable resource management.

Costa Rica took many steps on this impressive road to success. One key step was the decision to pay landholders to conserve forest on private land, in a novel government program called *Pago por Servicios Ambientales* (PSA)—Payment for Environmental Services.

Nature provides ecosystem services (pp. 4, 116–117), such as air and water purification, climate regulation, and nutrient cycling. For example, forests in Costa Rica's mountains capture rainfall and provide clean drinking water for farms, towns, and cities below. Ecosystem services are vital for our lives and our society, but historically we have tended to take them for granted, and rarely do we pay for them in the marketplace. This has contributed to the loss of these services as we degrade the natural systems that provide them. For this reason, many economists have proposed creating financial incentives for conserving ecosystem services.

In Costa Rica, which had lost more than three-quarters of its forest, political leaders adopted this approach in Forest Law 7575, passed in 1996. Since then, the Costa Rican government has been paying farmers and ranchers to preserve forest on their land, replant cleared areas, allow forest to regenerate naturally, and establish sustainable forestry systems. Payments are designed to be competitive with potential profits from farming or cattle ranching, and in recent years these payments have averaged $78/hectare (ha)/yr ($32/acre/yr).

The PSA program recognizes four ecosystem services that forests provide:

- Watershed protection: Forests cleanse water by filtering pollutants, and they conserve water and reduce soil erosion by slowing runoff.
- Biodiversity: Tropical forests such as Costa Rica's are especially rich in life.
- Scenic beauty: Appreciation of scenic beauty encourages recreation and ecotourism, which bring money to the economy.
- Carbon sequestration: Forests pull carbon dioxide from the atmosphere, thereby slowing global warming.

To fund the PSA program, Costa Rica's government sought money from people and companies who benefited from these services. For watershed protection, irrigators, bottlers, municipal water suppliers, and utilities that generate hydropower all made voluntary payments into the program, and a tariff on water users was added in 2005. For biodiversity and scenery, the country targeted ecotourism, while international lending agencies provided loans and donations. Because carbon dioxide is emitted when fossil fuels are burned, the nation used a 3.5% tax on fossil fuels to help fund the program. It also sought to sell carbon offsets in global carbon trading markets (p. 510).

Costa Rican landholders rushed to sign up for the PSA program. The agency administering it, *Fondo Nacional de Financiamiento Forestal* (FONAFIFO), signed landowners to contracts and sent agents to advise them on forest conservation and to monitor compliance.

Caribbean Sea

COSTA RICA

San José

■ Forested area

(a) 1940　　　　　　　　　(b) 1987　　　　　　　　　(c) 2005

FIGURE 6.1 Forest cover in Costa Rica decreased between 1940 and 1987, but it increased thereafter.
Data from FONAFIFO.

By 2015, FONAFIFO had paid 172 billion colónes ($325 million in today's U.S. dollars) to nearly 15,000 landholders and had registered more than 1 million ha (2.5 million acres)—20% of the nation's land area.

Deforestation slowed in Costa Rica in the wake of the program; forest cover rose by 10% in the decade after 1996. Policymakers, economists, and environmental advocates cheered the PSA program's apparent success in safeguarding forests and the ecosystem services they provide.

However, some observers argued that forest loss had been slowing for other reasons and that the program itself was having little effect. They contended that payments were being wasted on people who had had no plans to cut down their trees. Critics also lamented that large wealthy landowners used the program more than low-income small farmers. All these concerns were borne out by researchers (see **THE SCIENCE BEHIND THE STORY**, pp. 142–143).

In response, the government modified its policies, making the program more accessible to small farmers and targeting the payments to locations where forest is most at risk and environmental assets are greatest.

In recent years forested area in Costa Rica has been rising (**FIGURE 6.1**). From a low of 17% in 1983, forest cover has risen to more than 53% today. The nation has thrived economically while protecting its environment. Since the PSA program began, Costa Ricans have enjoyed an increase in real, inflation-adjusted per capita income of more than 60%—a rise in wealth surpassing the vast majority of nations.

Many factors have contributed to Costa Rica's success in building a wealthier society while protecting its ecological assets. Back in 1948, Costa Rica abolished its army and shifted funds from the military budget into health and education. With a stable democracy and a healthy and educated citizenry, the stage was set for well-managed development, including innovative advances in conservation. In 1970, the nation began creating one of the world's finest systems of national parks, which today covers fully one-quarter of its territory. Ecotourism at the parks brings wealth into the country: Each year more than 2 million tourists from around the world inject over $2 billion into Costa Rica's economy.

As a result, Costa Ricans understand the economic value of protecting their natural capital. With the economic value of nature so clear, an ethic of conservation has grown and flourished. In today's Costa Rica, an ethical appreciation for nature and an economic appreciation for nature go hand in hand, pointing the way toward truly sustainable development.

Culture, Worldview, and the Environment

Costa Rica's program of payment for environmental services has inspired similar approaches throughout the world. From Mexico to Australia to Tanzania to China, these programs are succeeding as people everywhere begin to better appreciate the contributions of ecological systems to human economies. In the United States, the federal government issues payments for ecological services through its Conservation Reserve Program (p. 230), which pays farmers to retain natural vegetation to prevent erosion, conserve soil, enhance wildlife habitat, and reduce water pollution. All such programs seek to defuse a dilemma facing many rural landholders, who often feel short-term economic pressure to clear forest for agriculture even though they may feel an ethical concern for the forest and its flora and fauna. Such trade-offs between ethical concerns and short-term economic benefits arise frequently in today's world.

Ethics and economics involve values

Environmental science entails a firm understanding of the natural sciences. To address environmental problems, however, we also need to understand how people perceive, value, and relate to their environment philosophically and pragmatically.

Ethics and economics are two very different disciplines, but each deals with questions of what we value and how those values influence our decisions and actions. To find sustainable solutions to environmental problems, we must aim to comprehend not only how natural systems work but also how values shape human behavior.

Culture and worldview influence our decisions

Every action we take affects our environment. Whether we are growing food, building homes, manufacturing products, or fueling vehicles, we meet our needs by extracting resources and altering our surroundings. In deciding how to manipulate our environment to meet our needs, we rely on rational assessments of costs and benefits, but our decisions are also influenced by our culture and our worldview (**FIGURE 6.2**).

(a) A Costa Rican guide shows a Hercules beetle to ecotourists

(b) A pineapple farmer makes a living from his crops

FIGURE 6.2 Differences in culture, wealth, and personal circumstance may lead people to interact differently with their environment. In Costa Rica's national parks **(a)**, ecotourism provides visitors exploration, recreation, education, and inspiration, while providing quality employment for Costa Ricans, who may see economic advantage in preserving forest as a result. In contrast, a Costa Rican farm worker **(b)** needs agriculture that produces enough to live on, and may favor clearing forest if the short-term economic benefits outweigh the costs.

Culture can be defined as the ensemble of knowledge, beliefs, values, and learned ways of life shared by a group of people. Culture, together with personal experience and personal circumstance, influences each person's perception of the world and his or her place within it—the person's **worldview.** A worldview reflects beliefs about the meaning, operation, and essence of the world.

People with different worldviews can study the same situation yet draw dramatically different conclusions. For example, two Costa Rican ranchers owning identical landholdings may make different decisions about how to manage them. One might opt to receive payments and conserve forest, whereas the other might opt to clear every hectare for cattle grazing.

Many factors shape our worldviews

Many factors shape people's worldviews and perception of their environment. Among the most influential are spiritual beliefs and political ideology. For instance, one's views on the proper role of government may guide whether one wants government to intervene in a market economy to protect environmental quality. Shared cultural experience is another factor. For example, early European settlers in the Americas, facing the struggles of frontier life, viewed their environment as a hostile force because inclement weather and wild animals frequently destroyed crops and killed livestock. Many people still view nature as a hostile adversary to overcome.

As you progress through your course and through your life, you will encounter scientific data on the environmental impacts of our choices (where to live, how to make a living, what to eat, what to wear, how to spend our leisure time, and so on). You will also see that culture and worldviews play critical roles in such choices. Acquiring scientific understanding is vital in our search for sustainable solutions, but we also need to consider ethics and economics, because these disciplines help us understand how and why we value the things we value.

Environmental Ethics

Ethics is a branch of philosophy that involves the study of good and bad, of right and wrong. The term *ethics* can also refer to the set of moral principles or values held by a person or a society. Ethicists help clarify how people judge right from wrong by elucidating the criteria that people use in making these judgments. Such criteria are grounded in values—for instance, promoting human welfare, maximizing individual freedom, or minimizing pain and suffering.

People of different cultures or worldviews may differ in their values and thus may disagree about actions they consider to be right or wrong. This is why some ethicists are **relativists** who believe that ethics do and should vary with social context. However, different societies show a remarkable extent of agreement on what moral standards are appropriate. For this reason, many ethicists are **universalists** who maintain that there exist objective notions of right and wrong that hold across cultures and contexts. For both relativists and universalists, ethics is a prescriptive pursuit; rather than simply describing behavior, it prescribes how we *ought* to behave.

Ethical standards help us judge right from wrong

Ethical standards are the criteria that help differentiate right from wrong. One classic ethical standard is the *categorical imperative* proposed by German philosopher Immanuel Kant, which advises us to treat others as we would prefer to be treated ourselves. In Christianity this standard is called the "Golden Rule," and most of the world's religions teach this same lesson. Another ethical standard is the *principle of utility*, elaborated by British philosophers Jeremy Bentham and John Stuart Mill. The utilitarian principle holds that something is right when it produces the greatest practical benefits for the most people. We all employ ethical standards as we make countless decisions in our everyday lives.

We value things in two ways

People ascribe value to things in two main ways. One way is to value something for the pragmatic benefits it brings us if we put it to use. This is termed **instrumental value** (or *utilitarian value*). The other way is to value something for its intrinsic worth, to feel that something has a right to exist and is valuable for its own sake. This notion is termed **intrinsic value,** or *inherent value*.

A person may ascribe instrumental value to a forest because we can harvest timber from it, hunt game in it, use it for recreation, and drink clean water it has captured and filtered. A person may perceive intrinsic value in a forest because it provides homes for countless organisms that the person feels have an inherent right to live. A forest, an animal, a lake, or a mountain can have both intrinsic and instrumental value. However, different people may emphasize different types of value.

Paying money for ecosystem services is a utilitarian approach that attempts to quantify instrumental values by assigning market prices to them. For people who tend to view nature through the lens of intrinsic values, this approach may make them uneasy. Indeed, scientists who share the goal of conserving natural amenities sometimes disagree on the means of doing so. For instance, Stanford University biologist Gretchen Daily is a key proponent of assigning market values to ecosystem services, with the utilitarian goal of engaging market forces to assist in their conservation. Yet in 2006 a student in her department, Douglas McCauley, authored a commentary in the scientific journal *Nature* that argued eloquently against such an approach. McCauley warned that the "commodification of nature" distracted from the environment's intrinsic value, which he maintained was infinite and priceless. "Nature conservation must be framed as a moral issue," McCauley wrote. "We will make more progress in the long run by appealing to people's hearts rather than to their wallets."

Environmental ethics pertains to people and the environment

The application of ethical standards to relationships between people and nonhuman entities is known as **environmental ethics.** This branch of ethics arose once people began to perceive environmental change brought by industrialization. Our interactions with our environment can give rise to ethical questions that are difficult to resolve. Consider some examples:

1. Is the present generation obligated to conserve resources for future generations? If so, how much should we conserve?

2. Can we justify exposing some communities to a disproportionate share of pollution? If not, what actions are warranted to prevent this?

3. Are humans justified in driving species to extinction? If destroying a forest would drive extinct a little-known insect species but would create jobs for 10,000 people, would that action be ethically admissible? What if it were an owl species? What if only 100 jobs would be created? What if it were a species harmful to people, such as a mosquito or a noxious bacterium or virus?

The first question is central to the notion of sustainability (p. 15) and to the pursuit of sustainable development (p. 153), which lie at the heart of environmental science. From an ethical perspective, sustainability means treating future generations as we would prefer to be treated ourselves. The second question goes to the heart of environmental justice (pp. 137–139). The third set of questions involves intrinsic values (but also instrumental values) and typifies questions that arise in debates over endangered species management, habitat protection, and conservation biology (Chapter 11).

We have expanded our ethical consideration

Answers to the three sets of questions above depend partly on what ethical standard(s) a person adopts. They also depend on the breadth and inclusiveness of the person's domain of ethical concern. A person who ascribes intrinsic value to insects and feels responsibility for their welfare would answer the third set of questions very differently from a person whose domain of ethical concern ends with human beings.

Many traditional non-Western cultures have long granted nonhuman entities intrinsic value and ethical standing. Australian Aborigines view their landscape as sacred and alive. Many native cultures across the Americas feature ethical systems that encompass both people and aspects of their environment.

As the history of Western cultures (European and European-derived societies) has progressed, people have granted intrinsic value and extended ethical consideration to more and more people and things. Today, concern for the welfare of domesticated animals is evident in the great care many people provide for their pets. Animal rights advocates voice concern for animals that are hunted, raised in pens, or used in laboratory testing. Most people now accept that wild animals merit ethical consideration. Increasing numbers of people today see intrinsic value in whole natural communities. Some go further and suggest that all of nature—living and nonliving things alike—should be ethically recognized.

What has helped broaden our ethical domain in these ways? Rising economic prosperity has played a role, by making us less anxious about our day-to-day survival. Science has also played a role, by demonstrating that people do not stand apart from nature but rather are part of it. For example, ecology makes clear that organisms are interconnected and that what affects plants, animals, and ecosystems also affects people. Evolutionary biology shows that human beings, as one species out of millions, have evolved subject to the same pressures as other organisms.

We can simplify our continuum of attitudes toward the natural world by dividing it into three ethical perspectives: anthropocentrism, biocentrism, and ecocentrism (**FIGURE 6.3**).

Anthropocentrism People who have a human-centered view of our relationship with the environment display **anthropocentrism.** An anthropocentrist denies, overlooks, or devalues the notion that nonhuman things have intrinsic value. An anthropocentrist evaluates the costs and benefits of actions solely according to their impact on people. For example, if cutting down a Costa Rican forest for farming or ranching would provide significant economic benefits while doing little harm to aesthetics or human health, the anthropocentrist would conclude this was worthwhile, even if it would destroy many plants and animals. Conversely, if protecting the forest would provide greater economic, spiritual, or other benefits to people, an anthropocentrist would favor its protection. In the anthropocentric perspective, anything not providing benefit to people is considered to be of negligible value.

Biocentrism **Biocentrism** ascribes intrinsic value to certain living things or to the biotic realm in general. In this perspective, human life and nonhuman life both have ethical standing. A biocentrist might oppose clearing a forest if this would destroy countless plants and animals, even if it would increase food production and generate economic growth for people. Some biocentrists advocate equal consideration for all living things, whereas others grant some types of organisms more consideration than others.

Ecocentrism **Ecocentrism** judges actions in terms of their effects on whole ecological systems, which consist of living and nonliving elements and their interrelationships. An ecocentrist values the well-being of entire species, communities, or ecosystems over the welfare of a given individual. Implicit in this view is that preserving systems generally protects their components, whereas protecting components may not safeguard the entire system. An ecocentrist would respond to a proposal to clear forest by broadly assessing the potential impacts on water quality, air quality, wildlife populations, soil structure, nutrient cycling, and ecosystem services. Ecocentrism is a more holistic perspective than biocentrism or anthropocentrism. It encompasses a wider variety of entities at a larger scale and seeks to preserve the connections that tie them together into functional systems.

Environmental ethics has ancient roots

Environmental ethics arose as an academic discipline in the 1970s, but people have contemplated our ethical relations with nature for thousands of years. The ancient Greek philosopher Plato argued that humanity had a moral obligation to our environment, writing, "The land is our ancestral home and we must cherish it even more than children cherish their mother."

Some ethicists and theologians have pointed to the religious traditions of Christianity, Judaism, and Islam as sources of anthropocentric hostility toward the environment. They point out biblical passages such as, "Be fruitful and multiply, and fill the earth and subdue it; and have dominion over the fish of the sea and over the birds of the air and over every living thing that moves upon the earth" (Genesis 1:28). Such wording, according to many scholars, has encouraged animosity toward and disregard of nature.

Others interpret sacred texts of these religions to encourage benevolent human stewardship over nature. Consider the directive, "You shall not defile the land in which you live" (Numbers 35:34). If one views the natural world as God's creation, then under this interpretation it would be considered sinful to degrade that creation. Indeed, many people holding the stewardship interpretation see it as one's ethical duty to act as a responsible steward of the world in which we live. In today's world, Pope Francis has become an especially powerful voice for the notion that humanity is ethically obliged to care for our environment.

FIGURE 6.3 We can categorize people's ethical perspectives as anthropocentric, biocentric, or ecocentric.

The industrial revolution inspired reaction

As industrialization spread in the 19th century, it amplified human impacts on the environment. In this period of social and economic transformation, agricultural economies became industrial ones, machines enhanced or replaced human and animal labor, and many people moved from farms to cities. Population rose dramatically, consumption of natural resources accelerated, and pollution intensified as we burned coal to fuel railroads, steamships, ironworks, and factories.

Some writers of the time drew attention to the drawbacks of industrialization. British critic John Ruskin worried that although people prized the material benefits that nature provided, they no longer appreciated its spiritual and aesthetic benefits. Motivated by such concerns, a number of citizens' groups sprang up in 19th-century England, forerunners of today's environmental organizations.

In the United States during the 1840s, a philosophical movement called *transcendentalism* flourished, espoused in New England by philosophers Ralph Waldo Emerson and Henry David Thoreau and by poet Walt Whitman. Like Ruskin, the transcendentalists objected to materialism, and they saw natural entities as symbols or messengers of deeper truths. Although Thoreau viewed nature as divine, he also observed the natural world closely and came to understand it in the manner of a scientist; indeed, he can be considered one of the first ecologists. His book *Walden*, in which he recorded observations and thoughts while living at Walden Pond away from the bustle of urban Massachusetts, remains a classic of American literature.

Conservation and preservation arose with the 20th century

One admirer of Emerson and Thoreau was **John Muir** (1838–1914), a Scottish immigrant to the United States who made California's Yosemite Valley his wilderness home. Although Muir chose to live in isolation in his beloved Sierra Nevada for long stretches of time, he also became politically active and won fame as a tireless advocate for the preservation of wilderness (**FIGURE 6.4**).

Muir was motivated by the rapid deforestation and environmental degradation he witnessed throughout North America and by his belief that the natural world should be treated with the same respect that we give to cathedrals. Today he is associated with the **preservation ethic,** which holds that we should protect the natural environment in a pristine, unaltered state. Muir argued that nature deserved protection for its own sake (an ecocentrist argument resting on the notion of intrinsic value), but he also maintained that nature promoted human happiness (an anthropocentrist argument based on instrumental value). "Everybody needs beauty as well as bread," he wrote in his 1912 book *The Yosemite*, "Places to play in and pray in, where nature may heal and give strength to body and soul alike."

Some of the factors that motivated Muir also inspired **Gifford Pinchot** (1865–1946), the first professionally trained American forester (**FIGURE 6.5**). Pinchot founded what would become the U.S. Forest Service and served as its chief in

FIGURE 6.4 A pioneering advocate of the preservation ethic, John Muir helped establish the Sierra Club, a leading environmental organization. Here Muir (**right**) is shown with President Theodore Roosevelt in Yosemite National Park in 1903. After this wilderness camping trip with Muir, the president expanded protection of areas in the Sierra Nevada.

President Theodore Roosevelt's administration. Like Muir, Pinchot opposed the deforestation and unregulated economic development that occurred during their lifetimes. However, Pinchot took a more anthropocentric view of how and why we should value nature. He espoused the **conservation ethic,** which holds that people should put natural resources to use but that we have a responsibility to manage them wisely. The conservation ethic uses a utilitarian standard and holds that we should allocate resources to provide the greatest good to the greatest number of people for the longest time. Whereas preservation aims to preserve nature for its own sake and for our aesthetic and spiritual benefit, conservation promotes the prudent, efficient, and sustainable extraction and use of natural resources for the good of present and future generations.

FIGURE 6.5 Gifford Pinchot was a leading proponent of the conservation ethic. This ethic holds that we should use natural resources in ways that ensure the greatest good for the greatest number of people for the longest time.

Pinchot and Muir came to represent different branches of the American environmental movement. Their contrasting ethical approaches often pitted them against one another on policy issues of the time, and they each left legacies that reverberate today. Yet despite their differences, both Muir and Pinchot represented reactions against a prevailing "development ethic," which holds that people should be masters of nature and which promotes economic development without regard to its negative consequences.

weighing the ISSUES

Preservation and Conservation

Do you identify more with the preservation ethic, the conservation ethic, both, or neither? Think of a forest, wetland, or other important natural resource in your region. Give an example of a situation in which you might adopt a preservation ethic and an example of one in which you might adopt a conservation ethic. Are there conditions under which you'd follow neither, but instead adopt a "development ethic"?

Today these schools of thought have spread globally, and people worldwide are wrestling with how to balance preservation, conservation, and economic development. In Costa Rica, leaders and citizens felt that development had gone too far and that precious ecosystem services were being degraded and lost, threatening the country's future. The nation responded by establishing an extensive system of national parks, eventually protecting 24% of its land area—one of the highest percentages of any nation in the world. Beyond this preservationist policy, many conservationist policies were implemented to encourage the sustainable use of soil, water, and forests. Costa Rica's program to pay for ecological services mixes these approaches, encouraging the preservation of forests within lands actively used for agricultural production to meet overall goals of conservation and sustainable development.

Aldo Leopold's land ethic inspires many people

As a young forester and wildlife manager, **Aldo Leopold** (1887–1949; **FIGURE 6.6**) began his career in the conservationist camp after graduating from Yale Forestry School, which Pinchot had helped found just as Roosevelt and Pinchot were advancing conservation on the national stage. As a forest manager in Arizona and New Mexico, Leopold embraced the government policy of shooting predators, such as wolves, to increase populations of deer and other game animals.

At the same time, Leopold followed the advance of ecological science. He eventually ceased to view certain species as "good" or "bad" and instead came to see that healthy ecological systems depend on protecting all their interacting parts. Drawing an analogy to mechanical maintenance, he wrote, "to keep every cog and wheel is the first precaution of intelligent tinkering."

It was not just science that pulled Leopold from an anthropocentric perspective toward a more holistic one. One day he shot a wolf, and when he reached the animal, Leopold was transfixed by "a fierce green fire dying in her eyes." At that moment, Leopold perceived intrinsic value in the wolf, and the experience remained with him for the rest of his life, helping to lead him

FIGURE 6.6 Aldo Leopold, a wildlife manager, author, and philosopher, articulated a new relationship between people and the environment. In his essay "The Land Ethic" he called on people to embrace their environment in their ethical outlook.

to an ecocentric ethical outlook. Years later, as a University of Wisconsin professor, Leopold argued that people should view themselves and "the land" as members of the same community and that we are obligated to treat the land in an ethical manner.

Leopold intended the land ethic to help guide decision making. "A thing is right," he wrote, "when it tends to preserve the integrity, stability, and beauty of the biotic community. It is wrong when it tends otherwise." Leopold died before seeing his seminal 1949 essay "The Land Ethic" and what would become his best-known book, *A Sand County Almanac*, in print, but today many view him as the most eloquent philosopher of environmental ethics.

Environmental justice seeks fair treatment for all people

Our society's domain of ethical concern has been expanding from rich to poor, and from majority races and ethnic groups to minority ones. This ethical expansion involves applying a standard of fairness and equality, and it has given rise to the environmental justice movement. **Environmental justice** involves the fair and equitable treatment of all people with respect to environmental policy and practice, regardless of their income, race, or ethnicity.

The struggle for environmental justice has been fueled by the recognition that poor people tend to be exposed to a greater share of pollution, hazards, and environmental degradation than are richer people (**FIGURE 6.7**, p. 138). Environmental justice advocates also note that racial and ethnic minorities tend to suffer more than their share of exposure to most hazards. Indeed, studies repeatedly document that poor and nonwhite communities each tend to bear heavier

(a) Migrant farm workers in Colorado

(b) Homes near a coal-fired power plant

(c) Children in New Orleans after Hurricane Katrina

FIGURE 6.7 Environmental justice efforts are inspired by the fact that the poor are often exposed to more hazards than are the rich. For example, Latino farm workers **(a)** may experience health risks from pesticides, fertilizers, and dust. Low-income white Americans in Appalachia **(b)** may suffer air pollution from nearby coal-fired power plants. African-American communities in New Orleans **(c)** were most susceptible to flooding and were devastated by Hurricane Katrina in 2005.

burdens of air pollution, lead poisoning, pesticide exposure, toxic waste exposure, and workplace hazards. This is thought to occur because lower-income and minority communities often have less access to information on environmental health risks, less political power with which to protect their interests, and less money to spend on avoiding or alleviating risks.

A protest in the 1980s by residents of Warren County, North Carolina, in opposition to a proposed toxic waste dump in their community helped to ignite the environmental justice movement. The state had chosen to site the dump in the North Carolina county with the highest percentage of African Americans. Warren County residents lost their battle and the dump was established—but the protest raised the profile of environmental justice concerns and inspired countless efforts elsewhere.

Like African Americans, Native Americans have encountered many environmental justice issues. For instance, uranium mining on lands of the Navajo nation in the Southwest employed many Navajo in the 1950s and 1960s. Although uranium mining had been linked to health problems and premature death, neither the mining industry nor the U.S. government provided the miners information or safeguards against radiation and its risks. Many Navajo families built homes and bread-baking ovens out of waste rock from the mines, not realizing it was radioactive. Lung cancer began to appear among Navajo miners in the 1960s. A later generation of Americans perceived this as negligence and discrimination, and they sought justice through the Radiation Exposure Compensation Act of 1990, a federal law compensating Navajo miners who suffered health effects from unprotected work in the mines.

Likewise, low-income white residents of the Appalachian region have long been the focus of environmental justice concerns. Mountaintop coal-mining practices (pp. 530, 639) in this economically neglected region provide jobs to local residents but also destroy forests, pollute water, bury streams, and cause flooding. Low-income residents of affected Appalachian communities continue to have little political power to voice complaints over the impacts of these mining practices.

Today the world's economies have grown, but the gaps between rich and poor have widened. And despite much progress toward racial equality, significant inequities remain. Environmental laws have proliferated, but minorities and the poor still suffer substandard environmental conditions. Yet today more people are fighting environmental hazards in their communities and winning.

One ongoing story involves Latino farm workers in California's San Joaquin Valley. These workers harvest much of

weighing the
ISSUES

Environmental Justice

Consider the place where you grew up. Where were the factories, waste dumps, and polluting facilities located, and who lived closest to them? Who lives nearest them in the town or city that hosts your campus? Do you think the concerns of environmental justice advocates are justified? If so, what could be done to ensure that poor communities do not suffer more hazards than wealthy ones?

the U.S. food supply of fruits and vegetables yet suffer some of the nation's worst air pollution. Industrial agriculture generates pesticide emissions, dairy feedlot emissions, and windblown dust from eroding farmland, yet this pollution was not being regulated. Valley residents enlisted the help of organizations including the Center on Race, Poverty, and the Environment, a San Francisco–based law firm focusing on environmental justice. Together they persuaded California regulators to enforce Clean Air Act provisions and convinced California legislators to pass new legislation regulating agricultural emissions.

Environmental justice is a key component in pursuing the environmental, economic, and social goals of sustainability and sustainable development (pp. 15, 153). As we explore environmental issues from a scientific standpoint throughout this book, we will also encounter the social, political, ethical, and economic aspects of these issues, and the concept of environmental justice will come up again and again.

Questions of environmental justice, like questions of how to value ecosystem services, intertwine ethical issues with economic ones—and friction often develops between people's ethical concerns and their economic desires. Addressing ethical, economic, and environmental concerns together in a mutually productive way is a primary goal of the modern drive for sustainable development (pp. 153–154).

Economics and the Environment

An **economy** is a social system that converts resources into *goods* (material commodities manufactured for and bought by individuals and businesses) and *services* (work done for others as a form of business). **Economics** is the study of how people decide to use potentially scarce resources to provide goods and services that are in demand. The word *economics* and the word *ecology* come from the same Greek root, *oikos*, meaning "household." Economists traditionally have studied the household of human society, and ecologists the broader household of all life.

Economies rely on goods and services from the environment

Our economies and our societies exist within the natural environment and depend on it in important ways. Economies receive inputs (such as natural resources and ecosystem services) from the environment, process them, and discharge outputs (such as waste) into the environment (**FIGURE 6.8**). The material inputs and the waste-absorbing capacity that Earth can provide are ultimately finite.

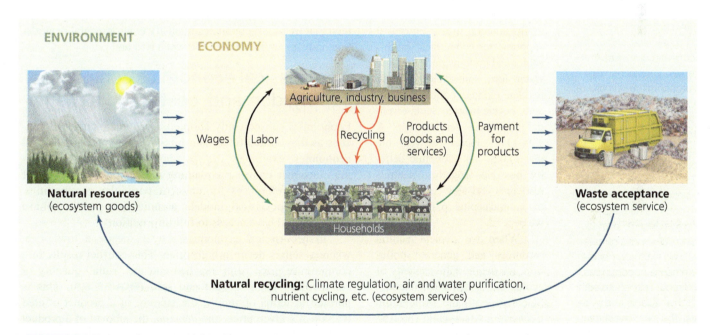

FIGURE 6.8 Economies exist within the natural environment, receiving resources from it, discharging waste into it, and benefiting from ecosystem services. Conventional neoclassical economics has focused on processes of production and consumption between households and businesses (**tan box in middle**), viewing the environment merely as an external factor. In contrast, environmental and ecological economists emphasize that economies exist within the natural environment and depend on all that it offers.

DATA Q Follow each arrow in this figure to make sure you understand what it indicates. Then answer these questions:
 • When you work at a job, what do you give and what do you receive?
• When you buy a product, what do you give and what do you receive?
• Describe three things that the environment provides to the economy, basing your answers on this illustration.

Go to **Interpreting Graphs & Data** on **Mastering**EnvironmentalScience°

These interactions are readily apparent, yet traditional economic schools of thought have long overlooked their importance. Many mainstream economists still adhere to a worldview that largely ignores the environment (and instead considers only the tan box in the middle of Figure 6.8). This conventional view, which continues to drive many policy decisions, implies that natural resources are free and limitless and that wastes can be endlessly absorbed at no cost. Newer schools of thought, however, recognize that economies exist within the environment and rely on its natural resources and ecosystem services. This is why Costa Rica and other nations have taken steps to better protect their natural assets.

Natural resources (p. 4) are the substances and forces that sustain our society and our everyday lives: the fresh water we drink; the trees that provide our lumber; the rocks that provide our metals; and the energy from the sun, wind, water, and fossil fuels. We can think of natural resources as "goods" produced by nature. Environmental systems also naturally function in a manner that supports our economies. Earth's ecological systems purify air and water, form soil, cycle nutrients, regulate climate, pollinate plants, and recycle waste. Such essential ecosystem services (pp. 4, 116–117) support the very life that makes our economic activity possible. Together, nature's resources and services make up the natural capital (p. 6) on which we depend.

When we deplete natural resources and generate pollution, we degrade the capacity of ecological systems to function. Scientists with the Millennium Ecosystem Assessment (pp. 15–16) concluded in 2005 that 15 of 24 ecosystem services they surveyed globally were being degraded or used unsustainably. The degradation of ecosystem services can weaken economies. In Costa Rica, rapid forest loss up through the 1980s was causing soil erosion, water pollution, and biodiversity loss that increasingly threatened the country's economic potential. Low-income small farmers were the first to feel these impacts. Indeed, across the world, ecological degradation is harming poor and marginalized people before wealthy ones, the Millennium Ecosystem Assessment found. As a result, restoring ecosystem services stands as a prime avenue for alleviating poverty.

The relationships among economic and environmental conditions are only now becoming widely recognized. Let's briefly examine how economic thought has evolved, tracing the path that is finally beginning to lead economies to become more compatible with the natural systems on which they depend.

Adam Smith proposed an "invisible hand"

Economics shares a common intellectual heritage with ethics, and practitioners of both study the relationship between individual action and societal well-being. Early philosophers had long believed that individuals acting in their own self-interest harm society. However, Scottish philosopher Adam Smith (1723–1790) argued that self-interested economic behavior can benefit society, as long as the behavior is constrained by the rule of law and private property rights within a competitive marketplace. Known today as a founder of **classical economics,** Smith felt that when people pursue their own economic self-interest under these conditions, the marketplace will behave as if guided by "an invisible hand" to benefit society as a whole.

Smith's philosophy remains a pillar of free-market thought today, and many credit it for the tremendous gains in material wealth that industrialized nations have achieved. Others contend that free-market policies intensify environmental degradation and worsen inequalities between rich and poor.

Neoclassical economics considers supply, demand, costs, and benefits

Economists subsequently adopted more quantitative approaches as they aimed to explain human behavior. **Neoclassical economics** examines the psychological factors underlying consumer choices, explaining market prices in terms of consumer preferences for units of particular commodities. Standard neoclassical models assume that people behave rationally and have access to full information.

In neoclassical economics, buyers desire a low price, whereas sellers desire a high price. This conflict results in a compromise price being reached and the "right" quantity of commodities being bought and sold (**FIGURE 6.9**). This is phrased in terms of *supply*, the amount of a product offered for sale at a given price, and *demand*, the amount of a product people will buy at a given price if free to do so. Theoretically, the market moves toward an equilibrium point, a price at which supply equals demand. By applying similar reasoning to environmental issues, economists can determine "optimal" levels of resource use or pollution control. For instance, reducing pollution emitted by a car or factory is often cost-effective at first, but as pollution is reduced, it becomes more and more costly to eliminate each remaining amount. At some point the cost per unit of reduction rises to match the benefits, and this break-even point is the optimal level of pollution reduction.

To evaluate an action or decision, neoclassical economists use **cost-benefit analysis,** which compares the estimated costs

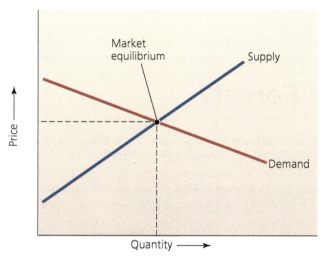

FIGURE 6.9 We can show economic fundamentals in a supply-and-demand graph. The demand curve indicates the quantity of a given good (or service) that consumers desire at each price, and the supply curve indicates the quantity produced at each price. In theory, the market automatically moves toward an equilibrium point, at which supply equals demand.

of a proposed action with the estimated benefits. If benefits exceed costs, the action should be pursued; if costs exceed benefits, it should not. Given a choice of actions, the one with the greatest excess of benefits over costs should be chosen.

This reasoning seems eminently logical, but problems arise when not all costs and benefits can be easily identified, defined, or quantified. It may be simple to quantify the dollar value of bananas grown or cattle raised on a tract of Costa Rican land cleared for agriculture, yet difficult to assign monetary value to the complex ecological costs of clearing the forest. Because monetary benefits are usually more easily quantified than environmental costs, benefits tend to be over-represented in cost-benefit analyses. As a result, environmental advocates often feel such analyses are predisposed toward economic development and against environmental protection.

Neoclassical economics has environmental consequences

Today's market systems operate largely in accord with the principles of neoclassical economics. These systems have generated unprecedented material wealth for market-oriented societies, yet four fundamental assumptions of neoclassical economics often contribute to environmental degradation.

Replacing resources One assumption of neoclassical economics is that natural resources and human resources (such as workers or technologies) are either infinite or largely substitutable and interchangeable. This implies that once we have depleted a resource, we will always be able to find some replacement for it. As a result, ecosystem goods and services are treated as free gifts of nature, endlessly abundant and resilient, and the market imposes no penalties for depleting them.

It is true that many resources can be replaced. Our societies have transitioned from manual labor to animal labor to

steam-driven power to fossil fuel power, and we are beginning a transition to renewable power sources such as wind and solar energy. However, Earth's material resources are ultimately limited. Nonrenewable resources (such as fossil fuels) can be depleted. Many renewable resources (such as soils, fish stocks, and forest products) can be used up if we exploit them faster than they are replenished. Even seemingly inexhaustible resources can become so polluted that we can no longer use them (such as contaminated water supplies).

External costs A second assumption of neoclassical economics is that all costs and benefits associated with an exchange of goods or services are borne by individuals engaging directly in the transaction. In other words, it is assumed that the costs and benefits are "internal" to the transaction, experienced by the buyer and seller alone.

However, many transactions affect other members of society. When a landholder fells a forest, people nearby suffer poorer water quality, dirtier air, and less wildlife. When a factory, power plant, or mining operation pollutes the air or water, it harms the health of people who live nearby. In such cases, people who are not involved in degrading the environment end up paying the costs. Costs of a transaction that affect people other than the buyer or seller are known as **external costs** (**FIGURE 6.10**). Often whole communities suffer external costs while certain individuals enjoy private gain. External costs commonly include the following:

- Health problems, stress, or anxiety experienced by people downstream or downwind from a pollution source
- Declines in resources, such as fewer fish in a stream
- Aesthetic damage, such as from air pollution, clear-cutting, or strip mining
- Declining real estate values, lost tourism revenue, higher health care expenses, and more

FIGURE 6.10 People commonly suffer external costs from air pollution. Here, residents of an Indonesian town cycle through smoke from fires set to clear forest for oil palm plantations.

Do Payments Help Preserve Forest?

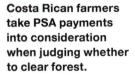

Costa Rican farmers take PSA payments into consideration when judging whether to clear forest.

Costa Rica's program to pay for ecosystem services has garnered international praise and has inspired other nations to implement similar programs. But have Costa Rica's payments actually been effective in preventing forest loss? A number of research teams have sought to answer this surprisingly difficult question by analyzing data from the PSA program.

Some early studies were quick to credit the PSA program for saving forests. A 2006 study conducted for FONAFIFO, the agency administering the program, concluded that PSA payments in the central region of the country had prevented 108,000 ha (267,000 acres) of deforestation—38% of the area under contract. Indeed, deforestation rates fell as the program proceeded; rates of forest clearance in 1997–2000 were half what they were in the preceding decade.

However, some researchers hypothesized that PSA payments were not responsible for this decline and that deforestation would have slowed anyway because of other factors. To test this hypothesis, a team led by G. Arturo Sanchez-Azofeifa of the University of Alberta and Alexander Pfaff of Duke University worked with FONAFIFO's payment data, as well as data on land use and forest cover from satellite surveys. They layered these data onto maps using a geographic information system (GIS) (p. 115), then explored the patterns revealed.

In 2007 in the journal *Conservation Biology*, they reported that only 7.7% of PSA contracts were located within 1 km of regions where forest was at greatest risk of clearance. PSA contracts were only slightly more likely to be near such a region than far from it. This meant, they argued, that PSA contracts were not being targeted to regions where they could have the most impact.

Moreover, because enrollment was voluntary, most landowners applying for payments likely had land unprofitable for agriculture and were not actually planning to clear forest for this purpose (**FIGURE 1**). In a 2008 paper, these researchers compared lands under PSA contracts with similar lands not under contracts. PSA lands experienced no forest loss, whereas the deforestation rate on non-PSA lands was 0.21%/yr. However, their analyses indicated that PSA lands stood only a

0.08%/yr likelihood of being cleared in the first place, suggesting that the program prevented only 0.08%/yr of forest loss, not 0.21%/yr. Other research was bearing this out; at least two studies found that many PSA participants, when interviewed, said they would have retained their forest even without the PSA program.

These researchers argued that Costa Rica's success in halting forest loss was likely due to other factors. In particular, Forest Law 7575, which had established the PSA system, had also banned forest clearing nationwide. This top-down government mandate, assuming it was enforceable, in theory made the PSA payments unnecessary. However, the PSA program made the mandate far more palatable to legislators, and Forest Law 7575 might never have passed had it not included the PSA payments.

Despite the PSA program's questionable impact on preserving existing forest, scientific studies show that it has been effective in regenerating new forest. In Costa Rica's Osa Peninsula, Rodrigo Sierra and Eric Russman of the University of Texas at Austin found in 2006 that PSA farms had five times more regrowing forest than did non-PSA farms. Interviews with farmers indicated that the program encouraged them to let land grow back into forest if they did not soon need it for production.

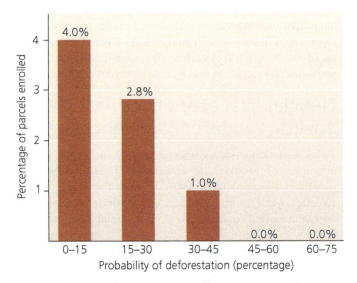

FIGURE 1 In areas at greater risk of deforestation, lower percentages of land parcels were enrolled in the PSA program. This is because land more profitable for agriculture was less often enrolled. *Data from Pfaff, A., et al., 2008.* Payments for environmental services: Empirical analysis for Costa Rica. *Working Papers Series SAN08-05, Terry Sanford Institute of Public Policy, Duke University.*

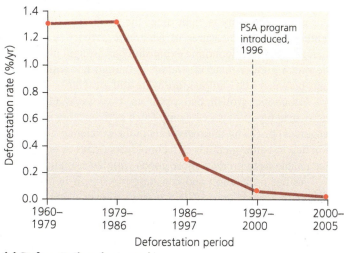

(a) Deforestation decreased

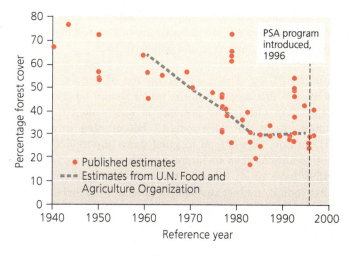

(b) Forest cover began increasing

FIGURE 2 Forest recovery was underway in Costa Rica before the PSA program began. Deforestation rates **(a)** had already dropped steeply, indicating that other factors were responsible. Forest cover **(b)** also began increasing shortly before the program's initiation (the graph gathers together estimates from many sources). *Data from (a) Sanchez-Azofeifa, G.A., et al., 2007. Costa Rica's payment for environmental services program: Intention, implementation, and impact.* Conservation Biology *21: 1165–1173. (b) Kleinn, C., et al., 2002. Forest area in Costa Rica: A comparative study of tropical forest cover estimates over time.* Environmental Monitoring and Assessment *73: 17–40.*

In the nation's northern Caribbean plain, a team led by Wayde Morse of the University of Idaho combined satellite data with on-the-ground interviews, finding that PSA payments plus the clearance ban had reduced deforestation rates from 1.43%/year to 0.10%/year and that the program encouraged forest regrowth still more. Meanwhile, dissertation work by Rodrigo Arriagada indicated that the regeneration of new forest seemed to be the PSA program's major effect at the national level as well.

Most researchers today hold that Costa Rica's forest recovery results from a long history of conservation policies and economic developments. Indeed, deforestation rates had been dropping steeply before the PSA program was initiated (**FIGURE 2**). There are several major reasons:

- Earlier policies (tax rebates and tax credits for timber production) encouraged forest cover.
- The creation of national parks fed a boom in ecotourism, so Costa Ricans saw how conserving natural areas could bring economic benefits.
- Falling market prices for meat discouraged ranching.
- After an economic crisis roiled Latin America in the 1980s, Costa Rica ended subsidies that had encouraged ranchers and farmers to expand into forested areas.

To help the PSA program make better use of its money, most researchers today feel that PSA payments should be targeted. Instead of paying equal amounts to anyone who applies,

FONAFIFO should prioritize applicants, or pay more money, in regions that are ecologically most valuable or that are at greatest risk of deforestation.

In a 2008 paper in the journal *Ecological Economics*, Tobias Wünscher of Bonn, Germany, and colleagues modeled and tested seven possible ways to target the payments, using data from Costa Rica's Nicoya Peninsula. They found that all seven approaches would improve on the existing program. The best approach paid some landowners more than others while selecting sites based on the perceived value of their ecological services and the apparent risk of deforestation. This approach more than doubled the economic efficiency of the program.

To determine how much money to offer each landowner, Wünscher's team suggested using auctions, in which applicants for PSA funds put in bids stating how much they were requesting. Because applicants have outnumbered available contracts 3 to 1, FONAFIFO could favor the lower bids to keep costs low, while the auction system could make differential payments politically acceptable.

The Costa Rican government is responding to suggestions from researchers by aiming PSA payments toward regions of greater environmental value and by making the program more accessible to small low-income farmers in undeveloped regions. The government has also raised the payment amounts considerably. Researchers—and other nations—are watching closely to see how the program develops.

If market prices do not take the social, ecological, or economic costs of environmental degradation into account, then taxpayers bear the burden of paying them. When economists ignore external costs, this creates a false impression of the consequences of our choices, and people continue to be unjustly subjected to the impacts of activities in which they did not participate. External costs are one reason that governments develop environmental policy (p. 162).

Discounting The third assumption of neoclassical economics is that an event in the future should be granted less value than one in the present. In economic terminology, future effects are "discounted." **Discounting** is meant to reflect how people tend to grant more importance to present conditions than to future conditions. Just as you might rather have an ice cream cone today than be promised one next month, market demand is greater for goods and services that are received sooner. Economists quantify this by assigning discount rates when calculating costs and benefits. For example, applying a 5% annual discount rate to forestry decisions means that a stand of trees whose timber is worth $500,000 on the market today would drop in perceived value by 5% each year. From the perspective of today's market, having the timber in 10 years would be worth only $299,368. By this logic, the more quickly the trees are cut, the more they are worth.

Discounting encourages policymakers to play down the long-term consequences of decisions. Many environmental problems unfold gradually, yet discounting discourages us from addressing resource depletion, pollution buildup, and other cumulative impacts. Instead, discounting shunts the costs of dealing with such problems onto future generations. Discounting has emerged as a flashpoint in the debate over how to respond to climate change. Economists agree that climate change will impose major costs on society, but they differ on how much to discount future effects—and so they differ on how much we should invest today to battle climate change (see **THE SCIENCE BEHIND THE STORY**, pp. 146–147).

Growth The fourth assumption of neoclassical economics pertains to **economic growth,** which can be defined as an increase in an economy's production and consumption of goods and services. Neoclassical economics assumes that economic growth is essential for maintaining social order, because a growing economy can alleviate the discontent of the poor by creating opportunities for poor people to become wealthier. A rising tide raises all boats, as the saying goes; if we make the overall economic pie larger, then each person's slice can become larger (even if some people still have much smaller slices than others).

To the extent that economic growth is a means to an end—a path to greater human well-being—it is a good thing. However, when growth becomes an end in itself it may no longer be the best route toward well-being. Sociologists have coined a word for the way that consumption and material affluence often fail to bring people contentment: *affluenza.* Moreover, people in poverty may feel poorer and less happy as the gap between rich and poor widens, even if they gain

more wealth themselves. Critics of the growth paradigm maintain that endless growth cannot be sustained, because resources to support growth are ultimately limited.

How sustainable is economic growth?

Our global economy is eight times the size it was just a half-century ago. All measures of economic activity are greater than ever before. Economic expansion has brought many people much greater material wealth (although not equally, and gaps between rich and poor are wide and growing).

Economic growth can occur in two ways: (1) by an increase in inputs to the economy (such as more labor or natural resources) or (2) by improvements in the efficiency of production due to better methods or technologies (ideas or equipment that enable us to produce more goods with fewer inputs).

As our population and consumption rise, it is becoming clearer that we cannot sustain growth forever using the first approach. Nonrenewable resources are finite in quantity, and renewable resources can also be exhausted if we overexploit them (as is happening with many fisheries today). As for the second approach, we have used technological innovation to push back the limits on growth time and again. More efficient technologies for extracting minerals, fossil fuels, and groundwater allow us to mine these resources more fully with less waste. Better machinery in our factories speeds our manufacturing. We continue to make computer chips more powerful while also making them smaller (**FIGURE 6.11**). In these ways, we are producing more goods and services with relatively fewer resources.

Can we conclude, then, that human ingenuity and technology will allow us to overcome all environmental constraints and continue economic growth indefinitely? Answering "yes" are people sometimes referred to as **Cornucopians.** (In Greek mythology, a *cornucopia*—literally "horn of plenty"—is a magical goat's horn that overflowed with grain, fruit, and flowers.) Responding "no" are people called **Cassandras,** named after the mythical princess of Troy with the gift of prophecy, whose dire predictions were not believed.

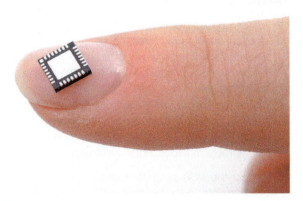

FIGURE 6.11 The miniaturization of computer chips is a striking advance in efficiency. It allows us to store and use far more information with far less input of raw materials.

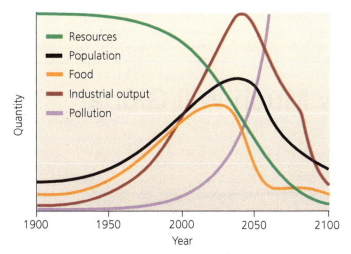

(a) Projection based on status quo policies

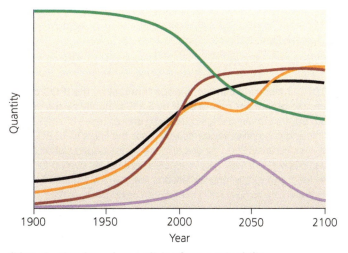

(b) Projection based on policies for sustainability

FIGURE 6.12 **Environmental scientists have used data to project future trends in resource availability, human population, food production, industrial output, and pollution.** Shown in **(a)** is a projection for a world in which society pursues policies typical of the 20th century, but allowing for twice as many resources to be discovered and extracted as were accessible at the time. In this projection, population and production rise until declining resources cause them to fall suddenly, while pollution continues to rise. In contrast, **(b)** shows a scenario with policies aimed at sustainability. In this projection, population levels off below 8 billion, production and resource availability stabilize, and pollution declines to low levels. *Data from Meadows, D., et al., 2004.* Limits to growth: The 30-year update. *White River Junction, VT: Chelsea Green Publishing. Used by permission.*

DATA Q
- For each of the five lines of data, describe how the pattern differs in the two graphs. That is, what happens to (1) resources, (2) population, (3) food, (4) industrial output, and (5) pollution in the projection for sustainability, part **(b)**, compared with the projection for the status quo, part **(a)**?
- Now suppose the researchers created a third projection, one that simulated a world in which we used resources even faster and more intensively than we do today. Draw a graph showing how you predict each of the five trend lines would appear.

Go to **Interpreting Graphs & Data** on **Mastering**EnvironmentalScience®

The Cassandra view has been articulated most famously by a group of researchers who published a series of books including *The Limits to Growth* (1972), *Beyond the Limits* (1992), and *Limits to Growth: The Thirty-Year Update* (2004). Starting from the premise that Earth's natural capital is finite, and using calculations of resource availability and consumption, they ran simulation models to predict how our economies would fare in the future. Model runs that used data matching our society's current consumption patterns consistently predicted economic collapse as resources become scarce (**FIGURE 6.12a**). The team experimented to see what parameters would produce model runs predicting a sustainable civilization (**FIGURE 6.12b**). They used these results to make recommendations for achieving sustainability.

Cornucopians, such as the economist Julian Simon and the statistician Bjorn Lomborg, counter that Cassandras underestimate the human capacity to innovate and the degree to which technologies can expand our access to resources. They also argue that market forces help us avoid resource depletion. As resources become scarce, prices rise, and individuals and firms gain incentive to turn to different resources, shift to different products, or reuse and recycle. All of these actions will alleviate pressure on the resource. As Simon put it in his 1981 book, *The Ultimate Resource:*

> The natural world allows, and the developed world promotes through the marketplace, responses to human needs and shortages.... The main fuel to speed our progress is our stock of knowledge, and the brake is our lack of imagination. The ultimate resource is people—skilled, spirited, and hopeful people who will exert their wills and imaginations for their own benefit, and so, inevitably, for the benefit of us all.

What accounts for this divergence in views between Cornucopians and Cassandras? It may be because we are living in a unique time of transition. Throughout our long history as a species, we have lived in a world with low numbers of people. In such an "empty world," we could always rely on being able to exploit more resources. Today, however, the human population is so large that it is straining Earth's systems and depleting its resources. In this new "full world," we are encountering limits. Because we are still learning what those limits are, people have a wide diversity of viewpoints.

As we will see with many issues, both Cornucopians and Cassandras make valid points. Cassandras have often underestimated our ability to innovate and adapt—yet ultimately, nonrenewable resources are finite, and renewable resources can be exploited only at limited rates. If our population and consumption continue to grow and we do not enhance reuse and recycling, we will continue depleting our natural capital, putting greater and greater demands on our capacity to innovate.

weighing the
ISSUES

Cornucopian or Cassandra?

Would you consider yourself more of a Cornucopian or a Cassandra? Which aspect(s) of each point of view do you share? Do you think our society would be better off if everyone were a Cornucopian or if everyone were a Cassandra—or do we need both?

Ethics and Economics: How Much Will Global Climate Change Cost?

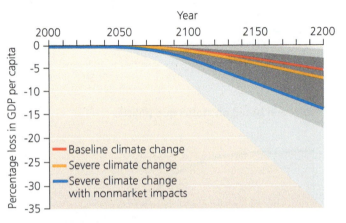

British economist Sir Nicholas Stern

How much will it cost our society if we do nothing about global climate change? The first major effort to address this question came when the British government commissioned economist Nicholas Stern to lead a team to assess the economic costs of a changing climate. But once Stern's report was published in 2006, a debate ensued that had as much to do with ethics as with economics. The dispute centered on discounting (p. 144), an economic practice heavy with ethical implications.

To produce the *Stern Review on the Economics of Climate Change*, Stern's research team surveyed the burgeoning scientific literature on the impacts of rising temperatures, changing rainfall patterns, rising sea level, and increasing storminess (see Chapter 18). It then estimated the economic consequences of these climatic changes and tried to put a price tag on the global cost. The report concluded that without action to forestall it, climate change would cause annual losses in global Gross Domestic Product (GDP; see p. 150) of 5–20% by the year 2200 (**FIGURE 1**).

Stern's team also calculated that by paying just 1% of GDP annually, starting now, our society could stabilize atmospheric greenhouse gas concentrations and prevent most of these future economic losses. (Two years later, with climate change growing worse, he revised this estimate upward to 2%.) The bottom line is this: Spending a relatively small amount of money today will save us much larger expenses in the future. This conclusion caught the attention of governments worldwide; for the first time, leading economists were advancing a strong economic argument for tackling climate change immediately.

FIGURE 1 Baseline climate change forecast by the IPCC could decrease global per capita GDP by 5.3% annually by the year 2200. Severe climate change could bring annual losses of 7.3%—and adding nonmarket values (p. 148) raises this figure to 13.8%. Gray-shaded areas show ranges of future values judged statistically to be 95% likely; darkest gray indicates where ranges overlap for all three lines, and lightest gray for just one line. *Data from HM Treasury, 2007. Stern review on the economics of climate change. London, U.K.*

The *Stern Review*'s conclusions depend partly on how one chooses to weigh future impacts versus current impacts. Stern used two discount factors. One accounted for the likelihood that people in the future will be richer than we are today and thus better able to handle economic costs. The other considered whether the future should be discounted simply because it is the future. This latter discount factor (called a *pure time discount factor*) is essentially an ethical issue because it assigns explicit values to the welfare of future versus current generations.

The *Stern Review* used a pure time discount rate of 0.1%. This means that an impact occurring next year is judged 99.9% as important as one occurring this year. It means that the welfare of a person born 100 years from now is valued at 90% of one born today. This discount rate treats current and

Environmental and ecological economists devise strategies for sustainability

The Cornucopian view has long held sway over mainstream economics, but today many economists are concluding that growth may be unsustainable if we do not reduce our demand for resources and make resource use far more efficient. Economists in the field of **environmental economics** feel we can modify neoclassical economic principles to address environmental challenges, and thereby attain sustainability within our current economic systems. Environmental economists were the first to develop methods to tackle the problems of external costs and discounting.

future generations *nearly* equally. Future generations are down-weighted only because of the (very small) possibility that our species could go extinct (in which case, there would be no future generations to be concerned about).

Many economists viewed this discount rate as too low. Yale University economist William Nordhaus proposed a discount rate starting at 3% and falling to 1% in 300 years. He maintained that such numbers are more objective because they reflect how people value things, as revealed by prices we pay in the marketplace. Indeed, when economists assess capital investments or construction projects (say, development of a railroad, dam, or highway) they typically choose discount rates close to those Nordhaus suggests. Nordhaus argued that the *Stern Review*'s near-zero discount rate overweighted the future, forcing people today to pay too much to address hypothetical future impacts.

Responding to Nordhaus and other critics, Stern's team argued that discount rates of 3% or 1% may be useful for assessing development projects but are too high for long-term environmental problems that directly affect human well-being. A 3% rate means that a person born in 1995 is valued only half as much as a person born in 1970. It means a grandchild is judged to be worth far less than a grandparent simply because of the dates they were born. For various reasons, Stern argued, the market should not be used to guide ethical decisions.

Stern's group also published sensitivity analyses that examined how their conclusions would vary under different discount rates. These showed that the report's main message—that it's cheaper to deal with climate change now than later—was robust across all discount rates up to at least 1.5%.

At the end of the day, the choice of a discount rate is an ethical decision on which well-intentioned people may differ. As governments, businesses, and individuals begin to invest in addressing climate change, the debate over the *Stern Review* reveals how ethics and economics remain intertwined.

Since the *Stern Review*, many more economic assessments have been made. In 2015, scientists for the U.S. Environmental Protection Agency summarized the peer-reviewed literature and calculated predicted economic impacts on the United States. Of the impacts included in their review, they estimated that reducing greenhouse gas emissions would help the nation avoid $235–334 billion in annual expected costs by the year 2050, and $1.3–1.5 *trillion* in such costs by 2100 (TABLE 1). This review used a 3% discount rate; the savings would have appeared even greater had a lower discount rate been used.

Today more and more of us are feeling the effects of climate change. Hurricane Sandy, flooding in Miami, drought in California, and other climate-related disasters are all driving home the same message: The sooner we address climate change, the better off we'll be.

TABLE 1 **Monetary Benefits of Reducing Greenhouse Gas Emissions to Minimize Climate Change in the United States**

TYPE OF IMPACT AVOIDED	YEARLY SAVINGS IN 2050	YEARLY SAVINGS IN 2100
Health	$199 billion	$1243 billion
Infrastructure	$1–7 billion	$8–19 billion
Electricity	$10–34 billion	Not estimated
Water Resources	$5–55 billion	$14–186 billion
Agriculture and Forestry	$11–13 billion	$7–13 billion
Ecosystems	$9–26 billion	$3–15 billion
TOTAL	**$235–334 billion**	**$1275–1476 billion**

Source: U.S. Environmental Protection Agency, 2015. *Climate change in the United States: Benefits of global action.* EPA 430-R-15-001.

Economists in the field of **ecological economics** feel that sustainability requires more far-reaching changes. They stress that in nature, every population has a carrying capacity (p. 65) and systems generally operate in self-renewing cycles, not in a linear or progressive manner. Ecological economists maintain that societies, like natural populations, cannot surpass environmental limitations. Many of these economists advocate economies that neither grow nor shrink, but rather are stable. Such a **steady-state economy** is intended to mirror natural ecological systems. Critics of steady-state economies assert that to halt growth will dampen our quality of life. Proponents respond that technological advances would continue, behavioral changes (such as greater use of recycling) would enhance sustainability, and wealth and happiness would rise.

Attaining sustainability will certainly require the reforms pioneered by environmental economists and may require the fundamental shifts in values, strategy, and behavior advocated by ecological economists. One approach they each take is to assign monetary values to ecosystem goods and services, to better integrate them into traditional cost-benefit analyses.

We can assign monetary value to ecosystem goods and services

Ecosystems provide us essential resources and life-support services, including arable soil, waste treatment, clean water, and clean air. Yet we often abuse the very ecological systems that sustain us. Why is this? From the economist's perspective, people overexploit natural resources and systems largely because the market assigns these entities no quantitative monetary value—or assigns values that underestimate their true worth.

Ecosystem services are said to have **nonmarket values**, values not usually included in the price of a good or service (**FIGURE 6.13**). For example, the aesthetic and recreational pleasure we obtain from natural landscapes is something of real value. Yet because we do not generally pay money for this, its value is hard to quantify and appears in no traditional measures of economic worth. Or consider Earth's water cycle (pp. 118–119): Rain fills our reservoirs with drinking water, rivers give us hydropower and flush away our waste, and water evaporates, purifying itself of contaminants and later falling as rain. This natural cycle is vital to our very existence, yet because we do not pay for it, markets impose no financial penalties when we disturb it.

In Costa Rica and elsewhere, environmental and ecological economists have sought ways to assign market values to ecosystem services. One technique, **contingent valuation**, uses surveys to determine how much people are willing to pay to protect or restore a resource. Such an approach was used to assess a proposal to develop a mine near world-famous Kakadu National Park in Australia in the 1990s. To determine how much the land was "worth" economically if preserved undeveloped, researchers interviewed citizens

(a) **Use value:** The worth of something we use directly

(b) **Existence value:** The worth of knowing that something exists, even if we never experience it ourselves

FIGURE 6.13 **Accounting for nonmarket values such as those shown here may help us make better environmental and economic decisions.**

(c) **Option value:** The worth of something we might use later

(d) **Aesthetic value:** The worth of something's beauty or emotional appeal

(e) **Scientific value:** The worth of something for research

(f) **Educational value:** The worth of something for teaching and learning

(g) **Cultural value:** The worth of something that sustains or helps define a culture

and asked how much they would be willing to pay to prevent mine development. On average, respondents said their households would pay $80 to $143 per year to prevent the predicted impacts. Multiplying these figures by the number of households in Australia, the researchers found that preservation was "worth" $435 to $777 million annually to Australia's population. These numbers exceeded the $102 million in annual economic benefits expected from mine development, so the researchers concluded that it was best to preserve the land undeveloped.

Because contingent valuation relies on survey questions and not actual expenditures, critics point out that people may volunteer idealistic (inflated) values, knowing that they will not actually have to pay the price they name. As a result, many researchers prefer to use methods that measure people's preferences as revealed by data on actual behavior. For example, to gauge how much people value parks, researchers may calculate the amount of money, time, or effort people expend to travel to parks. Economists may compare housing prices for similar homes in different settings to infer the dollar value of landscapes, views, or peace and quiet. They may also calculate how much it costs to restore natural systems that have been damaged, to replace their functions with technology, or to clean up pollution.

In Costa Rica, Taylor Ricketts of Stanford University, working with Gretchen Daily and others, studied native bees at a coffee plantation. By carefully measuring bee pollination (pp. 78, 220) and comparing the resulting coffee production in areas near forest and far from forest, this team calculated that forests were providing the farm with pollination services worth $60,000 per year.

Researchers have even set out to calculate the total economic value of all the services that oceans, forests, wetlands, and other systems provide across the world. Teams headed by ecological economist Robert Costanza have combed the scientific literature and evaluated hundreds of studies that estimated dollar values for 17 major ecosystem services (**FIGURE 6.14**). The researchers reanalyzed the data using multiple valuation techniques to improve accuracy, then multiplied average estimates for each ecosystem by the global area it occupied. Their initial analysis in 1997 was groundbreaking, and in 2014 they updated their

research. The 2014 study calculated that Earth's biosphere in total provides more than $125 trillion worth of ecosystem services each year, in 2007 dollars. This is equal to $143 trillion in 2014 dollars, an amount that exceeds the global annual monetary value of goods and services created by people!

Costanza also joined Andrew Balmford and 17 other colleagues to compare the benefits and costs of preserving natural systems intact versus converting wild lands for agriculture, logging, or fish farming. After reviewing many studies, they reported in the journal *Science* in 2002 that a global network of nature reserves covering 15% of Earth's land surface and 30% of the ocean would be worth $4.4 to $5.2 trillion. This amount is 100 times greater than the amount those areas would be worth were they to be converted for direct exploitative human use. This demonstrates, as stated in their research paper, that "conservation in reserves represents a strikingly good bargain."

Such research has sparked debate. Some ethicists argued that we should not put dollar figures on services such as clean air and water, because they are priceless and we would perish without them. Others say that arguing for conservation purely on economic grounds risks not being able to justify it whenever it fails to deliver clear economic benefits. However, backers of the research counter that valuation does not argue for making decisions on monetary grounds

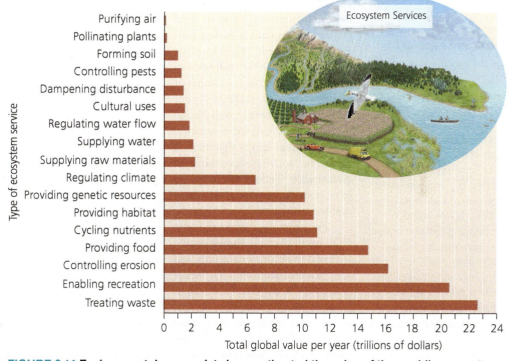

FIGURE 6.14 Environmental economists have estimated the value of the world's ecosystem services at more than $143 trillion (in 2014 dollars). This amount is an underestimate because it does not include ecosystems and services for which adequate data were unavailable. Shown are subtotals for each ecosystem service in 2007 dollars. *Data from Costanza, R., et al., 2014. Changes in the global value of ecosystem services.* Global Env. Change *26: 152–158.*

DATA Q Which three ecosystem services provide the greatest benefit to us, in dollar value, according to the data in this bar chart?

Go to **Interpreting Graphs & Data** on **Mastering**EnvironmentalScience®

alone, but instead clarifies and quantifies values that we already hold implicitly.

In 2010, researchers wrapped up a large international effort to summarize and assess attempts to quantify the economic value of natural systems. *The Economics of Ecosystems and Biodiversity* study has published a number of fascinating reports that you can download online. Regarding measuring, or valuation, of nature's economic worth, this effort concludes:

> Valuation is seen not as a panacea, but rather as a tool to help recalibrate the faulty economic compass that has led us to decisions that are prejudicial to both current well-being and that of future generations. The invisibility of biodiversity values has often encouraged inefficient use or even destruction of the natural capital that is the foundation of our economies.

We can measure progress with full cost accounting

If assigning market values to ecosystem services helps to give us a fuller and truer picture of costs and benefits, then we can take a similar approach to measuring the economic progress we make as a society. For decades, policymakers and the public have assessed each nation's economy by calculating its **Gross Domestic Product (GDP),** the total monetary value of final goods and services a nation produces each year. Governments regularly use GDP to make policy decisions that affect billions of people. However, GDP is a poor measure of economic well-being. It does not account for nonmarket values. It also lumps together all economic activity, desirable and undesirable. Thus, GDP can rise in response to economic activities that harm society.

For example, crime can boost GDP because crime forces people to invest in security measures and to replace stolen items. Oil spills (**FIGURE 6.15**) increase GDP because they require cleanups, which cost money and increase the production of goods and services. Natural disasters such as hurricanes, tornados, and earthquakes can boost GDP because of all the costs incurred in emergency measures and cleanup and rebuilding operations. War can augment GDP because of the economic activity involved in manufacturing weapons and servicing armies. Pollution causes GDP to

FIGURE 6.15 Oil spills like this one along the coast of Santa Barbara County, California, in 2015, actually increase GDP. The many monetary transactions resulting from cleanup efforts mean that pollution can increase GDP—one of many reasons GDP is not a reliable indicator of societal well-being.

FAQ

Does having more money make a person happier?

This age-old question has long been debated in the realm of philosophy. In recent years, though, social scientists have conducted serious research on the issue. So far, studies have found a surprising degree of consensus: In general, we become happier as we get wealthier, but once we gain a moderate level of wealth (roughly $50,000–$90,000 in yearly income), attaining further money no longer increases our happiness. Apparently, reaching a basic level of financial security alleviates day-to-day economic worries, but once those worries are taken care of, our happiness revolves around other aspects of our lives (such as family, friends, and the satisfaction of helping others). Research on happiness can help us guide our personal life decisions. It also suggests that enhancing a society's happiness might best be achieved by raising many people's incomes by a little, rather than by raising some people's income by a lot.

increase when the polluting substance is manufactured and again when society pays to clean up the pollution.

Environmental economists have developed indicators meant to differentiate desirable from undesirable economic activity and to better reflect our well-being. One such alternative to the GDP is the **Genuine Progress Indicator (GPI).** To calculate GPI, we begin with conventional economic activity and add to it positive contributions not paid for with money, such as volunteer work and parenting. We then subtract negative impacts, such as crime and pollution (**FIGURE 6.16a**).

GPI can differ strikingly from GDP: **FIGURE 6.16b** compares these indices for the United States across 50 years. On a per-person basis, the nation's GDP rose greatly, but its GPI remained flat for 30 years. This discrepancy suggests that Americans have been spending more and more money but that their quality of life is not improving.

The GPI is an example of **full cost accounting** (also called *true cost accounting*) because it aims to account for all costs and benefits. A variety of full cost accounting indicators have been devised. The Index of Sustainable Economic Welfare (ISEW) gave rise to the GPI. Net Economic Welfare (NEW) adjusts GDP by adding the value of leisure time and personal transactions while deducting costs of environmental degradation. The Human Development Index assesses standard of living, life expectancy, and education. Several U.S. states are beginning to use the GPI to measure progress and help guide policy. Since Maryland's governor embraced the approach in 2010, that state's GPI has grown faster than its GDP.

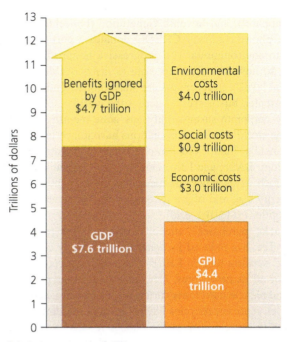

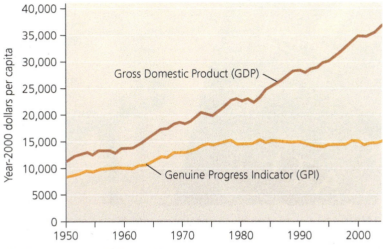

(a) Components of GPI

(b) Change in U.S. GDP vs. GPI

FIGURE 6.16 Full cost accounting indicators such as the GPI attempt to measure progress and well-being more effectively than pure economic indicators such as the GDP. We see in **(a)** how the GPI (**orange bar at right**) adds to the GDP (**reddish brown bar at left**) benefits such as volunteering and parenting (**upward gold arrow**). The GPI then subtracts external environmental costs such as pollution, social costs such as divorce and crime, and economic costs such as borrowing and the gap between rich and poor (**downward gold arrow**). Shown are values for the United States in 2004, the most recent year available. We see in **(b)** that per capita U.S. GDP has increased dramatically since 1950, yet per capita U.S. GPI leveled off after 1975. *Data from Talberth, J., C. Cobb, and N. Slattery, 2007.* The Genuine Progress Indicator 2006: A tool for sustainable development. *Oakland, CA: Redefining Progress. By permission of John Talberth, Ph.D. All data are adjusted for inflation by using year-2000 dollars.*

 DATA • What was the ratio of GDP to GPI in 1950? (Divide GDP by GPI using the values shown on the part **(b)** graph for that year.)
• What was this ratio in the year you were born?
• What was this ratio in 2004?
• How has the ratio of GDP to GPI changed through time, and what does this indicate to you?

Go to Interpreting Graphs & Data on Mastering EnvironmentalScience°

Critics of full cost accounting argue that the approach is subjective and too easily driven by ideology. Proponents respond that making a subjective attempt to measure progress is better than misapplying a more objective indicator such as the GDP to quantify well-being—something it was never meant to do.

Today, attempts are gaining ground to measure happiness (rather than economic output) as the prime goal of national policy. The small Asian nation of Bhutan pioneered this approach with its measure of Gross National Happiness. Another indicator is the Happy Planet Index, which measures how much happiness we gain per amount of resources we consume. By this measure, Costa Rica was calculated to be the top-performing nation in the world.

Costa Rica is also one of five nations working with the World Bank (p. 175) in the WAVES program (Wealth Accounting and Valuation of Ecosystem Services) to implement full cost accounting methods. Together they are addressing questions such as how much economic benefit the nation's

forests, national parks, and other natural amenities generate through tourism and watershed protection.

Markets can fail

When markets do not take into account the positive outside effects on economies (such as ecosystem services) or the negative side effects of economic activity (external costs), economists call this **market failure.** Traditionally, we have tried to counteract market failure by using government intervention. Government can restrain individual and corporate behavior through laws and regulations. It can tax harmful activities. It can also design economic incentives that use market mechanisms to promote fairness, resource conservation, and economic sustainability. Paying for the conservation of ecosystem services, as Costa Rica does, is one way of deploying economic incentives toward policy goals. (We will examine all these approaches in Chapter 7 in our discussion of environmental policy.)

FIGURE 6.17 Ecolabeling gives consumers information on the environmental impact of products, enabling us to promote sustainable business practices through our purchasing decisions. Fair-trade chocolate is just one example of the many ecolabeled products now widely available.

Ecolabeling empowers consumers

Another way to mitigate market failure is to help consumers gain better information with which to make purchasing decisions. With **ecolabeling,** sellers who use sustainable practices in growing, harvesting, or manufacturing products advertise this fact on their labels, hoping to win approval from buyers (**FIGURE 6.17**). Examples include labeling recycled paper, organic foods (pp. 257–261), dolphin-safe tuna, fair-trade and shade-grown coffee, and sustainably harvested lumber (p. 318). When labeling information is accurate, each of us as consumers can provide businesses and industries a powerful incentive to switch to more sustainable processes when we favor ecolabeled products.

In a similar vein, individuals who invest money in the stock market can pursue **socially responsible investing,** which entails investing in companies that have met criteria for environmental or social sustainability. A fast-growing trend, by 2014 U.S. investors had sunk $6.6 trillion into socially responsible investing, representing one-sixth of all investment.

Businesses are responding to sustainability concerns

As more consumers and investors express preferences for sustainable products and services, many industries, businesses, and corporations are "greening" their operations. By finding ways to enhance energy efficiency, reduce toxic substances, increase the use of recycled materials, and minimize greenhouse gas emissions, businesses often discover that they reduce costs and increase profit.

Some companies have cultivated an eco-conscious image from the start, such as Ben & Jerry's (ice cream), Patagonia

(outdoor apparel), and Seventh Generation (household products). The phone company CREDO donates a portion of its proceeds to environmental and progressive nonprofit groups according to how its customers vote to distribute the funds. At the local level, entrepreneurs are starting thousands of sustainably oriented businesses all across the world.

Today corporate sustainability has gone mainstream, and some of the world's largest corporations have joined in, including McDonald's, Intel, Ford Motor Company, Toyota, IKEA, Dow, DuPont, BASF, and IBM. Google uses renewable energy to meet much of its power demand, has taken many steps to reduce energy use, and offers its employees sustainable transportation options (**FIGURE 6.18**).

Starbucks purchases "ethically sourced" fair-trade and organic coffee, has reduced its water and electricity consumption, and is seeking to gain LEED green-building certification (p. 343) for all its new stores. Hewlett-Packard runs programs to reuse and recycle used toner cartridges, electronics, and plastics; and Dell reuses or recycles 98% of its nonhazardous waste. Sprint aims to recycle 90% of the phones it sells by crediting customers who turn in old phones. Nike, Inc., collects more than 1.5 million used sneakers each year, recycling the materials to create synthetic surfaces for basketball courts, tennis courts, and running tracks. Microsoft became carbon-neutral by charging each of its divisions with monitoring and offsetting its greenhouse gas emissions.

Wal-Mart provides the highest-profile example of corporate greening efforts. Advocates of sustainability had long criticized the world's largest retailer for its environmental and social impacts. The company responded by launching a quest to sell organic and sustainable products, reduce packaging and use recycled materials, enhance fuel efficiency in its truck fleet, reduce energy use in its stores, power itself with renewable energy, cut carbon dioxide emissions, and preserve 1 acre of natural land for every acre developed. It

FIGURE 6.18 Google offers employees free bicycles to ride to work. Google is just one of thousands of corporations working to make its operations more sustainable.

also developed a "sustainability index" that rates products it carries to help inform eco-conscious consumers. Since then, Wal-Mart has met some of its goals and fallen short on others, while saving money through efficiency measures. Many observers remain skeptical of Wal-Mart's commitment to sustainability and point to unfulfilled promises. Yet, if the corporation delivers on its stated goals, the social and environmental benefits could be substantial. Wal-Mart's vast global reach and its ability to persuade suppliers to alter their ways to retain its business mean that the changes it enacts tend to have far-reaching impacts.

Of course, corporations exist to make money for their shareholders, so they cannot be expected to pursue goals that do not turn a profit. Moreover, many corporate greening efforts are more rhetoric than reality, pursued mostly for public relations purposes. Such corporate **greenwashing** can mislead consumers into thinking a company is acting more sustainably than it actually is. The bottled water industry presents a stark example of greenwashing. Advertising with words such as "pure" and "natural" and images of forests and alpine springs leads us to believe that bottled water is cleaner and healthier for us to drink. In reality, bottled water is often less safe than tap water, the plastic bottles are a major source of waste, considerable amounts of oil are burned to transport the bottles, and the industry depletes aquifers in local communities (pp. 396–399).

In the end, it is up to all of us in our roles as consumers to encourage trends in sustainability by rewarding those businesses that truly promote sustainable solutions. This is one way that each of us can express the ethical values we cherish through the economic system in which we live.

Sustainable Development

Today's search for sustainable solutions centers on **sustainable development,** economic progress that maintains resources for the future. The United Nations defines sustainable development as development that "meets the needs of the present without sacrificing the ability of future generations to meet their own needs." Sustainable development is an economic pursuit shaped by policy and informed by science. It is also an ethical pursuit because it asks us to manage our resource use so that future generations can enjoy similar access to resources.

Sustainable development involves environmental protection, economic well-being, and social equity

Economists use the term **development** to describe the use of natural resources for economic advancement (as opposed to simple subsistence, or survival). Development involves making purposeful changes intended to improve the quality of human life. The construction of homes, schools, hospitals, power plants, factories, and transportation networks represents an example of development. In the past, "sustainable development" might have been viewed as an oxymoron—a phrase that contradicts itself. Advocates of

development felt that protecting the environment threatened people's economic needs, whereas advocates of environmental protection held that development degrades the environment, thereby jeopardizing the very improvements for human life that were intended. Today, however, people increasingly perceive how we all depend on a healthy and functional natural environment.

We also now recognize that society's poorer people tend to suffer the most from environmental degradation. As a result, advocates of environmental protection, economic development, and social justice began working together toward common goals. This cooperation gave rise to the modern drive for sustainable development, which seeks ways to promote social justice, economic well-being, and environmental quality at the same time (**FIGURE 6.19**). Governments, businesses, industries, and organizations pursuing sustainable development aim to satisfy a **triple bottom line,** a trio of goals including economic advancement, environmental protection, and social equity.

Programs that pay for ecosystem services are one example of a sustainable development approach that seeks to satisfy a triple bottom line. Costa Rica's PSA program aims to enhance its citizens' well-being by conserving the country's natural assets while compensating affected landholders for any economic losses. The intention is to achieve a win-win-win result that pays off in economic, social, and environmental dimensions.

Of course, designing sustainable solutions to complex problems is never simple, and people differ in what they mean by "sustainable development." Proponents of a school of thought called *weak sustainability* feel that we can allow natural capital to decline as long as human-made capital increases to compensate for it. In contrast, proponents of *strong sustainability* insist that human-made capital cannot substitute for natural capital and that we must not allow natural capital to diminish.

FIGURE 6.19 Sustainable development occurs when social, economic, and environmental goals overlap.

Sustainable development is global

Sustainable development has blossomed as an international movement. The United Nations, the World Bank, and other global organizations (p. 175) sponsor conferences, fund projects, publish research, and facilitate collaboration across borders among governments, businesses, and NGOs.

The Earth Summit at Rio de Janeiro, Brazil, in 1992 was the world's first major gathering focused on sustainable development. With representatives from more than 200 nations, this conference gave rise to several notable achievements, including the Convention on Biological Diversity (p. 291) and the Framework Convention on Climate Change (p. 506). Ten years later, nations reconvened in Johannesburg, South Africa, at the 2002 World Summit on Sustainable Development. Then in 2012, the world returned to Rio de Janeiro for the Rio+20 conference (p. 174).

In 2015, world leaders met at the United Nations and adopted 17 **Sustainable Development Goals** for humanity (TABLE 6.1). Each broad goal for sustainable development has a number of specific underlying targets—169 in all—that may be met by implementing concrete strategies.

Many of the Sustainable Development Goals were given a 2030 target date, and we are making better progress on some than on others. We still have a long way to go to resolve the many challenges facing humanity. Pursuing solutions that meet a triple bottom line of environmental, economic, and social goals can help pave the way for a truly sustainable world.

TABLE 6.1 U.N. Sustainable Development Goals

- End poverty in all its forms everywhere
- End hunger, achieve food security, and promote sustainable agriculture
- Promote health and well-being for all
- Ensure quality education for all
- Achieve gender equality and empower women and girls
- Ensure water and sanitation for all
- Ensure access to affordable, reliable, and sustainable energy sources
- Promote jobs and sustainable economic growth
- Build resilient infrastructure, promote sustainable industry, and foster innovation
- Reduce inequality within and among nations
- Make cities safe, resilient, and sustainable
- Ensure sustainable consumption and production
- Take urgent action to combat climate change and its many impacts
- Conserve marine resources
- Protect and restore terrestrial ecosystems and halt biodiversity loss
- Promote peaceful, inclusive, just institutions
- Renew partnerships for sustainable development

Adapted from the United Nations Division for Sustainable Development (www.un.org/sustainabledevelopment).

closing
THE LOOP

As our global society begins to chart a path toward long-term sustainability, we must pay heed to people's economic needs and to their ethical values. Science can tell us a great deal about how to conserve resources and maintain ecosystem services, but to achieve a sustainable civilization we must integrate this scientific knowledge with a keen understanding of ethical and economic concerns.

The nation of Costa Rica provides one useful model of a pathway toward sustainable development. A series of decisions by its political leaders has enabled the nation to make impressive progress in social, economic, and environmental dimensions. By paying farmers and ranchers to preserve and restore forest on private land, for example, the country's citizenry reaps the rewards of a cleaner and healthier environment, which in turn has enhanced economic progress. As Costa Rica's leaders use research-based feedback to refine and improve the PSA program, that program should be able to accomplish its goals more efficiently and effectively. The government, businesses, and people of Costa Rica recognize how economic health depends on environmental protection, and seem poised to build on their success so far.

Across the world, policymakers, economists, businesspeople, scientists, and everyday people are pulling their societies in a sustainable direction in all kinds of ways. Ecolabeling, ecotourism, the valuation of ecosystem services, corporate sustainability efforts, and alternative means of measuring progress are all helping to bring economic approaches to bear on environmental protection and resource conservation. As our economic approaches reorient us toward sustainability, modern economics renews some of its historic ties to ethics. And as people's ethical concerns expand outward to encompass more cultures, creatures, and even nonliving entities, our societies are increasingly pursuing the concept of distributional equity, or equal treatment, the aim of environmental justice. One type of distributional equity is equity among generations. Concern among today's generations for the well-being of tomorrow's generations is the basis of sustainability and sustainable development.

REVIEWING Objectives

You should now be able to:

Describe how culture and worldview influence the choices people make

Culture and personal experience influence a person's worldview. Factors such as religion, political ideology, and shared experiences shape our perspectives on the environment and the choices we make. (pp. 132–133)

Discuss the nature and historical expansion of environmental ethics in Western culture

Environmental ethics applies ethical standards to relationships between people and aspects of their environments. We value things for utilitarian reasons or for their own sake. Anthropocentrism values people above all else, whereas biocentrism values all life, and ecocentrism values ecological systems. The industrial revolution in Western culture inspired philosophical reactions that fed into environmental ethics. (pp. 133–136)

Compare and contrast major approaches in environmental ethics

The preservation ethic values preserving natural systems intact, whereas the conservation ethic promotes responsible long-term use of resources. Environmental justice seeks equal treatment for people of all income levels, races, and ethnicities. (pp. 136–139)

Explain how our economies exist within the environment and how economies rely on ecosystem services

Economies depend on the ecological systems around them for natural resources and ecosystem services. The depletion of resources and the degradation of ecosystem services threaten our economic well-being. (pp. 139–140)

Identify principles of classical and neoclassical economics and summarize their implications for the environment

Classical economics proposes that individuals acting for their own economic good can benefit society as a whole, and provides a philosophical basis for free-market capitalism. Neoclassical economics focuses on supply and demand and quantifies costs and benefits, but four of its assumptions tend to worsen environmental impacts. Conventional economic theory has promoted infinite economic growth, with little regard to resource depletion or environmental impact. (pp. 140–141, 144–145)

Describe aspects of environmental economics and ecological economics

Environmental economists advocate reforming economic practices to promote sustainability; ecological economists advocate going further by pursuing a steady-state economy. Assigning monetary value to ecosystem goods and services can help reduce external costs and make market prices reflect full costs and benefits. Full cost accounting indicators aim to measure economic progress and human well-being more effectively than GDP. (pp. 146–151)

Discuss how individuals and businesses can help move our economic system in a sustainable direction

Consumer choice in the marketplace, facilitated by ecolabeling, can encourage businesses to modify their operations to pursue more sustainable practices. (pp. 152–153)

Define sustainable development and the "triple bottom line," and discuss how sustainable development is pursued worldwide

Sustainable development promotes people's economic advancement while using resources to satisfy today's needs without compromising future needs. The United Nations has played a key role in promoting sustainable development globally. Advocates of sustainable development pursue a "triple bottom line" of environmental, economic, and social goals in a coordinated way. (pp. 153–154)

TESTING Your Comprehension

1. What does the study of ethics encompass? Describe and differentiate instrumental value and intrinsic value. What is environmental ethics?

2. Compare and contrast anthropocentrism, biocentrism, and ecocentrism. Explain how individuals with each perspective might evaluate the development of a shopping mall atop a wetland in your town or city.

3. Differentiate the preservation ethic from the conservation ethic. Explain the contributions of John Muir and Gifford Pinchot in the history of environmental ethics.

4. Describe Aldo Leopold's land ethic. How did Leopold define the "community" to which ethical standards should be applied?

5. Explain the concept of environmental justice. Give an example of an inequity relevant to environmental justice that you believe exists in your city, state, or country.

6. Name and describe two key contributions that the environment makes to the economy.

7. Describe four ways in which critics hold that neoclassical economic approaches can worsen environmental problems.

8. Compare and contrast the views of neoclassical economists, environmental economists, and ecological economists, particularly regarding the issue of economic growth.

9. What are ecosystem services? Give several examples. Describe ways in which some economists have assigned monetary values to ecosystem services.

10. Define sustainable development. What is meant by the triple bottom line? Why is it important to pursue sustainable development?

SEEKING Solutions

1. Describe your worldview as it pertains to your relationship with your environment. How do you think your culture has influenced your worldview? How do you think your personal experience has affected it? How do you think your worldview shapes your decisions?

2. Explore the U.S. EPA's online tool for environmental justice data at www.epa.gov/ejscreen. Launch the EJScreen Tool and go to a map of your own city. Click on "Map Data" and then "Map EJ Indexes." Now click on "Demographic Indicators" and see which areas of your city have the greatest densities of minority residents and of low-income residents. ("Demographic Index" averages these two variables.) Next click on "Environmental Indicators," and map each of the listed indicators one by one. Describe what patterns you see. Are any environmental indicators most severe in the neighborhoods with high percentages of minority or low-income people? Which indicators show such a pattern?

   What do you think accounts for the patterns you see? Do you perceive any problems revealed by these data? What could be done to alleviate such problems?

3. Do you think that a steady-state economy is a practical alternative to our current approach that prioritizes economic growth? Why or why not?

4. Do you think we should attempt to quantify and assign market values to ecosystem services? Why or why not? What consequences might this have?

5. **CASE STUDY CONNECTION** Suppose you are a Costa Rican farmer who needs to decide whether to clear a stand of forest or apply to receive payments to preserve it through the PSA program. Describe all the types of information you would want to consider before making your decision. Now, describe what you think each of the following people would recommend to you if you were to go to them for advice: (a) a preservationist, (b) a conservationist, (c) a neoclassical economist, and (d) an ecological economist.

6. **THINK IT THROUGH** You have just returned from serving in the U.S. Peace Corps in Costa Rica, where you worked closely with farmers, foresters, ecologists, and policymakers on issues related to Costa Rica's PSA program. You have now been hired as an adviser on natural resource issues to the governor of your state. Think about the condition of the forests, soil, and water in your state. Given what you learned in Costa Rica, would you advise your governor to institute a program to pay citizens to conserve ecosystem services? Why or why not? Describe what policies you would advocate.

CALCULATING Ecological Footprints

The population of the United States grew from 152,271,000 to 292,892,000 between 1950 and 2004, but the nation's GDP grew still faster. Taking these two trends together, Figure 6.16b (p. 151) shows how per capita GDP rose over this half-century. During this period, the per capita Genuine Progress Indicator (GPI) grew as well, but more slowly than GDP.

Given the values of the major components of the GPI for the United States in 1950 and 2004, and the population figures above, calculate and enter the per capita rates for these components, as well as the overall GPI values. Refer to Figure 6.16a (p. 151) to check how to calculate the GPI.

COMPONENTS OF GPI	U.S. TOTAL IN 1950 (TRILLIONS OF DOLLARS)	PER CAPITA IN 1950 (THOUSANDS OF DOLLARS)	U.S. TOTAL IN 2004 (TRILLIONS OF DOLLARS)	PER CAPITA IN 2004 (THOUSANDS OF DOLLARS)
GDP	1.153		7.589	
Benefits	1.041		4.746	
Environmental costs	0.407		3.990	
Social and economic costs	0.476		3.926	
GPI				

Data from Talberth, J., C. Cobb, and N. Slattery, 2007. *The Genuine Progress Indicator 2006: A tool for sustainable development*. Oakland, CA: Redefining Progress.

1. How many times greater was the GDP in 2004 than in 1950? By how many times did the GPI increase between 1950 and 2004? What does this comparison tell you?

2. By how many times, respectively, did benefits, environmental costs, and social and economic costs increase between 1950 and 2004? How are trends in each of these components driving the overall trend between the GPI and the GDP? Which component has gotten the worst over the years? How would you account for these trends?

3. There are many ways to define and measure the benefits and costs that go into the GPI. How do you think a person with a biocentric worldview would measure these differently from a person with an anthropocentric worldview? Whose GPI for the year 2005 would likely be higher?

4. Now consider your own life. Very roughly, what would you estimate are the values of the benefits, environmental costs, and social and economic costs you experience? What could you do to help improve these trends in your own personal accounting?

MasteringEnvironmentalScience®

Students Go to **MasteringEnvironmentalScience** for assignments, the etext, and the Study Area with practice tests, videos, current events, and activities.

Instructors Go to **MasteringEnvironmentalScience** for automatically graded activities, current events, videos, and reading questions that you can assign to your students, plus Instructor Resources.

CHAPTER 7

Environmental Policy

Making Decisions and Solving Problems

Fracking the Marcellus Shale

Dimock, Pennsylvania

UNITED STATES

> **Hydraulic fracturing and horizontal drilling have opened up a new era of energy security, job growth, and economic strength.**
> American Petroleum Institute director Erik Milito

> **There's no safe way to put toxic chemicals into the ground and control them.**
> New York schoolteacher Elizabeth Bouiss

When the men from Cabot Oil and Gas Corporation came to the small town of Dimock in rural Pennsylvania, many of Dimock's 1500 residents were happy to sign on to the contracts the company offered. In exchange for the right to drill for natural gas on their land, Cabot would pay them royalties on sales of the gas extracted from drilling pads placed on their property. To some in this small community, the gas payments and the potential for jobs promised economic security.

Soon the new drilling sites around Dimock were producing the largest quantities of natural gas from anywhere in the Marcellus Shale, the vast gas-bearing rock formation that underlies portions of Pennsylvania, New York, West Virginia, and Ohio (**FIGURE 7.1**, p. 160). Money and jobs from the gas boom kept Dimock economically afloat, even as other towns in the region reeled from recession and cut funding for schools and basic services.

Yet despite the economic gains, some Dimock residents began having second thoughts about drilling. Their once-quiet community was now experiencing noise, light, and air pollution; heavy truck traffic; and toxic wastewater spills. Soon, many people's drinking water began to turn brown, gray, or cloudy with sediment, and chemical smells began wafting from their water wells. On New Year's Day, 2009, Norma Fiorentino's well exploded. Methane had built up in her well water, and a spark from a motorized pump set off a potentially lethal blast.

Residents blamed the drilling technique that Cabot Oil and Gas was using: hydraulic fracturing. Citizens who could no longer drink their own well water appealed for help. Retired schoolteacher Victoria Switzer approached Cabot, local political leaders, and the Pennsylvania Department of Environmental Protection (DEP) but was turned away by them all. Then she went to the news media, and the story got out. Documentary filmmaker Josh Fox came to town and filmed residents setting their methane-contaminated tap water on fire. His 2010 film *Gasland* won numerous awards, and Dimock became Ground Zero in the burgeoning national debate over hydraulic fracturing.

Across the United States, virtually all the easily accessible oil and natural gas has already been discovered and extracted. To extract more, we've needed to develop ever more powerful technology to reach petroleum deposits deeper underground, deeper underwater, and at lower concentrations.

 Hauling fracking wastewater—one of many jobs created by shale gas extraction

Upon completing this chapter, you will be able to:

- Describe environmental policy and discuss its societal context

- Explain the role of science in policymaking

- Discuss the history of U.S. environmental policy and summarize major U.S. environmental laws

- List institutions that influence international environmental policy and describe how nations handle transboundary issues

- Compare and contrast the different approaches to environmental policymaking

Actor and anti-fracking activist Mark Ruffalo holds up water from a Dimock, Pennsylvania, well during a protest at New York's capitol building in Albany.

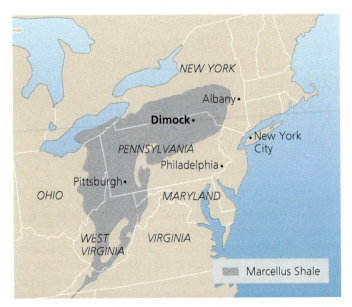

FIGURE 7.1 The Marcellus Shale formation underlies large portions of Pennsylvania, New York, Ohio, and West Virginia.

In formations such as the Marcellus Shale, natural gas is locked up in tiny bubbles dispersed throughout the shale rock. The technique of hydraulic fracturing is now making this **shale gas** accessible.

Hydraulic fracturing (also called *hydrofracking*, or simply *fracking*) involves drilling deep into the earth and then angling the drill horizontally once a shale formation is reached. An electric charge sets off targeted explosions that perforate the drilling pipe and create fractures in the shale. Drillers then pump a slurry of water, sand, and chemicals down the pipe under great pressure. The sand lodges in the fractures and holds them open, while some of the liquids return to the surface as wastewater. Natural gas trapped in the shale migrates into the fractures and rises through the drilling pipe to the surface, where it is collected in tanks (**FIGURE 7.2**).

In this way, fracking boosts the amount of natural gas that can be extracted from a rock formation. In recent years, fracking has allowed us to extract so much additional gas from the Marcellus Shale, the Barnett Shale in Texas, the Haynesville Shale in Louisiana, the Fayetteville Shale in Arkansas, and other rock formations that it has ignited a boom in domestic U.S. natural gas production. This boom has employed thousands of people and has driven down the price of natural gas. Expanded use of domestically produced natural gas has reduced the United States' reliance on coal for electricity. Because natural gas is cleaner-burning than coal, burning it in place of coal reduces the greenhouse gas emissions that drive climate change.

For these reasons—and also because of the powerful political influence of oil and gas corporations—policymakers have encouraged fracking. They have removed many regulatory constraints that would normally apply; fracking has been exempted from seven major federal environmental laws that protect public health, including the National Environmental Policy Act and the Safe Drinking Water Act. As a result, gas companies do not need to report the chemical additives they plan to use in fracking, nor do they need to test for chemical compounds drawn up in fracking wastewater. (Much of this wastewater is radioactive because drillers add radioisotopes [p. 24] as tracers to the fracking fluids they inject and also because naturally occurring radioisotopes are brought up from deep underground.) Consequently, neither regulators nor policymakers nor scientists nor homeowners have access to data to make fully informed judgments about potential health or environmental effects of fracking.

In such a climate of uncertainty, people in places like Dimock are left to wonder, worry, and argue. Victoria Switzer and several dozen families in town brought lawsuits against Cabot, but many of their neighbors became angry with them for doing so. Residents whose water was not affected—and those who felt that the benefits of receiving gas payments outweighed the health risks from pollution—blamed townspeople who spoke out about water quality for bringing media attention to Dimock and driving down property values.

With outside influences disrupting the community and causing neighbor to turn against neighbor, some hoped that government could help by enforcing measures to protect public welfare and to ensure that people are treated equally and justly. After recognizing that for some Dimock families the water was undrinkable, Pennsylvania's DEP eventually fined Cabot and required the company to pay for clean drinking water to be hauled in from outside town. However, once Tom Corbett, an avid supporter of fracking, was elected as Pennsylvania's governor, the DEP allowed Cabot to end the water shipments.

As publicity built and activists complained that Pennsylvania's government was not protecting its citizens, the U.S. Environmental Protection Agency (EPA) stepped in, sending federal researchers to run tests of Dimock's water. Results showed elevated levels of several chemicals that could threaten health in 5 out of 64 wells tested, but the EPA stated that Cabot was addressing these impacts. Some residents and scientists suspected that industry and politicians were influencing the EPA's conclusions. The EPA denied such influence and embarked on a nationwide study of the health and environmental impacts of hydraulic fracturing on drinking water, issued in 2015.

The debates in Dimock have been occurring throughout Pennsylvania and across the nation as fracking has spread. People from Ohio to California are experiencing impacts on their water, land, and air because of fracking, and are trying to weigh these against the economic promises. State governments have approached hydraulic fracturing in various ways. Pennsylvania and New York offer a study in contrast. Both neighboring states sit atop the Marcellus Shale, which holds some of the richest gas deposits in America. Pennsylvania's political leaders welcomed the gas industry with open arms, exempting it from regulations. New York's leaders, in contrast, chose to ban fracking entirely. Vermont and Maryland have also banned fracking, whereas Ohio, Texas, Louisiana, Wyoming, and other states have encouraged it.

People differ in their views on the proper role of government, but most would agree that public policy should protect people's equality of opportunity, promote their economic advancement, and safeguard them from undue harm. As we discuss environmental policy, we will see how society is struggling to balance these aims in the case of natural gas extraction.

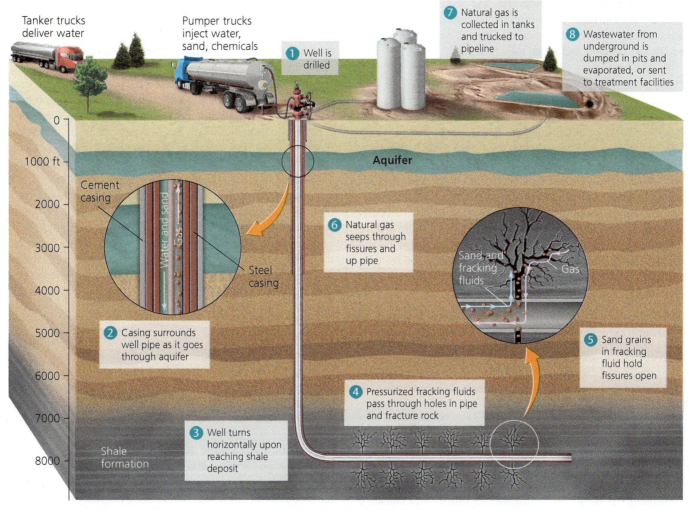

FIGURE 7.2 Hydraulic fracturing is used to extract natural gas trapped in shale deposits deep underground. ❶ A well is drilled with ❷ protective casing and ❸ is turned horizontally on reaching a shale deposit. ❹ Pressurized fluids fracture the rock, and ❺ sand lodges in the cracks, holding them open for ❻ natural gas to seep into them and rise through the pipe to ❼ tanks at the surface where the gas is collected in tanks. ❽ Wastewater also rises to the surface and is piped to treatment facilities or to wastewater pits, where it is left to evaporate.

Environmental Policy: An Overview

When a society recognizes a problem, its leaders may try to resolve the problem using policy. **Policy** consists of a formal set of general plans and principles intended to address problems and guide decision making in specific instances. **Public policy** is policy made by governments, including those at the local, state, federal, and international levels. Public policy consists of laws, regulations, orders, incentives, and practices intended to advance societal well-being. **Environmental policy** is policy that pertains to our interactions with our environment. Environmental policy generally aims to regulate resource use or reduce pollution to promote human welfare and/or protect natural systems.

Forging effective policy requires input from science, ethics, and economics. Science (Chapter 1) provides information and analyses needed to identify and understand problems and devise potential solutions. Ethics and economics (Chapter 6) offer criteria by which to assess problems and help clarify how society might address them.

The ongoing debate over hydraulic fracturing across the United States illustrates how science, economics, and ethics each inform and motivate policymaking. Science has enabled the technological advances that allow us to find and extract fossil fuels—and science also helps us understand the impacts of fossil fuel extraction and use, including air pollution, water pollution, and health effects. In economic terms, hydraulic fracturing for shale gas helps us exploit a valuable fuel resource that powers our economy and can substitute for dirtier fuel sources. For communities near drilling sites, shale gas extraction supplies a short-term boost in jobs and income but also introduces health risks and long-term costs of pollution cleanup. Ethically, fracking can pose problems if drilling by private companies pollutes water and air on which the public relies. Such conflicts commonly result from development, and environmental policy tries to address them by balancing the benefits of economic advancement against the costs of impacts on human health, ecological systems, and social well-being.

Environmental policy addresses issues of fairness and resource use

Because market capitalism is driven by incentives for short-term economic gain, it provides businesses and individuals little motivation to minimize environmental impacts, seek long-term social benefits, or equalize costs and benefits among parties. Market prices often do not reflect the value of environmental contributions to economies or the full costs imposed on the public by private parties when their actions degrade the environment (Chapter 6). Such market failure (p. 151) has traditionally been viewed as justification for government involvement. Governments typically intervene in the marketplace for several reasons:

- To provide social services, such as national defense, health care, and education
- To provide "safety nets" (for the elderly, the poor, victims of natural disasters, and so on)
- To eliminate unfair advantages held by single buyers or sellers
- To manage publicly held resources
- To minimize pollution and other threats to health and quality of life

Environmental policy aims to protect people's health and well-being, to safeguard environmental quality and conserve natural resources, and to promote equity or fairness in people's use of resources.

The tragedy of the commons When publicly accessible resources are open to unregulated exploitation, they tend to become overused, damaged, or depleted. So argued environmental scientist Garrett Hardin in his 1968 essay "The Tragedy of the Commons." Basing his argument on an age-old scenario, Hardin explained that in a public pasture (or "common") open to unregulated grazing, each person who grazes animals will be motivated by self-interest to increase the number of his or her animals in the pasture. Because no single person owns the pasture, no one has incentive to expend effort taking care of it. Instead, each person takes what he or she can until the resource is depleted and overgrazing causes the pasture's food production to collapse, harming everyone. This scenario, known as the **tragedy of the commons,** pertains to many types of resources held and used in common by the public: forests, fisheries, clean air, clean water—even global climate.

When shared resources are being depleted or degraded, it is in society's interest to develop guidelines for their use. In Hardin's example of a common pasture, guidelines might require pasture users to help restore and manage the resource or might limit the number of animals each person can graze. These two concepts—management and restriction of use—are central to environmental policy today.

Public oversight through government is a standard way to alleviate the tragedy of the commons, but this dilemma can also be addressed in other ways. One is a bottom-up cooperative approach, in which users of the resource band together and cooperate to prevent overexploitation. Indeed, many traditional societies over the centuries have developed ways to manage resources cooperatively and sustainably at the community level. Another approach is privatization, in which the resource is subdivided and allotments are sold into private ownership, so that each owner has incentive to conserve his or her portion of the resource in ways that maximize its productivity.

The cooperative approach may work if the resource is localized and enforcement is simple, but often these conditions do not hold. Privatization may work if property rights can be clearly assigned (as with land), but it tends not to work with resources such as air or water. Privatization also opens the door to short-term profit taking at the long-term expense of the resource. Thus, in many cases public oversight and regulation by democratic government are likely the best ways to avoid the tragedy of the commons.

Free riders A second reason we develop policy for publicly held resources is the predicament of the **free rider.** Let's say a community on a river suffers from water pollution that emanates from 10 different factories. The problem could in theory be solved if every factory voluntarily agreed to reduce its own pollution. However, once they all begin reducing their pollution, it becomes tempting for any one of them to stop doing so. A factory that avoids the efforts others are making would in essence get a "free ride." If enough factories take a free ride, the whole collective endeavor will collapse. Because of the free-rider problem, private voluntary efforts are often less effective than efforts mandated by public policy.

External costs Environmental policy also aims to promote fairness by dealing with external costs (p. 141), harmful impacts suffered by people not involved in the actions that created them. For example, a factory that discharges waste into a river imposes external costs (water pollution, health problems, reduced fish populations, aesthetic impacts) on downstream users of the river. Likewise, shale gas drilling operations may impose a variety of external costs on people living nearby by polluting drinking water (**FIGURE 7.3**), polluting air, causing noise pollution and light pollution, and even causing earthquakes (pp. 40–41). Contamination of drinking water by methane from natural gas, or by the many chemicals used in fracking fluids in the drilling process, can disrupt people's lives and affect their health (see **THE SCIENCE BEHIND THE STORY**, pp. 164–165).

When governments take action to force industries to protect water quality, clean up their pollution, pay fees, or reimburse residents for damage, this helps to "internalize" costs. The costs are paid by the companies, which will likely pass them on to consumers by raising the prices of their products. Higher market prices, in turn, may reduce demand for the products, and consumers may instead favor less-expensive products that impose fewer costs on society.

These goals of environmental policy—to protect resources against the tragedy of the commons and to promote fairness

FIGURE 7.3 Pollution from shale gas drilling creates external costs. This man in Dimock, Pennsylvania, noticed contamination of his drinking water once fracking began nearby.

by eliminating free riders and addressing external costs—are reflected in today's diversity of approaches to environmental policy. As an example, the **polluter-pays principle** specifies that a party responsible for pollution should be held responsible for covering the costs of its impacts. This principle helps protect resources such as clean water and air, promotes just treatment of all parties, and helps shift external costs into the market prices of goods and services.

Various factors can obstruct environmental policy

If environmental policy brings clear benefits, then why are environmental laws and regulations often challenged? One reason is the perception that environmental protection equates to economic sacrifice (see FAQ, p. 140). Businesses and individuals often view regulations as restrictive, bureaucratic, or costly. Landowners may fear that zoning (p. 337) or protections for endangered species (p. 289) will restrict how they can use their land. Developers complain of time and money lost in obtaining permits; reviews by government agencies; and required environmental controls, monitoring, and mitigation.

Another hurdle for environmental policy stems from the nature of environmental problems, which often develop gradually over long time periods. In contrast, human behavior is geared toward addressing short-term needs, and this is reflected in our social institutions. Businesses usually opt for short-term financial gain. The news media focus coverage on new and sudden events. Politicians often act in their short-term interest because they depend on re-election every few years. For all these reasons, environmental policy goals that attract wide public support may nonetheless be obstructed.

Policy in general can be held up for a variety of reasons—even if a majority of people favor it. Translating any given policy idea into reality is long, hard work. The checks and balances in a constitutional democracy seek to ensure that new policy is implemented only after it has gone through extensive review and discussion—and in general, this is a very good thing.

However, a number of other factors often hinder the making of policy that a majority of citizens would favor. In democracies such as the United States, each person has a political voice and can make a difference—yet money tends to wield influence. People, organizations, industries, or corporations with enough wealth to buy access to power can exert disproportionate influence over policymakers. Elected officials can become reliant on campaign contributions from wealthy donors, and money also funds lobbyists who seek to influence policymakers on behalf of clients with financial interests in certain political outcomes.

Another reason that advances in policy can be stalled is the movement of individuals between government and the private sector, known as the **revolving door.** Some individuals employed in industry gain political influence when they take jobs with the government agencies responsible for regulating their industries. Conversely, businesses often hire former government officials who had regulated their industries. As an example, after Pennsylvania regulators tried to strengthen oversight of wastewater from gas drilling, three of the regulators left their government posts and went to work for the gas industry. As another example, before becoming U.S. vice president, Dick Cheney was chairman and CEO of Halliburton, a leading energy services company that helped to pioneer hydraulic fracturing. After taking office in 2001, Cheney convened an energy task force that met 40 times with industry officials, yet only once with environmental advocates, and never publicly revealed its proceedings. Many policies friendly to fossil fuel industries followed, including language inserted into the Energy Policy Act of 2005 that exempted hydraulic fracturing from the Safe Drinking Water Act.

Defenders of the revolving door assert that corporate executives who take government jobs regulating their own industry bring with them an intimate knowledge of the industry that makes them highly qualified and likely to benefit society with well-informed policy. Critics contend that taking a job regulating your former employer is a clear conflict of interest that undermines the effectiveness of the regulatory process.

Science informs policy but is sometimes disregarded

Economic interests, ethical values, and political ideology all influence the policy process, but environmental policy that is effective is generally informed by scientific research. For

weighing the
ISSUES

Internalizing External Costs

Imagine that we were to use policy to internalize all the external costs of gasoline (pollution, health risks, climate change, impacts from oil drilling and transport, etc.) and that as a result, gas prices rise to $13 per gallon. What effects do you think this would have on the choices we make as consumers (such as driving behavior and types of vehicles purchased)? What influence might it have on the types of vehicles produced and the types of energy sources developed? What effects might it have on our taxes and our health insurance premiums? In the long run, do you think that internalizing external costs in this way would end up costing society more money or saving society money? What factors might be important in determining the outcome?

Does Fracking Contaminate Drinking Water?

A Pennsylvania homeowner sets fire to her tap water

When the U.S. Environmental Protection Agency (EPA) issued its long-awaited assessment on the potential impacts of hydraulic fracturing on drinking water in 2015, everyone rushed to interpret the news. The gas industry proclaimed that the report showed fracking to be safe. Environmental advocates countered that the report documented many instances of water contamination. What the EPA report actually said, in carefully phrased language, was that it "did not find evidence that [fracking has] led to widespread, systemic impacts on drinking water resources in the United States." Of course, not finding evidence is not proof that no impacts exist.

In reality, the EPA and the scientists whose research it summarized were under so many constraints that they could not collect data adequate to answer the questions they'd set out to address. In particular, most researchers were unable to obtain data on water quality from sites *before* gas drilling began. (When such information exists, it is generally collected by the gas industry and not made available to independent researchers.) Without having these baseline data, it is hard to know whether chemicals found in drinking water *after* drilling began actually resulted from the drilling.

Consequently, researchers have taken other approaches, such as comparing water quality near and far from drilling sites. Dr. Robert Jackson of Duke University became interested in this approach after watching Pennsylvanians setting their methane-rich tap water on fire in the documentary film *Gasland*. His research team visited private wells that draw drinking water from aquifers that lie above the Marcellus Shale formation in Pennsylvania. With homeowners' permission, researchers collected water samples from 59 wells located within 1 km of active gas drilling sites and 82 wells more than 1 km away from the nearest drilling site. The researchers took the water samples to the lab and analyzed them for their chemical constituents.

Jackson's team found no evidence that well water was being contaminated by chemicals from the fracking process or by salty fluids rising up from deep underground as a result of fracking. The chemistry of samples near and far from drilling sites was similar, and these data were also similar to water samples collected from nearby regions by other researchers in past years.

However, when the researchers tested for methane, the main component of natural gas, they found that concentrations averaged six times higher in wells near natural gas drilling sites than in wells far from drilling sites (**FIGURE 1**). Methane dissolved in water is not considered toxic, but methane is an asphyxiating and flammable gas. Eleven samples from near drilling sites exceeded the level at which explosions are deemed a risk, and experts recommend remediation. "When the methane concentrations are that high," Jackson told one science reporter, "the water can bubble like champagne." Levels of ethane and propane were also high only in the samples near drilling sites. This research made a splash when it was published in 2013 in the prestigious scientific journal *Proceedings of the National Academy of Sciences* (*PNAS*).

In 2015 a research group led by Donald Siegel of Syracuse University produced results that challenged Jackson's conclusions. Siegel's group tested more than 11,300 well water samples from northeast Pennsylvania—far more than any previous study—and found that samples near drilling sites contained no more methane than samples far from drilling sites. Siegel's team obtained so many samples by cooperating with Chesapeake Energy Corporation, which supplied them with data it had collected. However, soon after this paper's publication in the journal *Environmental Science and Technology*, reporters discovered that two of its authors had failed to reveal potential conflicts of interest. Siegel had been paid by Chesapeake Energy to conduct the study, and one of his coauthors was a former employee of the company. For many, this called into question the reliability of the paper.

A vital question is where the methane in people's drinking water is coming from. Jackson's team asked whether the gas in the samples they had studied was *thermogenic* methane formed by heat and pressure deep underground or *biogenic* methane produced at shallow depths by microorganisms. They analyzed ratios of isotopes (p. 23) of carbon and hydrogen in the methane

instance, when deciding whether to regulate a substance that may pose a public health risk, regulatory agencies such as the Environmental Protection Agency (EPA) comb the scientific literature for information and may commission new studies to research unresolved questions. When crafting a bill to reduce pollution, a legislator may use data from scientific studies to quantify the cost of the pollution or the predicted benefits of its reduction. In today's world, a nation's strength depends on its commitment to science. This is why governments devote a portion of their tax revenue to fund scientific research.

Unfortunately, sometimes policymakers allow ideology, rather than science, to determine policy on scientific

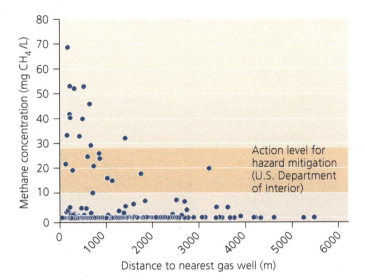

FIGURE 1 **Methane concentrations were far higher at water wells near gas drilling sites (left end of graph) than at wells far from them (right end of graph).** *Data from Jackson, Robert B., et al., 2013. Increased stray gas abundance in a subset of drinking water wells near Marcellus shale gas extraction. Proc. Natl. Acad. Sci. USA 110(28): 11250–11255.*

DATA Q Let's say you are going to sample two new water wells for methane, one just 250 m from a gas drilling site and one 5000 m from a drilling site. Based on the data shown here, predict which well would most likely have methane concentrations that are above the "Action level for hazard mitigation." Explain how you can conclude this.

Go to **Interpreting Graphs & Data** on **Mastering**EnvironmentalScience®

wells are apparently allowing gas pulled up from deep below to escape near the surface.

If methane can leak from drilling holes into drinking water, then what about the cocktails of chemicals that make up fracking fluids, some of which are toxic or carcinogenic? A 2013 study in Texas examined samples from 95 private drinking water wells near drilling sites over the Barnett Shale formation and compared them to samples far from drilling sites and to historical data collected by the state of Texas decades earlier. Researchers Brian Fontenot, Kevin Schug, and others from the University of Texas at Arlington found levels of dissolved solids and several heavy metals to be much greater near the drilling sites, but they could not tie the difference definitively to drilling.

A 2015 study in *PNAS* was the first to find clear evidence of aquifer pollution by fracking chemicals after they had migrated laterally, along with methane, through 1–3 km of rock above the Marcellus Shale. This research, by Garth Llewellyn of Appalachia Consulting, Susan Brantley of Pennsylvania State University, and others, used sophisticated chemical analysis to identify numerous compounds in water samples from three Pennsylvania homes and then match these to chemicals from nearby fracking operations. The researchers inferred that the contamination resulted either from underground migration from a borehole or from a leak of flowback fluid at a gas pad.

Today researchers are increasingly studying air pollution resulting from fracking. Methane is a powerful greenhouse gas (p. 482), and methane that escapes from drilling operations enters the atmosphere and contributes to global climate change. Anthony Ingraffea of Cornell University is a leading researcher on methane leaks and air pollution. In 2014 he led a team that analyzed compliance reports from drillers to the Pennsylvania DEP for 41,000 oil and gas wells drilled between 2000 and 2012. They found instances of "loss of well integrity" in 1.9% of the wells. Wells using hydraulic fracturing showed 57% more integrity problems than wells not using fracking, as well as six times more problems with cement or casing. Wells in northeast Pennsylvania (where Dimock is located) had the worst records, with nearly 10% of the wells showing loss of structural integrity. The resulting methane leaks can potentially reach air, water, or both.

Researchers across the spectrum agree that more research is needed into the potential sources of air and water pollution from fracking, so that industry can improve its safeguards and government can protect people's health while enabling energy development.

as well as ratios of methane to other hydrocarbon gases. Their results showed that the methane from well water near drilling sites was thermogenic gas originating deep underground. In fact, its chemistry was consistent with that of the Marcellus Shale underlying the region, suggesting that it resulted from drilling.

Still, the Marcellus Shale lies 1–2 km below the surface, whereas drinking water wells extend at most only 200 m underground. How could methane from so far down reach these shallow wells? A series of scientific studies by Jackson and others has now shown that leaks from faulty or damaged steel and concrete casings that surround the drilling boreholes of gas

matters. Politicians may ignore scientific consensus on well-established matters such as evolution, vaccination, or climate change if it suits their political needs or if they are motivated chiefly by political or religious ideology. Some may reject scientific advice if this helps to please campaign contributors or powerful constituencies. For these reasons, taxpayer-funded

science often is suppressed or distorted for political ends. As a result, we cannot simply take for granted that science will play a role in policy. Each of us can help ensure that our representatives in government make proper use of our scientific knowledge by re-electing those who do and voting out of office those who do not.

U.S. Environmental Law and Policy

The United States provides a good focus for understanding environmental policy in constitutional democracies worldwide, for several reasons. First, the United States has pioneered innovative environmental policy. Second, U.S. policies serve as models—of both success and failure—for other nations and for international bodies. Third, the United States exerts a great deal of influence on the affairs of other nations. Fourth, understanding U.S. policy at the federal level helps us to understand it at local, state, and international levels.

Federal policy arises from the three branches of government

Federal policy in the United States results from actions of the three branches of government—legislative, executive, and judicial—established under the U.S. Constitution. Statutory law, or **legislation,** is created by Congress, which consists of the Senate and the House of Representatives. For instance, to deal with water pollution across the United States, Congress passed the Federal Water Pollution Control Acts of 1965 and 1972, and then in 1977 passed the Clean Water Act. These laws regulated the discharge of wastes, especially from industry, into rivers and streams, and thereby improved water quality markedly across the nation.

Bills introduced by legislators may be shepherded from subcommittee through committee and on to passage by Congress (**FIGURE 7.4**). If a bill passes through all of these steps, it may become law if the president, the head of the executive branch, signs it. Legislation is thereby enacted (approved) or vetoed (rejected) by the president, whose veto may be overridden by a two-thirds vote of Congress.

A bill can die in countless fashions along the way, however, and only a small proportion of bills ever become law. For instance, legislators from New York, Pennsylvania, and Colorado in 2009 and several years since have introduced a bill to restore Safe Drinking Water Act regulations on hydraulic fracturing and to require the gas industry to disclose all chemicals it uses in fracking (chemicals known to number more than 1000). Amid opposition from the industry, this bill for the so-called FRAC Act (Fracturing Responsibility and Awareness to Chemicals Act) has so far been unable to secure enough votes to make it out of committee.

Once a law is enacted, its implementation and enforcement is assigned to an administrative agency within the executive branch. Administrative agencies (**TABLE 7.1**) are sometimes nicknamed the "fourth branch" of government. They create **regulations,** specific rules or requirements intended to help achieve the objectives of the more broadly written statutory law. For instance, in 2015 the U.S. Department of the Interior issued new regulations on hydraulic fracturing on public lands, intended to better control practices on the 100,000 oil and gas wells drilled on federally managed lands.

Administrative agencies also monitor compliance with laws and regulations and enforce them when they are violated. For instance, the EPA is an administrative agency that

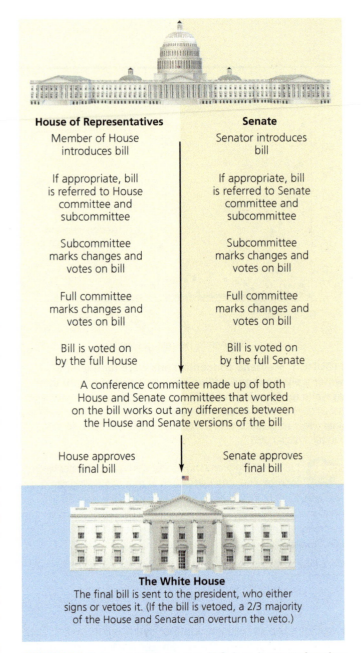

House of Representatives	Senate
Member of House introduces bill	Senator introduces bill
If appropriate, bill is referred to House committee and subcommittee	If appropriate, bill is referred to Senate committee and subcommittee
Subcommittee marks changes and votes on bill	Subcommittee marks changes and votes on bill
Full committee marks changes and votes on bill	Full committee marks changes and votes on bill
Bill is voted on by the full House	Bill is voted on by the full Senate

A conference committee made up of both House and Senate committees that worked on the bill works out any differences between the House and Senate versions of the bill

House approves final bill	Senate approves final bill

The White House
The final bill is sent to the president, who either signs or vetoes it. (If the bill is vetoed, a 2/3 majority of the House and Senate can overturn the veto.)

FIGURE 7.4 Before a bill becomes U.S. law, it must clear hurdles in both legislative bodies. If the bill passes the House and Senate, a conference committee works out differences between versions before the bill is sent to the president. The president may then sign or veto the bill.

regulates some aspects of fossil fuel extraction and waste handling under various laws, including the Clean Water Act.

The president may also issue **executive orders,** specific legal instructions for government agencies. In 2012 President Barack Obama issued an executive order requiring oil and gas companies to publicly disclose the chemicals they use in hydraulic fracturing. The order required disclosure only after drilling is complete. Today gas companies report chemicals they use in fracking after the fact at an industry-funded website called FracFocus. However, more than 10% of chemicals are kept unreported as commercial trade secrets, raising questions over whether the site truly conforms to the executive order.

TABLE 7.1 Federal Administrative Agencies That Influence Environmental Policy

EXECUTIVE OFFICE OF THE PRESIDENT

Council on Environmental Quality

DEPARTMENT OF AGRICULTURE (USDA)

Natural Resources Conservation Service
U.S. Forest Service (USFS)

DEPARTMENT OF COMMERCE

Bureau of the Census
National Marine Fisheries Service
National Oceanic and Atmospheric Administration (NOAA)

DEPARTMENT OF DEFENSE

Army Corps of Engineers

DEPARTMENT OF ENERGY (DOE)

Energy Efficiency and Renewable Energy
Energy Information Administration (EIA)
Federal Energy Regulatory Commission (FERC)

DEPARTMENT OF HEALTH AND HUMAN SERVICES

Centers for Disease Control and Prevention (CDC)
Food and Drug Administration (FDA)

DEPARTMENT OF THE INTERIOR

Bureau of Indian Affairs
Bureau of Land Management (BLM)
Bureau of Reclamation
Minerals Management Service
National Park Service (NPS)
U.S. Fish and Wildlife Service (USFWS)
U.S. Geological Survey (USGS)

DEPARTMENT OF LABOR

Occupational Safety and Health Administration (OSHA)

INDEPENDENT AGENCIES

Consumer Product Safety Commission
Environmental Protection Agency (EPA)
National Aeronautics and Space Administration (NASA)
Nuclear Regulatory Commission (NRC)

Source: U.S. General Services Administration, Washington, D.C.

The judiciary, consisting of the Supreme Court and various lower courts, is charged with interpreting law. This is necessary because social norms, societal conditions, and technologies change over time, and because Congress writes laws broadly to apply to varied circumstances throughout the nation. Decisions rendered by the courts make up a body of law known as *case law.* Previous rulings serve as *precedents,* or legal guides, for later cases, steering judicial decisions through time. The judiciary is an important arena for environmental policy. Grassroots environmental advocates and organizations use lawsuits to help level the playing field with large corporations and agencies—as some residents of Dimock, Pennsylvania, have in their dispute with Cabot Oil and Gas. Conversely, the courts hear complaints from businesses and individuals challenging the constitutional validity of environmental laws they believe to be infringing on their

rights. Individuals and organizations also lodge suits against government agencies when they feel the agencies are failing to enforce their own regulations.

Courts interpret the constitutionality of policy

The U.S. Constitution lays out several principles that have come to be especially relevant to environmental policy. One of these is from the Fifth Amendment, which ensures, in part, that private property shall not "be taken for public use without just compensation." Courts have interpreted this clause, known as the *takings clause*, to ban not only the literal taking of private property but also what is known as regulatory taking. A **regulatory taking** occurs when the government, by means of a law or regulation, deprives a property owner of all or some economic uses of his or her property. Many people cite the takings clause in opposing regulations that restrict development on privately owned land. For example, some would contend that zoning regulations (p. 337) that prohibit a landowner from allowing gas drilling in a residential neighborhood deprive the landowner of an economically valuable use of the land and, therefore, may violate the Fifth Amendment.

In a landmark case in 1992, the U.S. Supreme Court ruled that a state land use law intended to "prevent serious public harm" violated the takings clause. *Lucas v. South Carolina Coastal Council* involved a developer named Lucas, who in 1986 purchased beachfront property in South Carolina for $975,000. In 1988, before Lucas began to build, South Carolina's legislature passed a law banning construction on eroding beaches (**FIGURE 7.5**). A state agency classified the Lucas property as an eroding beach and prohibited

FIGURE 7.5 Should government be able to restrict development in areas where erosion, storms, and flooding pose risks to life and property, such as on this beach? If so, does this constitute a taking of private property rights, and should the property owner be compensated? These questions were addressed in *Lucas v. South Carolina Coastal Council*.

residential construction there. Lucas contested this decision, and the Supreme Court ruled in his favor, declaring that the state law deprived Lucas of all economically beneficial uses of his land. As a result, Lucas was allowed to build homes on the land. Today, regulatory taking remains a contentious area of law—a key issue in the sensitive balance between private rights and the public good.

State and local governments also make environmental policy

The structure of the federal government is mirrored at the state level with legislatures, governors, judiciaries, and agencies. States, counties, and municipalities all generate and enforce environmental policy of their own. Policymakers at these various levels frequently interact to address issues. For instance, after monitoring by inspectors for Pennsylvania's Department of Environmental Protection began showing increasing problems with gas wells (**FIGURE 7.6**), Pennsylvania's legislature in 2012 passed a bill updating regulations on the natural gas industry, tightening safety standards, and imposing an "impact fee" on drillers to help recover some of the costs that drilling imposes. Pennsylvania had been the largest gas-producing state not to tax the industry, but as criticism of hydraulic fracturing grew and as state financial needs intensified, the fee became more desirable. Pennsylvania's law allows county and municipal governments to decide whether to impose the fee. Proceeds help to fund various state and local programs. Governor Corbett signed the bill into law, but he had also asked legislators to prevent municipalities from regulating drilling. The legislators refused, and in the resulting compromise, the law forced towns to permit drilling in all areas, including residential zones, but allowed towns to use their zoning standards to regulate noise, lighting, and structures.

Such disagreements between policymakers at different levels have flared up in other states as well. The 125,000 citizens of Denton, Texas, have seen many impacts of hydraulic fracturing for shale gas in their region, which lies above the Barnett Shale formation. In 2014, Denton's residents voted to ban fracking within city limits. However, the Texas state legislature wanted to continue encouraging fracking across the state, so it passed a law nullifying Denton's ban and limiting the power of municipalities in Texas to restrict fracking.

Ideally, states and localities can act as laboratories for experimenting with novel policy concepts, so that ideas that succeed may be adopted elsewhere. Because the neighboring Marcellus Shale states of Pennsylvania and New York have pursued different approaches to hydraulic fracturing, people throughout the United States will be able to compare results from their two approaches.

In New York, the state maintained a moratorium on fracking until 2012, when Governor Andrew Cuomo and the Department of Environmental Conservation (DEC) proposed banning fracking in the watersheds of New York City and Syracuse while allowing it in the rest of the state. More than 140 upstate New York towns responded by using their zoning

(a) Monitoring water quality

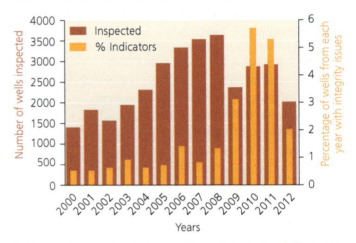

(b) Inspections and problems at Pennsylvania gas drilling sites

FIGURE 7.6 State and local governments administer environmental policy. Technicians **(a)** take water samples from a home to give regulators from Pennsylvania's Department of Environmental Protection (DEP) evidence on impacts of fracking. DEP staff also inspect gas drilling sites and cite drillers for violations, as shown by data **(b)** on inspections and well integrity issues. *Data from Ingraffea, A., et al., 2014. Assessment and risk analysis of casing and cement impairment in oil and gas wells in Pennsylvania, 2000–2012.* Proc. Natl. Acad. Sci. USA *111: 10955–10960.*

DATA What happened to the percentage of wells documented as having integrity problems after 2008? Suggest two different hypotheses for this pattern. (The researchers don't know the explanation but suggested two possibilities.)

Go to **Interpreting Graphs & Data** on **Mastering**EnvironmentalScience®

powers to enact local bans on fracking, setting up potential struggles among the governor, the legislature, and the towns to determine policy. Cuomo retreated from his stance in the face of widespread anti-fracking protests, and held off on a decision as his administration assessed new scientific reports and as polls showed his state's citizens split evenly on whether to allow fracking. Then in December 2014 the state health commissioner released a long-awaited report that found "significant public health risks" with fracking, and Cuomo responded by issuing a ban on fracking across the state. Environmental advocates and many residents of

New York City (whose water comes from regions above the Marcellus Shale) celebrated the decision. But many rural communities whose leaders desperately desire economic development were upset, and 15 towns threatened to secede from New York and join Pennsylvania.

For political leaders in New York, Pennsylvania, and 30 other states across the country, the benefits of shale gas development have been alluring. Development means jobs for rural communities and a boost in economic output. Moreover, because natural gas burns more cleanly than coal, it has gained wide appeal. Some experts view natural gas as a "bridge" between fossil fuels and cleaner renewable energy sources, but others think pursuing shale gas will simply commit us more deeply to fossil fuels and delay our transition to renewable sources. Policymakers must weigh all these considerations along with the potential risks to health and water supplies.

State laws cannot violate principles of the U.S. Constitution, and if state and federal laws conflict, federal laws take precedence. Federal policymakers may influence environmental policy at the level of the states by:

- Supplanting state law to force change. (This is uncommon.)

- Providing financial incentives to encourage change. (This can be effective if federal funding is adequate and if states need the money.)

- Following an approach of *cooperative federalism*, whereby a federal agency sets national standards and then works with state agencies to achieve them in each state. (This is most common.)

Sometimes relations between federal and state governments grow testy. When the U.S. EPA came in to test Dimock's water, the head of Pennsylvania's DEP complained that this was undue interference in state affairs. The EPA subsequently backed away from taking action in Dimock, as it also has in communities in Texas, Wyoming, and elsewhere where scientific evidence showing groundwater pollution from fracking had been collected.

In recent years, pressure to weaken federal oversight and hand over power to the states has grown, but political scientists argue that retaining strong federal control over environmental policy is a good idea for several reasons:

- Citizens of all states should have equitable protection against environmental and health impacts.

- Dealing with environmental problems often requires an "economy of scale" in resources, such that one strong national effort is far more efficient than 50 state efforts.

- Many issues involve "transboundary" disputes that cross state lines, and nationwide efforts minimize disputes among states.

As we proceed through our discussion of federal policy, keep in mind that a great deal of environmental policy is created and administered at the state and local levels.

The first wave of U.S. environmental policy promoted development

Distinct types of environmental policy were created in three different periods of U.S. history. Laws enacted during the first period, from the 1780s to the late 1800s, accompanied the westward expansion of the nation and were intended mainly to promote settlement and the extraction and use of the continent's abundant natural resources (**FIGURE 7.7**). Among these early laws were the General Land Ordinances of 1785 and 1787, by which the new federal government gave itself the right to manage the lands it was expropriating from Native Americans. As the young country accrued

(a) Settlers in Nebraska, circa 1860

(b) Loggers felling an old-growth tree, Washington

FIGURE 7.7 Early U.S. environmental policy promoted settlement and natural resource extraction.
The Homestead Act of 1862 allowed settlers **(a)** to claim 65 ha (160 acres) of public land by paying $16, living there for 5 years, and farming or building a home. The timber industry was allowed to cut the nation's ancient forests **(b)** with little policy to encourage conservation.

additional territory through purchase, treaties with European powers and Native American nations, and military conquest, these laws created a grid system for surveying these lands and readying them for private ownership. From 1785 to the 1870s, the government promoted settlement in the Midwest and West and doled out millions of acres to its citizens and to railroad companies, encouraging settlers, entrepreneurs, and land speculators to move west.

Western settlement was meant to provide U.S. citizens with means to achieve prosperity while relieving crowding in eastern cities. It expanded the geographic reach of the United States at a time when the young nation was still jostling with European powers for control of the continent. It also wholly displaced the millions of Native Americans whose ancestors had inhabited these lands for millennia. U.S. environmental policy of this era reflected a perception that the vast western lands were inexhaustible in natural resources.

The second wave of U.S. environmental policy encouraged conservation

In the late 1800s, as the continent became more populated and its resources became increasingly exploited, public policy regarding natural resources began to shift. Reflecting the emerging conservation and preservation ethics (pp. 136–137) in American society, laws of this period aimed to alleviate some of the environmental impacts of westward expansion.

In 1872, Congress designated Yellowstone as the world's first national park. In 1891, Congress authorized the president to create forest reserves to prevent overharvesting and protect forested watersheds. In 1903, President Theodore Roosevelt created the first national wildlife refuge. These acts launched the creation of a national park system, national forest system, and national wildlife refuge system that still stand as global models (pp. 310, 319). These developments reflected a new understanding that the continent's resources were exhaustible and required legal protection.

Land management policies continued through the 20th century, targeting soil conservation in the wake of the Dust Bowl (p. 224) and wilderness preservation with the Wilderness Act of 1964 (p. 319), which sought to preserve pristine lands "untrammeled by man, where man himself is a visitor who does not remain."

The third wave responded to pollution

Further social changes in the 20th century gave rise to the third major period of U.S. environmental policy. In a more densely populated nation driven by technology, heavy industry, and intensive resource consumption, Americans found themselves better off economically but living amid dirtier air, dirtier water, and more waste and toxic chemicals. Events in the 1960s and 1970s triggered greater awareness of environmental problems, bringing about a profound shift in public policy.

A landmark event was the 1962 publication of *Silent Spring*, a book by American scientist and writer Rachel

FIGURE 7.8 **Scientist and writer Rachel Carson revealed the effects of DDT and other pesticides in her 1962 book,** *Silent Spring.*

Carson (**FIGURE 7.8**). *Silent Spring* awakened the public to the ecological and health impacts of pesticides and industrial chemicals (p. 363). The book's title refers to Carson's warning that pesticides might kill so many birds that few would be left to sing in springtime.

Ohio's Cuyahoga River (**FIGURE 7.9**) also drew attention to pollution hazards. The Cuyahoga was so polluted with oil and industrial waste that the river actually caught fire near Cleveland a number of times in the 1950s and 1960s. This spectacle, coupled with an oil spill off the Pacific coast near Santa Barbara, California, in 1969, moved the public to prompt Congress and the president to better safeguard the environment and public health.

FIGURE 7.9 **Ohio's Cuyahoga River was so polluted with oil and waste that the river caught fire multiple times in the 1950s and 1960s and would burn for days at a time.**

A number of leaders in government and academia were influential in responding to the growing public desire for environmental protection. Secretary of the Interior Stewart Udall helped shape many key environmental laws and oversaw the creation of more than 100 federal parks and refuges for conservation and public use. Legal scholar Joseph Sax developed the **public trust doctrine,** which holds that natural resources such as air, water, soil, and wildlife should be held in trust for the public and that government should protect them from exploitation by private parties. Wisconsin Senator Gaylord Nelson founded Earth Day in 1970, a now-annual event, which galvanized public support for action to address pollution problems.

With the help of such leaders, public demand for a cleaner environment during this period inspired a number of major laws that underpin modern U.S. environmental policy (TABLE 7.2, p. 172). These laws helped to clean up air and water, protect rare and endangered species, and control hazardous waste and toxic substances. You will encounter most of them in later chapters of this book, and they have already helped to shape the quality of your life.

Historians suggest that major advances in environmental policy occurred in the 1960s and 1970s because (1) environmental problems became readily apparent and were directly affecting people's lives; (2) people could visualize policies to deal with the problems; and (3) the political climate was ripe, with a supportive public and leaders who were willing to act. In addition, photographs from NASA's space program allowed humanity to see, for the first time ever, images of Earth from space (see photos on pp. 2 and 661). It is hard for us today to comprehend the power those images had at the time, but they revolutionized many people's worldviews by making them aware of the finite nature of our planet.

Today, largely because of policies enacted since the 1960s, our health is better protected and the nation's air and water are considerably cleaner. Thanks to the many citizens who worked tirelessly in grassroots efforts, and to the policymakers who listened and chose to make a difference in people's lives, we now enjoy a cleaner environment where industrial chemicals, waste disposal, and resource extraction are more carefully regulated. Much remains to be done, but all of us alive today owe a great deal to the dedicated people who inspired policy to tackle pollution during this period.

NEPA and the EIS process grant citizens input

One of the foremost U.S. environmental laws is the **National Environmental Policy Act (NEPA),** drafted by Indiana University political scientist Lynton Caldwell and signed into law by Republican President Richard Nixon in 1970. NEPA created an agency called the Council on Environmental Quality and required that an **environmental impact statement (EIS)** be prepared for any major federal action that might significantly affect environmental quality. An EIS summarizes results from studies that assess the impacts on the environment that could result from a development project undertaken or funded by the federal government.

The EIS process forces government agencies and businesses that contract with them to evaluate environmental impacts using a cost-benefit approach (p. 140) before proceeding with a new dam, highway, or building project. The EIS process rarely halts development projects, but it serves as an incentive to lessen environmental damage. NEPA also grants ordinary citizens input in the process by requiring that EISs be made publicly available and that policymakers solicit and consider public comment on them.

State governments have adopted EIS processes as well. Examining New York's experience with regard to hydraulic fracturing provides an example of how the process works. In 1992 New York developed a Generic EIS (GEIS) that set parameters for permitting oil and gas wells in general. In 2009, as demand grew for shale gas extraction, a Draft Supplemental Generic EIS (SGEIS) for fracking was prepared and released for public review. The public submitted more than 13,000 comments. In 2010 Governor David Paterson ordered the Department of Environmental Conservation (DEC) to conduct further review, and in 2011 the DEC released a Revised Draft SGEIS. This time, a record-setting 67,000 public comments were received. The DEC also proposed regulations, which drew 180,000 public comments. After reviewing all this public input, the Final SGEIS was prepared, laying out parameters for how the DEC should review individual fracking applications. If New York's statewide ban on fracking is ever overturned in the future, the DEC would use these parameters and assess drilling proposals one by one with EIS processes, then would issue permits for those that meet the terms of the GEIS and the SGEIS.

Creation of the EPA marked a shift in environmental policy

Six months after signing NEPA into law, Nixon issued an executive order calling for a new integrated approach to environmental policy. "The Government's environmentally related activities have grown up piecemeal over the years," the president told Congress. "The time has come to organize them rationally and systematically." Nixon's order moved elements of agencies regulating water quality, air pollution, solid waste, and other issues into the newly created **Environmental Protection Agency (EPA).** The order charged the EPA with conducting and evaluating research, monitoring environmental quality, setting and enforcing standards for pollution

FAQ

Isn't the EPA an advocate for the environment?

Like all administrative agencies, the EPA is part of the executive branch and operates in line with the policies of the presidential administration in power at the time. As such, the EPA under one president may function very differently from the EPA under another president. Indeed, sometimes the agency may impede environmental regulations! The EPA employs many dedicated scientists who carry out careful research and make scientifically informed policy recommendations. They advise administrators appointed by the president, however, and policy decisions are ultimately made by these politically appointed administrators.

TABLE 7.2 **Major U.S. Environmental Protection Laws, 1963–1980**

Clean Air Act
1963;
amended 1970 and 1990

Sets standards for air quality, restricts emissions from new sources, enables citizens to sue violators, funds research on pollution control, and established an emissions trading program for sulfur dioxide. As a result, the air we breathe today is far cleaner (pp. 454–457).

Resource Conservation and Recovery Act
1976

Sets standards and permitting procedures for the disposal of solid waste and hazardous waste (p. 613). Requires that the generation, transport, and disposal of hazardous waste be tracked "from cradle to grave."

Endangered Species Act
1973

Seeks to protect species threatened with extinction. Forbids destruction of individuals of listed species or their critical habitat on public and private land, provides funding for recovery efforts, and allows negotiation with private landholders (pp. 289-291).

Clean Water Act
1977

Regulates the discharge of wastes, especially from industry, into rivers and streams (p. 408). Aims to protect wildlife and human health, and has helped to clean up U.S. waterways.

Safe Drinking Water Act
1974

Authorizes the EPA to set quality standards for tap water provided by public water systems and to work with states to protect drinking water sources from contamination.

Soil and Water Conservation Act
1977

Directs the U.S. Department of Agriculture to survey and assess soil and water conditions across the nation and prepare conservation plans. Responded to worsening soil erosion and water pollution on farms and rangeland as production intensified.

Toxic Substances Control Act
1976

Directs the EPA to monitor thousands of industrial chemicals and gives it power to ban those found to pose too much health risk (p. 378). However, the number of chemicals continues to increase far too quickly for adequate testing.

CERCLA ("Superfund")
1980

Funds the Superfund program to clean up hazardous waste at the nation's most polluted sites (p. 625). Costs were initially charged to polluters but most are now borne by taxpayers. The EPA continues to progress through many sites that remain. Full name is the Comprehensive Environmental Response Compensation and Liability Act.

levels, assisting the states in meeting the standards, and educating the public.

Since its inception, the EPA has played a central role in environmental policy. Industries whose practices are regulated by the EPA complain constantly of expense and bureaucracy, but studies that examine how regulations can reduce external costs show that EPA actions bring hundreds of billions of dollars of net benefits to the American public every year (p. 182). EPA regulations are frequently challenged by industry-backed lawsuits, and EPA administrators are continually pressured by policymakers, many of whom rely for campaign contributions on industries regulated by the EPA. As a result, environmental advocates often feel the EPA falls short in carrying out its mission of protecting the public from environmental hazards.

The EPA's 2015 report on fracking hazards to drinking water is a case in point; the report was hampered by the EPA's inability to obtain adequate data from an industry unwilling to provide information, and the report ended up addressing only some of the hazards. In this case the EPA also found itself at cross-purposes with the Obama administration, which was enthusiastically promoting shale gas as a way to expand domestic production of energy that is cleaner than coal. Residents of Dimock and elsewhere whose water had been polluted went to Washington, D.C., in 2015 to complain to the EPA's Scientific Advisory Board that the EPA had abandoned them and downplayed fracking's pollution risks in its report, and the Scientific Advisory Board agreed.

The social context for policy evolves

In the 1980s Congress strengthened, broadened, and elaborated on the laws of the 1970s. Major amendments were made to the Clean Water Act in 1987 and to the Clean Air Act in 1990. But the political climate in the United States soon changed. Although public support for the goals of environmental protection remained high, many people began to feel that the legislative and regulatory means used to achieve these goals too often imposed economic burdens on businesses or individuals. Increasingly, attempts were made at the federal level to roll back or weaken environmental laws. These attempts began with the administration of President Ronald Reagan and continued with the George W. Bush administration and with most congressional sessions since 1994.

For example, as policymakers sought to encourage natural gas extraction, hydraulic fracturing won exemptions from the Safe Drinking Water Act and at least six more of the nation's most fundamental environmental laws. Fracking remains exempted from key aspects of the National Environmental Policy Act, the Clean Air Act, the Clean Water Act, the Resource Conservation and Recovery Act, the Superfund Act, and the Emergency Planning and Community Right to Know Act.

As advocates of environmental protection watched their hard-won gains eroding, many began to feel that new perspectives and strategies were needed. In a provocative 2004 essay titled "The Death of Environmentalism," political consultants Michael Shellenberger and Ted Nordhaus argued that environmental advocates needed to appeal to people's core values and not simply offer technical policy fixes. They needed to stop labeling problems as "environmental" and start showing people why these problems are actually human issues that lie at the heart of our quality of life. They needed to be more responsive to people's needs and to articulate a positive, inspiring vision for the future.

Today in the United States, legal protections for public health and environmental quality remain strong in some areas but have eroded in others. Past policies restricting toxic substances such as lead and DDT have improved public health, but scientists and regulators cannot keep up with the flood of new chemicals being introduced by industry. And as energy issues move to the fore, people continue to experience impacts from fossil fuel use and extraction while striving to find a path toward cleaner energy.

Environmental policy advances today on the international stage

Amid the heightened partisanship of U.S. politics today, environmental policy has gotten caught in the political crosshairs. Environmental issues have come to be identified as a predominantly Democratic concern—despite the fact that some of the greatest early conservationists were Republicans and that the words *conservative* and *conservation* share the same root and original meaning. As a result of ideological gridlock, significant bipartisan advances now rarely occur, and U.S. leadership in environmental policy internationally has waned.

However, many other nations are forging ahead with environmental policy. Germany has used policy to make impressive strides with solar energy (pp. 577–578). Sweden maintains a thriving society while promoting progressive environmental policies (pp. 549–550). Small developing nations such as Costa Rica are bettering their citizens' lives while protecting and restoring their natural capital (p. 131). Even China, despite becoming the world's biggest polluter, is also taking the world's biggest steps toward renewable energy, reforestation, and pollution control (p. 508).

FAQ

If something is harmful, wouldn't the government have made it illegal?

This assumption vastly overestimates the power of government. In fact, we are surrounded by risky or hazardous things that remain perfectly legal. Some (such as junk food, cigarettes, and alcohol) persist because they are popular. Others (such as fossil fuel pollution) persist because mitigating the problem is costly, complicated, or grand in scale. Still others (such as many toxic chemicals) go unregulated because they are released to consumers more quickly than scientists can determine their health effects. In many cases, financially valuable products or practices that harm health or the environment have politically powerful constituencies. Corporations and industries lobby policymakers to shield their products or practices from regulation, fearing that regulation could adversely affect sales and profits. Often these powerful societal, economic, and political pressures cause health and environmental hazards to go unaddressed by government.

FIGURE 7.10 Concerns over climate change are driving environmental policy today. Here, protesters demonstrate near the Arc de Triomphe in Paris during the United Nations Climate Change Conference held in the city in December 2015.

In all nations, the pressing issue of global climate change (Chapter 18) has come to dominate discussion of environmental policy (**FIGURE 7.10**). A series of international conferences (p. 507) in recent years has brought together representatives of the world's nations to grapple with how to reduce the greenhouse gas emissions that drive climate change.

Worldwide, we have now embarked on a fourth wave of environmental policy, one focused on sustainability and sustainable development (pp. 15, 153–154). This approach aims to safeguard natural systems while raising living standards for the world's people. In 2012, world leaders met in Rio de Janeiro, Brazil, at the United Nations–sponsored Rio+20 conference, to explore the latest strategies for promoting economic vitality and social equity while safeguarding environmental quality. This conference built on the 1992 Earth Summit at Rio de Janeiro and the 2002 World Summit on Sustainable Development in Johannesburg, South Africa, each of which brought together leaders from 200 nations.

International Environmental Policy

Environmental systems pay no heed to political boundaries, so environmental problems often are not restricted to the confines of particular nations. Consider climate change: It is a global issue because carbon pollution from any one nation spreads through the atmosphere and oceans, affecting all nations. Because one nation's laws have no authority in other nations, international policy is vital to solving transboundary problems. However, international law is often more nebulous in its origins and weaker in its authority than national law. As environmental scientist Hilary French has noted, "The world is still composed of nation-states that view themselves as sovereign, meaning that international law has the force of moral suasion, but few if any real teeth."

Globalization makes international institutions vital

We live in an era of rapid and profound change. **Globalization** describes the process by which the world's societies have become more interconnected, linked by diplomacy, commercial trade, and communication technologies in countless ways. Globalization has brought us many benefits by facilitating the spread of ideas and technologies that empower individuals and enhance our lives. Billions of people enjoy a degree of access to news, education, arts, and science that we could barely have imagined in the past, and billions also now live under governments that are more democratic. A better awareness of other cultures promotes peace and understanding, and warfare is on the decline.

Yet as globalization proceeds, societies and ecological systems are altered at unprecedented rates and scales. People move organisms from one continent to another, allowing invasive species to affect ecosystems everywhere. Multinational corporations operate outside the reach of national laws and rarely have incentive to conserve resources or limit pollution while moving from nation to nation. Our biggest environmental challenges are now global in scale (such as climate change, ozone depletion, overfishing, and biodiversity loss)—yet we lack an adequate global legal framework to address these issues effectively. For all these reasons, in today's globalizing world the institutions that do shape international law and policy play increasingly vital roles.

International law includes customary law and conventional law

International law known as **customary law** arises from long-standing practices, or customs, held in common by most cultures. In contrast, international law known as **conventional law** arises from *conventions*, or *treaties* (written contracts among nations), into which nations enter. One example is the United Nations Framework Convention on Climate Change, a 1994 treaty that established a framework for agreements to reduce greenhouse gas emissions that contribute to global climate change. The Kyoto Protocol (a *protocol* is an amendment or addition to a convention) later specified the agreed-upon details of the emissions limits (p. 506). **TABLE 7.3** shows a selection of major environmental treaties.

Treaties such as the Montreal Protocol and the Kyoto Protocol are truly global in scope and have been ratified by most of the world's nations. However, treaties are also signed among pairs or groups of nations. The United States, Mexico, and Canada entered into the North American Free Trade Agreement (NAFTA) in 1994 (**FIGURE 7.11**). NAFTA eliminated trade barriers such as tariffs on imports and exports, making goods cheaper to buy. Yet NAFTA also threatened to

TABLE 7.3 Major International Environmental Treaties

CONVENTION OR PROTOCOL	YEAR IT CAME INTO FORCE	NATIONS THAT HAVE RATIFIED IT	STATUS IN U.S.
CITES: Convention on International Trade in Endangered Species of Wild Fauna and Flora (p. 291)	1975	175	Ratified
Ramsar Convention on Wetlands of International Importance	1975	159	Ratified
Montreal Protocol, of the Vienna Convention for the Protection of the Ozone Layer (p. 468)	1989	196	Ratified
Basel Convention on the Control of Transboundary Movements of Hazardous Wastes and Their Disposal (p. 623)	1992	172	Signed but not ratified
Convention on Biological Diversity (p. 291)	1993	168	Signed but not ratified
Stockholm Convention on Persistent Organic Pollutants (p. 379)	2004	152	Signed but not ratified
Kyoto Protocol, of the UN Framework Convention on Climate Change (p. 506)	2005	184	Signed but not ratified

FIGURE 7.11 The North American Free Trade Agreement (NAFTA) eliminated trade barriers to make goods cheaper. However, some U.S. manufacturing jobs moved to Mexico (such as to this garment factory in Tehuacan), where wages are lower and health and environmental regulations are more lax.

undermine protections for workers and the environment by steering economic activity to areas where regulations were most lax. Side agreements were negotiated to try to address these concerns, and NAFTA's impacts on jobs and on environmental quality in the three nations have been complex. Many U.S. jobs moved to Mexico, but fears that pollution would soar and regulations would be gutted largely did not come to pass—indeed, some sustainable products and practices spread from nation to nation.

Today these issues are being debated anew with the proposed Trans-Pacific Partnership trade agreement among Pacific Rim nations. Deliberations recur with each proposed trade agreement, as we try to find ways to gain the benefits of free trade while avoiding environmental damage and economic harm to working people.

Several organizations shape international environmental policy

In this age of globalization, a number of international organizations influence the policy and behavior of nations by providing funding, applying political or economic pressure, and directing media attention.

The United Nations Founded in 1945 and including representatives from virtually all nations of the world, the **United Nations (UN)** seeks to maintain peace, security, and friendly relations among nations; to promote respect for human rights and freedoms; and to help nations cooperate to resolve global economic, social, cultural, and humanitarian challenges. Headquartered in New York City, the United Nations plays an active role in shaping international environmental policy by sponsoring conferences, coordinating treaties, and publishing research. An agency within it, the United Nations Environment Programme (UNEP), promotes sustainability with research and outreach activities that provide information to policymakers and scientists.

The World Bank Established in 1944 and based in Washington, D.C., the **World Bank** is one of the largest sources of funding for economic development. It shapes policy by funding dams, irrigation infrastructure, and other major development projects. In fiscal year 2015, the World Bank provided

$42.5 billion in loans and support for projects designed to benefit low-income people in developing countries.

Despite its admirable mission, the World Bank is often criticized for funding unsustainable projects that cause environmental impacts, such as dams that flood valuable forests and farmland to provide electricity. Providing for the needs of growing human populations in poor nations while minimizing damage to the ecological systems on which people depend can be a tough balancing act. Environmental scientists agree that the concept of sustainable development must be the guiding principle for such efforts.

The European Union

The **European Union (EU)** seeks to promote Europe's unity and its economic and social progress and to "assert Europe's role in the world." The EU can sign binding treaties on behalf of its member nations and can enact regulations that have the same authority as national laws. The EU's European Environment Agency addresses waste management, noise pollution, water pollution, air pollution, habitat degradation, and natural hazards. The EU also seeks to remove trade barriers among member nations. It has classified some nations' environmental regulations as barriers to trade, arguing that the stricter environmental laws of some northern European nations limit the import and sale of environmentally harmful products from other member nations.

The World Trade Organization

Based in Geneva, Switzerland, the **World Trade Organization (WTO)** represents multinational corporations. It promotes free trade by reducing obstacles to international commerce and enforcing fairness among nations in trading practices. The WTO has authority to impose financial penalties on nations that do not comply with its directives. These penalties can sometimes affect environmental policy.

Like the EU, the WTO has interpreted some national environmental laws as unfair barriers to trade. For instance, in 1995 the U.S. EPA issued regulations requiring cleaner-burning gasoline in U.S. cities, following congressional amendments of the Clean Air Act. Brazil and Venezuela filed a complaint with the WTO, claiming the new rules discriminated against the petroleum they exported to the United States, which did not burn as cleanly. The WTO agreed, ruling that even though the dirtier-burning South American gasoline posed a threat to human health in the United States, the EPA's rules were an illegal trade barrier. The ruling forced the United States to weaken its regulations of gasoline. Not surprisingly, critics have frequently charged that the WTO aggravates environmental problems.

weighing the ISSUES

Trade Barriers and Environmental Protection

If Canada has stricter laws for environmental protection than Mexico, and if these laws restrict Mexico's ability to export its goods to Canada, then by WTO policy Canada's laws could be overruled in the name of free trade. Do you think this is fair? Now consider that Canada is wealthier than Mexico and that Mexico could use an economic boost. Does this affect your response?

Nongovernmental organizations

A great number of **nongovernmental organizations (NGOs)**—non-profit, mission-driven organizations not overseen by any government—have become international in scope and exert influence over policy. Those that advocate for aspects of environmental protection are known as environmental NGOs (ENGOs). Groups such as the Nature Conservancy focus on conservation objectives on the ground (such as purchasing and managing land and habitat for rare species) without becoming politically involved. Other groups, such as Conservation International, the World Wildlife Fund, Greenpeace, and Population Connection, attempt to shape policy through research, education, lobbying, or protest. NGOs apply more funding and expertise to environmental problems—and conduct more research intended to solve them—than do many national governments.

Approaches to Environmental Policy

When most of us think of environmental policy, what comes to mind are major laws, such as the Clean Water Act, or government regulations, such as those limiting what an industry can dump into a water supply. However, environmental policy is diverse.

Policy can follow three approaches

Environmental policy can use a variety of strategies, which in turn can be categorized into three major approaches (**FIGURE 7.12**).

Lawsuits in the courts Before the legislative push of recent decades, most environmental policy questions in the United States were addressed with lawsuits in the courts. Individuals suffering external costs from pollution would sue polluters, one case at a time, much as some residents of Dimock, Pennsylvania, decided to do against Cabot Oil and Gas. The courts sometimes punished polluters by ordering them to stop their operations or pay damages to the affected individuals. However, as industrialization proceeded and population grew denser, pollution became harder to avoid and judges became reluctant to hinder industry.

In 1970, the U.S. Supreme Court heard the case *Boomer v. Atlantic Cement Company*. The court ruled that residents of Albany, New York, were suffering pollution from a nearby cement plant, yet the justices refused to shut down the plant. Instead, they allowed the plant to continue operating once it paid financial compensation to the residents. The court had calculated that the economic costs to the company of controlling its pollution were greater than the economic costs of the pollution to the residents, and the court based its decision on an attempt to minimize overall costs. In handing down this ruling, the justices essentially let the market decide between right and wrong. To many people, rulings like these showed that lawsuits in themselves were no longer a viable avenue

PROBLEM
Pollution from factory harms people's health

FIGURE 7.12 Three major policy approaches exist to resolve environmental problems. To address pollution from a factory, we might ① seek damages through lawsuits, ② limit pollution through legislation and regulation, or ③ reduce pollution using market-based strategies.

SOLUTIONS
Three policy approaches

① **Lawsuits in the courts:** People can sue factories

② **Command-and-control policy:** Governments can regulate emissions

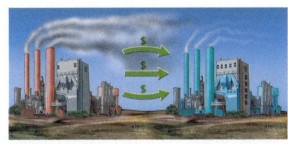

③ **Economic policy tools:** Policy can create incentives; a factory that pollutes less (right) will outcompete one that pollutes more (left) through permit trading, avoiding green taxes, or selling ecolabeled products

for preventing pollution. People began to view legislation and regulation as more effective means of protecting public health and safety.

Command-and-control policy Most environmental laws and regulations of recent decades use a **command-and-control** approach, in which a regulating agency such as the EPA prohibits certain actions—or sets rules, standards, or limits—and threatens punishment for violations. This simple and direct approach to policymaking has brought citizens of the United States and other nations cleaner air, cleaner water, safer workplaces, healthier neighborhoods, and many other advances. The relatively safe, healthy, comfortable lives most of us enjoy today owe much to the command-and-control environmental policy of the past several decades.

Even in plain financial terms, putting health and quality of life aside, command-and-control policy has been effective. Each year the White House Office of Management and Budget analyzes U.S. regulatory policy to calculate the economic costs and benefits of regulations. These analyses have consistently revealed that benefits outweigh costs and that environmental regulations have been most beneficial of all. You can explore some of these data in *Calculating Ecological Footprints* (p. 182).

Economic policy tools Despite the successes of command-and-control policy, many people have grown disenchanted with its top-down nature and have come to view government mandates as restrictions on their freedom. Moreover, whereas command-and-control policy dictates particular solutions to problems, many people feel that private entities competing

and innovating in a free market may produce new or better solutions at lower cost. As a result, political scientists, economists, and policymakers today are exploring policy approaches that aim to channel the innovation and economic efficiency of the market in directions that benefit the public. Such economic policy tools use financial incentives to encourage desired outcomes, discourage undesired outcomes, and set market dynamics in motion to achieve goals in an economically efficient manner.

All three of these approaches aim to "internalize" external costs suffered by the public by building these costs into market prices. Each approach has strengths and weaknesses, and each is best suited to different conditions. The approaches may also be used together. For instance, government regulation is often needed to frame market-based efforts, and citizens can use the courts to ensure that regulations are enforced. Let's now explore several types of economic policy tools: taxes, subsidies, and emissions trading.

Green taxes discourage undesirable activities

In taxation, money passes from private parties to the government, which uses it to pay for services to benefit the public. Taxing undesirable activities helps to internalize external costs by making these costs part of the normal expense of doing business. A tax on an environmentally harmful activity or product is called a **green tax.** When a business pays a green tax, it is essentially reimbursing the public for environmental damage it causes. One example is the impact fee that Pennsylvania's state government imposed on shale gas drillers; the fee was meant to recover some of the external costs of drilling, such as damage to roads caused by increased heavy truck traffic.

Under an effective program of green taxation, a firm owning a drilling site that pollutes groundwater, surface water, soil, or air would pay taxes on the amount of pollution it discharges—the more pollution, the higher the tax payment. This would give firms a financial incentive to reduce pollution while allowing them the freedom to decide how to do so. One polluter might choose to invest in pollution control technology if this is more affordable than paying the tax. Another polluter might instead choose to pay the tax—funds the government could use to reduce pollution in some other way.

Green taxes have yet to gain wide support in the United States, although similar "sin taxes" on cigarettes and alcohol are long-accepted tools of U.S. social policy. Taxes on pollution are more common in Europe, where many nations have adopted the polluter-pays principle (p. 163). Today there is widespread debate about whether and how to implement carbon taxes—taxes on gasoline, coal-based electricity, and fossil-fuel-intensive products according to the carbon emissions they produce—to fight global climate change (p. 509).

Green taxation provides incentive to reduce pollution not merely to some level specified in a regulation but to still lower levels. However, businesses generally pass on their tax expenses to consumers, and these increased costs may affect low-income consumers disproportionately more than high-income ones.

Subsidies promote targeted activities

Another type of economic policy tool is the **subsidy,** a government giveaway of money or resources that is intended to support or promote an industry or activity. Subsidies take many forms, and one is the *tax break*, which relieves the tax burden on an industry, firm, or individual. Because a tax break deprives the government's treasury of funds it would otherwise collect, it has the same financial effect as a direct giveaway of money.

Subsidies can promote environmentally sustainable activities. Examples include the Conservation Reserve program that pays American farmers to conserve areas of natural vegetation around active farmland (p. 230), Costa Rica's program of payment for ecological services (pp. 131–132), and Germany's feed-in tariff system for solar power (pp. 577–578). However, subsidies often are used to prop up activities that result in environmental impacts. The U.S. Forest Service

spends $35 million of taxpayer money each year building roads in national forests to allow private timber corporations access to cut trees, which the companies then sell at a profit (**FIGURE 7.13**). Under the General Mining Act of 1872 (pp. 641–642), mining companies extract $500 million to $1 billion in minerals from U.S. public lands each year without paying a penny in royalties to the taxpayers who own these lands. Since this law was enacted, the U.S. government has given away nearly $250 billion of mineral resources, and mining activities have polluted more than 40% of watersheds in the West. The 140-year-old law still allows mining companies to buy public lands for $5 or less per acre.

Fossil fuels have been a major recipient of subsidies over the years. From 1950 to 2010, the U.S. government gave $594 billion of its citizens' money to oil, gas, and coal corporations (most of this in tax breaks), according to one recent compilation (**FIGURE 7.14**). In comparison, just $171 billion was granted to renewable energy, and most of these subsidies went to hydropower and to corn ethanol, which is not widely viewed as a sustainable fuel (pp. 566–567). In recent years, global fossil fuel subsidies have outpaced renewable energy subsidies four to eight times, according to the International Energy Agency.

In 2009, President Obama and other leaders of the Group of 20 (G-20) nations resolved to gradually phase out their collective $300 billion of annual fossil fuel subsidies. Doing so would hasten a shift to cleaner renewable energy sources and accomplish half the greenhouse gas emissions cuts needed to hold global warming to 2°C. However, since that time, fossil fuel subsidies have *grown*, not shrunk, rising to $493 billion in 2014. A prime reason is that consumers have grown used to artificially low subsidized prices for gasoline and electricity

FIGURE 7.13 When companies extract timber from U.S. national forests, taxpayers pay the costs of building and maintaining access roads. "By absorbing these costs, the federal government shields the timber industry from the true cost of doing business," the nonprofit group Taxpayers for Common Sense concludes.

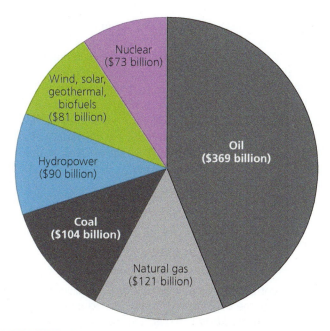

FIGURE 7.14 The well-established fossil fuel industries (slices in shades of gray) receive the vast majority of U.S. energy subsidies. Cumulative data for the United States from 1950 to 2010 are shown. Apportionments remain similar today. *Data from Management Information Services, Inc., 2011. 60 years of energy incentives: Analysis of federal expenditures for energy development. Washington, D.C.: Management Information Services.*

DATA How many dollars in subsidies have gone to fossil fuels (oil, coal, and natural gas) for every dollar that has gone to renewable energy (excluding hydropower)?

Go to **Interpreting Graphs & Data** on **Mastering**EnvironmentalScience®

and might punish policymakers who try to lift the subsidies. Economists are now trying to design politically palatable means of shifting subsidies from fossil fuels to renewable energy sources or toward rebates to taxpayers.

Emissions trading can produce cost-effective results

Another approach that employs market dynamics to achieve policy goals is **emissions trading.** In an emissions trading system, a government creates a market in permits for the emission of pollutants, and companies, utilities, or industries then buy and sell the permits among themselves. In a **cap-and-trade** emissions trading system, the government first caps the overall amount of pollution it will allow, then grants or auctions off permits to polluters that allow them each to emit a certain fraction of that amount. As polluters trade these permits, the government progressively lowers the cap of overall emissions allowed (see Figure 18.33, p. 509).

Suppose, for example, you own an industrial plant with permits to release 10 units of pollution, but you find that you can make your plant more efficient and release only 5 units

instead. You now have a surplus of permits, which might be valuable to some other plant owner who is having trouble reducing pollution or who wants to expand production. In such a case, you can sell your extra permits. Doing so generates income for you and meets the needs of the other plant, while the total amount of pollution does not rise. By providing firms an economic incentive to reduce pollution, emissions trading can lower expenses for both industry and the public relative to a conventional regulatory system.

The United States pioneered the cap-and-trade approach with its program to reduce sulfur dioxide emissions, established by the 1990 amendments to the Clean Air Act (p. 454). Since then, sulfur dioxide emissions from sources in the program have declined by 67%, acid rain has been reduced, and air quality and visibility have improved (see Figure 17.30, p. 471). The 67% reduction in pollution was greater than the amount actually required by the legislation, offering evidence that cap-and-trade systems can sometimes cut pollution more effectively than command-and-control regulation. Moreover, the cuts were attained at much less cost than was predicted and with no apparent effect on electricity supply or economic growth. Similar cap-and-trade programs have shown success with smog in the Los Angeles basin and with nitrogen oxides among northeastern states.

Although cap-and-trade programs can reduce pollution, they do allow hotspots of pollution to occur around plants that buy permits to pollute more. Moreover, large firms can hoard permits, deterring smaller new firms from entering the market, thereby suppressing competition. Nonetheless, emissions trading shows promise for safeguarding environmental quality while granting industries the flexibility to lessen their impacts in ways that are economically palatable.

In recent years, emissions trading programs have begun to address the greenhouse gas emissions that drive global climate change. In the European Union Emission Trading Scheme, each participating nation allocated permits to its industries, which by trading them have established a market in which the price of a permit fluctuates according to supply and demand. In the United States, the Chicago Climate Exchange pioneered this approach, and carbon trading programs are running in California and among northeastern states (p. 509).

Emissions trading programs need to be carefully designed, however. In the European Union Emission Trading Scheme, nations allocated too many permits, causing their price to fall as supply outstripped demand. Because of this overallocation, the program has largely failed to provide industries adequate financial incentive to reduce emissions. (We will examine carbon trading markets further in Chapter 18.)

weighing the
ISSUES

A License to Pollute?

Some environmental advocates oppose emissions trading because they view it as giving polluters "a license to pollute." How do you feel about emissions trading as a means of reducing air pollution? Would you favor command-and-control regulation instead? What advantages and disadvantages do you see in each approach?

Market incentives also operate at the local level

You may have already taken part in transactions involving financial incentives as policy tools. Many municipalities charge residents for waste disposal according to the amount of waste they generate. Some cities place taxes or disposal fees on items whose safe disposal is costly, such as tires and motor oil. Others give rebates to residents who buy water-efficient toilets and appliances, because rebates can cost a city less than upgrading its wastewater treatment system. Likewise, power utilities may offer discounts to customers who buy high-efficiency lightbulbs and appliances, because doing so is less costly than expanding the generating capacity of their plants.

The creative use of economic policy tools is growing at all levels, while command-and-control regulation and legal action in the courts continue to play vital roles in environmental policymaking. As a result, we have a variety of effective policy strategies available to us as we seek sustainable solutions to our society's challenges.

closing
THE LOOP

Environmental policy is a problem-solving tool that seeks to safeguard and manage natural resources and to protect the public's health and well-being. This can often be challenging amid pressures from wealthy and powerful private interests. Command-and-control legislation and regulation remain the most common approaches to policymaking, but legal precedent as established by lawsuits retains influence, and innovative market-based policy tools are increasingly being used. When considering hydraulic fracturing for shale gas from the Marcellus Shale and elsewhere, policymakers often feel torn: They want to encourage fracking to gain the benefits of cleaner-burning domestic energy, but they also want to protect people against impacts on their water and their health. In Dimock, Pennsylvania, most residents who sued Cabot Oil and Gas have reached nondisclosed settlement terms, while a few continue to fight. Pennsylvania regulators have stepped up monitoring and enforcement efforts in the wake of public pressure and media attention, but the DEP's efforts remain limited by funding. Elsewhere across the United States, the debate over fracking is growing, as scientific evidence for pollution builds while the EPA struggles to regulate the industry effectively and while Americans reap the benefits of expanded domestic energy production.

Fracking for shale gas is expanding internationally as well. As it does, nations are looking to America's experience to inform their own policymaking regarding fracking. Meanwhile, the International Energy Association is urging drillers to show greater transparency and environmental sensitivity than they have thus far. Historically, the United States has often led the way with environmental policy, but many environmental issues span political boundaries and require international cooperation. By integrating the fundamentals of environmental policy introduced in this chapter with your knowledge of the natural sciences, you will be well equipped to develop your own creative solutions to many of our society's most challenging problems.

REVIEWING Objectives

You should now be able to:

Describe environmental policy and discuss its societal context

Environmental policy aims to protect natural resources and amenities from degradation or depletion and to promote equitable treatment of people. It addresses the tragedy of the commons, free riders, and external costs. In a democracy everyone has a voice, yet wealth tends to buy influence. (pp. 161–163)

Explain the role of science in policymaking

Data from scientific research are vital for informing policy. However, policymakers may sometimes ignore or distort science for political ends, so we in the public need to remain vigilant. (pp. 163–165)

Discuss the history of U.S. environmental policy and summarize major U.S. environmental laws

The legislative, executive, and judicial branches, along with administrative agencies and state and local governments, all play roles in U.S. environmental policy. U.S. environmental policy came in three waves. The first promoted frontier expansion and resource extraction. The second aimed to mitigate impacts of the first through conservation. The third targeted pollution and gave us much of today's environmental policy architecture: NEPA; the EPA; the EIS process; and major laws such as the Clean Air Act, Clean Water Act, and Endangered Species Act. Today a fourth wave of environmental policy is building internationally around sustainable development and global climate change. (pp. 166–174)

List institutions that influence international environmental policy and describe how nations handle transboundary issues

In our globalizing world, many environmental issues cross political boundaries. International policy includes customary law and conventional law. The United Nations, World Bank, European Union, World Trade Organization, and nongovernmental organizations all help shape international policy. (pp. 174–176)

Compare and contrast the different approaches to environmental policymaking

Lawsuits have provided a traditional approach to resolving environmental disputes, but today legislation from Congress and regulations from administrative agencies make up most U.S. federal policy. These top-down approaches are referred to as command-and-control. Economic policy tools include green taxes to discourage undesirable activities, subsidies to encourage targeted activities, and market-based approaches such as emissions trading that harness market dynamics to achieve policy goals. (pp. 176–180)

TESTING Your Comprehension

1. Describe two major justifications for environmental policy. Now discuss three problems that environmental policy commonly seeks to address.

2. What is *the tragedy of the commons*? Explain how this concept might apply to an unregulated industry that is a source of water pollution.

3. Outline the primary responsibilities of the legislative, executive, and judicial branches of the U.S. government. What is the "fourth branch" of the U.S. government?

4. What is meant by a *regulatory taking*?

5. Summarize how the first, second, and third waves of environmental policy in U.S. history differed from one another. Describe two current priorities in international environmental policy.

6. What did the National Environmental Policy Act accomplish? Briefly describe the origin and mission of the U.S. Environmental Protection Agency.

7. What is the difference between customary law and conventional law? What challenges do transboundary environmental problems present?

8. Why are environmental regulations sometimes considered to be unfair barriers to trade?

9. Compare and contrast the three major approaches to environmental policy: lawsuits, command-and-control, and economic policy tools.

10. Explain how each of the following work: a green tax, a subsidy, and an emissions trading system.

SEEKING Solutions

1. Reflect on the causes for the transitions in U.S. history from one type of environmental policy to another. Now peer into the future, and think about how life and society might be different in 25, 50, or 100 years. What would you predict about the environmental policy of the future, and why? What issues might future policy address? Do you predict we will have more or less environmental policy?

2. Compare the roles of the United Nations, the World Bank, the European Union, the World Trade Organization, and nongovernmental organizations. If you could gain the support of just one of these institutions for a policy you favored, which would you choose? Why?

3. Consider the main approaches to environmental policy— lawsuits, command-and-control laws and regulations, and economic policy tools. Describe an advantage and a disadvantage of each. Do you think any one approach is most effective? Could we do with just one approach, or does it help to have more?

4. Think of one environmental problem that you would like to see solved. From what you've learned about policy-making in this chapter, describe how you think you could best create policy to address this problem.

5. **CASE STUDY CONNECTION** You have just been elected to Congress as the representative from the Pennsylvania district that includes the town of Dimock. What policy approaches would you choose to pursue in a search for solutions to the debate over hydraulic fracturing that has been dividing your constituents? Give reasons for your choices.

6. **THINK IT THROUGH** You are the mayor of a small Texas town above the Barnett Shale, and a gas company has expressed interest in drilling in your town. Some of your town's residents are eager to have jobs they believe that hydraulic fracturing for shale gas will bring. Others are fearful that leaks of methane and fracking fluids from drilling could contaminate the air and the water supply. Some of your town's landowners are looking forward to receiving payments from the gas company for use of their land, whereas others dread the prospect of noise and pollution. If the company receives too much local opposition in your town, it says it will drill elsewhere instead. What information would you seek from the gas company, from your state regulators, and from scientists and engineers before deciding whether support for hydraulic fracturing is in the best interest of your town? How would you make your decision? How might you try to address the diverse preferences of your town's residents?

CALCULATING Ecological Footprints

Critics of command-and-control policy often argue that regulations are costly to business and industry, yet cost-benefit analyses (p. 140) have repeatedly shown that regulations bring citizens more benefits than costs, overall. Each year the U.S. Office of Management and Budget assesses costs and benefits of major federal regulations of administrative agencies.

Results from the most recent report, covering the decade from 2005 to 2015, are shown below. This decade includes periods of both Republican and Democratic control of the presidency and of Congress. Subtract costs from benefits, and enter these values for each agency in the third column. Divide benefits by costs, and enter these values in the fourth column.

Costs and Benefits of Major U.S. Federal Regulations, 2005–2015
(average values from ranges of estimates, in billions of dollars)

AGENCY	BENEFITS	COSTS	BENEFITS MINUS COSTS	BENEFIT: COST RATIO
Department of Energy	22.7	7.7	15.0	2.9
Department of Health and Human Services	26.8	3.1		
Department of Transportation	25.8	12.4		
Environmental Protection Agency (EPA)	474.0	41.5		
Other departments	71.6	21.0		
Total	**620.9**	**85.7**		

Data from U.S. Office of Management and Budget, 2016. *2015 report to Congress on the benefits and costs of federal regulations and agency compliance with the Unfunded Mandates Reform Act.* Washington, D.C.: OMB.

1. For how many of the agencies shown do regulations exert more costs than benefits? For how many do regulations provide more benefits than costs?

2. Which agency's regulations have the greatest excess of benefits over costs? Which agency's regulations have the greatest ratio of benefits to costs?

3. What percentage of total benefits from regulations comes from EPA regulations? Most of the benefits and costs from EPA regulations are from air pollution rules resulting from the Clean Air Act and its amendments. Judging solely by these data, would you say that Clean Air Act legislation has been a success or a failure for U.S. citizens? Why?

MasteringEnvironmentalScience®

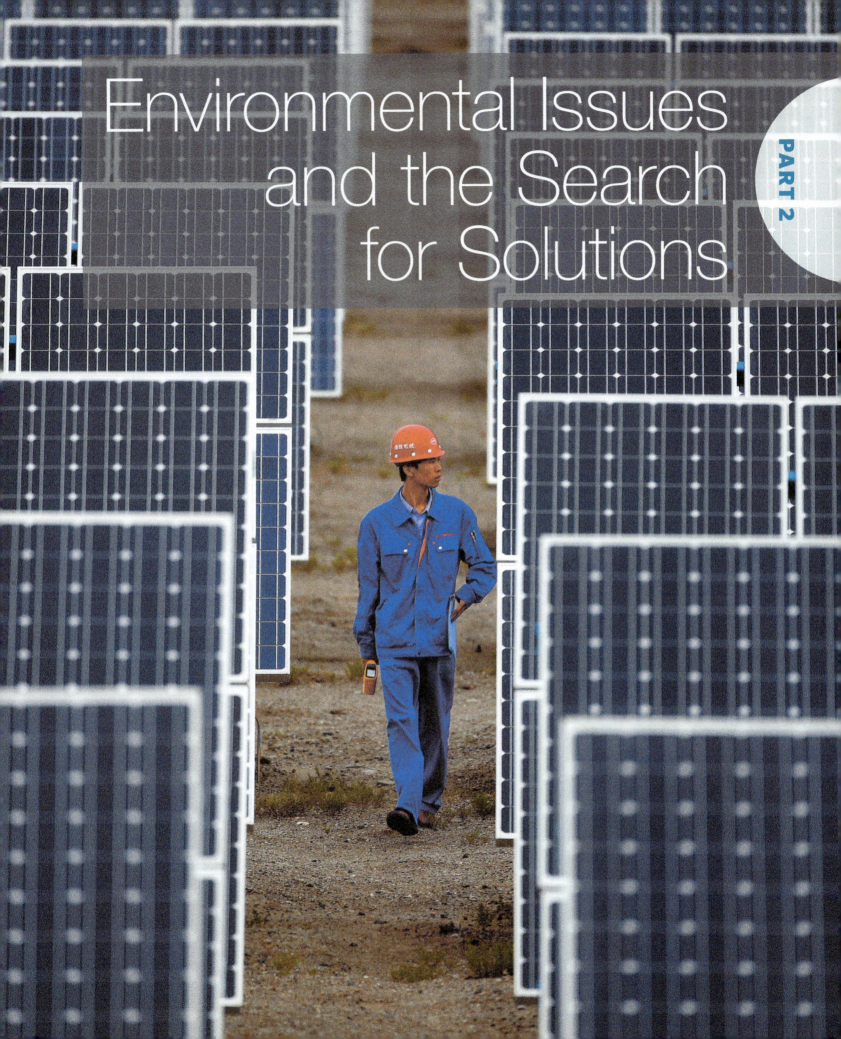

Environmental Issues and the Search for Solutions

PART 2

CHAPTER 8

Human Population

Crowded street in Shanghai, one of China's largest cities

Upon completing this chapter, you will be able to:

- Describe the scope of human population growth

- Discuss divergent views on population growth

- Explain how human population, affluence, and technology affect the environment

- Explain the fundamentals of demography

- Describe the concept of demographic transition

- Explain how family planning, the status of women, and affluence affect population growth

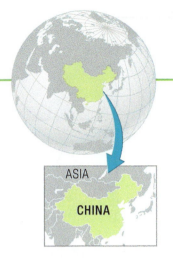

ASIA
CHINA

Will China's New "Two-Child Policy" Defuse Its Population "Time Bomb"?

> **We don't need adjustments to the family-planning policy. What we need is a phaseout of the whole system.**
> Gu Baochang, Chinese demographer at People's University, Beijing, referring to the nation's "one-child" policy in 2013

> **As you improve health in a society, population growth goes down. Before I learned about it, I thought it was paradoxical.**
> Bill Gates, Founder, Microsoft Corporation

The People's Republic of China is the world's most populous nation, home to one-fifth of the more than 7 billion people living on Earth. It is also the site of one of the most controversial social experiments in history.

When Mao Zedong founded the country's current regime six decades ago, roughly 540 million people lived in a mostly rural, war-torn, impoverished nation. Mao's policies encouraged population growth, and by 1970 improvements in food production, food distribution, and public health allowed China's population to swell to 790 million people. At that time, Chinese women gave birth to an average of 5.8 children in their lifetimes.

However, the country's burgeoning population and its industrial and agricultural development were eroding the nation's soils, depleting its water, and polluting its air. Realizing that the nation might not be able to continue to feed its people, Chinese leaders decided in 1970 to institute a population control program that prohibited most Chinese couples from having more than one child.

The program began with outreach efforts encouraging people to marry later and have fewer children. Along with these efforts, the Chinese government increased the accessibility of contraceptives and abortion. Fertility declined with these initiatives, and by 1975 China's annual population growth rate had dropped from 2.8% to 1.8%.

In 1979, the government began rewarding one-child families with government jobs and better housing, medical care, and access to schools. Families with more than one child, meanwhile, were subjected to costly monetary fines, employment discrimination, and social scorn. The one-child program applied mostly to families in urban areas. Many farmers and ethnic minorities in rural areas were exempted, because success on the farm often depends on having multiple children. The experiment was a success in slowing population growth: The nation's growth rate is now down to 0.5%, and Chinese women now have only an average of 1.7 children in their lifetimes.

However, the one-child policy also produced a population with a shrinking labor force, increasing numbers

Will the two-child policy balance China's skewed sex ratio?

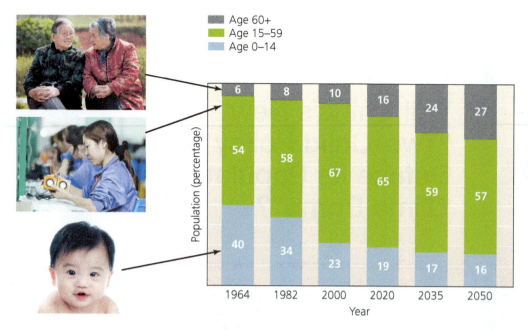

FIGURE 8.1 China's restrictive reproductive policies are leading to a shrinking workforce and rising numbers of older citizens. Values in the figure represent the percentage of the Chinese population in each age group. *Data from Population Reference Bureau, 2004.* China's Population: New Trends and Challenges.

of older people, and too few women. These unintended consequences led some demographers to question whether China's one-child policy simply traded one population problem—overpopulation—for other population problems.

The rapid reduction in fertility that resulted from this policy drastically changed China's age structure. Once consisting predominantly of young people, China's population has shifted, such that the numbers of children and older people are more even (**FIGURE 8.1**). This means there will be relatively fewer workers for China's growing economy and proportionately larger numbers of older people relying on governmental support and services.

The shrinking workforce caused by the one-child policy may now slow the growth of the thriving Chinese economy it helped to produce. Many employers in China are struggling to find workers and must pay wages up to 35% higher than they did a few years ago to keep their employees. Although this is a welcome development for workers, it threatens China's ability to produce goods as cheaply as it once did, which may induce companies to relocate from China to other nations where labor costs are lower.

The growing number of older Chinese individuals poses problems because the Chinese government lacks the resources to fully support them. And because tradition dictates that Chinese sons care for their parents and grandparents in old age, this puts a heavy economic burden on the millions of only children produced under the one-child policy.

Modern China also has too few women. Chinese culture has traditionally valued sons because they carry on the family name, assist with farm labor in rural areas, and care for aging parents. Daughters, in contrast, will most likely marry and leave their parents, as the traditional culture dictates. Thus, when faced with being limited to just one child, many Chinese couples preferred a son to a daughter. Tragically, this has led in some instances to selective abortion and the killing of female infants. This has caused a highly unbalanced ratio of young men and women in China, leading to the social instability that arises when large numbers of young men are unable to find brides and remain longtime bachelors.

Until recently, Chinese authorities attempted to address this looming population "time bomb" of an aging population with skewed ratios of men and women by occasionally loosening the one-child policy and giving some citizens greater control of their reproductive choices. For example, Chinese authorities announced in 2013 that if either member of a married couple is an only child, the couple would be allowed to have a second child—but only 1.5 million of the 11 million citizens eligible for this exemption applied for it. Faced with the prospect of continued population issues, the Chinese government announced in October 2015 that the former one-child policy would immediately become a two-child policy, and couples would be permitted to have two children without penalty.

It is unclear, however, if Chinese couples, used to the material wealth and urban lifestyle many enjoy, will embrace the opportunity to grow their families—and accept the costs of raising a second child—now that it is allowed. In 2015, 60% of Chinese women of childbearing age were over 35, so many may not wish to (or be able to) conceive a second child. In fact, an earlier survey from 2008 by China's family-planning commission reported that only 19% of the people they surveyed wished to have a second child. It may therefore be that China's relaxing of the one-child policy came too late to defuse the pending time bomb the nation may experience as China's population grays in the midst of its rapid industrialization.

China's reproductive policies have long elicited intense criticism worldwide from people who oppose government intrusion into personal reproductive choices, and such intrusion continues today, albeit with a higher allowable family size. The policy, however, has proven highly effective in slowing population growth rates and aiding the economic rise of modern China. As other nations become more crowded and they seek to emulate China's economic growth, might their governments also feel forced to turn to drastic policies that restrict individual freedoms? In this chapter, we examine human population dynamics in China and worldwide, consider their causes, and assess their consequences for the environment and human society.

Our World at Seven Billion

China receives a great deal of attention on population issues because of its unique reproductive policies and its status as the world's most populous nation. But China is not alone in dealing with population issues. India, China's neighbor, is also a population powerhouse, and is only slightly less populous than China. India lacks China's stringent reproductive policies, though, and soon will overtake China and possess the world's largest population (**FIGURE 8.2**).

Many of the world's poorer nations continue to experience substantial population growth. Some of these nations are ill-equipped to handle such growth, and this leads to stresses on society, the environment, and people's well-being. In our world of more than 7.3 *billion* people, one of our greatest challenges in this century is finding ways to slow human population growth without requiring measures such as those used in China, but rather by establishing conditions for all people that lead them to desire to have fewer children.

The human population is growing rapidly

Our global population grows by more than 88 million people each year. This means that we add 2.8 people to the planet *every second*. Take a look at **FIGURE 8.3** and note just how recently and suddenly this increase came about.

It took until after 1800—which is recent when viewed in terms of human history—for our population to reach 1 billion. Yet we reached 2 billion by 1930 and 3 billion in just 30 more years. Our population added its next billion in just 15 years, and it has taken only 12 years to add each of the next three installments of a billion people. It is expected that the globe will be home to more than 9.7 billion people by 2050.

What accounts for our unprecedented growth? As you may recall, exponential growth—the increase in a quantity by a fixed percentage per unit time—accelerates increase in population size, just as compound interest accrues in a savings account (Chapter 3; p. 64). The reason for this pattern is that a fixed percentage of a small number makes for a small increase, but that the same percentage of a large number produces a large increase. Thus, even if the growth *rate* remains steady, population *size* will increase by greater increments with each successive generation.

For much of the 20th century, the growth rate of the human population rose from year to year. This rate peaked at 2.1% during the 1960s and has declined to 1.2% since then. Although 1.2% may sound like a small percentage of increase, exponential growth endows small numbers with large consequences. A hypothetical population starting with one man and one woman that grows at 1.2% gives rise to a population of 2939 after 40 generations and 112,695 after just 60 generations. Note that this 1.2% rate is the global average, and rates of

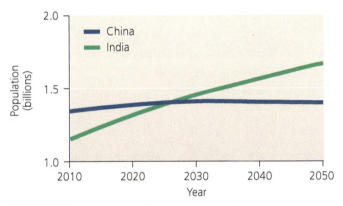

FIGURE 8.2 India will likely soon surpass China as the most populous nation. China's rate of growth is now lower than India's as a result of China's aggressive population policies. *Data from U.S. Census Bureau International Database, www.census.gov/population/international/data/idb/.*

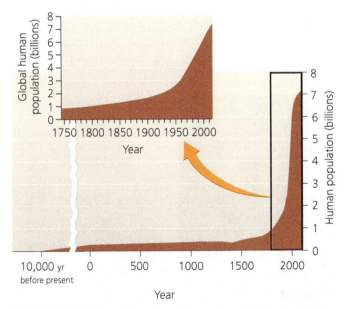

FIGURE 8.3 We have risen from fewer than 1 billion in 1800 to more than 7.3 billion today. Viewing global human population size over a long timescale **(bottom graph)** and growth since the industrial revolution **(inset top graph)** shows that nearly all growth has occurred in just the past 200 years. *Data from U.S. Census Bureau.*

FAQ

How big is a billion?

Human beings have trouble conceptualizing huge numbers. As a result, we often fail to recognize the true magnitude of a number such as 7 billion. Although we know that a billion is bigger than a million, we tend to view both numbers as impossibly large and therefore similar in size. For example, guess (without calculating) how long it would take a banker to count out $1 million if she did it at a rate of a dollar a second for 8 hours a day, 7 days a week. Now guess how long it would take to count $1 billion at the same rate. The difference between your estimate and the answer may surprise you. Counting $1 million would take a mere 35 days, whereas counting $1 billion would take 95 years! Living 1 million seconds takes only 12 days, while living for 1 billion seconds requires more than 31 years. You couldn't live for 7 billion seconds—that would take 221 years. Examples like these can help us appreciate the *b* in *billion*.

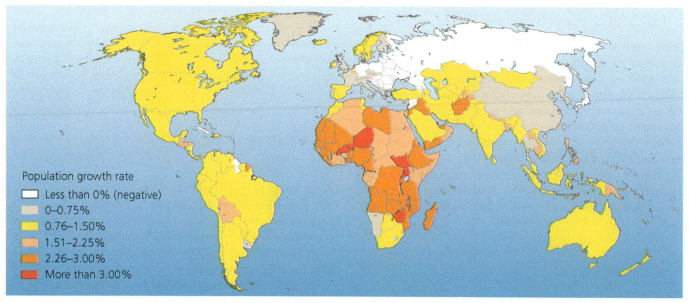

FIGURE 8.4 Population growth rates vary greatly from place to place. Populations are growing fastest in poorer nations, while populations are beginning to decrease in some highly industrialized nations. Shown are rates of natural increase as of 2015. *Data from U.S. Census Bureau International Database, www.census.gov/population/international/data/idb/.*

 DATA
- Which world region has the highest population growth rates?
- Which world region has the lowest population growth rates?

Go to **Interpreting Graphs & Data** on **Mastering**EnvironmentalScience•.

annual growth vary greatly from region to region (**FIGURE 8.4**) and are typically higher in developing nations.

At a 1.2% annual growth rate, a population doubles in size in just 58 years. We can roughly estimate doubling times with a handy rule of thumb. Just take the number 70 and divide it by the annual percentage growth rate: 70/1.2 = 58.3. Had China not instituted its original one-child policy, and had its growth rate remained at 2.8%, it would have taken only 25 years to double in size.

Is population growth a problem?

Our spectacular growth in numbers has resulted largely from technological innovations, improved sanitation, better medical care, increased agricultural output, and other factors that have brought down death rates. These improvements have been particularly successful in reducing **infant mortality rates,** the frequency of children dying in infancy. Birth rates have not declined as much, so births have outpaced deaths for many years now. Thus, our population explosion has arisen from a very good thing—our ability to keep more of our fellow human beings alive longer! Why, then, do so many people view population growth as a problem?

Let's start with a bit of history. At the outset of the industrial revolution (p. 5), population growth was universally regarded as a good thing. For parents, high birth rates meant more children to support them in old age. For society, it meant a greater pool of labor for factory work. However, British economist **Thomas Malthus** (1766–1834) had a different view. Malthus (**FIGURE 8.5a**) argued that unless population

growth were controlled by laws or other social strictures, the number of people would eventually outgrow the available food supply. Malthus's most influential work, *An Essay on the Principle of Population,* published in 1798, argued that if society did not limit births (through abstinence and contraception, for instance), then rising death rates would reduce the population through war, disease, and starvation.

In our day, biologists Paul (**FIGURE 8.5b**) and Anne Ehrlich of Stanford University have been called "neo-Malthusians" because they too have warned that our

(a) Thomas Malthus   (b) Paul Ehrlich

FIGURE 8.5 Thomas Malthus and Paul Ehrlich each argued that runaway population growth would surpass food supply and lead to disaster.

population may grow faster than our ability to produce and distribute food. In his best-selling 1968 book, *The Population Bomb,* Paul Ehrlich predicted that population growth would unleash famine and conflict that would consume civilization by the end of the 20th century.

Although human population quadrupled in the past 100 years—the fastest rate at which it has ever grown (see Figure 8.3)—Ehrlich's forecasts have not fully materialized. This is due, in part, to the way we have intensified food production in recent decades (pp. 237–238). Population growth has indeed contributed to famine, disease, and conflict—but as we shall see, enhanced prosperity, education, and increasing gender equality have also helped to reduce birth rates.

Does this mean we can disregard the concerns of Malthus and Ehrlich? Some Cornucopians (p. 144) say yes. Under the Cornucopian view that many economists hold, population growth poses no problem if new resources can be found or created to replace depleted ones. In contrast, environmental scientists recognize that not all resources can be replaced. Once a species has gone extinct, for example, we cannot replicate its exact functions in an ecosystem or know what benefits it might have provided us. Thus, the environmental scientists argue, population growth is indeed a problem if it depletes resources, stresses social systems, and degrades the natural environment, such that our quality of life declines.

Population is one of several factors that affect the environment

One widely used formula gives us a handy way to think about population and other factors that affect environmental quality. Nicknamed the **IPAT model**, it represents how our total impact (*I*) on the environment results from the interaction among population (*P*), affluence (*A*), and technology (*T*):

$$I = P \times A \times T$$

Increased population intensifies impact on the environment as more individuals take up space, use resources, and generate waste. Increased affluence magnifies environmental impact through the greater per capita resource consumption that generally has accompanied enhanced wealth. Technology that enhances our abilities to exploit minerals, fossil fuels, old-growth forests, or fisheries generally increases impact, but technology to reduce smokestack emissions, harness renewable energy, or improve manufacturing efficiency can decrease impact.

We might also add a sensitivity factor (*S*) to the equation to denote how vulnerable a given environment is to human pressures:

$$I = P \times A \times T \times S$$

For instance, the arid lands of western China are more sensitive to human disturbance than the moist regions of southeastern China. Plants grow more slowly in the arid west, making the land more vulnerable to deforestation and soil degradation. Thus, adding an additional person to western China has more environmental impact than adding one to southeastern China.

We could refine the IPAT equation further by adding terms for the effects of social institutions such as education, laws and their enforcement, stable and cohesive societies, and ethical standards that promote environmental well-being. Such factors all affect how population, affluence, and technology translate into environmental impact.

Impact can be thought of in various ways, but we can generally boil it down either to pollution or to resource consumption. The depletion of resources by ever-larger populations has been a focus of scientists and philosophers since Malthus's time. Today, researchers calculate that humanity is appropriating for its own use nearly one-quarter of Earth's terrestrial net primary production (see **THE SCIENCE BEHIND THE STORY**, pp. 190–191)—an extraordinary percentage for one species among the millions on Earth.

One reason our population has kept growing, despite limited resources, is that repeatedly we have developed technology—the *T* in the IPAT equation—to increase efficiency, alleviate our strain on resources, and allow the human population to expand further. For instance, we have used technological advances to increase global agricultural production faster than our population has risen (p. 237).

Modern-day China shows how all elements of the IPAT formula can combine to cause tremendous environmental impact in little time. China is the world's fastest-growing economy over the past two decades, and is "demonstrating what happens when large numbers of poor people rapidly become more affluent," in the words of Earth Policy Institute president Lester Brown. Although millions of Chinese people are increasing their material wealth and their resource consumption, the country is battling unprecedented environmental challenges brought about by its rapid economic development. Intensive agriculture has expanded into western China, and farming in this arid region has caused farmland to erode and blow away, much like the Dust Bowl tragedy that befell the U.S. heartland in the 1930s (pp. 223–224). China has overpumped aquifers and has drawn so much water for irrigation from the Yellow River that the once-mighty waterway now dries up in many stretches. Although China is reducing its air pollution from industry and charcoal-burning homes, the country faces new threats to air quality from rapidly rising numbers of automobiles. The air in Beijing is so polluted, for example, that simply breathing it on a daily basis damages the lungs to the same extent as smoking 40 cigarettes. As the world's other industrializing countries strive to attain the material prosperity that industrialized nations enjoy, they too may soon face many of the same challenges as China.

Demography

We exist within our environment as one species out of many. As such, all the principles of population ecology (Chapter 3) that apply to birds, frogs, and passenger pigeons apply to humans as well. The application of principles from population ecology to the study of statistical change in human populations is the focus of **demography.**

Can We Map Our Population's Environmental Impact?

Dr. Helmut Haberl of Austria's Institute for Social Ecology

Burgeoning numbers of people are making heavy demands on Earth's natural resources and ecosystem services. How can we quantify and map the environmental impacts our expanding population is exerting?

One way is to ask the following question: "Of all the biomass that Earth's plants can produce, what proportion do human beings use (for food, clothing, shelter, etc.) or otherwise prevent from growing?" To answer this question, nine environmental scientists led by Helmut Haberl of the Institute for Social Ecology in Austria teamed up to measure our consumption of net primary production (NPP; p. 111), the net amount of energy stored in plant matter as a result of photosynthesis. NPP was seen as a useful measure of human impact, because our overuse of NPP diminishes resources for other species; alters habitats, communities, and ecosystems; and threatens our future ability to derive ecosystem services.

Haberl's team began with a well-established model that maps how vegetation varies with climate across the globe and used it to produce a detailed world map of "potential NPP"— vegetation that would exist if there were no human influence. The team then gathered data for the year 2000 on crop harvests, timber harvests, grazing pressure, and other human uses of vegetation from various global databases from the United Nations Food and Agriculture Organization (FAO) and other sources. They also gathered data on how people affect vegetation indirectly, such as through fires, erosion and soil degradation, and other changes due to land use. To calculate the proportion of NPP that people appropriate, the researchers divided the amounts used up in these impacts by the total "potential" amount.

When all the data crunching was done, Haberl's group concluded that people harvest 12.5% of global NPP and that our modifications associated with land use reduce it 9.6% further and fires 1.7% further still (**FIGURE 1**). This makes us responsible for using up fully 23.8% of the planet's NPP—a staggeringly large amount for just a single species! Half of this use occurred on farmed land, where 83.5% of NPP was used. In urban areas, 73.0% of NPP was consumed; on grazing land, 19.4%; and in forests, 6.6%.

To determine how human use of NPP varies across regions of the world, Haberl's group layered the data sets atop one another in a geographic information systems (GIS) approach (p. 115). Again, they calculated the proportion of NPP

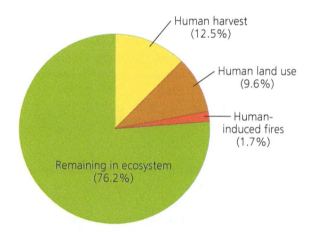

FIGURE 1 Humanity uses or causes Earth to lose 23.8% of the planet's net primary production. Direct harvesting (of crops, timber, etc.) accounts for most of this, and land use impacts and fire also contribute. *Data from Haberl, H., et al., 2007. Quantifying and mapping the human appropriation of net primary production in Earth's terrestrial ecosystems. Proc. Natl. Acad. Sci. U.S.A. 104: 12942–12947, Table 1.*

Earth has a carrying capacity for us

Environmental factors set limits on our population growth (Chapter 3), and the environment has a carrying capacity (p. 65) for our species, just as it does for every other. From an evolutionary standpoint, we happen to be a particularly successful species, however, and so we have repeatedly increased this carrying capacity by developing technology to overcome the natural limits on our population growth.

Environmental scientists who have tried to pin a number to the human carrying capacity have come up with wildly differing estimates. The most conservative estimates range from 1 to 2 billion people living prosperously in a healthy environment to 33 billion living in extreme poverty in a degraded

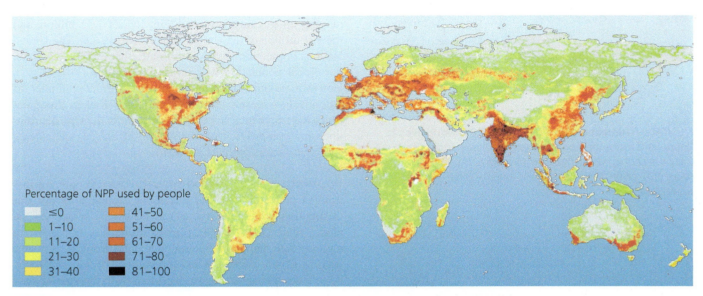

FIGURE 2 The proportion of Earth's net primary production that people appropriate varies from region to region. Regions that are densely populated or intensively farmed exert the heaviest impact. *Source: Haberl, H., et al., 2007. Quantifying and mapping the human appropriation of net primary production in Earth's terrestrial ecosystems. Proc. Natl. Acad. Sci. U.S.A. 104: 12942–12947, Fig 1b. © 2007 National Academy of Sciences, U.S.A. By permission.*

Percentage of NPP used by people
≤0
1–10
11–20
21–30
31–40
41–50
51–60
61–70
71–80
81–100

that we appropriate and produced a global map (**FIGURE 2**). The researchers published their results in 2007 in the *Proceedings of the National Academy of Sciences of the USA.*

In their global map of NPP consumption, densely populated and heavily farmed regions such as India, eastern China, and Europe show the greatest proportional use of NPP. The influence of population is clear. For instance, although people in southern Asia consume very little per capita, dense populations here result in a 63% use of NPP. In contrast, in sparsely inhabited regions of the world (such as the boreal forest, Arctic tundra, Himalayas, and Sahara Desert), humans consume almost no NPP. In North America, NPP use is heaviest in the East, Midwest, and Great Plains. In general, the map shows heavy appropriation of NPP in areas where population is dense relative to the area's vegetative production.

The map does not fully show the effects of resource consumption due to affluence. Wealthy societies tend to import food, fiber, energy, and products from other places, and this consumption can drive environmental degradation in poorer regions. For instance, North Americans and Europeans import timber logged from the Amazon basin, as well as soybeans and beef grown in areas where Amazonian forest was cleared. Through global trade, we redistribute the products we gain from the planet's NPP. As a result, the environmental impacts of our consumption are often felt far from where we consume products.

By showing areas of high and low impact, maps like the one produced in this project can help us to make better decisions and minimize our impacts on ecosystems and ecosystem services. Environmental scientists who have commented on the published paper agree and say the team's data should give us pause. As Jonathan Foley of the University of Wisconsin and his colleagues put it, "Ultimately, we need to question how much of the biosphere's productivity we can appropriate before planetary systems begin to break down. 30%? 40%? 50%? More? . . . Or have we already crossed that threshold?"

world of intensive cultivation and without wild natural areas. As our population climbs well beyond 7 billion, we may yet continue to find ways to increase our carrying capacity and preserve at least some of our current quality of life. Given our knowledge of population ecology and logistic growth (p. 65), however, we have no reason to assume that human numbers on Earth can keep growing indefinitely.

Demography is the study of human population

Just as population ecologists study such characteristics in other organisms, **demographers** study population size; density; distribution; age structure; sex ratio; and rates of birth, death, immigration, and emigration of people. Each of these

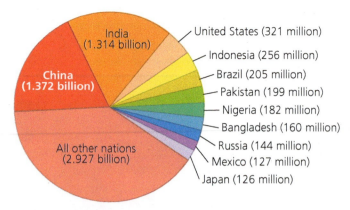

FIGURE 8.6 Almost one in five people in the world lives in China, and more than one of every six live in India. Three of every five people live in one of the 11 nations that have populations above 100 million. *Data from Population Reference Bureau, 2015 World population data sheet.*

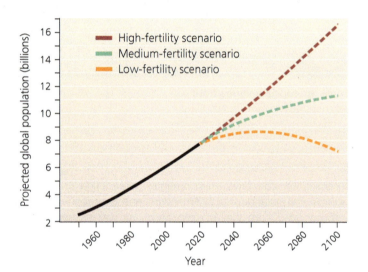

characteristics is useful for predicting population dynamics and environmental impacts.

Population size Our global human population of more than 7.3 billion is spread among 200 nations with populations ranging up to China's 1.37 billion, India's 1.31 billion, and the 321 million of the United States (**FIGURE 8.6**). The United Nations Population Division estimates that by the year 2050, the global population will surpass 9.7 billion (**FIGURE 8.7**). However, population size alone—the absolute number of individuals—doesn't tell the whole story. Rather, a population's environmental impact depends on its density, distribution, and composition (as well as on affluence, technology, and other factors outlined earlier).

Population density and distribution People are distributed unevenly over our planet. In ecological terms, our distribution is clumped (p. 61) at all spatial scales. At the global scale (**FIGURE 8.8**), population density is highest in regions with temperate, subtropical, and tropical climates, such as China, Europe, Mexico, southern Africa, and India. Population density is lowest in regions with extreme-climate

FIGURE 8.7 The United Nations predicts world population growth. In the latest projection, population is estimated to reach 9.7 billion for 2050 and around 10.5 billion in 2100 using a medium-fertility scenario. In the high-fertility scenario, women on average have 0.5 child more than in the medium scenario. In the low-fertility scenario, women have 0.5 child fewer than in the medium scenario. *Adapted by permission from Population Division of the Department of Economic and Social Affairs of the United Nations Secretariat, 2015. World population prospects: The 2015 revision. esa.un.org/unpd/wpp, © United Nations, 2015.*

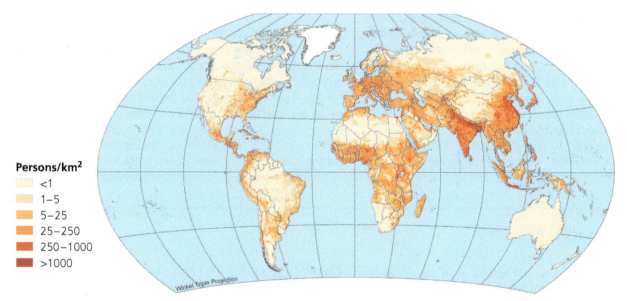

FIGURE 8.8 Human population density varies significantly from one region to another. Arctic and desert regions have the lowest population densities, whereas areas of India, Bangladesh, and eastern China have the densest populations. *Source: Center for International Earth Science Information Network–CIESIN–Columbia University, 2015. Gridded Population of the World, Version 4 (GPWv4): Population Density. Palisades, NY: NASA Socioeconomic Data and Applications Center (SEDAC).*

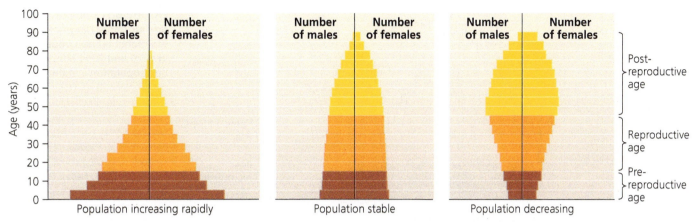

FIGURE 8.9 Age structure diagrams show numbers of individuals of different age classes in a population. A diagram like that on the left is weighted toward young age classes, indicating a population that will grow quickly. A diagram like that on the right is weighted toward old age classes, indicating a population that will decline. Populations with balanced age structures, like the one shown in the middle diagram, will remain relatively stable in size.

biomes, such as desert, rainforest, and tundra. It is highest along seacoasts and rivers. At more local scales, we cluster together in cities and towns.

This uneven distribution means that certain areas bear more environmental impact than others. Just as the Yellow River experiences pressure from Chinese cities and farms, the world's other major rivers, from the Nile to the Danube to the Ganges to the Mississippi, all receive greater impact because of the accompanying high population density. At the same time, some areas with low population density are sensitive (a high *S* value in our revised IPAT model) and thus especially vulnerable to impact. Deserts and arid grasslands, for instance, are easily degraded by development that commandeers too much water.

Age structure Age structure (p. 64) describes the relative numbers of individuals of each age class within a population. Data on age structure are especially valuable to demographers

trying to predict future dynamics of human populations. A population made up mostly of individuals past reproductive age will tend to decline over time. In contrast, a population with many individuals of reproductive age or pre-reproductive age is likely to increase. A population with an even age distribution will likely remain stable as births keep pace with deaths.

Age structure diagrams, often called population pyramids, are visual tools scientists use to illustrate age structure (**FIGURE 8.9**). The width of each horizontal bar represents the number of people in each age class. A pyramid with a wide base denotes a large proportion of people who have not yet reached reproductive age—and this indicates a population soon capable of rapid growth. In this respect, a wide base of a population pyramid is like an oversized engine on a rocket—the bigger the booster, the faster the increase.

As an example, compare age structures for the nations of Canada and Nigeria (**FIGURE 8.10**). Nigeria's large concentration of individuals in young age groups portends a great deal

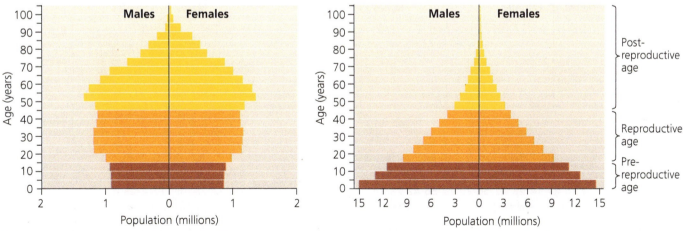

(a) Age structure diagram of Canada in 2015

(b) Age structure diagram of Nigeria in 2015

FIGURE 8.10 Canada (a) shows a fairly balanced age structure, whereas Nigeria (b) shows an age distribution heavily weighted toward young people. Nigeria's population growth rate (2.5%) is over six times greater than Canada's (0.4%). *Data from U.S. Census Bureau International Database, www.census.gov/population/international/data/idb/.*

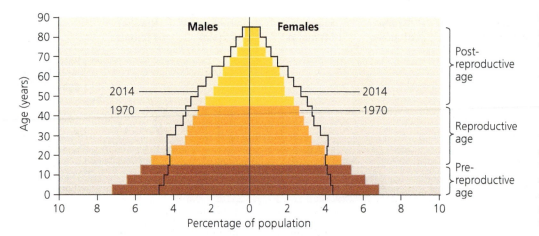

FIGURE 8.11 Comparing the age structure diagrams of the global human population in 1970 and 2014 shows how our population is aging. The more balanced age distribution also predicts slower growth in coming decades than what occurred after 1970. *Adapted by permission from Population Division of the Department of Economic and Social Affairs of the United Nations Secretariat, 2013.* World population prospects: The 2012 revision. *esa.un.org/unpd/wpp/,* © *United Nations, 2013.*

of reproduction. This is one of several reasons why Nigeria's population growth rate is much greater than Canada's.

Today, populations are aging in many nations, and the global population is "grayer" than in the past (**FIGURE 8.11**). The global median age today is 28, but it is predicted to be 38 by the year 2050. Population aging is pronounced in the United States, where the "baby boom" generation is now beginning to reach retirement age. Changing age distributions have caused concerns in some nations that have declining numbers of workers and strong social welfare programs for retirees (which are supported by current workers), such as the Social Security program in the United States. Despite the long-term benefits associated with smaller populations, many policymakers find it difficult to let go of the notion that population growth increases a nation's economic, political, and military strength. So while China and India struggle to get their population growth under control, some national governments are offering financial and social incentives that encourage their own citizens to have more children. These incentives include extended maternity and paternity leave, subsidized child care, and tax breaks for larger families.

By causing dramatic reductions in the number of children born since 1970, China's former one-child policy virtually guaranteed that the nation's population age structure would change. Indeed, in 1970 the median age in China was 20; by 2050 it is predicted to be 45. Today there are 130 million Chinese people older than 65, but that number is expected to triple by 2050 (**FIGURE 8.12**). Although graying populations have benefits–such as increased charitable contributions to society by retirees–the dramatic shift in age structure will challenge China's economy, health care system, families, and military forces because fewer working-age people will be available to support social programs to assist the rising number of older people.

Sex ratios The ratio of males to females also can affect population dynamics. Note that population pyramids give data on sex ratios by representing numbers of males and females on opposite sides of each diagram. The naturally occurring sex ratio at birth in human populations features a slight preponderance of males; for every 100 female infants born, about 106 male infants are born. This phenomenon is likely an evolutionary adaptation (p. 49) to account for the

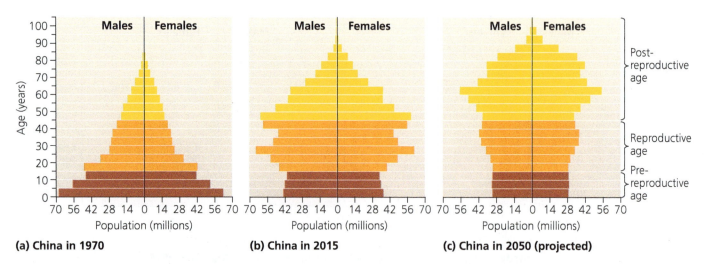

(a) China in 1970   **(b) China in 2015**   **(c) China in 2050 (projected)**

FIGURE 8.12 As China's population ages, older people will outnumber the young. China's "one child" policy was highly successful in reducing birth rates but also in significantly changing China's age structure. Population pyramids show the predicted graying of the Chinese population from **(a)** 1970 to **(b)** 2015 to **(c)** what is predicted for 2050. *Data from U.S. Census Bureau International Database, www.census.gov/population/international/data/idb.*

fact that males are slightly more prone to death during any given year of life. A higher birth rate of males tends to ensure that the ratio of men to women will be approximately equal when people reach reproductive age. Thus, a slightly uneven sex ratio at birth may be beneficial to some species. However, a greatly distorted ratio can lead to problems.

In recent years, demographers have witnessed an unsettling trend in China: The ratio of newborn boys to girls has become heavily skewed. Today, roughly 116 boys are born for every 100 girls. Some provinces have reported sex ratios as high as 138 boys for every 100 girls. The leading hypothesis for these unusual sex ratios is that many parents, having learned the sex of their fetuses by ultrasound, are selectively aborting female fetuses.

Recall that Chinese culture has traditionally valued sons over daughters. Sociologists maintain that this cultural gender preference, combined with the government's one-child policy, has led some couples to selectively abort female fetuses or to abandon or kill female infants. The Chinese government reinforced this gender discrimination when, in 1984, it exempted rural peasants from the one-child policy if their first child was a girl but not if the first child was a boy. Chinese authorities are hoping that the liberalization of their reproductive policies instituted in 2015 may help combat this skewed sex ratio, as parents will no longer be constrained to having only one child and therefore may be more embracing of daughters.

China's skewed sex ratio may further lower population growth rates. However, it has the undesirable social consequence of leaving large numbers of Chinese men single. Without the anchoring effect a wife and family provide, many of these men leave their native towns and find work elsewhere as migrant workers. Living as bachelors far from home, these men often engage in more risky sexual activity than their married counterparts. Researchers speculate that this could lead to higher incidence of HIV infection in China in coming decades, as tens of millions of bachelors find work as migrant workers.

Population change results from birth, death, immigration, and emigration

Rates of birth, death, immigration, and emigration determine whether a population grows, shrinks, or remains stable. The formula for measuring population growth (p. 64) also pertains to people: Birth and immigration add individuals to a population, whereas death and emigration remove individuals. Technological advances have led to a dramatic decline in human death rates, widening the gap between birth rates and death rates and resulting in the global human population expansion.

Falling infant mortality rates have played an especially large role in population growth. Throughout much of human history, parents needed to have larger families as insurance against the likelihood that one or more of their children would die during infancy. Poor nutrition, disease, exposure to hostile elements, and limited medical care claimed the lives of many infants in their first year of life. As societies have industrialized and become more affluent, infant mortality rates have plummeted as a result of better nutrition, prenatal care, and the presence of medically trained practitioners during birth.

As shown in **FIGURE 8.13**, infant mortality rates vary widely around the world and are closely tied to a nation's level of industrialization. China, for example, saw its infant mortality rate drop from 47 children per 1000 live births in 1980 to 16 children per 1000 live births in 2013 as the nation industrialized and prospered. Many other industrializing nations enjoyed similar success in reducing infant mortality during this time period.

weighing the
ISSUES

China's Reproductive Policy

Consider the benefits as well as the problems associated with a reproductive policy such as China's. Do you think a government should be able to enforce strict penalties for citizens who fail to abide by such a policy? If you disagree with China's policy, what alternatives can you suggest for dealing with the resource demands of a rapidly growing population?

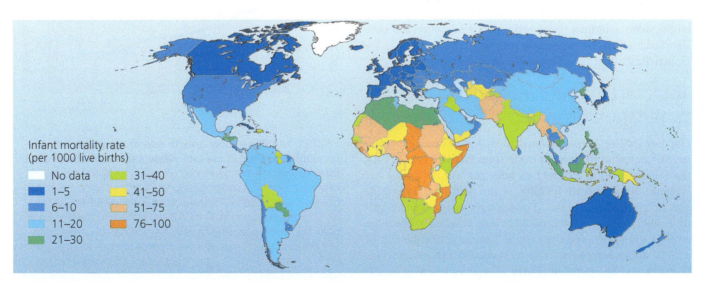

Infant mortality rate
(per 1000 live births)

No data	31–40
1–5	41–50
6–10	51–75
11–20	76–100
21–30	

FIGURE 8.13 Infant mortality rates are highest in poorer nations, such as in sub-Saharan Africa, and lowest in wealthier nations. Industrialization brings better nutrition and medical care, which greatly reduce the number of children dying in their first year of life. *Data from Population Reference Bureau,* 2015 World population data sheet.

TABLE 8.1 Trends in China's Population Growth

MEASURE	1950	1970	1990	2015
Total fertility rate	5.8	5.8	2.2	1.7
Rate of natural increase (% per year)	1.9	2.6	1.4	0.5
Doubling time (years)	37	27	49	140
Population (billions)	0.56	0.83	1.15	1.37

Data from China Population Information and Research Center; and Population Reference Bureau, *2015 World population data sheet.*

For most of the past 2000 years, China's population was relatively stable. The first significant increases resulted from enhanced agricultural production and a powerful government during the Qing Dynasty in the 1800s. Population growth began to outstrip food supplies by the 1850s, and quality of life for the average Chinese peasant began to decline. Over the next 100 years, China's population grew slowly, at about 0.3% per year, amid food shortages and political instability. Population growth rates rose again following Mao's establishment of the People's Republic in 1949, and they have declined since the advent of the one-child policy (**TABLE 8.1**).

In recent decades, falling growth rates in many countries have led to an overall decline in the global growth rate (**FIGURE 8.14**). This decline has come about, in part, from a steep drop in birth rates. Note, however, that although the rate of growth is slowing, the absolute size of the population continues to increase. Even though our percentage increases are getting smaller year by year, these are percentages of ever-larger numbers, so we continue to add more than 88 million people to the planet each year.

Immigration and emigration play increasingly important roles in population change in our modern, international world. People regularly relocate from one nation to another to improve their economic opportunities or to flee conflict or environmental degradation in their home nation. Such migration can have significant effects on population growth in nations, such as the United States, that accept large numbers of immigrants.

Total fertility rate influences population growth

One key statistic demographers calculate to examine a population's potential for growth is the **total fertility rate (TFR)**, the average number of children born per woman during her lifetime. **Replacement fertility** is the TFR that keeps the size of a population stable. For humans, replacement fertility roughly equals a TFR of 2.1. (Two children replace the mother and father, and the extra 0.1 accounts for the risk of a child dying before reaching reproductive age.) If the TFR drops below 2.1, population size in a given country (in the absence of immigration) will shrink.

Factors such as industrialization, improved women's rights (pp. 199–204), and quality health care have driven the

TFR downward in many nations in recent years. All these factors have come together in Europe, where the TFR has dropped from 2.6 to 1.4 in the past half-century. Nearly every European nation now has a fertility rate below the replacement level, and populations are declining in 15 of 45 European nations. In 2015, Europe's overall annual **rate of natural increase** (also called the *natural rate of population change*)—change due to birth and death rates alone, excluding migration—was between 0.0% and 0.1%. Worldwide by 2015, 84 countries had fallen below the replacement fertility of 2.1. These low-fertility countries make up a sizeable portion of the world's population and include China (with a TFR of 1.7). **TABLE 8.2** shows total fertility rates of major continental regions.

Many nations are experiencing the demographic transition

Many nations with lowered birth rates and TFRs are experiencing a common set of interrelated changes. In countries with good sanitation, effective health care, and reliable food supplies, more people than ever are living long lives. As a result, over the past 50 years the life expectancy for the average person globally has increased from 46 to 71 years as the

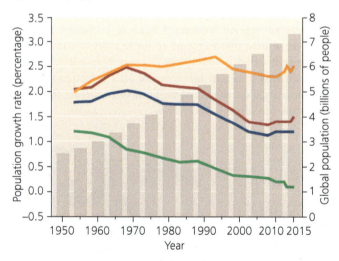

FIGURE 8.14 The annual growth rate of the global human population peaked in the late 1960s and has declined since then. Growth rates of developed nations have fallen since 1950, whereas those of developing nations have fallen since the global peak in the late 1960s. For the world's least developed nations, growth rates began to fall in the 1990s. Although growth rates are declining, global population size is still growing about the same amount each year, because smaller percentage increases of ever-larger numbers produce roughly equivalent additional amounts. *Data from Population Division of the Department of Economic and Social Affairs of the United Nations Secretariat, 2011. World population prospects: The 2010 revision. esa.un.org/unpd/wpp. © United Nations, 2011. Data updates for 2011–2015 from Population Reference Bureau, 2011–2015 World population data sheets.*

TABLE 8.2 Total Fertility Rates for Major Continental Regions

REGION	TOTAL FERTILITY RATE (TFR)
Africa	4.7
Australia and South Pacific	2.5
Latin America and Caribbean	2.1
Asia	2.2
North America	1.8
Europe	1.4

Data from Population Reference Bureau, *2015 World population data sheet.*

worldwide death rate has dropped from 20 deaths per 1000 people to 8 deaths per 1000 people. **Life expectancy** is the average number of years that an individual in a particular age group is likely to continue to live, but often people use this term to refer to the average number of years a person can expect to live from birth. Much of the increase in life expectancy is due to reduced rates of infant mortality. Societies going through these changes are generally those that have undergone urbanization and industrialization and have generated personal wealth for their citizens.

To make sense of these trends, demographers developed a concept called the **demographic transition.** This is a model of economic and cultural change first proposed in the 1940s and 1950s by demographer Frank Notestein to explain the declining death rates and birth rates that have occurred in Western nations as they industrialized. Notestein argued that nations move from a stable pre-industrial state of high birth and death rates to a stable post-industrial state of low birth and death rates (FIGURE 8.15). Industrialization, he proposed, causes these rates to fall by first decreasing mortality and then lessening the need for large families. Parents thereafter choose to invest in quality of life rather than quantity of children. Because death rates fall before birth rates fall, a period of net population growth results. Thus, under the demographic transition model, population growth is seen as a temporary phenomenon that occurs as societies move from one stage of development to another.

weighing the ISSUES

What Are the Consequences of Low Fertility?

In the United States, Canada, and almost every European nation, the total fertility rate is now at or below the replacement fertility rate (although some of these nations are still growing because of immigration). What economic or social consequences do you think might result from below-replacement fertility rates? Would you rather live in a society with a growing population, a shrinking population, or a stable population? Why?

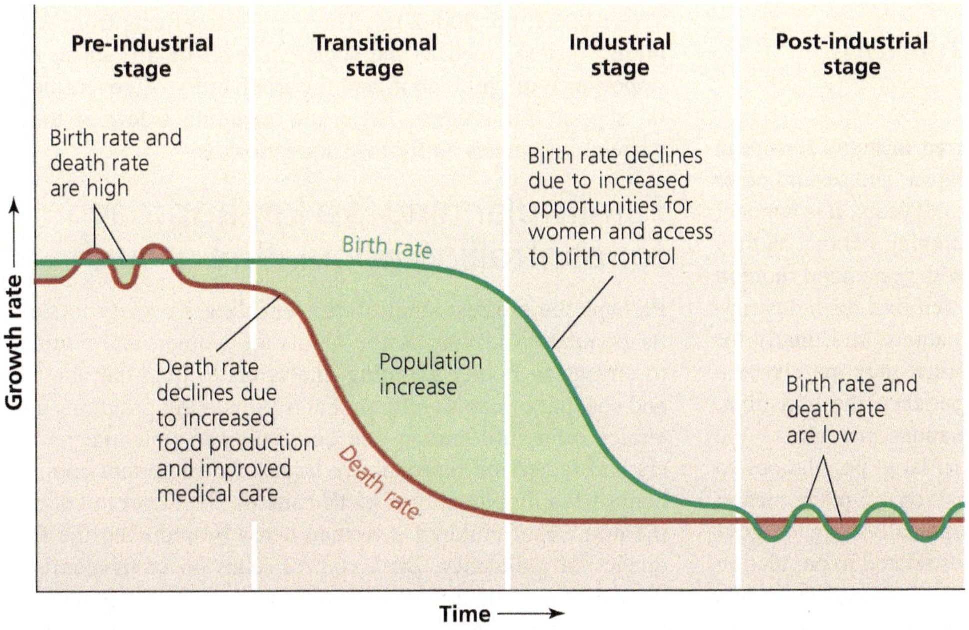

FIGURE 8.15 The demographic transition models a process that has taken some populations from a pre-industrial stage of high birth rates and high death rates to a post-industrial stage of low birth rates and low death rates. In this diagram, the wide green area between the two curves illustrates the gap between birth and death rates that causes rapid population growth during the middle portion of this process. *Adapted from Kent, M., and K. Crews, 1990.* World population: Fundamentals of growth. *By permission of the Population Reference Bureau.*

DATA Q
- In which stage of the demographic transition does population increase the most?
- Is growth greatest at the beginning or end of this stage?

Go to **Interpreting Graphs & Data** on MasteringEnvironmentalScience®.

The pre-industrial stage

The first stage of the demographic transition model is the **pre-industrial stage,** characterized by conditions that have defined most of human history. In pre-industrial societies, both death rates and birth rates are high. Death rates are high because disease is widespread, medical care is rudimentary, and food supplies are unreliable and difficult to obtain. Birth rates are high because people must compensate for infant mortality by having several children and because effective methods of preventing pregnancy are not available. In this stage, children are valuable as workers who can help meet a family's basic needs. Populations within the pre-industrial stage are not likely to experience much growth, which is why the human population grew relatively slowly until the industrial revolution.

Industrialization and falling death rates
Industrialization initiates the second stage of the demographic transition, known as the **transitional stage.** This transition from the pre-industrial stage to the industrial stage is generally characterized by declining death rates due to increased food production and improved medical care. Birth rates in the transitional stage remain high, however, because people have not yet grown used to the new economic and social conditions. As a result, population growth surges.

The industrial stage and falling birth rates
The third stage in the demographic transition is the **industrial stage.** Industrialization increases opportunities for employment outside the home, particularly for women. Children become less valuable, in economic terms, because they do not help meet family food needs as they did in the pre-industrial stage. If couples are aware of this, and if they have access to birth control, they may choose to have fewer children. Birth rates fall, closing the gap with death rates and reducing population growth.

The post-industrial stage
In the final stage, the **post-industrial stage,** both birth and death rates have fallen to low and stable levels. Population sizes stabilize or decline slightly. The society enjoys the fruits of industrialization without the threat of runaway population growth. The United States is an example of a nation in this stage, although the U.S. population is growing faster than most other post-industrial nations because of a relatively high immigration rate.

Is the demographic transition a universal process?

The demographic transition has occurred in many European countries, the United States, Canada, Japan, and several other developed nations over the past 200 to 300 years. It is a model that may or may not apply to all developing nations as they industrialize now and in the future. On the one hand, note in Figure 8.14 (p. 196) how growth rates fell first for industrialized nations, then for less developed nations, and finally for least developed nations, suggesting that it may merely be a matter of time before all nations experience the transition. On the other hand, some developing nations may already be suffering too much from the impacts of large populations to replicate the developed world's transition, a phenomenon called **demographic fatigue.** Demographically fatigued governments face overwhelming challenges related to population growth, including educating and employing swelling ranks of young people. When these stresses are coupled with large-scale environmental degradation or disease epidemics, the society may never complete the demographic transition.

Many nations in sub-Saharan Africa are experiencing such challenges today, given their large populations and the stunningly high prevalence of HIV/AIDS. Because it removes young and productive members of society, AIDS undermines the ability of poorer nations to develop. Nations lose billions of dollars in productivity when large numbers of its citizens are battling the disease, and treatment puts a huge burden on health care systems. Children orphaned by AIDS further strain social safety nets, requiring interventions to prevent the cycle of poverty and disease from claiming yet another generation. Improved public health efforts (including sex education, contraceptives, and intravenous drug abuse policies) have slowed HIV transmission rates in many nations, however, offering hope that these nations may escape the "trap" of demographic fatigue.

Although increased prosperity for all people is a noble goal, it does come with environmental cost. Environmental scientists estimate that for people of all nations to attain the material standard of living that North Americans now enjoy, we would need the natural resources of four-and-a-half more planet Earths. Hence, whether developing nations (which include the vast majority of the planet's people) pass through the demographic transition is one of the most important and far-reaching questions for the future of our civilization and Earth's environment.

Population and Society

Demographic transition theory links the quantitative study of how populations change with the societal factors that influence (and are influenced by) population dynamics. There are many factors that affect fertility in a given society. They include public health issues, such as people's access to contraceptives and the rate of infant mortality. They also include cultural factors—such as the level of women's rights, the relative acceptance of contraceptive use, and even cultural influences like television programs (see **THE SCIENCE BEHIND THE STORY**, pp. 200–201). There are also effects from economic factors, such as the society's level of affluence, the importance of child labor, and the availability of governmental support for retirees. Let's now examine a few of these societal influences on fertility more closely.

Family planning is a key approach for controlling population growth

Perhaps the greatest single factor enabling a society to slow its population growth is the ability of women and couples to engage in **family planning,** the effort to plan the number and spacing of one's children. Family-planning programs and clinics offer information and counseling to potential mothers and fathers on reproductive issues. An important component of family planning is **birth control,** the effort to control the number of children a woman bears by reducing the frequency of pregnancy. Birth control relies on **contraception,** the deliberate attempt to prevent pregnancy despite engaging in sexual intercourse. Common methods of modern contraception in use today include condoms, spermicides, hormonal treatments (birth control pill/hormone injection), intrauterine devices (IUDs), and permanent sterilization through tubal ligation or vasectomy. Many family-planning organizations aid clients by offering free or discounted contraceptives.

Worldwide in 2015, 56% of women aged 15–49 reported using modern contraceptives, with rates of use varying widely among nations. China and the United Kingdom, at 84%, had the highest rate of contraceptive use of any nations. Eight European nations showed rates of contraceptive use of 70%

or more, as did Brazil, Canada, Colombia, Costa Rica, Cuba, Micronesia, New Zealand, Nicaragua, Paraguay, Puerto Rico, South Korea, Thailand, and Uruguay. At the other end of the spectrum, 12 African nations had rates at or below 10%.

Low usage rates for contraceptives in some societies are caused by limited availability, especially in rural areas. In others, low usage may be due to religious doctrine or cultural influences that hinder family planning, denying counseling and contraceptives to people who might otherwise use them. This can result in family sizes that are larger than the parents desire and lead to elevated rates of population growth.

In a physiological sense, access to family planning (and the civil rights to demand its use) gives women control over their **reproductive window,** the period of their life, beginning with sexual maturity and ending with menopause, in which they may become pregnant. A healthy woman can potentially bear up to 25 children within this window (**FIGURE 8.16a**), but she may choose to delay the birth of her first child to pursue education and employment. She may also use contraception to delay her first child, space births within the window, and "close" her reproductive window after achieving her desired family size (**FIGURE 8.16b**).

Family-planning programs are working around the world

Data show that funding and policies that encourage family planning can lower population growth rates in all types of nations, even those that are least industrialized. No other nation has pursued a sustained population control program as intrusive as China's, but some rapidly growing nations have implemented programs that are less restrictive.

India was the first nation to implement population control policies. However, when some policymakers in India introduced forced sterilization in the 1970s, the resulting outcry brought down the government. Since then, India's efforts have been more modest and far less coercive, focusing on family planning and reproductive health care. This has greatly reduced rates of growth in India, but India will nonetheless likely overtake China and become the world's most populous nation in several decades because of China's more aggressive population initiatives.

The government of Thailand has reduced birth rates and slowed population growth as well. In the 1960s, Thailand's growth rate was 2.3%, but in 2015 it was 0.4%. This decline was achieved without a one-child policy, resulting instead from an education-based approach to family planning and the increased availability of contraceptives. Brazil, Cuba, Iran, Mexico, and many other developing countries have instituted active programs to reduce their population growth; these entail setting targets and providing incentives, education, contraception, and reproductive health care.

Empowering women reduces fertility rates

Today, many social scientists and policymakers recognize that for population growth to slow and stabilize, women in societies worldwide should be granted equality in both

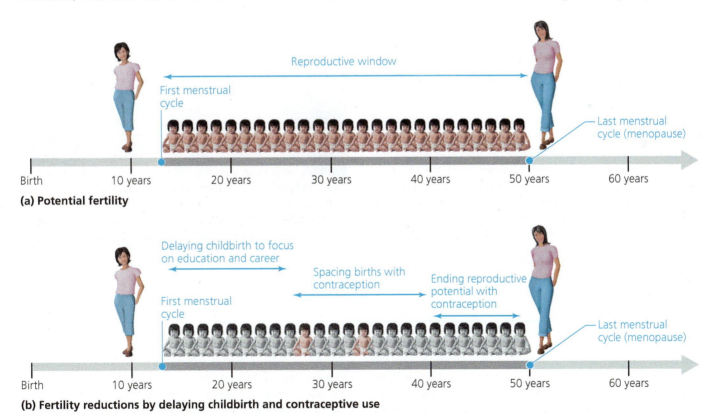

(a) Potential fertility

(b) Fertility reductions by delaying childbirth and contraceptive use

FIGURE 8.16 Women can potentially have very high fertility within their "reproductive window" but can choose to reduce the number of children they bear. They may do this by delaying the birth of their first child, or by using contraception to space pregnancies or to end their reproductive window at the time of their choosing.

Did Soap Operas Reduce Fertility in Brazil?

Eliana La Ferrara,
Bocconi University

Over the past 50 years, the South American nation of Brazil experienced the second-largest drop in fertility among developing nations with large populations—second only to China. In the 1960s, the average woman in Brazil had six children. Today, Brazil's total fertility rate is 1.9 children per woman, which is lower than that of the United States. Brazil's drastic decrease in fertility is interesting because, unlike China, it occurred without intrusive governmental policies to control its citizens' reproduction.

Brazil accomplished this, in part, by providing women equal access to education and opportunities to pursue careers outside the home. Women now make up 40% of the workforce in Brazil and graduate from college in greater numbers than men. In 2010, Brazilians elected a woman, Dilma Rousseff, as their nation's president.

The Brazilian government also provides family planning and contraception to its citizens free of charge. Eighty percent of married women of childbearing age in Brazil currently use contraception, a rate higher than that in the United States or Canada. Universal access to family planning has given women control over their desired family size and has helped reduce fertility across all economic groups, from the very rich to the very poor. It is interesting to note that induced abortion is not used in Brazil as it is in China; the procedure is illegal except in rare circumstances.

As Brazil's economy grew with industrialization, people's nutrition and access to health care improved, greatly reducing infant mortality rates. Increasing personal wealth promoted materialism and greater emphasis on career and possessions over family and children. The nation also urbanized as people flocked to growing cities such as Río de Janeiro and São Paulo. This brought about the fertility reductions that typically occur when people leave the farm for the city.

It turns out, however, that Brazil may also have had a rather unique influence affecting its fertility rates over the past several decades—soap operas (**FIGURE 1**). Brazilian soap operas, called *telenovelas* or *novelas*, are a cultural phenomenon and are watched religiously by people of all ages, races, and incomes. Each *novela* follows the activities of several fictional families, and these TV shows are wildly popular because they have characters, settings, and plot lines with which everyday Brazilians can identify.

Telenovelas do not overtly address fertility issues, but they do promote a vision of the "ideal" Brazilian family. This family is typically middle- or upper-class, materialistic, individualistic, and full of empowered women. By challenging existing cultural and religious values through their characters, *novelas* had, and continue to have, a profound impact on Brazilian society. In essence, these programs provided a model family for Brazilians to emulate—with small family sizes being a key characteristic.

In a 2012 paper in the *American Economic Journal: Applied Economics*, a team of researchers from Bocconi University in Italy, George Washington University, and the Inter-American Development Bank (based in Washington, D.C.) analyzed various parameters to investigate statistical relationships between

FIGURE 1 *Telenovelas* **are a surprising force for promoting lower fertility rates.** Here, residents gather outside a café in Río de Janeiro to watch the popular program *Avenida Brasil*.

decision-making power and access to education and job opportunities. In addition to providing a basic human right, empowerment of women would have many benefits with respect to fertility rates: Studies show that where women are freer to decide whether and when to have children, fertility rates fall, and children are better cared for, healthier, and better educated.

Another benefit of equal rights for women is the ability to make reproductive decisions themselves. In many societies, men restrict women's decision-making abilities, including decisions about how many children a woman will bear. Fertility rates have dropped most noticeably in nations where women have gained improved access to contraceptives and to family planning education. These differences are seen

telenovelas and fertility patterns in Brazil from 1965 to 2000. Rede Globo, the network that has a virtual monopoly on the most popular *novelas*, increased the number of areas that received its signal in Brazil over those 35 years (**FIGURE 2**), and it now reaches 98% of Brazilian households. By combining data on Rede Globo broadcast range with demographic data, the researchers were able to compare changes in fertility patterns over time in areas of Brazil that received access to *novelas* with areas of Brazil that did not.

The team, led by Dr. Eliana La Ferrara, found that women in areas that received the Globo signal had significantly lower fertility than those in areas not served by Rede Globo. They also found that fertility declines were age-related, with substantial reductions in fertility occurring in women aged 25–44, but not in younger women (**FIGURE 3**). The authors hypothesized that this effect was likely due to the fact that women between 25 and 44 were closer in age to the main female characters in *novelas*, who typically had no children or only a single child. The depressive effect on fertility among women in areas served by Globo was therefore attributed to wider spacing of births and earlier ending of reproduction by women over 25, rather than to younger women delaying the birth of their first child.

The researchers determined that access to television alone did not depress fertility. For example, comparisons of fertility rate in areas with access to a different television network, Sistema

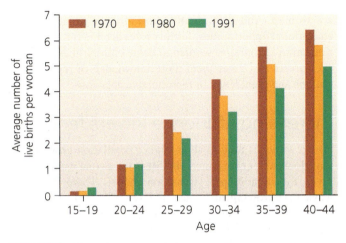

FIGURE 3 Fertility declines among Brazilian women between 1970 and 1991 were most pronounced in later age classes. The authors attribute some of this decline to women in those age classes emulating the low fertility of lead female characters in *novelas*. *Source: La Ferrara, E., et al., 2012. Soap operas and fertility: Evidence from Brazil. Am. Econ. J. Appl. Econ. 4: 1–31.*

Brasileiro de Televisão, found no relationship. The study authors concluded that this was likely due to the reliance of Sistema Brasileiro de Televisão on programming imported from other nations, with which everyday Brazilians did not connect as they did with *novelas* from Rede Globo.

Television's ability to influence fertility is not limited to Brazil. A 2014 study found that in the United States, tweets and Google searches for terms such as "birth control" increased significantly the day following the airing of new episodes of MTV's *16 and Pregnant*. By correlating geographic patterns in viewership with fertility data, the study authors concluded that MTV's *Teen Mom* series may have been responsible for reducing teenage births by up to 20,000 per year.

The factors that affect human fertility can be complex and vary greatly from one society to another. As this is a correlative study (p. 13), it does not prove causation between watching *telenovelas* and reduced fertility. It does show, however, that effects on fertility can come from intentional factors, such as a government increasing the availability of birth control, and other times can come from unexpected and unintentional factors—such as popular television shows.

FIGURE 2 The Globo television network expanded over time and now reaches nearly all households in Brazil. Fertility declines were correlated with the availability of Globo, and its *novelas*, over the time periods in the study. *Source: La Ferrara, E., et al., 2012. Soap operas and fertility: Evidence from Brazil. Am. Econ. J. Appl. Econ. 4: 1–31.*

when comparing nations that have similar cultures and levels of economic development but very different approaches to family planning—such as Bangladesh and Pakistan. When both nations were faced with rapid population growth due to high fertility in the 1970s (with TFR in both nations hovering around 7), Bangladesh instituted a government-supported program to improve access to contraception and reproductive

counseling to its citizens in an effort to reduce its rate of population growth. Pakistan took a far less aggressive and coordinated approach, which made access to family planning by Pakistani women far less reliable than that for Bangladeshi women. After 40 years of differing approaches to reproductive issues, the results are striking. While Bangladesh's TFR in 2015 had fallen to 2.3, Pakistan's TFR was 3.8 children per woman—one of the highest in southern Asia. Expanding educational opportunities for women is an important component of equal rights. In many nations girls are discouraged from pursuing an education or are kept out of school altogether. Worldwide, more than two-thirds of people who cannot read are women. And data clearly show that as women receive educational opportunities, fertility rates decline (**FIGURE 8.17**). Education encourages women to delay childbirth as they pursue careers and gives them more knowledge of reproductive options and greater say in reproductive decisions.

Increasing affluence lowers fertility

Poorer societies tend to show higher population growth rates than do wealthier societies (**FIGURE 8.18**), as one would expect given the demographic transition model. There are many ways that growing affluence and reducing poverty lead to lower rates of population growth.

Historically, people tended to conceive many children, which helped ensure that at least some would survive. Today's improved medical care in wealthy nations has reduced infant mortality rates, making it less necessary to bear multiple children. Increasing urbanization has also driven TFR down; whereas rural families need children to contribute to farm labor, in urban areas children are usually excluded from the labor market, are required to go to school, and impose significant economic costs on their families. Moreover, if a government provides some form of social security, parents need fewer children to support them in their old age. Finally, with greater educational opportunities and changing roles in society, women tend to move into the labor force. When women are able to focus on their careers, this often acts to delay the birth of their first child until later in life or enables them to choose to not have children.

Economic factors are tied closely to population growth. Poverty exacerbates population growth, and rapid population growth worsens poverty. This connection is important because a great majority of the next billion people to be added to the global population will be born into nations in Africa, Asia, and Latin America that have emerging,

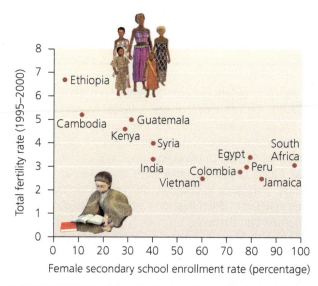

FIGURE 8.17 Increasing female literacy is strongly associated with reduced birth rates in many nations. *Data from McDonald, M., and D. Nierenberg, 2003.* Linking population, women, and biodiversity. State of the world 2003. *Washington, D.C.: Worldwatch Institute.*

DATA Q Is the relationship between total fertility rate and the rate of enrollment of girls in secondary school positive (as variable 1 increases, so does variable 2), negative (as variable 1 increases, variable 2 decreases), or is there no obvious relationship (increases in variable 1 are not correlated with changes in variable 2)?

Go to **Interpreting Graphs & Data** on **Mastering**EnvironmentalScience*.

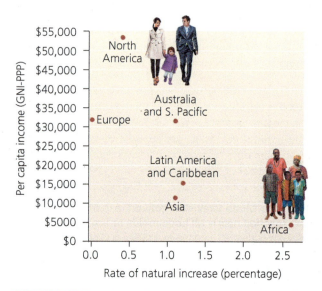

FIGURE 8.18 Poverty and population growth show a fairly strong correlation, despite the influence of many other factors. Regions with the lowest per capita incomes tend to have the most rapid population growth. Per capita income is here measured in GNI PPP, or "gross national income in purchasing power parity." GNI PPP is a measure that standardizes income among nations by converting it to "international" dollars, which indicate the amount of goods and services one could buy in the United States with a given amount of money. *Data from Population Reference Bureau, 2015 World population data sheet.*

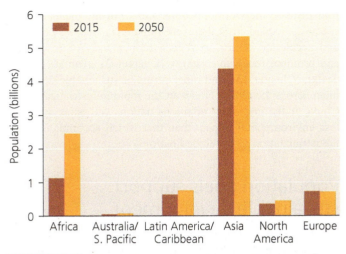

FIGURE 8.19 **Africa will experience the greatest population growth of any region in coming decades.** The vast majority of future population growth will occur in developing regions. The highly industrialized regions of Europe and North America are predicted to experience only minor population change. *Data from Population Reference Bureau, 2015 World population data sheet.*

- Which region will add the most people between 2015 and 2050?
- Which will increase by the greatest percentage during this time period?
- Which regions will experience very little population growth during this period?
- Propose one explanation for why the fastest-growing region will increase faster than other regions.

Go to **Interpreting Graphs & Data** on **Mastering**EnvironmentalScience°.

developing economies (**FIGURE 8.19**). This is unfortunate from a social standpoint, because in some cases these countries, many of which already are strained by rapid population growth, will be unable to provide for them. It is also unfortunate from an environmental standpoint, because poverty often results in environmental degradation. People who depend on agriculture and live in areas of poor farmland, for instance, may need to farm even if doing so degrades the soil and is not sustainable. Poverty also drives people to cut forests and to deplete biodiversity. For example, impoverished settlers and miners hunt large mammals for "bush meat" in Africa's forests, including the great apes that are now heading toward extinction.

Expanding wealth can increase the environmental impact per person

Poverty can lead people into environmentally destructive behavior, but ironically wealth can produce even more severe and far-reaching environmental impacts. The affluence of a society such as the United States, Japan, or France is built on levels of resource consumption unprecedented in human history. Much of this chapter has dealt with numbers of people rather than with the amount of resources each member of the population consumes or the amount of waste each

member produces. The environmental impact of human activities, however, depends not only on the number of people involved but also on the way those people live. Recall the *A* (for affluence) in the IPAT equation. Affluence and consumption are spread unevenly across the world, and wealthy societies generally consume resources from regions far beyond their own.

We have explored the concept of the ecological footprint, the cumulative amount of Earth's surface area required to provide the raw materials a person or population consumes and to dispose of or recycle the waste produced. Individuals from affluent societies leave considerably larger per capita ecological footprints (see Figure 1.12, p. 15). In this sense, the addition of 1 American to the world has as much environmental impact as the addition of 3.4 Chinese, 8 Indians, or 14 Afghans. This fact reminds us that the "population problem" does not lie solely with nations in the developing world but is relevant to people everywhere.

As population is rising, so is consumption, and some environmental scientists have calculated that we are already living beyond the planet's means to support us sustainably. One recent analysis concludes that humanity's global ecological footprint surpassed Earth's capacity to support us in 1971 and that our species is now living 50% beyond its means (**FIGURE 8.20**, p. 204). This is what is known as our overshoot (see Figure 1.5, p. 6). In this analysis, our ecological footprint can be compared to the amount of biologically productive land and sea available to us—an amount termed **biocapacity.** For any given area, if the footprint is greater than the biocapacity, there is an "ecological deficit." If the footprint is less than the biocapacity, there is an "ecological reserve." Because our footprint exceeds our biocapacity by 50% worldwide, we are running a global ecological deficit, gradually draining our planet of its natural capital and its long-term ability to support our civilization.

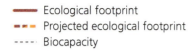

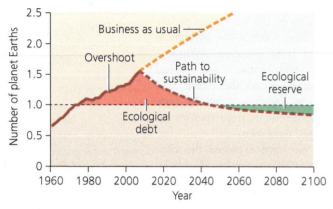

FIGURE 8.20 The global ecological footprint of the human population is estimated to be 50% greater than what Earth can bear. If population and consumption continue to rise (**orange dashed line**), we will increase our ecological deficit, or degree of overshoot, until systems give out and populations crash. If, instead, we pursue a path to sustainability (**red dashed line**), we can eventually repay our ecological debt and sustain our civilization. *Adapted from WWF, 2008.* Living planet report 2008. *Gland, Switzerland: WWF International.*

The richest one-fifth of the world's people possesses more than 80 times the income of the poorest one-fifth (**FIGURE 8.21**). The richest one-fifth also uses 86% of the world's resources. That leaves only 14% of global resources—energy, food, water, and other essentials—for the remaining four-fifths of the world's people, and all of the planet's other species, to share. It is therefore imperative that we continue and accelerate efforts to reduce human uses of resources, and promote renewable energy (Chapter 21), "smart" urban design (Chapter 13), and other forms of sustainable development across the globe. This way, the rapid industrialization of China, India, and other populous nations can occur with far less environmental impact than that which accompanied the industrialization of developed nations.

Population goals support sustainable development

The factors that influence fertility are complex and interacting, so initiatives to slow population growth must be diverse, flexible, and culturally specific. This call for a multifaceted approach was echoed at the milestone 1994 United Nations conference on population and development in Cairo, Egypt. The conference marked a transition away from older notions of top-down command-and-control (p. 177) population policy geared toward pushing contraception and lowering populations to preset targets. Instead, the conference organizers urged governments to offer better education and health care and to address social needs that affect population from the bottom up (such as alleviating poverty, disease, and sexism). Although nations learn from one another how to deal with rising (or shrinking) populations, it has become clear that there is no "one size fits all" solution to population issues, and programs must be as unique as the nations they are serving.

Average per U.S. resident:
- Economic activity: $55,860
- Footprint: 8.2 hectares
- CO_2: 17.0 metric tons

(a) A family living in the United States

Average per India resident:
- Economic activity: $5640
- Footprint: 1.2 hectares
- CO_2: 1.7 metric tons

(b) A family living in India

FIGURE 8.21 Material wealth varies widely from nation to nation. A typical U.S. family **(a)** may own a large house with a wealth of material possessions. A typical family in a developing nation such as India **(b)** may live in a much smaller home with far fewer material possessions. Compared with the average resident of India, the average U.S. resident shares in 10 times more economic activity, has an ecological footprint 8 times higher, and emits 10 times more carbon dioxide. Economic activity is calculated by dividing the gross national income (GNI) of each country by its population. One hectare = 2.471 acres. *Data for ecological footprints are for 2012 and are from Global Footprint Network, www.footprintnetwork.org; data for economic activity and carbon dioxide emissions are from World Bank, 2015, data.worldbank.org.*

closing
THE LOOP

China has demonstrated that it is possible to rapidly and drastically slow population growth, but its one-child policy created demographic problems by rapidly "graying" the Chinese population and raised important human rights issues. China's neighbor India, also a population powerhouse, started employing population control initiatives at roughly the same time as China, and its initial draconian policies caused public outcry. India's subsequent relaxed policies, along with the societal changes brought about by industrialization, have acted to reduce fertility rates, but not to the extent of its neighbor to the north—which means that India will soon surpass China as the world's most populous nation.

Today's human population is larger than at any time in the past. Our growing population and our growing consumption affect the environment and our ability to meet the needs of all the world's people. However, there are at least two reasons to be encouraged. First, although global population is still rising, the rate of growth has decreased nearly everywhere, and some countries are even seeing population declines. Most developed nations have passed through the demographic transition, showing that it is possible to lower death rates while stabilizing population and creating more prosperous societies. Second, progress has been made in expanding rights for women worldwide. Although there is still a long way to go, women are obtaining better education, more economic independence, and more ability to control their reproductive decisions. Aside from the clear ethical progress these developments entail, they are also helping to slow population growth.

Human population cannot continue to rise forever. The question is how it will stop rising: Will it be through the gentle and benign process of the demographic transition as is happening in India, through restrictive governmental intervention such as China's policies, or through the miserable Malthusian checks of disease and social conflict caused by overcrowding and competition for scarce resources? How we answer this question today will determine not only the quality of the world in which we live but also the quality of the world we leave to our children and grandchildren.

REVIEWING Objectives

You should now be able to:

Describe the scope of human population growth

The human population currently stands at more than 7 billion and increases by 1.2% annually. It is predicted to rise to around 9.7 billion people by 2050. (pp. 185–188)

Discuss divergent views on population growth

Because growing populations can deplete resources, strain food supplies, and stress social systems, Thomas Malthus and Paul Ehrlich warned that overpopulation could greatly harm humanity. The contrasting Cornucopian view holds that innovation will replace consumed resources and thus that population growth can continue. (pp. 188–189)

Explain how human population, affluence, and technology affect the environment

The IPAT model summarizes how environmental impact (I) results from interactions among population size (P), affluence (A), and technology (T): $I = P \times A \times T$. Rising population and rising affluence may each increase consumption and environmental impact. Technology has frequently worsened environmental degradation, but it can also help lessen our impacts. (p. 189)

Explain the fundamentals of demography

Demography applies principles of population ecology to the statistical study of human populations. Demographers study size, density, distribution, age structure, and sex ratios of populations, as well as rates of birth, death, immigration, and emigration. (pp. 189–196)

Describe the concept of demographic transition

The demographic transition model explains why population growth slows as nations experience the process of industrialization. Economic and societal factors in agrarian societies that favored large family sizes are replaced with those that favor smaller families with industrialization and urbanization. The demographic transition may or may not proceed to completion in all of today's developing nations. (pp. 196–198)

Explain how family planning, the status of women, and affluence affect population growth

Family-planning programs, reproductive education, and access to modern contraceptives have reduced population growth in many nations. Fertility rates also tend to fall in societies that grant women equal rights to men, as women delay childbirth to pursue education and employment. Poorer societies tend to show faster population growth than do wealthier societies (pp. 198–204)

TESTING Your Comprehension

1. What is the approximate current human global population? How many people are being added to the population each day?

2. Why has the human population continued to grow despite environmental limitations? Do you think this growth is sustainable? Why or why not?

3. Contrast the views of environmental scientists with those of Cornucopian economists and policymakers regarding whether population growth is a problem. Name several reasons why population growth is commonly viewed as a problem.

4. Explain the IPAT model. How can technology either increase or decrease environmental impact? Provide at least two examples.

5. Describe how demographers use size, density, distribution, age structure, and sex ratio of a population to estimate how it may change in the future. How does each

of these factors help determine the impact of human populations on the environment?

6. What is the total fertility rate (TFR)? Why is the replacement fertility for humans approximately 2.1? How is Europe's TFR affecting its rate of natural increase?

7. How does the demographic transition model explain the increase in population growth rates in recent centuries? How does it explain the recent decrease in population growth rates in many countries?

8. Why have fertility rates fallen in many countries?

9. Why are the empowerment of women and the pursuit of gender equality viewed as important to controlling population growth? Describe the aim of family-planning programs.

10. Why do poorer societies have higher population growth rates than wealthier societies? How does poverty affect the environment? How does affluence affect the environment?

SEEKING Solutions

1. The World Bank estimates that half the world's people survive on less than $2 per day. How do you think this situation affects the political stability of the world? Explain your answer.

2. Apply the IPAT model to the example of China provided in the chapter. How do population, affluence, technology, and ecological sensitivity affect China's environment? Now consider your own country or your own state. How do population, affluence, technology, and ecological sensitivity affect your environment? How can we minimize the environmental impacts of growth in the human population?

3. Do you think that all of today's developing nations will complete the demographic transition and come to enjoy a permanent state of low birth and death rates? Why or why not? What steps might we as a global society take to help ensure that they do? Now think about developed nations such as the United States and Canada. Do you think these nations will continue to lower and stabilize their birth and death rates in a state of prosperity? What factors might affect whether they do so?

4. **CASE STUDY CONNECTION** China's reduction in birth rates is leading to significant change in the nation's age structure. Review Figure 8.12, which shows that the population is growing older, leading to the top-heavy population pyramid for the year 2050. If you were tasked with maximizing the contributions of the growing number of retirees to Chinese society, what sorts of programs would you devise?

5. **THINK IT THROUGH** India's prime minister puts you in charge of that nation's population policy. India has a population growth rate of 1.4% per year, a TFR of 2.3, a 47% rate of contraceptive use, and a population that is 68% rural. What policy steps would you recommend, and why?

6. **THINK IT THROUGH** Now suppose that you have been tapped to design population policy for Germany. Germany is losing population at an annual rate of 0.3%, has a TFR of 1.5, a 62% rate of contraceptive use, and a population that is 73% urban. What policy steps would you recommend, and why?

CALCULATING Ecological Footprints

A nation's population size and the affluence of its citizens each influence its resource consumption and environmental impact. As of 2015, the world's population passed 7.3 billion. In 2015, the average per capita income was $15,030 per year, and the latest estimate for the world's average ecological footprint was 2.8 hectares (ha) per person. The sampling of data in the table will allow you to explore patterns in how population, affluence, and environmental impact are related.

NATION	POPULATION (MILLIONS OF PEOPLE)	AFFLUENCE (PER CAPITA INCOME)[1]	PERSONAL IMPACT (PER CAPITA FOOTPRINT, IN HA/PERSON)	TOTAL IMPACT (NATIONAL FOOTPRINT, IN MILLIONS OF HA)
Belgium	11.1	$39,260	7.4	82.1
Brazil	194.3	$11,720	3.1	
China	1350.4	$9,210	3.4	
Ethiopia	87.0	$1,140	1.0	
India	1259.7	$3,840	1.2	
Japan	127.6	$36,320	5.0	
Mexico	116.1	$16,630	2.9	
Russia	143.2	$22,760	5.7	
United States	313.9	$50,610	8.2	2574.0

[1]Measured in GNI PPP (gross national income in purchasing power parity), a measure that standardizes income among nations by converting it to "international" dollars, the amount of goods and services one could buy in the United States with a given amount of money.

Sources: Population and affluence data are from Population Reference Bureau, *2012 and 2013 World population data sheets*. Footprint data are from Global Footprint Network, www.footprintnetwork.org/ecological_footprint_nations/. All data are for 2012.

1. Calculate the total impact (national ecological footprint) for each country.

2. Draw a graph illustrating per capita impact (on the *y* axis) versus affluence (on the *x* axis). What do the results show? Explain why the data look the way they do.

3. Draw a graph illustrating total impact (on the *y* axis) in relation to population (on the *x* axis). What do the results suggest to you?

4. Draw a graph illustrating total impact (on the *y* axis) in relation to affluence (on the *x* axis). What do the results suggest to you?

5. You have just used three of the four variables in the IPAT equation. Now give one example of how the *T* (technology) variable could potentially increase the total impact of the United States and one example of how it could potentially decrease the U.S. impact.

MasteringEnvironmentalScience®

Students Go to **MasteringEnvironmentalScience** for assignments, the etext, and the Study Area with practice tests, videos, current events, and activities.

Instructors Go to **MasteringEnvironmentalScience** for automatically graded activities, current events, videos, and reading questions that you can assign to your students, plus Instructor Resources.

CHAPTER 9

The Underpinnings of Agriculture

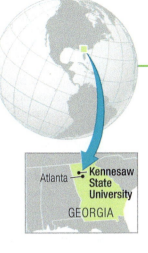

Farm to Table (and Back Again) at Kennesaw State University

> **There are two spiritual dangers in not owning a farm. One is the danger of supposing that breakfast comes from the grocery, and the other that heat comes from the furnace.**
>
> Conservationist and philosopher Aldo Leopold

> **The nation that destroys its soil destroys itself.**
>
> U.S. President Franklin D. Roosevelt

It's not surprising to see phrases such as "think globally, eat locally" and "farm to table" when you're dining at a trendy restaurant, but would you expect to see them used at your campus dining hall?

Believe it or not, campus dining services around the country are now among the industry leaders in culinary sustainability, a pursuit that embraces the use of fresh, healthy, locally produced foods to provide diners delicious, nutritious meals. One leader in sustainable dining is Kennesaw State University (KSU) in suburban Atlanta, Georgia. In 2009, the university opened The Commons, a dining facility that serves more than 5000 students each day and was granted gold-level certification as a sustainable building by the Leadership in Energy and Environmental Design (LEED) program (p. 343). The next year, KSU launched a Farm-to-Campus program when it acquired a plot of farmland just off campus.

Today, the university runs three farms on 27 hectares (67 acres) of land near the campus that grow thousands of pounds of produce each year, supplying many of the fruits and vegetables served to students in The Commons. But KSU does even more, and is working to create a fully closed loop system. Uneaten food waste from The Commons is fed into an aerobic digestion system behind the facility, where it is broken down to generate a nutrient-rich liquid. This liquid compost is trucked back to the campus farms and applied there as a fertilizer to nourish the soil and help grow new crops. As KSU's first director of Culinary and Hospitality Services put it, "We go beyond farm-to-campus. We embrace farm-to-campus and back-to-farm operations."

Kennesaw State's ambitious program to mesh sustainable agriculture with campus dining got its start when the university committed to make sustainability a prime consideration in construction and operation of The Commons. KSU's architects and engineers designed a facility that minimizes energy use, water consumption, and waste generation. Illuminated with floor-to-ceiling windows and high-efficiency lighting, The Commons offers nine themed food stations and a rotating menu of 200–300 items. Foods are prepared to order or in small batches according to demand, drastically reducing the amount of leftover food. A "trayless" approach to food service further reduces food waste and the amount of water used in dishwashing,

Kennesaw State University's Hickory Grove Farm

Tending to crops on a college farm

as diners without trays tend to eat less and use fewer plates. The Commons has special dishwashing systems designed for water and energy efficiency, napkin dispensers that reduce paper waste, hydration stations for refilling reusable water bottles, and a recycling and composting program that diverts 44,000 pounds of waste from the landfill each month. The facility even generates biodiesel from its used cooking oil (p. 567) to fuel university vehicles.

With an award-winning green building as the anchor of its program, KSU set about creating an agricultural system that could supply diners with fresh, healthy, local produce. The three campus farms grow dozens of items—everything from tomatoes to cucumbers to melons to apples—using practices that protect soil quality by minimizing the use of chemical pesticides and synthetic fertilizers. Indeed, soil quality is enhanced by the application of nutrient-rich compost and liquid from the digester. Sixty free-range chickens produce 300 organic eggs per week, and apiaries house 48 honeybee colonies that produce 30 gallons of honey each year.

On campus, herbs, lettuce, and shiitake mushrooms are grown in an herb garden and greenhouse behind The Commons and in 10 hydroponic stations scattered throughout the dining area. These are each watered with rainwater collected in barrels on the roof of The Commons. There is even a grist mill on site that grinds fresh grits and cornmeal. In addition, the university has cultivated relationships with local farms and meat producers, sourcing locally produced food whenever possible.

KSU's culinary program is receiving wide recognition, and in 2013 Kennesaw State became the first educational institution to win the National Restaurant Association's Innovator of the Year award, beating out competitors such as Walt Disney Parks and Resorts and the U.S. Air Force. That year, the university opened its Institute for Culinary Sustainability and Hospitality, the first

such degree program in the United States. Its students learn the culinary, business, and scientific skills required to implement sustainable food practices in restaurants, hotels, hospitals, and schools.

Today at KSU, the links between farm and campus continue to grow. A farmers' market has been established on campus; and a student group, Students for Environmental Sustainability, has launched a permaculture project next to the science building. Student volunteers and interns work at the farms, and new classes in organic farming and beekeeping are using the farms as well.

Kennesaw State has enjoyed unusual success in linking campus and farm to enhance students' dining experiences, but food service operations on many campuses across North America today are embracing fresh locally farmed food as they pursue culinary sustainability in various ways. Michigan State University engages students in organic farming through teaching, research, and production. At California State University, Chico, more than 35 students are employed at the school's 800-acre farm, supplying fruits, vegetables, meat, and dairy products to the dining halls, farmers' markets, and local restaurants. Many dozens of schools—from Hampshire College to Clemson University to Dickinson College to Case Western Reserve University—provide their communities with food through community-supported agriculture (CSA) programs (p. 262).

Altogether, about 70% of America's largest colleges and universities now have a campus farm or garden from which their dining halls can source food directly, according to a recent survey of more than 300 major institutions. Meanwhile, more and more campus dining halls are gaining LEED certification as green buildings. Collectively, these efforts are reducing the ecological footprints of American colleges and universities and pointing the way to a more sustainable food future—all while supplying students with healthy and delicious meals.

The Changing Face of Agriculture

Most of us don't think about agriculture from day to day; after all, most people today live in cities and suburbs, not on farms or ranches. Yet agriculture provides our most basic daily needs, from the food we eat to the clothes we wear. When we eat food from a college dining hall, a grocery store, or a restaurant, that food originated with the efforts of farmers or ranchers working the land to produce crops or raise animals. When we buy shirts, pants, socks, or linens, many of these originated with cotton crops grown and tended by farmers. Our lives depend directly on agriculture—and agriculture is responsible for some of our biggest impacts on the environment, as well. For these reasons, it's important to recognize where our food and fiber come from, to understand how they are produced, and to devise ways of reducing their environmental impacts and making their production sustainable.

Several factors underpin agriculture

We can define **agriculture** as the practice of raising crops and livestock for human use and consumption. We obtain most of our food and fiber from **cropland,** land used to raise plants for human use, and from **rangeland,** or pasture, land used for grazing livestock.

Growing crops and raising animals require inputs of resources. Crops require soil, sunlight, water, nutrients, and mechanisms for pollination. Livestock require food and water. These resources are the factors that underpin agriculture—and making agriculture sustainable means safeguarding their availability and quality. As the human population has grown, so have the amounts of these resources we devote to agriculture. Today we commandeer more than 1 out of every 3 acres of land on Earth to produce food and fiber for ourselves. Rangeland covers 26% of Earth's land surface, and cropland covers 12%. The percentages are even greater in the United States, where nearly half the land is devoted to agriculture. Here, rangeland covers 27% and cropland covers 19% of our land area.

Agriculture led to modern societies

During most of the human species' 200,000-year existence, we were hunter-gatherers, depending on wild plants and animals for our food and fiber. Then about 10,000 years ago, as the climate warmed and glaciers retreated, people in some cultures began to raise plants from seed and to domesticate animals.

Agriculture most likely began as hunter-gatherers brought wild fruits, grains, and nuts back to their encampments. Some of these foods fell to the ground, were thrown away, or survived passage through the digestive system. The plants that grew from these seeds likely produced fruits larger and tastier than those in the wild, because they sprang from seeds of fruits that people had selected in the first place because they were especially large and delicious. As these plants bred with similar ones nearby, they gave rise to subsequent generations of plants with large and flavorful fruits.

Eventually people realized they could guide this evolutionary process, and our ancestors began intentionally planting seeds from plants whose produce was most desirable. This practice of selective breeding (p. 51) has generated the many hundreds of crops we enjoy today, all of which are artificially selected versions of wild plants. People followed the same process of selective breeding with animals, creating livestock from wild species.

Evidence from archaeology suggests that agriculture was invented independently by different cultures in at least 5 areas of the world, and possibly in 10 or more (FIGURE 9.1). The earliest widely accepted evidence for plant and animal domestication comes from the "Fertile Crescent" region of the Middle East at least 10,500 years ago. Wheat and barley originated in this region, as did rye, peas, lentils, onions, garlic, carrots, grapes, and domesticated goats and sheep. In China, domestication began 9500 years ago, leading to the varieties of rice, millet, and pigs we know today. Agriculture in Africa (coffee, yams, sorghum, and more) and the Americas (corn, beans, squash, potatoes, llamas, and more) developed in several regions at least 4500–7000 years ago, and likely as much as 10,000 years ago.

In one mainstream scholarly view, once our ancestors learned to cultivate crops and raise animals, they began to settle in more permanent camps and villages, often near water sources. In a self-reinforcing cycle of positive feedback (p. 104), the need to harvest crops kept people sedentary, and once they were sedentary, it made sense to plant more crops. As food supplies became more abundant, carrying capacities (p. 65) increased and populations rose. Population increase, in turn, promoted the intensification of agriculture.

Moreover, the ability to grow excess farm produce enabled some people to live off the food that others produced. This eventually led to the development of professional specialties, commerce, technology, densely populated urban centers, social stratification, and politically powerful elites. For better or worse, the advent of agriculture may be largely responsible for the civilization we know today.

Industrial agriculture dominates today

For thousands of years, the work of cultivating, harvesting, storing, and distributing crops was performed by human and animal muscle power, along with hand tools and simple machines—an approach known as **traditional agriculture.** In the past, most traditional farmers produced only enough food for their own

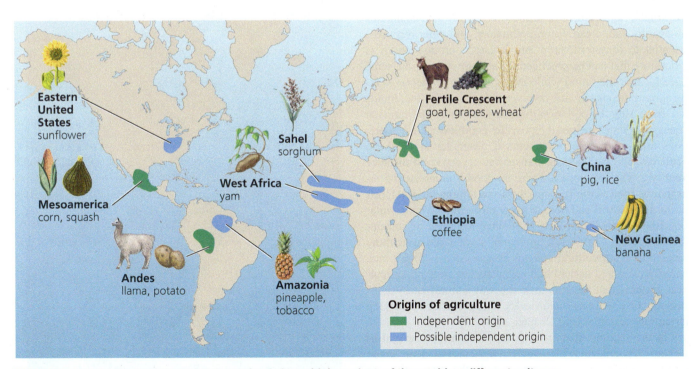

FIGURE 9.1 Agriculture originated independently in multiple regions of the world as different cultures domesticated plants and animals from wild species. *Data from Diamond, J., 1997.* Guns, germs, and steel. *New York, NY: W.W. Norton; and Goudie, A., 2000.* The human impact, *5th ed. Cambridge, MA: MIT Press.*

subsistence. Today many have integrated into market economies and produce excess food to sell, using teams of animals for labor and significant quantities of irrigation water and fertilizer.

The industrial revolution (p. 5) introduced large-scale mechanization and fossil fuel combustion to agriculture, just as it did to industry. Many farmers in industrializing societies replaced horses and oxen with machinery that provided faster and more powerful means of cultivating, harvesting, transporting, and processing crops. Such **industrial agriculture** also boosted yields by intensifying irrigation and introducing synthetic fertilizers, while the advent of chemical pesticides reduced herbivory by crop pests and competition from weeds. The use of machinery created a need for highly organized approaches to farming, leading large-scale farmers to plant vast areas with single crops in straight orderly rows. Such **monocultures** ("one type") are distinct from the **polycultures** ("many types") typical of traditional agriculture, such as Native American farming systems that mixed maize, beans, squash, and peppers in the same fields.

Industrial agriculture spread from developed nations to developing nations with the advent of the Green Revolution (see Chapter 10; p. 238). Beginning around 1950, the Green Revolution introduced new technology, crop varieties, and farming practices to the developing world. These advances dramatically increased yields and helped millions avoid starvation. Yet despite its successes, industrial agriculture is exacting a price. The intensive cultivation of monocultures using pesticides, irrigation, and chemical fertilizers has many consequences, among them the degradation of soil, water, and pollinators that we rely on for our terrestrial food supply.

Sustainable agriculture reduces environmental impacts

Sustainable agriculture describes agriculture that maintains the healthy soil, clean water, pollinators, and other resources essential to long-term crop and livestock production. It is agriculture that can be practiced in the same way in the same place far into the future while maintaining high yields. Making agriculture sustainable involves reducing fossil fuel inputs and decreasing the pollution these inputs cause. The farms at Kennesaw State University provide good examples; there, crops are grown with few chemical pesticides or synthetic fertilizers, and harvests are transported a very short distance directly to The Commons, where the food is consumed. Subsequently, food waste is composted and returned to the farms to nourish the soil and help to grow more food. This circular recycling-oriented approach is another hallmark of sustainable agriculture. In overall approach, sustainable agriculture seeks to treat agricultural systems as ecosystems. To this end, many farmers and agricultural scientists are creating agricultural systems that better mimic the way natural ecosystems function. (We will explore a variety of approaches for making agriculture sustainable in Chapter 10.)

In the sections that follow, we will examine several of the most basic factors that underpin agriculture—soil, water, nutrients, and pollinators. We will study the threats to these vital resources, and we will see how farmers, ranchers, scientists, and policymakers are working to conserve them.

Soil: A Foundation of Agriculture

Soil supports all of Earth's terrestrial ecosystems, and healthy soil is vital for agriculture. When we abuse soil, we hamper its ability to sustain our production of food and fiber. Because every one of us relies directly on agriculture for the meals we eat and the clothing we wear, the quality of our lives is closely tied to the quality of our soil.

Soil is a living system

Soil is a multifaceted system consisting of disintegrated rock, organic matter, water, gases, nutrients, and microorganisms. Although it is derived from rock, soil is shaped by living organisms (**FIGURE 9.2**). By volume, soil consists roughly of

FIGURE 9.2 Soil, a complex mixture of organic and inorganic components, is full of living organisms. Most soil organisms decompose organic matter. Some, such as earthworms, also help to aerate the soil.

50% mineral matter and up to 5% organic matter. The rest consists of the space between soil particles (pore space) taken up by air or water. The organic matter in soil includes living and dead microorganisms as well as decaying material from plants and animals. The composition of a region's soil strongly influences the character of its ecosystems. In fact, because soil is composed of living and nonliving components that interact in complex ways, soil itself meets the definition of an ecosystem (pp. 59, 110).

Soil supports agriculture

Agriculture relies on healthy soil in several ways (**FIGURE 9.3**). Crop plants depend on soil that contains organic matter to provide the nutrients they need for growth. Plants also need soil with a structure and texture suitable for roots to penetrate deeply, which assists with uptake of water and nutrients, allowing for proper growth. Plants need soil that holds water and dissolved nutrients in a way that makes them accessible to plant roots. And plants also depend on living organisms in the soil. In particular, many fungi form symbiotic mutualistic associations with plant roots, called *mycorrhizae* (p. 78). The dense network of fungal tissue in the soil helps draw up water and minerals, some of which are passed to the plant, and the plant provides to the fungus the carbohydrates produced by photosynthesis. Wheat and many other crops rely on mycorrhizae for proper growth. Animal agriculture also depends on soil with all these characteristics; because animals subsist on plants, the livestock we raise rely indirectly on healthy soil.

Healthy soil has sustained agriculture for thousands of years. When people first began farming, they were able to take advantage of deep, nutrient-rich soils that had built up over vast spans of time. Today we face the challenge of producing immense amounts of food from soil that has been farmed many times, while also conserving its fertility for the future.

Productive soil is a renewable natural resource; once depleted, soil may renew itself over time. However, renewal generally occurs very slowly. Most farming and grazing practiced so far has depleted soils faster than they form, so it is imperative for our civilization's future that we develop sustainable methods of working with soil.

Soil forms slowly

The formation of soil plays a key role in terrestrial primary succession (p. 83), which begins when the lithosphere's parent material is exposed to the effects of the atmosphere, hydrosphere, and biosphere. **Parent material** is the base geologic material in a particular location. It may be hardened lava or volcanic ash; rock or sediment deposited by glaciers; wind-blown dunes; sediments deposited by rivers, in lakes, or in the ocean; or **bedrock,** the mass of solid rock that makes up Earth's crust. Parent material is broken down by **weathering,** the physical, chemical, and biological processes that convert large rock particles into smaller particles. Physical weathering results from wind, rain, freezing, and thawing. Chemical weathering occurs as water or gases chemically alter rock. Biological weathering involves living things; for example, lichens (p. 83) produce acid that eats away at rock, and trees' roots rub against rock.

Once weathering has produced fine particles, biological activity contributes to soil formation through the deposition, decomposition, and accumulation of organic matter. As plants, animals, and microbes die or deposit waste, this material is incorporated amid the weathered rock particles, mixing with minerals. For example, the deciduous trees of temperate forests drop their leaves each fall, making leaf litter available to the detritivores and decomposers (p. 78) that break it down and incorporate its nutrients into the soil. In decomposition, complex organic molecules are broken down into simpler ones that plants can take up through their roots.

FAQ

Isn't soil just lifeless dirt?

Don't let appearances fool you—healthy soil is a complex ecological system full of life! A single teaspoon of soil can contain millions of bacteria and thousands of fungi, algae, and protists. Soil provides habitat for earthworms, insects, mites, millipedes, centipedes, nematodes, sow bugs, and other invertebrates, as well as for burrowing mammals, amphibians, and reptiles. These organisms improve the nutrient content of soil as well as its texture for retaining water and helping plants' roots to grow. In general, where we find soil rich in life, we find thriving ecological communities aboveground and excellent conditions for agriculture.

FIGURE 9.3 Crop plants such as wheat depend on healthy soil for nutrients, organic matter, water retention, and proper root growth.

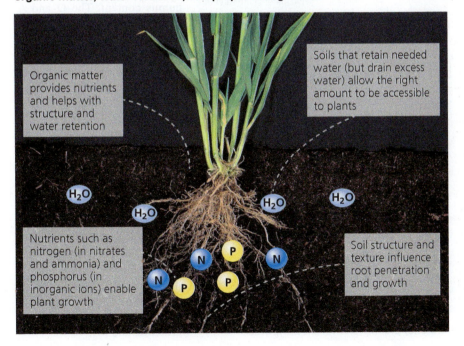

Organic matter provides nutrients and helps with structure and water retention

Soils that retain needed water (but drain excess water) allow the right amount to be accessible to plants

Nutrients such as nitrogen (in nitrates and ammonia) and phosphorus (in inorganic ions) enable plant growth

Soil structure and texture influence root penetration and growth

Partial decomposition of organic matter creates **humus,** a dark, spongy, crumbly mass of material made up of complex organic compounds. Soils with high humus content hold moisture well and are productive for plant life. For example, soils on the Kennesaw State University farms have an abundance of well-developed humus because they are regularly resupplied with organic material collected from dining hall operations.

Weathering and the accumulation and transformation of organic matter are influenced by five main factors:

- *Climate:* Soil forms faster in warm, wet climates, because heat and moisture speed most physical, chemical, and biological processes.
- *Organisms:* Plants and decomposers add organic matter to soil.
- *Topography:* Hills and valleys affect exposure to sun, wind, and water, and they influence how soil moves.
- *Parent material:* Its attributes influence properties of the soil.
- *Time:* Soil formation can take decades, centuries, or millennia.

Although soil is a renewable resource, it forms so slowly that for all practical purposes we cannot regain soil once it has been lost. Because forming just 1 inch of soil can easily require hundreds or thousands of years, we would be wise to conserve the soil we have.

A soil profile consists of horizons

As wind, water, and organisms move and sort the fine particles that weathering creates, distinct layers of soil eventually develop. Each layer is known as a **soil horizon,** and the cross-section as a whole, from surface to bedrock, is known as a **soil profile.**

The simplest way to categorize soil horizons is to recognize A, B, and C horizons corresponding respectively to topsoil, subsoil, and parent material. However, soil scientists often recognize at least three additional horizons (**FIGURE 9.4**). Soils vary by location, and few soil profiles contain all six horizons, but any given soil contains at least some of them.

Generally, the degree of weathering and the concentration of organic matter decrease as one moves downward in a soil profile. Minerals are transported downward as a result of **leaching,** the process whereby minerals suspended or dissolved in liquid are transported to another location. Soil that undergoes leaching is a bit like coffee grounds in a drip filter. When it rains, water infiltrates the soil, dissolves some of its components, and carries them downward. Minerals commonly leached from the E horizon include iron, aluminum, and silicate clay. In some soils, minerals may be leached so rapidly that plants are deprived of nutrients. Minerals that are leached from soils may enter groundwater, and some can pose human health risks when the affected water is extracted for drinking.

A crucial horizon for agriculture and ecosystems is the A horizon, or **topsoil.** Topsoil consists mostly of inorganic mineral components, with organic matter and humus from above mixed in. Topsoil is the portion of the soil that is most nutritive for plants, and it takes its loose texture, dark coloration, and strong water-holding capacity from its humus content. The O and A horizons are home to most of the organisms that give life to soil. Topsoil is vital for agriculture, but agriculture practiced unsustainably over time will deplete organic matter, reducing the topsoil's fertility and ability to hold water.

Soils differ in quality

The six horizons shown in Figure 9.4 depict an idealized soil, but soils display great variety. Scientists classify soils—and farmers judge their quality for farming—based on properties such as color, texture, structure, and pH.

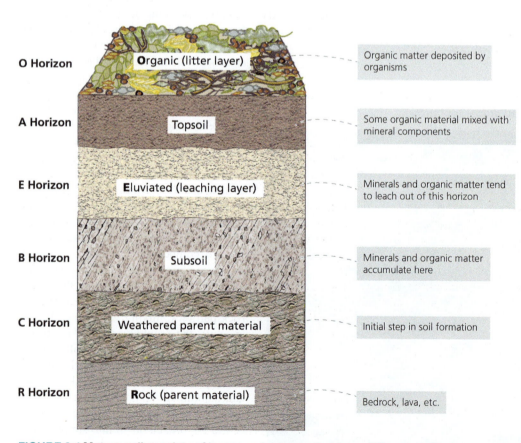

O Horizon — **O**rganic (litter layer)	Organic matter deposited by organisms
A Horizon — Topsoil	Some organic material mixed with mineral components
E Horizon — **E**luviated (leaching layer)	Minerals and organic matter tend to leach out of this horizon
B Horizon — Subsoil	Minerals and organic matter accumulate here
C Horizon — Weathered parent material	Initial step in soil formation
R Horizon — **R**ock (parent material)	Bedrock, lava, etc.

FIGURE 9.4 Mature soil consists of layers, or horizons, that have different attributes.

Color A soil's color can indicate its composition and its fertility. Black or dark brown soils are usually rich in organic matter, whereas a pale color often indicates a history of leaching or low organic content.

Texture Soil texture is determined by the size of particles (**FIGURE 9.5**). **Clay** particles are less than 0.002 mm in diameter; **silt,** 0.002–0.05 mm; and **sand,** 0.05–2 mm. Sand grains, as any beachgoer knows, are large enough to see individually and do not adhere to one another. Clay particles, in contrast, readily adhere to one another and give clay a sticky feeling when moist. Silt is intermediate, feeling powdery when dry and smooth when wet. Soil with an even mixture of the three particle sizes is known as **loam.**

Soils with large particles are porous and allow water to pass through quickly—so crops planted in sandy soils require frequent irrigation. Conversely, soils with very fine particles have small pore spaces because particles pack closely together,

making it difficult for water and air to pass through—so in clay soils, water infiltrates slowly and less oxygen is available to soil life. For these reasons, the best soils for plant growth and agriculture tend to be silty soils with medium-sized pores, or loamy soils with a mix of pore sizes.

Structure Soil structure is a measure of the "clumpiness" of soil. An intermediate degree of clumpiness is generally best for plant growth. Soil can be compacted by excessive foot traffic or by repeated plowing; this compaction reduces the soil's ability to absorb water and inhibits the penetration of plants' roots.

pH Plants can die in soils that are too acidic or too alkaline, so soils of intermediate pH values (p. 27) are best for most plants. Soil pH influences the availability of nutrients for plants' roots. For instance, acids from organic matter may leach some nutrients from the sites of exchange between plant roots and soil particles.

Regional soil differences affect agriculture

Soil characteristics vary from place to place, and they are affected by climate and other variables. For example, it may surprise you to learn that soils of the Amazon rainforest are much less productive than soils in Iowa or Kansas. This is because the enormous amount of rain that falls in tropical regions such as the Amazon basin readily leaches minerals and nutrients out of the topsoil and E horizon, below the reach of plants' roots. At the same time, warm temperatures in the Amazon speed the decomposition of leaf litter and the uptake of nutrients by plants, so only small amounts of humus remain in the thin topsoil layer.

As a result, when tropical rainforest is cleared for farming, cultivation quickly depletes the soil's fertility. This is why the traditional form of agriculture in tropical forested areas is *swidden* agriculture, in which the farmer cultivates a plot for one to a few years and then moves on to clear another plot, leaving the first to grow back to forest. Plots are often burned before planting, in which case the practice is called **slash-and-burn** agriculture (**FIGURE 9.6a**, p. 216). At low population densities this process can be sustainable, but with today's dense human populations, soils are often not allowed enough time to regenerate—and increasingly, farmed plots are converted to pasture for ranching.

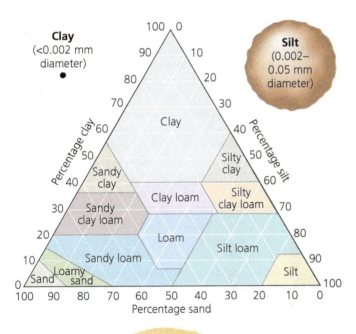

FIGURE 9.5 The texture of soil depends on its mix of particle sizes. Using this diagram, scientists classify soil texture according to the proportions of sand, silt, and clay. After measuring the percentage of each particle size in a soil sample, a scientist can trace the appropriate white lines inward from each side of the triangle to determine texture. Loam is generally best for plant growth, although some plants grow better in other textures.

 What type of soil contains 20% clay, 60% silt, and 20% sand?

Go to **Interpreting Graphs & Data** on **Mastering**EnvironmentalScience®.

FAQ

What is "slash-and-burn" agriculture?

Soils of tropical rainforests are not well suited for cultivating crops because they contain relatively low levels of plant nutrients. Instead, most nutrients are tied up in the forest's lush vegetation. When farmers cut tropical rainforest for agriculture, they enrich the soil by burning the plants on site. The nutrient-rich ash is tilled into the soil, providing sufficient fertility to grow crops. Unfortunately, the nutrients from the ash are usually depleted in just a few years. At this point, farmers move deeper into the forest and slash and burn another swath of land, causing additional impacts to these productive and biologically diverse ecosystems.

(a) Slash-and-burn agriculture on nutrient-poor soil in the tropics

(b) Industrial agriculture on rich topsoil in Iowa

FIGURE 9.6 Regional soil differences affect how people farm. In tropical forested areas such as Indonesia **(a)**, farmers pursue swidden agriculture by the slash-and-burn method because tropical rainforest soils **(inset)** are nutrient-poor and easily depleted. On the Iowa prairie **(b)**, less rainfall means fewer nutrients are leached from the topsoil, and more organic matter accumulates, forming a thick, dark topsoil layer **(inset)**.

As a result, agriculture has degraded the soils of many tropical areas.

In contrast, on the grasslands of North America, which have been almost entirely converted to agriculture (**FIGURE 9.6b**), there is less rainfall and therefore less leaching, so nutrients remain within reach of plants' roots. Plants return nutrients to the topsoil as they die, maintaining its fertility. This creates the thick, rich topsoil of temperate grasslands, which can be farmed repeatedly with minimal loss of fertility as long as farmers guard against loss of soil.

Water for Agriculture

Just as soil is a crucial resource for farming and ranching, so is water. Plants require water for growth, and we have long provided our crops and our livestock with supplemental water when needed to boost production.

Irrigation boosts productivity

The artificial provision of water beyond that which crops receive from rainfall is known as **irrigation.** Some crops, such as rice and cotton, use large amounts of water and generally require irrigation, whereas others, such as beans and wheat, need relatively little water. The amount of water a crop requires also is influenced by the rate of evaporation and by the soil's ability to hold water and make it available to plant roots. If the climate is dry, or if too much water evaporates or runs off before it can infiltrate the soil, crops may require irrigation. By irrigating crops, people can farm in arid regions (**FIGURE 9.7**) and may maintain high yields in times of drought.

Fully 70% of all fresh water that people withdraw from rivers, lakes, and underground aquifers is used for irrigation. Irrigated acreage has increased dramatically worldwide along with the adoption of industrial farming methods, reaching almost 400 million ha (nearly 1 billion acres), twice the area of Mexico. In some cases, withdrawing water for irrigation has depleted groundwater and dried up rivers and lakes. (We will examine irrigation further in Chapter 15.)

Salinization and waterlogging are easier to prevent than to correct

It's possible to overwater, however. **Waterlogging** occurs when overirrigation saturates the soil and causes the water table to rise to the point that water drowns plant roots, depriving them of access to gases and essentially suffocating them.

A more frequent problem is **salinization,** the buildup of salts in surface soil layers. In dryland areas where precipitation and humidity are low, the evaporation of water from the soil's A horizon may pull water containing dissolved salts up from lower horizons. As the water evaporates at the surface, those salts remain, often turning the soil surface white. Irrigation in arid areas generally hastens salinization, and irrigation water often contains some dissolved salt in the

(a) Flood-and-furrow irrigation of cotton in the southern California desert near the Colorado River

(b) Center-pivot irrigation in the southern California desert

FIGURE 9.7 Irrigating water-thirsty crops in arid regions causes us to lose a great deal of water to evaporation.

Sustainable approaches to irrigation maximize efficiency

One of the most effective ways to reduce water use in agriculture is to better match crops and climate. Many arid regions have been converted into productive farmland through extensive irrigation, often with the support of government subsidies that make irrigation water artificially inexpensive. Some farmers in these areas cultivate crops that require large amounts of water, such as rice and cotton. This leads to extensive water loss from evaporation in the arid climate. Choosing other crops that require far less water could enable these areas to remain agriculturally productive while greatly reducing water use.

Another approach is to embrace technologies that improve efficiency in water use. Currently, irrigation efficiency worldwide is low, as plants end up using only about 40% of the water that we apply. The rest evaporates or soaks into the soil away from plant roots (**FIGURE 9.8a**). Drip irrigation systems that target water directly toward plant roots

(a) Conventional irrigation

(b) Drip irrigation

FIGURE 9.8 Irrigation methods vary in their water use.
Conventional methods **(a)** are inefficient, because most water is lost to evaporation and runoff. In drip irrigation systems **(b)**, hoses drip water directly into soil near plants' roots, so that much less is wasted.

first place, which introduces new salt to the soil. Salinization inhibits crop production; it currently lowers yields on roughly one-fifth of all irrigated cropland globally, costing more than $11 billion each year.

Once salinization has occurred, one way to alleviate it is to stop irrigating and wait for rain to flush salts from the soil. However, in dryland areas where salinization is most often a problem, precipitation is rarely adequate to flush soils. A better solution may be to plant salt-tolerant plants, such as barley, that can be used as food or pasture. A third option is to bring in large quantities of less-saline water (if available) to flush the soil. However, using too much water may cause waterlogging.

Because remedying salinization once it has occurred is expensive and difficult, preventing it in the first place is better. The best way to prevent salinization is to avoid planting crops that require a great deal of water in dryland areas. A second way is to irrigate with water low in salt content. A third way is to irrigate efficiently, supplying no more water than a crop requires.

through hoses or tubes can increase efficiencies to more than 90% (**FIGURE 9.8b**). Such drip irrigation systems are used on the Kennesaw State University campus farms that supply The Commons with much of its produce, greatly reducing water use. In addition, rainwater is gathered in barrels atop the roof of the Commons; rainwater harvesting is another technique for making good use of water. As systems for drip irrigation and rainwater harvesting become more affordable, more farmers, gardeners, and homeowners are turning to them.

Nutrients for Plants

Along with water and soil, nutrients are vital for plant growth, and thus for cropland agriculture. Plants require nitrogen, phosphorus, and potassium to grow, as well as smaller amounts of more than a dozen other nutrients. Plants remove these nutrients from soil as they grow, as does leaching. If farmland soil is depleted of nutrients, crop yields decline. For this reason, farmers enhance nutrient-limited soils by adding **fertilizer,** substances that contain essential nutrients.

(a) Inorganic fertilizer

(b) Organic fertilizer

FIGURE 9.9 Two main types of fertilizer exist. Inorganic fertilizer **(a)** consists of synthetically manufactured granules. Organic fertilizer **(b)** includes substances such as compost crawling with natural decomposers like earthworms.

Fertilizers boost crop yields but can be overapplied

There are two main types of fertilizers (**FIGURE 9.9**). **Inorganic fertilizers** are mined or synthetically manufactured mineral supplements. **Organic fertilizers** consist of the remains or wastes of organisms and include animal manure; crop residues; charcoal; fresh vegetation; and **compost,** a mixture produced when decomposers break down organic matter such as food and crop waste in a controlled environment.

One of the highlights of the culinary sustainability efforts at Kennesaw State University is its closed-loop system for recycling wastes. Uneaten food and scraps from food preparation are placed in a large aerobic digester tank outside the dining hall. Over time, the food items break down inside the digester, generating roughly 1900 L (500 gal) of nutrient-rich "compost tea" each day. This liquid is trucked to the nearby campus farms, where it is used as an organic fertilizer.

Historically, people relied on organic fertilizers to replenish soil nutrients, but during the latter half of the 20th century many farmers in industrialized nations and nations experiencing the Green Revolution embraced inorganic fertilizers (**FIGURE 9.10**). Inorganic fertilizers have boosted our global food production, but their overapplication has triggered increasingly severe pollution problems. Because inorganic fertilizers are generally more susceptible to leaching and runoff than are organic fertilizers, they more readily contaminate surface water bodies and groundwater supplies.

Nutrients from inorganic fertilizers can have impacts far beyond the boundaries of the fields on which they are applied (**FIGURE 9.11**). For instance, nitrogen and phosphorus runoff from farms and other inland sources spurs phytoplankton blooms in the Chesapeake Bay (p. 104), the Gulf of Mexico (p. 406), and other marine and coastal regions, creating oxygen-depleted "dead zones" that kill fish and shellfish.

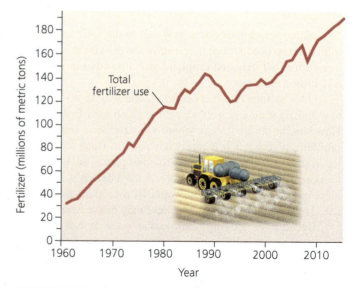

FIGURE 9.10 Use of synthetic, inorganic fertilizers has risen sharply over the past half-century. Today, usage stands at more than 190 million metric tons annually. *Data from International Fertilizer Industry Association and UN Food and Agriculture Organization (FAO).*

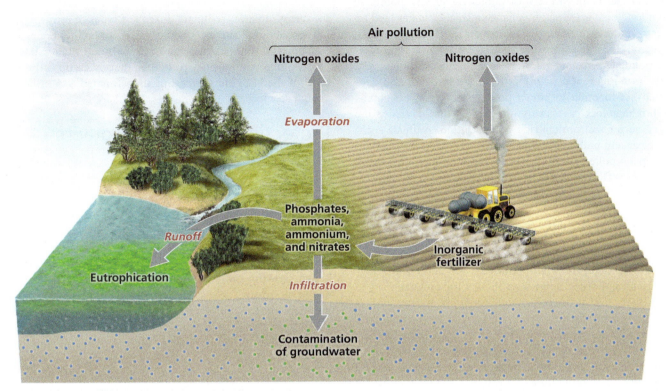

FIGURE 9.11 Overapplication of fertilizers has impacts beyond the farm field, because nutrients not taken up by plants end up elsewhere. Nitrates can leach into aquifers and contaminate drinking water. Runoff of phosphates and nitrogen compounds can alter the ecology of waterways by eutrophication. Compounds such as nitrogen oxides may pollute the air.

Such eutrophication (pp. 109, 406) occurs at countless river mouths, lakes, and ponds throughout the world. Moreover, nitrates are readily leached through soil and contaminate groundwater. Components of some nitrogen fertilizers can even volatilize (evaporate) into the air, contributing to photochemical smog (p. 462) and acid deposition (p. 469). Through these processes, unnatural amounts of nitrates and phosphates spread through ecosystems and can pose human health risks, including cancer and methemoglobinemia, a disorder that can asphyxiate and kill infants. Indeed, human inputs of nitrogen have modified the nitrogen cycle and now account for one-half the total nitrogen flux on Earth (p. 124).

Sustainable fertilizer use involves targeting and monitoring nutrients

Sustainable approaches to fertilizing crops with inorganic fertilizers target the delivery of nutrients to plant roots and avoid the overapplication of fertilizer. Farmers using drip irrigation systems can add fertilizer to irrigation water, thereby releasing it only above plant roots. Growers practicing no-till farming or conservation tillage (p. 226) often inject fertilizer along with seeds, concentrating it near the developing plant. Farmers can also avoid overapplication by regularly monitoring soil nutrient content and applying fertilizer only when nutrient levels are too low. These types of approaches are examples of **precision agriculture,** which involves using technology to precisely monitor crop conditions, crop needs, and resource use,

to maximize production while minimizing waste of resources. In addition, by planting buffer strips of vegetation along field edges and watercourses, growers can help to capture nutrient runoff before it enters streams and rivers.

Sustainable agriculture embraces the use of organic fertilizers, because they can provide some benefits that inorganic fertilizers cannot. Organic fertilizers provide not only nutrients but also organic matter that improves soil structure, nutrient retention, and water-retaining capacity. For example, *biochar* is a term for charcoal (wood heated in the absence of oxygen) used as a soil amendment. Biochar was mixed into soil by Amazonian people for centuries to increase soil fertility and crop productivity, and today it is also being considered as a means of carbon sequestration (p. 538).

The use of organic fertilizers is not without cost, though. When manure is applied in amounts needed to supply sufficient nitrogen for a crop, it may introduce excess phosphorus, which can run off into waterways. Accordingly, sustainable approaches do not rely solely on organic fertilizers but integrate them with the targeted use of inorganic fertilizer.

Pollination

Like all plants, crops need to be pollinated to set seed and produce fruit. **Pollination** (p. 78) is the process by which male sex cells of a plant (pollen) fertilize female sex cells of a plant (ova, or egg cells). Pollination can occur in different ways. Plants such as grasses and conifer trees are pollinated thanks to the wind. Millions of minuscule pollen grains are blown long distances, and by chance a small number land on the female parts of other plants of their species. In contrast, the many kinds of plants that sport showy flowers are typically pollinated by animals, such as hummingbirds, bats, and insects (see Figure 4.8, p. 78). Flowers are evolutionary adaptations that function to attract pollinators; the sweet smells and bright colors of flowers are signals that advertise the sugary nectar and protein-rich pollen within. Animals seeking nectar and pollen are drawn to flowers, and they end up transferring pollen from flower to flower as they visit them. This enables flowering plants to reproduce, set seed, and create fruits.

Many crops rely on pollinators

Our staple grain crops, such as wheat and corn, are derived from grasses and are wind-pollinated, but many fruit, vegetable, and nut crops depend on insects for pollination. The insects that pollinate our crops are among the most vital (yet least appreciated) factors contributing to our food production. They provide more than $150 billion of ecosystem services each year, and without them we would have trouble producing enough food to feed our population. Pollinators are the unsung heroes of agriculture.

The most complete survey on pollination to date, undertaken by tropical bee biologist Dave Roubik, documented 800 types of cultivated plants that rely on bees and other insects for pollination. An estimated 73% of these types are pollinated by bees, 19% by flies, 5% by wasps, 5% by beetles, and 4% by moths and butterflies. Bats pollinate 6.5% and birds 4%. Overall, native species of bees in the United States alone are estimated to provide $3 billion of pollination services each year to crop agriculture.

The European honeybee (*Apis mellifera*)—introduced to America from the Old World many decades ago—underpins modern agriculture like no other pollinator. Farmers and orchardists regularly hire beekeepers to bring hives of domesticated honeybees to their fields and orchards when it is time to pollinate crops (**FIGURE 9.12**).

FIGURE 9.12 Beekeepers bring hives of honeybees to farmers' crops when it is time for flowers to be pollinated.

At Kennesaw State University's farms, honeybees from the farms' apiaries pollinate dozens of types of crops, ensuring productive harvests while also producing honey for consumption on campus. Across the United States, honeybees pollinate more than 100 crops that make up fully one-third of the U.S. diet, providing an estimated $15 billion in annual services.

Protecting pollinators protects agriculture

When we harm pollinating insects we harm our crop yields, but by encouraging pollinators we can enhance our yields. As one example of many, the U.S. Great Basin states are a world center for the production of alfalfa seed, and alfalfa flowers are pollinated mostly by native alkali bees that live in the soil as larvae. In the mid-20th century, many farmers began plowing the land and increasing chemical pesticide use in an effort to boost yields. These measures killed vast numbers of the soil-dwelling bees, and alfalfa seed production plummeted in areas where alkali bees were lost. However, farmers who realized the importance of the bees began transplanting divots of soil containing bee larvae to establish and manage new populations near their crops. By encouraging these native bees, these farmers were able to raise yields of alfalfa seed from 300–600 lb/acre to 1000–2400 lb/acre.

Today many pollinating insects are in trouble. Mostly as a result of pesticide use and habitat loss, populations of native pollinators are dwindling and species are disappearing from region after region. At the same time, European honeybees are suffering dramatic declines, and scientists are rushing to learn why and to help us respond (p. 245). (We will examine these losses, assess the risks and impacts, and explore solutions in our discussion of pollinator conservation in Chapter 10.) Finding ways to conserve pollinators is a crucial step to attaining sustainable agriculture.

Conserving Agricultural Resources

Today we face challenges with each of the key resources necessary for agriculture: soil, water, nutrients, and pollinators. If we are to feed the world's rising human population, we will need to modify our diets or increase agricultural production (or both)—and do so sustainably, without degrading our resources. We cannot simply keep expanding farming and grazing into new areas, because land suitable and available for agriculture is running out. When we pursue farming or grazing on unsuitable lands, it can turn grasslands into deserts; remove ecologically precious forests; diminish biodiversity; encourage invasive species; pollute soil, air, and water with toxic chemicals; and allow fertile soil to be blown and washed away. Instead, we must find ways to improve the efficiency of food production in areas already under cultivation, while pursuing agricultural methods that exert less impact on natural systems.

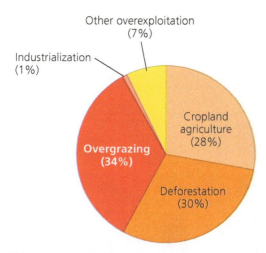

Industrialization
(1%)

Other overexploitation
(7%)

Cropland
agriculture
(28%)

Overgrazing
(34%)

Deforestation
(30%)

(a) Causes of soil degradation

FIGURE 9.13 Most of the world's soil degradation results from overgrazing by livestock, deforestation, and cropland agriculture. *Data from Wali, M.K., et al., 1999. Assessing terrestrial ecosystem sustainability: Usefulness of regional carbon and nitrogen models.* Nature and Resources 35: 21–33.

(b) Farmer with degraded soil

Damage to land and soil makes conservation vital

Each year, our planet gains more than 80 million people yet loses 5–7 million ha (12–17 million acres, about the size of West Virginia) of productive cropland to land degradation. **Land degradation** refers to a general deterioration of land that diminishes its productivity and biodiversity, impairs the functioning of its ecosystems, and reduces the services these ecosystems provide to us. Land degradation is caused by the cumulative impacts of unsustainable agriculture, deforestation, and urban development. It is manifested in processes such as soil erosion, nutrient depletion, water scarcity, salinization, waterlogging, chemical pollution, changes in soil structure and pH, and loss of organic matter from the soil. Through its impacts on agriculture, land degradation directly affects as much as one-third of the world's people.

Within the broad problem of land degradation, agricultural concerns often focus on what is known as **soil degradation,** the process by which soils deteriorate in quality and decline in productivity. Soil degradation results primarily from forest removal, cropland agriculture, and overgrazing of livestock (**FIGURE 9.13a**). In many places throughout the world, especially in drier regions, it has become more difficult to raise crops and graze livestock as soil continues to degrade (**FIGURE 9.13b**). Over the past 50 years, scientists estimate that soil degradation has reduced potential rates of food production by 13% on cropland and 4% on rangeland. By midcentury, there will likely be 2 billion more mouths to feed (p. 192), so it is imperative that we learn to practice agriculture in sustainable ways that maintain the integrity of our soil. As we shall see, scientists, policymakers, farmers, and ranchers are making great strides in discovering and implementing ways to conserve soil and other agricultural resources.

Erosion threatens ecosystems and agriculture

Erosion is the removal of material from one place and its transport to another by the action of wind or water (**FIGURE 9.14**). When eroded material is deposited at a new

FIGURE 9.14 Water erosion can readily remove soil from areas where soil is exposed, such as farmland.

location, this is called **deposition.** Erosion and deposition are natural processes, and in the long run deposition helps to create new soil. For example, flowing water may deposit freshly eroded sediment rich in nutrients across river valleys and deltas, helping to form rich and productive soils. This is why floodplains are excellent for farming.

However, erosion can be a problem for ecosystems and for agriculture because it tends to occur much more quickly than soil is formed. Erosion also tends to remove topsoil, the most valuable soil layer for living things. And when eroded soils are carried out to sea, their nutrients are lost to terrestrial systems. Windy regions with sparse plant cover experience the most wind erosion, whereas areas with steep slopes, intense precipitation, and sparse vegetative cover suffer the most water erosion.

Although erosion is a natural process, people have made land more vulnerable to erosion in three ways:

- Overcultivating fields through poor planning or excessive tilling (plowing)
- Grazing rangeland with more livestock than the land can support
- Clearing forests on steep slopes or with large clear-cuts

Erosion can be difficult to detect and measure, even when it is having substantial consequences. For example, losing just a penny's thickness of surface soil may be hard to notice, yet this translates to a loss of fully 12 tons per hectare (5 tons/acre) of valuable topsoil. One study determined that at erosion rates typical for the United States, U.S. croplands lose about 2.5 cm (1 in.) of topsoil every 15–30 years, reducing corn yields by 4.7–8.7% and wheat yields by 2.2–9.5%.

To minimize erosion, we can erect physical barriers that capture soil. In the long term and across large areas, however, the growth of vegetation is what prevents soil loss. Vegetation slows wind and water flow while plant roots hold soil in place and take up water.

Soil erosion is a global issue

More than 19 billion ha (47 billion acres) of the world's croplands suffer from erosion and other forms of soil degradation resulting from human activity. U.S. farmlands lose roughly 5 tons of soil for every ton of grain harvested. In the past decade, China lost an area of arable farmland the size of Indiana. In Kazakhstan, wind eroded tens of millions of hectares after industrial crop agriculture was imposed on land better suited for grazing. In Africa, soil degradation in coming decades could reduce crop yields by half.

People are the primary cause of erosion today, and we have accelerated it to unnaturally high rates. In a 2004 study, geologist Bruce Wilkinson analyzed prehistoric erosion rates from the geologic record and compared these with modern rates. He concluded that human activities move over 10 times

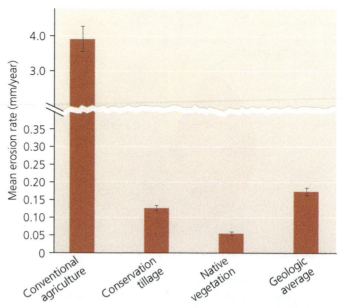

FIGURE 9.15 Erosion rates from conventional (industrial) agriculture are high. They greatly exceed rates in fields farmed under conservation tillage (p. 226), rates in areas covered by native vegetation, and rates averaged over the geologic record. *Data from Montgomery, D.R., 2007. Soil erosion and agricultural sustainability. Proc. Natl. Acad. Sci. USA 104: 13268–13272.*

DATA How many times less is the erosion rate from conservation tillage agriculture than the erosion rate from conventional agriculture?

Go to **Interpreting Graphs & Data** on **Mastering**EnvironmentalScience®.

more soil than all natural processes combined. A 2007 study by soil scientist David Montgomery found an even greater degree of impact (**FIGURE 9.15**). Montgomery's study also pointed toward a solution, revealing that land farmed under conservation approaches erodes less than land under conventional farming.

Desertification reduces productivity of arid lands

Much of the world's population lives and farms in *drylands*, arid and semi-arid environments that cover about 40% of Earth's land surface. Precipitation in these regions is meager, so drylands are prone to desertification. **Desertification** describes a form of land degradation in which more than 10% of productivity is lost as a result of erosion, soil compaction, forest removal, overgrazing, drought, salinization, climate change, water depletion, and other factors. Most such degradation results from wind and water erosion (**FIGURE 9.16**). Severe desertification can expand existing desert areas and create new ones (**FIGURE 9.17**). This process has occurred most dramatically in areas of the Middle East that have been inhabited, farmed, and grazed for long periods—including

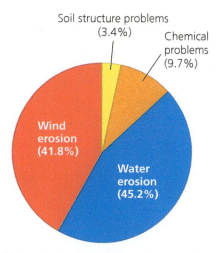

FIGURE 9.16 Soil degradation on drylands is due primarily to erosion by water and wind. *Data from UN Environment Programme, 2002.* Tackling land degradation and desertification. *Washington, D.C., and Rome, Italy: Global Environment Facility and International Fund for Agricultural Development.*

the Fertile Crescent region, where agriculture first originated (p. 211).

By some estimates, desertification endangers the food supply or well-being of more than 1 billion people in more than 100 countries and costs tens of billions of dollars in income each year. China alone loses $6.5 billion annually from desertification. In China's western reaches, desert areas are expanding because of overgrazing from more than 400 million goats, sheep, and cattle. In the Sistan Basin along the border of Iran and Afghanistan, a formerly moist oasis that supported a million livestock recently turned barren in just five years, and windblown sand buried more than 100 villages. In Kenya, overgrazing and deforestation fueled by rapid population growth have left 80% of its land vulnerable to desertification.

Everywhere, soil degradation forces ranchers to crowd onto poorer land and farmers to reduce the fallow periods during which land lies unplanted and can regain nutrients. In a positive feedback cycle (p. 104), both of these actions worsen soil degradation further.

Desertification is expected to grow worse as climate change alters rainfall patterns, making some areas drier. The United Nations estimates that tens of millions of people could be displaced, and suggests that industrialized nations fund reforestation projects in drylands of the developing world to slow desertification while gaining carbon credits in emissions trading programs (pp. 308, 509).

As a result of desertification, gigantic dust storms from denuded land in China now sometimes blow across the Pacific Ocean to North America, and dust storms from Africa's Sahara Desert blow across the Atlantic Ocean to the Caribbean Sea (see Figure 17.11c, p. 453). Such massive dust storms occurred in the United States during the early 20th century, when desertification shook American agriculture and society to their very roots.

The Dust Bowl prompted the United States to fight erosion

Before the large-scale cultivation of North America's Great Plains, native prairie grasses of this temperate grassland region held soils in place. In the late 19th and early 20th centuries, many homesteading settlers arrived in Oklahoma, Texas, Kansas, New Mexico, and Colorado, hoping to make a living there as farmers. Between 1879 and 1929, cultivated area in the region soared from 5 million ha (12 million acres) to 40 million ha (100 million acres). Farmers grew abundant wheat and ranchers grazed many thousands of cattle, sometimes expanding onto unsuitable land and causing erosion by removing native grasses and altering soil structure.

FIGURE 9.17 Severe desertification can cause desert areas to expand. In this photo, immense sand dunes moving in from the Gobi Desert are burying rural communities in northwestern China.

(a) Kansas dust storm, 1930s

FIGURE 9.18 Drought and poor agricultural practices devastated millions of U.S. farmers in the 1930s in the Dust Bowl. The photo **(a)** shows towering clouds of dust approaching homes near Dodge City, Kansas. The map **(b)** shows the Dust Bowl region. Eroded soil from this region blew eastward all the way to the Atlantic Ocean.

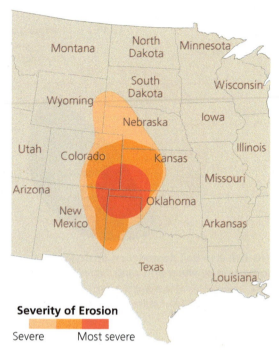

Severity of Erosion

Severe   Most severe

(b) Dust Bowl region

In the early 1930s, a drought worsened the ongoing human impacts, and the region's strong winds began to erode millions of tons of topsoil (**FIGURE 9.18**). Massive dust storms traveled up to 2000 km (1250 mi), blackening rain and snow as far away as New York and Washington, D.C. Some areas lost 10 cm (4 in.) of topsoil in just a few years. The most affected region in the southern Great Plains became known as the **Dust Bowl,** a term now also used for the historical event itself. The "black blizzards" of the Dust Bowl forced many thousands of farmers off their land, driving them into poverty and westward as refugees, and adding to the economic hardship of America's Great Depression.

In response, the U.S. government, along with state and local governments, increased support for research into soil conservation. The U.S. Congress passed the Soil Conservation Act of 1935, which established the Soil Conservation Service (SCS). This agency worked closely with farmers to develop conservation plans for individual farms, using science to assess the land's resources and problems, and collaborating with landowners to ensure that the plans harmonized with landowners' objectives. The teams formed by the SCS to combat erosion included soil scientists, forestry experts, engineers, economists, and biologists. These teams were among the earliest examples of interdisciplinary environmental problem solving.

The first director of the SCS, Hugh Hammond Bennett, was an innovator and evangelist for soil conservation. Under his dynamic leadership, the agency promoted soil conservation practices through county-based conservation districts. Organized by the states but operating with federal direction, authorization, and funding, these districts continue today to implement soil conservation programs and empower local residents to plan and set priorities.

In 1994 the SCS was renamed the Natural Resources Conservation Service (NRCS), and its responsibilities were expanded to include water quality protection and pollution control. The NRCS and most state universities employ agricultural extension agents (**FIGURE 9.19**), trained experts who assist farmers by providing information on new research and techniques and by helping them apply this knowledge to implement conservation measures on their land.

FIGURE 9.19 Agricultural extension agents assist farmers by providing information on new research and techniques. In this photo, an extension agent from the National Resources Conservation Service inspects potato plants grown by a Montana farmer.

(a) Crop rotation

(b) Contour farming

(c) Terracing

(d) Intercropping

(e) Shelterbelts

(f) No-till farming

FIGURE 9.20 Farmers have adopted various strategies to conserve soil. Rotating crops **(a)** such as soybeans and corn helps restore soil nutrients and reduce impacts of pests. Contour farming **(b)** reduces erosion on hillsides. Terracing **(c)** minimizes erosion in mountainous areas. Intercropping **(d)** reduces soil loss and maintains soil fertility. Shelterbelts **(e)** protect against wind erosion. In **(f),** soybeans grow through stubble remaining from a wheat crop, in no-till agriculture.

Farmers conserve soil and resources in many ways

Through a diversity of efforts since the Dust Bowl, scientists, policymakers, and resource managers have learned how to better protect soil quality, and today millions of farmers in America and worldwide are practicing conservation measures. A number of well-tested farming methods can reduce erosion and alleviate the impacts of conventional cultivation on soil and other resources (**FIGURE 9.20**). Some of these approaches have been promoted since the Dust Bowl, some have become popular more recently, and some have been practiced for centuries. These approaches are now widely applied in many places around the world.

Crop rotation **Crop rotation** describes the process in which farmers alternate the type of crop grown in a given field from one season or year to the next (FIGURE 9.20a). Many U.S. farmers who grow wheat or corn alternate these crops with soybeans from year to year. Soybeans are legumes, plants with specialized bacteria on their roots that fix nitrogen (p. 124), revitalizing the soil with nutrients. In this way, crop rotation returns nutrients to the soil.

Rotating crops also reduces insect pests; if an insect is adapted to feed and lay eggs on one crop, then planting a different type of crop will leave its offspring with nothing to eat. This helps to break cycles of disease associated with continuous planting of the same crop. Finally, crop rotation helps to minimize the erosion that occurs when farmers let fields lie fallow (unplanted).

Contour farming Water running down a hillside with little plant cover can carry soil away, so farmers have developed methods for cultivating slopes. **Contour farming** (FIGURE 9.20b) is a process in which furrows are plowed sideways across a hillside, perpendicular to its slope and following the natural contours of the land. In contour farming, the side of each furrow acts as a small dam that slows runoff and captures eroding soil. Contour farming is most effective on gradually sloping land with crops that grow well in rows.

Terracing On very steep terrain, the most effective method for reducing erosion is terracing (FIGURE 9.20c). Terraces are level platforms, sometimes with raised edges, that are cut into steep hillsides to contain water from irrigation or precipitation. **Terracing** transforms slopes into a series of steps like a staircase, enabling farmers to cultivate hilly land without losing huge amounts of soil to water erosion. Terracing also helps farmers conserve and make maximum use of water. Farmers have used terracing for centuries in mountainous regions, such as the foothills of the Himalayas and the Andes. Terracing is labor-intensive to establish but in the long term is likely the only sustainable way to farm in mountainous terrain.

Intercropping Farmers also minimize erosion by **intercropping,** planting different crops in alternating bands or other spatially mixed arrangements (FIGURE 9.20d). Intercropping helps slow erosion by providing more ground cover than does a single crop. Like crop rotation, intercropping reduces vulnerability to insects and disease and, when a nitrogen-fixing legume is included, replenishes the soil with nutrients.

Shelterbelts A widespread technique to reduce erosion from wind is to establish **shelterbelts,** or windbreaks (FIGURE 9.20e). These are rows of trees or tall shrubs that are planted along the edges of fields to slow the wind. On North America's Great Plains, fast-growing species such as poplars are often used. Shelterbelts can be combined with intercropping; mixed crops are planted in rows surrounded by or interspersed with rows of trees that provide fruit, wood, and wildlife habitat, as well as protection from wind.

Conservation tillage Turning the earth by tilling (plowing, disking, harrowing, or chiseling) aerates the soil and works weeds and old crop residue into the soil to nourish it. However, tilling also leaves the surface bare, allowing wind and water to erode away precious topsoil. **Conservation tillage** encompasses an array of approaches that reduce the amount of tilling relative to conventional farming. **No-till** farming (FIGURE 9.20f)—the ultimate form of conservation tillage—eliminates tilling altogether. Rather than plowing after each harvest, farmers practicing conservation tillage leave crop residues atop their fields, keeping the soil covered with plant material.

Often, temporary *cover crops* are planted to hold the soil in place between times that main food crops are growing. To plant the next food crop, the grower cuts a thin, shallow groove into the soil surface using a "no-till drill" mounted on a tractor. This device drops seeds into the furrow and closes the furrow over the seeds, minimizing disturbance to the soil. Often a localized dose of fertilizer is added to the soil along with the seed. By planting seeds of the new crop through the residue of the old, less soil erodes away, organic material accumulates, and the soil soaks up more water—all of which encourages better plant growth.

By adding organic matter to the soil, no-till farming captures carbon that otherwise would make its way to the atmosphere and contribute to climate change (Chapter 18), and instead stores it in the soil. Moreover, because no-till farming reduces tractor use, the farmer burns less gasoline. For all these reasons, conservation tillage has been on the rise.

Today across the United States, nearly one-quarter of farmland is under no-till cultivation, and more than 40% is farmed using conservation tillage. Forty percent of soybeans, 21% of corn, and 18% of cotton receive no-till treatment. All these conservation approaches are having positive effects: According to U.S. government figures, soil erosion on cropland in the United States declined from 3.05 billion tons in 1982 to 1.72 billion tons in 2010.

weighing the
ISSUES

How Would You Farm?

You are a farmer who owns land on both sides of a steep ridge. You want to plant a sun-loving crop on the sunny, but very windy, south slope of the ridge and a crop that needs a great deal of irrigation on the north slope. What farming techniques might best conserve your soil? What factors might you want to know about before you decide to commit to one or more methods?

Grazing practices affect soil quality

We have focused in this chapter largely on the cultivation of crops, but raising livestock (discussed further in Chapter 10) is a major component of agriculture and likewise exerts a toll on soils and ecosystems. People across the globe tend more than 3.4 billion cattle, sheep, and goats, most of which graze on grasses on the open range. As long as livestock populations do not exceed the rangeland's carrying capacity (p. 65) and do not consume grass faster than it can regrow, grazing can be sustainable. Moreover, human use of rangeland does not

FIGURE 9.21 Overgrazing occurs when livestock deplete plants faster than they can regrow, reducing or eliminating the vegetation that covers the soil.

necessarily exclude its use by wildlife or its continued functioning as a grassland ecosystem. However, grazing too many livestock that destroy too much of the plant cover impedes plant regrowth. Without adequate regeneration of plant biomass, the result is **overgrazing** (FIGURE 9.21).

When livestock remove too much plant cover or churn up the soil with their hooves, soil is exposed and made vulnerable to erosion. In a positive feedback cycle, soil erosion makes it difficult for vegetation to regrow, a problem that perpetuates the lack of cover and gives rise to more erosion (FIGURE 9.22). Moreover, non-native weedy plants that are unpalatable or poisonous to livestock may invade and outcompete native vegetation in the new, modified environment. Too many livestock trampling the ground can also compact soil and alter its structure. Soil compaction makes it more difficult for water to infiltrate, for soil to be aerated, for plants' roots to expand, and for roots to conduct cellular respiration (p. 31). All of these effects further decrease plant growth and survival.

Worldwide, overgrazing causes as much soil degradation as cropland agriculture does, and it causes more desertification. Degraded rangeland costs an estimated $23 billion per year in lost productivity. Grazing exceeds the sustainable supply of grass in India by 30% and in parts of China by up to 50%. To relieve pressure on rangelands, both nations are now beginning to feed crop residues to livestock.

Subsidies and conservation measures each influence ranching

In the United States, range managers assess the carrying capacity of rangelands and inform livestock owners of these limits, so that herds are rotated from site to site as needed to conserve grass cover and soil integrity. Managers also can establish and enforce limits on grazing on publicly owned land. Most U.S. rangeland is federally owned and managed by the Bureau of Land Management (BLM). The BLM is the nation's largest landowner, with more than 100 million ha (248 million acres), mostly in 12 western states (see Figure 12.10, p. 310). Ranchers can be granted rights to graze livestock on BLM lands for inexpensive fees; a grazing permit in 2016 was just $2.11 per month per "animal unit" (one horse, one cow plus calf, five sheep, or five goats).

As a result of these low fees, American ranchers have traditionally had little incentive to conserve rangelands. Because most grazing has taken place on public lands leased from the government and because U.S. taxpayers have subsidized grazing, a classic "tragedy of the

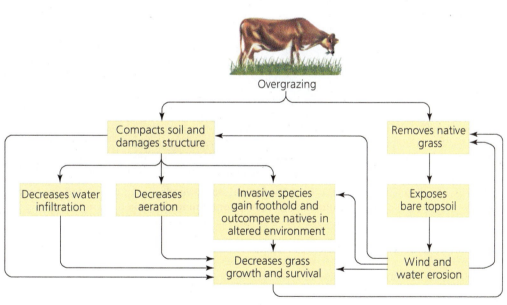

FIGURE 9.22 Overgrazing has ecological consequences. When grazing by livestock exceeds the carrying capacity of rangelands and their soil, this can set in motion a series of consequences and positive feedback loops (p. 104) that degrade soils and grassland ecosystems.

DATA Q As in most charts like this one, arrows lead from causes to consequences.
• What is the immediate cause of exposure of bare topsoil?
• What is the immediate consequence of exposing bare topsoil?
• How many immediate consequences of wind and water erosion are shown in this diagram?

Go to **Interpreting Graphs & Data** on **Mastering**EnvironmentalScience®.

How Can Ranchers and Scientists Work Together to Restore Grasslands?

Burning brush to reestablish grass cover in the Malpai Borderlands

In the high desert of southern Arizona and New Mexico, scientists and cattle ranchers trying to heal the scars left by decades of overgrazing found they had to contend with a creature even more damaging than a hungry steer: Smokey Bear.

You might think wildfires would be a natural enemy of grasslands, but researchers in the Malpai Borderlands realized that people's efforts to suppress fire had done far more harm. Before large numbers of settlers and ranchers arrived more than a century earlier, this semi-arid western landscape thrived under an ecological cycle common to many grasslands. Trees such as mesquite grew near creeks. Hardy grasses such as black grama covered the drier plains. Periodic wildfires, usually sparked by lightning, burned back shrubs and trees and kept grasslands open. Deer, rabbits, and bighorn sheep grazed on the grasses but rarely ate enough to deplete the range. Fed by seasonal rains, new grasses sprouted without being overeaten or crowded out by larger plants.

By the 1990s, however, those grasslands had become scarce. Ranchers had brought large cattle herds to the area in the late 1800s. The cows chewed through vast expanses of grass, trampled the soil, and scattered mesquite seed into areas where grasses had dominated. Ranchers fought wildfires to keep their herds safe, and the federal government joined in the firefighting efforts.

The Malpai's ranching families found themselves struggling to make a living. Decades of photos taken by ranchers and by University of Arizona botanist Raymond Turner show how soil had eroded and how trees and brush had overgrown the grass. The ranchers knew their cattle were part of the problem, but they also suspected that firefighting efforts were to blame.

In 1993, a group of ranchers launched an innovative plan. They formed the Malpai Borderlands Group, designating about 325,000 ha (800,000 acres) of land for protection and study. Ranchers enlisted government agencies and scientists to study the region's ecology, bring back grasses, and restore native animal species. They sought to return periodic fire to the landscape by conducting prescribed burns (p. 314) and by allowing natural fires to run their course (**FIGURE 1**).

FIGURE 1 The Malpai Borderlands Group reinstated fire as a landscape process, conducting controlled burns. Monitoring indicates fire has improved ecological conditions in the region.

commons" situation (p. 162) has developed, and overgrazing has resulted in extensive environmental impacts across the American West.

For this reason, ranchers and environmental advocates have traditionally been at loggerheads. In recent years, however, some ranchers and environmental advocates have teamed up to preserve ranchland against what each of them views as a common threat—the encroaching housing developments of suburban sprawl (p. 334). Although developers often pay high prices for ranchland, many ranchers do not want to see the loss of the wide-open spaces and the ranching lifestyle that they cherish. Today increasing numbers of ranchers are working cooperatively with government agencies, environmental scientists, and even environmental advocates to find ways to raise livestock more sustainably and safeguard the health of the land (see **THE SCIENCE BEHIND THE STORY**, above).

The group's research efforts have centered on the Gray Ranch, a 121,000-ha (300,000-acre) parcel in the heart of the borderlands. At McKinney Flats within the Gray Ranch, scientists led by Denver Zoo researcher Charles Curtin divided rangeland into four study areas of about 890 ha (2200 acres) each. Each area is further divided into four "treatments," or areas with varying land management techniques:

- In Treatment 1, land is burned and grazed.
- In Treatment 2, land is burned but not grazed.
- In Treatment 3, land is grazed but not burned.
- In Treatment 4, land is neither grazed nor burned.

Treatments 1 and 3, which allow grazing, also feature small fenced-off areas that prevent animals from eating grass. These "exclosures" allow scientists to make precise side-by-side comparisons of how grazing affects grasses. Scientists measure rainfall in each area and monitor soils for degradation and erosion. Teams of wildlife and vegetation specialists monitor each treatment for the distribution and abundance of birds, insects, animals, and vegetation (**FIGURE 2**).

By comparing areas where fire is suppressed to those where fire can burn, researchers have documented how the suppression of fire leads to more brush and trees and less grass. Conversely, when fire burns an area, woody plants such as mesquite trees are damaged and their seed production is disrupted, and grass subsequently returns to replace the trees. Such changes follow both natural fires and carefully monitored, deliberately set controlled burns.

All told, the landscape-level (p. 114) McKinney Flats project may be the largest replicated terrestrial ecological experiment in North America. It has helped spur other research initiatives, and the Malpai Borderlands Group hosts an annual scientific conference. Each year ranchers hear results of scientific research into soils, grasslands, forests, fire, and climate on both the U.S. and the Mexican sides of the border, as well as studies of jackrabbits, prairie dogs, bison, bighorn sheep, and an array of the region's endangered plants and animals. Ranchers and researchers have even found ways to assist threatened species and ranching at the same time, like creating ponds to serve an endangered frog and thirsty cattle.

FIGURE 2 Researchers measure vegetation on research plots in the Malpai Borderlands.

Scientists have also helped ranchers develop a cycle of "grassbanking," in which cattle are allowed to graze on shared plots of land while other areas recover. Ranchers must also work with the weather. Scientists have found that controlled burns or grassbanking should track with cycles of rain and drought to bring back grass.

Because of such research, controlled burns are now a regular part of the Malpai landscape. Ranchers have burned more than 100,000 ha (250,000 acres) since 1994, and natural fires are often allowed to run their course with little or no intervention. Damaged areas have been reseeded with native grasses.

Agency grants and personnel have also helped ranchers to reduce water runoff and soil erosion in this arid region where water is precious. Engineering rocks and brush to slow water flow along 33 miles of streams and gullies is helping water infiltrate into the soil and remain available to plants, cattle, and wildlife.

Scientists increasingly believe that sustainable ranching, if managed properly, can help restore grasslands damaged by overgrazing. "We cannot assume rangelands will recover on their own," Curtin wrote in one study on the Malpai Borderlands. "Conservation of grazed lands requires restoring and sustaining natural processes."

Policy can promote conservation measures in agriculture

Governments have long sought ways to encourage agricultural production, but some long-standing policies have worsened land degradation. Many nations spend billions of dollars in subsidies to fund practices that are not economically and environmentally sustainable, such as growing water-thirsty crops in desert regions. Such subsidies may cause growers to farm in areas or in ways they would not choose to farm without the guarantee of financial support. In the United States, roughly one-fifth of the income of the average U.S. farmer comes from subsidies. Proponents of such subsidies stress that the uncertainties of weather make profits and losses from farming unpredictable from year to year. To persist, these proponents say, farmers need some way of being compensated in bad years. This may be the case, but subsidies can encourage people to cultivate land that would otherwise not be farmed; to produce more food than is needed, driving down prices for other producers; and to practice methods that damage or deplete soil, water, nutrients, and pollinators. Opponents of environmentally destructive subsidies suggest that a better model is for farmers to buy insurance to protect themselves against production shortfalls.

In recent years, more and more public policy has sought to lessen the environmental impacts and external costs (pp. 141, 162) of agriculture.

FIGURE 9.23 Buffer strips of natural vegetation between farmed fields reduce erosion and improve water quality. The Conservation Reserve Program pays farmers to establish these areas.

In theory, the marketplace should discourage people from farming and grazing using intensive methods that degrade the land they own if such practices are not profitable in the long run. But land degradation often unfolds gradually, whereas farmers and ranchers generally cannot afford to go without profits in the short term, even if they know conservation is in their long-term interests. For this reason, we have increasingly developed public policy to encourage conservation measures. In current U.S. policy, financial incentives are used as a means to influence agricultural land use.

Every five to six years, the U.S. Congress has passed comprehensive legislation that guides agricultural policy. The 2014 Farm Bill funded 15 programs that encourage the conservation of soil, grasslands, wetlands, wildlife habitat, and other natural resources on agricultural lands. For example, the Wetlands Reserve Program offers payments to landowners who protect, restore, or enhance wetland areas on their property. Many of the provisions promoting soil conservation require farmers to adopt soil conservation plans and practices before they can receive government subsidies.

The **Conservation Reserve Program,** first established in the 1985 Farm Bill, pays farmers to stop cultivating damaged and highly erodible cropland, and instead to place it in conservation reserves planted with grasses and trees (**FIGURE 9.23**). Lands under the Conservation Reserve Program now cover an area nearly the size of Virginia, and the U.S. Department of Agriculture (USDA) estimates that each dollar invested in the program saves nearly 1 ton of topsoil. Besides reducing erosion, the Conservation Reserve Program generates income for farmers, improves water quality, and provides habitat for wildlife.

Farmers apply for the program, and the U.S. Farm Service Agency uses an "environmental benefits index" to select those who will be awarded contracts. This index includes the predicted benefits to wildlife, water quality, air quality, and the farm through reduced erosion, as well as benefits likely to endure beyond the contract period, with all these balanced against the cost of payments. Contracts generally run for 10–15 years, and annually the federal government pays farmers about $2 billion for the conservation of these lands. The most recent (2014) farm bill re-authorized the Conservation Reserve Program, but capped the amount of land it could cover by 2017 at 10 million ha (24 million acres). The actual area in reserves tends to vary with market prices for crops. Since 2007, the area in conservation reserves has decreased by 35% in response to higher food prices as farmers have withdrawn lands from the program and planted them with crops or pasture grasses.

Internationally, the United Nations promotes soil conservation and sustainable agriculture through a variety of programs led by the Food and Agriculture Organization (FAO). As just one example, FAO's Farmer-Centered Agricultural Resource Management Program (FARM) supports innovative approaches to resource management and sustainable agriculture in eight Asian nations. This program studies success stories and tries to help farmers elsewhere duplicate successful efforts. Rather than following a top-down, government-controlled approach, the FARM program calls on the creativity of local communities to educate and encourage farmers to conserve soils and secure their food supply.

weighing the
ISSUES

Soil, Subsidies, and Sustainability

Do you think that financial incentive programs such as the Conservation Reserve Program and the Wetlands Reserve Program are a good use of taxpayers' money? Are financial incentives more effective than government regulation for promoting certain land use goals? Do you think they can help lead us toward agriculture that is truly sustainable?

closing
THE LOOP

Even if we have never set foot on a farm, milked a cow, or shorn a sheep, we depend on agriculture for our daily needs, including food and clothing. It is thus important for all of us to help ensure that the world's agricultural systems are sound and sustainable. This means safeguarding the quality and availability of resources on which crops and livestock rely, including soil, water, nutrients, and pollinators. When students at Kennesaw State University enter their dining hall, they actually know where a good portion of their food comes from, and how it has been grown. By sourcing produce locally at its campus farms, The Commons at KSU provides wholesome, delicious food produced in sustainable ways. With a minimum of fossil fuel consumption, chemical use, and processing, Kennesaw State supplies its students with food choices that have a light footprint, thus protecting the region's soil, water, and other agricultural resources.

Today many colleges and universities are managing campus farms, serving organic food, running trayless dining halls, composting food scraps, reusing waste oil as biodiesel, and even growing food for the communities around them. Chances are that you, as a college student, can take part in such activities and play a role in helping to reduce the ecological footprint of modern agriculture. Experiences as a student on campus can ready you for entry into a society struggling to meet the challenges of feeding a global population of 7 billion people and climbing. With the help of science, technology, policy, and good old-fashioned experience on the farm and the ranch, our society is finding ways to boost yields while conserving vital resources and reducing agriculture's environmental impacts. In light of growth in population and consumption, we will likely need still wider adoption of conservation measures if we are to achieve sustainable agriculture and feed the 9 billion people expected to crowd our planet by midcentury.

REVIEWING Objectives

You should now be able to:

Outline broad developments in the history of agriculture

At least 10,000 years ago, people began producing crop plants and domesticating animals by selective breeding. Agriculture originated independently in multiple cultures across the world. Over the past century or so, industrial agriculture has been replacing traditional agriculture. As a result, today we are producing more food than ever before but also face widespread impacts such as soil degradation and chemical pollution. (pp. 210–212)

Explain the importance of soils to agriculture

Productive agriculture requires healthy soil. Soil is a renewable resource, but it forms extremely slowly. (pp. 212–214)

Discuss the fundamentals of soil science, including soil formation and soil properties

Soil is a complex system full of organisms that decompose organic matter. Weathering helps form soil; and soil formation is also influenced by climate, organisms, topography, parent material, and time. Soil profiles consist of layers (such as the A horizon, or topsoil) with characteristic properties. Soil can be characterized by color, texture, structure, and pH. Soil properties differ among regions, affecting plant growth and the choice of farming approaches. (pp. 212–216)

Describe how farmers supply water to crops, and explain why sustainable alternatives are important

Conventional irrigation boosts crop yields but uses water inefficiently. Overirrigation can cause salinization and waterlogging, which lower crop yields. Farmers can conserve water by using efficient techniques and choosing crops to match soils and climates. (pp. 216–218)

Describe how farmers supply nutrients to crops, and assess sustainable alternatives

Growers can use inorganic or organic fertilizers to supplement nutrients for crops. Overapplying fertilizers can lead to pollution, as leaching and run-off transport nutrients that affect ecosystems and human health. Fertilizer use can be made more sustainable by targeting fertilizers directly to plants and monitoring when they are needed. (pp. 218–219)

Explain the importance of pollinators for crop success

Many crops require pollination by insects or other animals to set seed and form fruit. As such, conserving healthy populations of pollinators is crucial to farming productivity. (pp. 219–220)

Analyze the causes and impacts of soil erosion and land degradation, and discuss solutions

Some agricultural practices cause erosion, which results in increasingly poorer crop yields. As industrial agriculture intensifies, soils become degraded, and we have been losing millions of acres of productive cropland each year. Soil degradation is the major component of land degradation globally, and desertification affects large areas in the world's arid regions. Farming techniques such as crop rotation, contour farming, intercropping, terracing, shelter-belts, and conservation tillage help to reduce soil erosion and boost crop yields. Overgrazing can degrade soil and affect native ecosystems, but herd rotation, grazing limits, and cooperation with scientists researching grazing techniques and impacts can all contribute to more sustainable grazing. (pp. 220–230)

Summarize major policy approaches for conservation in agriculture

Some existing policies and subsidies worsen land degradation—for example, government payments to farmers that encourage overcultivation. Policies for conservation include the Conservation Reserve Program and other programs funded in U.S. farm bills. (pp. 229–230)

TESTING Your Comprehension

1. Explain how each of us depends on healthy soil for our day-to-day well-being. What characteristics of soil are important to crop growth and agriculture, and how?

2. Compare and contrast the methods used in traditional and industrial agriculture. How does sustainable agriculture differ from industrial agriculture?

3. What processes are most responsible for the formation of soil? Name the five primary factors thought to influence soil formation.

4. List and describe the major horizons in a typical soil profile. How are soil horizons created? How is organic matter distributed in a typical soil profile?

5. Explain how overirrigation can damage soils and reduce crop yields. How can irrigation practices be made more sustainable?

6. Explain the differences between inorganic and organic fertilizers. How do fertilizers boost crop growth? How does excess fertilizer added to soil sometimes end up in water supplies and in the atmosphere?

7. What was the Dust Bowl? What impacts did it have on the United States? Describe innovations in soil conservation introduced since that time.

8. Name three human activities that can promote soil erosion. Describe six farming methods that help reduce erosion. Explain how each approach works.

9. Describe the effects of overgrazing on soil. What policies can be linked to the practice of overgrazing? What conditions characterize sustainable grazing practices?

10. Explain the goals and methods of the Conservation Reserve Program.

SEEKING Solutions

1. Select two specific techniques or approaches described in this chapter that you think are especially effective in sustaining agricultural resources—for instance, in conserving soil, conserving water, or minimizing nutrient pollution—and describe how these techniques or approaches accomplish the goal of sustaining resources.

2. How do you think a farmer can best help to conserve soil? How do you think a scientist can best help to conserve soil? How do you think a national government can best help to conserve soil? Give an example of each.

3. What pollinating insects exist in your area? Do a bit of research online to find several species of bees or other insects that pollinate crops in your region. Describe them and explain the benefits they bring. Now research three steps that could be taken (by yourself, homeowners, farmers, or others) to help conserve or increase populations of these beneficial insects.

4. **CASE STUDY CONNECTION** List several steps that Kennesaw State University has taken with its dining hall and its campus farm to pursue sustainable food services. Has your own college or university taken any of these same steps? If so, which ones? Has your campus taken any additional steps toward food and dining sustainability? Describe three actions that you think your campus could and should take in regard to these issues.

5. **THINK IT THROUGH** You are a land manager with the U.S. Bureau of Land Management (BLM, p. 310) and have just been put in charge of 500,000 acres of public grasslands that have been degraded by decades of overgrazing. Soil is eroding, creating large gullies. Invasive weeds are replacing native grasses. Shrubs are encroaching on grassland areas because fire was suppressed. Environmental advocates want an end to ranching on the land. Ranchers want grazing to continue. What steps would you take to assess the land's condition? What steps would you take to begin restoring its soil and vegetation? Would you allow grazing? If so, would you set limits on it? How might you decide what those limits should be?

6. **THINK IT THROUGH** You are the head of an international granting agency that assists farmers with soil conservation and sustainable agriculture. You have $10 million to disburse. Your agency's staff has decided that the funding should go to (1) farmers in an arid area of Africa that is prone to salinization, (2) farmers in a fast-growing area of Indonesia where swidden agriculture is practiced, (3) farmers in Argentina practicing no-till agriculture, and (4) farmers in a dryland area of Mongolia undergoing desertification. What types of projects would you recommend funding in each of these areas, how would you apportion your funding among them, and why?

CALCULATING Ecological Footprints

In the United States, approximately 5 tons of topsoil are lost for every ton of grain harvested. Erosion rates vary greatly with soil type, topography, tillage method, and crop type. For simplicity, let us assume that the 5:1 ratio applies to all plant crops and that a typical diet includes 1 pound of plant material or its derived products (sugar, for example) per day. In the first two columns of the table, calculate the annual topsoil losses associated with growing this food for you and for other groups, assuming the same diet.

	PLANT PRODUCTS CONSUMED (LB)	SOIL LOSS (LB) AT 5:1 RATIO	SOIL LOSS (LB) AT 3:1 RATIO	SOIL LOSS PREVENTED (LB) AT 3:1 RELATIVE TO 5:1 RATIO
You	365	1825	1095	730
Your class				
Your state				
United States				

1. Improved soil conservation measures reduced erosion in the United States by 41% from 1982 to 2010. If additional measures were again able to reduce the current rate of soil loss by this percentage, the ratio of soil lost to grain harvested would fall from 5:1 to about 3:1. Calculate the soil losses associated with food production at a 3:1 ratio, and record your answers in the third column of the table.

2. Calculate the amount of topsoil hypothetically saved by the additional conservation measures in Question 1, and record your answers in the fourth column of the table.

3. Define a "sustainable" rate of soil loss. Describe how you might determine whether a given farm was practicing sustainable use of soil.

MasteringEnvironmentalScience®

CHAPTER 10

Making Agriculture Sustainable

AUSTRALIA

Kojonup

Can Organic Farming and GMOs Coexist?

> " **Worrying about starving future generations won't feed them. Food biotechnology will.**
>
> Advertising campaign of the Monsanto Company
>
> **Industrial agriculture has destroyed diverse sources of food, and it has stolen food from other species . . . using huge quantities of fossil fuels and water and toxic chemicals in the process.**
>
> Vandana Shiva, Director of the Research Foundation for Science, Technology, and Natural Resource Policy, India "

Steve Marsh and Michael Baxter are neighbors and had been friends since childhood. For years, the two farmers had tended adjacent parcels of land in the small rural town of Kojonup in Western Australia. But now, flanked by their lawyers, they stared each other down in a courtroom in the full glare of the media. Their conflict captured the world's attention: It was the first time a farmer had brought a lawsuit against another farmer for "contamination" of organic fields by genetically modified crops.

Despite their legal feud, both men shared a love of farming and a commitment to the idea of sustainable agriculture. Both had tried to do the right thing for their families, society, and the environment as they made their living from the land. Their courtroom battle was a microcosm of the global debate over how we should pursue agriculture.

Marsh had turned to organic farming on his 477 hectares (ha; 1200 acres) of land because he wanted to safeguard the health of his family and the integrity of his soil and water by eliminating many of the pollution sources common to conventional industrial agriculture. He raised sheep and grew wheat, rye, and oats without using chemical pesticides or synthetic fertilizers, and he had worked diligently to obtain official certification as an organic farmer.

Baxter grew canola in the conventional way on his 900 ha (2200 acres): planting vast stands of the crop in monocultural fields, using chemical pesticides to ward off insects and weeds, and applying synthetic fertilizers to provide extra nutrients to the plants. Like most conventional farmers, he was proud to be producing large quantities of crops to help feed people inexpensively and efficiently.

In 2010, Baxter began planting Roundup Ready canola, a genetically modified (GM) crop engineered by the Monsanto Company. Like other Roundup Ready crops, this one was genetically engineered to resist the lethal effects of Monsanto's best-selling weed-killer, Roundup. With Roundup Ready crops, a farmer simply sprays Roundup throughout a field, and it kills weeds without killing the crop.

But after Baxter harvested his GM canola, Marsh discovered that some of its seeds, stalks, and leaves had blown into Marsh's fields. Under Australian law, farm produce can be labeled organic only if it is 100%

An organic produce market in Australia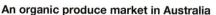

Upon completing this chapter, you will be able to:

- **Explain the challenge of feeding a growing human population**
- **Identify the goals, methods, and consequences of the Green Revolution**
- **Discuss how we raise animals for food, and assess the impacts that result**
- **Describe reasons and approaches for preserving crop diversity**
- **Discuss threats to pollinators and identify potential solutions**
- **Explore strategies for pest and weed management**
- **Describe the science behind genetic engineering**
- **Compare the benefits and costs of genetically modified foods, and assess the public debate over them**
- **Analyze the nature, growth, and potential of organic agriculture**
- **Contrast conventional industrial, organic, and biotech approaches to agriculture, and summarize potential pathways toward sustainable agriculture**

A farmer tends a field of GM canola in flower

free of any genetically engineered material. Consequently, inspectors de-certified the crops on 70% of Marsh's land, and as a result he was no longer able to sell his harvest as certified organic. Without the higher revenue farmers gain from organic sales, Marsh realized that he would be unable to recoup the costly investments he'd made to convert and maintain his fields to meet organic standards. Reluctantly, he judged that a lawsuit was his best remaining option.

In 2014, Marsh lost his case before Western Australia's Supreme Court. The court declared that Baxter had done nothing illegal. It said the problem was that Australia's organic certification agency held to a zero-tolerance policy and de-certified Marsh's crop even though the influx of GM material had occurred by accident. The court also ordered Marsh to pay Baxter $804,000 in legal costs. Marsh's lawyers appealed the decision, but in 2015 the Court of Appeal ruled against him. Two of that court's three judges agreed with the original ruling, but one dissented, asserting that Baxter had in fact been negligent in allowing seeds from his GM crop to reach his neighbor's land. Encouraged by the dissent, Marsh appealed to Australia's High Court, the highest court in the nation. In 2016, however, the High Court decided not to hear the case, closing off Marsh's final judicial avenue.

The protracted legal battle reverberated across Australia, which has the most land in certified organic agriculture (mostly rangeland) of any nation in the world. Thus far, only two GM crops (canola and cotton) have been approved at the national level for planting, and Australia's states vary in their policies, with some states banning these GM crops and others allowing them. Western Australia passed a law restricting them in 2003, but a new governing party in 2010 lifted the restrictions (allowing farmers like Baxter to plant GM canola). Its leaders now aim to repeal the law entirely.

Australia is the only nation to require organic farmers to meet a zero-tolerance threshold for GM-free purity, but across the world, organic farmers face the same dilemma. They pay costs up front to protect their fields from windblown pollen and seeds from GM plants, or from incursion by drifting pesticide spray from conventional industrial farming. Yet they still risk costs later if such incursion, despite their best efforts, renders their crops unsaleable as organic. A survey of 268 organic farmers across the United States in 2013 revealed that on average they pay $6500–$8500 each year in upfront preventive costs and that 16% of them had lost sales as a result of contamination, averaging losses of $4500.

Nonorganic conventional farmers who do not grow GM crops also fear such contamination. They know that Monsanto sends agents to farms to look for the presence of the company's patented GM crops on the land of farmers who did not pay for the seed. Monsanto has sued farmers for growing patented crops without payment, even if the plants appeared by accident as a result of windblown pollen or seed from neighbors.

Today farmers are increasingly fighting back in court to protect their investments. Wheat farmers in Kansas and Washington sued Monsanto in 2013 for letting its experimental GM wheat leak into the U.S. supply, because this caused millions of dollars of losses to farmers when several Asian nations that restrict GM food cut their imports of U.S. wheat as a result. In 2015, farmers sued multinational agribusiness corporation Syngenta for marketing a new strain of GM corn before getting full approval from all international buyers. The corn had gotten mixed into the U.S. supply, and China refused to import U.S. corn, resulting in perhaps $1 billion in losses.

Amid such controversies, people on all sides are trying to do their best for agriculture and our food supply. Supporters of conventional industrial agriculture point to its high yields and say theirs is the only model that can feed the world. Supporters of organic agriculture say that conventional agriculture threatens future yields with its pollution and that protecting natural resources is the best way to safeguard agriculture. Supporters of genetic engineering say that science and technology can produce novel and efficient solutions to enhance nutrition, reduce pollution, and fight hunger and poverty. In today's world, agriculture is complex and fast changing. In this chapter, we explore the various pathways available to us in the pursuit of sustainable farming and grazing.

The Race to Feed the World

As the human population continues to grow, we can expect our numbers to swell past 9 billion by the middle of this century. For every four people living today, there will be five in 2050. Feeding 2 billion additional mouths while protecting the soil, water, and ecological resources that underpin agriculture will require the large-scale embrace of farming and grazing practices that are more sustainable than those we currently use.

The industrialization of agriculture (p. 212) has boosted our production of food and fiber to extraordinary levels. By intensifying our inputs of energy and resources and by managing production efficiently at immense scales, we have increased our agricultural output beyond what any farmer or rancher of the past would have imagined. Yet our high-input industrial agriculture also has brought pollution and resource depletion on unprecedented scales. The environmental impacts of high-input industrial agriculture now threaten to roll back our gains, calling into question whether the industrial model can be sustained in the long term. An alternative vision is one of low-input agriculture, such as organic agriculture conducted at more local scales. This model reduces impacts but sacrifices economic efficiency; as a result, many people wonder if it can produce enough food, at low enough cost, to feed the world's billions. Alongside the debate over which model is more sustainable, the quickly evolving science and technology of genetic engineering is offering intriguing solutions while also raising questions and anxieties.

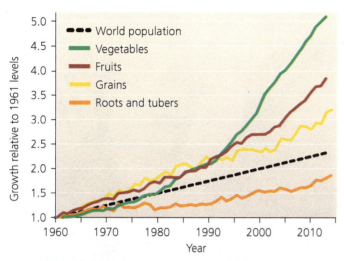

FIGURE 10.1 Global production of most foods has risen more quickly than world population. This means that we have produced more food per person each year. The data lines show cumulative increases relative to 1961 levels (for example, a value of 2.0 means twice the 1961 amount). Food is measured by weight. *Data from UN Food and Agriculture Organization (FAO).*

We are producing more food per person

Over the past half-century, our ability to produce food has grown faster than our global population (**FIGURE 10.1**). We have increased food production by devoting more fossil fuel energy to agriculture; intensifying our use of irrigation, fertilizers, and pesticides; planting and harvesting more frequently; cultivating more land; and developing (mostly through traditional crossbreeding, partly through genetic engineering) more productive crop and livestock varieties.

Improving people's quality of life by producing more food per person is a monumental achievement of which humanity can be proud. However, ensuring that our food production can be sustained depends on conserving soil, water, and biodiversity by using careful agricultural practices. Today many of the world's soils are in decline, and most of the planet's arable land has already been claimed (Chapter 9). Hence, even though agricultural production has outpaced population growth so far, we have no guarantee that this will continue.

We face undernutrition, overnutrition, and malnutrition

Despite our rising food production, nearly 800 million people worldwide do not have enough to eat. These people suffer from **undernutrition,** receiving fewer calories than the minimum dietary energy requirement. As a result, every 5 seconds, somewhere in the world, a child starves to death. In most cases, poverty limits the amount of food people can buy. One out of every seven of the world's people lives on less than $1.25 per day, and one out of three lives on less than $2 per day, according to World Bank estimates. Political obstacles, regional conflict and wars, and inefficiencies in distribution contribute significantly to hunger as well.

Most people who are undernourished live in the developing world. However, hunger is also a problem in the United States, where the U.S. Department of Agriculture (USDA) has classified 49 million Americans as "food insecure," lacking the income required to reliably procure sufficient food. Agricultural scientists and policymakers worldwide pursue a goal of **food security,** the guarantee of an adequate, safe, nutritious, and reliable food supply available to all people at all times.

The good news is that globally the number of people suffering from undernutrition has been falling since the 1960s. The percentage of people who are undernourished has fallen even more (**FIGURE 10.2**). We still have a long way to go to eliminate hunger, but these positive trends are encouraging.

Although nearly 800 million people lack access to adequate food, many others consume more than is healthy and suffer from **overnutrition,** receiving too many calories each day. Overnutrition is a problem in developed nations such as the United States, where food is abundant, junk food is cheap, and people tend to lead sedentary lives with little exercise. As a result, more than one in three U.S. adults are obese, according to the Centers for Disease Control and Prevention. Across the world as a whole, the World Health Organization estimates that 39% of adults are overweight and that of these, one-third are obese. Excessive weight can lead to heart disease, diabetes, stroke, some types of cancer, and other health problems. The growing availability of highly processed foods (which are often calorie-rich, nutrient-poor, and inexpensive) suggests that overnutrition will remain a challenge.

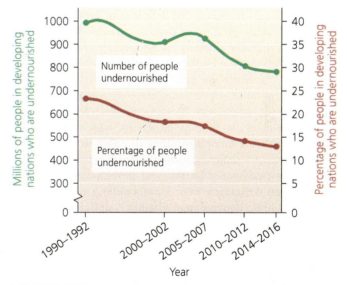

FIGURE 10.2 The number and the percentage of people in the developing world who suffer undernutrition have each been declining. *Data from Food and Agriculture Organization of the United Nations, 2015. The state of food insecurity in the world, 2015. Rome: FAO.*

DATA Explain how the percentage of undernourished people can have decreased slightly between 2000–2002 and 2005–2007, while the absolute number of undernourished people increased slightly during the same period.

Go to **Interpreting Graphs & Data** on **Mastering**EnvironmentalScience®.

FIGURE 10.3 Millions of children suffer from forms of malnutrition, such as kwashiorkor and marasmus.

FIGURE 10.4 Norman Borlaug helped launch the Green Revolution. The high-yielding, disease-resistant wheat that he bred helped boost agricultural productivity in developing countries.

Just as the *quantity* of food a person eats is important for health, so is the *quality* of food. **Malnutrition,** a shortage of nutrients the body needs, occurs when a person fails to obtain a complete complement of essential proteins (p. 27), lipids (p. 28), vitamins, or minerals. Malnutrition can lead to disease (**FIGURE 10.3**). For example, people who eat a diet high in starch but deficient in protein or essential amino acids (p. 27) can develop *kwashiorkor*. Children who have recently stopped breast-feeding are most at risk for developing kwashiorkor, which causes bloating of the abdomen, deterioration and discoloration of hair, mental disability, immune suppression, developmental delays, anemia, and reduced growth. Protein deficiency together with a lack of calories can lead to *marasmus,* which causes wasting or shriveling among millions of children in the developing world. A deficiency of iron in the diet can lead to anemia, which causes fatigue and developmental disabilities; iodine deficiency can cause swelling of the thyroid gland and brain damage; and vitamin A deficiency can lead to blindness.

The Green Revolution boosted agricultural production

The desire for greater quantity and quality of food for the world's growing population led in the mid- and late-20th century to the **Green Revolution** (p. 212), which introduced new technology, crop varieties, and farming practices to the developing world. Agricultural scientists had realized that farmers could not go on forever converting additional land to increase production, so they devised methods and technologies to increase crop yields on existing cultivated land. As a result, yields rose dramatically in industrialized nations. For instance, the average acre of U.S. corn field raised its corn output fivefold during the 20th century. Many people viewed such growth in production and efficiency as key to ending starvation in developing nations.

The transfer of technology and practices to the developing world that marked the Green Revolution began in the 1940s, when American agricultural scientist Norman Borlaug introduced Mexico's farmers to a specially bred type of wheat (**FIGURE 10.4**). This strain of wheat produced large seed heads, was resistant to diseases, was short in stature to resist wind, and produced high yields. Within two decades of planting this new crop, Mexico tripled its wheat production and began exporting wheat. The stunning success of this program inspired others. Borlaug—who won the Nobel Peace Prize for his work—took his wheat to India and Pakistan and helped transform agriculture there as well.

Soon many developing countries were doubling, tripling, or quadrupling their yields using selectively bred strains of wheat, rice, corn, and other crops from industrialized nations. When Borlaug died in 2009 at age 95, he was widely celebrated as having saved more lives than anyone in history—estimated to be as many as a billion.

Industrialized agriculture has brought mixed consequences

Along with the new grains, developing nations imported the methods of industrial agriculture. They began applying large amounts of synthetic fertilizers and chemical pesticides on their fields, irrigating crops generously with water, and using more machinery powered by fossil fuels. From 1900 to 2000, people increased energy inputs into agriculture by 80 times while expanding the world's cultivated area by just one-third.

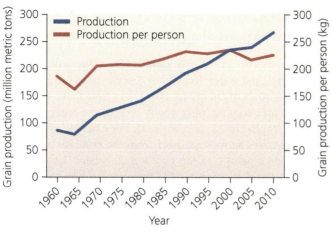

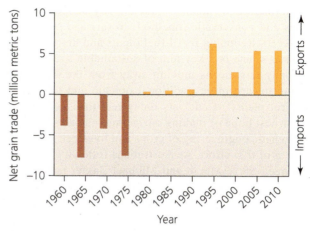

(a) Production and per-capita production rose

(b) Imports turned to exports

FIGURE 10.5 Green Revolution technology enabled India to boost its grain production. India's production grew faster than its population **(a)**, so that grain per person increased from the 1960s to the 1990s. As a result **(b)**, India was able to stop importing grain and begin exporting it to other nations. *Data from UN Food and Agriculture Organization (FAO).*

This high-input agriculture succeeded in producing more corn, wheat, rice, and soybeans from each hectare of land. Intensified agriculture saved millions in India from starvation in the 1970s and turned that nation into a net exporter of grain (**FIGURE 10.5**).

The environmental and social impacts of these developments have been mixed. On the positive side, intensifying the use of already-cultivated land reduced pressures to convert additional lands for cultivation. Between 1961 and 2013, global food production more than tripled and per-person food production rose 48%, while area converted for agriculture increased only 11%. In this way, the Green Revolution helped preserve biodiversity and natural ecosystems by preventing a great deal of deforestation and habitat destruction. On the negative side, the intensified use of fossil fuels, water, inorganic fertilizers, and synthetic pesticides has worsened pollution, topsoil erosion, and soil and water quality (Chapter 9).

In today's conventional industrial agriculture, crops are planted in monocultures, large expanses of single crop types (p. 212; **FIGURE 10.6**). This makes planting and harvesting efficient and thereby increases output. However, monocultures also reduce biodiversity over large areas, because many fewer wild organisms are able to live in monocultures than in native habitats or in traditional small-scale polycultures. Moreover, when all plants in a field are of the same species (and thus genetically similar), they are susceptible to viral diseases, fungal pathogens, or insect pests that can multiply and spread quickly from plant to plant. For this reason, monocultures bring risks of catastrophic failure.

Today, yields are declining in some regions targeted by the Green Revolution, likely due to soil degradation from the heavy use of fertilizers, pesticides, and irrigation. Moreover, wealthier farmers with larger holdings of land were best positioned to invest in Green Revolution technologies. As a result, many low-income farmers who could not afford these technologies were driven out of business and moved to cities,

adding to the immense migration of poor rural people to urban areas of the developing world (p. 333).

How can we achieve sustainable agriculture?

Most experts feel that to sustain our population in the long run we will need to begin raising crops and livestock in ways that are less polluting and less resource-intensive. **Sustainable agriculture** (p. 212) consists of farming and grazing that maintain the healthy soil, clean water, pollinators, genetic diversity, and other resources needed for the production of crops and livestock over the long term. The

FIGURE 10.6 Monocultures improve the efficiency of planting and harvesting but are susceptible to outbreaks of pests. Armyworms (**inset**) attack corn fields like this one.

world's farmers and ranchers have already adopted many strategies and conservation methods to this end (explored in Chapter 9). Reducing fossil fuel inputs and the pollution these inputs cause is a key aim of sustainable agriculture, and to many this means moving away from the industrial model and toward more traditional organic and low-input models. Yet plenty of analysts today feel that technology offers our best hope of making agriculture sustainable, and that the genetic engineering of crops and livestock is a vital component of this effort. As Australians following the dispute between Steve Marsh and Michael Baxter have seen, competing visions exist for agriculture in our world today, and there are valid arguments on all sides.

Raising Animals for Food

Food from cropland agriculture makes up a large portion of the human diet, but most of us also eat animal products. As our population has grown, consuming meat and other animal products has come to have significant environmental, social, and economic impacts. How we respond to demand for animal products affects our society's ecological footprint and our quest for sustainable agriculture.

Consumption of animal products is growing

As wealth and global commerce have increased, so have humanity's production and consumption of meat, milk, eggs, and other animal products (**FIGURE 10.7**). Worldwide, more than 30 billion domesticated animals are raised and slaughtered for food each year. Global meat production has increased more than fivefold since 1950, and per capita meat consumption has doubled. The Food and Agriculture Organization of the United Nations (FAO) predicts that as developing nations become wealthier, per capita meat consumption could nearly double again by the year 2050.

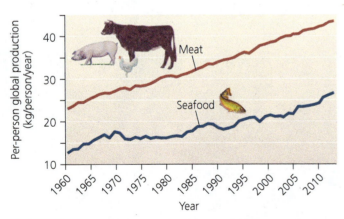

FIGURE 10.7 Per-person production of meat from farmed animals and of seafood has risen steadily worldwide. *Data from UN Food and Agriculture Organization (FAO).*

Our food choices have environmental consequences

What we choose to eat has ramifications for how we use energy, land, and water. Recall our discussions of trophic levels and pyramids of energy (p. 80). Every time that one organism consumes another, only about 10% of the energy moves from one trophic level up to the next. For example, if we feed grain to a cow and then eat beef from the cow, we lose most of the grain's energy to the cow's metabolism. Energy was used up as the cow converted the grain to tissue as it grew and as the cow conducted cellular respiration (p. 31) on a constant basis to maintain itself. For this reason, eating meat is far less energy-efficient than relying on a vegetarian diet, and it leaves a far greater ecological footprint.

In contrast, if we eat lower on the food chain (a more vegetarian diet with fewer animal products), we put a greater proportion of the sun's energy to use as food for ourselves. The lower on the food chain we eat, the smaller is our ecological footprint, and the more of us Earth can support.

In addition, some animals convert grain feed into milk, eggs, or meat more efficiently than others (**FIGURE 10.8**).

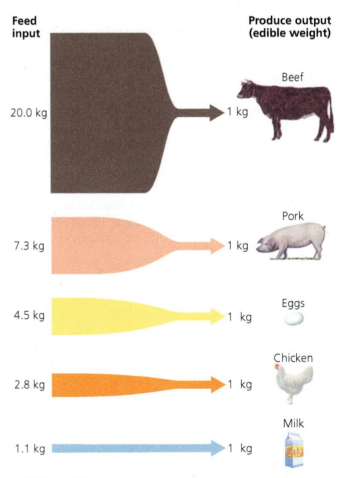

FIGURE 10.8 Producing different animal food products requires different amounts of animal feed. Twenty kilograms of feed must be provided to cattle to produce 1 kg of beef. *Data from Smil, V., 2001. Feeding the world: A challenge for the twenty-first century. Cambridge, MA: MIT Press.*

Scientists have calculated relative energy-conversion efficiencies for different types of animals and have used these to infer the area of land (**FIGURE 10.9a**) and weight of water (**FIGURE 10.9b**) required to produce 1 kg (2.2 lb) of edible protein from milk, eggs, chicken, pork, and beef. The research illustrates that producing eggs and chicken meat requires the least land and water, whereas producing beef requires the most. These differences in efficiency demonstrate that when we choose what to eat, we are also choosing indirectly how to use resources such as land and water.

Eating meat and animal products has additional environmental consequences. The FAO recently estimated that livestock in the United States account for 55% of the nation's soil erosion, 37% of pesticide applications, 50% of antibiotics consumed, and one-third of the nitrogen and phosphorus pollution in U.S. waterways.

Our diets also have consequences for climate change, because livestock are a major source of greenhouse gases that lead to global warming (p. 482). **FIGURE 10.9c** shows the amount of emissions (in carbon dioxide equivalents) released in the production of 1 kg (2.2 lb) of edible protein from milk, eggs, chicken, pork, and beef. Again, beef exerts the largest footprint. Livestock release methane and nitrous oxide in their metabolism and waste. Nitrous oxide is also released from certain crops used to feed animals, and from fertilizers applied to those crops. Carbon dioxide is released to the atmosphere when forests are cleared for ranching or for growing feed, as well as when fossil fuels are burned to grow feed or transport animals. The FAO reports that across the world, livestock agriculture contributes 5% of our carbon dioxide emissions, 44% of our methane emissions, and 53% of our nitrous oxide emissions. Altogether, this represents 14.5% of the emissions driving climate change—a larger share than automobile transportation! On a brighter note, the FAO also concludes that we can reduce greenhouse gas emissions from livestock by 30% by the widespread adoption of clean and efficient technologies and practices already being used by the most resourceful producers.

Feedlots have benefits and costs

In traditional agriculture, livestock are kept by farming families near their homes or are grazed on open grasslands by nomadic herders or sedentary ranchers. These traditions survive today, but the advent of industrial agriculture and the rising demand for meat, milk, and eggs has brought an additional method of raising animals for food. **Feedlots,** also known as *factory farms* or *concentrated animal feeding operations*, are huge pens designed to deliver energy-rich

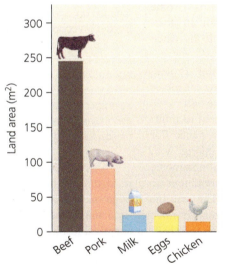

(a) Land required to produce 1 kg of protein

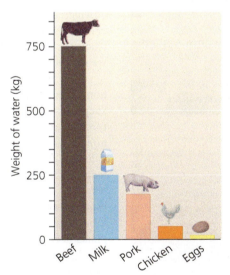

(b) Water required to produce 1 kg of protein

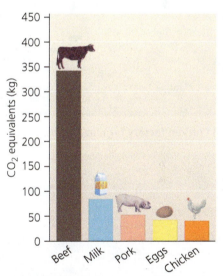

(c) Greenhouse gas emissions released in producing 1 kg of protein

FIGURE 10.9 Producing different types of animal products requires different amounts of (a) land and (b) water—and releases different amounts of (c) greenhouse gas emissions. Raising cattle for beef exerts the greatest impacts in all three ways. *Data (a, b) from Smil, V., 2001. Feeding the world: A challenge for the twenty-first century. Cambridge, MA: MIT Press; and (c) from FAO, 2015. Global Livestock Environmental Assessment Model (GLEAM).*

DATA Answer the following in terms of protein, pound for pound.
- How many times more land does it take to produce beef than chicken?
- How many times more water does beef require, compared with chicken?
- How many times more greenhouse gas emissions does beef release, relative to chicken?

Go to **Interpreting Graphs & Data** on **Mastering**EnvironmentalScience°.

food to animals living at high densities (FIGURE 10.10). Today nearly half the world's pork and most of its poultry come from feedlots.

Feedlots allow for economic efficiency and greater food production, and this makes animal products affordable to more people. However, feedlot animals are generally fed grain grown on cropland. One-third of the world's cropland is now devoted to growing feed for animals, and 45% of global grain production goes to feed livestock and poultry. This elevates the price of staple grains and can endanger food security for the poor.

For environmental quality, feedlots offer one very significant benefit: Taking cattle and other livestock off pastureland and concentrating them in feedlots reduces grazing impacts across large areas of the landscape (p. 227). Animals that are densely concentrated in feedlots will not contribute to overgrazing.

At the same time, intensified animal production at feedlots can result in intensive pollution. Livestock produce prodigious amounts of manure and urine, and the highly concentrated waste from feedlots can pollute surface water and groundwater. Rich in nitrogen and phosphorus, livestock waste is a common cause of eutrophication (pp. 109, 406). This waste may also release bacterial and viral pathogens that can sicken people, including *Salmonella, Escherichia coli, Giardia, Microsporidia, Pfiesteria*, and pathogens that cause diarrhea, botulism, and parasitic infections.

The crowded conditions under which animals are kept in feedlots necessitate heavy use of antibiotics to control disease. Overuse of antibiotics can cause microbes to evolve resistance to the antibiotics (just as pests evolve resistance to pesticides; p. 247). This makes the drugs less effective, and leads feedlot managers to increase dosages and switch to ever-stronger antibiotics. Hormones are administered to livestock as well, and metals are added to feed to spur growth. Livestock excrete most of these chemicals, which end up in wastewater and may be transferred up the food chain in downstream ecosystems. Some chemicals that remain in livestock are transferred to us when we eat their meat.

The Environmental Protection Agency and state agencies keep a watchful eye on U.S. feedlots, regulating them to minimize environmental impacts. For example, wastewater along with manure may be stored in lagoons, where it undergoes treatment similar to that of municipal wastewater (p. 410). The resulting sludge may be applied to farm fields as fertilizer (or injected into the ground where plants need it), reducing the need for chemical fertilizers.

(a) Chicken factory farm in Arkansas

(b) Cattle feedlot in Colorado

FIGURE 10.10 Most meat eaten in the United States comes from animals raised in feedlots or factory farms. These facilities house thousands of animals such as (a) chickens or (b) cattle at high densities. The animals are dosed liberally with antibiotics to control disease.

weighing the
ISSUES

Feedlots and Animal Rights

Animal rights advocates denounce factory farming because they argue that it mistreats animals. Chickens, pigs, and cattle are crowded together in small pens their entire lives, fattened up, and slaughtered. Should we concern ourselves with the quality of life of the animals that constitute part of our diet? Do you think animal rights concerns are as important as environmental concerns? Are conditions at feedlots a good reason for being vegetarian?

We use aquaculture to raise seafood

Part of the typical human diet consists of aquatic organisms. However, wild fish populations are plummeting throughout the world's oceans as increased demand and new technologies lead us to overexploit marine fisheries (pp. 433–438). As a result, raising fish and shellfish on "fish farms" may be one of the best ways to meet our growing demand for these foods (FIGURE 10.11).

The cultivation of aquatic organisms for food in controlled environments, called **aquaculture,** is now being pursued with more than 220 freshwater and marine species, ranging from fish to shrimp to clams to seaweeds (FIGURE 10.12). Many aquatic species are raised in open water in large, floating net-pens. Others are raised in ponds or holding tanks.

FIGURE 10.11 People practice many types of aquaculture. Here, fish-farmers tend their animals at a fish farm in China.

Aquaculture is our fastest-growing type of food production; global output has increased 10-fold in just the past 30 years. Most widespread in Asia, aquaculture today produces more than $160 billion worth of food and provides half of the fish and two-thirds of the shellfish that people eat.

Aquaculture brings benefits and costs

Aquaculture increases food supplies and helps ensure reliable protein sources, thus improving people's food security. Aquaculture also helps reduce fishing pressure on wild fish stocks, many of which, as we've discussed, are overharvested and declining. Reducing fishing pressure also lessens bycatch (p. 435), the unintended catch of nontarget organisms that results from commercial fishing. Aquaculture consumes fewer fossil fuels and provides a safer work environment than does commercial fishing. Fish farming can produce up to 10 times more fish per unit area than is harvested from waters of the continental shelf and up to 1000 times more than from the open ocean.

When practiced on a small scale by families or villages, as in much of the developing world, community-based aquaculture can be sustainable, and its compatibility with other activities can make it an effective path toward sustainable agriculture (p. 263). For instance, uneaten fish scraps make excellent fertilizers for crops, and food waste can be fed to fish.

At large scales, industrialized aquaculture produces ample amounts of food but also exerts environmental impacts. Dense concentrations of farmed animals can incubate disease, which necessitates antibiotic treatment, results in expense, and can reduce food security. Industrial-scale shrimp farming along tropical coastlines has destroyed large areas of valuable mangrove forest and has polluted coastal waters. Indeed, aquaculture can produce remarkable amounts of waste, both from the farmed organisms and from feed that goes uneaten and decomposes in the water. As with feedlot livestock, commercially farmed fish often are fed grain, which is energy-inefficient and can reduce food supplies for people. Farmed fish are also sometimes fed fish meal made from wild ocean fish such as herring and anchovies, whose harvest may place additional stress on wild populations. For all these reasons, industrial-scale aquaculture can leave a large ecological footprint.

If farmed aquatic organisms escape into ecosystems where they are not native, they may spread disease to native stocks or may outcompete native organisms for food or habitat. This is happening with several Asian carp species farmed in the central United States that have escaped and whose growing populations are altering ecosystems and harming fisheries. Such concerns delayed government approval of genetically engineered Atlantic salmon, which in 2015 became the first GM animal approved for sale as food. The GM salmon created by AquaBounty Technologies are engineered to grow especially fast (**FIGURE 10.13**). Such GM fish could help reduce fishing

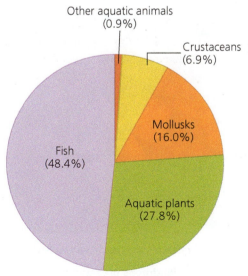

FIGURE 10.12 Aquaculture involves many types of fish, but also a wide diversity of other marine and freshwater organisms. *Data from UN Food and Agriculture Organization (FAO).*

Other aquatic animals (0.9%)
Crustaceans (6.9%)
Mollusks (16.0%)
Fish (48.4%)
Aquatic plants (27.8%)

FIGURE 10.13 Transgenic salmon (top) grow faster than wild salmon of the same species. They often reach a larger size; these two fish are the same age.

pressures on wild stocks, but critics contend that they might also—if they escape into the wild—outcompete their wild relatives, interbreed with them, or spread disease to them. In response, AquaBounty has promised to cultivate only sterile females and to raise the fish in freshwater tanks isolated from rivers and oceans.

Preserving Crop Diversity and Pollinators

Industrial agriculture puts pressure on many of the ecological resources that support our food production. Among these resources are the genetic diversity of crop plants and the populations of pollinating insects that help plants reproduce.

Crop diversity provides insurance

In a modern industrial monoculture, all crop plants in a field are of the same species and genetically similar. This places all our eggs in one basket, such that any single catastrophe might potentially wipe out an entire crop. Monocultures also have narrowed the human diet by reducing the diversity of crops we grow—both the types of crops and the diversity of genetic variants within each crop type. Although expanded international trade provides most individual people access to a wider diversity of foods than were available in the past, for humanity as a whole fully 90% of the food consumed now comes from just 15 crop species and 8 livestock species—a drastic reduction in diversity from earlier times. One can find examples with every food type in every region of the world. Only 30% of the maize varieties that grew in Mexico in the 1930s exist today. The number of wheat varieties in China dropped from 10,000 in 1949 to 1000 by the 1970s. In the United States, apples and other fruit and vegetable crops have decreased in diversity by 90% in less than a century.

Mass-market forces contribute to the trend toward lesser diversity because commercial food processors prefer items to be uniform in size and shape for convenience, and because consumers are often wary of what they perceive as unusual-looking food products.

Preserving the integrity of diverse native crop variants gives us a bulwark against the potential failure of our homogenized commercial crops. Every crop has a wild ancestral species from which it was domesticated, and most crops have had a complex evolutionary history, with people creating many *landraces* (variants adapted to local conditions) over the centuries. The wild relatives of crop plants and their local landraces contain a diversity of genes that we may someday need to introduce into our commercial crops (through cross-breeding or genetic engineering) to confer resistance to disease or pests or to meet other unforeseen challenges.

For instance, all the world's maize (corn) arose from a wild grass species from the highlands of Mexico. Generations of Mexican farmers have selectively bred varieties of maize, creating hundreds of local landraces (**FIGURE 10.14a**). When researchers 15 years ago presented evidence that genetically engineered corn might be interbreeding with these landraces,

the world's agricultural scientists refocused attention on the importance of preserving genetic diversity of the Mexican landraces, as a form of insurance for the world's corn.

Likewise, the potato blight that devastated Ireland in the 1840s and forced the emigration of so many Irish farmers to America occurred because every potato in Ireland was derived from just one or a few strains. These strains originated in the Andes Mountains in South America, where potatoes were first domesticated. Today thousands of amazingly diverse potato varieties survive in the Andes (**FIGURE 10.14b**), cultivated in backyard gardens and serving as a reservoir of genetic diversity that can protect the world's potato crops. Safeguarding regions and cultures that maintain a wealth of crop diversity is one way to conserve the genetic resources so vital to our long-term success with agriculture.

Seed banks are living museums

Another way to preserve genetic assets for agriculture is to collect and store seeds from diverse crop varieties. This is the work of **seed banks,** institutions that preserve seed types as

(a) A sampling of maize varieties from Mexico

(b) A sampling of potato varieties from the Andes

FIGURE 10.14 Local landraces preserve genetic diversity for crop plants. Mexico hosts numerous varieties of maize **(a)** bred by farmers over centuries, whereas Peru and other Andean nations host thousands of types of potatoes **(b)** showing striking variety.

FIGURE 10.15 The "doomsday seed vault" in Arctic Norway stores seed samples as insurance against global agricultural catastrophe.

a kind of living museum of genetic diversity. These facilities keep seed samples in cold, dry conditions to keep them viable, and they plant and harvest them periodically to renew the stocks.

Examples of seed banks include the Millennium Seed Bank in the United Kingdom; the U.S. National Seed Storage Laboratory at Colorado State University; Seed Savers Exchange in Iowa; Native Seeds/SEARCH in Tucson, Arizona; and the Wheat and Maize Improvement Center in Mexico. In total, 1400 such facilities house 1–2 million distinct types of seeds worldwide.

The most renowned seed bank is the so-called doomsday seed vault established in 2008 on the island of Spitsbergen in Arctic Norway. The internationally funded Svalbard Global Seed Vault (**FIGURE 10.15**) is storing millions of seeds from around the world (spare sets from other seed banks) as a safeguard against global agricultural calamity—"an insurance policy for the world's food supply." This secured, refrigerated facility is built deep into a mountain in an area of permanently frozen ground. The site has no tectonic activity, has little natural radiation or humidity, and is high enough above sea level to stay dry even if climate change melts all the planet's ice. The doomsday seed vault is an admirable effort, but we would be well advised not to rely on it to save us. Far better to manage our agriculture wisely and sustainably so that we never need to break into the vault!

Bee declines make pollinator conservation urgent

Just as genetic diversity in crops is a resource on which agriculture relies, so are healthy populations of the insects and other animals that pollinate many of our crops (pp. 219–220). Farmers in the United States alone gain

an estimated $15 billion per year in pollination services from the introduced European honeybee that beekeepers have domesticated for use with crops (see Figure 9.12, p. 220), and more than $3 billion per year from many of the nation's 4000 species of wild native bees. However, scientific data indicate that populations of honeybees and of wild native bees are declining steeply across North America. As bees disappear, we lose their services and risk lowered crop yields.

Scientists studying the pressures on bees, butterflies, and other pollinators are concluding that they are suffering a "perfect storm" of stresses—many of which result from industrial agriculture. A direct source of mortality is the vast arsenal of chemical insecticides we apply to crops, lawns, and gardens to kill pest insects. All insects are vulnerable to these poisons, so when we try to control pests, we also end up killing beneficial insects such as bees. Pollinators have suffered loss of habitat and flower resources for decades, but this has grown worse recently as weed-killers like Roundup have eliminated weeds from farm monocultures, depriving pollinators of a diversity of nectar and pollen sources. Bees are also being attacked by novel parasites and pathogens that, like many invasive species (pp. 86, 287), have been moved around the world by human travel and trade. In particular, two accidentally introduced parasitic mites have swept through honeybee populations in recent years, decimating hives and pushing beekeepers toward financial ruin.

Researchers are finding that these multiple sources of stress seem to interact and cause more damage than the sum of their parts. For example, pesticide exposure and difficulty finding food might weaken a bee's immune system, making it more vulnerable to parasites. Any or all of these factors may possibly play a role in what is being called **colony collapse disorder,** a mysterious malady that for the past decade has destroyed up to one-third of all honeybees in the United States each year (**FIGURE 10.16**).

Fortunately, we have a number of solutions at hand to help restore bee populations. By retaining or establishing

FIGURE 10.16 Honeybees have been dying off in huge numbers in recent years. Chemical insecticides, introduced parasites and pathogens, and loss of habitat and flower resources are likely all to blame.

FIGURE 10.17 Farmers can help conserve pollinating insects by leaving buffer strips of flowering native plants along the edges of their cultivated fields.

wildflowers and flowering shrubs in or near farm fields, farmers can provide bees a refuge and a source of diverse food resources (**FIGURE 10.17**). Organic farmers such as Steve Marsh often plant such buffer strips to protect against incursion of pesticides or GM material from neighbors, and conventional farmers sometimes plant buffer strips to protect streams and receive subsidies under conservation programs (p. 230). Encouraging flowers on highway rights-of-way can provide resources to pollinators while beautifying roadsides, and the U.S. Federal Highway Administration is working with the Xerces Society, a pollinator conservation non-profit group, to restore and create pollinator habitat along the nation's highways. In addition, farmers can decrease their use of chemical insecticides by using biocontrol or integrated pest management (pp. 248–249). Homeowners can help pollinators by reducing or eliminating the use of pesticides, tending gardens of flowering plants, and providing nesting sites for bees. These solutions and others have been embraced in a national strategy put forth in 2015 by the Obama administration's Pollinator Health Task Force.

Controlling Pests and Weeds

Although pollinating insects are beneficial for agriculture, many other organisms can weaken or destroy our crops and livestock. Some insects feed on crop plants, some pathogens attack livestock or crops, and some plants compete with crop plants for sun, water, and nutrients. Anyone who has ever planted a crop or raised animals has struggled with these natural adversaries, and any pursuit of sustainable agriculture will need to find safe and effective ways of limiting losses to them.

"Pests" and "weeds" hinder agriculture

What people term a **pest** is any organism that damages crops or livestock. What we term a **weed** is any plant that competes with our crops. These are subjective terms that we define

entirely by our own economic interests. There is nothing inherently malevolent about a pest or a weed. These organisms are simply growing and behaving naturally, adapted to survive and reproduce, like any other animal or plant. They just happen to affect our farm productivity in doing so.

Throughout the history of agriculture, the insects, mites, rats, fungi, and viruses that attack our crops have taken advantage of the ways we cluster food plants in agricultural fields. Pests pose an especially great threat to monocultures, because a pest adapted to specialize on a crop can move easily from plant to plant (see Figure 10.6). From the perspective of an insect that feeds on corn, grapes, or apples, encountering a grain field, vineyard, or orchard is like discovering an endless buffet. In a natural ecosystem, each organism's population is kept in check by its predators, competitors, parasites, and pathogens—but in an industrial monoculture, the abundance of one type of food and the lack of other habitats can allow a pest population to flourish, unhindered by its natural enemies.

We have developed thousands of chemical pesticides

Because industrial monocultures limit the ability of natural enemies to control pest populations, farmers have felt the need to introduce some type of pest control to produce food economically at a large scale. In the past half-century, most farmers have turned to chemicals to suppress pests and weeds. In that time we have developed thousands of synthetic chemicals to kill insects (*insecticides*), plants (*herbicides*), and fungi (*fungicides*). Such poisons are collectively termed **pesticides.**

All told, nearly 400 million kg (900 million lb) of active ingredients from conventional pesticides are applied in the United States each year—almost 3 pounds per person. Four-fifths of this total is applied on agricultural land (**FIGURE 10.18**). Since 1960, pesticide use has risen fourfold worldwide. Usage in industrialized nations has leveled off, but it continues to rise in the developing world. Today more than $44 billion is expended annually on pesticides across the world.

FIGURE 10.18 Synthetic chemical pesticides are widely applied to crops in conventional industrial agriculture.

Pesticides boost food production but also have negative impacts

Chemical pesticides have helped to greatly increase our food production. They help us control pests in our homes, and they continue to protect millions of people in developing nations from insect-borne diseases such as malaria. However, exposure to synthetic pesticides can have health consequences for people (Chapter 14). Especially vulnerable are farm workers, who experience high levels of pesticide exposure, and children, whose brains and bodies are growing and developing. We all encounter and ingest chemical pesticide residues when we eat non-organic produce we buy at the grocery store, and pesticides leave residues and breakdown products in the soil, water, and air, affecting organisms and ecosystems in many ways. Although not every pesticide has effects on people, in the United States and most other nations, new chemical products are not thoroughly tested for health effects before being brought to market (p. 378), so a pesticide may cause health impacts yet still be widely used.

An additional problem is that pesticides commonly kill nontarget organisms, including the predators and parasites of the pests they are meant to target. When these valuable natural enemies are eliminated, pest populations become even harder to control. Moreover, as just discussed, chemical pesticides affect pollinating insects. For instance, the use of Roundup in North America has been so widespread that milkweed has gone from being abundant in and near farm fields to being scarce. Monarch butterflies rely on milkweed for food, so as Roundup has eliminated milkweed from the landscape, monarch populations have plummeted (pp. 287–288).

Today a new class of chemical insecticide has become common: **neonicotinoids.** Seed companies treat crop seed with neonicotinoids and they become systemic in the plant, dispersing throughout its tissues as it grows, and ending up in leaves, stems, flowers, fruit, and pollen. "Neonics" make a plant toxic to insects; they kill insects that feed on the plant (as intended), but also harm bees pollinating the plant and predators that eat insects that feed on the plant (unintended consequences). Neonicotinoids are also applied to plants chemically, and they may enter soil, water, and even plants that grow from treated soil, continuing to kill a diversity of organisms that pose no threat to crops.

Pests evolve resistance to pesticides

Despite the toxicity of chemical pesticides, their effectiveness tends to decline with time as pests evolve resistance to them. Recall from our discussion of natural selection (pp. 48–50) that organisms within populations vary in their traits. Because most insects, weeds, and microbes can occur in huge numbers, it is likely that a small fraction of individuals may by chance already have genes that enable them to metabolize and detoxify a given pesticide (p. 366). These individuals will survive exposure to the pesticide, whereas individuals without these genes will not (**FIGURE 10.19**).

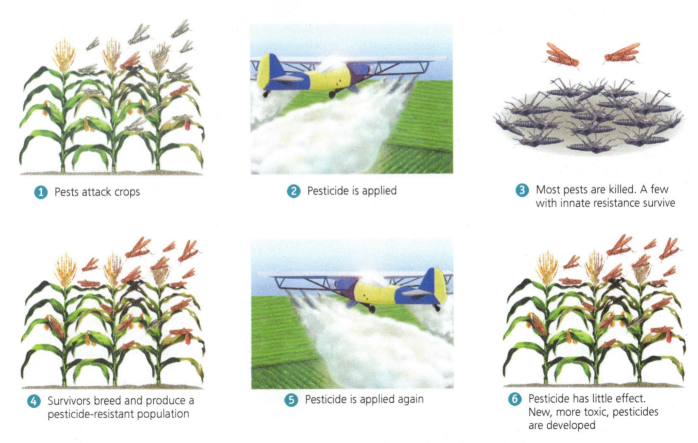

1 Pests attack crops

2 Pesticide is applied

3 Most pests are killed. A few with innate resistance survive

4 Survivors breed and produce a pesticide-resistant population

5 Pesticide is applied again

6 Pesticide has little effect. New, more toxic, pesticides are developed

FIGURE 10.19 Through the process of natural selection, crop pests often evolve resistance to the poisons we apply to kill them.

Let's say an insecticide application kills 99.99% of the insects in a field. That sounds like a successful application, but it means that 1 in 10,000 insects survives. If an insect that is genetically resistant to an insecticide survives and mates with other resistant individuals, the genes for insecticide resistance will be passed on to their offspring. As resistant individuals become more prevalent in the pest population, insecticide applications will cease to be effective and the population will grow.

In many cases, industrial chemists are caught up in an evolutionary arms race (p. 77) with the pests they battle, racing to intensify or retarget the toxicity of their chemicals while the armies of pests evolve ever-stronger resistance to their efforts. Because we seem to be stuck in this cyclical process, it has been nicknamed the "pesticide treadmill." Today among arthropods (insects and their relatives) alone, there are more than 10,000 known cases of resistance by nearly 600 species to more than 330 insecticides. Hundreds more weed species and plant diseases have evolved resistance to herbicides and other pesticides. Many species, including insects such as the green peach aphid, Colorado potato beetle, and diamondback moth, have evolved resistance to multiple chemicals.

Biological control pits one organism against another

Because of pesticide resistance, toxicity to nontarget organisms, and human health risks from some synthetic chemicals, agricultural scientists have sought alternatives to chemical pesticides. The most obvious alternative has been to battle pests and weeds with organisms that eat or infect them. This strategy, called **biological control**—often referred to as *biocontrol*—operates on the principle that "the enemy of one's enemy is one's friend." Biological control is essentially an attempt to reestablish the restraining influence that predators and parasites exert over populations in nature. For example, parasitoid wasps (p. 77) are natural enemies of many caterpillars. These wasps lay eggs on a caterpillar, and the larvae that hatch from the eggs feed on the caterpillar, eventually killing it. Parasitoid wasps are frequently used as biocontrol agents and have often succeeded in controlling pests and reducing chemical pesticide use.

One classic case of successful biological control is the introduction of the cactus moth, *Cactoblastis cactorum*, from Argentina to Australia in the 1920s to control invasive prickly pear cactus that was overrunning rangeland. Larvae of the moth devour cactus tissue, and within just a few years, the species managed to free millions of hectares of rangeland from the cactus (**FIGURE 10.20**).

A widespread modern biocontrol effort has been the use of **Bacillus thuringiensis (Bt),** a naturally occurring soil bacterium that produces a protein that kills many caterpillars and some fly and beetle larvae. Farmers spray Bt spores on their crops to protect against insect attack. In addition, scientists have managed to isolate the gene responsible for the

(a) Before cactus moth introduction

(b) After cactus moth introduction

FIGURE 10.20 Photos from the 1920s show an Australian ranch before (a) and after (b) introduction of the cactus moth. Larvae of this moth were used to clear invasive non-native prickly pear cactus from millions of hectares of rangeland.

bacterium's poison and engineer it into crop plants so that the plants produce the poison (p. 251).

Biocontrol agents themselves can become pests

When a pest is not native to the region where it is damaging crops, scientists may consider introducing a natural enemy (a predator, parasite, or pathogen) of the pest from its native range, expecting that the enemy will attack it. Alternatively, scientists may consider importing a biocontrol agent from abroad that the pest has never encountered, reasoning that the pest has not evolved ways to avoid the biocontrol agent. In either case, this involves introducing an animal or microbe from a foreign ecosystem into a new ecological context. This is risky, because no one can know for certain what effects the biocontrol agent might have. In some cases biocontrol agents have turned invasive and become pests themselves. When this happens, biocontrol organisms are more difficult to manage than chemical controls, because they cannot be "turned off" once they are set loose.

Following the cactus moth's success in Australia, for example, the moth was introduced in other countries to control non-native prickly pear. However, moths introduced to Caribbean islands spread to Florida on their own and are now eating their way through rare native cacti in the southeastern United States. If these moths reach Mexico and the southwestern United States, they could decimate many native and economically important species of prickly pear.

In Hawai'i, wasps and flies have been introduced to control pests at least 122 times over the past century, and biologists Laurie Henneman and Jane Memmott suspected that some of these might be harming native Hawaiian caterpillars that were not pests. They sampled parasitoid wasp larvae from 2000 caterpillars of various species in a remote mountain swamp far from farmland. In this wilderness preserve, they found that fully 83% of the parasitoids were biocontrol agents that had been intended to combat lowland agricultural pests.

Because of concerns about unintended impacts, researchers study biocontrol proposals carefully before putting them into action, and government regulators must approve these efforts. If biological control works as planned, it can be a permanent, effective, and environmentally benign solution. Yet there will never be a sure-fire way of knowing in advance whether a given biocontrol program will work as planned.

Integrated pest management combines biocontrol and chemical methods

Because both biocontrol and chemical control methods pose risks, agricultural scientists and farmers began developing more sophisticated strategies, trying to combine the best attributes of each approach. **Integrated pest management (IPM)** incorporates numerous techniques, including biocontrol, use of chemicals when needed, close monitoring of populations, habitat alteration, crop rotation, transgenic crops, alternative tillage methods, and mechanical pest removal.

IPM has become popular in many parts of the world that are embracing sustainable agriculture. Indonesia stands as an exemplary case (**FIGURE 10.21**). This nation had subsidized pesticide use heavily for years but came to understand that pesticides were actually making pest problems worse. They were killing the natural enemies of the brown planthopper, which began to devastate rice fields as its populations exploded. Concluding that pesticide subsidies were costing money, causing pollution, and decreasing yields, the Indonesian government in 1986 banned the import of 57 pesticides, slashed pesticide subsidies, and promoted IPM. International experts helped teach Indonesian rice farmers about IPM, and collaborative groups of farmers traded information and experimented with new approaches. Within just 4 years, pesticide production

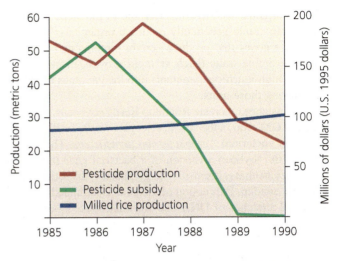

FIGURE 10.21 Once Indonesia threw its weight behind integrated pest management in 1986, pesticide production and imports were reduced, pesticide subsidies were phased out, and yields of rice increased. Data from the World Bank.

fell by half, pesticide imports fell by two-thirds, and subsidies were phased out (saving taxpayers $179 million annually). Rice yields rose 13%.

Genetically Modified Food

Although industrial agriculture has been feeding an ever-greater number and proportion of the world's people, relentless population growth and intensified environmental impacts are demanding further innovation. A new set of potential solutions began to arise in the 1980s and 1990s as advances in genetics enabled scientists to directly alter the genes of organisms, including crop plants and livestock. The genetic modification of organisms that provide us food and fiber holds promise to enhance nutrition and the efficiency of agriculture while lessening impacts on ecosystems. However, many people fear that genetic engineering may pose risks that are not yet fully understood, and the widening role of biotechnology in agriculture has strengthened the influence of multinational corporations over farmers and our food supply. For these reasons, agricultural biotechnology has inspired anxiety and protest among many consumer advocates, small farmers, and critics of big business.

Genetic engineering creates GMOs

Genetic engineering refers to any process whereby scientists directly manipulate an organism's genetic material in the laboratory by adding, deleting, or changing segments of DNA (pp. 27–28). **Genetically modified foods** are foods derived from **genetically modified organisms (GMOs),** organisms that are genetically engineered. Genetic engineering uses recombinant DNA (DNA patched together from the DNA of multiple organisms). **FIGURE 10.22** shows the steps typically followed to create genetically modified organisms. The goal is to place genes that produce certain proteins and code for certain desirable traits (such as rapid growth, disease resistance, or high nutritional content) into the genomes of organisms lacking those traits.

For instance, to create Roundup Ready soybeans, the most abundant GM crop in the world, scientists worked with a strain of the soil bacterium *Agrobacterium tumefaciens*. They transferred to the soybean a version of a bacterial gene that codes for an enzyme that is insensitive to glyphosate, the active ingredient in Roundup. Expression of the gene is regulated by three additional stretches of DNA that were introduced—one from

Agrobacterium, one from a type of petunia, and one derived from a common plant virus, the cauliflower mosaic virus.

An organism that contains DNA from another species is called a **transgenic** organism, and a gene transferred between them is called a **transgene.** The creation of transgenic organisms is one type of **biotechnology,** the material application of biological science to create products derived from organisms. Biotechnology has helped us develop medicines, clean up pollution, understand the causes of cancer and other diseases, dissolve blood clots after heart attacks, and make better beer and cheese. **TABLE 10.1** shows a selection of notable developments in GM foods. The stories behind them illustrate both the promises and pitfalls of food biotechnology.

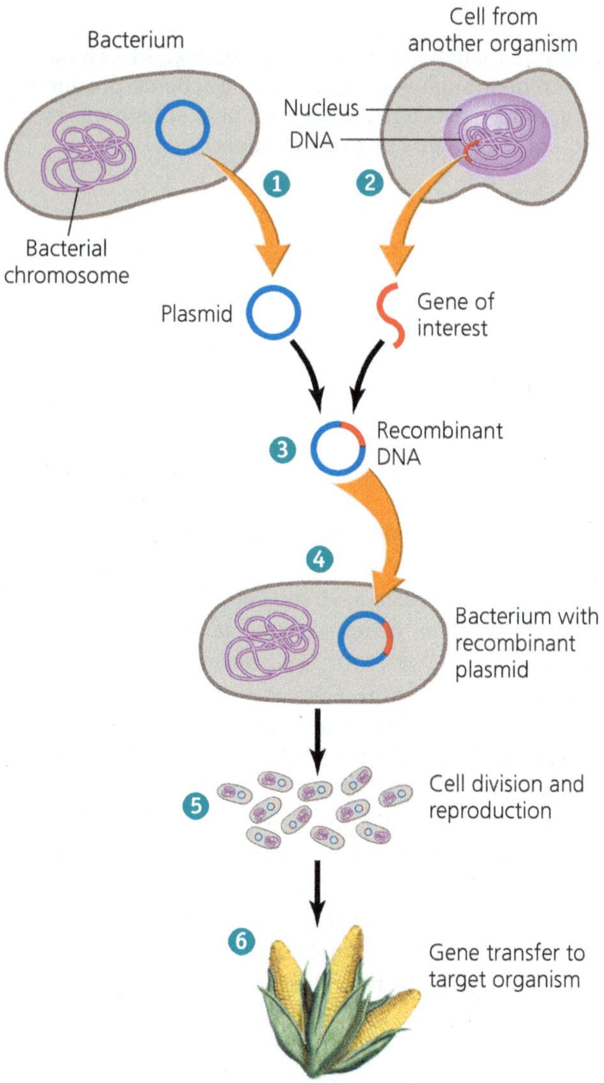

Genetic engineering is both like—and unlike—traditional breeding

In principle, the genetic alteration of plants and animals by people is nothing new; through artificial selection (pp. 50–51), we have been influencing the genetic makeup of our livestock and crop plants for thousands of years. Our ancestors altered the gene pools of domesticated plants and animals through selective breeding by preferentially mating individuals with favored traits (fast growth, higher yields, disease resistance, etc.) so that offspring would inherit those traits (p. 211).

The techniques geneticists use to create engineered organisms, however, differ from traditional selective breeding in several ways. First, selective breeding mixes genes from individuals of the same or similar species, whereas scientists creating recombinant DNA routinely mix genes of organisms as different as viruses and crops, or spiders and goats. Second, selective breeding involves whole organisms living in the field, whereas genetic engineering works with genetic material in the lab. Third, traditional breeding selects from combinations of genes that come together on their own, whereas genetic engineering creates novel combinations directly in a more controlled way. Thus, in traditional breeding, people use a process of selection (pp. 48–51) acting on random mutations, whereas in genetic engineering, scientists use a process of mutation (p. 49) that is nonrandom and precisely directed.

Biotechnology is transforming the products around us

In just over three decades, GM foods have gone from science fiction to mainstream business. When recombinant DNA technology was first being developed in the 1970s, scientists collectively regulated and monitored their own research. Once the scientific community declared itself confident in the 1980s that the technique was safe, industry leaped at the chance to develop hundreds of applications, from improved medicines (such as hepatitis B vaccine and insulin for diabetes) to designer plants and animals (including glow-in-the-dark pet goldfish).

Since then, GM crops have been adopted and planted across the world with remarkable speed (**FIGURE 10.23**). Worldwide, four of every five soybean plants are now transgenic, as are three of every four cotton plants, one of every

FIGURE 10.22 To create recombinant DNA, scientists follow several steps. First they isolate plasmids ❶, small circular DNA molecules, from a bacterial culture. DNA containing a gene of interest is then removed ❷ from another organism. Scientists insert this gene into the plasmid to form ❸ recombinant DNA. This recombinant DNA enters ❹ new bacteria that reproduce, generating many copies ❺ of the desired gene. The gene is then transferred to individuals of the target plant or animal ❻ and will be expressed in the genetically modified organism as a desirable trait, such as rapid growth or insect resistance.

TABLE 10.1 Several Notable Examples of Genetically Modified Food Technologies

CROP	DESCRIPTION AND STATUS	CROP	DESCRIPTION AND STATUS
Golden rice 	Engineered to produce beta-carotene to fight vitamin A deficiency in Asia and the developing world. May offer only moderate nutritional enhancement despite years of work. Currently undergoing field trials in Asia.	**Bt cotton** 	Engineered with genes from bacterium *Bacillus thuringiensis* (Bt) that kills insects. Has increased yield, decreased insecticide use, and boosted income for 14 million small farmers in India, China, and other nations.
Virus-resistant papaya 	Resistant to ringspot virus and grown in Hawai'i. In 2011 became the first biotech crop approved for consumption in Japan.	**Roundup-Ready alfalfa** 	One of many crops engineered to tolerate Monsanto's Roundup herbicide (glyphosate). Because the crop can withstand it, the chemical can be applied in great quantities to kill weeds. Unfortunately, many weeds are evolving resistance to glyphosate as a result. Planted in the U.S., 2005–2007, GM alfalfa was then banned because a lawsuit forced the USDA to better assess its environmental impact. Reapproved in 2011.
GM salmon 	Engineered for fast growth, large size. Would be the first GM animal approved for sale as food. To prevent fish from breeding with wild salmon and spreading disease to them, company AquaBounty promised to make their fish sterile and raise them in inland pens.	**Roundup-Ready sugar beet** 	Tolerant of Monsanto's Roundup herbicide (glyphosate). Swept to dominance (95% of U.S. crop) in just 2 years. As with alfalfa, a lawsuit forced more environmental review, after it had already become widespread. Reapproved in 2012.
Biotech potato 	Resistant to late blight, the pathogen that caused the 1845 Irish Potato Famine and that still destroys $7.5 billion of potatoes each year. Being developed by European scientists, but struggling with EU regulations on research.	**Biotech soybean** 	The most common GM crop in the world, covering nearly half the cropland devoted to biotech crops. Engineered for herbicide tolerance, insecticidal properties, or both. Like other crops, soybeans may be "stacked" with more than one engineered trait.
Bt corn 	Engineered with genes from bacterium *Bacillus thuringiensis* (Bt) that kills insects. One of many Bt crops developed.	**Sunflowers and superweeds**	Research on Bt sunflowers suggests that transgenes might spread to their wild relatives and turn them into vigorous "superweeds" that compete with the crop or invade ecosystems. This is most likely to occur with crops like squash, canola, and sunflowers that can breed with their wild relatives.

FIGURE 10.23 GM crops have spread with remarkable speed since their commercial introduction in 1996. They now are planted on more than 10% of the world's cropland. *Data from the International Service for the Acquisition of Agri-biotech Applications (ISAAA).*

 In the past 5 years, have GM crops been growing faster in industrialized nations or in developing nations?
- If current trends continue, which group of nations will have more GM crops in 2020?
- Can you estimate how much more this group of nations might have?

Go to **Interpreting Graphs & Data** on **Mastering**EnvironmentalScience®.

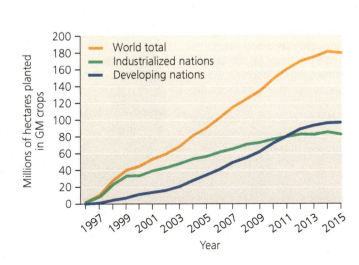

three corn plants, and one of every four canola plants. Globally in 2015, nearly 18 million farmers grew GM crops on 180 million ha (445 million acres) of farmland—nearly 12% of all cropland in the world.

In the United States today, roughly 90% of corn, soybeans, cotton, and canola consist of genetically modified strains, and close to half of these crops are engineered for more than one trait. It is conservatively estimated that more than 70% of processed foods in U.S. stores contain GM ingredients. Thus, it is extremely likely that you consume GM foods on a daily basis.

Most GM crops today are like the Roundup Ready canola that Australian farmer Michael Baxter planted—engineered to tolerate herbicides, so that farmers can apply herbicides to kill weeds without having to worry about killing their crops. Other crops are engineered to resist insect attack. Some are modified for both traits. Tolerance of herbicides and resistance to pests enables large-scale commercial farmers to grow crops more efficiently, and this is largely why sales of GM seeds to farmers have risen so quickly.

Soybeans account for half the world's GM crops (**FIGURE 10.24a**). Of the 28 nations growing GM crops in 2015, five (the United States, Brazil, Argentina, India, and Canada) accounted for 90% of production, with the United States alone growing 40% of the global total (**FIGURE 10.24b**). However, more than half the world's GM crops are now grown in developing nations.

Many people view biotechnology as a promising avenue toward sustainable agriculture. Such proponents argue that GM crops offer economic benefits to farmers while reducing environmental impacts. Many other people are skeptical and concerned. What, then, does scientific research tell us about the benefits and the risks of genetic engineering?

What are the benefits of GM crops?

Genetically engineered foods have long been promoted as a way to assist poor people in developing nations. If foods can be made more nutritious, then this can help us fight malnutrition. If crops can be made tolerant of drought or of salinized soils

(p. 216), then small farmers can more easily produce crops on marginal land. However, most of these noble intentions have not yet come to pass. This is largely because the corporations that develop GM varieties cannot easily profit from selling seed to small farmers in developing nations. Instead, most biotech crops thus far have been engineered for insect resistance and herbicide tolerance, which improve efficiency for large-scale industrial farmers who can afford the technology.

Regardless, proponents of genetic engineering maintain that GM foods bring substantial environmental and social benefits and assist the pursuit of sustainable agriculture in several major ways:

- By increasing production of food and fiber while lowering costs, GM crops enhance food security and reduce poverty and hunger for millions of people.

- By raising yields on existing farmland, GM crops alleviate pressure to clear forests and convert natural lands for agriculture, thus helping to conserve biodiversity, habitat, and ecosystem services. By one estimate, adoption of GM crops saved 152 million ha (375 million acres) of land from being cleared between 1996 and 2014—an area larger than Texas, Florida, and California put together.

- Crops engineered for drought tolerance help conserve water by reducing the need for irrigation.

- Crops engineered for better nutrition—such as golden rice (p. 251)—could help fight malnutrition and enhance children's health in developing nations.

- GM crops that reduce fossil fuel use during cultivation help to lower greenhouse gas emissions and address climate change. For example, herbicide-tolerant crops free farmers from having to plow to control weeds, and no-till farming (p. 226) can reduce erosion, increase carbon storage, and cut down on fossil fuel use. In 2014, GM crops were estimated to have reduced carbon dioxide emissions by the equivalent of taking 12 million cars off the road for a year.

- Crops engineered for insect resistance allow farmers to eliminate or reduce use of chemical insecticides.

FIGURE 10.24 So far, genetic engineering has mainly involved common crops grown in industrialized nations. Of the world's GM crops **(a)**, soybeans are the most common. Of global acreage planted in GM crops **(b)**, the United States devotes the most area. *Data from the International Service for the Acquisition of Agri-biotech Applications (ISAAA).*

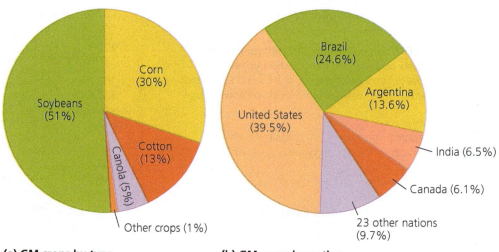

(a) GM crops by type

(b) GM crops by nation

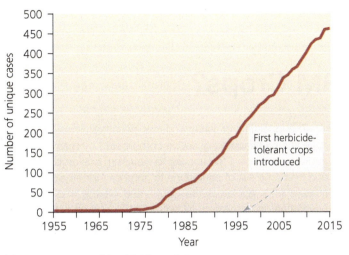

(a) Known cases of herbicide resistance

FIGURE 10.25 Weeds are evolving resistance to herbicides.
Documented cases of herbicide resistance have surpassed **(a)** 450 biotypes involving 250 species of plants. In little over a decade, weed resistance to glyphosate **(b)** spread across North America.
Data from Heap, I. International Survey of Herbicide-Resistant Weeds. Spring, 2016. www.weedscience.com.

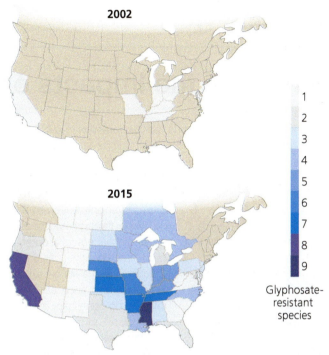

Glyphosate-resistant species

(b) Spread of glyphosate resistance

For reasons like these, millions of farmers like Michael Baxter feel that they are farming responsibly and contributing toward sustainable agriculture. Scientific research has largely supported most of these assertions. In 2014, researchers conducted a **meta-analysis,** an effort that gathers results from all scientific studies on a particular research question and statistically analyzes their data for significant patterns or trends. This meta-analysis calculated that, on average, researchers to that point had found that the adoption of GM crops increased yields by 22% and boosted farmers' profits by 68% while reducing chemical pesticide use by 37% (see **THE SCIENCE BEHIND THE STORY**, pp. 254–255).

To understand the effects on pesticide use, however, we need to consider insecticides and herbicides separately. Farmers who adopt insect-resistant Bt crops (pp. 248, 251) tend to use fewer insecticides because their crops do not need them; the insect resistance is already built into the plant. In India and China, Bt cotton has enabled more than 15 million farmers to increase their yields while reducing or eliminating their application of chemical insecticides. In contrast, herbicide-tolerant GM crops tend to result in *more* use of chemical herbicides, because farmers often apply more herbicide (to make sure weeds are killed) if their crops can withstand heavier applications.

A 2012 study by Washington State University scientist Charles Benbrook, using USDA data, calculated that as GM crops of all types expanded in the United States between 1996 and 2011, insecticide use declined by 56 million kg (123 million lb), but herbicide use increased by 239 million kg (527 million lb). This rise in herbicide use is accelerating, according to Benbrook's analysis, and the prime reason is that weeds are evolving resistance to herbicides, causing farmers

to apply even more of them. Just as insects can evolve resistance to insecticides (see Figure 10.19), weeds can evolve resistance to the chemicals used to kill them. Worldwide so far, more than 450 varieties of 250 weed species are known to have evolved resistance to herbicides (**FIGURE 10.25a**).

One of the most widely used herbicides is Monsanto's Roundup, which contains glyphosate as its active ingredient. As we saw in our Central Case Study at the beginning of this chapter, besides producing Roundup, Monsanto engineers seeds for crops that survive spraying with Roundup. For years farmers have loved these Roundup Ready crops because they reduce labor and boost yields: Simply spray fields with Roundup, and the weeds die while the crops thrive. However, widespread use of Roundup and other glyphosate-based herbicides is resulting in the evolution of resistance (**FIGURE 10.25b**). Scientists have confirmed glyphosate resistance in 32 weed species so far, and farmers report they believe the actual number is higher. As glyphosate-resistant weeds fill their fields, farmers douse their plants with more chemicals, including additional herbicides that are more powerful and more polluting.

What are the risks of GM crops?

Many people worry about health impacts from eating GM foods, but this widespread public anxiety has not been corroborated by research. A 2013 review of 1783 scientific studies found that "the scientific research conducted so far has not detected any significant hazards directly connected with the use of GE [genetically engineered] crops." Committees from many major organizations representing the world's scientists, including the U.S. National Academy of Sciences, the

What Are the Impacts of GM Crops?

**Dr. Wilhelm Klümper (left)
and Dr. Matin Qaim (right)**

Most of us are confronted with such a deluge of information—and misinformation—about genetic engineering and GM crops that it can be hard to know what to believe. Even for a trained scientist, it's not easy to wade through hundreds of research papers in a complex and fast-changing field and come up with a simple answer to a question such as, "What impacts do genetically modified crops have?"

In cases like this, scientists often turn to meta-analyses (p. 253). The prefix *meta* means beyond, or encompassing and transcending, or at a higher level—and a *meta-analysis* is essentially an analysis of analyses. For instance, if your doctor wants to prescribe a treatment for your medical condition, he or she probably won't rely on just one research paper, because different studies may come up with different results (through error, by chance, or because of varying conditions). Instead, your doctor would prefer to assess the entirety of the research literature to come up with the best information—and consulting a published meta-analysis can help him or her do so.

The question of what impacts genetically modified crops have is a multifaceted question complicated by many variables, and we can expect the answer to vary from one crop to another and from one set of conditions to another. Nonetheless, policymakers and regulators are confronted on the one hand with advocates demanding bans on all GMOs and on the other with agronomic experts urging them to speed and strengthen support for crop biotechnology—so they need general answers to help inform them before making decisions. Thus, in recent years several researchers have tried to make broad assessments of the scientific research on genetically modified foods.

The most widely cited work so far is that of Wilhelm Klümper and Matin Qaim of Georg-August-University in Göttingen, Germany, which was published in the open-access journal *PLOS ONE* in November 2014. They found 24,000 potential studies in total and screened them for suitability until arriving at 147 that compared GM crops to non-GM ones and met all requirements of the meta-analysis. Klümper and Qaim then used variants on the statistical technique known as regression (p. 62) to test for patterns in types of impacts the studies had documented.

Klümper and Qaim's major results can be seen in **FIGURE 1**. According to their analysis, research so far finds that GM crops produce yields that average 22% higher than non-GM crops. The analysis also found that GM crops require 37% less chemical pesticide use than non-GM crops. Thus, the analysis indicated that GM crops help farmers produce more food while applying fewer pesticides.

American Association for the Advancement of Science, the American Medical Association, the European Commission, the World Health Organization, and others, have reviewed the literature and concluded that GM foods present no clear evidence of danger.

Most scientists who harbor concerns over GMOs focus on potential ecological impacts. Many conventional crops can interbreed with their wild relatives (domesticated rice can breed with wild rice, for instance), so there seems little reason to believe that transgenic crops would not also occasionally breed with wild strains. Some scientists are concerned that in such an event, transgenes may be transferred from crops to other plants. This could in theory "contaminate" the genomes of local landraces of crops. In addition, if genes for glyphosate resistance are transferred to other plants, the genes from Roundup Ready "supercrops" could end up creating "superweeds."

In the first confirmed case of transgene escape, GM oilseed rape (a relative of canola) was found hybridizing with wild mustard. In another case, creeping bentgrass engineered

for use on golf courses—a GM plant not yet approved by the USDA—pollinated wild grass up to 21 km (13 mi) away from its test growing site. In the region of Mexico where corn was first domesticated, years of scientific debate and controversy have accompanied the apparent contamination of local landraces of maize by transgenic corn imported from the United States. Most scientists think that more transgenes will inevitably make their way from GM crops into both conventional crops and wild plants, but the consequences of this are open to debate.

Because biotechnology is rapidly changing, and because the large-scale introduction of GMOs into our environment is recent, we cannot yet know everything about the consequences. Many experts feel we should proceed with caution, adopting the **precautionary principle,** the idea that one should not undertake a new action until its ramifications are well understood. Others feel that enough research has been done to allow for informed choices, and that it is time to double our bet on biotechnology because it appears to offer more benefits than risks.

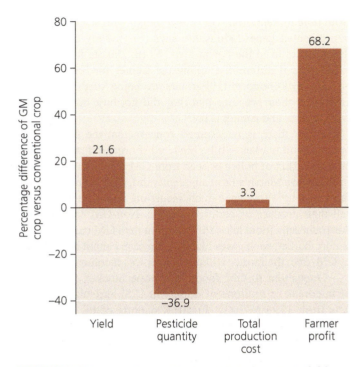

FIGURE 1 GM crops showed higher yields, lower pesticide use, and higher farmer profits than conventional non-GM crops. *Data from Klümper, W., and M. Qaim, 2014. A meta-analysis of the impacts of genetically modified crops. PLOS ONE 9(11) e111629: 1–7.*

The analysis also showed that GM crops increase farmers' production costs by 3%. GM seeds are more expensive, but this is balanced by the fact that farmers save money on pesticides. The 3% higher overall production costs, in turn, are far surpassed by the 22% yield gains. Because farmers have so much more crop to sell, their profits rise by 68% with GM crops, according to the analysis.

In exploring the data further, the researchers found that the pesticide savings were entirely due to reductions in insecticide use due to insect-resistant crops such as Bt corn and cotton (p. 251). Insect-resistant crops showed a 42% reduction in chemical application, whereas herbicide-tolerant crops such as Roundup Ready canola showed no significant difference in chemical application compared with non-GM crops. Yields were also substantially greater with insect-resistant crops than with herbicide-tolerant ones.

Critics took issue with the study's methods. Some complained that Klümper and Qaim had included results from conference presentations and studies by advocacy organizations and said they should have included only peer-reviewed studies published in respected journals. However, the analysis had found that GM crop yield gains were actually greater in peer-reviewed papers; had they included only these, the results would have been even more favorable to GM crops. Some critics faulted Klümper and Qaim for including mostly surveys of farmers rather than experimental field trials. Qaim responded that both types of studies are useful and that surveys have the advantage of reflecting results in real-world conditions on working farms, rather than idealized settings on smaller-scale research plots.

Further scientific research on the impacts of biotechnology in agriculture will continue to be published at a rapid pace. Only time will tell if future studies modify the big picture painted by Klümper and Qaim: that GM crops can bring greater food production and economic benefits to farmers, with less chemical pollution.

Public debate over GMOs continues

Science helps inform us about genetic engineering, but ethical and economic concerns have largely driven the public debate. For many people, the idea of "tinkering" with the food supply seems dangerous or morally wrong. Because every person relies on food for survival and cannot choose *not* to eat, the genetic modification of dietary staples such as corn and rice essentially forces people to consume biotech products or to go to unusual effort to avoid them.

The perceived lack of control over one's own food has driven concern that the global food supply is being dominated by a handful of large corporations that develop GM technologies, among them Monsanto, Syngenta, Bayer CropScience, Dow, DuPont, and BASF. Critics say these multinational corporations threaten the independence and well-being of the family farmer and that government regulators generally side with big business rather than small farmers. Critics of biotechnology (**FIGURE 10.26**) also voice concern that much of

FIGURE 10.26 Grassroots opposition to GM foods is widespread. At this demonstration in the Philippines, protestors maintained that Monsanto's genetically engineered corn increases costs to farmers and harms the environment while enriching the corporation.

the research into the impacts of GMOs is funded, conducted, or influenced by the corporations that stand to benefit if their products are approved.

In 2013, biotech companies appeared to gain even more power when a rider in a budget bill passed by the U.S. Senate stripped courts of the ability to revoke USDA approval of any GM crop found to be unsafe. Dubbed the "Monsanto Protection Act" by its critics, it inspired a groundswell of opposition from food safety advocates and citizens who viewed it as a violation of the government's system of checks and balances. The provision expired after six months.

So far, GM crops have not lived up to their promise of feeding the world's hungry. Crops with traits that might benefit poor farmers of developing countries (such as increased nutrition, drought tolerance, and salinity tolerance) have not been widely commercialized, likely because corporations have little economic incentive to do so. Whereas the Green Revolution was largely a public venture, the "gene revolution" promised by genetic engineering is largely driven by the financial interests of corporations selling proprietary products.

Indeed, corporations patent the transgenes they develop and go to great lengths to protect their investments. Thousands of North American farmers have found themselves at the mercy of the Monsanto Company. This first came to light in Canada, after Monsanto's private investigators took seed samples from canola plants grown by Saskatchewan farmer Percy Schmeiser and charged him with violating Canada's law that makes it illegal for farmers to reuse patented seed or grow the seed without a contract with the company. Schmeiser maintained that pollen from Monsanto's Roundup Ready canola used by his neighbors blew onto his land and pollinated his non-GM canola. When Schmeiser harvested his seed and replanted it the next year, many of the plants that grew contained Monsanto's patented herbicide-tolerance gene. Monsanto sued Schmeiser, and the courts sided with Monsanto, ordering the farmer to pay the corporation $238,000.

In 2008, Schmeiser and Monsanto reached an out-of-court settlement whereby Monsanto agreed to pay cleanup costs of the contamination. But the company continued suing other farmers. Monsanto has launched 150 or more such lawsuits against several hundred farmers and farm companies and has reached out-of-court settlements with several thousand farmers, resulting in payments of tens of millions of dollars. Most of these farmers were sued for saving seeds from one harvest to another—something farmers have done from time immemorial, yet is now illegal with seeds that contain patented genes. Monsanto says it is merely demanding that farmers heed the patent laws. North Dakota farmer Tom Wiley sees it differently, telling the Center for Food Safety, an anti-GMO nonprofit, "Farmers are being sued for having GMOs on their property that they did not buy, do not want, will not use, and cannot sell."

Today there is widespread concern that the burgeoning organic food market will be hindered if organic farms experience an influx of pollen or seed from GM plants, as Australian farmer Steve Marsh's farm apparently did. Like Marsh, organic farmers who experience such incursion might be unable to sell their produce as certified organic. Non-GM canola from Australia sells for at least $50/ton more than GM canola on the export market, so in cases like this, if a crop cannot be certified as GM-free, the farmer will receive far less income for it.

Opposition to GM foods in Europe blocked the import of hundreds of millions of dollars in U.S. agricultural products from 1998 to 2003. In 2013, exports of the $8 billion U.S. wheat crop were threatened after unapproved GM wheat plants were found on an Oregon farm, an apparent vestige of Monsanto field trials. Japan and South Korea suspended purchases of U.S. wheat in response.

Given such developments, the future of GM foods seems likely to hinge on social, economic, legal, and political factors as well as scientific ones—and these factors vary in different nations. In Europe and Japan, consumers have expressed widespread unease about genetic engineering. In contrast, most American consumers remain unaware that the majority of the food they eat now contains GM products.

Many nations label GM foods

More than 60 nations require that foods with genetically engineered ingredients be labeled so that consumers know what they are buying. In contrast, the United States does not label GM foods, despite the fact that polls consistently show that a large majority of Americans would like their food to be labeled. A petition requesting labeling sent to the U.S. Food and Drug Administration in 2011 garnered more than 1 million signatures. Since then, Vermont, Connecticut, and Maine have passed GMO labeling laws. In California and several other states, ballot measures to mandate labeling of GM food have been defeated amid well-funded opposition from food, biotech, and pesticide industries. In 2015, the U.S. House of Representatives passed a bill that would prevent states from requiring GMO labeling and would weaken government's ability to regulate GMOs. Opponents nicknamed the bill the "Denying Americans the Right-to-Know (DARK) Act." As of this writing it has not passed the Senate.

Proponents of labeling argue that consumers have a right to know what's in the food they buy. Opponents argue that labeling implies that labeled foods are dangerous, whereas research has not shown that to be the case. They also say that labeling will entail expense and that consumers wishing to avoid GM foods can do so by buying certified organic foods. In nations where labeling has been allowed, stores have ended up eliminating some GM foods from their shelves because of consumer avoidance. If this were to occur in the United States, it could pressure food producers to avoid using GM ingredients and farmers to avoid growing GM crops.

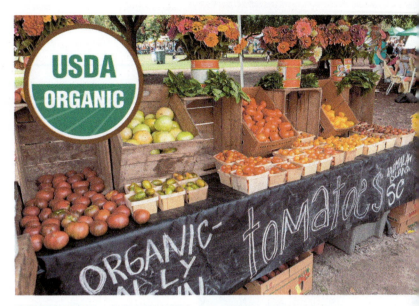

FIGURE 10.27 Look for the USDA organic logo to ascertain whether a product is certified organic under the National Organic Program.

Organic Agriculture

Concerns over genetic engineering, chemical pesticides, and other aspects of high-input industrial agriculture have led many people to support agriculture that involves fewer fossil fuel inputs and less pollution. *Low-input agriculture* describes farming and ranching that use lesser amounts of pesticides, fertilizers, growth hormones, antibiotics, water, and fossil fuel energy than in industrial agriculture. This approach seeks to reduce the costs of food production by allowing nature to provide ecosystem services (such as pest control, pollination, and fertilizer) that farmers using industrial methods must pay for themselves. Food-growing practices that use no synthetic fertilizers, pesticides, hormones, or antibiotics—but instead rely on biological approaches such as composting (p. 609) and biocontrol—are termed **organic agriculture.**

Organic approaches reduce inputs and pollution

The bounty of organic agriculture is increasingly available to us (**FIGURE 10.27**). But what exactly is meant by the term *organic*? In 1990, the U.S. Congress passed the Organic Food Production Act to establish national standards for organic products and facilitate their sale. Under this law, the USDA in 2000 issued criteria by which crops and livestock can be officially certified as organic (**TABLE 10.2**). These standards went into effect in 2002 as part of the National Organic Program. California, Washington, and Texas established stricter state guidelines for labeling foods organic, and today many U.S. states and more than 80 nations have laws spelling out organic standards.

For farmers, organic farming can bring many benefits: lower input costs, enhanced income from higher-value produce, and reduced chemical pollution and soil degradation (see **THE SCIENCE BEHIND THE STORY**, pp. 258–259).

TABLE 10.2 USDA Criteria for Certifying Crops and Livestock as Organic

For crops to be considered organic . . .

- The land must be free of prohibited substances for 3 years.
- Crops must not be genetically engineered.
- Crops must not be irradiated to kill bacteria.
- Sewage sludge cannot be used.
- Organic seeds and planting stock are preferred.
- Farmers must not use synthetic fertilizers. Only crop rotation, cover crops, animal or crop wastes, or approved synthetic materials are allowed.
- Most conventional pesticides are prohibited. Pests, weeds, and diseases should be managed with biocontrol, mechanical practices, or approved synthetic substances.

For livestock to be considered organic . . .

- Mammals must be raised organically from the last third of gestation; poultry, from the second day of life.
- Feed must be 100% organic, although vitamin and mineral supplements are allowed.
- Dairy cows must receive 80% organic feed for 9 months, followed by 3 months of 100% organic feed.
- Hormones and antibiotics are prohibited; vaccines are permitted.
- Animals must have access to the outdoors.

Adapted from the National Organic Program, 2002. *Organic production and handling standards.* Washington, DC: U.S. Department of Agriculture.

How Productive Is Organic Farming?

Organic farming puts fewer synthetic chemicals into the soil, air, and water than conventional industrial farming does. But can organic farming produce large enough crop yields to feed the human population? The world's two longest-running field experiments on the topic—in Switzerland and in Pennsylvania—suggest that the answer is yes.

Let's first visit Switzerland, where 1 in every 8 hectares of agricultural land is managed organically (the sixth-highest rate among nations) and where people consume more organic food per capita than anywhere else in the world. Back in 1977, Swiss researchers established experimental farms at Therwil, near the city of Basel. At the research site,

Swiss scientist Dr. Paul Mäder visits Pennsylvania's Rodale Institute

wheat, potatoes, and other crops are grown in plots cultivated in different treatments:

- Conventional farming using chemical pesticides, herbicides, and inorganic fertilizers
- Conventional farming that also uses organic fertilizer (cattle manure)
- Organic farming using only manure, mechanical weeding, and plant extracts to control pests
- Organic farming that also adds natural boosts, such as herbal extracts in compost

Researchers record crop yields at harvest each year. They analyze the soil regularly, measuring nutrient content, pH, structure, and other variables. They also measure the biological diversity and activity of microbes and invertebrates in the soil. Such indicators of soil quality help researchers assess the potential for long-term productivity.

In 2002, Paul Mäder and colleagues from two Swiss research institutes reported in the journal *Science* results from 21 years of data. Long-term studies are rare in agriculture and in ecology. They are highly valuable, because they can reveal slow processes or subtle effects that get swamped out by year-to-year variation in shorter-term studies. Over the study's 21 years, the organic fields yielded 80% of what the conventional fields produced. Organic potato crops averaged just 58–66% of conventional yields because of nutrient deficiency and disease, but organic crops of winter wheat produced 90% of conventional yields.

Although the organic plots produced 20% less on average, they received 35–50% less fertilizer than the conventional fields and 97% fewer pesticides. Thus, Mäder's team concluded, the organic plots were highly efficient and represent "a realistic alternative to conventional farming."

How can organic fields produce decent yields without relying on synthetic chemicals? The answer lies in the soil. Mäder's team found that soil in the organic plots had better structure, better supplies of some nutrients, and much more microbial activity and invertebrate biodiversity than soil in conventional fields supplemented with manure (**FIGURE 1**). Organic fields outperformed conventional fields without manure still more.

Studies are continuing at the Swiss plots today, producing new research results. As one example, Jens Leifeld and two colleagues at a Zurich research institute analyzed soil carbon content after 27 years. They found that soil carbon had decreased in all treatments, but that conventional plots had suffered the greatest decline. Conventional plots supplemented with manure, however, did just as well as the organic plots in retaining soil carbon.

Organic farming has proved even more successful in Pennsylvania, where the Rodale Institute has compared organic and conventional fields of corn and soybeans in a large-scale experiment running since 1981 on its 330-acre farm. In 2011 it released results from 30 years' worth of data (**FIGURE 2**).

Averaged across the 30 years, yields of organically grown crops equaled yields of conventionally grown crops. Moreover, the organic crops required 30% less energy input, and raising them released 35% fewer greenhouse gas emissions. Because of lower energy inputs and higher crop prices for organic produce, farming organically resulted in three times the profit for the farmer.

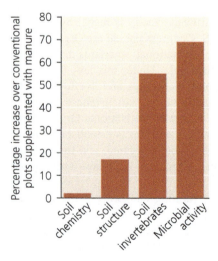

FIGURE 1 Organic fields developed better soil quality than conventional fields supplemented with manure. Values for soil chemistry (6 variables), structure (3 variables), invertebrates (5 variables), and microbial activity (6 variables) were compared. *Data from Mäder, P., et al., 2002. Soil fertility and biodiversity in organic farming.* Science *296: 1694–1697.*

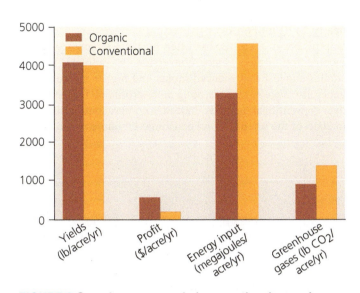

FIGURE 2 **Organic crops equaled conventional crops in yields while producing more profit, requiring less energy input, and releasing fewer greenhouse gas emissions.** *Data from Rodale Institute, 2011.* The farming systems trial: Celebrating 30 years. *Kutztown, PA: Rodale Institute.*

FIGURE 3 **Organically grown corn did better than conventionally grown corn during periods of drought in the Rodale Institute experiment.** Better-quality soil (shown in left half of the central inset) was part of the reason.

As with the Swiss experiment, the secret lies in the soil. At the Rodale Institute's farm over the years, the soil of the organic fields became visibly darker and better textured than the conventional fields' soil. This helped corn crops in the organic fields to outperform those in the conventional fields during times of drought (**FIGURE 3**).

Given such differences in soil quality, researchers expect that organic fields should perform better and better relative to conventional fields as time goes by—in other words, that they are more sustainable. Moreover, the striking yield results of the Rodale experiment suggest that perhaps organic agriculture can feed the world's people every bit as reliably as today's conventional industrial agriculture.

Several other long-term experiments comparing organic and conventional farming systems are currently running in the United States (**TABLE 1**). Their results so far parallel those found in Pennsylvania and Switzerland: Organic yields approach or match conventional yields, while reducing pollution, building better soil quality, and producing higher profits for farmers.

TABLE 1   Five Additional Long-Term Studies of Organic vs. Conventional Farming

STUDY AND LOCATION	YEAR STARTED	KEY RESULTS SO FAR
University of California, Davis: Davis, California	1988	Organic beans and safflower show yields comparable to those of conventional plots. Crops that require more nitrogen show lower yields in organic plots.
University of Minnesota: Lamberton, Minnesota	1989	Organic and conventional corn yields are equivalent. Soil quality has increased in organic plots. Organic plots are more resilient to stress.
University of Wisconsin–Madison: Arlington, Wisconsin	1989	Organic corn and soybeans yield ~80% of conventional plots in rainy years when managing weeds is difficult. Forage crops and milk production are equivalent between plot types. Better technology improves organic yields.
USDA Agricultural Research Service: Beltsville, Maryland	1996	Organic plots accumulate more nitrogen and soil organic carbon, and can improve soil quality even more than no-till farming.
Iowa State University: Greenfield, Iowa	1998	Organic yields are equivalent to or greater than conventional yields for corn and soybeans. Organic farming improves soil quality and produces higher profits for farmers.

Summarized from Delate, K., et al. 2015. A review of long-term organic comparison trials in the U.S. *Sustainable Agriculture Research* 4(3): 5–14.

Australian farmer Steve Marsh was experiencing all these benefits on his organic farm before running into conflict with his neighbor. In many cases more pests attack organic crops because of the lack of chemical pesticides—indeed, Michael Baxter accused Marsh of causing pests to spill over onto his property—but in many cases, biocontrol methods (p. 248) can help keep pests in check. Moreover, the lack of synthetic chemicals in organic farming maintains soil quality and encourages pollinating insects. Surveys reveal that farmers who adopt organic techniques do so primarily because they want to practice stewardship toward the land and to safeguard their family's health.

Farmers face obstacles to adopting organic methods, however. Foremost among these are the risks and costs of shifting to new methods, particularly during the transition period. Australian farmers like Steve Marsh need to meet standards for three years before their products can be certified, and U.S. farmers face the same requirement.

Many consumers favor organic food out of concern about health risks posed by the pesticides, hormones, and antibiotics used in conventional industrial agriculture. Consumers also buy organic products out of a desire to improve environmental quality. The main obstacle for consumers is price. Organic products tend to be 10–30% more expensive than conventional ones, and some (such as milk) can cost twice as much. However, enough consumers are willing to pay more for organic products that grocers and other businesses are making them more widely available.

Organic agriculture is on the rise

Just a decade or two ago, few farmers grew organic food, few consumers demanded it, and the only place to buy it was in specialty stores. Today that has changed; about 80% of Americans buy organic food at least occasionally, most retail groceries offer it, and Americans are the world's eighth-highest per-person consumers of organic food. U.S. consumers now spend more than $35 billion annually on organic food, amounting to 5% of all food sales (**FIGURE 10.28**). Worldwide, sales of organic food remain less than 1% of all food sales, but they grew 4.5 times between 2000 and 2014, when sales surpassed $80 billion.

Production is increasing along with demand. Although organic agriculture takes up just 1% of agricultural land worldwide (44 million ha, or 108 million acres, in 2014), this area is rapidly expanding. Two-thirds of this area is in developed nations, and nearly two-thirds is grazing land. In the United States, the number of certified operations has more than tripled since the 1990s, while acreage devoted to organic crops and livestock has quadrupled: today 2.2 million ha (5.4 million acres) are under organic management, representing 0.6% of U.S. agricultural land. Europe boasts far more: 11.6 million ha (28.7 million acres) as of 2014, most in nations of the European Union, where 5.7% of the agricultural area is in organic production.

Government initiatives are assisting the growth of organic farming. The European Union supports farmers financially during conversion. This is an example of a subsidy (p. 178) aimed at reducing the external costs (pp. 141, 162) of industrial agriculture. The United States offers no such subsidy, which may explain why U.S. organic production lags behind that of Europe. However, the 2014 Farm Bill (p. 230) did expand U.S. funding for research on organic agriculture and to assist growers with the costs of certification, totaling nearly $170 million over five years. Government support is helpful, because conversion often means a temporary loss in income for farmers. Once conversion is complete, though, studies suggest that reduced inputs and higher market prices generally make organic farming more profitable for the farmer than conventional methods. Indeed, a comprehensive summary in 2015 of the scientific and economic literature found that on

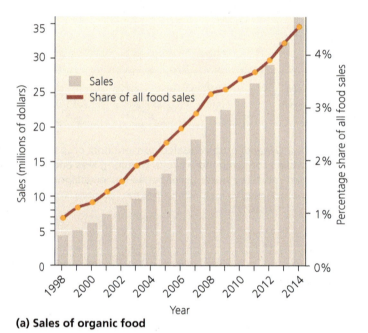

(a) Sales of organic food

(b) Extent of organic agriculture

FIGURE 10.28 Organic agriculture is growing. Sales of organic food in the United States **(a)** have increased rapidly, both in total dollar amounts (**bars**) and as a percentage of the overall food market (**line**). Since the 1990s **(b)**, U.S. acreage devoted to organic crops and livestock has quadrupled, and certified operations have more than tripled. Data (a) from Organic Trade Association's 2016 Organic Industry Survey, and previous years; (b) from USDA Economic Research Service.

average across the world, organic farming is 22–35% more profitable than conventional farming. Data from the USDA also support this; recent figures for corn, soy, wheat, and milk indicate that although organic yields are somewhat lower than conventional yields, today's price premiums are great enough that organic approaches are more profitable for the farmer.

Can organic agriculture feed the world?

Although the benefits of organic approaches in reducing pollution are clear, people have long wondered whether organic agriculture can produce enough food for the world's growing population. In recent years a number of researchers have reviewed the growing literature of published scientific studies comparing yields of organic agriculture and conventional industrial agriculture. These meta-analyses—analyzed summaries of many studies (p. 253)—have found that organic agriculture, on average, tends to produce yields of roughly 80% of those of conventional agriculture (FIGURE 10.29). The gap between organic and conventional yields is highly variable, however, and depends on many factors.

So if yields from organic agriculture average 20% less than those of conventional industrial agriculture, can we feed the world with organic agriculture? The mainstream view has long been "no," and critics point out that lower yields mean that more land would need to be converted to agriculture to produce the same amount of food. However, increasing numbers of scientists, farmers, and other experts are more optimistic. They point out that organic yields have been rising as farmers gain more experience, and could rise still further if organic agriculture were to receive the research funding that goes toward conventional agriculture (the USDA sends only 2% of its research funding to organic agriculture). In addition, they say, by protecting soil quality, organic farming keeps more acreage productive for farming over the long term than does conventional farming. Moreover, experts argue that overall food production is not what limits our ability to feed the world. Only 43% of the global grain crop actually feeds people directly; the rest goes to livestock feed, biofuels (p. 566), and processed products like high-fructose corn syrup. Politics and the logistics of transport limit food distribution, and one-third of all food intended for our consumption ends up going to waste. Thus, by reducing waste, eating less meat, dealing with distribution problems, and prioritizing food for people, the world could easily be fed by organic agriculture even with lower yields, supporters maintain.

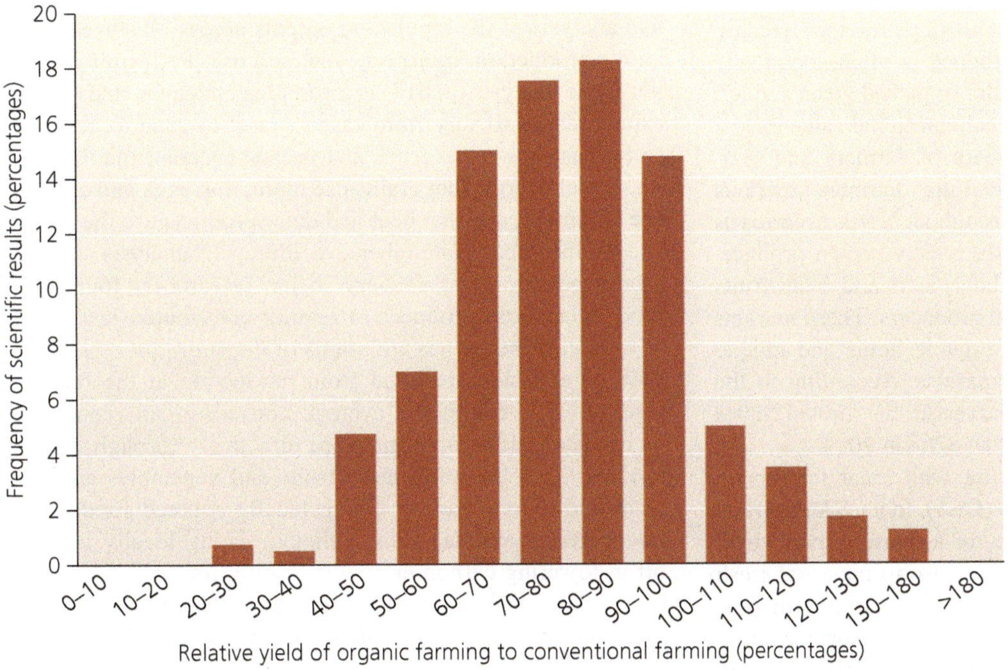

Sustainable Food Production

Growing adequate food and fiber for our burgeoning global population while maintaining the integrity of the environmental systems that support our agriculture is a tremendous challenge. Sustainable agriculture, like sustainability itself, involves a triple bottom line of social, economic, and environmental dimensions (p. 153). Sustainable agriculture consists of agriculture that produces enough to provide food security to society, that is profitable enough to provide farmers and ranchers a viable living, and that conserves resources adequate to support future agriculture.

FIGURE 10.29 On average, crop yields on organic farms were about 80% of yields on conventional industrial farms. This graph shows the results of a meta-analysis (p. 253); each bar represents the frequency with which scientific results (for organic yields relative to conventional yields) fell into each interval shown on the x axis. *Data from de Ponti, T., 2012. The crop yield gap between organic and conventional agriculture. Agricultural Systems 108: 1–9.*

DATA Q What percentage of the relative yield results surveyed by the meta-analysis fell between 40% and 50%? What percentage of the results showed organic yields to be greater than conventional yields? Explain how you determined this.

Go to **Interpreting Graphs & Data** on **Mastering**EnvironmentalScience®.

We can consider multiple paths toward sustainable agriculture

A variety of approaches exist as pathways to sustainable agriculture. On one end of the spectrum is conventional industrial agriculture. It produces high yields thanks to intensive inputs of fossil fuels, pesticides, and fertilizers applied to monocultures. At the other end of the spectrum is organic agriculture, which rejects chemical pesticides and fertilizers and takes a low-input approach, accepting lower yields but protecting natural resources that support agriculture in the long term. Into the debate between the high-input industrial approach and the low-input organic approach has stepped biotechnology. Using genetic engineering to increase crop yields while reducing environmental impacts is one strategy that can theoretically lead toward sustainable food production. Genetic engineering has tremendous potential for the future if we can harness and direct it toward safely serving people's needs, especially those of small farmers and consumers.

Regardless of the paths taken, any effort to make agriculture truly sustainable will need to address the adverse environmental and social impacts we now struggle with. These include soil degradation; overuse of water; loss of crop diversity, pollinators, and natural habitats; overuse of chemical pesticides and fertilizers; reliance on fossil fuels; and greenhouse gas emissions and climate change.

Locally supported agriculture is growing

Many proponents of sustainable agriculture call attention to the fossil fuel energy we use to transport food. Those who tally "food miles" say the average food product sold in a U.S. supermarket travels at least 1600 km (1000 mi) between the farm and the grocery, burning large amounts of petroleum. Because of the travel time, supermarket produce is often chemically treated to preserve freshness and color or picked green and left to ripen in transit, affecting flavor and nutritional value.

In response, increasing numbers of farmers and consumers are supporting local agriculture. Farmers' markets (**FIGURE 10.30**) are springing up throughout North America as people rediscover the joys of fresh, locally grown produce. At **farmers' markets,** consumers buy meats and fresh fruits and vegetables in season from local producers. These markets generally offer a wide choice of organic items and unique local varieties not found in supermarkets. According to the USDA, the number of farmers' markets in the United States grew from 1755 in 1994 to more than 8268 in 2014.

Some consumers are partnering with local farmers in **community-supported agriculture (CSA).** In a CSA program, consumers pay farmers in advance for a share of their yield, usually a weekly delivery of produce. Consumers get fresh seasonal produce, while farmers get a guaranteed income stream up front to invest in their crops—a welcome alternative to taking out loans and being at the mercy of the weather.

Farmers' markets and community-supported agriculture help strengthen local economies while giving the consumer access to fresher foods. They also often cut down on food miles. However, the number of miles a given item has traveled is not in itself a reliable measure of fossil fuel consumption. In many cases, shipping a food item far away as part of a large shipment by barge or rail may result in *less* fossil fuel use per item than transporting the item a short distance in small quantities by truck.

FIGURE 10.30 Farmers' markets are flourishing as consumers rediscover the benefits of buying fresh, locally grown produce.

To determine which alternative involves less energy use overall, one needs to conduct a **life-cycle analysis,** a quantitative analysis of the inputs and outputs across all stages of an item's production, transport, sale, and use. Performing a full life-cycle analysis (p. 618) is a complex endeavor, and we can expect results to vary from case to case. In general, research so far has shown that fruits and vegetables consume the least fossil fuel energy, that grains use more, that eggs and chicken use still more, and that beef and dairy products use the most.

In the most comprehensive life-cycle analysis of U.S. food production and delivery so far, researchers found that food miles from producer to retailer contributed just 4–5% of total greenhouse gas emissions of the entire process. Fully 83% of emissions resulted from production at the farm or feedlot. As a result, the average consumer can reduce his or her ecological footprint more effectively through dietary choices (such as eating more fruits and vegetables and less meat and dairy) than by eating locally sourced food, these researchers maintain. Thus, although eating locally is helpful in lowering carbon emissions, eating fewer animal products makes a far bigger difference. In Calculating Ecological Footprints (p. 266), you will work with some of these data yourself.

Sustainable agriculture mimics natural ecosystems

The best approach for making an agricultural system sustainable is to mimic the way a natural ecosystem functions. Ecosystems operate in cycles and are internally stabilized with negative feedback loops (p. 104). In this way they provide a useful model for agriculture.

One example comes from Japan, where rice farmers are reviving ancient traditions and finding them superior to modern industrial methods. Takao Furuno is one such farmer. Starting 25 years ago, he and his wife added a crucial element to their rice paddies: the crossbred aigamo duck. Each spring after they plant rice, the Furunos release hundreds of aigamo ducklings into their paddies (**FIGURE 10.31**). The ducklings eat weeds that compete with the rice, and they eat insects and snails that attack the rice. The ducklings also fertilize the rice plants with their waste and oxygenate the water by paddling. Noting his success, scientists have worked with Furuno to determine what he does that works so well. They have found that in paddies that have ducks, rice plants grow larger and yield far more rice. Once the rice grains form, the ducks are removed from the paddies (because they would eat the rice grains) and kept in sheds, where they are fed waste grain. The ducks mature, lay eggs, and can be sold at market.

Besides ducks, the Furunos raise fish in the paddies, and these provide food and fertilizer as well. They also let the aquatic fern *Azolla* cover the water surface. This plant fixes nitrogen; feeds the ducks; and provides habitat for insects, plankton, and aquatic invertebrates, which provide additional food for the fish and ducks. Because fast-growing *Azolla* can double its biomass in 3 days, surplus plant matter is harvested and used as cattle feed. The end result is a productive ecosystem in which pests and weeds are transformed into resources and which yields multiple types of organic food. From only 2 ha of paddies and 1 ha of organic vegetables, the Furunos annually produce 7 tons of rice, 300 ducks, 4000 ducklings, and enough vegetables to feed 100 people. At this rate—twice the productivity of the region's conventional farmers—just 2% of Japan's people could supply the nation's food needs.

Takao Furuno wrote a book to popularize the "aigamo method," and an estimated 75,000 Asian farmers are now using it, increasing their yields 20–50% and regaining huge amounts of time they used to spend manually weeding. All across the world, treating agricultural systems as ecosystems is a key aspect of sustainable agriculture.

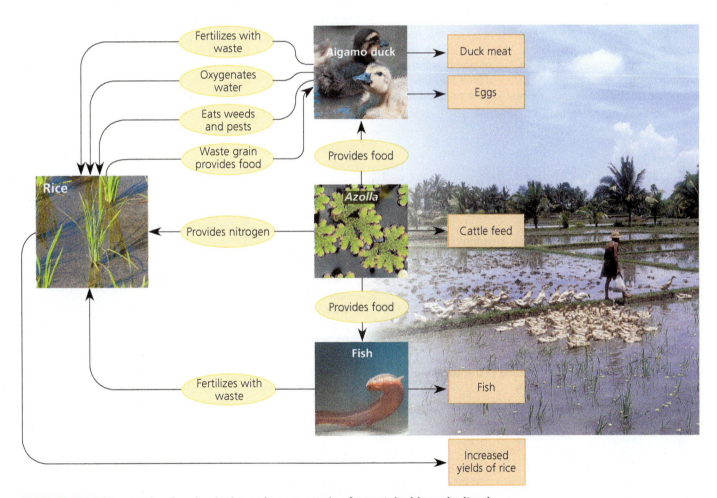

FIGURE 10.31 Aigamo rice-farming in Japan is an example of a sustainable agricultural system. Ducks, fish, and a water fern help create an ecosystem with the rice that makes each of them more productive. In this diagram, benefits that one organism provides to another are shown in ovals, with arrows pointing in the direction the benefits flow. In the rectangles are types of food that result as outputs from the system.

closing
THE LOOP

The high yields and efficient production of industrial agriculture have helped to provide food security and relieve pressures for land conversion. At the same time, industrial agriculture as practiced today emits so much pollution and consumes so many resources that it calls into question its own long-term sustainability. Organic farming represents an alternative pathway toward sustainable agriculture by putting resource conservation first and scaling down the intensity of farming to lessen its impacts. Biotechnology, meanwhile, scales up the technological aspects of agriculture through genetic engineering, seeking to find novel solutions to produce more food at less expense and with less environmental impact. Amid debates among these three camps, the U.S. Department of Agriculture holds that all means of farming should coexist and be supported and that we should try to take advantage of the best attributes of each.

Many organic producers are skeptical of the policy of coexistence urged by the USDA. They look at the situation of organic farmer Steve Marsh in Australia and wonder how organic farmers can protect the purity of their harvests if governments do not place tougher restrictions on GM crops. In Australia, the debate has continued long past the end of the legal battle between Marsh and Baxter. Supporters of Marsh (such as the Safe Food Foundation) are urging policymakers to pass laws protecting non-GM farmers, while supporters of Baxter (such as the Pastoralists and Graziers Association) argue that coexistence works and that policymakers should repeal policies restricting GM crops. Meanwhile, in the wake of the Marsh–Baxter dispute, further reports are surfacing in Australia of stray GM canola plants found growing on non-GM farms and on roadsides along routes where trucks transport canola seed.

The societal and scientific debates worldwide over whether and how different methods of farming can coexist are nuanced and complex. What is certain is that if our planet is to support 9 billion people by midcentury, we must find ways to shift to agricultural techniques that do not further degrade the soil, water, pollinators, and other resources and ecosystem services that support our food production. Approaches such as biological pest control, organic production, pollinator conservation, preservation of native crop diversity, sustainable aquaculture, and likely some degree of careful and responsible genetic engineering may all be parts of the game plan we will need to set in motion to work together toward a sustainable future.

REVIEWING Objectives

You should now be able to:

Explain the challenge of feeding a growing human population

Our food production has outpaced our population growth thus far, yet nearly 800 million people still go hungry each year. Undernutrition, overnutrition, and malnutrition all challenge the goal of food security, as population and resource consumption continue to grow. (pp. 236–238)

Identify the goals, methods, and consequences of the Green Revolution

The Green Revolution aimed to enhance agricultural productivity in developing nations. Scientists used selective breeding to develop crop strains that grew quickly, were more nutritious, or were resistant to disease or drought. The increased efficiency and higher yields fed more people while reducing the amount of natural land converted for farming. However, the resulting expanded use of fossil fuels, chemical fertilizers, and synthetic pesticides has increased pollution and soil degradation. (pp. 238–239)

Discuss how we raise animals for food, and assess the impacts that result

As consumption of animal products has increased, we have turned to intensive production through feedlots and aquaculture. Feedlots create waste and pollution, but also relieve pressure on lands that could otherwise be overgrazed. Aquaculture provides food security and relieves pressures on wild fish stocks but, at industrial scales, can give rise to pollution, habitat loss, and other impacts. Eating animal products leaves a considerably greater ecological footprint than eating plant products. (pp. 240–244)

Describe reasons and approaches for preserving crop diversity

Protecting diversity of native crop varieties provides insurance against failure of major crops. Seed banks preserve rare and local varieties of seed, acting as storehouses for genetic diversity. (pp. 244–245)

Discuss threats to pollinators and identify potential solutions

Many pollinators are dwindling due to habitat loss, chemical pesticides, parasites, and disease. Conserving pollinating insects through habitat provision and pesticide reduction is vital to our food security. (pp. 245–246)

Explore strategies for pest and weed management

We kill "pests" and "weeds" with synthetic chemicals that can pollute and pose health hazards. Pests and weeds can evolve resistance to chemical pesticides or herbicides, leading us to design more toxic poisons. Biological control uses natural enemies of pests against them, whereas integrated pest management combines various techniques and minimizes the use of synthetic chemicals. (pp. 246–249)

Describe the science behind genetic engineering

Genetic engineering uses recombinant DNA to move genes for desirable traits from one type of organism to another. The process is both like, and unlike, traditional selective breeding. (pp. 249–251)

Compare the benefits and costs of genetically modified foods, and assess the public debate over them

Biotech crops offer several major benefits for agriculture and society. These have not yet reached their potential but could help make our agriculture sustainable. There is no clear evidence for human health effects, but GM crops may have ecological impacts, including the spread of transgenes and an increase in herbicide pollution. Many people have ethical qualms about altering food through genetic engineering. Development of biotech foods has been controlled by multinational corporations, which many critics view as a threat to small farmers. Debate continues over whether to label GM foods in the marketplace. (pp. 252–257)

Analyze the nature, growth, and potential of organic agriculture

Organic agriculture, because it reduces chemical and fossil fuel inputs, exerts fewer environmental impacts than industrial agriculture. Organic produce makes up a small part of the market but is growing rapidly. Scientific studies show that organic agriculture is productive and is a realistic alternative to industrial agriculture. (pp. 257–261)

Contrast conventional industrial, organic, and biotech approaches to agriculture, and summarize pathways toward sustainable agriculture

We can work toward sustainable agriculture in a variety of ways. Conventional industrial agriculture produces high yields, but at the cost of pollution. Organic agriculture protects resources but generally has somewhat lower yields. Biotechnology and genetic engineering aim to offer novel ways to enhance production while minimizing environmental impact. Locally supported agriculture (e.g., farmers' markets and CSAs) minimizes food miles and fossil fuel use. Mimicking natural ecosystems is a key approach to making agriculture sustainable. (pp. 261–263)

TESTING Your Comprehension

1. What kinds of techniques have people used to increase agricultural production? How did Norman Borlaug help inaugurate the Green Revolution?

2. Name several positive and negative environmental consequences of feedlot operations. Why is beef an inefficient food from the perspective of energy consumption?

3. What are some economic benefits of aquaculture? What are some negative environmental impacts?

4. Why is it important to preserve the genetic diversity of crops and their landraces and wild relatives? How can we preserve populations of honeybees and of native pollinators?

5. Explain in your own words how pesticide resistance occurs.

6. Explain the concept of biocontrol. List several components of a system of integrated pest management (IPM).

7. How is a transgenic organism engineered? How is genetic engineering different from traditional agricultural breeding? How is it similar?

8. Describe several reasons why many people support the development of genetically modified organisms, and name several uses of such organisms that have been developed so far.

9. Describe the scientific concerns held by opponents of GM crops. Describe some of their other concerns.

10. Define *organic agriculture*, and describe what standards a farmer in the United States must meet to gain organic certification. What factors are driving the growth of organic agriculture?

SEEKING Solutions

1. Assess several ways in which high-input industrial agriculture can be beneficial for people and for environmental quality, as well as several ways in which it can be detrimental. Now suggest several ways in which we might modify industrial agriculture to mitigate its impacts on people and the environment.

2. What factors make for an effective biological control strategy of pest management? What risks are involved in biocontrol? If you had to decide whether to use biocontrol against a particular pest, what questions would you want to have answered before you decide?

3. People who view GM foods as solutions to world hunger and insecticide overuse often want to speed their development and approval. Others adhere to the precautionary principle and want extensive testing for health and environmental safety. How much caution do you think is warranted before a new GM crop is introduced? Describe what you feel the ideal process should be.

4. **CASE STUDY CONNECTION** Put yourself in the seat of one of the judges in Australia's long-running legal battle between Steve Marsh and Michael Baxter. If you had to decide whom to favor and what damages to award, what questions would you ask Marsh? What questions would you ask Baxter? How do you expect you might rule in the case? Give reasons for your answers.

5. **THINK IT THROUGH** You are a farmer in your local area, and you have 500 acres to farm as you wish. What crops and animals would you raise, what farming approaches would you pursue, and why? Would you farm organically or by conventional industrial methods? Would you choose to grow genetically modified crops? How would you manage for pests and weeds? Give reasons for each of your answers.

6. **THINK IT THROUGH** You are a USDA official and must decide whether to allow the commercial planting of a new genetically modified strain of cabbage that is tolerant to a best-selling herbicide and has twice the vitamin content of regular cabbage. What questions would you ask of scientists before deciding whether to approve the new crop? What scientific data would you want to see? Would you also consult nonscientists? Would you consider ethical, economic, and social factors?

CALCULATING Ecological Footprints

Many people who want to reduce their ecological footprint have focused on how much energy is expended (and how many climate-warming greenhouse gases are emitted) in transporting food from its place of production to its place of sale. The typical grocery store item is shipped by truck, air, and/or sea for many hundreds of miles before reaching the shelves, and this transport consumes oil. This concern over "food-miles" has helped drive the "locavore" movement to buy and eat locally sourced food.

However, food's transport from producer to retailer, as measured by food-miles, is just one source of carbon emissions in the overall process of producing and delivering food.

In 2008, environmental scientists Christopher Weber and H. Scott Mathews conducted a thorough life-cycle analysis (p. 262) of U.S. food production and delivery. By filling in the table below, you will get a better idea of how our dietary choices contribute to climate change.

FOOD TYPE	TOTAL EMISSIONS[1] ACROSS LIFE CYCLE	EMISSIONS[1] FROM DELIVERY[2]	PERCENTAGE EMISSIONS FROM DELIVERY[2,3]
Fruits and vegetables	0.85	0.10	11.8
Cereals and carbohydrates	0.90	0.07	
Dairy products	1.45	0.03	
Chicken/fish/eggs	0.75	0.03	
Red meat	2.45	0.03	
Beverages	0.50	0.04	

[1]Emissions are measured in metric tons of carbon dioxide equivalents, per household per year.
[2]"Delivery" means transport from producer to retailer.
[3]"Percentage emissions from delivery" is calculated by dividing "Emissions from delivery"/"Total emissions across life cycle" and then multiplying by 100 (to convert proportion to percentage).
Source: Weber, C.L., and H.S. Mathews, 2008. Food-miles and the relative climate impacts of food choices in the United States. *Environmental Science and Technology* 42: 3508–3513.

1. Which type of food is responsible for the most greenhouse gas emissions across its whole life cycle? Which type is responsible for the least emissions?

2. What is the range of values for percentage of emissions from transport of food to retailers (delivery)? Weber and Mathews found that 83% of food's total emissions came from its production process on the farm or feedlot. What do these numbers tell you about how you might best reduce your own footprint with regard to food?

3. After measuring mass, energy content, and dollar value for each food type, the researchers calculated emissions per kilogram, calorie, and dollar. In every case, red meat produced the most emissions, followed by dairy products and chicken, fish, and eggs. They then calculated that shifting one's diet from meat and dairy to fruits, vegetables, and grains for just one day per week would reduce emissions as much as eating 100% locally (cutting food-miles to zero) all the time. Knowing all this, how would you choose to reduce your own food footprint? By how much do you think you could reduce it?

MasteringEnvironmentalScience®

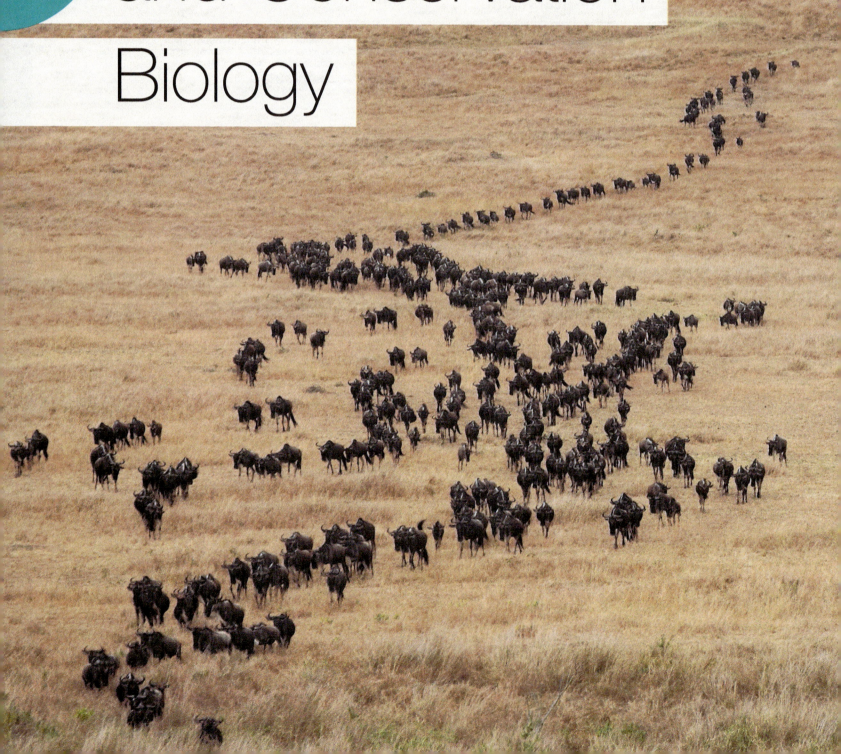

CHAPTER 11

Biodiversity and Conservation Biology

Wildebeest crossing the vast plains of the Serengeti

central
CASE STUDY

Will We Slice through the Serengeti?

> **Construction of the road will be a huge relief for us. We will sell . . . maize and horticultural products to our colleagues in Arusha and they will bring us cows and goats.**
> Bizare Mzazi, a small farmer outside Serengeti National Park

> **If we construct this road, all our rhinos will disappear. . . . We should strive to conserve our heritage for future generations.**
> Sirili Akko, executive officer of the Tanzania Association of Tour Operators

It's been called the greatest wildlife spectacle on Earth. Each year more than 1.2 million wildebeest migrate across the vast plains of the Serengeti in East Africa, along with more than 700,000 zebras and hundreds of thousands of antelope. The herds can stretch as far as the eye can see. Packs of lions track the procession and pick off the weak and the unwary, while hungry crocodiles wait in ambush at river crossings. After bearing their calves in the wet season, the wildebeest journey north to find fresh grass. The great herds spend the dry season at the northern end of the Serengeti ecosystem, and then head back south to complete their annual cyclical journey.

This epic migration, with its dramatic interplay of predators and prey, has cycled on for millennia. Yet today, the entire phenomenon may be threatened by a proposal to build a commercial highway across the Serengeti, slicing straight across the animals' migratory route.

Before examining the highway proposal, let's step back for a broad view of the Serengeti. The people native to this region, the Maasai, are semi-nomadic herders. Because the Maasai subsist on their cattle and have lived at low population densities, wildlife thrived here long after it had declined in other parts of Africa.

When East Africa was under colonial rule, the British created game reserves to conserve wildlife for their own hunting. After Tanzania, Kenya, and other African nations gained independence in the mid-20th century, the British reserves became the basis for today's national protected areas. Serengeti National Park was established in Tanzania in 1951, and the Maasai Mara National Reserve was later created just across the border in Kenya. These two protected areas, together with several adjacent ones, encompass the Serengeti ecosystem. This 30,000-km² (11,500-mi²) region is one of the last places on the planet where an ecosystem remains nearly intact and functional over a vast area.

Today 2 million people from around the world visit Tanzania and Kenya each year, most of them ecotourists who visit the parks and protected areas. Serengeti National Park alone receives 800,000 annual visitors. Tourism injects close to $3 billion into these nations' economies and creates jobs for tens of thousands of local people. Because the region's people see that functional ecosystems full of wildlife bring foreign dollars into their communities, many support the parks. Indeed, East Africa has been at the forefront of

A Maasai man herding his cattle ▲

Upon completing this chapter, you will be able to:

- Characterize the scope of biodiversity on Earth
- Specify the benefits that biodiversity brings us
- Discuss today's extinction crisis in geologic context
- Evaluate the primary causes of biodiversity loss
- Assess the science and practice of conservation biology
- Analyze efforts to conserve threatened and endangered species
- Compare and contrast conservation efforts above the species level

community-based conservation (p. 296), in which local people act as stewards managing their own natural resources, often in collaboration with international conservationists.

However, most people living in northern Tanzania remain desperately poor. Farmers, villagers, and townspeople along the shores of Lake Victoria feel isolated from the rest of Tanzania by a poor road system. Walled off by Serengeti National Park to their east (which does not allow commercial truck traffic on its few dirt roads), these people have little access to outside markets to buy and sell goods. In response, Tanzania's president Jakaya Kikwete promised to build a paved commercial highway across the Serengeti, connecting Lake Victoria communities with cities to the east and ports on the Indian Ocean. The World Bank and the German government offered to finance the $480-million project, and Chinese contractors stood ready to build it.

Around the world, conservationists reacted with alarm. The proposed highway would slice right through the middle of the wildebeest migration path (**FIGURE 11.1**). Scientists predicted that the road and its vehicles would physically block migration routes and kill countless animals in collisions. A highway would also provide access for poaching (the illegal killing of wildlife for meat or body parts) and would allow an entry corridor for exotic plant species that could invade the ecosystem. A highway would encourage human settlement right up to the park boundary, making the park an island of habitat hemmed in by agriculture, housing, and commerce. And by boosting development, a highway could spur the towns on Lake Victoria to grow into large cities, creating demand for still-larger transportation corridors in the future. For all these reasons, experts predicted that the highway would diminish animal populations and possibly destroy the migration spectacle.

Such an outcome could devastate tourism, so the region's tourism operators opposed the highway. So did most Kenyans, who feared that the highway would prevent migratory animals from reaching Kenya's Maasai Mara Reserve. In 2010 a Kenyan nongovernmental organization, the African Network for Animal Welfare, sought to stop the highway with a lawsuit in the East Africa Court of Justice, a body that adjudicates international matters in the region. This court in 2011 issued an injunction halting the road project, and in 2014 after hearing appeals from the Tanzanian government, it issued a new ruling reinforcing its prohibition of the project.

Meanwhile, international pressure rose on Tanzania to abandon its plans. Highway opponents proposed an alternative route that would wrap around the Serengeti's southern end, passing through more towns and serving five times as many people along the way. The World Bank and the German government offered to help fund this alternative route instead.

Then in 2016, shortly after John Magufuli took office as Tanzania's president, a new plan kicked off debate. The nation of Uganda wanted to export oil from its recently developed oilfields eastward to ports on the Indian Ocean. Three routes for an oil pipeline were proposed, and the most direct option would cut straight across Serengeti National Park.

Such an oil pipeline (accompanied by a road) would create a barrier to migratory animals just as a highway would. It would bring all the other potential impacts of a highway as well, along with risks of oil spills. So when news of the pipeline project became public, scientists and conservationists again rushed to express opposition to a development corridor through the park. The nonprofit group Serengeti Watch said it would sue to stop the proposal in the East Africa Court of Justice.

Faced with the opposition, Magufuli's government announced that any pipeline route would avoid passing through game reserves and national parks. As this book went to press, a southern route around the Serengeti began to look more likely, but the issue remained unresolved.

Today in Tanzania, poaching is on the rise and animal populations are falling. The Serengeti is one of our planet's last intact large ecosystems, so impacts here have global ramifications for biological diversity on Earth. We would all be impoverished if the Serengeti's biodiversity were lost, so we must hope that Africans can find ways to improve their standard of living while conserving their wildlife and natural systems. East Africa has helped to pioneer win-win solutions in conservation thus far, so there is hope that it will show the way yet again.

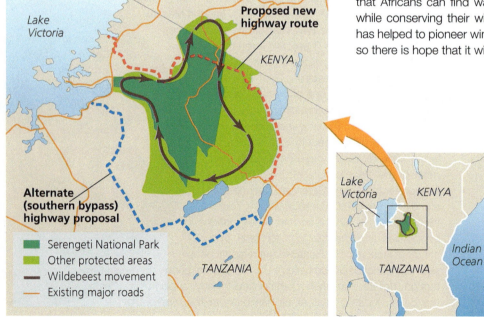

Lake Victoria

Proposed new highway route

KENYA

Alternate (southern bypass) highway proposal

■ Serengeti National Park
■ Other protected areas
— Wildebeest movement
— Existing major roads

TANZANIA

Lake Victoria

KENYA

Indian Ocean

TANZANIA

FIGURE 11.1 A proposed highway would slice through Serengeti National Park. It would increase commerce and connect Tanzanian people on each side, but it would also cut across the migration route for wildebeest and other animals. Highway opponents suggest an alternate route around the park's southern edge.

Life's Diversity on Earth

Rising human population and resource consumption are putting ever-greater pressure on the flora and fauna of our planet. We are diminishing the ultimate source of our civilization's wealth and happiness: Earth's diversity of life, the very quality that makes our planet unique in the known universe. Thankfully, many people around the world are working tirelessly to save threatened animals, plants, and ecosystems in efforts to stem the loss of our planet's priceless biological diversity.

Biodiversity encompasses multiple levels

Biological diversity, or **biodiversity,** is the variety of life across all levels of biological organization, and includes diversity in species, genes, populations, communities, and ecosystems (p. 59). Biodiversity is a concept as multifaceted as life itself, and biologists use different working definitions according to their own aims and philosophies. Yet scientists agree that the concept applies across major levels in the organization of life (**FIGURE 11.2**). The level that people find easiest to visualize and that we use most commonly is species diversity.

Species diversity A **species** is a set of individual organisms that share certain unique characteristics and can breed with one another and produce fertile offspring (p. 48). Species form by the process of speciation (p. 52) and may disappear by extinction (p. 57). **Species diversity** describes the number or variety of species found in a particular region. One component of species diversity is *species richness*, the number of species. Another component of species diversity is *evenness* or *relative abundance*, the extent to which species differ in numbers of individuals (greater evenness means they differ less).

Biodiversity exists below the species level in the form of *subspecies*, populations of a species that occur in different geographic areas and differ from one another in slight ways. Subspecies arise by the same processes that drive speciation but result when divergence stops short of forming separate species. As an example, the black rhinoceros diversified into about eight subspecies, each inhabiting a different part of Africa. The eastern black rhino, which is native to Kenya and Tanzania, differs slightly in its attributes from each of the other subspecies.

Genetic diversity Scientists designate subspecies when they recognize substantial genetically based differences among individuals from different populations of a species. However, all species consist of individuals that vary genetically from one another to some degree, and this variation is another important component of biodiversity. **Genetic diversity** encompasses the differences in DNA composition (p. 27) among individuals, and these differences provide the raw material for adaptation to local conditions. In the long term, populations with more genetic diversity may be more likely to persist, because their variation better enables them to cope with environmental change.

Populations with little genetic diversity are vulnerable to environmental change if they lack genetic variants to help them adapt to novel conditions. Populations with low genetic diversity may also show less vigor, be more vulnerable to disease, and suffer inbreeding depression, which occurs when genetically similar parents mate and produce weak or defective offspring. Scientists have sounded warnings over low genetic diversity in species that have dropped to low population sizes, including American bison, elephant seals, and the cheetahs of the East African plains. Diminished genetic diversity in our crop plants is a prime concern to humanity (p. 244).

Ecosystem diversity **Ecosystem diversity** refers to the number and variety of ecosystems (pp. 59, 110), but biologists may also refer to the diversity of communities (pp. 58, 78) or habitats (p. 59). Scientists may also consider the geographic arrangement of habitats, communities, or ecosystems across a landscape, including the sizes and shapes of patches and the connections among them (p. 114). Under any of these concepts, a seashore of beaches, forested cliffs, offshore coral reefs, and ocean waters would hold far more biodiversity than the same acreage of a

FIGURE 11.2 The concept of biodiversity encompasses multiple levels in the hierarchy of life.

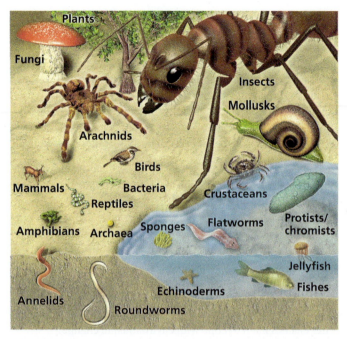

FIGURE 11.3 Some groups contain more species than others.
This illustration shows organisms scaled in size to the number of
species known so far from each group, giving a visual sense
of their species richness. *Data from Roskov, Y., et al. (eds.), 2016.
Species 2000 & ITIS catalogue of life, 26 May 2016. Digital resource at
www.catalogueoflife.org/col. Leiden, the Netherlands: Species 2000: Naturalis.*

The Serengeti region holds a diversity of habitats, including savanna (p. 96), grassland, hilly woodlands, seasonal
wetlands, and rock outcroppings. This habitat diversity contributes to the rich diversity of species in the region.

Biodiversity is unevenly distributed

Some groups of organisms include more species than others.
For example, in numbers of species, insects show a staggering predominance over all other forms of life (**FIGURE 11.3**;
FIGURE 11.4). Among insects, about 40% are beetles, and
beetle species alone outnumber all non-insect animal species
and all plant species. No wonder the British biologist J.B.S.
Haldane famously quipped that God must have had "an inor-
dinate fondness for beetles."

In some groups, large numbers of species formed rapidly as populations spread into a variety of environments and
adapted to local conditions. Other groups diversified because
of a tendency to become subdivided by barriers that promote
speciation (p. 52). Still other groups accumulated species
through time because of low rates of extinction.

Biodiversity is also greater in some places than in others. For instance, species richness generally increases as one
approaches the equator (**FIGURE 11.5**). This pattern of variation with latitude, called the *latitudinal gradient in species
richness*, is one of the most obvious patterns in ecology, yet
one of the most difficult for scientists to explain. Hypotheses
abound for the cause of the latitudinal gradient. Some focus
on geographic area, arguing that the tropics provide more
room for speciation. Some focus on solar energy, arguing that more plant growth makes areas nearer the equator

monocultural cornfield. A mountain slope whose vegetation changes with elevation from desert to forest to alpine
meadow would hold more biodiversity than a flat area the
same size consisting of only desert, forest, or meadow.

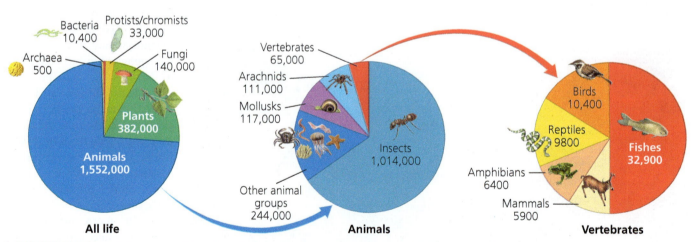

FIGURE 11.4 Most species are animals. Nearly two-thirds of animals are insects (whereas vertebrates
make up only 4%). *Data from Roskov, Y., et al. (eds.), 2016. Species 2000 & ITIS catalogue of life, 26 May 2016. Digital
resource at www.catalogueoflife.org/col. Leiden, the Netherlands: Species 2000: Naturalis.*

DATA Q
- What percentage of vertebrate species do mammal species make up?
- What percentage of animal species are mammals?
- What percentage of the world's total species are mammals?
- How many insect species exist for every mammal species on Earth?

Go to **Interpreting Graphs & Data** on **Mastering**EnvironmentalScience®

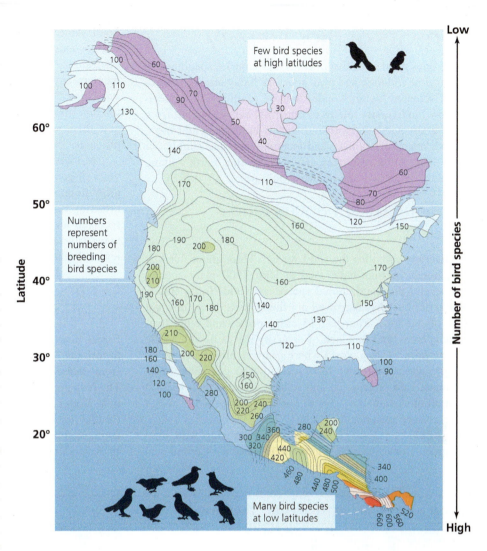

FIGURE 11.5 Species richness tends to increase toward the equator. Birds show a clear latitudinal gradient in species richness. Regions of arctic Canada and Alaska are home to just 30–100 breeding species of birds, whereas areas of Costa Rica and Panama host more than 600. *Adapted from Cook, R.E., et al., 1969. Variation in species density in North American birds. Systematic Biology 18: 63–84. (Originally published as Systematic Zoology.) By permission of Oxford University Press.*

more productive. Others focus on the relative stability of tropical climates, arguing that this discourages small numbers of generalist species from dominating ecosystems and instead allows numerous specialist species to coexist. Still others focus on history, arguing that polar and temperate regions are relatively species-poor because glaciation events repeatedly forced organisms out of these regions and toward tropical latitudes.

Regardless of the explanation, the latitudinal gradient influences the species diversity of Earth's biomes (pp. 92–98). Tropical dry forests and rainforests support far more species than tundra and boreal forests, for instance. Tropical biomes typically show more evenness as well, whereas in high-latitude biomes with low species richness, particular species greatly outnumber others. For example, Canada's boreal forest is dominated by immense expanses of black spruce, whereas Panama's tropical forest contains hundreds of tree species, none of which greatly outnumber others.

For any given region, species diversity tends to increase with diversity of habitats, because each habitat supports a somewhat different set of organisms. Thus, ecotones (transition zones where habitats intermix; p. 114) tend to support high biodiversity. Human disturbance often creates ecotones or patchwork combinations of habitats. This increases habitat diversity locally; so in moderately disturbed areas, species diversity tends to rise. However, at larger scales, human disturbance decreases diversity because it replaces regionally unique habitats with homogenized disturbed habitats, causing species reliant on the regionally unique habitats to disappear. Moreover, species that rely on large expanses of habitat disappear when those habitats are fragmented by human disturbance. Within a given habitat, diversity varies according to the attributes of the habitat. Structurally diverse habitats tend to allow for more ecological niches (pp. 60, 75) and thereby support greater species richness and evenness. For instance, forests generally support more diversity than grasslands, and coral reefs support more diversity than the open ocean.

Many species await discovery

Despite centuries of exploration and study by professional biologists and amateur naturalists, Earth is home to so much life that humanity remains profoundly ignorant of the number of species that exist. So far, scientists have identified and described about 1.8 million species of plants, animals, and microorganisms. However, estimates for the total number that actually exist range from 3 million to 100 million, with the most widely accepted estimates in the neighborhood of 14 million.

Our knowledge of species numbers is incomplete for several reasons. First, many species are tiny and easily overlooked. These include bacteria, nematodes (roundworms), fungi, protists, and soil-dwelling arthropods. Second, many organisms are difficult to identify; sometimes, organisms thought to be of the same species turn out to be different species once examined more closely. This is frequently the case with microbes, fungi, and small insects, but also sometimes with organisms as large as birds, trees, and whales. Third, some areas of Earth remain little studied. We have barely sampled the ocean depths, hydrothermal vents (p. 31), or the tree canopies and soils of tropical forests. There remain many frontiers on our planet to explore!

Benefits of Biodiversity

These days most of us live in cities and suburbs, spend nearly all our time indoors, and pass hours each day staring at electronic screens. It's no wonder that we often fail to appreciate how biodiversity relates to our lives! Yet we benefit from biodiversity and it supports our society in fundamental ways. Indeed, our cities, homes, and technology would simply not exist without the resources and services that Earth's living species provide us—and neither would we.

Biodiversity enhances food security

Biodiversity provides the food we eat. Throughout our history, human beings have used at least 7000 plant species and several thousand animal species for food. Today industrial agriculture has narrowed our diet. Globally, we now get 90% of our food from just 15 crop species and eight livestock species, and this lack of diversity leaves us vulnerable to crop failures. In a world where nearly 1 billion people go hungry, we can improve food security (the guarantee of an adequate, safe, nutritious, and reliable food supply; p. 237) by finding sustainable ways to harvest or farm wild species and rare crop varieties.

TABLE 11.1 shows a selection of promising wild food resources from just one region of the world—Central and South America. Plenty of additional new or underused food sources exist there and elsewhere worldwide. As examples, the babassu palm of the Amazon produces more vegetable oil than any other plant. The serendipity berry generates a sweetener 3000 times sweeter than table sugar. Some salt-tolerant grasses and trees are so hardy that farmers can irrigate them with salt water to produce animal feed and other products.

Moreover, the wild relatives of our crops hold reservoirs of genetic diversity that can help protect the crops we grow in monocultures by providing helpful genes for cross-breeding or genetic engineering (p. 244). We have already received tens of billions of dollars' worth of disease resistance from the wild relatives of potatoes, wheat, corn, barley, and other crops.

Organisms provide drugs and medicines

People have made medicines from plants and animals for centuries, and about half of today's pharmaceuticals are derived from chemical compounds from wild plants (TABLE 11.2). The rosy periwinkle of Madagascar produces compounds that treat Hodgkin's disease and a deadly form of leukemia. The Pacific yew of North America's Pacific Northwest produces a compound that forms the basis for the anti-cancer drug taxol. In Australia, a rare species of cork, *Duboisia leichhardtii*,

---

TABLE 11.1 Potential New Food Sources*

Amaranths
(three species of Amaranthus)

Grain and leafy vegetable; livestock feed; rapid growth, drought resistant

Capybara
(Hydrochoeris hydrochaeris)

World's largest rodent; meat esteemed; easily ranched in open habitats near water

Buriti palm
(Mauritia exuosa)

"Tree of life" to Amerindians; vitamin-rich fruit; pith as source for bread; palm heart from shoots

Vicuna
(Lama vicugna)

Threatened species related to llama; source of meat, fur, and hides; can be profitably ranched

Maca
(Lepidium meyenii)

Cold-resistant root vegetable resembling radish, with distinctive flavor; near extinction

Chachalacas
(Ortalis, many species)

Tropical birds; adaptable to human habitations; fast-growing

*The wild species shown here are just some of the many plants and animals that could supplement our food supply.
Source: Adapted from Wilson, E.O., 1992. *The diversity of life.* Cambridge, MA: Belknap Press.

TABLE 11.2 Natural Plant Sources of Pharmaceuticals*

Pineapple
(*Ananas comosus*)

Drug: Bromelain
Application: Controls tissue inflammation

Pacific yew
(*Taxus brevifolia*)

Drug: Taxol
Application: Anticancer agent (especially ovarian cancer)

Autumn crocus
(*Colchicum autumnale*)

Drug: Colchicine
Application: Anticancer agent

Velvet bean
(*Mucuna deeringiana*)

Drug: L-Dopa
Application: Parkinson's disease suppressant

Yellow cinchona
(several species of *Cinchona*)

Drug: Quinine
Application: Antimalarial agent

Common foxglove
(*Digitalis purpurea*)

Drug: Digitoxin
Application: Cardiac stimulant

*Shown are just a few of the many plants that provide chemical compounds of medical benefit.
Source: Adapted from Wilson, E.O., 1992. *The diversity of life.* Cambridge, MA: Belknap Press.

provides hyoscine, a compound that physicians use to treat cancer, stomach disorders, and motion sickness. Aspirin was derived from chemicals found in willows and in meadowsweet. Each year, pharmaceutical products owing their origin to wild species generate up to $150 billion in sales and save millions of human lives.

The world's biodiversity holds a still-greater treasure chest of medicines yet to be discovered. For this reason, people working for pharmaceutical companies engage in **bioprospecting,** searching for organisms that might provide new drugs, medicines, foods, or other valuable products. A recent international survey highlighted animals that show particular promise yet may be lost to extinction before we can profit from what they have to offer (TABLE 11.3, p. 276). We already lost such an opportunity in two species of gastric brooding frogs discovered in the rainforests of Queensland, Australia (see top photo in Table 11.3). Females of these bizarre frogs raised their young inside their stomachs, where in any other animal, stomach acids would soon destroy them! Apparently the young frogs exuded substances that neutralized their mother's acid production. Any such substance could be of immense use for treating human stomach ulcers, which affect 25 million Americans. Sadly, both frog species went extinct in the 1980s, taking their medical secrets with them forever. With every species that goes extinct, we lose one more potential opportunity to find cures and treatments.

Biodiversity provides ecosystem services

Contrary to popular opinion, some things in life can indeed be free—as long as we protect the ecological systems that provide them. Forests provide clean air and water, and they buffer hydrologic systems against flooding and drought. Native crop varieties provide insurance against disease and other stresses. Wildlife can attract tourism that boosts economies. Intact ecosystems provide these and other valuable processes, known as ecosystem services (pp. 4, 116), for all of us, free of charge. Maintaining these ecosystem services is one clear benefit of protecting biodiversity.

According to scientists, biodiversity helps to:

- Provide food, fuel, fiber, and shelter.
- Purify air and water.
- Detoxify and decompose wastes.
- Stabilize Earth's climate.
- Moderate floods, droughts, and temperatures.
- Cycle nutrients and renews soil fertility.
- Pollinate plants, including many crops.
- Control pests and diseases.
- Maintain genetic resources for crop varieties, livestock breeds, and medicines.
- Provide cultural and aesthetic benefits.

TABLE 11.3 Major Types of Animals at Risk That Offer Potential Medical Uses

SPECIES AT RISK	POTENTIAL MEDICAL USES
Amphibians 30% of all species are threatened with extinction. 	• Antibiotics; chemicals for pain-killers, heart disease, and high blood pressure • Natural adhesives for treating tissue damage • Ability to regenerate organs and tissues could suggest how we might, too. • "Antifreeze" compounds that allow frogs to survive freezing might help us preserve organs for transplants.
Sharks Overfishing has reduced populations of most species. Some risk extinction. 	• Squalamine from sharks' livers could lead to novel antibiotics, appetite-suppressants, drugs to shrink tumors, and drugs to fight vision loss. • Study of salt glands is helping address kidney diseases.
Horseshoe crabs Overfishing is sharply diminishing populations. 	• A number of antibiotics are being developed. • The compound T140 may treat AIDS, arthritis, and several cancers. • Cells from blood can help detect cerebral meningitis in people.
Bears Nine species are at risk of extinction. 	• An acid from bears' gallbladders treats gallstones and liver disease, and prevents bile buildup during pregnancy. • While hibernating, bears build bone mass. If we learn how, we might apply this knowledge to treat osteoporosis and hip fractures. • Hibernating bears excrete no waste for months. Learning how could help treat renal disease.
Cone snails Most live in coral reefs, which are threatened ecosystems. 	• One compound may prevent death of brain cells from head injuries or strokes. • Another is a painkiller 1000 times more potent than morphine. So far just a few hundred of the 70,000–140,000 compounds these snails produce have been studied.

Source: Adapted from Chivian, E., and A. Bernstein, 2008. *Sustaining life: How human health depends on biodiversity.* New York, NY: Oxford University Press.

In these ways, organisms and ecosystems support vital processes that people cannot replicate or would need to pay for if nature did not provide them. The economic value of 17 of the world's ecosystem services has been estimated at more than $143 trillion per year (p. 149)—more than the gross domestic product of all national economies combined.

Biodiversity helps maintain ecosystem function

Ecological research demonstrates that biodiversity tends to enhance the stability of communities and ecosystems. Research has also found that biodiversity tends to increase the resilience (p. 82) of ecological systems—their ability to withstand disturbance, recover from stress, or adapt to change. Thus, the loss of biodiversity can diminish a natural system's ability to function and to provide services to our society.

Will the loss of a few species really make much difference in an ecosystem's ability to function? Consider a metaphor first offered by Paul and Anne Ehrlich (p. 188): The loss of one rivet from an airplane's wing—or two, or three—may not cause the plane to crash. But as rivets are removed the structure will be compromised, and eventually the loss of just one more rivet will cause it to fail. Keeping this metaphor in mind, we would be wise to preserve as many components of our ecosystems as possible to make sure these systems continue to function.

Research shows that removing a keystone species (p. 80) such as a top predator can significantly alter an ecological system. Think of lions, leopards, and cheetahs on the Serengeti—or wolves, mountain lions, and grizzly bears at Yellowstone National Park (a place sometimes called "America's Serengeti"). These predators prey on herbivores that consume many plants; thus the removal of a top predator can have consequences that multiply as they cascade down the food chain.

Likewise, losing an "ecosystem engineer" (p. 81) has major effects. For example, elephants normally eat and trample many young plants, helping to maintain the open structure of Africa's savannas. Scientists have found that when elephants are removed (as by illegal hunting), the landscape fills in with scrubby vegetation, converting the savanna into a dense scrub forest and affecting countless other species.

Ecosystems are complex, however, and it is difficult to predict which species may be most influential. Thus, many people prefer to apply the precautionary principle (p. 254) in the spirit of Aldo Leopold (p. 137), who advised, "To keep every cog and wheel is the first precaution of intelligent tinkering."

Biodiversity boosts economies through tourism and recreation

Biodiversity can generate income for people through tourism, particularly in developing countries in the tropics that boast impressive species diversity. When wealthier people travel to observe wildlife and explore natural areas, they create economic opportunities for area residents. Visitors spend money at local businesses, hire local people as guides, and support parks that employ local residents. Ecotourism (p. 67) can thereby bring jobs and income to areas that might otherwise suffer poverty.

The parks and wildlife of Kenya and Tanzania are prime examples. Ecotourism brings in fully a quarter of all foreign money entering Tanzania's economy each year. Leaders and citizens in both nations who recognize biodiversity's economic benefits have managed their parks and reserves diligently. Ecotourism is a vital source of income for nations such as Costa Rica, with its tropical forests; Australia, with its Great Barrier Reef; and Belize, with its caves, rainforests, and coral reefs. The United States, too, benefits from ecotourism; its national parks draw millions of visitors from around the world.

Popular sites for ecotourism can sometimes become victims of their own success, however. Excessive development of infrastructure for tourism can damage an area's natural assets, and too many visitors to natural areas can degrade the outdoor experience and disturb wildlife. Anyone who has been to Yosemite, the Grand Canyon, or the Great Smoky Mountains on a crowded summer weekend can attest to this. Still, ecotourism commonly provides a powerful financial incentive for nations, states, and local communities to preserve natural areas and reduce impacts on the landscape and on native species.

FIGURE 11.6 An Indonesian girl gazes into a flower of *Rafflesia arnoldii*, the largest flower in the world. Biophilia holds that human beings have an instinctive love and fascination for nature and a deep-seated desire to affiliate with other living things.

People value connections with nature

Not all of biodiversity's benefits to people can be expressed in the hard numbers of economics or the practicalities of food and medicine. Some scientists and philosophers argue that people find a deeper value in biodiversity. Harvard University biologist Edward O. Wilson has popularized the notion of **biophilia,** asserting that human beings share an instinctive love for nature and feel an emotional bond with other living things (**FIGURE 11.6**). Wilson and others cite as evidence of biophilia our affinity for parks and wildlife, our love for pets, the high value of real estate with a view of natural landscapes, and our interest in hiking, bird-watching, fishing, hunting, backpacking, and similar outdoor pursuits.

In a 2005 book, writer Richard Louv added that as children in recent years have been increasingly deprived of outdoor experiences and direct contact with wild organisms, they suffer what he calls "nature-deficit disorder." Louv argues that this alienation from biodiversity and nature damages childhood development and may lie behind much of the emotional stress, angst, and anxiety felt by young people in developed nations today.

Do we have ethical obligations toward other species?

Aside from all of biodiversity's pragmatic benefits, many people feel that living organisms simply have an inherent right to exist. Human beings are part of nature, and like any other animal we need to use resources and consume other organisms to survive. However, we also have conscious reasoning ability and can make deliberate decisions. Our ethical sense has developed from this intelligence and ability to choose. As our society's sphere of ethical consideration has widened over time, and as more of us take up biocentric or ecocentric worldviews (p. 135), more people have come to feel that other organisms have intrinsic value and an inherent right to exist. In this view, the conservation of biodiversity is justified on ethical grounds alone.

Biodiversity Loss and Extinction

Despite our society's expanding ethical breadth and despite the many clear benefits that biological diversity brings us, the future of life remains far from secure. In today's fast-changing world, every corner of our planet has been touched in some manner by human impact, and biological diversity is rapidly being lost.

Human disturbance creates winners and losers

We affect ecosystems and landscapes in many ways, and this creates both "winners" and "losers" among the world's plants and animals. The creatures that benefit from the changes we make—and those that are harmed—each tend to show predictable sets of attributes, because our many impacts share common themes. In general when we alter natural systems we tend to make each area more similar to

TABLE 11.4 Characteristics of Winning and Losing Species

WINNERS tend to be	LOSERS tend to be
• Generalists	• Specialists
• Geographically widespread	• Limited to a small range
• Users of open, early successional habitats	• Users of mature, dense habitats
• Able to cope with rapidly changing conditions	• Needing stable unchanging conditions
• Small and fast-reproducing (*r*-selected)	• Large and slow-reproducing (*K*-selected)
• Low on the food chain	• High on the food chain
• Not in need of large areas of habitat	• Needing large areas of habitat
• Mainland species	• Island species

WINNER: The house mouse (*Mus musculus*) is a small, fast-reproducing, generalist mammal that thrives by living and feeding in and near our buildings.

LOSER: The tiger (*Panthera tigris*) is a large, slow-reproducing mammal that needs large areas of mature habitat full of prey and free of people.

other areas. This is because we spread into diverse natural environments and then shape them to our own species' particular tastes and needs. We also tend to make landscapes more open in structure, by clearing vegetation away to make room for farms, pastures, towns, and cities. And of course we frequently leave pollution in our wake. Because the general nature of these changes is similar across regions and cultures, certain types of organisms tend to do well and other types tend not to do well.

As shown in **TABLE 11.4**, "winning" species tend to be generalists that can cope with disturbance and changing conditions; tend to use open spaces, edges, or early successional habitats; tend to be low on the food chain; tend to be small and fast-reproducing; and tend not to need large areas of habitat. In contrast, "losing" species tend to be those that specialize on certain habitats or resources; those that use mature densely vegetated habitats; those that are high on the food chain; those that are large and slow-reproducing; and those that need large areas of habitat. Geographically widespread species stand a much better chance to succeed in a changing world undergoing human impact than species limited to small areas, and mainland species tend to do better than island species.

Biodiversity loss involves population declines

One major facet of our world's loss of biodiversity involves declining population sizes of species. As a population shrinks, it encounters two problems. First, it loses genetic diversity; and second, its geographic range tends to become smaller as the species disappears from parts of its range. Both problems make a population vulnerable to further declines. In the Serengeti, scientific studies have documented recent population declines among large mammals (see **THE SCIENCE BEHIND THE STORY**, pp. 280–281). Such declines mean that species diversity, genetic diversity, and ecosystem diversity are all being lost.

To quantify and summarize such change at the global scale, scientists at the World Wildlife Fund and the United Nations Environment Programme (UNEP) developed the *Living Planet Index*. This index expresses how large the average population size of a species is now, relative to its size in the baseline year

of 1970. The most recent compilation summarized trends from populations of 1562 terrestrial species, 757 freshwater species, and 910 marine species that are sufficiently monitored to provide reliable data. Between 1970 and 2010, the Living Planet Index fell by 52%—meaning that on average, population sizes became 52% smaller than they were just four decades earlier (**FIGURE 11.7**). This decline has been driven primarily by biodiversity losses in tropical regions. In the temperate zones (where forests are now regrowing, pollution is being better controlled, and ecological restoration is taking hold in industrialized countries) the index decreased by 36%. In tropical regions (where deforestation is rife) it fell by 56%.

Extinction is irreversible

When a population has declined to a very low level, extinction becomes a possibility. **Extinction** (p. 57) occurs when the last member of a species dies and the entire species ceases to exist. The disappearance of a particular population from a

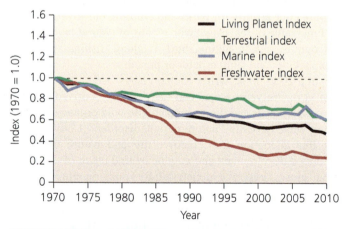

FIGURE 11.7 Populations of vertebrate animals are one-half the size today that they were just 40 years ago. Between 1970 and 2010, the Living Planet Index, an indicator of the status of global biodiversity, fell by 52%. Its index values summarize trends for 10,380 populations of 3038 vertebrate species. The index for terrestrial species fell by 39%; for marine species, by 39%; and for freshwater species, by 76%. *Data from WWF, 2014. Living planet report 2014. Gland, Switzerland: WWF International.*

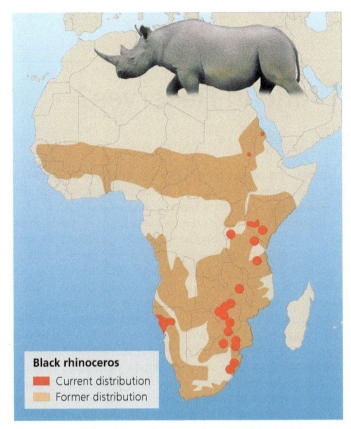

FIGURE 11.8 The black rhinoceros has disappeared from most of its range across Africa. *Based on data from Deon Furstenburg, wildliferanching.com, and other sources.*

Black rhinoceros
- ■ Current distribution
- ■ Former distribution

Human impact is responsible for most extirpation and extinction today, but these processes also occur naturally, albeit at a much slower rate. If species did not naturally go extinct, our world would be filled with dinosaurs, trilobites, ammonites, and the millions of other creatures that vanished from Earth during the immense span of time before humans appeared. Paleontologists estimate that roughly 99% of all species that ever lived are now extinct.

Most extinctions preceding the appearance of human beings occurred singularly for independent reasons, at a pace referred to as the **background extinction rate.** By studying traces of organisms preserved in the fossil record (p. 56), scientists infer that for mammals and marine animals, each year, on average, 1 species out of every 1–10 million has vanished.

Earth has experienced five mass extinction events

Extinction rates rose far above this background rate at several discrete points in Earth's history. In the past 440 million years, our planet has experienced five major **mass extinction events** (p. 58). Each event eliminated more than one-fifth of life's families (p. 56) and at least half its species (**TABLE 11.5**). The most severe episode occurred at the end of the Permian period (see **APPENDIX E**). At this time, about 250 million years ago, close to 90% of all species went extinct. The best-known episode occurred 66 million years ago at the end of the Cretaceous period, when evidence points to an asteroid impact (and possibly volcanism) that brought an end to the dinosaurs and many other groups.

If current trends continue, our modern era, known as the Quaternary period, may see the extinction of more than half of all species. Although similar in scale to previous mass extinctions, today's ongoing mass extinction is different in two primary respects. First, we are causing it. Second, we will suffer as a result.

given area, but not the entire species globally, is referred to as "local extinction" or **extirpation.** Extirpation is an erosive process that can, over time, lead to extinction. The black rhinoceros has been extirpated from most of its historic range across Africa (**FIGURE 11.8**), but as a species it is not yet extinct. However, at least three of its subspecies are extinct.

TABLE 11.5 Mass Extinctions

EVENT	DATE (MILLIONS OF YEARS AGO [MYA])	CAUSE	TYPES OF LIFE MOST AFFECTED	PERCENTAGE OF LIFE DEPLETED
Ordovician	440 mya	Unknown	Marine organisms; terrestrial record is unknown	>20% of families
Devonian	360 mya	Unknown	Marine organisms; terrestrial record is unknown	>20% of families
Permo-Triassic	250 mya	Possibly volcanism	Marine organisms; terrestrial record is less known	>50% of families; 80–95% of species
End-Triassic	200 mya	Unknown	Marine organisms; terrestrial record is less known	20% of families; 50% of genera
Cretaceous-Paleogene	66 mya	Likely asteroid impact	Marine and terrestrial organisms, including dinosaurs	5% of families; >50% of species
Current	Beginning 0.01 mya	Human impacts	Large animals, specialized organisms, island organisms, organisms harvested by people	Ongoing

Why Is Wildlife Declining in African Reserves?

Zebras in Serengeti National Park

Tanzania and Kenya have some of the world's most famous parks and protected areas, with the greatest variety and density of large mammals to be found anywhere. The parks are sizeable, well managed, and well funded. Yet even these places of refuge are not immune to pressures from rising human population, development, and resource extraction.

For several decades, biologists and park managers have censused wildlife in and around the parks and reserves. In recent years, researchers have been analyzing these long-term data sets to assess population trends for large mammals of the East African savanna. These studies are finding that most animals are declining in number—inside the parks and reserves as well as outside them.

In Tanzania, several government agencies and nonprofit groups have collaborated to census mammals by airplane. Aerial surveys began in Serengeti National Park in the 1970s and expanded to other parks in the 1980s. In 2006, Chantal Stoner and six colleagues from Tanzania and from the University of California, Davis, compiled and analyzed data on 25 species over a 10-year period (roughly 1990–2000) from eight regions, each centered on a major park.

Across the eight regions in wet and dry seasons for all species, this team found that population declines outnumbered increases by more than 10 to 1 (**FIGURE 1**). In the Serengeti region, five species declined while two increased. Population

declines were greater outside parks and in areas that received less protection. However, even within the boundaries of well-protected reserves, many species decreased in number.

In Kenya, researchers are seeing patterns similar to those in Tanzania. In 2009, conservation biologist David Western and two colleagues reviewed 30 years of data from aerial surveys conducted from 1977 to 2007 across Kenya's rangelands, which make up three-quarters of the nation's land area. In most locations, Western's group found downward trends in the populations of most species. Moreover, the populations were trending downward both inside and outside the parks (**FIGURE 2**). Many species, including wildebeest, zebras, and some antelope, migrate into and out of parks, so factors that affect their populations outside the parks will affect numbers counted within park boundaries as well. Cycles of rainfall and drought affect grazing mammals, but statistical analysis of the data indicated that the long-term declines transcended short-term effects due to rain or drought.

Researchers have given particular attention to the Maasai Mara National Reserve, on the Kenyan side of the border adjacent to Serengeti National Park. A decade ago Dutch scientist Wilber Ottichilo analyzed aerial survey data and concluded that nonmigratory mammals had declined by 58% from 1977 to 1997 within the reserve and by a similar amount outside the reserve. Resident wildebeest had fared even worse, showing an 81% decline.

FIGURE 1 Across protected areas in Tanzania, more animals have decreased than have increased. *Data are combined from six locations, from Stoner, C., et al., 2006. Changes in large herbivore populations across large areas of Tanzania. African Journal of Ecology 45: 202–215.*

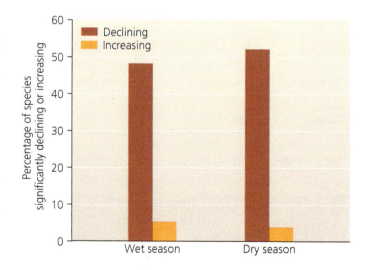

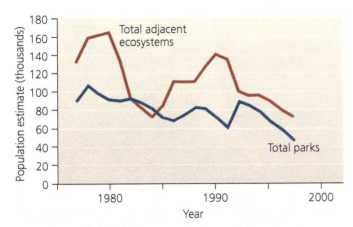

FIGURE 2 Wildlife populations have declined across Kenya, both inside and outside parks. *Data from Western, D., et al., 2009. The status of wildlife in protected areas compared to non-protected areas of Kenya. PLoS ONE 4(7): e6140.*

More recently, Joseph Ogutu of the International Livestock Research Institute in Nairobi, Kenya, and his colleagues extended the analysis to 2009 and found that most species had continued declining, both on and off the reserve, and were now at just one-third of their 1977 population sizes (**FIGURE 3**). As elsewhere, drought appeared to account only for short-term fluctuations and not long-term declines.

So why are all these animals declining? For decades East Africa has had one of the world's fastest rates of human population growth. Kenya and Tanzania have five times as many people today as they did 50 years ago, and this has intensified pressures on all resources, including wildlife and ecosystems. Researchers have attributed the wildlife declines to several main factors:

- Settlements have increased as nomadic Maasai herders have become sedentary and as people have migrated in from elsewhere.
- Farmers convert grasslands to crops (especially wheat). This destroys habitat for antelope, wildebeest, and the predators that follow them.
- Livestock compete with wild grazing animals for food on the grasslands.
- Local residents kill animals for food for their own subsistence, while criminal gangs kill animals and export "bush meat," elephant tusks, and rhino horns to rich consumers abroad.

All these factors harm wild animals directly outside the reserves, reducing their populations regionally. Because many animals move seasonally in and out of protected areas, these factors thereby indirectly lessen their numbers inside the reserves.

As a result of studies like these, researchers have concluded that the mere act of setting aside parks is not adequate to successfully conserve wildlife and ecosystems. Instead, we need to view the big picture and consider animals' needs across the landscape, as well as how human impacts might spill into reserves. Conservation success thus requires linking reserves with corridors of habitat that animals can use, and it requires working with local people living near protected areas. Community-based conservation (p. 296) encourages people to be stewards of the natural assets in the areas where they live, and this approach shows promise for sustaining East Africa's celebrated wildlife populations.

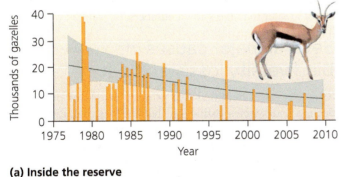

(a) Inside the reserve

(b) Outside the reserve

FIGURE 3 Thomson's gazelles, the most abundant species at Kenya's Maasai Mara National Reserve, decreased by 59% (a) inside the reserve, and by 77% (b) outside the reserve. *Data from Ogutu, Joseph, et al., 2011. Continuing wildlife population declines and range contraction in the Mara region of Kenya during 1977–2009. Journal of Zoology 285: 99–109.*

We are setting the sixth mass extinction in motion

In the past few centuries alone, we have recorded hundreds of instances of species extinction caused by people. Sailors documented the extinction of the dodo on the Indian Ocean island of Mauritius in the 17th century, for example, and today only a few body parts of this unique bird remain in museums. Among North American birds in the past two centuries, we have driven into extinction the Carolina parakeet, great auk, Labrador duck, and passenger pigeon (p. 60); almost certainly the Bachman's warbler and Eskimo curlew; and likely the ivory-billed woodpecker (**FIGURE 11.9**). Several more species, including the whooping crane, Kirtland's warbler, and California condor (p. 291), teeter on the brink of extinction.

FAQ

If a mass extinction is happening, why don't I notice species going extinct all around me?

There are two reasons that most of us don't personally sense the scale of biodiversity loss. First, if you live in a town or city, the plants and animals you see from day to day are generalist species that thrive in disturbed areas. In contrast, the species most in trouble are those that rely on less-disturbed habitats, and you may need to go farther afield to find them.

Second, a human lifetime is very short! The loss of populations and species over the course of our lifetime may seem a slow process to us, but on Earth's timescale it is sudden—almost instantaneous. Because each of us is born into a world that has already lost many species, we don't recognize what's already vanished. Likewise, our grandchildren won't appreciate what we lose in our lifetimes. Each human generation experiences just a portion of the overall phenomenon, so we have difficulty sensing the big picture. Nonetheless, researchers and naturalists who spend their time outdoors observing nature see biodiversity loss around them all the time—and that's precisely why they feel so passionate about preventing it.

People have been hunting species to extinction for thousands of years. Archaeological evidence shows that in case after case, a wave of extinction followed close on the heels of human arrival on islands and continents (**FIGURE 11.10**). After Polynesians reached Hawai'i, half its birds went extinct. Birds, mammals, and reptiles vanished following human arrival on many other oceanic islands, including large island masses such as New Zealand and Madagascar. Dozens of species of large vertebrates died off in Australia after people arrived roughly 50,000 years ago. North America lost 33 genera of large mammals (such as camels, lions, horses, and giant ground sloths) after people arrived more than 13,000 years ago.

Today, species loss is accelerating as our population growth and resource consumption put increasing strain on habitats and wildlife. In 2005, scientists with the Millennium Ecosystem Assessment (pp. 15–16) calculated that the current global extinction rate is 100 to 1000 times greater than the background extinction rate. They projected that the rate would increase 10-fold or more in future decades.

To monitor threatened and endangered species, the International Union for Conservation of Nature (IUCN) maintains the **Red List,** a regularly updated list of species facing high risks of extinction. As of 2016 the Red List reported that 22% (1197) of mammal species, 13% (1375) of bird species, 20% (944) of reptile species, 31% (1994) of amphibian species, and 16% (2073) of fish species were threatened with extinction. For most other groups, scientists do not yet have enough data to make accurate global assessments. In the United States alone over the past 500 years, 236 animal species and 33 plant species are known to have gone extinct. For all these figures, the actual numbers of species are without doubt greater than the known numbers.

FIGURE 11.9 The ivory-billed woodpecker was one of North America's most majestic birds. It lived in old-growth forests of the southeastern United States. Forest clearing and timber harvesting eliminated the mature trees it needed for food, shelter, and nesting, and this symbol of the South appeared to go extinct. Recent fleeting, controversial observations raised hopes that the species persists, but proof has been elusive.

Several major causes of biodiversity loss stand out

Scientists have identified five primary causes of population decline and species extinction: habitat loss, pollution, overharvesting, invasive species, and climate change. Each of these causes is intensified by human population growth and by our increasing per capita consumption of resources.

Habitat loss Habitat loss is the single greatest threat to biodiversity today. A species' habitat is the specific environment in which it lives (p. 59). Because organisms have adapted to their habitats over thousands or millions of years of evolution, any sudden, major change in their habitat will likely render it less suitable for them. Species lose their habitats when those habitats are destroyed outright, but also when they are

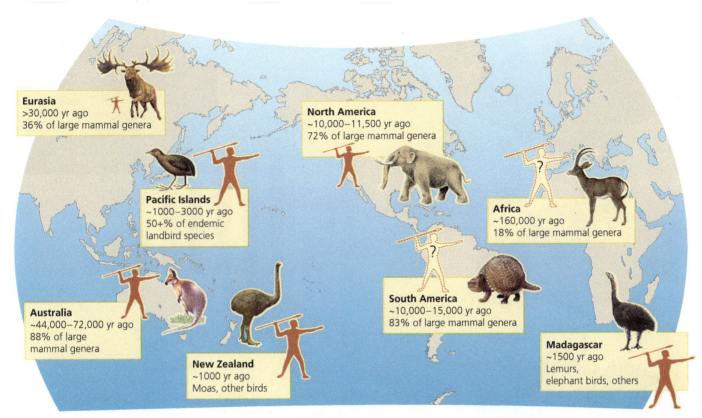

FIGURE 11.10 This map shows when humans arrived in each region, and the extent of extinctions that followed. One extinct animal from each region is illustrated. Larger human hunter icons indicate more evidence and certainty that hunting (as opposed to climate change or other factors) was a primary cause of extinctions. Data for South America and Africa are so far too sparse to be conclusive. *Adapted from Barnosky, A.D., et al., 2004. Assessing the causes of late Pleistocene extinctions on the continents. Science 306: 70–75; and Wilson, E.O., 1992. The diversity of life. Cambridge, MA: Belknap Press.*

altered through more subtle processes, including fragmentation and other forms of degradation.

Many human activities alter, degrade, or destroy habitat. Urban sprawl supplants diverse natural ecosystems, driving many species from their homes (**FIGURE 11.11**). Farming replaces diverse natural communities with simplified ones of only a few plant species. Grazing modifies the structure and species composition of grasslands, and can lead to desertification (p. 222). Clearing forests removes the food and shelter that forest-dwelling organisms need to survive. Damming rivers creates reservoirs upstream while affecting water conditions and floodplain communities downstream.

Habitat loss occurs most commonly through gradual, piecemeal degradation, such as **habitat fragmentation** (**FIGURE 11.12**, p. 284). When farming, logging, road building, or development intrude into an unbroken expanse of forest or grassland, this breaks up a continuous area of habitat into fragments, or patches.

As habitat fragmentation proceeds across a landscape, animals and plants requiring the habitat disappear from one fragment after another. Fragmentation can also prevent animals from moving from place to place, and this is the concern of opponents of the proposed highway and oil pipeline through the Serengeti. In response to habitat fragmentation, conservationists try to link

FIGURE 11.11 Development of land for housing is one way in which habitat is altered or destroyed.

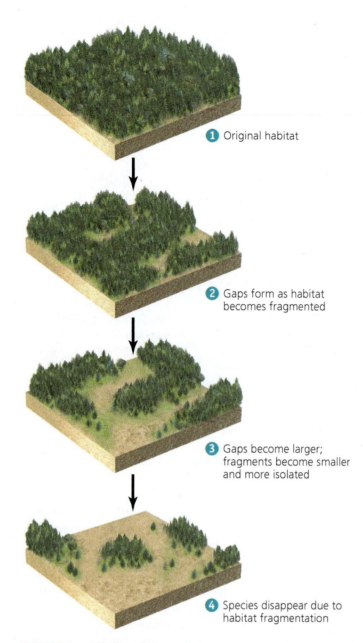

① Original habitat

② Gaps form as habitat becomes fragmented

③ Gaps become larger; fragments become smaller and more isolated

④ Species disappear due to habitat fragmentation

FIGURE 11.12 Habitat fragmentation occurs as human impact creates gaps that expand and eventually come to dominate the landscape, stranding islands of habitat. As habitat becomes fragmented, fewer populations can persist, and numbers of species in the fragments decline.

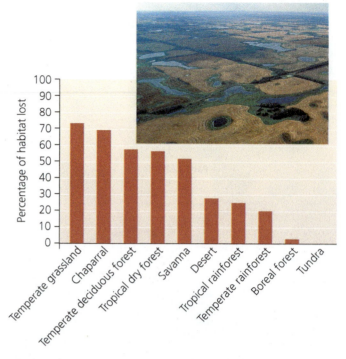

FIGURE 11.13 Human impact has caused habitat loss in all the world's biomes. Shown are percentages of original area that were fully and directly converted for human use through 1990. The photo shows "prairie pothole" wetlands (lakes and ponds once abundant in the northern Great Plains) that have been partly drained for farming. *Data adapted from Millennium Ecosystem Assessment, 2005.* Ecosystems and human well-being: Biodiversity synthesis. *Washington, D.C.: World Resources Institute.*

fragments together with corridors of habitat along which animals can travel. (In Chapter 12 we will learn more about the fragmentation of forests, the effects on wildlife, and potential solutions.)

Habitat loss affects nearly every biome (**FIGURE 11.13**). More than half of the world's temperate forests, grasslands, and shrublands had been converted by 1950 (mostly for agriculture). Today habitat is being lost most rapidly in tropical rainforests, tropical dry forests, and savannas. Within most biomes, wetlands (p. 391) are especially threatened. More than half the wetlands of the lower 48 U.S. states and Canada have been drained for agriculture (see photo, Figure 11.13).

Habitat loss is the primary source of population declines in more than 80% of threatened mammals and birds, according to UNEP data. For example, the prairies native to North America's Great Plains are today almost entirely converted to agriculture. Less than 1% of original prairie habitat remains. As a result, grassland bird populations have declined by an estimated 82–99%.

Of course, our habitat alteration benefits some species. Animals such as starlings, raccoons, house sparrows, pigeons, gray squirrels, rats, mosquitoes, and cockroaches thrive among us in farmland, towns, and cities. However, these "winning" species that do well in our midst tend to be weedy generalists that are in little danger of disappearing any time soon. The concern is that far too many other species are "losing" as a result of the way we alter habitats.

weighing the ISSUES

Habitat Fragmentation in Your Region

Examine satellite imagery of the region where you live using an online application such as Google Earth or Google Maps. How much of the area around you is developed with roads and buildings? How much is forested or in natural areas? Are the natural areas connected to one another, or are they fragmented? Discuss how easy or difficult it would be for an animal such as a deer or a bear to move from one habitat patch to another. Do you see ways in which natural areas could be better connected using corridors? Now zoom in or zoom out a bit on the map, and address these same questions at a different landscape scale.

Pollution Pollution can harm organisms in many ways. Air pollution degrades forest ecosystems and affects the atmosphere and climate. Noise pollution and light pollution impinge on the behavior and habitat use of animals. Water pollution impairs fish and amphibians. Agricultural runoff containing fertilizers, pesticides, and sediments harms many terrestrial and aquatic species. Heavy metals, polychlorinated biphenyls (PCBs), endocrine-disrupting compounds, and other toxic chemicals poison people and wildlife. Plastic garbage in the ocean can strangle, drown, or choke marine creatures. The effects of oil spills on wildlife are dramatic and well known. We examine all these impacts in other chapters of this book. However, although pollution can cause extensive damage to wildlife and ecosystems, it tends to be less significant as a cause of population-wide decline than public perception holds it to be, and it is far less influential than habitat loss.

Overharvesting We have always hunted and harvested animals and plants from nature, but our growth in population and consumption is now leading us to remove many species at faster rates than they are able to reproduce. Forests around the world are being lost (p. 305), and tree species prized for their wood, such as teak and mahogany, are disappearing fastest. In Africa, gorillas and other primates killed for "bush meat" could soon face extinction. In the oceans, many fish stocks are now overharvested (p. 434). Decades of whaling drove the Atlantic gray whale extinct and have left several other whales threatened or endangered. Thousands of sharks are killed each year simply for their fins, which are used in soup. Indeed, today our oceans contain only 10% of the large animals they once did (p. 437), and this has far-reaching impacts on marine food webs.

For most species, hunting or harvesting will not in itself pose a threat of extinction, but for some species it can. Animals that are large in size, are long-lived, and raise few young in their lifetimes (K-selected species; p. 67) are most vulnerable to hunting. For example, people have long killed elephants to extract their tusks for ivory (**FIGURE 11.14**). By 1989, 7% of African elephants were being slaughtered each year, so the world's nations enacted a global ban on the commercial trade of ivory. Elephant numbers recovered following the ban, but since 2007 **poaching** (the illegal killing of wildlife for meat or body parts) has increased to all-time highs, driven by high black-market prices for ivory paid by wealthy overseas buyers. From 2010 through 2015, more than 170,000 African elephants were killed—enough to send populations downward and threaten the species' future.

With the illegal global trade in wildlife products surpassing $20 billion per year, poaching has led to steep population declines for many animals. Rhinoceros populations have crashed as poachers slaughter rhinos for their horns, which are ground into powder and sold to ultra-wealthy Asian consumers as cancer cures, party drugs, or hangover treatments, even though the horns have no such properties and consist of the same material as our fingernails. Across Asia, tigers are threatened by poaching and habitat loss; body parts from one tiger can fetch a poacher $15,000 on the black market, where they are sold as aphrodisiacs in Asian countries. Today half the world's tiger subspecies are extinct, and most of the remaining animals are crowded onto just 1% of the land they occupied historically.

In much of Africa today, protecting wildlife has become a dangerous job. Poaching is now conducted with brutal efficiency by organized crime syndicates using helicopters, night-vision goggles, and automatic weapons. Park rangers are heavily armed, yet are routinely outgunned in firefights with poachers, and many courageous rangers have lost their lives. In this way the demand for luxury goods by wealthy foreign consumers in Asia, Europe, and America has grave consequences for Africans living in regions like the Serengeti.

Today scientists are assisting efforts to curb poaching. Researchers in the field are tracking elephants and ivory

FIGURE 11.14 Poachers slaughter elephants to sell their tusks for ivory. Despite the ban on ivory trade, poaching is at an all-time high. Here, Kenyan officials at Maasai Mara National Reserve prepare to set fire to tusks confiscated from poachers, in an effort to discourage the trade.

TABLE 11.6 Invasive Species

European gypsy moth
(Lymantria dispar)

Introduced to Massachusetts in the hope it could produce silk. The moth failed to do so, and instead spread across the eastern United States, where its outbreaks defoliate trees over large regions every few years.

Asian long-horned beetle
(Anoplophora glabripennis)

Since the 1990s, has repeatedly arrived in North America in imported lumber. These insects burrow into wood and can kill the majority of trees in an area. Chicago, Seattle, Toronto, New York, and other cities have cleared thousands of trees to eradicate these invaders.

European starling
(Sturnus vulgaris)

Introduced to New York City in the 1800s by Shakespeare devotees intent on bringing every bird mentioned in Shakespeare's plays to America. Outcompeting native birds for nest holes, within 75 years starlings became one of North America's most abundant birds.

Emerald ash borer
(Agrilus planipennis)

Discovered in Michigan in 2002, this wood-boring insect reached 12 U.S. states and Canada by 2010, killing millions of ash trees in the upper Midwest. Billions of dollars will be spent in trying to control its spread.

Cheatgrass
(Bromus tectorum)

After introduction to Washington state in the 1890s, cheatgrass spread across the western United States. It crowds out other plants, uses up the soil's nitrogen, and burns readily. Fire kills many native plants, but not cheatgrass, which grows back stronger without competition.

Sudden oak death
(Phytophthora ramorum)

This disease has killed more than 1 million oak trees in California since the 1990s. The pathogen (a water mold) was likely introduced via infected nursery plants. Scientists are concerned about damage to eastern U.S. forests if it spreads to oaks there.

Brown tree snake
(Boiga irregularis)

Nearly every native forest bird on the South Pacific island of Guam has disappeared, eaten by these snakes, which arrived from Asia as stowaways on ships and planes after World War II. Guam's birds had not evolved with snakes, and had no defenses against them.

Nile perch
(Lates niloticus)

A large fish from the Nile River. Introduced to Lake Victoria in the 1950s, it proceeded to eat its way through hundreds of species of native cichlid fish, driving a number of them extinct. People value the perch as food, but it has radically altered the lake's ecology.

Kudzu
(Pueraria montana)

A Japanese vine that can grow 30 m (100 ft) in a single season, the U.S. Soil Conservation Service introduced kudzu in the 1930s to help control erosion. Kudzu took over forests, fields, and roadsides throughout the southeastern United States.

Polynesian rat
(Rattus exulans)

One of several rat species that have followed human migrations across the world. Polynesians transported this rat to islands across the Pacific, including Easter Island (pp. 8–9). On each island it caused ecological havoc, and has driven extinct birds, plants, and mammals.

by putting radio collars on animals, satellite-tracking tusks with microchips, and flying drones (unmanned surveillance aircraft) over parks to capture real-time video of poachers. Meanwhile, researchers in the lab are conducting genetic analyses to expose illegal hunting and wildlife trade. For instance, forensic DNA testing can reveal the geographic origins of elephant ivory, helping authorities to focus on hotspots of illegal activity (see **THE SCIENCE BEHIND THE STORY**, pp. 294–295).

Invasive species Another major cause of biodiversity loss today is invasive species (**TABLE 11.6**). When non-native species are introduced to new environments, most perish, but the few that survive may do very well, especially if they find themselves freed from the predators and parasites that had attacked them back home or from the competitors that had limited their access to resources. Once released from the limiting factors (p. 65) of predation, parasitism, and competition, an introduced species may proliferate and become invasive, displacing native species. Invasive species cause billions of dollars in economic damage each year, and can sometimes push native species toward extinction.

Some introductions are accidental. Examples include animals that escape from the pet trade; weeds whose seeds cling to our socks as we travel from place to place; and aquatic organisms, such as zebra mussels (p. 73), transported in the ballast water of ships. If a highway is built through the Serengeti, ecologists fear that it would introduce weed seeds from passing vehicles. Park managers already are concerned about several American plants, such as datura, parthenium, and prickly poppy, that have spread rapidly in this way across other African grasslands and that are toxic to native herbivores.

Other introductions are intentional. In Lake Victoria, west of the Serengeti, the Nile perch was introduced as a food fish to supply people much-needed protein (see Table 11.6). It soon multiplied and spread throughout the vast lake, however, preying on and driving extinct dozens of native species of cichlid fish from one of the world's most spectacular evolutionary radiations of animals. The Nile perch is providing people food, but at a significant ecological cost. Throughout history, people everywhere have brought food crops and animals with them as they colonized new places, and today we continue international trade in exotic pets and ornamental plants, often heedless of the ecological consequences.

Species native to islands are especially vulnerable to introduced species. Island species have existed in isolation for millennia with relatively few parasites, predators, and competitors; as a result, they have not evolved the defenses necessary to resist invaders that are better adapted to these pressures. For instance, Hawaii's native plants and animals have been under siege from invasive organisms such as rats, pigs, and cats, and this has led to a number of extinctions (Chapter 3).

Some of the most devastating invasive species are microscopic pathogens that cause disease. Hawai'i offers examples; the malaria and avian pox transmitted by introduced mosquitoes are killing off the islands' native birds, which lack immunity to these foreign diseases. Indeed, some scientists classify disease separately as an additional major cause of biodiversity loss.

FIGURE 11.15 The polar bear became the first species listed under the Endangered Species Act as a result of climate change. As Arctic warming melts the sea ice from which they hunt seals, polar bears must swim farther and farther for food.

Climate change Our manipulation of Earth's climate system (Chapter 18) is having global impacts on biodiversity. As we warm the atmosphere with emissions of greenhouse gases from fossil fuel combustion, we modify long-term climate patterns and we increase the frequency of extreme weather events (such as droughts and storms) that put stress on populations.

In the Arctic, where temperatures have warmed the most, melting sea ice and other impacts are threatening polar bears and people alike (**FIGURE 11.15**). Across the world, warming temperatures are forcing many organisms to shift their geographic ranges toward the poles and higher in altitude. Some species will not be able to adapt. Mountaintop organisms cannot move farther upslope to escape warming temperatures, so many may perish. Trees may not be able to move toward the poles fast enough. As ranges shift, animals and plants encounter new communities of prey, predators, and parasites to which they are not adapted. All in all, scientists predict that a rise in global temperature of 1.5–2.5°C (2.7–4.5°F) could put 20–30% of the world's plants and animals at increased risk of extinction.

A mix of causes threatens many species

For many species, multiple factors are conspiring to cause declines. The monarch butterfly, once familiar to every American schoolchild, is today in precipitous decline (**FIGURE 11.16**, p. 288). On its breeding grounds in the United States and Canada, industrial agriculture and chemical herbicides have eliminated most of the milkweed plants monarchs depend on. Our highly efficient monocultures leave no natural habitat remaining, while insecticides intended for crop pests also kill monarchs and other beneficial insects. In winter the entire monarch population of eastern and central North America

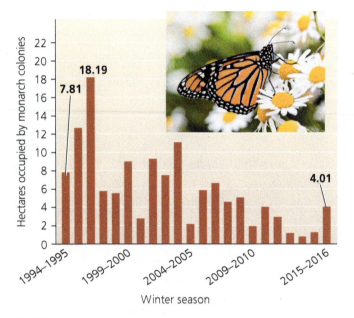

(a) Male golden toad from Monteverde, Costa Rica

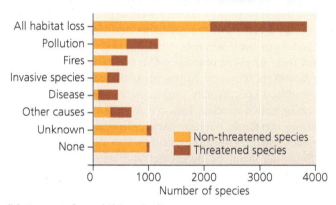

(b) Causes of amphibian declines

FIGURE 11.16 The once-abundant monarch butterfly has undergone alarming declines due to herbicides, pesticides, and habitat loss. Annual surveys on its Mexican wintering grounds show the population occupying many fewer hectares of forest than in the past. *Data from MonarchWatch, collected by the Monarch Butterfly Biosphere Reserve and World Wildlife Fund Mexico.*

 How much area did monarchs occupy in 2015–2016, as a proportion of the area occupied in 1994–1995? As a proportion of the area occupied in 1996–1997?

Go to **Interpreting Graphs & Data** on **Mastering**EnvironmentalScience®

FIGURE 11.17 The world's amphibians are vanishing. The golden toad **(a)** is one of at least 170 species of amphibians that have suddenly gone extinct in recent years. This brilliant orange toad of Costa Rican cloud forests disappeared due to drought, climate change, and/or disease. Habitat loss **(b)** is the main reason for amphibian declines, but many declines remain unexplained. *Data from IUCN, 2008. Global amphibian assessment.*

DATA
- What is the second greatest known cause of amphibian declines, after habitat loss?
- What is the greatest cause for threatened species?
- What is a greater cause for non-threatened species, fires or pollution?

Go to **Interpreting Graphs & Data** on **Mastering**EnvironmentalScience®

migrates south and funnels into a single valley in Mexico, where the butterflies cluster by the millions in groves of tall trees. Here some people are illegally logging these forests while others fight to save the trees, the butterflies, and the ecotourism dollars they bring to the community.

Reasons for the decline of a population or species can be complex and difficult to determine. The worldwide collapse of amphibians provides an example of a "perfect storm" of bewildering factors. In recent years entire populations of frogs, toads, and salamanders have vanished without a trace. More than 40% of the 7200 known species of amphibians are in decline, 30% are threatened, and at least 170 species studied just years or decades ago are thought to be extinct (**FIGURE 11.17a**). As these creatures disappear before our eyes, scientists are racing to discover why, and studies implicate a wide array of causes (**FIGURE 11.17b**). These include habitat destruction, chemical pollution, invasive species, climate change, and a disease called chytridiomycosis caused by the fungal pathogen *Batrachochytrium dendrobatidis*. Biologists suspect that multiple factors are interacting and multiplying the effects.

As researchers learn more, they are designing responses to amphibian declines. An IUCN conservation action plan recommends that we protect and restore habitat, crack down

on illegal harvesting, enhance disease monitoring, and establish captive breeding programs.

Today many people are striving to save vanishing species. The search for solutions to our biodiversity crisis is dynamic and inspiring, and scientists are developing innovative approaches to sustain Earth's diversity of life.

Conservation Biology: The Search for Solutions

The urge to act as responsible stewards of natural systems, and to use science as a tool in this endeavor, sparked the rise of **conservation biology.** This scientific discipline is devoted

(a) Sampling insects in Madagascar

(b) Drawing blood from a Seychelles Magpie Robin

(c) Checking camera traps in Africa

FIGURE 11.18 Conservation biologists use many approaches to study the loss, protection, and restoration of biodiversity, seeking to develop scientifically sound solutions.

to understanding the factors, forces, and processes that influence the loss, protection, and restoration of biological diversity. It arose as biologists became increasingly alarmed at the degradation of the natural systems they had spent their lives studying. Conservation biologists choose questions and pursue research with the aim of developing solutions to such problems as habitat degradation and species loss (**FIGURE 11.18**). Conservation biology is thus an applied and goal-oriented science, with implicit values and ethical standards.

Conservation biology responds to biodiversity loss

Conservation biologists integrate ecology and evolution as they use field data, lab data, theory, and experiments to study our impacts on other organisms. They also design, test, and implement responses to these impacts. These researchers address the challenges facing biological diversity at all levels, from genes to species to ecosystems.

At the genetic level, conservation geneticists study genetic attributes of organisms to infer the status of their populations. If two populations of a species are genetically distinct, they may have different ecological needs and may require different types of management. Moreover, as a population dwindles, genetic variation is lost.

Conservation geneticists investigate how small a population can become and how much genetic variation it can lose before running into problems such as inbreeding depression (p. 271), whereby genetic similarity causes parents to produce weak or defective offspring. By determining a population's minimum viable population size, conservation geneticists help wildlife managers decide how vital it may be to increase the population.

Studies of genes, populations, and species inform conservation efforts with habitats, communities, ecosystems, and landscapes. As landscape ecologists know, organisms may be distributed across a landscape in a metapopulation, or network of subpopulations (p. 115). Because small and isolated subpopulations are most vulnerable to extirpation, conservation biologists pay special attention to them. By examining how organisms disperse from one habitat patch to another, and how their genes flow among subpopulations, conservation biologists try to learn how likely a population is to persist or succumb in the face of environmental change.

Endangered species are a focus of conservation efforts

The primary legislation for protecting biodiversity in the United States is the **Endangered Species Act.** Enacted in 1973, the Endangered Species Act (ESA) offers protection to species that are judged to be **endangered** (in danger of becoming extinct in the near future) or **threatened** (likely to become endangered soon). The ESA forbids the government and private citizens from taking actions that destroy individuals of these species or the habitats that are critical to their survival. The ESA also forbids trade in products made from threatened and endangered species. The aim is to prevent extinctions and enable declining populations to recover. As of 2016, there were 1229 species in the United States listed as endangered and 367 more listed as threatened. For nearly 75% of these species, government agencies are running recovery plans to protect them and stabilize or increase their populations.

The ESA has had a number of successes. Following the 1973 ban on the pesticide DDT (p. 363) and years of effort by

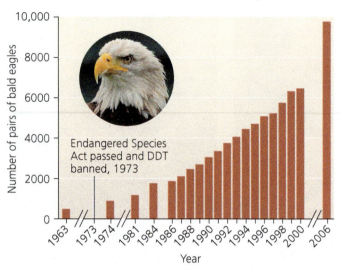

FIGURE 11.19 **The bald eagle's recovery is a success story.** America's national symbol was close to extinction in the lower 48 states in the 1960s. Following protection under the Endangered Species Act and a ban on DDT, its numbers began to rebound. In 2007, the bald eagle was declared recovered and was removed from the Endangered Species List. *Data from U.S. Fish and Wildlife Service, based on annual volunteer surveys. Data after 2000 are scarce because surveys were discontinued once it became clear the eagle was recovering.*

wildlife managers, the bald eagle, peregrine falcon, brown pelican, and other birds have recovered and are no longer listed as endangered (**FIGURE 11.19**). Intensive management programs with other species, such as the red-cockaded woodpecker (see Figure 12.15, p. 313), have held populations steady in the face of continued pressure on habitat. Overall, roughly 40% of declining populations have been stabilized. For every listed species that has gone extinct, three have recovered enough that they have been removed from the endangered species list.

These successes have come despite the fact that the U.S. Fish and Wildlife Service and the National Marine Fisheries Service, the agencies that administer the ESA, are perennially underfunded. Federal authorization for spending under the ESA expired in 1992; since then Congress has failed to re-authorize the law. The ESA remains in force, but Congress appropriates funds for its administration year by year. As a result, today a number of species are judged by scientists to be in need of ESA protection but have not been added to the endangered species list because government funding is inadequate to help recover them. Such species are said to be "warranted but precluded"—meaning that their listing is warranted by scientific research but precluded by lack of resources. This has led some environmental advocacy groups to sue the government for failing to enforce the law. Dedicated and well-meaning Fish and Wildlife Service staff are frequently caught in a no-win situation, battling lawsuits at the same time as they suffer budget cuts.

Polls repeatedly show that most Americans support protecting endangered species. Yet some opponents feel that the ESA can imperil people's livelihoods. This has been a common perception in the Pacific Northwest, where protection for the northern spotted owl and the marbled murrelet—birds that rely on old-growth forest—have slowed timber harvesting, causing loggers to fear for their jobs. In addition, many landowners worry that federal officials will restrict the use of private land on which threatened or endangered species are found. This has led to a practice described as "shoot, shovel, and shut up," among landowners who want to conceal the presence of such species on their land.

In fact, however, the ESA has stopped few development projects—and a number of its provisions and amendments promote cooperation with landowners. A landowner and the government can agree to a *habitat conservation plan*, an arrangement that grants the landowner an "incidental take permit" to harm some individuals of a species if he or she voluntarily improves habitat for the species. Similarly, in a *safe harbor agreement*, the government agrees not to mandate additional or different management requirements if the landowner pursues actions that assist a species' recovery.

Recent efforts to conserve the greater sage grouse (**FIGURE 11.20**) exemplify the cooperative public-private approach. This species had been judged "warranted but precluded," but cattle ranchers and the oil and gas industry opposed listing the species as endangered because restrictions on land use across vast sagebrush regions of the West could complicate ranching and drilling activities. As a result, federal agencies embarked on a wide campaign with ranchers and the energy industry across 13 western states to design voluntary agreements to lessen impacts on sage grouse populations. In 2015 the bird was denied listing, as federal officials said the collaborative agreements were adequate to conserve the

FIGURE 11.20 **This male greater sage grouse is doing a courtship display in its sagebrush habitat.** Sage grouse have declined steeply throughout the western United States. Their listing as an endangered species would have complicated efforts to ranch and drill for oil and gas on these lands, so the U.S. government instead designed collaborative agreements with ranchers and the oil and gas industry to try to conserve the bird without listing it.

species. This announcement produced celebration and relief in many quarters but also criticism both from development advocates, who felt the agreements were too restrictive, and from environmental advocates, who judged that the strategy would fail to save the species.

International treaties promote conservation

On the global stage, the United Nations has facilitated international treaties to protect biodiversity. The 1973 **Convention on International Trade in Endangered Species of Wild Fauna and Flora (CITES)** protects rare species by banning the international transport of their body parts. The 1990 global ban on the ivory trade may be this treaty's biggest accomplishment so far. When nations enforce it, CITES can protect rhinos, elephants, tigers, and other species whose body parts are traded internationally.

In 1992, leaders of many nations agreed to the **Convention on Biological Diversity,** a treaty that aims to help nations conserve biodiversity, use biodiversity in a sustainable manner, and ensure the fair distribution of its benefits. By providing funding and incentives for conservation and by promoting public education and scientific cooperation, this treaty has produced many success stories. It has helped African nations gain economic benefits from ecotourism at wildlife preserves such as Serengeti National Park, has prompted nations worldwide to protect more area in reserves, has enhanced global markets for sustainable crops such as shade-grown coffee, and has encouraged movement away from pesticide-intensive farming practices. Yet the treaty's overall goal—"to achieve, by 2010, a significant reduction of the current rate of biodiversity loss at the global, regional and national level"—was not met.

Today the Convention's signatory nations aim to achieve 20 new biodiversity targets by 2020. Goals include:

- Cutting the loss of natural habitats in half—and, where feasible, bringing this loss close to zero
- Conserving 17% of land areas and 10% of marine and coastal areas
- Restoring at least 15% of degraded areas
- Alleviating pressures on coral reefs

Captive breeding, reintroduction, and cloning are being pursued

In the effort to save species at risk, zoos and botanical gardens have become centers for **captive breeding,** in which individuals are bred and raised in controlled conditions with the intent of reintroducing their progeny into the wild. The IUCN counts 65 plant and animal species that now exist *only* in captivity or cultivation.

Reintroducing species into areas they used to inhabit is expensive and resource-intensive, but it can pay big dividends. In 2010 the first of 32 black rhinos was translocated from South Africa to Serengeti National Park to help restore a former population (**FIGURE 11.21**). This followed similar reintroduction projects elsewhere in Africa.

FIGURE 11.21 We can reestablish populations and rescue species by reintroducing them to areas where they were extirpated. Black rhinos have been helicoptered in to Serengeti National Park from other areas where populations are increasing.

A prime example of captive breeding and reintroduction is the program to save the California condor, North America's largest bird (**FIGURE 11.22**). Although condors are harmless scavengers of dead animals, people in the early 20th century used to shoot them for sport. Condors also collided with electrical wires and succumbed to lead poisoning after scavenging carcasses of animals killed with lead shot. By 1982, only

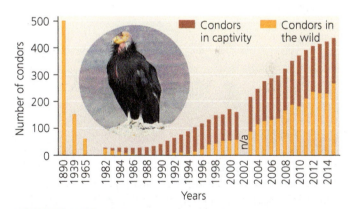

FIGURE 11.22 California condors are being bred in captivity and released to the wild, gradually rebuilding their population.

 DATA
- In 2015, were there more condors in the wild or in captivity?
- Describe how the ratio of wild to captive condors has changed over the years.
- Roughly how many condors are alive today compared with the number alive in the 1980s?
- How large is today's condor population relative to the wild condor population in 1890?

Go to **Interpreting Graphs & Data** on **Mastering**EnvironmentalScience®

22 condors remained, and biologists made the wrenching decision to take all the birds into captivity.

Today the collaborative program between the Fish and Wildlife Service and several zoos has boosted condor numbers. As of 2015 there were 167 birds in captivity and 268 birds living in the wild. Condors have been released at sites in California, Arizona, and Baja California (Mexico), where they thrill people lucky enough to spot the huge birds soaring through the skies. Unfortunately, many of these long-lived birds still die of lead poisoning, and wild populations will likely not become sustainable until hunters convert from lead shot to nontoxic shot made of copper or steel. California has now banned lead shot for all hunting, effective in 2019. This may give condors a fighting chance, while helping to stop the accumulation of a highly toxic substance in the environment.

One new idea for saving species from extinction is to create individuals by cloning them. In this technique, DNA from an endangered species is inserted into a cultured egg without a nucleus, and the egg is implanted into a female of a closely related species that acts as a surrogate mother. Several mammals have been cloned in this way, with mixed results. Some scientists even talk of re-creating extinct species from DNA recovered from preserved body parts. Indeed, in 2009 a subspecies of Pyrenean ibex (a type of mountain goat) was cloned from cells taken from the last surviving individual, which had died in 2000. The cloned baby ibex died shortly after birth, however. Even if cloning can succeed from a technical standpoint, it is not an adequate response to biodiversity loss. Cloning does nothing to protect genetic diversity, and without ample habitat and protection in the wild, having cloned animals in a zoo does little good.

Forensics can help to protect species

To counter poaching and other illegal harvesting, scientists have a new tool at their disposal. **Forensic science,** or *forensics*, involves the scientific analysis of evidence to make an identification or answer a question relating to a crime or an accident. Conservation biologists are now using forensics to protect species at risk. By analyzing DNA from organisms or their tissues sold at market, researchers can often determine the species or subspecies of organism—and sometimes its geographic origin. This information can help detect illegal activity and enhance law enforcement. A prime example is the analysis of illegal ivory shipments to determine the origin of African elephants that were killed to obtain their tusks (see The Science behind the Story, pp. 294–295).

Another example involves whales. To determine whether whales that are legally protected by international agreement are being illicitly killed, researchers have been genetically analyzing samples of whale meat sold in Asian markets (**FIGURE 11.23**). By comparing their genetic data with data from known whale populations, conservation geneticists C. Scott Baker, Stephen Palumbi, Frank Cipriano, and their colleagues have identified the species and geographic origin of hundreds of market samples and have documented numerous instances of illegal hunting. They even found that some meat labeled as whale meat came instead from dolphins, orcas, porpoises, sheep, and horses! These researchers calculated that four times more whales were being killed in the Sea of Japan than were being reported, and that Japanese and South Korean ships

(a) Whale meat for sale

(b) Genetic testing of market samples

FIGURE 11.23 Genetic testing can reveal the origin of illegally killed wildlife. Whale meat is widely sold **(a)** in Japanese and Korean markets. Dr. C. Scott Baker and his colleagues run genetic analyses in Asian hotel rooms **(b)** on samples from markets to detect meat illegally harvested from protected species and populations.

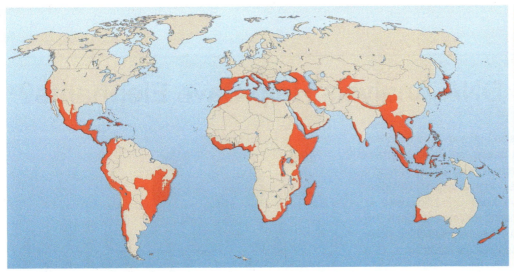

(a) The world's biodiversity hotspots

(b) Ring-tailed lemur

FIGURE 11.24 Biodiversity hotspots are priority areas for habitat preservation. Highlighted in color **(a)** are the 34 hotspots mapped by Conservation International, a nongovernmental organization. (Only 15% of the highlighted area is actually habitat; most is developed.) These regions are home to species such as the ring-tailed lemur **(b)**, a primate endemic to Madagascar that has lost more than 90% of its forest habitat as a result of human population growth and resource extraction. *Data from Conservation International.*

were killing so many minke whales that they would eventually wipe out this region's population. The scientists used their findings to urge that the international community monitor catches more closely.

Several strategies help to protect habitats, communities, and ecosystems

Scientists know that protecting species does little good if the larger systems they rely on are not also sustained. Yet no law or treaty exists to protect communities or ecosystems. For these reasons, conservation biologists pursue several strategies for conserving ecological systems on broader scales.

Umbrellas and flagships As with endangered species under the ESA, particular species can often serve as tools to conserve habitats, communities, and ecosystems. Such species are called *umbrella species* because they act as a kind of umbrella to protect many others. Umbrella species often are large animals that roam great distances, like the Serengeti's lions and wildebeest. Because such animals require large areas, meeting their habitat needs helps meet those of thousands of less charismatic animals, plants, and fungi that might never elicit as much public interest.

Environmental advocacy organizations often use large charismatic vertebrates as *flagship species* to promote biodiversity conservation. For instance, the symbol of the World Wide Fund for Nature (in North America, the World Wildlife Fund) is the panda. An endangered animal requiring large stands of undisturbed bamboo forest, the panda's lovable appearance helps solicit public support for conservation efforts that protect far more than just the panda.

Biodiversity hotspots To prioritize regions for conservation efforts, scientists have mapped biodiversity hotspots (**FIGURE 11.24a**). A **biodiversity hotspot** is a region that supports an especially great number of species that are endemic (p. 57), found nowhere else in the world (**FIGURE 11.24b**). To qualify as a hotspot, a region must harbor at least 1500 endemic plant species (0.5% of the world's total plant species). In addition, a hotspot must have already lost 70% of its habitat to human impact and be at risk of losing more.

The ecosystems of the world's biodiversity hotspots together once covered 15.7% of the planet's land surface. Today, because of habitat loss, they cover only 2.3%. This small amount of land is the exclusive home for half the world's plant species and 42% of terrestrial vertebrate species. The hotspot concept motivates us to focus on these valuable areas, where the greatest number of unique species can be protected.

Parks and protected areas A prime way to conserve habitats, communities, ecosystems, and landscapes is to set aside areas of undeveloped land to be protected in parks and preserves. Some preserves are privately owned; The Nature Conservancy purchases lands that protect important habitats and seeks to connect them with other preserved lands so that ecological processes can function across broad regions. Most protected areas are publicly owned and managed. Currently we have set aside 13% of the world's land area in national parks, state parks, provincial parks, wilderness areas, biosphere reserves, and other protected areas. Many of these lands are managed for recreation, water quality protection, or other purposes (rather than for biodiversity), and many suffer from illegal logging, poaching, and resource extraction. Yet these areas offer animals and plants a degree of protection from human persecution, and some are large enough to

Can Forensic DNA Analysis Help Save Elephants?

As any television buff knows, forensic science is a crucial tool in solving mysteries and fighting crime. In recent years, conservation biologists have been using forensics to unearth secrets and catch bad guys in the multi-billion-dollar illegal global wildlife trade. One such detective story centers on the poaching of Africa's elephants for ivory.

Each year, tens of thousands of elephants are slaughtered illegally by poachers, simply for their tusks (**FIGURE 1**). Customs agents and law enforcement authorities manage to discover and confiscate tons of tusks being shipped internationally in the ivory trade—51 tons in 2013 alone. Yet only a small percentage of tusks are found and confiscated, and poachers are rarely apprehended, so the organized international crime syndicates that run these lucrative operations have been largely unhindered thus far.

Confiscated tusks being destroyed in Kenya, to discourage poaching

Enter conservation biologist Samuel Wasser of the University of Washington in Seattle. By bringing the tools of genetic analysis, he and his students and colleagues have been shedding light on where elephants are being killed and where tusks are being shipped, thereby helping law enforcement efforts. In 2015, Wasser's team published a summary of nearly 20 years of work in the journal *Science*, revealing two major "poaching hotspots" in Africa.

The researchers began by accumulating 1350 reference samples of DNA from elephants at 71 locations across 29 African nations. Two subspecies of African elephant exist—savanna elephants, which live in open savannas, and forest elephants, which live in the forests of West and Central Africa. Of the 1350 genetic samples (from tissues or dung), 1001 came from savanna elephants and 349 came from forest elephants.

Wasser's teams sequenced DNA from the reference samples and compiled data on 16 highly variable stretches of DNA. For

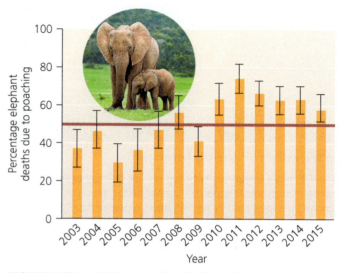

FIGURE 1 Since 2010, more than half of African elephant deaths have been due to poaching, a level scientists believe is unsustainable. Values above the horizontal line in this graph are thought to cause declines in the population. *Data from MIKE (Monitoring the Illegal Killing of Elephants), 2016.* Trends in levels of illegal killing of elephants in Africa to 31 December 2015.

each of the 71 geographic locations, they measured frequencies of alleles (different versions of genes) in these variable DNA stretches. By this process they created a map of allele frequencies for 71 geographic areas for both types of elephants; this would act as a kind of reference library with which they could compare any samples from tusks confiscated from the ivory trade.

Working with law enforcement officials, Wasser was able to access 20% of all ivory seizures made internationally between 1996 and 2005, 28% made between 2006 and 2011, and 61% made between 2012 and 2014 (**FIGURE 2**). Taking samples from these tusks and sequencing the DNA, Wasser's teams were then able to compare the results to their library of 71 locations and look for matches. In this way, with the help of sophisticated statistical techniques, they were able to determine the geographic origin of each of the tusks they tested, within an estimated distance of about 300–400 km (185–250 mi).

preserve entire natural systems that otherwise would be fragmented, degraded, or destroyed. Increasingly, areas of ocean are also being protected in marine reserves and marine protected areas (p. 439).

Serengeti National Park and the adjacent Maasai Mara National Reserve are two of the world's largest and most

celebrated parks, but Tanzania and Kenya have each set aside a number of other protected areas. Some of the best known include (in Kenya) Amboseli, Tsavo, Mount Kenya, and Lake Nakuru National Parks, and (in Tanzania) Ngorongoro Conservation Area, Selous Game Reserve, Kilimanjaro National Park, and Gombe Stream National Park. Altogether

FIGURE 2 Dr. Samuel Wasser worked with law enforcement officials to obtain samples of confiscated ivory.

For instance, ivory seized in the Philippines from 1996 to 2005 all appeared to come from forest elephants in an area of eastern Democratic Republic of Congo—just one small portion of their large geographic range. This region was difficult to patrol, however, due to its remoteness and to warfare occurring at the time, so little could be done with the information.

In contrast, when customs agents seized 6.5 tons of tusks in Singapore in 2002, Wasser's team determined that their DNA matched known samples from savanna elephants in Zambia,

indicating that many more elephants were being killed there than Zambia's government had realized. The Zambian government responded; it replaced its wildlife director and began imposing harsher sentences on poachers and ivory smugglers.

For shipments seized between 2006 and 2014, the genetic research indicated a surprisingly clear pattern of two main poaching hotspots. Most forest elephant tusks seized originated from a small region of West-Central Africa where two protected areas overlap the boundaries of four nations (FIGURE 3a). As for savanna elephant tusks, most came from animals in southern Tanzania and northern Mozambique during the early portion of the period. Later in the period, tusks originated from throughout Tanzania (FIGURE 3b), pointing to a shift northward and an increase in poaching in the parks of central and northern Tanzania.

In most cases, shipments seized in ports such as Hong Kong, Malaysia, Taiwan, and Sri Lanka were labeled with their shipping origin (often a coastal port in Kenya, Tanzania, Togo, or other African countries). With the additional help of Wasser's research, authorities could learn the entire route of the ivory shipments, from where the elephants were killed to where the tusks were exported to where they were imported. Combined with other data on poaching collected by international survey efforts, the genetic information from DNA forensic studies is painting a clearer picture of the crime network threatening elephants, and is giving law enforcement authorities more and better information with which to work.

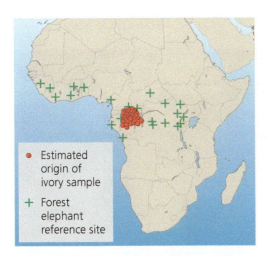

(a) Origin of tusks confiscated in Hong Kong

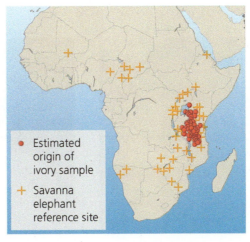

(b) Origin of tusks confiscated in Uganda

FIGURE 3 Genetic analysis of confiscated ivory reveals where elephants were killed. For example, analysis of a shipment confiscated in Hong Kong in 2013 (a) shows that it came from forest elephants killed in a small area of West-Central Africa. Likewise, tusks from a shipment confiscated in Uganda in 2013 (b) were found to have come from various areas within Tanzania. *Adapted from Wasser, S., et al., 2015. Genetic assignment of large seizures of elephant ivory reveals Africa's major poaching hotspots. Science 349: 84–87.*

roughly 25% of Tanzania's land area and 12% of Kenya's land area is protected.

Alas, setting aside land is not always enough to ensure effective conservation. In Kenya and Tanzania, pressures from outside the reserves (settlements, hunting, competition with livestock, and habitat loss to farming) are reducing

populations of migratory wildlife within the reserves (see The Science behind the Story, pp. 280–281). Similar situations occur in North America: Despite the large size of Yellowstone National Park (which is larger than the state of Delaware), elk, bears, bison, and wolves roam seasonally in and out of the park, sometimes coming into conflict with ranchers. As a result, conservationists have tried to find ways to protect animals and habitats across the Greater Yellowstone Ecosystem, the broader region over which the animals roam.

Moreover, as global warming (Chapter 18) drives species toward the poles and upward in elevation (p. 499), this can force them out of protected areas. A major challenge today is to link protected areas across the landscape with corridors of habitat so that species like wildebeest can move in response to climate change. (We will explore parks and protected areas more fully in Chapter 12.)

Ecological restoration Protecting natural areas before they become degraded is the best way to safeguard biodiversity and ecological systems. However, in some cases we can restore degraded natural systems to a semblance of their former condition through the practice of ecological restoration (p. 90). Ecological restoration aims not simply to bring back populations of animals and plants but to reestablish the processes—the cycling of matter and the flow of energy—that make an ecosystem function. By restoring complex natural systems such as the Illinois prairie (p. 91), the Florida Everglades (p. 91), or the southeastern longleaf pine forest (p. 313), restoration ecologists aim to re-create functioning systems that filter pollutants, cleanse water and air, build soil, and recharge groundwater, providing habitat for native wildlife and services for people.

In Kenya, efforts are being made to restore the Mau Forest Complex, Kenya's largest remaining forested area and a watershed that provides water for the Maasai Mara Reserve and for the people of the region (**FIGURE 11.25**). Over the years so much forest in this densely populated area has been destroyed by agriculture, settlement, and timber extraction

that the water supply for the Serengeti's wildlife and for millions of Kenyan people is now threatened. Kenya's government is working with international agencies and with U.S. funding to replant and protect areas of the forest.

Community-based conservation is growing

Helping people, wildlife, and ecosystems all at the same time is the focus of many current efforts in conservation biology. In the past, conservationists from industrialized nations, in their zeal to preserve ecosystems in developing nations, often neglected the needs of people in the areas they wanted to protect. Developing nations came to view this as a kind of neocolonialism. Today, in contrast, many international conservation biologists, including those hailing from developing nations, actively engage local people in efforts to protect land and wildlife—a cooperative approach called **community-based conservation.** A quarter of the world's protected areas are now being managed using community-based conservation. In several African nations, the African Wildlife Foundation funds community-based conservation programs to help communities conserve elephants, lions, rhinos, gorillas, and other animals.

In East Africa, conservationists and scientists began working with the Maasai and other people of the region years ago, understanding that to conserve animals and ecosystems, local people need to be stewards of the land and feel invested in conservation (**FIGURE 11.26**). This has proven challenging because the parks and reserves were created on land historically used by local people. Residents were forcibly relocated; by some estimates 50,000 Maasai were evicted to create Serengeti National Park. In the view of many local people,

FIGURE 11.25 Ecological restoration attempts to restore communities to their original state prior to human disturbance. Here people plant trees in an effort to restore the Mau Forest Complex in Kenya.

FIGURE 11.26 Scientists and conservation advocates work cooperatively with local people to conserve wildlife. Biologist Alayne Cotterill of the conservation research group Living with Lions works with Maasai warriors to monitor lion populations near the Serengeti.

the parks were a government land grab, and laws against poaching deprive them of a right to kill wildlife. As human population grew in the region, conflicts between people and wildlife increased. Ranchers worried that wildebeest and buffaloes might spread disease to their cattle. Farmers lost produce when elephants ate their crops. And the economic benefits of ecotourism were not being shared with all people in the region.

In response, proponents of conservation have tried to reallocate tourist dollars to local villages and to transfer some authority over wildlife management to local people. In the regions around the Maasai Mara Reserve, the Kenya Wildlife Service and international nonprofits have been helping farmers and ranchers build strong electric fences to keep wildlife away from their crops and livestock. These efforts are reducing conflicts between people and wildlife and are fostering more favorable attitudes toward conservation, but they also raise new challenges. For instance, in some cases elephants

are thriving on reserves, but because they are now hemmed in by fences, they become overcrowded and overgraze the vegetation, destroying trees and endangered plants the reserves were meant to protect.

Working cooperatively with local people to make conservation beneficial for them requires patience, investment, and trust on all sides. Setting aside land for preservation may deprive local people of access to exploitable resources, but it also helps ensure that those resources can be sustainably managed and will not be used up or sold to foreign corporations. If tourism revenues are adequately distributed, people gain direct economic benefits from conserving wildlife and thus may be willing to work alongside government agencies and international conservationists. Community-based conservation has not always been successful, but in a world of rising human population, sustaining biodiversity will require locally based management that sustainably meets people's needs.

closing
THE LOOP

Data from scientists worldwide confirm what any naturalist who has watched the habitat change in his or her hometown already knows: From amphibians to zebras, biological diversity is being lost rapidly within our lifetimes. This erosion of biodiversity threatens to result in a mass extinction event equivalent to those of the geologic past. Habitat alteration, pollution, overharvesting, invasive species, and climate change are the primary causes of biodiversity loss. This loss matters, because human society cannot function without biodiversity's pragmatic benefits. Conservation biologists are conducting research that guides efforts to save endangered species, protect their habitats, recover populations, and preserve and restore natural ecosystems.

Areas such as East Africa—where proposals for a highway and an oil pipeline through the Serengeti threaten to disrupt the annual wildebeest migration and have myriad consequences

for other species—have outsized importance globally for biodiversity and its conservation. Tanzania and Kenya are home to exceptionally rich species diversity. Both nations have invested heavily in protected areas to try to safeguard wildlife populations and functioning ecosystems, and both nations have been pioneers in community-based conservation. However, these countries also face severe economic and demographic challenges as people in their swelling populations attempt to rise up from poverty amid widespread land degradation in rural regions. East Africa's people desire economic development, and it can be challenging to achieve such development while protecting natural resources. The debates over the Serengeti highway and pipeline proposals bring these issues into stark relief, but there is hope that if local people can obtain economic benefits from ecotourism, they will be inspired to help protect the Serengeti ecosystem and achieve a win-win solution for economic development and biodiversity conservation.

REVIEWING Objectives

You should now be able to:

Characterize the scope of biodiversity on Earth

Biodiversity can be considered at three levels: species diversity, genetic diversity, and ecosystem diversity. Some taxonomic groups (such as insects) are more diverse than others, and diversity is unevenly distributed across habitats, biomes, and regions of the world. Roughly 1.8 million species have been described so far, but scientists agree that the world holds millions more. (pp. 271–273)

Specify the benefits that biodiversity brings us

Wild species are sources of food, medicine, and economic development; and biodiversity supports functioning ecosystems and the services they provide us. In addition, many people feel we have a psychological need to connect with the natural world, as well as an ethical duty to preserve nature. (pp. 274–277)

Discuss today's extinction crisis in geologic context

Most populations of wild species are declining, and today's extinction rate is far higher than the natural background rate. Earth has experienced five mass extinction events in the past 440 million years, and human impact is now initiating a sixth mass extinction. (pp. 277–282)

Evaluate the primary causes of biodiversity loss

Habitat loss (through destruction, alteration, or fragmentation) is the main cause of current biodiversity loss. Pollution, overharvesting, invasive species, and climate change are also important causes. Many organisms (such as amphibians) are facing impacts from a mix of factors. (pp. 282–288)

Assess the science and practice of conservation biology

Conservation biologists study biodiversity loss and seek ways to protect and restore biodiversity. These scientists integrate research at the genetic, population, species, ecosystem, and landscape levels. (pp. 288–289)

Analyze efforts to conserve threatened and endangered species

The U.S. Endangered Species Act has been largely effective, despite debate over its merits and limitations in funding. Internationally, CITES and the Convention on Biological Diversity aim to safeguard biodiversity. Modern recovery strategies include captive breeding and reintroduction programs, and forensics can help us trace products from illegally poached animals. (pp. 289–295)

Compare and contrast conservation efforts above the species level

Species cannot truly be conserved without conserving habitats, communities, and ecosystems. Charismatic species are often used in popular appeals to conserve habitats and ecosystems. Biodiversity hotspots help prioritize regions globally for conservation. Parks and protected areas conserve biodiversity at the landscape level. Ecological restoration efforts are restoring degraded ecosystems. Community-based conservation empowers people to invest in conserving their local species and ecosystems. (pp. 293–297)

TESTING Your Comprehension

1. What is biodiversity? Describe three levels of biodiversity.

2. Define the term *ecosystem services*. Give three examples of ecosystem services that people would have a hard time replacing if these were lost.

3. What is the relationship between biodiversity and food security? Between biodiversity and pharmaceuticals? Give three examples of benefits of biodiversity conservation for food supplies and medicine.

4. List three reasons why biodiversity conservation is important.

5. What are the five primary causes of biodiversity loss? Give one specific example of each.

6. List three invasive species, and describe their impacts.

7. Describe one successful accomplishment of the U.S. Endangered Species Act. Now describe one reason some people have criticized it.

8. Explain how captive breeding can help with endangered species recovery, and give an example. Now explain why cloning could never be, in itself, an effective response to species loss.

9. Name two reasons that a large national park like Serengeti or Yellowstone might not be enough to effectively conserve a population of a threatened species. What solutions exist to address these reasons?

10. Explain the notion of community-based conservation. Why have conservation advocates been turning to this approach? What challenges exist in implementing it?

SEEKING Solutions

1. Many arguments have been advanced regarding the importance of preserving biodiversity. Which argument do you think is most compelling, and why? Which argument do you think is least compelling, and why?

2. Some people declare that we shouldn't worry about endangered species because extinction has always occurred. How would you respond to this view?

3. Compare the approach of setting aside protected areas with the approach of community-based conservation.

What are the advantages and disadvantages of each? Can we—and should we—follow both approaches?

4. **CASE STUDY CONNECTION** You are an advisor to the new president of Tanzania, who is seeking to develop a formal policy on potential highways and pipelines through Serengeti National Park. Given what you know from our Central Case Study, what would be your preliminary advice to the president, and why? To improve your advice, what further information would you seek to

learn from the region's residents? From scientists and conservationists? From tourism operators? From nations and international bodies that might help fund a project?

5. **THINK IT THROUGH** You are a legislator in a nation with no endangered species act. You want to introduce a law to protect your nation's biodiversity. Consider the U.S. Endangered Species Act, as well as international efforts such as CITES and the Convention on Biological Diversity. What strategies would you write into your legislation? How would your proposed law be similar to and different from each of these efforts? Explain your answers.

6. **THINK IT THROUGH** As a resident of your community, you attend a town meeting called to discuss the proposed development of a shopping mall and condominium complex. The development would eliminate a 100-acre stand of forest, the last sizeable forest stand in your town. The developers say the forest loss will not matter because plenty of 1-acre stands still exist, scattered throughout the area. Consider the development's possible impacts on the community's biodiversity, children, and quality of life. What will you choose to tell your fellow citizens and the town's decision-makers at this meeting, and why?

CALCULATING Ecological Footprints

Research shows that much of humanity's footprint on biodiversity comes from our use of grasslands for livestock grazing and forests for timber and other resources. Grasslands and forests contribute different amounts to each nation's biocapacity (pp. 6, 203), depending on how much of these habitats each nation has. Likewise, the per capita biocapacity and per capita ecological footprints of each nation vary further according to their populations. When footprints are equal to or below biocapacity, then resources are being used sustainably. When footprints surpass biocapacity, then resources are being used unsustainably.

In the table, fill in the proportion of each nation's per capita footprint accounted for by use of grazing land and forest land. Then fill in the proportion of each nation's per capita biocapacity provided by grazing land and forest land.

FOOTPRINTS AND BIOCAPACITIES IN HECTARES PER PERSON	KENYA	TANZANIA	UNITED STATES	CANADA
Footprint from grazing land	0.27	0.36	0.19	0.42
Footprint from forest land	0.28	0.24	0.86	0.74
Total ecological footprint	0.95	1.19	7.19	6.43
Percentage footprint from grazing and forest	58			
Biocapacity of grazing land	0.27	0.39	0.26	0.23
Biocapacity of forest land	0.02	0.13	1.56	8.27
Total biocapacity	0.53	1.02	3.86	14.92
Percentage biocapacity from grazing and forest				57

Data from WWF, 2012. *Living planet report 2012*. Gland, Switzerland: WWF International.

1. In which nations is grazing land being used sustainably? In which nations is forest land being used sustainably?

2. Do forest use and grazing make up a larger part of the footprint for temperate-zone industrialized nations such as Canada and the United States or for tropical-zone developing nations such as Kenya and Tanzania? What do you think accounts for this difference between these two types of nations? What else besides land use of forests and grasslands contributes to an ecological footprint?

3. The Living Planet Index declined 52% between 1970 and 2010 (p. 278), but its temperate and tropical components differed. During this period the index for temperate regions decreased by 36%, whereas the index for tropical regions declined by 56%. Based on this information, do you predict that biodiversity loss has been steepest in Kenya and Tanzania or in Canada and the United States? Explain your answer.

MasteringEnvironmentalScience®

Students Go to **MasteringEnvironmentalScience** for assignments, the etext, and the Study Area with practice tests, videos, current events, and activities.

Instructors Go to **MasteringEnvironmentalScience** for automatically graded activities, current events, videos, and reading questions that you can assign to your students, plus Instructor Resources.

CHAPTER 12

Forests,
Forest Management,
and Protected Areas

Certified Sustainable Paper in Your Textbook

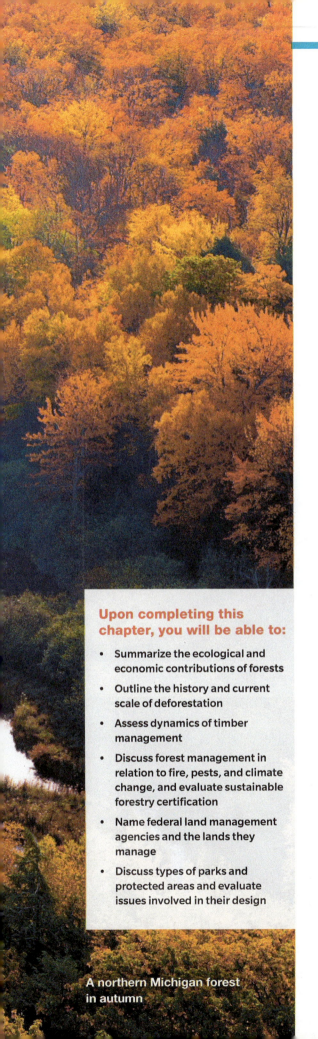

> " FSC is the high bar in forest certification, because its standards protect forests of significant conservation value and consistently deliver meaningful improvements in forest management on the ground.
> Kerry Cesareo, deputy director, WWF-US Forests Program

> Our commitment to sustainability remains a shared value of the company.
> Doug Osterberg, CEO, Appleton Coated paper company "

Upper Peninsula of MICHIGAN

WISCONSIN — Combined Locks

As you turn the pages of this textbook, you are handling paper made from trees that were grown, managed, harvested, and processed using certified sustainable practices. The trees cut to make this book's paper were selected for harvest based on a sustainable management plan designed to avoid depleting the forests of their mature trees or degrading the ecological functions the forests perform. The logs were then transported to mills to create pulp and then paper.

If you were to trace the paper in this book back to its origin, you might find yourself standing in a mixed forest of aspen, birch, beech, maple, spruce, and pine in northern Wisconsin or in the Upper Peninsula of Michigan. These sparsely populated regions near the Canadian border, flanked by Lake Michigan and Lake Superior, remain heavily forested, despite supplying timber to our society for nearly 200 years and continuing to supply us with much of the wood required to make paper.

Once harvested, the trees are shipped to a nearby pulp mill where the timber is chipped and fed into a digester, which cooks the wood chips with chemicals to break down the wood's molecular bonds. To make white paper, lignin (the polymer that gives wood its sturdiness) is separated out and discarded, leaving only the wood's cellulose fibers. Millworkers bleach, wash, and screen the cellulose fibers and mix them with water and dye.

A paper-making mill such as the one in the riverside village of Combined Locks, Wisconsin, will pour such a mixture onto a moving mat, where heavy rollers press it into thin sheets. The moist sheets of newly formed paper are then dried on heated rollers and made ready to receive any of a variety of coatings. The paper is wound into immense reels, later to be cut into sheets. In this way, the mill at Combined Locks, owned by the company Appleton Coated, produces about 400,000 tons of paper each year, employing nearly 600 people. The mill recycles chemicals and water used in the process, and it combusts discarded lignin to help power the mill.

Each stage in the paper-making process for this textbook, from the harvest in the northern forests to the pulp and paper production, has been examined by independent third-party inspectors from the Forest Stewardship Council® (FSC®) to ensure that practices meet the FSC's strict criteria for sustainable forest management and paper production. The Forest Stewardship Council is an organization that

A worker tending to equipment in a paper mill

A northern Michigan forest in autumn

certifies forests, companies, and products that meet sustainability standards. To become FSC-certified, harvesting operations in timber-producing regions such as Wisconsin and Michigan are required to protect rare species and sensitive habitats, safeguard water sources, control erosion, minimize pesticide use, and maintain the diversity of the forest and its ability to regenerate after harvesting. FSC certification is the best way for consumers of forest products to know that they are supporting sustainable practices that protect forests.

Your textbook's paper gained FSC certification because Pearson Education, the publisher of this book, is striving to follow sustainable practices. Pearson supported FSC paper for this book at the request of your authors and editors because our entire team feels that an environmental science textbook should walk its talk. So, as you flip through this book, you can feel satisfied that you are doing a small part to help safeguard the world's forests by supporting sustainable forestry practices.

Forest Ecosystems and Forest Resources

A **forest** is any ecosystem with a high density of trees. Forests provide habitat for countless organisms; help maintain the quality of soil, air, and water; and play key roles in our planet's biogeochemical cycles (p. 117). Forests also provide humanity with wood for fuel, construction, paper production, and more.

There are many types of forests

Most of the world's forests occur as boreal forest, a biome that stretches across much of Canada, Scandinavia, and Russia; or as tropical rainforest, a biome in Latin America, Indonesia, Southeast Asia, and equatorial Africa. Temperate deciduous forests, temperate rainforests, and tropical dry forests also cover large regions (pp. 92–98).

Within each forest biome, the plant community varies from region to region because of differences in soil and

(a) Maple-beech-birch forest, Michigan's Upper Peninsula

(b) Oak-hickory forest, West Virginia

(c) Ponderosa pine forest, northern Arizona

(d) Redwood forest, coastal northern California

FIGURE 12.1 Shown are 4 of the 23 forest types found in the continental United States.

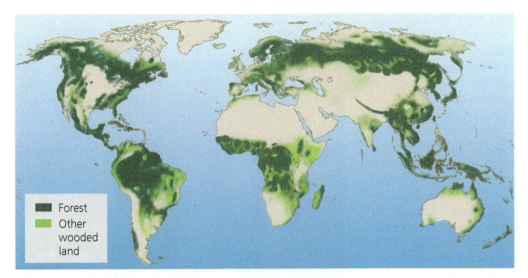

FIGURE 12.2 **Forests cover 31% of Earth's land surface.** Most widespread are boreal forests in the north and tropical forests near the equator. "Other wooded land" supports trees at sparser densities. *Data from Food and Agriculture Organization of the United Nations, 2010.* Global forest resources assessment 2010.

Forest

Other wooded land

climate. As a result, ecologists and forest managers find it useful to classify forests into **forest types,** categories defined by their predominant tree species (**FIGURE 12.1**). The eastern United States contains 10 forest types, ranging from spruce-fir to oak-hickory to longleaf–slash pine. The western United States has 13 forest types, ranging from Douglas fir and hemlock–sitka spruce forests of the moist Pacific Northwest to ponderosa pine and pinyon-juniper woodlands of the drier interior.

Northern Wisconsin and Michigan have a diverse mix of forest types because this region lies where the temperate deciduous forest biome of the eastern United States merges into the boreal forest of Canada. In this region, spruce-fir forest intermixes with forests of aspen, birch, maple, and beech, and areas of white, red, and jack pine.

Altogether, forests currently cover 31% of Earth's land surface (**FIGURE 12.2**) and occur on all continents except Antarctica.

Forests are ecologically complex

Because of their structural complexity and their capacity to provide many niches for organisms, forests comprise some of the richest ecosystems for biodiversity (**FIGURE 12.3**). The leaves, fruits, and seeds of trees, shrubs, and forest-floor plants furnish food and shelter for an immense diversity of insects, birds, mammals, and other animals. Plants are also colonized by an extensive array of fungi and microbes, in both parasitic and mutualistic relationships (pp. 77–78).

In a forest's **canopy**—the upper level of leaves and branches in the treetops—beetles, caterpillars, and other

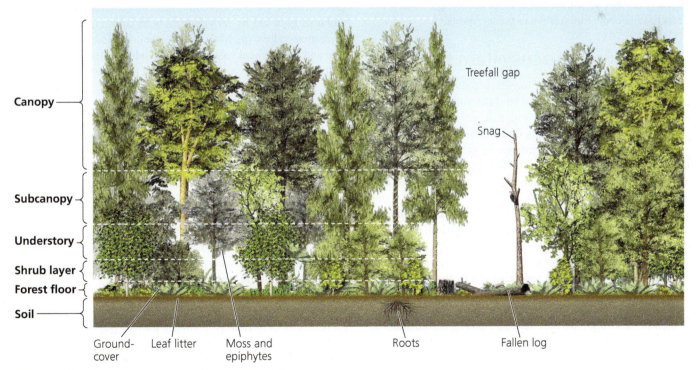

Canopy

Subcanopy

Understory

Shrub layer

Forest floor

Soil

Treefall gap

Snag

Ground-cover Leaf litter Moss and epiphytes Roots Fallen log

FIGURE 12.3 A mature forest is complex in its structure.

leaf-eating insects abound, providing food for birds such as warblers and tanagers, while arboreal mammals from squirrels to sloths to monkeys consume fruit and leaves. Animals also live and feed in the **subcanopy** (the middle portion of a forest beneath the tree crowns of the canopy), in shrubs and small trees of the **understory** (the shaded lower level of a forest), and among groundcover plants on the forest floor. Other animals use the bark, branches, and trunks of trees, or the vines, lichens, and mosses that grow on them.

Cavities in trunks and limbs provide nest and shelter sites for a wide variety of animals. Dead and dying trees (called **snags**) are especially valuable; insects that break down the wood provide food for woodpeckers, which create cavities that other animals may later use. Fallen trees create openings called *treefall gaps*, letting sunlight through past the canopy and down to the forest floor, encouraging the growth of early successional plants and creating habitat diversity within the forest. Much of a forest's biodiversity resides on the forest floor, where decaying logs and the fallen leaves and branches of the leaf litter nourish the soil. A multitude of soil organisms helps decompose plant material and cycle nutrients (p. 212).

In tropical rainforests (p. 95), particularly tall individual trees called emergent trees protrude here and there above the canopy. Epiphytes, plants specialized to grow atop other plants, add to the biomass and species diversity at all levels of a rainforest. Epiphytes include many ferns, mosses, lichens, orchids, and bromeliads.

Forests with a greater diversity of plants tend to host a greater diversity of organisms overall. As forests change by the process of succession (p. 83), their species composition changes along with their structure. In general, old-growth forests host more biodiversity than young forests, because older forests contain more structural diversity, microhabitats, and resources for more species. Old-growth forests also are home to more species that today are threatened, endangered, or declining, because old forests have become rare relative to young forests.

Forests provide ecosystem services

Besides hosting biodiversity, forests supply us with many vital ecosystem services (pp. 4, 116; **FIGURE 12.4**). As plants grow, their roots stabilize the soil and help to prevent erosion. Trees' roots draw minerals up from deep soil layers and deliver them to surface soil layers where other plants can use them. Plants also return organic material to the topsoil when they die or drop their leaves. When rain falls, leaves and leaf litter slow runoff by intercepting water. This helps water soak into the ground to nourish roots and recharge aquifers, thereby preventing flooding, reducing soil erosion, and helping keep streams and rivers clean.

Forest plants also filter pollutants and purify water as they take it up from the soil and release it to the atmosphere in transpiration (p. 118). Plants draw carbon dioxide from the air for use in photosynthesis (p. 30), release the oxygen that we breathe, and regulate moisture and precipitation. Indeed, because trees remove carbon dioxide from the air and store carbon in their tissues, forests are viewed as a vital safeguard against global climate change (p. 308; Chapter 18).

FIGURE 12.4 Forests provide us a diversity of ecosystem services, as well as resources that we can harvest.

By performing all these ecological functions, forests are indispensable for our survival. Forests also enhance our health and quality of life by providing us with cultural, aesthetic, and recreation values (p. 148). People seek out forests for adventure and for spiritual solace alike—to admire beautiful trees, to observe wildlife, to enjoy clean air, and for many other reasons.

Forests provide us valuable resources

Ecosystem services alone make forests priceless to our society, but forests also provide many economically valuable resources. Among these are plants for medicines, dyes, and fibers; animals, plants, and fungi for food; and, of course, wood from trees. For millennia, wood has fueled the fires to cook our food and keep us warm. It has built the homes that keep us sheltered. It built the ships that carried people and cultures between continents. And it gave us paper, the medium of the first information revolution. Forests and their resources have helped our society achieve the standard of living we now enjoy.

In recent decades, industrial harvesting has allowed us to extract more timber than ever before, supplying all these needs of a rapidly growing human population and its expanding economy. Most commercial timber extraction today takes place in Canada, Russia, and other nations with large expanses of boreal forest, and in tropical nations with large areas of rainforest, such as Brazil and Indonesia. In the United States, most logging takes place in pine plantations of the South and conifer forests of the West.

Forest Loss

When trees are removed more quickly than they can regrow, the result is **deforestation,** the clearing and loss of forests. Deforestation has altered landscapes across much of our planet. In the time it takes you to read this sentence, 2 hectares (ha; 5 acres) of tropical forest will have been cleared. As we alter, fragment, and eliminate forests, we lose biodiversity, accelerate climate change, and disrupt the ecosystem services that support our societies.

Agriculture and demand for wood put pressure on forests

To make way for agriculture and to extract wood products, people have been clearing forests for millennia. This has fed our civilization's growth, but the loss of forests also exerts damaging impacts as our population grows. Deforestation leads to biodiversity loss, soil degradation, and desertification (p. 222). It also adds carbon dioxide to the atmosphere, contributing to climate change.

In 2015, the Food and Agriculture Organization (FAO) of the United Nations released its latest *Global Forest Resources Assessment.* In this report, researchers combined remote sensing data from satellites, analysis from forest experts, questionnaire responses, and statistical modeling to form a comprehensive picture of the world's forests. The assessment concluded that we are eliminating 7.6 million ha (18.8 million acres) of forest each year. Subtracting annual regrowth from this amount makes for an annual net loss of 3.3 million ha (8.2 million acres)—an area about the size of Maryland. This rate (for the period 2010–2015) is lower than deforestation rates for earlier years. In the 1990s, the world was losing 1.8% of its forest each year; today it is losing just 0.8% each year.

However, some other research efforts have calculated higher rates of forest loss. In 2013, a large research team analyzed satellite data, published their work in the journal *Science*, and worked with staff from Google to create interactive maps of forest loss and gain across the world. They found much more deforestation than the FAO had: From 2000 to 2012, this team calculated annual losses of 19.2 million ha and annual gains of 6.7 million ha, for a net loss per year of 12.5 million ha (30.9 million acres, larger than New York State).

We deforested much of North America

Deforestation for timber and farmland propelled the expansion of the United States and Canada westward across the North American continent. The vast deciduous forests of the East were cleared by the mid-1800s, making way for countless small farms. Timber from these forests built the cities of the Atlantic seaboard. Cities of the upper Midwest, such as Chicago, Detroit, and Milwaukee, were constructed from timber felled in the vast pine and hardwood forests of Wisconsin and Michigan.

As a farming economy shifted to an industrial one, wood was used to stoke the furnaces of industry. Logging operations moved south to the Ozarks of Missouri and Arkansas, and then to the pine woodlands and bottomland hardwood forests of the South, which were logged and converted to pine plantations. Once mature trees were removed from these areas, timber companies moved west, cutting the continent's biggest trees in the Rocky Mountains, the Sierra Nevada, the Cascade Mountains, and the Pacific Coast ranges. Exploiting forest resources helped American society to develop, but we were not harvesting forests sustainably. Instead, we were depleting our store of renewable resources for the future.

By the 20th century, very little **primary forest**—natural forest uncut by people—remained in the lower 48 U.S. states, and today even less is left (**FIGURE 12.5**). Nearly all the large

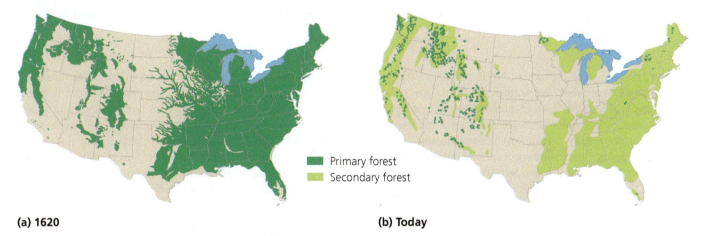

(a) 1620   (b) Today

Primary forest
Secondary forest

FIGURE 12.5 Areas of primary (uncut) forest have been dramatically reduced. When Europeans first colonized North America **(a)**, much of what is now the United States was covered in primary forest (**dark green**). Today, nearly all this primary forest is gone **(b)**, cut for timber and to make way for agriculture. Much of the landscape has become reforested with secondary forest (**pale green**). Sources: (a) U.S. Forest Service; and (b) Hansen, M.C., et al., 2013. High-resolution global maps of 21st-century forest cover change. Science 342: 850–853, and George Draffan, Endgame Research (www.endgame.org).

(a) Cattle on burned and cleared land

FIGURE 12.6 Tropical forests are being lost. Many are burned to farm soybeans or to create grazing land for cattle, as seen here **(a)** in Brazil. Most of the soybeans and beef are exported to consumers in wealthier nations. Africa and Latin America are losing forests at rapid rates **(b)**, whereas Europe and the United States are slowly gaining secondary forest. In Asia, tree plantations are increasing, but natural primary forests are still being lost. *Data from Food and Agriculture Organization of the United Nations, 2015.* Global forest resources assessment 2015. *By permission.*

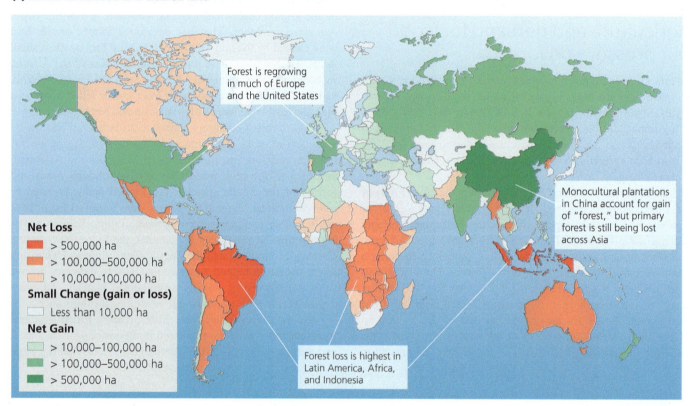

Forest is regrowing in much of Europe and the United States

Monocultural plantations in China account for gain of "forest," but primary forest is still being lost across Asia

Net Loss
- > 500,000 ha
- > 100,000–500,000 ha
- > 10,000–100,000 ha

Small Change (gain or loss)
- Less than 10,000 ha

Net Gain
- > 10,000–100,000 ha
- > 100,000–500,000 ha
- > 500,000 ha

Forest loss is highest in Latin America, Africa, and Indonesia

(b) Change in forest area by region

oaks and maples found in eastern North America today, and even most redwoods of the California coast, are *second-growth* trees: trees that have sprouted and grown to partial maturity after old-growth trees were cut. Second-growth trees characterize **secondary forest,** which contains smaller trees than does primary forest. The species composition, structure, and nutrient balance of a secondary forest may differ markedly from the primary forest that it replaced.

Forests today are being cleared most rapidly in developing nations

Uncut primary forests still remain in many developing countries. These nations are in the position the United States and Canada enjoyed a century or two ago: having a resource-rich frontier that they can develop. Today's powerful industrial technologies allow these nations to exploit their resources and push back their frontiers even faster than occurred in North America. As a result, deforestation is rapid today in places such as Indonesia, Africa, Central America, and South America (**FIGURE 12.6**). In these regions, developing nations are striving to expand settlement for their burgeoning populations and to boost their economies by extracting natural resources. Moreover, people in many of these societies harvest fuelwood for their daily cooking and heating needs (p. 563). In contrast, parts of Europe and the United States are gaining forest as they recover from past deforestation. This is not compensating for the loss of tropical forests, however, because tropical forests are home to far more biodiversity than the temperate forests of North America and Europe.

The South American nation of Brazil, home to most of the vast Amazon rainforest, illustrates success in reducing

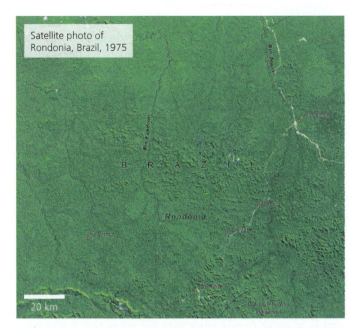

FIGURE 12.7 Deforestation of Amazonian rainforest has been rapid in recent decades. Satellite images of the state of Rondonia in Brazil show extensive clearing resulting from settlement in the region.

deforestation but also the continuing pressures on tropical forests. Not long ago, Brazil was losing forests faster than any other country (**FIGURE 12.7**). Its government was promoting settlement on the forested frontier and was subsidizing an expansion of large-scale cattle ranching and soybean farming to meet demand from consumers in the United States and Europe. More recently, however, Brazil reduced deforestation significantly, while advancing economically and politically as a stable democracy, thereby becoming a global model for fighting forest loss while 80% of its forest remained. However, in 2012 Brazil's legislature weakened the nation's Forest Code, which had helped to slow deforestation. President Dilma Rousseff vetoed some aspects of the legislation in an effort to balance conservation and development interests, but regulation and enforcement were cut and large dam and road projects were initiated. With a now-declining economy prompting people to clear forests illegally for short-term financial gain, deforestation rose markedly.

Developing nations often are desperate enough for economic development and foreign capital that they impose few restrictions on logging. Often they allow their timber to be extracted by foreign multinational corporations, which pay fees for a **concession,** or right to extract the resource. Once a concession is granted, the corporation has little or no incentive to manage forest resources sustainably. Local people may receive temporary employment from the corporation, but once the timber is gone they no longer have the forest and the ecosystem services it provided. As a result, most economic benefits are short term and are reaped not by local residents but by the foreign corporation. Much of the wood extracted in developing nations is exported to Europe and North America. In this way, our consumption of high-end furniture and other wood products can drive forest destruction in poorer nations.

Throughout Southeast Asia and Indonesia today, vast swaths of tropical rainforest are being cut to establish plantations of oil palms (**FIGURE 12.8**). Oil palm fruit

FIGURE 12.8 Oil palm plantations are replacing primary forest across Southeast Asia and Indonesia. Since 1950, the immense island of Borneo (maps at bottom) has lost most of its forest. *Data from Radday, M., WWF-Germany, 2007. Designed by Hugo Ahlenius, UNEP/GRID-Arendal. Extent of deforestation in Borneo 1950–2001, and projection towards 2020; http://maps.grida.no/go/graphic/extent-of-deforestation-in-borneo-1950–2005-and-projection-towards-2020.*

produces palm oil, which has come to be a common ingredient in snack foods, soaps, and cosmetics, and now is beginning to be used as a biofuel. In Indonesia, the world's largest palm oil producer, oil palm plantations have displaced more than 8 million ha (20 million acres) of rainforest. On top of this, clearing for plantations encourages further development and eases access for people to enter the forest and conduct logging illegally.

weighing the ISSUES

Logging Here or There

Suppose you are an activist protesting the logging of old-growth trees near your hometown. Now let's say you know that if the protest is successful, the company will move to a developing country and cut its primary forest instead. Would you still protest the logging in your hometown? Would you pursue any other approaches?

The palm oil boom represents a conundrum for environmental advocates. Many people eager to fight climate change had urged the development of biofuels (p. 564) to replace fossil fuels. Yet grown at the large scale that our society is demanding, monocultural plantations of biofuel crops such as oil palms are causing severe environmental impacts by displacing natural forests.

Forest loss worsens climate change

Because trees absorb carbon dioxide from the air during photosynthesis and then store carbon in their tissues, forests serve as a major reservoir for carbon. Scientists estimate that the world's forests store about 296 billion metric tons of carbon in living tissue, which is more than the atmosphere contains. When plant matter is burned or when plants die and decompose, carbon dioxide is released—and thereafter less vegetation remains to soak it up. Carbon dioxide is the primary greenhouse gas contributing to global climate change (p. 482). Therefore, when we destroy forests, we accelerate climate change. The more forests we preserve or restore, the more carbon we keep out of the atmosphere, and the better we can address climate change.

Solutions are emerging

New solutions are being proposed to address deforestation in developing nations. Some conservation proponents are pursuing community-based conservation projects (p. 296) that empower local people to act as stewards of their forest resources. In other cases, conservation organizations are buying concessions and using them to preserve forest rather than to cut it down. In such a **conservation concession,** the nation receives money *and* keeps its natural resources intact. The South American nation of Surinam entered into such an agreement with the nongovernmental organization Conservation International and virtually halted logging while gaining $15 million.

In Indonesia, one former supplier of this textbook's paper, NewPage Corporation, helped to fund a project called POTICO (Palm Oil, Timber, Carbon Offsets) that aimed to reduce deforestation and illegal logging. In this still-evolving project, the nonprofit World Resources Institute (WRI) is working with palm oil companies that own concessions to clear primary rainforest and is steering them instead toward land that is already logged and degraded. WRI then protects the forests that were slated for conversion or allows the forests to become certified to the FSC forest management standard. Because primary forests store more carbon than oil palm plantations, these land swaps can reduce Indonesia's greenhouse gas emissions and qualify for credit via carbon offsets (p. 510).

Carbon offsets and other financial incentives are central to emerging international plans to curb deforestation and climate change. Forest loss accounts for at least 12% of the world's greenhouse gas emissions—nearly as much as all the world's vehicles emit. Across a decade's worth of international climate conferences (p. 507), negotiators have outlined a proposed program called **Reducing Emissions from Deforestation and Forest Degradation** (REDD, soon changed to REDD+ as the program expanded in scope), whereby wealthy industrialized nations pay poorer developing nations to conserve forest. The aim is to make forests more valuable when saved than when cut down. Under this plan, poor nations gain income while rich nations receive carbon credits to offset their emissions in an international cap-and-trade system (pp. 507, 509). Although the REDD+ plan has not yet been formally agreed to, leaders of rich nations have proposed to transfer $100 billion per year to poor nations by 2020, and the approach gained crucial momentum at the Paris Climate Conference in 2015.

Forest Management

Our demand for forest resources and amenities is rising, so we need to take care in managing forests. *Foresters* are professionals who manage forests through the practice of **forestry.** Foresters must balance our society's demand for forest products against the central importance of forests as ecosystems. Today, sustainable forest management practices are spreading as informed consumers demand sustainably produced products. Just as your textbook uses FSC-certified paper from sustainably managed forests, more and more paper, lumber, and other forest products are now made using certified sustainable practices. In this way, consumer choice is influencing the ways forests are managed.

Forest management is one type of resource management

Debates over how to manage forest resources reflect broader questions about how to manage natural resources in general. Resources such as fossil fuels and many minerals are nonrenewable, whereas resources such as the sun's energy are perpetually renewable (p. 4). Between these extremes lie resources that are renewable if they are not exploited too rapidly. These include timber, as well as soils, fresh water, rangeland, wildlife, and fisheries.

Resource management describes our use of strategies to manage and regulate the harvest of renewable resources. Sustainable resource management involves harvesting these resources in ways that do not deplete them. Resource managers help to conserve soil resources with farming practices that fight erosion, to safeguard the supply and quality of surface water and groundwater, to encourage low-impact grazing to manage rangeland sustainably, and to protect fisheries and wildlife from overharvesting. Resource managers are guided by research in the natural sciences and by social, political, and economic factors.

Resource managers follow several strategies

A key question in managing resources is whether to focus strictly on the resource of interest or to look more broadly at the environmental system of which it is a part. Taking a broader view often helps avoid degrading the system and thereby helps sustain the resource in the long term.

Maximum sustainable yield A guiding principle in resource management has been **maximum sustainable yield.** Its aim is to achieve the maximum amount of resource extraction without depleting the resource from one harvest to the next. Recall the logistic growth curve (see Figure 3.17, p. 65), which reflects how limiting factors slow exponential population growth and then cap it at a carrying capacity. The logistic curve indicates that a population grows most quickly when it is at an intermediate size—specifically, at one-half of carrying capacity. A fisheries manager aiming for maximum sustainable yield will therefore prefer to keep fish populations at intermediate levels so that they rebound quickly after each harvest. Doing so should result in the greatest amount of fish harvested over time while sustaining the population indefinitely (**FIGURE 12.9**).

This management approach, however, keeps the fish population at only half its carrying capacity—well below the size it would attain in the absence of fishing. Suppressing population size in this way will likely affect other species and alter the food web dynamics of the community. From an ecological point of view, management for maximum sustainable yield may set in motion significant ecological changes.

In forestry, maximum sustainable yield argues for cutting trees shortly after they go through their fastest stage of growth. Because trees often increase in biomass most quickly at an intermediate age, trees are generally cut long before they have grown as large as they would in the absence of harvesting. This practice maximizes timber production over time, but it also alters forest ecology and eliminates habitat for species that depend on mature trees.

Ecosystem-based management Because of these dilemmas, more and more managers espouse **ecosystem-based management,** which aims to minimize impact on the ecosystems and ecological processes that provide the resource. Many certified sustainable forestry plans protect certain forested areas, restore ecologically important habitats, and consider patterns at the landscape level (p. 114), allowing timber harvesting while preserving the functional integrity of the forest ecosystem. This means fostering the area's

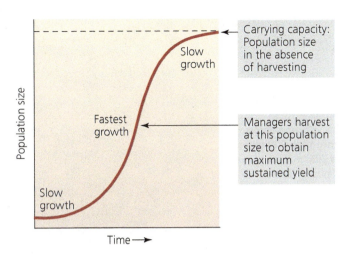

FIGURE 12.9 Maximum sustainable yield maximizes the amount of resource harvested while sustaining the harvest in perpetuity. For a wildlife population or fisheries stock that grows according to logistic growth, managers aim to keep the population at half the carrying capacity, because populations grow fastest at intermediate sizes.

ecological processes, including succession (p. 83), in which the forest community naturally changes over time. Ecosystems are complex, however, so it can be challenging to determine how best to implement this type of management. As a result, ecosystem-based management has come to mean different things to different people.

Adaptive management Some management actions will succeed, and some will fail. A wise manager will try new approaches if old ones are ineffective. **Adaptive management** involves systematically testing different approaches and aiming to improve methods through time. For managers, it entails monitoring the results of one's practices and adjusting them as needed, based on what is learned. Adaptive management is intended to be a fusion of science and management, because it explicitly tests hypotheses about how best to manage resources. This process is time-consuming and complex but can be highly effective.

Adaptive management was featured in the Northwest Forest Plan, a 1994 plan crafted by the administration of President Bill Clinton to resolve disputes between loggers and preservationists over the last remaining old-growth temperate rainforests in the continental United States. This plan sought to allow limited logging to continue in the Pacific Northwest, with adequate protections for old-growth-dependent species such as the northern spotted owl, and to let science guide management.

Fear of a "timber famine" inspired national forests

The United States began formally managing forest resources a century ago, in response to rampant deforestation. The depletion of forests throughout the eastern United States had

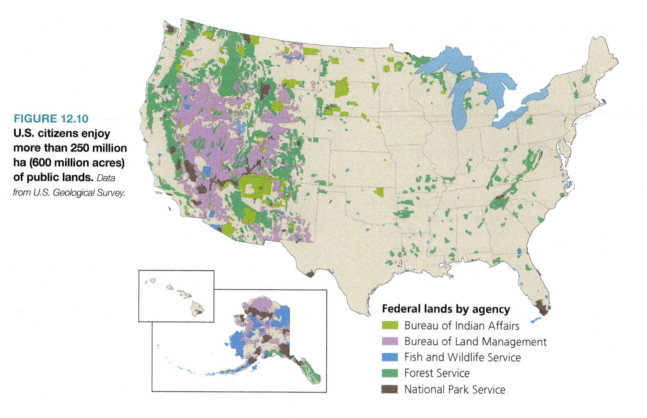

FIGURE 12.10
U.S. citizens enjoy more than 250 million ha (600 million acres) of public lands. *Data from U.S. Geological Survey.*

Federal lands by agency
- Bureau of Indian Affairs
- Bureau of Land Management
- Fish and Wildlife Service
- Forest Service
- National Park Service

prompted widespread fear of a "timber famine." This led the federal government to form a system of forest reserves: public lands set aside to grow trees, produce timber, and protect water quality. Today the U.S. **national forest** system consists of 77 million ha (191 million acres), managed by the U.S. Forest Service and covering more than 8% of the nation's land area (**FIGURE 12.10**).

The U.S. Forest Service was established in 1905 under the leadership of Gifford Pinchot (p. 136) during the Progressive Era, a time of social reform when people began using science to inform public policy. In line with Pinchot's conservation ethic (p. 136), the Forest Service aimed to manage the forests for "the greatest good of the greatest number in the long run." Pinchot believed that the nation should extract and use resources from its public lands, but that wise and careful management of timber resources was imperative.

We extract timber from private and public lands

Today almost 90% of the timber harvesting in the United States takes place on private land owned by the timber industry or by small landowners (**FIGURE 12.11**). Timber companies pursue maximum sustainable yield on their land, to obtain maximal profits year after year.

Private timber companies also extract timber from the U.S. national forests and from publicly held state forests. On the national forests, U.S. Forest Service employees conduct timber sales and build roads to provide access for logging companies. The Forest Service sells timber below the costs it incurs for marketing and administering the harvest and for

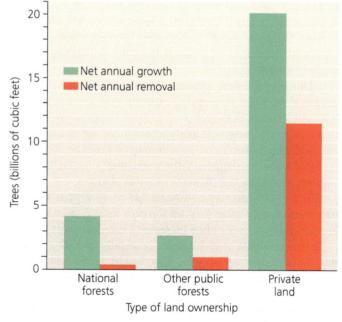

FIGURE 12.11 In the United States today, trees are growing faster than they are being removed. However, forests that regrow after logging often differ substantially from the forests that were removed. "Private land" here combines land owned by the timber industry and by small landholders. *Data from Oswalt, S.N., et al., 2014. Forest resources of the United States, 2012. For. Serv. Gen. Tech. Rep. WO-91. Washington, D.C.: USDA Forest Service.*

DATA For which of the three types of land is the ratio of growth to removal greatest?

Go to **Interpreting Graphs & Data** on MasteringEnvironmentalScience®

building access roads, while the companies go on to sell the timber they harvest for profit. In this way, taxpayers subsidize private timber harvesting on public land (p. 178). These subsidies also tend to inflate harvest levels beyond what would occur in a free market.

On the U.S. national forests, private timber extraction increased in the 1950s as the nation underwent a postwar economic boom, paper consumption rose, and the growing population moved into newly built suburban homes. Harvests then began to decrease in the 1980s as economic trends shifted, public concern over clear-cutting grew, and management philosophy evolved. Today, regrowth is outpacing removal on these lands by 11 to 1 (see Figure 12.11). At present, in an average year, about 2% of U.S. forest acreage is cut for timber. Overall, timber harvesting in the United States and other developed nations has remained stable for the past half-century, while it has more than doubled in developing nations.

Note, however, that even when the regrowth of trees outpaces their removal, the character of a forest may change. Once primary forest is replaced by younger secondary forest or by single-species plantations, the resulting community may be very different, and generally it is less ecologically valuable.

Plantation forestry has grown

Today's timber industry focuses on production from plantations of fast-growing tree species planted in single-species monocultures (p. 239). All trees in a stand are planted at the same time, so the stands are **even-aged;** that is, all trees are the same age (**FIGURE 12.12**). Stands are cut after a certain number of years (called the rotation time), and the land is replanted with seedlings. Such plantation forestry is growing, and 7% of the world's forests are now plantations. One-quarter of these feature non-native tree species.

Ecologists and foresters alike view plantations as akin to crop agriculture. Because there are few tree species and little variation in tree age, plantations do not offer habitat to many forest organisms. For instance, stands of red pine planted in northern Wisconsin and Michigan host far less biodiversity than the more-diverse forests of multiple tree species that surround them. Even-aged single-species plantations lack the structural complexity that characterizes a mature natural forest as seen in Figure 12.3 (p. 303). Plantations are also vulnerable to outbreaks of pest species such as bark beetles, as we shall soon see. For all these reasons, some harvesting methods aim to maintain **uneven-aged** stands, in which a mix of ages (and often a mix of tree species) creates greater structural diversity and makes the stand more similar to a natural forest.

We harvest timber by several methods

Timber companies choose from several methods to harvest trees. In the simplest method, **clear-cutting,** all trees in an area are cut at once (**FIGURE 12.13**). Clear-cutting is cost-efficient, and to

FIGURE 12.12 Even-aged management, with all trees of equal age, can be seen in the foreground in a plantation that is regrowing after having been clear-cut. Uneven-aged management maintains a mix of tree ages, as seen in the more mature forest in the background.

some extent it can mimic natural disturbance events such as fires, tornadoes, or windstorms. However, the ecological impacts of clear-cutting are considerable. An entire ecological community is removed, soil erodes away, and sunlight penetrates to ground level, changing microclimatic conditions. As a result, new types of plants replace those of the original forest. Clear-cutting essentially sets in motion a process of succession (p. 83) in which the resulting climax community may be quite different from the

FIGURE 12.13 Clear-cutting is cost-efficient for timber companies but has ecological consequences. These include soil erosion, water pollution, and altered community composition.

original climax community. For example, when we clear-cut an eastern U.S. oak-hickory forest, the forest that will regrow may be dominated by maples, beeches, and tulip trees.

Concerns about clear-cutting led foresters and the timber industry to develop alternative harvesting methods. Clear-cutting (**FIGURE 12.14a**) remains the most widely practiced method, but alternative approaches involve cutting some trees while leaving others standing. In the **seed-tree** approach (**FIGURE 12.14b**), small numbers of mature and vigorous seed-producing trees are left standing so that they can reseed the logged area. In the **shelterwood** approach (also represented by Figure 12.14b), small numbers of mature trees are left in place to provide shelter for seedlings as they grow. These three methods all lead to largely even-aged stands of trees.

In contrast, **selection systems** (**FIGURE 12.14c**) allow uneven-aged stand management, because only some trees are cut at any one time. In single-tree selection, widely spaced trees are cut one at a time, whereas in group selection, small patches of trees are cut. The stand's overall rotation time may be the same as in an even-aged approach, because multiple harvests are made, but the stand remains mostly intact between harvests. Selection systems can maintain much of a forest's structural diversity, but they still have ecological impacts, because moving trucks and machinery over a network of roads and trails to access individual trees compacts the soil and disturbs the forest floor. Timber companies often resist selection methods because they are expensive, and selection methods are unpopular with loggers because they pose more safety risks than clear-cutting.

All timber-harvesting methods disturb soil, alter habitat, and affect plants and animals. All methods modify forest structure and composition. Most methods speed runoff, raise flooding risk, and increase soil erosion, thereby degrading water quality. When steep hillsides are clear-cut, landslides

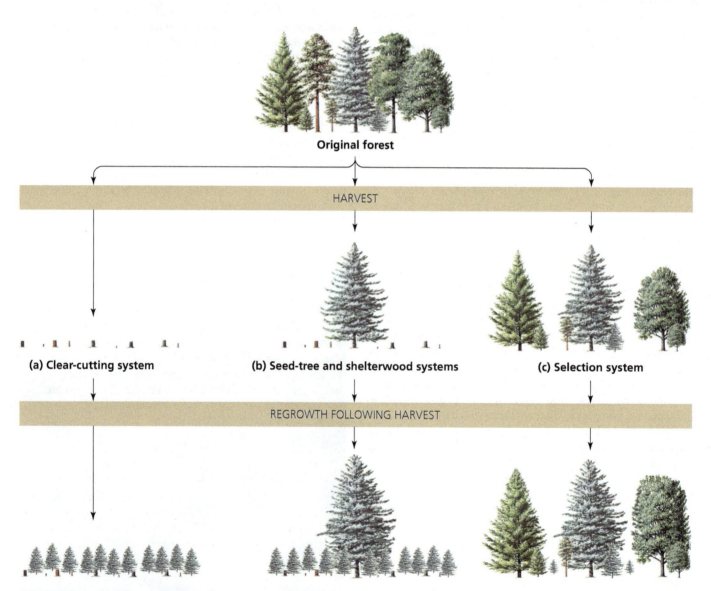

FIGURE 12.14 Foresters have devised several methods to harvest timber. In clear-cutting **(a)**, all trees are cut, extracting a great deal of timber inexpensively but leaving a vastly altered landscape. In seed-tree systems and shelterwood systems **(b)**, a few large trees are left in clear-cuts to reseed the area or provide shelter for seedlings. In selection systems **(c)**, a minority of trees is removed at any one time, while most are left standing.

can result. Finding ways to minimize these impacts is important, because timber harvesting is necessary to obtain the wood products that all of us use.

Forest management has evolved over time

For the past half-century, the U.S. Forest Service has nominally been guided by the policy of **multiple use,** meaning that the national forests were to be managed for recreation, wildlife habitat, mineral extraction, and various other uses. In practice, however, timber production was often the primary use. In recent decades, as people became more aware of the impacts of logging and as development spread across the landscape, citizens began to urge that public forests be managed for recreation, wildlife, and ecosystem integrity, as well as for timber.

In 1976 the U.S. Congress passed the **National Forest Management Act.** This act mandated that every national forest draw up plans for renewable resource management based on the concepts of multiple use and maximum sustainable yield and subject to public input under the National Environmental Policy Act (p. 171). Guidelines specified that these plans:

- Consider environmental factors as well as economic ones, such that profit alone does not determine harvesting decisions.

- Provide for diverse ecological communities and preserve regional diversity of tree species.

- Ensure research and monitoring of management practices.

- Allow increased harvests only if sustainable.

- Ensure that timber is extracted only where impacts on tree regeneration and soil, watershed, fish, wildlife, recreation, and aesthetic resources can be assessed and minimized.

Following passage of the National Forest Management Act, the U.S. Forest Service developed new programs to manage wildlife, non-game animals, and endangered species. Forest Service scientists and managers pushed for ecosystem-based management and launched ecological restoration programs to recover plant and animal communities that had been lost or degraded (**FIGURE 12.15**). Timber-harvesting methods were brought more in line with ecosystem-based management goals. A set of approaches dubbed **new forestry** called for timber cuts that mimicked natural disturbances. For instance, "sloppy clear-cuts" that leave a variety of trees standing were intended to mimic the changes a forest might experience if hit by a severe windstorm.

Another policy milestone that signaled a shift toward conservation occurred in 2001, when President Bill Clinton issued an executive order that became known as the **roadless rule.** The roadless rule put 23.7 million ha (58.5 million acres)—31% of national forest land and 2% of total U.S. land—off-limits to road construction or maintenance (and thus to logging). The roadless rule received strong popular support, including a record 4.2 million public comments from citizens to the federal government.

The administration of President George W. Bush marked a change in policy direction. In 2004, the Bush administration

FIGURE 12.15 Ecosystem-based management is practiced in this longleaf pine forest in the southeastern United States. Foresters and biologists restore and nurture mature longleaf pine trees while burning and removing brush from the understory. This habitat is home to unique species such as the endangered red-cockaded woodpecker (inset photo), whose nest and roost trees are identified by bands painted on the trunks.

freed forest managers from many requirements of the National Forest Management Act, granting them more flexibility in managing forests but loosening environmental protections and restricting public oversight. In 2005, the Bush administration repealed the roadless rule, inviting states to decide how national forests within their boundaries should be managed. Some states responded favorably, whereas others sued the administration, asking that the roadless rule be reinstated. Following a series of court rulings, the Obama administration reinstated most of the roadless policy but also negotiated with some states to allow them to develop their own plans.

Fire can hurt or help forests

Another area of policy debate involves how to handle wildfire. For more than a century, the Forest Service and other agencies suppressed fire whenever and wherever it broke out. Yet scientific research clearly shows that many species and ecological communities depend on fire. Some plants have seeds that germinate only in response to fire, and researchers studying tree rings have documented that North America's grasslands and pine woodlands burned frequently. (Burn marks in a tree's growth rings reveal past fires, giving scientists an accurate history of fire events extending back hundreds or even thousands of years.) Ecosystems dependent on fire are adversely affected when fire is suppressed: Grasslands are

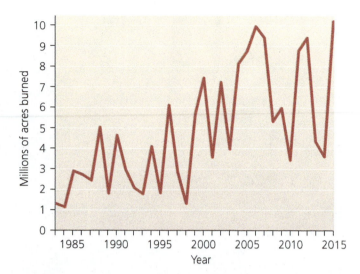

FIGURE 12.16 **Wildfires have become larger in the United States.** Fuel buildup from decades of fire suppression has contributed to this trend. *Data from National Interagency Fire Center.*

invaded by shrubs, and pine woodlands become cluttered with hardwood understory. Invasive plants move in, and animal diversity and abundance decline.

Over time, suppressing frequent low-intensity fires leads to occasional catastrophic fires that damage forests, destroy property, and threaten human lives. This is because fire suppression allows limbs, logs, sticks, and leaf litter to accumulate on the forest floor, producing kindling for a catastrophic fire. Severe fires have become more numerous in recent years (**FIGURE 12.16**), and fire control expenses now eat up more than half of the U.S. Forest Service's budget. At the same

time, increased residential development alongside forested land—in the **wildland-urban interface**—is placing more homes in fire-prone situations (**FIGURE 12.17**).

To reduce fuel loads, protect property, and improve the condition of forests, land management agencies now burn areas of forest intentionally with low-intensity fires under carefully controlled conditions (**FIGURE 12.18**). Such **prescribed fire** clears away fuel loads, nourishes the soil with ash, and encourages the vigorous growth of new vegetation. Because prescribed burning (also called controlled burning) is time intensive and sometimes misunderstood by politicians and the public, prescribed fires are conducted on only a small proportion of land (just more than 2 million acres per year). As a result, vast areas of American forests remain vulnerable to catastrophic fires.

In the wake of major fires in California in 2003, the U.S. Congress passed the Healthy Forests Restoration Act. Although this legislation encouraged some prescribed fire, it primarily promoted the physical

weighing the ISSUES

How to Handle Fire?

A century of fire suppression has left vast areas of North American forests vulnerable to catastrophic wildfires. Prescribed fire helps to alleviate this risk, yet we will never have adequate resources to conduct careful prescribed burning over all these lands. Can you suggest solutions to help protect people's homes in the wildland-urban interface while improving the ecological condition of forests? Should people who choose to live in homes in fire-prone areas pay a premium for insurance against fire damage? Should homeowners in fire-prone areas be fined if they don't adhere to fire-prevention maintenance guidelines? Should new home construction be allowed in areas known to be fire-prone?

FIGURE 12.17 **Habitual suppression of fire has led to catastrophic wildfires that damage forests and threaten homes.** To avoid these unnaturally severe fires, ecologists suggest we allow natural fires to burn when we can and conduct prescribed burns to reduce fuel loads and restore forest ecosystems.

FIGURE 12.18 **Prescribed fire helps to promote forest health and prevent larger damaging fires.** Forest Service workers are shown here conducting a carefully controlled, low-intensity burn in a pine forest.

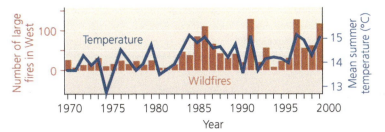

(a) Wildfires correlate with hotter summer temperatures

FIGURE 12.19 Forests face intensifying threats as temperatures rise. The number of large wildfires in the western United States **(a)** has risen along with the region's summer temperatures. Warm summers and mild winters favor bark beetles **(b)**, which are killing vast areas of trees throughout the West, especially in even-aged plantations. *Source: (a) Westerling, A.L., et al., 2006. Warming and earlier spring increase western U.S. forest wildfire activity. Science 313: 940–943, Fig 1A. Reprinted with permission from AAAS.*

(b) Mountain pine beetles kill more trees in a warmer climate

removal of small trees, underbrush, and dead trees by timber companies. The removal of dead trees, or snags, following a natural disturbance (such as fire, windstorm, insect damage, or disease) is called **salvage logging.** From a short-term economic standpoint, salvage logging may seem to make good sense. However, ecologically, snags have immense value; the insects that decay them provide food for wildlife, and many animals depend on holes in snags for nesting and roosting. Removing timber from recently burned land can also cause soil erosion, impede forest regeneration, and promote additional wildfire (see THE SCIENCE BEHIND THE STORY, pp. 316–317).

Climate change and pest outbreaks are altering forests

Global climate change (Chapter 18) is now worsening wildfire risk by bringing warmer weather to North America and drier weather to the American West (**FIGURE 12.19a**). Scientific climate models predict further warming and drying (p. 502).

Pest insects such as bark beetles, which feed within the bark of conifer trees, are adding to the wildfire risk. These beetles attract one another to weakened trees and attack en masse, eating tissue, laying eggs, and bringing with them a small army of fungi, bacteria, and other pathogens (**FIGURE 12.19b**). Bark beetle infestations can wipe out vast areas of trees with amazing speed. Since the 1990s, beetle outbreaks have devastated tens of millions of acres of forest in western North America, killing tens of *billions* of conifer trees and leaving them as fodder for fires.

Scientists studying beetles and their impacts say there are two primary reasons for today's unprecedented outbreaks. The first reason is that past forest management has resulted in even-aged forests across large regions, and many trees in these forests are now at a prime age for beetle infestation. Most at risk are plantation forests dominated by single species that the beetles prefer. The second reason is climate change. Milder winters allow beetles to overwinter farther north, and warmer summers speed up their feeding and reproduction. In Alaska, beetles have switched from a two-year life cycle to a one-year cycle. In parts of the Rocky Mountains, they now produce two broods per year instead of one. Meanwhile, droughts like those that have plagued the western and southern United States in recent years have stressed and weakened trees, making them vulnerable to attack.

As climate change interacts with pests, diseases, and management strategies, our forest systems are being altered in profound ways. Already many dense, moist forests devastated by beetles have been replaced by drier woodlands, shrublands, or grasslands. Further changes to our forests could create novel types of ecosystems not seen today.

Sustainable forestry is gaining ground

We can help address all of today's challenges to our forests by pursuing sustainable forestry practices. Such practices reduce impacts on land, water, and wildlife; conserve resources

FAQ

Aren't all forest fires destructive?

No. Fire is a natural process that helps to maintain the health of many forests and grasslands. When allowed to occur naturally, low-intensity fires generally burn moderate amounts of material, return nutrients to the soil, and promote lush growth of new vegetation. However, each time we suppress a fire, we allow more and more dead wood, dried grass, and leaf litter to accumulate, unburned. This material becomes kindling that can eventually feed a truly damaging fire that grows too big and too hot to control. This is why many land managers today conduct carefully controlled prescribed burns and also allow some natural fires to run their course. By doing so, they aim to prevent large catastrophic fires and help return our fire-dependent ecosystems to a healthier and safer condition.

Does Salvage Logging Help or Hurt Forests?

It's not often that a scientific paper throws an entire college into turmoil and lands a graduate student in a congressional hearing to face hostile questioning from federal lawmakers. But such is the political sensitivity of salvage logging.

When a fire burns a forest, should the killed trees be cut and sold for timber? Proponents of salvage logging say yes: We should not let economically valuable wood go to waste. But many scientists counter that the burned wood is often more valuable left in place—for erosion control, wildlife habitat (snags provide holes for cavity-dwelling animals and food for insects and birds), and organic material to enhance the soil and nurse the growth of future trees.

Logging a fire-damaged tree in a national forest

Many proponents of salvage logging argue that forests regenerate best after a fire if they are logged and replanted with seedlings. Moreover, they maintain, salvage logging reduces fire risk by removing woody debris that could fuel the next fire. When the Biscuit Fire consumed 200,000 ha (500,000 acres) in Oregon in 2002, foresters from the College of Forestry at Oregon State University (OSU) made these arguments in support of plans to log portions of the burned area.

Meanwhile, OSU forestry graduate student Daniel Donato and five other OSU researchers were setting up research plots in areas burned by the Biscuit Fire to test whether salvage logging really does reduce fire risk and help seedlings regenerate. They measured seedling growth and survival and the amount of woody debris in a number of study plots on burned land before (2004) and after (2005) salvage logging took place and on burned land that was not logged.

The researchers found that conifer seedlings sprouted naturally in the burned areas at densities exceeding what foresters aim for when they replant sites manually. This suggested that manual planting of seedlings may be unnecessary. In contrast, natural seedling densities in logged areas were only 29% as high (**FIGURE 1**). This indicated that salvage logging was hindering seedling survival, presumably because logging disturbs the soil and crushes many seedlings.

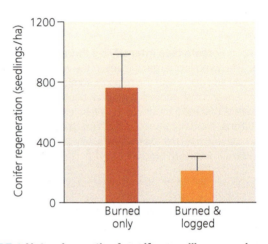

FIGURE 1 Natural growth of conifer seedlings was lower in areas that underwent salvage logging. *Data from Donato, D.C., et al., 2006. Post-wildfire logging hinders regeneration and increases fire risk. Science 311: 352, Fig 1A. Reprinted with permission from AAAS.*

Donato's team also found that salvage logging more than tripled the amount of woody debris on the ground relative to unlogged sites (**FIGURE 2**). The research team suggested that the best strategy after fire may be to leave the site alone and leave dead trees standing, so that seedlings regenerate safely.

These conclusions directly contradicted what some OSU forestry professors had argued following the 2002 fire. When they learned that the prestigious journal *Science* had accepted the Donato team's paper for publication, they took the unusual step of asking the journal's editors to reconsider their decision. Claiming that *Science*'s peer review process had failed, Professor John Sessions and others tried through back channels to stop publication of the paper—actions that were widely condemned as an attempt at censorship.

The paper's publication in 2006 unleashed a torrent of bizarre events. U.S. Congressmen Greg Walden of Oregon and Brian Baird of Washington felt the paper threatened legislation they had sponsored to accelerate salvage logging. Walden and Baird called the 29-year-old Donato and others before a hearing of the House of Representatives' Committee on Resources and grilled them before a packed crowd in Medford, Oregon (**FIGURE 3**).

The Bureau of Land Management (BLM) then suspended the team's research grant, in what many viewed as a response

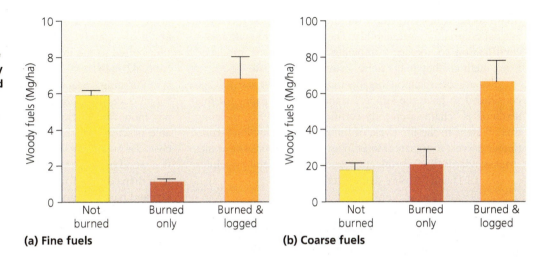

FIGURE 2 Burned sites that were salvage-logged (orange bar) contained more fine (a) and coarse (b) woody debris than unlogged burned sites or sites that did not burn. *Data from Donato, D.C., et al., 2006. Post-wildfire logging hinders regeneration and increases fire risk. Science 311: 352, Fig 1B. Reprinted with permission from AAAS.*

(a) Fine fuels

(b) Coarse fuels

to political pressure. This highly unusual action drew media attention, and the BLM reinstated the funding.

A heated debate roiled for months in the OSU College of Forestry. The college receives 12% of its funding from taxes on timber sales, leading many to suggest that the college is open to influence from industry. E-mail correspondence was subpoenaed and showed the college's dean, Hal Salwasser, collaborating with timber industry representatives to refute the paper. As publicity built, the college's reputation suffered, and a faculty committee on academic freedom criticized Salwasser for "significant failures of leadership." The dean admitted mistakes, survived a no-confidence vote of the faculty, and pledged to make reforms.

In the pages of *Science* and elsewhere, scientific criticisms of the study's methods were largely rebutted, yet many felt that the study's conclusions stretched beyond what its short-term data could support. Scientists on both sides of the debate agreed that long-term research was needed to fully assess the effects of salvage logging on forest regeneration and fire risk.

Two studies by different OSU forestry scientists soon provided the first such long-term data. Jeffrey Shatford and colleagues documented widespread natural regrowth of conifers across areas of Oregon and northern California that had burned 9–19 years earlier. And Jonathan Thompson and colleagues examined satellite data, aerial photography, and government records for regions within the Biscuit Fire area that had burned in a previous fire 15 years earlier. They found that of the regions burned in that 1987 fire, those that were salvage-logged burned more severely in 2002 than regions that were not logged. In a paper in the *Proceedings of the National Academy of Sciences*, Thompson's team concluded that salvage logging increases the risk of severe fires, even when debris is removed and seedlings are manually planted.

Then in 2015, Donato, by then at the Washington Department of Natural Resources, wrapped up a re-examination of the Biscuit Fire study sites with several colleagues, 10 years after the fire. They found that over this decade, the differences lessened between sites that were salvage-logged and those that were not. As young trees regrew after the fire, these trees began contributing woody material to the soil, and the sites became more similar. They also found that trees, especially Douglas fir, regrew best near the forested edges of areas denuded by fire, while areas in the center of burned regions grew fewer trees and more knobcone pine, which is a species that sets seed after fires. The team found that the sites were still actively changing, and that many more years would go by before the new forests matured and stabilized.

As more long-term studies on the impacts of salvage logging are conducted, we should become better able to manage our forests in an age of increasingly frequent wildfire.

FIGURE 3 Daniel Donato, at the time an OSU graduate student, was called to testify in a congressional hearing.

rather than depleting them; and enable people to make a living from timber harvesting over the long term.

Any company can claim that its timber-harvesting practices are sustainable, but how is the purchaser of wood products to know whether they really are? Organizations such as the Forest Stewardship Council (FSC) examine practices and rate them against criteria for sustainability. They grant **sustainable forest certification** to forests, companies, and products produced using methods they judge to be sustainable (**FIGURE 12.20**).

Several certification organizations exist, but the FSC is considered to have the strictest standards. The paper for this textbook is "chain-of-custody certified" by the FSC, meaning that all steps in the life cycle of the paper's production—from timber harvest to transport to pulping to production—have met strict standards. FSC certification rests on 10 general principles (**TABLE 12.1**) and 56 more-detailed criteria.

FIGURE 12.20 Logs from trees harvested using practices that meet FSC guidelines are marked with the FSC logo to certify that the wood was sustainably harvested.

TABLE 12.1 Ten Principles of Forest Stewardship Council (FSC) Certification

To receive FSC certification, forest product companies must:
- Comply with all laws and treaties.
- Show uncontested, clearly defined, long-term land rights.
- Recognize and respect indigenous peoples' rights.
- Maintain or enhance long-term social and economic well-being of forest workers and local communities, and respect workers' rights.
- Use and share benefits derived from the forest equitably.
- Reduce environmental impacts of logging and maintain the forest's ecological functions.
- Continuously update an appropriate management plan.
- Monitor and assess forest condition, management activities, and social and environmental impacts.
- Maintain forests of high conservation value.
- Promote restoration and conservation of natural forests.

Source: Forest Stewardship Council.

The number of FSC-certified forests, companies, and products is growing quickly. More than 6% of the world's forests managed for timber production are now FSC-certified. This totaled more than 187 million ha (462 million acres) in 81 nations as of 2016. More than 1350 operations are certified, and nearly 30,000 companies or projects are chain-of-custody certified. Some of the growth of certification is tied to the increasing construction of green buildings (pp. 342–344).

Pursuing sustainable forestry practices is often more costly for producers, but producers recoup these costs when consumers pay more for certified products. And in the long term, sustainable practices conserve the resource base, thus holding down costs for everyone. When we ask businesses if they carry certified wood or paper, we make them aware of our preferences and help drive demand for these products. For example, consumer demand has led Home Depot and other major retailers to carry sustainable wood, and these retailers' purchasing decisions are influencing timber-harvesting practices around the world. You can look for the logos of certifying organizations on forest products where they are sold. If certification standards are kept strong, then we as consumers can exercise choice in the marketplace and thereby help to promote sustainable forestry practices.

Parks and Protected Areas

As our world fills with more people consuming more resources, the sustainable management of forests and other ecosystems becomes ever more important. So does our need to preserve functional ecosystems by setting aside tracts of undisturbed land to remain forever undeveloped.

Preservation has been part of the American psyche ever since John Muir rallied support for saving scenic lands in the Sierras (p. 136). For ethical reasons as well as pragmatic ecological and economic ones, Americans and people worldwide have chosen to set aside areas of land in perpetuity to be protected from development. Today nearly 13% of the world's land area has been granted some degree of protection in various types of parks and reserves.

Why create parks and reserves?

People establish parks and preserves because these areas offer us:

- Inspiration from enormous or unusual scenic features.
- Hiking, fishing, hunting, kayaking, bird-watching, and other recreation.
- Revenue from ecotourism (pp. 67, 277).
- Health, peace of mind, exploration, wonder, and spiritual solace.
- Utilitarian benefits and ecosystem services (such as how forested watersheds provide cities clean drinking water and a buffer against floods).
- Refuges for biodiversity, helping to maintain populations, habitats, communities, and ecosystems.

Federal parks and reserves began in the United States

The striking scenery of the American West persuaded the U.S. government to create the world's first **national parks,** public lands protected from resource extraction and development but open to nature appreciation and recreation (**FIGURE 12.21**). Yellowstone National Park was established in 1872, followed by Sequoia, General Grant (now Kings Canyon), Yosemite, Mount Rainier, and Crater Lake National Parks. The Antiquities Act of 1906 gave the U.S. president authority to declare selected public lands as *national monuments*, which may later become national parks.

The National Park Service was created in 1916 to administer the growing system of parks and monuments, which today numbers 408 units totaling more than 34 million ha (84 million acres) and includes national historic sites, national recreation areas, national seashores, and other areas (see Figure 12.10). The parks receive more than 280 million reported recreation visits per year—almost one per U.S. resident.

Because America's national parks are open to everyone and showcase the nation's natural beauty in a democratic way, writer Wallace Stegner famously called them "the best idea we ever had." A visitor from the Upper Peninsula of Michigan can easily take his or her family for the weekend to Pictured Rocks National Lakeshore, where they can camp along sandstone cliffs on the shore of Lake Superior. They can head across the lake to remote Isle Royale National Park to hike and canoe through lands that are home to moose and wolves. Or they can cross Lake Michigan and enjoy climbing immense sand dunes at Sleeping Bear Dunes National Lakeshore.

Another type of federal protected area in the United States is the **national wildlife refuge** (**FIGURE 12.22**, p. 320). The system of national wildlife refuges, begun in 1903 by President Theodore Roosevelt, now totals more than 560 refuges comprising 39 million ha (96 million acres; see Figure 12.10), plus tens of millions of acres of ocean and islands owned or co-managed in marine refuges and monuments. There are wildlife refuges within an easy drive of nearly every major American city.

The U.S. Fish and Wildlife Service administers the national wildlife refuges, which serve as havens for wildlife but also in many cases encourage hunting, fishing, wildlife observation, photography, environmental education, and other public uses. Some wildlife advocates find it ironic that hunting is allowed at many refuges, but hunters have long been in the forefront of the conservation movement and have traditionally supplied the bulk of funding for land acquisition and habitat management. Many refuges are managed for waterfowl, but managers increasingly consider non-game species and work to maintain and restore habitats and ecosystems.

Wilderness areas are established on federal lands

In response to the public's desire for undeveloped areas of land, in 1964 the U.S. Congress passed the Wilderness Act, which allowed some areas of existing federal lands to be designated

(a) Crater Lake National Park, Oregon

(b) Arches National Park, Utah

(c) Great Smoky Mountains National Park, North Carolina and Tennessee

FIGURE 12.21 The awe-inspiring beauty of America's national parks draws millions of people for recreation and wildlife-watching.

FIGURE 12.22 National wildlife refuges offer safe havens for wildlife. For example, thousands of sandhill cranes spend the winter at Bosque del Apache NWR in New Mexico.

as **wilderness areas.** These areas are off-limits to development but are open to hiking, nature study, and other low-impact public recreation (**FIGURE 12.23**).

Congress declared that wilderness areas were needed "to assure that an increasing population, accompanied by expanding settlement and growing mechanization, does not occupy and modify all areas … leaving no lands designated for preservation and protection in their natural condition." Despite these words, some preexisting extractive land uses, such as grazing and mining, were allowed to continue within some wilderness areas as a political compromise so the act could be passed.

Wilderness areas are established within national forests, national parks, national wildlife refuges, and land managed by the Bureau of Land Management (BLM). They are overseen by the agencies that administer those areas. In Michigan's Upper

FIGURE 12.23 Wilderness areas are preserved and protected from development. Here visitors enjoy kayaking on Lake Superior at Pictured Rocks National Lakeshore in Michigan's Upper Peninsula.

Peninsula, for instance, several wilderness areas are designated in the Hiawatha and Ottawa National Forests, one in Pictured Rocks National Lakeshore, and one in Seney National Wildlife Refuge. Overall, the nation has more than 760 wilderness areas totaling 44 million ha (110 million acres). These cover 4.5% of U.S. land area (2.7% if Alaska is excluded).

Not everyone supports setting land aside

The restriction of activities in wilderness areas has generated some opposition to U.S. land protection policies, especially among citizens and policymakers in western states. When these states came into existence, the federal government retained jurisdiction over much of their acreage. Idaho, Oregon, and Utah control less than half the land within their borders, and in Nevada 80% of the land is federally managed. Some western state governments have sought to obtain land from the federal government and to facilitate resource extraction and development on it. They have been supported by the industries that extract timber, minerals, and fossil fuels, as well as by farmers, ranchers, trappers, and mineral prospectors at the grassroots level. Both wealthy advocates of privatization and people who make their living off the land have supported efforts to secure local private control of public lands, limit government regulation, encourage the extraction of resources, and promote motorized recreation on public lands. These advocates have driven debates over national park policy, such as whether recreational activities that disturb wildlife, such as snowmobiles and jet-skis, should be allowed.

Similar debates take place wherever parks are established. In East Africa, many Maasai people living near Serengeti National Park resent the park because it displaced their families from their traditional land. Many people living near the park want a highway to be built through it to provide them better mobility and access to trade (pp. 269–270).

Parks and protected areas regularly bring substantial economic benefits to people who live nearby, through ecotourism. However, individuals who do not have jobs related to parks and tourism may feel that a park restricts their economic opportunities. For this reason, proponents of protected areas increasingly try to ensure that the economic benefits of preserving natural land are spread equitably among people in nearby communities.

Groups of indigenous people frequently oppose government actions to set aside land. As with the Maasai in Africa, Native Americans were forced from their land while the U.S. government imposed rules of property ownership that were foreign to most Native American cultures. Some sites used today for recreation in the national parks are sites long sacred to Native cultures. The huge basalt rock formation at Devil's Tower National Monument in Wyoming is sacred to many Plains tribes, yet it is also popular with technical rock climbers. Native Americans view rock climbing as a desecration of the site. As a compromise, the National Park Service asks climbers not to climb the tower in June, when Native Americans travel there for religious ceremonies—an accommodation the U.S. Supreme Court upheld after a lengthy court battle.

However, protected areas sometimes serve the interests of indigenous people. In Brazil and other Latin American nations, some rainforest parks and reserves encompass regions occupied by indigenous tribes. The tribes are thereby granted some protection from conflict with miners, farmers, and settlers, and they can continue their traditional way of life.

Many agencies and groups protect land

Efforts to set aside land at the national level are paralleled at regional and local levels. Each U.S. state has agencies that manage resources on public lands, as do many counties and municipalities. When Mackinac Island, the nation's second national park, was transferred to the state of Michigan, it became the first officially designated state park in the nation. Another example is Adirondack State Park in New York. In the 19th century, the state of New York established this park to protect land in a mountainous area where streams converge to form the Hudson River, which thereafter flows south past Albany to New York City. Seeing the need for river water to power industries, keep canals filled, and provide drinking water, the state made a far-sighted decision that has paid dividends through the years. Today these parks are among nearly 7000 state parks across the United States, along with regional parks, county parks, and others.

Private nonprofit groups also preserve land. **Land trusts** are local or regional organizations that purchase land to preserve in its natural condition. The Nature Conservancy, the world's largest land trust, uses science to select areas and ecosystems in greatest need of protection. Smaller land trusts

are diverse in their missions and methods. Nearly 1700 local and state land trusts in the United States together own 870,000 ha (2.1 million acres) and have helped preserve an additional 5.6 million ha (13.9 million acres), including scenic areas such as Big Sur on the California coast, Jackson Hole in Wyoming, and Maine's Mount Desert Island. Moreover, thousands of local volunteer groups (often named "Friends of" an area) have organized from the grassroots to help care for protected lands.

Parks and reserves are increasing internationally

Many nations have established protected areas and are benefiting from ecotourism as a result—from Kenya and Tanzania (Chapter 11) to Costa Rica (Chapter 6) to Belize to Ecuador to India to Australia. Worldwide area in protected parks and reserves has increased nearly sevenfold since 1970, and today the world's 158,000 protected areas cover about 12.7% of the planet's land area. However, parks in developing nations do not always receive funding adequate to manage resources, provide for recreation, and protect wildlife from poaching and trees from logging. As a result, many of the world's protected areas are merely "paper parks"—protected on paper but not in reality.

Some types of protected areas are designated or partly managed by the United Nations. **Biosphere reserves** are tracts of land with exceptional biodiversity that couple preservation with sustainable development (p. 153) to benefit local people (**FIGURE 12.24**). Each biosphere reserve consists of (1) a core area that preserves biodiversity; (2) a buffer zone that allows local activities and limited development; and (3) an outer

Biodiversity protection

Sustainable agriculture, small settlements

Core area

Buffer zone

Transition zone

Limited development; research; education; ecotourism

FIGURE 12.24 Biosphere reserves couple preservation with sustainable development. Each biosphere reserve includes three zones. In the photo shown here from a transition zone, women at the Maya Biosphere Reserve in Guatemala process and sell Maya nuts harvested from rainforest trees. FSC certification in the transition zone here helped prevent illegal logging.

transition zone where agriculture, human settlement, and other land uses are pursued sustainably.

The Maya Biosphere Reserve in Guatemala is an example. Rainforest here is protected in core areas, and in the transition zone timber harvesting takes place in concessions, some of which are FSC-certified. A study by the nonprofit Rainforest Alliance found that FSC certification here gave people economic incentives to conserve the forest. The study found that rates of deforestation and wildfire were much lower in the FSC-certified areas where sustainable logging occurred than in the core area that was supposed to be fully protected.

World heritage sites are another type of international protected area. Nearly 1000 sites in more than 150 countries are listed for their natural or cultural value. One such site is a mountain gorilla reserve shared by three African countries. This reserve, which integrates national parklands of Rwanda, Uganda, and the Democratic Republic of Congo, is also an example of a transboundary park, a protected area overlapping national borders. A North American example of a transboundary park is Waterton–Glacier National Parks on the Canadian–U.S. border. Some transboundary reserves function as "peace parks," helping to ease tensions by acting as buffers between nations with historical boundary

disputes. This is the case with Peru and Ecuador as well as Costa Rica and Panama, and many people hope that peace parks might also help resolve conflicts between Israel and its neighbors.

Beyond all these efforts on land, the importance of conserving the oceans' natural resources is leading us to establish protected areas and reserves in marine waters (p. 440). Today 1.6% of the world's ocean area and 7.2% of coastal waters fall within designated protected areas.

Economic incentives can help preserve land

Innovative economic strategies can facilitate international efforts to protect natural lands. One strategy is the conservation concession discussed earlier (p. 308). Another is the **debt-for-nature swap,** in which a conservation organization raises money and offers to pay off a portion of a developing nation's international debt in exchange for a promise by the nation to set aside reserves, fund environmental education, and better manage protected areas. The U.S. government committed itself to a program of debt-for-nature swaps through its 1998 Tropical Forest Conservation Act. As of 2013, deals

(a) Fragmentation from clear-cuts in British Columbia

(c) Wood thrush

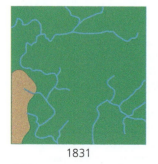

1831

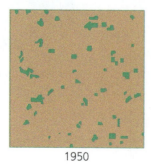

1882

1950

(b) Fragmentation of wooded area (green) in Cadiz Township, Wisconsin

FIGURE 12.25 Forest fragmentation has ecological consequences. Fragmentation results from **(a)** clear-cutting and **(b)** agriculture and residential development. Fragmentation affects forest-dwelling species such as the **(c)** wood thrush, whose nests are parasitized by cowbirds that thrive in surrounding open country. *Source: (b) Curtis, J.T., 1956. The modification of mid-latitude grasslands and forests by man. In Thomas, W.L., Jr. (Ed.), Man's role in changing the face of the earth. ©1956. Used by permission of the publisher, University of Chicago Press.*

had been struck with 14 developing nations through this act, enabling more than $326 million in their funds intended for debt payments to go to conservation efforts instead. In the largest such deal yet, the U.S. government forgave Indonesia $30 million of debt while two conservation groups paid Indonesia $2 million. In return, Indonesia promised to preserve forested areas that are home to the Sumatran tiger and other species.

Habitat fragmentation makes preserves more vital

Protecting large areas of land has taken on new urgency now that scientists understand the risks posed by habitat fragmentation (p. 283). Often, it is not the outright destruction of forests and other habitats that threatens species and ecosystems but rather their fragmentation (see **THE SCIENCE BEHIND THE STORY**, pp. 324–325). Expanding agriculture, spreading cities, highways, logging, and other impacts routinely divide large contiguous expanses of habitat into small, disconnected ones (see Figure 11.12, p. 284). Forests are being fragmented everywhere these days, as a result of logging (**FIGURE 12.25a**) and other factors, including agriculture and residential development (**FIGURE 12.25b**). Even in areas where forest cover is increasing (such as the northeastern United States), regrowing forests are becoming fragmented into ever-smaller parcels.

When forests are fragmented, many species suffer. Bears, mountain lions, and other animals that need large areas of habitat may disappear. For other species, the problem may lie with **edge effects,** impacts that result because the conditions along a fragment's edge differ from conditions in the interior. Birds that thrive in the interior of forests may fail to reproduce near the edge of a fragment (**FIGURE 12.25c**). Their nests often are attacked by predators and parasites that favor open habitats or travel along habitat edges. Because of edge effects, avian ecologists judge forest fragmentation to be a main reason why populations of many North American songbirds are declining. A 2015 research study in the journal *Science* concluded that across the world, 70% of forested area now lies within 1 km (0.62 mi) of an edge. As a result, the vast majority of the world's forested area is now being directly influenced by human activities, modified microclimates, and the incursion of non-forest species.

Insights from islands warn us of habitat fragmentation

In assessing the impacts of habitat fragmentation on populations, ecologists and conservation biologists have leaned on concepts from **island biogeography theory.** Introduced by ecologist Robert MacArthur and biologist E.O. Wilson (p. 277) in 1963, this theory explains how species come to be distributed among oceanic islands. Since MacArthur and Wilson's work, researchers have applied it widely on land to "habitat islands"—patches of one habitat type isolated within "seas" of others.

Island biogeography theory explains how the number of species on an island results from a balance between the number added by immigration and the number lost through extirpation. It predicts an island's species richness based on the island's size and its distance from the mainland:

- The farther an island lies from a continent, the fewer species tend to find and colonize it. Thus, remote islands host few species because of low immigration rates (**FIGURE 12.26a**).
- Larger islands have higher immigration rates because they present fatter targets for dispersing organisms to encounter (**FIGURE 12.26b**).
- Larger islands have lower extinction rates because more space allows for larger populations, which are less likely to drop to zero by chance (**FIGURE 12.26c**).

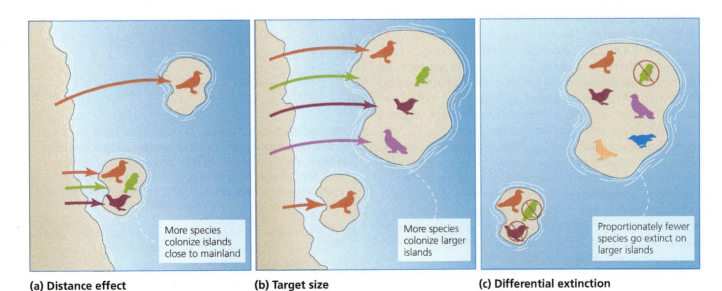

(a) Distance effect

More species colonize islands close to mainland

(b) Target size

More species colonize larger islands

(c) Differential extinction

Proportionately fewer species go extinct on larger islands

FIGURE 12.26 Islands that are larger or closer to a mainland tend to support more species.

What Happens When a Forest Is Fragmented?

Dr. Thomas Lovejoy in the Amazon

What happens to animals, plants, and ecosystems when we fragment a forest? A massive experiment smack in the middle of the Amazon rainforest is helping scientists learn the answers.

Stretching across 1000 km² (386 mi²), the Biological Dynamics of Forest Fragments Project (BDFFP) is the world's largest and longest-running research study on forest fragmentation. For more than 30 years, hundreds of researchers have muddied their boots in the rainforest here, publishing more than 700 scientific research papers, graduate theses, and books.

The story begins back in the 1970s as conservation biologists debated the SLOSS question (p. 326) and how to apply island biogeography theory to forested landscapes being fragmented by development. Biologist Thomas Lovejoy decided some good, hard data were needed. He conceived a huge experimental project to test ideas about forest fragmentation and established it in the heart of the biggest primary rainforest on the planet—South America's Amazon rainforest.

Farmers, ranchers, loggers, and miners were streaming into the Amazon then, and deforestation was rife. If scientists could learn how large fragments had to be to retain their species, it would help them work with policymakers to preserve forests in the face of development pressures.

Lovejoy's team of Brazilians and Americans worked out a deal with Brazil's government: Ranchers could clear some forest within the study area if they left square plots of forest standing as fragments within those clearings. By this process, 11 fragments of three sizes (1 ha [2.5 acres], 10 ha [25 acres], and 100 ha [250 acres]) were left standing, isolated as "islands" of forest, surrounded by "seas" of cattle pasture (**FIGURE 1**). Each fragment was fenced to keep cattle out. Then, 12 study plots (of 1, 10, 100, and 1000 ha) were established within the large expanses of continuous forest still surrounding the pastures; these would serve as control plots against which the fragments could be compared.

Besides comparing treatments (fragments) and controls (continuous forest), the project also surveyed populations in fragments before and after they were isolated. These data on trees, birds, mammals, amphibians, and invertebrates showed declines in the diversity of most groups (**FIGURE 2**).

As researchers studied the plots over the years, they found that small fragments lost more species, and lost them faster,

FIGURE 1 Experimental forest fragments of 1, 10, and 100 ha were created in the BDFFP.

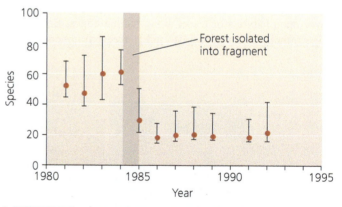

FIGURE 2 Species richness of understory birds declined in this 1-ha forest plot after it was isolated as a fragment in 1984. Error bars show statistical uncertainty around mean estimates. *Data from Ferraz, G., et al., 2003. Rates of species loss from Amazonian forest fragments. Proc. Natl. Acad. Sci. U S A 100: 14069–14073. © 2003 National Academy of Sciences. By permission.*

DATA Q On average, about how many bird species were present in this forest plot before fragmentation? How many were present after fragmentation?

Go to **Interpreting Graphs & Data** on MasteringEnvironmentalScience®

than large fragments—just as island biogeography theory predicts. To slow down species loss by 10 times, researchers found that a fragment needs to be 1000 times bigger. Even 100-ha fragments were not large enough for some animals, and lost half their species in less than 15 years. Monkeys died out because they need large ranges. So did colonies of army ants, and the birds that follow them to eat insects scared up by the ants as they swarm across the forest floor.

Fragments distant from continuous forest lost more species, but data revealed that even very small openings can stop organisms adapted to deep interior forest from dispersing to recolonize fragments. Many understory birds would not traverse cleared areas of only 30–80 m (100–260 ft). Distances of just 15–100 m (50–330 ft) were insurmountable for some bees, beetles, and tree-dwelling mammals.

Soon, a complication ensued: Ranchers abandoned many of the pastures because the soil was unproductive, and these areas began filling in with secondary forest. As this young forest grew, it made the fragments less like islands. However, this led to new insights. Researchers learned that secondary forest can act as a corridor for some species, allowing them to disperse from mature forest and recolonize fragments where they had disappeared. By documenting which species did this, scientists learned which may be more resilient to fragmentation.

The secondary forest habitat also introduced new species—generalists adapted to disturbed areas. Frogs, leaf-cutter ants, and small mammals and birds that thrive in second-growth areas soon became common in adjacent fragments. Open-country butterfly species moved in, displacing interior-forest butterflies.

The invasion of open-country species illustrated one type of edge effect. There were more: Edges receive more sunlight, heat, and wind than interior forest, which can kill trees adapted to the dark, moist interior. Tree death results in the flux of carbon dioxide to the atmosphere, which accelerates climate change. Sunlight also promotes growth of vines and shrubs that create a thick, tangled understory along edges. As BDFFP researchers documented these impacts, they found that many edge effects extended far into the forest (**FIGURE 3**). Small fragments essentially became "all edge."

The results on edge effects are relevant for all of Amazonia, because forest clearance and road construction create an immense amount of edge. A satellite image study in the 1990s estimated that for every 2 acres of land deforested, 3 acres were brought within 1 km of a road or pasture edge.

BDFFP scientists emphasize that impacts across the Amazon will be more severe than the impacts revealed by their experiments. This is because most real-life fragments (1) are not protected from hunting, logging, mining, and fires; (2) do not have secondary forest to provide connectivity; (3) are not near large tracts of continuous forest that provide recolonizing species and maintain humidity and rainfall; and (4) are not square in shape, and thus feature more edge. Scientists say the years of data argue for preserving numerous tracts of Amazonian forest that are as large as possible.

The BDFFP has inspired other large-scale, long-term experiments on forest fragmentation elsewhere in the world; such studies are now running in Kansas, South Carolina, Australia, Borneo, Canada, France, and the United Kingdom. These projects are providing data showing how smaller area, greater isolation, and more edge are affecting species, communities, and ecosystems. Thus far, they indicate that fragmentation decreases biodiversity by 13–75% while altering nutrient cycles and other ecological processes—and that the effects magnify with time. Insights from these projects are helping scientists and policymakers strategize how best to conserve biodiversity amid continued pressures on forests.

Ironically, today the BDFFP study site is itself threatened by forest fragmentation. A Brazilian government agency has settled colonists just outside the site and has proposed settling 180 families inside it. Development is proceeding up a new highway from Manaus, a city of 1.7 million people, to Venezuela, and wherever roads are built and people settle, logging, hunting, mining, and burning follow. BDFFP researchers can now hear chainsaws and shotgun blasts from their study plots. "It would be tragic," says leading BDFFP scientist William Laurance, "to see a site that's given us so much information be lost so easily."

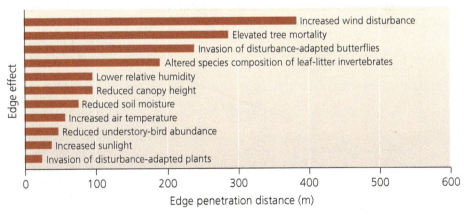

FIGURE 3 Edge effects can extend far into the interior of forest fragments. *Data from studies summarized in Laurance, W.F., et al., 2002. Ecosystem decay of Amazonian forest fragments: A 22-year investigation.* Conservation Biology 16: 605–618, Fig. 3. Adapted by permission of John Wiley & Sons, Inc.

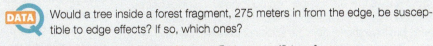

Would a tree inside a forest fragment, 275 meters in from the edge, be susceptible to edge effects? If so, which ones?

Go to **Interpreting Graphs & Data** on Mastering EnvironmentalScience®

Together, the latter two trends give large islands more species than small islands. Large islands also contain more species because they tend to possess more habitats than smaller islands. Very roughly, the number of species on an island is expected to double as island size increases 10-fold. This phenomenon is illustrated by **species-area curves** (**FIGURE 12.27**).

These patterns hold up for terrestrial habitat islands as well, such as forests fragmented by logging and road building. Small "islands" of forest lose diversity fastest, starting with large species that were few in number to begin with. One of the first researchers to show this experimentally was University of Michigan graduate student William Newmark, who in 1983 examined historical records of mammal sightings in North American national parks. The parks, increasingly surrounded by development, are islands of natural habitat isolated by farms, ranches, roads, and cities.

Newmark found that many parks were missing a few species they had held previously. The red fox and river otter had vanished from Sequoia and Kings Canyon National Parks, for example, and the white-tailed jackrabbit and spotted skunk no longer lived in Bryce Canyon National Park. In all, 42 species had disappeared. As island biogeography theory predicted, smaller parks lost more species than larger parks. Species were disappearing because the parks were too small to sustain their populations, Newmark concluded, and because the parks had become too isolated to be recolonized by new arrivals.

Reserve design has consequences for biodiversity

Because habitat fragmentation is a central issue in biodiversity conservation, and because there are limits on how much land can feasibly be set aside, conservation biologists have argued heatedly about whether it is better to make reserves large in size and few in number or many in number but small in size. This so-called **SLOSS** debate (for "**s**ingle **l**arge **o**r **s**everal **s**mall") is complex, but it seems clear that large species that

FIGURE 12.28 We can connect two areas of forest with a corridor. Shown here is a wildlife overpass in Canada's Banff National Park that allows animals to cross safely over the Trans-Canada Highway.

roam great distances, such as wildebeest and zebras (Chapter 11), benefit more from the "single large" approach to reserve design. In contrast, creatures such as insects that live as larvae in small areas may do just fine in a number of small isolated reserves, if they can disperse as adults by flying from one to another. The SLOSS debate was one motivation for establishing the Amazonian forest fragmentation project described in our **Science behind the Story** (pp. 324–325).

A related issue is how effectively **corridors** of protected land allow animals to travel between islands of habitat (**FIGURE 12.28**, and also see Figure 5.11, p. 114). In theory, connections between fragments provide animals access to more habitat and encourage gene flow to maintain populations in the long term. Many land managers now try to join new reserves to existing reserves for these reasons.

Climate change threatens protected areas

Today global climate change (Chapter 18) threatens our investments in protected areas. As temperatures become warmer, species ranges shift toward cooler climes: toward the poles and upward in elevation (p. 499). In a landscape of fragmented habitat, some organisms may be unable to move from one fragment to another. Species we had hoped to protect in parks may, in a warming world, become trapped in them. High-elevation species face the most risk from climate change, because they have nowhere to go once a mountaintop becomes too warm or dry. For this reason, corridors to allow movement from place to place become still more important. In response to these challenges, conservation biologists are now looking beyond parks and protected areas as they explore strategies for saving biodiversity.

FIGURE 12.27 Larger islands hold more species. This species-area curve shows that the number of amphibian and reptile species on Caribbean islands increases with island area. The increase is not linear, but logarithmic; note the scales of the axes. *Data from MacArthur, R.H, and E.O. Wilson. The theory of island biogeography.* © 1967 Princeton University Press, 1995 renewed PUP. Reprinted by permission of Princeton University Press.

closing
THE LOOP

Forests are ecologically vital and economically valuable, yet we continue to lose them around the world, as natural forests are cut at a faster rate than forests are regrowing. We all use forest products, and this textbook is no exception, because the paper in any book requires trees. But there are different ways to harvest trees, and different ways to manage forests. The paper in this book comes from trees in forests that are sustainably managed and that are certified as such by the Forest Stewardship Council. Sustainable forest certification provides harvesters and the companies that use and produce forest products economic incentives for promoting responsible conservation practices on forested lands.

In the United States and other areas of the world, forest management reflects trends in land and resource management in general. Early emphasis on resource extraction evolved into policies of sustainable yield and multiple use as land and resource availability declined and as the public became more aware of environmental degradation. Public forests today are managed not only for timber production but also for recreation, wildlife habitat, and ecosystem integrity.

Meanwhile, public support for the preservation of natural lands has led to the establishment of parks and protected areas worldwide. As development spreads across the landscape, fragmenting habitats and subdividing populations, scientists trying to conserve species, communities, and ecosystems are thinking and working at the landscape level.

REVIEWING Objectives

You should now be able to:

Summarize the ecological and economic contributions of forests

Forests are ecologically complex, support a wealth of biodiversity, store carbon, and contribute ecosystem services. Forests also provide us timber and other economically important products and resources. (pp. 302–305)

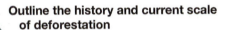

Outline the history and current scale of deforestation

We have lost forests as a result of timber harvesting and clearance for agriculture. The United States deforested much of its land as settlement, farming, and industrialization proceeded. Today deforestation is occurring most rapidly in developing nations. (pp. 305–308)

Assess dynamics of timber management

Resource managers have long managed for maximum sustainable yield and have begun to implement ecosystem-based management and adaptive management. The U.S. national forests were established to conserve timber and allow its sustainable extraction, although most U.S. timber today comes from private lands. Plantation forestry, featuring single-species, even-aged stands, is widespread and growing. Harvesting methods include clear-cutting and other even-aged techniques, as well as selection strategies that maintain uneven-aged stands that more closely resemble natural forest. (pp. 308–313)

Discuss forest management in relation to fire, pests, and climate change, and evaluate sustainable forestry certification

Foresters are now managing for recreation, wildlife habitat, and ecosystem integrity. However, increasingly severe fires are posing challenges. Fire suppression encourages eventual catastrophic fires. One solution is to reduce fuel loads by conducting prescribed burns. Today climate change and outbreaks of bark beetles are also affecting forests. Certification of sustainable forest products allows consumer choice in the marketplace to influence forestry practices. (pp. 313–318)

Name federal land management agencies and the lands they manage

The U.S. Forest Service, National Park Service, Fish and Wildlife Service, and Bureau of Land Management manage U.S. national forests, national parks, national wildlife refuges, and BLM land, respectively. (pp. 310, 319–320)

Discuss types of parks and protected areas and evaluate issues involved in their design

Public demand for preservation and recreation led to the creation of national parks, reserves, and wilderness areas. States, municipalities, and private land trusts all manage protected areas at the state, regional, and local levels. At the international level, biosphere reserves are one of several types of protected lands. Because habitat fragmentation affects wildlife, conservation biologists are using island biogeography theory to learn how best to design systems of parks and reserves. However, climate change is now posing new threats to protected areas. (pp. 318–326)

TESTING Your Comprehension

1. Name at least two reasons why natural primary forests contain more biodiversity than single-species forestry plantations.

2. Describe three ecosystem services that forests provide.

3. Name several major causes of deforestation. Where is deforestation most severe today?

4. Compare and contrast maximum sustainable yield, ecosystem-based management, and adaptive management. How may pursuing maximum sustainable yield sometimes affect populations and communities?

5. Compare and contrast the major methods of timber harvesting. Name an advantage and a disadvantage of each method.

6. Describe several ecological impacts of logging. How has the U.S. Forest Service responded to public concern over these impacts?

7. Are forest fires a bad thing? Explain your answer.

8. Name at least three reasons that people have created parks and reserves. How do national parks differ from national wildlife refuges? What is a wilderness area?

9. What percentage of Earth's land is protected? Describe one type of protected area that has been established outside North America.

10. Give two examples of how forest fragmentation affects animals. How does island biogeography theory help us design reserves?

SEEKING Solutions

1. People in industrialized nations are fond of warning people in developing nations to stop destroying rainforest. People of developing nations often respond that this is hypocritical, because the industrialized nations became wealthy by deforesting their land and exploiting its resources in the past. What would you say to the president of a developing nation, such as Indonesia or Brazil, in which a great deal of forest is being cleared?

2. What might you tell an opponent of parks and preserves to help him or her understand why a wilderness hiker wants scenic land in Utah federally protected? What might you tell a wilderness hiker to help him or her understand why the park opponent disapproves of the protection? How might you help them find common ground?

3. Consider the impacts that climate change may have on species' ranges. If you were trying to preserve an endangered mammal that occurs in a small area and you had generous funding to acquire land to help restore its population, how would you design a protected area for it? Would you use corridors? Would you include a diversity of elevations? Would you design few large reserves or many small ones? Explain your answers.

4. **CASE STUDY CONNECTION** You are the publisher of a magazine with a circulation of 1 million readers. Recently you have begun receiving queries from readers about whether you use certified sustainable paper. As you decide whether to switch to FSC-certified paper, what information would you want to learn (for instance, in terms of your readership, finances, and effects on forests)? Examine Table 12.1 and select three items from the table that you might want to feature in your magazine to impress on readers why such a switch

makes a positive difference. Why do you think each of these three items might resonate with your readers?

5. **THINK IT THROUGH** You have just become the supervisor of the Hiawatha National Forest in the Upper Peninsula of Michigan. Timber companies are requesting to cut as many trees as you will let them, and environmentalists want no logging at all. Ten percent of your forest is old-growth primary forest, and the remaining 90% is secondary forest. Your forest managers are split among preferring maximum sustainable yield, ecosystem-based management, and adaptive management. What management approach(es) will you take? Will you allow logging of all old-growth trees, some, or none? Will you allow logging of secondary forest? If so, what harvesting strategies will you encourage? What would you ask your scientists before deciding on policies on fire management and salvage logging?

6. **THINK IT THROUGH** You run a nonprofit environmental advocacy organization and are trying to save a tract of tropical forest in a poor developing nation. You have worked in this region for years and care for the local people, who want to save the forest and its animals but also need to make a living and use the forest's resources. The nation's government plans to sell a concession to a multinational timber corporation to log the entire forest unless your group can work out some other solution. Describe what solution(s) you would try to arrange. Consider the range of options discussed in this chapter, including government-protected areas, private protected areas, biosphere reserves, forest management techniques, carbon offsets, FSC-certified sustainable forestry, and more. Explain reasons for your choice(s).

CALCULATING Ecological Footprints

We all rely on forest resources. The average North American consumes 225 kg (500 lb) of paper and paperboard each year. Using the estimates of paper and paperboard consumption for each region of the world, calculate the per capita consumption for each region using the population data in the table. Note: 1 metric ton = 2205 pounds.

	POPULATION (MILLIONS)	TOTAL PAPER CONSUMED (MILLIONS OF METRIC TONS)	PER CAPITA PAPER CONSUMED (POUNDS)
Africa	1136	8	16
Asia	4351	190	
Europe	741	92	
Latin America	618	27	
North America	353	76	
Oceania	39	4	
World	7238	400	122

Data are for 2014, from Population Reference Bureau and UN Food and Agriculture Organization (FAO).

1. How much paper would be consumed if everyone in the world used as much paper as the average North American?

2. How much paper would North Americans save each year if they consumed paper at the rate of Europeans?

3. North Americans have been reducing their per-person consumption of paper and paperboard by nearly 5% annually in recent years as a result of a shift to online activity, recycling, and reducing packaging of some products. Name three specific things you personally could do to reduce your own consumption of paper products.

4. Describe three ways in which consuming FSC-certified paper rather than conventional paper can reduce the environmental impacts of paper consumption.

MasteringEnvironmentalScience®

Students Go to **MasteringEnvironmentalScience** for assignments, the etext, and the Study Area with practice tests, videos, current events, and activities.

Instructors Go to **MasteringEnvironmentalScience** for automatically graded activities, current events, videos, and reading questions that you can assign to your students, plus Instructor Resources.

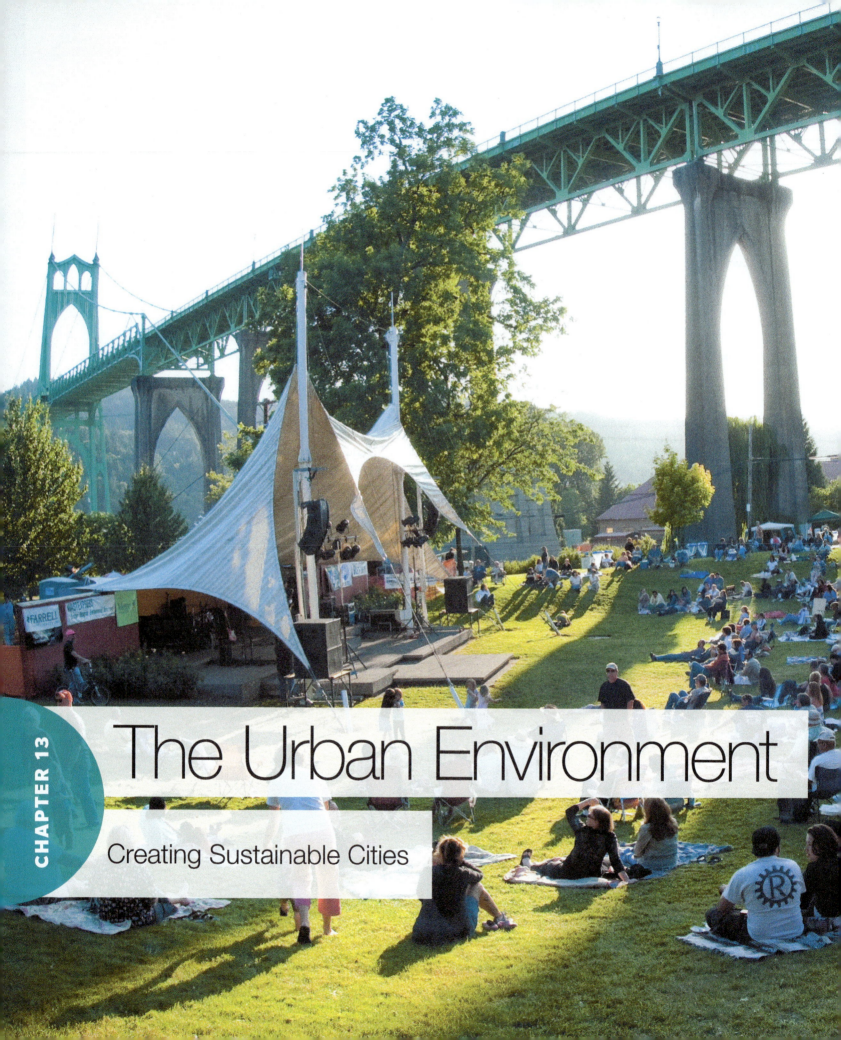

The Urban Environment

Creating Sustainable Cities

Managing Growth in Portland, Oregon

Pacific Ocean

Portland

OREGON

> " **Sagebrush subdivisions, coastal condomania, and the ravenous rampage of suburbia in the Willamette Valley all threaten to mock Oregon's status as the environmental model for the nation.**
> Oregon Governor Tom McCall, 1973
>
> **We have planning boards. We have zoning regulations. We have urban growth boundaries and 'smart growth' and sprawl conferences. And we still have sprawl.**
> Environmental scientist Donella Meadows, 1999 "

With fighting words, Oregon governor Tom McCall challenged his state's legislature in 1973 to take action against runaway sprawling development, which many Oregonians feared would ruin the communities and landscapes they loved. McCall was echoing the growing concerns of state residents that farms, forests, and open space were being gobbled up and paved over.

Foreseeing a future of subdivisions, strip malls, and traffic jams engulfing the pastoral Willamette Valley, Oregon acted. The state legislature passed Senate Bill 100, a sweeping land use law that would become the focus of acclaim, criticism, and careful study for years afterward by other states and communities trying to manage their own urban and suburban growth.

Oregon's law required every city and county to draw up a comprehensive land use plan in line with statewide guidelines that had gained popular support from the state's electorate. As part of each land use plan, each metropolitan area had to establish an urban growth boundary (UGB), a line on a map separating areas desired to be urban from areas desired to remain rural. Development for housing, commerce, and industry would be encouraged within these urban growth boundaries but severely restricted beyond them. The intent was to revitalize city centers; prevent suburban sprawl; and protect farmland, forests, and open landscapes around the edges of urbanized areas.

Residents of the area around Portland, the state's largest city, established a new regional planning entity to apportion land in their region. The Metropolitan Service District, or Metro, represents 25 municipalities and 3 counties. Metro adopted the Portland-area urban growth boundary in 1979 and has tried since then to focus growth in existing urban centers and to build communities where people can walk, bike, or take mass transit between home, work, and shopping. These policies have largely worked as intended. Portland's downtown and older neighborhoods have thrived, regional urban centers are becoming denser and more community oriented, mass transit has expanded, and development has been limited on land beyond the UGB. By the year 2000, Portland was attracting international attention for its "livability."

To many Portlanders today, the UGB remains the key to maintaining quality of life in city and countryside alike. In the view of its critics, however, the "Great Wall of Portland" is an elitist and intrusive government regulatory tool.

Upon completing this chapter, you will be able to:

- Describe the scale of urbanization
- Define sprawl and discuss its causes and effects
- Outline city and regional planning and land use strategies
- Evaluate transportation options, urban parks, and green buildings
- Analyze environmental impacts and advantages of urban centers
- Discuss urban ecology and the pursuit of sustainable cities

Concert below the St. John's Bridge in Portland, Oregon

Mount Hood overlooking downtown Portland

In 2004, Oregon voters approved a ballot measure that threatened to eviscerate their state's land use rules. Ballot Measure 37 required the state to compensate certain landowners if government regulation had decreased the value of their land. For example, regulations prevented landowners outside UGBs from subdividing their lots and selling them for housing development. Under Measure 37, the state had to pay these landowners to make up for theoretically lost income, or else allow them to ignore the regulations. Because state and local governments did not have enough money to pay such claims, the measure was on track to gut Oregon's zoning, planning, and land use rules.

Landowners filed more than 7500 claims for payments or waivers affecting 295,000 hectares (ha; 730,000 acres). Although the measure had been promoted to voters as a way to protect the rights of small family landowners, most claims were filed by large developers. Neighbors suddenly found themselves confronting the prospect of massive housing subdivisions, gravel mines, strip malls, or industrial facilities being developed next to their homes—and many who had voted for Measure 37 began to have misgivings.

The state legislature, under pressure from opponents and supporters alike, settled on a compromise: to introduce a new ballot measure. Oregon's voters passed Ballot Measure 49 in 2007. It restricts development outside the UGB that is on a large scale or that impacts sensitive natural areas, but it protects the rights of small landowners to gain income from their property by developing small numbers of homes.

In 2010, Metro finalized a historic agreement with its region's three counties to determine where urban growth will be allowed over the next 50 years. Metro and the counties apportioned more than 121,000 ha (300,000 acres) of undeveloped land into "urban reserves" open for development and "rural reserves" where farmland and forests would be preserved. Boundaries were precisely mapped to give clarity and direction for landowners and governments alike.

People are confronting similar issues in communities throughout North America, and debates and negotiations like those in Oregon will determine how our cities and landscapes will change in the future.

Our Urbanizing World

In 2007, we passed a turning point in human history. For the first time ever, more people were living in urban areas (cities and suburbs) than in rural areas. As we undergo this historic shift from the countryside into towns and cities—a process called **urbanization**—two pursuits become ever more important. One is to make our urban areas more livable by meeting residents' needs for a safe, clean, healthy urban environment and a high quality of life. The other is to make urban areas sustainable by creating cities that can prosper in the long term while minimizing ecological impacts.

Industrialization has driven urbanization

Since 1950, the world's urban population has multiplied by more than five times, whereas the rural population has not even doubled. Urban populations are growing because the human population overall is growing (Chapter 8) and because more people are moving from farms to cities than are moving from cities to farms.

This shift from country to city began long ago. Agricultural harvests that produced surplus food freed a proportion of citizens from farm life and allowed the rise of specialized manufacturing professions, class structure, political hierarchies, and urban centers (p. 211). The industrial revolution (p. 5) spawned technological innovations that created jobs and opportunities in urban centers for people who were no longer needed on farms. Industrialization and urbanization bred further technological advances that increased production efficiencies, both on the farm and in the city, and economic opportunities continued to grow faster in cities. This process of positive feedback continues today.

United Nations demographers project that in the year 2050, two of every three people will live in urban areas.

Between now and then, they estimate, the urban population will increase by 63%, whereas the rural population will decline by 4%. Trends differ between developed and developing nations, however (**FIGURE 13.1**). In developed nations such as the United States, urbanization has slowed because

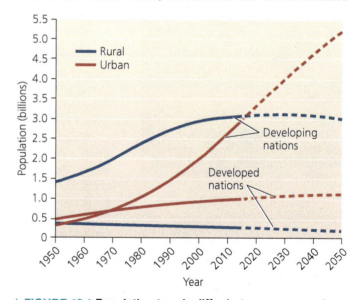

FIGURE 13.1 Population trends differ between poor and wealthy nations. In developing nations, urban populations are growing quickly, and rural populations will soon begin declining. Developed nations are already largely urbanized, so their urban populations are growing slowly, whereas rural populations are falling. Solid lines in the graph indicate past data, and dashed lines indicate future projections. *Data from UN Population Division, 2015. World urbanization prospects: The 2014 revision. By permission.*

DATA Beginning in what decade will the majority of people in developing nations be living in urban areas?

Go to **Interpreting Graphs & Data** on **Mastering**EnvironmentalScience®

three of every four people already live in cities, towns, and **suburbs,** the smaller communities that ring cities. Back in 1850, the U.S. Census Bureau classified only 15% of U.S. citizens as urban-dwellers. That percentage now stands at 80%. Most U.S. urban-dwellers reside in suburbs; fully half the U.S. population today is suburban.

In contrast, today's developing nations, where most people still reside on farms and in rural villages, are urbanizing rapidly. As industrialization diminishes the need for farm labor while increasing urban commerce and jobs, rural people are streaming to cities. Sadly, across the globe, wars, conflict, and ecological degradation are also driving millions of people from the countryside into urban centers. For all these reasons, most fast-growing cities today are in the developing world. In cities such as Delhi, India; Lagos, Nigeria; and Karachi, Pakistan, population growth often exceeds economic growth, and the result is overcrowding, pollution, and poverty. United Nations demographers estimate that urban areas of developing nations will absorb nearly all of the world's population growth from now on.

Environmental factors influence the location of urban areas

Real estate agents use the saying, "Location, location, location" to stress how a home's setting determines its value. Location is vital for urban centers as well. Think of any major city, and chances are it's situated along a major river, seacoast, railroad, or highway—some corridor for trade that has driven economic growth (**FIGURE 13.2**).

Well-located cities often serve as linchpins in trading networks, funneling in resources from agricultural regions, processing them, manufacturing products, and shipping those products to other markets. Portland got its start in the mid-19th century as pioneers arriving by the Oregon Trail settled where the Willamette River joins the Columbia River, just upriver from where the Columbia flows into the Pacific Ocean. With this strategic location for trade, Portland grew as it received, processed, and shipped overseas the produce from farms of the river valleys and as it imported products shipped in from other ports.

Another example of this geographic pattern is Chicago, which grew with extraordinary speed in the 19th and early 20th centuries. At that time, railroads funneled through it the resources from the vast lands of the Midwest and West on their way to consumers and businesses in the populous cities of the East. Chicago became a center for grain processing, livestock slaughtering, meatpacking, and much else.

Today, powerful technologies and cheap transportation enabled by fossil fuels have allowed cities to thrive even in resource-poor regions. The Dallas–Fort Worth area prospers from—and relies on—oil-fueled transportation by interstate highways and a major international airport. Southwestern cities such as Los Angeles, Las Vegas, and Phoenix flourish in desert regions by appropriating water from distant sources. Whether such cities can sustain themselves as oil and water become increasingly scarce in the future is an important question.

(a) St. Louis, Missouri

(b) Fort Worth, Texas

FIGURE 13.2 Cities tend to develop along trade corridors. St. Louis **(a)** is situated on the Mississippi River near its confluence with the Missouri River, where river trade drove its growth in the 19th and early 20th centuries. Fort Worth, Texas **(b)**, grew in the late 20th century as a result of the interstate highway system and a major international airport.

In recent years, many cities in the southern and western United States have undergone growth spurts as large numbers of people have moved there in search of warmer weather, more space, new economic opportunities, or places to retire. Between 1990 and 2015, the population of the Denver metropolitan area grew by more than 70%; that of the Dallas–Fort Worth and Houston metropolitan areas each grew by more than 75%; that of the Atlanta area grew by more than 90%; that of the Phoenix region grew by 104%; and that of the Las Vegas metropolitan area grew by a whopping 148%.

weighing the
ISSUES

What Made Your City?

Consider the town or city in which you live, or the major urban center located nearest you. Why do you think it developed where it did? What physical, social, or environmental factors may have aided its growth? Do you think it will prosper in the future? Why or why not?

People increasingly moved to suburbs

American cities grew rapidly in the 19th and early 20th centuries as a result of immigration from abroad and increased trade as the nation expanded westward. The bustling economic activity of downtown districts held people in cities despite growing crowding, poverty, and crime. However, by the mid-20th century, many affluent city-dwellers were choosing to move outward to cleaner, less crowded suburban communities. These people were pursuing more space, better economic opportunities, cheaper real estate, less crime, and better schools for their children.

As affluent people moved out into the expanding suburbs, jobs followed. This hastened the economic decline of downtown districts, and American cities stagnated. Chicago's population declined to 80% of its peak because so many residents moved to its suburbs. Philadelphia's population fell to 76% of its peak, Washington, D.C.'s to 71%, and Detroit's to just 55%.

Portland followed this trajectory: Its population growth stalled in the 1950s to 1970s as crowding and deteriorating economic conditions drove city-dwellers to the suburbs. But the city bounced back. Subsequent policies to revitalize the city center helped restart Portland's growth (**FIGURE 13.3**).

The exodus to the suburbs in 20th-century America was aided by the rise of the automobile, an expanding road network, and inexpensive and abundant oil. Millions of people could now commute by car to downtown workplaces from new homes in suburban "bedroom communities." By facilitating long-distance transport, fossil fuels and highway networks also made it easier for businesses to import and export resources, goods, and waste. The federal government's development of the interstate highway system was pivotal in promoting these trends.

Today, technology and social trends are reinforcing the spread of suburban areas and are driving people still farther out to **exurbs,** regions beyond the suburbs, generally inhabited by affluent individuals seeking even more space than the suburbs provide. In our age of the Internet, handheld devices, jet travel, and video-conferencing, being located in a city's downtown on a river or seacoast is no longer so vital to success. As globalization continues to connect distant societies, businesses and individuals can more easily communicate from far-flung locations.

In most ways, suburbs and exurbs have delivered the qualities people have sought in them. The wide spacing of homes, with each one on its own plot of land, gives families room and privacy. However, by allotting more space to each person, suburban and exurban growth has spread human impact across the landscape. Natural areas have disappeared as housing developments are constructed. Our extensive road networks ease travel, but people find themselves needing to climb into a car to get anywhere. People commute longer distances to work and spend more time stuck in traffic. The expanding rings of suburbs and exurbs surrounding cities have grown larger than the cities themselves, and towns are merging into one another. These aspects of growth inspired a new term: *sprawl.*

Sprawl

The term *sprawl* has become laden with meanings and suggests different things to different people, but we can begin our discussion by giving **sprawl** a simple, nonjudgmental definition: the spread of low-density urban, suburban, or exurban development outward from an urban center.

Urban areas spread outward

The spatial growth of urban and suburban areas is clear from maps and satellite images of rapidly spreading cities such as Las Vegas (**FIGURE 13.4**). Another example is Chicago, whose metropolitan area now spreads over a region 40 times the size of the city. All in all, houses and roads supplant more than 2700 ha (6700 acres) of U.S. land every day.

Sprawl results from development approaches that place homes on spacious lots in residential tracts that spread over large areas but are far from urban centers and commercial amenities (**FIGURE 13.5**). Such approaches allot each person more space than in cities. For example, the average resident of Chicago's suburbs takes up 11 times more space than a resident of the city. As a result, the outward spatial growth of suburbs and exurbs across the landscape generally outpaces growth in numbers of people.

In fact, many researchers define *sprawl* as the physical spread of development at a rate that exceeds the rate of population growth. For instance, the population of Phoenix grew 12 times larger between 1950 and 2000, yet its land area grew 27 times larger. Between 1950 and 1990, the population of 58 major U.S. metropolitan areas rose by 80%, but the land area they covered rose by 305%. Even in 11 metro areas where

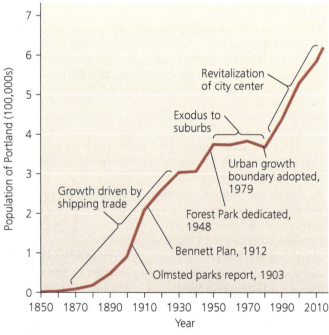

FIGURE 13.3 Portland grew, stabilized, and then grew again. Jobs in the shipping trade helped to boost Portland's economy and population in the 1890s–1920s. City residents began leaving for the suburbs in the 1950s–1970s, but policies to enhance the city center revitalized Portland's growth. *Data from U.S. Census Bureau.*

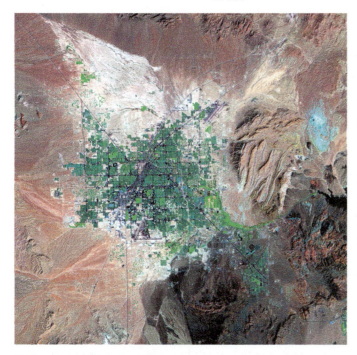

(a) Las Vegas, Nevada, 1986

(b) Las Vegas, Nevada, 2013

FIGURE 13.4 Satellite images show the rapid urban and suburban expansion referred to as sprawl. Las Vegas, Nevada, is one of the fastest-growing cities in North America. Between 1986 **(a)** and 2013 **(b)**, its population and its developed area each tripled.

population declined between 1970 and 1990 (for instance, Rust Belt cities such as Detroit, Cleveland, and Pittsburgh), the amount of land covered increased.

Sprawl has several causes

There are two main components of sprawl. One is human population growth—there are simply more of us alive each year (Chapter 8). The other is per capita land consumption—each person is taking up more land than in the past. The amount of sprawl is a function of the number of people added to a region times the amount of land each person occupies.

A study of U.S. metropolitan areas between 1970 and 1990 found that these two factors contribute about equally to sprawl but that cities vary in which is more influential. The Los Angeles metro area increased in population density by 9% between 1970 and 1990, becoming the nation's most densely populated metro area. Increasing density should be a good recipe for preventing sprawl, yet L.A. grew in size by a whopping 1021 km² (394 mi²) because of an overwhelming influx of new people. In contrast, the Detroit metro area lost 7% of its population between 1970 and 1990, yet it expanded in area by 28%. In this case, sprawl clearly was caused solely by increased per capita land consumption.

Each person is taking up more space these days in part because of factors mentioned earlier: Better highways, relatively inexpensive gasoline, telecommunications, and the Internet have fostered movement away from city centers by freeing businesses from dependence on the centralized infrastructure a major city provides and by giving workers greater flexibility to live where they desire. Given a choice, most

people desire space and privacy and prefer living in less congested, more spacious, more affluent communities.

Economists and politicians have encouraged the unbridled spatial expansion of cities and suburbs. The conventional assumption has been that growth is always good and that attracting business, industry, and residents will enhance a community's economic well-being, political power, and cultural influence. Today, this assumption is being challenged as growing numbers of people feel the negative effects of sprawl on their lifestyles. Citizens

FIGURE 13.5 Sprawl is characterized by the spread of development across large areas of land. This requires people to drive cars to reach commercial amenities or community centers.

and leaders in Portland and many other metropolitan areas are instead now seeking growth in wealth, influence, and prestige without spatially expanding across the landscape.

What is wrong with sprawl?

To some people, the word *sprawl* evokes strip malls, traffic jams, homogeneous commercial development, and tracts of cookie-cutter houses encroaching on farmland, ranchland, or forests. For other people, sprawl is simply the collective result of choices made by millions of well-meaning individuals trying to make a better life for their families. What can scientific research tell us about the impacts of sprawl?

Transportation Most studies show that sprawl constrains transportation options, essentially forcing people to own a vehicle, drive it most places, drive greater distances, and spend more time in it. Sprawling communities suffer more traffic accidents and have few or no mass transit options. Across the United States during the 1980s and 1990s the average length of work trips rose by 36%, and total vehicle miles driven increased three times faster than population growth. An automobile-oriented culture encourages congestion and increases dependence on oil.

Pollution By promoting automobile use, sprawl increases pollution. Carbon dioxide emissions from vehicles contribute to climate change (Chapter 18) and air pollutants that contain nitrogen and sulfur lead to tropospheric ozone, urban smog, and acid precipitation (Chapter 17). Runoff of water from roads and parking lots may be polluted by motor oil that has leaked from vehicles and road salt applied in cold-climate regions to combat ice. Paved areas produce 16 times more runoff than do naturally vegetated areas, and polluted runoff that reaches waterways can pose risks to ecosystems and human health.

Health In addition to the health impacts of pollution, some research suggests that sprawl promotes physical inactivity because driving cars largely takes the place of walking during daily errands. Physical inactivity increases obesity and high blood pressure, which can lead to other ailments. A 2003 study found that people from the most-sprawling U.S. counties show higher blood pressure and weigh 2.7 kg (6 lb) more for their height than people from the least-sprawling U.S. counties.

Land use As more land is developed, less is left as forests, fields, farmland, or ranchland. Of the estimated 1 million ha (2.5 million acres) of U.S. land converted each year, roughly 60% is agricultural land and 40% is forest. These lands provide vital resources, recreation, aesthetic beauty, wildlife habitat, and air and water purification. Sprawl generally diminishes all these amenities. Perhaps of most concern, more and more children these days are growing up without the ability to roam through woods, fields, and open space, which used to be a normal part of childhood. Being deprived of access to regular experience with nature as a child, many experts argue, can have psychological and emotional consequences for the individual and for society (p. 277).

Economics Sprawl drains tax dollars from communities and funnels money into infrastructure for new development on the fringes of those communities. Funds that could be spent maintaining and improving downtown centers is instead spent on extending the road system, water and sewer system, electricity grid, telephone lines, police and fire service, schools, and libraries. The costs of extending public infrastructure are generally not charged to developers but are paid by taxpayers of the community. In theory, fees on developers or property taxes on new homes and businesses could pay back the public investment, but studies find that in most cases existing taxpayers have ended up subsidizing new development.

Creating Livable Cities

To respond to the challenges that sprawl presents, architects, planners, developers, and policymakers are trying to revitalize city centers and to plan and manage how urbanizing areas develop. They aim to make cities safer, cleaner, healthier, and more pleasant for their residents.

Planning helps to create livable urban areas

How can we design cities to maximize their efficiency, functionality, and beauty? These are the questions central to **city planning** (also known as **urban planning**). City planners advise policymakers on development options, transportation needs, public parks, and other matters.

Washington, D.C., is the earliest example of city planning in the United States. President George Washington hired French architect Pierre Charles L'Enfant in 1791 to design a capital city for the new nation on undeveloped land along the Potomac River. L'Enfant laid out a baroque-style plan of diagonal avenues cutting across a grid of streets, with plenty of space allotted for majestic public monuments, and the city was built largely according to his plan (**FIGURE 13.6**). A century later, as the city became crowded and dirty, a special commission in 1901 undertook new planning efforts to beautify the city while staying true to the intentions of L'Enfant's original plan. These planners imposed a height restriction on new buildings, which kept the magnificent government edifices and monuments from being crowded and dwarfed by modern skyscrapers. This preserved the spacious, stately feel of the city.

City planning in North America came into its own at the turn of the 20th century, as urban leaders sought to beautify and impose order on fast-growing, unruly cities. Landscape architect Daniel Burnham's 1909 *Plan of Chicago* was perhaps the grandest effort of this time, and it was largely implemented

(a) The L'Enfant plan, 1791

(b) Washington, D.C., today

FIGURE 13.6 Washington, D.C., is a prime example of early city planning. The 1791 plan **(a)** for the new U.S. capital laid out splendid diagonal avenues cutting across gridded streets, allowing space for the magnificent public monuments **(b)** that still grace the city today.

over the following years and decades. Burnham's plan expanded Chicago's city parks and playgrounds, improved neighborhood living conditions, streamlined traffic systems, and cleared industry and railroads from the shore of Lake Michigan to provide public access to the lake.

Portland gained its own comprehensive plan just three years later, in 1912. Edward Bennett's *Greater Portland Plan* recommended rebuilding the harbor; dredging the river channel; constructing new docks, bridges, tunnels, and a waterfront railroad; superimposing wide radial boulevards on the old city street grid; establishing civic centers downtown; and greatly expanding the number of parks. Voters approved the plan by a two-to-one margin, but they defeated a bond issue that would have paid for park development. As the century progressed, other major planning efforts were conducted, and some ideas, such as establishing public squares downtown, came to fruition.

In today's world of sprawling metropolitan areas, **regional planning** has become at least as important as city planning. Regional planners deal with the same issues as city planners, but they work on broader geographic scales and must coordinate their work with multiple municipal governments. In some places, regional planning has been institutionalized in formal government bodies; the Portland area's Metro is the epitome of such a regional planning entity. When Metro and its region's three counties in 2010 announced their collaborative plan apportioning undeveloped land into "urban reserves" and "rural reserves," it marked a historic accomplishment in regional planning. The agreement enables homeowners, farmers, developers, and policymakers to feel informed and secure knowing what kinds of land uses lie in store on and near their land over the next half-century.

Zoning is a key tool for planning

One tool that planners use is **zoning,** the practice of classifying areas for different types of development and land use (**FIGURE 13.7**). For instance, to preserve the cleanliness and tranquility of residential neighborhoods, industrial facilities

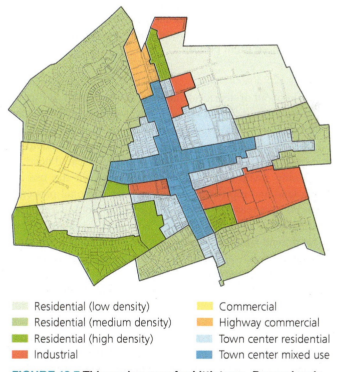

Residential (low density)	Commercial
Residential (medium density)	Highway commercial
Residential (high density)	Town center residential
Industrial	Town center mixed use

FIGURE 13.7 This zoning map for Littletown, Pennsylvania, shows several typical zoning patterns. Public and institutional uses are clustered together in "mixed-use" areas in the center of town and along major roads. Industrial zones tend to be located away from most residential areas. Residential zones vary in the density of homes allowed.

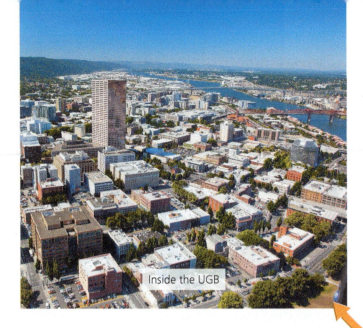

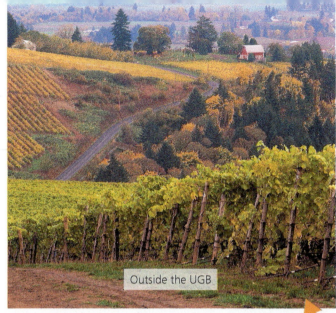

Inside the UGB

Outside the UGB

FIGURE 13.8 The Portland region's urban growth boundary separates areas earmarked for high-density urban development (left photo) from rural areas where development is more restricted (right photo).

may be kept out of districts zoned for residential use. By specifying zones for different types of development, planners can guide what gets built where. Zoning also gives home buyers and business owners security, because they know in advance what types of development can and cannot be located nearby.

Zoning involves government restriction on the use of private land and represents a constraint on personal property rights. For this reason, some people consider zoning a regulatory taking (p. 167) that violates individual freedoms. Most people defend zoning, however, saying that government has a proper and useful role in setting limits on property rights for the good of the community.

When Oregon voters passed Ballot Measure 37 in 2004, this shackled government's ability to enforce zoning regulations with landowners who bought their land before the regulations were enacted. However, many Oregonians soon began witnessing new development they did not condone, so in 2007 they passed Ballot Measure 49 to restore public oversight over development. The passage of Oregon's Measure 37 spawned similar ballot measures in other U.S. states, but voters defeated most of these. In general, people have supported zoning over the years because the common good it produces for communities is widely perceived to outweigh the restrictions on private use.

Urban growth boundaries are now widely used

Planners in Oregon put limits on sprawl by containing growth largely within existing urbanized areas. They did this by establishing **urban growth boundaries,** lines on a map that separate areas zoned to be high-density and urban from areas intended to remain low-density and rural. Oregon's urban growth boundaries (UGBs) have aimed to revitalize downtowns; protect working farms, orchards, ranches, and forests; and ensure urban-dwellers some access to open space (**FIGURE 13.8**). UGBs also save taxpayers money by reducing the amounts that municipalities need to pay for infrastructure, compared with the amounts required by sprawling areas. Since Oregon instituted its policies, many other states, regions, and cities have adopted UGBs—from Boulder, Colorado, to Lancaster, Pennsylvania, to many California communities.

In most ways, the Portland region's urban growth boundary has worked as intended. It has lowered prices for land outside the UGB while raising prices within it. It has restricted development outside the UGB, preserving farms and forests. It has increased the density of new housing inside the UGB by more than 50% as homes are built on smaller lots and as multistory apartments replace low-rise structures. Neighborhoods have thrived, and downtown employment has grown as businesses and residents invest anew in the central city. Through it all, Portland has been able to absorb considerable immigration while avoiding rampant sprawl.

Nonetheless, the Portland region's urbanized area grew by 101 km² (39 mi²) in the decade after its UGB was established, because 146,000 people were added to the population. Rising population pressure has led Metro to enlarge the UGB three dozen times since its establishment. In addition, UGBs tend to increase housing prices within their boundaries, and in Portland, housing has become far less affordable. Today in the city, demand for housing exceeds supply, rents are soaring, and low- and middle-income people are being forced out of neighborhoods they have lived in for years as these neighborhoods experience **gentrification,** a transformation to conditions that cater to wealthier people. These trends suggest that relentless population growth may thwart even the best anti-sprawl efforts and that livable cities may fall victim to their own success if they are in high demand as places to live.

"Smart growth" and "new urbanism" aim to counter sprawl

As more people feel the impacts of sprawl on their everyday lives, efforts to manage growth are springing up throughout North America. Oregon's Senate Bill 100 was one of the first, and since then dozens of states, regions, and cities have adopted similar land use policies. Urban growth boundaries and other approaches from these policies have coalesced under the concept of **smart growth** (TABLE 13.1).

Proponents of smart growth aim to rejuvenate the older existing communities that so often are drained and impoverished by sprawl. Smart growth means "building up, not out"—focusing development and economic investment in existing urban centers and favoring multistory shop-houses and high-rises.

A related approach among architects, planners, and developers is **new urbanism.** This approach seeks to design neighborhoods on a walkable scale, with homes, businesses, schools, and other amenities all nearby for convenience. The aim is to create functional neighborhoods in which families can meet most of their needs close to home without using a car. Trees, green spaces, a mix of architectural styles, and creative street layouts add to the visual interest of new-urbanist developments. These developments mimic the traditional urban neighborhoods that existed until the advent of suburbs.

New-urbanist neighborhoods are generally served by public transit systems. In **transit-oriented development,** compact communities in the new-urbanist style are arrayed around stops on a rail line, enabling people to travel most places they need to go by train and foot alone. Several lines of the Washington, D.C., Metro system have been developed in this manner.

Among the 600 communities in the new-urbanist style across North America are Seaside, Florida; Kentlands in Gaithersburg, Maryland; Addison Circle in Addison, Texas; Mashpee Commons in Mashpee, Massachusetts; Harbor Town in Memphis, Tennessee; Celebration in Orlando, Florida; and Orenco Station, west of Portland.

Transit options help cities

Traffic jams on roadways cause air pollution, stress, and countless hours of lost personal time. They cost Americans an estimated $74 billion each year in fuel and lost productivity. To encourage more efficient urban transportation, policymakers can raise fuel taxes, charge trucks for road damage, tax inefficient modes of transport, and reward carpoolers with carpool lanes. But a key ingredient in any planner's recipe for improving the quality of urban life is to give residents alternative transportation options.

Bicycle transportation is one key option (FIGURE 13.9). Portland has embraced bicycles like few other American cities, and today 6% of its commuters ride to work by bike (the national average is 0.5%). The city has developed nearly 400 miles of

TABLE 13.1 Ten Principles of "Smart Growth"

- Mix land uses.
- Take advantage of compact building design.
- Create a range of housing opportunities and choices.
- Create walkable neighborhoods.
- Foster distinctive, attractive communities with a strong sense of place.
- Preserve open space, farmland, natural beauty, and critical environmental areas.
- Strengthen existing communities, and direct development toward them.
- Provide a variety of transportation choices.
- Make development decisions predictable, fair, and cost-effective.
- Encourage community and stakeholder collaboration in development decisions.

Source: U.S. Environmental Protection Agency.

FIGURE 13.9 Bicycles provide a healthy alternative to transportation by car. This Portland bicycle lot accommodates riders who commute downtown by bike, and is conveniently located at a streetcar stop.

bike lanes and paths, 5000 public bike racks, and special markings at intersections to protect bicyclists. Amazingly, all this infrastructure was created for the typical cost of just 1 mile of urban freeway. Portland also has a bike-sharing program similar to programs in cities such as Montreal, Toronto, Denver, Minneapolis, Miami, San Antonio, Boston, and Washington, D.C.

Other transportation options include **mass transit** systems: public systems of buses, trains, subways, or *light rail* (smaller rail systems powered by electricity) that move large numbers of passengers while easing traffic congestion, taking up less space than road networks, and emitting less pollution than cars. A 2005 study calculated that each year rail systems in U.S. metropolitan areas save taxpayers $67.7 billion in costs for congestion, consumer transportation, parking, road maintenance, and accidents—far more than the $12.5 billion that governments spend to subsidize rail systems each year. As long as an urban center is large enough to support the infrastructure necessary, rail systems are cheaper, more energy-efficient, and less polluting than roadways choked with cars (**FIGURE 13.10**).

In Portland, buses, light rail, and streetcars together carry 100 million riders per year (**FIGURE 13.11**). The nation's most-used train systems are the extensive "heavy rail" systems in America's largest cities, such as New York City's subways, Washington, D.C.'s Metro, the T in Boston, and the San Francisco Bay area's BART. Each of these rail systems carries more than one-fourth of its city's daily commuters.

In general, however, the United States lags behind most nations in mass transit. Many countries, rich and poor alike, have extensive and accessible bus systems that ferry citizens within and between towns and cities cheaply and effectively. And whereas Japan, China, and many European nations have developed entire systems of modern high-speed "bullet" trains (**FIGURE 13.12**), the United States has only one such train, Amtrak's *Acela Express*. This train connects Boston and Washington, D.C., via New York, Philadelphia, and Baltimore, and it travels more slowly than most bullet trains.

The United States chose instead to invest in road networks for cars and trucks largely because (relative to most other nations) its population density was low and gasoline was cheap. As energy costs and population rise, however, mass transit becomes increasingly appealing, and citizens begin to clamor for train and bus systems in their communities. Americans may eventually see more high-speed rail; the 2009 stimulus bill passed by Congress set aside $8 billion for developing high-speed rail, and the Obama administration identified 10 potential corridors for development of such trains.

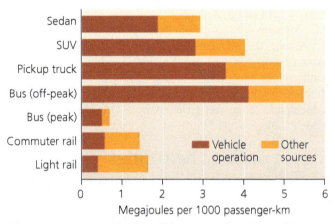

(a) Energy consumption for different modes of transit

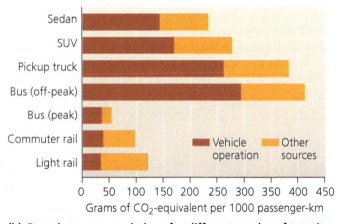

(b) Greenhouse gas emissions for different modes of transit

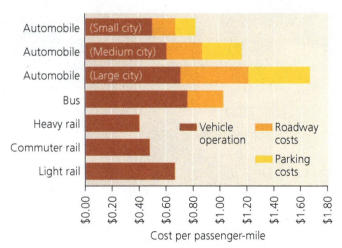

(c) Operating costs for different modes of transit

FIGURE 13.10 Rail transit consumes less energy (a), emits fewer greenhouse gases (b), and costs less (c) than automobile transit. Bus transit is highly efficient in places and at times of high use ("peak" in figure), but much less so when and where use is low ("off-peak" in figure). Data presented for light rail are averages of systems in Boston and San Francisco. *Data from (a, b) Chester, M., and Horvath, A., 2009. Environmental assessment of passenger transportation should include infrastructure and supply chains. Environmental Research Letters 4: 024008 (8 pp.); and (c) Litman, T., 2005. Rail transit in America: A comprehensive evaluation of benefits. © 2005 T. Litman.*

DATA Let's contrast the impacts of traveling by automobile with those of traveling by rail.
- In part **(a)**, compare the energy consumed by driving an SUV with the energy consumed by taking commuter rail.
- In part **(b)**, compare the greenhouse gases emitted by driving a pickup truck with those produced by riding light rail.
- In part **(c)**, automobile traffic creates two types of operating costs that are *not* created by rail traffic. What are they?

Go to **Interpreting Graphs & Data** on **Mastering**EnvironmentalScience®

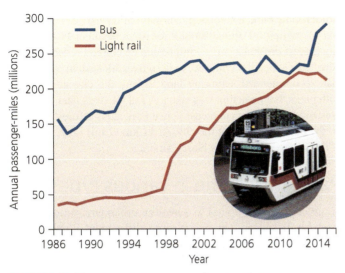

FIGURE 13.11 Ridership has grown on Portland's buses and on its MAX light rail system.

Establishing mass transit is not easy, however. Once a road system is in place and lined with businesses and homes, it is difficult and expensive to replace or complement it with a mass transit system. Strong and visionary political leadership may be required. Such was the case in Curitiba, Brazil. Faced with an influx of immigrants from outlying farms in the 1970s, city leaders led by Mayor Jaime Lerner undertook an aggressive planning process so that they could direct growth rather than being overwhelmed by it. They established a large fleet of public buses and reconfigured Curitiba's road system to maximize its efficiency. Today this metropolis of 2.5 million people has an outstanding bus system that is used each day by three-quarters of its population. The 340 bus routes, 250 terminals, and 1900 buses accompany measures to encourage bicycles and pedestrians. All of this has resulted in a steep drop in car use, despite the city's rapidly growing population.

FIGURE 13.12 "Bullet trains" of high-speed rail systems in Europe and Asia can travel at 150–220 mph. This Chinese train is speeding through the city of Qingdao.

Urban residents need park lands

City-dwellers often desire some sense of escape from the noise, commotion, and stress of urban life. Natural lands, public parks, and open space provide greenery, scenic beauty, freedom of movement, and places for recreation. These lands also keep ecological processes functioning by helping to regulate climate, purify air and water, and provide wildlife habitat. The animals and plants of urban and suburban parks and natural lands also serve to satisfy our natural affinity for contact with other organisms (sometimes called biophilia; p. 277). In the wake of urbanization and sprawl, protecting natural lands and establishing public parks have become more important as many of us come to feel increasingly disconnected from nature.

America's city parks arose in the late 19th century as urban leaders yearning to make crowded and dirty cities more livable established public spaces using aesthetic ideals borrowed from European parks, gardens, and royal hunting grounds. The lawns, shaded groves, curved pathways, and pastoral vistas of many American city parks originated with these European ideals, as interpreted by America's leading landscape architect, Frederick Law Olmsted. Olmsted designed Central Park in New York City (**FIGURE 13.13**) and many other urban parks.

East Coast cities such as New York, Boston, and Philadelphia developed parks early on, but western cities were not far behind. In San Francisco, William Hammond Hall transformed 2500 ha (1000 acres) of the peninsula's dunes into

FIGURE 13.13 Central Park in New York City was one of America's first city parks, and it remains one of the largest and finest.

FIGURE 13.14 The High Line Park was created thanks to a visionary group of Manhattan citizens. They pushed to make a park out of an abandoned elevated rail line.

FIGURE 13.15 Forest preserves wind through the suburbs of Chicago. This regional greenbelt system (visible in the satellite image) features 40,000 ha (100,000 acres) of woodlands, fields, marshes, and prairies, making up the largest holding of locally owned public conservation land in the United States.

Golden Gate Park, a verdant playground of lawns, trees, gardens, and sports fields. Portland's quest for parks began in 1900, when city leaders created a parks commission and hired Frederick Law Olmsted's son, John Olmsted, to design a park system. His 1904 plan recommended acquiring land to ring the city generously with parks, but no action was taken. A full 44 years later, citizens pressured city leaders to create Forest Park along a forested ridge on the northwest side of the city. At 11 km (7 mi) long, it is one of the largest city parks in North America.

Park lands come in various types

Large city parks are vital to a healthy urban environment, but even small spaces can make a big difference. Playgrounds give children places to be active outdoors and interact with their peers. Community gardens allow people to grow vegetables and flowers in a neighborhood setting.

Greenways are strips of land that connect parks or neighborhoods, often run along rivers, streams, or canals, and provide access to walking trails. They can protect water quality, boost property values, and serve as corridors for the movement of wildlife. Across North America, the Rails-to-Trails Conservancy has helped to convert 35,500 km (22,000 mi) of abandoned railroad rights-of-way into greenways for walking, jogging, and biking.

One newly developed and instantly popular linear park along an old rail line is The High Line Park in Manhattan in New York City (**FIGURE 13.14**). An elevated freight line running above the city's streets was going to be demolished, but a group of citizens saw its potential for a park, and they pushed the idea until city leaders came to share their vision. More than 13,000 people per day now use the 23-block-long High Line for recreation or commuting to work.

The concept of the corridor is sometimes implemented on a large scale. **Greenbelts** are long and wide corridors of park lands, often encircling an entire urban area. One example is the system of forest preserves that stretches through Chicago's suburbs like a necklace (**FIGURE 13.15**). In Canada, cities including Toronto, Ottawa, and Vancouver employ greenbelts as urban growth boundaries, containing sprawl while also preserving open space for city residents.

Green buildings bring benefits

Although we need park lands, we spend most of our time indoors, so the buildings in which we live and work affect our health. Moreover, buildings consume 40% of our energy and 70% of our electricity, contributing to the greenhouse gas emissions that drive climate change. As a result, there is a thriving movement in architecture and construction to design and build **green buildings,** structures meant to minimize the ecological footprint of their construction and operation.

Green buildings are built from sustainable materials, limit their use of energy and water, minimize adverse health impacts, control pollution, and recycle waste (**FIGURE 13.16**). Constructing or renovating buildings using new efficient technologies is probably the most effective way that cities can reduce energy consumption and greenhouse gas emissions.

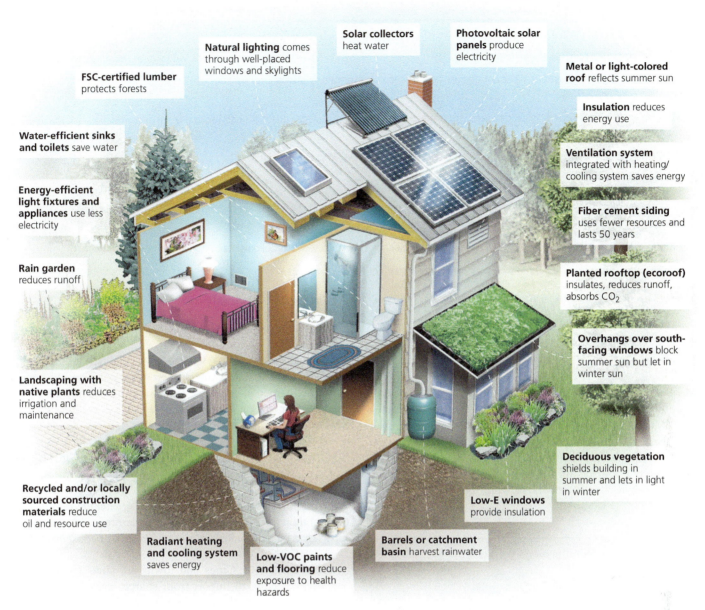

Natural lighting comes through well-placed windows and skylights

Solar collectors heat water

Photovoltaic solar panels produce electricity

FSC-certified lumber protects forests

Metal or light-colored roof reflects summer sun

Water-efficient sinks and toilets save water

Insulation reduces energy use

Energy-efficient light fixtures and appliances use less electricity

Ventilation system integrated with heating/cooling system saves energy

Fiber cement siding uses fewer resources and lasts 50 years

Rain garden reduces runoff

Planted rooftop (ecoroof) insulates, reduces runoff, absorbs CO_2

Landscaping with native plants reduces irrigation and maintenance

Overhangs over south-facing windows block summer sun but let in winter sun

Deciduous vegetation shields building in summer and lets in light in winter

Recycled and/or locally sourced construction materials reduce oil and resource use

Low-E windows provide insulation

Radiant heating and cooling system saves energy

Low-VOC paints and flooring reduce exposure to health hazards

Barrels or catchment basin harvest rainwater

FIGURE 13.16 A green building incorporates design features to minimize its ecological footprint.

The U.S. Green Building Council promotes these efforts by running the **Leadership in Energy and Environmental Design (LEED)** certification program. New buildings or renovation projects apply for certification and, depending on their performance, may be granted silver, gold, or platinum status (**TABLE 13.2**, p. 344).

LEED certification brings enough prestige to a building that architects, contractors, and building owners are often willing to pay the extra expense that green building techniques add to construction. The added cost is generally less than 3% for a LEED Silver–certified building and 10% for a LEED Platinum–certified building, but because construction is expensive, these small percentages can represent large sums of money. Today, LEED certification is booming. Portland now features several dozen LEED-certified buildings, as well as the nation's first LEED Gold–certified sports arena (the

Moda Center, where the Trailblazers basketball team plays). The savings on energy, water, and waste at the refurbished Moda Center paid for the cost of its LEED upgrade after just one year.

Schools, colleges, and universities are leaders in sustainable building. In Portland, the Rosa Parks Elementary School was built with locally sourced and nontoxic materials, uses 24% less energy and water than comparable buildings, and diverted nearly all of its construction waste from the landfill. Schoolchildren learn about renewable energy in their own building by watching a display of the electricity produced by its photovoltaic solar system. Portland State University, the University of Portland, Reed College, and Lewis and Clark College are just a few of the many colleges and universities nationwide that are constructing green buildings as part of their campus sustainability efforts (Chapter 24).

TABLE 13.2 Green Building Approaches for LEED Certification of New Construction

APPROACHES THAT ARE REWARDED	MAXIMUM POINTS*
ENERGY: Monitor energy use; use efficient design, construction, appliances, systems, and lighting; use clean, renewable energy sources	37
THE SITE: Build on previously developed land; minimize erosion, runoff, and water pollution; use regionally appropriate landscaping; integrate with transportation options	21
INDOORS: Improve indoor air quality; provide natural daylight and views; improve acoustics	17
MATERIALS: Use local or sustainably grown, harvested, and produced products; reduce, reuse, and recycle waste	14
WATER USE: Use efficient appliances inside; landscape for water conservation outside	11
POINTS ABOVE MAY TOTAL UP TO 100. THEN, UP TO 10 BONUS POINTS MAY BE AWARDED FOR:	
INNOVATION: New and innovative technologies and strategies to go beyond LEED requirements	6
THE REGION: Addressing environmental concerns most important for one's region	4

*Out of 110 possible points, 40 are required for LEED certification, 50 for silver, 60 for gold, and 80 for platinum levels.

Urban Sustainability

Most of our efforts to make cities safer, cleaner, healthier, and more beautiful are also helping to make them more sustainable. A sustainable city is one that can function and prosper over the long term, providing generations of residents a good quality of life far into the future. In part, this entails minimizing the city's impacts on the natural systems and resources that nourish it. It also entails viewing the city itself as an ecological system (see **THE SCIENCE BEHIND THE STORY**, pp. 346–347). Urban centers exert both positive and negative environmental impacts. The extent and nature of these impacts depend on how we use resources, produce goods, transport materials, and deal with waste.

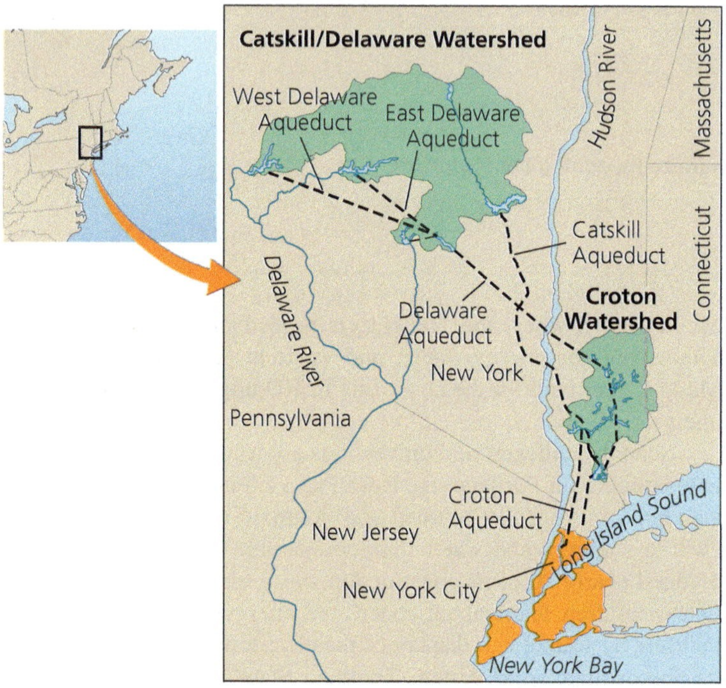

Urban resource consumption brings a mix of environmental effects

You might guess that urban living has a greater environmental impact than rural living. However, the picture is not so simple; instead, urbanization brings a complex mix of consequences.

Resource sinks Cities and towns are sinks (p. 117) for resources, having to import from source areas beyond their borders nearly everything they need to feed, clothe, and house their inhabitants and power their commerce. Urban and suburban areas rely on large expanses of land elsewhere to supply food, fiber, water, timber, metal ores, and mined fuels. Urban centers also need areas of natural land to provide ecosystem services, including purification of water and air, nutrient cycling, and waste treatment. Indeed, for their day-to-day survival, major cities such as New York, Boston, San Francisco, and Los Angeles depend on water they pump in from faraway watersheds (**FIGURE 13.17**).

The long-distance transportation of resources and goods from countryside to city requires fossil fuel use and thereby has considerable environmental impact. However, imagine that all the world's 3.9 billion urban residents were instead spread evenly across the landscape. What would the transportation requirements be, then, to move all those resources and goods around to all those people? A world without cities would likely require *more* transportation to provide people the same degree of access to resources and goods.

Efficiency Once resources arrive at an urban center, the concentration of people allows goods and services to be

FIGURE 13.17 New York City pipes in its drinking water from reservoirs in two upstate watersheds. The city acquires, protects, and manages watershed land to minimize pollution of these water sources.

delivered efficiently. For instance, providing electricity for densely packed urban homes and apartments is more efficient than providing electricity to far-flung homes in the countryside. The density of cities facilitates the provision of many social services that improve quality of life, including medical services, education, water and sewer systems, waste disposal, and public transportation.

More consumption Because cities draw resources from afar, their ecological footprints are much greater than their actual land areas. For instance, urban scholar Herbert Girardet calculated that the ecological footprint of London, England, extends 125 times larger than the city's actual area. By another estimate, cities take up only 2% of the world's land surface but consume more than 75% of its resources.

However, the ecological footprint concept is most meaningful when used on a per-person basis. So, in asking whether urbanization intensifies resource consumption, we must ask whether the average urban- or suburban-dweller has a larger footprint than the average rural-dweller. The answer is yes, but urban and suburban residents also tend to be wealthier than rural residents, and wealth correlates with resource consumption. Thus, although urban and suburban citizens tend to consume more than their rural counterparts, the reason could simply be that they tend to be wealthier.

Urbanization preserves land

Because people pack together densely in cities, more land outside cities is left undeveloped. Indeed, this is the very idea behind urban growth boundaries. If cities did not exist, and if instead all 7 billion of us were evenly spread across the planet's land area, no large blocks of land would be left uninhabited, and we would have much less room for agriculture, wilderness, biodiversity, or privacy. The fact that half the human population is concentrated in discrete locations helps allow room for natural ecosystems to continue functioning and provide the ecosystem services on which all of us, urban and rural, depend.

Urban centers suffer and export pollution

Just as cities import resources, they export wastes, either passively through pollution or actively through trade. In so doing, urban centers transfer the costs of their activities to other regions—and mask the costs from their own residents. Citizens of Indianapolis, Columbus, or Buffalo may not recognize that pollution from nearby coal-fired power plants worsens acid precipitation hundreds of miles to the east. New York City residents may not realize how much garbage their city produces if it is shipped elsewhere for disposal.

However, not all waste and pollution leave the city. Urban-dwellers are exposed to heavy metals, industrial compounds, and chemicals from manufactured products that accumulate in soil and water. Airborne pollutants cause smog and acid precipitation (Chapter 17). Fossil fuel combustion releases greenhouse gases as well as pollutants that pose health risks.

City residents suffer thermal pollution as well, because cities tend to have ambient temperatures that are several degrees higher than those of surrounding areas. This **urban heat island effect** (**FIGURE 13.18**) results from the concentration of heat-generating buildings, vehicles, factories, and people. It also results from the way that buildings and dark paved surfaces absorb sunlight throughout the day and then release the energy slowly as heat. In addition, buildings block cooling wind currents, and impervious surfaces cause rainwater to run off into sewers rather than evaporating and cooling the air. Because much of the energy a city gathers in the daytime is released as heat at night, this warms the

FAQ

Aren't cities bad for the environment?

Stand in the middle of a big city and look around. You see concrete, cars, and pollution. Environmentally bad, right? Not necessarily. The widespread impression that urban living is less sustainable than rural living is largely a misconception. Consider that in a city you can walk to the grocery store instead of driving. You can take the bus or the train. Police, fire, and medical services are close at hand. Water and electricity are easily supplied to your entire neighborhood, and waste is easily collected. In contrast, if you live in the country, resources must be used to transport all these services for long distances, or you need to burn gasoline traveling to reach them. By clustering people together, cities distribute resources efficiently while also preserving natural lands outside the city.

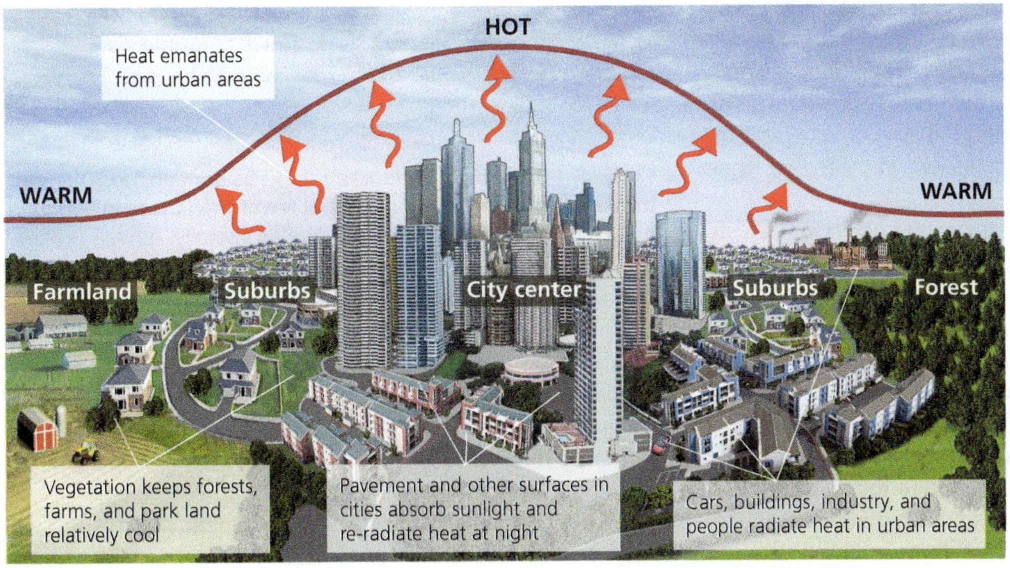

HOT

Heat emanates from urban areas

WARM

WARM

Farmland   Suburbs   City center   Suburbs   Forest

Vegetation keeps forests, farms, and park land relatively cool

Pavement and other surfaces in cities absorb sunlight and re-radiate heat at night

Cars, buildings, industry, and people radiate heat in urban areas

FIGURE 13.18 Cities produce urban heat islands, creating temperatures warmer than surrounding areas.

How Do Baltimore and Phoenix Function as Ecosystems?

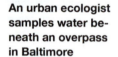

An urban ecologist samples water beneath an overpass in Baltimore

Researchers in urban ecology examine how ecosystems function in cities and sub-urbs, how natural systems respond to urbanization, and how people inter-act with the urban environment. Today, Baltimore and Phoenix are centers for urban ecology.

These two cities are very different: Baltimore is an Atlan-tic port city on the Chesapeake Bay with a long history, whereas Phoenix is a young and fast-growing southwestern metropolis sprawling across the desert. Each was selected by the U.S. National Science Foundation as a research site in its prestigious Long Term Eco-logical Research (LTER) program, which funds multidecade ecological research. Since 1997, hundreds of researchers have studied Baltimore and Phoenix explicitly as ecosystems, examining nutrient cycling, biodiversity, air and water quality, environmental health threats, and more.

Research teams in both cities are combining old maps, aer-ial photos, and new remote sensing satellite data to reconstruct the history of landscape change. In Phoenix, one group showed how urban development spread across the desert in a "wave of advance," affecting soils, vegetation, and microclimate as it went. In Baltimore, mapping showed that development frag-mented the forest into smaller patches over the past 100 years, even while the overall amount of forest remained the same.

The study regions designated for each city encompass both heavily urbanized central city areas and rural and natural areas on the urban fringe. To measure the impacts of urbanization, many research projects compare conditions in these two types of areas.

The Baltimore project's scientists can see ecological effects of urbanization just by comparing the urban lower end of their

site's watershed with its less developed upper end. In the lower end, pavement, rooftops, and compacted soil prevent rainfall from infiltrating the soil, so water runs off quickly into streams. The rapid flow cuts streambeds deeply into the earth while leav-ing the surrounding soil drier. As a result, wetland-adapted trees and shrubs are vanishing, replaced by dry-adapted upland trees and shrubs.

The fast flow of water also worsens pollution. In natural areas, streams and wetlands filter pollution by breaking down nitrogen compounds. But in urban areas, where wetlands dry up and run-off from pavement creates flash floods, streams lose their filtering ability. In Baltimore, the resulting pollution ends up in the Chesa-peake Bay, which suffers eutrophication and a large hypoxic dead zone (pp. 104, 108). Baltimore-project scientists studying nutrient cycling (p. 117) found that urban and suburban watersheds suffer far more nitrate pollution than natural forests, yet also found that the filtering ability of urban park lands helps keep pollution much lower than in agricultural landscapes (**FIGURE 1**).

Baltimore research also reveals that applying salt to icy roads in winter has environmental impacts. Road salt makes

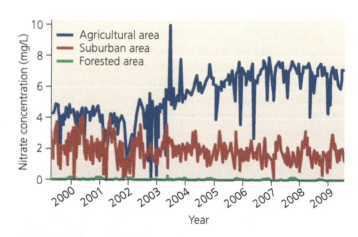

FIGURE 1 **Streams in Baltimore's suburbs contain more nitrates than streams in nearby forests, but fewer than those in agricul-tural areas, where fertilizers are applied liberally.** *Data from Baltimore Ecosystem Study, www.lternet.edu/research/keyfindings/urban-watersheds.*

nighttime air and interferes with patterns of convective circula-tion that would otherwise cool the city. To minimize the urban heat island effect, we can plant more vegetation and we can paint rooftops pale colors to reflect sunlight.

Urban residents also suffer noise pollution and light pol-lution. **Noise pollution** consists of undesired ambient sound. Excess noise degrades one's surroundings aesthetically, can

cause stress, and at intense levels (such as with prolonged exposure to the sounds of leaf blowers, lawn mowers, and jackhammers) can harm hearing. The glow of **light pollution** from city lights may impair sleep and obscures the night sky, impeding the visibility of stars.

These various forms of pollution and the health risks they pose are not evenly shared among urban residents.

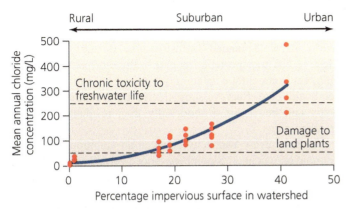

FIGURE 2 Salt concentrations in Baltimore-area streams are high enough to damage plants in the suburbs and to kill aquatic animals in urban areas. *Adapted from Kaushal, S.S., et al., 2005. Increased salinization of fresh water in the northeastern United States. Proc. Natl. Acad. Sci. USA 102: 13517–13520, Fig 2. ©2005 National Academy of Sciences, U.S.A. By permission.*

DATA At what percentage of impervious (paved) surface would you expect to begin seeing damage to land plants as a result of salty road runoff entering streams? At what percentage would you expect to begin seeing chronic toxicity to freshwater life?

its way into streams, which become up to 100 times saltier, even in summer, when salt is not applied. Such high salinity kills organisms (**FIGURE 2**), degrades habitat and water quality, and impairs streams' ability to remove nitrate.

To study contamination of groundwater and drinking water, researchers are using isotopes (p. 23) to trace where salts in the most polluted streams are coming from. Baltimore is now improving water quality substantially with a $900-million upgrade of its sewer system.

Urbanization also affects species and ecological communities. Cities and suburbs facilitate the spread of non-native species, because people introduce exotic ornamental plants and because urbanization's impacts on the soil, climate, and landscape favor weedy generalist species over more specialized native ones. In Baltimore, non-native plant species are most abundant in urban areas. In Phoenix's dry climate, pollen from some non-native plants causes allergy problems for city residents.

Community ecologists studying the wild animals and plants that persist within Phoenix are finding that urbanization alters relationships among them. Compared with natural landscapes, cities offer steady and reliable food resources—think of people's bird feeders, or food scraps from dumpsters.

Growing seasons are extended and seasonal variation is buffered in cities, as well. The urban heat island effect (p. 345) raises nighttime temperatures and makes temperatures more similar year-round. Buildings and ornamental vegetation shelter animals from extreme conditions, and irrigation in yards and gardens provides water. In a desert city like Phoenix, watering boosts primary productivity and lowers daytime temperatures. Together, all these changes lead to higher population densities of animals but lower species diversity as generalists thrive and displace specialists.

Urban ecologists in Phoenix and Baltimore are also studying social and demographic aspects of the urban environment. Some research measures how natural amenities affect property values. One study found that proximity to a park increases a home's property values—unless crime is pervasive. If the robbery rate surpasses 6.5 times the national average (as it does in Baltimore), then proximity to a park begins to depress property values.

Other studies focus on environmental justice concerns (pp. 137–139). These studies have repeatedly found that sources of industrial pollution tend to be located in neighborhoods that are less affluent and that are home primarily to people of racial and ethnic minorities. Phoenix researchers mapped patterns of air pollution and toxic chemical releases and found that minorities and poor citizens are exposed to a greater share of these hazards. And as a result, they suffer from higher rates of childhood asthma.

In Baltimore, researchers found a more complex pattern. Toxic release sites were more likely to be in working-class white neighborhoods than in African American neighborhoods. This, the researchers concluded, was a result of historical inertia. In the past, living close to one's workplace—the factories that release toxic chemicals—was something people preferred, and white workers claimed the privilege of living near their workplaces.

Whether addressing the people, natural communities, or changing ecosystems of the urban environment, studies on urban ecology like those in Phoenix and Baltimore will be vitally informative in our ever more urban world.

Those who receive the brunt of the pollution are often those who are too poor to live in cleaner areas. Environmental justice concerns (pp. 137–139) center on the fact that a disproportionate number of people living near, downstream from, or downwind from factories, power plants, and other polluting facilities are people who are poor and, often, people of racial or ethnic minorities.

Urban centers foster innovation

Cities promote a flourishing cultural life and, by mixing together diverse people and influences, spark innovation and creativity. The urban environment can promote education and scientific research, and cities have long been viewed as engines of technological and artistic inventiveness.

This inventiveness can lead to solutions to societal problems, including ways to reduce environmental impacts. For instance, research into renewable energy sources is helping us develop ways to replace fossil fuels. Technological advances have helped us reduce pollution. Wealthy and educated urban populations provide markets for low-impact goods, such as organic produce. Recycling programs help reduce the solid waste stream. Environmental education is helping people choose their own ways to live cleaner, healthier, lower-impact lives. All these phenomena are facilitated by the education, innovation, science, and technology that are part of urban culture.

Urban ecology helps cities move toward sustainability

Cities that import all their resources and export all their wastes have a linear, one-way metabolism. Linear models of production and consumption tend to destabilize environmental systems and are ultimately not sustainable. Proponents of sustainability for cities stress the need to develop circular systems, akin to systems found in nature, which recycle materials and use renewable sources of energy.

Researchers in the field of **urban ecology** hold that cities can be viewed explicitly as ecosystems and that the fundamentals of ecosystem ecology and systems science (Chapter 5) apply to the urban environment. Major urban ecology projects are ongoing in Baltimore and Phoenix, where researchers are studying these cities as ecological systems (see The Science Behind the Story, pp. 346–347).

Planners and visionary leaders have come up with designs for entire "eco-cities" built from scratch that follow cyclical, sustainable patterns of resource use and waste recycling. So far, none of these efforts have come to fruition, but existing cities across the world are adopting ecologically sustainable strategies by maximizing the efficient use of resources, recycling waste and wastewater, and developing green technologies. Urban agriculture that recycles organic waste and produces locally consumed food is thriving in many places, from Portland to Cuba to Japan. Curitiba, Brazil, shows the kind of success that can result when a city invests in well-planned infrastructure. Besides its highly effective bus transportation network, the city provides recycling, environmental education, job training for the poor, and free health care. Surveys show that its citizens are unusually happy and better off economically than residents of other Brazilian cities.

In 2007, New York City unveiled an ambitious plan that then-mayor Michael Bloomberg hoped would make it "the first environmentally sustainable 21st-century city." PlaNYC was a 132-item program to reduce greenhouse gas emissions, improve mass transit, plant trees, clean up polluted land and rivers, and enhance access to park land. The aim of PlaNYC was to make New York City a better place to live as it accommodates 1 million more people by 2030 (**FIGURE 13.19**). Under the plan as of 2014, the city had improved energy efficiency in 174 city buildings; planted 950,000 trees; opened or renovated 234 school playgrounds; established 129 community gardens; upgraded wastewater treatment; and acquired 36,000 acres to protect upstate water supplies. It had expanded recycling, installed solar panels, cleaned up polluted sites, installed bike lanes and racks and launched a bike-sharing program, retrofitted ferries to reduce pollution, introduced electric vehicles, and converted hundreds of taxis to hybrid vehicles. So far, such actions have improved air quality and reduced greenhouse gas emissions by 19%. In 2015, Mayor Bill de Blasio continued the program under a new name, OneNYC, while adding new dimensions to address economic equity.

New York City's Sustainability Goals
- Reduce greenhouse gas emissions 80% by 2050
- Send zero waste to landfills by 2030
- Enhance parks
- Convert contaminated land to safe, beneficial use
- Manage water resources and alleviate flooding
- Achieve the best air quality of all large U.S. cities

FIGURE 13.19 New York City is making impressive strides in urban sustainability under the OneNYC (formerly PlaNYC) program. The newly completed One World Trade Center tower that now dominates the skyline has many energy-saving and water-saving green building features. *Goals from City of New York, Office of the Mayor, 2015. One New York: The plan for a strong and just city.*

Steps toward livability enhance sustainability

Most steps being taken to make cities more livable also help to make them more sustainable. Planning and zoning are pursuits that specifically entail a long-term vision. By projecting further into the future than political leaders or businesses generally do, planning and zoning are powerful forces for sustaining urban communities. The principles and practices of smart growth and new urbanism cut down on energy consumption, helping us to address the looming challenge of climate change (Chapter 18). Moving transportation away from cars and toward mass transit reduces gasoline consumption and carbon emissions. Parks offer ecosystem services while promoting residents' health. And green buildings bring a diversity of health and environmental benefits.

Successes from Portland to Curitiba to New York City show how we can make cities more sustainable. Indeed, because they affect the environment in many positive ways and can encourage efficient resource use, urban centers are a key element in achieving progress toward global sustainability.

closing
THE LOOP

As the human population shifts from rural to urban lifestyles, the nature of our environmental impact changes. As predominantly urban- and suburban-dwellers, our impacts are less direct but more far-reaching. Goods must be delivered to us over long distances, requiring the use of more resources. Limiting the waste of those resources by making urban and suburban areas more efficient will be vital for our future. Fortunately, the innovative cultural environment that cities foster is helping us develop sustainable solutions.

Portland, Oregon, is among those American cities that have been enhancing the quality of life for their citizens while making strides toward environmental sustainability. However, as people stream to Portland in droves, the city risks becoming a victim of its own success. Growth forecasts estimate that the number of households in Portland will jump by 44–57% as early as 2035, and that households in its three-county region will rise by 56–74%. As density increases inside the urban growth boundary, new challenges—such as rising rents, increased traffic, and parking congestion in residential neighborhoods—are beginning to strain the smart-growth vision that has worked so well. Portland's leaders have been engaging citizens in a planning process to design solutions to keep their city "prosperous, healthy, equitable, and resilient," and they have recently completed a Comprehensive Plan to help guide decision making through 2035.

Portland is just one of many urban centers whose residents are seeking to enhance their quality of life while protecting the quality of their environment. Both city and regional planning have improved many urban areas, and we should be encouraged about such progress. Ongoing experimentation in our urbanizing world will help us determine how to continue creating better and more sustainable communities in which to live.

REVIEWING Objectives

You should now be able to:

Describe the scale of urbanization

The world's population has become predominantly urban. Our ongoing shift from rural to urban living is driven largely by industrialization and is proceeding fastest in the developing world. The location and growth of cities have always been influenced by environmental factors, but the geography of urban areas changes as cities decentralize and suburbs grow and expand. (pp. 332–334)

Define sprawl and discuss its causes and effects

Sprawl covers large areas of land with low-density development. Both population growth and increased per capita land use contribute to sprawl. Sprawl

results from the home-buying choices of individuals who prefer suburbs to cities, but has been facilitated by government policy and technological developments. Sprawl may lead to negative impacts involving transportation, pollution, health, land use, natural habitat, and economics. (pp. 334–336)

Outline city and regional planning and land use strategies

City planning and regional planning, along with zoning, are key tools for improving the quality of urban life. Urban growth boundaries, "smart growth," and "new urbanism" attempt to re-create compact and vibrant urban spaces. (pp. 336–339)

Evaluate transportation options, urban parks, and green buildings

Mass transit systems can enhance the efficiency and sustainability of urban areas. Urban park lands provide recreation, soothe the stress of urban life, and keep people in touch with natural areas. Green buildings minimize their ecological footprints by using sustainable materials, limiting the use of energy and water, minimizing health impacts on their occupants, controlling pollution, and recycling waste. (pp. 339–344)

Analyze environmental impacts and advantages of urban centers

Cities are resource sinks with high per capita resource consumption, and they create substantial waste and pollution. However, cities also maximize efficiency, help preserve natural lands, and foster innovation that can lead to solutions for environmental problems. (pp. 344–348)

Discuss urban ecology and the pursuit of sustainable cities

Linear modes of consumption and production are unsustainable, and more circular modes will be needed to create truly sustainable cities. Many cities worldwide are taking steps to decrease their ecological footprints. Most steps taken for urban livability also enhance sustainability. (pp. 346–349)

TESTING Your Comprehension

1. What factors lie behind the shift of population from rural areas to urban areas? What types of nations are experiencing the fastest urban growth today, and why?

2. Why have so many city-dwellers in the United States and other developed nations moved into suburbs?

3. Give two definitions of *sprawl*. Describe five negative impacts that have been suggested to result from sprawl.

4. What are city planning and regional planning? Contrast planning with zoning. Give examples of some of the suggestions made by early planners such as Daniel Burnham and Edward Bennett.

5. How are some people trying to prevent or slow sprawl? Describe some key elements of "smart growth." What effects, positive and negative, do urban growth boundaries tend to have?

6. Describe several apparent benefits of rail transit systems. What is a potential drawback?

7. How are parks thought to make urban areas more livable? Give three examples of types of parks or public spaces.

8. What is a green building? Describe several features a LEED-certified building may have.

9. Describe the connection between urban ecology and sustainable cities. List three actions a city can take to enhance its sustainability.

10. Name two positive effects of urban centers on the natural environment.

SEEKING Solutions

1. Describe the causes of the spread of suburbs and outline the environmental, social, and economic impacts of sprawl. Overall, do you think the spread of urban and suburban development that is commonly labeled *sprawl* is predominantly a good thing or a bad thing? Do you think it is inevitable? Give reasons for your answers.

2. Would you personally want to live in a neighborhood developed in the new urbanist style? Why or why not? Would you like to live in a city or region with an urban growth boundary? Why or why not?

3. All things considered, do you feel that cities are a positive thing or a negative thing for environmental quality? How much do you think we may be able to improve the sustainability of our urban areas?

4. **CASE STUDY CONNECTION** After you earn your college degree, you decide to settle in the Portland, Oregon, region, where you are being offered three equally desirable jobs in three very different locations. If you accept the first, you will live in downtown Portland, amid commercial and cultural amenities but where population density is high and growing. If you take the second, you will live in one of Portland's suburbs where you have more space but where commute times are long and sprawl may soon surround you for miles. If you select the third, you will live in a rural area outside the urban growth boundary with plenty of space and scenic beauty but few cultural amenities. You are a person who aims to live in an ecologically sustainable way. Where would you choose to live? Why? What considerations will you factor into your decision?

5. **THINK IT THROUGH** You are the facilities manager on your campus, and your school's administration has committed funds to retrofit one existing building with sustainable green construction techniques so that it earns LEED certification. Consider the various buildings on your campus, and select one that you feel is unhealthy

or that wastes resources in some way and that you would like to see retrofitted. Describe for an architect three specific ways in which green building techniques might be used to improve this particular building.

6. **THINK IT THROUGH** You are the president of your college or university, and students are clamoring for you to help create the world's first fully sustainable campus. Consid-ering how people enhance livability and sustainability in cities, what lessons might you try to apply to your college or university? You are scheduled to give a speech to the campus community about your plans and will need to name five specific actions you plan to take to pursue a sustainable campus. What will they be, and what will you say about each choice to describe its importance?

CALCULATING Ecological Footprints

One way to reduce your ecological footprint is with alternative transportation. Each gallon of gasoline is converted during combustion to approximately 20 pounds of carbon dioxide (CO_2), which is released into the atmosphere. The table lists typical amounts of CO_2 released per person per mile through various modes of transportation, assuming typical fuel efficiencies.

For an average North American person who travels 12,000 miles per year, calculate and record in the table the CO_2 emitted yearly for each transportation option, and the reduction in CO_2 emissions that one could achieve by relying solely on each option.

MODE OF TRANSPORT	CO_2 PER PERSON PER MILE	CO_2 PER PERSON PER YEAR	CO_2 EMISSION REDUCTION	YOUR ESTIMATED MILEAGE PER YEAR	YOUR CO_2 EMISSIONS PER YEAR
Automobile (driver only)	0.825 lb	9900 lb	0		
Automobile (2 persons)	0.413 lb				
Bus	0.261 lb				
Walking	0.082 lb				
Bicycle	0.049 lb				
				Total = 12,000	

1. Which transportation option provides the most miles traveled per unit of carbon dioxide emitted?

2. Clearly, it is unlikely that any of us will walk or bicycle 12,000 miles per year. In the last two columns, estimate what proportion of the 12,000 annual miles you think that you actually travel by each method, and then calculate the CO_2 emissions that you are responsible for generating over the course of a year. Which transportation option accounts for the most emissions for you?

3. How could you reduce your CO_2 emissions? How many pounds of emissions do you think you could realistically eliminate over the course of the next year by making changes in your transportation decisions?

MasteringEnvironmentalScience®

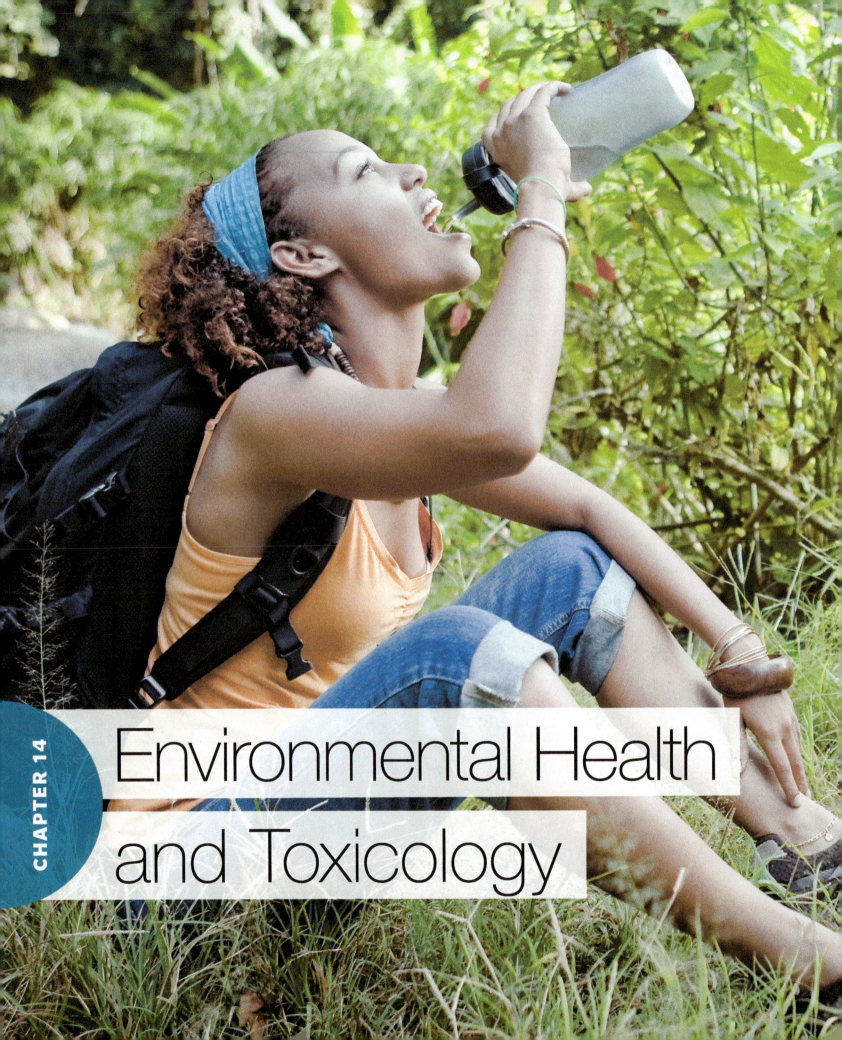

Environmental Health
and Toxicology

Many consumers are embracing "BPA free" water bottles due to concerns about the safety of bisphenol A.

Poison in the Bottle: How Safe Is Bisphenol A?

> **This chemical is harming snails, insects, lobsters, fish, frog, reptiles, birds, and rats, and the chemical industry is telling people that because you're human, unless there's human data, you can feel completely safe.**
> Dr. Frederick vom Saal, BPA researcher

> **There is no basis for human health concerns from exposure to BPA.**
> The American Chemistry Council

How is it that a chemical found to alter reproductive development in animals gets used in baby bottles? How can it be that a substance linked to breast cancer, prostate cancer, and heart disease is routinely used in food and drink containers? The chemical bisphenol A (BPA for short) has been associated with everything from neurological effects to miscarriages. Yet it's in hundreds of products we use every day, and there's a better than 9 in 10 chance that it is coursing through your body right now.

To understand how chemicals that may pose health risks come to be widespread in our society, we need to explore how scientists and policymakers study toxic substances and other environmental health risks—and the vexing challenges these pursuits entail.

Chemists first synthesized BPA, an organic compound (p. 27) with the chemical formula $C_{15}H_{16}O_2$, in 1891. As they began producing plastics in the 1950s, chemists found BPA to be useful in creating epoxy resins used in lacquers and coatings. Epoxy resins containing BPA were soon being used to line the insides of metal food and drink cans and the insides of pipes for our water supply, as well as in enamels, varnishes, adhesives, and even dental sealants for our teeth.

Chemists also found that linking BPA molecules into polymers (p. 27) helped create polycarbonate plastic, a hard, clear type of plastic that soon found use in water bottles, food containers, eating utensils, eyeglass lenses, CDs and DVDs, laptops and other electronics, auto parts, sports equipment, baby bottles, and children's toys. With so many uses, BPA has become one of the world's most-produced chemicals; each year we make about half a kilogram (1 lb) of BPA for each person on the planet, and 9 kg (20 lb) per person in the United States!

Unfortunately, BPA leaches out of its many products and into our food, water, air, and bodies. Fully 93% of Americans carry detectable concentrations in their urine, according to the latest National Health and Nutrition Examination Survey conducted by the Centers for Disease Control and Prevention (CDC). Because most BPA passes through the body within hours, these data suggest that we are receiving almost continuous exposure. Babies and children have higher relative exposure to BPA because they eat more for their body weight and metabolize the chemical less effectively.

What, if anything, is BPA doing to us? To address such questions, scientists run experiments on laboratory animals, administering known doses of the substance and

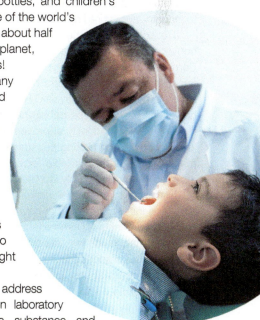

Dental sealants may contain BPA. ▲

353

Behavioral issues

Cardiovascular disease

Type 2 diabetes

Obesity

Asthma

Hypertension

Sperm quality

Male sexual performance

Male genital abnormalities

Blood hormone levels

Kidney functioning

Thyroid function

Immune function

Egg development and maturation

FIGURE 14.1 Studies have linked elevated blood/urine BPA concentrations to numerous health impacts in humans. Although these correlative studies do not conclusively prove that BPA causes each observed ailment, they indicate topics for further research.

measuring the health impacts that result. Hundreds of studies with rats, mice, and other animals have shown many apparent effects of BPA, including a wide range of reproductive abnormalities. Recent studies suggest humans suffer health impacts from BPA as well (see **THE SCIENCE BEHIND THE STORY**, pp. 356–357).

Many of these effects occur when BPA is present at extremely low doses—much lower than the exposure levels set so far by regulatory agencies for human safety. Scientists say this is because BPA mimics the female sex hormone estrogen; that is, it is structurally similar to estrogen and can induce some of its effects in animals (see Figure 14.10, p. 365). Hormones such as estrogen function at minute concentrations, so when a synthetic chemical similar to estrogen reaches the body in a similarly low concentration, it can fool the body into responding.

In reaction to research involving animals, a growing number of researchers, doctors, and consumer advocates are calling on governments to regulate BPA and for manufacturers to stop using it. The chemical industry insists that BPA is safe, pointing to industry-sponsored research that finds no health impacts.

To sort through the debate, several expert panels have convened to assess the fast-growing body of scientific studies. Some panels have found typical BPA exposure to be a cause for concern, whereas others have concluded such exposure is not a meaningful health risk. Regardless of their conclusion, however, most of these panels have indicated a need for the development of federally approved testing protocols for studies of hormone-mimicking substances. Such guidelines would make more studies available for consideration by expert panels, who can evaluate only studies that meet established federal guidelines for toxicological research. As existing guidelines are designed for substances that have "traditional" toxicity profiles, such as increasing adverse effects with increasing exposure to the toxin, new guidelines need to be developed for hormone-mimicking substances that have unconventional toxicity profiles and exert effects at very low doses. Indeed, dealing with substances like BPA is forcing us toward a challenging paradigm shift in the way we assess environmental health risks.

As governments continue to consider differing regulatory approaches for products containing BPA, studies suggesting human health impacts of the chemical are now emerging. A 2013 review found 91 studies that examined the relationship between the level of BPA in research participants' urine or blood and a variety of health problems (**FIGURE 14.1**). Although these studies are correlative (p. 13), they collectively suggest that exposure to elevated BPA levels may be harmful to humans.

In light of a growing body of research, some governments have taken steps to regulate the use of BPA in consumer products. Canada, for example, has banned BPA completely. In many other nations, including the United States, its use in products for babies and small children has been restricted. Accordingly, concerned parents can now more easily find BPA-free items for their infants and children, but the rest of us remain exposed through most food cans, many drink containers, and thousands of other products.

In the face of mounting public concern about the safety of BPA, many companies are voluntarily choosing to remove it from their products, even in the absence of regulation by the U.S. government. WalMart and Toys "R" Us, for example, decided to stop carrying children's products with BPA several years before the U.S. Food and Drug Administration (FDA) banned BPA use in baby bottles in 2012. Campbell's has announced that it is transitioning away from the use of BPA in its soup can liners, and food giants ConAgra, Nestlé, and Heinz have also pledged to remove BPA from their food packaging. There is precedent for such efforts, because BPA was voluntarily phased out of can liners in Japan starting in the late 1990s.

Although we don't yet know everything there is to know about BPA, it isn't likely to be among our greatest environmental health threats. However, it provides a timely example of how we as a society assess health risks and decide how to manage them. As scientists and government regulators assess BPA's potential risks, their efforts give us a window on how hormone-disrupting chemicals are challenging the way we appraise and control the environmental health risks we face.

Environmental Health

Examining the impacts of human-made chemicals such as BPA is just one aspect of the broad field of **environmental health,** which assesses environmental factors that influence our health and quality of life. These factors include wholly natural aspects of the environment over which we have little or no control, as well as anthropogenic (human-caused) factors. Practitioners of environmental health seek to prevent adverse effects on human health and on the ecological systems that are essential to our well-being.

We face four types of environmental hazards

Many environmental health hazards exist in the world around us. We can categorize them into four main types: physical, chemical, biological, and cultural. For each type of hazard, there is some amount of risk that we cannot avoid—but there is also some amount of risk that we *can* avoid by taking precautions. Much of environmental health consists of taking steps to minimize the risks of encountering hazards and to lessen the impacts of the hazards we do encounter.

Physical hazards **Physical hazards** arise from processes that occur naturally in our environment and pose risks to human life or health. Some are ongoing natural phenomena, such as ultraviolet (UV) radiation from sunlight (**FIGURE 14.2a**). Excessive exposure to UV radiation damages DNA in cells and has been tied to skin cancer, cataracts, and immune suppression. We can reduce these risks by shielding our skin from intense sunlight with clothing and sunscreen and avoiding excessive sun exposure.

Other physical hazards include discrete events such as earthquakes, volcanic eruptions, fires, floods, blizzards, landslides, hurricanes, and droughts. We can do little to predict the timing of a natural disaster such as an earthquake, and nothing to prevent one. However, we can minimize risk by preparing ourselves. Scientists can map geologic faults to determine areas at risk of earthquakes, engineers can design buildings to resist damage, and governments and individuals can create emergency plans to prepare for a quake's aftermath.

Some common practices make us more vulnerable to certain physical hazards. Clear-cutting hillsides makes landslides more likely, for example, and channelizing rivers promotes flooding in some areas while preventing it in others (pp. 396–397). We can reduce risk from such hazards by improving our forestry and flood control practices and by carefully regulating development in areas that are prone to landslides or flooding.

Chemical hazards **Chemical hazards** include many of the synthetic chemicals that humanity manufactures, such as pharmaceuticals, disinfectants, and pesticides (**FIGURE 14.2b**). Some substances produced naturally by organisms (such as venoms) also can be hazardous, as can many substances that we find in nature and then process for our use (such as hydrocarbons, lead, and asbestos). Following our overview of environmental health, much of this chapter

(a) Physical hazard

(b) Chemical hazard

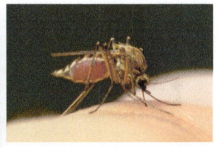

(c) Biological hazard

(d) Cultural hazard

FIGURE 14.2 Environmental health hazards come in four types. The sun's ultraviolet radiation is an example of a physical hazard **(a)**. Chemical hazards **(b)** include both synthetic and natural chemicals. Biological hazards **(c)** include diseases and the organisms that transmit them. Cultural or lifestyle hazards **(d)** include the behavioral decisions we make, such as smoking, as well as the socioeconomic constraints forced on us.

THE SCIENCE
behind the story

Did an Error Cleaning Mouse Cages Alert Us to the Dangers of Bisphenol A?

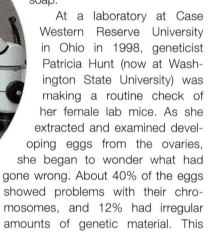

Dr. Patricia Hunt, Washington State University

Of the many studies documenting health impacts of BPA on lab animals, one of the first came about because a lab assistant reached for the wrong soap.

At a laboratory at Case Western Reserve University in Ohio in 1998, geneticist Patricia Hunt (now at Washington State University) was making a routine check of her female lab mice. As she extracted and examined developing eggs from the ovaries, she began to wonder what had gone wrong. About 40% of the eggs showed problems with their chromosomes, and 12% had irregular amounts of genetic material. This dangerous condition, called *aneuploidy*, can lead to miscarriages or birth defects in mice and people alike.

A bit of sleuthing revealed that a lab assistant had mistakenly washed the lab's plastic mouse cages and water bottles with an especially harsh soap. The soap damaged the cages so badly that parts of them seemed to have melted.

The cages were made from polycarbonate plastic, which contains BPA. Hunt knew that BPA mimics estrogen and that some studies had linked the chemical to reproductive abnormalities in mice, such as low sperm counts and early sexual development. Other research indicated that BPA leaches out of plastic into water and food when the plastic is treated with heat, acidity, or harsh soap.

Hunt wondered whether the chemical might be adversely affecting the mice in her lab. Deciding to re-create the accidental cage-washing incident in a controlled experiment, Hunt instructed researchers in her lab to wash polycarbonate cages and water bottles using varying levels of the harsh soap. They then compared mice kept in damaged cages with plastic water bottles to mice kept in undamaged cages with glass water bottles.

The developing eggs of mice exposed to BPA through the deliberately damaged plastic showed significant problems during meiosis, the division of chromosomes during egg formation—just as they had in the original incident (**FIGURE 1**). In contrast, the eggs of mice in the control cages were normal.

In another round of tests, Hunt's team gave sets of female mice daily oral doses of BPA over 3, 5, and 7 days. They observed the same meiotic abnormalities in these mice, although at lower levels (**FIGURE 2**). The mice given BPA for 7 days were most severely affected.

Published in 2003 in the journal *Current Biology*, Hunt's findings set off a new wave of concern over the safety of BPA. The findings were disturbing because sex cells of mice and of people divide and function in similar ways. "We have observed meiotic defects in mice at exposure levels close to or even below those considered 'safe' for humans," the

will focus on chemical health hazards and the ways we study and regulate them.

Biological hazards **Biological hazards** result from ecological interactions among organisms (**FIGURE 14.2c**). When we become sick from a virus, bacterial infection, or other pathogen, we are suffering parasitism (p. 77). This is what we call **infectious disease.** Some infectious diseases are spread when pathogenic microbes attack us directly. With others, infection occurs through a **vector,** an organism (such as a mosquito) that transfers the pathogen to the host. Infectious diseases such as malaria, cholera, tuberculosis, and influenza (flu) are major environmental health hazards, especially in developing nations with widespread poverty and few resources for health care. As with physical and chemical hazards, it is impossible for us to avoid risk from biological

agents completely, but through monitoring, sanitation, and medical treatment we can reduce the likelihood and impacts of infection.

Cultural hazards Hazards that result from our place of residence, the circumstances of our socioeconomic status, our occupation, or our behavioral choices can be thought of as **cultural hazards.** We can minimize or prevent some of these cultural or lifestyle hazards, whereas others may be beyond our control. For instance, choosing to smoke cigarettes, or living or working with people who smoke, greatly increases our risk of lung cancer (**FIGURE 14.2d**). Choosing to smoke is a personal behavioral decision, but exposure to secondhand smoke in the home or workplace may be beyond one's control. The influences of personal choices and "forced" decisions on health can also apply to diet and nutrition, workplace

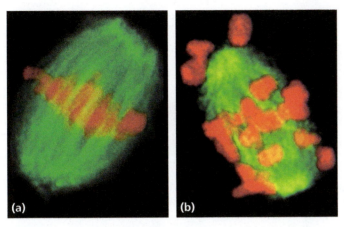

FIGURE 1 In normal cell division (a), chromosomes (red) align properly. Exposure to BPA causes abnormal cell division **(b)**, whereby chromosomes scatter and are distributed improperly and unevenly between daughter cells.

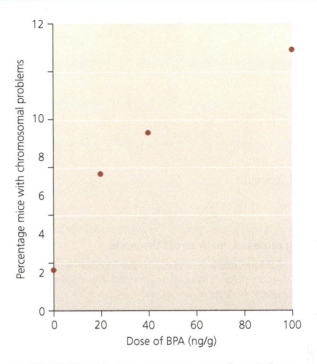

FIGURE 2 In this dose-response experiment, the percentage of mice showing chromosomal problems during cell division rose with increasing doses of BPA. In the United States and Europe, regulators have set safe intake levels for people at doses of 50 ng/g of body weight per day. *Data from Hunt, P. A., et al., 2003. Bisphenol A exposure causes meiotic aneuploidy in the female mouse.* Current Biology *13: 546–553.*

DATA Q Using the figure, predict the percentage of mice in the study that would likely suffer chromosomal problems when exposed to a BPA dosage of 70 ng/g. What would this percentage be?

Go to **Interpreting Graphs & Data** on Mastering EnvironmentalScience*

research paper stated. "Clearly, the possibility that BPA exposure increases the likelihood of genetically abnormal offspring is too serious to be dismissed without extensive further study."

Since that time, hundreds of other studies of BPA at low doses have documented harmful effects in lab animals, including reproductive disorders related to estrogen mimicry and other maladies ranging from thyroid problems to liver damage to elevated anxiety. Epidemiological studies are also beginning to shed light on the potentially far-reaching impacts of BPA on human health (see the opening Case Study, pp. 353–354). As a result, more and more scientists are urging regulators to adopt a precautionary position and restrict BPA based on this diverse body of evidence.

hazards, and drug use. As advocates of environmental justice (pp. 137–139) argue, health factors such as living near toxic waste sites or working with pesticides without proper training and safeguards are often correlated with socioeconomic deprivation. In general, the fewer economic resources or political clout one has, the harder it is to avoid cultural hazards and environmental health risks.

Noninfectious disease is affected by genes, environment, and lifestyle

Despite all our technological advances, we still find ourselves battling disease, which causes the vast majority of human deaths worldwide (**FIGURE 14.3**, p. 358). We've seen

that infectious diseases are caused by a pathogenic organism infecting a host. **Noninfectious diseases,** such as cancer and heart disease, develop without the action of a foreign organism. You don't "catch" noninfectious diseases—people develop them through a combination of their genetics, coupled with environmental and lifestyle factors. For instance, whether a person develops lung cancer depends not only on his or her genes but also on environmental conditions, such as the individual's exposure to airborne cancer-inducing chemicals, and to lifestyle choices, such as whether or not he or she chooses to smoke.

More than half the world's deaths result from noninfectious diseases (see Figure 14.3a), but the incidence of such diseases can be lessened by wider adoption of healthy lifestyles. In the United States, lifestyle trends are altering the prevalence of noninfectious disease. Over the past 25 years,

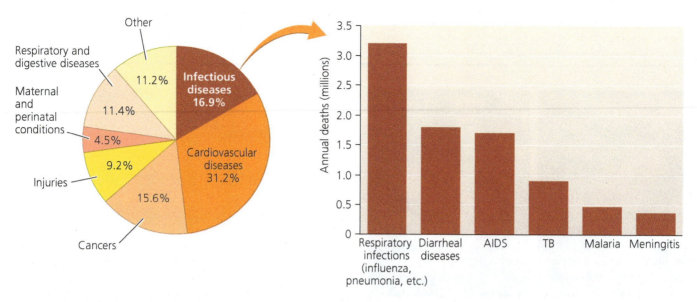

(a) Leading causes of death across the world

(b) Leading causes of death by infectious diseases

FIGURE 14.3 Infectious diseases are the second-leading cause of death worldwide. Six types of diseases account for 80% of all deaths from infectious disease. *Data from World Health Organization, 2015. Geneva, Switzerland: WHO, http://www.who.int.*

DATA AIDS is a well-known infectious disease, but respiratory and diarrheal diseases claim far more lives every year than AIDS. According to the figure, how many times more lives were lost to respiratory infections and diarrheal diseases than to AIDS?

Go to **Interpreting Graphs & Data** on **Mastering**EnvironmentalScience*.

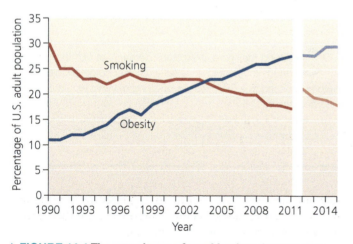

FIGURE 14.4 The prevalence of smoking has decreased in the United States in recent decades, but obesity is on the rise. Data from 1990–2011 is not directly comparable to data from 2012–2015 for both smoking and obesity due to methodology changes implemented after collecting the 2011 data. *Data from United Health Foundation, 2015. America's health rankings, 2015 edition. Minnetonka, MN: United Health Foundation.*

DATA The population of the United States in mid-2015 was approximately 320 million people. Using the figure, estimate the number of Americans in 2015 who were (a) obese and (b) smokers.

Go to **Interpreting Graphs & Data** on **Mastering**EnvironmentalScience*

the percentage of Americans who smoke cigarettes has decreased while obesity rates have increased (**FIGURE 14.4**). This has reduced the prevalence of factors that induce lung cancer but increased those that induce heart disease and type 2 diabetes. Obesity is typically due to low levels of physical activity coupled with a diet high in calories and fat, but other factors can be involved. Studies have found that exposing pregnant mice to high levels of BPA increases the production of fat cells in her developing embryos. This causes prenatally exposed mice to store more fat as adults than mice that were not exposed to BPA in the womb, even if both groups of mice have identical diets.

Infectious disease has long plagued humanity

Infectious diseases have ravaged human populations throughout history, sometimes claiming huge numbers of lives in massive epidemics. Examples of such diseases include cholera, bubonic plague, tuberculosis, malaria, smallpox, and various strains of flu—just to name a few.

Infectious diseases spread when a pathogenic organism enters a host through the skin, via the respiratory system, or by the consumption of contaminated food or water. Once established in the host, the pathogen uses energy from the host's tissues and the favorable internal conditions inside the body (a warm, wet environment that is ideal for bacteria and viruses) to produce huge numbers of offspring. Because these offspring will need new hosts to survive and reproduce, the pathogen helps to expel its offspring by inducing vomiting or diarrhea (for pathogens that inhabit the digestive tract, such as *Salmonella*) or coughing and sneezing (for those that inhabit the respiratory system, such as influenza viruses) in the host. These pathogen-containing body fluids or aerosols are then taken in by caregivers, or contaminate local food and water sources, completing the cycle of infection.

As shown in Figure 14.3a, infectious diseases account for about 17% of deaths worldwide that occur each year, but can be responsible for up to 50% of annual deaths in some poorer nations. Further, although infectious disease accounts for fewer deaths than noninfectious disease, infectious disease robs society of more "years of life" (a measure of the difference between the actual age at which a person dies and the age at which a person would die if he or she were able to lead a healthy lifestyle), because it tends to strike people of all ages, including the very young. The World Health Organization (WHO) estimates that although infectious diseases are responsible for fewer than 20% of deaths worldwide each year, they are responsible for more than 40% of the "years of life" lost to death worldwide each year—showing how many young lives are tragically cut short by infectious disease. Fortunately, modern advances have reduced mortality from infectious disease in the human population but, as these data show, we still have much work to do.

We fight infectious disease with diverse approaches

Although infectious disease has troubled human civilizations since the dawn of time, it wasn't until the late 1800s that scientists firmly established the connection between diseases and pathogenic viruses, bacteria, and protists. This discovery gave rise to antibiotics, chemicals that combat disease by killing pathogens. It also produced societal approaches that help to minimize the spread of disease. These include sterilizing drinking water, providing sanitary facilities (such as indoor plumbing or communal latrines) that prevent fecal contamination of drinking water sources, ensuring nutritious diets for all people to strengthen the immune system, and providing medical care that identifies infected individuals and breaks the cycle of infection through treatment, quarantining, and early intervention (**FIGURE 14.5**).

FIGURE 14.5 The infectious disease cholera ravaged London in the 19th century. Feces, both animal and human, were ever-present in city streets; the disease was spread through direct contact with feces and contamination of local drinking water sources.

Collectively, these approaches illustrate the ways we can combat infectious disease in the modern world. Besides providing food security (p. 237), this means ensuring access to safe drinking water and improving sanitation for all people, especially poor individuals. In recent years, we have made slow but steady progress in providing adequate drinking water and sanitation to the world's people, which helps reduce the incidence of diseases such as cholera and dysentery that are spread through drinking water contaminated with human or animal feces. For example, in 2015, 91% of people across the globe had access to safe drinking water—an increase of 2.6 billion people since 1990. Some 2.1 billion people gained access to sanitary facilities from 1990 to 2015, resulting in 68% of the global population using sanitary facilities. Areas currently unserved by safe drinking water and advanced sanitation are typically rural areas in poorer nations (**FIGURE 14.6**, p. 360).

Another important tool in fighting infectious disease is expanding access to health care. In developing nations, this includes opening clinics, immunizing children against diseases, providing prenatal and postnatal care for mothers and babies, and making generic and inexpensive pharmaceuticals available.

Education campaigns play a vital role in rich and poor nations alike. Public service announcements and government programs that educate the public about proper hygiene and safe food-handling procedures can minimize the incidence of disease. And as some infectious diseases are spread by sexual contact, such as HIV/AIDS, education on sex and reproductive health is helping men and women reduce risk from disease and avoid unwanted pregnancies.

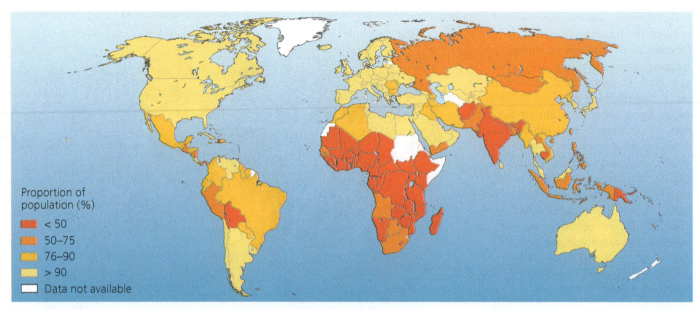

FIGURE 14.6 **We are gradually improving access to improved sanitation for the world's people, but access in some regions remains inadequate.** In much of central Africa, for example, less than half of the people have access to sanitary facilities that reduce the spread of disease. *Data from World Health Organization, 2015. Geneva, Switzerland: WHO, http://www.who.int.*

Infectious disease remains a concern

Although we have made progress in combating disease around the world, we still must remain vigilant because potential epidemics are constantly emerging. Recent examples include severe acute respiratory syndrome (SARS) in 2003, the H5N1 avian flu ("bird flu") starting in 2004, the H1N1 swine flu that spread across the globe in 2009–2010, and the outbreak of Ebola in West Africa in 2014 (**FIGURE 14.7**).

FIGURE 14.7 **Disease can spread rapidly in our highly mobile, internationalized world.** The Ebola outbreak in 2014 claimed more than 6000 lives as it spread among several nations in West Africa and spurred many nations to adopt protocols for evaluating travelers for signs of the disease at airports, ports, and border crossings to limit the spread of the virus.

Diseases like influenza are caused by pathogens that mutate readily, giving rise to a variety of strains of the disease with slightly different genetics. As a pathogen's genes determine its virulence (a measure of how fast a disease spreads and the harm it does to infected individuals), these frequent mutations make it more likely that a highly virulent strain may arise and threaten a global pandemic at any time. In addition to natural strains of diseases, there is growing concern over bioterrorism, the intentional genetic manipulation of a pathogenic organism to increase its virulence and/or transmission between humans for use as a weapon. To complicate matters further, some long-identified pathogens, such as those causing tuberculosis and strains of malaria, are evolving resistance to our antibiotics, in the same way that some pests have evolved resistance to our pesticides (pp. 247–248).

The spread of disease today is much easier than in the past. In our world of dense human populations and extensive international travel, novel diseases (and new strains of old diseases) that emerge in one location are more likely to spread rapidly to other locations than they were in the past. A pathogen in its human host can now hop continents in a matter of hours by airplane. The changes people produce in the environment can also cause diseases to spread. Human-induced global warming (Chapter 18), for example, is causing tropical diseases such as malaria, dengue, cholera, and yellow fever to expand into temperate climates where they formerly could not thrive. Humans can also indirectly promote the spread of disease when they inadvertently aid disease vectors. For instance, clearing forests for agriculture can increase mosquito populations by providing them farm ponds in which they can breed and by driving away mosquito predators such as frogs, toads, and bats.

Environmental health experts continue to work, along with governments and nongovernmental organizations such as the WHO, to implement programs around the world that act to reduce the incidence of infectious disease and prevent widespread outbreaks. But given that the pathogens that infect us will not stop evolving, we cannot afford to let up on our vigilance in acting proactively to head off epidemics before they begin.

Toxicology is the study of chemical hazards

Although most indicators of human health are improving as the world's wealth increases, our modern society is exposing us to more and more synthetic chemicals. Some of these substances pose threats to human health, but figuring out which of them do—and how, and to what degree—is a complicated scientific endeavor. **Toxicology** is the science that examines the effects of poisonous substances on humans and other organisms. Toxicologists assess and compare substances to determine their **toxicity,** the degree of harm a chemical substance can inflict. A toxic substance, or poison, is called a **toxicant,** but any chemical substance may exert negative impacts if we ingest or expose ourselves to enough of it. Conversely, if the quantity is small enough, a toxicant may pose no health risk at all. These facts are often summarized in the catchphrase, "The dose makes the poison." In other words, a substance's toxicity depends not only on its chemical properties but also on its quantity.

In recent decades, our ability to produce new chemicals has expanded, concentrations of chemical contaminants in the environment have increased, and public concern for health and the environment has grown. These trends have driven the rise of **environmental toxicology,** which deals specifically with toxic substances that come from or are discharged into the environment. Toxicologists generally focus on human health, using other organisms as models and test subjects. Environmental toxicologists study animals and plants to determine the ecological impacts of toxic substances and to see whether other organisms can serve as indicators of health threats that could soon affect people.

Many environmental health hazards exist indoors

Modern Americans spend roughly 90% of their lives indoors. Unfortunately, the spaces inside homes and workplaces, just like the outdoors, can be rife with environmental hazards (pp. 472–474).

Cigarette smoke and radon are leading indoor hazards (p. 472) and are the top two causes of lung cancer in developed nations. Cigarette smoke contains substances that can harm the respiratory system and induce cancer. **Radon** is a highly toxic radioactive gas that is colorless and undetectable without specialized kits (**FIGURE 14.8**). Radon seeps up from the ground in areas with certain types of bedrock and can accumulate in basements and homes with poor air circulation.

Homes and offices can have problems with toxic compounds produced by mold, which can flourish in wall spaces when moisture levels are high. **Asbestos,** used in the past as insulation in walls and other products, is dangerous when inhaled. Long-term exposure to asbestos scars the lung tissue, impairs lung function, and leads to a disorder called **asbestosis.**

FIGURE 14.8 People can determine their exposure to radon gas with in-home testing. Air samples are collected in specialized collectors like the one shown, and then mailed to a laboratory for analysis.

Lead poisoning is another indoor health hazard. **Lead** is a heavy metal, and when ingested, it can cause damage to the brain, liver, kidney, and stomach; learning problems and behavioral abnormalities in children; anemia; hearing loss; and even death. Lead poisoning among U.S. children has greatly declined in recent years, however, as a result of education campaigns and the U.S. federal government mandating the phaseout of lead-based paints and leaded gasoline (p. 7) in the 1970s. Today lead poisoning can result from drinking water that has passed through the lead pipes common in older homes or from ingesting or inhaling lead-containing dust produced by the slow wearing-away of leaded paint.

One recently recognized hazard is a group of chemicals known as **polybrominated diphenyl ethers (PBDEs).** These compounds provide fire-retardant properties and are used in a diverse array of consumer products, including computers, televisions, plastics, and furniture. PBDEs are emitted during production and disposal of products in which they are used and may also release into the air at very slow rates throughout the lifetime of these products. These chemicals persist and accumulate in living tissue, and their abundance in the environment and in people in the United States is doubling every few years.

Like BPA, PBDEs appear to act as hormone disruptors; lab testing with animals shows them to affect thyroid hormones. Animal testing also suggests that PBDEs affect the development of the brain and nervous system and may cause cancer. Concern about PBDEs rose after a study showed that concentrations in the breast milk of Swedish mothers had increased exponentially from 1972 to 1997. U.S. studies also show rising concentrations in breast milk. The European Union decided in 2003 to ban PBDEs, and industries in Europe phased them out. As a result, concentrations in breast milk of European mothers have fallen substantially. In the United States, however, there has so far been little movement to address the issue. The dangers posed by fire retardants such as PBDEs have caused some to question the stringent flammability standards that often make the use of such chemicals necessary. As stated by Linda Birnbaum of the National Institute of Environmental Health Sciences in an interview on PBDEs, "I don't question the need for flame retardants in airplanes, but do we need them in nursing pillows and babies' strollers?"

Risks must be balanced against rewards

The job of toxicologists and other scientists who study environmental health hazards is to learn as much as they can about the hazards, but then the rest of us need to take this information and weigh it against any benefits we obtain from exposing ourselves to the hazards. With most hazards, there is some trade-off between risk and reward, and we must judge as best we can how these compare. In regard to BPA, its usefulness for many purposes means that despite its health risks, we may as a society choose to continue using it.

As we review the impacts of toxic substances throughout this chapter, it is important to keep in mind that artificially produced chemicals have played a crucial role in giving us the standard of living we enjoy today. These chemicals have helped create the industrial agriculture that produces much of our food (pp. 238–239), the medical advances that protect our health and prolong our lives, and many of the modern materials and conveniences we use every day. It is appropriate to remember these benefits as we examine some of the unfortunate side effects of these advances and as we search for better alternatives.

Toxic Substances and Their Effects on Organisms

Our environment contains countless natural substances that may pose health risks. These include petroleum, oozing naturally from the ground; radon gas, seeping up from bedrock; and **toxins,** toxic chemicals manufactured in the tissues of living organisms. For example, toxins can be chemicals that plants use to ward off herbivores or that insects use to defend themselves from predators. In addition, we are exposed to many synthetic (artificial, or human-made) chemicals, some of which also have toxic properties.

Synthetic chemicals are all around us—and in us

Tens of thousands of synthetic chemicals have been manufactured (**TABLE 14.1**), and synthetic chemicals surround us in our daily lives. Each year in the United States, we manufacture or import 113 kg (250 lb) of chemical substances for every man, woman, and child. Many of these substances find their way into soil, air, and water, as revealed by researchers who monitor environmental quality. For instance, scientists at the U.S. Geological Survey's National Water-Quality Assessment Program (NAWQA) have carried out systematic surveys for synthetic chemicals in U.S waterways and aquifers since

TABLE 14.1 Estimated Numbers of Chemicals in Commercial Substances

TYPE OF CHEMICAL	ESTIMATED NUMBER
Chemicals in commerce	100,000
Industrial chemicals	72,000
New chemicals introduced per year	2000
Pesticides (21,000 products)	600
Food additives	8700
Cosmetic ingredients (40,000 products)	7500
Human pharmaceuticals	3300

Data are for the 1990s, from Harrison, P., and F. Pearce, 2000. *AAAS atlas of population and environment.* Berkeley, CA: University of California Press.

the 1980s. A 2002 study found that 80% of U.S. streams contain at least trace amounts of 82 wastewater contaminants, including antibiotics, detergents, drugs, steroids, plasticizers, disinfectants, solvents, perfumes, and other substances. A 2006 study of groundwater detected 42 volatile organic compounds (VOCs, p. 455) in 18% of wells and 92% of aquifers tested throughout the nation, although fewer than 2% of samples violated federal health standards for drinking water. VOCs are emitted from products such as gasoline, paints, and plastics, and they come from many sources, including urban runoff; engine exhaust; industrial emissions; wastewater; and leaky storage tanks, landfills, and septic systems.

The pesticides we use to kill insects and weeds on farms, lawns, and golf courses are some of the most widespread synthetic chemicals. A 2006 NAWQA study concluded that pesticides are regularly present in streams and groundwater nationwide, finding traces of at least one pesticide in every stream that was tested. The data showed that concentrations were seldom high enough to pose health risks to people, but they were often high enough to affect aquatic life (**FIGURE 14.9**). Pesticide contamination is most severe in the farming states of the Midwest and Great Plains.

As a result of all this exposure, every one of us carries traces of hundreds of industrial chemicals in our bodies. The U.S. government's latest National Health and Nutrition Examination Survey (the one that found 93% of Americans showing traces of BPA in their urine; p. 353) gathered data on 148 foreign compounds in Americans' bodies. Among these were several toxic persistent organic pollutants restricted by international treaty (p. 379). Depending on the pollutant, these were detected in 41% to 100% of the people tested. Smaller-scale surveys have found similar results.

Our exposure to synthetic chemicals begins in the womb, as substances our mothers ingested while pregnant were transferred to us. A 2009 study by the non-profit Environmental Working Group found 232 chemicals in the umbilical cords of 10 newborn babies it tested. Nine of the 10 umbilical cords contained BPA, leading researchers to note that we are born "pre-polluted."

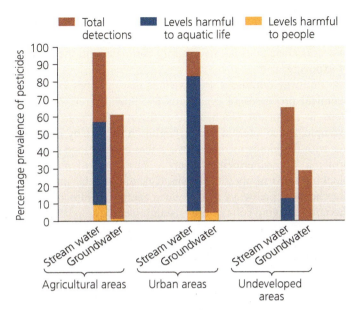

FIGURE 14.9 Nearly all U.S. streams and most aquifers in agricultural and urban areas contain pesticides throughout the year. Fewer than 10% of tested samples violate human health standards, but most violate standards for aquatic life. *Data from Gilliom, Robert J., et al., 2006.* Pesticides in the nation's streams and ground water, 1992–2001. *Circular 1291, National Water-Quality Assessment Program, U.S. Geological Survey.*

All this should not necessarily be cause for alarm. Not all synthetic chemicals pose health risks, and relatively few are known with certainty to be toxic. However, of the roughly 100,000 synthetic chemicals on the market today, very few have been thoroughly tested. For the vast majority, we simply do not know what effects, if any, they may have.

Silent Spring began the public debate over synthetic chemicals

It was not until the 1960s that people began to seriously consider the risks of exposure to pesticides. One event behind this growing awareness was the publication of Rachel Carson's 1962 book *Silent Spring*. At the time, pesticides were indiscriminately sprayed over residential neighborhoods and public areas, on an assumption that the chemicals would do no harm to people. Carson brought together a diverse collection of scientific studies, medical case histories, and other data to demonstrate that the insecticide DDT in particular, and artificial pesticides in general, were hazardous to people, wildlife, and ecosystems. Most consumers had no idea that the store-bought chemicals they used in their houses, gardens, and crops might be toxic.

The chemical industry challenged Carson's book vigorously, attempting to discredit the author's science and personal reputation. Carson suffered from cancer as she finished *Silent Spring* (p. 170), and she lived only briefly after its publication. However, the book was a best-seller and helped generate significant social change in views and actions toward the environment. The use of DDT was banned in the United States in 1973 and is now illegal in a number of nations.

Despite its damaging effects, DDT is still manufactured today because some developing countries with tropical climates use it to control disease vectors, such as mosquitoes that transmit malaria. In these countries, malaria represents a greater health threat than do the toxic effects of the pesticide. New technologies are promising to reduce the need for pesticides to control malaria, however. In 2012, researchers reported that they genetically modified a type of bacteria found in the digestive tract of mosquitoes so that it produced a protein that impairs the hatching of the malarial parasite from its egg sacs in the mosquito's gut. The bacteria were introduced into mosquitoes by mixing them in a sugar solution for mosquitoes to drink. Although 90% of mosquitoes that didn't drink the solution were later found to contain malarial parasites, only 20% of the mosquitoes that drank the solution later contained parasites. The use of genetically modified organisms is not itself without risk (pp. 253–256), but such research may provide non-chemical options for reducing human mortality from malaria and speed the phaseout of DDT around the world.

weighing the ISSUES

A Circle of Poison?

Although many nations have banned the use of DDT, the compound is still manufactured in India and exported to developing nations that lack such bans. How do you feel about this? Is it unethical for a company to sell a substance that has been deemed toxic by so many nations? Or would it be unethical *not* to sell DDT to tropical nations if they desire it for improving public health, such as controlling mosquitoes that transmit malaria or the Zika virus?

Not all toxic substances are synthetic

Although many toxicologists focus on synthetic chemicals, toxic substances also exist naturally in the environment around us and in the foods we eat. Thus, it would be a mistake to assume that all artificial substances are unhealthy and that all natural substances are healthy. In fact, the plants and animals we eat contain many chemicals that can cause us harm. Recall that plants produce toxins to ward off animals that eat them. In domesticating crop plants, we have selected for strains with reduced toxin content, but we have not eliminated these dangers. Furthermore, when we consume animal meat, we ingest toxins the animals obtained from plants or animals they ate. Scientists are actively debating just how much risk natural toxicants pose, and it is clear that more research is required on these questions.

Toxic substances come in different types

Toxicants can be classified based on their particular effects on health. The best known are **carcinogens,** which are substances or types of radiation that cause cancer. In cancer, malignant cells grow uncontrollably, creating tumors, damaging the body, and often leading to death. Cancer frequently has a genetic component, but a wide variety of environmental factors are thought to raise the risk of cancer. Indeed, in 2010 the President's Cancer Panel concluded that the prevalence of environmentally induced cancer has been "grossly underestimated." In our society today, the greatest number of cancer cases is thought to result from carcinogens contained in cigarette smoke. Polycyclic aromatic hydrocarbons (PAHs; p. 27) make up some of the carcinogens found in cigarette smoke. PAHs also occur in charred meats and are released from the combustion of coal, oil, and natural gas.

Carcinogens can be difficult to identify because there may be a long lag time between exposure to the agent and the detectable onset of cancer—up to 15–30 years in the case of cigarette smoke. Moreover, as with all risks, only a portion of people exposed to a carcinogen will eventually get cancer. Cancer is a leading cause of death that kills millions and leaves few families untouched. Two of every five Americans are diagnosed with cancer at some time in their lives, and one of every five dies from it. Thus, the study of carcinogens has played a large role in shaping the way that toxicologists pursue their work.

Mutagens are substances that cause genetic mutations in the DNA of organisms (p. 49). Although most mutations have little or no effect, some can lead to severe problems, including cancer and other disorders. If mutations occur in an individual's sperm or egg cells, then the individual's offspring suffer the effects.

Chemicals that cause harm to the unborn are called **teratogens.** Teratogens that affect development of human embryos in the womb can cause birth defects. One example of a teratogen is the drug thalidomide, developed in the 1950s to aid sleeping and to prevent nausea during pregnancy. Tragically, the drug caused severe birth defects in thousands of babies whose mothers were prescribed thalidomide. Even a single dose during pregnancy could result in limb deformities and organ defects. Thalidomide was banned in the 1960s once scientists recognized its connection with birth defects. Ironically, today the drug shows promise in treating a wide range of diseases, including Alzheimer's disease, AIDS, and various types of cancer.

Other chemical toxicants known as **neurotoxins** assault the nervous system. Neurotoxins include venoms produced by animals such as snakes and stinging insects, heavy metals such as lead and mercury, pesticides, and some chemical weapons developed for use in war. A famous case of neurotoxin poisoning occurred in Japan, where a chemical factory dumped mercury waste into Minamata Bay between the 1930s and 1960s. Thousands of people there ate fish contaminated with the mercury and soon began suffering from slurred speech, loss of muscle control, sudden fits of laughter, and in some cases death.

The human immune system protects our bodies from disease. Some toxicants weaken the immune system, reducing the body's ability to defend itself against bacteria, viruses, allergy-causing agents, and other attackers. Others, called **allergens,** overactivate the immune system, causing an immune response when one is not necessary. One hypothesis for the increase in asthma in recent years is that allergenic synthetic chemicals are more prevalent in our environment.

Allergens are not universally considered toxicants, however, because they affect some people but not others and because one's response does not necessarily correlate with the degree of exposure.

Pathway inhibitors are toxicants that interrupt vital biochemical processes in organisms by blocking one or more steps in important biochemical pathways. Rat poisons, for example, cause internal hemorrhaging in rodents by interfering with the biochemical pathways that create blood clotting proteins. Some herbicides, such as atrazine, kill plants by blocking steps in photosynthesis. Cyanide kills by interrupting chemical pathways that produce energy in mitochondria, thereby depriving cells of life-sustaining energy.

Most recently, scientists have recognized **endocrine disruptors,** toxicants that interfere with the *endocrine system.* The endocrine system consists of chemical messengers (*hormones*) that travel through the bloodstream at extremely low concentrations and have many vital functions. They stimulate growth, development, and sexual maturity, and they regulate brain function, appetite, sex drive, and many other aspects of our physiology and behavior. Some hormone-disrupting toxicants affect an animal's endocrine system by blocking the action of hormones or accelerating their breakdown. Others are so similar to certain hormones in their molecular structure and chemistry that they "mimic" the hormone by interacting with receptor molecules just as the actual hormone would (**FIGURE 14.10**).

BPA is one of many chemicals that appear to mimic the female sex hormone estrogen and bind to estrogen receptors. Many plastic products also contain another class of hormone-disrupting chemical, called **phthalates.** Used to soften plastics and enhance fragrances, phthalates are used widely in children's toys (**FIGURE 14.11a**), perfumes and cosmetics (**FIGURE 14.11b**), and other items. Health research on phthalates has linked them to birth defects, breast cancer, reduced sperm counts, and other reproductive effects. The European Union and nine other nations have banned phthalates, California and Washington enacted bans for children's toys, and the United States in 2008 banned six types of phthalates in toys. Still, across North America many routes of exposure remain. Like BPA, phthalates show how a substance can be a carcinogen, a mutagen, and an endocrine disruptor all at the same time.

Organisms have natural defenses against toxic substances

Although synthetic toxicants are new, organisms have long been exposed to natural toxicants. Mercury, arsenic, cadmium, and other harmful substances are found naturally in the environment. Some organisms produce biological toxins to avoid predators or capture prey. Examples include venom in poisonous snakes and spiders, toxins in sea urchins, and the natural insecticide pyrethrin found in chrysanthemums. Over time, organisms able to tolerate these harmful substances have gained an evolutionary advantage.

Skin, scales, and feathers are the first line of defense against toxic substances because they resist uptake from the

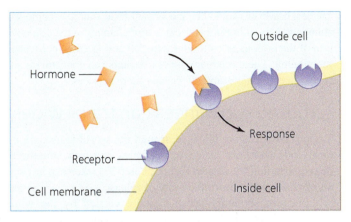

(a) Normal hormone binding

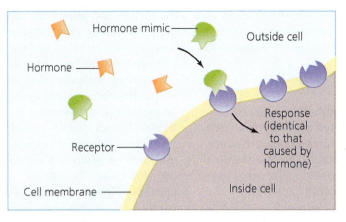

(b) Hormone mimicry

FIGURE 14.10 Many endocrine-disrupting substances mimic the chemical structure of hormone molecules. Like a key similar enough to fit into another key's lock, the hormone mimic binds to a cellular receptor for the hormone, causing the cell to react as though it had encountered the hormone.

surrounding environment. However, toxicants can circumvent these barriers and enter the body from vital activities such as eating, drinking, and breathing. Once inside the organism, they are distributed widely by the circulatory and lymph systems in animals, and by the vascular system in plants.

Organisms possess biochemical pathways that use enzymes to detoxify harmful chemicals once they enter the body. Some pathways break down, or metabolize, toxic substances to render them inert. Other pathways make toxic substances water soluble so they are easier to excrete through the urinary system. In humans, many of these pathways are found in the liver, so this organ is disproportionately affected by intake of harmful substances, such as excessive alcohol.

Some toxic substances cannot be effectively detoxified or made water soluble by detoxification enzymes. Instead, the body sequesters these chemicals in fatty tissues and cell membranes to keep them away from vital organs. Heavy metals, dioxins, and some insecticides (including DDT) are stored in body tissue in this manner.

Defense mechanisms for natural toxins have evolved over millions of years. For the synthetic chemicals that are so prevalent in today's environment, however, organisms have not had long-term exposure, so the impacts of these toxic substances can be severe and unpredictable.

Individuals vary in their responses to hazards

Some of the defenses described above have a genetic basis. As a result, individuals may respond quite differently to identical exposures to hazards because they happen to have different combinations of genes. Poorer health also makes an individual more sensitive to biological and chemical hazards.

(a) Exposure through toys

(b) Exposure through cosmetics

FIGURE 14.11 Many soft plastic children's toys and many cosmetics contain hormone-disrupting phthalates. Phthalates have been banned in Europe and in some products in the United States, but exposure to them remains widespread in the United States through many consumer goods.

Sensitivity also can vary with sex, age, and weight. Because of their smaller size and rapidly developing organ systems, younger organisms (for example, fetuses, infants, and young children) tend to be much more sensitive to toxicants than are adults. Regulatory agencies such as the U.S. Environmental Protection Agency (EPA) typically set human chemical exposure standards for adults and extrapolate downward for infants and children. However, many scientists contend that these linear extrapolations often do not offer adequate protection to fetuses, infants, and children.

FAQ

Do individual organisms survive exposure to a toxic chemical because they are "mutated" by the chemical and develop defenses to the toxicant?

When a population of organisms is exposed to a toxicant, such as a pesticide, a few individuals often survive while the vast majority of the population is killed. These individuals survive because they *already* possess genes (which others in the population do not) that code for enzymes that counteract the toxic properties of the toxicant. Because the effects of these genes are expressed only when the pesticide is applied, many people think the toxicant "creates" detoxification genes by mutating the DNA of a small number of individuals. This is not the case. The genes for detoxifying enzymes were present in the DNA of resistant individuals from birth, but their effects were seen only when pesticide exposure caused selective pressure (p. 50) for resistance to the toxic substance.

The type of exposure can affect the response

The risk posed by a hazard often varies according to whether a person experiences high exposure for short periods of time, known as **acute exposure,** or low exposure over long periods of time, known as **chronic exposure.** Incidences of acute exposure are easier to recognize, because they often stem from discrete events, such as accidental ingestion, an oil spill, a chemical spill, or a nuclear accident. Toxicity tests in laboratories generally reflect acute toxicity effects. However, chronic exposure is more common—and more difficult to detect and diagnose. Chronic exposure often affects organs gradually, as when smoking causes lung cancer or when alcohol abuse leads to liver or kidney damage. Arsenic in drinking water or pesticide residues on food also pose chronic risk. Because of the long time periods involved, relationships between cause and effect may not be readily apparent.

Toxic Substances and Their Effects on Ecosystems

When toxicants concentrate in environments and harm the health of many individuals, populations (p. 48) of the affected species become smaller. This decline in population can then affect other species. For instance, species that are prey of the organism affected by toxicants could experience population growth because predation levels are lower. Predators of the poisoned species, however, would decline as their food source became less abundant. Cascading impacts can cause changes in the composition of the biological community (p. 58) and threaten ecosystem functioning. There are many ways toxicants can concentrate and persist in ecosystems and affect ecosystem services.

Airborne substances can travel widely

Toxic substances are released around the world from agricultural, industrial, and domestic activities and may sometimes be redistributed by air currents (Chapter 17), exerting impacts on ecosystems far from their site of release.

Because so many substances are carried by the wind, synthetic chemicals are ubiquitous worldwide, even in seemingly pristine areas. Scientists who travel to the most remote alpine lakes in the wilderness of British Columbia find them contaminated with industrial toxicants, such as polychlorinated biphenyls (PCBs), which are by-products of chemicals used in transformers and other electrical equipment and as hydraulic fluids. These chemicals enter the air, soil, or water when the equipment in which they are housed burns, leaks, or corrodes.

Earth's polar regions are particularly contaminated, because natural patterns of global atmospheric circulation (p. 451) tend to move airborne chemicals toward the poles (**FIGURE 14.12**). Thus, although we manufacture and apply synthetic substances

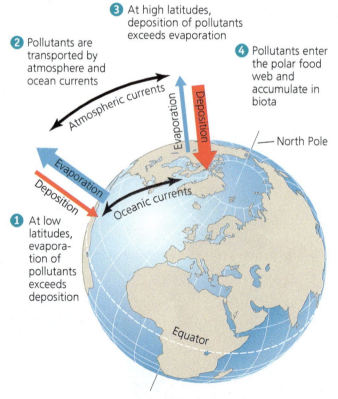

③ At high latitudes, deposition of pollutants exceeds evaporation

② Pollutants are transported by atmosphere and ocean currents

④ Pollutants enter the polar food web and accumulate in biota

Atmospheric currents

Evaporation

Deposition

— North Pole

Evaporation

Deposition

Oceanic currents

① At low latitudes, evaporation of pollutants exceeds deposition

Equator

FIGURE 14.12 Air and water currents direct pollutants to the poles. In the process of "global distillation," pollutants that evaporate and rise high into the atmosphere at lower latitudes are carried toward the poles by atmospheric currents, while ocean currents carry pollutants deposited in the ocean toward the poles. This process exposes polar organisms to unusually concentrated levels of toxic substances.

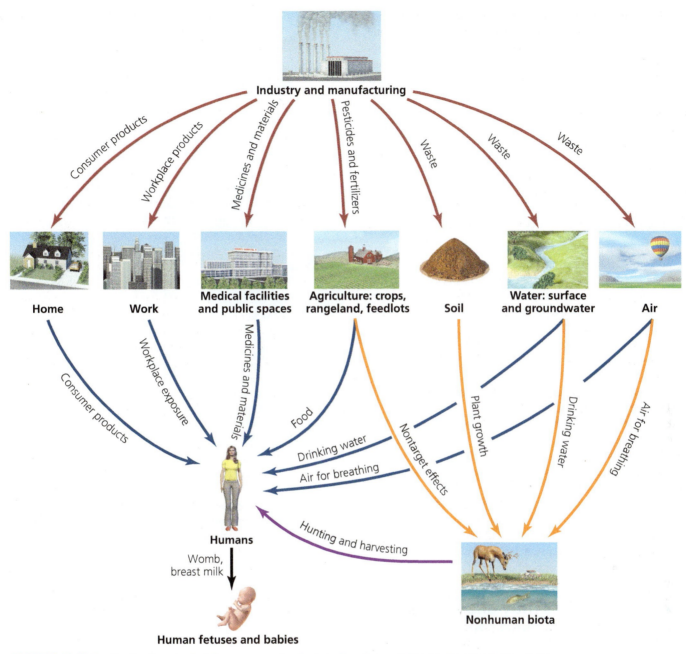

FIGURE 14.13 Synthetic chemicals take many routes in traveling through the environment. People take in only a tiny proportion of these compounds, and many compounds are harmless. However, people receive small amounts of toxicants from many sources, and developing fetuses and babies are particularly sensitive.

mainly in temperate and tropical regions, contaminants are strikingly concentrated in the tissues of Arctic polar bears, Antarctic penguins, and people living in Greenland.

Effects can also occur over relatively shorter distances. Pesticides, for example, can be carried by air currents to sites far from agricultural fields in a process called pesticide drift. The Central Valley of California is the world's most productive agricultural region, and the region's frequent winds often blow airborne pesticide spray—and dust particles containing pesticide residue—for long distances. In the nearby mountains of the Sierra Nevada, research has associated pesticide drift from the Central Valley with population declines in four species of frogs.

Toxic substances may concentrate in water

Toxic substances are not evenly distributed in the environment, and they move about in specific ways (**FIGURE 14.13**). Water running off from land often transports toxicants from large areas and concentrates them in small volumes of surface water. The NAWQA findings on water quality reflect this concentrating effect. Wastewater treatment plants also add toxins, pharmaceuticals, and detoxification products from humans to waterways. If chemicals persist in soil, they can leach into groundwater and contaminate drinking water supplies.

Many chemicals are soluble in water and enter organisms' tissues through drinking or absorption. For this reason, aquatic animals such as fish, frogs, and stream invertebrates are effective indicators of pollution. If scientists find low concentrations of pesticides harming frogs, fish, and invertebrates, they view this as a warning that people could be next. The contaminants that wash into streams and rivers also flow and seep into the water we drink and drift through the air we breathe. Once concentrated in waters, toxic substances can move long distances and affect a variety of ecosystems (p. 388).

Some toxicants persist in the environment

A toxic substance that is released into the environment may degrade quickly and become harmless, or it may remain unaltered and persist for many months, years, or decades. The rate at which a given substance degrades depends on its chemistry and on factors such as temperature, moisture, and sun exposure. The *Bt* toxin (p. 248) used in biocontrol and genetically modified crops has a very short persistence time, whereas chemicals such as DDT and PCBs persist for decades. Atrazine, one of our most widely used herbicides, is highly variable in its persistence, depending on environmental conditions.

Persistent synthetic chemicals exist in our environment today because we have designed them to persist. The synthetic chemicals used in plastics, for instance, are used precisely because they resist breakdown. Sooner or later, however, most toxicants degrade into simpler compounds called **breakdown products.** Often these are less harmful than the original substance, but sometimes they are just as toxic as the original chemical, or more so. For instance, DDT breaks down into DDE, a highly persistent and toxic compound in its own right. Atrazine produces a large number of breakdown products whose effects have not been fully studied.

Toxic substances may accumulate and move up the food chain

Within an organism's body, some toxic substances are quickly excreted, and some are degraded into harmless breakdown products. Others persist intact in the body. Substances that are fat soluble or oil soluble (including organic compounds such as DDT and DDE) are absorbed and stored in fatty tissues. Substances such as methylmercury (CH_3Hg^+) may be stored in muscle tissue. Such persistent toxicants accumulate in an animal's body in a process termed **bioaccumulation,** such that the animal's tissues have a greater concentration of the substance than exists in the surrounding environment.

Toxic substances that bioaccumulate in an organism's tissues may be transferred to other organisms as predators consume prey, resulting in a process called **biomagnification** (**FIGURE 14.14**). When one organism consumes another, the predator takes in any stored toxicants and stores them in its own body. Thus bioaccumulation takes place on all trophic levels. Moreover, each individual predator consumes many individuals from the trophic level beneath it, so with each

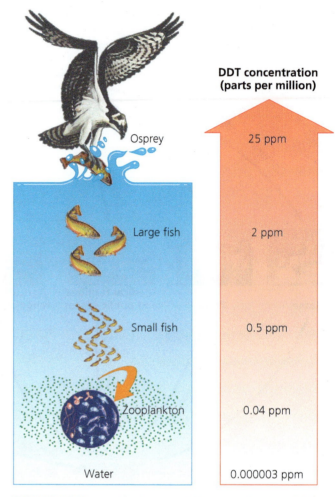

FIGURE 14.14 In a classic case of biomagnification, DDT becomes highly concentrated in fish-eating birds such as ospreys. Organisms at the lowest trophic level take in fat-soluble compounds such as DDT from water. As animals at higher trophic levels eat organisms lower on the food chain, each organism passes its load of toxicants up to its consumer, such that organisms on all trophic levels bioaccumulate the substance in their tissues.

step up the food chain, concentrations of toxicants become magnified.

The process of biomagnification occurred throughout North America with DDT. Top predators, such as birds of prey, ended up with high concentrations of the pesticide because concentrations became magnified as DDT moved from water to algae to plankton to small fish to larger fish and finally to fish-eating birds.

Biomagnification of DDT caused populations of many North American birds of prey to decline precipitously from the 1950s to the 1970s. The peregrine falcon was nearly wiped out in the eastern United States, and the bald eagle, the U.S. national bird, was virtually eliminated from the lower 48 states. Eventually scientists determined that DDT was causing these birds' eggshells to grow thinner, so that eggs were breaking in the nest and killing the embryos within. In a remarkable environmental success story, populations of all these birds have rebounded (pp. 289–290) since the United States banned DDT.

Unfortunately, DDT continues to impair wildlife in parts of the world where it is still used. In addition, mercury bioaccumulates in some commercially important fish species, such as tuna. Polar bears of Svalbard Island in Arctic Norway show extremely high levels of PCB contamination from biomagnification as a result of the global distillation process shown in Figure 14.12. Polar bear cubs suffer immune suppression, hormone disruption, and high mortality—and because the cubs receive PCBs in their mothers' milk, contamination persists and accumulates across generations.

In all these cases, biomagnification affects ecosystem composition and functioning. When populations of top predators such as eagles or polar bears are reduced, species interactions (pp. 74–78) change, and effects cascade through food webs (p. 80).

Toxic substances can threaten ecosystem services

Toxicants can alter the biological composition of ecosystems and the manner in which organisms interact with one another and their environment. In so doing, harmful compounds can threaten the ecosystem services (pp. 4, 116) provided by nature. For example, pesticide exposure has been implicated as a factor in the recent declines in honeybee populations (pp. 245–246), affecting the ecosystem service of pollination they provide to wild plants and agricultural crops.

Nutrient cycling is one of the many services that healthy, functioning ecosystems provide. Decomposers and detritivores in the soil (p. 78) break down organic matter and replenish soils with nutrients for plants to use. When soils are exposed to pesticides or antifungal agents, nutrient cycling rates are altered. This can make nutrients less available to producers, affecting their growth and causing impacts that cascade throughout the ecosystem.

Studying Effects of Hazards

Determining health effects of particular environmental hazards is a challenging job, especially because any given person or organism has a complex history of exposure to many hazards throughout life. Scientists rely on several different methods with people and with wildlife, ranging from correlative surveys to manipulative experiments (p. 13).

Wildlife studies integrate work in the field and lab

Scientists study the impacts of environmental hazards on wild animals to help conserve animal populations and also to understand potential risks to people. Just as placing canaries in coal mines helped miners determine whether the air was safe for them to breathe, studying how wild animals respond to pollution and other hazards can help us detect environmental health threats before they do us too much harm.

Often wildlife toxicologists work in the field with animals to take measurements, document patterns, and generate hypotheses, before heading to the laboratory to run controlled manipulative experiments to test their hypotheses. The work of two of the pioneers in the study of endocrine disruptors illustrates the approaches embraced in wildlife studies.

Biologist Louis Guillette studied alligators in Florida (**FIGURE 14.15a**) and discovered that many showed bizarre reproductive problems. Females had trouble producing viable eggs, young alligators had abnormal gonads, and male hatchlings had too little of the male sex hormone testosterone while female hatchlings had too much of the female sex hormone

(a) Louis Guillette taking blood sample from alligator

(b) Tyrone Hayes in lab with frog

FIGURE 14.15 Wildlife studies examine the effects of toxic substances in the environment. Researchers Louis Guillette **(a)** and Tyrone Hayes **(b)** found that alligators and frogs, respectively, show reproductive abnormalities that they attribute to endocrine disruption by pesticides.

estrogen. Because certain lakes received agricultural runoff that included insecticides such as DDT and dicofol and herbicides such as atrazine, Guillette hypothesized that chemical contaminants were disrupting the endocrine systems of alligators during their development in the egg. Indeed, when Guillette and his team compared alligators in polluted lakes with those in cleaner lakes, they found the ones in polluted lakes to be suffering far more problems. Moving into the lab, the researchers found that several contaminants detected in alligator eggs and young could bind to receptors for estrogen and reverse the sex of male embryos. Their experiments showed that atrazine appeared to disrupt hormones by inducing production of aromatase, an enzyme that converts testosterone to estrogen.

Following Guillette's work, researcher Tyrone Hayes (**FIGURE 14.15b**) found similar reproductive problems in frogs and attributed them to atrazine. In lab experiments, male frogs raised in water containing very low doses of the herbicide became feminized and hermaphroditic, developing both testes and ovaries. Hayes then moved to the field to look for correlations between herbicide use and reproductive impacts in the wild. His field surveys showed that leopard frogs across North America experienced hormonal problems in areas of heavy atrazine usage. His work indicated that atrazine, which kills plants by blocking biochemical pathways in photosynthesis, can also act as an endocrine disruptor.

Human studies rely on case histories, epidemiology, and animal testing

In studies of human health, we gain much knowledge by directly studying sickened individuals. Medical professionals have long treated victims of poisonings, so the effects of common poisons are well known. This process of observation and analysis of individual patients is known as a **case history** approach. Case histories have advanced our understanding of human illness, but they do not always help us infer the effects of rare hazards, new hazards, or chemicals that exist at low environmental concentrations and exert minor, long-term effects. Case histories also tell us little about probability and risk, such as how many extra deaths we might expect in a population due to a particular cause.

For such questions, which are common in environmental toxicology, we need **epidemiological studies,** large-scale comparisons among groups of people, usually contrasting a group known to have been exposed to some hazard and a group that has not. Epidemiologists track the fate of all people in the study for a long period of time (often years or decades) and measure the rate at which deaths, cancers, or other health problems occur in each group. The epidemiologist then analyzes the data, looking for observable differences between the groups, and statistically tests hypotheses accounting for differences. When a group exposed to a hazard shows a significantly greater degree of harm, it suggests that the hazard may be responsible. For example, epidemiologists have tracked asbestos miners for evidence of asbestosis, lung cancer, and mesothelioma (cancer of the cells that line the body's internal organs). Survivors of the Chernobyl and Fukushima nuclear disasters have been monitored for thyroid cancer and other illnesses (p. 560). Canadian epidemiologists are now tracking people for impacts of BPA exposure.

The epidemiological process is akin to a natural experiment (p. 13) in which the experimenter studies groups of research participants made available by some event that has occurred. A similar approach was followed by anthropologist Elizabeth Guillette, the wife of biologist Louis Guillette, to study the effects of pesticide exposure on child development in the Yaqui Valley of Mexico (see **THE SCIENCE BEHIND THE STORY**, pp. 372–373). The advantages of epidemiological studies are their realism and their ability to yield relatively accurate predictions about risk. Drawbacks include the need to wait a long time for results and an inability to address future effects of new hazards, such as products just coming to market. In addition, participants in epidemiological studies encounter many factors that affect their health besides the one under study. Epidemiological studies measure a statistical association between a health hazard and an effect, but they do not confirm that the hazard *causes* the effect.

To establish causation, manipulative experiments are needed. However, subjecting people to massive doses of toxic substances in a lab experiment would clearly be unethical. This is why researchers have traditionally used nonhuman animals as test subjects. Foremost among these animal models have been laboratory strains of rats, mice, and other mammals (**FIGURE 14.16**). Because of shared evolutionary history, the bodies of all mammals function similarly, so substances that harm mice and rats are reasonably likely to harm us. Some people feel the use of animals for testing is unethical, but animal testing enables scientific and medical advances that would be impossible or far more difficult otherwise. Still, new techniques (with human cell cultures,

FIGURE 14.16 Animal testing is used to study toxic substances in the laboratory. Tests with specially bred strains of mice, rats, and other animals allow researchers to study the toxicity of substances, develop safety guidelines, and make medical advances in ways they could not achieve without these animals.

bacteria, or tissue from chicken eggs) are being devised that may soon replace some live-animal testing.

Dose-response analysis is a mainstay of toxicology

The standard method of testing with lab animals in toxicology is **dose-response analysis.** Scientists quantify the toxicity of a substance by measuring the strength of its effects or the number of animals affected at different doses. The **dose** is the amount of substance the test animal receives, and the **response** is the type or magnitude of negative effects the animal exhibits as a result. The response is generally quantified by measuring the proportion of animals exhibiting negative effects. The data are plotted on a graph, with dose on the x axis and response on the y axis (**FIGURE 14.17a**). The resulting curve is called a **dose-response curve.**

Once they have plotted a dose-response curve, toxicologists can calculate a convenient shorthand gauge of a substance's toxicity: the amount of the substance it takes to kill half the population of study animals used. This lethal dose for 50% of individuals is termed the LD_{50}. A high LD_{50} indicates low toxicity, and a low LD_{50} indicates high toxicity.

If the experimenter is interested in nonlethal health effects, he or she may want to document the level of toxicant at which 50% of a population of test animals is affected in some other way (for instance, the level of toxicant that causes 50% of lab mice to lose their hair). Such a level is called the effective dose–50%, or ED_{50}.

Some substances can elicit effects at any concentration, but for others, responses may occur only above a certain dose, or threshold. Such a **threshold dose** (**FIGURE 14.17b**) might be expected if the body's organs can fully metabolize or excrete a toxicant at low doses but become overwhelmed at high concentrations. It might also occur if cells can repair damage to their DNA from mutagenic chemicals only up to a certain point.

Sometimes a response may *decrease* as a dose increases. Toxicologists are finding that some dose-response curves are U-shaped, J-shaped, or shaped like an inverted U (**FIGURE 14.17c**). Such counterintuitive curves contradict toxicology's traditional assumption that "the dose makes the poison." These unconventional dose-response curves often occur with endocrine disruptors, likely because the hormone system is geared to respond to minute concentrations of substances (normally, hormones in the bloodstream). Because the endocrine system responds to minuscule amounts of chemicals, it may be vulnerable to disruption by contaminants that are dispersed through the environment and that reach our bodies in very low concentrations. In research with BPA, a number of studies with lab animals have found unconventional dose-response curves of this sort.

Researchers generally give lab animals much higher doses relative to body mass than people would receive in the environment. This is so that the response is great enough to be measured and so that differences between the effects of small and large doses are evident. Data from a range of doses give shape to the dose-response curve. Once the data from animal tests

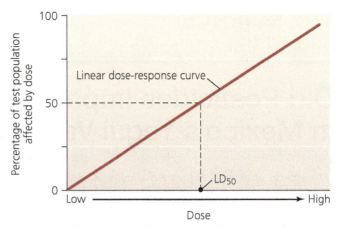

(a) Linear dose-response curve

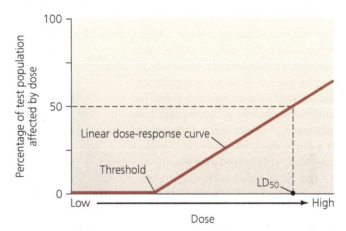

(b) Dose-response curve with threshold

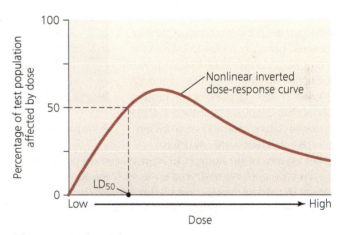

(c) Unconventional dose-response curve

FIGURE 14.17 Dose-response curves show that organisms' responses to toxicants may sometimes be complex. In a classic linear dose-response curve **(a)**, the percentage of animals killed or otherwise affected by a substance rises with the dose. The point at which 50% of the animals are killed is labeled the lethal dose–50, or LD_{50}. For some toxic substances, a threshold dose **(b)** exists, below which doses have no measurable effect. Some substances, in particular endocrine disruptors, show unconventional, nonlinear dose-response curves **(c)** that are U-shaped, J-shaped, or shaped like an inverted U.

Did Pesticides Impair Child Development in Mexico's Yaqui Valley?

With spindly arms and big, round eyes, one set of pictures shows the sorts of stick figures drawn by young children everywhere. Next to them is another group of drawings, mostly disconnected squiggles and lines. Both sets of pictures are intended to depict people. The main difference identified between the two groups of young artists was this: long-term pesticide exposure.

Children's drawings are not a typical tool of toxicology, but anthropologist Elizabeth Guillette was interested in the effects of pesticides on children and wanted to try new methods. She devised tests to measure childhood development based on techniques from anthropology and medicine. Searching for a study site, Guillette found the Yaqui valley region of northwestern Mexico.

The Yaqui valley is farming country, worked for generations by the indigenous group that gives the region its name. Synthetic pesticides arrived in the area in the 1940s. Some Yaqui embraced the agricultural innovations, spraying their farms in the valley to increase their yields. Yaqui farmers in the surrounding foothills, however, generally chose to bypass the chemicals and to continue following more traditional farming practices. Although differing in farming techniques, Yaqui in the valley and foothills continued to share the same culture, diet, education system, income levels, and family structure.

A Yaqui family

At the time of the study, in 1994, valley farmers planted crops twice a year, applying pesticides up to 45 times from planting to harvest. A previous study conducted in the valley in 1990, focusing on areas with the largest farms, had indicated high levels of multiple pesticides in the breast milk of mothers and in the umbilical cord blood of newborn babies. In contrast, foothill families avoided chemical pesticides in their gardens and homes.

The researchers were interested in determining if exposure to neurotoxins in pesticides interfered with cognitive function and coordination in children. Processing and retaining information, as well as effectively using one's motor skills, occurs though the actions of innumerable interacting nerves, or neurons. For example, committing a reading passage to memory requires the neurons in the eye to effectively transmit the image of the words on the page to neurons in your short-term memory in the brain, which then transfer this information to other

neurons in long-term memory. If neurotoxins are present in an organism's system, however, they may interfere with these transfers, impairing learning and memory. Similar results would be expected for tasks involving coordination or fine motor skills, because they similarly require high levels of interconnected neural activity. Long-term, elevated dosages of other neurotoxins, such as the heavy metal mercury, have been shown to interfere with mental functioning and coordination in humans, so similar effects with long-term exposure to neurotoxic pesticides would be reasonable.

To understand how pesticide exposure affects childhood development, Guillette and fellow researchers studied 50 preschoolers aged 4 to 5, of whom 33 were from the valley and 17 from the foothills. Each child underwent a half-hour exam, during which researchers showed a red balloon, promising to give the balloon later as a gift, and using the promise to evaluate long-term memory. Each child was then put through a series of physical and mental tests, such as catching a ball, dropping raisins into a bottle cap, drawing a picture of a person (as a measure of perception), and repeating a short string of numbers (to test short-term memory). The researchers also measured each child's height and weight. When all tests were completed, each child was asked what he or she had been promised and received a red balloon.

Although the two groups of children did not differ in height and weight, they differed markedly in other measures of development. Valley children had greater difficulty than foothill children when attempting to catch a ball or drop raisins into the bottle cap. Each group did fairly well repeating numbers, but valley children showed poor long-term memory. At the end of the test, all but one of the foothill children remembered that they had been promised a balloon, and 59% remembered it was red. However, of the valley children only 27% remembered the color of the balloon, only 55% remembered they'd be getting a balloon, and 18% were unable to remember anything about a balloon.

The children's drawings exhibited the most dramatic difference between valley and foothill children (**FIGURE 1**). The researchers determined each drawing could earn 5 points, with 1 point each for a recognizable feature: head, body, arms, legs, and facial features. Foothill children drew pictures that looked

FIGURE 1 Data from children in Mexico's Yaqui valley offer a startling example of apparent neurological effects of pesticide poisoning. Young children from foothills areas where pesticides were not commonly used drew recognizable figures of people. Children of the same age from valley areas where pesticides were heavily used drew less-recognizable figures. *Adapted from Guillette, E.A., et al., 1998. An anthropological approach to the evaluation of preschool children exposed to pesticides in Mexico. Environmental Health Perspectives 106: 347–353.*

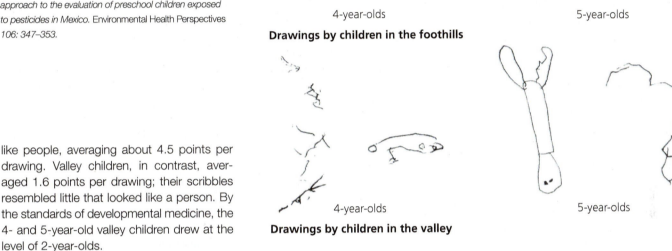

4-year-olds   5-year-olds

Drawings by children in the foothills

4-year-olds   5-year-olds

Drawings by children in the valley

like people, averaging about 4.5 points per drawing. Valley children, in contrast, averaged 1.6 points per drawing; their scribbles resembled little that looked like a person. By the standards of developmental medicine, the 4- and 5-year-old valley children drew at the level of 2-year-olds.

Some scientists greeted Guillette's study skeptically, pointing out that its sample size was too small to be meaningful. Others said that factors the researchers missed, such as different parenting styles or unknown health problems, could be to blame. Prominent toxicologists argued that because the researchers lacked time and money to take blood or tissue samples to check for pesticides or other toxic substances, the study results couldn't be tied to agricultural chemicals.

Subsequent studies have, however, better illuminated potential effects of toxic chemicals on cognitive development in children. Shortly after the publication of Guillette's study, a paper published in the *New England Journal of Medicine* by Joseph and Sandra Jacobson showed that exposure to endocrine-disrupting PCBs in the womb permanently stunted cognitive performance in children. The Jacobsons tracked children born to mothers who had high levels of the endocrine disruptor PCB during pregnancy from eating PCB-laden fish from Lake Michigan. The study found that even at age 11, cognitive issues in children exposed to PCBs in utero remained. PCB-exposed children had lower IQ scores and poorer memory and attention span than those not exposed, even when important factors that affect cognitive development, such as socioeconomic status, were considered. Those exposed in the womb to the highest levels of PCBs were three times more likely to have lower average IQ and two times more likely to be delayed in reading comprehension.

In the intervening decades, other studies have found negative effects on cognitive ability in children from toxic substances that only temporarily reside in the body. For example, a 2015 study from France studied 287 children and found that those with higher levels of two pyrethroid pesticide metabolites (a metabolite is a chemical produced when the body partially metabolizes a pesticide in the liver to make it more easily excreted through the urinary system) in their urine scored significantly lower on cognitive tests, particularly in the areas of memory and verbal comprehension, than children with lower levels of the metabolite. Metabolizing and excreting this type of pesticide typically occurs within two days after exposure, showing that children do not have to be permanently impaired by exposure to a toxicant in the womb but may suffer temporary impacts on cognition from exposure to chemicals that are ubiquitous in our environment.

These results are significant because in modern America, educational systems are increasingly adopting "high stakes" testing in schools with children of all ages. Poor performance on such tests can have profoundly negative consequences for students (such as failing to advance to the next grade), teachers (poor performance evaluation if student test scores are considered), and even entire school systems (reduced resources if funding is tied to test scores). As this study and others show, our environment may be rife with chemicals that impair, even if only temporarily, children's ability to effectively process and recall information. This calls into question the validity of putting heavy weight on the results of individual assessments, and illustrates the need for additional study of this issue, particularly in agricultural and urban areas where pesticides are heavily utilized.

are plotted, researchers can extrapolate downward to estimate responses to still-lower doses from a hypothetically large population of animals. This way, they can come up with an estimate of, say, what dose causes cancer in 1 mouse in 1 million. A second extrapolation is required to estimate the effect on humans, with our greater body mass. Because these two extrapolations stretch beyond the actual data obtained, they introduce uncertainty into the interpretation of what doses are safe for people.

Chemical mixes may be more than the sum of their parts

It is difficult enough to determine the impact of a single hazard, but the task becomes astronomically more difficult when multiple hazards interact. Chemical substances, when mixed, may act together in ways that cannot be predicted from the effects of each in isolation. Mixed toxicants may sum each other's effects, cancel out each other's effects, or multiply each other's effects. Whole new types of impacts may arise when toxicants are mixed together. Interactive impacts that are greater than the simple sum of their constituent effects are called **synergistic effects.**

With Florida's alligators, lab experiments have indicated that the DDT breakdown product DDE can either help cause sex reversal or inhibit it, depending on the presence of other chemicals. Mice exposed to a mixture of nitrate, atrazine, and aldicarb have been found to show immune, hormone, and nervous system effects that were not evident from exposure to each of these chemicals alone.

Traditionally, environmental health has tackled effects of single hazards one at a time. In toxicology, the complex experimental designs required to test interactions, and the sheer number of chemical combinations, have meant that single-substance tests have received priority. This approach is changing, but the interactive effects of most chemicals are unknown.

Endocrine disruption poses challenges for toxicology

As today's emerging understanding of endocrine disruption leads toxicologists to question their assumptions, unconventional dose-response curves are presenting challenges for scientists studying toxic substances and for policymakers trying to set safety standards for them. Knowing the shape of a dose-response curve is crucial if one is using it to predict responses at doses below those that have been tested. Because so many novel synthetic chemicals exist in very low concentrations over wide areas, many scientists suspect that we may have underestimated the dangers of compounds that exert impacts at low concentrations.

Scientists first noted endocrine-disrupting effects decades ago, but the idea that synthetic chemicals might be altering the hormones of animals was not widely appreciated until the 1996 publication of the book *Our Stolen Future*, by Theo Colburn, Dianne Dumanoski, and J.P. Myers. Like *Silent Spring*, this book integrated scientific work from various fields and presented a unified view of the hazards posed by endocrine-disrupting chemicals.

Today, thousands of studies have linked hundreds of substances to effects on reproduction, development, immune function, brain and nervous system function, and other hormone-driven processes. Evidence is strongest so far in nonhuman animals, but many studies suggest impacts on humans (see Figure 14.1, p. 354). Some researchers argue that the sharp rise in breast cancer rates (one in eight U.S. women today develops breast cancer) may be due to hormone disruption, because an excess of estrogen appears to feed tumor development in older women. Other scientists attribute male reproductive problems to elevated BPA exposure. For example, studies found that workers in Chinese factories that manufactured BPA had elevated rates of erectile dysfunction and reduced sperm counts when compared to workers in factories manufacturing other products.

Much of the research into hormone disruption has brought about strident debate. This is partly because scientific uncertainty is inherent in any developing field. Another reason is that negative findings about chemicals pose an economic threat to the manufacturers of those chemicals, who stand to lose many millions of dollars in revenue if their products were to be banned or restricted in the United States.

Risk Assessment and Risk Management

Policy decisions on whether to ban chemicals or restrict their use generally follow years of rigorous testing for toxicity. Likewise, strategies for combating disease and other health threats are based on extensive scientific research. However, policy and management decisions also incorporate economics and ethics—and all too often the decision-making process is heavily influenced by pressure from powerful corporate and political interests. The steps between the collection and interpretation of scientific data and the formulation of policy involve assessing and managing risk.

We express risk in terms of probability

Exposure to an environmental health threat does not invariably produce some given harmful effect. Rather, it causes some probability of harm, some statistical chance that damage will result. To understand a health threat, a scientist must know more than just its identity and strength. He or she must also know the chance that one will encounter it, the frequency with which one may encounter it, the amount of substance or degree of threat to which one is exposed, and one's sensitivity to the threat. Such factors help determine the overall risk posed by a particular threat.

Risk can be measured in terms of **probability,** a quantitative description of the likelihood of a certain outcome. The probability that some harmful outcome (for instance, injury, death, environmental damage, or economic loss) will result from a given action, event, or substance expresses the **risk** posed by that phenomenon.

Our perception of risk may not match reality

Every action we take and every decision we make involves some element of risk, some (generally small) probability that things will go wrong. We try in everyday life to behave in ways that minimize risk, but our perceptions of risk do not always match statistical reality (**FIGURE 14.18**). People often worry unduly about negligibly small risks yet happily engage in other activities that pose high risks. For instance, most people perceive flying in an airplane as a riskier activity than driving a car, but according to a 2016 report by the National Safety Council, a person's chance of dying from an automobile accident is many times higher than dying from an airplane crash. Psychologists agree that this difference between perception and reality stems from the fact that we feel more at risk when we are not controlling a situation and safer when we are "at the wheel"—regardless of the actual risk involved.

This psychology may help account for people's anxiety over nuclear power, toxic waste, and pesticide residues on foods—environmental hazards that are invisible or little understood and whose presence in our lives is largely outside our personal control. In contrast, people are more ready to accept and ignore the risks of smoking cigarettes, overeating, and not exercising—voluntary activities statistically shown to pose far greater risks to health.

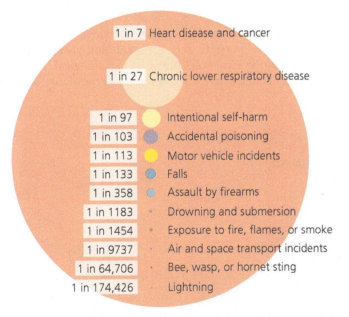

FIGURE 14.18 Our perceptions of risk do not always match the reality of risk. Listed here are several leading causes of death in the United States, along with a measure of the risk each poses. The larger the area of the circle in the figure, the greater the risk of dying from that cause. *Data are for 2013, from* Injury Facts, 2016. Itasca, IL: National Safety Council.

DATA Q People tend to view car travel as being safer than airplane travel, but a person is how many times more likely to die from a car accident than from an airplane crash?

Go to **Interpreting Graphs & Data** on **Mastering**EnvironmentalScience®.

Risk assessment analyzes risk quantitatively

The quantitative measurement of risk and the comparison of risks involved in different activities or substances together are termed **risk assessment.** Risk assessment is a way to identify and outline problems. In environmental health, it helps ascertain which substances and activities pose health threats to people or wildlife and which are largely safe.

Assessing risk for a chemical substance involves several steps. The first steps involve the scientific study of toxicity we examined above—determining whether a substance has toxic effects and, through dose-response analysis, measuring how effects vary with the degree of exposure. Subsequent steps involve assessing the individual's or population's likely extent of exposure to the substance, including the frequency of contact, the concentrations likely encountered, and the length of encounter.

To assess risk from a widely used substance such as BPA, teams of scientific experts may be convened to review hundreds of studies so that regulators and the public can benefit from informed summaries. In 2008, for example, the government's National Toxicology Program of the National Institute of Environmental and Health Sciences (National Institutes of Health [NIH]) convened a panel to review the literature pertaining to the safety of BPA, and the regulatory challenges faced with endocrine disruptors such as BPA were highlighted in this review. Initially, the panel deemed 80 studies appropriate for informing policy on regulating BPA, 70% of which were from academic laboratories (many of which found adverse effects of BPA on organisms) rather than industry laboratories (which typically found no effects of BPA). Shortly thereafter, the panel received a 93-page letter from the American Chemistry Council contending that many of the studies had flaws that made them unsuitable for informing regulatory policy. One common criticism advanced in the letter was that academic studies did not follow Good Laboratory Practice (GLP), which is part of the protocols adopted by regulatory agencies around the world to evaluate potentially toxic substances. After considering these concerns, the panel eliminated many academic studies from consideration, reducing academic studies to a mere 30% of the studies being considered. Given only this pool of studies to consider, the panel failed to rate any impacts of BPA as of "Concern for Adverse Impact" or "Serious Concern for Adverse Impact."

Scientists argued that as a federal agency regulating potentially toxic substances, the panel is expected to consider both GLP and non-GLP studies in its deliberations. Also, they contended, very rigid GLP methods were not always appropriate for chemicals such as BPA that show unusual dose-response curves. To aid efforts to standardize research protocols and produce more studies that would qualify as suitable for consideration, the National Institute of Environmental and Health Sciences sponsored a meeting in 2009 of BPA researchers at which common protocols were established. Further, the agency devoted $30 million to BPA research to stimulate studies using these protocols, generate additional information, and better inform future panels about potential health impacts of BPA on humans. Because ratings like this heavily influence regulatory decisions, initiatives like these can broaden the scientific studies used to evaluate threats to public health.

Risk management combines science and other social factors

Accurate risk assessment is a vital step toward effective **risk management,** which consists of decisions and strategies to minimize risk (**FIGURE 14.19**). In most nations, risk management is handled largely by federal agencies. In the United States, these include agencies such as the FDA, the EPA, and the CDC. In risk management, scientific assessments of risk are considered in light of economic, social, and political needs and values. Risk managers assess costs and benefits of addressing risk in various ways with regard to both scientific and nonscientific concerns before making decisions on whether and how to reduce or eliminate risk.

In environmental health and toxicology, comparing costs and benefits (pp. 140–141) can be difficult because the benefits are often economic, whereas the costs often pertain to health. Moreover, economic benefits are generally known, easily quantified, and of a discrete and stable amount, whereas health risks are hard-to-measure probabilities, often involving a small percentage of people likely to suffer greatly and a large majority likely to experience little effect. When a government agency bans a pesticide, it may mean considerable economic loss for the manufacturer and potential economic loss for the farmer, whereas the benefits accrue less predictably over the long term to some percentage of factory workers, farmers, and the general public. Because of the lack of equivalence in the way costs and benefits are measured, risk management frequently tends to stir up debate.

In the case of BPA, eliminating plastic linings in our food and drink cans could do more harm than good, because the linings help prevent metal corrosion and the contamination of food by pathogens. Alternative substances exist for most of BPA's uses, but replacing BPA with alternatives will entail economic costs to industry, and these costs are passed on to consumers in the prices of products. Such complex considerations can make risk management decisions difficult even if the science of risk assessment is fairly clear. This may help account for the observed hesitancy of U.S. regulatory agencies, so far, to issue stringent restrictions on the uses of BPA. But the issue is far from settled, because both the FDA and the EPA continue to review options for managing risk from BPA.

Philosophical and Policy Approaches

Because we cannot know a substance's toxicity until we measure and test it, and because there are so many untested chemicals and combinations, science will never eliminate the many uncertainties that accompany risk assessment. In such a world of uncertainty, there are two basic philosophical approaches to categorizing substances as safe or dangerous (**FIGURE 14.20**).

One approach is to assume that substances are harmless until shown to be harmful. We might nickname this the "innocent-until-proven-guilty" approach. Because thoroughly testing every existing substance (and combination of substances) for its effects is a hopelessly long, complicated, and expensive pursuit, the innocent-until-proven-guilty approach has the virtue of facilitating technological innovation and economic activity. However, it has the disadvantage

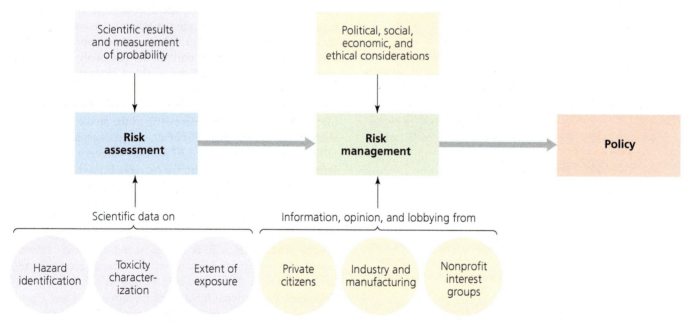

FIGURE 14.19 The first step in addressing risks from an environmental hazard is risk assessment. Once science identifies and measures risks, then risk management can proceed. In risk management, economic, political, social, and ethical issues are considered in light of the scientific data from risk assessment.

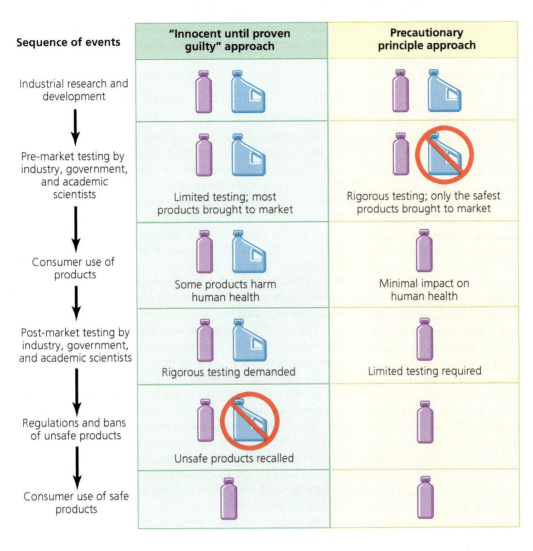

Sequence of events	"Innocent until proven guilty" approach	Precautionary principle approach
Industrial research and development		
Pre-market testing by industry, government, and academic scientists	Limited testing; most products brought to market	Rigorous testing; only the safest products brought to market
Consumer use of products	Some products harm human health	Minimal impact on human health
Post-market testing by industry, government, and academic scientists	Rigorous testing demanded	Limited testing required
Regulations and bans of unsafe products	Unsafe products recalled	
Consumer use of safe products		

FIGURE 14.20 Two main approaches can be taken to introduce new substances to the market. In one approach, substances are "innocent until proven guilty"; they are brought to market relatively quickly after limited testing. Products reach consumers more quickly, but some fraction of them (**blue bottle in diagram**) may cause harm to some fraction of people. The other approach is to adopt the precautionary principle, bringing substances to market cautiously, only after extensive testing. Products that reach the market should be safe, but many perfectly safe products (**purple bottle in diagram**) will be delayed in reaching consumers.

of putting into wide use some substances that may later turn out to be dangerous.

The other approach is to assume that substances are harmful until shown to be harmless. This approach follows the precautionary principle (p. 254). This more cautious approach should enable us to identify troublesome toxicants before they are released into the environment, but it may also impede the pace of technological and economic advance.

These two approaches are actually two ends of a continuum of possible approaches. The two endpoints differ mainly in where they lay the burden of proof—specifically, whether product manufacturers are required to prove a product is safe or whether government, scientists, or citizens are required to prove a product is dangerous.

Philosophical approaches are reflected in policy

This choice of philosophical approach has direct implications for policy, and nations vary in how they blend the two approaches when it comes to regulating synthetic substances. European nations have recently embarked on a policy course that largely incorporates the precautionary principle, whereas the United States largely follows an innocent-until-proven-guilty approach. For instance, compounds in cosmetics require no FDA review or approval before being sold to the public.

In the United States, several federal agencies apportion responsibility for tracking and regulating synthetic chemicals. The FDA, under the Food, Drug, and Cosmetic Act of 1938 and its subsequent amendments, monitors foods and food additives, cosmetics, drugs, and medical devices. The EPA regulates pesticides under the Federal Insecticide, Fungicide, and Rodenticide Act of 1947 (FIFRA) and its amendments. The Occupational Safety and Health Administration (OSHA) regulates workplace hazards under a 1970 act. Several other agencies regulate other substances.

weighing the
ISSUES

The Precautionary Principle

Industry's critics say chemical manufacturers should bear the burden of proof for the safety of their products before they hit the market. Industry's supporters say that mandating more safety research will hamper the introduction of products that consumers want, increase the price of products as research costs are passed on to consumers, and cause companies to move to nations where standards are more lax. What do you think? Which approach should U.S. government regulators embrace?

EPA regulates industrial chemicals

The widespread regulation of chemicals in the United States began with the **Toxic Substances Control Act of 1976 (TSCA)**, which directed the EPA to monitor thousands of industrial chemicals, ranging from PCBs to lead to BPA. The act gave the agency power to regulate these substances and ban those that posed excessive risk, but the law had several weaknesses.

For example, chemicals that were already in use in 1976 were simply grandfathered into the program, and chemicals introduced after 1976 were not tested extensively for toxicity *prior* to their use in products. They were only tested *after* they were suspected of causing harm, and banning dangerous chemicals was prohibitively difficult.

As our knowledge of chemical hazards expanded in recent decades, it became clear that the TSCA was in dire need of update. Legislators, scientists, and public health advocates lobbied for change and these efforts were realized in 2016 with the passage of the **Frank R. Lautenberg Chemical Safety for the 21st Century Act.** The act improves upon TSCA by directing the EPA to review the potential toxicity of *all* industrial chemicals currently in use and by subjecting new chemicals to rigorous safety testing *before* they are used in products—provisions which will greatly improve chemical safety for consumers.

In its regulation of pesticides under FIFRA, the EPA is charged with "registering" each new pesticide that manufacturers propose to bring to market. The registration process involves risk assessment and risk management. The EPA first asks the manufacturer to provide information, including results of safety assessments the company has performed according to EPA guidelines. The EPA examines the company's research and all other relevant scientific research. It examines the product's ingredients and how the product will be used and tries to evaluate whether the chemical poses risks to people, other organisms, or water or air quality. The EPA then approves, denies, or sets limits on the chemical's sale and use. It also must approve language used on the product's label.

Because the registration process takes economic considerations into account, critics say it allows hazardous chemicals to be approved if the economic benefits are judged to outweigh the hazards. Here the challenges of weighing intangible risks involving human health and environmental quality against the tangible and quantitative numbers of economics become apparent.

FAQ

If the government allows a product to be sold in stores, isn't it safe?

Just because a product is available to the public doesn't mean it poses no risk to consumers. Some products, such as firearms and cigarettes, are recognized to be hazardous and have restrictions placed on their purchase. Medicines, cosmetics, and some types of food are tested for safety prior to release, but other potentially dangerous products, such as endocrine disrupting chemicals in plastics, are not similarly tested. The Lautenberg Act should improve the safety of the chemicals in the products we use, but with more than 80,000 chemicals in need of assessment, these efforts will take time. Hence, it would be a mistake to assume that the products you use today have been thoroughly tested prior to their arrival on store shelves.

Toxicants are regulated internationally

The European Union took the world's boldest step toward testing and regulating manufactured chemicals. In 2007, the EU's **REACH** program went into effect (*REACH* stands for Registration, Evaluation, Authorisation, and Restriction of Chemicals). REACH largely shifts the burden of proof for testing chemical safety from national governments to industry and requires that chemical substances produced or imported in amounts of more than 1 metric ton per year be registered with a new European Chemicals Agency. This agency evaluates industry research and decides whether the chemical seems safe and should be approved, whether it is unsafe and should be restricted, or whether more testing is needed. It is expected that REACH will require the registration of about 30,000 substances.

The REACH policy also aims to help industry by giving it a single streamlined regulatory system and by exempting it from having to file paperwork on substances under 1 metric ton. By requiring stricter review of major chemicals already in use, exempting chemicals made only in small amounts, and providing financial incentives for innovating new chemicals, the EU hopes to help European industries research and develop safer new chemicals and products while safeguarding human health and the environment. REACH differs markedly from TSCA (**TABLE 14.2**), illustrating how the approach that Europe pursued was different from the approach pursued in the United States until 2016.

In an impacts assessment in 2003, EU commissioners estimated that REACH will cost the chemical industry and

---

TABLE 14.2 American vs. European Approaches to Chemical Regulation

TSCA (UNITED STATES)	REACH (EUROPEAN UNION)
• Government bore burden of proof to show harm	• Industry bears burden of proof to show safety
• Few data on new chemicals were required from industry	• More data on new chemicals are required from industry
• Chemicals in use before 1976 were not regulated	• Chemicals in use before 1981 bear scrutiny like that directed toward newer chemicals
• Prioritizing problems was hampered by lack of data	• Problems are prioritized using data on risk
• Industry was allowed to keep trade secrets from the public	• Database will allow public access to chemical information

Source: Adapted from Schwarzman, M.R., and M.P. Wilson, 2009. New science for chemicals policy. *Science* 326: 1065–1066.

TABLE 14.3 **The "Dirty Dozen" Persistent Organic Pollutants (POPs) Targeted by the Stockholm Convention**

TOXICANT	DESCRIPTION	TOXICANT	DESCRIPTION
Aldrin	Insecticide to kill termites and crop pests	Furans	By-product of processes that release dioxins; also present in commercial mixtures of PCBs
Chlordane	Insecticide to kill termites and crop pests	Heptachlor	Broad-spectrum insecticide
DDT	Insecticide to protect against insect-spread disease; still applied in some countries to control malaria	Hexachloro-benzene	Fungicide for crops; released by chemical manufacture and processes that release dioxins and furans
Dieldrin	Insecticide to kill termites, textile pests, crop pests, and disease vectors	Mirex	Household insecticide; fire retardant in plastics, rubber, and electronics
Dioxins	By-product of incomplete combustion and chemical manufacturing; released in metal recycling, pulp and paper bleaching, auto exhaust, tobacco smoke, and wood and coal smoke	PCBs	Industrial chemical used in heat-exchange fluids, electrical transformers and capacitors, paints, sealants, and plastics
Endrin	Pesticide to kill rodents and crop insects	Toxaphene	Insecticide to kill crop insects and livestock parasites

Data from United Nations Environment Programme (UNEP), 2001.

chemical users 2.8–5.2 billion euros (U.S. $3.8–7.0 billion) over 11 years but that the health benefits to the public would be roughly 50 billion euros (U.S. $67 billion) over 30 years. Changes in the program since then have made the predicted cost:benefit ratio even more favorable.

The world's nations have also sought to address chemical pollution with international treaties. The Stockholm Convention on Persistent Organic Pollutants (POPs) came into force in 2004 and has been ratified by 172 nations. POPs are toxic chemicals that persist in the environment,

bioaccumulate and biomagnify up the food chain, and often can travel long distances. The PCBs and other contaminants found in polar bears are a prime example. Because contaminants often cross international boundaries, an international treaty seemed the best way to deal fairly with such transboundary pollution. The Stockholm Convention aims first to end the use and release of 12 POPs shown to be most dangerous, a group nicknamed the "dirty dozen" (TABLE 14.3). It sets guidelines for phasing out these chemicals and encourages transition to safer alternatives.

closing
THE LOOP

International agreements such as REACH and the Stockholm Convention indicate that governments may act to protect the world's people, wildlife, and ecosystems from toxic substances and other environmental hazards. At the same time, solutions often come more easily when they do not arise from government regulation alone. Consumer choice exercised through the market can often be an effective way to influence industry's decision making, but this requires consumers to have full information from scientific research regarding the risks involved. Once scientific results are in, a society's philosophical approach to risk management will determine what policy decisions are made.

All of these factors have come into play regarding regulation of BPA in consumer products. Although some nations have banned the chemical, many others have only restricted its use in children's products or chosen not to restrict BPA at all. But growing consumer concern over the presence of BPA, brought about by media attention, has spurred some companies to remove BPA from their products, even in the absence of governmental regulation in the United States.

It is important to remember, however, that synthetic chemicals, while exposing people to some risk, have brought us innumerable modern conveniences, a larger food supply, and medical advances that save and extend human lives. The lining of cans that contain BPA, for example, can affect human health by leaching BPA into foods but also serves a beneficial function by preventing corrosion and contamination of canned goods. A safer and happier future, one that safeguards the well-being of both people and the environment, therefore depends on knowing the risks that some hazards pose, assessing these risks, and having means in place to phase out harmful substances and replace them with safer ones whenever possible.

REVIEWING Objectives

You should now be able to:

Explain the goals of environmental health and identify major environmental health hazards

The study of environmental health assesses environmental factors that affect human health and quality of life. Environmental health threats include physical, chemical, biological, and cultural hazards that occur both indoors and outside. Disease, both infectious and noninfectious, remains a major threat to human health and is being addressed with a diversity of approaches. (pp. 355–362)

Describe the types of toxic substances in the environment, the factors that affect their toxicity, and the defenses that organisms have against them

Toxicant types include carcinogens, mutagens, teratogens, allergens, pathway inhibitors, neurotoxins, and endocrine disruptors. The toxicity of a substance may be influenced by the nature of exposure (acute or chronic) and individual variation in the strength of the organism's defenses, such as detoxifying enzymes, against the toxin. (pp. 362–366)

Explain the movements of toxic substances and how they affect organisms and ecosystems

Toxic substances may travel long distances through the atmosphere, waterways, or groundwater. Some toxic substances bioaccumulate and move up the food chain, poisoning consumers at high trophic levels through the process of biomagnification, and impairing ecosystem services. (pp. 366–369)

Discuss the approaches used to study the effects of toxic chemicals on organisms

Scientists use wildlife toxicology, case histories, epidemiology, animal testing, and dose-response analysis to assess the toxicity of chemicals. Toxicity is affected by dosage, but some chemicals have unconventional dose-response curves and synergistic interactions with other chemicals. (pp. 369–374)

Summarize risk assessment and risk management

Risk assessment involves quantifying and comparing risks involved in different activities or substances. Risk management integrates science with political, social, and economic concerns to design strategies to minimize risk. (pp. 374–376)

Compare philosophical approaches to risk and how they relate to regulatory policy

An innocent-until-proven-guilty approach assumes that a chemical substance is safe unless shown to be harmful after release to the public, whereas a precautionary approach assumes that a substance may be harmful unless proven safe by its manufacturer prior to sale. (pp. 376–379)

TESTING Your Comprehension

1. What four major types of health hazards are examined by practitioners of environmental health?

2. In what way is disease the greatest hazard that people face? What kinds of interrelationships must environmental health experts study to learn about how diseases affect human health?

3. Where does most exposure to radon, asbestos, lead, and PBDEs occur?

4. When did concern over the effects of pesticides start to grow in the United States? Describe the argument presented by Rachel Carson in *Silent Spring*. What policy resulted from the book's publication? Where and how is DDT still used?

5. List and describe the general categories of toxic substances described in this chapter.

6. How do toxic substances travel through the environment, and where are they most likely to be found? Describe and contrast the processes of bioaccumulation and biomagnification.

7. What are epidemiological studies, and how are they most often conducted?

8. Why are animals used in laboratory experiments in toxicology? Explain the dose-response curve. Why is a substance with a high LD_{50} considered safer than one with a low LD_{50}?

9. What factors may affect an individual's response to a toxic substance? Why is chronic exposure to toxic agents often more difficult to measure and diagnose than acute exposure? What are synergistic effects, and why are they difficult to measure and diagnose?

10. How do scientists identify and assess risks from substances or activities that may pose health threats?

SEEKING Solutions

1. Describe some environmental health hazards you may be living with indoors. How may you have been affected by indoor or outdoor hazards in the past? How could you best deal with these hazards in the future?

2. Do you feel that laboratory animals should be used in experiments in toxicology? Why or why not?

3. Why has research on endocrine disruption spurred so much debate? What steps do you think could be taken to help establish greater consensus among scientists, industry, regulators, policymakers, and the public?

4. **CASE STUDY CONNECTION** You work for a public health organization and have been asked to educate the public about BPA and to suggest ways to minimize exposure to the chemical. You begin by examining your lifestyle and finding ways to use alternatives to BPA-containing products. Create a list of five ways you are exposed daily to BPA, and then list approaches that would avoid or minimize these exposures. Do these steps require more time and/or money? What are some costs of embracing these changes? What would you tell an interested person about BPA as it relates to human health?

5. **THINK IT THROUGH** In your public speaking class, you have been asked to create a presentation that supports either the policy approach of the United States or that of the European Union concerning the study and management of the risks of synthetic chemicals. Which position would you advocate? Explain the major points you would raise in your presentation to support your position.

6. **THINK IT THROUGH** You are the parent of two young children, and you want to minimize the environmental health risks your kids are exposed to. Name five steps that you could take in your household and in your daily life that would minimize your children's exposure to environmental health hazards.

CALCULATING Ecological Footprints

In 2007, the last year the EPA gathered and reported data on pesticide use (pp. 246–247), Americans used 1.13 billion pounds of pesticide active ingredients, and global use totaled 5.21 billion pounds. In that same year, the U.S. population was 302 million, and the world's population was 6.63 billion. In the table, calculate your share of pesticide use as a U.S. citizen in 2007 and the amount used by (or on behalf of) the average citizen of the world.

1. What is the ratio of your annual pesticide use to the world's per capita average?

2. In 2007, the average U.S. citizen had an ecological footprint of 8.0 hectares, and the average world citizen's footprint was 2.7 hectares (Chapter 1). Compare the ratio of pesticide use with the ratio of the overall ecological footprints. How do these differ, and how would you account for the difference?

3. Does the per capita pesticide use for a U.S. citizen seem reasonable for you personally? Why or why not? Do you find this figure alarming or of little concern? What else would you like to know to assess the risk associated with this level of pesticide use?

Annual Pesticide Use

	POUNDS OF ACTIVE INGREDIENTS
You	
Your class	
Your state	
United States	1.13 billion
World (total)	5.21 billion
World (per capita)	

MasteringEnvironmentalScience®

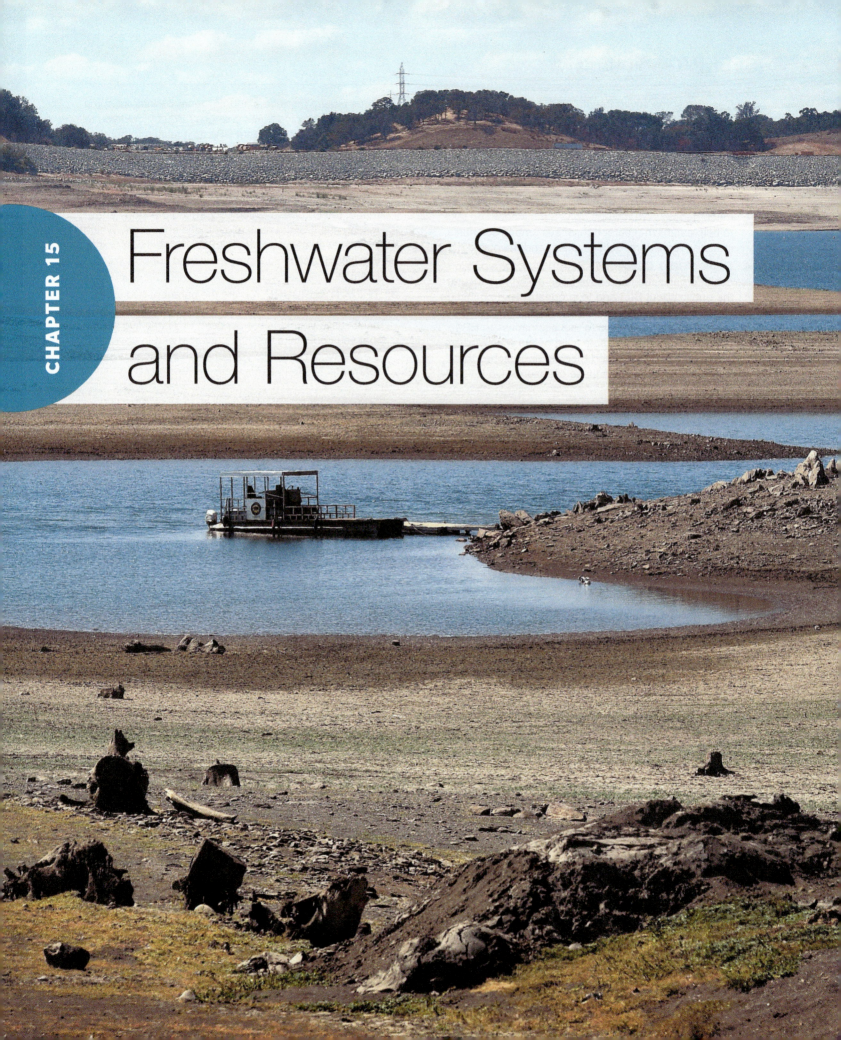

CHAPTER 15

Freshwater Systems and Resources

Conserving Every Drop in California

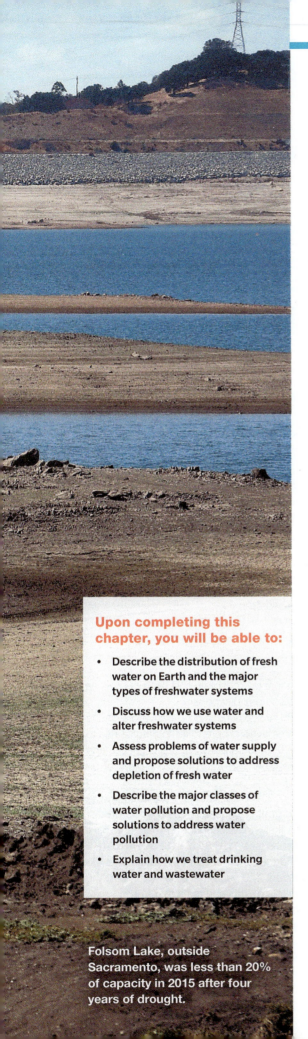

Folsom Lake, outside Sacramento, was less than 20% of capacity in 2015 after four years of drought.

Upon completing this chapter, you will be able to:

- Describe the distribution of fresh water on Earth and the major types of freshwater systems

- Discuss how we use water and alter freshwater systems

- Assess problems of water supply and propose solutions to address depletion of fresh water

- Describe the major classes of water pollution and propose solutions to address water pollution

- Explain how we treat drinking water and wastewater

> **It's a different world. We have to act differently.**
> Jerry Brown, Governor of California, when announcing mandatory water restrictions in 2015

> **A grower, like myself, who has so much emotion and so much of his passion tied up in dirt and production and making things grow, this is a heartbreaker.**
> Bill Diedrich, California almond farmer, surveying his drought-withered trees

As Governor Jerry Brown stood in a field of dry grass near the shore of Lake Tahoe in April 2015, California was enduring its fourth straight year of drought, and it was clear that water-use patterns had to change if the Golden State was to survive and prosper. "This is the new normal," Brown said, referring to his state's current drought, "and we'll have to deal with it."

At that press conference, the governor announced mandatory water restrictions that required cities and towns in California to reduce their water use by 25%—an aggressive target intended to help the United States' most populous state and the world's eighth largest economy deal with one of its worst droughts in recorded history.

Record-low levels of precipitation coupled with record-high heat have kept California in drought conditions since 2012 and made water scarcity one of the greatest challenges faced by the state in recent memory. The drought has largely been driven by the lack of intense "atmospheric river" precipitation events, which can supply up to 50% of the state's annual precipitation. In these events, moist air from the Pacific rushes inland over California, to the Sierra Nevada Mountains. As the air rises over the mountains, it cools, resulting in torrential rains and heavy mountain snow, which eventually supplies water to rivers during snowmelt in spring. Since 2011, however, a high-pressure atmospheric system in the eastern Pacific has interfered with these events, resulting in reduced precipitation. The Sierra Nevada snowpack in 2014, for example, was a mere 18% of its normal volume after years of reduced snowfall coupled with elevated temperatures. According to a National Aeronautics and Space Administration (NASA) study, the accumulated precipitation deficit in 2015 was 50 cm (20 in.)—the equivalent of an entire year's average rainfall.

The drought has had devastating effects on California's agricultural sector, which uses 80% of California's water and supplies half the United States' produce, including 90% of its tomatoes, 95% of its broccoli, and 99% of its pistachios and almonds (**FIGURE 15.1a**, p. 384). With water supplies for irrigation threatened by drought, some farmers are extracting groundwater at unsustainable rates to support their crops and as a result are depleting groundwater reserves. The drought is having severe economic impacts, costing the state an estimated $2.2 billion in agricultural losses in 2015 alone.

Detail from a water conservation sign in California ▲

(a) Pistachio trees killed by drought

(b) Lake Oroville reservoir in 2014

(c) The endangered giant kangaroo rat (*Dipodomys ingens*)

(d) Billboard advertising water conservation in California

FIGURE 15.1 California's historic drought has touched many aspects of the state. Crops deprived of water, such as these pistachio trees in the San Joaquin Valley **(a)**, have withered in the dry heat. Lake Oroville **(b)**, like many other reservoirs in California, currently holds only a small fraction of its usual volume. Populations of some species of wildlife, such as the giant kangaroo rat **(c)**, have reached critically low levels. Billboards in California promote water-conserving strategies **(d)** to aid citizens in reducing water use.

Reservoirs that supply cities with water are draining, threatening the water supply for millions of people (**FIGURE 15.1b**). In some cases, water levels in reservoirs have fallen tremendously, revealing the remnants of towns submerged more than 75 years ago when nearby dams were first constructed. Low water levels in California reservoirs have also reduced hydroelectric power (pp. 570–573) generation by 60% since 2012, forcing utilities to turn to other energy sources to meet demands for power.

Humans aren't the only casualties of drought, though. The U.S. Forest Service estimated in 2015 that the drought had led to the death of more than 66 million trees in California alone, many killed when trees weakened by drought were infested by parasitic beetles. So many dead trees have raised concerns about catastrophic wildfires that could burn entire forests.

In addition, endangered giant kangaroo rats (*Dipodomys ingens*) are starving in California's Carrizo Plain outside Los Angeles as the grasslands they occupy slowly turn to desert (**FIGURE 15.1c**). With only 5% of the kangaroo rat population surviving, cascading effects are seen on rat predators, such as the endangered San Joaquin fox, coyotes, and birds of prey, as one of their primary food sources disappears.

And Chinook salmon are experiencing massive die-offs in northern California as water levels drop and temperatures rise to lethal levels. Populations of smelt are also at critical levels, and their scarcity is depriving other fish and predatory birds of an important food source.

The need for drastic action to reduce water demand in California is particularly urgent because climate models predict that California is highly likely to experience similar long-term droughts in coming decades (see **THE SCIENCE BEHIND THE STORY**, pp. 386–387). Further, analyses of the state's past climate indicate that the recent 150 years have been some of the wettest in the past 1000 years. This suggests that the "normal" levels of precipitation seen in California (and on which California has based its allowable water withdrawals from rivers) were, in fact, abnormally high.

In his speech, Governor Brown said, "As Californians, we have to pull together and save water in every way we can."

His words, it seems, found fertile ground with the residents of his state (**FIGURE 15.1d**). More timid approaches to water conservation proposed in the previous year had yielded only modest improvements, but the ambitious nature of the new restrictions spurred action. Data from the state's water utilities indicated that in May 2015, one month after the statewide restrictions were mandated, residential water use across the state dropped by 29% compared with May 2013, with some regions averaging reductions of more than 38% (**FIGURE 15.2**).

Abnormally warm temperatures in the Pacific Ocean in 2015 predicted a strong El Niño event (p. 423) for late 2015 and early 2016, and such events have historically caused elevated levels of rain and snow in California. While the event did bring much-needed precipitation in northern and eastern California, replenishing reservoirs and building snowpack in the Sierra Nevada mountains, southern California ended the winter with lower than average levels of precipitation.

The El Niño winter of 2015–2016 provided California a reprieve from stifling drought, but with the state's population expected to grow and the threat of severe drought ever-looming, it is clear the state will be dealing with water supply issues for the foreseeable future. Much work remains to be done in California to develop sustainable, long-term water conservation programs. But one of the most important hurdles that has already been overcome is convincing Californians that saving water is important to the future of their state, and is a movement worth embracing.

FIGURE 15.2 Following statewide mandatory restrictions on water use, water use across California declined significantly. Water savings are reported by hydrologic region, which are regional boundaries based on river watersheds and major bodies of water. *Data from California State Water Resources Control Board, May 2015 Statewide Conservation Progress, www.waterboards.ca.gov.*

Freshwater Systems

"Water, water, everywhere, nor any drop to drink." The well-known line from the poem *The Rime of the Ancient Mariner* describes the situation on our planet well. Water may seem abundant, but water that we can drink is quite rare (**FIGURE 15.3**). About 97.5% of Earth's water resides in the oceans and is too salty to drink or to use in watering the vast majority of plants we grow for food. Only 2.5% is considered **fresh water,** water that is relatively pure, with few dissolved

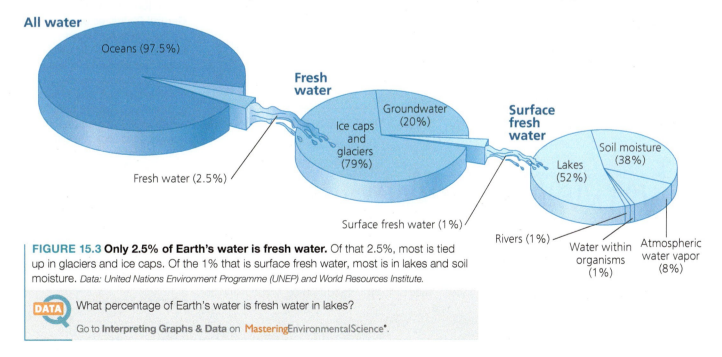

FIGURE 15.3 Only 2.5% of Earth's water is fresh water. Of that 2.5%, most is tied up in glaciers and ice caps. Of the 1% that is surface fresh water, most is in lakes and soil moisture. *Data: United Nations Environment Programme (UNEP) and World Resources Institute.*

DATA What percentage of Earth's water is fresh water in lakes?

Go to **Interpreting Graphs & Data** on **Mastering**EnvironmentalScience®.

Are We Destined for a Future of "Megadroughts" in the United States?

**Benjamin I. Cook,
NASA**

As the state of California struggles to reduce water use in agriculture, industry, and homes to deal with its current water shortages, a new study suggests that much more of the same is in store for the American Southwest and the Central Great Plains for the remainder of this century.

This was the sobering conclusion of a team of researchers led by Benjamin Cook, research scientist at NASA Goddard Institute for Space Studies, advanced in a paper published in the journal *Science Advances* in 2015. In an effort to frame the current drought in the western United States in the proper context, the group compared the climate of the Southwest and Central Plains over the past 1000 years, gleaned from natural archives of climate such as tree rings, with predictions of their climates over the next 100 years based on computer simulations. "We are the first to do this kind of quantitative comparison between

the projections and the distant past," said co-author Jason Smerdon, "and the story is a bit bleak."

Estimates of past climates in these regions used the North American Drought Atlas (NADA), a reconstruction of climates based on data from tens of thousands of samples of tree rings from the United States, Canada, and Mexico. Trees are a natural archive of climate data, because they grow at varying rates depending on moisture, temperature, and other factors. Using data from the recent past, scientists have determined the relationship between tree growth patterns and climatic factors, and they then use this relationship to reconstruct past climates for which they have data on tree growth from tree rings.

To predict the future climate of the southwestern United States and the Central Plains, the researchers used 17 climate models and ran simulations based on a "business as usual" scenario, in which the growth in greenhouse gas emissions followed current trends, or based on a "moderate reduction" scenario, in which growth in greenhouse gas emissions was more modest.

The study used three measures of drought, all of them indications of levels of soil moisture available to plants. These were soil moisture content from surface to 30-cm depth; soil moisture content from surface to 2-m depth; and the Palmer Drought Severity Index (PDSI), a measure of the difference between soil

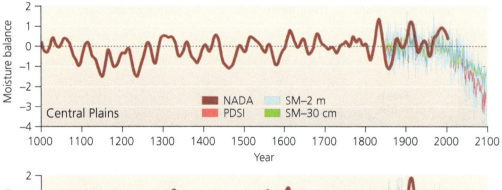

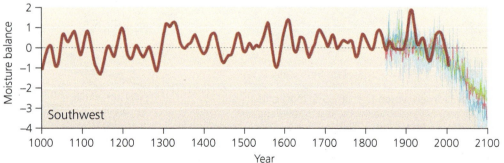

FIGURE 1 Soil moisture levels for the Southwest and Central Plains, as predicted by the North American Drought Atlas (NADA) and computer climate models. Negative values for moisture balance indicate drier soils (drought) and positive values indicate wetter soils. The predicted drought conditions in the late 21st century are unprecedented in the last 1000 years. (PDSI = Palmer Drought Severity Index, SM-30 cm = Soil moisture to 30 cm depth, and SM-2 m = Soil moisture to 2 m depth. The gray-shaded areas represent the variability in model PDSI values across computer climate models). *Source: Cook, B.I., et al., 2015, Science Advances 1(1): e1400082.*

moisture supply (from precipitation) and soil moisture demand (from evaporation and uptake by plants). In the PDSI, negative values mean drier conditions—drought—and positive values mean wetter conditions.

When the researchers combined data from the past, present, and future, the results were stunning—and troubling. The climate models predicted unprecedented levels of drought in both regions through 2100, with high levels of agreement between the 17 climate models and the three measures of soil moisture (**FIGURE 1**). The study concluded that the drying of soils was not being driven by drastic reductions in precipitation but rather by increased evaporative pressure, that is, warmer temperatures leading to higher rates of evaporation and elevated rates of soil water uptake by plants. The models also concluded that human-induced climate change, and not natural variability in climate, would be the driving force for drier soils in the Southwest and Central Plains. A comparison with the past indicated that these future conditions would be far worse than those seen during the "Medieval megadrought period" from A.D. 1100 to 1300. This extended drought is thought to have led to the decline of the ancient Pueblos, the Anasazi, who lived along the Colorado Plateau. Smerdon described their results as follows: "Even when selecting for the worst megadrought-dominated period, the 21st Century projections make the megadroughts seem like quaint walks through the Garden of Eden."

Comparing the probability of extended droughts in the latter half of the 20th century (1950–2000) with the probability of such droughts in the latter half of the 21st century (2050–2099) provided further cause for concern. According to the simulations, the probability that the two regions would experience a decade-long drought essentially doubled, and the chance that they would experience a multidecade drought increased from around 10% to more than 80% (**FIGURE 2**).

If greenhouse gases are reduced, the chance of a multidecade drought drops to 60–70% for the Central Plains but remains above 80% for the Southwest, providing yet another incentive for combating global climate change. Most important, studies like these help us to prepare for the future by providing an idea of what to expect. In the words of study co-author Toby Ault, "The time to act is now. The time to start planning for adaptation is now. We need to assess what the rest of this century will look like for our children and grandchildren."

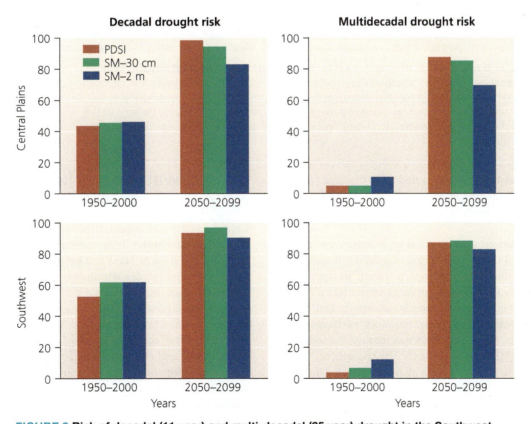

FIGURE 2 **Risk of decadal (11 year) and multi-decadal (35 year) drought in the Southwest and Central Plains in the late 20th century and late 21st century for three measures of soil moisture.** Due to human-induced climate change, both regions are far more likely to have long-term drought in the future than in the past. *Source: Cook, B.I., et al., 2015, Science Advances 1(1): e1400082.*

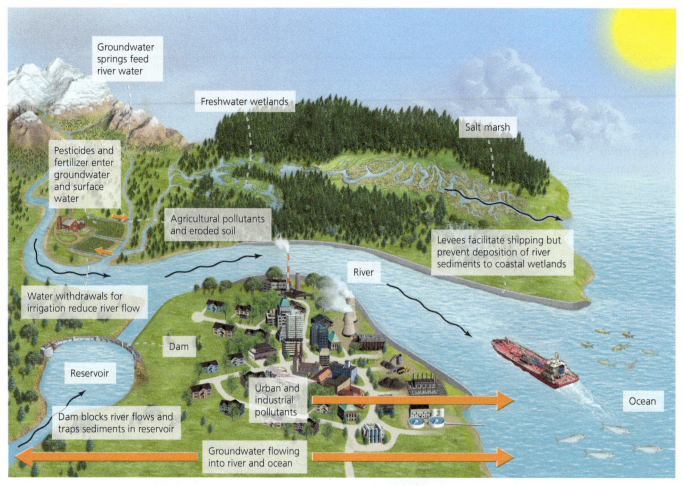

FIGURE 15.4 Water flows through freshwater systems and marine and coastal aquatic systems that interact extensively with one another. People affect the components of the system by constructing dams and levees, withdrawing water for human use, and introducing pollutants. Because the systems are closely connected, these impacts can cascade through the system and cause effects far from where they originated. In the figure, orange arrows indicate inputs into water bodies and black lines indicate the direction of water flow.

Labels in figure:
- Groundwater springs feed river water
- Freshwater wetlands
- Salt marsh
- Pesticides and fertilizer enter groundwater and surface water
- Agricultural pollutants and eroded soil
- Levees facilitate shipping but prevent deposition of river sediments to coastal wetlands
- River
- Water withdrawals for irrigation reduce river flow
- Dam
- Reservoir
- Urban and industrial pollutants
- Ocean
- Dam blocks river flows and traps sediments in reservoir
- Groundwater flowing into river and ocean

salts. Because most fresh water is tied up in glaciers, ice caps, and underground aquifers, a little more than 1 part in 10,000 of Earth's water is fresh water that is easily accessible for human use.

Water is renewed and recycled as it moves through the water cycle (pp. 118–119). Precipitation falling from the sky either sinks into the ground or acts as runoff to form rivers, which carry water to the oceans or large inland lakes. As they flow, rivers can interact with ponds, wetlands, and coastal aquatic ecosystems. Underground aquifers exchange water with rivers, ponds, lakes, and the ocean through the sediments on the bottoms of these water bodies. The movement of water in the water cycle creates a web of interconnected freshwater and marine aquatic systems (**FIGURE 15.4**) that exchange water, organisms, sediments, pollutants, and other dissolved substances. What happens in one system therefore affects other systems—even those that are far away. Let's examine the freshwater components of the interconnected system, beginning with groundwater. Marine and coastal components of the system will be examined subsequently (Chapter 16), but note that all these systems interact extensively.

Groundwater plays key roles in the hydrologic cycle

Liquid fresh water occurs either as surface water or groundwater. **Surface water** is water located atop Earth's surface (such as in a river or lake) and **groundwater** is water beneath the surface that resides within pores in soil or rock. Any precipitation reaching Earth's land surface that does not evaporate, flow into waterways, or get taken up by organisms infiltrates (sinks into) the surface. Groundwater flows slowly beneath the surface from areas of high pressure to areas of low pressure and can remain underground for a long time, in some cases for thousands of years. Groundwater makes up one-fifth of Earth's freshwater supply and plays a key role in meeting human water needs.

Groundwater is contained within **aquifers:** porous, spongelike formations of rock, sand, or gravel that can hold water (**FIGURE 15.5**). An aquifer's upper layer contains pore spaces partly filled with water. In the lower layer, the spaces are completely filled with water. The boundary between these two zones is the **water table.** Picture a sponge resting partly submerged in a tray of water; the lower part of the sponge is saturated, whereas the upper portion contains plenty of air in its pores. Any area where water infiltrates Earth's surface and reaches an aquifer below is known as a **recharge zone.**

The earth underground consists of layers of materials with different textures and densities. When a porous, water-bearing layer of rock, sand, or gravel is trapped between upper and lower layers of less permeable substrate (often clay), we have a **confined aquifer,** or *artesian aquifer.* In such a situation, the water is under great pressure. In contrast, an **unconfined aquifer** has no imperme-able upper layer to confine it, so its water is under less pressure and can be readily recharged by surface water.

The largest known aquifer is the Ogallala Aquifer, which lies under the Great Plains of the United States. Water from this massive aquifer has enabled American farmers to create the most boun-tiful grain-producing region in the world. However, unsustainable water withdraw-als are threatening the long-term use of the aquifer for agriculture (**FIGURE 15.6**).

Surface water converges in river and stream ecosystems

Surface water is vital for our survival and for the planet's ecological systems. Groundwater and surface water interact, and water can flow from one type of sys-tem to the other. Surface water becomes groundwater by infiltration. Groundwater becomes surface water through springs (and human-drilled wells), often keeping streams flowing or wetlands moist when surface conditions are otherwise dry.

Water that falls from the sky as rain, emerges from springs, or melts from snow or a glacier and then flows over the land surface is called **runoff.** As it flows downhill, runoff converges where the land dips lowest, forming streams, creeks, or brooks. These small watercourses may merge into rivers, whose water eventually reaches a lake or ocean. A smaller river flowing into a larger one is a *tributary.* The area of land drained by a *river sys-tem*—a river and all its tributaries—is that river's **watershed** (p. 103), also called a *drainage basin.*

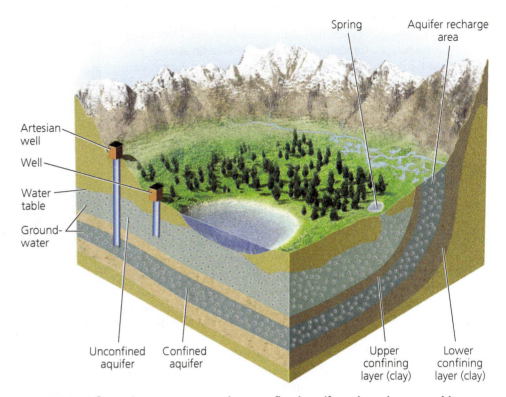

FIGURE 15.5 Groundwater may occur in unconfined aquifers above impermeable layers or in confined aquifers under pressure between impermeable layers. Water may rise naturally to the surface at springs, wetlands, and through the wells we dig. Artesian wells tap into confined aquifers to mine water under pressure.

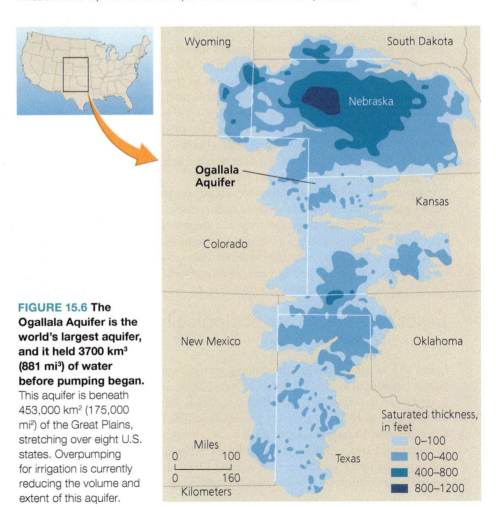

FIGURE 15.6 The Ogallala Aquifer is the world's largest aquifer, and it held 3700 km³ (881 mi³) of water before pumping began. This aquifer is beneath 453,000 km² (175,000 mi²) of the Great Plains, stretching over eight U.S. states. Overpumping for irrigation is currently reducing the volume and extent of this aquifer.

Saturated thickness, in feet
- 0–100
- 100–400
- 400–800
- 800–1200

Landscapes determine where rivers flow, but rivers shape the landscapes through which they run as well. A river that runs through a steeply sloped region and carries a great deal of sediment may flow as an interconnected series of watercourses called a *braided river* (FIGURE 15.7a). In flatter regions, most rivers are *meandering rivers* (FIGURE 15.7b). In a meandering river, the force of water rounding a bend gradually eats away at the outer shore, eroding soil from the bank.

(a) Braided river in Nebraska

FAQ

Is groundwater found in huge underground caverns?

As one of the "out of sight" elements of the water cycle, it's sometimes difficult for people to visualize how water exists underground. Many incorrectly assume that groundwater is always found in large, underground caves—essentially lakes beneath Earth's surface. This is not the case. If you look at soil under a microscope, you'll see there are small pores between the particles of minerals and organic matter that compose the soil. Many types of rock, such as limestone and sandstone, have relatively large pores between the particles of minerals that compose the rock. So when we extract groundwater with wells, we are simply sucking water out of the pores between soil particles or the pores within rocks in the portion of the soil beneath the water table.

Meanwhile, sediment is deposited along the inside of the bend, where water currents are weaker. Over time, river bends become exaggerated in shape, forming oxbows (FIGURE 15.7c). If water erodes a shortcut from one end of the loop to the other, pursuing a direct course, the oxbow is cut off and remains as an isolated, U-shaped water body called an *oxbow lake*.

Over thousands or millions of years, a meandering river may shift from one course to another, back and forth over a large area, carving out a flat valley. Areas nearest a river's course that are flooded periodically are said to be within the river's **floodplain.** Frequent deposition of silt from flooding makes floodplain soils especially fertile. As a result, agriculture thrives in floodplains, and *riparian* (riverside) forests are productive and species-rich. A river's meandering is often driven by large-scale flooding events that scour new channels during periods of high flow.

(b) Meandering river in Colorado

Rivers and streams host diverse ecological communities. Algae and detritus support many types of invertebrates, from water beetles to crayfish. Insects as diverse as dragonflies, mayflies, and mosquitoes develop as larvae in streams and rivers before maturing into adults that take to the air. Fish and amphibians consume aquatic invertebrates and plants, and birds such as kingfishers, herons, and ospreys dine on fish and amphibians.

Lakes and ponds are ecologically diverse systems

Lakes and ponds are bodies of standing surface water. The largest lakes, such as North America's Great Lakes, are sometimes known as inland seas. Lake Baikal in Asia is the world's deepest lake, at 1637 m (just over 1 mile) deep. The Caspian Sea is the world's largest body of fresh water, covering nearly as much land area as Montana or California. Although lakes and ponds can vary greatly in size, scientists have described several zones common to these waters (FIGURE 15.8).

(c) Oxbow along a river in Arizona

FIGURE 15.7 Rivers are classified according to their flow across the landscape. The Platte River in Nebraska **(a)** shows an example of a braided river, whereas the East River in Colorado **(b)** meanders across the landscape. The Horseshoe Bend portion of the Colorado River in Arizona **(c)** demonstrates an oxbow.

Around the nutrient-rich edges of a water body, the water is shallow enough that aquatic plants grow from the mud and reach above the water's surface. This region, named the *littoral zone*, abounds in invertebrates—such as insect larvae, snails, and crayfish—that fish, birds, turtles, and amphibians feed on. The *benthic zone* extends along the bottom of the lake or pond, from shore to the deepest point. Many invertebrates live in the mud, feeding on detritus or on one another. In the open portion of a lake or pond, far from shore, sunlight penetrates shallow waters of the *limnetic zone*. Because light enables photosynthesis, the limnetic zone supports phytoplankton (algae, protists, and cyanobacteria), which in turn support zooplankton, both of which are eaten by fish. Below the limnetic zone lies the *profundal zone*, the volume of open water that sunlight does not reach. This zone lacks photosynthetic life and is lower in dissolved oxygen than upper waters.

Ponds and lakes change over time as streams and runoff bring them sediment and nutrients. **Oligotrophic** lakes and ponds, which are low in nutrients and high in oxygen, may slowly transition to the high-nutrient, low-oxygen conditions of **eutrophic** water bodies (pp. 108–109). Eventually, water bodies

may fill in completely by the process of aquatic succession (p. 86). These changes occur naturally, but eutrophication can also result from human-caused nutrient pollution.

Freshwater wetlands include marshes, swamps, bogs, and vernal pools

Wetlands are systems in which the soil is saturated with water and which generally feature shallow standing water with ample vegetation. There are many types of wetlands, and most are enormously rich and productive. In *freshwater marshes*, shallow water allows plants such as cattails and bulrushes to grow above the water surface. *Swamps* also consist of shallow water rich in vegetation, but they occur in forested areas.

In cypress swamps of the southeastern United States, cypress trees grow in standing water. In northern forests, swamps are created when beavers dam streams with limbs from trees they have cut, flooding wooded areas upstream.

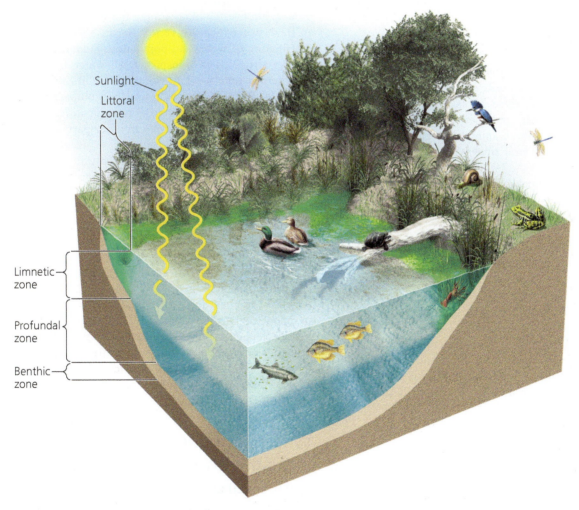

FIGURE 15.8 In lakes and ponds, emergent plants grow along the shoreline in the littoral zone. The limnetic zone is the layer of open, sunlit water, where photosynthesis takes place. Sunlight does not reach the deeper profundal zone. The benthic zone, at the bottom of the water body, often is muddy, rich in detritus and nutrients, and low in oxygen.

FIGURE 15.9 Bogs are freshwater wetlands characterized by abundant growth of mosses and the accumulation of non-decomposed plant remains called peat.

Bogs are ponds covered with thick floating mats of vegetation and can represent a stage in aquatic succession (**FIGURE 15.9**).

Some wetlands are seasonal, being wet only at some times of the year. *Vernal pools* are an example. In many parts of North America, these pools form in early spring from rain and snowmelt, and then dry up once weather becomes warmer. Numerous animals have evolved to take advantage of the ephemeral time windows in which they exist each year.

Wetlands are extremely valuable habitat for wildlife and also provide important ecosystem services (p. 116) by slowing runoff, reducing flooding, recharging aquifers, and filtering pollutants.

Despite the vital roles played by wetlands, people have drained and filled them, especially for agriculture. Many wetlands are lost when people divert and withdraw water, channelize rivers, and build dams. Southern Canada and the United States, for example, have lost more than half their wetlands since European colonization.

Wetlands and other aquatic systems are affected by people when they modify the flow of rivers and introduce pollutants that alter water's chemical, biological, and physical properties. Let's now take a closer look at such impacts on freshwater ecosystems.

Human Activities Affect Waterways

Fresh water is one of the world's most precious resources. Not only do we need it to keep our bodies hydrated and healthy, but we also require huge quantities of water for our homes, farms, and factories. Although water is a limited resource, it is also a renewable resource as long as we manage our use

sustainably. Unfortunately, people are withdrawing water at unsustainable levels and are depleting many sources of surface water and groundwater. Already, one-third of the world's people are affected by chronic water shortages.

Additionally, people have intensively engineered freshwater waterways with dams, levees, and diversion canals to satisfy demands for water supplies, transportation, and flood control. An estimated 60% of the world's largest 227 rivers (and 77% of those in North America and Europe), for example, have been strongly or moderately affected by artificial dams, levees, and diversions. We will now examine the many ways fresh water use and other human activities affect freshwater systems.

Fresh water and human populations are unevenly distributed across Earth

World regions differ in the size of their human populations and, as a result of climate and other factors, possess varying amounts of groundwater, surface water, and precipitation. Hence, not every human around the world has equal access to fresh water (**FIGURE 15.10**). Because of the mismatched distribution of water and population, human societies have always struggled to transport fresh water from its source to where people need it. Southern California is one such example. Population growth has long outpaced water supply in this arid region, requiring municipalities in the region to "pipe in" water from rivers in other parts of the state to meet demand.

Fresh water is distributed unevenly in time as well as space. For example, India's monsoon storms can dump half of a region's annual rain in just a few hours. Rivers have seasonal differences in flow because of the timing of rains and snowmelt. For this reason, people build dams to store water from wetter months that can be used in drier times of the year, when river flow is reduced.

As if the existing mismatches between water availability and human need were not enough, global climate change (Chapter 18) has and will continue to worsen conditions in many regions by altering precipitation patterns, melting glaciers, causing early-season runoff, and intensifying droughts and flooding. A 2009 study found that one-third of the world's 925 major rivers experienced reduced flow from 1948 to 2004, with the majority of the reduction attributed to effects of climate change.

Water supplies households, industry, and especially agriculture

We all use water at home for drinking, cooking, cleaning, and irrigation of lawns and gardens. Most mining, industrial, and manufacturing processes require water. Farmers and ranchers use water to irrigate crops and water livestock. Globally, we

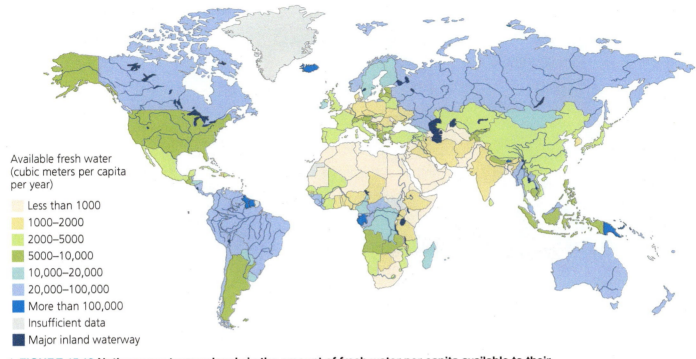

Available fresh water
(cubic meters per capita
per year)

- Less than 1000
- 1000–2000
- 2000–5000
- 5000–10,000
- 10,000–20,000
- 20,000–100,000
- More than 100,000
- Insufficient data
- Major inland waterway

FIGURE 15.10 Nations vary tremendously in the amount of fresh water per capita available to their citizens. For example, with over 100,000 cubic meters per capita per year, Iceland, Papua New Guinea, Gabon, and Guyana each have more than 100 times more water per person than do many Middle Eastern and North African countries. *Data from Harrison, P., and F. Pearce, 2000. AAAS atlas of population and the environment, edited by the American Association for the Advancement of Science, © 2000 by the American Association for the Advancement of Science.*

DATA Q Compare the water availability per person in the developing regions of Africa, Asia, and Latin America and the Caribbean. Which region has the most water per person?

Go to **Interpreting Graphs & Data** on **Mastering**EnvironmentalScience®.

allot about 70% of our annual use of fresh water to agriculture. Industry accounts for roughly 20%, and residential and municipal uses for only 10%.

Why do we allocate 70% of our water use to agriculture? Our rapid population growth requires us to feed and clothe more people each year, and the intensification of agriculture that occurred during the Green Revolution (p. 238) to keep up with that demand required significant increases in irrigation. As a result, we withdraw 70% more water for irrigation today than we did 50 years ago and have doubled the amount of land under irrigation. This expansion of irrigated land has helped food and fiber production to keep up with population growth, but many irrigated areas are using water unsustainably, threatening their long-term productivity.

When we remove water from an aquifer or surface water body and do not return it, this is called **consumptive use.** Our primary consumptive use of water is for irrigation, which is the water applied to crops (p. 216). In contrast, **nonconsumptive use** of water does not remove, or only temporarily removes, water from an aquifer or surface water body. Using water to generate electricity at hydroelectric dams is an example of nonconsumptive use; water is taken in, passed through dam machinery to turn turbines, and released

downstream. Consumptive uses of water are typically of the greatest concern in water-poor regions, as they act to reduce water levels in aquifers, lakes, and rivers.

Excessive water withdrawals can drain rivers and lakes

In many places we are withdrawing surface water at unsustainable rates to meet our many demands for fresh water. As a result, many of the world's major rivers regularly run dry before reaching the sea. The Colorado River often fails to reach the Gulf of California after the many withdrawals of its water in the arid western United States and Mexico. This reduction in flow threatens the future of the cities and farms that depend on the river. And it has drastically altered the ecology of the river and its delta, changing plant communities, wiping out populations of fish and invertebrates, and devastating fisheries.

The Colorado's plight is not unique. Several hundred miles to the east, the Rio Grande also frequently runs dry, the victim of overextraction by both Mexican and U.S. farmers in times of drought. China's Yellow River also often fails to reach the sea. Even the river that has nurtured human

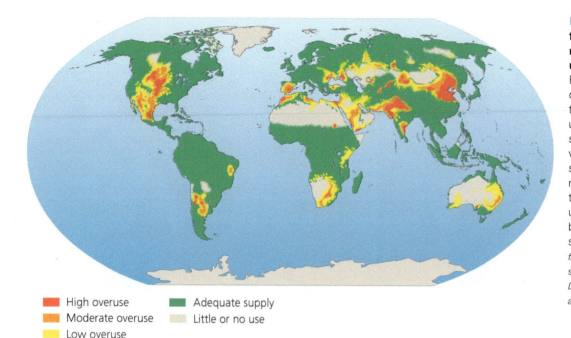

■ High overuse   ■ Adequate supply
■ Moderate overuse   Little or no use
■ Low overuse

civilization longer than any other, the Nile in Egypt, now peters out before reaching its mouth.

Worldwide, roughly 15–35% of water withdrawals for irrigation are thought to be unsustainable. In areas where agriculture is demanding more fresh water than can be sustainably supplied, *water mining*—withdrawing water faster than it can be replenished—is taking place. In these areas, aquifers are being depleted or surface water is being piped in from other regions (**FIGURE 15.11**). This is occurring today in California's central agricultural regions, where farmers are mining groundwater to grow crops as supplies of surface waters have dwindled due to drought.

Nowhere are the effects of surface water depletion so evident as in the Aral Sea. Once the fourth-largest lake on Earth, just larger than Lake Huron, it lost more than four-fifths of its volume in just 45 years (**FIGURE 15.12**). This dying inland sea, on the border of present-day Uzbekistan and Kazakhstan,

is the victim of poor irrigation practices. The former Soviet Union instituted industrial cotton farming in this dry region by flooding the land with water from the two rivers that supplied the Aral Sea its water. For a few decades this boosted Soviet cotton production, but it shrank the Aral Sea, and the irrigated soil became salty and waterlogged. Today 60,000 fishing jobs are gone, winds blow pesticide-laden dust up from the dry lake bed (**FIGURE 15.13**), and little cotton grows on the blighted soil.

Groundwater can also be depleted

Groundwater is more easily depleted than surface water because most aquifers recharge very slowly. Today we are mining groundwater, extracting 160 km³ (5.65 trillion ft³) more water each year than returns to the ground. This is a problem because one-third of Earth's human population—including

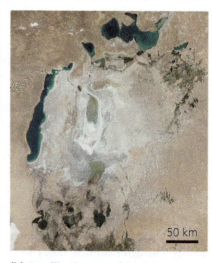

FIGURE 15.12 The Aral Sea in central Asia was once the world's fourth largest lake. However, it has been shrinking **(a, b)** because so much water was withdrawn to irrigate cotton crops.

(a) Satellite image of Aral Sea, 1987   (b) Satellite image of Aral Sea, 2015

FIGURE 15.13 Ships were stranded along the former shoreline of the Aral Sea because the waters receded so far and so quickly. Today, restoration efforts are beginning to reverse the decline in the northern portion of the sea, and waters there are slowly rising.

99% of the rural population of the United States—relies on groundwater to meet its needs for water.

As aquifers are mined, water tables drop. Groundwater becomes more difficult and expensive to extract, and eventually it may be depleted. In parts of Mexico, India, China, and many Asian and Middle Eastern nations, water tables are falling 1–3 m (3–10 ft) per year. In California, the overpumping of groundwater has dropped water tables across the state. In some locations in California, the water table is more than 3 m (10 ft) lower in the soil than it was a mere decade ago (**FIGURE 15.14**).

When groundwater is overextracted in coastal areas, saltwater from the ocean can intrude into inland aquifers, making well water undrinkable. This occurs when overpumping of groundwater near the coast greatly reduces the volume of fresh groundwater flowing from rocks under the land into connected rocks that are beneath the ocean. Saline groundwater, the seawater that exists in rocks beneath the ocean, can then penetrate into aquifers beneath the land. Typically, saline groundwater does not penetrate far inland because it is constantly being "pushed back" by fresh groundwater flowing into the sea. But when fresh groundwater flows are reduced by overextraction, saline groundwater experiences less resistance to inland movements and can penetrate into coastal aquifers. This causes coastal wells to draw up saline groundwater instead of the desired fresh groundwater. This problem has occurred in California, Florida, India, the Middle East, and many other locations where groundwater is heavily extracted.

Moreover, as aquifers lose water, they can become less capable of supporting overlying strata, and the land surface above may subside. For this reason, cities from Venice to Bangkok to Beijing are slowly sinking. Mexico City's downtown has sunk over 10 m (33 ft) since the time of Spanish colonialism; streets are buckled, old buildings lean at angles,

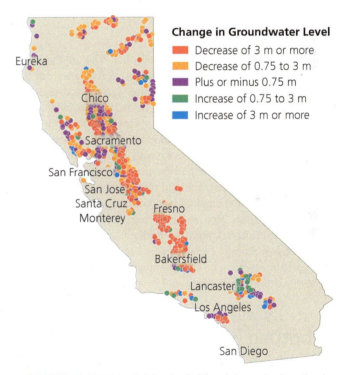

Change in Groundwater Level

- 🟥 Decrease of 3 m or more
- 🟧 Decrease of 0.75 to 3 m
- 🟪 Plus or minus 0.75 m
- 🟩 Increase of 0.75 to 3 m
- 🟦 Increase of 3 m or more

FIGURE 15.14 Water tables in California's agricultural valleys have dropped precipitously in the past decade. This has been predominantly driven by overextraction of aquifers for agriculture. *Data from Groundwater Information Center, California Department of Water Resources, http://www.water.ca.gov/groundwater/MAP_APP/.*

DATA Q The area between which two cities in California saw the consistently greatest decrease in water table in the decade examined in the figure? What do you think was responsible for reducing groundwater levels in this area more than in other parts of the state?

Go to **Interpreting Graphs & Data** on **Mastering**EnvironmentalScience®.

(a) 16th-century chapel in Mexico City

(b) Sinkhole in Florida

FIGURE 15.15 When too much groundwater is withdrawn, the land may weaken and subside. This can cause buildings to lean, as seen in Mexico City **(a)**. Large areas of land may sometimes collapse suddenly in sinkholes, as seen here in Florida **(b)**.

and underground pipes break so often that 30% of the system's water is lost to leaks (**FIGURE 15.15a**). Subsidence due to groundwater overextraction caused some agricultural regions in California to sink by 9 m (30 ft) from 1925 to 1977, and many areas continue to sink today as overextraction continues.

Sometimes land subsides suddenly, creating **sinkholes,** areas where the ground gives way with little warning, occasionally swallowing homes and businesses (**FIGURE 15.15b**). Once the ground subsides, soil and rock becomes compacted, losing the porosity that enabled it to hold water. Recharging a depleted aquifer thereafter becomes more difficult. Estimates suggest that compacted aquifers under California's Central Valley have lost storage capacity equal to that of 40% of the state's surface reservoirs. This compaction can also lead to land slowly sinking over time.

Groundwater supplies our bottled water

These days, our groundwater is being withdrawn for a new purpose: to be packaged in plastic bottles and sold on supermarket shelves. Bottled water is booming business. The average American drinks 36 gallons of bottled water a year, and in 2015 sales topped $14 billion in the United States and $160 billion worldwide. Most people who buy bottled water do so for portability and convenience, or because they believe it will taste better or be safer and healthier than tap water. However, in blind taste tests people think tap water tastes just as good, and chemical analyses show that bottled water is no safer or healthier than tap water (see **THE SCIENCE BEHIND THE STORY**, pp. 398–399). Bottled water also exerts substantial ecological impact. A 2009 study

calculated the energy costs of bottled water to be 1000–2000 times greater than the energy costs of tap water, contributing greenhouse gases to the atmosphere. Also, roughly three of four plastic water bottles are not recycled after use in the United States, generating close to 1.5 million tons of plastic waste annually.

As a result of the environmental impacts of bottled water, as of 2015 more than 90 colleges and universities in the United States and Canada have banned the sale of bottled water or restricted the use of plastic water bottles on campus. Similarly, major U.S. cities, including New York City, San Francisco, and Seattle, prohibit using government funds to purchase bottled water, in part due to the cost savings of drinking tap water instead of bottled water.

People build levees to control floods

Among the reasons we control the movement of fresh water, flood prevention ranks high. People have always been attracted to riverbanks for their water supply and for the flat topography and fertile soil of floodplains. But if one lives in a floodplain, one must be prepared to face flooding. **Flooding** is a normal, natural process that occurs when snowmelt or heavy rain swells the volume of water in a river so that water spills over the river's banks. In the long term, floods are immensely beneficial to both natural systems and human agriculture, because floodwaters build and enrich soil by spreading nutrient-rich sediments over large areas.

In the short term, however, floods can do tremendous damage to the farms, homes, and property of people who

Legend:
- Major rivers
- Federal aqueduct projects
- State aqueduct projects
- Local aqueduct projects

Redding

Sacramento River

Sacramento

San Francisco

San Joaquin–Sacramento Delta

San Joaquin River

Fresno

Colorado River

Los Angeles

San Diego

FIGURE 15.16 California's aqueducts redistribute water throughout the state. Water is sent to urban areas in southern California and agricultural regions in central California from rivers in the north of the state, and from the Colorado River to the east. The California aqueduct system is a collection of aqueduct systems administered by the federal government, the state of California, and local entities.

choose to live in floodplains. To protect against floods, communities and governments have built **levees** (also called *dikes*) along banks of rivers to hold water in main channels. These structures prevent flooding at most times and places but can sometimes worsen flooding because they force water to stay in channels and accumulate, building up enormous energy and leading to occasional catastrophic overflow events.

We divert surface water to suit our needs

People have long diverted water from rivers and lakes to farm fields, homes, and cities. California has one of the most extensive diversion schemes for fresh water in the world (**FIGURE 15.16**). Aqueducts, artificial river channels that are sometimes called *canals,* channel water from rivers in northern California and the Sierra Nevada Mountains in eastern California to the state's central agricultural valleys and urban centers on the coast. The state-supported California State Water Project uses 1100 km (700 mi) of aqueducts and pipelines to transport water from wetter northern California to the more arid, highly populated regions of southern California. Along the way, the system joins the 800-km (500-mi) federal Central Valley Project aqueduct system, which helps to supply farmers in the central agricultural valleys with water for irrigation (**FIGURE 15.17a**). Agricultural and urban areas in southern California are also supplied with water diverted from the Colorado River on the state's eastern border with Arizona. Although the predominant force driving the movement of water in these aqueducts is gravity flow, in some locations water must be pumped over mountains as high as 600 m (2000 ft).

weighing the
ISSUES

Reaching for Water

The diversion of water from Mono Lake to Los Angeles is not the only controversial diversion of fresh water in the United States. The rapidly growing Las Vegas metropolitan area is exceeding its allotment of water from the Colorado River and has proposed a $3.5 billion project that would divert groundwater from 450 km (280 mi) away to meet the growing demand in Nevada's largest city. Do you think such diversions are ethically justified? If rural communities and wetland ecosystems at the diversion site in eastern Nevada are destroyed by this project, is this an acceptable cost given the economic activity generated in Las Vegas? How else might cities like Las Vegas meet their future water needs?

(a) Aqueduct in California's agricultural Central Valley region

(b) The receding shoreline of Mono Lake

FIGURE 15.17 Aqueducts function as man-made rivers to redistribute water in California (a). Mono Lake **(b)** in central California currently contains about half the volume of water it did in 1941, as streams that once fed the lake are now diverted to supply water to Los Angeles.

Is Your Bottled Water as Safe as You Think It Is?

Billions of dollars are spent worldwide on bottled water each year.

Which is safer and healthier for you to drink, tap water or bottled water? If you said bottled water, your answer is like most people's. After all, bottled water is delivered to us in single-use, sterilized containers adorned with images of bubbling springs and pristine mountain streams—a far cry from the impression most people have of the purity of the tap water that flows from their faucets. But are our impressions accurate?

Before we answer that question, it is important to note that tap water and bottled water are regulated very differently. Municipalities that provide tap water to their residents need to submit regular reports to the Environmental Protection Agency (EPA) describing their water sources, treatment methods, and contaminants. In contrast, bottled water is classified as a "food" and so is regulated by the Food and Drug Administration (FDA). For example, bottlers are not required to list the source of the water, how it was purified, and the safety testing protocols used at the bottling plant, details your local water company must supply to customers and the EPA. This "lighter" regulation by the FDA results in very uneven levels of information across brands of bottled water, with some providing a great deal of information to consumers and others offering very little.

So, to find out what's in bottled water, scientists have to do some detective work. In one study of the safety of bottled water, researchers from the nonprofit Environmental Working Group in 2008 sent samples of 10 major brands of bottled water to the University of Iowa's Hygienic Laboratory for analysis. The lab's chemists ran a battery of tests and detected 38 chemical pollutants, including traces of heavy metals, radioactive isotopes, caffeine and pharmaceuticals from wastewater pollution, nitrate and ammonia from fertilizer, and various industrial compounds such as solvents and plasticizers. Each brand contained eight contaminants on average, and two brands had levels of chemicals that exceeded legal limits in California and industry safety guidelines. Another two brands showed the

chemical composition of standard municipal water treatment—including chlorine, fluoride, and other by-products of disinfection—indicating that these brands were likely just tap water that was filtered, bottled, and sold at tremendous markup.

But as concern over endocrine-disrupting chemicals increased, researchers began to test if the bottles that give bottled water its impression of purity may actually be adding chemicals to our bodies that interfere with hormonal signals. Endocrine disruptors such as bisphenol A (BPA), phthalates, and other compounds that leach out of plastics can exert a wide array of health impacts, even at very low doses (Chapter 14).

The human body uses hormones (chemical messengers) to signal receptors in specific cells to "turn on" or "turn off" during embryonic development, sexual maturation, and other physiological processes. Much of the study of endocrine disruptors has focused on the hormones that affect sexual development. Some endocrine-disrupting chemicals, such as BPA, mimic the female hormone estrogen and act to "turn on" receptors for estrogen—these are called estrogenic chemicals. Other endocrine-disrupting chemicals turn on receptors for male hormones, such as testosterone; these are called androgenic chemicals. Still other endocrine-disrupting chemicals block receptors for hormones, preventing natural hormones from reaching the receptors they are attempting to signal. These chemicals are classified as antiestrogenic (blocks female hormones) or antiandrogenic (blocks male hormones). Whether these chemicals are blocking receptors when they should be open, or binding to receptors that should not be turned on at that time, endocrine-disrupting chemicals act to disrupt vital physiological processes and impair human health and development.

In 2009, researchers Martin Wagner and Jörg Oehlmann of Johann Wolfgang Goethe University in Frankfurt, Germany, tested 20 brands of bottled water for estrogenic chemicals. They compared nine brands packaged in glass bottles, nine brands packaged in plastic bottles (PET, or polyethylene terephthalate, the type of plastic with a #1 symbol on the bottom), and two brands packaged in "Tetra Pak" paperboard boxes with an inner plastic coating. They placed samples (as well as tap-water samples as controls) in a "yeast estrogen screen," a standard test-tube screening procedure that uses yeast cells engineered with genes to change color when exposed to estrogen-mimicking compounds.

The researchers detected estrogenic contamination in 60% of the samples. Both Tetra Pak brands and seven of nine plastic

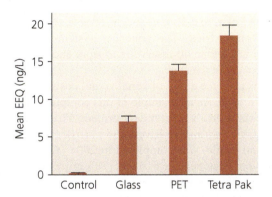

FIGURE 1 Estrogenic activity (as determined by a yeast cell culture test, and measured as "estradiol equivalent concentrations [EEQ]") was highest in bottled water from Tetra Pak containers, followed by PET plastic containers, and then glass containers. A negative control showed no appreciable estrogenic potency. *Source: Wagner, M., and J. Oehlmann, 2009. Endocrine disruptors in bottled mineral water: Total estrogenic burden and migration from plastic bottles. Environmental Science and Pollution Research 16: 278–286.*

brands contained hormone-mimicking substances that apparently leached from the packaging. So did three of the glass-bottled brands, presumably from contamination at the bottling plant. The strength of the estrogenic activity was highest in Tetra Pak brands, followed by plastic containers and then glass bottles (**FIGURE 1**). The researchers' results, published in the journal *Environmental Science and Pollution Research*, caused

people to question their assumptions regarding the purity and safety of bottled water.

But is bottled water safer than tap water? In a follow-up study published in 2013 in the journal *PLOS One*, Wagner and Oehlmann, along with two collaborators, shed light on this question. Using methods like those in their previous study, they tested 18 brands of bottled water from Europe for antiestrogenic and antiandrogenic properties. They found that 13 of 18 brands inhibited estrogen activity, with inhibition rates of up to 60% (**FIGURE 2**). Sixteen brands inhibited androgen activity, with inhibition rates reaching more than 90%. Tap water, on the other hand, did not elicit significant antiestrogenic and antiandrogenic activity.

Subsequent analyses of the chemicals present in the samples found that antagonistic activity in the samples correlated with concentrations of fumarates, a class of chemicals used in plastics manufacturing. Although fumarates are known to have antiestrogenic properties, they are not antiandrogenic, leading the investigators to conclude that the observed inhibition of androgens was due to the action of an unidentified chemical or chemicals.

As these studies show, bottled water—by its very nature of being housed in plastic bottles—may pose a greater health risk from hormone-mimicking chemicals than does your local tap water. So, then, what should we drink? Groups like EWG recommend drinking filtered tap water instead of bottled water. Carry it around with you in a refillable, stainless steel or glass-lined water bottle, and you'll stay hydrated—all the while avoiding harmful chemicals.

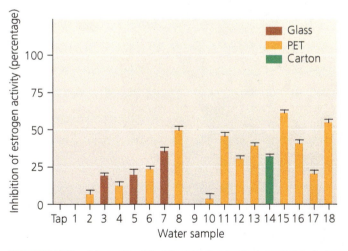
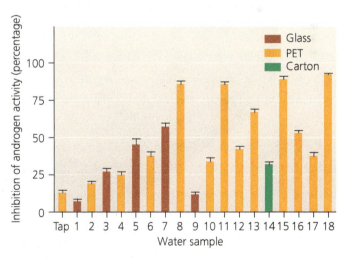

FIGURE 2 Most brands of bottled water tested contained chemicals that inhibited the activity of estrogenic and androgenic hormones. Tap water had little or no inhibitory activity. *Source: Wagner, M., et al. 2013. Identification of putative steroid receptor antagonists in bottled water: Combining bioassays and high-resolution mass spectrometry. PLoS One 8(8): e72472.*

In the past, large-scale diversion projects have enabled politically strong yet water-poor regions to forcibly appropriate water from communities too weak to keep it for themselves. For example, the city of Los Angeles grew by commandeering water from the rural Owens Valley 350 km (220 mi) away. In so doing, it turned the environment of this region to desert, creating dustbowls and destroying its economy. Then in 1941, city leaders in Los Angeles decided to divert streams feeding into Mono Lake, over 565 km (350 mi) away in northern California. As the lake level fell 14 m (45 ft) over 40 years, salt concentrations doubled, and aquatic communities suffered (**FIGURE 15.17b**).

We have erected thousands of dams

A **dam** is any obstruction placed in a river or stream to block its flow. Dams create **reservoirs,** artificial lakes that store water for human use. We build dams to prevent floods, provide drinking water, facilitate irrigation, and generate electricity (pp. 570–573).

Worldwide, we have erected more than 45,000 large dams (greater than 15 m, or 49 ft, high) across rivers in more than 140 nations. We have built tens of thousands of smaller dams. Only a few major rivers in the world remain undammed and free-flowing. These run through the tundra and taiga of Canada, Alaska, and Russia and in remote regions of Latin America and Africa. The aqueduct systems in California alone use more than 40 reservoirs to store and then transport water throughout the state.

Dams produce a mix of benefits and costs, as illustrated in **FIGURE 15.18**. As an example of this complex mix, we can consider the world's largest dam project. The Three Gorges Dam on China's Yangtze River, 186 m (610 ft) high and 2.3 km (1.4 mi) wide, was completed in 2008 (**FIGURE 15.19a**). Its reservoir stretches for 616 km (385 mi; as long as Lake Superior). This project provides flood control, enables boats and barges to travel farther upstream, and generates enough hydroelectric power to replace dozens of large coal or nuclear plants.

However, the Three Gorges Dam cost $39 billion to build, and its reservoir flooded 22 cities and the homes of

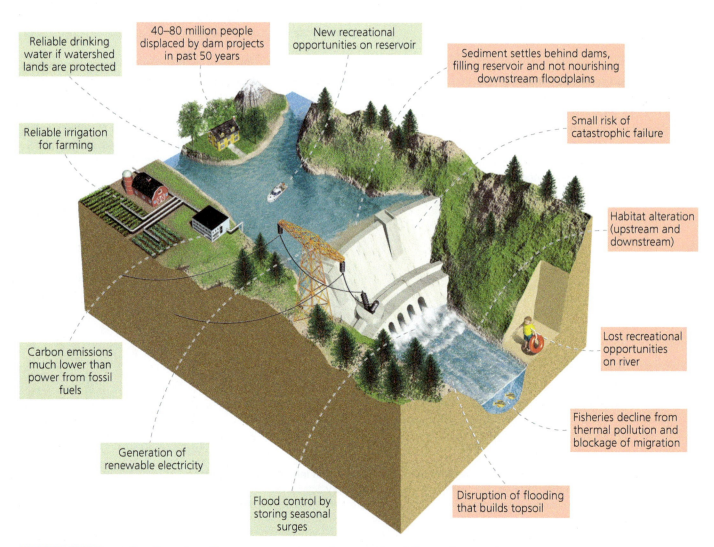

Reliable drinking water if watershed lands are protected

40–80 million people displaced by dam projects in past 50 years

New recreational opportunities on reservoir

Sediment settles behind dams, filling reservoir and not nourishing downstream floodplains

Reliable irrigation for farming

Small risk of catastrophic failure

Habitat alteration (upstream and downstream)

Carbon emissions much lower than power from fossil fuels

Lost recreational opportunities on river

Generation of renewable electricity

Fisheries decline from thermal pollution and blockage of migration

Flood control by storing seasonal surges

Disruption of flooding that builds topsoil

FIGURE 15.18 Damming rivers has diverse consequences for people and the environment. The generation of clean and renewable electricity is one of several major benefits (**green boxes**) of hydroelectric dams. Habitat alteration is one of several negative impacts (**orange boxes**).

(a) The Three Gorges Dam in Yichang, China

(b) Displaced people in Sichuan Province, China

FIGURE 15.19 China's Three Gorges Dam (a), completed in 2008, is the world's largest dam. More than 1.2 million people were displaced and whole cities were leveled for its construction, as shown here **(b)** in Sichuan Province.

1.24 million people, requiring the largest resettlement project in China's history (**FIGURE 15.19b**). The rising water submerged 10,000-year-old archaeological sites, productive farmlands, and wildlife habitat. The reservoir slows the river's flow so that suspended sediment settles behind the dam. Because the river downstream is deprived of sediment, the tidal marshes at the Yangtze's mouth are eroding away. This has left the city of Shanghai with a degraded coastal environment and less coastal land to develop. Many scientists worry that the Yangtze's many pollutants will also be trapped in the Three Gorges Dam reservoir, eventually making the water undrinkable. To avoid such a buildup of pollutants, the Chinese government plans to spend $5 billion building hundreds of sewage treatment and waste disposal facilities. On top of all these worries, earthquakes in southern China in 2008 and again in 2012 raised fears that a future quake could damage the dam, perhaps even leading to its collapse. Such a dam failure would be catastrophic, as it could generate a wall of water that breaches downstream dams and creates a domino effect of dam failures. This occurred in 1975, when record flooding along the Yangtze caused breaches at 62 separate dams, and the subsequent flooding claimed 26,000 lives.

Some dams are being removed

People who feel that the costs of some dams outweigh their benefits are pushing for such dams to be dismantled. By removing dams and letting rivers flow free, they say, we can restore riparian ecosystems, reestablish economically valuable fisheries, and revive river recreation such as fly-fishing and rafting. Increasingly, private dam owners and the Federal Energy Regulatory Commission (FERC), the U.S. government agency charged with renewing licenses for dams, have agreed. Many aging dams are in need of costly repairs or have

outlived their economic usefulness, and roughly 400 dams have been removed in the United States in the past decade.

The drive to remove dams first gathered steam in 1999 with the dismantling of the Edwards Dam on Maine's Kennebec River. FERC had determined that the environmental benefits of removing the dam outweighed the economic benefits of relicensing it. Within a year after the 7.3-m (24-ft) high, 279-m (917-ft) long dam was removed, large numbers of 10 species of migratory fish, including salmon, sturgeon, shad, herring, alewife, and bass, ventured upstream and began using the 27-km (17-mi) stretch of river above the dam site. Some property owners along the former reservoir who had opposed the dam's removal had a change of heart once they saw the healthy and vibrant river that now ran past their property.

In 2014, the world's largest dam removal project was completed when the last section of the 64-m (210-ft) Glines Canyon Dam on the Elwah River in Washington State was demolished. Built in 1914 to supply power for local wood mills, the dam decimated local fisheries and imperiled the livelihood of Native Americans who had long harvested the river's salmon and shellfish by preventing salmon from migrating upriver to spawn. Calls for the dam's removal began in the 1980s and finally came to fruition when the first section of the dam was removed in 2011. Sediments once held behind the dam have flooded downstream, rebuilding riverbanks, beaches, and estuaries. Habitats for shellfish and small fish are being created in and around the river's mouth. And as salmon migrate upriver to spawn in the Elwah's tributaries, it is hoped that an entire functional ecosystem will emerge along with their return.

In California, a movement is underway to remove the O'Shaughnessy Dam on the Tuolumne River in Yosemite National Park. Such a removal would drain the Hetch Hetchy Valley, which is currently submerged beneath the dam's

(a) Hetch Hetchy Valley before dam (1907)

(b) Hetch Hetchy Valley after dam

FIGURE 15.20 Calls for the removal of the O'Shaughnessy Dam on the Tuolumne River in the Hetch Hetchy Valley region of Yosemite National Park have increased as more and more dams have been retired in the United States. Proponents of dam removal seek to restore valley to the natural ecosystem **(a)** that existed before the dam's reservoir **(b)** submerged the valley.

reservoir (**FIGURE 15.20**) and restore the native plants and animals that occupied the area prior to the dam's construction in 1913. This action is strongly opposed by the city of San Francisco, however, as the reservoir provides drinking water for the city.

Wetlands are affected by human manipulations of waterways

From California to the Aral Sea, wetlands are being lost as we divert and withdraw water, channelize rivers, build dams, and otherwise engineer natural waterways. Coupled with the extensive draining of wetlands for agriculture (p. 230), these activities impair wetlands worldwide. As wetlands disappear, we lose the many ecosystem services they provide us, such as filtering pollutants, harboring wildlife, controlling floods, and helping to maintain drinking water supplies. One such example is the Sacramento–San Joaquin Delta that lies between Sacramento and San Francisco. This estuary (p. 425), which is fed by runoff originating in the Sierra Nevada Mountains that empties into San Francisco Bay, has been extensively altered by levees, water diversions, and dredging to create agricultural land and urban areas, and to protect against local flooding. These manipulations have degraded these wetlands so extensively that populations of several species of fish in the delta are currently at their lowest levels in recorded history. An ambitious plan has been proposed to restore these wetlands by increasing water flow to the delta and transporting water to the nearby California and Central Valley aqueducts through tunnels that will run beneath the delta. The plan is controversial, however, and its ultimate fate undetermined as of 2016.

Solutions to Depletion of Fresh Water

Population growth, expansion of irrigated agriculture, and industrial development doubled our annual use of fresh water in the past 50 years. We now use an amount equal to 10% of total global runoff. The hydrologic cycle makes fresh water a renewable resource, but if we take more than a lake, river, or aquifer can provide, we must reduce our use, find another water source, or be prepared to run out of water.

To address shortages of fresh water, we aim either to increase supply or to reduce demand. We can increase supply temporarily through more intensive extraction, but this is generally not sustainable. Diversions may solve supply problems in one area while causing shortages in others, as seen with California's aqueducts. In contrast, strategies for reducing demand include conservation and efficiency measures. Lowering demand is more difficult politically in the short term but may be necessary in the long term. In the developing world, international aid agencies are increasingly funding demand-based solutions over supply-based solutions, because demand-based solutions offer better economic returns and cause less ecological and social damage.

Desalination "makes" more fresh water

A supply strategy with some potential for sustainability is to generate fresh water by **desalination,** or **desalinization,** the removal of salt from seawater or other water of marginal quality. One method of desalination mimics the hydrologic cycle by heating and evaporating ocean water and then condensing the vapor—essentially distilling fresh water. Another method forces water through membranes with tiny pores to filter out salts; the most common such process is reverse osmosis. The process converts saline water with up to 35,000 parts per million (ppm) of dissolved salts to fresh water with less than 1000 ppm dissolved salts.

More than 20,000 desalination facilities are operating worldwide. However, desalination is expensive, requires large inputs of fossil fuel energy, kills aquatic life at water intakes, and generates concentrated salty waste. As a result, large-scale desalination is pursued mostly in wealthy oil-rich nations where water is extremely scarce. In Saudi Arabia, for example, desalination produces half the nation's drinking water. As

California continues to struggle with supplies of fresh water, more eyes are turning westward to the sprawling Pacific Ocean and see desalination as a viable option in the parched state. In late 2015, a desalination plant began supplying San Diego County in southern California with 190 million L (50 million gal) of water a day—roughly 7% of the county's daily demand. The $1 billion plant, which is the largest in the United States, is contracted to provide fresh water to the county for the next 30 years at a cost of at least $110 million per year. While proponents argue that the plant provides the area a reliable water source for drought-prone California, critics contend that the contract will force residents to purchase water from the plant at costs far higher than other water sources, and many times higher than the costs of reducing water use through conservation.

FAQ

Can't we just use desalination to fulfill our demand for water?

Given the seemingly endless supply of water in Earth's oceans, many people assume that desalination is the answer to our world's water crises. So why aren't we eagerly utilizing this technology everywhere?

Simply put, we lack the abundant, clean energy sources needed to make the widespread use of desalination economically viable and environmentally sustainable. For example, the United States withdraws more than 700 billion liters (185 billion gallons) of fresh water every day for use in food production, industry, and public supplies. Diverting the energy necessary to supply even a tiny fraction of this quantity from desalination would drastically increase the demand for energy, causing prices for electricity, natural gas, and other fuels to skyrocket. Using fossil fuels as an energy source for desalination would also drastically increase U.S. emissions of air pollutants and greenhouse gases. Due to these constraints, it is unlikely that desalination will be widely embraced in the United States unless we are able to find abundant, environmentally friendly energy sources.

Agricultural demand can be reduced

Because most water is used for agriculture, it makes sense to look first to agriculture for ways to decrease demand. Farmers can improve efficiency by lining irrigation canals to prevent leaks, leveling fields to minimize runoff, and adopting efficient irrigation methods. Low-pressure spray irrigation squirts water downward toward plants, and drip irrigation systems target individual plants and introduce water directly onto the soil (see Figure 9.8b, p. 217). Both methods reduce water lost to evaporation and runoff. Experts estimate that drip irrigation (in which as little as 10% of water is wasted) could reduce water withdrawals while raising yields that would produce $3 billion in extra annual income for farmers of the developing world.

Fortunately, there is room for efficiency improvement in irrigation. One commonly used method of irrigation, "flood and furrow" irrigation (p. 217), ends up wasting much of the water that is applied to cropland from seepage and evaporation, and leads to the soil becoming waterlogged and laden with salts (pp. 216–217). Were we to embrace more efficient approaches, such as drip irrigation, we could maintain high crop yields with smaller inputs of water, enhancing sustainability. Such inefficient irrigation is possible because many national governments subsidize irrigation to promote agricultural self-sufficiency, drastically lowering water costs for farmers.

Choosing crops to match the land and climate in which they are farmed can also save huge amounts of water. Currently, crops that require a great deal of water, such as cotton, rice, and alfalfa, are often planted in arid areas with government-subsidized irrigation. As a result of the subsidies, the true cost of water is not part of the costs of growing the crop. Eliminating subsidies and growing crops in climates with adequate rainfall could greatly reduce water use. In light of California's problems with water supplies, many scientists are calling for a re-examination of agriculture in California's Central Valley, which grows crops in its arid environment using water diverted from wetter regions. Agriculture consumes 80% of the water used in California, and decisions to grow crops such as almond trees, which alone use 10% of all the water consumed in California, are now facing close scrutiny.

In addition, selective breeding (pp. 50–51) and genetic modification (pp. 249–250) can create crop varieties that produce high yields with less water.

We can lower residential and industrial water use

In our households, we can reduce water use by installing low-flow faucets, showerheads, washing machines, and toilets. Automatic dishwashers, studies show, use less water than does washing dishes by hand. Catching rain runoff from your roof in a barrel—*rainwater harvesting*—will reduce the amount you need to use from the hose. And if your city allows it, you can use *gray water*—the wastewater from showers and sinks—to water your yard. Better yet, you can replace a water-intensive lawn with native plants adapted to your region's natural precipitation patterns. **Xeriscaping**, landscaping using plants adapted to arid conditions, has become increasingly popular in the U.S. Southwest (**FIGURE 15.21**). Many residents of southern California have accepted rebates from their local water districts of up to $3.75 per square foot to rip out water-intensive grass. Some are choosing to replace their typical lawns with low-demand native plants, while

FIGURE 15.21 Xeriscaping lets homeowners and businesses reduce water consumption by landscaping with attractive, drought-tolerant plants.

others are embracing artificial turf like that used on athletic fields. Each of us can significantly cut our daily water use by reexamining aspects of our daily lives.

Industry and municipalities can take water-saving steps as well. Manufacturers are shifting to processes that use less water and in doing so are reducing their costs. Las Vegas is one of many cities that are recycling treated municipal wastewater for irrigation and industrial uses. Governments in Arizona and in England are capturing excess runoff and pumping it into aquifers. Finding and patching leaks in pipes has saved some cities and companies large amounts of water—and money. Boston and its suburbs reduced water demand by 30% over 17 years by patching leaks, retrofitting homes with efficient plumbing, auditing industry, and promoting conservation to the public. This program enabled Massachusetts to avoid an unpopular $500 million river diversion scheme.

Market-based approaches to water conservation are being debated

Economists who want to use market-based strategies to achieve sustainable water use have suggested ending government subsidies of inefficient practices and instead letting water become a commodity whose price reflects the true costs of its extraction. Others worry that making water a fully priced commodity would make it less available to the world's poor and increase income disparity. Because industrial use of water can be 70 times more profitable than agricultural use, market forces alone might favor uses that would benefit wealthy and industrialized people, companies, and nations at the expense of the rural poor.

Similar concerns surround another potential solution, the privatization of water supplies. During the 1990s, many public water systems were partially or wholly privatized, as governments transferred construction, maintenance, management, or ownership to private companies. This was done to enhance efficiency, but firms have little incentive to allow equitable access to water for rich and poor alike. Already in some developing countries, rural residents without access to public water supplies find themselves forced to buy water from private vendors and pay up to 12 times more than those connected to public supplies.

Other experiences indicate that decentralization of control over water, from the national level to the local level, can help conserve water. In Mexico, the effectiveness of irrigation systems improved dramatically once they were transferred from public ownership to the control of 386 local water user associations.

Regardless of how demand is addressed, the shift from supply-side to demand-side solutions is paying dividends. In Europe, a new focus on demand (through government mandates and public education) has decreased public water consumption, and industries are becoming more water-efficient. From 1980 to 2010, water conservation efforts enabled the United States to decrease its water consumption by 16%, even while its population grew 31%.

Nations often cooperate to resolve water disputes

We've seen that nations have unequal access to freshwater supplies, and there are fears that scarcity of this vital resource can lead to conflict. For example, a 2012 report by the intelligence agencies of the U.S. government concluded that in the subsequent decade, many nations vital to American interests will experience political and economic instability due to water shortages, making freshwater supplies one of the greatest threats to U.S. national security interests. In the face of such instability, one recourse would be to commandeer the water resources of other nations by force, sparking international military conflict. **FIGURE 15.22** shows some waterways for which conflict is a concern in coming decades.

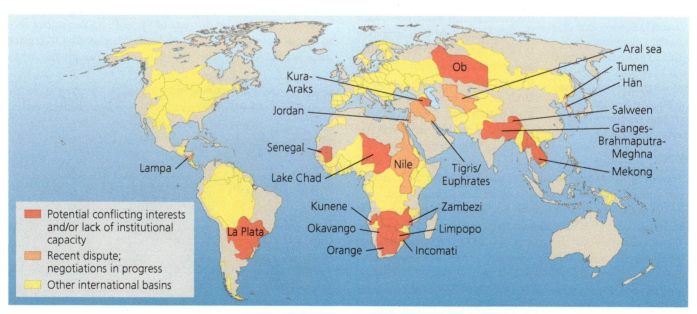

FIGURE 15.22 Water basins that cross national boundaries (yellow) have the potential for conflict if water supplies become scarce. Basins with higher potential for conflict (**red**) are found in regions with growing populations, but negotiations are underway on several international basins to prevent conflict (**orange**).

A total of 261 major rivers (whose watersheds cover 45% of the world's land area) cross national borders, and transboundary disagreements are common. Water is already a key element in the hostilities among Israel, the Palestinian people, and neighboring nations. The United States has its share of conflicts over water. Water allocations within California and from the Colorado River have long been a source of conflict between farms and growing cities. The states of Georgia, Alabama, and Florida are currently embroiled in disputes over water withdrawals from shared rivers.

Yet on the positive side, so far many nations have cooperated to resolve water disputes. India has struck agreements to co-manage transboundary rivers with Pakistan, Bangladesh, Bhutan, and Nepal. In Europe, nations along the Rhine and Danube rivers have signed water-sharing treaties.

Freshwater Pollution and Its Control

We have seen that people affect aquatic systems by withdrawing too much water and by altering the systems' natural processes by engineering waterways with dams, diversions, and levees. However, people also affect aquatic ecosystems and threaten human health when they introduce toxic substances and disease-causing organisms into surface waters and groundwater.

Developed nations have made admirable advances in cleaning up water pollution over the past few decades. Still, the World Commission on Water recently concluded that more than half the world's major rivers remain "seriously depleted and polluted, degrading and poisoning the surrounding ecosystems, threatening the health and livelihood of people who depend on them." Levels of impairment are similar in U.S. waterways. In 2013, the Environmental Protection Agency (EPA) reported that 55% of the 2000 U.S. streams and rivers sampled in 2008–2009 were in poor condition to support aquatic life. The largely invisible pollution of groundwater, meanwhile, has been termed a "covert crisis." Preventing pollution is easier and more effective than correcting it later, so many of our current solutions to pollution problems embrace preventative strategies rather than "end-of-pipe" treatment and cleanup.

Water pollution comes from point and non-point sources

Water pollution—changes in the chemical, physical, or biological properties of waters caused by human activities—comes in many forms and can have diverse impacts on aquatic ecosystems and human health.

Most forms of water pollution are not conspicuous to the human eye, so scientists and technicians measure water's chemical properties (such as pH, nutrient concentrations, and dissolved oxygen concentration); physical characteristics (such as temperature and turbidity—the density of suspended particles in a water sample); and biological properties (such

as the presence of harmful microorganisms or the species diversity in aquatic ecosystems).

Some water pollution is emitted from **point sources**—discrete locations, such as a factory or sewer pipe. In contrast, pollution from **non-point sources** is cumulative, arising from multiple inputs over larger areas, such as farms, city streets, and residential neighborhoods (**FIGURE 15.23**, p. 406). The U.S. Clean Water Act (p. 172) addressed point-source pollution with some success by targeting industrial discharges. As a result, water quality in the United States today suffers most from non-point-source pollution, resulting from countless common activities such as applying fertilizers and pesticides to farms and lawns, applying salt to roads in winter, and leaking automobile oil. To minimize non-point-source pollution of drinking water, governments limit development on watershed land surrounding reservoirs.

Water pollution takes many forms

Water pollution comes in many forms that can impair waterways and threaten people and organisms that drink or live in affected waters.

Toxic chemicals Our waterways have become polluted with toxic organic substances of our own making, including pesticides, petroleum products, and other synthetic chemicals (pp. 362–363). Many of these can poison animals and plants, alter aquatic ecosystems, and cause an array of human health problems, including cancer. In addition, toxic metals such as arsenic, lead, and mercury also damage human health and the environment, as do acids from acid precipitation (p. 469) and from acid drainage from mining sites (p. 634). Issuing and enforcing more stringent regulations on industry can help reduce releases of many toxic chemicals. We can also modify our industrial processes and our purchasing decisions to rely less on these substances.

Pathogens and waterborne diseases Disease-causing organisms (pathogenic viruses, protists, and bacteria) can enter drinking water supplies when these are contaminated with human waste from inadequately treated sewage or with animal waste from feedlots (p. 241). Specialists monitoring water quality can tell when water has been contaminated by such waste when they detect fecal coliform bacteria, which live in the intestinal tracts of people and other vertebrates. These bacteria are usually not pathogenic themselves, but they serve as indicators of fecal contamination, alerting us that the water may hold other pathogens that can cause ailments such as giardiasis, typhoid, or hepatitis A.

Biological pollution by pathogens causes more human health problems than any other type of water pollution. In the United States, an estimated 20 million people fall ill each year from drinking water contaminated with pathogens. Worldwide, the United Nations estimates that 3800 children die every day of diseases associated with unsafe drinking water, such as cholera, dysentery, and typhoid fever.

Although many people today still lack reliable access to safe drinking water and sanitation/sewer facilities, we are making slow but steady progress worldwide in supplying

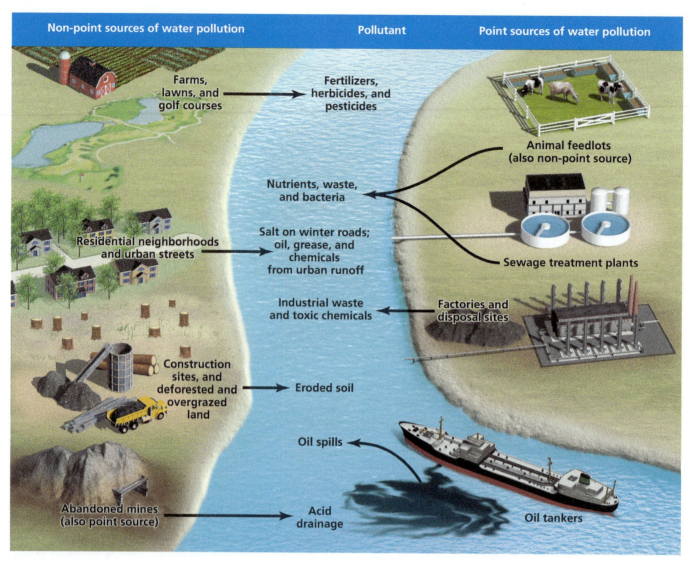

Non-point sources of water pollution	Pollutant	Point sources of water pollution

Farms, lawns, and golf courses → Fertilizers, herbicides, and pesticides

Animal feedlots (also non-point source)

Nutrients, waste, and bacteria

Residential neighborhoods and urban streets → Salt on winter roads; oil, grease, and chemicals from urban runoff

Sewage treatment plants

Industrial waste and toxic chemicals ← Factories and disposal sites

Construction sites, and deforested and overgrazed land → Eroded soil

Oil spills

Abandoned mines (also point source) → Acid drainage

Oil tankers

FIGURE 15.23 Point-source pollution (on right) comes from discrete facilities or locations, usually from single outflow pipes. Non-point-source pollution (such as runoff from streets, residential neighborhoods, lawns, and farms; **on left**) originates from numerous sources spread over large areas.

people with these services (p. 359). We reduce the risks posed by waterborne pathogens by using chemical or other means to disinfect drinking water (p. 409) and by treating wastewater (pp. 409–410). Other measures to lessen health risks include public education to encourage personal hygiene and government enforcement of regulations to ensure the cleanliness of food production, processing, and distribution.

Nutrient pollution The Chesapeake Bay's dead zone shows how nutrient pollution from fertilizers and other sources can lead to eutrophication and hypoxia in surface waters (pp. 108–109). When excess nitrogen and/or phosphorus enters a water body, it fertilizes algae and aquatic plants, boosting their growth. Algae then spread and cover the water's surface, depriving underwater plants of sunlight. As algae die off, bacteria consume them. Because this decomposition requires oxygen, the increased bacterial activity drives down levels of dissolved oxygen. These levels can drop too low to support fish and shellfish, leading to dramatic changes in aquatic ecosystems.

A "dead zone"—an area of very low dissolved oxygen levels—appears annually in the northern Gulf of Mexico, fueled by nutrients from Midwest farms carried by the Mississippi and Atchafalaya rivers. The low-oxygen conditions have adversely affected marine life and reduced catches of shrimp and fish to half of what they were in the 1980s, affecting people whose livelihoods depended on seafood harvests.

Eutrophication (**FIGURE 15.24**) is a natural process, but nutrient input from runoff from farms, golf courses, lawns, and sewage can dramatically increase the rate at which it occurs. We can reduce nutrient pollution by treating wastewater, reducing fertilizer application, using phosphate-free detergents, and planting vegetation and protecting natural areas to increase nutrient uptake.

(a) Oligotrophic water body

(b) Eutrophic water body

FIGURE 15.24 Pollution of freshwater bodies by excess nutrients accelerates the process of eutrophication. An oligotrophic water body **(a)** with clear water and low nutrient content may eventually become a eutrophic water body **(b)** with high nutrient content and abundant algae and plant life.

Biodegradable Wastes Introducing large quantities of biodegradable materials into waters decreases dissolved oxygen levels, too. When human wastes, animal manure, paper pulp from paper mills, or yard wastes (grass clippings and leaves) enter waterways, bacterial decomposition escalates as organic material is metabolized. This lowers dissolved oxygen levels in the water, just as in waters receiving elevated inputs of plant nutrients. **Wastewater** is water affected by human activities and can be a source of biodegradable wastes. It includes water from toilets, showers, sinks, dishwashers, and washing machines; water used in manufacturing or industrial cleaning processes; and stormwater runoff. The widespread practice of treating wastewater to remove organic matter has greatly reduced impacts from biodegradable wastes in rivers in developed nations. Oxygen depletion remains a major problem in some developing nations, however, where wastewater treatment

is less common, and around 90% of wastewater is released untreated into waterways.

Sediment Eroded soils can be carried to rivers by runoff and transported long distances by river currents (**FIGURE 15.25**). Clear-cutting, mining, clearing land for development, and cultivating farm fields all expose soil to wind and water erosion (p. 221). Some water bodies, such as the Colorado River and China's Yellow River, are naturally sediment-rich, but many others are not. When a clear-water river receives a heavy influx of eroded sediment, aquatic habitat changes dramatically, and fish adapted to clear water may be killed. We can reduce sediment pollution by better managing farms and forests and avoiding large-scale disturbance of vegetation.

Thermal pollution Water's ability to hold dissolved oxygen decreases as temperature rises, so some aquatic organisms may not survive when human activities raise water temperatures. When we withdraw water from a river and use it to cool an industrial facility, we transfer heat from the facility back into the river where the water is returned. People also raise water temperatures by removing streamside vegetation that shades water.

Too little heat can also cause problems. On the Mississippi and many other dammed rivers, water at the bottoms of reservoirs is colder than water at the surface. When dam operators release water from the depths of a reservoir, downstream water temperatures drop suddenly. In some river systems, these pulses of cold water have favored cold-loving invasive fish species over native species adapted to normal river temperatures.

FIGURE 15.25 Sediments wash into the Pacific Ocean from a river in Costa Rica. Farming, construction, and other human activities can cause elevated levels of soil to enter waterways, affecting water quality and aquatic wildlife.

Groundwater pollution is a difficult problem

Most pollution control efforts focus on surface water. Yet groundwater sources once assumed to be pristine are regularly polluted by industry and agriculture. Groundwater pollution is hidden from view and difficult to monitor; it can be out-of-sight, out-of-mind for decades until widespread contamination of drinking supplies is discovered.

Groundwater pollution is also more difficult to address than surface water pollution. Rivers flush their pollutants fairly quickly, but groundwater retains its contaminants until they decompose, which in the case of persistent pollutants can be many years or decades. The long-lived pesticide DDT, for instance, is found widely in U.S. aquifers even though it was banned more than 35 years ago. Moreover, chemicals break down much more slowly in aquifers than in surface water or soils. Decomposition is slower in groundwater because it is not exposed to sunlight, contains fewer microbes and minerals, and holds less dissolved oxygen and organic matter. For example, concentrations of the herbicide alachlor decline by half after 20 days in soil, but in groundwater this takes almost four years.

Further, when groundwater is overdrafted, as is occurring in California today, pollutants concentrate in the remaining groundwater, making it potentially unusable for irrigation or domestic purposes.

There are many sources of groundwater pollution

Some chemicals that are toxic at high concentrations, including aluminum, fluoride, nitrates, and sulfates, occur naturally in groundwater. After all, groundwater is in contact with rock for thousands of years, and during that time all kinds of compounds, both toxic and benign, may leach into the water.

However, groundwater pollution resulting from human activity is widespread. Industrial, agricultural, and urban wastes—from heavy metals to petroleum products to solvents to pesticides—can leach through soil and seep into aquifers. Pathogens and other pollutants can enter groundwater through improperly designed wells and from the pumping of liquid hazardous waste below ground (p. 624). Leakage of carcinogenic pollutants (such as chlorinated solvents and gasoline) from underground tanks of oil and industrial chemicals also poses a threat to groundwater. Across the United States, the EPA has embarked on a nationwide cleanup program to unearth and repair leaky tanks. After more than 15 years of work, by 2016 the EPA had confirmed leaks from 510,000 tanks, and had completed cleanups on more than 430,000 of them.

The leaking of radioactive compounds from underground tanks is also a source of groundwater pollution. Currently, 67 of the 177 underground storage tanks at the Hanford Nuclear Reservation in Washington have been confirmed to be leaking radioactive waste into the soil (**FIGURE 15.26**). The site is the most radioactively contaminated area in the United States and stores 60% of the United States' high-level radioactive waste.

FIGURE 15.26 Leaky underground storage tanks are a major source of groundwater pollution. More than one-third of the underground tanks housing radioactive waste at the Hanford Nuclear Reservation in Washington are leaking, threatening groundwater and the nearby Columbia River.

It has been storing wastes since the 1940s, and billions of dollars have been spent on remediation efforts at the facility. Its cleanup has experienced delays and cost overruns, however, and the radioactive material in aging underground tanks is not scheduled to be completely removed until 2047.

Agriculture contributes to groundwater pollution in several ways. Pesticides were detected in most of the shallow aquifer sites tested in the United States in the 1990s, although levels generally did not violate EPA safety standards for drinking water. Nitrate from fertilizers has leached into aquifers in California's agricultural regions and in other agricultural regions of the United States. Nitrate in drinking water has been linked to cancers, miscarriages, and "blue-baby" syndrome, which reduces the oxygen-carrying capacity of infants' blood.

Legislative and regulatory efforts have helped to reduce pollution

As numerous as our freshwater pollution problems may seem, it is important to remember that many were worse a few decades ago, when, for example, the Cuyahoga River repeatedly caught fire (p. 170). Citizen activism and government response during the 1960s and 1970s in the United States resulted in legislation such as the Federal Water Pollution Control Act of 1972 (later amended and renamed the Clean Water Act in 1977). These acts made it illegal to discharge pollution from a point source without a permit, set standards for industrial wastewater, set standards for contaminant levels in surface waters, and funded construction of sewage treatment plants. Thanks to such legislation, point-source pollution in the United States was reduced, and rivers and lakes became notably cleaner.

In the past several decades, however, enforcement of water quality laws has grown weaker, as underfunded and understaffed state and federal regulatory agencies faced great challenges when enforcing existing laws. A comprehensive investigation by the *New York Times* in 2009 revealed that violations of the Clean Water Act have risen and that documented violations now number more than 100,000 per year (to say nothing of undocumented instances). The EPA and the states act on only a tiny percentage of these violations, the *Times* found. As a result, 1 in 10 Americans have been exposed to unsafe drinking water—for the most part unknowingly, because many pollutants cannot be detected by smell, taste, or color.

The Great Lakes of Canada and the United States represent an encouraging success story in fighting water pollution. In the 1970s these lakes, which hold 18% of the world's surface fresh water, were badly polluted with wastewater, fertilizers, and toxic chemicals. Algal blooms fouled beaches, and Lake Erie was pronounced "dead." Today, efforts of the Canadian and U.S. governments have paid off. According to Environment Canada, releases of seven toxic chemicals are down by 71%, municipal phosphorus has decreased by 80%, and chlorinated pollutants from paper mills are down by 82%. Levels of PCBs and DDE are down by 78% and 91%, respectively. Bird populations are rebounding, and Lake Erie is now home to the world's largest walleye fishery. The Great Lakes' troubles are by no means over—sediment pollution is still heavy, algal blooms still plague Lake Erie, and fish are not always safe to eat. However, the progress so far shows how conditions can improve when citizens push their governments to take action.

We treat our drinking water

Technological advances as well as government regulation have improved our control of pollution. The treatment of drinking water is a widespread and successful practice in developed nations today. Before being sent to your tap, water from a reservoir or aquifer is treated with chemicals to remove particulate matter; passed through filters of sand, gravel, and charcoal; and/or disinfected with small amounts of an agent such as chlorine to combat pathogenic bacteria. The U.S. EPA sets standards for more than 90 drinking water contaminants, which local governments and private water suppliers are obligated to meet.

We treat our wastewater

Wastewater treatment is also now a mainstream practice. Wastewater includes water that carries sewage; water from showers, sinks, washing machines, and dishwashers; water used in manufacturing or industrial cleaning processes; and stormwater runoff. Natural systems can process moderate amounts of wastewater, but the large and concentrated amounts that our densely populated areas generate can harm ecosystems and pose health threats. Thus, attempts are now widely made to treat wastewater before it is released into the environment.

In rural areas, **septic systems** are the most popular method of wastewater disposal. In a septic system, wastewater runs from the house to an underground septic tank, inside which solids and oils separate from water. The clarified water proceeds downhill to a drain field of perforated pipes laid horizontally in gravel-filled trenches underground. Microbes decompose pollutants in the wastewater these pipes emit. Periodically, solid waste from the septic tank is pumped out and taken to a landfill.

In more densely populated areas, municipal sewer systems carry wastewater from homes and businesses to centralized treatment locations. There, pollutants are removed by physical, chemical, and biological means (**FIGURE 15.27**, p. 410). At a treatment facility, **primary treatment,** the physical removal of contaminants in settling tanks or clarifiers, removes about 60% of suspended solids. Wastewater then proceeds to **secondary treatment,** in which water is stirred and aerated so that aerobic bacteria degrade organic pollutants. Roughly 90% of suspended solids may be removed after secondary treatment. Finally, the clarified water is treated with chlorine, and sometimes ultraviolet light, to kill bacteria. Most often, the treated water, called *effluent*, is piped into rivers or the ocean following primary and secondary treatment. However, many municipalities are recycling "reclaimed" water for lawns and golf courses, for irrigation, or for industrial purposes such as cooling water in power plants.

As water is purified throughout the treatment process, the solid material removed is termed *sludge*. Sludge is sent to digesting vats, where microorganisms decompose much of the matter. The result, a wet solution of "biosolids," is then dried and disposed of in a landfill, incinerated, or used as fertilizer on cropland. Methane-rich gas created by the decomposition process is sometimes burned to generate electricity, helping to offset the cost of treatment. Each year about 6 million dry tons of sludge are generated in the United States.

weighing the
ISSUES

Sludge on the Farm

It is estimated that up to half the biosolids from sewage sludge produced each year are used as fertilizer on farmland. This practice makes productive use of the sludge, increases crop output, and conserves landfill space, but many people have voiced concern over accumulation of toxic metals, proliferation of dangerous pathogens, and odors. Do you feel that this practice represents an efficient use of resources or an unnecessary risk? What further information would you want to know to inform your decision?

Constructed wetlands can aid treatment

Long before people built the first wastewater treatment plants, natural wetlands were filtering and purifying water. Recognizing this, engineers have begun manipulating wetlands and even constructing new wetlands to employ as tools to cleanse wastewater. Generally in this approach, wastewater that has gone through primary or secondary treatment at

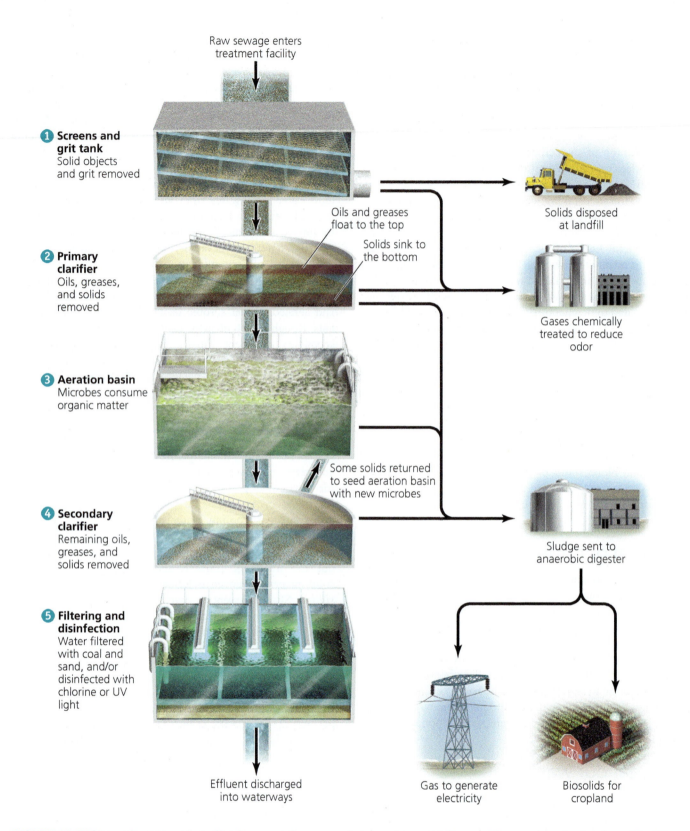

1 Screens and grit tank
Solid objects and grit removed

Raw sewage enters treatment facility

Oils and greases float to the top

Solids sink to the bottom

Solids disposed at landfill

2 Primary clarifier
Oils, greases, and solids removed

Gases chemically treated to reduce odor

3 Aeration basin
Microbes consume organic matter

4 Secondary clarifier
Remaining oils, greases, and solids removed

Some solids returned to seed aeration basin with new microbes

Sludge sent to anaerobic digester

5 Filtering and disinfection
Water filtered with coal and sand, and/or disinfected with chlorine or UV light

Effluent discharged into waterways

Gas to generate electricity

Biosolids for cropland

FIGURE 15.27 Shown here is a generalized process from a modern, environmentally sensitive wastewater treatment facility. Wastewater initially passes through screens to remove large debris and into grit tanks to let grit settle **1**. It then enters tanks called primary clarifiers **2**, in which solids settle to the bottom and oils and greases float to the top for removal. Clarified water then proceeds to aeration basins **3**, which oxygenate the water to encourage decomposition by aerobic bacteria. Water then passes into secondary clarifier tanks **4** for removal of further solids and oils. Next, the water may be purified **5** by chemical treatment with chlorine, passage through carbon filters, and/or exposure to ultraviolet light. The treated water (called *effluent*) may then be piped into natural water bodies, used for urban irrigation, flowed through a constructed wetland, or used to recharge groundwater. In addition, most treatment facilities use anaerobic bacteria to digest sludge removed from the wastewater. Biosolids from digesters may be sent to farm fields as fertilizer, and gas from digestion may be used to generate electric power.

a conventional facility is pumped into the wetland, where microbes living amid the algae and aquatic plants decompose the remaining pollutants. Water cleansed in the wetland can then be released into waterways or allowed to percolate underground.

One of the first constructed wetlands was established in Arcata, a town on northern California's scenic Redwood Coast (**FIGURE 15.28**). This 35-hectare (ha; 86-acre) engineered wetland system was built in the 1980s after residents objected to a $50 million system plan to build a large treatment plant and pump treated wastewater into the ocean. In the wetland system that was built instead, oxidation ponds send partially treated wastewater to the wetland, where plants and microbes continue to perform secondary treatment. The project cost just $7 million, and the site also serves as a haven for wildlife and human recreation. In fact, the Arcata Marsh and Wildlife Sanctuary has brought the town's waterfront back to life; more than 100,000 people visit each year, and more than 300 species of birds have been observed there. The practice of treating wastewater with artificial wetlands is growing fast; today more than 500 artificially constructed or restored wetlands in the United States are performing this service.

FIGURE 15.28 The Arcata Marsh and Wildlife Sanctuary is the site of an artificially engineered, constructed wetland system. The wetlands help treat this northern California city's wastewater, and upland areas around the marsh are open to the public for recreation.

closing
THE LOOP

Citizen action, government legislation and regulation, new technologies, economic incentives, and public education are all helping us to confront a rising challenge of our new century: ensuring adequate quantity and quality of fresh water for ourselves and for the planet's ecosystems. Faced with a multiyear drought and prospects for longer, more severe droughts in coming decades, California is finally enacting stringent measures to reduce water use in the state—and providing a model for other states to follow.

Although the water issues confronting the Golden State are intense, there are many signs of hope. Farmers are embracing drip irrigation techniques, enabling them to conserve water while continuing to grow half the nation's produce. Citizens are reducing water use in their homes and embracing xeriscaping, saving huge quantities of water. Such conservation measures have saved one water district in southern California a stunning 3.7 billion liters (1 billion gallons) of water a day! These water conservation efforts have not depressed economic growth, as the state's economy grew faster than the national average during its historic drought.

California is not alone. Water depletion has become a serious concern in many areas of the developing world and in arid regions of developed nations. Water pollution, meanwhile, continues to take a toll on the health, economies, and societies of nations both rich and poor. Better regulation has improved water quality in the United States and other developed nations, and there is reason to hope that we may yet attain sustainability in our water use. Potential solutions are numerous, and the issue is too important to ignore.

REVIEWING Objectives

You should now be able to:

Describe the distribution of fresh water on Earth and the major types of freshwater systems

Of all the water on Earth, only about 1% is readily available for our use. Fresh water occurs as groundwater and surface waters, and common freshwater ecosystems include rivers and streams, lakes and ponds, and wetlands. (pp. 385–392)

Discuss how we use water and alter freshwater systems

Water from groundwater and surface waters is used in agriculture, industry, and homes. Overextraction of water from these sources can imperil their long-term use and endanger ecosystems. Humans also modify freshwater systems by controlling floods with levees, diverting waters with canals, and creating reservoirs with dams. (pp. 392–402)

Assess problems of water supply and propose solutions to address depletion of fresh water

Solutions to shortages of fresh water include increasing supplies of fresh water, through processes such as desalination, or reducing demand for water by farms, businesses, and homes. Making water reflect its true costs, by eliminating government subsidies and/or privatizing water utilities, has also been offered as a water-conserving approach. (pp. 402–405)

Describe the major classes of water pollution and propose solutions to address water pollution

Water pollution stems from point sources and non-point sources, and is caused by toxic chemicals, microbial pathogens, excessive nutrients, biodegradable wastes, sediment, and thermal pollution. Legislation and regulation have succeeded in improving water quality in developed nations in recent decades. (pp. 405–409)

Explain how we treat drinking water and wastewater

Municipalities treat drinking water by filtering and disinfection in a multi-step process. Septic systems help treat wastewater in rural areas, while cities and towns use municipal wastewater treatment facilities, which treat wastewater physically, biologically, and chemically in a series of steps. (pp. 409–411)

TESTING Your Comprehension

1. Compare and contrast the main types of freshwater ecosystems. Name and describe the major zones of a typical pond or lake.
2. Why are sources of fresh water unreliable for some people and plentiful for others?
3. Describe three benefits and three costs of damming rivers. What particular environmental, health, and social concerns has China's Three Gorges Dam and its reservoir raised?
4. Why do the Colorado, Rio Grande, Nile, and Yellow rivers now slow to a trickle or run dry before reaching their deltas?
5. Why are water tables dropping around the world? What are some negative impacts of falling water tables?
6. Name three major types of water pollutants, and provide an example of each. Explain which classes of water pollutants you think are most important in your local area.
7. Define *groundwater*. Why do many scientists consider groundwater pollution a greater problem than surface water pollution?
8. What are some sources of groundwater pollution that come from human activities?
9. Describe how drinking water is treated. How does a septic system work?
10. Describe and explain the major steps in the process of wastewater treatment. How can artificial wetlands aid such treatment?

SEEKING Solutions

1. How can we lessen agricultural demand for water? Describe some ways we can reduce household water use. How can industrial uses of water be reduced?
2. Describe three ways in which your own actions contribute to water pollution. Now describe three ways in which you could diminish these impacts.
3. Have the provisions of the Clean Water Act been effective? Discuss some of the methods we can adopt, in addition to "end-of-pipe" solutions, to prevent water pollution.
4. **CASE STUDY CONNECTION** California's governor has put you in charge of water policy for the state. Aquifers beneath the state have been overpumped, and many wells have run dry. Agricultural production in the state is down, and farmers are clamoring for you to do something. Meanwhile, more water is needed for burgeoning urban populations in cities such as Los Angeles and San Francisco. What policies would you consider to restore California's water supply? Would you try to take steps to increase supply, decrease demand, or both? Explain why you would choose such policies.

5. THINK IT THROUGH Having solved the water depletion problem in your state, your next task is to deal with pollution of the groundwater that provides your state's drinking water supply. Recent studies have shown that one-third of the state's groundwater has levels of pollutants that violate EPA standards for human health, and citizens are fearful for their safety. What steps would you consider taking to safeguard the quality of your state's groundwater supply, and why?

6. THINK IT THROUGH The Army Corps of Engineers is proposing building a dam on the river near your town and some local residents are voicing opposition to the plan. You are city manager, and the City Council has asked you to present them with a brief overview of the costs and benefits of such a dam. Which three benefits would you highlight? Which three costs? Why did you choose these particular benefits and costs as most important?

CALCULATING Ecological Footprints

One of the single greatest personal uses of water is for showering. Showerheads installed in homes and apartments built before 1992 dispense at least 5 gallons of water per minute, but low-flow showerheads produced after that year dispense just 2.5 gallons per minute. Given an average daily shower time of 8 minutes, calculate the amounts of water used and saved over the course of a year with old standard versus low-flow showerheads, and record your results in the table.

	ANNUAL WATER USE WITH STANDARD SHOWERHEADS (GALLONS)	ANNUAL WATER USE WITH LOW-FLOW SHOWERHEADS (GALLONS)	ANNUAL WATER SAVINGS WITH LOW-FLOW SHOWERHEADS (GALLONS)
You			
Your class			
Your state			
United States			

1. The EPA is currently promoting showerheads that produce still lower flows of 2 gallons per minute (gpm). How much water would you save per year by using a 2-gpm showerhead instead of a 2.5-gpm showerhead?

2. How much water would you be able to save annually by shortening your average shower time from 8 minutes to 6 minutes? Assume you use a 2.5-gpm showerhead.

3. Compare your answers to those of Questions 1 and 2. Do you save more water by showering 8 minutes with a 2-gpm showerhead or 6 minutes with a 2.5-gpm showerhead?

4. Can you think of any factors that are *not* being considered in this scenario of water savings? Explain.

MasteringEnvironmentalScience®

CHAPTER 16

Marine and Coastal Systems and Resources

NEWFOUNDLAND

Atlantic Ocean

MASSACHUSETTS

central
CASE STUDY

Collapse of the Cod Fisheries

> **Either we have sustainable fisheries, or we have no fishery.**
> David Anderson, Canadian Fisheries Minister (1997–1999)

> **So the human management side of this is critical. We don't want to shoot ourselves in the foot as stocks begin to rebuild. We've got to nurse them along.**
> George Rose, Centre for Fisheries Ecosystems Research, Memorial University of Newfoundland

Atlantic cod has had a profound impact on human civilization. Europeans exploring the coasts of North America 500 years ago discovered that they could catch these abundant fish merely by dipping baskets over the railings of their ships. The race that ensued to harvest this resource helped colonize the New World. Starting in the early 1500s, schooners captured millions of cod, and the fish became a dietary staple in cultures on both sides of the Atlantic. If you enjoy traditional British "fish and chips," then you've probably eaten cod, as it is commonly used in this dish.

Since then, cod fishing has been the economic engine for hundreds of communities in coastal New England and eastern Canada. Massachusetts honored the fish by naming Cape Cod after it and by erecting a carved wooden cod statue in its statehouse. In many Canadian coastal villages, cod fishing has been a way of life for generations. So it came as a shock when the cod all but disappeared, and governments had to step in and close the fisheries.

The Atlantic cod (*Gadus morhua*) is a type of groundfish, a name given to fish that live or feed on the bottom—other groundfish species include halibut, pollock, and haddock. Adult cod eat smaller fish and invertebrates, commonly grow 60–70 cm long, and can live 20 years. A mature female cod can produce several million eggs each breeding season. Atlantic cod inhabit cool ocean waters on both sides of the North Atlantic and occur in 24 discrete populations, called stocks. One stock inhabits the Grand Banks off Newfoundland, and another lives on Georges Bank off Massachusetts (**FIGURE 16.1**, p. 416).

The Grand Banks provided ample fish for centuries. With advancing technology, however, ships became larger and more effective at finding fish. By the 1960s, massive industrial trawlers (the majority from Europe) were vacuuming up unprecedented numbers of groundfish off the Canadian coast. In 1977, Canada exercised its legal right to the waters 200 nautical miles from shore, kicked out foreign fishing fleets, and claimed most of the Grand Banks for itself. Then, the Canadian fleet developed the same industrial technologies and revved up its fishing industry like never before.

Then came the crash. Catches dwindled in the 1980s because too many fish had been harvested and because bottom-trawling (fishing by dragging weighted nets across the seafloor, p. 435) had destroyed huge expanses of the cod's underwater habitat. By 1992 the situation was dire: Scientists reported that mature cod were at just 10% of their long-term abundance. Canadian Fisheries Minister John Crosbie announced a two-year ban on commercial cod fishing off Labrador and Newfoundland, where the $700 million fishery supplied income to 16% of the province's workforce. To compensate fishers, the government offered weekly payments for 10 weeks along with training for new job skills and incentives for early retirement. Over the next two years, 40,000 fishers and processing-plant workers lost their jobs, and some coastal communities faced economic ruin.

Cod stocks did not rebound by 1994, so the government extended the moratorium, enacted bans on all other major cod fisheries, and scrambled to offer more compensation to displaced fishers, eventually spending more than $4 billion. In 1997–1998, Canada partially reopened some fisheries, but data soon confirmed that the stocks were not recovering. In

Upon completing this chapter, you will be able to:

- Identify physical, geographic, and chemical aspects of the marine environment

- Explain how the oceans influence and are influenced by climate

- Describe major types of marine ecosystems

- Assess impacts from marine pollution

- Review the state of ocean fisheries and identify reasons for their decline

- Evaluate marine protected areas and reserves as solutions for conserving biodiversity

Massachusetts cod fishermen haul in their catch.

415

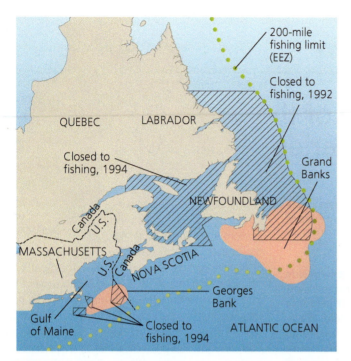

FIGURE 16.1 Populations of Atlantic cod inhabit areas of the northwestern Atlantic Ocean, including the Grand Banks and Georges Bank, regions of shallow water that are especially productive for groundfish. Portions of these and other areas have been closed to fishing in recent years because cod populations have collapsed after being overfished.

2003, the cod fisheries were closed indefinitely, to both commercial and recreational fishing.

Then in 2009, a portion of the Grand Banks off the southeastern coast of Newfoundland was reopened to cod fishing after data showed the stock was recovering slightly. Some dreamed of a comeback for the fishery, but others thought the decision to reopen the area to fishing was ill-advised. Allowable

cod harvests were set beyond what a scientific review board had recommended, and many feared fishing would decimate the stock yet again. Fishing continues today in this limited area, and researchers and resource managers are monitoring populations closely to see how they fare over time.

Across the border in U.S. waters, cod stocks had collapsed in the Gulf of Maine and on Georges Bank. In 1994, the National Marine Fisheries Service (NMFS) closed three prime fishing areas on Georges Bank. Over the next several years, NMFS designed a number of regulations meant to protect and restore the fishery, but these steps were too little, too late. A 2005 report revealed that the cod were not recovering, and further restrictions were enacted. As of 2008, managers announced that the Gulf of Maine stock was 58% of what was necessary for sustainability, and the Georges Bank stock was only 12% as large as it needed to be.

There has been some good news in the cod fisheries, however, since the bans initiated in 1992 (Canada) and 1994 (United States). A 2011 study reported that Grand Banks cod populations were at 34% of historical levels, after hovering at around 5% of historical levels for the previous 20 years. Research on cod populations reveals only part of the story, however. When cod were driven to low numbers by overharvesting, populations of forage fish, such as capelin, increased ninefold as a result of reduced cod predation. This put into motion a cycle in which these forage fish then preyed on and outcompeted young cod, further slowing cod recovery. Populations of forage fish are now in decline, however, because they have outstripped their plankton food supply and are now themselves being harvested by fishermen; this in turn gives cod a better opportunity to rebound. And there are hopeful signs in the Georges Bank as well: Seafloor invertebrates have begun to recover in the absence of trawling. Spawning stock of haddock and yellowtail flounder has risen. Sea scallops have increased in biomass 14-fold. These modest but encouraging recoveries show us that protecting oceans and creating no-fishing areas give us hope that depleted fisheries can one day recover and thrive.

The Oceans

It's been said that our planet "Earth" should more properly be named "Ocean." After all, ocean water covers most of our planet's surface. The oceans are an important component of Earth's interconnected aquatic systems (p. 388). The vast majority of rivers empty into oceans (a small number of rivers empty into inland seas), so the oceans receive most of the inputs of water, sediments, pollutants, and organisms carried by freshwater systems.

The oceans touch and are touched by virtually every environmental system and every human endeavor. They shape our planet's climate, teem with biodiversity, provide us resources, and facilitate our transportation and commerce. Even if you live in a landlocked region far from the coast, the oceans affect you. The oceans provide the fish you eat, the crude oil you need for your car—or that's needed to power your public transportation—and they influence your weather, wherever you live in the world.

Oceans cover most of Earth's surface

The world's five oceans—Pacific, Atlantic, Indian, Arctic, and Southern—are all connected, making up a single vast body of water. This one "world ocean" covers 71% of Earth's surface and contains 97.5% of its water (see Figure 15.3, p. 385). The oceans take up most of the hydrosphere, influence the atmosphere and lithosphere, and encompass much of the biosphere (p. 58). Let's first briefly survey the physical and chemical makeup of the oceans—for although they may look homogeneous from a beach, boat, or airplane, marine systems are actually very complex and dynamic.

Seafloor topography can be rugged

You might not realize that our planet's longest mountain range is under water: The Mid-Atlantic Ridge (p. 33) runs the length of the Atlantic Ocean. Even though most maps depict oceans as smooth swaths of blue, when we examine what's beneath the waves we see that the geology of the

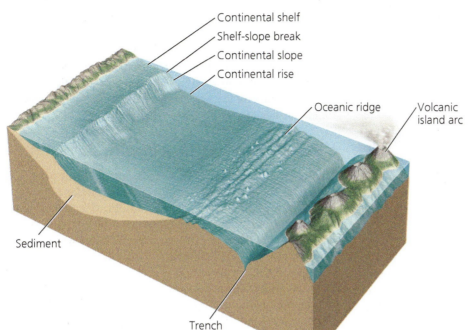

Continental shelf
Shelf-slope break
Continental slope
Continental rise
Oceanic ridge
Volcanic island arc
Sediment
Trench

FIGURE 16.2 A stylized bathymetric profile shows key geologic features of the submarine environment. Shallow water exists around the edges of continents over the continental shelf, which drops off at the shelf-slope break. The steep continental slope gives way to the more gradual continental rise, all of which are underlain by sediments from the continents. Vast areas of seafloor are flat abyssal plain. Seafloor spreading occurs at oceanic ridges, and oceanic crust is subducted in trenches (p. 34). Volcanic activity along trenches may give rise to island chains such as the Aleutian Islands. Features on the left side of this diagram are more characteristic of the Atlantic Ocean, and features on the right side of the diagram are more characteristic of the Pacific Ocean.

ocean floor is anything but uniform. Underwater volcanoes shoot forth enough magma to build islands above sea level, such as the Hawaiian Islands (p. 39). Steep canyons as large as Arizona's Grand Canyon lie just offshore of some continents. The lowest spot in the oceans—the Mariana Trench in the South Pacific—is deeper than Mount Everest is high, by more than a mile.

Stylized maps (**FIGURE 16.2**) that reflect bathymetry (the measurement of ocean depths) and topography (physical geography, or the shape and arrangement of landforms) illustrate underwater geographic features. In bathymetric profile, gently sloping **continental shelves** sit beneath the shallow waters bordering the continents. Continental shelves vary tremendously in width but average 80 km (50 mi) wide, with an average slope of just 1.9 m/km (10 ft/mi). These shelves drop off at the *shelf-slope break*, where the *continental slope* angles more steeply downward to the deep ocean basin below.

Some island chains, such as the Florida Keys, are formed by reefs (pp. 426–427) and lie atop the continental shelf. Others, such as the Aleutian Islands, which curve across the North Pacific from Alaska toward Russia, are volcanic in origin. The Aleutians are also the site of a deep trench that, like the Mariana Trench, formed at a convergent tectonic plate boundary, where one slab of crust dives beneath another in the process of subduction (p. 34).

Wherever reefs, volcanism, or other processes create physical structure underwater, life thrives. Marine animals make use of physical structure as habitat, and as a result topographically complex areas often make for productive fishing grounds. Georges Bank and the Grand Banks are examples; they are essentially huge underwater mounds formed during the ice ages, when glaciers dumped debris at their southernmost extent. As climate warmed and the glaciers retreated, sea level rose, and these hilly areas were submerged in the ocean's salty water and eventually became prime habitat for cod.

Ocean water contains high concentrations of dissolved salts

Ocean water contains approximately 96.5% H_2O by mass. Most of the remainder consists of ions from dissolved salts (**FIGURE 16.3**). Ocean water is salty primarily because ocean

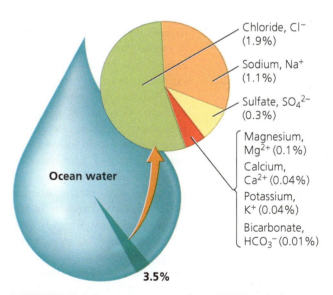

Chloride, Cl^- (1.9%)

Sodium, Na^+ (1.1%)

Sulfate, SO_4^{2-} (0.3%)

Magnesium, Mg^{2+} (0.1%)
Calcium, Ca^{2+} (0.04%)
Potassium, K^+ (0.04%)
Bicarbonate, HCO_3^- (0.01%)

Ocean water

3.5%

FIGURE 16.3 Ocean water consists of 3.5% salt, by mass. Most of this salt is NaCl in solution, so sodium and chloride ions are abundant. A number of other ions and trace elements are also present.

DATA Q If you had a beaker containing one kilogram (1000 grams) of seawater, how many grams of salt are in the beaker? Of this salt, how many grams are from negatively charged ions?

Go to **Interpreting Graphs & Data** on **Mastering**EnvironmentalScience®.

FAQ

When fresh water is scarce, why don't we irrigate crops with seawater?

Freshwater resources are often strained by agricultural irrigation (pp. 216–218), the largest human use of water. Some agriculture occurs in coastal areas, where huge volumes of seawater are available a short distance away, yet farmers do not use it to water their crops, even in times of extreme drought. Why?

The crops we grow for food, such as wheat and broccoli and strawberries, are terrestrial plants and are evolutionarily adapted (pp. 48–51) to use fresh water to grow. If the cells of a terrestrial plant, such as corn, are placed in a solution of seawater, the water inside the cells will be "pulled out" into the surrounding saltwater by the process of diffusion. This occurs because there are fewer dissolved substances (such as sugars, proteins, and salts) within the cell than there are dissolved substances (such as salts) in the seawater. Accordingly, the concentration of water inside the cell is therefore greater than the concentration of water in the seawater.

Water always diffuses from areas of higher concentration to lower concentration, so it flows from inside the cell to outside the cell (this diffusion is called osmosis when it occurs across cell membranes). With continued exposure to saltwater, the plant will dehydrate and die. Marine organisms have evolved physiological mechanisms to combat or replace such water loss from their cells, so they are able to survive in seawater, unlike the terrestrial plants that supply us so much of our food.

basins are the final repositories for runoff that collects salts from weathered rocks and carries them, along with sediments, to the ocean. Wind also blows salts from the land out to sea. Whereas the water in the ocean evaporates, the salts do not, and they accumulate in ocean basins. If we were able to evaporate all the water from the oceans, their basins would be left covered with a layer of dried salt 63 m (207 ft) thick.

The salinity of ocean water generally ranges from 33,000 to 37,000 parts per million, varying from place to place because of differences in evaporation, precipitation, and freshwater runoff (for comparison, fresh water has less than 500 parts per million salinity) from land and glaciers. Coastal waters are often less saline because of the influx of freshwater runoff. Salinity in surface ocean waters near the equator is low because this region has a great deal of precipitation, which is relatively salt free. In contrast, surface ocean water salinity is high at latitudes roughly 30–35 degrees north and south, because at these latitudes evaporation exceeds precipitation, concentrating salts in seawater.

Besides dissolved salts, nutrients such as nitrogen and phosphorus occur in seawater in trace amounts (well under 1 part per million) and play essential roles in nutrient cycling (p. 117) in marine ecosystems. Another aspect of ocean chemistry is dissolved gas content. Roughly 36% of the gas dissolved in seawater is oxygen, which is produced by photosynthetic plants, bacteria, and phytoplankton (p. 74) and enters by diffusion from the atmosphere. Oxygen concentrations are highest in the upper layer of the ocean, reaching 13 ml/L of water. Marine organisms depend on oxygen dissolved in water, just as terrestrial organisms rely on oxygen in air, so most marine life, including fish such as cod, live in these well-oxygenated waters. If dissolved oxygen is depleted from ocean waters, a hypoxic "dead zone" may ensue (pp. 104, 112–113), killing animals or forcing them to leave (see **THE SCIENCE BEHIND THE STORY**, pp. 420–421). Another gas that is soluble in ocean water is carbon dioxide (CO_2). As we pump excess CO_2 into the atmosphere by burning fossil fuels, more CO_2 diffuses into the oceans. We shall soon see (pp. 432–433) how this affects ocean pH (p. 27), turning the water more acidic and posing problems for marine life.

Solar energy stratifies ocean water

Sunlight warms the ocean's surface but does not penetrate deeply, so ocean water is warmest at the surface and becomes colder with depth. Surface waters in tropical regions receive more solar radiation and therefore are warmer than surface waters in temperate or polar regions. Warmer water is less dense than cooler water, but water also becomes denser as it gets saltier. This occurs because as salt dissolves in water, it increases solution mass (the mass of water and its dissolved salts) more than it increases solution volume. These relationships give rise to different layers of water: Heavier (colder and saltier) water sinks, whereas lighter (warmer and less salty) water remains nearer the surface. Waters of the surface zone are heated by sunlight and stirred by wind such that they are of similar density down to a depth of about 150 m (490 ft). Below this zone lies the **pycnocline,** a region in which density increases rapidly with depth. The pycnocline contains about 18% of ocean water by volume, compared to the surface zone's 2%. The remaining 80% lies in the deep zone beneath the pycnocline. The dense water in this deep zone is sluggish and unaffected by winds, storms, sunlight, and temperature fluctuations.

Despite the daily heating and cooling of surface waters, ocean temperatures are much more stable than temperatures on land. Midaltitude oceans experience yearly temperature variation of only around 10°C (18°F), and tropical and polar oceans have more stable temperatures than land at those latitudes. The reason for this stability is that water has a high **heat capacity,** a measure of the heat required to increase temperature by a given amount. It takes more energy to increase the temperature of water than it does to increase the temperature of air. High heat capacity enables ocean water to absorb a tremendous amount of heat from the air. In fact, just the top 2.6 m (8.5 ft) of the oceans holds as much heat as the entire atmosphere! By absorbing heat and releasing it to the atmosphere, the oceans help regulate Earth's climate (Chapter 18). They also influence climate by moving heat from place to place via the ocean's surface circulation, a system of currents that move in the pycnocline and the surface zone.

Surface water flows horizontally in currents

Earth's ocean is composed of vast, riverlike flows driven by density differences, heating and cooling, gravity, and wind. Surface **currents** flow horizontally within the upper 400 m (1300 ft) of water for great distances and in long-lasting patterns across the globe (**FIGURE 16.4**). Warm-water currents carry water heated by the sun from equatorial regions, while cold-water currents carry

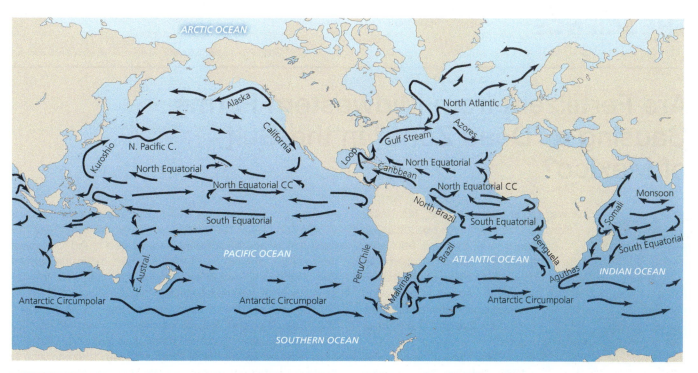

FIGURE 16.4 The upper waters of the oceans flow in surface currents, long-lasting and predictable global patterns of water movement. Warm- and cold-water currents interact with the planet's climate system, and people have used them for centuries to navigate the oceans. *Adapted from Rick Lumpkin (NOAA/AOML).*

DATA If you released a special buoy that traveled on the surface ocean currents shown above into the Pacific Ocean from the southeastern coast of Japan, would it likely reach the United States or Australia first? On what currents would it be carried?

Go to **Interpreting Graphs & Data** on **Mastering**EnvironmentalScience®.

water cooled in high-latitude regions or from deep below. Some surface currents are very slow. Others, like the Gulf Stream, are rapid and powerful. From the Gulf of Mexico, the Gulf Stream flows up the U.S. Atlantic coast and past the eastern edges of Georges Bank and the Grand Banks at nearly 2 m/sec, or more than 4 mph. Averaging 70 km (43 mi) across, the Gulf Stream continues across the North Atlantic, bringing warm water to Europe and moderating that continent's climate (p. 422), which otherwise would be much colder.

Besides influencing climate, ocean currents have aided navigation and shaped human history. Currents helped carry Polynesians to Easter Island (p. 8), Darwin to the Galápagos (p. 49), and Europeans to the New World. Currents transport heat, nutrients, pollution, and the larvae of cod and many other marine species from place to place. Currents in the Pacific Ocean have even transported debris from the tsunami that devastated eastern Japan in 2011 all the way to the western coast of the United States (**FIGURE 16.5**).

Vertical movement of water affects marine ecosystems

Surface winds and heating also create vertical currents in seawater, which move water from the surface to the deep ocean. Where horizontal surface currents diverge from one another,

FIGURE 16.5 Currents carried a 20-m (65-ft), 188-ton dock from Japan to the Oregon coast. The dock was dislodged by the 2011 tsunami in Japan and washed ashore in Oregon more than a year later. The algae and invertebrates that had attached to the dock were removed by state wildlife officials to prevent them from becoming invasive species (pp. 86–90) in the western United States.

Are Fertilizers from Midwestern Farms Causing a "Dead Zone" in the Gulf of Mexico?

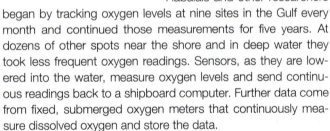

Dr. Nancy Rabalais, LUMCON

She was prone to seasickness, but Nancy Rabalais cared too much about the Gulf of Mexico to let that stop her. Leaning over the side of an open boat idling miles from shore, she hauled a water sample aboard—and helped launch efforts to breathe life back into the Gulf's "dead zone."

Since that first expedition in 1985, Rabalais, her colleague and husband Eugene Turner, and fellow scientists at the Louisiana Universities Marine Consortium (LUMCON) and Louisiana State University have made great progress in unraveling the mysteries of the region's hypoxia—and in getting it on the political radar screen.

Rabalais and other researchers began by tracking oxygen levels at nine sites in the Gulf every month and continued those measurements for five years. At dozens of other spots near the shore and in deep water they took less frequent oxygen readings. Sensors, as they are lowered into the water, measure oxygen levels and send continuous readings back to a shipboard computer. Further data come from fixed, submerged oxygen meters that continuously measure dissolved oxygen and store the data.

The team also collected hundreds of water samples, using lab tests to measure levels of nitrogen, salt, bacteria, and phytoplankton. LUMCON scientists logged hundreds of miles in their ships, regularly monitoring more than 70 sites in the Gulf. They also donned scuba gear to view firsthand the condition of shrimp, fish, and other sea life. Such a range of long-term data allowed the researchers to build a "map" of the dead zone, tracking its location and its consequences.

In 1991, Rabalais made that map public, earning immediate headlines. That year, her group mapped the size of the zone at more than 10,000 km² (about 4000 mi²). Bottom-dwelling shrimp were stretching out of their burrows, straining for oxygen. Many fish had fled. The bottom waters, infused with sulfur from bacterial decomposition, smelled like rotten eggs.

The group's years of monitoring also enabled them to explain and predict the dead zone's emergence. As rivers rose each spring (and as fertilizers were applied in the Midwestern farm states), oxygen would start to disappear in the northern Gulf. The hypoxia would last through the summer or fall, until seasonal storms mixed oxygen into hypoxic areas.

The source of the problem, Rabalais said, lay back on land. The Mississippi and Atchafalaya rivers draining into the Gulf were polluted with agricultural runoff, and the nutrient pollution from fertilizers spurred algal blooms whose decomposition by bacteria snuffed out oxygen in wide stretches of ocean water (pp. 108–109). This work had clearly demonstrated the interconnections between freshwater aquatic systems and the Gulf, and how pollutants from farm fields in the upper Midwest could exert effects far away at the mouth of the Mississippi River.

Over time, monitoring linked the dead zone's size to the volume of river flow and its nutrient load. The 1993 flooding of the Mississippi created a zone much larger than the year before, whereas a drought in 2000 brought low river flows, low nutrient loads, and a small dead zone (**FIGURE 1**). Similar relationships between river flow and dead zone size have been seen ever since. In 2005, the dead zone was predicted to be large, but Hurricanes Katrina and Rita stirred oxygenated surface water into the depths, decreasing the dead zone that year.

Many Midwestern farming advocates and some scientists, such as Derek Winstanley, chief of the Illinois State Water Survey, challenged the findings. They argued that the Mississippi naturally carries high loads of nitrogen from runoff and that Rabalais's team had not ruled out upwelling in the Gulf as a source of nutrients.

But sediment analyses showed that Mississippi River mud contained many fewer nitrates early in the century, and Rabalais and Turner found that silica residue from phytoplankton blooms increased in Gulf sediments between 1970 and 1989, paralleling rising nitrogen levels. In 2000, a federal integrative assessment team of dozens of scientists laid the blame for the dead zone on nutrients from fertilizers and other sources in the fresh waters emptying into the Gulf.

Then in 2004, while representatives of farmers and fishermen debated political fixes, Environmental Protection Agency water quality scientist Howard Marshall suggested that to alleviate the dead zone we'd be best off reducing phosphorus

FIGURE 1 The map in (a) shows dissolved oxygen concentrations in bottom waters of the Gulf of Mexico off the Louisiana coast in 2015. The darkest areas indicate the lowest oxygen levels, with regions considered hypoxic (<2 mg/L) are outlined in black. The dead zone forms to the west of the mouth of the Mississippi River because prevailing currents carry nutrients in that direction. The graph in (b) shows that the size of the hypoxic zone (shown by bars) is correlated with the amount of nitrogen pollution entering from the Mississippi River (shown by line), with nutrient delivery being higher in wetter years and lower in years with Midwest drought. The dead zone in 2015 covered more than 16,000 km² (6400 mi²) of the Gulf, and scientists and policymakers aim to reduce its size to 5000 km² (1930 mi²). *Hypoxic zone data from Nancy Rabalais, LUMICON, and R. Eugene Turner, LSU. Nitrogen flux data from USGS, toxics.usgs.gov/hypoxia/mississippi/index.html.*

DATA Q In which year was the hypoxic zone largest, and how many square kilometers was it? In which year was nitrogen flux the highest, and what was its approximate value that year?

Go to **Interpreting Graphs & Data** on **Mastering**EnvironmentalScience®.

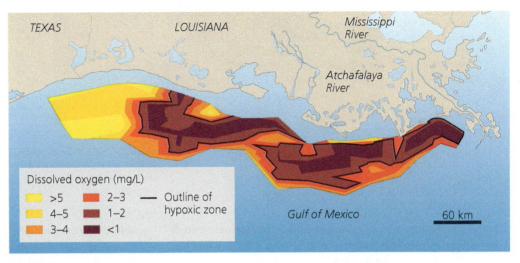

(a) Dissolved oxygen at ocean bottom

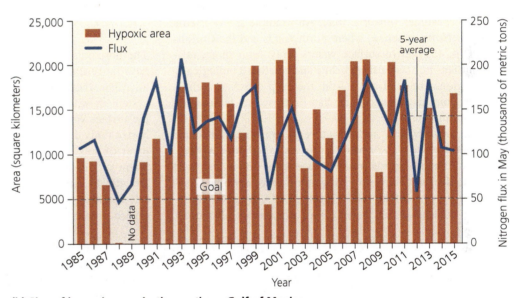

(b) Size of hypoxic zone in the northern Gulf of Mexico

pollution from industry and sewage treatment. His reasoning was this: Phytoplankton need both nitrogen and phosphorus, but there is now so much nitrogen in the Gulf that phosphorus has become the limiting factor on phytoplankton growth.

Since then, research has supported this contention, and scientists now propose that nitrogen and phosphorus should be managed jointly. Moreover, recent research indicates that a federally mandated 30% reduction in nitrogen in the river will not be adequate to eliminate the dead zone. Scientists also maintain that large-scale restoration of wetlands along the river and at the river's delta would best filter pollutants before they reach the Gulf.

All this research is guiding a federal plan to reduce farm runoff, clean up the Mississippi, restore coastal wetlands, and shrink the Gulf's dead zone. It has also led to a better understanding of hypoxic zones around the world.

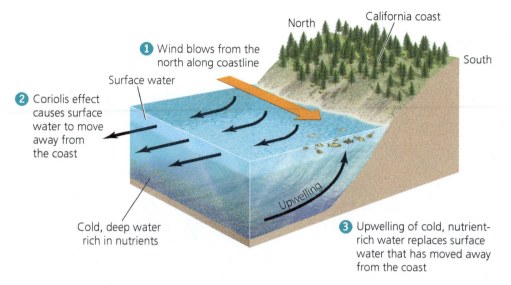

North
California coast
South

1 Wind blows from the north along coastline

Surface water

2 Coriolis effect causes surface water to move away from the coast

Cold, deep water rich in nutrients

Upwelling

3 Upwelling of cold, nutrient-rich water replaces surface water that has moved away from the coast

FIGURE 16.6 Upwelling is the movement of bottom waters upward. This often brings nutrients up to the surface, creating productive areas for marine life. For example, north winds blow along the California coastline **1**, while the Coriolis effect (p. 451) draws wind and water away from the coast **2**. Water is then pulled up from the bottom **3** to replace the water that moves away from shore.

cold, deep waters are pulled to the surface in a process called **upwelling.** Upwelled water is rich in nutrients brought up from the ocean bottom, so upwellings are often sites of high primary productivity (p. 111) and lucrative fisheries. Upwellings will occur where strong winds blow away from or parallel to coastlines (**FIGURE 16.6**). An example is the Pacific coast of North America, where north winds and the Coriolis effect (p. 451) move surface waters away from shore, raising nutrient-rich water from below and creating a biologically rich region. The cold water from the deep ocean also chills the air along the coast, giving San Francisco its famous fog and cool summers.

In areas where surface currents converge, or come together, surface water sinks—a process called **downwelling.** Downwelling transports warm surface water rich in dissolved gases to deeper waters, providing an influx of dissolved oxygen for deep-water life and "burying" CO_2 from the atmosphere in deep ocean waters. Vertical currents also occur

within the deep zone of the ocean, where differences in density can lead to rising and falling convection currents, similar to those in molten rock (p. 32) and in air (p. 451).

Ocean currents affect Earth's climate

The horizontal and vertical movements of ocean water can have far-reaching effects on climate both regionally and globally. The **thermohaline circulation** is a worldwide current system in which warmer, lower-salinity water moves along the surface and colder, saltier water (which is denser) moves horizontally deep beneath the surface (**FIGURE 16.7**). One segment of this worldwide conveyor-belt system includes the warm surface water in the Gulf Stream that flows across the Atlantic Ocean to Europe. On reaching Europe, this water releases heat to the air, keeping Europe warmer than it otherwise would be given its latitude. The now-cooler water

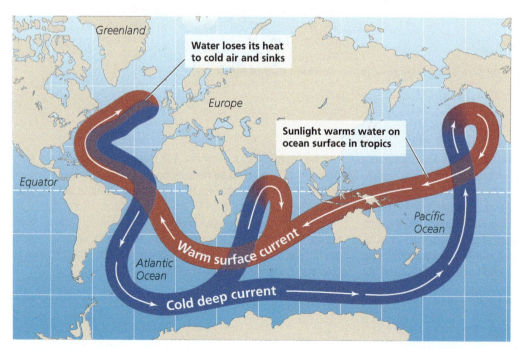

Greenland

Water loses its heat to cold air and sinks

Europe

Sunlight warms water on ocean surface in tropics

Equator

Pacific Ocean

Warm surface current

Atlantic Ocean

Cold deep current

FIGURE 16.7 As part of the oceans' thermohaline circulation, warm surface currents carry heat from equatorial waters northward toward Europe, where they warm the atmosphere. The water then cools and sinks, forming the North Atlantic Deep Water (NADW). Scientists debate whether rapid melting of Greenland's ice sheet could interrupt this heat flow and cause Europe to cool dramatically.

becomes saltier through evaporation, and thus becomes denser, and sinks, creating a region of downwelling known as the **North Atlantic Deep Water (NADW).**

Scientists hypothesize that interrupting the thermohaline circulation could trigger rapid climate change. If global warming (Chapter 18) causes much of Greenland's ice sheet to melt, the resulting freshwater runoff into the North Atlantic would make surface waters less dense (because fresh water is less dense than saltwater). This could potentially stop the NADW formation and shut down the northward flow of warm water, causing Europe to cool rapidly. Some data suggest that the thermohaline circulation in this region is already slowing, but other researchers maintain that Greenland will not produce enough runoff to cause a shutdown this century. A 2015 study indicated that global warming is, however, slowing the flow of this current, potentially affecting climate in Europe.

Another interaction between ocean currents and the atmosphere that influences climate is the **El Niño–Southern Oscillation (ENSO),** a systematic shift in atmospheric pressure, sea surface temperature, and ocean circulation in the tropical Pacific Ocean. Under normal conditions, prevailing winds blow from east to west along the equator, from a region of high pressure in the eastern Pacific to one of low pressure in the western Pacific, forming a large-scale convective loop in the atmosphere (**FIGURE 16.8a**). The winds push surface waters westward, causing water to "pile up" in the western Pacific. As a result, water near Indonesia can be 50 cm (20 in.) higher and 8°C warmer than water near South America, elevating the risk of coastal flooding in the Pacific. The westward-moving surface waters allow cold water to rise up from the deep in a nutrient-rich upwelling along the coast of Peru and Ecuador.

El Niño conditions are triggered when air pressure decreases in the eastern Pacific and increases in the western Pacific, weakening the equatorial winds and allowing the warm water to flow eastward toward South America (**FIGURE 16.8b**). This suppresses upwelling along the Pacific coast of the Americas, shutting down the delivery of nutrients that support marine life and fisheries. This phenomenon was called El Niño (Spanish for "little boy" or "Christ Child") by Peruvian fishermen because the arrival of warmer waters usually occurred shortly after Christmas. El Niño events alter weather patterns around the world, creating rainstorms and floods in areas that are generally dry (such as southern California, p. 385) and causing drought and fire in regions that are typically moist (such as Indonesia).

Coastal industries such as Peru's anchovy fisheries are devastated by El Niño events, and the 1982–1983 El Niño alone caused over $8 billion in economic losses worldwide. Food shortages in Asia caused by the 1997–1998 El Niño plunged an estimated 15% of the people in some nations into poverty. The El Niño event of 2015–2016 was especially strong, with massive heat waves afflicting India and drought ravaging Southeast Asia.

La Niña events are the opposite of El Niño events; in a La Niña event, unusually cold waters rise to the surface and extend westward in the equatorial Pacific when winds blowing to the west strengthen, and weather patterns are affected in opposite ways. ENSO cycles are periodic but irregular, occurring every

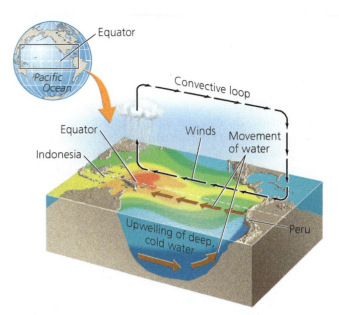

(a) Normal conditions

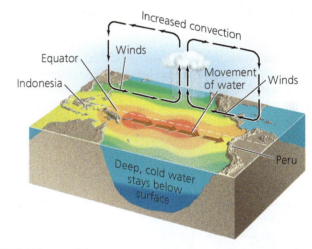

(b) El Niño conditions

FIGURE 16.8 El Niño conditions occur every 2 to 8 years, causing marked changes in weather patterns. In these diagrams, red and orange colors denote warmer water, and blue and green colors denote colder water. Under normal conditions **(a)**, prevailing winds push warm surface waters toward the western Pacific. Under El Niño conditions **(b)**, winds weaken, and the warm water flows back across the Pacific toward South America, like water sloshing in a bathtub. *Adapted from National Oceanic and Atmospheric Administration, Tropical Atmospheric Ocean Project.*

2–8 years. Scientists are exploring whether warming air and sea temperatures due to climate change (Chapter 18) may be increasing the frequency and strength of these cycles.

Marine and Coastal Ecosystems

With their variation in topography, temperature, salinity, nutrients, and sunlight, marine and coastal environments feature a variety of ecosystems. Regions of ocean water differ greatly, and some zones support more life than others. The uppermost 10 m (33 ft) of water absorbs 80% of solar energy, so nearly all of the oceans' primary productivity occurs in the top layer, or **photic zone.** Generally, the warm, shallow waters of continental shelves are most biologically productive and support the greatest species diversity. Habitats and ecosystems occurring between the ocean's surface and floor are termed **pelagic,** whereas those that occur on the ocean floor are called **benthic.** Most marine and coastal ecosystems are powered by solar energy, with sunlight driving photosynthesis by phytoplankton in the photic zone. Yet even the darkest ocean depths host life.

As we survey marine and coastal ecosystems, keep in mind that they are part of a web of interconnected freshwater and marine aquatic systems (see Figure 15.4, p. 388) that exchange water, organisms, sediments, pollutants, and other dissolved substances with one another. Hence, the topics we discuss in this chapter are greatly influenced by those in freshwater ecosystems, and vice versa.

Intertidal zones undergo constant change

Where the ocean meets the land, **intertidal,** or *littoral,* ecosystems (**FIGURE 16.9**) spread between the uppermost reach of the high tide and the lowest limit of the low tide. **Tides** are the periodic rising and falling of the ocean's height at a given location, caused by the gravitational pull of the moon and sun. High and low tides occur roughly 6 hours apart, so intertidal organisms spend part of each day submerged in water, part of the day exposed to air and sun, and part of the day being lashed by waves. These organisms must also protect themselves from marine predators at high tide and terrestrial predators at low tide.

The intertidal environment is a tough place to make a living, but it is home to a remarkable diversity of organisms. Life abounds in the crevices of rocky shorelines, which provide shelter and pools of water (tide pools) during low tides. Sessile (stationary, non-swimming) animals such as anemones, mussels, and barnacles live attached to rocks, filter-feeding on plankton in the water that washes over them. Urchins, sea slugs, chitons, and limpets eat intertidal algae or scrape food from the rocks. Sea stars (starfish) creep slowly along, preying on the filter-feeders and herbivores. Crabs clamber around the rocks, scavenging detritus. Rocky intertidal regions such as these line the shoreline of many parts of Newfoundland and New England onshore of the Grand Banks and Georges Bank, respectively.

The rocky intertidal zone is so diverse because environmental conditions such as temperature, salinity, and moisture change dramatically from the high to the low reaches. This environmental variation gives rise to horizontal bands dominated by different sets of organisms arrayed according to their habitat needs, competitive abilities, and adaptation to exposure. Sandy intertidal areas, such as those of Cape Cod, host less biodiversity, yet plenty of organisms burrow into the sand at low tide to await the return of high tide, when they emerge to feed.

Supratidal zone (splash zone)

Level of high tide

Intertidal zone

Level of low tide

Subtidal zone

FIGURE 16.9 The rocky intertidal zone stretches along rocky shorelines between the lowest and highest reaches of the tides. The intertidal zone provides niches for a diversity of organisms, including sea stars (starfish), crabs, sea anemones, corals, chitons, mussels, nudibranchs (sea slugs), and sea urchins. Areas higher on the shoreline are exposed to the air more frequently and for longer periods, so organisms that tolerate exposure best specialize in the upper intertidal zone. The lower intertidal zone is exposed less frequently and for shorter periods, so organisms less tolerant of exposure thrive in this zone.

FIGURE 16.10 Salt marshes occur in temperate intertidal zones where the substrate is muddy. Tidal waters flow in channels called tidal creeks amid flat areas called benches, sometimes partially submerging the salt-adapted grasses.

FIGURE 16.11 Mangrove forests line tropical and subtropical coastlines. Mangrove trees, with their unique roots, are adapted for growing in saltwater and provide habitat for many fish, birds, crabs, and other animals.

Salt marshes line temperate shorelines

Along many of the world's coasts at temperate latitudes, **salt marshes** occur where the tides wash over gently sloping sandy or silty substrates. Rising and falling tides flow into and out of channels called tidal creeks and at highest tide spill over onto elevated marsh flats (**FIGURE 16.10**). Marsh flats grow thick with salt-tolerant grasses, as well as rushes, shrubs, and other herbaceous plants.

Salt marshes boast very high primary productivity and provide critical habitat for shorebirds, waterfowl, and many commercially important fish and shellfish species. Salt marshes also filter pollution and stabilize shorelines against storm surges, an unusual rise in ocean level caused by offshore storms. However, because people desire to live and do business along coasts, we have altered or destroyed vast expanses of salt marshes to make way for coastal development. When salt marshes are destroyed, we lose the ecosystem services they provide. When Hurricane Katrina struck the Gulf Coast in 2005, for instance, the flooding was made worse because vast areas of salt marshes had vanished during the preceding decades as a result of engineering of the Mississippi River, development, and subsidence from oil and gas drilling.

Mangrove forests line coasts in the tropics and subtropics

In tropical and subtropical latitudes, mangrove forests replace salt marshes along gently sloping sandy and silty coasts. **Mangroves** are among the few types of trees that are salt tolerant, and they have rather unique roots. Some of their roots curve upward like snorkels to get oxygen from the air, while others curve downward like stilts to support the tree in changing water levels (**FIGURE 16.11**). Fish, shellfish, crabs, snakes, and other organisms thrive among the root networks, and birds feed and nest in the dense foliage of these coastal forests. Besides serving as nurseries for fish and shellfish

that people harvest, mangroves also provide wood for fuel and construction, natural dyes, and chemical compounds that have medicinal properties.

From Florida to Mexico to the Philippines, half the world's mangrove forests have been destroyed as people have developed coastal areas. Shrimp farming in particular has greatly impacted mangroves, as mangrove forests are often destroyed to build large pens along the coastline in which shrimp are raised in aquaculture (p. 242). When mangroves are removed, coastal areas lose the ability to slow runoff, filter pollutants, and retain soil. As a result, offshore systems such as coral reefs and eelgrass beds are more readily degraded. Moreover, mangrove forests protect coastal communities against storm surges and tsunamis. The 2004 Indian Ocean tsunami (p. 42) devastated areas where mangroves had been removed because of shrimp farming and coastal development but caused less damage where mangroves were intact.

Fresh water meets saltwater in estuaries

Many salt marshes and mangrove forests occur in or near **estuaries,** water bodies where rivers flow into the ocean, mixing fresh water with saltwater. Estuaries are biologically productive ecosystems that experience daily and seasonal fluctuations in salinity as tides come in and go out and river flows vary. Sheltered from crashing surf, the shallow water of estuaries nurtures eelgrass beds and other plant life, producing abundant food and resources. For shorebirds and for many commercially important shellfish species such as oysters and scallops, estuaries provide critical habitat. For fishes such as salmon, which spawn in streams and mature in the ocean, estuaries provide a transitional zone where young fish make the passage from fresh water to saltwater.

Estuaries everywhere have been affected by coastal development, water pollution, habitat alteration, and overfishing. The Chesapeake Bay demonstrates many of the challenges faced by estuaries. Bay waters suffer from seasonal oxygen depletion from cultural eutrophication (p. 108), and its once-thriving oyster populations, which naturally filter the water of pollutants, have been decimated by overharvesting (Chapter 5). Coastal ecosystems such as estuaries have borne the brunt of human impact because two-thirds of Earth's people choose to live within 160 km (100 mi) of the ocean.

Kelp forests harbor many organisms

Along many temperate coasts, large brown algae, or **kelp,** grow from the floor of continental shelves, reaching up toward the sunlit surface. Some kelp reaches 60 m (200 ft) in height and can grow 45 cm (18 in.) per day. Dense stands of kelp form underwater "forests" (**FIGURE 16.12**). Kelp forests supply shelter and food for invertebrates and fish, which in turn provide food for predators such as seals and sharks. Kelp forests also absorb wave energy and protect shorelines from erosion. Some types of kelp are eaten by people, and kelp provides alginates, chemical compounds that serve as thickeners in consumer products such as cosmetics, paints, paper, soaps, and ice cream.

Coral reefs are treasure troves of biodiversity

Shallow subtropical and tropical waters are home to coral reefs. A reef is an underwater outcrop of rock, sand, or other material. A **coral reef** is a mass of calcium carbonate composed of the shells of tiny marine animals known as **corals.** A coral reef may occur as an extension of a shoreline; along a *barrier island* paralleling a shoreline; or as an *atoll,* a ring around a submerged island.

Corals are tiny invertebrate animals related to sea anemones and jellyfish. They remain attached to rock or existing reef and capture passing food with stinging tentacles. Corals also derive nourishment from symbiotic algae known as **zooxanthellae,** which inhabit their bodies and produce food

(a) Coral reef community

Bleaching is evident in the whitened regions of this coral

(b) Bleached coral

FIGURE 16.13 Coral reefs provide food and shelter for a tremendous diversity (a) of fish and other creatures. Today these reefs face multiple stresses from human impacts. Many corals have died as a result of coral bleaching **(b)**, in which corals lose their zooxanthellae.

FIGURE 16.12 "Forests" of tall brown algae known as kelp grow from the floor of the continental shelf. Numerous fish and other creatures eat kelp or find refuge among its fronds.

through photosynthesis (and provide the diversity of vibrant colors in reefs). Most corals are colonial, and the surface of a coral reef consists of millions of densely packed individuals. As corals die, their shells remain part of the reef and new corals grow atop them. This accumulation of coral shells enables the reef to persist and grow larger over time.

Like kelp forests, coral reefs protect shorelines by absorbing wave energy. They also host tremendous biodiversity (**FIGURE 16.13a**). This is because coral reefs provide complex physical structure (and thus many habitats) in shallow nearshore waters, which are regions of high primary productivity. If you have ever gone diving or snorkeling over a coral reef, you will have noticed the staggering diversity of anemones, sponges, hydroids, tubeworms, and other sessile invertebrates; the innumerable mollusks, flatworms, sea stars, and urchins; and the many fish species that find food and shelter in reef nooks and crannies.

It is their promotion of biodiversity that makes the alarming decline in coral reefs worldwide particularly disturbing. Many reefs have fallen victim to "coral bleaching," a process that occurs when zooxanthellae die or abandon the coral, thereby depriving the coral of nutrition. Corals lacking zooxanthellae lose color and frequently die, leaving behind ghostly white patches in the reef (**FIGURE 16.13b**). Once large areas of coral die, species that hide within the reef are exposed to higher levels of predation, and their numbers decline. Without living coral—the foundation of the ecosystem on which so many species rely—biological diversity on reefs declines as organisms flee or perish. Coral bleaching is thought to occur when coral are strongly stressed, with common stressors including increased sea surface temperatures associated with global climate change, and exposure to elevated levels of pollutants.

weighing the ISSUES

Coastal Development

A developer wants to build a large marina on an estuary in your coastal town. The marina would boost the town's economy but eliminate its salt marshes. What consequences would you expect for property values? For water quality? For wildlife? As a homeowner living adjacent to the marshes, how would you respond? Do you think that developers or town officials should pay to upgrade homeowner's insurance against damage from storm surges for coastal residents when a protective salt marsh is destroyed?

Another threat to coral comes from nutrient pollution in coastal waters, which promotes the growth of algae that are smothering reefs in the Florida Keys and in many other regions. In addition, coral reefs sustain damage when divers use cyanide to stun fish in capturing them for food or for the pet trade, a common practice in the waters of Indonesia and the Philippines. And as global climate change accelerates, the oceans are becoming increasingly more acidic as excess carbon dioxide from the atmosphere reacts with seawater to form carbonic acid. The resulting acidification threatens to deprive corals of the carbonate ions they need to produce their structural parts (pp. 432–433).

A few coral species thrive in waters outside the tropics and build reefs on the ocean floor at depths of 200–500 m (650–1650 ft). These little-known reefs, which occur in cold-water areas off the coasts of Norway, Spain, the British Isles, and elsewhere, are only now being studied by scientists. Already, however, many have been badly damaged by bottom-trawling (p. 435)—the same practice that has drastically degraded the benthic habitats of groundfish such as the Atlantic cod. Norway and other countries are now beginning to protect some of these deep-water reefs.

Open-ocean ecosystems vary in their biodiversity

Biological diversity in pelagic regions of the open ocean is highly variable in its distribution. Near the surface, primary production (p. 111) and animal life are concentrated in regions of nutrient-rich upwelling. Microscopic phytoplankton constitute the base of the marine food chain in the pelagic zone and produce up to half of the atmosphere's oxygen. These photosynthetic algae, protists, and cyanobacteria feed zooplankton (p. 74), which in turn become food for fish, jellyfish,

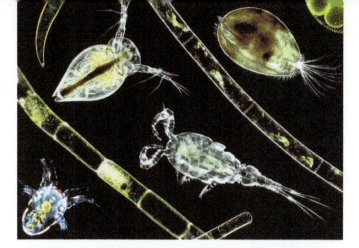

FIGURE 16.14 The uppermost reaches of ocean water contain billions upon billions of phytoplankton—tiny photosynthetic algae, protists, and bacteria that form the base of the marine food chain. This part of the ocean is also home to zooplankton, small animals and protists that dine on phytoplankton.

whales, and other free-swimming animals (**FIGURE 16.14**). Predators at higher trophic levels include larger fish, sea turtles, and sharks. Fish-eating birds such as puffins, petrels, and shearwaters feed at the surface of the open ocean, returning periodically to nesting sites on islands and coastlines.

In the little-known deep-water ecosystems, animals have adapted to tolerate extreme water pressures and to live in the dark without food from autotrophs. Many of these often bizarre-looking creatures scavenge carcasses or organic detritus that fall from above. Others are predators, and still others attain food from symbiotic mutualistic (p. 77) bacteria.

Ecosystems form around hydrothermal vents (p. 31) on the bottom of the deep ocean, where heated water spurts from the seafloor, carrying minerals that precipitate to form rocky structures. Tubeworms, shrimp, and other creatures in these systems use symbiotic bacteria to derive their energy from chemicals in the heated water rather than from sunlight, a process called *chemosynthesis* (p. 31). They manage to thrive within amazingly narrow zones between scalding-hot and icy-cold water.

Marine Pollution

People have long made the oceans a sink for waste and pollutants. Even into the mid-20th century, it was common for coastal U.S. cities to dump trash and untreated sewage along their shores. The Clean Water Act (p. 172) reduced the discharge of point-source pollution (p. 405) into the oceans. But a great deal of pollution is non-point-source pollution that comes from countless small things we all do and from items we throw away far inland: The oceans are downstream from everywhere. As long as rivers flow to the sea, much of our pollution on land sooner or later ends up in the oceans.

Accordingly, oil, plastic, toxic chemicals, and excess nutrients all eventually make their way into the oceans. Ocean-based sources, such as sewage and trash from cruise ships and abandoned fishing gear from fishing boats, add to the ocean's pollution inputs. The volumes of trash added to Earth's oceans is stunning, and the scope of the problem can be gauged by the amount picked up by volunteers who trek beaches in the Ocean Conservancy's annual International Coastal Cleanup. Over 25 years, more than 65 million kg (144 million lb) of waste have been collected from the world's beaches as part of this program. As the average human has a mass of around 63 kg (140 lb), this is the mass equivalent of more than *1 million* people.

Plastic debris endangers marine life

Plastics are a particularly dangerous type of trash for the oceans. Plastic bags and bottles, fishing nets and line, and small particles of plastic and other trash can harm marine creatures. Organisms can become entangled and drown in large pieces of plastic debris, or they may die as a result of ingesting material they cannot digest or expel (FIGURE 16.15a). Ocean currents concentrate floating plastic and debris in **gyres,** areas of the ocean where currents converge (see THE SCIENCE BEHIND THE STORY, pp. 430–431). The North Pacific Gyre contains the **Great Pacific Garbage Patch,** an area larger than Texas, in which tiny pieces of floating plastic outnumber organisms by a 6 to 1 margin.

Because plastic is designed not to break down, it can drift for decades before washing up on beaches. Plastic does degrade slowly in seawater and sunlight but, in doing so, breaks down into smaller and smaller bits that become more and more numerous. This can actually make things worse: The oceans are now filled with uncountable trillions of tiny pellets of plastic floating just under the surface like confetti (FIGURE 16.15b). Some larger pieces of plastic sink, littering the bottom of the ocean where the low-temperature, low-light conditions are not favorable for degradation. In the North Sea off Norway, for example, underwater surveys have found more than 100 pieces of plastic per square kilometer on the ocean floor.

Floating plastics are particularly harmful to marine life. Organisms mistake the small pieces of plastic for plankton or fish eggs, ingest them, and suffer injury or death (FIGURE 16.15c). A 2010 study found that the average

(a) Plastics entangling wildlife

(b) Tiny bits of debris from ocean gyres

(c) Organisms ingesting plastics

(d) Preventing oceanic plastic pollution

FIGURE 16.15 **Plastics affect oceanic life.** Like thousands of marine animals every year, this loggerhead sea turtle **(a)** became entangled in a discarded fishing net. Pieces of plastic like this sample from the "Great Pacific Garbage Patch" are also dangerous to organisms **(b)**. Plastics are often mistaken for food items and ingested by wildlife, such as by this dead albatross **(c)**. Prevention is the best approach for addressing oceanic plastic pollution. Here, booms capture trash in the Los Angeles River **(d)** to prevent it from entering the Pacific Ocean.

plankton-eating fish in the Great Pacific Garbage Patch had 2.1 pieces of plastic in its digestive system. Fish are not the only organisms affected. Seabird chicks are killed when parents unknowingly feed them pieces of plastic when regurgitating food. One study attributed 40% of premature deaths of albatross chicks in the Pacific islands of Midway to this cause. Some 267 species are affected by marine plastic debris, leading to the death of an estimated 100,000 marine mammals, such as seals and dolphins, and 1 million seabirds each year.

FAQ

Is the "Great Pacific Garbage Patch" a huge mass of floating debris?

The vast majority of plastics that accumulate in the ocean are quite small and are dispersed over relatively large areas, so they are difficult to detect with the naked eye. Hence, a person will not see dense expanses of floating debris when sailing through the Great Pacific Garbage Patch (or viewing pictures of it), and this can incorrectly lead him or her to believe the problem of oceanic debris is exaggerated.

But even though the plastic and other debris in these gyres are typically small (usually only several millimeters in length), they still pose threats to wildlife from ingestion and toxicity: Smaller pieces of plastic are more dangerous to marine life than larger pieces in some ways, because they closely resemble food items. So although a floating tire or fishing net certainly provides a striking visual impact, it's important to realize that the oceanic plastic pollution we cannot easily see is often of greatest concern.

Plastics, of course, can have toxic effects on organisms. Plastics contain harmful substances, such as bisphenol A and phthalates (Chapter 14), which leach into ocean water or the digestive tracts of animals (if the plastic is ingested). Pieces of plastic in the ocean can also chemically "grab" and concentrate persistent organic pollutants (POPs; p. 379) such as DDT and PCBs, thereby contaminating the plastic and increasing its toxicity to any organism that ingests it.

Floating pieces of plastic can serve as "rafts" and transport species over long distances they could not travel on their own. Seaweeds, sessile invertebrates, and other creatures have become invasive species when carried to new habitats on floating plastic. For example, two species of bryozoans (a predatory, sessile invertebrate that forms polyps on seaweed and stones) are reaching the Florida coast from the Caribbean on floating plastics and are adversely affecting native species when polyps damage ecologically important seaweed species.

Because plastics take 500–1000 years to degrade at sea and there is no viable way to collect the small bits of plastics that litter the oceans, preventing their entry into the oceans is the only key to remedying oceanic plastic pollution (**FIGURE 16.15d**). In 2006, the U.S. Congress responded to ocean pollution by passing the Marine Debris Research, Prevention, and Reduction Act, aiding efforts to keep plastics out of marine waters. In 2015, Congress strengthened these efforts by passing a ban on the sale and distribution of products containing tiny plastic "microbeads." Added to some toothpastes and shower gels to act as tiny "scrubbers," trillions of these plastic particles were entering U.S. waters on a daily basis. These and other efforts are ecologically wise but also convey significant economic benefits. In Asia alone, the costs of plastic pollution on fisheries, tourism, and other industries is an estimated $1 billion a year.

Oil pollution comes from spills of all sizes

Oil pollution is another important source of marine pollution. Although large oil spills occur infrequently, their impacts can be staggering near the spill site. The danger that oil spills pose to fisheries, economies, and ecosystems became clear in 1989, when the oil tanker *Exxon Valdez* ran aground in Prince William Sound, spilling hundreds of thousands of barrels of oil and causing an ecological disaster along the Alaskan coast.

The dangers of oil pollution made headlines again in 2010, when British Petroleum's *Deepwater Horizon* offshore drilling platform exploded and sank into the Gulf of Mexico off the Louisiana coast (pp. 534–535). Oil gushed from the platform's underwater well, was spread widely by ocean currents, and washed up on coastal areas across the northern Gulf of Mexico. The economic and ecological impacts of the spill were visited on hundreds of miles of water, sediments, and shoreline along the coasts of Louisiana, Mississippi, Alabama, and Florida. Five years after the accident, populations of oysters, crabs, and sea turtles remained at low levels, and large numbers of marine mammals were beaching themselves on Gulf shores, suggesting that the impacts of this accident may continue to be felt for some time.

About 30% of our crude oil and much of our natural gas come from seafloor deposits. Most offshore oil and gas is concentrated in petroleum-rich regions such as the North Sea and the Gulf of Mexico, but energy companies extract smaller amounts from diverse locations, among them the Grand Banks and adjacent Canadian waters. Proposals to drill for oil and gas in Georges Bank and the Gulf of Maine have been stalled by both the U.S. and Canadian governments, in large part because any spilled oil could damage the region's fisheries.

Given the severity of the *Deepwater Horizon* and other oil spills, it may be surprising to learn that about half of the petroleum entering the world's oceans in a given year originates from natural seeps in the ocean bottom. The impacts on organisms from these petroleum releases are less severe than the impacts from spills, however, because the seepage from the ocean floor is more diffuse. The spillage that results from an accident involving an oil tanker or oil rig, in contrast, is concentrated in a relatively small area.

Much of the oil entering the ocean from sources other than seeps accumulates in waters from innumerable, widely spread, small non-point sources. Shipping vessels and recreational boats can leak oil as they ply ocean waters. Motor oil from vehicles on roads and parking lots is washed into streams by rains and carried to the sea. Spills from oil tankers account for 12% of oil pollution in an average year, and 3% comes from leakage that occurs during the extraction of oil by offshore oil rigs. Although the *Exxon Valdez* spill was catastrophic, the good news is that the amount of oil spilled from tankers worldwide has decreased over the past three decades,

Can We Predict the Oceans' "Garbage Patches"?

Nikolai Maximenko, International Pacific Research Center at the University of Hawai'i at Manoa

In 1997, while sailing across a little-traveled portion of the Pacific Ocean as he returned from a recreational yacht race, Captain Charles Moore encountered a huge floating mass of debris he described as a "soupy" collection of items including tires, plastics, chemical drums, coat hangers, fishing nets, and other items. This was the first recorded visual confirmation of what is now called the "Great Pacific Garbage Patch" and brought firmly into the public eye the issue of plastic pollution in the oceans. Knowing where floating debris will accumulate in the oceans is important to marine biologists and oceanographers, but short of stumbling upon these areas in the ocean, as Captain Moore did, is there any way to know where to look for them? Thanks to the work of Nikolai Maximenko and his collaborators, the answer is yes.

Maximenko, a senior researcher of the International Pacific Research Center at the University of Hawai'i at Manoa, studies the movements of currents in the oceans. In particular, his work examines *Ekman drift*, wind-driven currents in the water's upper layers first modeled by scientist Walfrid Ekman in the early 20th century. Although the large-scale ocean currents that are driven by pressure gradients and the Coriolis force (collectively called *geostrophic currents*) are fairly well described, the smaller-scale movements associated with Ekman currents are not. This lack of information hinders researchers' ability to predict movements of floating material in the oceans—such as plastics and other debris.

Data gathered by the Global Drifter Program of the U.S. National Oceanic and Atmospheric Administration (NOAA) are helping scientists better understand these currents. Since 1979, the program has deployed more than 12,000 buoys in the world's oceans and tracked their movements to determine patterns in ocean currents. Each buoy (**FIGURE 1**) is composed of a 30- to 40-cm (12- to 16-in.) floating drifter, which contains sensors for detecting ocean temperature, salinity, and other properties, as well as a transmitter to send its information to satellites overhead. The drifter is tethered to a type of anchor, called a subsurface drogue, which hangs at a depth of 15 m (49 ft).

The drogue keeps the drifter upright on the surface and provides an underwater "sail" for currents to push. Drifters typically operate for about 400 days before they stop transmitting, and the program aims to maintain an array of 1250 drifters at any given time in the world's oceans.

In 2008, Maximenko partnered with Peter Niiler of the Scripps Institution of Oceanography to use data from the Global Drifter program to produce a more detailed map of surface ocean currents. The team combined data on drifter movements with satellite altimetry and wind currents, producing a highly detailed map of both geostrophic and Ekman currents. As the issue of oceanic plastic pollution gained attention, Maximenko saw an opportunity to use his map to predict the areas of the ocean where floating debris is likely to accumulate. To do this, he partnered with his colleague, Jan Hafner, to create a computer model that predicts the movements of drifters over long time periods in the world's oceans. He then ran a simulation in which he uniformly distributed drifters in the oceans and saw where they concentrated over time.

The simulation's results, published in 2012 in *Marine Pollution Bulletin*, revealed that oceanic debris was likely to accumulate in portions of five subtropical gyres including the area of the Great Pacific Garbage Patch. The simulation also revealed that 70% of the drifters remained at sea after 10 years, showing the lengthy life span of floating debris in the oceans. One of the model's predictions, which were first released in 2008

FIGURE 1 A researcher prepares to deploy a buoy for the Global Drifter Program. The cylindrical sea anchor (or drogue) extends vertically in the water column after deployment and is pushed along by shallow ocean currents.

FIGURE 2 Maximenko's computer simulation predicts where plastics and trash accumulate based on oceanic currents. They include the North and South Atlantic Gyres, Indian Ocean Gyre, and the North and South Pacific Gyres. The North Pacific Gyre off the coast of California is home to the "Great Pacific Garbage Patch." *Source: International Pacific Research Center, IPRC Climate, Vol. 8, No. 2, 2008.*

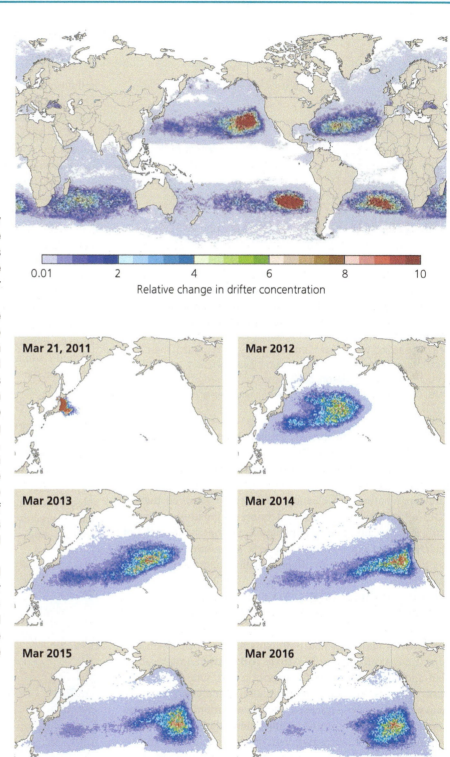

(**FIGURE 2**), was verified when a 2010 study found high concentrations of plastic in the North Atlantic Gyre where Maximenko's model predicted it. The remaining "garbage patches" have all since been verified at their predicted locations.

Maximenko's model was also put to use to predict the fate of the debris washed into the Pacific by the tsunami that struck Japan in 2011 (p. 21). The original debris model (**FIGURE 3**) predicted how ocean currents would push debris across the Pacific. In 2012, the team updated their model to include the effects of winds pushing floating debris at faster rates than heavier debris. The model's predictions closely match observations of debris movements in the Pacific; the model also aided efforts to prepare for the arrival of debris, some of which could harbor marine life that act as invasive species (pp. 86–90) in the United States.

Thanks to the work of Maximenko and his colleagues, we now have a greater understanding of the movements of plastics and other debris in the oceans. Armed with this knowledge, we will be better able to properly assess and, we hope, reduce the impacts of plastics on marine life.

FIGURE 3 Computer models predicted the fate of debris from the 2011 tsunami in northern Japan. After reaching the west coast of North America, debris was expected to remain off the coast of California in the North Pacific Gyre. *Source: Nikolai Maximenko, International Pacific Research Center. Press release, April 5, 2011.*

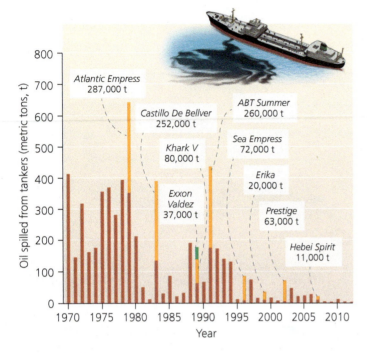

FIGURE 16.16 **Less oil is being spilled into ocean waters today in large tanker spills, thanks in part to regulations imposed on the oil-shipping industry and improved spill response techniques.** The bar chart shows cumulative quantities of oil spilled worldwide from nonmilitary spills of more than 7 metric tons. Larger spills are identified by vessel name, and spill amounts from these events are indicated with orange or green bars. *Data from International Tanker Owners Pollution Federation Ltd.*

in part because of an increased emphasis on spill prevention and response (**FIGURE 16.16**).

To help deal with the problem of catastrophic oil spills, the U.S. Oil Pollution Act of 1990 created a $1 billion oil pollution prevention and cleanup fund. It also required that by 2015 all oil tankers in U.S. waters be equipped with double hulls as a precaution against puncture. Further, in the wake of the *Deepwater Horizon* spill, the U.S. government is evaluating tighter regulations on offshore drilling operations.

Toxic pollutants can contaminate seafood

Aside from the harm that pollutants such as petroleum and plastic can do to marine life, toxic pollutants can make some fish and shellfish unsafe for people to eat. One prime concern today is mercury contamination. Mercury is a toxic heavy metal (p. 364) released into the environment from coal combustion (p. 457), mine tailings, and other sources. After settling onto land and water, mercury bioaccumulates in animals' tissues and biomagnifies as it makes its way up the food chain (p. 368). As a result, fish and shellfish at high trophic levels can contain dangerously elevated levels of mercury. Eating seafood high in mercury is particularly dangerous for young children and for pregnant or nursing mothers, because of the neurological damage it causes.

Since seafood is an important part of a healthy diet, nutritionists do not advocate avoiding seafood entirely. However,

people in at-risk groups should avoid fish high in mercury (such as swordfish, shark, and albacore tuna) while continuing to eat seafood low in mercury (such as catfish, salmon, and canned light tuna).

Excess nutrients cause eutrophication and algal blooms

Pollution from fertilizer runoff or other nutrient inputs can create low-oxygen "dead zones" in coastal ecosystems, as we saw with the Chesapeake Bay (Chapter 5) and the Gulf of Mexico (pp. 420–421). The release of excess nutrients into surface waters can spur unusually rapid growth of phytoplankton, causing eutrophication (p. 109) in freshwater and saltwater systems.

Excessive nutrient concentrations sometimes give rise to population explosions among some species of marine algae. These dinoflagellate algae produce powerful toxins that attack the nervous systems of vertebrates. Blooms of these algae are known as **harmful algal blooms.** Some dinoflagellates produce reddish pigments that discolor surface waters, and blooms of these species are nicknamed **red tides** (**FIGURE 16.17**). Harmful algal blooms can cause illness and death among zooplankton, birds, fish, marine mammals, and people as their toxins are passed up the food chain. They also cause economic loss for communities that rely on beach tourism and fishing. Reducing nutrient runoff into coastal waters can lessen the frequency of these outbreaks.

Climate change is altering ocean chemistry

Elevated levels of carbon dioxide in the atmosphere from fossil fuel combustion can pollute ocean water and change its chemical properties—much in the way excess plant nutrients or toxic substances change the chemical properties of seawater and affect marine organisms. The oceans absorb carbon dioxide (CO_2) from the atmosphere, as we first saw in our study of the carbon

FIGURE 16.17 **Red tides are a type of harmful algal bloom in which the algae produce pigment that turns the water red.**

cycle (see Figure 5.17, p. 120). As our civilization pumps excess carbon dioxide into the atmosphere by burning fossil fuels for energy and removing vegetation from the land, the buildup of atmospheric CO_2 is causing the planet to grow warmer, setting in motion many changes and consequences (Chapter 18).

The oceans have soaked up roughly a third of the excess CO_2 that we've added to the atmosphere, and this has slowed global climate change. However, there are two concerns. The first concern is that the ocean's surface water may soon become saturated with as much CO_2 as it can hold. Once it reaches this limit, then climate change will accelerate as the oceans will no longer remove large amounts of carbon dioxide from the atmosphere.

The second concern is that as ocean water soaks up CO_2, it becomes more acidic. As **ocean acidification** proceeds, sea creatures such as coral, snails, and mussels have difficulty forming shells because the carbonate ions (CO_3^{2-}) they need to create them become less available with increasing acidity, and elevated acidity levels can cause the shells of sea creatures to even begin dissolving.

Chemistry tests in the lab show that coral shells, for example, begin to erode faster than they are built once the carbonate ion concentration falls below 200 micromoles/kg of seawater. Researchers studying coral reefs in the field are finding the same thing: Reefs are growing only in waters with greater than 200 micromoles/kg of carbonate ion availability. A 2007 study used historical data and computer simulations to model how the distribution of coral reefs around the world would vary at differing levels of atmospheric CO_2 and, hence, differing levels of ocean acidity. The results predicted that as atmospheric CO_2 levels rise, coral reefs will shrink in distribution, diversity, and density (**FIGURE 16.18**). By the time atmospheric CO_2 levels pass 500 ppm, little area of ocean will be left with conditions to support coral reefs.

Because of this threat to coral reefs, scientists are intensifying the study of coral responses to warmer and acidified ocean waters to better inform efforts to conserve ecologically important reef habitats. Research published in 2015, for example, revealed that individual coral colonies vary greatly in their responses to heat and acidity, suggesting some reefs may be more resilient than others to changes in water temperature and pH. In another study, scientists found that an endangered Caribbean coral was able to sustain growth under stressful conditions by increasing its feeding rate, indicating that some reefs may be able to better persist under increasingly stressful conditions as long as food is plentiful.

Emptying the Oceans

As severe as the impacts of oceanic pollution on marine organisms can be, most scientists concur that the more worrisome dilemma is overharvesting. Sadly, the old cliché that "there are always more fish in the sea" may not prove to be true. As we have discussed, the oceans today have been overfished, and as with the groundfish of the Northwest Atlantic, many stocks have been severely depleted.

The oceans and their biological resources have met human needs for thousands of years, but today we are placing

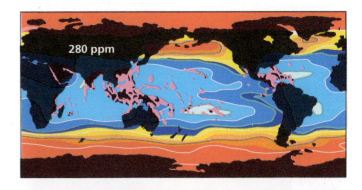

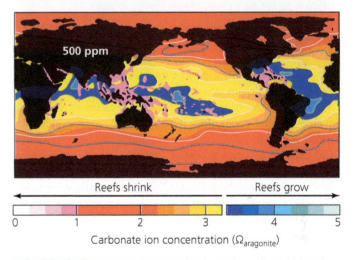

FIGURE 16.18 Increased atmospheric carbon dioxide levels have decreased the number of ocean areas that support coral reefs. Most of Earth's tropical and subtropical oceans were suitable for the growth of coral reefs (blue colors; **top panel**) before people began emitting carbon dioxide into the atmosphere. In 2007, when concentrations were 380 ppm atmospheric CO_2 (**middle panel**), fewer regions are suitable. When the planet reaches 500 ppm (**bottom panel**), very few areas of the ocean will have conditions suitable for coral reefs. *Adapted from Hoegh-Guldberg, O., et al., 2007. Coral reefs under rapid climate change and ocean acidification. Science 318: 1737–1742. Fig 4. Reprinted with permission from AAAS.*

unprecedented pressure on marine resources. More than half the world's marine fish populations are fully exploited, meaning that we cannot harvest them more intensively without depleting them, according to the United Nations Food and Agriculture Organization (FAO). An additional 28% of marine fish populations are overexploited and already being driven toward extinction. Only one-fifth of the world's marine

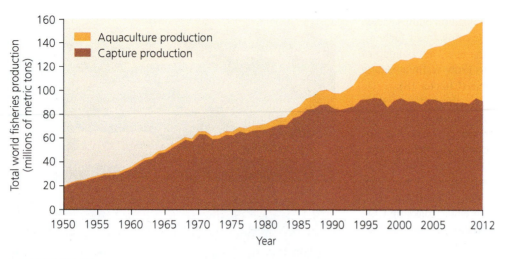

FIGURE 16.19 After rising for decades, world fisheries capture production has stalled for the past 20 years. Many scientists fear that a global decline is imminent if conservation measures are not taken. Note how the growth in aquaculture has enabled seafood production to increase, despite flat hauls from capture fisheries in recent decades. *Data from the Food and Agriculture Organization of the United Nations, 2014. The state of world fisheries and aquaculture 2014. Fig. 1.*

DATA Q Approximately what percentage of the total global production of fisheries in 2012 came from aquaculture? How does this compare to 1980? Offer an explanation for the difference in aquaculture production in these two years.

Go to **Interpreting Graphs & Data** on **Mastering**EnvironmentalScience°.

fish populations can yield more than they are already yielding without being driven into decline.

Total global fisheries catch, after decades of increases, leveled off after about 1988 (**FIGURE 16.19**), despite increased fishing effort. In 2008, the FAO concluded that "the maximum wild capture fisheries potential from the world's oceans has probably been reached." Fishery collapses such as those off Newfoundland and New England are ecologically devastating and take a severe economic toll on human communities that depend on fishing. A comprehensive 2006 study in the journal *Science* predicted that if current trends continue, populations of *all* ocean species that we fish for today will collapse by the year 2048. If fisheries collapse as predicted, we will lose not only an important food source for people but also the ecosystem services healthy fisheries provide. Productivity will be reduced, ecosystems will become more sensitive to disturbance, and the filtering of water by vegetation and organisms (such as oysters) will decline, making harmful algal blooms, dead zones, fish kills, and beach closures more common. Aquaculture (raising fish in tanks or pens; see Figure 10.11) is booming and is helping to relieve pressure on wild stocks, but fish farming comes with its own set of environmental dilemmas (pp. 242–244). To prevent such a collapse of the world's fisheries, it is vital that we turn immediately to more sustainable fishing practices.

Industrialized fishing facilitates overharvesting

It is tempting to blame Earth's problems exclusively on modern civilization. But scientists are learning that people began depleting some marine populations centuries or millennia ago. Steller's sea cow, a relative of the manatee, which was once abundant in the Pacific Ocean near Alaska, was hunted to extinction in 1768 by Native Americans and European whalers for its meat, oil, and blubber. These depletions of marine populations tended to be localized, however, as people long ago lacked the technology to intensively harvest organisms and exploit fisheries, especially those in the open ocean.

Mostly sustainable harvesting started changing in the 19th century, when many species of whales were nearly hunted to extinction as part of the commercial whaling industry. Even the cod fishery in the North Atlantic was periodically overharvested in some areas during this century, as fishers introduced new approaches and technology that increased fish catches. Continued technological advances in the 20th century, however, took commercial fishing to what were previously unimaginable levels, and many fisheries, such as the cod fishery, finally buckled and broke under the intense pressure.

Today's industrialized commercial fishing fleets employ massive ships and powerful new technologies to find and capture fish in great volumes. Some vessels even process and freeze their catches while at sea. And because no portion of the ocean is unreachable by modern fleets, our impacts are much more rapid and intensive than in the past.

The modern fishing industry uses several methods to capture fish at sea that are highly efficient but also environmentally damaging:

- In *purse seining*, vessels deploy large nets, some as long as a kilometer (0.6 mi), around schools of fish near the surface (**FIGURE 16.20a**). The nets are suspended in the upper water column by floating buoys on the net top and weights on the net bottom. Once the school of fish is encircled, the net is drawn shut like a laundry bag (or the hood on a sweatshirt) by the purse line, a rope running through the top of the net.

- Some vessels set out long driftnets that span large expanses of water (**FIGURE 16.20b**). These chains of transparent nylon mesh nets are arrayed to drift with currents to capture passing fish, and they are held vertical by floats at the top and weights at the bottom. *Driftnetting* usually targets species that traverse open water in immense schools, such as herring, sardines, and mackerel. Specialized forms of driftnetting are used for sharks, shrimp, and other animals.

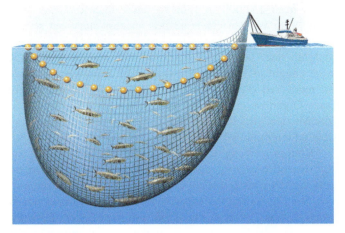

(a) Purse seining

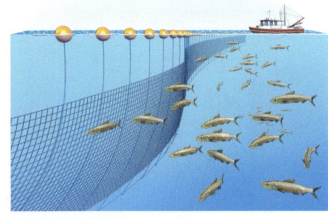

(b) Driftnetting

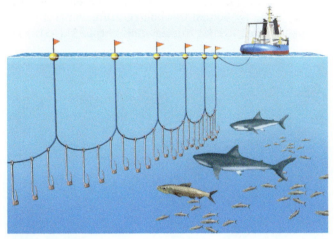

(c) Longlining

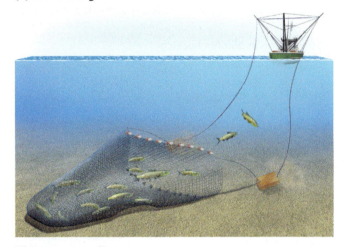

(d) Bottom-trawling

FIGURE 16.20 Commercial fishing fleets use several methods of capture. The illustrations above are schematic for clarity and do not portray the immense scale that these technologies can attain; for instance, industrial trawling nets can be large enough to engulf multiple jetliners.

- *Longline fishing* (**FIGURE 16.20c**) involves setting out extremely long lines (up to 80 km [50 mi] long) with up to several thousand baited hooks spaced along their lengths. Tuna and swordfish are among the species targeted by longline fishing.

- *Trawling* entails dragging immense cone-shaped nets through the water, with weights at the bottom and floats at the top. Trawling in open water captures pelagic fish, whereas *bottom-trawling* (**FIGURE 16.20d**) involves dragging weighted nets across the floor of the continental shelf to catch groundfish and other benthic organisms, such as scallops.

Industrialized fishing kills nontarget animals and damages ecosystems

Modern fishing practices catch more than just the species they target. *Bycatch*, the accidental capture of animals, accounts for the deaths of millions of fish, sharks, marine mammals, and birds each year. The impact of bycatch can be substantial.

A 2011 report from NOAA reported that 17% of all commercially harvested fish were captured unintentionally.

Purse seining and driftnetting capture dolphins, seals, and sea turtles, as well as countless nontargeted fish. Most of these creatures end up drowning (mammals and turtles need to surface to breathe) or dying from air exposure on deck (fish suffocate when kept out of the water). Driftnetting is now banned in international waters because of excessive bycatch, but the practice continues in many national waters.

Similar bycatch problems exist with longline fishing, which kills turtles; sharks; and albatrosses, magnificent seabirds with wingspans up to 3.6 m (12 ft). Several methods are being developed to limit bycatch from longline fishing (such as using flagging to scare birds away from the lines), but an estimated 300,000 seabirds of various species die each year when they become caught on hooks while trying to ingest bait.

The scale of bycatch and solutions to address it are illustrated by the story of dolphins and tuna. Several species of dolphins that associate with yellowfin tuna become caught in purse seines set for the tuna in the tropical Pacific. In purse seining, boats surround schools of tuna with a net and draw

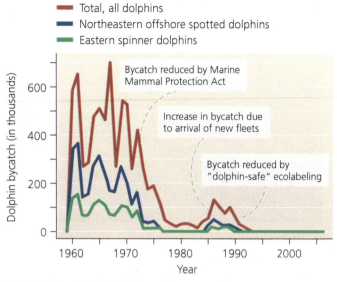

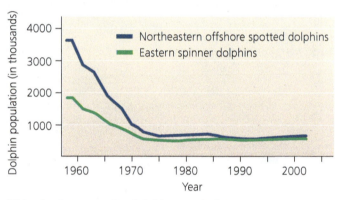

(a) Reduction in dolphin bycatch

(b) Lack of recovery by dolphin populations

FIGURE 16.21 Bycatch has severely decreased dolphin populations. Bycatch of dolphins by tuna fleets **(a)** declined first as a result of regulations following the 1972 U.S. Marine Mammal Protection Act and later when "dolphin-safe" ecolabeling encouraged fleets to adopt methods that reduced bycatch. However, despite this success, populations of the two dolphin species most affected **(b)** have not rebounded, possibly because there are fewer fish for them to eat. *Data from (a) National Oceanic and Atmospheric Administration, www.noaa.gov, and (b) Wade, P.R., et al., 2007. Depletion of spotted and spinner dolphins in the eastern tropical Pacific: Modeling hypotheses for their lack of recovery. Mar. Ecol. Prog. Ser. 343: 1–14.*

the net in, trapping both tuna and dolphins. Hundreds of thousands of dolphins were being needlessly killed each year throughout the 1960s. The U.S. Marine Mammal Protection Act of 1972 forced U.S. fleets to modify their gear and fishing practices to allow dolphins to escape from fishing nets. Dolphin bycatch dropped greatly as a result (**FIGURE 16.21a**).

However, as other nations' ships began catching tuna, dolphin bycatch rose again. Because U.S. fleets were operating under more restrictions, the U.S. government required that tuna imported from foreign fleets also minimize dolphin bycatch, and it supported ecolabeling efforts (p. 152) to label tuna as "dolphin-safe" if its capture used methods designed to avoid bycatch. These measures helped reduce dolphin

deaths from 133,000 in 1986 to fewer than 2000 per year since 1998. We can celebrate this success story but should also recognize that there is little accountability for the many "dolphin-safe" labels, that sharks and other animals continue to be caught as bycatch, and that dolphin populations have not recovered (**FIGURE 16.21b**). Further, in 2015 the World Trade Organization dealt such labeling a blow when it sided with Mexico, which argued that U.S. rules regarding labeling violated international trade laws.

Bottom-trawling not only results in bycatch but also can destroy entire ecosystems. The weighted nets crush organisms in their path and leave long swaths of damaged sea bottom. Bottom-trawling is especially destructive to structurally complex areas, such as reefs, that provide shelter and habitat for animals. In recent years, underwater photography has begun to reveal the extent of structural and ecological disturbance done by bottom-trawling (**FIGURE 16.22**). Bottom-trawling is often likened to clear-cutting (pp. 311–312) and

(a) Before trawling at Georges Bank

(b) After trawling at Georges Bank

FIGURE 16.22 Bottom-trawling causes severe structural damage to reefs and benthic habitats, and it can decimate underwater communities and ecosystems. A photo of an untrawled location **(a)** on the seafloor of Georges Bank shows a vibrant and diverse benthic community. A photo of the same site after trawling **(b)** shows a flattened expanse of sea bottom with only scarce biological diversity and productivity.

strip mining (p. 635). In heavily fished areas, the bottom may be damaged multiple times. At Georges Bank, it is estimated that the average expanse of ocean floor has been trawled three times. Bottom-trawling here is known to destroy young cod as bycatch, and this is thought to be a main reason why the Georges Bank cod stock is not recovering. Reducing bycatch is also an important part of restoring Grand Banks populations; bycatch of cod while fishing for other species in the Grand Banks rose from 600 metric tons in 2006 to 1100 metric tons in 2009, slowing the recovery of cod populations.

Fisheries collapse quickly under intensive harvest

Throughout the world's oceans, today's industrialized fishing fleets are depleting marine populations with astonishing speed. In a 2003 study, Canadian fisheries biologists Ransom Myers and Boris Worm analyzed data from FAO archives and found the same pattern for region after region: In just a decade after the arrival of industrialized fishing, catch rates dropped precipitously, with 90% of large-bodied fish and sharks eliminated. Populations then stabilized at 10% of their former levels. This means, Myers and Worm concluded, that the oceans

today contain only one-tenth of the large-bodied animals they once did. We can see the effects of large-scale industrialized fishing in the catch records of groundfish from the Northwest Atlantic for both the Grand Banks (**FIGURE 16.23a**) and Georges Bank cod fisheries (**FIGURE 16.23b**).

As we have seen (pp. 80–82), when animals at high trophic levels are removed from a food web, the proliferation of their prey can alter the nature of the entire community. Many scientists conclude that most marine communities may have been very different prior to industrial fishing.

Several factors mask the collapse of large fisheries

Although industrialized fishing has depleted fish stocks in region after region, the overall global catch has remained roughly stable for two decades (see Figure 16.19). This can be explained by several factors. Fishing fleets are now traveling longer distances to reach less-fished portions of the ocean. They also are fishing in deeper waters; average depth of catches was 150 m (495 ft) in 1970 and 250 m (820 ft) in 2000. Moreover, fleets are spending more time fishing and are setting out more nets and lines—expending increasing effort just to catch

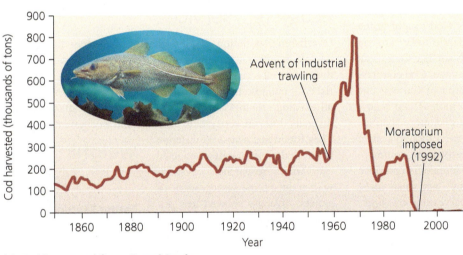

(a) Cod harvested from Grand Banks

(b) Cod harvested from Georges Bank

FIGURE 16.23 In the North Atlantic off the coast of Newfoundland, commercial catches of Atlantic cod have fluctuated over time. Catches increased with intensified fishing by industrial trawlers in the 1960s and 1970s **(a)**, but the fishery subsequently crashed, and moratoria imposed in 1992 and 2003 have not brought it back. A similar pattern is seen in the cod catches at Georges Bank **(b)**; industrial fishing produced 30 years of high catches, followed by a collapse and the closure of some areas to fishing. Note also that in each case, there is one peak before 1977 and one after 1977. The first peak and decline resulted from foreign fishing fleets, whereas the second peak and decline resulted from Canadian and U.S. fleets, respectively, after they laid claim to their 200-mile exclusive economic zones. *Data from (a) Millennium Ecosystem Assessment, 2005. Ecosystems and human well-being: biodiversity synthesis. Washington, D.C.: World Resources Institute. Used with permission; and (b) O'Brien, L., et al., 2008. Georges Bank Atlantic cod. An assessment of 19 Northeast groundfish stocks through 2007. Woods Hole, MA: Northeast Fisheries Science Center.*

the same number of fish. For example, a 2010 study showed that British trawlers were working 17 times harder just to catch the same number of cod and other fish as 120 years ago.

More powerful technology also helps explain large catches despite declining stocks. Today's Japanese, European, Canadian, and U.S. fleets can reach almost any spot on the globe with vessels that attain speeds of 80 kph (50 mph). They boast an array of military technologies developed for locating enemy submarines, including advanced sonar mapping equipment, satellite navigation, and thermal sensing systems. Some fleets rely on airplanes to find schools of commercially valuable fish, such as bluefin tuna. Technology cannot continue indefinitely to increase catches and mask the collapse of fisheries, however, because at some point simply not enough fish will remain to be harvested, regardless of the technology used to find and extract them.

We are "fishing down the food chain"

Numbers of fish do not tell the whole story of fisheries depletion. Analyses of fisheries data reveal, in case after case, that as fishing increases, the fish that are caught decline in size and age. This is not only because fishers prefer to take large fish but also because under intense fishing pressure, few fish escape being caught for very many years, so that few have a chance to grow to a large size. Cod caught in the Northwest Atlantic today are, on average, much smaller than those caught decades ago. Cod continue to grow as they age and can live for 20 years or more. But it is now rare to find a cod more than 10 years of age, even though cod of this age formerly were common. Because large female fish produce far more young than small ones, the intense harvesting of larger fish makes it harder for populations to recover once they have been depleted.

FIGURE 16.24 When stocks of desirable fish crash, fishers switch to less desirable species. The once little-desired slimehead (*Hoplostethus atlanticus*) is now captured extensively and marketed as the far more appealing "orange roughy."

called "orange roughy" was called "slimehead" by fishermen because of its unique mucus canals (**FIGURE 16.24**). Similarly, the toad-colored "toothfish" that fishermen once threw overboard has found new life as "Chilean sea bass," even though the species is not biologically classified as a sea bass. With intensive harvest, populations of orange roughy and Chilean sea bass are now overfished in many areas, forcing harvesters to continue to "fish *further* down the food chain."

Loss of marine biodiversity erodes ecosystem services

Overfishing, pollution, habitat change, and other factors that deplete biodiversity can threaten the ecosystem services we derive from the oceans. In the 2006 study that predicted a global fisheries collapse by 2048 (p. 434), the study's authors analyzed all existing scientific literature to summarize the effects of biodiversity loss on ecosystem function and ecosystem services. They found that across 32 different controlled experiments conducted by various researchers, systems with reduced species diversity or genetic diversity showed less primary and secondary production and were less able to withstand disturbance.

The team also found that when biodiversity was reduced, habitats that serve as nurseries for fish and shellfish were also affected, threatening future populations of these species. Moreover, when biodiversity was lower, so was the amount of water purification by filter-feeding organisms (such as oysters), which clean waters of excess nutrients, sediments, and phytoplankton that can lead to eutrophication, harmful algal blooms, dead zones, fish kills, and beach closures.

Marine Conservation

Because we bear responsibility and stand to lose a great deal if valuable marine ecological systems collapse, scientists have been working to develop solutions to the problems that

In addition, as particular species become too rare to fish profitably, fleets begin targeting other species that are more abundant. Generally this means shifting from large, desirable species to smaller, less desirable ones. Time and again, fleets have depleted popular food fish (such as cod) and shifted to species of lower value (such as capelin, a smaller fish eaten by cod). Because this often entails catching species at lower trophic levels, this phenomenon has been termed "fishing down the food chain."

Some of these species were undesirable ones that fishermen formerly threw back when fishing for more marketable species but that underwent "image makeovers" to aid their sale to consumers. For example, the species of fish now

threaten the oceans. Many have begun by reconsidering the strategies used traditionally in fisheries management.

Fisheries management has been based on maximum sustainable yield

Fisheries managers conduct surveys, study fish population biology, and monitor catches. They then use that knowledge to regulate the timing of harvests, the scale of harvests, and the techniques used to catch fish. The goal is to allow for maximal harvests while keeping fish available for the future—the concept of maximum sustainable yield (p. 309). Recall that this means keeping populations at about half the level they would otherwise achieve, because populations growing by logistic growth grow fastest at this size (pp. 65–66). If data indicate that current yields are unsustainable, managers might limit the number or biomass of fish that can be harvested, restrict the type of gear fishers can use, or completely close fisheries.

The science of estimating population sizes of species is imprecise and can lead to significant changes in maximum sustained yield. For example, the cod population in the Gulf of Maine was sampled in 2008 and its biomass estimated at 34 million kg (74.9 million lb). Allowable harvests by fishermen were set based on this survey. A subsequent survey three years later, however, concluded that the cod biomass in the Gulf was only 12 million kg (24.6 million lb). This nearly one-third reduction in cod biomass was not explainable by fish harvests between 2008 and 2011, so resource managers concluded that the 2008 survey overestimated cod populations by putting too much weight on a few large trawls, which were not indicative of true cod numbers. In light of the new assessment of cod biomass, fishing quotas were immediately lowered. This forced fishermen in the area to quickly adapt to new, smaller levels of allowable catch, which was financially challenging for many small fishing operations. Similar events unfolded in the Georges Bank at the same time.

Despite efforts to restrict fish harvests to sustainable levels, many fish and shellfish stocks around the world have crashed. Thus, many scientists and resource managers feel it is time to shift the focus away from individual species and toward viewing marine resources as elements of larger ecological systems. This perspective means considering the impacts of fishing practices on habitat quality, species interactions, and other factors that may have indirect or long-term effects on populations. One key aspect of such ecosystem-based management (p. 309) is to set aside areas of ocean where systems can function without human interference.

We can protect unique and biodiverse areas in the ocean

Hundreds of **marine protected areas (MPAs)** have been established, most of them along the coastlines of developed countries. Much like our national parks, these areas restrict some human activities within their boundaries (such as oil drilling), but nearly all MPAs allow fishing or other extractive activities. As one report from the National Resources Defense Council put it, MPAs "are dredged, trawled, mowed for kelp, crisscrossed with oil pipelines and fiber-optic cables, and swept through with fishing nets." Currently, slightly less than 3% of the world's oceans are designated as MPAs (**FIGURE 16.25**).

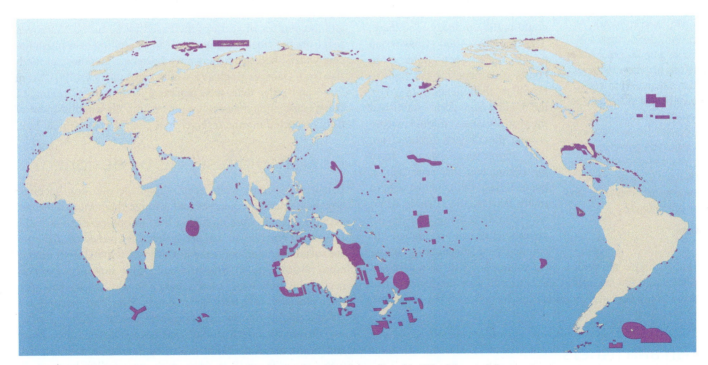

FIGURE 16.25 Marine protected areas are found across the globe. Roughly 3% of the world's oceans are afforded some sort of protection (highlighted in purple), but critics contend that for many, these protections do little to prevent biodiversity loss, habitat destruction, and ecosystem degradation. *Source: IUCN and UNEP-WCMC (October 2013). The World Database on Protected Areas (WDPA); available at: www.protectedplanet.net.*

Because of the lack of true refuges from fishing pressure, many scientists want to establish areas where fishing is prohibited. Such no-take areas have come to be called **marine reserves.** Designed to preserve ecosystems intact, marine reserves are also intended to improve fisheries. Scientists argue that marine reserves can act as production factories for fish for surrounding areas, because fish larvae produced inside reserves will disperse outside and stock other parts of the ocean. By serving both purposes, proponents maintain, marine reserves are a win-win proposition for conservationists and fishers alike.

Perhaps not surprisingly, many commercial and recreational fishers dislike the idea of no-take reserves, just as most were opposed to the Canadian groundfish moratoria and the Georges Bank closures. Nearly every marine reserve that has been established or proposed has met with opposition from people and businesses who use the area for fishing or recreation.

Clearly, when marine reserves are established, it pays to be sensitive to the concerns of people of the area. In 2006, President George W. Bush established the Papahanaumokuakea Marine National Monument around the northwestern Hawaiian Islands—at 362,000 km² (140,000 mi²), an area larger than all U.S. national parks combined. Fishing was made off-limits here except for native Hawaiians, who were given fishing rights.

weighing the
ISSUES

Preservation at Sea

Despite the many benefits of coastal and oceanic ecosystems, very little of the world's ocean is protected, even in limited capacity. Why do you think it is taking so long for the preservation ethic we use commonly in terrestrial habitats to make the leap to the oceans? What challenges do you see with preserving areas of coastal ocean? Do you think local people should be given fishing rights in marine reserves? What could be done to better protect marine ecosystems, fisheries, and local fishing cultures?

In contrast, consider the Chagos Archipelago, a group of 55 islands in the Indian Ocean controlled by the United Kingdom. The indigenous people of these islands were forcibly evicted from the main island, Diego Garcia, between 1967 and 1973 so that the United States could build a military base there. The roughly 4000 Chagossian people have been fighting in courts ever since for the right to return to their islands. In 2010, the United Kingdom established the world's largest marine reserve around the Chagos Archipelago, and indigenous people from the region fear that with fishing banned, they will never be able to return because there would be no way to make a living. The British government has attempted to counter these concerns by noting that if the people return, the fishing bans will be reassessed and that in the meantime, the bans will save the region's fish from being depleted by industrial fleets from other countries.

Reserves can work for both fish and fishers

Data from marine reserves around the world have indicated that reserves can work as win-win solutions that benefit ecosystems, fish populations, and fishing economies. A comprehensive review of data from marine reserves in 2001 revealed that just one to two years after their establishment, marine reserves, on average, increased species diversity by 23%; the density of organisms within the reserve by 91%; the total biomass of organisms by 192%; and the average size of each organism by 31%.

That year, 161 prominent marine scientists signed a "consensus statement" summarizing the effects of marine reserves. Besides boosting fish biomass, total catch, and record-sized fish, the report stated, marine reserves yield several benefits. Within reserve boundaries, for example, mortality and habitat destruction are reduced; the chance of biodiversity loss is lessened; and the reserve produces rapid and long-term increases in the abundance, diversity, and productivity of marine organisms. Outside the reserve, benefits are seen when individuals of protected species spread outside reserves into nearby waters and when larvae of species protected within reserves "seed the seas" outside reserves.

The consensus statement was backed up by research from reserves around the world. At Apo Island in the Philippines, biomass of large predators increased eightfold inside a marine reserve, and fishing improved outside the reserve. At two coral reef sites in Kenya, commercially fished species and keystone species were up to 10 times more abundant in the protected area as in the fished area. At Leigh Marine Reserve in New Zealand, snapper increased 40-fold, and spiny lobsters increased by 5–11% yearly. Spillover from this reserve improved fishing and ecotourism, and local residents who once opposed the reserve came to support it. On Georges Bank, once commercial trawling was halted in 1994, populations of many organisms began to recover. As benthic invertebrates began to come back, numbers of groundfish such as haddock and yellowtail flounder rose inside the closed areas, and the number of scallops increased by 14 times. Moreover, fish from the closure areas appear to be spilling over into adjacent waters, because fishers have been catching more groundfish from Georges Bank as a whole since the late 1990s. From these and other data sets, increasing numbers of scientists, fishers, and policymakers are advocating the establishment of fully protected marine reserves as a central management tool.

How should reserves be designed?

If marine reserves work in principle, the question becomes how best to design reserves and arrange them into connected networks. Studies are modeling how to optimize the size and spacing of reserves so that ecosystems are protected, fisheries are sustained, and people are not overly excluded from marine areas (**FIGURE 16.26**). Scientists are asking how large reserves need to be, how many there need to be, and where they need to be placed to take best advantage of ocean currents and levels of available nutrients that promote productivity.

Of several dozen studies that have estimated how much area of the ocean *should be* protected in no-take reserves, estimates range from 10 to 65%, with most falling between 20 and 50%. But prohibiting fishing in such large expanses of the ocean would affect people who make a living by fishing and economically damage nations that sell fish on the international

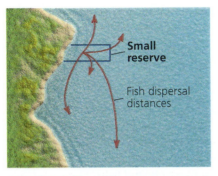
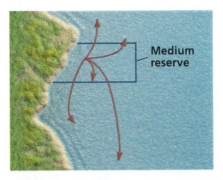
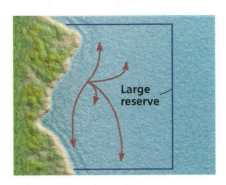

FIGURE 16.26 Marine reserves of different sizes may have varying effects on ecological communities and fisheries. Young and adult fish and shellfish of different species can disperse different distances, as indicated by the red arrows in the figure. A small reserve (**left panel**) may fail to protect animals because too many disperse out of the reserve. A large reserve (**right panel**) may protect fish and shellfish very well but will provide relatively less "spillover" into areas where people can legally fish. Thus medium-sized reserves (**middle panel**) may offer the best hope of preserving species and ecological communities while also providing adequate fish to people. *Adapted from Halpern, B.S., and R.R. Warner, 2003. Matching marine reserve design to reserve objectives. Proceedings of the Royal Society of London B 270: 1871–1878, Fig. 1. Used by permission of The Royal Society and the author.*

market. It would also be very difficult to enforce such protections in international waters without the cooperation of the fishers. Hence, most scientists feel that involving fishers directly in the planning process for designating protected areas is crucial for coming up with answers to all these questions. If marine reserves can be made to work and to be accepted, then they may well seed the seas and help lead us toward solutions to one of our most pressing environmental problems.

closing
THE LOOP

The story of saving the cod fisheries in the Grand Banks and Georges Bank illustrates the complex challenges faced in conserving marine environments and aquatic biodiversity when an economically important species, such as cod, is at stake. But as we have seen, the cod fishery off the coast of Newfoundland and New England is not the only one in jeopardy of collapsing: Overfishing is decimating stocks of numerous species across the globe.

Overharvesting is not the only danger faced by marine environments, however, as pollution, ocean acidification, and destructive fishing practices degrade coastal and open-ocean ecosystems. Because many of these impacts come from non-point sources, minimizing their inputs into the

world's oceans has proven problematic. Climate change is altering the temperature and chemistry of the world's oceans, endangering ecologically important habitats such as coral reefs and the biodiversity those reefs support.

The good news is that we have solutions that can save these critical ecosystems; we just need the collective will to see them implemented and maintained. Closing fisheries to harvest or reducing allowable catch levels, as was done for the cod fishery of North America, is one approach that can aid recovery, as can the establishment of marine reserves, where ecosystems can function with minimal impact from human activities. The challenges we face in saving our oceans, and the species that occupy them, is great—but it's a battle worth fighting to protect 71% of our planet.

REVIEWING Objectives

You should now be able to:

Identify physical, geographic, and chemical aspects of the marine environment

The five major oceans, with their complex underwater topography, contain 97% of Earth's water and cover nearly three-fourths of its surface. Ocean waters move horizontally in currents, and vertical currents transport water, heat, and dissolved substances from surface waters to the deep ocean, and vice versa. (pp. 416–422)

Explain how the oceans influence and are influenced by climate

The thermohaline circulation redistributes heat around the world and shapes regional climate. El Niño and La Niña events occur periodically and alter climate. It is hypothesized that global warming could alter existing circulation patterns, which in turn would affect Earth's climate. (pp. 422–423)

Describe major types of marine ecosystems

Marine and coastal ecosystems include intertidal zones, salt marshes, mangrove forests, estuaries, kelp forests, coral reefs, and pelagic and deep-water open-ocean systems. Many of these systems are highly productive, are rich in biodiversity, and can suffer heavy impacts from human activities. (pp. 423–427)

Assess impacts from marine pollution

Plastic debris harms marine life and can accumulate in ocean regions where currents converge. Large oil spills from offshore oil rigs or oil tankers can profoundly affect marine ecosystems. Toxic levels of mercury can bioaccumulate in some large fish species. Nutrient pollution can lead to dead zones and harmful algal blooms in the ocean. Absorption of excess carbon dioxide from the atmosphere by the oceans leads to the acidification of waters, which hinders corals, oysters, and other creatures that create carbonate shells. (pp. 427–433)

Review the state of ocean fisheries and identify reasons for their decline

Three-fourths of the world's marine fish populations are exploited or overexploited, and the global fish catch has stagnated since the late 1980s despite increased fishing effort and improved technologies. Industrial fishing approaches have overexploited fisheries and resulted in high rates of bycatch and degradation of marine habitats. Traditional fisheries management has not stopped declines, so many scientists feel that ecosystem-based management is needed. (pp. 433–438)

Evaluate marine protected areas and reserves as solutions for conserving biodiversity

We have established fewer protected areas in the oceans than we have on land. Most marine protected areas allow some level of fishing to occur but no-take marine reserves that prohibit fish harvests can protect ecosystems, boost fish populations, and make fisheries sustainable. (pp. 438–441)

TESTING Your Comprehension

1. What proportion of Earth's surface do oceans cover? What is the average salinity of ocean water? How are density, salinity, and temperature related in each layer of ocean water?

2. What factors drive ocean currents? Give an example of how a surface current affects climate in a particular region.

3. Where in the oceans are productive areas of biological activity likely to be found?

4. Describe three kinds of ecosystems found near coastal areas and the types of life they support.

5. Why are coral reefs biologically valuable? How are they being degraded by human impact? What is causing the disappearance of mangrove forests and salt marshes?

6. Discuss three ways plastics affect marine life.

7. What is causing ocean acidification? What consequences do scientists expect ocean acidification to bring about?

8. Describe an example of how overfishing can lead to ecological damage and fishery collapse.

9. Name three industrial fishing practices, and explain how they result in bycatch and marine habitat degradation.

10. How does a marine reserve differ from a marine protected area? Why do many fishers oppose marine reserves? Explain why many scientists say no-take reserves will be good for fishers.

SEEKING Solutions

1. What benefits do you derive from the oceans? How do your choices affect the oceans? Give specific examples.

2. We have been able to reduce the amount of oil we spill into the oceans, but other petroleum-based products such as plastic continue to litter our oceans and shorelines. Discuss some ways that we can reduce this impact on the marine environment.

3. Describe the trends in global fish capture from 1950 to 1990 and from 1990 to 2012, and explain several factors that account for these trends.

4. **CASE STUDY CONNECTION** You make your living fishing for cod in the Georges Bank, just as your parents and grandparents did, and as most of your neighbors do in your small coastal village. After years of regulated fish harvests failed to recover the fishery, the government has decided to completely close the fishery, and scientists are pushing for a permanent marine reserve to be established on your former fishing grounds. You have no desire to move away from your village, so what steps will you take now? Will you protest the closure? What compensation will you ask of the government if it prevents you from fishing? Will you work with scientists to establish a reserve that improves fishing in the future, or will you oppose their attempts to create a reserve? What data and what assurances will you ask of them?

5. **THINK IT THROUGH** You are mayor of a coastal town where some residents are employed as commercial fishers and others make a living serving ecotourists who come to snorkel and scuba dive at the nearby coral reef. In recent years, several fish stocks have crashed, and ecotourism is dropping off as fish disappear from the increasingly degraded reef. Scientists are urging you to help establish a marine reserve around portions of the reef, but most commercial and recreational fishers are opposed to this idea. What steps would you take to restore your community's economy and environment?

6. **THINK IT THROUGH** You operate an aquaculture (p. 242) operation that raises oysters for sale to seafood restaurants. Ocean acidification poses a major threat to your business by robbing oysters of the ions they need to grow shells, potentially leading to slower growth rates and higher mortality in your oysters. You approach your U.S. senator about the issue when she makes a campaign stop in your area to see if the federal government can help. How would you lobby your senator to aid your business? Would you request financial help from the government or advocate aggressive action to combat climate change? Or both? What arguments would you make to convince the government to support aquaculture?

CALCULATING Ecological Footprints

The relationship between the ecological goods and services that individuals use and the amount of land area needed to provide those goods and services is relatively well developed. People also use goods and services from Earth's oceans, where the concept of area is less useful. It is clear, however, that our removal of fish from the oceans has an impact, or an ecological footprint.

The table shows data on the mean annual per-person consumption from ocean fisheries for North America, China, and the world as a whole. Using this consumption data, calculate the amount of fish each consumer group would consume per year, given the annual per capita consumption rates, for each of these three regions. Record your results in the table.

Annual Consumption

CONSUMER GROUP	NORTH AMERICA (21.8 KG PER PERSON)	CHINA (35.1 KG PER PERSON)	WORLD (18.9 KG PER PERSON)
You			
Your class			
Your state			
United States			
World			

Data: Food and Agriculture Organization of the United Nations, 2014. *The state of world fisheries and aquaculture 2014.*

1. Calculate the ratio of North America's per capita fish consumption rate to that of the world. Compare this ratio to the ratio of the per capita ecological footprints for the United States, Canada, and Mexico (see Figure 1.12, p. 15) versus the world average footprint of 2.7 hectares (ha)/person/year. Can you account for similarities and differences between these ratios?

2. The population of China has grown at an annual rate of 1.1% since 1987, while over the same period fish consumption in China has grown at an annual rate of 8.9%. Speculate on the reasons behind China's rapidly increasing consumption of fish.

3. What ecological concerns do the combined trends of human population growth and increasing per capita fish consumption raise for you? What role might you play in contributing to these concerns or to their solutions?

MasteringEnvironmentalScience®

CHAPTER 17

The Atmosphere, Air Quality, and Pollution Control

Clearing the Air in L.A. and Mexico City

> **I left L.A. in 1970, and one of the reasons I left was the horrible smog. And then they cleaned it up. That was one of the greatest things the government has ever done for me. You have beautiful days now. It's a much, much nicer place to live.**
>
> Actor and comedian Steve Martin, speaking to *Los Angeles Magazine*

> **This city can be a model for others.**
>
> Mexico City Mayor Miguel Ángel Mancera

For Americans, Los Angeles has long symbolized air pollution. Smog—that unhealthy mix of air pollutants resulting from fossil fuel combustion—has blanketed the city for decades. Exhaust from millions of automobiles clogging L.A.'s freeways regularly becomes trapped by the mountains that surround the city, and the region's warm sunshine turns the pollutants to smog. In response, Los Angeles has worked hard to fight air pollution and has succeeded in improving its air quality notably, thanks to policy efforts and new technologies. Today L.A. still suffers the nation's worst smog, but its skies are clearer than in decades.

One city that looked to Los Angeles's efforts as it planned its own responses to smog is Mexico City, the capital of Mexico and one of the world's largest metropolises. Not long ago, Mexico City suffered the most polluted air in the world. On days of poor air quality throughout the 1990s, residents wore face masks on the streets, teachers kept students inside at recess, and outdoor sports events were canceled. Children drawing pictures would use brown crayons to color the sky. Each year thousands of deaths and tens of thousands of hospital visits were blamed on pollution. Mexican novelist Carlos Fuentes called his capital "Makesicko City."

As in Los Angeles, traffic generates most of the pollution in Mexico City, where motorists in nearly 7 million cars sputter across miles of urban sprawl. And like L.A., Mexico City lies in a valley surrounded by mountains, vulnerable to temperature inversions that trap pollutants over the city. Moreover, at Mexico City's high altitude—2240 m (7350 ft) above sea level—solar radiation is intense, which worsens smog formed by the interaction of pollutants with sunlight. Mexico City environmental chemist Armando Retama likens his hometown to "a casserole dish with a lid on top."

Despite the challenges facing the world's sixth-largest metropolis, Mexico City's 21 million people fought back and made notable improvements in air quality. A succession of mayors took bold action to clean up the air, and in recent years Mexico City has been enjoying a renaissance. As the smog began to clear, revealing beautiful views of the snow-capped peaks that ring the valley, Mexico City became an international model for other cities seeking to fight pollution.

Efforts began in the 1990s, when city leaders shut down an oil refinery in the city and pushed factories

Upon completing this chapter, you will be able to:

- Describe the composition, structure, and function of Earth's atmosphere

- Relate weather and climate to atmospheric conditions

- Identify major outdoor air pollutants and outline the scope of air pollution

- Assess strategies and solutions for control of outdoor air pollution

- Explain stratospheric ozone depletion and identify steps taken to address it

- Describe acid deposition, discuss its consequences, and explain how we are addressing it

- Characterize the scope of indoor air pollution and assess solutions

Testing emissions at a vehicle inspection station in Los Angeles

Mexico City on a smoggy day

and power plants to shift to cleaner-burning natural gas. Policymakers reduced sources of key pollutants by mandating that lead be removed from gasoline, that the sulfur content of diesel fuel be lowered, and that pollution control technologies such as catalytic converters be phased in for new vehicles.

Levels of many pollutants fell, but as the city's population continued to grow, smog from automobile traffic persisted. In response, city officials stepped up vehicle emissions testing and upgraded taxis and city vehicles to cleaner models. To monitor air quality, 34 sampling stations were set up across the city, sending real-time data to city engineers.

In 2007 Mayor Marcelo Ebrard accelerated efforts as part of a 15-year sustainability plan he launched aiming to make Mexico City "the greenest city in the Americas." New lines were added to the subway system, and 800 exhaust-spewing minibuses were replaced with fuel-efficient buses. More than 450,000 people now use these buses each day, cutting carbon dioxide emissions by an estimated 80,000 tons per year.

In 2010 Ebrard introduced a bicycle-sharing program to free short-distance commuters from dependence on cars. With 1000 bikes at rental stations throughout the city, people can rent a bike cheaply at one location and drop it off at another. In Ebrard's most popular initiative, every Sunday morning the city's main boulevard, the Paseo de la Reforma, is closed to car traffic, creating a safe and pleasant community space for pedestrians, bikers, joggers, and skateboarders.

After 2012, Ebrard's successor as mayor, Miguel Ángel Mancera, built on these efforts by expanding the bicycle program and putting electric buses and taxis on the roads. Mancera also rolled out a car-sharing program, aiming to remove 40,000 vehicles from circulation. Private entrepreneurs are now getting in on the act; recent university graduates have launched companies providing car-sharing and carpooling services, some using electric vehicles.

All these changes paid off with cleaner air. In 1991, Mexico City's air was deemed hazardous to breathe on all but 8 days of the year. By 2010–2015, most pollutants had been slashed by more than 75%, and the air was meeting health standards on one of every two days.

However, plenty of hurdles remain. Rampant development is challenging efforts to plan for sustainable growth. New highway construction is inducing more people to drive. The typical driver still spends three hours a day stuck in traffic that averages just 13 mph. Worst of all, smog still contributes to an estimated 4000 deaths each year. In 2016, a monthlong spell of hot, windless weather brought back foul air conditions that the city had not suffered in over a decade. Angry residents complained that the city's politicians had become overconfident and were failing to follow through on further reforms. In response, Mexico's president, Enrique Peña Nieto, urged leaders to buckle down anew and tighten emissions testing on vehicles. The hope is that Mexico City will continue to follow in the footsteps of Los Angeles, making steady progress toward cleaner air.

L.A. and Mexico City typify cities of developed and developing nations today. Nations that are industrializing as they try to build wealth for their citizens are confronting the same air quality challenges that plagued the United States and other wealthy nations a generation and more ago. We will examine the solutions sought in Los Angeles, Mexico City, and elsewhere as we learn about Earth's atmosphere and how to reduce the pollutants we release into it.

The Atmosphere

Every breath we take reaffirms our connection to the **atmosphere,** the layer of gases that envelops our planet. The atmosphere moderates our climate, provides oxygen, helps to shield us from meteors and hazardous solar radiation, and transports and recycles water and nutrients.

Earth's atmosphere consists of 78% nitrogen (N_2) and 21% oxygen (O_2) by volume of dry air. The remaining 1% is composed of argon (Ar) and minute concentrations of other gases (**FIGURE 17.1**). The atmosphere also contains water vapor (H_2O) in concentrations that vary with time and place from 0% to 4%.

Over our planet's long history, the atmosphere's composition has changed. When Earth was young, our atmosphere was dominated by carbon dioxide (CO_2), nitrogen, carbon monoxide (CO), and hydrogen (H_2), but about 2.7 billion years ago, oxygen began to build up with the emergence of microbes that emitted oxygen by photosynthesis (p. 30). Today, human activity is altering the quantities of some atmospheric gases, such as carbon dioxide, methane (CH_4), and ozone (O_3). Before exploring how our pollutants change the air we breathe and how we strive to control pollution, we will begin with an overview of Earth's atmosphere.

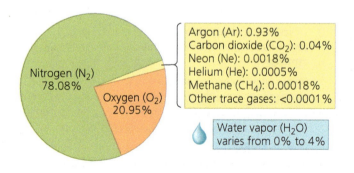

FIGURE 17.1 Earth's atmosphere consists of nitrogen, oxygen, argon, and a mix of gases at dilute concentrations.

The atmosphere is layered

The atmosphere that stretches so high above us and seems so vast is actually just 1/100 of Earth's diameter—a thin coating like the fuzzy skin of a peach. It consists of four layers that differ in temperature, density, and composition (**FIGURE 17.2**).

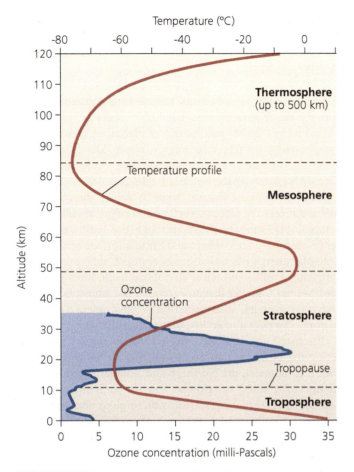

FIGURE 17.2 The atmosphere is layered. Temperature (**red line**) drops with altitude in the troposphere, rises with altitude in the stratosphere, drops in the mesosphere, and rises in the thermosphere. The tropopause separates the troposphere from the stratosphere. Ozone (**blue area**) is densest in the lower stratosphere. *Adapted from Jacobson, M.Z., 2002. Atmospheric pollution: History, science, and regulation. Cambridge, U.K.: Cambridge University Press; and Parson, E.A., 2003. Protecting the ozone layer: Science and strategy. Oxford, U.K.: Oxford University Press.*

The bottommost layer, the **troposphere,** blankets Earth's surface and provides us the air we breathe. Movement of air within the troposphere drives our planet's weather. Although it is thin (averaging 11 km or 7 mi high) relative to the atmosphere's other layers, the troposphere contains three-quarters of the atmosphere's mass. This is because gravity pulls mass downward, making air denser near Earth's surface. Tropospheric air gets colder with altitude, dropping to roughly −52°C (−62°F) at the top of the troposphere. At this point, temperatures stabilize, marking a boundary called the *tropopause*. The tropopause acts like a cap, limiting mixing between the troposphere and the atmospheric layer above it, the stratosphere.

The **stratosphere** extends 11–50 km (7–31 mi) above sea level. Similar in composition to the troposphere, the stratosphere is much drier and less dense. Its gases experience little vertical mixing, so once substances (including pollutants) enter it, they tend to remain for a long time. The stratosphere warms with altitude, because its ozone and oxygen absorb the sun's ultraviolet (UV) radiation (p. 30). Most of the atmosphere's ozone concentrates in a portion of the stratosphere roughly 17–30 km (10–19 mi) above sea level, a region we call the **ozone layer.** By absorbing and scattering incoming UV radiation, the ozone layer greatly reduces the amount of this radiation that reaches Earth's surface. UV light can damage living tissue and induce mutations in DNA, and life has evolved to rely on the protective presence of the ozone layer.

Above the stratosphere lies the mesosphere, where temperatures decrease with altitude and where incoming meteors burn up. Above this, the thermosphere extends upward to an altitude of 500 km (300 mi). Still higher, the atmosphere merges into space in a region called the exosphere.

Pressure, humidity, and temperature vary within the atmosphere

Air moves dynamically within the lower atmosphere as a result of differences in the physical properties of air masses. Among these properties are pressure and density, relative humidity, and temperature.

Gravity pulls gas molecules toward Earth's surface, causing air to be most dense near the surface and less dense as altitude increases. **Atmospheric pressure,** which measures the force per unit area produced by a column of air, also decreases with altitude, because at higher altitudes there are fewer molecules being pulled down by gravity (**FIGURE 17.3**). At sea level, atmospheric pressure averages

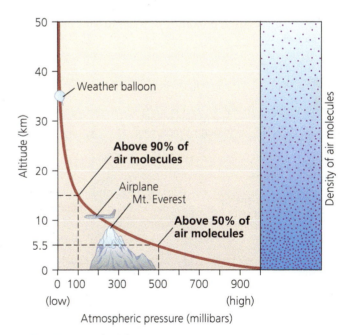

FIGURE 17.3 As one climbs higher through the atmosphere, gas molecules become less densely packed, and atmospheric pressure decreases. One needs to be only 5.5 km (3.4 mi) high to be above half the planet's air molecules. *Adapted from Ahrens, C.D., 2007. Meteorology today, 8th ed., Figure 1.9. © 2007. Belmont, CA: Brooks/Cole. By permission of Cengage Learning.*

14.7 lb/in.² or 1013 millibars (mb). Mountain climbers trekking to Mount Everest, the world's highest mountain, can look up and view their destination from Kala Patthar, a nearby peak, at roughly 5.5 km (18,000 ft) in elevation. At this altitude, pressure is 500 mb, and half the atmosphere's air molecules are above the climber, whereas half are below. A climber who reaches Everest's peak at 8.85 km (29,035 ft) in elevation, where the "thin air" is just over 300 mb, stands above two-thirds of the molecules in the atmosphere! When we fly on a commercial jet airliner at a typical cruising altitude of 11 km (36,000 ft), we are above 80% of the atmosphere's molecules.

Another property of air is **relative humidity,** the ratio of water vapor that air contains at a given temperature to the maximum amount it *could* contain at that temperature. Average daytime relative humidity in June in the desert at Phoenix, Arizona, is only 31% (meaning that the air contains less than a third of the water vapor possible at its temperature), whereas on the tropical island of Guam, relative humidity rarely drops below 88%.

People are sensitive to changes in relative humidity because we perspire to cool our bodies. When humidity is high, the air is already holding nearly as much water vapor as it can, so sweat evaporates slowly and the body cannot cool itself efficiently. This is why high humidity makes it feel hotter than it actually is. Conversely, low humidity speeds evaporation and makes it feel cooler than it actually is.

The temperature of air also varies with location and time. At the global scale, temperature varies over Earth's surface because the sun's rays strike some areas more directly than others. At more local scales, temperature varies because of topography, plant cover, proximity of water to land, and many other factors.

Solar energy heats the atmosphere, helps create seasons, and causes air to circulate

An enormous amount of energy from the sun constantly bombards the upper atmosphere—over 1000 watts/m², thousands of times more than the total output of electricity generated by human society. Of this solar energy, about 70% is absorbed by the atmosphere and planetary surface, while the rest is reflected back into space (see Figure 18.2, p. 481).

Sunlight is most intense when it shines directly overhead and meets the planet's surface at a perpendicular angle. At this angle, sunlight passes through the least amount of energy-absorbing atmosphere and Earth's surface receives the most solar energy per unit area. In contrast, solar energy that approaches Earth's surface at an oblique angle loses intensity as it traverses a longer distance through the atmosphere. This is why, on average, solar radiation is most intense near the equator and weakest near the poles (**FIGURE 17.4**).

Because Earth is tilted on its axis (an imaginary line connecting the poles, running perpendicular to the equator) by about 23.5 degrees, the Northern and Southern Hemispheres end up being tilted toward the sun for half of each year, resulting in the seasons (**FIGURE 17.5**). Regions near the equator experience about 12 hours each of sunlight and darkness per day throughout the year. Near the poles, in contrast, day length varies greatly between summer and winter, and seasonality is pronounced.

On the surface of our planet, land and water absorb solar energy and then emit thermal infrared radiation, which warms the air and causes some water to evaporate. As a result, air near Earth's surface tends to be warmer and moister than air at higher altitudes. These differences set into motion a

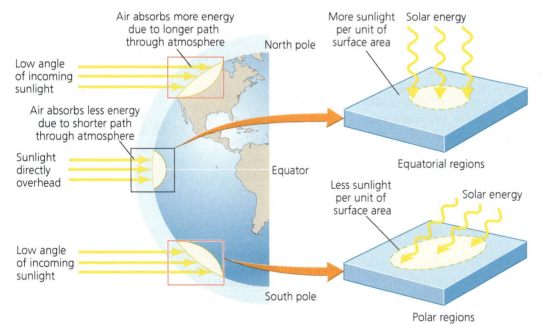

FIGURE 17.4 Because of Earth's curvature, polar regions receive less solar energy than equatorial regions. One reason is that sunlight gets spread over a larger area when striking the surface at an angle. Another reason is that sunlight approaching at a lower angle near the poles must traverse a longer distance through the atmosphere, causing more energy to be absorbed or reflected. These patterns represent year-round averages; the latitude at which radiation approaches the surface perpendicularly varies with the seasons (see Figure 17.5).

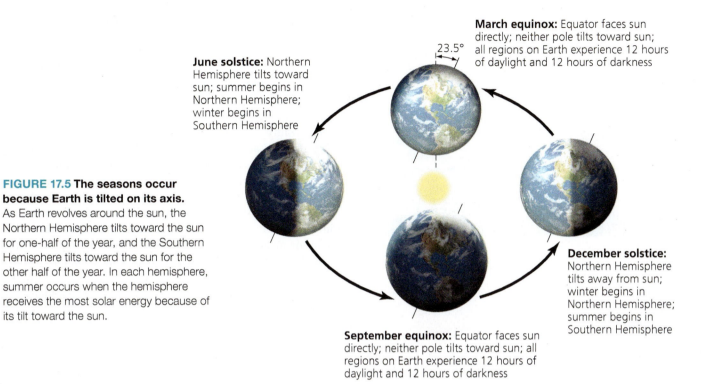

June solstice: Northern Hemisphere tilts toward sun; summer begins in Northern Hemisphere; winter begins in Southern Hemisphere

23.5°

March equinox: Equator faces sun directly; neither pole tilts toward sun; all regions on Earth experience 12 hours of daylight and 12 hours of darkness

December solstice: Northern Hemisphere tilts away from sun; winter begins in Northern Hemisphere; summer begins in Southern Hemisphere

September equinox: Equator faces sun directly; neither pole tilts toward sun; all regions on Earth experience 12 hours of daylight and 12 hours of darkness

FIGURE 17.5 The seasons occur because Earth is tilted on its axis. As Earth revolves around the sun, the Northern Hemisphere tilts toward the sun for one-half of the year, and the Southern Hemisphere tilts toward the sun for the other half of the year. In each hemisphere, summer occurs when the hemisphere receives the most solar energy because of its tilt toward the sun.

process of **convective circulation** (**FIGURE 17.6**). Warm air, being less dense, rises and creates vertical currents. As air rises into regions of lesser atmospheric pressure, it expands and cools, causing moisture to condense and fall as rain. Once the air cools, it descends and becomes denser, replacing warm air that is rising. The descending air picks up heat and moisture near ground level and prepares to rise again, continuing the process. Convective circulation patterns occur in ocean waters (p. 422), in magma beneath Earth's surface (pp. 35–36), and even in a simmering pot of soup. Convective circulation influences both weather and climate.

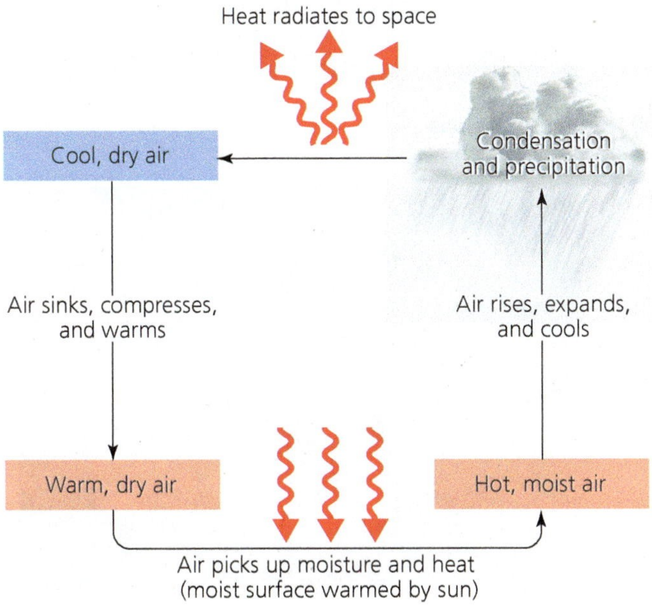

The atmosphere drives weather and climate

Weather and climate each involve the physical properties of the troposphere, such as temperature, pressure, humidity, cloudiness, and wind. **Weather** specifies atmospheric conditions in a location over short time periods, typically minutes, hours, days, or weeks. In contrast, **climate** describes typical patterns of atmospheric conditions in a location over long periods of time, typically years, decades, centuries, or millennia. Mark Twain once noted the distinction by remarking, "Climate is what we expect; weather is what we get." For example, Los Angeles has a climate characterized by reliably warm, dry summers and mild, rainy winters, yet on occasional autumn days, dry Santa Ana winds blow in from the desert and bring extremely hot weather.

Air masses interact, producing weather

Weather can change quickly when air masses with different physical properties meet. The boundary between air masses that differ in temperature and moisture (and therefore density) is called a **front.** The boundary along which a mass of warmer, moister air replaces a mass of colder, drier air is

Heat radiates to space

Cool, dry air

Condensation and precipitation

Air sinks, compresses, and warms

Air rises, expands, and cools

Warm, dry air

Hot, moist air

Air picks up moisture and heat (moist surface warmed by sun)

FIGURE 17.6 Convective circulation helps to drive weather. Air heated near Earth's surface picks up moisture and rises. Once aloft, this air cools and moisture condenses, forming clouds and precipitation. Cool, drying air begins to descend, compressing and warming in the process. Warm, dry air near the surface begins the cycle anew.

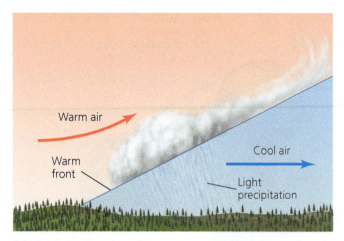

(a) Warm front

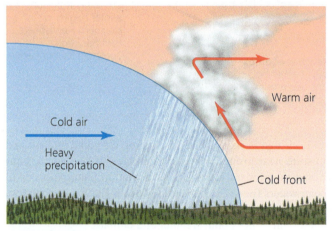

(b) Cold front

FIGURE 17.7 Fronts occur where air masses meet. When a warm front approaches **(a)**, warmer air rises over cooler air, causing light or moderate precipitation as moisture in the warmer air condenses. When a cold front approaches **(b)**, colder air pushes beneath warmer air, and the warmer air rises, resulting in condensation and heavy precipitation.

termed a **warm front** (**FIGURE 17.7a**). Some of the warm, moist air along the leading edge of a warm front usually rises over the cooler air mass that is blocking its progress. As it rises, the warm air cools and the water vapor within condenses, forming clouds and light rain. A **cold front** (**FIGURE 17.7b**) is the boundary along which a colder, drier air mass displaces a warmer, moister air mass. The colder air, being denser, tends to wedge beneath the warmer air. The warmer air rises, expands, then cools to form clouds and thunderstorms. Once a cold front passes through, the sky usually clears, and temperature and humidity drop.

Adjacent air masses may also differ in atmospheric pressure. A **high-pressure system** contains air that descends because it is cool and then spreads outward as it nears the ground. High-pressure systems typically bring fair weather. In a **low-pressure system,** warmer air rises, drawing air inward toward the center of low atmospheric pressure. The rising air expands and cools, and clouds and precipitation often result.

Inversions affect air quality

Under most conditions, air in the troposphere becomes cooler as altitude increases. Because warm air rises, vertical mixing results (**FIGURE 17.8a**). Occasionally, however, a layer of cool air may form beneath a layer of warmer air. This departure from the normal temperature profile is known as a **temperature inversion,** or thermal inversion (**FIGURE 17.8b**). The band of air in which temperature rises with altitude is called an **inversion layer** (because the normal direction of temperature change is inverted). The cooler air at the bottom of the inversion layer is denser than the warmer air above, so it resists vertical mixing and remains stable. Temperature inversions can occur in different ways, sometimes involving cool air at ground level and sometimes involving an inversion layer higher above the ground. One common type of inversion (shown in Figure 17.8b) occurs in mountain valleys where slopes block morning sunlight, keeping ground-level air within the valley shaded and cool.

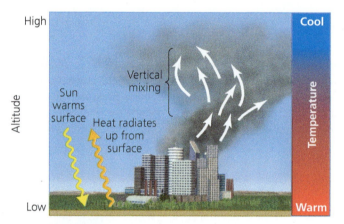

(a) Normal conditions

(b) Temperature inversion

FIGURE 17.8 Temperature inversions trap air and pollutants. Under normal conditions **(a)**, air becomes cooler with altitude and air of different altitudes mixes, dispersing pollutants upward. In a temperature inversion **(b)**, dense cool air remains near the ground, and air warms with altitude within the inversion layer. Little mixing occurs, and pollutants are trapped.

Vertical mixing allows pollutants in the air to be carried upward and diluted, but temperature inversions trap pollutants near the ground. As a result, cities such as Los Angeles and Mexico City suffer their worst pollution when inversions prevent pollutants from being dispersed. As noted earlier, both of these metropolitan areas are encircled by mountains that promote inversions, interrupt air flow, and trap pollutants. Los Angeles experiences inversions most often when a "marine layer" of air cooled by the ocean moves inland. In Mexico City in 1996, a persistent temperature inversion sparked a five-day crisis in which pollution killed at least 300 people and sent 400,000 to hospitals. Desperate for a solution, some people proposed dynamiting a hole in the mountains and installing immense fans to blow out the air!

Large-scale circulation systems produce global climate patterns

At large geographic scales, convective air currents contribute to broad climate patterns (**FIGURE 17.9**). Near the equator, solar radiation sets in motion a pair of convective cells known as *Hadley cells*. Here, where sunlight is most intense, surface air warms, rises, and expands. As it does so, it releases moisture, producing the heavy rainfall that gives rise to tropical rainforests. After releasing much of its moisture, this air diverges and moves in currents heading north and south. The air in these currents cools and descends at about 30 degrees latitude north and south. Because the descending air is now dry, the regions around 30 degrees latitude are quite arid, giving rise to deserts. Two additional pairs of convective cells, *Ferrel cells* and *polar cells*, lift air and create precipitation around 60 degrees latitude north and south and cause air to descend at 30 degrees latitude and in the polar regions.

Together these three pairs of convective cells create wet climates near the equator, arid climates near 30° latitude, moist regions near 60° latitude, and dry conditions near the poles. These patterns, combined with temperature variation, help explain why biomes tend to be arrayed in latitudinal bands (see Figure 4.16, p. 92).

The Hadley, Ferrel, and polar cells interact with Earth's rotation to produce global wind patterns (see Figure 17.9). As Earth rotates on its axis, regions of the planet's surface near the equator move west to east more quickly than regions near the poles. As a result, from the perspective of an Earth-bound observer, air currents of the convective cells that flow north or south appear to be deflected from a straight path. This deflection is called the **Coriolis effect,** and it results in the curving global wind patterns in Figure 17.9. Near the equator lies a region with few winds known as the *doldrums*. Between the equator and 30° latitude, *trade winds* blow from east to west. From 30° to 60° latitude, *westerlies* blow from west to east. For centuries, people made use of these patterns to facilitate ocean travel by wind-powered sailing ships.

The atmosphere interacts with the oceans to affect weather, climate, and the distribution of biomes. Winds and convective circulation in ocean water together maintain ocean currents (p. 419). Trade winds weaken periodically, leading to El Niño conditions (p. 423). And oceans and atmosphere sometimes interact to create violent storms that can threaten life and property.

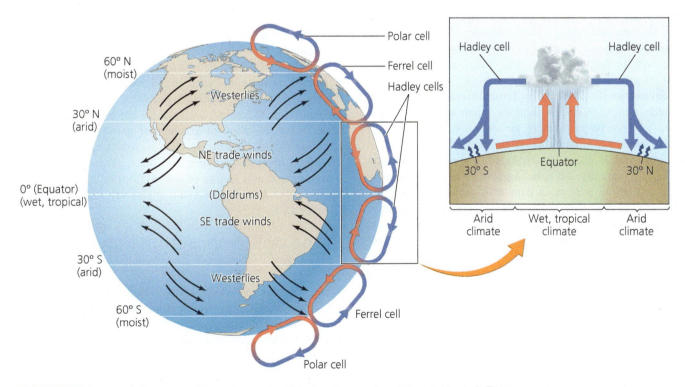

FIGURE 17.9 Large-scale convective cells create global patterns of moisture and wind. These cells give rise to a wet climate in tropical regions, arid climates around 30° latitude, moist climates around 60°, and dry climates near the poles. Surface air movement of these cells interacts with the Coriolis effect to create global wind currents.

(a) Satellite image of a hurricane

(b) Photograph of a tornado

FIGURE 17.10 Hurricanes and tornadoes are cyclonic storms that pose hazards to life and property.

Storms pose hazards

Hurricanes (**FIGURE 17.10a**) form when warm, moisture-laden air over tropical oceans rises and winds rush into these areas of low pressure. In the Northern Hemisphere, these winds turn counterclockwise because of the Coriolis effect. In other regions, such cyclonic storms are called *cyclones* or *typhoons*. The powerful convective currents of these storms draw up immense amounts of water vapor. As the warm, moist air rises and cools, water condenses (because cool air cannot hold as much water vapor as warm air) and falls heavily as rain. In North America, the Gulf Coast and Atlantic Coast are most susceptible to hurricanes.

Tornadoes (**FIGURE 17.10b**) form when a mass of warm air meets a mass of cold air and the warm air rises quickly, setting a powerful convective current in motion. If high-altitude winds are blowing faster and in a different direction from low-altitude winds, the rising column of air may begin to rotate. Eventually the spinning funnel of rising air may lift up soil and objects in its path with winds up to 500 km per hour (310 mph). In North America, tornadoes are most apt to form in the Great Plains and the Southeast, where cold air from Canada and warm air from the Gulf of Mexico frequently meet.

Understanding how the atmosphere functions can help us predict violent storms and warn people of their approach. Such knowledge also helps us comprehend how our pollution of the atmosphere affects climate, ecological systems, economies, and human health.

Outdoor Air Quality

Throughout history, we have made the atmosphere a dumping ground for our airborne wastes. Whether from simple wood fires or modern coal-burning power plants, we have generated **air pollutants,** gases and particulate material added to the atmosphere that can affect climate or harm people or other living things. At the same time, our efforts to control **air pollution,** the release of air pollutants, have brought some of our best successes in confronting environmental problems thus far.

In recent decades, public policy and improved technologies have helped us reduce most types of **outdoor air pollution** (often called *ambient air pollution*) in industrialized nations. However, outdoor air pollution remains a problem, particularly in industrializing nations and in urban areas. Scientists estimate that each year 3.3 million people die prematurely as a result of health problems caused by outdoor air pollution. Moreover, we face an enormous air pollution issue in our emission of greenhouse gases (p. 482), which contribute to global climate change. Addressing our release of carbon dioxide, methane, and other gases that warm the atmosphere stands as one of our civilization's primary challenges. (We discuss this issue separately and in depth in Chapter 18.)

Some pollution is from natural sources

When we think of outdoor air pollution, we tend to envision smokestacks belching smoke from industrial plants. However, natural processes also pollute the air. Some of these natural impacts are made worse by human activity and land use policies.

Fires from burning vegetation emit soot and gases. Worldwide, more than 60 million hectares (ha; 150 million acres, an area the size of Texas) of forest and grassland burn in a typical year (**FIGURE 17.11a**). Fires occur naturally, but human influence can make them more severe. In regions

(a) Natural fire in California

(b) Mount Saint Helens eruption, 1980

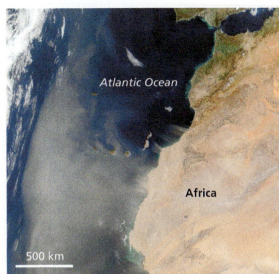

Atlantic Ocean

Africa

500 km

(c) Dust storm blowing dust from Africa to the Americas

FIGURE 17.11 Wildfire, volcanoes, and dust storms are three natural sources of air pollution.

like the Los Angeles basin, residential development has encroached into chaparral ecosystems (p. 98) that are naturally fire-prone, often resulting in costly damage when fires occur. Across North America, the suppression of fire has allowed fuel to build up in forests and eventually feed highly destructive fires (p. 314). In the tropics, many farmers clear forest for farming and grazing using a "slash-and-burn" approach (p. 215). Today climate change (Chapter 18) is leading to drought in many regions, including the Los Angeles basin, and worsening fires as a result.

Volcanic eruptions (pp. 38–39) release large quantities of particulate matter, as well as sulfur dioxide and other gases, into the troposphere (**FIGURE 17.11b**). In 2012, residents of Mexico City went on alert as Popocatepetl, a volcano just 70 km (45 mi) from the city, let loose a series of moderate eruptions, adding to the region's pollution challenges. Ash from volcanic eruptions near populated areas can ground airplanes, destroy car engines, and pose respiratory health dangers. Major eruptions may also blow matter into the stratosphere, where it can circle the globe for months or years. Sulfur dioxide reacts with water and oxygen and then condenses into fine droplets, called aerosols (p. 483), which reflect sunlight back into space and thereby cool the atmosphere and surface. The 1991 eruption of Mount Pinatubo in the Philippines ejected nearly 20 million tons of ash and aerosols and cooled global temperatures by 0.5°C (0.9°F).

Winds sweeping over arid terrain can send huge amounts of dust aloft. Dust storms occur naturally, but they are made worse by unsustainable farming and grazing practices that strip vegetation from the soil and lead to desertification (p. 222). Continental-scale dust storms took place in the United States in the 1930s, when soil from the drought-plagued Dust Bowl states blew eastward to the Atlantic (p. 224). Today, trade winds blow soil across the Atlantic

Ocean from Africa to the Americas (**FIGURE 17.11c**). Strong westerlies sometimes lift soil from deserts in Mongolia and China and blow it all the way across the Pacific Ocean to North America.

We create outdoor air pollution

Human activity produces many air pollutants. As with water pollution, anthropogenic (human-caused) air pollution can emanate from point sources or non-point sources (p. 405). A point source describes a specific location from which large quantities of pollutants are discharged (such as a coal-fired power plant). Non-point sources are more diffuse, consisting of many small, widely spread sources (such as millions of automobiles).

Pollutants released directly from a source are termed **primary pollutants.** Ash from a volcano, sulfur dioxide from a power plant, and carbon monoxide from an engine are all primary pollutants. Often primary pollutants react with one another, or with constituents of the atmosphere, and form other pollutants; the resulting compounds are called **secondary pollutants.** Examples include ozone formed from pollutants in urban smog or the acids in acid rain, formed when certain primary pollutants react with water and oxygen.

Because substances differ in how readily they react in air and in how quickly they settle to the ground, pollutants differ in their **residence time,** the amount of time a pollutant spends in the atmosphere. Pollutants with brief residence times exert localized impacts over short time periods. Most particulate matter and most pollutants from automobile exhaust stay aloft only hours or days, which is why air quality in a city like Mexico City or Los Angeles changes from day to day. In contrast, pollutants with long residence times can exert impacts regionally or globally for long periods, even

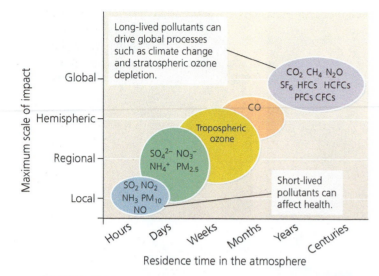

FIGURE 17.12 **Substances with short residence times affect air quality locally, whereas those with long residence times affect air quality globally.** *Source: United Nations Environment Programme, 2007. Global environment outlook (GEO-4), FIGURE 2.1. Data from European Environment Agency 1995; EPA Center for Airborne Organics 1997.*

centuries (**FIGURE 17.12**). The pollutants that drive climate change and those that deplete Earth's ozone layer (two separate phenomena!—see **FAQ**, p. 468) are each able to cause these global and long-lasting impacts because they persist in the atmosphere for so long.

The Clean Air Act addresses pollution

To address air pollution in the United States, Congress has passed a series of laws, most notably the **Clean Air Act.** First enacted in 1963, the Clean Air Act has been amended multiple times, chiefly in 1970 and 1990. This body of legislation funds research into pollution control, sets standards for air quality, and encourages emissions standards for automobiles and for stationary point sources such as industrial plants. It also imposes limits on emissions from new sources, funds a nationwide air quality monitoring system, and enables citizens to sue parties violating the standards. The 1990 amendments introduced an emissions trading program for sulfur dioxide.

Under the Clean Air Act, the U.S. Environmental Protection Agency (EPA) sets nationwide standards for (1) emissions of several key pollutants and (2) concentrations of major pollutants in ambient air. It is largely up to the states to monitor emissions and air quality and to develop, implement, and enforce regulations within their borders. States submit implementation plans to the EPA for approval, and if a state's plans are not adequate, the EPA can take control of enforcement. If a region fails to clean up its air, the EPA can prevent it from receiving federal money for transportation projects.

Agencies monitor emissions

State and local agencies monitor and report to the EPA emissions of six major pollutants, profiled below. Across the United States in 2014, human activity polluted the air with 88 million tons of these six monitored pollutants. Carbon monoxide was the most abundant pollutant by mass (**FIGURE 17.13**).

Carbon monoxide **Carbon monoxide (CO)** is a colorless, odorless gas produced primarily by the incomplete combustion of fuel. Vehicles and engines account for most CO emissions in the United States. Other sources include industrial processes, waste combustion, and residential wood burning. Carbon monoxide is hazardous because it binds to hemoglobin in red blood cells, which in turn prevents the hemoglobin from binding with oxygen. Carbon monoxide poisoning induces nausea, headaches, fatigue, heart and nervous system damage, and potentially death.

Sulfur dioxide **Sulfur dioxide (SO_2)** is a colorless gas with a pungent odor. Most emissions result from the combustion of coal for electricity generation and industry. During combustion, elemental sulfur (S), a contaminant in coal, reacts with oxygen (O_2) to form SO_2. Once in the atmosphere, SO_2 may react to form sulfur trioxide (SO_3) and sulfuric acid (H_2SO_4), which may then settle back to Earth in acid deposition (p. 469).

Nitrogen oxides **Nitrogen oxides (NO_x)** are a family of compounds that include nitric oxide (NO) and nitrogen dioxide (NO_2). Most U.S. emissions of nitrogen oxides result when nitrogen and oxygen from the atmosphere react at high temperatures during combustion in vehicle engines. Fossil

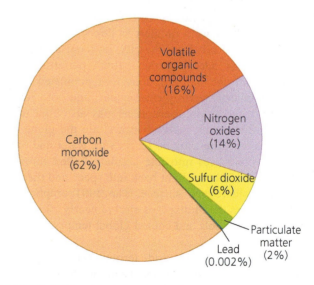

FIGURE 17.13 **In 2014, the United States emitted 88 million tons of the six major pollutants whose emissions are monitored by the EPA and state agencies.** These figures omit pollutants from dust and wildfires. *Data from U.S. EPA.*

fuel combustion in industry and at electrical utilities accounts for most of the rest. NO_x emissions contribute to smog, acid deposition, and stratospheric ozone depletion.

Volatile organic compounds   **Volatile organic compounds (VOCs)** are carbon-containing chemicals emitted by vehicle engines and a wide variety of solvents, industrial processes, household chemicals, and consumer items. Examples range from benzene to acetone to formaldehyde. One common group of VOCs consists of hydrocarbons (p. 27) such as methane (CH_4, the primary component of natural gas), propane (C_3H_8, used as a portable fuel), butane (C_4H_{10}, found in cigarette lighters), and octane (C_8H_{18}, a component of gasoline). Human activities account for about half the VOC emissions in the United States. The remainder comes from natural sources; for example, plants produce isoprene and terpenes, compounds that generate a bluish haze that has given the Blue Ridge Mountains their name. VOCs can react to produce secondary pollutants, as occurs in urban smog.

Particulate matter   **Particulate matter** is composed of solid or liquid particles small enough to be suspended in the atmosphere. Particulate matter includes primary pollutants such as dust and soot, as well as secondary pollutants such as sulfates and nitrates. Scientists classify particulate matter by the size of the particles. Smaller particles are more likely to get deep into the lungs and cause respiratory damage and heart problems. PM_{10} pollutants consist of particles less than 10 microns in diameter (one-seventh the width of a human hair), whereas $PM_{2.5}$ pollutants consist of still-finer particles less than 2.5 microns in diameter. Most PM_{10} pollution is from road dust, whereas most $PM_{2.5}$ pollution results from combustion.

Lead   **Lead** (Pb) is a heavy metal that enters the atmosphere as a particulate pollutant. The lead-containing compounds tetraethyl lead and tetramethyl lead, when added to gasoline, improve engine performance. However, exhaust from the combustion of leaded gasoline emits airborne lead, which can be inhaled or can be deposited on land and water. When lead enters the food chain, it accumulates in body tissues and can cause central nervous system malfunction and many other ailments (p. 361). Since the 1980s, most developed nations have phased out leaded gasoline (p. 7), and today most developing nations are following suit. In developed nations, the main remaining source of atmospheric lead is industrial metal smelting.

We have reduced pollutant emissions

Since passage of the Clean Air Act of 1970, the United States has reduced emissions of each of the six monitored pollutants substantially (**FIGURE 17.14a**). These dramatic

reductions in emissions have occurred despite significant increases in the nation's population, energy consumption, miles traveled by vehicle, and gross domestic product (**FIGURE 17.14b**). Likewise, most other industrialized nations have taken their own steps to reduce emissions and have attained similar results.

We have achieved this success in controlling pollution as a result of policy steps and technological developments, each motivated by grassroots social demand for cleaner air. In factories, power plants, and refineries, technologies such as baghouse filters, electrostatic precipitators, and

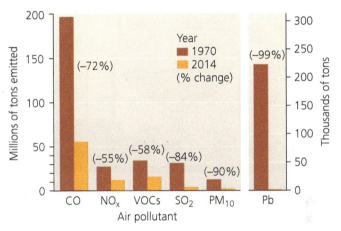

(a) Declines in six major pollutants

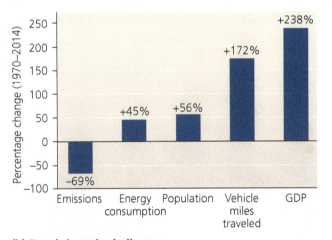

(b) Trends in major indicators

FIGURE 17.14 U.S. emissions have declined sharply since 1970. We have achieved reductions **(a)** in the six major pollutants tracked by the EPA, despite increases **(b)** in U.S. energy consumption, population, vehicle miles traveled, and gross domestic product. *Data from U.S. EPA.*

DATA Q   By what percentage has population increased since 1970? By what percentage have emissions decreased? Using these two amounts, calculate the change in emissions per person.

Go to **Interpreting Graphs & Data** on **Mastering**EnvironmentalScience®

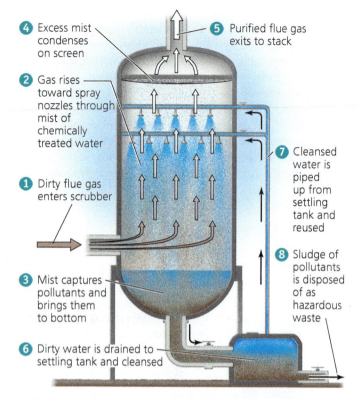

④ Excess mist condenses on screen

⑤ Purified flue gas exits to stack

② Gas rises toward spray nozzles through mist of chemically treated water

① Dirty flue gas enters scrubber

⑦ Cleansed water is piped up from settling tank and reused

③ Mist captures pollutants and brings them to bottom

⑧ Sludge of pollutants is disposed of as hazardous waste

⑥ Dirty water is drained to settling tank and cleansed

FIGURE 17.15 Scrubbers typically remove at least 90% of particulate matter and gases such as sulfur dioxide. Scrubbers and other pollution control devices come in many designs. In this spray-tower wet scrubber, polluted air rises through a chamber while nozzles spray a mist of water mixed with lime or other active chemicals to capture pollutants and wash them out of the air.

scrubbers (FIGURE 17.15) have been installed to chemically convert or physically remove airborne pollutants before they are emitted from smokestacks. Meanwhile, cleaner-burning motor vehicle engines and automotive technologies such as catalytic converters have cut down on pollution from automobile exhaust. In a **catalytic converter** (FIGURE 17.16), engine exhaust reacts with metals that convert hydrocarbons, CO, and NO_X into carbon dioxide, water vapor, and nitrogen gas. Such pollution control technologies were initially resisted by industry because of added expense, but came to be widely adopted once policy measures were in place encouraging or mandating them. Other policies have been influential, as well. Phaseouts of leaded gasoline caused lead emissions to plummet, and the EPA's Acid Rain Program and its emissions trading system (p. 471), along with clean coal technologies (p. 538), have reduced SO_2 and NO_X emissions.

The reduction of outdoor air pollution in the United States since 1970 has resulted in notably cleaner air throughout the nation and represents one of America's greatest accomplishments in safeguarding human health and environmental quality. It demonstrates how seemingly intractable problems can be tackled and addressed within a democratic system when government and industry are informed by science and are responsive to the public's demands. The EPA estimates that between 1970 and 1990 alone, clean air regulations and the resulting technological advances in pollution control saved the lives of 200,000 Americans. Similar advances in pollution control have been made by most other wealthy nations in recent years.

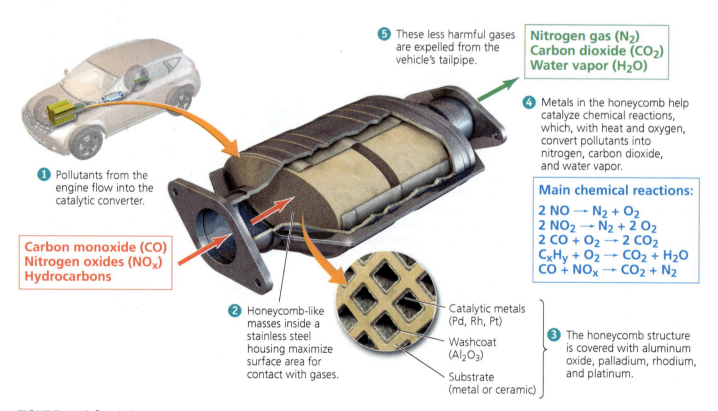

⑤ These less harmful gases are expelled from the vehicle's tailpipe.

Nitrogen gas (N_2)
Carbon dioxide (CO_2)
Water vapor (H_2O)

④ Metals in the honeycomb help catalyze chemical reactions, which, with heat and oxygen, convert pollutants into nitrogen, carbon dioxide, and water vapor.

① Pollutants from the engine flow into the catalytic converter.

Carbon monoxide (CO)
Nitrogen oxides (NO_X)
Hydrocarbons

Main chemical reactions:

$2\ NO \rightarrow N_2 + O_2$
$2\ NO_2 \rightarrow N_2 + 2\ O_2$
$2\ CO + O_2 \rightarrow 2\ CO_2$
$C_xH_y + O_2 \rightarrow CO_2 + H_2O$
$CO + NO_x \rightarrow CO_2 + N_2$

② Honeycomb-like masses inside a stainless steel housing maximize surface area for contact with gases.

Catalytic metals (Pd, Rh, Pt)

Washcoat (Al_2O_3)

Substrate (metal or ceramic)

③ The honeycomb structure is covered with aluminum oxide, palladium, rhodium, and platinum.

FIGURE 17.16 Catalytic converters improve air quality by filtering pollutants from vehicle exhaust.

Air quality has improved

As a result of emissions reductions, air quality has improved markedly in industrialized nations. In the United States, the EPA and the states monitor outdoor air quality by measuring the concentrations of six **criteria pollutants,** pollutants judged to pose substantial risk to human health. For each of these, the EPA has established *national ambient air quality standards,* which are maximum concentrations allowable in ambient outdoor air. The six criteria pollutants include four of the six pollutants whose emissions are monitored—carbon monoxide, sulfur dioxide, particulate matter, and lead—and also nitrogen dioxide and tropospheric ozone.

Nitrogen dioxide (NO_2) is a highly reactive, reddish brown, foul-smelling gas that contributes to smog and acid deposition. Along with nitric oxide (NO), NO_2 belongs to the family of compounds called nitrogen oxides (NO_x). Nitric oxide reacts readily in the atmosphere to form NO_2, which is both a primary and secondary pollutant.

Although ozone in the stratosphere shields us from the dangers of UV radiation, ozone from human activity accumulates low in the troposphere. **Tropospheric ozone (O_3),** also called *ground-level ozone,* is a secondary pollutant, created by the reaction of nitrogen oxides and volatile carbon-containing chemicals in the presence of sunlight. A major component of photochemical smog (pp. 462–463), this colorless gas poses health risks due to the instability of the O_3 molecule. This triplet of oxygen atoms will readily split into a molecule of oxygen gas (O_2) and a free oxygen atom. The oxygen atom may then participate in reactions that can injure living tissues and cause respiratory problems. Tropospheric ozone is the pollutant that most frequently exceeds its national ambient air quality standard.

Across the United States, more than 4000 monitoring stations take hourly or daily air samples to measure pollutant concentrations. The EPA compiles these data and calculates values on its Air Quality Index (AQI) for each site. Each of six pollutants—CO, SO_2, NO_2, O_3, PM_{10}, and $PM_{2.5}$—receives an AQI value from 0 to 500 that reflects its current concentration. AQI values below 100 indicate satisfactory air conditions, and values above 100 indicate unhealthy conditions. The highest AQI value from the pollutants on a particular day is reported as the overall AQI value for that day, and these values are made available online and reported in weather forecasts.

The actions of scientists, policymakers, industrial leaders, and everyday citizens have made air quality today far better than it was a generation or two ago (**FIGURE 17.17**). However, there remains plenty of room for improvement. Concerns over new pollutants are emerging, greenhouse gas emissions are altering the climate, and people in low-income communities suffer disproportionately from living near hotspots of pollution. In fact, many Americans live in areas where pollution continues to reach unhealthy levels. Despite enormous improvement over the past two decades, residents of Los Angeles County, for instance, still regularly breathe air that violates national ambient air quality standards for five of the six criteria pollutants. People in four adjacent southern

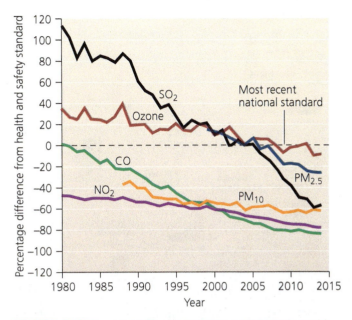

FIGURE 17.17 Concentrations of criteria pollutants in ambient air across the United States have steadily fallen. All now average below their standards for health and safety set by the EPA. Lead (not shown) has plummeted from 900% of its standard in 1980. Still, local hotspots of pollution occur, and U.S. standards are sometimes more lax than those set by European nations or the World Health Organization. These data omit pollutants from dust and wildfires. *Data from U.S. EPA.*

California counties breathe air that violates four of the standards. Altogether, as of 2014, 57 million Americans lived in counties that violated the national ambient air quality standard for at least one criteria pollutant. Still, L.A. and other metropolises are making perceptible headway toward cleaner air for their citizens (**FIGURE 17.18**, p. 458).

Toxic pollutants pose health risks

We are also reducing emissions of **toxic air pollutants,** substances known to cause cancer; reproductive defects; or neurological, developmental, immune system, or respiratory problems in people and other organisms. Under the 1990 Clean Air Act, the EPA regulates 187 toxic air pollutants produced by a variety of activities, including metal smelting, sewage treatment, and industrial processes. These pollutants range from the heavy metal mercury (from coal-burning power plant emissions and other sources) to VOCs such as benzene (a component of gasoline) and methylene chloride (found in paint stripper). Based on monitoring at 300 sites across the United States, scientists estimate that toxic air pollutants cause cancer in 1 out of every 25,000 Americans (40 cancer cases per 1 million people). Health risks are highest in urban and industrialized areas such as the Los Angeles region, but nationwide the EPA estimates that Clean Air Act regulations on facilities such as chemical plants, waste incinerators, dry cleaners, and coke ovens have helped to reduce emissions of toxic air pollutants since 1990 by more than 42%.

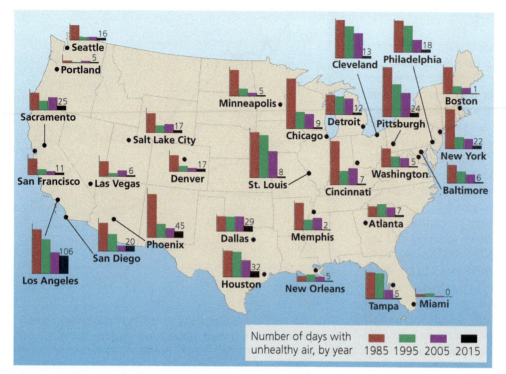

Number of days with unhealthy air, by year — 1985 1995 2005 2015

FIGURE 17.18 In most U.S. cities, air is becoming cleaner. This map shows numbers of days with unhealthy air from years spanning four decades for 29 metropolitan areas, according to the Air Quality Index (AQI). The AQI combines data on CO, NO_2, SO_2, O_3, and particulate matter. Days with AQI values over 100, shown here, indicate unhealthy air and national ambient air quality standard violations. *Data from U.S. EPA.*

DATA Q Locate where you live on the map. How does your city or the nearest major city to you compare to others in its air quality? Has it improved its air quality in recent years? Now explore one of the EPA websites that lets you browse information on the air you breathe: www.airnow.gov, www3.epa.gov/aircompare, or www3.epa.gov/air/emissions/where.htm. What factors do you think influence the quality of your region's air? Propose three steps for reducing air pollution in your region.

Go to **Interpreting Graphs & Data** on **Mastering**EnvironmentalScience®

Rural areas also confront pollution challenges

Although we tend to focus on pollution in cities, air quality is also a rural issue. In rural areas, people suffer from drift of airborne pesticides from farms, as well as industrial pollutants that drift far from cities, factories, and power plants. Air pollution also emanates from feedlots (p. 241) where cattle, hogs, or chickens are raised. The huge numbers of animals densely concentrated at feedlots and the voluminous amounts of waste they produce generate methane, hydrogen sulfide, and ammonia. These gases create objectionable odors, and ammonia contributes to nitrogen deposition. Studies show that people working at and living near feedlots have high rates of respiratory illness.

Indeed, some of the worst air quality in the United States occurs in certain rural regions, including California's Central Valley (the nation's agricultural fruit basket) and areas near natural gas extraction sites, where fumes from extraction pollute the air.

Air pollution remains severe in industrializing nations

Although the United States and other industrialized nations have improved their air quality, outdoor air pollution is worsening in many industrializing countries. In these societies, proliferating factories and power plants are emitting more pollutants as governments encourage economic growth. Additionally, more citizens own and drive automobiles. At the same time, many people continue to burn traditional sources of fuel, such as wood, charcoal, and coal, for cooking and home heating. As a result, only one person out of eight who live in cities that report on their air quality enjoys air that meets the health guidelines of the World Health Organization (WHO). Half the world's urban population breathes air polluted at levels at least 2.5 times beyond the WHO's standards.

Mexico embodies these trends, and despite the progress in its capital, residents of Mexico City and many other Mexican cities and towns continue to suffer a variety of health impacts from polluted air (see **THE SCIENCE BEHIND THE STORY**, pp. 460–461).

Currently, people in the vast sprawling cities of India, China, and other Asian nations suffer the world's worst air quality. Most Asian nations have fueled their rapid industrial development with coal, the most-polluting fossil fuel. Power plants and factories often use outdated, inefficient, heavily polluting technology because it is cheaper and quicker to build. In addition, car ownership is skyrocketing. Urban air quality is nearly as bad in eastern European nations such as Poland and Bulgaria, where old Soviet-era factories still pollute the air and where many people burn coal for home heating. In countless cities in these regions of the world, the haze often becomes thick enough to obscure the sun.

In China's capital of Beijing in winter 2013, pollution became so severe that airplane flights were cancelled and people wore face masks to breathe (**FIGURE 17.19**). Levels of particulate matter were literally off the charts; a monitor atop the U.S. Embassy designed to measure the Air Quality Index detected record-breaking readings beyond the maximum value of 500, up to 755. During this "airpocalypse," a fire at a factory went unnoticed for three hours because the pollution was so thick

FIGURE 17.19 At Tiananmen Square in Beijing, China, children wear face masks during an "airpocalypse" gripping the city.

FIGURE 17.20 Delhi, India, was rated by the World Health Organization as having the poorest air quality in the world.

that no one could see the smoke! Thousands of people suffered ill health as the pollution soared 30 times past the WHO's safe limits. In 2014 and in 2015, Beijing's "airmageddon" returned, and conditions became so bad that the national government and the official state media were finally forced to admit the problem and begin a public discussion about solutions.

In many Chinese cities, air pollution is worse than in Beijing, and across China, the health impacts of air pollution are enormous. A 2013 international research report blamed outdoor air pollution for 1.2 million premature deaths in China each year. Another 2013 study found that residents of polluted northern China die, on average, five years earlier than residents of southern China, where the air is cleaner. Moreover, prevailing westerly winds carry some of China's pollution across the Pacific Ocean to North America! A 2014 study quantified Chinese pollution reaching the U.S. West Coast and estimated that it adds at least one extra smoggy day per year to Los Angeles's total.

In 2015, China's citizens were transfixed by an online video that went viral, titled *Under the Dome*. This powerful 104-minute documentary, produced and narrated by a well-known Chinese investigative reporter, Chai Jing, confirmed for China's people what they already know: They spend their daily lives trapped under a dome of dangerously polluted air. The Chinese government sent mixed messages, banning the video while also proclaiming a commitment to fighting pollution. Many commentators drew a parallel with *Silent Spring*, Rachel Carson's 1962 book on DDT that galvanized the grassroots environmental movement in the United States (p. 170).

China's government is now striving to reduce pollution. It has closed down some heavily polluting factories and mines, phased out some subsidies for polluting industries, and installed pollution controls in power plants. It subsidizes efficient electric heaters for homes to replace dirty, inefficient coal stoves. It has mandated cleaner formulations for gasoline and

diesel and has raised standards for fuel efficiency and emissions for cars above what the United States requires. In Beijing, mass transit is being expanded, many buses run on natural gas, and heavily polluting vehicles are kept out of the central city. China is also aggressively developing cleaner wind, solar, and nuclear power to substitute for coal-fired power.

As bad as air pollution is in China, it is even worse in India's cities, and sometimes in other Asian metropolises such as Karachi, Jakarta, Bangkok, and Singapore. Delhi, India, has the world's worst air quality (**FIGURE 17.20**), according to 2014 WHO data for $PM_{2.5}$ pollutants. In this WHO survey, 13 of the 20 most polluted cities on Earth were in India. Because India's cities struggle with many challenges of urban poverty, there have been few efforts to tackle air pollution, despite an estimated toll of 1.5 million premature deaths each year.

Across India, China, and other industrializing nations of Asia, pollution from autos, industry, agriculture, and wood-burning stoves has resulted in a persistent 2-mile-thick layer of pollution that hangs over southern Asia throughout the dry season each December through April. Dubbed the Asian Brown Cloud, or Atmospheric Brown Cloud, this massive layer of brownish haze is estimated to reduce the sunlight reaching Earth's surface in southern Asia by 10–20%; promote flooding in some areas and drought in others by altering the monsoon; decrease rice yields by 5–10%; speed the melting of Himalayan glaciers by depositing dark soot that absorbs sunlight; and contribute to many thousands of deaths each year.

Smog poses health risks

Now let's take a closer look at one of the most prevalent types of air pollution: smog. As we saw in our Central Case Study, **smog** is a general term for a mixture of air pollutants that can accumulate as a result of fossil fuel combustion, generally over industrial regions or urban areas with heavy automobile traffic.

What Are the Health Impacts of Mexico City's Air Pollution?

Dr. Lilian Calderón-Garcidueñas

"I know I'm inhaling poison," a 38-year-old candy vendor named Guadalupe told a reporter amid the fumes of a traffic-choked intersection in Mexico City. "But there is nothing I can do."

For as long as we have polluted our air, people have felt effects on their health. But identifying and quantifying those impacts poses a challenge for scientists. For researchers wanting to understand pollution's health impacts—and to design solutions for people like Guadalupe—what better place to go than Mexico City, long home to some of the world's worst air pollution?

A key first step is to determine what's actually in the air. One researcher who has led the way is Mario Molina, the Nobel Prize–winning chemist who helped discover the cause of stratospheric ozone depletion and who appears in this chapter's other **Science behind the Story** feature (pp. 466–467). Molina stepped away from scholarly work at U.S. universities to return to his hometown of Mexico City and help address its pollution issues. In 2003 and 2006, Molina organized intensive air-sampling projects in Mexico City involving hundreds of scientists.

Nearly 200 research publications later, these efforts have clarified many aspects of the city's pollution. One study used machines that could identify and record individual particles in real time. Its data indicated that metal-rich particulates from trash incinerators were peaking in the morning, whereas smoke from fires outside the city blew in during the afternoon. Other researchers discovered that volatile organic carbons control the amount of tropospheric ozone formed in smog (not nitrogen oxides, as was expected). City officials responded by targeting VOC emissions for reduction, while also taking steps to discourage automobile traffic (**FIGURE 1**).

Few people today understand Mexico City's air pollution in more detail than Armando Retama, the city's director of atmospheric monitoring. But he may grasp its impacts best when he leaves town. "I can breathe better. I'm not all dry. My eyes aren't irritated. My skin doesn't crack," he says. "We have chronic symptoms that we aren't aware of."

Most health impacts of urban pollution affect the respiratory system. At high altitudes like Mexico City's, the "thin air" forces people to breathe deeply to obtain enough oxygen. This means they pull more air pollutants into their lungs than people at lower elevations. As result, respiratory problems are commonplace. Many studies have confirmed that Mexico City residents show reduced lung function in comparison with people from less-polluted areas and that respiratory problems become worse and emergency room visits become more numerous when pollution is severe.

Most studies have looked at short-term exposure, but in 2007 a research team led by Isabelle Romieu of Mexico's National Institute of Public Health examined the effects of growing up amid polluted air. Her team measured lung function in 3170 eight-year-old children from 39 Mexico City schools across 3 years and correlated this with their exposure to tropospheric ozone, nitrogen dioxide, and particulate matter. The children's ability to inhale and exhale deeply improved as they matured, but children from more-polluted areas lagged behind those from cleaner areas, indicating smaller, weaker lungs.

Romieu and her colleagues also showed in a series of studies that the city's pollution worsens asthma in children, particularly those with certain genetic profiles. In 2008 her team analyzed data from 200 asthmatic and healthy children and found that children in areas with more traffic and pollutants coughed, wheezed, and used respiratory medication more often.

FIGURE 1 During a resurgence of smog in March, 2016, Mexico City commuters wore face masks to guard against air pollutants and took advantage of free mass transit when authorities restricted car traffic.

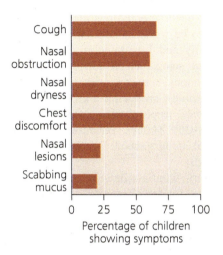

FIGURE 2 Mexico City children show respiratory symptoms from air pollution. These data are from 174 Mexico City children. Of 27 similar children from less-polluted areas outside Mexico City, none showed any of these conditions. *Data from Calderón-Garcidueñas, L., et al., 2003. Respiratory damage in children exposed to urban pollution. Pediatric Pulmonology 36: 148–161.*

Another Mexican researcher, Lilian Calderón-Garcidueñas, has led several studies comparing chest X-ray films and medical records of Mexico City children with those of similar children from nonpolluted locations. Her team found hyperinflation (chronic overexpansion) and other problems with the lungs of the Mexico City youth. Mexico City children also reported many respiratory problems, whereas rural children did not (**FIGURE 2**).

Air pollution harms the heart and the cardiovascular system, too. Multiple studies reveal that pollution can affect heart rate, blood pressure, blood clotting, blood vessels, and atherosclerosis. Epidemiological studies (p. 370) show that pollution correlates with emergency room admissions for heart attacks, chest pain, and heart failure, as well as death from heart-related causes.

Part of the reason smog affects the cardiovascular system is that tiny particulates can work their way into the bloodstream, causing the heart to reduce blood flow or go out of rhythm. The heart mounts an inflammatory response against pollutant particles laden with dead bacteria in the blood, but because the pollution is persistent, the inflammation becomes chronic and stresses the heart, even from an early age. Even young people are at risk. One

Mexican research team analyzed the hearts of 21 people from Mexico City who had died at an early age, and found that pollution exacts a toll before age 18.

Besides affecting the heart and lungs, air pollution can affect the brain. Recent research shows it can damage children's brain tissue in ways similar to Alzheimer's disease. In one study, Calderón-Garcidueñas and her colleagues used brain scans and found that 56% of youth in Mexico City had lesions on the prefrontal cortex, whereas fewer than 8% did in a region with clean air. In another study, they compared 20 children aged 7–8 from Mexico City with 10 similar children from a Mexican city with clean air, measuring their cognitive skills and scanning their brains with magnetic resonance imaging (MRI). Results showed that the Mexico City children performed more poorly on most tests of reasoning, knowledge, and memory. The differences in cognition were consistent with differences in the volume of white matter in key portions of the brain, as revealed by the MRIs.

All these impacts can lead to higher rates of death. Studies by a U.S. and Mexican research team in Mexico City in the late 1990s confirmed this by comparing death certificate records against air pollution measurements. The team found that death rates rose on the day of and the day after severe pollution episodes, especially in response to particulate matter (**FIGURE 3a**). They also found that infant mortality was significantly higher in the days following strong pollution episodes (**FIGURE 3b**).

The extensive research showing a diversity of health impacts from air pollution in Mexico City has caught the attention of city leaders. These scientific findings have helped motivate them to work hard to clean up their city's air.

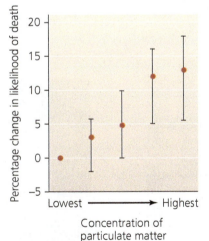

(a) Death rates increase with pollution intensity

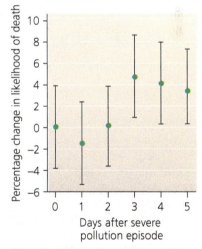

(b) Infant mortality rates are higher after a pollution episode

FIGURE 3 Rates of (a) death and (b) infant mortality each increase in Mexico City with exposure to air pollution. *Data from (a) Borja-Aburto, V., et al., 1997. Ozone, suspended particulates, and daily mortality in Mexico City. American Journal of Epidemiology 145: 258–268; and (b) Loomis, D., et al., 1999. Air pollution and infant mortality in Mexico City. Epidemiology 10: 118–123.*

Since the onset of the industrial revolution, cities have suffered a type of smog known as **industrial smog** (**FIGURE 17.21a**). When coal or oil is burned, some portion is completely combusted, forming CO_2; some is partially combusted, producing CO; and some remains unburned and is released as soot (particles of carbon). Moreover, coal contains contaminants such as mercury and sulfur. Sulfur reacts with oxygen to form sulfur dioxide, which can undergo a series of reactions to form sulfuric acid and other compounds. These substances, along with soot, are the main components of industrial smog.

America's most severe industrial smog event occurred in the small town of Donora, Pennsylvania, in 1948 (**FIGURE 17.21b**). Donora is located in a mountain valley, and one day after air had cooled during the night, the morning sun did not reach the valley floor to warm and disperse the cold air. The resulting temperature inversion trapped smog containing particulate matter emissions from a steel and wire factory. Twenty-one people died, and more than 6000 people—nearly half the town—became ill.

The world's worst industrial smog crisis was even more catastrophic: In 1952, in London, England, a high-pressure system settled over the city for several days, trapping sulfur dioxide and particulate matter emitted from factories and coal-burning stoves and creating foul conditions that killed 4000 people—and by some estimates up to 12,000. In the wake of "killer smog" episodes such as those in London and Donora, governments of developed nations began regulating industrial emissions and greatly reduced industrial smog. However, in industrializing regions such as China, India, and eastern Europe, coal burning and lax pollution control continue to result in industrial smog that poses significant health risks.

Most smog pollution in urban areas today results largely from automobile exhaust. In Mexico City, vehicles contribute 31% of VOCs, 50% of sulfur dioxide, and 82% of nitrogen oxides. Some of these emissions react with sunlight, so pollution tends to be worst in cities with sunny climates, such as Mexico City and Los Angeles. Such cities suffer from **photochemical smog,** which forms when sunlight drives chemical reactions between primary pollutants and atmospheric compounds, producing a mix of more than 100 different chemicals, tropospheric ozone often being the most abundant (**FIGURE 17.22a**). Because it also includes NO_2, photochemical smog generally appears as a brownish haze (**FIGURE 17.22b**).

Hot, sunny, windless days in urban areas provide perfect conditions for the formation of photochemical smog. On a typical weekday, exhaust from morning traffic releases NO and VOCs into a city's air. Sunlight then promotes the production of ozone and other secondary pollutants, leading pollution typically to peak in midafternoon. Photochemical smog irritates people's eyes, noses, and throats, and over time can lead to asthma, lung damage, heart problems, decreased resistance to infection, and even cancer.

We can take steps to reduce smog

Smog afflicts countless American cities and suburbs, from Atlanta to Newark to Baltimore to Houston to Salt Lake

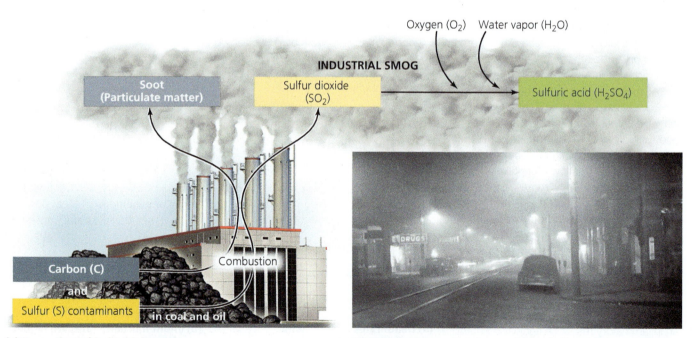

(a) Formation of industrial smog

(b) Donora, Pennsylvania, at midday in its 1948 smog event

FIGURE 17.21 Industrial smog results from fossil fuel combustion. When coal or oil is burned in a power plant or factory, soot (particulate matter of carbon) is released, and sulfur contaminants give rise to sulfur dioxide, which may react with atmospheric gases to produce further compounds **(a)**. Carbon monoxide and carbon dioxide are also emitted. Under certain weather conditions, industrial smog can blanket whole regions, as it did in Donora, Pennsylvania **(b)**, shown in the daytime during its deadly 1948 smog episode.

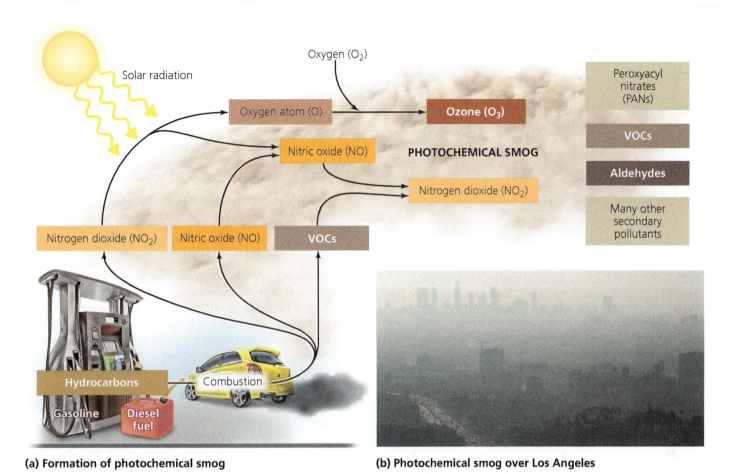

(a) Formation of photochemical smog

(b) Photochemical smog over Los Angeles

FIGURE 17.22 Photochemical smog results when pollutants from automobile exhaust react amid exposure to sunlight. Nitrogen dioxide, nitric oxide, and VOCs initiate a series of chemical reactions **(a)** that produce a toxic brew of secondary pollutants, including ozone, peroxyacyl nitrates (PANs), aldehydes, and others. Photochemical smog is common over many urban areas, especially those with hilly topography or frequent inversions, such as **(b)** Los Angeles.

City. Mayors, city councils, and state and federal regulators everywhere are trying to devise ways to clear their air. Los Angeles's struggle with air pollution began in 1943, when the city's first major smog episode cut visibility to three blocks. With the city's then-predominant image as a clean and beautiful coastal haven at risk, civic leaders confronted the problem head-on. In fact, Los Angeles's quest to understand and solve its smog problem spurred much of the original research into how photochemical smog forms and how automobiles might burn fuel more cleanly.

Los Angeles's city and county officials began their air pollution control efforts by passing ordinances restricting emissions from power plants, oil refineries, and the petrochemical industry, then continued with efforts to cut emissions from motor vehicles. Because air pollution spreads from place to place, responsibility for pollution control soon moved from the city and county levels to the state and federal levels.

California took the lead among U.S. states in adopting pollution control technology and setting emissions standards for vehicles. In 1967 state leaders established the California Air Resources Board, the first state agency focused on regulating air quality. Today in California and 33 other states, drivers are required to have their vehicle exhaust inspected regularly. Inspection programs that require car owners to repair cars that emit excessive pollution have cut these emissions by 30%.

California's demands also helped lead the auto industry to develop less-polluting cars. A study by the nonprofit group Environment California concluded that a new car today generates just 1% of the smog-forming emissions of a 1960s-era car. For this reason, the air is cleaner, even with many more vehicles on the road. In Los Angeles, VOC pollution has declined by 98% since 1960, even though the city's drivers now burn 2.7 times more gasoline. L.A.'s peak smog levels have also decreased substantially since 1980 (**FIGURE 17.23**, p. 464).

Despite its progress, Los Angeles still suffers the worst tropospheric ozone pollution of any U.S. metropolitan area, according to 2016 rankings by the American Lung Association. L.A. residents breathe air exceeding California's health standard for ozone on more than 90 days per year. One recent study calculated that air pollution in the L.A. basin and the nearby San Joaquin Valley each year causes nearly 3900 premature

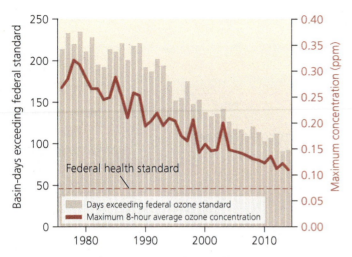

FIGURE 17.23 In the Los Angeles region, tropospheric ozone in photochemical smog has been reduced since the 1970s, thanks to public policy and improved automotive technology. Ozone pollution still violates the federal health standard, however. *Data from South Coast Air Quality Management District.*

DATA Q By roughly what percentage has Los Angeles reduced its ozone pollution since the late 1970s? Calculate changes in each data set shown. Do the two data sets show similar patterns?

Go to **Interpreting Graphs & Data** on **Mastering**EnvironmentalScience°

deaths and costs society $28 billion (due to hospital admissions, lost workdays, etc.).

Across the world, many cities struggle with photochemical smog and are working to develop solutions. One example is Tehran, the capital of Iran. Here, city leaders now require vehicle inspections, regulate traffic into the city center, and pay drivers to trade in old polluting cars for newer cleaner ones. Over the past decade or two, sulfur was reduced in diesel fuel, lead was removed from gasoline, and buses running on (cleaner-burning) natural gas hit the roads. To raise public awareness, 22 electronic billboards were installed around the city, displaying current pollutant levels. All these efforts helped reduce pollution, yet so many people continued to stream into the city and buy cars that pollution soon grew worse again. In response, officials lowered gasoline subsidies, rationed fuel, and began expanding the subway system.

Of all the world's cities, Mexico City is gaining attention today for its success in reducing smog—once the world's worst—even as its population, cars, and economic activity continue to grow. Regulations now require cars to have catalytic converters and get emissions tests. Some industrial facilities were forced to clean up their operations.

weighing the ISSUES

Smog-Busting Solutions

Does the city you live in, or the nearest major city to you, suffer from photochemical smog or other air pollution? How is this city responding? What policies do you think it should pursue? What benefits might your city enjoy from such policies? Would they bring any problems?

Under pressure from city leaders, the national oil company Pemex removed lead from gasoline, improved its refineries, imported cleaner gasoline, and removed pollutants from the liquefied petroleum gas that city residents use for cooking and heating. As a result of all these efforts, plus the expansion of the subway system, low-emission bus fleet, and bike-sharing and car-sharing programs, smog has been reduced since 1990 by more than half. Particulate matter is down by 70%, carbon monoxide by 74%, sulfur dioxide by 86%, and lead by 95%.

Should we regulate greenhouse gases as air pollutants?

Even as the world's nations struggle to reduce various sources of air pollution, they continue to release vast quantities of the greenhouse gases (p. 482) that are driving global climate change (Chapter 18). As a result, our emission of carbon dioxide and other gases that warm the lower atmosphere is arguably today's biggest air pollution problem. Industry and utilities generate much of these emissions, but all of us contribute by living carbon-intensive lifestyles. Each year the average U.S. vehicle driver releases close to 6 metric tons of carbon dioxide, 275 kg (605 lb) of methane, and 19 kg (41 lb) of nitrous oxide, all of them greenhouse gases that drive climate change.

In 2007 the U.S. Supreme Court ruled that the EPA has legal authority under the Clean Air Act to regulate carbon dioxide and other greenhouse gases as air pollutants. President Barack Obama urged Congress to address greenhouse gas emissions through bipartisan legislation. When Congress failed to do so, Obama instructed the EPA to develop regulations for these emissions. In 2011, the EPA introduced moderate carbon emission standards for cars and light trucks, and in 2012 it announced that it would limit carbon emissions for new coal-fired power plants and cement factories (but not existing ones). The EPA decided to phase in regulations gradually, beginning with the largest emitters.

The coal-mining and petrochemical industries objected, and these industries and several states sued to stop the regulations. A court of appeals unanimously upheld the EPA regulations in 2012. The automotive industry supported the EPA's regulations. U.S. automakers had begun investing in fuel-efficient vehicles, and preferred to have one set of federal emissions standards so as not to have to worry about meeting many differing state standards. The public also voiced strong support; 2.1 million Americans sent comments to the EPA in favor of its actions—a record number of public comments for any federal regulation.

In 2015, the EPA finalized and launched a regulatory plan for existing power plants, the Clean Power Plan. Under the plan, states would be allowed to choose how to reduce their plants' emissions—by upgrading technology, switching from coal to natural gas, enhancing efficiency, promoting renewable energy, or through carbon taxes or cap-and-trade programs. The plan aims to cut carbon dioxide emissions from power plants by 32% below 2005 levels by the year 2030. The EPA also estimates that SO_2 and NO_X will be reduced by 20% and that cleaner air will save 3600 lives. All told, the

EPA estimates that by 2030 the plan will bring total public health and climate benefits worth $54 billion each year.

A number of industries and states soon lined up to challenge the plan in court, and in 2016 the Supreme Court issued a stay of the plan in a controversial 5–4 ruling, preventing the EPA from enforcing the plan's requirements until the lawsuits are resolved. The EPA will no doubt continue to face formidable political opposition from emitting industries and from policymakers who fear that regulations will hamper economic growth. Yet if we were able to reduce emissions of other major pollutants sharply since 1970 while advancing our economy, we can hope to achieve similar results in reducing greenhouse gas emissions. Indeed, although U.S. carbon dioxide emissions rose by 51% from 1970 to 2005, they fell by 12% from 2005 to 2015 even as the economy grew. This decrease in emissions resulted from a shift from coal to cleaner-burning natural gas and from improved fuel efficiency in automobiles and other technologies.

Ozone Depletion and Recovery

Although ozone in the troposphere is a pollutant in photochemical smog, ozone is a highly beneficial gas in the stratosphere, where it forms the ozone layer (p. 447; see Figure 17.2). In this region of the stratosphere, concentrations of ozone reach only about 12 parts per million, but ozone molecules are so effective at absorbing the sun's ultraviolet radiation that even this diffuse concentration helps to protect life on Earth's surface from UV radiation's damaging impacts on tissues and DNA.

A generation ago, scientists discovered that our planet's stratospheric ozone was being depleted, posing a major threat to human health and the environment. Years of dynamic research by hundreds of scientists (see **THE SCIENCE BEHIND THE STORY**, pp. 466–467) revealed that certain airborne chemicals destroy ozone and that most of these **ozone-depleting substances** are human-made. Our subsequent campaign to halt degradation of the ozone layer stands as one of society's most successful efforts to address a major environmental problem.

Synthetic chemicals deplete stratospheric ozone

Researchers identifying ozone-depleting substances pinpointed primarily **halocarbons**—human-made compounds derived from simple hydrocarbons (p. 27) in which hydrogen atoms are replaced by halogen atoms, such as chlorine, bromine, or fluorine. In the 1970s, industry was producing more than 1 million tons per year of one type of halocarbon, **chlorofluorocarbons (CFCs).** CFCs were useful as refrigerants, as fire extinguishers, as propellants for aerosol spray cans, as cleaners for electronics, and for making polystyrene foam. Because CFCs rarely reacted with other chemicals, scientists surmised that they would be harmless to people and the environment.

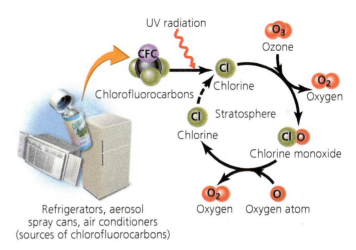

FIGURE 17.24 CFCs destroy ozone in a multistep process, repeated many times. A chlorine atom released from a CFC molecule in the presence of UV radiation reacts with an ozone molecule, forming one molecule of oxygen gas and one chlorine monoxide (ClO) molecule. The oxygen atom of the ClO molecule then binds with a stray oxygen atom to form oxygen gas, leaving the chlorine atom to begin the destructive cycle anew.

Unfortunately, the nonreactive qualities that made CFCs ideal for industrial purposes were having disastrous consequences for the ozone layer. Whereas reactive chemicals are broken down quickly in the troposphere, CFCs reach the stratosphere unchanged and can linger there for a century or more. In the stratosphere, intense UV radiation from the sun eventually breaks CFCs into their constituent chlorine and carbon atoms. In a two-step chemical reaction (**FIGURE 17.24**), each newly freed chlorine atom can split an ozone molecule and then ready itself to split more. During its long residence time in the stratosphere, each free chlorine atom can catalyze the destruction of as many as 100,000 ozone molecules!

The ozone hole appears each year

In 1985, researchers shocked the world when they announced that stratospheric ozone levels over Antarctica in the southern springtime had declined by nearly half during just the previous decade, leaving a thinned ozone concentration that was soon named the **ozone hole** (**FIGURE 17.25**, p. 468). During each Southern Hemisphere spring since then, ozone concentrations over this immense region have dipped to roughly half their historic levels.

Extensive scientific detective work has revealed why seasonal ozone depletion is so severe over Antarctica (and, to a lesser extent, the Arctic). During the dark and frigid Antarctic winter (June to August), temperatures in the stratosphere dip below −80°C (−112°F), enabling unusual high-altitude *polar stratospheric clouds* to form. Many of these icy clouds contain condensed nitric acid, which splits chlorine atoms off from compounds such as CFCs. The freed chlorine atoms accumulate in the clouds, trapped over Antarctica by wind currents that swirl in a circular *polar vortex* that prevents air from mixing with the rest of Earth's atmosphere.

THE SCIENCE
behind the story

Go to **Process of Science** on Mastering EnvironmentalScience®

How Did Scientists Discover Ozone Depletion and Its Causes?

In discovering the depletion of stratospheric ozone and coming to understand the roles of halocarbons and other substances, scientists have relied on historical records, field observations, laboratory experiments, computer models, and satellite technology.

The story starts back in 1924, when British scientist G.M.B. Dobson built an instrument that measured atmospheric ozone concentrations by sampling sunlight at ground level and comparing the intensities of wavelengths that ozone does and does not absorb. By the 1970s, the Dobson ozone spectrophotometer was being used by a global network of observation stations.

Meanwhile, atmospheric chemists were learning how stratospheric ozone is created and destroyed. Ozone and oxygen exist in a natural balance, with one occasionally reacting to form the other, and oxygen being far more abundant. Researchers found that certain chemicals naturally present in the atmosphere, such as hydroxyl (OH) and nitric oxide (NO), destroy ozone, making the ozone layer thinner than it would otherwise be. And nitrous oxide (N_2O) produced by soil bacteria can make its way to the stratosphere and produce NO, Dutch meteorologist Paul Crutzen reported in 1970. This last observation was important, because some human activities, such as fertilizer application, were increasing emissions of N_2O.

Following Crutzen's report, American scientists Richard Stolarski and Ralph Cicerone showed in 1973 that chlorine atoms can catalyze the destruction of ozone even more effectively than N_2O can. And two years earlier, British scientist James Lovelock had developed an instrument to measure trace amounts of atmospheric gases and found that virtually all the chlorofluorocarbons (CFCs) humanity had produced in the past four decades were still aloft, accumulating in the stratosphere.

This set the stage for the key insight. In 1974, American chemist F. Sherwood Rowland and his Mexican postdoctoral associate Mario Molina took note of all the preceding research and realized that CFCs were rising into the stratosphere, being broken down by UV radiation, and releasing chlorine atoms that ravaged the ozone layer (see Figure 17.24, p. 465). Molina and Rowland's analysis, published in the journal *Nature*, earned them the 1995 Nobel prize in chemistry jointly with Crutzen.

The paper also sparked discussion about setting limits on CFC emissions. Industry leaders attacked the research; DuPont's chairman of the board reportedly called it "a science fiction tale . . . a load of rubbish . . . utter nonsense." But measurements in the lab and in the stratosphere by numerous researchers soon confirmed that CFCs and other halocarbons were indeed depleting ozone. In response, the United States and several other nations banned the use of CFCs in aerosol spray cans in 1978. Other uses continued, however, and by the early 1980s global production of CFCs was again on the rise.

Then, a shocking new finding spurred the international community to take action. In 1985, Joseph Farman and colleagues

Releasing a high-altitude balloon equipped to measure stratospheric ozone

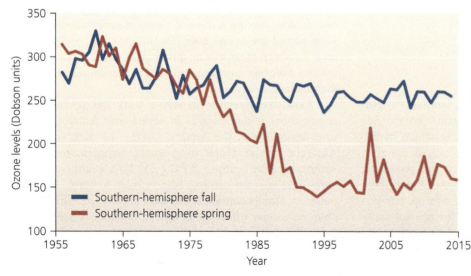

FIGURE 1 Data from Halley, Antarctica, show a decrease in springtime stratospheric ozone concentrations from the 1950s to 1990. Once ozone-depleting substances began to be phased out, ozone concentrations stopped declining. *Data from British Antarctic Survey.*

analyzed data from a British research station in Antarctica that had been recording ozone concentrations since the 1950s. Farman's team reported in *Nature* that springtime Antarctic ozone concentrations had plummeted by 40–60% just since the 1970s (**FIGURE 1**). Farman's team had beaten a group of National Aeronautics and Space Administration (NASA) scientists to the punch. The NASA scientists were sitting on reams of data from satellites showing a global drop in ozone levels, but they had not yet submitted their analysis for publication.

But why should an "ozone hole" be localized over Antarctica, of all places? And why only in the southern spring? To determine what was causing this odd phenomenon, atmospheric chemists Susan Solomon, James Anderson, Crutzen, and others mounted expeditions in 1986 and 1987 to analyze atmospheric gases using ground stations and high-altitude balloons and aircraft. From their data they figured out how the region's polar stratospheric clouds and circulating winds provide ideal conditions for chlorine from CFCs and other chemicals to set in motion the destruction of massive amounts of ozone.

By 1987, the mass of scientific evidence helped convince the world's nations to agree on the Montreal Protocol, which aimed to cut CFC production in half by 1998. Within two years, further scientific evidence and computer modeling showed that more drastic measures were needed. In 1990, the Montreal Protocol was strengthened to include a complete phaseout of CFCs by 2000, in the first of several follow-up agreements. Today, amounts of ozone-depleting substances in the stratosphere are beginning to level off.

As the ozone layer begins a long-term recovery, scientists continue their research. In 2009, a team led by A.R. Ravishankara of the National Oceanic and Atmospheric Administration determined that nitrous oxide (N_2O) had become the leading cause of ozone depletion (**FIGURE 2**). Its emissions are not regulated, so its impacts had come to surpass those that the remaining halocarbons were exerting. Ravishankara's team pointed out that regulating nitrous oxide, which is also a potent greenhouse gas, would help address climate change as well as speed ozone recovery. However, the picture has turned out to be more complicated, because two other greenhouse gases—carbon dioxide and methane—both help to increase amounts of stratospheric ozone, so our emission of these gases is actually helping the ozone layer to recover.

Today scientists are keeping close track of numerous measurements from satellites and from ground-based monitoring stations to gain a detailed chronicle of the ozone layer's slow recovery across the globe (**FIGURE 3**). As atmospheric concentrations of halocarbons continue to decline, the changing balance of gases such as nitrous oxide, methane, and carbon dioxide will play a larger role in determining the speed and extent of the recovery. If all goes well, scientists expect full recovery at midlatitudes by midcentury, followed by full recovery in the polar regions one or two decades later.

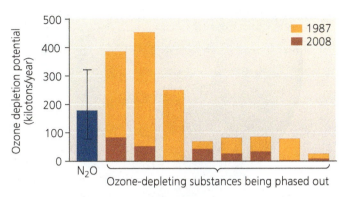

FIGURE 2 Because most ozone-depleting substances were phased out beginning in 1987, nitrous oxide (N_2O; left bar) has become the primary ozone-depleting substance we emit. It has less impact than CFCs and other halocarbons did in 1987 (**full-bar values**), but more impact than any other substance today (**lower portion of bars**). *Adapted from Ravishankara, A.R., et al., 2009. Nitrous oxide (N_2O): The dominant ozone-depleting substance emitted in the 21st century.* Science *326: 123–125, FIGURE 1. Reprinted by permission of AAAS and the author.*

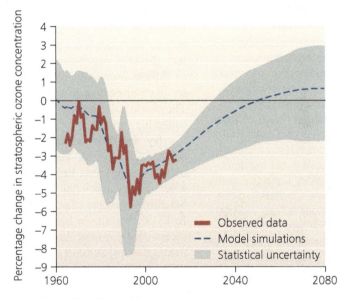

FIGURE 3 Globally, stratospheric ozone is starting to recover. Data averaged from multiple satellite and ground-based sources (**red line**) show a global decrease in observed ozone concentrations, followed by a gradual increase. Simulations from many models (**blue dashed line** shows mean; **gray shading** shows statistical uncertainty) indicate that recovery should be complete later this century. *Adapted from World Meteorological Organization, 2014. Scientific assessment of ozone depletion: 2014. Geneva, Switzerland: WMO, Global Ozone Research and Monitoring Project—Report #55.*

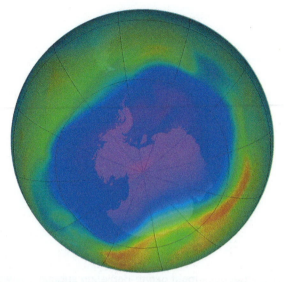

FIGURE 17.25 The "ozone hole" is a vast area of thinned ozone density in the stratosphere over the Antarctic region. It has reappeared seasonally each September in recent decades. This colorized satellite imagery of Earth's Southern Hemisphere from September 24, 2006, shows the ozone hole (**purple/blue**) at its maximal recorded extent to date.

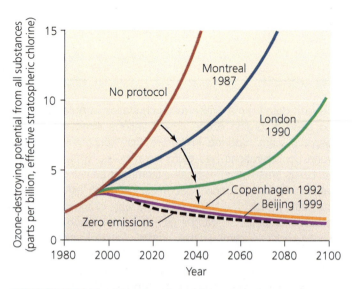

FIGURE 17.26 The Montreal Protocol reduced atmospheric concentrations of pollutants that destroy stratospheric ozone, and follow-up agreements in London, Copenhagen, and Beijing reduced them still more. In this graph, *y*-axis values give a collective measure of ozone-destructive potential from all substances, and most data indicate future projected values. *Data from Emmanuelle Bournay, UNEP/GRID-Arendal, http://maps.grida.no.*

In the Antarctic spring (starting in September), sunshine returns and dissipates the clouds. This releases the chlorine atoms, which begin destroying ozone. The solar radiation also catalyzes chemical reactions, speeding up ozone depletion as temperatures warm. The ozone hole lingers over Antarctica until December, when warmed summer air shuts down the polar vortex, allowing ozone-depleted air to diffuse away and ozone-rich air from elsewhere to stream in. The ozone hole vanishes until the following spring.

By the time scientists had determined most of this, plummeting ozone levels had become a serious international concern. Scientists worried that intensified UV exposure at Earth's surface would promote skin cancer, damage crops, and kill off ocean phytoplankton, the base of the marine food chain.

FAQ

Is the ozone hole related to global warming?

It is a common misconception that the ozone hole is related to global warming. In reality, stratospheric ozone depletion and global warming are completely different issues. Ozone depletion allows excess ultraviolet radiation from the sun to penetrate the atmosphere, but this does not significantly warm or cool the atmosphere. Conversely, global warming does not appreciably affect ozone loss. However, by coincidence many ozone-depleting substances banned by the Montreal Protocol also happen to be greenhouse gases that warm the atmosphere. Thus, although the Montreal Protocol was designed to combat ozone depletion, it is also helping us slow down climate change.

We addressed ozone depletion with the Montreal Protocol

Policymakers responded to the scientific concerns, and international efforts to restrict production of CFCs bore fruit in 1987 with the **Montreal Protocol.** In this treaty, the world's nations agreed to cut CFC production in half by 1998. Follow-up agreements deepened the cuts, advanced timetables for compliance, and added nearly 100 additional ozone-depleting substances (**FIGURE 17.26**). Most substances covered by these agreements have now been phased out, and industry has been able to shift to alternative chemicals. As a result, we have evidently halted the advance of ozone depletion and stopped the ozone hole from worsening (**FIGURE 17.27**)—a success that all humanity can celebrate.

Earth's ozone layer is not expected to recover completely until after 2060. Much of the 5.5 million tons of CFCs emitted into the troposphere has not yet diffused up into the stratosphere, so concentrations may not peak there until 2020.

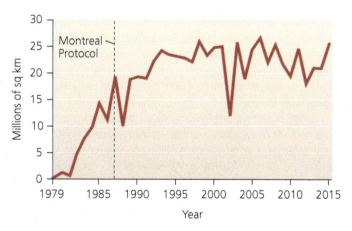

FIGURE 17.27 The Antarctic ozone hole grew quickly after its appearance, but phaseouts of ozone-depleting substances beginning in 1987 have halted its growth. *Data from NASA, reflecting averages from 7 Sept. to 13 Oct. each year.*

Because of this time lag and the long residence times of many halocarbons, we can expect many years to pass before our policies have the desired result.

One challenge in restoring the ozone layer is that nations can plead for some ozone-depleting substances to be exempt from the ban. For instance, the United States was allowed to continue using methyl bromide, a fumigant used to control pests on strawberries. Yet despite the remaining challenges, the Montreal Protocol and its follow-up amendments are widely considered our biggest success story in addressing a global environmental problem. The success has been attributed to several factors:

1. Informative scientific research developed rapidly, facilitated by new and evolving technologies.

2. Policymakers engaged industry in helping to solve the problem. Industry became willing to develop replacement chemicals in part because patents on CFCs were running out and firms wanted to position themselves to profit from next-generation chemicals.

3. Implementation of the Montreal Protocol after 1987 followed an adaptive management approach (p. 309), adjusting strategies in response to new scientific information, technological advances, and economic data.

Because of its success in addressing ozone depletion, the Montreal Protocol is widely viewed as a model for international cooperation to resolve other pressing global problems, from biodiversity loss (p. 291) to persistent organic pollutants (p. 379) to climate change (p. 506).

Addressing Acid Deposition

Just as stratospheric ozone depletion crosses political boundaries, so does another atmospheric pollution concern—**acid deposition,** the deposition of acidic (p. 26) or acid-forming pollutants from the atmosphere onto Earth's surface. As with ozone depletion, we are enjoying some success in addressing this challenge.

Fossil fuel combustion spreads acidic pollutants far and wide

Acid deposition often takes place by precipitation (commonly referred to as **acid rain**) but also may occur by fog, gases, or the deposition of dry particles. Acid deposition is one type of **atmospheric deposition,** which refers more broadly to the wet or dry deposition onto land of a wide variety of pollutants, including mercury, nitrates, organochlorines, and others.

Acid deposition originates primarily with the emission of sulfur dioxide and nitrogen oxides, largely through fossil fuel combustion by automobiles, electric utilities, and industrial facilities. Once airborne, these pollutants react with water, oxygen, and oxidants to produce compounds of low pH (p. 27), primarily sulfuric acid and nitric acid. Suspended in the troposphere, droplets of these acids may travel days or weeks for hundreds of kilometers (**FIGURE 17.28**). Depending on climate, 20–80% of all acidic compounds emitted into the atmosphere may fall in precipitation, with the remainder falling as dry deposition.

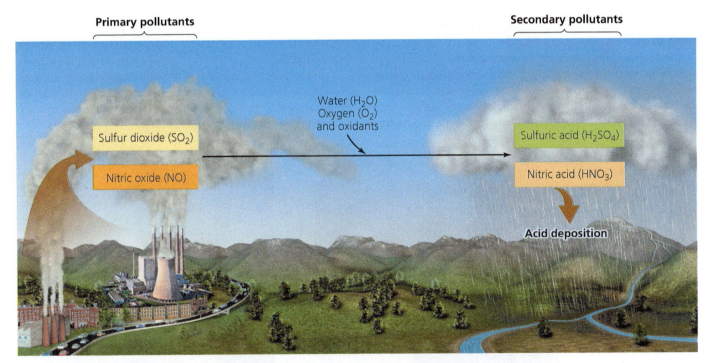

Primary pollutants

Secondary pollutants

Water (H_2O)
Oxygen (O_2)
and oxidants

Sulfur dioxide (SO_2)

Nitric oxide (NO)

Sulfuric acid (H_2SO_4)

Nitric acid (HNO_3)

Acid deposition

FIGURE 17.28 Acid deposition has impacts far downwind from where pollutants are released. Sulfur dioxide and nitric oxide emitted by industries, utilities, and vehicles react in the atmosphere to form sulfuric acid and nitric acid. These acidic compounds descend to Earth's surface in rain, snow, fog, and dry deposition.

Acid deposition has many impacts

Acid deposition has wide-ranging detrimental effects on ecosystems (TABLE 17.1). Acids leach nutrients such as calcium, magnesium, and potassium ions out of the topsoil, altering soil chemistry and harming plants and soil organisms. This occurs because hydrogen ions from acid precipitation take the place of calcium, magnesium, and potassium ions in soil compounds, and these valuable nutrients leach into the subsoil, where they become inaccessible to plant roots.

Acid precipitation also "mobilizes" toxic metal ions such as aluminum, zinc, mercury, and copper by chemically converting them from insoluble forms to soluble forms. Elevated soil concentrations of metal ions such as aluminum damage the root tissues of plants, hindering their uptake of water and nutrients. In some areas, acid fog with a pH of 2.3 (equivalent to vinegar, and more than 1000 times more acidic than normal rainwater) has enveloped forests for extended periods, killing trees. Animals are affected by acid deposition, too; populations of snails and other invertebrates typically decline, and this reduces the food supply for birds.

When acidic water runs off from land, it affects streams, rivers, and lakes. Thousands of lakes in Canada, Europe, the United States, and elsewhere have lost their fish because acid precipitation leaches aluminum ions out of soil and rock and into waterways. These ions damage the gills of fish and disrupt their salt balance, water balance, breathing, and circulation.

The severity of all these effects depends not only on the pH of the deposition but also on the acid-neutralizing capacity of the soil, rock, or water that receives the acidic input. Substrates differ naturally in their chemistry and pH, and regions with more alkaline soil, rock, or water have a greater capacity to buffer themselves against acid deposition. This also means that once calcium or similar ions are leached from a soil, the soil becomes more sensitive to acidification.

Besides altering natural ecosystems, acid precipitation damages crops, erodes stone buildings, and corrodes vehicles, causing billions of dollars in damage. And as acid precipitation erases the writing from tombstones and dissolves away features from ancient cathedrals in Europe, sacred temples in Asia, and revered monuments in Washington, D.C., it hastens the loss of cultural amenities that are beyond monetary value (FIGURE 17.29).

Because the pollutants leading to acid deposition can travel long distances, their effects may be felt far from their sources. For instance, much of the pollution from power plants and factories in Pennsylvania, Ohio, and Illinois travels east with prevailing winds and falls out in states such as New York, Vermont, and New Hampshire. As a result, regions of greatest acidification tend to be downwind from heavily industrialized source areas of pollution.

TABLE 17.1 Ecological Impacts of Acid Deposition

ACID DEPOSITION IN NORTHEASTERN FORESTS HAS . . .

- Accelerated leaching of base cations (ions such as Ca^{2+}, Mg^{2+}, NA^+, and K^+, which counteract acid deposition) from soil

- Allowed sulfur and nitrogen to accumulate in soil, where excess N can overfertilize native plants and encourage weeds

- Increased dissolved inorganic aluminum in soil, hindering plant uptake of water and nutrients

- Leached calcium from needles of red spruce, causing trees to die from wintertime freezing

- Increased mortality of sugar maples due to leaching of base cations from soil and leaves

- Acidified 41% of Adirondack, New York, lakes and 15% of New England lakes

- Diminished lakes' capacity to neutralize further acids

- Elevated aluminum levels in surface waters

- Reduced species diversity and abundance of aquatic life, affecting entire food webs

Adapted from Driscoll, C.T., et al., 2001. *Acid rain revisited*. Hubbard Brook Research Foundation. © 2001 C.T. Driscoll. Used with permission.

FIGURE 17.29 **Acid deposition corrodes statues and buildings.** Shown is an Egyptian obelisk known as Cleopatra's Needle, in Central Park, New York City, **(a)** before and **(b)** after significant acid deposition.

(a) Before acid rain damage

(b) After acid rain damage

We are addressing acid deposition

The Acid Rain Program established under the Clean Air Act of 1990 has helped fight acid deposition in the United States. This program set up an emissions trading system (p. 179) for sulfur dioxide. Coal-fired power plants were allocated permits for emitting SO_2 and could buy, sell, or trade these allowances. Each year the overall amounts of allowed pollution were decreased. (See p. 509 for further explanation of how such a system works.) The economic incentives created by this cap-and-trade program encouraged polluters to switch to low-sulfur coal, invest in technologies such as scrubbers, and devise other ways to become cleaner and more efficient. During the course of the cap-and-trade program, SO_2 emissions across the United States fell by 67% (**FIGURE 17.30**). As a result, average sulfate loads in precipitation across the eastern United States were 51% lower in 2008–2010 than in 1989–1991.

The Acid Rain Program also required power plants to reduce nitrogen oxide emissions, with the EPA allowing plants flexibility in how they implemented the reductions. Emissions of NO_X fell significantly as a result, and wet nitrogen deposition declined, as well. Thanks to the declines in SO_2 and NO_X, air and water quality improved throughout the eastern United States (**FIGURE 17.31**). This market-based program spawned similar cap-and-trade programs for other pollutants, including greenhouse gases (p. 482). The Los Angeles region adopted its own cap-and-trade program in 1994. The RECLAIM (Regional Clean Air Incentives Market) program has helped the L.A. basin reduce emissions of sulfur oxides and nitrogen oxides by more than 70%.

Many have attributed the success in reducing acid deposition at the national level to the Acid Rain Program, and the EPA has calculated that the program's economic benefits (in health care expenses avoided, for instance) outweighed its costs by 40 to 1. However, some experts maintain that pollution declined because cleaner fuels became less expensive and because simultaneous conventional regulation mandated emissions cuts. Indeed, during this time period European nations using command-and-control regulation (p. 177) instead of emissions trading reduced their SO_2 emissions by even more than the United States did. In 2011, emissions trading ended once the EPA issued its Cross-State Air Pollution Rule, which aimed to limit pollution drifting from upwind states into downwind states.

As with recovery of the ozone layer, there is a time lag before the positive consequences of emissions cuts kick in, so it will take years for acidified ecosystems to recover. Research indicates that soils across the northeastern United States are showing signs of recovery, but that they remain degraded and vulnerable. Scientists also point out that further

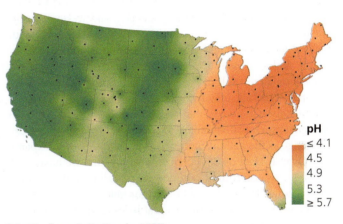

(a) pH of precipitation in 1990

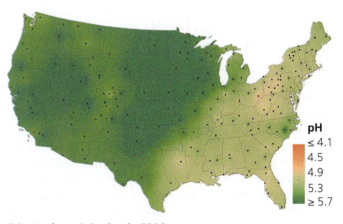

(b) pH of precipitation in 2014

FIGURE 17.31 Precipitation has become less acidic as air quality has improved under the Clean Air Act. Average pH values for precipitation rose between **(a)** 1990 and **(b)** 2014. Precipitation remains most acidic in regions near and downwind from areas of heavy industry. *Data from the National Atmospheric Deposition Program.*

DATA In the area where you live, how did the pH of precipitation change between 1990 and 2014? Has precipitation become more acidic or less acidic?

Go to **Interpreting Graphs & Data** on Mastering**Environmental**Science®

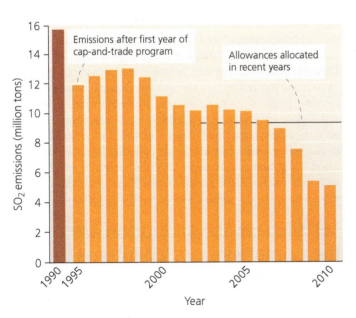

FIGURE 17.30 Sulfur dioxide emissions fell 67% in the wake of an emissions trading system. By 2010, emissions from U.S. power plants participating in this EPA program mandated by the 1990 Clean Air Act had dropped well below the amount allocated in permits (**black line**). *Data from U.S. EPA.*

pollution reductions are needed if we are to fully restore ecosystems in the Northeast and prevent further damage to property and infrastructure.

In the meantime, some researchers are jump-starting the process of ecosystem recovery. At Hubbard Brook Experimental Forest in New Hampshire, where scientists first studied acid deposition's effects in the United States, scientists used a helicopter to distribute 50 tons of a calcium-containing mineral called wollastonite over one watershed. Within three years of this experimental application, topsoil pH rose from 3.9 to 4.2. Sugar maples (one of the forest's key tree species that had been declining because of acid deposition) began producing healthier foliage, thicker root growth, more seeds, and more surviving seedlings. Over the next 50 years, scientists plan to evaluate the impact of calcium addition on the watershed's soil, water, and life, and compare these results to watersheds where calcium remains depleted.

Although the United States, Canada, and western Europe are beginning to recover from acid deposition after cutting sulfur emissions, acid deposition is becoming worse in industrializing nations. Today China emits the most sulfur dioxide of any nation as a result of coal combustion in power plants and factories that lack effective pollution control equipment. Not surprisingly, China has the world's worst acid rain problem. The government is beginning to tackle the issue, but it faces a challenge as the nation's industrial sector continues to expand.

Overall, data on acid deposition show that we have made advances in controlling outdoor air pollution, but that more can be done. The same can be said for indoor air pollution—a source of human health threats that is less familiar to most of us but statistically more dangerous.

Indoor Air Quality

Indoor air generally contains higher concentrations of pollutants than does outdoor air. As a result, the health impacts from **indoor air pollution** in workplaces, schools, and homes outweigh those from outdoor air pollution. The World Health Organization (WHO) attributes nearly 3.5 million premature deaths each year to indoor air pollution (compared with 3.3 million for outdoor air pollution). Indoor air pollution takes nearly 10,000 lives each day.

If this seems surprising, consider that the average American spends at least 90% of his or her time indoors. Then consider the dizzying array of consumer products in our homes and offices that play major roles in our daily lives. Many of these products are made of synthetic materials, and novel synthetic substances are not comprehensively tested for health effects before being brought to market (Chapter 14). Furniture, carpeting, cleaning fluids, insecticides, and plastics all exude volatile chemicals into the air.

Ironically, some attempts to be environmentally prudent during the energy crises of the 1970s (p. 540) worsened indoor air quality. To improve energy efficiency by reducing heat loss, building managers sealed off ventilation in buildings, and designers constructed new buildings with limited ventilation and with windows that did not open. These steps saved energy, but they also trapped stable, unmixed air—and pollutants—inside.

FIGURE 17.32 In the developing world, many people build fires indoors for cooking and heating, as in this Maasai home in Kenya. Indoor fires expose people to severe pollution from particulate matter and carbon monoxide.

Burning fuel causes indoor pollution in the developing world

Indoor air pollution has by far the greatest impact in the developing world, where poverty forces millions of people to burn wood, charcoal, animal dung, or crop waste inside their homes for cooking and heating, with little or no ventilation (**FIGURE 17.32**). In the air of such homes, WHO researchers have found that concentrations of particulate matter are commonly 20 times above U.S. EPA standards. As a result, people inhale dangerous amounts of soot, carbon monoxide, and other pollutants, which together increase risks of premature death by pneumonia, bronchitis, and lung cancer, as well as allergies, sinus infections, cataracts, asthma, emphysema, and heart disease. International health researchers estimate that indoor air pollution from burning fuelwood, dung, and coal is responsible for nearly 7% of all deaths each year. Many people are not aware of the health risks, and of those who are, many are too poor to have viable alternatives.

Tobacco smoke and radon are the primary indoor pollutants in industrialized nations

In industrialized nations, the primary indoor air health risks are cigarette smoke and radon (a naturally occurring radioactive gas). Smoking cigarettes irritates the eyes, nose, and throat; worsens asthma and other respiratory ailments; and greatly increases the risk of lung cancer and heart disease. Inhaling secondhand smoke (smoke inhaled by a nonsmoker who is nearby or shares

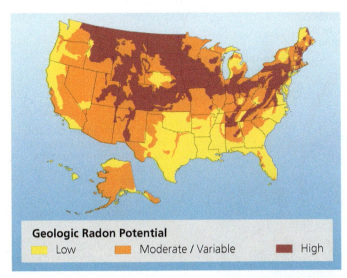

Geologic Radon Potential

Low	Moderate / Variable	High

FIGURE 17.33 One's risk from radon depends largely on underground geology. This map shows levels of risk across the United States. *Data from U.S. Geological Survey, 1993. Generalized geologic radon potential of the United States.*

an enclosed airspace with a smoker) causes many of the same problems. This hardly seems surprising when one considers that tobacco smoke is a brew of more than 4000 chemical compounds, over 250 of which are known or suspected to be toxic or carcinogenic. Smoking has become less prevalent in developed nations in recent years as a result of public education campaigns, and many public and private venues now ban smoking. Still, smoking is estimated in the United States alone to cause 160,000 lung cancer deaths per year, and secondhand smoke to cause 3000.

Radon is the second-leading cause of lung cancer in the developed world, responsible for an estimated 21,000 deaths per year in the United States and for 15% of lung cancer cases worldwide. Radon (p. 361) is a radioactive gas resulting from the natural decay of uranium in soil, rock, or water. It seeps up from the ground and can infiltrate buildings. Colorless and odorless, radon's presence can be impossible to predict without knowing an area's underlying geology (**FIGURE 17.33**). The only way to determine whether radon is entering a building is to sample air with a test kit. The EPA estimates that 6% of U.S. homes exceed its safety standard for radon. Since the 1980s, millions of U.S. homes have been tested for radon and close to a million have undergone radon mitigation. New homes are being built with radon-resistant features.

Many VOCs pollute indoor air

In our daily lives at home, we are exposed to many indoor air pollutants (**FIGURE 17.34**, p. 474). The most diverse are volatile organic compounds (p. 455). These airborne carbon-containing compounds are released by plastics, oils, perfumes, paints, cleaning fluids, adhesives, and pesticides. VOCs evaporate from furnishings, building materials, color film, carpets, laser printers, fax machines, and sheets of paper. Some products, such as chemically treated furniture, release large amounts of VOCs when new and progressively

less as they age. Other items, such as photocopying machines, emit VOCs each time they are used.

Although we are surrounded by products that emit VOCs, they are released in very small amounts. Studies find concentrations of VOCs in buildings to be nearly always less than 1 part per 10 million. This is, however, a much greater concentration than is found outdoors. Moreover, we experience instances of high exposure. The "new car smell" that fills the interiors of new automobiles comes from a complex mix of dozens of VOCs as they outgas from the newly manufactured plastic, metal, and leather components of the car. The smell diminishes with time, but some scientific studies warn of health risks from this mix and recommend that you keep a new car well ventilated.

The implications for human health of chronic exposure to VOCs are far from clear. Because they generally exist in low concentrations and because individuals regularly are exposed to mixtures of many different types, it is extremely difficult to study the effects of any one pollutant. An exception is formaldehyde, a VOC that has clear and known health impacts. Formaldehyde off-gasses from the glues and resins in pressed wood (such as plywood), insulation, and other products. It can irritate mucous membranes, can induce skin allergies, and is considered a likely carcinogen.

weighing the ISSUES

How Safe Is Your Indoor Environment?

Think about the amount of time you spend indoors. Name some potential indoor air quality hazards in your home, work, or school environment. Are these spaces well ventilated? What could you do to improve the safety of the indoor spaces you use?

Living organisms can pollute

The most widespread source of indoor air pollution in the developed world may be living organisms. Tiny dust mites can worsen asthma and cause allergies, as can dander (skin flakes) from pets. The airborne spores of some fungi, molds, and mildews can cause allergies, asthma, and other respiratory ailments. Some airborne bacteria can cause infectious disease, including legionnaires' disease. Of the estimated 10,000–15,000 annual U.S. cases of legionnaires' disease, 5–15% are fatal. Heating and cooling systems in buildings make ideal breeding grounds for microbes, providing moisture, dust, and foam insulation as substrates, along with air currents to carry the organisms aloft.

Microbes that induce allergic responses are thought to be a major cause of building-related illness, a sickness produced by indoor pollution. When the cause of such an illness is a mystery, and when symptoms are general and nonspecific, the illness is often called **sick building syndrome.** The U.S. Occupational Safety and Health Administration (OSHA) estimates that 30–70 million Americans have suffered ailments related to the building in which they live. We can reduce the prevalence of sick building syndrome by using low-toxicity construction materials and ensuring that buildings are well ventilated.

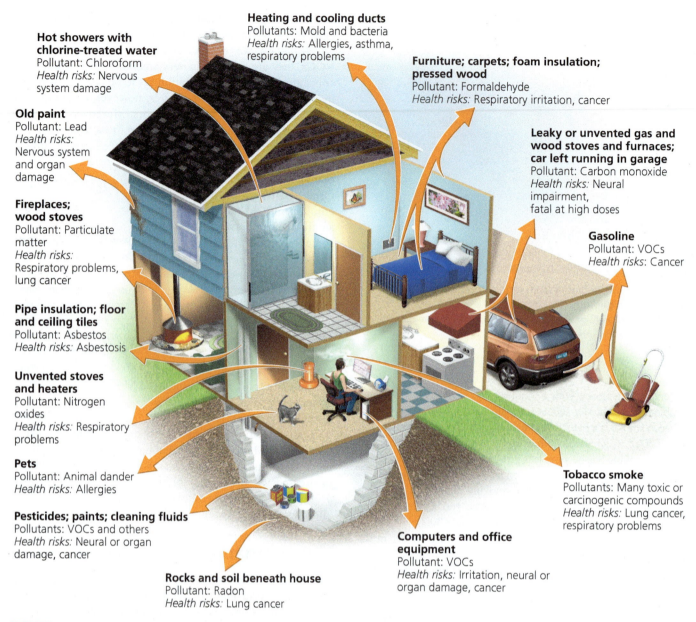

Hot showers with chlorine-treated water
Pollutant: Chloroform
Health risks: Nervous system damage

Old paint
Pollutant: Lead
Health risks: Nervous system and organ damage

Fireplaces; wood stoves
Pollutant: Particulate matter
Health risks: Respiratory problems, lung cancer

Pipe insulation; floor and ceiling tiles
Pollutant: Asbestos
Health risks: Asbestosis

Unvented stoves and heaters
Pollutant: Nitrogen oxides
Health risks: Respiratory problems

Pets
Pollutant: Animal dander
Health risks: Allergies

Pesticides; paints; cleaning fluids
Pollutants: VOCs and others
Health risks: Neural or organ damage, cancer

Heating and cooling ducts
Pollutants: Mold and bacteria
Health risks: Allergies, asthma, respiratory problems

Furniture; carpets; foam insulation; pressed wood
Pollutant: Formaldehyde
Health risks: Respiratory irritation, cancer

Leaky or unvented gas and wood stoves and furnaces; car left running in garage
Pollutant: Carbon monoxide
Health risks: Neural impairment, fatal at high doses

Gasoline
Pollutant: VOCs
Health risks: Cancer

Tobacco smoke
Pollutants: Many toxic or carcinogenic compounds
Health risks: Lung cancer, respiratory problems

Computers and office equipment
Pollutant: VOCs
Health risks: Irritation, neural or organ damage, cancer

Rocks and soil beneath house
Pollutant: Radon
Health risks: Lung cancer

FIGURE 17.34 The typical home contains many sources of indoor air pollution. Shown are common sources, the major pollutants they emit, and some of the health risks they pose.

We can enhance indoor air quality

Using low-toxicity materials, monitoring air quality, keeping rooms clean, and providing adequate ventilation are the keys to alleviating indoor air pollution in most situations. Remedies for fuelwood pollution in the developing world include drying wood before burning (which reduces the amount of smoke produced), cooking outside, shifting to less-polluting fuels (such as natural gas), and replacing inefficient fires with cleaner stoves that burn fuel more efficiently. Installing hoods, chimneys, or cooking windows can increase ventilation for little cost, alleviating most indoor smoke pollution.

In the industrialized world, we can try to avoid cigarette smoke, limit our exposure to new plastics and treated wood, and restrict our contact with pesticides, cleaning fluids, and other toxic substances by keeping them in garages or outdoor sheds. The EPA recommends that we test our homes and offices for radon, mold, and carbon monoxide. Because carbon monoxide is so deadly and so hard to detect, most U.S. states now require new homes to be equipped with alarms that sound if they detect dangerous levels of CO. In addition, keeping rooms and air ducts clean and free of mildew and other biological pollutants will reduce potential irritants and allergens. Most of all, keeping our indoor spaces well ventilated will minimize concentrations of the pollutants among which we live.

Progress is being made worldwide in alleviating the health toll of indoor air pollution. Researchers calculate that rates of premature death from indoor air pollution dropped nearly 40 percent from 1990 to 2010. With continued efforts, we should see additional progress in safeguarding people's health.

closing
THE LOOP

Air quality is vitally important for our health both indoors and outdoors, in cities and rural areas, and in developed and developing nations. Los Angeles was among the first cities to confront severe air pollution and take major steps to alleviate the problem. Many other cities and national governments of developed nations subsequently achieved cleaner air through public policy and technological advances. Now many cities of developing nations are following suit and taking steps to clean up their air. Mexico City has been a pioneer, and its success is making it a model for others. Urban areas in China, India, and elsewhere have a long way to go but are beginning to make progress in tackling outdoor air pollution, while in all nations the growing awareness of indoor air pollution is encouraging action and solutions.

Outdoor air pollution in any location is influenced not only by our emissions but also by natural sources of pollution and by atmospheric conditions. The more we understand about the science of the atmosphere, the better we can protect our health against pollution. Likewise, science has proven crucial to gaining a solid understanding of two other major air quality issues, ozone depletion and acid deposition. Policymakers responded quickly to scientific findings on stratospheric ozone depletion, and as a result, our global society appears to have dodged a bullet; today our planet's ozone layer is gradually on the mend. With acid deposition, we reacted to scientific research by launching policies and economic programs to reduce emissions of the acidic pollutants that lead to the problem, and ecosystems are now beginning to recover. As the world's less-wealthy nations industrialize, continued integration of science, policy, economics, and technology should help us achieve even cleaner air in the future.

REVIEWING Objectives

You should now be able to:

Describe the composition, structure, and function of Earth's atmosphere

The atmosphere consists of 78% nitrogen gas, 21% oxygen gas, and various other gases in minute concentrations. It includes four layers, across which temperature and other attributes vary. Ozone is concentrated in the stratosphere. The atmosphere moderates climate, provides oxygen, conducts and absorbs solar radiation, and transports and recycles nutrients and waste. (pp. 446–448)

Relate weather and climate to atmospheric conditions

The sun's energy heats the atmosphere; drives air circulation; and influences weather, climate, and the seasons. Weather is a short-term phenomenon, whereas climate is a long-term phenomenon. Fronts, pressure systems, and the interactions among air masses influence weather. Global convective cells create latitudinal climate zones. (pp. 448–452)

Identify major outdoor air pollutants and outline the scope of air pollution

Natural sources such as fires, volcanoes, and windblown dust pollute the atmosphere, and human activity can worsen some of their impacts. Most pollution is caused by people, however, and we emit pollutants from point and non-point sources. Major pollutants include carbon monoxide, sulfur dioxide, nitrogen oxides, VOCs, particulate matter, lead, nitrogen dioxide, and tropospheric ozone. Photochemical smog, created by chemical reactions of pollutants in the presence of sunlight, impairs visibility and human health widely in urban areas, whereas industrial smog from fossil fuel combustion remains a problem in urban and industrial areas of many developing nations. Today industrializing nations such as China and India are experiencing the world's worst air pollution. (pp. 452–463)

Assess strategies and solutions for control of outdoor air pollution

Thanks to public policy (such as the U.S. Clean Air Act) and to pollution control technologies (such as scrubbers and catalytic converters), pollutant emissions in the United States and many other wealthy nations have decreased substantially since 1970, and ambient air quality has improved in most respects. Today the U.S. EPA is taking early steps toward regulating greenhouse gases as pollutants because they drive climate change. Throughout the world, cities such as Los Angeles and Mexico City have taken bold steps to address photochemical smog. Governments of industrializing nations such as China are beginning to combat air pollution, as well. (pp. 454–465)

Explain stratospheric ozone depletion and identify steps taken to address it

CFCs and other persistent human-made compounds destroy ozone in the stratosphere; ozone depletion is most severe over Antarctica, where an "ozone hole" appears each spring. Thinning ozone concentrations pose dangers to life because they allow more ultraviolet radiation to reach Earth's surface. Fortunately, the

Montreal Protocol and its follow-up agreements have proven remarkably successful in reducing emissions of ozone-depleting substances. However, the long residence time of CFCs in the atmosphere accounts for a time lag between the protocol and full restoration of stratospheric ozone. (pp. 465–469)

and infrastructure. Regulation, cap-and-trade programs, and technology have all helped to reduce acid deposition in North America, and industrializing nations are beginning to tackle the problem, as well. (pp. 469–472)

Describe acid deposition, discuss its consequences, and explain how we are addressing it

Acid deposition results when pollutants such as SO_2 and NO react in the atmosphere to produce acids that are deposited on Earth's surface. Acid deposition may be wet or dry, and may occur a long distance from the source of pollution. Acid deposition damages soils, water bodies, plants, animals, ecosystems, and human property

Characterize the scope of indoor air pollution and assess solutions

Indoor air pollution causes more deaths and health problems worldwide than outdoor air pollution. Indoor burning of fuelwood is the developing world's primary indoor air pollution risk, whereas tobacco smoke and radon are the worst indoor pollutants in the developed world. VOCs and living organisms commonly pollute indoor air. Using low-toxicity materials, keeping spaces clean, monitoring air quality, and maximizing ventilation all help to enhance indoor air quality. (pp. 472–474)

TESTING Your Comprehension

1. About how thick is Earth's atmosphere? Name one characteristic of the troposphere and one characteristic of the stratosphere.

2. Where is the "ozone layer" located? Describe how and why stratospheric ozone is beneficial for people, whereas tropospheric ozone is harmful.

3. How does solar energy influence weather and climate? Describe how Hadley, Ferrel, and polar cells help to determine climate patterns and the location of biomes.

4. Describe a temperature inversion. Explain how inversions contribute to severe smog episodes such as the ones in London and in Donora, Pennsylvania.

5. How does a primary pollutant differ from a secondary pollutant? Give an example of each.

6. What has happened with the emissions of major pollutants in the United States in recent decades? What has

happened with concentrations of "criteria pollutants" in U.S. ambient air in recent decades? Name one health risk from toxic air pollutants.

7. How does photochemical smog differ from industrial smog? Give three examples of the health risks posed by the outdoor air pollutants in smog.

8. Explain how chlorofluorocarbons (CFCs) deplete stratospheric ozone. Why is this depletion considered a long-term international problem? What was done to address this problem?

9. Why are the effects of acid deposition often felt in areas far from where the primary pollutants are produced? List three impacts of acid deposition.

10. Name three common sources of indoor pollution and their associated health risks. For each pollution source, describe one way to reduce exposure to the source.

SEEKING Solutions

1. Name one type of natural air pollution, and discuss how human activity can sometimes worsen it. What potential solutions can you think of to minimize this human impact?

2. Explain how and why emissions of major pollutants have been reduced by well over 50% in the United States since 1970, despite increases in population, energy use, and economic activity. Describe at least two ways you think air quality might be further improved.

3. International action through a treaty has helped to halt further stratospheric ozone depletion, but other transboundary pollution issues, including acid deposition, have not yet been addressed as effectively. What types

of actions do you feel are appropriate for pollutants that cross political boundaries?

4. **CASE STUDY CONNECTION** Describe five ways in which Los Angeles or Mexico City has responded to its pollution challenges. What impacts have each of these responses had? Now consider a major city that is near where you live. Describe at least one approach used by L.A. or Mexico City that you feel would help address air pollution in your city, and explain why.

5. **THINK IT THROUGH** Suppose that you are the head of your county health department, and the EPA informs you that your county has failed to meet the national ambient

air quality standards for ozone, sulfur dioxide, and nitrogen dioxide. Your county is partly rural but is home to a city of 200,000 people and 10 sprawling suburbs. There are several large and aging coal-fired power plants, a number of factories with advanced pollution control technology, and no public transportation system. Describe at least three steps you would urge the county government to take to meet the air quality standards. Explain specifically what effects you would expect each of these steps to have, and why.

6. **THINK IT THROUGH** You have just taken a job at a medical clinic in your hometown. The nursing staff has asked you to develop a brochure for patients featuring tips on how to minimize health impacts from air pollution (both indoor and outdoor) in their daily lives. List the top five tips you will feature, and explain for each why you will include it in your brochure.

CALCULATING Ecological Footprints

"While only some motorists contribute to traffic fatalities, all motorists contribute to air pollution fatalities." So stated a writer for the Earth Policy Institute, pointing out that air pollution kills far more people than vehicle accidents. According to EPA data, emissions of nitrogen oxides in the United States in 2014 totaled 12.2 million tons. Nitrogen oxides come from fuel combustion in motor vehicles, power plants, and other industrial, commercial, and residential sources, but fully 7.2 million tons of the 2014 total came from vehicles. The U.S. Census Bureau estimates the nation's population to have been 318.9 million in 2014 and projects that it will reach 359.4 million in 2030. Considering these data, calculate the missing values in the table to the right (1 ton = 2000 lb).

	TOTAL NO_x EMISSIONS (lb)	NO_x EMISSIONS FROM VEHICLES (lb)
You		
Your class		
Your state		
United States	24.4 billion	14.4 billion

Data from U.S. EPA.

1. By what percentage is the U.S. population projected to increase between 2014 and 2030? Do you think that NO_x emissions will increase, decrease, or remain the same over that period of time? Why? (You may want to refer to Figure 17.14.)

2. Assume you are an average American driver. Using the 2014 emissions totals, how many pounds of NO_x emissions are you responsible for creating? How many pounds would you prevent if you were to reduce by half the vehicle miles you travel? What percentage of your total NO_x emissions would that be?

3. How might you reduce your vehicle miles traveled by 50%? What other steps could you take to reduce the NO_x emissions for which you are responsible?

MasteringEnvironmentalScience®

Students Go to **MasteringEnvironmentalScience** for assignments, the etext, and the Study Area with practice tests, videos, current events, and activities.

Instructors Go to **MasteringEnvironmentalScience** for automatically graded activities, current events, videos, and reading questions that you can assign to your students, plus Instructor Resources.

CHAPTER 18

Global
Climate Change

CASE STUDY

Rising Seas Threaten South Florida

FLORIDA

Atlantic
Ocean

Gulf of
Mexico   Miami •

> **Miami, as we know it today, is doomed. It's not a question of if. It's a question of when.**
> University of Miami geologist Dr. Harold Wanless

> **Miami Beach is not going to sit back and go underwater.**
> Philip Levine, mayor of Miami Beach

It happens now in Miami at least six times a year. Salty water bubbles up from drains, seeps up from the ground, fills the streets, and spills across lawns and sidewalks. Under the dazzling sun of a South Florida sky, floodwaters stall car traffic, creep into doorways, force businesses to close, and keep people from crossing the street. Employees struggle to get to work while tourists stand around, baffled.

The flooding is most severe in Miami Beach, the celebrated strip of glamorous hotels, clubs, shops, and restaurants that rises from a seven-mile barrier island just offshore from Miami. The carefree affluent image of Miami Beach, with its sun and fun, is increasingly jeopardized by the grimy reality of these unwelcome saltwater intrusions. By 2030, flooding is predicted to strike Miami and Miami Beach about 45 times per year—becoming no longer a curious inconvenience, but an existential threat.

These mysterious floods that seem to come out of nowhere are a recent phenomenon, so Miami-area residents are just now coming to realize that their coastal metropolis is slowly being swallowed by the ocean. The cause? Rising sea levels driven by global climate change.

The world's oceans rose 20 cm (8 in.) in the 20th century as warming temperatures expanded the volume of seawater and caused glaciers and ice sheets to melt, discharging water into the oceans. These processes are accelerating today, and scientists predict that sea level will rise another 26–98 cm (10–39 in.) or more in this century as climate change intensifies.

As sea levels rise, coastal cities across the globe—from Venice to Amsterdam to New York to San Francisco—are facing challenges. In the United States, scientists find that the Atlantic Seaboard and the Gulf Coast are especially vulnerable. The hurricane-prone shores of Florida, Louisiana, Texas, and the Carolinas are at risk, as are coastal cities such as Houston and New Orleans. From Cape Cod to Corpus Christi, millions of Americans who live in shoreline communities are beginning to suffer significant expense, disruption to daily life, and property damage as beaches erode, neighborhoods flood, aquifers are fouled, and storms strike with more force.

Perhaps nowhere in America is more vulnerable to sea-level rise than Miami and its surrounding communities in South Florida. Six million people live in this region, and three-quarters of them inhabit low-lying coastal areas that also hold most of the region's wealth and property. Experts calculate that Miami alone has more than $400 billion in assets at risk from sea-level rise—more than

Upon completing this chapter, you will be able to:

- Describe Earth's climate system and explain the factors that influence global climate

- Identify greenhouse gases and characterize human influences on the atmosphere and on climate

- Summarize how researchers study climate

- Outline current and expected future trends and impacts of climate change in the United States and around the world

- Suggest and assess ways we may respond to climate change

Bulldozing beach sand off a Fort Lauderdale boulevard after a storm surge

A motorist stranded in Miami floodwaters

FIGURE 18.1 In Miami, billions of dollars of property and infrastructure are located within just meters of the ocean.

any other city in the world (**FIGURE 18.1**). Hurricanes strike the area frequently, and every centimeter of sea-level rise makes the surge of seawater from a storm more expansive, costly, and dangerous.

South Florida is highly sensitive to sea-level change because its landscape is exceptionally flat; just 1 meter of sea-level rise would inundate more than a third of the region. A 4-m (13-ft) rise in sea level would submerge Miami and reduce the region to a handful of small islands.

The porous limestone bedrock that underlies South Florida also poses a challenge. Pockmarked with holes like Swiss cheese, this permeable rock lets water percolate through, like a sieve. This is why Miami's floods seem to come from out of nowhere; during the highest tides of the year, ocean water is forced inland, where it mixes with fresh water underground and is pushed up as a briny mixture through the limestone directly into yards and streets. As a result, Miami and its neighboring cities cannot simply wall themselves off from a rising ocean, because seawalls won't stop water from seeping up from below.

Moreover, as salt water moves inland, it contaminates the fresh drinking water of South Florida's Biscayne Aquifer. Fort Lauderdale and several other communities are already struggling with saltwater incursion. Florida is building desalination plants to convert seawater to drinking water, but desalination (p. 402) is expensive and consumes large amounts of energy.

Despite all these challenges, Miami Beach today is undergoing a building boom. Multi-million-dollar high-rise condos are sprouting one after another, and real-estate prices are soaring. Wealthy buyers are snapping up new buildings at record prices, even as the ocean begins lapping at their foundations. All along, politicians have been cheering on the building spree.

Many of Florida's top state-level politicians have long been in denial about climate change, but today Miami-area leaders and citizens are taking action to safeguard their region's future. Commissioners of Broward, Miami-Dade, Monroe, and Palm Beach Counties in 2010 adopted an agreement to work together on strategies to combat climate change and its effects in the region. This agreement, the Southeast Florida Regional Climate Change Compact, is garnering wide recognition as a model for regional cooperation on climate issues. Despite a lack of money from the state and federal governments, these policymakers are helping to build up dunes, raise building foundations, shift development inland, and stop subsidizing insurance for development in low-lying coastal areas.

In Miami Beach, Mayor Philip Levine won election to office in 2013 after a campaign ad showed him paddling a kayak through the streets of the South Beach neighborhood, promising to address flooding. "I wasn't swept into office," Levine is fond of saying. "I floated in." Under Levine, the city has raised some roadways 3 feet, and businesses are being urged to remodel their first floors. The city raised stormwater charges on residents and is spending $400 million installing a system of massive pumps to extract floodwater. Engineers expect these measures to get the city through the next couple of decades, but they recognize that more interventions will be needed later.

Only time will tell whether South Florida's communities will overcome their challenges and provide a shining example for other regions. In Miami and many other coastal cities, vast sums will be spent on pumps, drains, pipes, seawalls, and other engineering solutions, but ultimately these are only temporary fixes; they may buy time, but they cannot stop the water forever. In the long term, only reducing our emissions of greenhouse gases will halt sea level rise and the many other imminent consequences of global climate change.

Our Dynamic Climate

Climate influences virtually everything around us, from the day's weather to major storms, from crop success to human health, and from national security to the ecosystems that support our economies. In one way or another, the accelerating change in our climate today will affect each and every one of us for the remainder of our lives. If you are a student in your teens or twenties, climate change may well be *the* major event of your lifetime and the phenomenon that most shapes your future.

Climate change is also the fastest-developing area of environmental science. New scientific studies that refine our understanding of climate are published every week, and policymakers and businesspeople make decisions and take actions just as quickly in response. Because new developments are occurring so quickly, we urge you to explore beyond this book, with your instructor and on your own, the most recent information on climate change and the impacts it will have on your future.

What is climate change?

Climate describes an area's long-term atmospheric conditions, including temperature, precipitation, wind, humidity, barometric pressure, solar radiation, and other characteristics. *Climate* differs from *weather* (p. 449) in that weather specifies conditions over hours or days, whereas climate summarizes conditions over years, decades, or centuries.

Global climate change—generally referred to simply as **climate change**—describes an array of changes in aspects of Earth's climate, such as temperature, precipitation, and the frequency and intensity of storms. People often use the term *global warming* synonymously in casual conversation, but **global warming** refers specifically to an increase in Earth's average surface temperature. Global warming is only one aspect of global climate change, but warming does in turn drive other components of climate change. Some researchers point out that the terms "warming" and "change" are so mild-sounding as to be misleading, and that a more accurate term for what Earth is experiencing today would be "climate disruption."

Over the long term, our planet's climate varies naturally. However, today's disruptive changes are unfolding at an exceedingly rapid rate, and they are creating conditions humanity has never experienced. Earth's climate is a complex and finely tuned system on which life depends, and rapid and disruptive changes in the geologic past have resulted in mass extinctions (pp. 58, 279). Scientists agree that the climate disruption we are beginning to witness today is being driven by

human activities, notably fossil fuel combustion and deforestation. Understanding how and why today's climate is changing—and judging how we might respond—requires understanding how our planet's climate normally functions.

Three factors influence climate

Three natural factors exert the most influence on Earth's climate. The first is the sun. Without the sun, Earth would be dark and frozen. The second is the atmosphere. Without this protective layer of gases, Earth would be as much as 33°C (59°F) colder on average, and temperature differences between night and day would be far greater than they are. The third is the oceans, which store and transport heat and moisture.

The sun supplies most of our planet's energy. Earth's atmosphere, clouds, land, ice, and water together absorb about 70% of incoming solar radiation and reflect the remaining 30% back into space (**FIGURE 18.2**). The 70% that is absorbed powers everything from wind to waves to evaporation to photosynthesis.

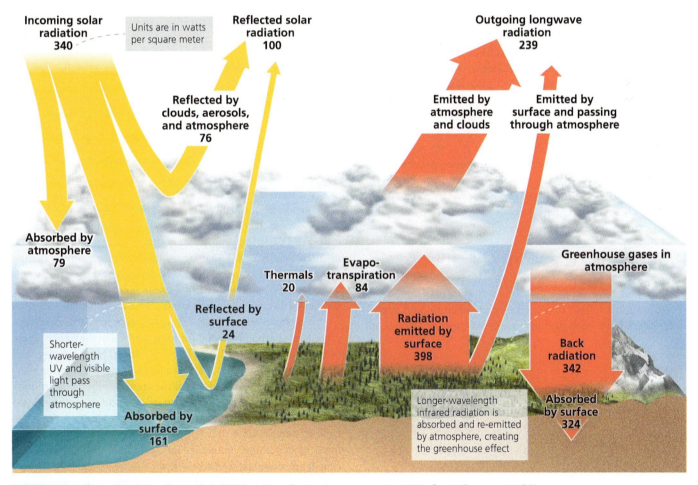

FIGURE 18.2 Our planet receives about 340 watts of energy per square meter from the sun, and it naturally reflects and emits this same amount. Earth absorbs nearly 70% of the solar radiation it receives, and reflects the rest back into space (**yellow arrows**). The radiation absorbed is then re-emitted (**orange arrows**) as infrared radiation, which has longer wavelengths. Greenhouse gases in the atmosphere absorb a portion of this long-wavelength radiation and then re-emit it, sending some back downward to warm the atmosphere and the surface by the greenhouse effect. *Data from Intergovernmental Panel on Climate Change (IPCC); Stocker, T.F., et al. (Eds.), 2013.* Climate change 2013: The physical science basis. Contribution of Working Group I to the fifth assessment report of the IPCC. *Cambridge, UK, and New York: Cambridge University Press.*

Greenhouse gases warm the lower atmosphere

As Earth's surface absorbs solar radiation, the surface increases in temperature and emits infrared radiation (p. 30), radiation with wavelengths longer than those of visible light. Atmospheric gases having three or more atoms in their molecules tend to absorb infrared radiation. These include water vapor (H_2O), ozone (O_3), carbon dioxide (CO_2), nitrous oxide (N_2O), and methane (CH_4), as well as halocarbons, a diverse group of mostly human-made gases (p. 465). All these gases are known as **greenhouse gases.** After absorbing radiation emitted from the surface of Earth, greenhouse gases re-emit infrared radiation in all directions. Some of this re-emitted energy is lost to space, but much of it travels back downward, warming the lower atmosphere (specifically the troposphere; p. 447) and the surface in a phenomenon known as the **greenhouse effect.**

Greenhouse gases differ in their ability to warm the troposphere and surface. *Global warming potential* refers to the relative ability of one molecule of a given greenhouse gas to contribute to warming. **TABLE 18.1** shows global warming potentials for several greenhouse gases. Values are expressed in relation to carbon dioxide, which is assigned a value of 1. For example, at a 20-year time horizon, a molecule of methane is 84 times more potent than a molecule of carbon dioxide. Yet because a typical methane molecule resides in the atmosphere for less time than a typical carbon dioxide molecule, methane's global warming potential is reduced at longer time horizons (it is 28 at a 100-year horizon).

Although carbon dioxide is less potent on a per-molecule basis than most other greenhouse gases, it is far more abundant in the atmosphere. Moreover, greenhouse gas emissions from human activity consist mostly of carbon dioxide; for this reason, carbon dioxide has caused nearly twice as much warming since the industrial revolution as have methane, nitrous oxide, and halocarbons combined.

Greenhouse gas concentrations are rising fast

The greenhouse effect is a natural phenomenon, and greenhouse gases have been present in our atmosphere throughout

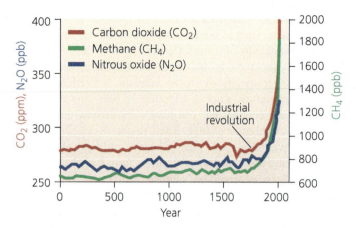

FIGURE 18.3 Since the start of the industrial revolution around 1750, global concentrations of carbon dioxide, methane, and nitrous oxide in the atmosphere have increased markedly. *Data from IPCC, 2013. Fifth assessment report.*

DATA Q By about what percentage has atmospheric carbon dioxide concentration increased since 1750?

Go to **Interpreting Graphs & Data** on **Mastering**EnvironmentalScience®

Earth's history. It's a good thing, too. Without the natural greenhouse effect, our planet would be too cold to support life as we know it. Thus, it is not the natural greenhouse effect that concerns scientists today but rather the anthropogenic (human-generated) intensification of the greenhouse effect. By increasing the concentrations of greenhouse gases over the past 250 years (**FIGURE 18.3**), we are intensifying the greenhouse effect beyond what our species has ever experienced.

We have boosted Earth's atmospheric concentration of carbon dioxide from roughly 278 parts per million (ppm) in the late 1700s to more than 400 ppm today (see Figure 18.3). The concentration of CO_2 in the atmosphere now is far higher than it has been in more than 800,000 years (four times longer than the human species has existed)—and it is likely the highest in the past *20 million* years. (We know this from research on ice cores and other sources of data on the ancient atmosphere, as we will see shortly.)

Why have atmospheric carbon dioxide levels risen so rapidly? Most carbon is stored for long periods in the upper layers of the lithosphere (p. 120). The deposition, partial decay, and compression of organic matter (mostly plants and phytoplankton) that grew in wetland or marine areas hundreds of millions of years ago led to the formation of coal, oil, and natural gas in buried sediments (p. 521). Over the past two centuries, we have extracted these fossil fuels from the ground and burned them in our homes, power plants, and automobiles, transferring large amounts of carbon from one reservoir (the underground deposits that stored the carbon for millions of years) to another (the atmosphere). This sudden flux of carbon from the lithosphere into the atmosphere is the main reason atmospheric CO_2 concentrations have increased so dramatically.

At the same time, people have cleared and burned forests to make room for crops, pastures, villages, and cities. Forests serve as a reservoir for carbon as plants conduct photosynthesis (p. 30) and store carbon in their tissues. When we

TABLE 18.1 Global Warming Potentials of Four Greenhouse Gases

GREENHOUSE GAS	RELATIVE HEAT-TRAPPING ABILITY (IN CO_2 EQUIVALENTS)	
	OVER 20 YEARS	OVER 100 YEARS
Carbon dioxide	1	1
Methane	84	28
Nitrous oxide	264	265
Hydrochlorofluorocarbon HFC-23	10,800	12,400

Data from IPCC, 2013. *Fifth Assessment Report.*

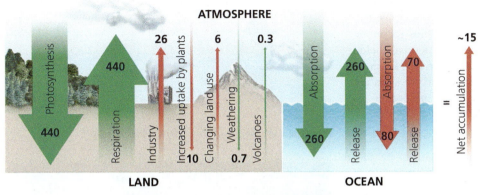

ATMOSPHERE

Photosynthesis 440 / 440 Respiration

Industry 26   Increased uptake by plants 10

Changing land use 6   Weathering 0.7

Volcanoes 0.3

Absorption 260 / Release 260

Absorption 70 / Release 80

Net accumulation ~15

=

LAND   **OCEAN**

FIGURE 18.4 Human activities are sending more carbon dioxide from Earth's surface to its atmosphere than is moving from the atmosphere to the surface. Shown are all current fluxes of CO₂, with arrows sized according to mass. Green arrows indicate natural fluxes, and red arrows indicate anthropogenic fluxes. *Adapted from IPCC, 2007.* Fourth assessment report.

DATA For every metric ton of carbon dioxide we emit due to changing land use (e.g., deforestation), how much do we emit from industry?

Go to **Interpreting Graphs & Data** on **Mastering**EnvironmentalScience®

clear forests, we reduce the biosphere's ability to remove carbon dioxide from the atmosphere. In this way, deforestation (p. 305) contributes to rising atmospheric CO₂ concentrations. **FIGURE 18.4** summarizes scientists' current understanding of the fluxes (both natural and anthropogenic) of carbon dioxide among the atmosphere, land, and oceans.

Methane concentrations are also rising—150% since 1750 (see Figure 18.3)—and today's atmospheric concentration is the highest by far in more than 800,000 years. We release methane by tapping into fossil fuel deposits, raising livestock that emit methane as a metabolic waste product, disposing of organic matter in landfills, and growing crops such as rice.

Human activities have also elevated atmospheric concentrations of nitrous oxide. This greenhouse gas, a by-product of feedlots, chemical manufacturing plants, auto emissions, and synthetic nitrogen fertilizers, has risen by 20% since 1750 (see Figure 18.3).

Among other greenhouse gases, ozone concentrations in the troposphere have risen roughly 42% since 1750, a result of photochemical smog (p. 462). The contribution of halocarbon gases to global warming has begun to slow because of the Montreal Protocol and subsequent controls on their production and use (p. 468).

Water vapor is the most abundant greenhouse gas in our atmosphere and contributes most to the natural greenhouse effect. Its concentrations vary locally, but because its global concentration has not changed, it is not thought to have driven industrial-age climate change.

Most aerosols exert a cooling effect

Whereas greenhouse gases exert a warming effect on the atmosphere, **aerosols** (p. 453), microscopic droplets and particles, can have either a warming or a cooling effect. Soot particles, or "black carbon aerosols," generally cause warming by absorbing

solar energy, but most other aerosols cool the atmosphere by reflecting the sun's rays. Sulfate aerosols produced by fossil fuel combustion may slow global warming, at least in the short term. When sulfur dioxide enters the atmosphere, it undergoes various reactions, some of which lead to acid deposition (p. 469). These reactions can form a sulfur-rich aerosol haze in the upper atmosphere that blocks sunlight. For this reason, aerosols released by major volcanic eruptions can cool Earth's climate for up to several years. This occurred in 1991 with the eruption of Mount Pinatubo in the Philippines (p. 453).

Radiative forcing expresses change in energy input

To measure the degree of impact that a given factor exerts on Earth's temperature, scientists calculate its **radiative forcing,** the amount of change in thermal energy that the factor causes. Positive forcing warms the surface, whereas negative forcing cools it. When scientists sum up the effects of all factors, they find that Earth is now experiencing overall radiative forcing of about 2.3 watts/m² (**FIGURE 18.5**). This means that our planet today is receiving

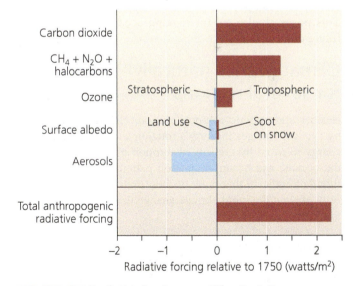

Carbon dioxide

CH₄ + N₂O + halocarbons

Ozone   Stratospheric — Tropospheric

Surface albedo   Land use — Soot on snow

Aerosols

Total anthropogenic radiative forcing

−2   −1   0   1   2
Radiative forcing relative to 1750 (watts/m²)

FIGURE 18.5 Radiative forcing quantifies the influence that aerosols, greenhouse gases, and other factors exert over Earth's energy balance. In this graph, radiative forcing is expressed as the warming or cooling effect that each factor has on temperature today relative to the year 1750, in watts/m². Red bars indicate positive forcing (warming), and blue bars indicate negative forcing (cooling). *Albedo* (p. 494) refers to the reflectivity of a surface. A number of more minor influences are not shown. *Data from IPCC, 2013.* Fifth assessment report.

and retaining 2.3 watts/m² more thermal energy than it is emitting into space. (By contrast, the pre-industrial Earth of 1750 was in balance, emitting as much radiation as it was receiving.) This extra amount is equivalent to the power converted into heat and light by 200 incandescent lightbulbs (or more than 900 compact fluorescent lamps) across a football field. For context, look back at Figure 18.2 and note that Earth is estimated naturally to receive and give off about 340 watts/m² of energy. Although 2.3 may seem like a small proportion of 340, heat from this imbalance accumulates, and over time this is enough to alter climate significantly.

Feedback complicates our predictions

As tropospheric temperatures increase, physics tells us that more water should evaporate from Earth's surface into the atmosphere, but scientists aren't yet sure how this will affect our climate. On one hand, more atmospheric water vapor could lead to more warming, which could cause more evaporation, in a positive feedback loop (p. 104) that would amplify the greenhouse effect. On the other hand, more water vapor could enhance cloudiness, which might, in a negative feedback loop (p. 104), slow global warming by reflecting more solar radiation back into space. In this second scenario, depending on whether low- or high-elevation clouds result, they might either shade and cool Earth (negative feedback) or else contribute to warming and accelerate evaporation and further cloud formation (positive feedback). We simply don't yet know which effect might predominate, or how the balance might vary from place to place. Because of feedback loops, minor changes in the atmosphere can potentially lead to major effects on climate. This poses challenges for making accurate predictions of future climate change.

Climate varies naturally for several reasons

Besides atmospheric composition, our climate is influenced by cyclical changes in Earth's rotation and orbit, variation in energy released by the sun, absorption of carbon dioxide by the oceans, and ocean circulation patterns. However, scientific data indicate that none of these four natural factors can fully explain the rapid climate change that we are experiencing today.

Milankovitch cycles In the 1920s, Serbian mathematician Milutin Milankovitch described three types of periodic changes in Earth's rotation and orbit around the sun. Over thousands of years, our planet wobbles on its axis, varies in the tilt of its axis, and experiences change in the shape of its orbit, all in regular long-term cycles of different lengths. These variations, known as **Milankovitch cycles,** alter the way solar radiation is distributed over Earth's surface (**FIGURE 18.6**). By modifying patterns of atmospheric heating, these cycles trigger long-term climate variation. This includes periodic episodes of glaciation during which global

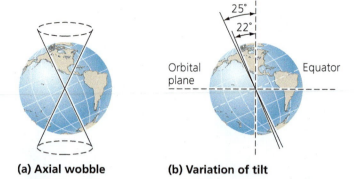

(a) Axial wobble　　　**(b) Variation of tilt**

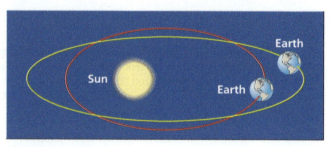

(c) Variation of orbit

FIGURE 18.6 There are three types of Milankovitch cycles: (a) an axial wobble that occurs on a 19,000- to 23,000-year cycle; **(b)** a 3-degree shift in the tilt of Earth's axis that occurs on a 41,000-year cycle; and **(c)** a variation in Earth's orbit from almost circular to more elliptical, which repeats every 100,000 years.

surface temperatures drop and ice sheets expand outward from the poles.

Solar output The sun varies in the amount of radiation it emits, over both short and long timescales. However, scientists are concluding that the variation in solar energy reaching our planet in recent centuries has simply not been great enough to drive significant temperature change on Earth's surface. Estimates place the radiative forcing of natural changes in solar output at only about 0.05 watts/m²—less than any of the anthropogenic causes shown in Figure 18.5.

Ocean absorption The oceans hold 50 times more carbon than the atmosphere holds. Oceans absorb carbon dioxide from the atmosphere when CO_2 dissolves directly in water and when marine phytoplankton use it for photosynthesis. However, the oceans are absorbing less CO_2 than we are adding to the atmosphere (see Figure 5.17, p. 120). Thus, carbon absorption by the oceans is slowing global warming but is not preventing it. Moreover, as ocean water warms, it absorbs less CO_2 because gases are less soluble in warmer water—a positive feedback effect (p. 104) that accelerates warming of the atmosphere.

Ocean circulation Ocean water exchanges heat with the atmosphere, and ocean currents move energy from place to place. In equatorial regions, the oceans receive more heat from

FAQ

The climate changes naturally, so why worry about climate change?

Earth's climate does indeed change naturally across very long periods of time, but there is nothing "natural" about today's climate disruption. We know that human beings are directly causing the unnaturally rapid change we are now witnessing. Moreover, humanity has never before experienced the sheer amount of change predicted for this century. In fact, the quantity by which the world's temperature is forecast to rise is greater than the amount of cooling needed to bring on an ice age! Greenhouse gas concentrations are already higher than they've been in more than 800,000 years, and they are rising. The human species, *Homo sapiens*, has existed for only 200,000 years, and our civilization arose only in the past few thousand years during an exceptionally stable period in Earth's climate history. Unless we reduce our emissions, we soon will be challenged by climate conditions our species has never lived through.

the sun and atmosphere than they emit. Near the poles, the oceans emit more heat than they receive. Because cooler water is denser than warmer water, the cooler water at the poles tends to sink, and the warmer surface water from the equator moves to take its place. This is one principle underlying global ocean circulation patterns (p. 422).

The oceans' thermohaline circulation system moves warm tropical water northward, providing Europe a far milder climate than it would otherwise have. Scientists are studying whether freshwater input from Greenland's melting ice sheet might shut down this warm-water flow (p. 423). Such an occurrence would plunge Europe into much colder conditions.

Multiyear climate variability results from the El Niño–Southern Oscillation (p. 423), which involves systematic shifts in atmospheric pressure, sea surface temperature, and ocean circulation in the tropical Pacific Ocean. These shifts overlie longer-term variability from a phenomenon known as the Pacific Decadal Oscillation. El Niño and La Niña events alter weather patterns in diverse ways, often promoting rainstorms and floods in dry regions and drought and fire in moist regions. This leads to impacts on wildlife, agriculture, and fisheries.

of indirect evidence that serve as proxies, or substitutes, for direct measurement.

For example, Earth's ice caps, ice sheets, and glaciers hold clues to climate history. In frigid areas near the poles and atop high mountains, snow falling year after year compresses into ice. Over the millennia, this ice accumulates to great depths, preserving within its layers tiny bubbles of the ancient atmosphere. Scientists can examine the trapped air bubbles by drilling into the ice and extracting long columns, or cores (**FIGURE 18.7**). The layered ice, accumulating season after season for thousands of years, provides a timescale. By studying the chemistry of the bubbles in each layer in these ice cores, scientists can determine atmospheric composition, greenhouse gas concentrations, temperature, snowfall, solar activity, and even (from trapped soot particles) the frequency of forest fires and volcanic eruptions during each time period. By extracting ice cores from Antarctica, scientists have now been able to go back in time 800,000 years, reading Earth's history across eight glacial cycles (see **THE SCIENCE BEHIND THE STORY**, pp. 486–487.)

Researchers also drill cores into beds of sediment beneath bodies of water. Sediments often preserve pollen grains and other remnants from plants that grew in the past (as described in the study of Easter Island; pp. 8–9). Because climate influences the types of plants that grow in an area, knowing what plants were present can tell us a great deal about the climate at that place and time.

Tree rings provide another proxy indicator. The width of each ring of a tree trunk cut in cross-section reveals how much the tree grew in a particular growing season. A wide ring means more growth, generally indicating a wetter year. Long-lived trees such as bristlecone pines can provide records of precipitation and drought going back several thousand years. Tree rings are also used to study fire history, because a charred ring indicates that a fire took place at the site in that year.

In arid regions such as the U.S. Southwest, packrat middens are a valuable source of climate data. Packrats are

Studying Climate Change

To comprehend any phenomenon that is changing, we must study its past, present, and future. Scientists monitor present-day climate, but they also have devised clever means of inferring past change and sophisticated methods to predict future conditions.

Proxy indicators tell us about the past

Evidence about **paleoclimate,** climate in the ancient past, is vital for giving us a baseline against which we can measure changes happening in our climate today. To understand paleoclimate, scientists decipher clues from thousands or millions of years ago by taking advantage of the record-keeping capacity of the natural world. **Proxy indicators** are types

FIGURE 18.7 Scientists drill deep into ancient ice sheets and extract cores of ice to learn the secrets of past climates.

What Can We Learn from the World's Longest Ice Core?

Inspecting an Antarctic ice core

In the most frigid reaches of our planet, snow falling year after year for millennia compresses into ice and stacks up into immense sheets that scientists can mine for clues to Earth's climate history. The ice sheets of Antarctica and Greenland trap tiny air bubbles, dust particles, and other proxy indicators (p. 485) of past conditions. By drilling boreholes and extracting ice cores, researchers can tap into these priceless archives.

Recently, researchers drilled and analyzed the deepest core ever. At a remote and pristine site in Antarctica named Dome C, they drilled down 3270 m (10,728 ft) to bedrock and pulled out more than 800,000 years' worth of ice. The longest previous ice core (from Antarctica's Vostok station) had gone back "only" 420,000 years.

Ice near the top of these cores was laid down most recently, and ice at the bottom is oldest, so by analyzing ice at intervals along the core's length, researchers can generate a timeline of environmental change. To date layers of the ice core, researchers first analyze deuterium isotopes (p. 23) to determine the rate of ice accumulation, referencing studies and models of how ice compacts over time. They then calibrate the timeline by matching recent events in the chronology (for example, major volcanic eruptions) with independent data sets from previous cores, tree rings, and other sources.

Dome C, a high summit of the Antarctic ice sheet, is one of the coldest spots on the planet, with an annual mean temperature of −54.5°C (−98.1°F). The Dome C ice core was drilled by the European Project for Ice Coring in Antarctica (EPICA), a consortium of researchers from 10 European nations.

In 2004, this team of 56 researchers published a paper in the journal *Nature*, reporting data on surface air temperature across 740,000 years. The researchers had obtained this temperature data by measuring the ratio of deuterium isotopes to normal hydrogen in the ice, a ratio that is temperature dependent.

From 2005 to 2008, five follow-up papers in the journals *Science* and *Nature* reported analyses of greenhouse gas concentrations from the EPICA ice core and extended the gas and temperature data back to cover all 800,000 years. By analyzing air bubbles trapped in the ice, the researchers quantified atmospheric concentrations of carbon dioxide and methane (red line and green line, respectively, in **FIGURE 1**).

These data show that by emitting these greenhouse gases since the industrial revolution, we have brought their atmospheric concentrations well above the highest levels they reached naturally at any time in the past 800,000 years. Today's carbon dioxide spike is too recent to show up in the ice core, but its concentration (surpassing 400 ppm in 2015) is far above previous maximum values (of ~300 ppm) shown in the red line of the figure. These data reveal that we as a society have brought ourselves deep into uncharted territory.

The EPICA results also confirm that temperature swings in the past were tightly correlated with concentrations of greenhouse gases (compare the top two data sets in Figure 1 with the temperature data set at bottom). This finding bolsters the scientific consensus that greenhouse gas emissions are causing our planet to warm today.

Also clear from the data is that temperature has varied with swings in solar radiation due to Milankovitch cycles (p. 484). The complex interplay of the Milankovitch cycles produces periodic

rodents that carry seeds and plant parts back to their middens, or dens, in caves and rock crevices sheltered from rain. In arid locations, plant parts may be preserved for centuries inside these middens, allowing researchers to study the past flora of the region.

To assess past ocean conditions, researchers gather data from coral reefs (p. 426). Living corals take in trace elements and isotope ratios (p. 23) from ocean water as they grow, and they incorporate these chemical clues, layer by layer, into growth bands in the reefs they build.

Proxy indicators generally provide us information about local or regional areas. To get a global perspective, scientists combine multiple records from various areas. Because the number of available indicators decreases the further back in time we go, estimates of global climate conditions for the recent past tend to be more reliable than those for the distant past.

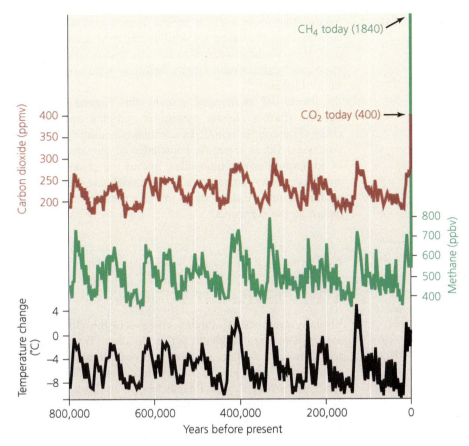

FIGURE 1 Data from the EPICA ice core reveal changes across 800,000 years.
Shown are atmospheric carbon dioxide concentration (**red line**), atmospheric methane concentration (**green line**), and deviations from today's average surface temperature (**black line**). Concentrations of CO_2 and methane rise and fall in tight correlation with temperature. Today's values for CO_2 and methane are included at the top right of the graph, for comparison. *Adapted by permission of Macmillan Publishers Ltd: Brook, E, 2008. Paleoclimate: Windows on the greenhouse. Nature 453: 291–292, Fig 1a. www.nature.com.*

Other findings from the ice core are not easily explained. Intriguingly, the early glacial cycles differ from the recent cycles (see the black line in Figure 1). In the recent cycles, glacial periods are long, whereas interglacial periods are brief, with a rapid rise and fall of temperature. Interglacials thus appear on the graph as tall, thin spikes. In older glacial cycles, the glacial and interglacial periods are of more equal duration, and the interglacials are not as warm. This shift in the nature of glacial cycles had been noted before by researchers working with oxygen isotope data from marine fossils. But why cycles should differ before and after the 450,000-year mark, no one knows.

Today polar scientists are searching for a site that might provide an ice core stretching back more than 1 million years. At that time, data from marine isotopes tell us that glacial cycles switched from a periodicity of roughly 41,000 years (reflecting the influence of planetary tilt) to intervals of about 100,000 years (more similar to orbital changes). An ice core that captures cycles on both sides of the 1-million-year divide might help clarify the influence of Milankovitch cycles or perhaps offer other explanations.

The intriguing patterns revealed by the Dome C ice core show that we still have plenty to learn about Earth's complex climate history. However, the clear relationship between greenhouse gases and temperature evident in the EPICA data suggests that if we want to prevent sudden global warming, we will need to reduce our society's greenhouse emissions.

temperature fluctuations on Earth resulting in periods of glaciation (when temperate regions of the planet are covered in ice) and in warm interglacial periods (periods between glaciations). The Dome C ice core spans eight glacial cycles.

Direct measurements tell us about the present

Today we measure temperature with thermometers, rainfall with rain gauges, wind speed with anemometers, and air pressure with barometers, using computer programs to integrate and analyze this information in real time. With these technologies and more, we document in detail the fluctuations in weather day-by-day and hour-by-hour across the globe.

We also measure the chemistry of the atmosphere and the oceans. Direct measurements of carbon dioxide concentrations in the atmosphere reach back to 1958, when scientist Charles Keeling began analyzing hourly air samples from a monitoring station at Hawaii's Mauna Loa Observatory. Here, unpolluted, well-mixed air from over vast stretches of

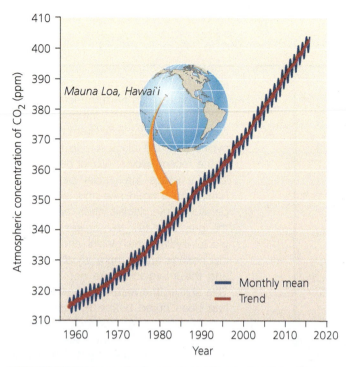

FIGURE 18.8 Atmospheric concentrations of carbon dioxide are rising steeply. Direct long-term measurements began in 1958, when Charles Keeling started collecting these data at Hawaii's Mauna Loa Observatory. *Data from National Oceanic and Atmospheric Administration, Earth System Research Laboratory, Global Monitoring Division, 2016.*

ocean blows across the top of Earth's most massive mountain. These data show that atmospheric CO_2 concentrations increased from 315 ppm in 1958 to just over 400 ppm today (**FIGURE 18.8**).

Direct measurements of climate variables such as temperature and precipitation extend back in time somewhat further. Precise and reliable thermometer measurements cover more than a century. Fishermen have recorded the timing of sea ice formation each year, and winemakers have kept meticulous records of precipitation and the length of the growing season. Accurate records of these types extend back, at most, a few hundred years.

Models help us predict the future

To understand how climate systems function and to predict future climate change, scientists simulate climate processes with sophisticated computer programs. **Climate models** are programs that combine what is known about atmospheric circulation, ocean circulation, atmosphere–ocean interactions, and feedback cycles to simulate climate dynamics (see **THE SCIENCE BEHIND THE STORY**, pp. 490–491). This requires manipulating vast amounts of data with complex mathematical equations—a task not possible until the advent of modern computers.

Climate modelers provide starting information to the model based on real data from the field, set up rules for

the simulation, and then let the program run. Researchers strive for accuracy by building in as much information as they can from what is understood about how the climate system functions. They then test the efficacy of a model by entering past climate data and running the model toward the present. If a model accurately reconstructs current climate, based on well-established data from the past, then we have reason to believe that it simulates climate mechanisms realistically and may accurately predict future climate.

Plenty of challenges remain for climate modelers, because Earth's climate system is complex and because many uncertainties remain in our understanding of feedback processes. Yet as scientific knowledge of climate improves, as computing power intensifies, and as we glean enhanced data from proxy indicators, climate models are improving in resolution and are allowing us to make predictions region by region across the world.

Current and Future Trends and Impacts

Virtually everyone is noticing changes in the climate these days. Miami-area residents suffer flooding. Texas ranchers and California farmers endure multiyear droughts. Coastal homeowners struggle to obtain insurance against hurricanes and storm surges. People from New York to Atlanta to Chicago to Los Angeles face one unprecedented weather event after another.

Extreme weather events are indeed part of a real pattern backed by a tremendous volume of scientific evidence. Climate change has already had numerous impacts on the physical properties of our planet, on organisms and ecosystems, and on human well-being (**FIGURE 18.9**). If we continue to emit greenhouse gases into the atmosphere, climate change will accelerate and the consequences will grow more severe.

Scientific evidence for climate change is extensive

For years, scientists have studied climate change in enormous breadth, depth, and detail. Researchers have been monitoring climate variables with everything from thermometers to satellites, building up detailed long-term databases. As a result, the scientific literature today is replete with many thousands of independent published studies, and we have gained a rigorous understanding of most aspects of climate change.

To make this vast and growing research knowledge accessible to policymakers and the public, the **Intergovernmental Panel on Climate Change (IPCC)** has taken up the task of periodically reviewing and summarizing all available data. This international panel consists of many hundreds of scientists and governmental representatives. Established in 1988 by the United Nations

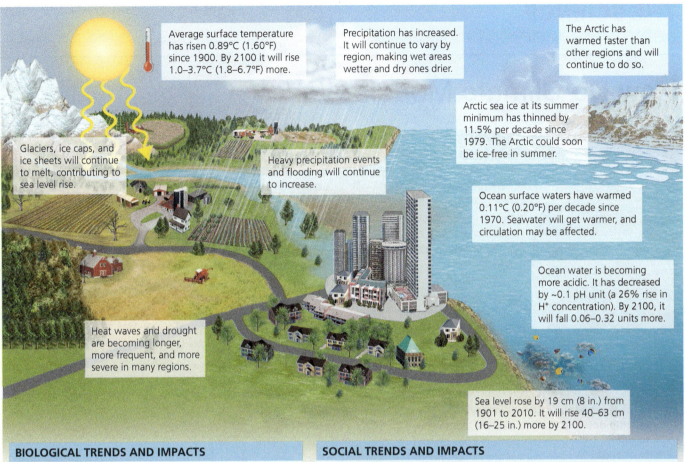

Average surface temperature has risen 0.89°C (1.60°F) since 1900. By 2100 it will rise 1.0–3.7°C (1.8–6.7°F) more.

Precipitation has increased. It will continue to vary by region, making wet areas wetter and dry ones drier.

The Arctic has warmed faster than other regions and will continue to do so.

Glaciers, ice caps, and ice sheets will continue to melt, contributing to sea level rise.

Heavy precipitation events and flooding will continue to increase.

Arctic sea ice at its summer minimum has thinned by 11.5% per decade since 1979. The Arctic could soon be ice-free in summer.

Ocean surface waters have warmed 0.11°C (0.20°F) per decade since 1970. Seawater will get warmer, and circulation may be affected.

Ocean water is becoming more acidic. It has decreased by ~0.1 pH unit (a 26% rise in H^+ concentration). By 2100, it will fall 0.06–0.32 units more.

Heat waves and drought are becoming longer, more frequent, and more severe in many regions.

Sea level rose by 19 cm (8 in.) from 1901 to 2010. It will rise 40–63 cm (16–25 in.) more by 2100.

BIOLOGICAL TRENDS AND IMPACTS

- Species ranges are shifting toward the poles and up in elevation.
- Seasonal timing (such as migration and breeding) is shifting.
- Some species are declining, and many could eventually face extinction.
- Ocean acidification is starting to affect marine life. Coral reefs could disappear, devastating marine ecosystems.
- Forests are being altered by drought, fire, and pest outbreaks.

SOCIAL TRENDS AND IMPACTS

- Ocean acidification could destroy major commercial fisheries.
- Droughts and flooding are leading to agricultural losses.
- Sea level rise will displace people and cause escalating expense.
- More intense storms are causing more property loss and loss of life.
- Farmers and foresters are struggling with altered growing seasons and disturbance regimes.
- Crop yields will likely fall in the dry tropics and subtropics.
- Impacts on biodiversity will cause losses of food, water, and ecosystem goods and services.
- Melting of mountain glaciers will reduce water supplies to millions.
- Economic costs will far outweigh benefits.
- Poorer nations and communities are suffering greater impacts.
- Increased warm-weather health hazards will outweigh decreased cold-weather hazards.

FIGURE 18.9 Current trends and future impacts of climate change are extensive. Shown are major physical, biological, and social trends and impacts (both observed and predicted) as reported by the IPCC. (Mean estimates are shown; the IPCC reports ranges and statistical probabilities as well.) *Data from IPCC, 2013.* Fifth assessment report.

Environment Programme and the World Meteorological Organization, the IPCC was awarded the Nobel Peace Prize jointly with former U.S. Vice President Al Gore in 2007 for its work in informing the world of the trends and impacts of climate change.

In 2013 and 2014, the IPCC released its *Fifth Assessment Report.* By summarizing thousands of scientific studies, this report documents observed trends in surface temperature, precipitation patterns, snow and ice cover, sea levels, storm intensity, and other factors (see Figure 18.9). It also predicts future changes in these phenomena after considering a range of potential scenarios for future greenhouse gas emissions. The report addresses impacts of climate disruption on wildlife, ecosystems, and society. Finally, it discusses strategies we might pursue in response.

How Do Climate Models Work?

Models are indispensable for scientists studying climate today—and they are increasingly vital for our society because they help us predict what conditions will confront us in the future. Yet to most of us, a climate model is a mysterious black box. So how *do* scientists create a climate model?

The colorful maps and data-rich graphs and charts that scientists generate from climate models are the end result, but the process begins when they put into the model a long series of mathematical equations. These equations describe how various components of Earth's systems function. Some equations are derived from physical laws such as those on the conservation of mass, energy, and momentum (p. 22). Others are derived from observational and experimental data on physics, chemistry, and biology, gathered from the field. Converted into computing language, these equations are integrated with information about Earth's landforms, hydrology, vegetation, and atmosphere (**FIGURE 1**).

Earth's climate system is mind-bogglingly complex, and modelers will never capture all the factors that influence climate. Yet as computers become more powerful and models more sophisticated, they are incorporating more and more of the factors that affect climate. To handle the complexity, most models consist of submodels, each handling a different component—ocean water, sea ice, glaciers, forests, deserts, troposphere, stratosphere. Years ago when models first coupled atmosphere and ocean components together, they were called "coupled" models. This is standard in today's far more complex general circulation models, or global climate models (which share the acronym GCM).

For a model to function, all its building blocks must be given equations to make them behave realistically in space and time. In the real climate system, time is continuous and spatial effects reach down to the level of molecules interacting with one another. But the virtual reality of climate models cannot be so detailed—there is simply not enough computer power available. Instead, modelers approximate reality by dividing time into periods (called time steps) and by dividing the Earth's surface into cells or boxes according to a grid (called grid boxes) (**FIGURE 2**).

Each grid box contains land, ocean, or atmosphere, much like a digital photograph is made up of discrete pixels of certain colors. The grid boxes are arrayed in a three-dimensional layer by latitude and longitude, or in equal-sized polygons.

The finer the scale of the grid, the greater resolution the model will have, and the better it will be able to predict results region by region. However, more resolution means more computing power is needed, and climate models already strain the most powerful supercomputing networks. Today's best climate models feature dozens of grid boxes piled up from the bottom

A researcher working with satellite data used to model climate change

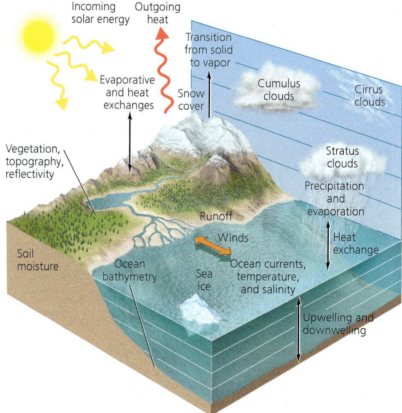

FIGURE 1 Climate models incorporate a diversity of natural factors and processes. Anthropogenic factors can then be added in.

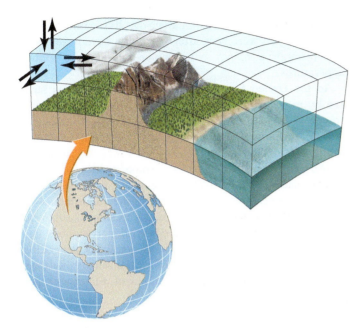

FIGURE 2 Climate models divide Earth's surface into a layered grid. Each grid box represents land, air, or water, and interacts with adjacent grid boxes via the flux of materials and energy. *Adapted from Bloom, A.J., 2010. Global climate change: Convergence of disciplines. Sunderland, MA: Sinauer Associates.*

of the ocean to the top of the atmosphere, with each grid box measuring a few dozen miles wide, and time measured in periods of just minutes.

Once the grid is established, the processes that drive climate are assigned to each grid box, with their rates parceled out among the time steps. The model lets the grid boxes interact through time by means of the flux of materials and energy into and out of each grid box.

Once modelers have input all this information, learned from our study of Earth and the climate system, they let the model run through time and simulate climate, from the past through the present and into the future. If the computer simulation accurately reconstructs past and present climate, then that enhances confidence that it will predict future climate accurately as well.

A number of studies have compared model runs that include only natural processes, model runs that include only human-generated processes, and model runs that combine both. Repeatedly these studies have found that the model runs that incorporate both human and natural processes are the ones that fit real-world climate observations the best (**FIGURE 3**). This supports the idea that human activities, as well as natural processes, are influencing our climate.

The major human influence on climate is our emission of greenhouse gases, and modelers need to select values to enter for future emissions if they want to predict future climate. Generally they will run their simulations multiple times, each time with a different future emission rate according to a specified scenario. Differences between the results from such scenarios tell us what influence these different emission rates would have. You can see results from such a comparison in Figure 18.26 (p. 503).

Throughout this chapter, you will see figures that show results of various models. In crafting its assessment reports, the IPCC consults nearly two dozen major models, considers the strengths and weaknesses of each, and presents the best summary its authors can muster.

Researchers are constantly testing and evaluating their models. They improve them by incorporating what is learned from new research and by taking advantage of what increasingly powerful computing technologies allow. As their work proceeds, we can expect increasingly precise and accurate predictions about future climate conditions.

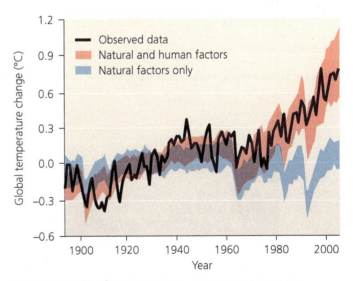

FIGURE 3 Models that incorporate both natural and anthropogenic factors predict observed climate trends best. *Adapted from Melillo, J.M., et al. (Eds.), 2014. Climate change impacts in the United States: The third national climate assessment. U.S. Global Change Research Program.*

DATA Q Why do you think the red-shaded area diverges from the blue-shaded area in the right-hand portion of the graph? Why does the observed data (black line) track with the red-shaded area rather than the blue-shaded area?

Go to **Interpreting Graphs & Data** on **Mastering**EnvironmentalScience®

To learn more, you may wish to download and explore the *Fifth Assessment Report* yourself. A series of three "working group" reports and a synthesis report are publicly accessible online at the IPCC's website. Meanwhile, scientists around the world continue to monitor, model, and analyze our changing climate. In response to the urgency of the situation, new research is being completed faster than ever.

Temperatures continue to rise

Average surface temperatures on Earth have risen by about 1.1°C (2.0°F) in the past 100 years (**FIGURE 18.10**). These include increases both in air temperature over land and in sea surface temperature. Most of this increase has occurred since the 1970s—and just since 1995, we have experienced the 19 warmest years on record since global measurement began 136 years ago! Since the 1960s, each decade has been warmer than the last. If you were born after 1976, you have never in your life lived through a year with average global temperatures lower than the 20th-century average—and if you were born after 1985, you have never even lived through a *month* with cooler-than-average global temperatures.

In the United States, temperatures in most areas have risen by more than 1 full degree Fahrenheit in just the past two decades (**FIGURE 18.11**). This may not sound like much, but it is exceedingly rapid change, and it is enough to cause countless impacts discussed below.

We can expect global surface temperatures to continue rising because we are still emitting greenhouse gases and because the greenhouse gases already in the atmosphere will continue warming the globe for decades to come. At the end of the 21st century, the IPCC predicts global temperatures will be 1.0–3.7°C (1.8–6.7°F) higher than today's, depending on how well we control our emissions. Unusually hot days and heat waves will become more frequent. Future changes

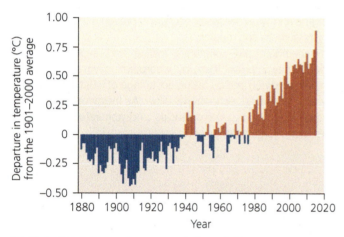

(a) Global temperature measured since 1880

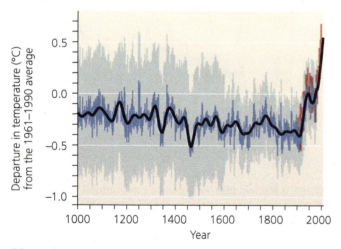

(b) Northern Hemisphere temperature, past 1000 years

FIGURE 18.10 Global temperatures have risen sharply in the past century. Data from thermometers **(a)** show changes in Earth's average surface temperature since 1880. Since 1976, every single year has been warmer than average. In **(b)**, proxy indicators (**blue line**) and thermometer data (**red line**) together show average temperatures in the Northern Hemisphere over the past 1000 years. The gray-shaded zone represents the 95% confidence range. *Data from **(a)** NOAA National Climatic Data Center; and **(b)** IPCC, 2001. Third assessment report.*

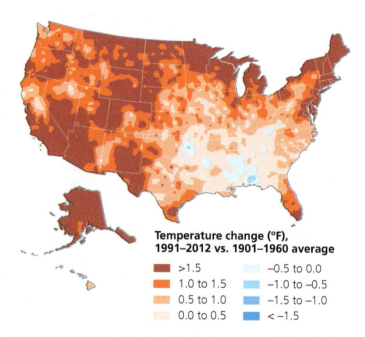

Temperature change (°F), 1991–2012 vs. 1901–1960 average

- ■ >1.5
- ■ 1.0 to 1.5
- ■ 0.5 to 1.0
- □ 0.0 to 0.5
- □ −0.5 to 0.0
- ■ −1.0 to −0.5
- ■ −1.5 to −1.0
- ■ < −1.5

FIGURE 18.11 Temperatures have risen across the United States. Most of the nation has warmed by more than 1 degree Fahrenheit (0.6°C) when the average from the period 1991–2012 is compared to the average from 1901–1960. *Data from NOAA National Climatic Data Center as presented in Melillo, J.M., et al. (Eds.), 2014. Climate change impacts in the United States: The third national climate assessment. U.S. Global Change Research Program.*

DATA Q By how much did the average temperature rise or fall where you live during the time frame illustrated? How does this compare with changes in other parts of the country?

Go to **Interpreting Graphs & Data** on **Mastering**EnvironmentalScience°

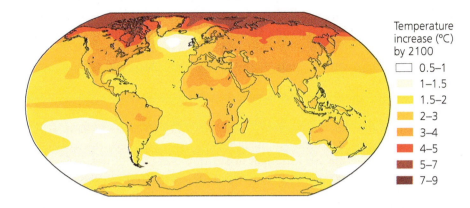

FIGURE 18.12 **Surface temperatures are projected to rise for the years 2081–2100, relative to 1986–2005.** Land masses are expected to warm more than oceans, and the Arctic will warm the most. This map was generated using an intermediate emissions scenario involving an average global temperature rise of 2.2°C (4.0°F). *Data from IPCC, 2013.* Fifth assessment report.

Temperature increase (°C) by 2100

- 0.5–1
- 1–1.5
- 1.5–2
- 2–3
- 3–4
- 4–5
- 5–7
- 7–9

in temperature are predicted to vary from region to region in ways that they already have (**FIGURE 18.12**). For example, polar regions will continue to experience the most intense warming.

Precipitation is changing

A warmer atmosphere speeds evaporation and holds more water vapor, and precipitation has increased worldwide by 2% over the past century. Yet some regions of the world are receiving more rain and snow than usual while others are receiving less. In the western United States, droughts have become more frequent and severe, harming agriculture, worsening soil erosion, reducing water supplies, and promoting wildfire. In parts of the eastern United States, heavy rain events have increased, leading to floods that have killed dozens of people, left thousands homeless, and inflicted billions of dollars in damage.

Future changes in precipitation are predicted to intensify regional changes that have already occurred (**FIGURE 18.13**). Many wet regions will receive more rainfall, increasing the risk of flooding, while many dry regions will become drier, worsening water shortages.

Extreme weather is becoming "the new normal"

The sheer number of extreme weather events in recent years—droughts, floods, tornadoes, hurricanes, snowstorms, cold snaps, heat waves—has caught everyone's attention,

and weather records are falling left and right. In the United States, 2012 was the hottest year ever recorded. The nation experienced a freakish heat wave in March, a severe summer drought that devastated agriculture across three-fifths of the country, and Superstorm Sandy, which inflicted $65 billion in damage along the Atlantic Coast.

Scientific data summarized by the U.S. Climate Extremes Index confirm that the frequency of extreme weather events in the United States has doubled since 1970. Moreover, a 2011 study by climate scientist James Hansen and others revealed that summer temperatures since the 1950s have become not only warmer but also more variable. As a result, extreme summers (some of them unusually cool, more of them unusually warm) have occurred more and more frequently.

Scientists are not the only ones to notice these trends. The insurance industry is finely attuned to such patterns, because insurers pay out money each time a major storm, drought, or flood hits. A major German insurer, Munich Re, calculated that since 1980, extreme weather events causing losses have increased by 50% in South America, have doubled in Europe, and have risen by 2.5 times in Africa, 4 times in Asia, and 5 times in North America.

For years, researchers have conservatively stated that although climate trends influence the probability of what the weather may be like on any given day, no single particular weather event can be directly attributed to climate change. In the aftermath of Superstorm Sandy, a metaphor spread across the Internet: When a baseball player takes artificial steroids and starts hitting more home runs, you can't attribute any

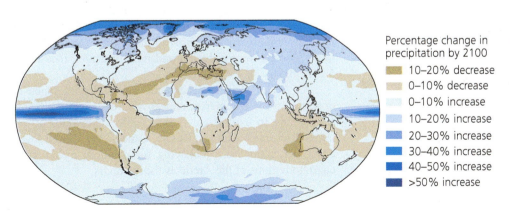

FIGURE 18.13 **Precipitation (June–Aug.) is projected to change for the years 2081–2100, relative to 1986–2005.** Browner shades indicate less precipitation; bluer shades indicate more. This map was generated using an intermediate emissions scenario involving an average global temperature rise of 2.2°C (4.0°F) by 2100. *Data from IPCC, 2013.* Fifth assessment report.

Percentage change in precipitation by 2100

- 10–20% decrease
- 0–10% decrease
- 0–10% increase
- 10–20% increase
- 20–30% increase
- 30–40% increase
- 40–50% increase
- >50% increase

one particular home run to the steroids, but you *can* conclude that steroids were responsible for the increase in home runs. Our greenhouse gas emissions are like artificial steroids that are supercharging our climate and increasing the instance of extreme weather events.

In 2012, research by Jennifer Francis of Rutgers University and Stephen Vavrus of the University of Wisconsin revealed a mechanism that may explain how and why global warming leads to more extreme weather. Warming has been greatest in the Arctic, and this has weakened the intensity of the Northern Hemisphere's polar jet stream. This *jet stream* is a high-altitude air current that blows west-to-east and meanders north and south, influencing much of the weather across North America and Eurasia. As the jet stream slows down, its meandering loops become longer. These long lazy loops move west to east more slowly, and may get stuck in a north–south orientation for long periods of time. Meteorologists call this an *atmospheric blocking pattern* because it blocks the eastward movement of weather systems (FIGURE 18.14). When this happens, a rainy system that would normally move past a

city in a day or two may instead be held in place for several days, causing flooding. Or dry conditions over a farming region might last two weeks instead of two days, resulting in drought. Hot spells last longer, and cold spells last longer, too.

Indeed, the record-breaking heat wave that roasted the eastern United States in March 2012 resulted after the jet stream became stuck in place (see FIGURE 18.14b). Atmospheric blocking patterns also were associated with the 2011 drought in Texas, the 2012 wildfires in Colorado, severe wintry weather in the eastern United States in 2014, floods and heat waves in Europe, and other extreme weather events.

Melting ice has far-reaching effects

As the world warms, mountaintop glaciers are disappearing (FIGURE 18.15). Between 1980 and 2015, the World Glacier Monitoring Service estimates that the world's glaciers, on average, each lost thickness equivalent to nearly 19 m (62 ft) in vertical height of water. Many glaciers on tropical mountaintops have disappeared already. In Glacier National Park in Montana, only 25 of 150 glaciers present at the park's inception remain, and scientists estimate that by 2030 even these will be gone.

Mountains accumulate snow in winter and release meltwater gradually during summer. One out of six people across the world live in regions that depend on mountain meltwater. As warming temperatures diminish mountain glaciers, summertime water supplies are declining for millions of people, and this will likely force whole communities to look elsewhere for water, or to move.

Warming temperatures are also melting vast amounts of polar ice. In the Arctic, the immense ice sheet that covers Greenland is melting faster and faster. In Antarctica, coastal ice shelves the size of Rhode Island have disintegrated as a result of contact with warmer ocean water, and research now suggests that the entire West Antarctic ice shelf may be on its way to unstoppable collapse, creating a 3-m (10-ft) rise in sea level.

One reason warming is accelerating in the Arctic is that as snow and ice melt, darker, less reflective surfaces (such as bare ground and pools of meltwater) are exposed, and Earth's *albedo*, or capacity to reflect light, decreases. As a result, more of the sun's rays are absorbed at the surface, and the surface warms. In a process of positive feedback, this warming causes more ice and snow to melt, which in turn causes more absorption of radiation and more warming (see Figure 5.2, p. 105).

As Arctic sea ice disappears, new shipping lanes are opening up for commerce, and governments and companies are rushing to exploit newly accessible underwater oil and mineral reserves. Already, Russia, Canada, the United States, and other nations are jockeying for position, trying to lay claim to regions of the Arctic as the ice melts.

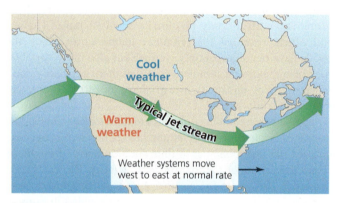

(a) Normal jet stream

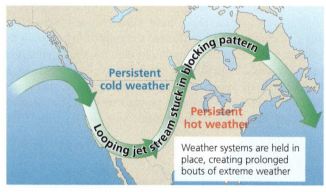

(b) Jet stream in March 2012

FIGURE 18.14 Changes in the jet stream can cause extreme weather events. When Arctic warming slows the jet stream, it departs from its normal configuration **(a)** and goes into a blocking pattern **(b)** that stalls weather systems in place, leading to extreme weather events. The blocking pattern shown here brought record-breaking heat to the eastern United States in March 2012.

(a) Jackson Glacier in 1911

(b) Jackson Glacier in 2009

FIGURE 18.15 Glaciers are melting rapidly as global warming proceeds. The Jackson Glacier in Glacier National Park, Montana, retreated substantially between **(a)** 1911 and **(b)** 2009. The graph **(c)** shows average declines in thickness of ice (water equivalent) in 40 of the world's major glaciers monitored since 1980. *Data from World Glacier Monitoring Service.*

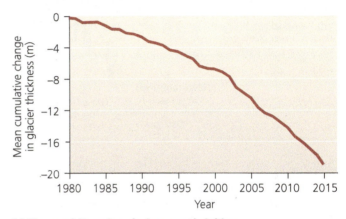

(c) The world's major glaciers are shrinking

Warmer temperatures in the Arctic are also causing *permafrost* (permanently frozen ground) to thaw. As ice crystals within permafrost melt, the thawing soil settles, destabilizing buildings, pipelines, and other infrastructure. When permafrost thaws, it also can release methane that has been stored for thousands of years. Because methane is a potent greenhouse gas, its release acts as a positive feedback mechanism that intensifies climate change.

Rising sea levels may affect hundreds of millions of people

As glaciers and ice sheets melt, increased runoff into the oceans causes sea levels to rise. Sea levels also are rising because ocean water is getting warmer, and water expands in volume as it warms. In addition, we are extracting groundwater from aquifers (for drinking and to apply to farmland) more quickly than groundwater is being replaced (p. 394), and as our wastewater enters rivers and as excess irrigation water runs off farmland, this further fuels sea level rise.

Worldwide, average sea levels have risen 24.1 cm (9.5 in.) in the past 135 years (**FIGURE 18.16**), reaching a rate of 3.4 mm/yr from 1993 to 2016. These numbers represent vertical rises in water level, and on most coastlines a vertical rise of a few inches translates into a great many feet of incursion inland.

Higher sea levels lead to beach erosion, coastal flooding, intrusion of saltwater into aquifers, and greater impacts from storm surges. A *storm surge* is a temporary and localized rise in sea level generated by a storm. The higher the sea level is to begin with, the farther inland a storm surge can reach.

Regions experience differing amounts of sea level change due to patterns of ocean currents or because land may be rising or subsiding naturally, depending on local geologic conditions. The United States is experiencing

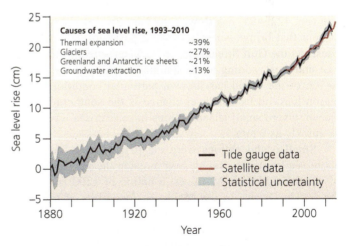

FIGURE 18.16 Global average sea level has risen more than 24 cm (9.5 in.) since 1880. Sea levels rise because water expands as it warms, glaciers and ice sheets are melting, and groundwater we extract eventually reaches the ocean. *Data from IPCC, 2013. Fifth assessment report; CSIRO; and NASA.*

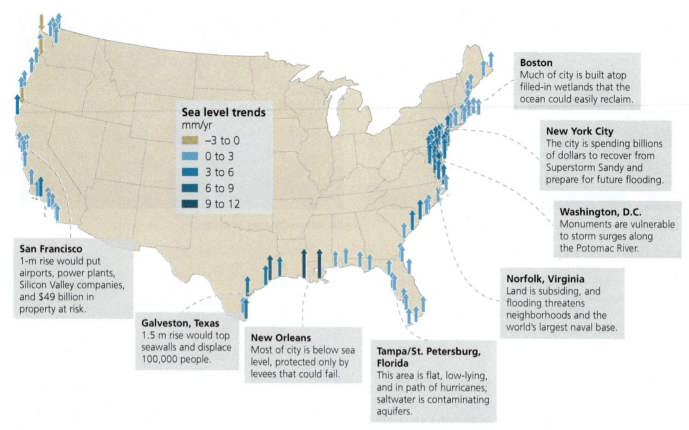

Sea level trends
mm/yr

- ⬆ −3 to 0
- ⬆ 0 to 3
- ⬆ 3 to 6
- ⬆ 6 to 9
- ⬆ 9 to 12

Boston
Much of city is built atop filled-in wetlands that the ocean could easily reclaim.

New York City
The city is spending billions of dollars to recover from Superstorm Sandy and prepare for future flooding.

Washington, D.C.
Monuments are vulnerable to storm surges along the Potomac River.

Norfolk, Virginia
Land is subsiding, and flooding threatens neighborhoods and the world's largest naval base.

San Francisco
1-m rise would put airports, power plants, Silicon Valley companies, and $49 billion in property at risk.

Galveston, Texas
1.5 m rise would top seawalls and displace 100,000 people.

New Orleans
Most of city is below sea level, protected only by levees that could fail.

Tampa/St. Petersburg, Florida
This area is flat, low-lying, and in path of hurricanes; saltwater is contaminating aquifers.

FIGURE 18.17 Sea level is rising at varying rates along the U.S. coast, putting many cities at risk of costly damage. Rates are highest (taller darker blue arrows) where land is subsiding along the Gulf Coast and the central Atlantic Seaboard. *Data from National Oceanic and Atmospheric Administration; city profiles adapted from Rising seas: A city-by-city forecast.* Rolling Stone, 20 June, 2013.

varying degrees of sea level rise (**FIGURE 18.17**), with the East Coast and the Gulf Coast most at risk. Miami and South Florida face rising seas in part because global warming's impact on the oceans' thermohaline circulation system (pp. 422, 485) has slowed the Gulf Stream, the massive current that brings water northward past Florida's Atlantic coast. As the Gulf Stream slows, its water spreads out to the east and west, adding to the amount reaching shore. Every region of the world has its own particular conditions. One region hard-hit by sea-level change is the southern Pacific Ocean, where small island nations have seen sea levels rise as quickly as 9 mm/yr.

Small island nations were the first to raise the alarm about sea level rise. Rising seas threaten the very existence of countries like the Maldives, a nation of 1200 islands in the Indian Ocean. In the Maldives, four-fifths of the land lies less than 1 m (39 in.) above sea level (**FIGURE 18.18a**). Already areas of the Maldives have flooded and salt water is contaminating drinking water supplies. Storms intensified by warmer water are eroding beaches and damaging the coral reefs that support the nation's tourism and fishing industries. Residents have evacuated several low-lying islands, and political leaders have looked to buy property in mainland nations in case their people one day need to abandon their homeland.

For these reasons, leaders of the Maldives have played a prominent role in international efforts to fight global warming. In 2009, Maldives President Mohamed Nasheed donned scuba gear and dove into the blue waters of Girifushi Island lagoon, followed by his entire cabinet. These officials held the world's first underwater cabinet meeting—part of a campaign to draw global attention to the impacts of climate change (**FIGURE 18.18b**). Nasheed then spoke out at international climate talks, where he pleaded with the United States, China, and other nations to reduce emissions of polluting gases that warm the atmosphere.

Today, people from island nations like the Maldives are no longer so alone in voicing concern over climate change. As major cities such as Miami come face-to-face with the challenges of sea-level rise, communities and nations around the world are hastening to come up with solutions.

In the United States, Superstorm Sandy demonstrated the havoc that storm surges can inflict. This massive storm—a hurricane until just before it made landfall—battered the eastern part of the nation in October 2012, causing $65 billion in damage and leaving 160 people dead and thousands homeless (**FIGURE 18.19**). In New Jersey, thousands of homes were destroyed, iconic boardwalks were washed away, and coastal communities were inundated with salt water and sand. In New York City, economic activity ground to a halt as tunnels and

(a) Malé, capital of the Maldives

(b) The "underwater cabinet meeting"

FIGURE 18.18 Rising sea levels threaten the Maldives. This nation's capital **(a)** is crowded onto an island averaging just 1.5 m (5 ft) above sea level. In 2009, the Maldives' president led his cabinet in an underwater meeting **(b)** to focus international attention on the plight of island nations vulnerable to sea-level rise.

FIGURE 18.19 Climate change contributes to the power and reach of devastating storms like Superstorm Sandy. The map shows areas in New York City flooded by the 2012 storm. The graph shows sea level rise in New York City in the past century. *Map data from* The New York Times *as adapted from federal agencies; graph data from Horton, R., et al., 2015. New York City panel on climate change 2015 report chapter 2: Sea level rise and coastal storms.* Ann. N.Y. Acad. Sci., *1336: 36–44.*

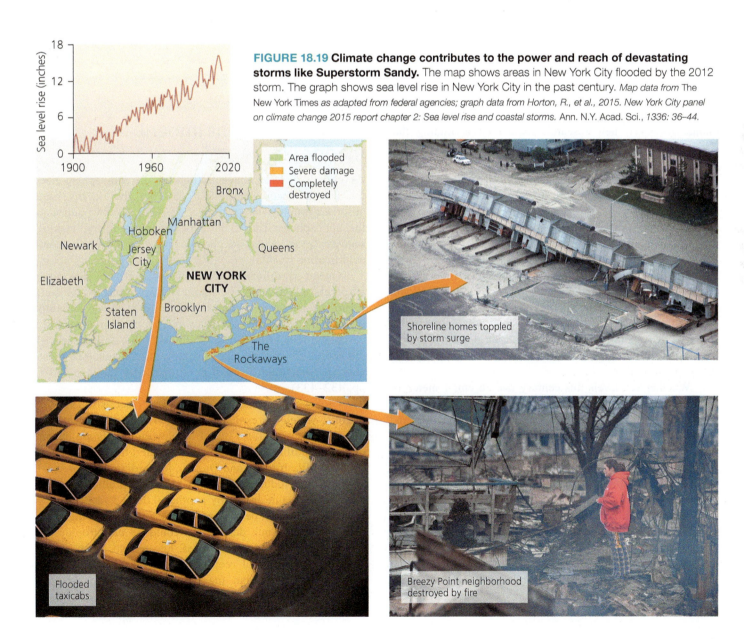

subway stations flooded and vehicles and buildings suffered damage. A fire broke out amid flooded homes in Queens and destroyed an entire neighborhood.

Superstorm Sandy was not directly and solely caused by global warming, but in a statistical sense it was facilitated and strengthened by it. Warmer ocean water boosts the chances of large and powerful hurricanes. A warmer atmosphere retains more moisture that a hurricane can dump onto land. A blocking pattern in the jet stream contributed to Sandy's energy. And higher sea levels magnify the damage caused by storm surges.

Seven years before Sandy, the United States was hit by an even costlier storm. Hurricane Katrina slammed into New Orleans and the Gulf Coast in 2005, killing more than 1800 people and inflicting $80 billion in damage. Outside New Orleans today, marshes of the Mississippi River delta continue to disappear as rising seas eat away at coastal vegetation, as dams upriver hold back silt that once maintained the delta, and as the land subsides due to petroleum extraction. All told, more than 2.5 million hectares (1 million acres) of Louisiana's coastal wetlands have vanished since 1940. Continued wetland loss will deprive New Orleans (much of which is below sea level, safeguarded only by levees) of protection against future storm surges.

The IPCC in its *Fifth Assessment Report* predicted that mean sea level worldwide will rise 26–82 cm (10–32 in.) higher by the year 2100, depending on our level of emissions. However, new research since then is finding that Greenland's ice is melting at an accelerating rate. If polar melting continues to accelerate, then sea levels will rise even more quickly than predicted by the IPCC. Indeed, many researchers are now predicting substantially faster and higher rises; estimates of 1 meter or more by the year 2100 are becoming common. Currently more than half of the U.S. population lives in coastal counties, and 3.7 million Americans live within 1 vertical meter of the high tide line. It has been estimated that a 1-m rise threatens 180 U.S. cities with losing an average of 9% of their land area. South Florida is judged most at risk (**FIGURE 18.20**). Here 2.4 million people, 1.3 million homes, and 1.8 million acres are vulnerable, according to experts with the Surging Seas project of Climate Central. In 107 South Florida towns, fully half the population is at risk.

Whether sea levels this century rise 26 cm, 1 meter, or more, hundreds of millions of people will be displaced or will need to invest in costly efforts to protect against high tides and storm surges. In the long term, if we do not cut greenhouse gas emissions, sea levels will keep rising. A 2015 study calculated that if we burn all the fossil fuels remaining in the world, the heated atmosphere would melt all the ice on Antarctica, half of it in just 1000 years, raising sea levels by about a foot per decade. If all the world's ice melted, the oceans would be 65 m (216 ft) higher. This would put the entire state of Florida under water, along with the rest of the Eastern Seaboard and Gulf Coast. The lower Mississippi River all the way up to Memphis would become a gigantic bay, and California's Central Valley, where much of our food is grown today, would become an inland sea.

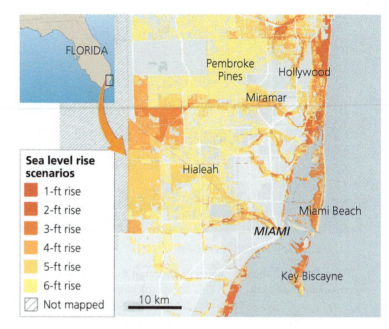

FIGURE 18.20 Miami, Florida, is one of many U.S. cities vulnerable to sea level rise. Shown are areas of the Miami region that would be flooded by rises in sea level of 1–6 feet. *Data from NOAA Coastal Flood Exposure Mapper, www.coast.noaa.gov/floodexposure/#/map.*

Acidifying oceans imperil marine life

As carbon dioxide concentrations in the atmosphere rise, the oceans absorb more CO_2. So far, the oceans have absorbed roughly one-quarter of the CO_2 we have added to the atmosphere. This is altering ocean chemistry, making seawater more acidic—a phenomenon referred to as **ocean acidification** (p. 433).

Ocean acidification threatens marine animals such as corals, clams, oysters, mussels, and crabs, which pull carbonate ions out of seawater to build their exoskeletons of calcium carbonate. As seawater becomes more acidic, carbonate ions become less available, and calcium carbonate begins to dissolve, jeopardizing the existence of these animals (see fuller discussion on this topic in Chapter 16, pp. 432–433).

Global ocean chemistry has already decreased by 0.1 pH unit, which corresponds to a 26% rise in acidity (hydrogen ion concentration). Initial impacts on corals are apparent, and acidified seawater has now killed billions of larval oysters in Washington and Oregon, jeopardizing the region's once-thriving industry. By 2100, scientists predict that seawater will decline in pH by another 0.06–0.32 units—very possibly enough to destroy most of our planet's living coral reefs (p. 427). Such destruction could be catastrophic for marine biodiversity and fisheries, because so many organisms depend on coral reefs for food and shelter. Indeed, ocean acidification and the potential loss of marine life threaten to become one of the most far-reaching impacts of global climate change.

Coral reefs face two additional risks from climate change: Warmer waters contribute to deadly coral bleaching (p. 426), and stronger storms physically damage reefs. All these factors concern residents of places like South Florida. South of Miami in the Florida Keys, coral reefs provide habitat for fish consumed locally and exported for profit, offer snorkeling and scuba diving sites for tourism, and protect coastlines from erosion by reducing wave intensity.

Climate change affects organisms and ecosystems

As the developing crisis with marine life shows, changes in Earth's physical systems have consequences for living things. Organisms are adapted to their environments, so changes to those environments affect them. As global warming proceeds, it is modifying biological phenomena that rely on temperature. In the spring, plants are now leafing out earlier, insects are hatching earlier, birds are migrating earlier, and animals are breeding earlier. These shifts can create mismatches in seasonal timing. For example, European birds known as great tits had evolved to raise their young when caterpillars peak in abundance. Now caterpillars are peaking earlier, but the birds have been unable to adapt, and fewer young birds are surviving.

Biologists are also recording spatial shifts in the ranges of organisms, as plants and animals move toward the poles or upward in elevation (i.e., toward cooler areas) as temperatures warm (FIGURE 18.21). Some organisms will not be able to cope; indeed, the IPCC has estimated that climate change could threaten as many as 20–30% of all plant and animal species with extinction. Trees may not be able to shift their distributions fast enough. Rare species may be forced out of preserves and into developed areas where they cannot survive. Animals and plants adapted to mountainous environments may be forced uphill until there is nowhere left to go.

Effects on plant communities make up an important component of climate change, because by drawing in carbon dioxide for photosynthesis, plants act as reservoirs for carbon. If an atmosphere richer in CO_2 enhances plant growth, then plants might in turn remove more CO_2 from the air, helping to mitigate carbon emissions, in a process of negative feedback. However, if climate change decreases plant growth (through drought, fire, or disease, for instance), then carbon flux to the atmosphere could increase, in a process of positive feedback. Free-Air CO_2 Enrichment (FACE) experiments are showing that extra carbon dioxide can both augment and diminish plant growth (see **The Science behind the Story,** Chapter 5, pp. 122–123).

In regions where precipitation and stream flow increase, erosion and flooding could pollute and alter aquatic systems. In regions where precipitation decreases, lakes, ponds, wetlands, and streams will shrink. The many impacts of climate change on ecological systems will tend to diminish the ecosystem goods and services on which our societies depend.

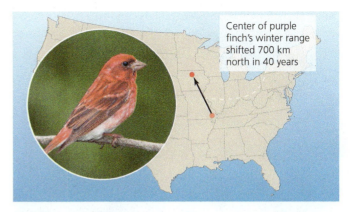

Center of purple finch's winter range shifted 700 km north in 40 years

(a) Birds are moving north

Pikas are disappearing from mountains after being forced upward

(b) Pikas are being forced upslope

FIGURE 18.21 Animal populations are shifting toward the poles and upward in elevation. Fully 177 out of 305 North American bird species have shifted their winter ranges significantly northward in the past 40 years, according to a 2009 analysis of Christmas Bird Count data by National Audubon Society researchers. The purple finch **(a)** has shown the greatest shift; its center of abundance moved 697 km (433 mi) north. Montane animals such as the pika **(b)**, a small mammal that lives at high elevations in western North America, are being forced upslope (into more limited habitat) as temperatures warm. Many pika populations have disappeared from mountains already.

Climate change affects society

Drought, flooding, storm surges, and sea level rise have already taken a toll on the lives and livelihoods of millions of people. The many consequences of climate change include impacts on agriculture, forestry, health, economics, and national security.

Agriculture For some crops in the temperate zones, moderate warming may slightly increase production because growing seasons become longer. Enriched atmospheric carbon dioxide for photosynthesis may also increase yields, although as mentioned above, the effects appear to be mixed. Moreover, some research shows that crops can become less nutritious when supplied with more carbon dioxide. Perhaps of most consequence, if rainfall continues to shift in space

FIGURE 18.22 Drought induced by climate change decreases crop yields. Withered corn fields like this one in Illinois were a common sight in 2012, when the U.S. government declared 1000 counties across 26 states to be disaster areas due to drought.

FIGURE 18.23 Climate change can contribute to humanitarian, geopolitical, and national security problems. Prolonged drought associated with climate change weakened agriculture and helped spark the civil war in Syria and the rise of the Islamic State, many experts feel. The resulting flow of refugees, in turn, put social and political strains on European nations receiving them.

and time, intensified droughts and floods will likely cut into agricultural productivity (**FIGURE 18.22**). Considering all factors together, the IPCC predicts that global crop yields will increase somewhat initially—but beyond a rise of 3°C (5.4°F), the IPCC expects crop yields to decline. This would worsen hunger in many of the world's developing nations.

Forestry In the forests that provide our timber and paper products, enriched atmospheric CO_2 may spur greater growth in the near term, but intensified drought, fire, and disease may eliminate these gains. Today's forest managers increasingly find themselves battling catastrophic fires, invasive species, and insect and disease outbreaks. Catastrophic fires are caused in part by decades of fire suppression, but are also promoted by longer, warmer, drier fire seasons (see Figure 12.19a, p. 315). Milder winters and hotter, drier summers are promoting outbreaks of bark beetles that are destroying millions of acres of trees (see Figure 12.19b, p. 315).

Health As climate change continues, we will face more heat waves—and heat stress can cause death, especially among older adults. A 1995 heat wave in Chicago killed at least 485 people, and a 2003 heat wave in Europe killed 35,000 people. A warmer climate also exposes us to other health problems:

- Respiratory ailments from air pollution, as hotter temperatures promote photochemical smog (p. 462)

- Expansion of tropical diseases, such as malaria and dengue fever, into temperate regions as vectors of infectious disease (such as mosquitoes) spread toward the poles

- Disease and sanitation problems when floods overcome sewage treatment systems

- Injuries and drowning from worsened storms

Health hazards from cold weather will decrease, but many researchers feel that the increase in warm-weather hazards will more than offset these gains.

National security The many impacts of climate change endanger the ability of nations to ensure social stability and protect their citizens from harm. The Pentagon, the White House, the U.S. Navy, the Council on Foreign Relations, and the Central Intelligence Agency have all concluded and publicly reported that climate change is contributing to political violence, war and revolution, humanitarian disasters, and refugee crises (**FIGURE 18.23**). "Climate change will affect the Department of Defense's ability to defend the Nation and poses immediate risks to U.S. national security," the U.S Defense Department stated bluntly in a 2014 report. The report described how storms, rising seas, and other impacts are "threat-multipliers," making small problems larger. Already, extreme weather events have damaged military installations, weakened the economies and infrastructure of allies and trading partners, disrupted flows of oil and gas, and strained emergency response abilities, while Arctic melting has set off a competitive race among nations to claim polar resources.

When environmental conditions worsen and people suffer as a result, some of them may leave their homes and become refugees, while others may turn to radical political ideologies and even terrorism. This is why national security experts have linked climate change and drought to the origins of the war in Syria, to conflicts elsewhere in the Middle East and Africa, and to the resulting refugee crisis in Europe. In the years ahead, the world's militaries and emergency responders will be devoting more and more of their efforts toward problems created or made worse by climate disruption.

Economics People will experience a variety of economic costs and benefits from the many impacts of climate change, but on the whole researchers predict that costs will substantially outweigh benefits, especially as climate change grows more severe. Climate change is also expected to widen the gap between rich and poor, both within and among nations. Poorer people have less wealth and technology with which to adapt to climate change, and they rely more on resources (such as local food and water) that are sensitive to climate disruption.

Economists have tried to quantify damages from climate change by totaling up its various external costs (pp. 141, 162). Their estimates for the **social cost of carbon,** the economic cost of damages resulting from each ton of carbon dioxide we emit, run the gamut from $10 to $350 per ton, depending on what costs are included and what discount rate (pp. 144, 146) is used. The U.S. government has calculated the social cost of carbon to be roughly $40 per ton ($37 in 2007 dollars, rising with inflation), and uses this official estimate to decide when and how to impose regulations that affect emissions. Other nations and many large corporations use their own estimates.

In terms of overall cost to society, the IPCC has estimated that climate change may cost 1–5% of gross domestic product (GDP) on average globally, with poor nations losing proportionally more than rich nations. The *Stern Review on the Economics of Climate Change* (see **The Science Behind the Story,** Chapter 6, pp. 146–147) concluded that climate change could cost the world 5–20% of GDP by the year 2200, but that investing just 1% of GDP starting now could enable us to avoid these future costs. A 2014 report titled *Risky Business* detailed hundreds of billions of dollars in likely damages from climate change. This report was issued by a politically diverse team of leading businesspeople and finance experts, chaired by former New York City Mayor Michael Bloomberg, former hedge fund manager Tom Steyer, and former U.S. Treasury Secretary Henry Paulson. Then in 2015 the U.S. Environmental Protection Agency (EPA) weighed in, calculating that reducing greenhouse gas emissions would help the United States avoid $235–334 billion in annual costs by the year 2050, and $1.3–1.5 trillion by 2100 (see Table 1, p. 147). Regardless of the precise numbers, many economists and policymakers have concluded that investing money now to fight climate change will save us a great deal more money in the future.

Impacts vary by region

Each of us will experience the impacts of climate change differently, depending on where we live. Temperature changes have been greatest in the Arctic (**FIGURE 18.24**). Here, ice sheets are melting, sea ice is thinning, storms are increasing,

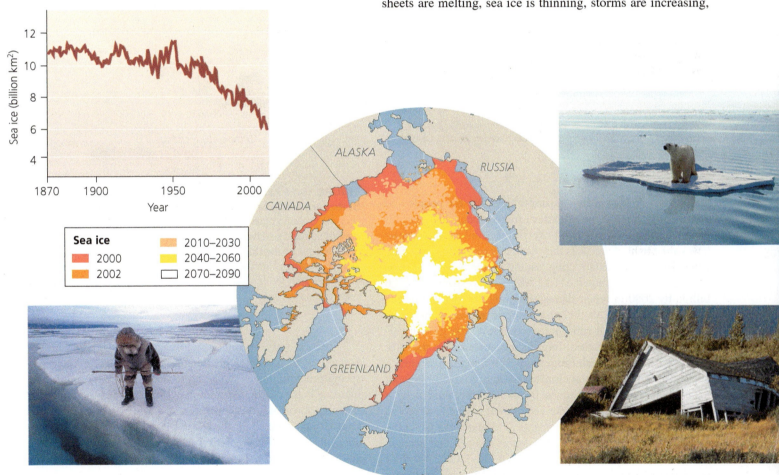

FIGURE 18.24 The Arctic has borne the brunt of climate change's impacts so far. As sea ice melts, it recedes from large areas. The map in the center of the figure shows the mean minimum summertime extent of sea ice for the recent past, present, and future. The graph (upper left) shows declines in sea ice averaged from six data sets. Inuit people find it difficult to hunt and travel in their traditional ways. Polar bears starve because they are less able to hunt seals. Structures are damaged as permafrost thaws beneath them. *Map data from National Center for Atmospheric Research and National Snow and Ice Data Center.*

TABLE 18.2 Some Predicted Impacts of Climate Change in the United States

- Average temperatures will rise 1.7–5.6°C (3–10°F) further by the year 2100.

- Droughts, flooding, and wildfire will worsen; dry areas will get drier and wet areas wetter.

- Extreme weather events will become more frequent. The costs they impose on society will grow.

- Sea level will rise an additional 0.3–1.2 m (1–4 ft) by 2100.

- Storm surges will continue to erode beaches and coastal wetlands, destroy real estate, and damage infrastructure.

- Health problems due to heat stress, disease, and pollution will rise. Some tropical diseases will spread north.

- Drought, fire, and pest outbreaks will continue to alter forests.

- Ocean acidification will affect marine ecosystems and fisheries.

- Although enhanced CO_2 and longer growing seasons favor crops, increased drought, heat stress, pests, and diseases will decrease most yields.

- Snowpack will decrease in the West; water shortages will worsen in many areas.

- Alpine ecosystems and barrier islands will begin to vanish.

- Melting permafrost will undermine Alaskan buildings and roads.

Adapted from Melillo, J.M., et al. (Eds.), 2014. *Climate change impacts in the United States: The third national climate assessment.* U.S. Global Change Research Program.

Reduced Emissions Scenario
Projected Temperature Change (°F)
End-of-Century (2071–2099 average)

Continued Emissions Scenario
Projected Temperature Change (°F)
End-of-Century (2071–2099 average)

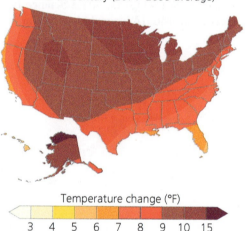

Temperature change (°F)

3 4 5 6 7 8 9 10 15

FIGURE 18.25 Average temperatures across the United States are predicted to rise by the end of this century. Even under a scenario of sharply reduced emissions (**top**), temperatures are predicted to rise by 3–4°F. Under a scenario of business-as-usual emissions (**bottom**), temperatures are predicted to rise by 7–11°F. *Data from Melillo, J.M., et al. (Eds.), 2014. Climate change impacts in the United States: The third national climate assessment. U.S. Global Change Research Program.*

and altered conditions are posing challenges for people and wildlife. As sea ice melts earlier, freezes later, and recedes from shore, it becomes harder for Inuit people and for polar bears alike to hunt the seals they each rely on for food. Permafrost is thawing, destabilizing buildings, roads, and bridges. A recent study estimates Alaska will need to spend $5.6–7.6 billion by 2080 to repair damage to public infrastructure due to climate change. As the strong Arctic warming melts ice caps and ice sheets, it contributes to sea level rise globally.

For the United States, impacts are assessed by the U.S. Global Change Research Program, which Congress created in 1990 to coordinate federal climate research. Its 2014 *National Climate Assessment,* produced by 300 experts guided by a 60-person committee, summarized current research, observed trends, and predicted future impacts of climate change on the United States (**TABLE 18.2**). This report showed that average U.S. temperatures have increased by 0.7–1.1°C (1.3–1.9°F) since record keeping began in 1895, with the vast majority of this rise occurring just since 1970. Temperatures are predicted to rise by another 1.7–5.6°C (3–10°F) by the end of this century (**FIGURE 18.25**). Extreme weather events have become more frequent, and the costs they impose on farmers, city-dwellers, coastal communities, and taxpayers across the country are escalating.

Impacts vary by region, and each region of the United States will face its own challenges. For instance, winter and spring precipitation is projected to decrease in the South but increase in the North. Drought may strike in some regions and flooding in others. Sea level rise will affect the Atlantic and Gulf Coasts more than the West Coast. You can learn more about the scientific predictions for your own region by consulting the 2014 report, which is publicly accessible online.

Are we responsible for climate change?

Scientists agree that today's global warming is a result of the well-documented recent increase in greenhouse gas concentrations in our atmosphere. They also agree that this rise results primarily from our combustion of fossil fuels for energy and secondarily from the loss of carbon-absorbing vegetation due to deforestation and other changes in land use.

Yet despite the overwhelming evidence for climate change and its impacts, many people, especially in the United States, have long tried to deny that it is happening. Most of these "climate skeptics" or "climate change deniers" now admit that the climate is changing but still express doubt that we are the cause. Indeed, while most of the world's nations moved forward to confront climate change through international dialogue, in the United States public discussion of climate change remained mired in outdated debates over whether the phenomenon was real and whether humans were to blame. These debates have been fanned by corporate interests, political think tanks, and a handful of scientists funded by fossil fuel industries, all of whom have aimed to cast doubt on the scientific consensus.

For instance, the oil corporation Exxon-Mobil funded attacks on climate science for years, methodically sowing doubt in the public discourse—after its own in-house scientists had done cutting-edge research back in the 1980s documenting climate change! In the 2010 book *Merchants of Doubt*, science historians Naomi Oreskes and Erik Conway reveal how some of the ideologically motivated individuals who cast doubt on climate science had previously done the same against scientific conclusions on the risks of tobacco smoke, DDT, ozone depletion, and acid rain.

All along, the views of climate change deniers have been amplified by the American news media, which seek to present two sides to every issue, even when the two sides' arguments are not equally supported by evidence. However, as data have mounted over the decades and as the economic and societal costs of climate change have grown clearer, more and more policymakers, corporate executives, military leaders, national security experts, heads of business and industry, and everyday citizens have concluded that climate change is escalating and is causing impacts to which we must respond.

Responding to Climate Change

From this point onward, our society will be focusing on how best to respond to the challenges of climate change. The good news is that everyone—not just leaders in government and business, but everyday people, and especially today's youth—can play a part in this vital search for solutions.

Shall we pursue mitigation or adaptation?

We can respond to climate change in two fundamental ways. One is to pursue actions that reduce greenhouse gas emissions, to lessen the severity of climate change. This strategy is called **mitigation** because the aim is to mitigate the problem, that is, to alleviate it or reduce its severity. Examples of mitigation include improving energy efficiency, switching to clean and renewable energy sources, preserving forests,

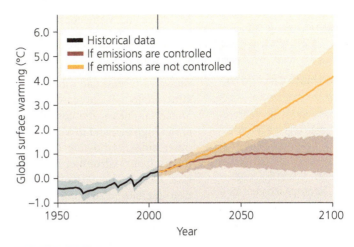

FIGURE 18.26 The sooner we stabilize our emissions, the less climate change we will cause. The dark red line shows the temperature change we can expect if we strongly limit our carbon emissions. The orange line shows the change expected if we fail to control emissions effectively. These data projections represent the averaged output of a large number of climate models, under two emissions scenarios studied by the IPCC. *Data from IPCC, 2013.* Fifth assessment report.

recovering landfill gas, and encouraging farm practices that protect soil quality. The sooner we begin reducing emissions, the lower the level at which they will peak, and the less we will alter climate (**FIGURE 18.26**).

The second type of response is to pursue strategies to cushion ourselves from the impacts of climate change. This strategy is called **adaptation** because the goal is to adapt to change. Installing elaborate pump systems, as Miami Beach is doing to pump out its floodwaters, is an example of adaptation (**FIGURE 18.27**). So is erecting seawalls, as residents of the

FIGURE 18.27 Miami Beach is trying to adapt to sea level rise by constructing an elaborate system of pumps and drainage pipes. Adaptation is a more costly and less effective response to climate change than mitigation, however.

Citizens desiring stronger action against climate change are now taking their grievances to court. In the Netherlands in 2015, a group named Urgenda and 900 citizens sued the Dutch government for "knowingly contributing" to global warming—and won. The court agreed that Dutch policy would not hold warming to the internationally set 2°C goal, and ordered the government to deepen emission cuts. In a U.S. lawsuit in 2016, the nonprofit Our Children's Trust and 21 young people demanded action on climate change, saying the federal government had "willfully ignored this impending harm." In Peru, a farmer sued a fossil-fuel power company, asking monetary compensation for impacts of glacier melt on his community.

Do you think lawsuits are an appropriate way to strengthen policy on climate change? To what degree is a government obligated to protect its citizens from climate change? Do you favor suing fossil-fuel companies for compensation for climate change impacts? What ethical or human rights issues, if any, do you think climate change presents? How could these best be resolved?

Maldives and many coastal cities have done. The Netherlands—a very low, flat, coastal nation—has so much experience with seawalls, levees, and other infrastructure to manage seawater that Dutch engineers and consultants are in high demand in Miami and other cities worldwide that find themselves confronting sea-level rise. Other examples of adaptation include restricting coastal development; adjusting farming practices to cope with drought; and modifying water management practices to deal with reduced river flows, glacial outburst floods, or salt contamination of groundwater.

Both adaptation and mitigation are necessary. Adaptation is needed because even if we could halt all our emissions right now, the greenhouse gas pollution already in the atmosphere would continue driving global warming until the planet's systems reach a new equilibrium, with temperature rising an estimated 0.6°C (1.0°F) more by the end of the century. Because this change is already locked in, it is wise to develop ways to minimize its impacts.

We also need to pursue mitigation, because if we do nothing to diminish climate change, it will eventually overwhelm any efforts we might make to adapt. To leave a sustainable future for our civilization and to safeguard the living planet that we know, we need to pursue mitigation. We will spend the remainder of our chapter examining approaches for the mitigation of climate change.

We are developing solutions in electricity generation

From cooking to heating to lighting, much of what we do each day depends on electricity. The generation of electricity produces the largest portion (40%) of U.S. carbon dioxide emissions. Fossil fuel combustion generates two-thirds of U.S. electricity, and coal accounts for most of the resulting emissions. There are two ways we can reduce the amount of fossil fuels we burn to generate electricity: (1) encourage conservation and efficiency (pp. 541–544) and (2) make greater use of nuclear power and renewable energy sources (Chapters 20 and 21).

Conservation and efficiency As individuals we all can make lifestyle choices to reduce electricity consumption. New energy-efficient technologies make it easier to conserve. Replacing standard lightbulbs with compact fluorescent lights reduces energy use for lighting by 40%. The U.S. EPA's Energy Star Program rates household appliances, lights, windows, fans, office equipment, and heating and cooling systems by their energy efficiency. Replacing an old washing machine with an Energy Star washing machine can cut your CO_2 emissions by 200 kg (440 lb) annually. Energy Star homes use highly efficient windows, ducts, insulation, and heating and cooling systems to reduce energy use and emissions by 30% or more. Such technological solutions also save consumers money by reducing utility bills.

Sources of electricity We can also reduce greenhouse gas emissions by switching to cleaner energy sources. Natural gas generates the same amount of energy as coal, with half the emissions. Cleaner still are alternatives to fossil fuels, including nuclear power, bioenergy, hydroelectric power, geothermal power, solar photovoltaic cells, wind power, and ocean sources. These energy sources give off no net emissions as they generate electricity. (We will examine these energy sources in Chapters 20 and 21.)

While our society begins to transition to clean and renewable alternatives, we are also trying to capture emissions before they leak to the atmosphere. **Carbon capture and storage** refers to technologies or approaches that remove carbon dioxide from emissions and then store it belowground under pressure in deep salt mines, depleted oil and gas deposits, or other underground reservoirs (see Figure 19.22, p. 538). However, we are still a long way from developing adequate technology and secure storage space to accomplish this without leakage. Moreover, it is questionable whether we will ever be able to sequester enough carbon to make a sizeable dent in our emissions. (Carbon capture and storage is discussed in Chapter 19, p. 538.)

Transportation solutions are at hand

Can you imagine life without a car? Most Americans probably can't—a reason why transportation is the second-largest source of U.S. greenhouse gas emissions. The average American family makes 10 trips by car each day, and U.S. taxpayers spend more than $200 million per day on road construction and repairs for the nation's 260 million registered automobiles.

Automotive technology The typical automobile is highly inefficient. Only 14% of the energy from fuel we pump into our tanks actually moves our cars down the road. More aerodynamic designs, increased engine efficiency, and improved tire design all can help make our vehicles more fuel-efficient. The vehicles of many nations are more fuel-efficient than those of the United States, and recent government mandates are now improving fuel efficiency in American vehicles (p. 543). Advancing technology is also bringing us alternatives to the traditional combustion-engine automobile. These include electric vehicles, gasoline-electric hybrid vehicles (p. 543), hydrogen fuel cells (p. 598), and alternative fuels such as compressed natural gas and biodiesel (p. 567).

Transportation choices We can make lifestyle choices that reduce our reliance on cars. Some people are choosing to live nearer to their workplaces. Others use mass transit, such as buses, subway trains, and light rail. Still others bike or walk to work or on errands. In the United States, public transportation currently serves 3–4% of passenger trips, reducing gasoline use by 4.2 billion gallons each year and saving 37 million metric tons of CO_2 emissions, the American Public Transportation Association estimates. If U.S. residents were to increase their use of mass transit to the levels of Canadians (7% of daily travel needs) or Europeans (10% of daily travel needs), the United States could cut its air pollution, its dependence on imported oil, and its contribution to climate change.

Unfortunately, reliable and convenient public transit is not yet available in many U.S. communities. Making automobile-based cities and suburbs more friendly to pedestrian and bicycle traffic and improving people's access to mass transit stand as central challenges for city and regional planners (p. 339).

We will need multiple strategies

Advances in agriculture, forestry, and waste management can help us mitigate climate change. In agriculture, sustainable management of cropland and rangeland enables soil to store more carbon. New techniques reduce methane emissions from rice cultivation and from cattle and manure and reduce nitrous oxide emissions from fertilizer. We can also grow renewable biofuel crops, although whether these decrease or increase emissions is an active area of research (p. 570). In forest management, preserving forests, reforesting cleared

areas, and pursuing sustainable forestry practices (p. 318) all help to absorb carbon from the air. Waste managers are cutting emissions by generating energy from waste in incinerators; capturing methane seeping from landfills; and encouraging recycling, composting, and the reuse of materials and products (pp. 609–618).

We should not expect to find a single "magic bullet" for mitigating climate change. Reducing emissions will require steps by many people and institutions across many sectors of our economy. The good news is that most reductions can be achieved using current technology and that we can begin implementing changes right away. Environmental scientists Stephen Pacala and Robert Socolow advise that we follow some age-old wisdom: When the job is big, break it into smaller parts. Pacala and Socolow have identified 15 strategies (**FIGURE 18.28**) that could each eliminate 1 billion tons of carbon per year by 2050 if deployed at a large scale. Achieving just 7 of these 15 aims would stabilize our emissions. If we achieve more, then we reduce emissions.

What role should government play?

Even if people agree on strategies and technologies to reduce emissions, they may disagree on what role government should play to encourage those strategies and technologies: Should it mandate change through laws and regulations? Should it impose no policies at all and hope that private enterprise will develop solutions on its own? Should it take the middle ground and design policies that give private entities financial incentives to reduce emissions? This debate has been

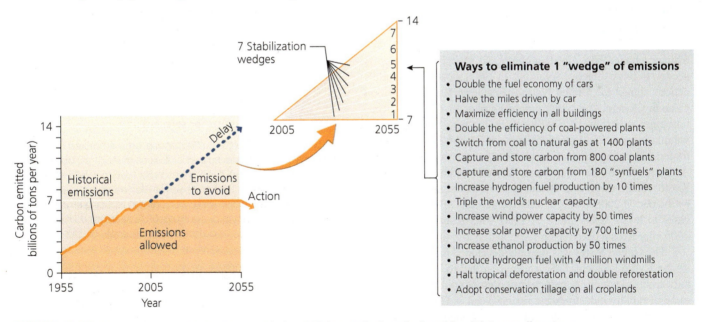

FIGURE 18.28 We can accomplish the large job of stabilizing emissions by breaking it into smaller steps. Environmental scientists Stephen Pacala and Robert Socolow began with a standard graph showing the doubling of CO_2 emissions that scientists expect to occur from 2005 to 2055. They added a flat line to represent the trend if emissions were held constant and then separated the graph into emissions allowed (below the line) and emissions to be avoided (the triangular area above the line). They then divided this "stabilization triangle" into seven equal-sized portions. Each of these "stabilization wedges" represents 1 billion tons of CO_2 emissions in 2055 to be avoided. Finally, they identified a series of strategies, each of which could take care of one wedge. If we accomplish just 7 of these strategies, we could halt our growth in emissions for the next half-century. *Adapted from Pacala, S., and R. Socolow, 2004. Stabilization wedges: Solving the climate problem for the next 50 years with current technologies. Science 305: 968–972.*

vigorous in the United States, where many business leaders and politicians have opposed all government action to address climate change, fearing that emissions reductions will impose economic costs on industry and consumers.

In 2007, the U.S. Supreme Court ruled that carbon dioxide was a pollutant that the Environmental Protection Agency (EPA) could regulate under the Clean Air Act (pp. 172, 454). When Barack Obama became president, he instead urged that Congress craft laws to address emissions. In 2009, the House of Representatives passed legislation to create a **cap-and-trade** system (pp. 179, 509) in which industries and utilities would compete to reduce emissions for financial gain, and under which emissions were mandated to decrease 17% by 2020. However, similar legislation did not pass in the Senate. As a result, responsibility for addressing emissions passed to the EPA, which is phasing in regulations on industry and utilities to spur energy efficiency retrofits and clean and renewable energy use.

In 2013 President Obama announced that, because of legislative gridlock, he would take steps to address climate change using the president's executive authority. His Climate Action Plan aimed to jumpstart renewable energy development, modernize the electrical grid, finance clean coal and carbon storage efforts, improve automotive fuel economy, protect and restore forests, and encourage energy efficiency. It also led to the Clean Power Plan (p. 464), under which the EPA proposed to regulate existing power plants. At the same time, the president sought to prepare the nation to adapt to the impacts of climate change and to engage with other countries to address greenhouse gas emissions.

The Kyoto Protocol sought to limit emissions

Climate change is a global problem, so global cooperation is needed to forge effective solutions. In 1992, most of the world's nations signed the **United Nations Framework Convention on Climate Change.** This treaty outlined a plan for reducing greenhouse gas emissions to 1990 levels by the year 2000 through a voluntary approach. Emissions kept rising, however, so nations forged a binding agreement to *require* emissions reductions. An outgrowth of the Framework Convention drafted in 1997 in Kyoto, Japan, the **Kyoto Protocol** mandated signatory nations, by the period 2008–2012, to reduce emissions of six greenhouse gases to levels below those of 1990. The treaty took effect in 2005 after Russia became the 127th nation to ratify it.

The United States was the only developed nation not to ratify the Kyoto Protocol. U.S. leaders objected to how it required industrialized nations to reduce emissions but did not require the same of rapidly industrializing nations such as China and India, whose emissions were rising quickly. Proponents of the Kyoto Protocol countered that the differential requirements were justified because industrialized nations had created the current problem and therefore should take the lead in resolving it.

As of 2012 (the close of the first commitment period), nations that signed the Kyoto Protocol had decreased their emissions by 10.6% from 1990 levels (**FIGURE 18.29**). However, much of this reduction was due to economic contraction in Russia and nations of the former Soviet Bloc following the

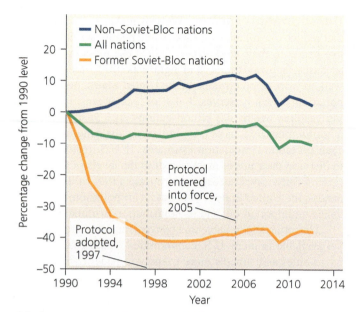

(a) Changes in emissions since 1990

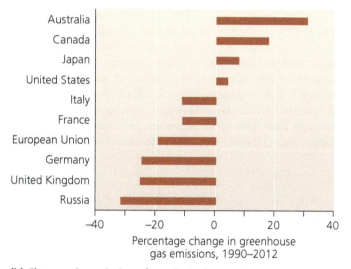

(b) Changes in emissions for selected nations

FIGURE 18.29 The Kyoto Protocol produced mixed results. Nations ratifying it reduced their emissions of six greenhouse gases by 10.6% by 2012 **(a)**, but this was largely due to unrelated economic contraction in the former Soviet-Bloc countries. A selection of major nations **(b)** shows varied outcomes in reducing emissions. The United States did not ratify the Protocol, Australia joined it late, and Canada left early. Values do not include influences of land use and forest cover. *Data from U.N. Framework Convention on Climate Change, 2015.*

DATA In part (b), compare the nations whose emissions increased with those whose emissions decreased. What difference(s) do you note between these two groups that might explain why their emissions trends differ?

Go to **Interpreting Graphs & Data** on **Mastering**EnvironmentalScience®

breakup of the Soviet Union. When these nations are factored out, the remaining signatories showed a 1.9% *increase* in emissions. Nations not parties to the accord, including China, India, and the United States, increased their emissions still more.

International climate negotiations seek a way forward

In recent years, representatives of the world's nations have met at a series of international conferences, trying to design a treaty to succeed the Kyoto Protocol (**FIGURE 18.30**). Delegates from European nations and small island nations have generally taken the lead, while China, India, and the United States have been reluctant to commit to emissions cuts.

At a contentious 2009 conference in Copenhagen, Denmark, nations endorsed a goal of limiting climate change to 2°C of warming, but failed to agree on commitments as the session ended in disarray. The process got back on track in Cancun, Mexico, in 2010, where nations made progress on a plan, nicknamed *REDD* (p. 308), to help tropical nations reduce forest loss, and where developed nations promised to fund developing nations to assist mitigation and adaptation

efforts. In Durban, South Africa, in 2011, nations agreed to a "road map" toward a legally binding international deal in 2015. And at the 2012 conference in Doha, Qatar, negotiators extended the Kyoto Protocol until 2020 (although a number of nations backed out of this extension, and the protocol now applies to only about 15% of the world's emissions).

At the 2015 climate conference in Paris, France, the world's nations made stronger commitments than ever before. European nations announced plans to further reduce their emissions. The United States committed to emissions cuts it hoped would be made possible by new regulations on coal-fired power plants and by switching from coal to natural gas. China pledged to cut back on coal-fired power and to establish a cap-and-trade program. Brazil promised to halt deforestation. And India agreed to slow its emissions growth and reforest its land while aggressively developing renewable energy.

The impressive progress at the Paris conference was made possible because all 197 nations were encouraged to bring their own particular solutions to the table. Success at Paris was also facilitated by an agreement a year earlier between U.S. President Obama and Chinese President Xi Jinping, which broke the impasse between these two largest polluting nations. In their joint announcement, Obama promised the United States would reduce carbon emissions by 28% by 2025, and Xi Jinping promised China would halt its emissions growth by 2030.

Ultimately, the historic conference in Paris will be judged by how well nations live up to their pledges. Moreover, the Paris Agreement brings the world only partway toward a goal of limiting warming to 2°C. This is because even if all nations fully meet the commitments they made in Paris, calculations indicate that the average global temperature by 2100 would still rise about 3.5°C (6.3°F) (**FIGURE 18.31**).

Nonetheless, the Paris conference produced progress that could conceivably lead to further action and deeper emissions cuts. Notably, individuals from the private sector became involved as never before. Microsoft founder and billionaire philanthropist Bill Gates led an effort to attract private

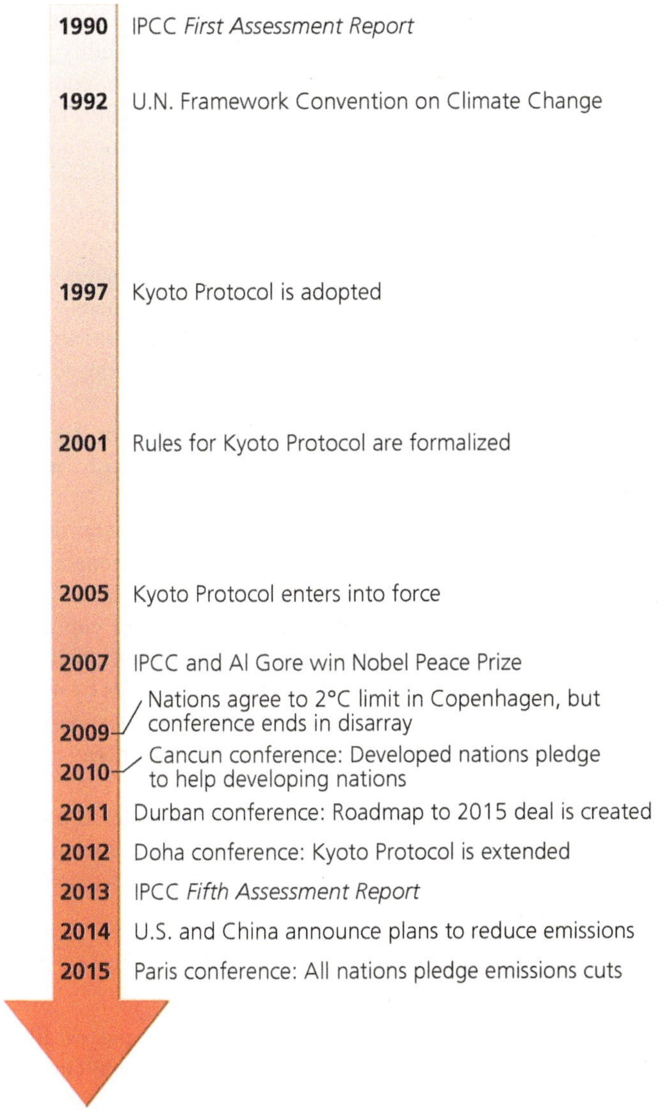

1990	IPCC *First Assessment Report*
1992	U.N. Framework Convention on Climate Change
1997	Kyoto Protocol is adopted
2001	Rules for Kyoto Protocol are formalized
2005	Kyoto Protocol enters into force
2007	IPCC and Al Gore win Nobel Peace Prize
2009	Nations agree to 2°C limit in Copenhagen, but conference ends in disarray
2010	Cancun conference: Developed nations pledge to help developing nations
2011	Durban conference: Roadmap to 2015 deal is created
2012	Doha conference: Kyoto Protocol is extended
2013	IPCC *Fifth Assessment Report*
2014	U.S. and China announce plans to reduce emissions
2015	Paris conference: All nations pledge emissions cuts

FIGURE 18.30 Negotiators have worked tirelessly for years to forge international agreements to reduce emissions. Shown is a timeline of highlights of climate change science and policy over the past quarter-century.

FIGURE 18.31 Commitments made at the 2015 Paris climate meeting would limit the global temperature increase to 3.5°C (6.3°F) by 2100. This is less than the 4.5°C (8.1°F) rise predicted in the absence of those commitments, but widely misses the formal target of limiting temperature rise to 2.0°C (3.6°F). *Data from Climate Interactive.*

investment into renewable energy from some of the wealthiest people in the world. Indeed, many experts now predict that success in mitigating climate change will come largely from business investment, technological advances, and economic incentives, as well as national, regional, and local government initiatives. Business and industry are accelerating investments in renewable energy and energy efficiency, and policymakers are looking to encourage private-sector efforts that can generate productive solutions through the marketplace.

Will emissions cuts hurt the economy?

Many U.S. policymakers have opposed mandates to reduce emissions because they fear this will hamper economic growth. China and India have long resisted emissions cuts under the same assumption. This is understandable, given that our current economies depend so heavily on fossil fuels. Yet nations such as Germany, France, and the United Kingdom have reduced their emissions since 1990 while enhancing their economies and providing their citizens high standards of living. Wealthy nations from Denmark to New Zealand to Hong Kong to Switzerland to Sweden emit less than half the greenhouse gases per person as the United States does.

Indeed, the United States reduced its carbon dioxide emissions by 12% from 2007 to 2015. This occurred as a result of efficiency measures and a shift from coal to natural gas. A recession also caused fossil fuel use to decline temporarily, yet despite the recession, the U.S. economy grew during this period overall, suggesting that cutting emissions need not hinder economic growth. Moreover, in 2015, for the first time in history, global carbon emissions were predicted to decline (after remaining stable in 2014), while the world economy continued to grow. These encouraging data from 2014 and 2015 suggested that we may perhaps have reached a historic turning point, at which economic growth becomes decoupled from greenhouse gas emissions.

Because resource use and per capita emissions are high in the United States, policymakers and industries often assume the United States has more to lose economically from restrictions on emissions than developing nations do. However, industrialized nations are also the ones most likely to *gain* economically from major energy transitions, because they are best positioned to invent, develop, and market new technologies to power the world in a post-fossil-fuel era. Germany, Japan, and China have realized this and are now leading the world in production, deployment, and sales of solar energy technology (**FIGURE 18.32**). If the United States does not act quickly to develop energy technologies for the future, then the future could belong to nations like China, Germany, and Japan.

States and cities are advancing climate change policy

In the absence of legislative action at the federal level to address climate change, state and local governments across the United States are advancing policies to limit emissions. Mayors from more than 1000 cities have signed on to the U.S. Mayors Climate Protection Agreement, committing their cities to pursue policies to "meet or beat" Kyoto Protocol guidelines. Former

FIGURE 18.32 China is racing to become the world's leader in renewable energy technology. Here, workers at a Chinese factory produce photovoltaic solar panels.

New York City mayor Michael Bloomberg launched a climate change panel in 2008 as part of his PlaNYC sustainability plan (p. 348). In 2015, city councilors in Portland, Oregon, voted to ban development of any new fossil fuel infrastructure. Local leaders in South Florida have focused thus far on adaptation measures but are starting to discuss mitigation policies as well.

A number of U.S. states have enacted targets or mandates for renewable energy production, seeking to boost cleaner alternatives to fossil fuels. The boldest state-level action so far has come in California. In 2006 that state's legislature worked with then-Governor Arnold Schwarzenegger to pass the Global Warming Solutions Act, which aims to cut California's greenhouse gas emissions 25% by the year 2020. This law established a cap-and-trade program for carbon emissions and followed earlier efforts in California to mandate higher fuel efficiency for automobiles.

Action is also being taken by nine northeastern states in the Regional Greenhouse Gas Initiative. In this effort, Connecticut, Delaware, Maine, Maryland, Massachusetts, New Hampshire, New York, Rhode Island, and Vermont run a joint cap-and-trade program for power plant emissions. From 2005 to 2013, these states cut their CO_2 emissions from power plants by 40%, even as their economies grew. It is estimated that investment of the auction proceeds will save nearly $3 billion in energy costs and eliminate 10 million tons of CO_2 emissions.

Voters in the state of Washington have considered a different path. In 2016 they were set to vote on a ballot measure introduced by political leaders to directly tax fossil fuel energy and then return the proceeds to taxpayers.

We can put a price on carbon

The policies being pursued or considered in places like California, Washington, and the states of the Regional Greenhouse Gas Initiative all involve "putting a price on carbon." **Carbon pricing** strategies are intended to compensate the public for the external costs (pp. 141, 162) we all suffer from fossil fuel emissions and climate change—those economic losses resulting from the various impacts of climate change on health, property, infrastructure, and more that we have surveyed in this chapter.

Carbon pricing lifts the burden of paying for these impacts off the shoulders of the public and shifts it to the entities responsible for emissions. It harnesses the financial incentives inherent in free-market capitalism while granting emitters the freedom to decide how they can best reduce emissions. In theory, once polluters are charged a price for polluting, market forces do the work of reducing pollution in an economically efficient way by allowing business, industry, or utilities flexibility in how they respond. Supporters of carbon pricing view this as the fairest, least expensive, and most effective approach to reducing emissions. Moreover, giving polluters a financial incentive to reduce emissions encourages them to innovate ways to cut emissions still further, even below a legally required amount. There are two primary approaches to carbon pricing: (1) carbon trading and (2) carbon taxation.

Carbon trading uses markets

In an emissions trading system (p. 179), a government sets up a market in permits for the emission of pollutants, and companies, utilities, or industries then buy and sell the permits among themselves (**FIGURE 18.33**). A **carbon trading** system is one in which permits are traded for the emission of carbon dioxide. The price of permits fluctuates freely in the market according to supply and demand, creating the same kinds of financial incentives as any other commodity that is bought and sold. In the approach known as cap-and-trade (p. 179), the government sets a cap on the amount of pollution it will allow, then gives, sells, or auctions permits to emitters that allow them to emit a certain fraction of the total amount. Over time the government lowers the cap to assure that total emissions decrease as intended. Emitters with too few permits to cover

their pollution must reduce their emissions, buy permits from other emitters, or pay for carbon offset credits (p. 510).

The world's largest cap-and-trade program is the European Union Emission Trading Scheme. This market began in 2005, but investors soon realized that national governments had allocated too many permits to their industries. The overallocation gave companies little incentive to reduce emissions, so permits lost their value and market prices collapsed. Europeans tried to address these problems by making emitters pay for permits and by taking other steps. Similar difficulties befell the world's first emissions trading program for greenhouse gas reduction, the Chicago Climate Exchange, which operated from 2003 to 2010 and involved several hundred corporations, institutions, and municipalities.

So far, California's cap-and-trade program and the Regional Greenhouse Gas Initiative each seem to be running effectively. In 2015, China announced that it will set up a national cap-and-trade program. Many observers are skeptical that China will be able to ensure transparency, prevent corruption, and minimize government interference—but if China succeeds, this could represent a major step toward controlling emissions globally. All these early experiments in carbon markets are providing lessons for how to set up effective and sustainable trading systems in the future.

Carbon emissions can be taxed

As the world's carbon trading markets show mixed results early in their growth, some economists and policymakers are saying that cap-and-trade systems are not effective enough, fail to produce results quickly, or leave too much to chance. Many of these critics would prefer that governments enact a **carbon tax** instead. A carbon tax is a type of green tax (p. 178) on the emission of carbon dioxide or on the carbon content of

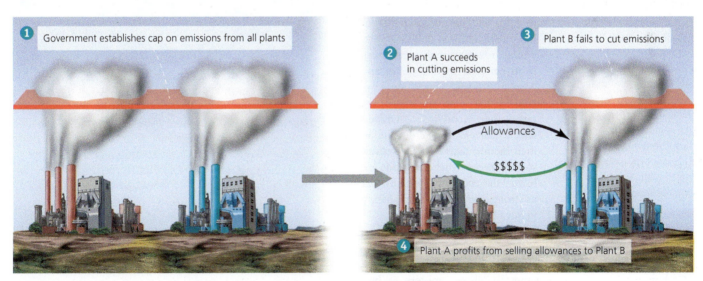

FIGURE 18.33 A cap-and-trade emissions trading system harnesses the power of market capitalism to achieve the goal of reducing emissions. In such a system, ❶ government first sets an overall cap on emissions. As polluting facilities respond, some will have better success reducing emissions than others. In this figure, ❷ Plant A succeeds in cutting its emissions well below the cap, whereas ❸ Plant B fails to cut its emissions at all. As a result, ❹ Plant B pays money to Plant A to purchase allowances that Plant A is no longer using. Plant A profits from this sale, and the government cap is met, reducing pollution overall. Over time, the cap can be progressively lowered to achieve further emissions cuts.

fossil fuels. Through carbon taxation, governments charge polluters a fee for each unit of CO_2 that results from their product or action. This gives polluters a powerful financial incentive to reduce emissions.

Carbon taxes can be implemented in several ways; they can be charged to energy producers, utilities, or motor vehicle users, and they can be scaled according to energy efficiency. Carbon taxes of various types have so far been introduced in roughly 40 nations and 20 cities and states. For instance, Boulder, Colorado, taxes electricity consumption; Montgomery County, Maryland, taxes power plants; and San Francisco Bay–area counties tax businesses for emissions.

The downside of a carbon tax is that polluters pass the cost along to consumers by charging higher prices for the products or services they sell. Proponents of carbon taxes have responded by proposing an approach called **fee-and-dividend.** In this approach, funds the government receives from polluters through the carbon tax, or "fee," are transferred as a tax refund, or "dividend," given to taxpayers. This way, if polluters pass their costs along to consumers, those consumers will be reimbursed for those costs by the tax refund they receive. In theory, the system should provide polluters a financial incentive to reduce emissions while imposing no financial burden on taxpayers and no drag on the economy. The fee-and-dividend approach is a type of **revenue-neutral carbon tax,** because there is no net transfer of revenue from taxpayers to the government. For this reason, the approach is gaining broad political appeal.

The Canadian province of British Columbia introduced a revenue-neutral carbon tax in 2008. In the five years that followed, the province reduced its emissions by 10% as fuel consumption declined 17% and the province's economy grew. The tax began at $10 per ton of CO_2 equivalent, was gradually raised to $30, and has been gathering more than a billion dollars a year. The tax revenues replace revenues from personal income taxes and corporate taxes, which are lowered by the same amount. Thus, rather than receiving a dividend check, citizens see a decline in their income taxes. The end result so far has been that emissions are lower, the economy remains just as strong, and taxpayers are happy to have lower income taxes.

Offsets help achieve carbon-neutrality

Carbon pricing programs generally allow emitters to buy **carbon offsets,** voluntary payments intended to enable another entity to help reduce the emissions that one is unable to reduce. The payment thus offsets one's own emissions. For example, a coal-burning power plant could pay for a reforestation project to plant trees that will soak up as much carbon dioxide as the coal plant emits. Or a university could fund the development of clean and renewable energy projects to make up for fossil fuel energy the university uses.

Carbon offsets are popular among utilities, businesses, universities, governments, and individuals trying to achieve **carbon-neutrality,** a condition in which no net carbon is emitted. In principle, carbon offsets are a powerful idea, but rigorous oversight is needed to make sure that offset funds achieve what they are intended for—and that offsets fund only emissions cuts that would not occur otherwise.

Should we engineer the climate?

What if all our efforts to reduce emissions are not adequate to rein in climate change? As climate disruption becomes more severe, some scientists and engineers are reluctantly considering drastic, assertive steps to alter Earth's climate in a last-ditch attempt to stop global warming—an approach called **geoengineering** (**FIGURE 18.34**).

One geoengineering approach would be to suck carbon dioxide out of the air. To achieve this we might enhance photosynthesis in natural systems by planting trees or by fertilizing ocean phytoplankton with nutrients like iron. A more high-tech method might be to design "artificial trees," structures that chemically filter CO_2 from the air.

A second geoengineering approach would be to block sunlight before it reaches Earth, thereby cooling the planet. We might deflect sunlight by injecting sulfates or other fine dust particles into the stratosphere, by seeding clouds with

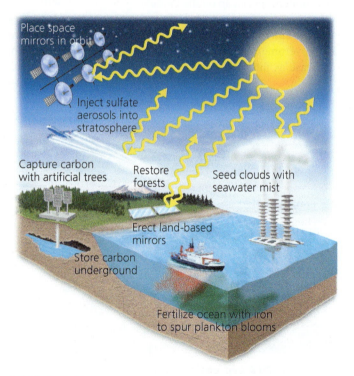

FIGURE 18.34 Geoengineering proposals seek to use technology to remove carbon dioxide from the air or reflect sunlight away from Earth. However, most geoengineering ideas would take years to develop, may not work well, or might cause undesirable side effects. Thus, they are not a substitute for reducing emissions.

seawater, or by deploying fleets of reflecting mirrors on land, at sea, or in space.

Scientists were long reluctant even to discuss the notion of geoengineering. The potential methods are technically daunting and would take years or decades to develop, and might also pose unforeseen risks. Moreover, blocking sunlight does not reduce greenhouse gas concentrations, so ocean acidification would continue. And any method would work only as long as society has money and ability to maintain it. In addition, many experts are wary of promulgating hope for easy technological fixes, lest politicians lose incentive to develop policy to reduce emissions.

However, as climate change intensifies, more scientists are becoming willing to contemplate geoengineering as a backup plan. Respected researchers and scientific institutions are beginning to assess the risks and benefits of geoengineering options, so that we can be ready to take well-informed action if climate change becomes severe enough to justify it.

You can address climate change

Government policies, corporate actions, international treaties, carbon pricing, technological innovations—and perhaps even geoengineering—will all play roles in addressing climate change. But in the end, the most influential factor may be the collective decisions of millions of regular people. Just as we each have an ecological footprint (p. 5), we each have a **carbon footprint** that expresses the amount of carbon we are responsible for emitting. To help reduce emissions, each of us can take steps in our everyday lives—from choosing energy-efficient appliances, to eating less meat, to deciding where to live and how to get to work.

College students are vital to driving the personal and societal changes needed to reduce carbon footprints and address climate change—through everyday lifestyle choices, educating others, and political activism. Today a groundswell of interest is sweeping across campuses, and many students are pressing their administrations to seek carbon-neutrality, divest from fossil fuels, and promote renewable energy (pp. 16–17, 649–655).

Global climate change may be the biggest challenge we face, but halting it would be our biggest victory. With concerted action, there is still time to avert the most severe impacts. Through outreach, education, innovation, and lifestyle choices, we have the power to turn the tables on climate change and help bring about a bright future for humanity and our planet.

closing
THE LOOP

Many factors influence Earth's climate, and human activities have come to play a major role. Climate change is well underway, and additional greenhouse gas emissions will intensify global warming and cause increasingly severe and varied impacts. Sea level rise and other consequences of global climate change are affecting locations worldwide from Miami to the Maldives, from Alaska to Bangladesh, and from New York to the Netherlands. As scientists and policymakers come to better understand anthropogenic climate change and its consequences, more and more of them are urging immediate action.

Policymakers at the international and national levels have struggled to take meaningful steps to slow greenhouse gas emissions, so increasingly, people at the local and regional levels are the ones making a difference. In South Florida, citizens and local leaders are investing time, thought, money, and creativity into finding solutions to rising sea levels. They are seeking to mitigate climate change by reducing greenhouse gas emissions and to adapt to climate change by building pumping systems, raising streets and foundations, and tailoring financial and insurance incentives to guide development toward upland areas. Like people anywhere who love their homes, residents of South Florida are girding themselves for a long battle to protect their land, communities, and quality of life while our global society inches its way toward emissions reductions. For all of us across the globe, taking steps to mitigate and adapt to climate change represents the foremost challenge for our future.

REVIEWING Objectives

You should now be able to:

Describe Earth's climate system and explain the factors that influence global climate

Solar radiation, Milankovitch cycles, ocean absorption, and ocean circulation all influence climate. Greenhouse gases warm the atmosphere by absorbing and re-emitting infrared radiation. Earth's climate changes naturally over time, but today human influence is changing it rapidly. Our planet is now experiencing radiative forcing of 2.3 watts/m² of thermal energy above what it was experiencing 250 years ago. (pp. 480–485)

Identify greenhouse gases and characterize human influences on the atmosphere and on climate

Greenhouse gases such as carbon dioxide, methane, water vapor, nitrous oxide, and ozone keep our atmosphere warm by exerting a natural greenhouse effect. As we burn fossil fuels and clear forests, we increase atmospheric concentrations of greenhouse gases, and this intensifies the greenhouse effect. Artificial greenhouse gases such as halocarbons are also being added. Aerosol pollution exerts a variable but slight cooling effect, yet overall, human emissions are warming the atmosphere, land surface, and ocean waters. (pp. 482–483)

Summarize how researchers study climate

Proxy indicators—such as data from ice cores, sediment cores, tree rings, packrat middens, and coral reefs—reveal information about past climate. Direct measurements of temperature, precipitation, and other conditions tell us about current climate. Climate models serve to predict future changes in climate. (pp. 485–491)

Outline current and expected future trends and impacts of global climate change in the United States and around the world

Temperatures have warmed by about 1.1°C (2.0°F) in the past century and are predicted to rise 1.0–3.7°C (1.8–6.7°F) more by 2100. Changes in precipitation are varying by region. Extreme weather events are becoming more frequent, likely due in part to modification of the jet stream. Melting glaciers will diminish water supplies, and melting ice sheets are adding to sea level rise. Sea level has risen 24.1 cm (9.5 in.) in the past 135 years, and its continued rise poses increasing risks of damage to islands and coasts. Of all impacts of our greenhouse gas emissions, ocean acidification may become the most far-reaching. All these changes exert impacts on organisms and ecosystems, as well as on agriculture, forestry, health, economics, and national security. These impacts vary regionally but will affect us all. (pp. 488–503)

Suggest and assess ways we may respond to climate change

Both mitigation and adaptation are necessary. We can reduce greenhouse gas emissions with multiple strategies, including conservation, energy efficiency, clean and renewable energy sources, new automotive technologies, and investment in mass transit. International efforts to design effective treaties to restrain climate change have, so far, fallen short of what is needed. However, renewable energy technologies present economic opportunities, and some states, cities, and businesses are acting to address emissions. Carbon trading and carbon taxation offer promising ways to harness market forces and financial incentives to help reduce emissions. (pp. 503–511)

TESTING Your Comprehension

1. What happens to solar radiation after it reaches Earth? How do greenhouse gases warm the lower atmosphere?

2. Why is carbon dioxide considered the main greenhouse gas? Why are carbon dioxide concentrations increasing in the atmosphere?

3. What evidence do scientists use to study the ancient atmosphere? Describe what a proxy indicator is, and give two examples.

4. Has simulating climate change with computer programs been effective in helping us predict climate? Briefly describe how these programs work.

5. List three major trends in climate that scientists have documented so far. Now list three future trends that they predict, along with their potential consequences.

6. Describe how rising sea levels, caused by global warming, can create problems for people. How is climate change affecting marine ecosystems?

7. How might a warmer climate affect agriculture? How is it affecting distributions of plants and animals? How might it affect human health?

8. What are the two largest sources of greenhouse gas emissions in the United States? How can we reduce emissions from these sources?

9. What roles have international treaties played in addressing climate change? Give two specific examples.

10. Describe two economic market-based approaches for reducing greenhouse gas emissions. Discuss advantages and disadvantages of each approach.

SEEKING Solutions

1. Some people argue that we need "more proof" or "better science" before we commit to substantial changes in our energy economy. How much certainty do you think we need before we should take action regarding climate change? How much certainty do you need in your own life before you make a major decision? Should nations and elected officials follow a different standard? Do you believe that the precautionary principle (pp. 254, 377) is an appropriate standard in the case of global climate change? Why or why not?

2. Suppose that you would like to make your own life-style carbon-neutral. You plan to begin by reducing the emissions you are responsible for by 25%. What three actions would you take first to achieve your goal?

3. How might your campus reduce its greenhouse gas emissions? Come up with three concrete proposals for ways to reduce emissions on your campus that you feel would be effective and feasible. How would you present these proposals to campus administrators to gain their support?

4. **CASE STUDY CONNECTION** You are the city manager for a coastal U.S. city that scientists predict will be hit hard by sea level rise, with risks and impacts trailing those in Miami by just a few years. You have just returned from a professional conference in Florida, where you toured Miami Beach and learned of the efforts being made there to adapt to climate change. What steps would you take to help your own city prepare for rising sea level? How would you explain the risks and impacts of climate change to your fellow city leaders to gain their support? Of the measures being taken in Florida communities, which would you choose to study closely, which would you want to begin right away, and which would be highest priority in the long run? Explain your choices.

5. **THINK IT THROUGH** You have just been elected governor of a medium-sized U.S. state. Polls show that the public wants you to take action to reduce greenhouse gas emissions, but does not want prices of gasoline or electricity to rise. Your state legislature will support you in your efforts as long as you remain popular with voters. One neighboring state has just passed ambitious legislation mandating steep greenhouse gas emissions reductions. Another neighboring state has recently joined a new regional emissions trading consortium. A third neighboring state has just established a revenue-neutral carbon tax. A fourth neighboring state is spending millions of dollars building seawalls around its coastal cities. What actions will you take in your first year as governor?

6. **THINK IT THROUGH** You have been appointed as the U.S. representative to an international conference to negotiate terms of an emissions reduction treaty that would build upon or replace the Paris Agreement. The U.S. government has instructed you to take a leading role in designing the new treaty and to engage constructively with other nations' representatives while protecting America's economic and political interests. What type of agreement will you try to shape? Describe at least three components that you would propose or agree to, and at least one that you would oppose.

CALCULATING Ecological Footprints

Global climate change is something to which we all contribute, because fossil fuel combustion plays such a large role in supporting the lifestyles we lead. Conversely, as individuals, each one of us can help to address climate change through personal decisions and actions in how we live our lives.

Several online calculators enable you to calculate your own personal carbon footprint, the amount of carbon emissions for which you are responsible. Go to www.nature.org/greenliving/carboncalculator/ or to www.carbonfootprint.com/calculator.aspx, take the quiz, and enter the relevant data in the table.

	CARBON FOOTPRINT (TONS PER PERSON PER YEAR)
World average	
U.S. average	
Your footprint	
Your footprint with three changes (see Question 3)	

1. How does your personal carbon footprint compare to that of the average U.S. resident? How does it compare to that of the average person in the world? Why do you think your footprint differs in the ways it does?

2. As you took the quiz and noted the impacts of various choices and activities, which one surprised you the most?

3. Think of three changes you could make in your lifestyle that would lower your carbon footprint. Now take the footprint quiz again, incorporating these three changes. Enter your resulting footprint in the table. By how much did you reduce your yearly emissions?

4. What do you think would be an admirable yet realistic goal for you to set as a target value for your own footprint? Would you choose to purchase carbon offsets to help reduce your impact? Why or why not?

MasteringEnvironmentalScience®

Students Go to **MasteringEnvironmentalScience** for assignments, the etext, and the Study Area with practice tests, videos, current events, and activities.

Instructors Go to **MasteringEnvironmentalScience** for automatically graded activities, current events, videos, and reading questions that you can assign to your students, plus Instructor Resources.

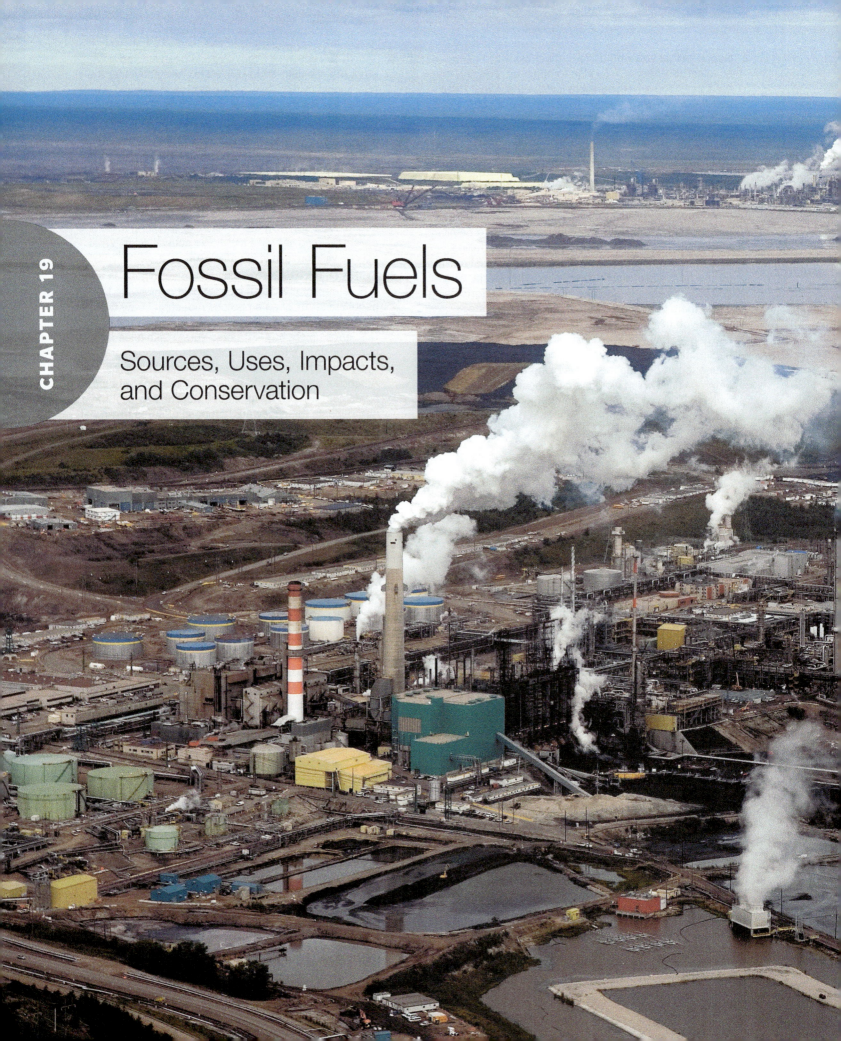

CHAPTER 19

Fossil Fuels

Sources, Uses, Impacts, and Conservation

Alberta's Oil Sands and the Keystone XL Pipeline

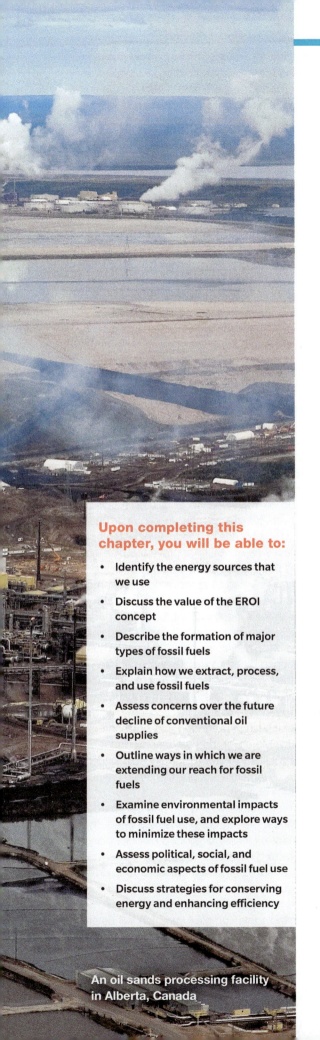

Upon completing this chapter, you will be able to:

- Identify the energy sources that we use
- Discuss the value of the EROI concept
- Describe the formation of major types of fossil fuels
- Explain how we extract, process, and use fossil fuels
- Assess concerns over the future decline of conventional oil supplies
- Outline ways in which we are extending our reach for fossil fuels
- Examine environmental impacts of fossil fuel use, and explore ways to minimize these impacts
- Assess political, social, and economic aspects of fossil fuel use
- Discuss strategies for conserving energy and enhancing efficiency

An oil sands processing facility in Alberta, Canada

> "The president's veto of the bipartisan Keystone bill represents a defeat for jobs, infrastructure, and the middle class."
> Senate Majority Leader Mitch McConnell (R-Kentucky)

> If Canada proceeds, and we do nothing, it will be game over for the climate.
> Climate scientist James Hansen

Everything relating to Canada's oil sands is huge. These fossil fuel deposits cover a region the size of Illinois, within boreal forests that span the width of the continent. Open pits miles wide are dug to extract the fuel. Inside these gargantuan pits, million-pound haul trucks with tires 14 feet wide and shovels five stories high look like ants when viewed from the rim of the pit. The economic value of the extracted oil is astounding. And the impacts of putting the oil to use are huge, as well: Burning all this fuel would alter the very climate of our planet.

Oil sands, also called **tar sands,** are layers of sand or clay saturated with a viscous, tarry type of petroleum called *bitumen*. Huge areas of these wet blackish deposits underlie a thinly populated region of northern Alberta, and the implications of mining them for oil are momentous. To some people the oil sands represent wealth and security, a key to maintaining our fossil-fuel-based lifestyle far into the future. To others they threaten appalling pollution and a severe disruption of Earth's climate.

To extract oil from oil sands, companies clear the forest and then strip-mine the land, creating open pits 215 m (400 ft) deep. The gooey deposits are mixed with hot water and chemicals to separate the bitumen from the sand, and the bitumen is removed and processed. Wastewater is dumped into toxic tailings lakes that are even larger than the mines. Where oil sands are located more deeply underground, hot water is injected down shafts to liquefy, separate, and extract the bitumen in place.

Mining for oil sands began in Alberta in 1967, but for many years it was hard to make money extracting these low-quality deposits. Rising oil prices after 2003 turned it into a profitable venture, and dozens of companies rushed in. Canadian oil sands became the source of up to 2.3 million barrels of oil per day, making up most of Canada's petroleum extraction, and each truckload that left a mine carried bitumen containing close to $20,000 in oil.

In 2015, a sharp downturn in world oil prices slowed the rush, but most companies continued mining, counting on profits from higher future prices to cover their short-term losses. Indeed, there is plenty left for the future; thanks to the oil sands, Canada boasts the world's third-largest proven reserves of oil, after Venezuela and Saudi Arabia.

Canada looked for buyers south of its border first, and the TransCanada Corporation built the Keystone

Protesting the Keystone XL pipeline at the U.S. White House

FIGURE 19.1 The Keystone pipeline system brings bitumen from Alberta's oil sands into the United States. The proposed Keystone XL extension (dashed line) set off a complex debate.

Pipeline to ship diluted bitumen into the United States. This pipeline system began operating in 2010, bringing oil from Alberta nearly 3500 km (2200 mi) to Illinois and Oklahoma (**FIGURE 19.1**). An extension later conveyed the oil to refineries on the Texas coast. TransCanada then proposed the Keystone XL extension, a 1400-km (875-mi) stretch of larger-diameter pipeline that would cut across the Great Plains to shave off distance and add capacity to the existing system. This proposed shortcut leg would also transport oil from the newly productive fields of the Bakken Formation in North Dakota and Montana.

The Keystone XL pipeline proposal soon met opposition from people living along the proposed route who were concerned about health, environmental protection, and property rights. It also faced nationwide opposition from advocates of action to address global climate change.

Pipeline proponents argued that the Keystone XL project would create jobs for workers in the U.S. heartland and would guarantee a dependable oil supply for decades to come. They stressed that buying oil from Canada—a stable, friendly, democratic neighbor—could help end U.S. reliance on oil-extracting nations such as Saudi Arabia and Venezuela that have authoritarian governments and poor human rights records.

Opponents of the pipeline extension expressed dismay at the destruction of boreal forest and anxiety about transporting oil above the continent's largest aquifer, where spills could contaminate drinking water for millions of people and irrigation water for America's breadbasket. They pointed out that most of the oil would be destined for export, and not for use in the United States. Most of all, they sought to avoid extracting a vast new source of fossil fuels whose combustion would release immense amounts of greenhouse gases and intensify climate change. By encouraging a source of oil that is energy-intensive to extract and that burns less cleanly than conventional fuels, they maintained, the United States would prolong fossil fuel dependence and worsen climate change, when it should instead be transitioning to clean renewable energy.

Under pressure from all sides, the administration of U.S. President Barack Obama walked a fine line. Because the proposed Keystone XL pipeline would cross an international border, it required a presidential permit from the U.S. Department of State. Because of concerns about damage to the ecologically sensitive Sandhills area of Nebraska and potential contamination of the Ogallala Aquifer, the State Department asked TransCanada to revise its proposed route to avoid the areas of concern. In 2014, the State Department released its final environmental impact statement for the new route. It concluded that encouraging oil sands extraction by building Keystone XL would increase greenhouse gas pollution—yet that if the pipeline was *not* built, Canada would likely find other ways to extract and sell its oil, including transporting it by rail, which presents its own set of risks (p. 533).

As the public debate intensified during the seven years of the project's review, tens of thousands of Americans protested against the pipeline in front of the White House. These protestors viewed the decision on Keystone XL as a crucial test of Obama's vow to deal with climate change. In contrast, business interests and many Republicans in Congress objected furiously to the administration's hesitancy to approve the pipeline, arguing that an economic opportunity was slipping by. In 2015 the Republican-led Congress voted to approve construction of Keystone XL. Obama vetoed the measure, maintaining that the decision legally rested with the executive branch.

In November 2015, President Obama decided against approving the Keystone XL pipeline, and told the nation that his administration had judged that the pipeline "would not serve the national interest of the United States." The pipeline's importance had been "overinflated," he stated, and its construction would not contribute meaningfully to the U.S. economy, would not lower gas prices for consumers, and would not enhance America's energy security. Moreover, he noted, approving it on the eve of global climate talks in Paris (p. 507) would undercut U.S. leadership just as America sought to gather nations together to address climate change.

Obama's decision on Keystone XL did not end the broader debate over the role of fossil fuels in our economy. Indeed, the divergent views on Canada's oil sands reflect our confounding relationship with fossil fuels. These energy sources power our civilization and have enabled our modern standard of living—yet as climate change worsens, we face the need to wean ourselves from them and shift to clean renewable energy sources. The way in which we handle this complex transition will determine a great deal about the quality of our lives and the future of our society and our planet.

Sources of Energy

Humanity has devised many ways to harness the renewable and nonrenewable forms of energy available on our planet (TABLE 19.1). We use these energy sources to heat and light our homes; power our machinery; fuel our vehicles; produce plastics, pharmaceuticals, and synthetic fibers; and provide the many comforts and conveniences to which we've grown accustomed.

Nature offers us a variety of energy sources

Most of Earth's energy comes from the sun. We can harness energy from the sun's radiation directly by using solar power technologies. Solar radiation also helps drive wind and the water cycle, enabling us to harness wind power and hydroelectric power. And, of course, sunlight drives photosynthesis (p. 30) and the growth of plants, from which we take wood and other biomass as a fuel source. When plants and other organisms die and are buried in sediments under particular conditions, their stored chemical energy may eventually be incorporated into **fossil fuels,** highly combustible substances formed from the remains of organisms from past geologic ages. We rely on three main fossil fuels, in the form of a solid (coal), liquid (oil), and gas (natural gas).

In addition, a great deal of energy emanates from Earth's core, making geothermal power available for our use. Energy also results from the gravitational pull of the moon and sun, which powers ocean tides. Finally, an immense amount of energy resides within the bonds among protons and neutrons in atoms, and this energy provides us with nuclear power. (We explore all these energy sources as alternatives to fossil fuels in Chapters 20 and 21.)

Energy sources such as sunlight, geothermal energy, and tidal energy are considered perpetually renewable because they are readily replenished, so we can keep using them without depleting them (p. 4). In contrast, energy sources such as coal, oil, and natural gas are considered nonrenewable. These nonrenewable fuels result from ongoing natural processes, but it takes so long for fossil fuels to form that, once depleted, they cannot be replaced within any time span useful to our civilization. It takes a thousand years for the biosphere to generate the amount of organic matter that must be buried to transform into a single day's worth of fossil fuels for our society. To replenish the fossil fuels we have depleted so far would take many millions of years. At our rising rate of consumption, we will use up Earth's easily accessible store of conventional fossil fuels in just decades. For this reason, and because fossil fuels exert severe environmental impacts, renewable energy sources increasingly are being developed as alternatives to fossil fuels.

We rely mostly on fossil fuels

Since the industrial revolution, fossil fuels have replaced biomass as our society's dominant source of energy. Global consumption of coal, oil, and natural gas has risen for years and is now at its highest level ever (FIGURE 19.2). We have favored fossil fuels as an energy source because their high energy content makes them efficient to burn, ship, and store. A single gallon of oil contains as much energy as a person would expend in nearly 600 hours of human labor.

TABLE 19.1 Energy Sources We Use

ENERGY SOURCE	DESCRIPTION	TYPE OF ENERGY
Coal	Fossil fuel extracted from ground (solid)	Nonrenewable
Oil	Fossil fuel extracted from ground (liquid)	Nonrenewable
Natural gas	Fossil fuel extracted from ground (gas)	Nonrenewable
Nuclear energy	Energy from atomic nuclei of uranium	Nonrenewable
Biomass energy	Energy stored in plant matter from photosynthesis	Renewable
Hydropower	Energy from running water	Renewable
Solar energy	Energy from sunlight directly	Renewable
Wind energy	Energy from wind	Renewable
Geothermal energy	Earth's internal heat rising from core	Renewable
Tidal and wave energy	Energy from tides and ocean waves	Renewable

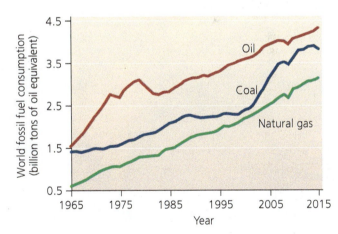

FIGURE 19.2 Annual global consumption of fossil fuels has risen greatly over the past 50 years. Oil remains our leading energy source. *Data from BP p.l.c., 2016.* Statistical review of world energy 2016.

DATA Q By roughly what percentage has the annual consumption of oil risen since the year you were born? Which fuel has risen fastest since you were born—oil, coal, or natural gas?

Go to **Interpreting Graphs & Data** on **Mastering**EnvironmentalScience®

We use fossil fuels for transportation, manufacturing, heating, and cooking, and also to generate **electricity,** a secondary form of energy that is convenient to transfer over long distances and apply to a variety of uses. Each type of fuel has its own mix of uses, and each contributes in different ways to our economies and our daily needs. For instance, oil is used mostly for transportation, whereas coal is used mostly to generate electricity.

Energy is unevenly distributed

Some regions of the globe have substantial reserves of coal, oil, or natural gas, whereas others have very few. Half the world's proven reserves of crude oil lie in the Middle East. The Middle East is also rich in natural gas, as is Russia. The United States possesses the most coal of any nation (**TABLE 19.2**).

Rates at which we consume energy also vary from place to place. A world map of per-person consumption shows that people in developed regions generally consume far more energy than do people of developing regions (**FIGURE 19.3**). Per person, the most industrialized nations use more than 50 times more energy than do the least industrialized nations. As an example, the United States has only 4.4% of the world's population, but it consumes nearly 18% of the world's energy.

Societies differ in how they use energy, as well. Industrialized nations apportion roughly one-third to transportation, one-third to industry, and one-third to all other uses. In contrast, industrializing nations devote more energy to subsistence activities such as growing and preparing food and heating homes. Moreover, people in developing countries often rely on manual or animal energy sources instead of automated ones. For instance, most rice farmers in Southeast Asia plant rice by hand, but industrial rice growers in California use airplanes. Because industrialized nations rely more on mechanized equipment and technology, they use more

TABLE 19.2 **Nations with the Largest Proven Reserves of Fossil Fuels**

COAL (% world reserves)		OIL (% world reserves)		NATURAL GAS (% world reserves)	
United States	26.6	Venezuela*	17.5	Iran	18.2
Russia	17.6	Saudi Arabia	15.7	Russia	17.3
China	12.8	Canada*	10.1	Qatar	13.1
Australia	8.6	Iran	9.3	Turkmenistan	9.4
India	6.8	Iraq	8.4	United States	5.6

*Most reserves in Venezuela and Canada consist of oil sands, which are included in these figures.
Data from BP p.l.c., 2016. *Statistical review of world energy 2016.*

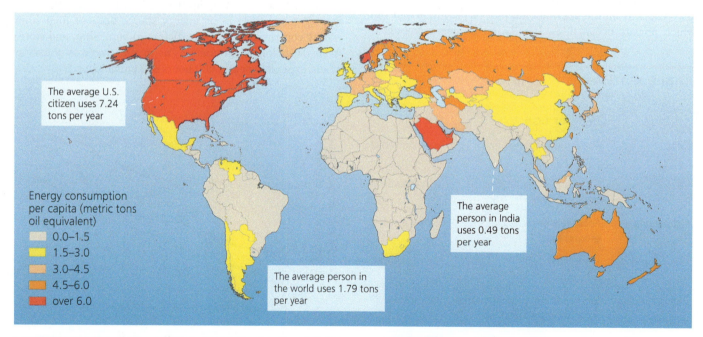

FIGURE 19.3 **People in wealthy industrialized nations tend to consume the most energy per person.** This map combines all types of energy, standardized to metric tons of "oil-equivalent" (the amount of fuel needed to result in the energy gained from burning one metric ton of crude oil). *Data from BP p.l.c., 2015.* Statistical review of world energy 2015.

DATA How many times more energy does the average U.S. citizen use than the average person in the world? How many times more energy does the average U.S. citizen use than the average citizen of India?

Go to **Interpreting Graphs & Data** on **Mastering**EnvironmentalScience®

fossil fuels. In the United States, oil, coal, and natural gas together supply 82% of energy demand (**FIGURE 19.4**).

It takes energy to make energy

We do not simply get energy for free. To harness, extract, process, and deliver energy, we need to invest substantial inputs of energy. For instance, mining oil sands in Alberta requires using powerful vehicles and heavy machinery, as well as constructing an immense infrastructure of roads, pipelines, waste ponds, storage tanks, water intakes, processing facilities,

housing for workers, and more—all requiring the use of energy. Natural gas must be burned to heat the water that is used to separate the bitumen from the sand. Processing and piping the oil away from the extraction site, and then refining it into products we can use, require further energy inputs. Thus, when evaluating an energy source, it is important to subtract costs in energy invested from the benefits in energy received. **Net energy** expresses the difference between energy returned and energy invested:

$$\text{Net energy} = \text{Energy returned} - \text{Energy invested}$$

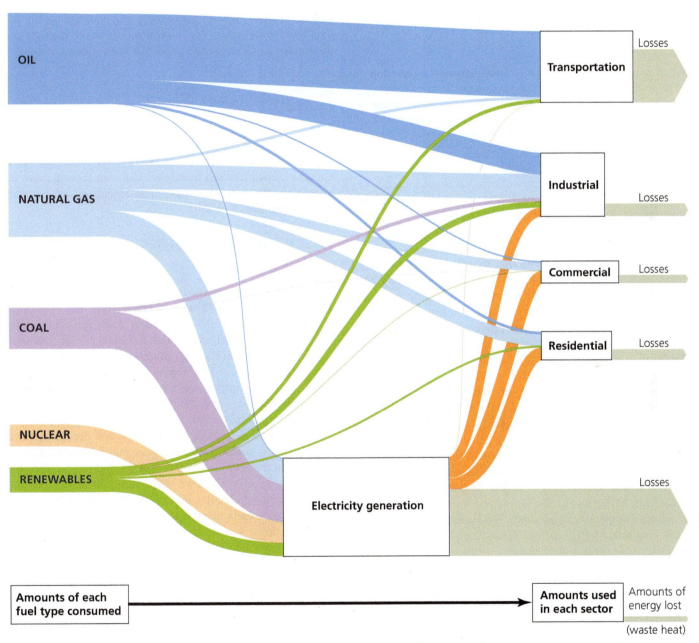

FIGURE 19.4 Total energy flow of the United States. Amounts are represented by the thickness of each bar. Energy consumed (domestic extraction plus imports minus exports) is shown on the left, and destinations of this energy are shown on the right. Portions of each energy source are used directly in the residential, commercial, industrial, and transportation sectors. Other portions are used to generate electricity, which in turn powers these sectors. The large amounts of energy lost as waste heat are shown on the right. *Adapted from Lawrence Livermore National Laboratory. Data are for 2015, from U.S. Energy Information Administration.*

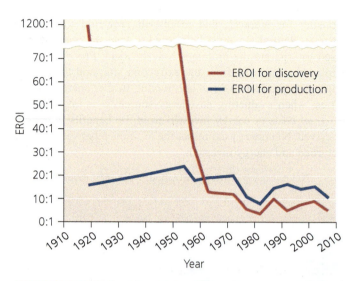

FIGURE 19.5 EROI values for the discovery and production of oil and gas in the United States have declined over the past century. *Data from Guilford, M., et al., 2011. A new long term assessment of energy return on investment (EROI) for U.S. oil and gas discovery and production. pp. 115–136 in Hall, C., and D. Hansen (Eds.), Sustainability, Special Issue, 2011, New studies in EROI (energy return on investment).*

When we assess energy sources, it is also useful to use a ratio often denoted as **EROI (energy returned on investment).** EROI ratios are calculated as follows:

$$EROI = \text{Energy returned}/\text{Energy invested}$$

Higher EROI ratios mean that we receive more energy from each unit of energy that we invest. Fossil fuels are widely used because their EROI ratios have historically been high. However, EROI ratios can change over time. Ratios rise as the technologies to extract and process fuels become more efficient, and they fall as resources are depleted and become harder to extract.

For example, EROI ratios for "production" (extraction and processing) of conventional oil and natural gas in the United States declined from roughly 24:1 in the 1950s to about 11:1 in recent years (**FIGURE 19.5**). This means that we used to be able to gain 24 units of energy for every unit of energy expended, but now we can gain only 11. EROI ratios for both the discovery and the production of oil and gas have declined because we found and extracted the easiest deposits first and now must work harder and harder to find and extract the remaining amounts. For the Alberta oil sands, EROI ratios are even lower, averaging around 4:1, because oil sands are a low-quality fuel that requires a great deal of energy to extract and process.

Where will we turn for energy?

Since the onset of the industrial revolution, abundant and inexpensive coal, oil, and natural gas have powered the astonishing advances of our civilization. These extraordinarily rich sources of energy have helped to bring us a standard of living our ancestors could scarcely have imagined. Yet because fossil fuel deposits are finite and nonrenewable, easily accessible supplies of the three main fossil fuels have dwindled, and EROI ratios have fallen.

In response, today we are devoting enormous amounts of money, energy, and technology to extend our reach for fossil fuels. We are using potent new extraction methods to free gas and oil tightly bound in rock layers. We are deploying powerful new machinery and techniques to squeeze more fuel from sites that were already extracted. We are drilling deeper underground, farther offshore, and into the Arctic seabed. And we are pursuing new types of fossil fuels, such as Alberta's oil sands.

There is, however, a different way we can respond to the ongoing depletion of conventional fossil fuel resources: We can hasten the development of clean and renewable energy sources to replace them. By transitioning to clean and renewable alternatives, we can gain energy that is sustainable in the long term while reducing pollution and its health impacts and the emission of greenhouse gases that drive climate change.

Fossil Fuels: Their Formation, Extraction, and Use

To grapple effectively with the energy issues we face, it is important to understand how fossil fuels are formed, how we locate deposits, how we extract these resources, and how our society puts them to use.

Fossil fuels are formed from ancient organic matter

The fossil fuels we burn today in our vehicles, homes, industries, and power plants were formed from the tissues of organisms that lived 100–500 million years ago. The energy these fuels contain originated with the sun and was converted to chemical-bond energy by photosynthesis. The chemical energy in these organisms' tissues then became concentrated as the tissues decomposed and their hydrocarbon compounds were altered and compressed (**FIGURE 19.6**).

Most organisms, after death, do not end up as part of a coal, gas, or oil deposit. A tree that falls and decays as a rotting log on the forest floor undergoes mostly aerobic decomposition; in the presence of air, bacteria and other organisms that use oxygen break down plant and animal remains into simpler carbon molecules that are recycled through the ecosystem. Fossil fuels are formed only when organic material is broken down in an anaerobic environment, one that has little or no oxygen. Such environments include the bottoms of lakes, swamps, and shallow seas. Over millions of years, organic matter that accumulates at the bottoms of such water bodies may be converted into crude oil, natural gas, or coal, depending on (1) the chemical composition of the material, (2) the temperatures and pressures to which it is subjected, (3) the presence or absence of anaerobic decomposers, and (4) the passage of time.

Because fossil fuels form only under certain conditions, they occur in isolated deposits. For example, Alberta's oil sands hold rich reserves of oil that surrounding regions do not. Geologists searching for fossil fuels drill cores and conduct

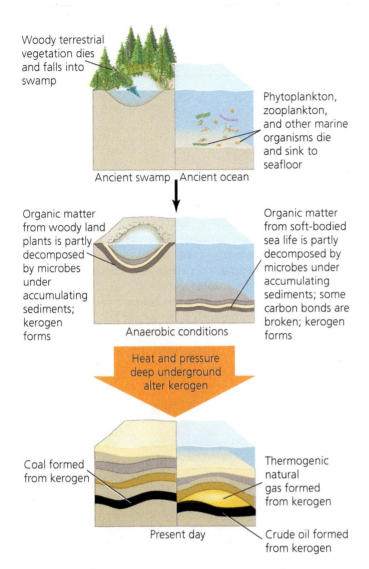

Woody terrestrial vegetation dies and falls into swamp

Phytoplankton, zooplankton, and other marine organisms die and sink to seafloor

Ancient swamp | Ancient ocean

Organic matter from woody land plants is partly decomposed by microbes under accumulating sediments; kerogen forms

Organic matter from soft-bodied sea life is partly decomposed by microbes under accumulating sediments; some carbon bonds are broken; kerogen forms

Anaerobic conditions

Heat and pressure deep underground alter kerogen

Coal formed from kerogen

Thermogenic natural gas formed from kerogen

Present day

Crude oil formed from kerogen

FIGURE 19.6 Fossil fuels form after organisms die and end up in oxygen-poor conditions. This can occur when trees fall into lakes and are buried by sediment or when phytoplankton and zooplankton drift to the seafloor and are buried (**top**). Organic matter that undergoes slow anaerobic decomposition deep under sediments forms kerogen (**middle**). Coal results when plant matter is compacted so tightly that little decomposition occurs (**bottom left**). Geothermal heating may turn kerogen into crude oil and natural gas (**bottom right**), which come to reside in porous rock layers beneath dense, impervious layers.

ground, air, and seismic surveys to map underground rock formations and predict where fossil fuel deposits might occur (see **THE SCIENCE BEHIND THE STORY**, pp. 522–523).

Coal **Coal** is a hard blackish substance formed from organic matter (generally woody plant material) compressed under very high pressure, creating dense, solid carbon structures (**FIGURE 19.7**). Coal typically results when water is squeezed out of such material as pressure and heat increase over time, and when little decomposition takes place. The proliferation 300–400 million years ago of swamps where organic material was buried created coal deposits in many regions of the world.

Coal varies from deposit to deposit in the amount of water, carbon, and potential energy it contains. Organic material that is broken down anaerobically but remains moist and near the surface is called *peat*. As peat decomposes further, as it is buried under sediments, as heat and pressure increase, and as time passes, moisture is squeezed out and carbon compounds compress more tightly together, forming coal. The more coal is compressed, the denser is its carbon content and the greater is its energy content per unit volume.

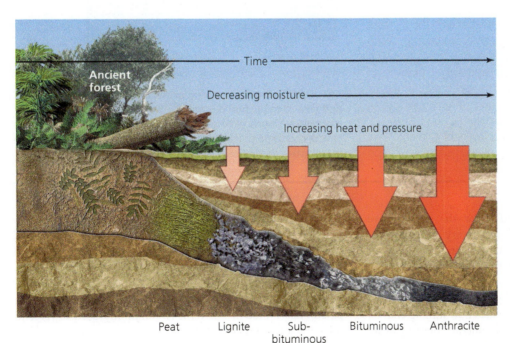

Time

Ancient forest

Decreasing moisture

Increasing heat and pressure

Peat | Lignite | Sub-bituminous | Bituminous | Anthracite

Types of coal

FIGURE 19.7 Coal forms as ancient plant matter turns to peat and then is compressed underground. Of the four classes of coal, anthracite coal is formed under the greatest pressure, where temperatures are high and moisture content is low. Anthracite coal has the densest carbon content and so contains the most potential energy.

How Do We Find and Estimate Fossil Fuel Deposits?

Drilling for oil or gas is risky business: Most wells are unproductive, and a company that doesn't pick its spots effectively could soon go bankrupt. So oil and gas companies turn to scientists to help them figure out where to drill.

The industry employs petroleum geologists who study underground rock formations to predict where deposits of oil and natural gas might lie. Because the organic matter that gave rise to fossil fuels was buried in sediments, geologists know to look for sedimentary rock that may act as a source. They also know that oil and gas tend to seep upward through porous rock until being trapped by impermeable layers.

Petroleum geologists study mapped seismic data to determine where oil or gas might be found.

To map subsurface rock layers, petroleum geologists first survey the landscape on the ground and from airplanes, studying rocks on the surface. Because rock layers often become tilted over geologic time, these strata may protrude at the surface, giving geologists an informative "side-on" view.

But to really understand what's deep beneath the surface, scientists conduct seismic surveys. In seismic surveying, a base station creates powerful vibrations at the surface by exploding dynamite, thumping the ground with a large weight, or using an electric vibrating machine (**FIGURE 1**). This sends seismic waves down and outward in all directions through the ground, just as ripples spread when a pebble is dropped into a pond.

As they travel, the waves encounter layers of different types of rock. Each time a seismic wave encounters a new type of rock with a different density, some of the wave's energy is reflected off the boundary. Other wave energy may be refracted, or bent, sending refraction waves upward.

As reflected and refracted waves return to the surface, devices called seismometers (also used to measure earthquakes) record their strength and timing. Scientists collect data from seismometers at multiple surface locations and run the data through computer programs. By analyzing how long it takes all the waves to reach the various receiving stations, and how strong they are at each site, researchers can triangulate and infer the densities, thicknesses, and locations of underlying geologic layers.

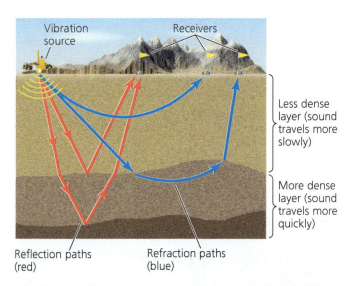

FIGURE 1 Seismic surveying provides clues to the location and size of fossil fuel deposits. Powerful vibrations are created, and receivers measure how long it takes seismic waves to reach other locations. Scientists interpret the patterns of wave reception to infer the densities, thicknesses, and locations of underlying rock layers.

To extract coal from deposits near the surface, we use strip mining, in which heavy machinery scrapes away huge amounts of earth. For deposits deep underground, we use subsurface mining, digging vertical shafts and blasting out networks of horizontal tunnels to follow seams, or layers, of coal. (Strip mining and subsurface mining are illustrated in Figure 23.7, p. 635.) We are also now mining coal on immense scales in the Appalachian Mountains, blasting away entire mountaintops in a process called mountaintop removal mining (pp. 530, 639).

Oil and natural gas The thick blackish liquid we know as **oil** consists of a mix of many types of hydrocarbon molecules (p. 27). The term **crude oil** refers specifically to oil extracted from the ground before it is refined. **Natural gas** is a gas consisting primarily of methane (CH_4) and lesser, variable amounts of other volatile hydrocarbons. Oil is also known as **petroleum,** although this term is commonly used to refer to oil and natural gas collectively.

Both oil and natural gas are formed from organic material (especially dead plankton) that drifted down

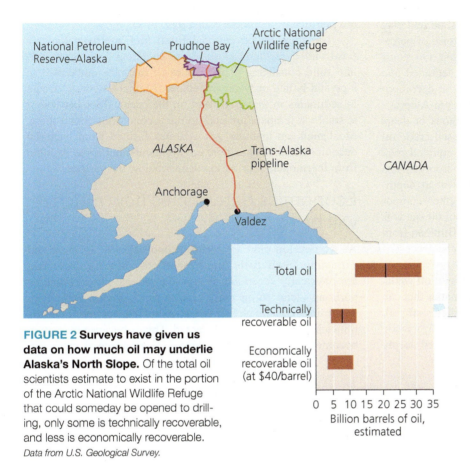

FIGURE 2 Surveys have given us data on how much oil may underlie Alaska's North Slope. Of the total oil scientists estimate to exist in the portion of the Arctic National Wildlife Refuge that could someday be opened to drilling, only some is technically recoverable, and less is economically recoverable.

Data from U.S. Geological Survey.

combined their results with a reanalysis of 2300 km (1400 mi) of seismic survey data that industry had collected in the 1980s.

After studying their resulting subsurface maps, USGS scientists concluded, with 95% certainty, that between 11.6 and 31.5 billion barrels of oil lay underneath the region of the refuge that Congress has debated opening for drilling. The mean estimate of 20.7 billion barrels is enough to supply the United States for 3 years at its current rate of consumption.

However, some portion of oil from any deposit is impossible to extract using current technology, so geologists estimate technically recoverable amounts of fuels. In its estimate for the Arctic Refuge, the USGS calculated technically recoverable oil to total 4.3–11.8 billion barrels, with a mean estimate of 7.7 billion barrels (just more than 1 year of U.S. consumption).

The portion of this oil that is economically recoverable depends on the costs of extracting it and the price of oil on the world market. USGS scientists calculated that at a price of $40 per barrel, 3.4–10.8 billion barrels would be economically worthwhile to recover. At higher prices, the economically recoverable amount would be closer to the technically recoverable amount.

Seismic surveying is similar to how we use sonar in water or how bats use echolocation as they fly. It is also used for finding coal deposits, salt and mineral deposits, and geothermal energy hotspots, as well as for studying faults, aquifers, and engineering sites.

Using data from such techniques, geologists with the U.S. Geological Survey (USGS) in 1998 assessed the subsurface geology of the Arctic National Wildlife Refuge on Alaska's North Slope to predict how much oil it may hold (**FIGURE 2**). Over three years, dozens of scientists conducted fieldwork and

In 2002, the USGS conducted similar analyses for the National Petroleum Reserve–Alaska, a vast parcel of tundra to the west of the Arctic Refuge and Prudhoe Bay that the U.S. government set aside 90 years ago as an emergency reserve of petroleum. USGS scientists estimated that this region contained 9.3 billion barrels of technically recoverable oil.

Across the world, petroleum geologists are using similar methods to try to determine how much oil remains to extract. Their work is of vital importance as the world's nations struggle to pursue well-informed energy policies.

through coastal marine waters millions of years ago and became buried in sediments on the ocean floor. This organic material was transformed by time, heat, and pressure into today's natural gas and crude oil. Natural gas may form directly, or it may form from coal or oil altered by heating. As a result, natural gas is often found above deposits of oil or seams of coal, and is often extracted along with those fuels.

Underground pressure tends to drive oil and natural gas upward through cracks and fissures in porous rock until they become trapped under a dense, impermeable rock layer. Oil and gas companies employ geologists to study rock formations to identify promising locations. Once a location is identified, a company conducts exploratory drilling, drilling small holes to great depths. If enough oil or gas is encountered, extraction may begin. Because oil and gas are under pressure while in the ground, they rise to the surface when a deposit is tapped. Once pressure is relieved and some portion has risen to the surface, the remainder will need to be pumped out.

Unconventional fossil fuels Besides the three conventional fossil fuels—coal, oil, and natural gas—other types of fossil fuels exist, often called "unconventional" because we are not (yet) using them as widely. Three examples of unconventional fossil fuels are oil sands, oil shale, and methane hydrate.

As we've seen, oil sands consist of moist sand and clay containing 1–20% bitumen, a thick and heavy form of petroleum that is rich in carbon and poor in hydrogen. Oil sands result from crude oil deposits that have been degraded and chemically altered by water erosion and bacterial decomposition. The leading scientific hypothesis to explain Alberta's oil sands is that geologic changes tens of millions of years ago as the Rocky Mountains were uplifted caused crude oil to migrate northeastward and upward until it saturated rock and soil in what is now northeastern Alberta. Microorganisms (which are abundant near the surface but absent at depth) began to consume the oil, particularly the lighter components, leaving degraded heavy bitumen. Oil from oil sands is extracted by two main methods (**FIGURE 19.8**). Bitumen from either process must then be chemically refined and processed to create synthetic crude oil (called *syncrude*).

The second type of unconventional fossil fuel, **oil shale,** is sedimentary rock (p. 35) filled with organic matter that can be processed into a liquid form of petroleum called **shale oil.** Oil shale is formed by the same processes that form crude oil but occurs when the organic matter was not buried deeply enough or subjected to enough heat and pressure to form oil. Oil shale is extracted using strip mines or subsurface mines. It can be burned directly like coal, or it can be processed in several ways. One way to process oil shale is to bake it in the presence of hydrogen and in the absence of air to extract liquid petroleum (a process called *pyrolysis*).

The third unconventional fossil fuel, **methane hydrate** (also called *methane clathrate* or *methane ice*), is an ice-like solid consisting of molecules of methane embedded in a crystal lattice of water molecules. Methane hydrate occurs in sediments in the Arctic and on the ocean floor because it is stable at temperature and pressure conditions found there. Most methane in these gas hydrates formed from bacterial decomposition in anaerobic environments, but some resulted from thermogenic formation deeper below the surface.

Economics determines how much will be extracted

As we develop more powerful technologies for locating and extracting fossil fuels, the proportions of these fuels that are physically accessible to us—the *technically recoverable* portions—tend to increase. However, whereas technology determines how much fuel *can* be extracted, economics determines how much *will* be extracted. This is because extraction becomes increasingly expensive as a resource is removed,

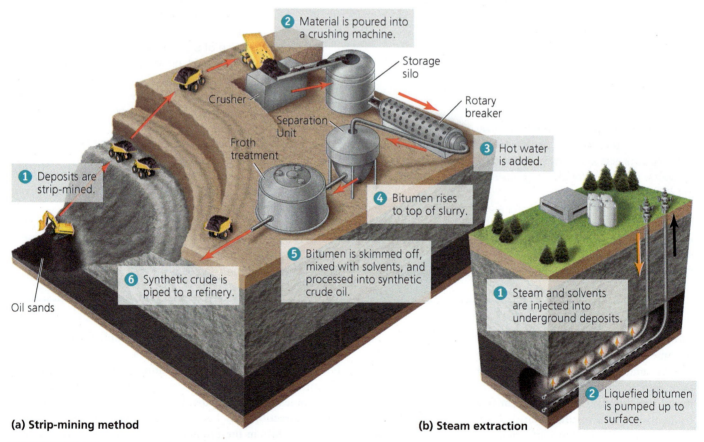

2 Material is poured into a crushing machine.

Storage silo

Crusher

Rotary breaker

Separation Unit

Froth treatment

3 Hot water is added.

1 Deposits are strip-mined.

4 Bitumen rises to top of slurry.

6 Synthetic crude is piped to a refinery.

5 Bitumen is skimmed off, mixed with solvents, and processed into synthetic crude oil.

Oil sands

1 Steam and solvents are injected into underground deposits.

2 Liquefied bitumen is pumped up to surface.

(a) Strip-mining method

(b) Steam extraction

FIGURE 19.8 Oil sands are extracted by two processes. Near-surface deposits of oil sands **(a)** are strip-mined. Deeper deposits of oil sands **(b)** are liquefied and extracted through well shafts.

so companies rarely find it profitable to extract the entire amount. Instead, a company will consider the costs of extraction (and other expenses), and balance these against the income it expects from sale of the fuel. Because market prices of fuel fluctuate, the portion of fuel from a given deposit that is *economically recoverable* fluctuates as well. As market prices rise, economically recoverable amounts approach technically recoverable amounts.

The amount of a fossil fuel that is technologically *and* economically feasible to remove under current conditions is termed its **proven recoverable reserve.** Proven recoverable reserves increase as extraction technology improves or as market prices of the fuel rise. Proven recoverable reserves decrease as fuel deposits are depleted by extraction or as market prices fall (making extraction unprofitable). Some examples of proven recoverable reserves are shown in Table 19.2 (p. 518).

Refining results in a diversity of fuels

Once we extract oil or gas, it must be processed and refined before we can use it as an energy source (**FIGURE 19.9**). Because crude oil is a complex mix of hundreds of types of hydrocarbons, we can create many types of petroleum products by separating its various components. The many hydrocarbon molecules in crude oil have carbon chains of different lengths (p. 27). Chain length affects a substance's chemical properties, and this has consequences for human use, such as whether a given fuel burns cleanly in a car engine. Through the process of **refining** at a refinery, hydrocarbon molecules are separated by size and are chemically transformed to create specialized fuels for heating, cooking, and transportation, and to create lubricating oils, asphalts, and the precursors of plastics and other petrochemical products.

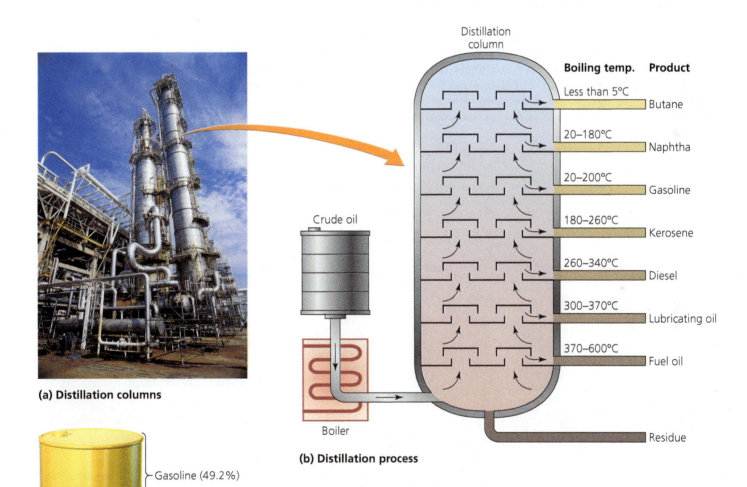

(a) Distillation columns

(b) Distillation process

(c) Typical output of refined oil

Gasoline (49.2%)

Diesel fuel and heating oil (25.0%)

Jet fuel (8.0%)

Liquefied petroleum gases (3.1%)

Heavy fuel oil (2.1%)

Other (12.7%)

FIGURE 19.9 The refining process results in a range of petroleum products. At oil refineries **(a)**, crude oil is boiled, causing its many hydrocarbon constituents to volatilize and proceed upward **(b)** through a distillation column. Constituents that boil at the hottest temperatures and condense readily once the temperature cools will condense at low levels in the column. Constituents that volatilize at cooler temperatures will continue rising through the column and condense at higher levels, where temperatures are cooler. In this way, heavy oils (generally those with hydrocarbon molecules with long carbon chains) are separated from lighter oils (generally those with short-chain hydrocarbon molecules). Shown in **(c)** are percentages of each major category of product typically generated from a barrel of crude oil. *Data (c) from U.S. Energy Information Administration.*

Fossil fuels have many uses

The amounts of a fossil fuel that a nation extracts, and the amounts of a fossil fuel that a nation consumes, each depend on many factors. **TABLE 19.3** shows amounts extracted and amounts consumed (as percentages of global extraction and consumption) by leading nations.

Each major type of fossil fuel has its own mix of uses. People have burned coal to cook food, heat homes, and fire pottery for thousands of years and in many cultures across the globe. Coal-fired steam engines helped drive the industrial revolution by powering factories, agriculture, trains, and ships, and coal fueled the furnaces of the steel industry. Today we burn coal largely to generate electricity. In coal-fired power plants, coal combustion converts water to steam, which turns turbines to create electricity (**FIGURE 19.10**). Coal provides 40% of the electrical generating capacity of the United States, and it has powered the rise of China's surging economy.

Like coal, natural gas is used to generate electricity in power plants, as well as to heat and cook in our homes, among many other uses. Converted to a liquid at low temperatures (liquefied natural gas, or LNG), it can be shipped long distances in refrigerated tankers. Versatile and clean-burning, natural gas emits half as much carbon dioxide per unit of energy released as coal and two-thirds as much as oil. For this reason, many energy experts have proposed natural gas as a climate-friendly "bridge fuel" that can help us transition from today's polluting fossil fuel economy toward a clean renewable energy economy. However, many other experts worry that investing in natural gas will simply delay our transition to renewables and instead deepen our reliance on fossil fuels.

The modern use of oil for energy began after the 1850s, when crude oil gushed from the world's first commercial oil wells drilled in Ontario and Pennsylvania. Over the next 40 years, drillers in Pennsylvania supplied half the world's oil and helped establish a fossil-fuel-based economy that would hold sway for decades to come. Today our global society extracts and consumes nearly 200 gallons of oil each year for every man, woman, and child. Most is used as fuel for vehicles, including gasoline for cars, diesel for trucks, and jet fuel for airplanes. Fewer homes burn oil for heating these days, but industry and manufacturing continue to use a great deal.

Over the past several decades, refining techniques and chemical manufacturing have greatly expanded our uses of petroleum to include a wide array of products and applications, from plastics to lubricants to fabrics to pharmaceuticals. In today's world, petroleum-based products are all around

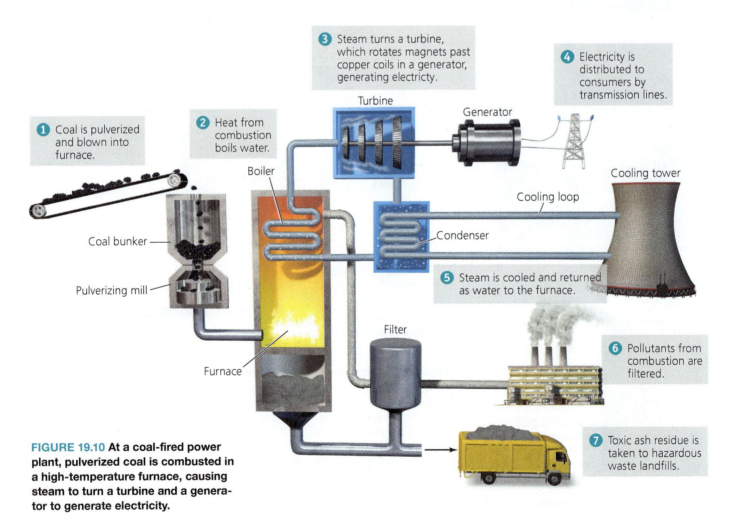

1 Coal is pulverized and blown into furnace.

2 Heat from combustion boils water.

3 Steam turns a turbine, which rotates magnets past copper coils in a generator, generating electricty.

4 Electricity is distributed to consumers by transmission lines.

5 Steam is cooled and returned as water to the furnace.

6 Pollutants from combustion are filtered.

7 Toxic ash residue is taken to hazardous waste landfills.

Turbine · Generator · Boiler · Coal bunker · Pulverizing mill · Furnace · Filter · Cooling loop · Cooling tower · Condenser

FIGURE 19.10 At a coal-fired power plant, pulverized coal is combusted in a high-temperature furnace, causing steam to turn a turbine and a generator to generate electricity.

TABLE 19.3 Top Extractors and Consumers of Fossil Fuels

EXTRACTION (% world extraction)		CONSUMPTION (% world consumption)	
COAL			
China	47.7	China	50.0
United States	11.9	India	10.6
India	7.4	United States	10.3
Australia	7.2	Japan	3.1
OIL			
United States	13.0	United States	19.7
Saudi Arabia	13.0	China	12.9
Russia	12.4	India	4.5
Canada	4.9	Japan	4.4
NATURAL GAS			
United States	22.0	United States	22.8
Russia	16.1	Russia	11.2
Iran	5.4	China	5.7
Qatar	5.1	Iran	5.5

Data from BP p.l.c., 2016. *Statistical review of world energy 2016.*

us in our everyday lives (**FIGURE 19.11**). Take a moment to explore Figure 19.11, and reflect on all the conveniences in your own life that result from petroleum products. The fact that petroleum is used to create so many items and materials we have come to rely on makes it vital that we take care to conserve our remaining oil reserves.

We are depleting fossil fuel reserves

Because fossil fuels are nonrenewable, the total amount available on Earth declines as we use them. Many scientists and oil industry analysts calculate that we have already extracted nearly half the world's conventional oil reserves. So far we have used up about 1.2 trillion barrels of oil, and most estimates hold that about 1.2 trillion barrels of proven reserves remain. Adding proven reserves of oil from oil sands in Canada and Venezuela brings the estimated total remaining to about 1.7 trillion barrels.

To assess how long this remaining oil will last, analysts calculate the **reserves-to-production ratio,** or R/P ratio, by dividing the amount of total remaining reserves by the annual rate of "production" (extraction and processing). At current levels of production (32 billion barrels globally per year), 1.7 trillion barrels would last about 53 more years. Applying the R/P ratio to natural gas, we find that the world's proven reserves of this resource would last 54 more years. For coal, the latest R/P ratio estimate is 110 years.

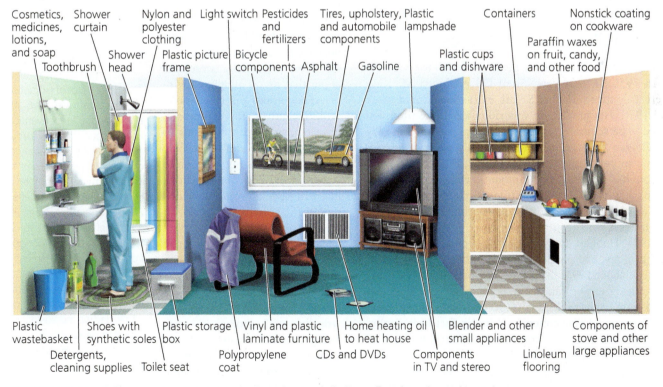

Cosmetics, medicines, lotions, and soap · Shower curtain · Toothbrush · Shower head · Nylon and polyester clothing · Plastic picture frame · Light switch · Bicycle components · Pesticides and fertilizers · Asphalt · Tires, upholstery, and automobile components · Gasoline · Plastic lampshade · Plastic cups and dishware · Containers · Nonstick coating on cookware · Paraffin waxes on fruit, candy, and other food

Plastic wastebasket · Detergents, cleaning supplies · Shoes with synthetic soles · Toilet seat · Plastic storage box · Vinyl and plastic laminate furniture · Polypropylene coat · CDs and DVDs · Home heating oil to heat house · Components in TV and stereo · Blender and other small appliances · Linoleum flooring · Components of stove and other large appliances

FIGURE 19.11 Petroleum products are everywhere in our daily lives. Petroleum is used to make many of the fabrics we wear, the materials we consume, and the plastics in countless items we use every day.

The actual number of years remaining for these fuels could turn out to be less than these figures suggest if our demand and production continue to increase. Alternatively, the actual number of years may end up being *more* than these figures suggest if we reduce demand and consumption by enhancing efficiency. The actual number of years may also turn out to be greater because proven recoverable reserves increase as new deposits are discovered, as extraction technology becomes more powerful, and as market prices rise.

For instance, hydraulic fracturing for natural gas in the Marcellus Shale and elsewhere in the United States has expanded the nation's proven reserves of natural gas considerably in recent years (Chapter 7). Likewise, hydraulic fracturing of the Bakken Formation—layers of shale and dolomite that underlie parts of North Dakota, Montana, and Canada—is allowing us to extract oil trapped tightly in this rock. By accessing this so-called *tight oil*, conventional oil held tightly in or near shale (which differs from shale oil, a petroleum liquid from specially processed oil shale), the United States has boosted its proven recoverable reserves of oil. In fact, the recent oil boom in North Dakota, along with increases in drilling deep offshore, enabled the United States, as of 2014, to become the world's largest extractor of oil.

Eventually, however, extraction of any nonrenewable resource will come to a peak and then decline. In general, extraction tends to decline once reserves are depleted halfway. If demand for the resource holds steady or rises while extraction declines, a shortage will result. With oil, this scenario has come to be nicknamed **peak oil.** Because we have already used roughly half of Earth's conventional oil reserves, many experts calculate that a peak oil crisis could begin soon.

Peak oil will pose challenges

To understand concerns about peak oil, let's turn back the clock to 1956. In that year, Shell Oil geologist M. King Hubbert calculated that U.S. oil extraction would peak around 1970. His prediction was ridiculed at the time, but it proved to be accurate; U.S. extraction peaked in that very year and then fell (**FIGURE 19.12a**). This peak in extraction came to be known as *Hubbert's peak*. Today U.S. oil extraction has reached this amount again, as hydraulic fracturing has enabled us to extract formerly inaccessible oil and as we have pursued various unconventional petroleum sources.

In 1974, Hubbert analyzed data on technology, economics, and geology, and predicted that worldwide oil extraction would peak about 1995. In fact, it grew past 1995, but many scientists using newer data today predict that global extraction will soon begin to decline (**FIGURE 19.12b**). Since about 2005, extraction of conventional oil has in fact been declining, and a variety of less-conventional petroleum sources have been compensating for this decline.

Discovery or development of wholly new sources of oil can delay the peak by boosting our overall proven reserves.

This is what is happening as we exploit Canada's oil sands. People who count on a series of new sources becoming available in the future tend to believe we can continue relying on oil for a long time. Indeed, R/P ratios for oil have risen over recent decades. However, at some point we will reach peak oil—as well as "peak gas" and "peak coal." The question is when—and whether we will be prepared to deal with the resulting challenges.

Predicting an exact date for peak oil is difficult because many companies and governments do not reveal their true data on oil reserves. Moreover, estimates differ about how much oil we can continue extracting from existing deposits. For these reasons, estimates vary for the timing of an oil extraction peak, but most studies predict dates before 2035.

Whenever an oil peak occurs, if demand for oil continues to rise as supply falls, the divergence of demand and supply would drive up oil prices dramatically, triggering major economic ripple effects that could profoundly affect our lives. Writer James Howard Kunstler has sketched a frightening scenario of a post-peak world during what he calls "the long emergency": Lacking cheap oil with which to transport goods long distances, today's globalized economy would collapse into isolated local economies. Large cities would need to run urban farms to feed their residents, and with less mechanized farming and fewer petroleum-based fertilizers and pesticides, we might feed only a fraction of the world's people. The American suburbs would be hit particularly hard because of their dependence on the automobile.

More optimistic observers argue that as oil supplies dwindle, rising prices will create powerful incentives for businesses, governments, and individuals to conserve energy and to develop alternative energy sources (Chapters 20 and 21)—and that this will save us from major disruptions.

If we discover and exploit enough new deposits to continue extracting more and more oil, we might postpone our day of reckoning for decades. If we do so, however, we will find ourselves wrestling with another concern: trying to avoid runaway climate change driven by greenhouse gas emissions from the combustion of all that additional oil!

FAQ

Hasn't "peak oil" been debunked? We're extracting more oil all the time, right?

An eventual peak is inevitable with any nonrenewable resource, so the notion of peak oil is sound. Its timing is a major question, though—and much depends on how one defines "oil." If one counts only conventional crude oil, then data indicate that we passed the global extraction peak back around 2005. If one also includes "unconventional" oil from difficult-to-access sources such as oil sands, deep-water offshore oil, polar oil, and "tight oil" freed by fracking, then extraction has been roughly flat since 2005. If one also lumps in various additional petroleum sources as "oil" (such as liquids condensed from natural gas), then extraction is still rising. In the big picture, conventional oil has already been declining for more than a decade, and today we increasingly rely on a host of petroleum sources that are more difficult and expensive to access.

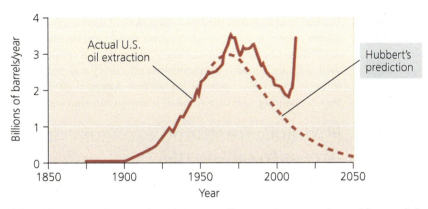

(a) Hubbert's prediction of peak in U.S. oil extraction, together with actual data

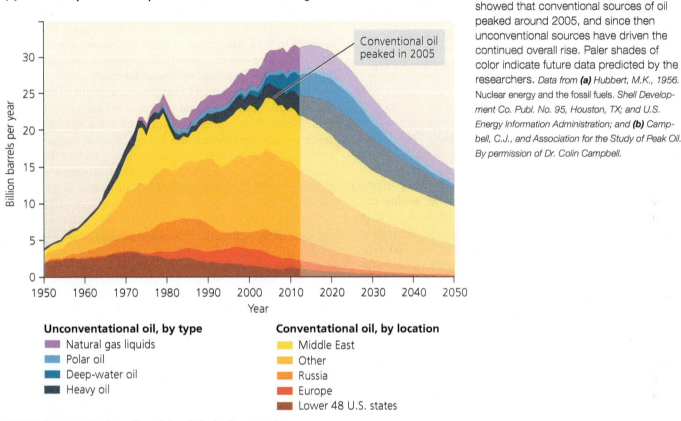

Unconventational oil, by type
- Natural gas liquids
- Polar oil
- Deep-water oil
- Heavy oil

Conventational oil, by location
- Middle East
- Other
- Russia
- Europe
- Lower 48 U.S. states

(b) Modern prediction of peak in global oil extraction

FIGURE 19.12 Peak oil describes a peak in extraction. U.S. oil extraction peaked in 1970 **(a)**, just as geologist M. King Hubbert had predicted. Drilling in Alaska and the Gulf of Mexico enhanced extraction during the subsequent decline, and today extraction has risen sharply as deep offshore drilling and hydraulic fracturing have made new deposits accessible. As for global oil extraction, many analysts calculate that it may soon peak. Shown in **(b)** is one projection, from a recent analysis by scientists at the Association for the Study of Peak Oil. This study showed that conventional sources of oil peaked around 2005, and since then unconventional sources have driven the continued overall rise. Paler shades of color indicate future data predicted by the researchers. *Data from* **(a)** *Hubbert, M.K., 1956. Nuclear energy and the fossil fuels. Shell Development Co. Publ. No. 95, Houston, TX; and U.S. Energy Information Administration; and* **(b)** *Campbell, C.J., and Association for the Study of Peak Oil. By permission of Dr. Colin Campbell.*

Reaching Further for Fossil Fuels

To stave off the day when supplies of oil, gas, and coal begin to decline, we are investing more and more money, energy, and technology into locating and extracting new fossil fuel deposits. We are extending our reach for fossil fuels in several ways:

- Mountaintop mining for coal
- Secondary extraction from existing wells
- Directional drilling
- Hydraulic fracturing for oil and gas
- Offshore drilling in deep waters
- Moving into ice-free waters of the Arctic
- Exploiting new "unconventional" fossil fuel sources

All these pursuits are expanding the amount of fossil fuel energy available to us. As we extend our mining and drilling efforts into less-accessible places to obtain fuel that is more difficult and more expensive to extract, we expand our proven reserves. However, we also reduce the EROI ratios of our fuels, intensify pollution, and worsen climate change. At a time when many people feel we should be transitioning to clean and renewable energy, our society instead appears to be doubling down on its bet on fossil fuels.

FIGURE 19.13 In mountaintop removal mining for coal, entire mountain peaks are leveled and fill is dumped into adjacent valleys, as shown in this aerial view spanning many square miles in West Virginia.

Mountaintop mining extends our reach for coal

Coal mining has long been an economic mainstay of the Appalachian region, but **mountaintop removal mining** has brought coal extraction to a whole new level (**FIGURE 19.13** and p. 639). In this method, entire mountaintops are blasted away to access coal seams running through them. The massive scale of mountaintop removal mining makes it economically efficient. However, it can cause staggering volumes of rock and soil to slide downslope, degrading or destroying entire hillsides, polluting or burying streams, and disrupting life for people living nearby.

Secondary extraction yields additional fuel

At a typical oil or gas well, as much as two-thirds of a deposit may remain in the ground after primary extraction, the initial drilling and pumping of oil or gas. So, companies may return and conduct secondary extraction using new technology or approaches to force the remaining oil or gas out by pressure. In secondary extraction for oil, solvents are injected, underground rocks are flushed with water or steam, or hydraulic fracturing is used. Because secondary extraction is more expensive than primary extraction, most deposits undergo secondary extraction only when market prices of oil and gas are high enough to make the process profitable.

Directional drilling reaches more fuel with less impact

Engineers have brought us a long way from the days when we simply sank a vertical shaft into the ground to suck up oil and gas. Today's **directional drilling** technology allows drillers to bore down vertically and then curve to drill horizontally. This enables them to follow horizontal layered deposits such as the

Bakken Formation or the Marcellus Shale. Directional drilling allows access to a large underground area (up to several thousand meters in radius) around each drill pad. As a result, fewer drill pads are needed. This enhances efficiency and profits for drillers, and it also reduces the surface footprint of drilling, thereby helping to lessen some of its environmental impacts.

Hydraulic fracturing expands our access to oil and gas

For oil and natural gas trapped tightly in shale or other rock, drillers now use hydraulic fracturing (see Figure 7.2, p. 161). **Hydraulic fracturing** (also called **fracking**) involves pumping chemically treated water under high pressure into deep layers of rock to crack them. Sand or small glass beads are injected to hold the cracks open as the water is withdrawn. Gas or oil then travels upward, with pressure and pumping, through the newly created system of fractures.

By unlocking formerly inaccessible deposits of shale gas and tight oil, fracking ignited a boom in oil and gas extraction in the United States. The new flow of oil reduced U.S. dependence on foreign imports and led by 2016 to a glut on the world market that brought oil prices down to less than one-quarter of their 2008 high. The new flow of natural gas has enabled many power plants to switch from coal to gas, helping to decrease U.S. carbon dioxide emissions since 2007.

Fracking has sparked debate everywhere it has occurred (**FIGURE 19.14**). Above the massive Marcellus Shale deposit, it

FIGURE 19.14 Hydraulic fracturing is expanding U.S. extraction of oil and natural gas, but is also sparking debates in communities where it takes place. This drill rig is fracking for shale gas on private land near homes in rural Pennsylvania.

is affecting the landscapes, economies, politics, and everyday lives of people in Pennsylvania, Ohio, New York, and neighboring states (see Chapter 7). The choices people face between financial gain and protecting their health, drinking water, and environment have been dramatized in popular films such as *Promised Land* and *Gasland*. Recently in North Dakota, fracking for oil supercharged the economy and drew young people from around the nation for high-paying jobs, but it also polluted the landscape, drew down water resources, and left many workers jobless once oil prices fell and rigs shut down.

We are drilling farther offshore

Roughly 35% of the oil and 10% of the natural gas extracted in the United States today come from offshore sites, primarily in the Gulf of Mexico and secondarily off southern California. The Gulf today is home to 90 drilling rigs and 3500 production platforms—each of them an engineering marvel built to withstand wind, waves, and ocean currents. Geologists estimate that most U.S. gas and oil remaining occurs offshore and that deep-water sites in the Gulf of Mexico alone may hold 59 billion barrels of oil.

We have been drilling in shallow water for several decades, but as oil and gas are depleted at shallow-water sites and as drilling technology improves, the industry is moving into deeper and deeper water. This poses risks, however, and our ability to drill in deep water has outpaced our capacity to deal with accidents. In the *Deepwater Horizon* oil spill of 2010 (pp. 429, 534–537), faulty equipment allowed natural gas accompanying an oil deposit to shoot up a well shaft. The gas ignited atop the platform, killing 11 workers and leading to the largest accidental oil spill in history. The fact that it took 86 days for the company that owned the platform, British Petroleum, to plug its leak demonstrates the challenge of addressing an emergency situation a mile or more beneath the surface of the sea. BP's Macondo well, where the accident took place, lay beneath 1500 m (5000 ft) of water. The deepest wells in the Gulf of Mexico are now twice that depth.

In 2008, responding to rising gasoline prices and seeking to lessen dependence on foreign oil, the U.S. Congress lifted a long-standing moratorium on offshore drilling along much of the nation's coastline. The Obama administration in 2010 followed through by designating vast areas open for drilling. These included most waters along the Atlantic coast from Delaware south to central Florida, a region of the eastern Gulf of Mexico, and most waters off Alaska's North Slope. However, just weeks after this announcement, the *Deepwater Horizon* spill occurred. Public reaction forced the Obama administration to backtrack, canceling offshore drilling projects it had approved and putting a hold on further approvals until new safety measures could be devised. In 2011, after weighing economic and environmental concerns, the administration issued a five-year plan that opened access to 75% of technically recoverable offshore oil and gas reserves while banning drilling offshore from states that did not want it. Drilling leases were expanded off Alaska and in the Gulf of Mexico, but areas along the East and West Coasts were not opened to drilling.

Melting ice is opening the Arctic

Today all eyes are on the Arctic. As global climate change melts the sea ice that covers the Arctic Ocean (pp. 494, 501), new shipping lanes are opening and nations and companies are jockeying for position, scrambling to lay claim to areas of ocean where fossil fuels might lie beneath the seafloor. Offshore drilling in Arctic waters, however, poses severe pollution and safety risks. Frigid temperatures, ice floes, winds, waves, and brutal storms make conditions challenging and accidents likely. If a spill were to occur, icebergs, pack ice, storms, cold, and wintertime darkness would hamper response efforts, while frigid water temperatures would slow the natural breakdown of oil.

So far, Royal Dutch Shell has been the only company to pursue offshore drilling in Alaskan waters—and it has met with one mishap after another. In 2012–2013 its *Kulluk* drilling rig ran aground while being towed south from Alaska during a storm. Another drilling ship nearly met the same fate in the Aleutian Islands, and the containment dome intended to control potential leaks was crushed during testing. Then in 2015 one of Shell's icebreakers, the *Fennica*, ran aground and had to be towed all the way to Portland, Oregon, for repairs. On its way out of the harbor, protesters rowing in kayaks and dangling daringly from a bridge blocked its path, gaining media attention for their cause. This followed protests in Seattle two months earlier over Shell's plans to house drilling rigs in Seattle's port (**FIGURE 19.15**). Meanwhile, Arctic storms halted Shell's drilling. That fall Shell gave up and withdrew from the Arctic after spending $7 billion in efforts to drill there.

We are exploiting new fossil fuel sources

Three sources of "unconventional" fossil fuels—oil sands, oil shale, and methane hydrate—are abundant, and together could theoretically supply our civilization for centuries. The oil sands

FIGURE 19.15 "Kayaktivists" in Seattle protested Shell's plans to drill for oil in the Arctic Ocean by blockading a drilling rig in 2015.

of Alberta—and those in Venezuela, which hold even more oil—are already increasing the amount of oil available to us.

The world's known deposits of oil shale may contain 3 trillion barrels of oil (more than all the world's conventional crude oil), but most of this will not easily be extracted. About 40% of global reserves are in the United States, mostly on public land in Colorado, Wyoming, and Utah. Shale oil is costly to extract, and its EROI is very low, with estimates ranging from 4:1 down to just 1.1:1. Historically, oil prices have been too low to make processing shale oil profitable, but each time oil prices rise, shale oil again attracts attention.

As for methane hydrate, scientists estimate there are enormous amounts of this substance on Earth, holding perhaps twice as much carbon as all known deposits of oil, coal, and natural gas combined. However, we do not yet know whether their extraction is safe and reliable. Japan recently extracted methane hydrate from the seafloor by sending down a pipe and lowering pressure within it so that the methane turned to gas and rose to the surface. Yet if extraction were to destabilize a seafloor methane hydrate deposit, a sudden release of gas could cause a landslide and tsunami and emit large amounts of methane, a potent greenhouse gas, into the atmosphere.

Oil sands, oil shale, and methane hydrate are abundant, but they are no panacea for our energy challenges. Their net energy values and EROI ratios are very low, because they are expensive to extract and process. These fuels also exert severe environmental impacts on land and water. Moreover, burning them would likely emit more greenhouse gases than our use of coal, oil, and natural gas currently does, worsening air pollution and climate change.

Addressing Impacts of Fossil Fuel Use

Our society's love affair with fossil fuels and the many petrochemical products we develop from them has helped to ease constraints on travel, lengthen our life spans, and boost our material standard of living beyond what our ancestors could have dreamed. Yet continued reliance on fossil fuels poses growing risks to human health; environmental quality; and social, political, and economic stability. The predicted impacts from climate change alone are so great that the International Energy Agency's chief economist recently joined a growing chorus of scientists in concluding that we will need to leave most fossil fuels in the ground if we are to avoid dangerous climate disruption and preserve the planet that we know. Let's now briefly survey the main impacts of fossil fuel use—beginning with extraction, continuing through emissions, and ending with social issues—and assess ways to minimize these impacts.

Extraction of fossil fuels brings many environmental impacts

The means by which we remove fossil fuel resources from the earth have direct impacts on landscapes and natural habitats (**FIGURE 19.16**), as well as a diversity of other consequences for ecological processes and environmental quality.

FIGURE 19.16 Oil sands extraction dramatically alters Alberta's forested landscape. Shown is a portion of the Suncor mine near Fort McMurray.

Coal The mining of coal exerts substantial impacts on natural systems and human well-being (p. 635). Subsurface coal mining has long posed health and safety risks to miners. Miners are vulnerable to accidents, and they breathe coal dust and hazardous gases in confined spaces, which often leads to respiratory diseases. Strip mining, for its part, destroys large swaths of habitat and causes extensive soil erosion. It also can cause chemical runoff into waterways through acid mine drainage (p. 634), whereby sulfide minerals in freshly exposed rock surfaces react with oxygen and rainwater to produce sulfuric acid, which can leach toxic metals from rocks. Most developed nations require mining companies to restore affected areas after mining. This reclamation is beneficial, but rarely is able to recreate the ecological communities that preceded mining (p. 641).

Mountaintop removal mining (pp. 530, 639) exerts impacts that exceed even those of strip mining. When countless tons of rock and soil are removed from the top of a mountain, material slides downhill, where immense areas of habitat can be degraded or destroyed and creek beds can be clogged and polluted, affecting wildlife, ecosystems, and local residents. (We explore all these issues of coal mining in Chapter 23.)

Oil and natural gas To drill for conventional oil or gas on land, road networks must be constructed and many sites may be explored in the course of prospecting. The extensive infrastructure needed to support a full-scale drilling operation typically includes access roads, transport pipelines, housing for workers, waste piles for excavated soil, and ponds to collect toxic sludge. These tend to pollute the soil, air, and water; fragment habitats; and disturb wildlife. These impacts can be readily seen at most drilling sites, and have been particularly well documented on the Alaskan tundra, where policymakers continue to debate whether to open the Arctic National Wildlife Refuge to drilling.

Fortunately, directional drilling (p. 530) is becoming widely used today. This provides much greater reach than traditional vertical drilling, so fewer drill pads are needed and less infrastructure is required, reducing the footprint on the landscape.

Hydraulic fracturing Extracting oil or natural gas by hydraulic fracturing exerts all the impacts of conventional drilling, as well as posing additional concerns. In fracking, chemicals are mixed with pressurized water and sand, a combination that is injected deep underground, giving rise to risks of water pollution that are not yet completely understood. One risk is that fracking fluids may leak out of drilling shafts and into aquifers that people rely on for drinking water. Another concern is that methane may contaminate groundwater used for drinking if it travels up the fractures or leaks through the shaft (see **The Science behind the Story**, Chapter 7, pp. 164–165).

Hydraulic fracturing also gives rise to air pollution as methane and volatile toxic components of fracking fluids seep up from drilling locations. Some of the unhealthiest air in America was recently found in a remote region of Wyoming near fracking operations. Many residents of areas near fracking sites have experienced polluted air and fouled drinking water; more research is needed to assess the extent of such pollution and to quantify the health risks.

In addition, hydraulic fracturing consumes immense volumes of fresh water. Injected water often returns to the surface laced with salts, radioactive elements such as radium, and toxic chemicals such as benzene from deep underground. This wastewater may be sent to sewage treatment plants that are not designed to handle the contaminants and that do not test for radioactivity. In Pennsylvania, millions of gallons of drilling waste from Marcellus Shale fracking sites have been sent to treatment plants, which then release their water into rivers that supply drinking water for people in Pittsburgh, Harrisburg, and other cities. In response to public outcry and government pressure, the oil and gas industry is beginning to reduce its consumption of clean water by reusing its wastewater in multiple injections.

Finally, fracking is now known to cause earthquakes (see **The Science behind the Story**, Chapter 2, pp. 40–41). Most have been minor, but this does raise questions about whether all social and economic costs have been considered.

Oil sands Impacts from the extraction of oil sands surpass those of conventional fossil fuel extraction. Because everything is cleared across huge areas when oil sands are mined (see Figure 19.16), this results in large-scale loss of boreal forest (in Alberta) or tropical forest (in Venezuela). Deforestation has severe consequences for species, habitats, ecosystems, soil, water, climate, and people (Chapter 12). Industry representatives counter that the area deforested in Alberta so far amounts to just 0.1% of Canada's vast boreal forest. They also point out that they are mandated to attempt restoration afterward. However, effective reclamation has not yet been demonstrated, and regions denuded by the very first oil sand mine in Alberta more than 30 years ago have still not recovered.

Three barrels of water are required to extract each barrel of oil from oil sands, and the toxic wastewater that results is discharged into huge reservoirs. The Syncrude mine's tailings reservoir near Fort McMurray, Alberta, for example, is so massive that it is held back by the world's second-largest dam. Migratory waterfowl land on water bodies like these and are killed as the oily water gums up their feathers and impairs their ability to insulate themselves.

Fuels can leak during transport and storage

Once fossil fuels are extracted, they must be transported for processing and refining, then transported again to consumers. Between transport episodes they may be stored for long periods. At each stage, there is the risk of leakage.

Rail transport Once mined, coal is generally transported by rail in open cars, which releases coal dust into the air. In the Pacific Northwest, clean energy advocates are opposing the transport of coal by train from the interior West where it is mined to coastal terminals, to be shipped to Asia. Pollution along the route is one concern; another is that this facilitates China's reliance on coal, a driver of global climate change.

Oil is also transported by rail, in pressurized tank cars. In recent years, as oil extraction in North America has soared, it has surpassed pipeline capacity, and a great deal of oil is being transported by rail instead. Tragically, a series of explosive derailments of trains carrying corrosive North Dakota crude oil has illustrated the dangers of carrying oil by train (**FIGURE 19.17**). Worst was the 2013 explosion in Lac-Mégantic, Quebec, which killed 47 people and destroyed the town's center. In the two following years the United States saw 10 major explosions, and in 2014 there were 141 tanker spills. The Obama administration responded with regulations to upgrade the safety of tanker cars; industry complained of additional cost while safety advocates said the steps were not strong enough. Ironically, while rejection of the Keystone XL pipeline may have averted a number of environmental and social impacts, it is also contributing to the surge in oil transport by rail.

FIGURE 19.17 Explosive crashes of trains carrying oil, such as this one in Lac-Mégantic, Quebec, are rising, as North American oil extraction has surpassed pipeline capacity.

Methane leaks Any time natural gas leaks into the air, its main component, methane, is added to the atmosphere. Methane is a powerful greenhouse gas (p. 483), so its leakage worsens global warming and climate change. Methane can leak at various points during the extraction, transport, and processing of natural gas—and it can leak from aging pipes that bring natural gas to consumers. A 2015 study of the Boston region found that 2.7% of natural gas piped to homes and businesses was escaping into the air—higher than the 1% previously assumed. By comparison, researchers estimate that 3.6–7.9% of natural gas extracted by shale gas fracking operations leaks into the air.

A great deal of methane also escapes to the atmosphere from oil drilling sites where gas accompanying an oil deposit is released to the air because pipelines for the gas do not exist. At many such sites, natural gas is flared off—in other words, the stream of gas is set on fire. This produces carbon dioxide, however, which is also a greenhouse gas.

Pipeline spills Oil is commonly conveyed by pipeline, and pipelines are subject to corrosion, vandalism, and equipment malfunction, any of which can result in spills that contaminate soil and water. Oil from Canada's oil sands (which is more corrosive than conventional crude oil) has spilled from a number of pipelines, fouling the Kalamazoo River in Michigan; a residential neighborhood of Mayflower, Arkansas; and other areas. People living along the initial proposed route of the Keystone XL pipeline worried that if oil were to spill from that pipeline, it would sink into the area's porous ground and quickly reach the region's shallow water table, contaminating the Ogallala Aquifer. This aquifer (p. 388) provides 2 million Americans with drinking water and irrigates a large portion of U.S. agriculture.

Marine oil spills Of the many ways in which fossil fuels pollute water, what comes to mind first for most people is how oil spilled from tanker ships or drilling platforms can

FIGURE 19.18 The explosion at BP's *Deepwater Horizon* drilling platform in 2010 unleashed the world's largest accidental oil spill. Here vessels try to put out the blaze.

foul coastal waters and beaches. In 2010, BP's *Deepwater Horizon* offshore drilling platform exploded and sank off the coast of Louisiana (**FIGURE 19.18**). Eleven workers were killed, and oil gushed out of a broken pipe on the ocean floor a mile beneath the surface at a rate of 30 gallons per second. Emergency shut-off systems failed, and BP engineers tried one solution after another; but as discussed earlier, the flow of oil and gas continued out of control for three months, spilling roughly 4.9 million barrels (206 million gallons) of oil. The crisis proved difficult to control because people had never had to deal with a spill so deep underwater. It revealed that offshore drilling presents serious risks of environmental impact that may be difficult to address.

As the oil spread through the Gulf of Mexico and washed ashore, the region suffered a wide array of impacts (**FIGURE 19.19**). Of the countless animals killed, most conspicuous were birds, which cannot regulate their body temperature once their feathers become coated with oil. However,

(a) Brown pelican coated in oil

(b) Beach cleanup

FIGURE 19.19 Impacts of the *Deepwater Horizon* spill were many. This brown pelican, coated in oil **(a)**, was one of countless animals killed. For months, volunteers and workers labored **(b)** to clean oil from the Gulf's beaches.

the underwater nature of the BP spill meant that unknown numbers of fish, shrimp, corals, and other marine animals also were killed, affecting ecosystems in complex ways. Plants in coastal marshes died, and the resulting erosion put New Orleans and other coastal cities at greater risk from storm surges and flooding. Gulf Coast fisheries, which supply much of the nation's seafood, were hit hard, with thousands of fishermen and shrimpers put out of work. Beach tourism suffered, and indirect economic and social impacts have lasted for years. Throughout this process, scientists have been studying aspects of the spill and its impacts on the region's people and natural systems (see **THE SCIENCE BEHIND THE STORY**, pp. 536–537).

The *Deepwater Horizon* spill was the largest accidental oil spill in world history, far eclipsing the spill from the *Exxon Valdez* tanker in 1989. In that event, oil from Alaska's North Slope, piped to the port of Valdez through the trans-Alaska pipeline, caused long-term damage to ecosystems and economies in Alaska's Prince William Sound after the tanker ran aground. Nearly 30 years later, a layer of oil remains in the sand of the region's beaches.

The *Exxon Valdez* event led U.S. policymakers to tighten regulation and improve spill-response capacity. However, if drilling moves into Arctic waters, our capacity to control accidents could be sorely tested by the harsh conditions. In U.S. Arctic waters open to oil and gas leasing, some sites are 1000 miles from the nearest Coast Guard station.

Fortunately, pollution from large spills has declined in recent decades (see Figure 16.16, p. 432), thanks to government regulations (such as requiring double-hulled ships) and improved spill-response efforts. And although large catastrophic oil spills have significant impacts on the marine environment, most water pollution from oil results from countless small non-point sources (see Figure 15.23, p. 406). Oil from automobiles, homes, industries, gas stations, and businesses runs off roadways and enters rivers and wastewater facilities, eventually being discharged into the ocean. Water pollution from industrial point sources has been greatly reduced in the United States following the Clean Water Act (Chapter 15), and many solutions exist to address non-point-source pollution.

Carbon emissions drive climate change

When we burn fossil fuels, we alter fluxes in Earth's carbon cycle (p. 120). We essentially remove carbon from a long-term reservoir underground and release it into the air. This occurs as carbon from the hydrocarbon molecules of fossil fuels unites with oxygen from the atmosphere during combustion, producing carbon dioxide (CO_2). Carbon dioxide is a greenhouse gas (p. 482), and CO_2 released from fossil fuel combustion warms our planet and drives changes in global climate (Chapter 18).

Because global climate change is beginning to have diverse, severe, and widespread ecological and socioeconomic impacts, carbon dioxide pollution (**FIGURE 19.20**) is becoming recognized as the single biggest negative consequence of

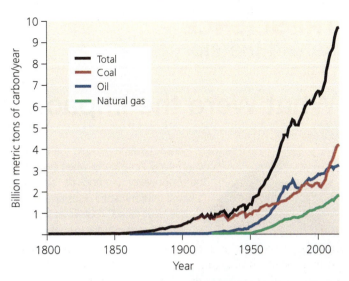

FIGURE 19.20 Emissions from fossil fuel combustion have risen dramatically as nations have industrialized and as population and consumption have grown. Here, global emissions of carbon from carbon dioxide are subdivided by source (coal, oil, or natural gas). *Data from Carbon Dioxide Information Analysis Center, Oak Ridge National Laboratory, U.S. Department of Energy, Oak Ridge, TN.*

 By what percentage have carbon emissions risen since the year your mother or father was born?

Go to **Interpreting Graphs & Data** on MasteringEnvironmentalScience®

fossil fuel use. Methane is also a potent greenhouse gas that drives climate warming. Switching to new fossil fuel sources may only make emissions worse; oil sands are estimated to generate 14–20% more greenhouse gas emissions than conventional oil, and shale oil is still more polluting.

Fossil fuel emissions also pose health risks

Besides modifying our climate, fossil fuel emissions affect human health. Combusting coal may emit mercury that can bioaccumulate in organisms' tissues, poisoning animals as it moves up food chains (p. 368) and posing health risks to people. Gasoline combustion in automobiles releases pollutants that irritate the nose, throat, and lungs, as well as cancer-causing hydrocarbons such as benzene and toluene. Gases such as hydrogen sulfide can evaporate from crude oil, irritate the eyes and throat, and cause asphyxiation. Crude oil also may contain trace amounts of poisons such as lead and arsenic. As a result, workers at drilling operations, at refineries, and in other jobs that entail frequent exposure to oil can develop serious health problems, including cancer.

The combustion of oil in vehicles and coal in power plants releases sulfur dioxide and nitrogen oxides, which contribute to smog (pp. 459–464) and acid deposition (p. 469). Air pollution from fossil fuel combustion is intensifying in developing nations that are industrializing, but it has been reduced in developed nations as a result of laws such as the U.S. Clean Air Act and government regulations to protect public health (Chapter 17). In these nations, public policy

What Were the Impacts of the Gulf Oil Spill?

President Barack Obama echoed the perceptions of many Americans when he called the *Deepwater Horizon* oil spill "the worst environmental disaster America has ever faced." But what has scientific research told us about the actual impacts of the Gulf oil spill?

We don't yet have all the answers, because the deep waters affected by the spill have been difficult for scientists to study. A great deal will remain unknown. Yet the intense and focused scientific response to the spill demonstrates the dynamic way in which science can assist society.

As the spill was taking place, government agencies called on scientists to help determine how much oil was leaking. Researchers eventually determined the rate reached 62,000 barrels per day. Using underwater imaging, aerial surveys, and shipboard water samples, researchers tracked the movement of oil up through the water column and across the Gulf. These data helped predict when and where oil might reach shore, thereby serving to direct prevention and cleanup efforts.

A scientist rescues an oiled Kemp's ridley sea turtle.

Meanwhile, as engineers struggled to seal off the well using remotely operated submersibles, researchers assisted government agencies in assessing the fate of the oil (**FIGURE 1**). These data would help inform studies of the oil's impacts on marine life and human communities.

Tracking movement of the oil underwater was challenging. University of Georgia biochemist Mandy Joye, who had studied natural seeps in the Gulf for years, documented that the leaking wellhead was creating a plume of oil the size of Manhattan. She also found evidence of low oxygen concentrations, or hypoxia (p. 104), because some bacteria consume oil and gas, depleting oxygen from the water and making it uninhabitable for fish and other creatures.

Joye and other researchers feared that the thinly dispersed oil might prove devastating to plankton (the base of the marine food chain) and to the tiny larvae of shrimp, fish, and oysters (the pillars of the fishing industry). Scientists taking water samples documented sharp drops in plankton during the spill, but it will take years to learn whether the impact on larvae will diminish populations of adult fish and shellfish. Studies on the condition of living fish in the region show gill damage, tail rot, lesions, and reproductive problems at much higher levels than is typical.

What was happening to life on the seafloor was a mystery, because only a handful of submersible vehicles in the world are able to travel to the crushing pressures of the deep sea. Luckily, a team of researchers led by Charles Fisher of Penn State University was scheduled to embark on a regular survey of deepwater coral across the Gulf of Mexico in late 2010. Using the three-person submersible *Alvin* and the robotic vehicles *Jason* and *Sentry*, the team found healthy coral communities at sites far away from the Macondo well but found dying corals and brittlestars covered in a brown material at a site 11 km from the Macondo well.

Eager to determine whether this community was contaminated by the BP oil spill, the research team added chemist Helen White of Haverford College and returned a month later, thanks to a National Science Foundation program that

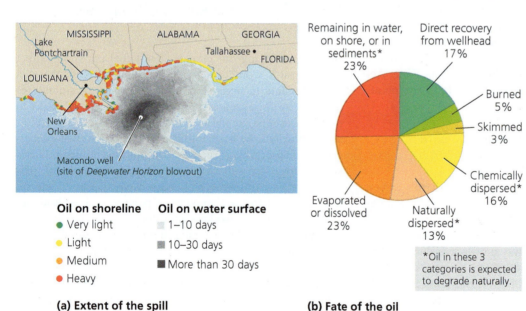

(a) Extent of the spill

Oil on shoreline
- Very light
- Light
- Medium
- Heavy

Oil on water surface
- 1–10 days
- 10–30 days
- More than 30 days

(b) Fate of the oil

Remaining in water, on shore, or in sediments* 23%
Direct recovery from wellhead 17%
Burned 5%
Skimmed 3%
Chemically dispersed* 16%
Naturally dispersed* 13%
Evaporated or dissolved 23%

*Oil in these 3 categories is expected to degrade naturally.

FIGURE 1 Scientists helped track oil from the *Deepwater Horizon* spill. The map **(a)** shows areas polluted by oil. The pie chart **(b)** gives a breakdown of the oil's fate. *Source: (a) NOAA Office of Response and Restoration; (b) NOAA.*

funds rapid response research. On this trip, chemical analysis of the brown material showed it to match oil from the BP spill.

Other questions revolve around impacts of the chemical dispersant that BP used to break up the oil, a compound called Corexit 9500. Work by biologist Philippe Bodin following the *Amoco Cadiz* oil spill in France in 1978 had found that Corexit 9500 appeared more toxic to marine life than the oil itself. BP threw an unprecedented amount of this chemical at the *Deepwater Horizon* spill, injecting a great deal directly into the path of the oil at the wellhead. This caused the oil to dissociate into trillions of tiny droplets that dispersed across large regions. Many scientists worried that this expanded the oil's reach, affecting more plankton, larvae, and fish.

Impacts of the oil on birds, sea turtles, and marine mammals were less difficult to assess, and hundreds of these animals were cleaned and saved by wildlife rescue teams. Officially confirmed deaths numbered 6104 birds, 605 turtles, and 97 mammals, but a much larger, unknown, number surely succumbed to the oil and were never found. What impacts this mortality may have on populations in coming years is unclear. (After the *Exxon Valdez* spill in Alaska in 1989, populations of some species rebounded, but populations of others have never come back.) Researchers are following the movements of marine animals in the Gulf with radio transmitters to try to learn what effects the oil may have had.

As images of oil-coated marshes saturated the media, researchers worried that widespread death of marsh grass would leave the shoreline vulnerable to severe erosion by waves. Fortunately, researchers found that oil did not penetrate to the roots of most plants and that oiled grasses were sending up new growth. Indeed, Louisiana State University researcher Eugene Turner said that loss of marshland from the oil "pales in comparison" with coastal wetlands lost each year in Louisiana due to other factors, including dredging, sea level rise, sinking of land due to the removal of oil and gas, and silt capture by dams on the Mississippi River.

The ecological damage caused by the spill had measurable consequences for people. The region's mighty fisheries were shut down, forcing thousands of fishermen out of work. The government tested fish and shellfish for contamination and reopened fishing once they were found to be safe, but consumers still balked at buying Gulf seafood. Beach tourism remained low all summer as visitors avoided the region. Together, losses in fishing and tourism totaled billions of dollars.

Scientists expect some impacts of the Gulf spill to be long-lasting. Oil from the similar *Ixtoc* blowout off Mexico's coast in 1979 still lies in sediments near dead coral reefs, and fishermen there say it took 15–20 years for catches to return to normal. After the *Amoco Cadiz* spill, it took seven years for oysters and other marine species to recover. In Alaska, oil from the 1989 *Exxon Valdez* spill remains embedded in beach sand today.

However, many researchers are hopeful about the Gulf of Mexico's recovery from the *Deepwater Horizon* spill. The Gulf's warm waters and sunny climate speed the natural breakdown of oil. In hot sunlight, volatile components of oil evaporate from the surface and degrade in the water, so that fewer toxic compounds reach marine life. In addition, bacteria that consume hydrocarbons thrive in the Gulf because some oil has always seeped naturally from the seafloor and because leakage from platforms, tankers, and pipelines is common. These microbes give the region a natural self-cleaning capacity.

Researchers continue to conduct a wide range of scientific studies (**FIGURE 2**). A consortium of federal and state agencies is coordinating research and restoration efforts in the largest ever Natural Resource Damage Assessment, a process mandated under the Oil Pollution Act of 1990. Answers to questions will come to light as long-term impacts become clear.

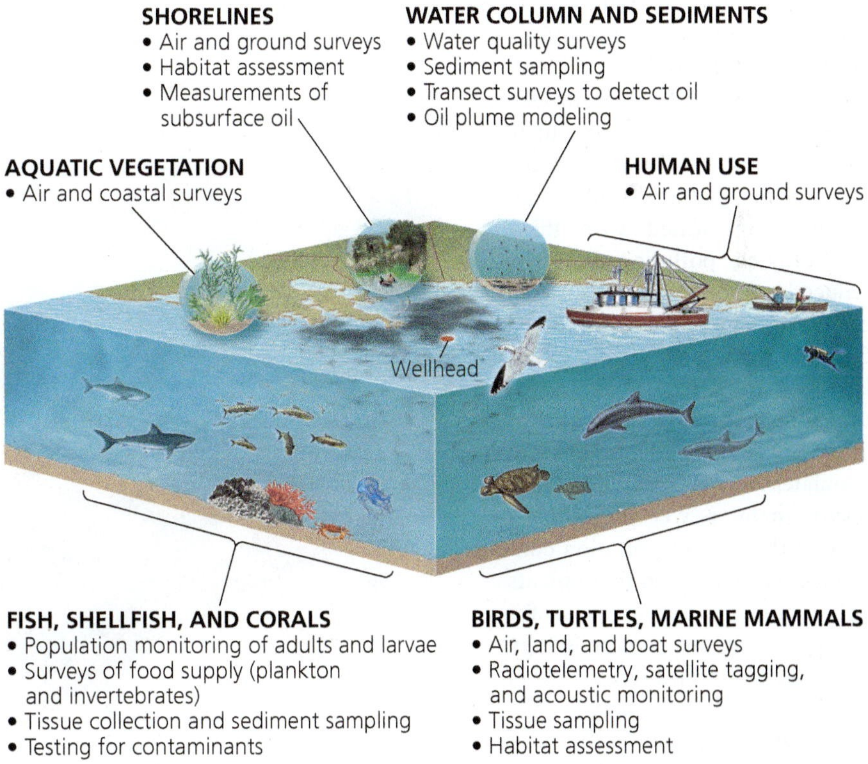

SHORELINES
- Air and ground surveys
- Habitat assessment
- Measurements of subsurface oil

WATER COLUMN AND SEDIMENTS
- Water quality surveys
- Sediment sampling
- Transect surveys to detect oil
- Oil plume modeling

AQUATIC VEGETATION
- Air and coastal surveys

HUMAN USE
- Air and ground surveys

Wellhead

FISH, SHELLFISH, AND CORALS
- Population monitoring of adults and larvae
- Surveys of food supply (plankton and invertebrates)
- Tissue collection and sediment sampling
- Testing for contaminants

BIRDS, TURTLES, MARINE MAMMALS
- Air, land, and boat surveys
- Radiotelemetry, satellite tagging, and acoustic monitoring
- Tissue sampling
- Habitat assessment

FIGURE 2 Scientific research is helping to assess damage to natural resources from the *Deepwater Horizon* oil spill. Thousands of researchers are surveying habitats, collecting samples and testing them in the lab, tracking wildlife, monitoring populations, and more.

has encouraged industry to develop and install technologies that reduce pollution, such as catalytic converters that cleanse vehicle exhaust (see Figure 17.16, p. 456).

Clean coal technologies aim to reduce air pollution from coal

At coal-fired power plants, scientists and engineers are seeking ways to cleanse coal exhaust of sulfur, mercury, arsenic, and other impurities. **Clean coal technologies** refer to an array of techniques, equipment, and approaches that aim to remove chemical contaminants during the generation of electricity from coal. Among these technologies are various types of scrubbers, devices that chemically convert or physically remove pollutants (see Figure 17.15, p. 456). Some scrubbers use minerals such as magnesium, sodium, or calcium in reactions to remove sulfur dioxide (SO_2) from smokestack emissions. Others use chemical reactions to strip away nitrogen oxides (NO_x), breaking them down into elemental nitrogen and water. Multilayered filtering devices are used to capture tiny ash particles.

Another "clean coal" approach is to dry coal that has high water content to make it cleaner-burning. We can also gain more power from coal with less pollution through gasification, in which coal is converted into a cleaner synthesis gas (called *syngas*) by reacting it with oxygen and steam at a high temperature. Syngas from coal can be used to turn a gas turbine and also to heat water to turn a steam turbine.

The U.S. government and the coal industry have each invested billions of dollars in clean coal technologies for new power plants. These efforts have helped to reduce air pollution from sulfates, nitrogen oxides, mercury, and particulate matter (pp. 455–457). If the many older plants that still pollute our air were retrofitted with these technologies, pollution could be reduced even more. This is a goal of the Obama administration's Clean Power Plan (pp. 464, 506), launched in 2015. However, the coal industry spends a great deal of money fighting regulations and mandates on its practices. As a result, many power plants have little in the way of pollution control technologies, and these plants will continue polluting our air for decades. Moreover, many energy analysts emphasize that "clean coal" technologies may make for *cleaner* coal but will never result in energy that is completely clean. Some argue that coal is an inherently dirty way of generating power and should be replaced outright with cleaner energy sources.

Can we capture and store carbon?

Even if clean coal technologies were able to remove every last contaminant from power plant emissions, coal combustion would still pump huge amounts of carbon dioxide into the air, intensifying the greenhouse effect and worsening climate change. This is why many current efforts focus on **carbon capture and storage** (CCS; p. 504). This approach involves capturing carbon dioxide emissions, converting the gas to a liquid form, and then sequestering (storing) it in the ocean or underground in a geologically stable rock formation (**FIGURE 19.21**).

Carbon capture and storage is being attempted at a variety of facilities. The world's first coal-fired power plant to approach zero emissions opened in 2008 in Germany. This plant removes its sulfate pollutants and captures its carbon dioxide, then compresses the CO_2 into liquid form, trucks it away, and injects it 900 m (3000 ft) underground into a depleted natural gas field.

In North Dakota, the Great Plains Synfuels Plant gasifies its coal, then sends half the CO_2 through a pipeline into Canada, where a Canadian oil company injects it into an oilfield to help it pump out the remaining oil. The North Dakota plant also captures, isolates, and sells seven other types of gases for various purposes.

The highest-profile effort at carbon capture and storage has suffered a rocky history. Beginning in 2003, the U.S. Department of Energy teamed up with seven energy companies to build a prototype of a near-zero-emissions coal-fired power plant. The *FutureGen* project (and its successor, the

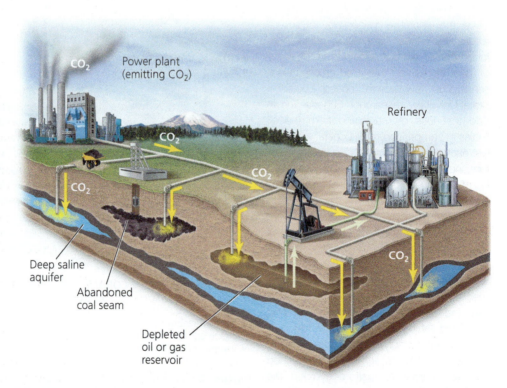

FIGURE 19.21 Carbon capture and storage schemes propose to inject liquefied carbon dioxide emissions underground. The CO_2 may be injected into depleted fossil fuel deposits, deep saline aquifers, or oil or gas deposits undergoing secondary extraction.

$1.65 billion *FutureGen 2.0*) aimed to design, construct, and operate a power plant that burns coal to generate electricity, captures 90% of its carbon dioxide emissions, and sequesters the CO_2 deep underground beneath layers of impermeable rock. It was hoped that this showcase project, located in downstate Illinois, could be a model for a new generation of power plants across the world, but financial challenges plagued it, and in 2015 the project was suspended.

At this point, carbon capture and storage remains too unproven to be the central focus of a clean energy strategy. We do not know how to ensure that carbon dioxide will stay underground once injected there. Injection might in some cases contaminate groundwater supplies or trigger earthquakes. Injecting carbon dioxide into the ocean would further acidify its waters (pp. 433, 498). Moreover, CCS is energy-intensive and decreases the EROI of coal, adding to its cost and the amount we consume. Finally, many renewable energy advocates fear that pursuing the CCS approach takes the burden off emitters and prolongs our dependence on fossil fuels rather than facilitating a shift to renewables.

We all pay external costs

The costs of addressing the many health and environmental impacts of fossil fuel extraction and use are generally not internalized in the market prices of fossil fuels. Instead, we all pay these external costs (pp. 141, 162) through medical expenses, costs of environmental cleanup, and impacts on our quality of life. Moreover, the prices we pay at the gas pump or on our monthly utility bill do not even cover the financial costs of fossil fuel extraction. Rather, fossil fuel prices have been kept inexpensive as a result of government subsidies and tax breaks to extraction companies. The profitable and well-established fossil fuel industries still receive far more financial support from taxpayers than do the young and struggling renewable energy sources (Figure 7.14, p. 179; Figure 21.7, p. 584). In this way, we all pay extra for fossil fuel energy through our taxes, generally without even realizing it.

Fossil fuel extraction has mixed consequences for local people

Wherever fossil fuels are extracted, people living nearby face the environmental, health, and social impacts of extractive development. When local people are allowed to participate in this development, however, they may also gain substantial financial benefits. In North America, communities where fossil fuel extraction takes place often experience a flush of high-paying jobs and economic activity, and for many people these economic benefits far outweigh other concerns. Perceptions can change with time, however. Economic booms generally prove temporary, lasting only as long as the resource holds out and market prices for fuel stay high. At some point all booms go bust, and once the jobs and money have left a community, its residents may be left with the legacy of a polluted environment for generations to come.

As an example, Fort McMurray has been the hub of Alberta's oil sands boom. This remote northern frontier town's population skyrocketed from 2000 in the 1960s to 80,000 as people flocked there looking for jobs. An additional 40,000 oilfield workers were being flown in and housed to work shifts, 3-weeks on, 10 days off, at the height of the recent boom. Most incoming residents have been men, averaging 32 years of age, and the city boasts the highest birth rate in Canada. Salaries are high, but so are rents and home prices. Like all boomtowns, Fort McMurray's population outgrew its infrastructure, and services became stretched thin. And like all boomtowns, it experienced a bust when the price of its principal resource fell. For Fort McMurray, the bust came in 2015 as world oil prices fell amid oversupply. Unable to make a profit on oil sands at the low prices, companies began cutting off contracts and orders and laying off workers. Across Alberta in 2015, more than 35,000 people in the energy sector lost their jobs. Still, after $200 billion of investment, many companies are in the game for the long haul, and are continuing to extract oil sands, hoping that prices will soon rise. As oil prices fluctuate in future years, Fort McMurray may see more booms and more busts.

Along the route the oil would have taken out of Alberta and through the United States via the proposed Keystone XL pipeline, TransCanada estimated that the project would have created 6500 construction jobs for two years plus 7000 one-year jobs for manufacturers of supplies. For landowners, the pipeline project would have had mixed consequences. TransCanada negotiated with thousands of landowners along the Keystone XL route, offering them money for the right to install the pipeline across their land. Many were happy to accept payments, but landowners who declined TransCanada's offers found their land rights taken away by **eminent domain**—the policy by which courts set aside private property rights to make way for projects judged to be for the public good. The landowner is paid an amount determined by a court to be fair and cannot appeal the decision.

In Alaska, the oil industry gains support for drilling by paying the Alaskan government a portion of its revenues. Since the 1970s, the state of Alaska has received more than $70 billion in oil revenues. One-quarter of these revenues are placed in the Permanent Fund, an investment fund that pays yearly dividends to all citizens. Since 1982, each Alaska resident has received annual payouts ranging from $331 to $2072.

Such distribution of revenue among citizens is unusual; in most parts of the world where fossil fuels are extracted, local residents suffer pollution without compensation. When multinational corporations pay developing nations for access to extract oil or gas, the money generally does not trickle down from government officials to the people who live where

weighing the
ISSUES

Clean Coal and Carbon Capture

Do you think we should spend billions of dollars to try to find ways to burn coal more cleanly and to sequester carbon emissions from fossil fuels? Or is our money better spent on developing new clean and renewable energy sources, even though they don't yet have enough infrastructure to generate power at the scale that coal can? What pros and cons do you see in each approach?

the extraction takes place. Moreover, oil-rich developing nations such as Ecuador, Venezuela, and Nigeria tend to have few environmental regulations, and governments may not enforce regulations if doing so would jeopardize losing the large sums of money associated with oil development.

In Ecuador, local people brought suit against Chevron for environmental and health impacts from years of oil extraction in the nation's rainforests. An Ecuadorian court in 2011 found the oil company guilty and ordered it to pay $9.5 billion for cleanup—the largest-ever such judgment. Chevron refused, and the court battle proceeded to the United States, where a judge threw out the ruling. The ongoing legal battle has now moved to other nations.

In Nigeria, the Shell Oil Company extracted $30 billion of oil from land of the native Ogoni people. Oil spills, noise, and gas flares caused chronic illness among them, but oil profits went to Shell and to Nigeria's military dictatorships, while the Ogoni remained in poverty with no running water or electricity. Ogoni activist Ken Saro-Wiwa worked for fair compensation to the Ogoni. After 30 years of persecution by Nigeria's government, he was arrested in 1994 on what many believed were trumped-up charges of involvement in four murders, given a trial universally regarded as a sham, and put to death by military tribunal.

Wherever in the world fossil fuel extraction comes to communities, people find themselves divided over whether the short-term economic benefits are worth the long-term health and environmental impacts. Today this debate is occurring in North Dakota and many parts of the West in response to oil and gas drilling, and in Pennsylvania, New York, and other states above the Marcellus Shale where the petroleum industry is fracking for gas (Chapter 7). The debate has gone on for years in Appalachia over mountaintop removal mining. And along the Gulf of Mexico, the oil and gas industry employs 100,000 people and helps fund local economies—yet still more people are employed in tourism, fishing, and service industries, all of which were hurt by the *Deepwater Horizon* spill. There are no easy answers, but impacts would be lessened if extraction industries were to put more health and environmental safeguards in place for workers and residents.

Dependence on foreign energy affects economic security

Putting all your eggs in one basket is always a risky strategy. Because virtually all our modern technologies and services depend in some way on fossil fuels, we are susceptible to supplies becoming costly or unavailable. Nations that lack adequate fossil fuel reserves of their own are especially vulnerable. For instance, nations such as Germany, France, South Korea, and Japan consume far more energy than they extract and thus rely almost entirely on imports (**FIGURE 19.22**). In the wake of its 1970 oil extraction peak, the United States began relying more on foreign supplies.

Such reliance means that seller nations can control energy prices, forcing buyer nations to pay more as supplies dwindle. This became clear in 1973, when the **Organization of Petroleum Exporting Countries (OPEC)** resolved to

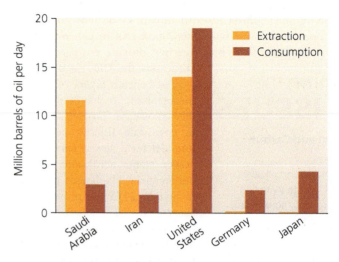

FIGURE 19.22 The United States, Germany, and Japan are among nations that consume more oil than they extract. Saudi Arabia and Iran extract more oil than they consume and are able to export oil to high-consumption countries. *Data from U.S. Energy Information Administration.*

DATA Q For every barrel of oil extracted in the United States, how many barrels are consumed in the United States?

Go to **Interpreting Graphs & Data** on **Mastering**EnvironmentalScience°

stop selling oil to the United States. The predominantly Arab nations of OPEC opposed U.S. support for Israel in the Arab–Israeli Yom Kippur War and sought to raise prices by restricting supply. OPEC's embargo created panic in the West and caused oil prices to skyrocket (**FIGURE 19.23**), spurring

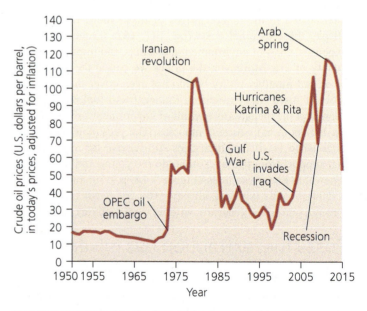

FIGURE 19.23 World oil prices have gyrated over the decades. Often this has resulted from political and economic events in oil-extracting countries, particularly in the Middle East. Prices shown are adjusted for inflation and expressed in 2015 dollars. *Data from BP p.l.c., 2016. Statistical review of world energy 2016.*

inflation. Fear of oil shortages drove Americans to wait in long lines at gas pumps. A similar supply shock occurred in 1979 in response to the Iranian revolution.

Today, experts attribute the recent fall in global oil prices to increased supply from hydraulic fracturing in the United States, followed by decisions by Saudi Arabia and OPEC to boost their own production, apparently hoping to keep prices low and drive new Western competitors out of business.

With the majority of world oil reserves located in the politically volatile Middle East, crises in this region are a constant concern for Western policymakers. The democratic street uprisings of the "Arab Spring" in 2011 put leaders of Western nations in an awkward position, because they had long supported many of the Middle East's autocratic rulers. These rulers had facilitated Western access to oil, even as they suppressed democracy in their own societies. The Arab Spring uprisings were part of a long history of events that have affected oil prices, stretching back through the two U.S.-led wars in Iraq, the Iran–Iraq war of the 1980s, and the 1973 OPEC embargo, and extending forward into recent events in Syria, negotiations with Iran over its nuclear program, and the rise of the terrorist Islamic State (ISIS).

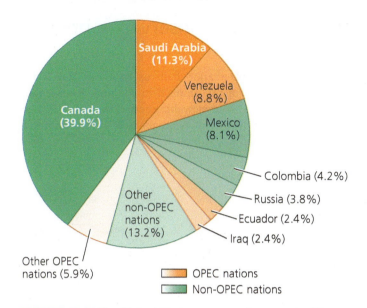

FIGURE 19.24 **The United States now receives most of its imported oil from non-OPEC nations and from non–Middle Eastern nations.** *Data from U.S. Energy Information Administration.*

weighing the
ISSUES

Drill, Baby, Drill?

Do you think the United States should encourage hydraulic fracturing by subsidizing it and loosening regulations? Do you think the United States should open more of its offshore waters to oil and gas extraction? In each case, what benefits and costs do you foresee? Would the benefits likely exceed the costs? How strongly should government regulate oil and gas extraction once drilling begins? Give reasons for your answers.

For U.S. leaders, the troubled history of the Middle East enhances the allure of using hydraulic fracturing to boost domestic oil and gas extraction. By supplying more of its own energy, the thinking goes, the United States gains leverage internationally and becomes less susceptible to foreign entanglements. U.S. proponents of the Keystone XL pipeline used this reasoning to argue for importing petroleum from the oil sands of Canada, a stable, friendly, democratic neighboring country. However, as President Obama and others pointed out, this oil would lessen U.S. reliance on Middle Eastern nations only if it were consumed at home, whereas the plan had always been to export it. In fact, in recent years the United States has already diversified its sources of imported petroleum considerably and now receives most from non–Middle Eastern nations, including Canada, Mexico, Venezuela, and Colombia (**FIGURE 19.24**).

Diversifying sources of foreign oil was one way in which the United States responded to the 1973 embargo. U.S. leaders also enacted conservation measures, funded research into renewable energy, and established an emergency stockpile (which today stores one month of oil) deep underground in salt caverns in Louisiana, called the Strategic Petroleum Reserve. They also called for secondary extraction at old wells and encouraged the development of additional domestic sources.

Since then, the desire to reduce reliance on foreign oil by boosting domestic supply has driven the expansion of offshore drilling into deeper and deeper water. It has repeatedly driven a proposal to open the Arctic National Wildlife Refuge in Alaska to oil extraction, despite arguments that drilling there would spoil America's last true wilderness while adding little to the nation's oil supply. Today it is driving the push to drill for oil in Arctic waters, despite the risks. As we extend our reach for fossil fuels, our society will continue to debate the complex mix of social, political, economic, and environmental costs and benefits.

Energy Efficiency and Conservation

Because fossil fuels are in limited supply, and because their use has health, environmental, political, and socioeconomic consequences, many people have concluded that fossil fuels are not a sustainable long-term solution to our energy needs. Many see a need to shift to clean and renewable sources of energy that exert less impact on climate and human health. As our society transitions to renewable energy sources, it will benefit us to extend the availability of our fossil fuel resources. We can do so by conserving energy and by improving energy efficiency.

Energy efficiency describes the ability to obtain a given amount of output while using less energy input. **Energy conservation** describes the practice of reducing wasteful or unnecessary energy use. In general, efficiency results from technological improvements, whereas conservation stems from behavioral choices. Because greater efficiency allows us to reduce energy use, efficiency is a primary means of conservation.

Efficiency and conservation bring benefits

Efficiency and conservation help us to waste less and to reduce our environmental impact. In addition, by extending the lifetimes of our nonrenewable energy supplies, efficiency and conservation help to alleviate many of the difficult individual choices and divisive societal debates related to fossil fuels, from oil sands development to Arctic drilling to hydraulic fracturing.

Americans use far more energy per person than people in most other nations (FIGURE 19.25a). Citizens of many European nations enjoy standards of living similar to those of U.S. citizens yet use less energy per capita. This indicates that Americans could reduce their energy consumption considerably without diminishing their quality of life. Indeed, per-person energy consumption *has* declined slightly in the United States over the past three decades (see data line in Figure 19.26a)—and this occurred during a period of sustained economic growth.

During this period the United States also cut in half its **energy intensity,** or energy use per dollar of Gross Domestic Product (GDP) (FIGURE 19.25b). Lower energy intensity indicates greater efficiency, and these data show that the United States now gets twice as much economic bang for its energy buck. Thus, although the United States continues to burn through more energy per dollar of GDP than most other industrialized nations, Americans have achieved tremendous gains in efficiency already and should be able to make further progress.

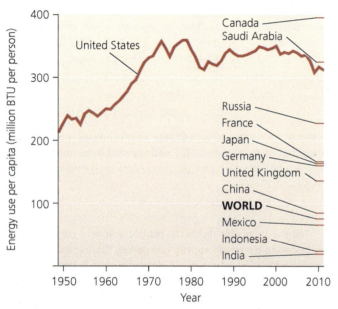

(a) Per capita energy consumption

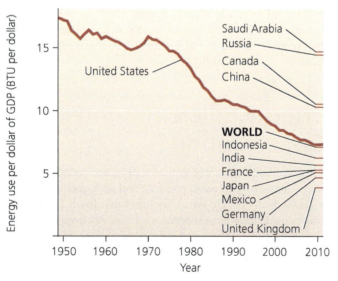

(b) Energy intensity

FIGURE 19.25 The United States trails other developed nations in energy efficiency but has made much progress. U.S. per-person energy use **(a)** has fallen slightly since 1979 but remains greater than that of most other nations. U.S. energy intensity **(b)** has fallen steeply and now approaches that of other developed nations. Energy intensity is energy use per inflation-adjusted 2005 dollars of GDP, using purchasing power parities, which control for differences among nations in purchasing power. *Data from U.S. Energy Information Administration.*

Personal actions and efficient technologies are two routes to conservation

As individuals, we can make conscious choices to reduce our energy consumption by driving less, dialing down thermostats, turning off lights when rooms are not in use, and investing in more energy-efficient devices and appliances. For any given individual or business, reducing energy consumption saves money while helping to conserve resources.

As a society, we can conserve energy by developing technologies and strategies to make energy-consuming devices and processes more efficient. Currently, more than two-thirds of the fossil fuel energy we use is simply lost, as waste heat, in automobiles and power plants (see Figure 19.4).

One way we can improve the efficiency of our power plants is through **cogeneration,** in which excess heat produced during the generation of electricity is captured and used to heat nearby workplaces and homes and to produce other kinds of power. Cogeneration can almost double the efficiency of a power plant. The same is true of coal gasification and combined cycle generation. In this process, coal is treated to create hot gases that turn a gas turbine, while exhaust from this turbine heats water to drive a conventional steam turbine.

In homes, offices, and public buildings, a significant amount of heat is needlessly lost in winter and gained in summer because of poor design and inadequate insulation (FIGURE 19.26). Improvements in design can reduce the energy required to heat and cool buildings. Such improvements may involve passive solar design (p. 584), better insulation, a building's location, the vegetation around it, and even the color of its roof (lighter colors keep buildings cooler by reflecting the sun's rays).

FIGURE 19.26 A thermogram reveals heat loss from buildings. It records energy in the infrared portion of the electromagnetic spectrum (p. 30). In this image, one house is uninsulated; its red color signifies warm temperatures where heat is escaping. Green shades signify cool temperatures, where heat is being conserved. Also note that in all houses, more heat is escaping from windows than from walls.

In most homes and businesses, several percent of electricity use is squandered to "vampire" power loss, energy that electronic devices in standby mode consume while switched off. However, recent government regulation has limited how much standby power devices can use, and this is reducing vampire power loss.

Many consumer products, from lightbulbs to appliances, have been reengineered through the years to enhance efficiency. Energy-efficient lighting, for example, can reduce energy use by 80%. Compact fluorescent bulbs are much more efficient than incandescent lightbulbs, which is why the United States and many other nations are phasing out incandescent bulbs.

Federal standards for energy-efficient appliances have already reduced per-person home electricity use below what it was in the 1970s. The U.S. EPA's Energy Star program (p. 504) labels refrigerators, dishwashers, and other appliances for their efficiency, enabling consumers to take energy use into account when shopping for them. Studies show that savings on utility bills readily offset the higher prices of energy-efficient appliances. The EPA estimates that if all U.S. households purchased energy-efficient appliances, the nation's annual energy expenditure would fall by $200 billion.

Automotive technology represents perhaps our best opportunity to conserve large amounts of fossil fuels fairly easily. We can accomplish this with vehicles such as electric cars, electric/gasoline hybrids, plug-in hybrids, or vehicles that use hydrogen fuel cells (p. 598). Among electric/gasoline hybrids, current U.S. models obtain fuel-economy ratings of up to about 50 miles per gallon (mpg)—twice that of the average American car. Many fully electric vehicles now obtain fuel-economy ratings of over 100 mpg. Automakers also possess

the means to enhance fuel efficiency for traditional gasoline-powered vehicles further by using lightweight materials, continuously variable transmissions, and more efficient engines.

Automobile fuel efficiency is a key to conservation

One of the ways in which U.S. leaders responded to the OPEC embargo of 1973 was to mandate an increase in the fuel efficiency of automobiles. Over the next three decades, however, as market prices for oil fell, many of the conservation initiatives of this time were abandoned. Without high market prices and an immediate threat of shortages, people lacked economic motivation to conserve, and U.S. policymakers repeatedly failed to raise the corporate average fuel efficiency (CAFE) standards, which set benchmarks for auto manufacturers to meet. The average fuel efficiency of new vehicles fell from 22.0 mpg in 1987 to 19.3 mpg in 2004, as sales of sport-utility vehicles increased relative to sales of cars.

In 2007, Congress mandated that automakers raise average fuel efficiency to 35 mpg by the year 2020. Since 2007, fuel economy has climbed substantially (**FIGURE 19.27**). Yet despite this advance, American automobiles continue to lag behind the vehicles of most other developed nations. When automakers requested a government bailout during the recent recession, President Obama forced them to agree to boost average fuel economies to 54.5 mpg by 2025. If this improvement comes to pass, it will enable a huge reduction in oil use. New technologies will add more than $2000 to the average price of a car, but drivers will save perhaps $6000 in fuel costs over the car's lifetime.

In 2009, Congress and the Obama administration took another major step to improve automobile fuel efficiency

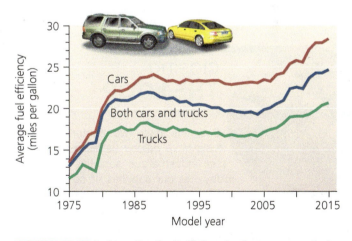

FIGURE 19.27 Automotive fuel efficiencies have responded to public policy. Fuel efficiency for automobiles in the United States rose dramatically in the late 1970s as a result of legislative mandates but then stagnated when no additional laws were enacted to improve fuel economy. Recent legislation is once again improving fuel efficiency. *Data from U.S. Environmental Protection Agency, 2015.* Light-duty automotive technology, carbon dioxide emissions, and fuel economy trends: 1975 through 2015.

while stimulating economic activity and saving jobs during a severe recession. The popular "Cash for Clunkers" program—formally named the Consumer Assistance to Recycle and Save (CARS) Act—paid Americans $3500 or $4500 each to turn in old vehicles and purchase newer, more fuel-efficient ones. The $3 billion program subsidized the sale or lease of 678,000 vehicles averaging 24.9 mpg that replaced vehicles averaging 15.8 mpg. It is estimated that 824 million gallons of gasoline are being saved as a result, preventing 9 million metric tons of greenhouse gas emissions and creating social benefits worth $278 million.

U.S. policymakers could do more to encourage oil conservation. The United States has kept taxes on gasoline extremely low, relative to most other nations. Americans pay two to three times *less* per gallon of gas than drivers in many European countries. In fact, gasoline in the United States is sold more cheaply than bottled water! As a result, U.S. gasoline prices do not account for the substantial external costs (pp. 141, 162) that oil extraction and consumption impose on society. Some experts estimate that if all costs to society were taken into account, the price of gasoline would exceed $13 per gallon. Instead, our artificially low gas prices diminish our economic incentive to conserve.

weighing the
ISSUES

More Miles, Less Gas

If you drive an automobile, what gas mileage does it get? How does it compare to the vehicle averages in Figure 19.27? If your vehicle's fuel efficiency were 10 mpg greater, and if you drove the same amount, how many gallons of gasoline would you no longer need to purchase each year? How much money would you save?

Do you think U.S. leaders should mandate further increases in the CAFE standards? Should the government raise taxes on gasoline sales as an incentive for consumers to conserve energy? What effects on economics, on health, and on environmental quality might each of these steps have?

The rebound effect cuts into efficiency gains

Energy efficiency is a vital pursuit, but it may not always save as much energy as we expect. This is because gains in efficiency from better technology may be partly offset if people engage in more energy-consuming behavior as a result. For instance, a person who buys a fuel-efficient car may choose to drive more because he or she feels it's okay to do so (or because he or she can afford to) now that less gas is being used per mile. This phenomenon is called the **rebound effect,** and studies indicate that it is widespread and significant. In some instances, the rebound effect may completely erase efficiency gains. As our society pursues energy efficiency, this will be an important factor to consider.

We need both conservation and renewable energy

Despite concerns over the rebound effect, energy efficiency and conservation efforts are vital to creating a sustainable future for our society. It is often said that reducing our energy use is equivalent to finding a new oil reserve. Some estimates hold that energy conservation and efficiency in the United States could save 6 million barrels of oil a day—nearly the amount gained from all offshore drilling, and considerably more than would be gained from Canada's oil sands. In fact, conserving energy is *better* than finding a new reserve because it alleviates health and environmental impacts while at the same time extending our future access to fossil fuels.

Yet regardless of how effectively we conserve, we will still need energy to power our civilization, and it will need to come from somewhere. Most energy experts maintain that the only sustainable way of guaranteeing ourselves a reliable long-term supply of energy is to ensure sufficiently rapid development of renewable energy sources (Chapters 20 and 21).

closing
THE LOOP

Over the past two centuries, fossil fuels have helped us build the complex industrialized societies we enjoy today. Yet our supplies of conventional fossil fuels are declining. We can respond to this challenge by expanding our search for new sources of fossil fuels—and paying ever-higher economic, health, and environmental costs. Or, we can encourage conservation and efficiency while developing alternative energy sources that are clean and renewable. The path we choose will have far-reaching consequences for human health and well-being, for Earth's climate, and for the stability and progress of our civilization.

The debate over the Canadian oil sands and the Keystone XL pipeline has been a microcosm of this larger conversation over our energy future. In all likelihood, the long-running debate over oil sands extraction and transport will have taken a few more turns by the time you read this book. We encourage you to research the latest developments, discuss them with your instructor, and wrestle thoughtfully with the issues.

Our modern energy situation is complex, but fortunately we are not caught in a simple trade-off between fossil fuels' economic benefits and their impacts on the environment, climate, and health. Instead, as renewable energy sources become increasingly well developed and economical, it becomes easier to envision freeing ourselves from a reliance on fossil fuels and charting a bright future for humanity and the planet with renewable energy.

REVIEWING Objectives

You should now be able to:

Identify the energy sources that we use

Many energy sources are available, but since the industrial revolution, nonrenewable fossil fuels—including coal, natural gas, and oil—have become our primary sources of energy. We are also developing alternative sources of clean and renewable energy. (pp. 517–519)

Discuss the value of the EROI concept

The concepts of *net energy* and *EROI* allow us to compare the amount of energy obtained from a source with the amount invested in its extraction and processing. This can help decide which energy sources are most efficient and productive. (pp. 519–520)

Describe the formation of major types of fossil fuels

Fossil fuels are formed very slowly as buried organic matter is chemically transformed by heat, pressure, and/or anaerobic decomposition. Coal results from organic matter that undergoes compression but little decomposition. Crude oil is a thick, liquid mixture of hydrocarbons formed under high temperature and pressure. Natural gas consists mostly of methane and can be formed in two ways. Oil sands contain bitumen, a tarry substance formed from oil degraded by bacteria, whereas shale oil and methane hydrate are other unconventional sources with future potential. (pp. 520–524)

Explain how we extract, process, and use fossil fuels

Scientists locate fossil fuel deposits by analyzing subterranean geology. Coal is mined underground and strip-mined from the surface, whereas we drill wells to pump out oil and gas. Oil sands may be strip-mined or dissolved underground and extracted through well shafts. Refineries separate components of crude oil to transform it into a wide variety of fuel types. Coal and natural gas are used today principally to generate electricity. Oil powers transportation and also is used to create a diversity of petroleum-based products. (pp. 522–527)

Assess concerns over the future decline of conventional oil supplies

Any nonrenewable resource can be depleted, and we have depleted nearly half the world's conventional oil. Once we pass the peak of oil extraction, the gap between rising demand and falling supply may pose immense challenges for society. (pp. 527–529)

Outline ways in which we are extending our reach for fossil fuels

The conventional fossil fuels that remain unexploited are harder to access and more expensive to extract, so we are applying technology and new approaches to extract them, including mountaintop removal mining, secondary extraction, directional drilling, hydraulic fracturing, deep offshore drilling, and Arctic exploration. We are also beginning to mine unconventional fossil fuels such as oil sands. (pp. 529–532)

Examine environmental impacts of fossil fuel use, and explore ways to minimize these impacts

In various ways, coal mining, oil and gas drilling, and oil sands extraction can alter ecosystems, pollute air and waterways, and pose health risks. Hydraulic fracturing magnifies some of these impacts. Fuels can also spill during transport, affecting soil, water, wildlife, and health. Emissions from fossil fuel combustion pollute air, pose human health risks, and drive global climate change. Public policy and advances in pollution control technology have reduced many of these emissions, but much more remains to be done. If we could safely capture carbon dioxide and sequester it underground, this would resolve a primary drawback of fossil fuels, but carbon capture and storage remains unproven so far. (pp. 532–539)

Assess political, social, and economic aspects of fossil fuel use

Fossil fuels impose a variety of external costs. Fossil fuel extraction creates jobs but leaves pollution. People living in areas of extraction experience a range of consequences. Today's societies are heavily reliant on fossil fuel energy, and nations that consume far more fossil fuels than they extract are especially vulnerable to supply restrictions. (pp. 539–541)

Discuss strategies for conserving energy and enhancing efficiency

Energy conservation involves both personal choices and efficient technologies. Efficiency in power plant combustion, lighting, consumer appliances, and automotive fuel efficiency all play roles in conserving energy. The rebound effect, however, can partly negate our conservation efforts. Conservation lengthens our access to fossil fuels and reduces environmental impact, but to build a sustainable society we will also need to shift to renewable energy sources. (pp. 541–544)

TESTING Your Comprehension

1. Name one reason that fossil fuels are our most prevalent source of energy today. Why are fossil fuels considered nonrenewable?

2. Describe how *net energy* differs from *energy returned on investment (EROI)*. Why are these concepts important when evaluating energy sources?

3. How are fossil fuels formed? How do environmental conditions determine which type of fossil fuel is formed in a given location? Why are fossil fuels often concentrated in localized deposits?

4. How do *technically recoverable* and *economically recoverable* amounts of a resource differ? Define the term *proven recoverable reserve*. Explain why the proven recoverable reserve of a given fuel could increase over time. Now explain why it could decrease.

5. Describe how coal is used to generate electricity.

6. How do we create petroleum products? Provide examples of several of these products.

7. Why do many experts think we are about to pass the global extraction peak for conventional oil? What consequences might there be for our society? What are two fundamental ways that we could respond to *peak oil*?

8. Describe three ways in which we are now extending our reach for fossil fuels. List several impacts (positive or negative) that these actions might have.

9. Give an example of a clean coal technology. Now describe how carbon capture and storage is intended to work.

10. Describe two specific examples of how technological advances can improve energy efficiency. Now describe one specific action you could take to conserve energy.

SEEKING Solutions

1. Describe three specific health or environmental impacts of fossil fuel extraction or consumption. For each impact, what steps could governments, industries, or individuals take to alleviate the impact?

2. Contrast the experiences of the Ogoni people of Nigeria with those of the citizens of Alaska. How have they been similar and different? Do you think businesses or governments should take steps to ensure that local people benefit from oil drilling operations? How could they do so?

3. Based on data in this chapter, explain two reasons why it is reasonable to expect that Americans should be able to use energy more efficiently in the future as the U.S. economy expands.

4. **CASE STUDY CONNECTION** Summarize the main arguments for and against the Keystone XL pipeline extension. What problems might the pipeline have helped solve? What problems might it have created? Do you personally think Canada should continue to develop Alberta's oil sands? Should the United States have approved construction of the Keystone XL pipeline extension? Give reasons for your answers.

5. **THINK IT THROUGH** You are the mayor of a rural Nebraska town along the route of the proposed Keystone XL pipeline, and a new U.S. presidential administration has reversed President Obama's decision and invited TransCanada to build the pipeline extension. Some of your town's residents are eager to have jobs they believe the pipeline will bring. Others are fearful that oil leaks from the pipeline could contaminate the water supply. Some of your town's landowners are looking forward to receiving payments from TransCanada for use of their land, whereas others dread the prospect of noise, pollution, and trees being cut on their property. If the company receives too much local opposition, it says it may move the pipeline route away from your town. What information would you seek from TransCanada, from your state regulators, and from scientists and engineers before deciding whether support for the pipeline is in the best interest of your town? How would you make your decision? How might you try to address the diverse preferences of your town's residents?

6. **THINK IT THROUGH** You are elected governor of the state of Florida as the federal government is debating opening new waters to offshore drilling for oil and natural gas. Drilling in Florida waters would create jobs for Florida citizens and revenue for the state in the form of royalty payments from oil and gas companies. However, there is always the risk of a catastrophic oil spill, with its ecological, social, and economic impacts. Would you support or oppose offshore drilling off the Florida coast? Why? What, if any, regulations would you insist be imposed on such development? What questions would you ask of scientists before making your decision? What factors would you consider in making your decision?

CALCULATING Ecological Footprints

Scientists at the Global Footprint Network calculate the energy component of our ecological footprint by estimating the amount of ecologically productive land and sea required to absorb the carbon released from fossil fuel combustion. This translates into 5.9 hectares (ha) of the average American's 8.2-ha ecological footprint. Another way to think about our footprint, however, is to estimate how much land would be needed to grow biomass with an energy content equal to that of the fossil fuel we burn.

Assume that you are an average American who burns about 6.1 metric tons of oil-equivalent in fossil fuels each year and that average terrestrial net primary productivity (p. 111) can be expressed as 0.0037 metric ton/ha/year. Calculate how many hectares of land it would take to supply our fuel use by present-day photosynthetic production.

	HECTARES OF LAND FOR FUEL PRODUCTION
You	1649
Your class	
Your state	
United States	

1. Compare the energy component of your ecological footprint calculated in this way with the 5.9 ha calculated using the method of the Global Footprint Network. Explain why results from the two methods may differ.

2. Earth's total land area is approximately 15 billion hectares. Compare this to the hectares of land for fuel production from the table.

3. In the absence of stored energy from fossil fuels, how large a human population could Earth support at the level of consumption of the average American, if all of Earth's area were devoted to fuel production? Do you consider this realistic? Provide two reasons why or why not.

MasteringEnvironmentalScience®

Conventional
Energy Alternatives

Upon completing this chapter, you will be able to:

- Discuss the reasons for seeking energy alternatives to fossil fuels

- Summarize the contributions to world energy supplies of conventional alternatives to fossil fuels

- Describe nuclear energy and explain how we harness it for electrical power

- Assess the benefits and drawbacks of nuclear power, and discuss the societal debate over this energy source

- Describe established and emerging sources and techniques involved in harnessing bioenergy, and assess bioenergy's benefits and shortcomings

- Outline the scale, methods, and impacts of hydroelectric power

Will Sweden Free Itself of Fossil Fuels?

> **Sweden will become one of the first fossil [fuel]-free . . . states in the world.**
> Swedish Prime Minister Stefan Löfven, 2015

> **If [Sweden] phases out nuclear power, then it will be virtually impossible for the country to keep its climate-change commitments.**
> Yale University economist William Nordhaus, 1997

Flashback to 1986: On the morning of April 28, alarms went off at a nuclear power plant in Sweden, as sensors detected unusually high levels of radiation. However, the radioactivity had not come from within the plant. Instead, it had spread through the atmosphere from the Soviet Union. The workers at this Swedish power plant had discovered the outside world's first evidence of the disaster at Chernobyl, more than 1200 km (750 mi) away in what is now the nation of Ukraine. Chernobyl's nuclear reactor had exploded two days earlier, the result of human error and unsafe engineering, but the Soviet government had not yet admitted it.

Fast-forward to 2011: On March 11, seismometers in Sweden and throughout the world recorded vibrations from the shaking of the massive Tohoku earthquake off the coast of Japan. Soon thereafter, a tsunami inundated the Japanese coast, destroying entire towns, killing more than 18,000 people, and disabling nuclear reactors at the Fukushima Daiichi power plant. For the next several weeks, the world would stand on edge as Japanese authorities frantically tried to control the leakage of radiation from the plant and avert further catastrophe.

The events at both Chernobyl and Fukushima had broad and long-lasting repercussions. Each altered the course of the world's use of nuclear energy, one of our main alternatives to fossil fuels. While people in Ukraine and Japan still struggle with the legacies of these events, the global debate over nuclear power affects all of us.

In Sweden in the days after Chernobyl, low levels of radioactive fallout rained down on the countryside, contaminating crops and cows' milk. For many Swedes, this confirmed the decision they had made collectively six years earlier, in a 1980 referendum, to phase out their country's nuclear power program, shutting down all nuclear plants by the year 2010.

But trying to phase out nuclear power proved difficult. Sweden had turned to nuclear power as it sought to reduce its use of coal, oil, and natural gas because of the health and environmental impacts of fossil fuels. Sweden became one of the few nations in the world to decrease its reliance on fossil fuels (**FIGURE 20.1**, p. 550). The nation has cut its fossil fuel use in half since the 1970s, while boasting a thriving economy that continues to provide its citizens with one of the highest living standards on Earth. Sweden accomplished this largely by replacing fossil-fuel-fired electricity with nuclear power—and

Using biogas to fuel vehicles in Kristianstad, Sweden ▲

549

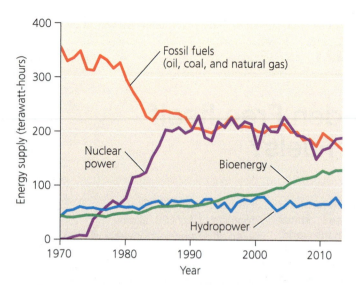

FIGURE 20.1 Sweden has greatly reduced its use of fossil fuels (mostly oil) since 1970. In their place, it has increased its use of nuclear power, hydroelectric power, and bioenergy. *Data from Swedish Energy Agency.*

DATA Q Based on the data in the graph, articulate a causal hypothesis regarding which energy source—nuclear power, hydropower, or bioenergy—was most responsible for taking the place of fossil fuels in Sweden. Explain the logic behind your hypothesis by referencing patterns in the data.

Go to **Interpreting Graphs & Data** on **Mastering**EnvironmentalScience®

as 2010 drew nearer, Sweden was still relying on 10 nuclear reactors for more than 30% of its energy supply and more than 40% of its electricity. If nuclear plants were to be shut down as planned, without reverting to fossil fuels, then the nation would need to boost its investment in alternative energy sources dramatically.

As a result, Sweden threw its weight behind research and development of renewable energy sources. Hydroelectric power was already supplying most of the rest of the nation's electricity and could not be expanded much more. Instead, the government hoped that energy from biomass sources and wind power could fill the gap. Manipulating market forces to make fossil fuels more expensive and renewable energy more affordable, Sweden boosted bioenergy and wind by applying a carbon tax (p. 509) to fossil fuels and by subsidizing renewable energy

through a certificate program in which producers and users of fossil fuel electricity were mandated to buy certificates from producers and users of renewable electricity.

Sweden soon built itself into an international leader in renewable energy, and today gets more than half its energy from renewable sources. Still, renewables took longer to develop than hoped, so policymakers repeatedly postponed the nuclear phaseout. In 2009, just before the original self-imposed deadline, the Swedish government announced that it was reversing its policy and would not phase out nuclear power after all. Instead, new power plants would be built as existing ones needed replacing.

In announcing this decision, Sweden's leaders said it would be fiscally and socially irresponsible to dismantle the nation's nuclear program without a ready replacement. Without an abundance of clean renewable power, a nuclear phaseout would mean a return to fossil fuels, or else would require cutting down immense areas of forest to combust biomass. Policymakers also cited Sweden's international obligations to hold down its carbon emissions under the Kyoto Protocol (p. 506). They reminded citizens that nuclear power is free of atmospheric pollution and is an effective way to minimize the greenhouse gas emissions that drive climate change.

Then Fukushima occurred. As the drama played out in Japan, anti-nuclear protestors all over the world staged demonstrations. Many national governments reassessed their commitments to nuclear power, ran safety checks of existing plants, and halted plans for new plants.

The Swedish government's shifting policies have reflected changing public sentiments. The majority of Swedes have always seen the clear benefits of fighting pollution and climate change by developing energy sources such as nuclear power, bioenergy, and hydropower, and they are proud of having moved so decisively away from fossil fuels. Yet each nuclear accident has eroded otherwise strong public support for nuclear power.

In 2015, Sweden's new prime minister, Stefan Löfven, announced that his government would aim one day to make Sweden completely free of fossil fuels. This proclamation was accompanied by an announcement of $550 million in new public funding toward renewable energy sources such as wind power, solar power, hydropower, and bioenergy. Time will tell the fate of nuclear power in Sweden, but that nation's citizens and leaders continue to support boosting the development of renewable energy sources to keep their nation a world leader in the shift away from fossil fuels.

Alternatives to Fossil Fuels

Fossil fuels drove the industrial revolution and helped to create the unprecedented material prosperity we enjoy today. Our global economy is powered largely by fossil fuels; more than 80% of our energy comes from oil, coal, and natural gas (**FIGURE 20.2a**). These three fuels also generate two-thirds of the world's electricity (**FIGURE 20.2b**). However, easily extractable supplies of these nonrenewable energy sources are in decline, and we are expending more and more energy and money to extract them (pp. 529–532). Moreover, the use of coal, oil, and natural gas drives global climate change and entails many other impacts to human health, society, and the environment (Chapters 17, 18, and 19).

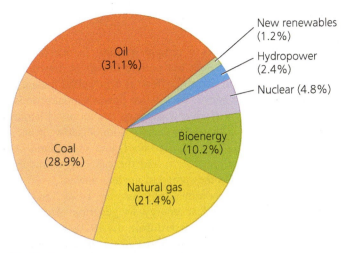

(a) World energy consumption, by source

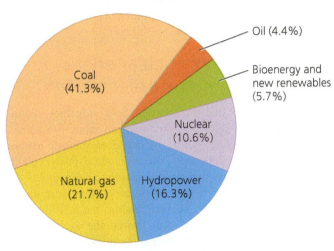

(b) World electricity generation, by source

FIGURE 20.2 Fossil fuels dominate the global energy supply. Together, oil, coal, and natural gas account for 81% **(a)** of the world's energy consumption. Nuclear power and hydroelectric power contribute substantially to global electricity generation **(b)**, but fossil fuels still power two-thirds of our electricity. *Data from International Energy Agency, 2015.* Key world energy statistics 2015. *Paris: IEA.*

For these reasons, most energy experts accept that we will need to shift from fossil fuels to energy sources that are less easily depleted and gentler on our health and environment. Developing alternatives to fossil fuels has the added benefit of helping to diversify an economy's mix of energy, thus lessening price volatility and dependence on foreign fuel imports.

We have developed a range of alternatives to fossil fuels (see Table 19.1, p. 517). Most of these energy sources are renewable, and most have less impact on health and the environment than oil, coal, or natural gas. Currently, most remain more expensive than fossil fuels, at least in the short term and when external costs (pp. 141, 162) are not included in market

prices. Yet as technologies develop and as we invest in infrastructure to better transmit power from renewable sources, prices will continue to fall, helping us to transition toward these alternative energy sources.

Nuclear power, bioenergy, and hydropower are conventional alternatives

Three alternative energy sources are currently the most developed and most widely used: nuclear power, hydroelectric power, and energy from biomass. These energy sources are well established, and each of them already plays a substantial role in our energy and electricity budgets. As a result, we can therefore describe nuclear power, hydropower, and biomass energy (bioenergy) as "conventional alternatives" to fossil fuels.

Each of these three conventional energy alternatives is generally considered to exert less environmental impact than fossil fuels but more impact than "new renewable" alternatives such as solar, wind, geothermal, and ocean power (Chapter 21). Yet as we will see, each of our conventional energy alternatives involves a unique and complex mix of benefits and drawbacks. Nuclear power is commonly considered a nonrenewable energy source, and hydropower and bioenergy are generally described as renewable, but the reality is more complicated.

Sweden has shown that it is possible for a wealthy and advanced economy to replace fossil fuels gradually with these conventional alternative sources while continuing to raise living standards for its citizens. Since 1970, Sweden has decreased its fossil fuel use from 81% to 30% of its national energy budget (see Figure 20.1). Today the three conventional alternatives—nuclear power, bioenergy, and hydropower—together provide Sweden with two-thirds of its energy and virtually all of its electricity.

Conventional alternatives provide much of our electricity

Across the globe, fuelwood and other bioenergy sources provide 10% of the world's energy, nuclear power provides about 5%, and hydropower provides about 2%. By comparison, the less established "new renewable" energy sources together account for just over 1% (see Figure 20.2a). Although their global contributions to our overall energy supply so far are minor, alternatives to fossil fuels do contribute greatly to our generation of electricity. Nuclear energy and hydropower together account for about 27% of the world's electricity generation (see Figure 20.2b).

Energy consumption patterns in the United States (**FIGURE 20.3a**, p. 552) are similar to those globally, except that the United States relies less on fuelwood and slightly more on fossil fuels and nuclear power than most other countries. A graph showing trends in energy consumption in the

United States over the past 65 years (**FIGURE 20.3b**) reveals two things. First, conventional alternatives play minor yet substantial roles in overall energy use. Second, use of conventional alternatives has been growing more slowly than use of fossil fuels.

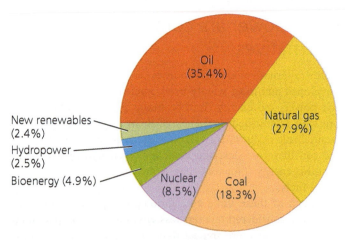

(a) U.S. energy consumption, by source

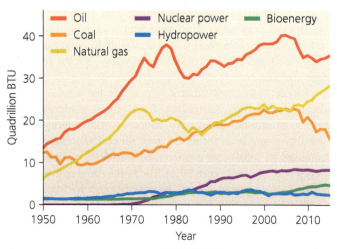

(b) U.S. energy consumption, 1950–2015

FIGURE 20.3 Fossil fuels predominate in the United States. Together, oil, natural gas, and coal account for 82% **(a)** of U.S. energy consumption. The line graph **(b)** shows amounts of each energy source used in the United States across the past 65 years. *Data from U.S. Energy Information Administration, 2016.*

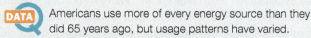 Americans use more of every energy source than they did 65 years ago, but usage patterns have varied.
- Since 1950, which energy source has grown the most, in absolute terms?
- Since 1950, which energy source has grown the most, in percentage terms?
- Which two energy sources grew fastest between 1950 and 1970?
- Which two energy sources grew fastest between 1970 and 1990?
- For each of the six energy sources shown, explain what has happened with their usage since 1990.

Go to **Interpreting Graphs & Data** on **Mastering**EnvironmentalScience®

Nuclear Power

Nuclear power occupies a unique position in our modern debate over energy. It does not pollute the air as does fossil fuel combustion, and thereby offers us a powerful means of combating climate change. Yet nuclear power's great promise has been clouded by nuclear weaponry, the thorny dilemma of radioactive waste disposal, and the long shadow of accidents at Three Mile Island, Chernobyl, and Fukushima. As a result, public safety concerns and the costs of addressing them have constrained the spread of nuclear power.

First developed commercially in the 1950s, **nuclear power** (the use of nuclear energy to generate electricity) experienced most of its growth during the 1970s and 1980s. The United States generates the most electricity from nuclear power of any nation—more than a quarter of the world's production—yet less than 20% of U.S. electricity comes from nuclear power. A number of other nations rely more heavily on nuclear power (**TABLE 20.1**). France leads the list, receiving 76% of its electricity from nuclear power.

Fission releases nuclear energy in reactors to generate electricity

Nuclear energy is the energy that holds together protons and neutrons (p. 23) in the nucleus of an atom. We harness this energy by converting it to thermal energy inside **nuclear reactors,** facilities contained within nuclear power plants. This thermal energy is then used to generate electricity.

The reaction that drives the release of nuclear energy inside nuclear reactors is **nuclear fission,** the splitting apart of atomic nuclei (**FIGURE 20.4**). In fission, the nuclei of large, heavy atoms, such as uranium or plutonium, are bombarded

TABLE 20.1 Top Producers of Nuclear Power

NATION	NUCLEAR POWER CAPACITY (gigawatts)	NUMBER OF REACTORS	ELECTRICITY FROM NUCLEAR POWER (%)
United States	99.2	99	19.5
France	63.1	58	76.3
Japan	42.4	48	0.5
China	26.8	31	3.0
Russia	25.4	35	18.6
South Korea	21.7	24	31.7
Canada	13.5	19	16.6
Ukraine	13.1	15	56.5
Germany	12.1	9	14.1
Sweden	9.6	10	34.3

Data from International Atomic Energy Agency.

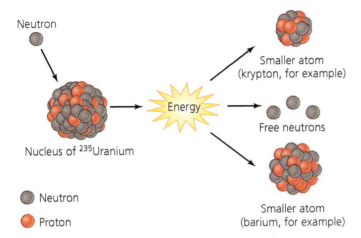

Neutron

Energy

Smaller atom
(krypton, for example)

Free neutrons

Nucleus of ^{235}Uranium

Smaller atom
(barium, for example)

● Neutron
● Proton

FIGURE 20.4 Nuclear fission drives modern nuclear power.
In nuclear fission, the nucleus of an atom of uranium-235 is
bombarded with a neutron. The collision splits the uranium atom
into smaller atoms and releases two or three neutrons, along with
energy in the form of heat, light, and radiation.

with neutrons. Ordinarily, neutrons move too quickly to
split nuclei when they collide with them, but if neutrons are
slowed down they can break apart nuclei. In a nuclear reactor, the neutrons bombarding uranium are slowed down with
a substance called a *moderator*, most often water or graphite.
Each split nucleus emits energy in the form of heat, light, and

radiation, and also releases neutrons. These resulting neutrons (two to three per nucleus in the case of uranium-235)
can in turn bombard other nearby uranium-235 (^{235}U) atoms,
resulting in a self-sustaining chain reaction.

If not controlled, this chain reaction becomes a runaway
process of positive feedback (p. 104)—the process that creates the explosive power of a nuclear bomb. Inside a nuclear
power plant, however, fission is controlled so that, on average, only one of the two or three neutrons emitted with each
fission event goes on to induce another fission event. To soak
up the excess neutrons produced when uranium nuclei divide,
control rods made of a metallic alloy that absorbs neutrons
are placed among the water-bathed fuel rods of uranium.
Engineers move these control rods into and out of the water
to maintain the fission reaction at the desired rate. In this way,
the chain reaction maintains a constant output of energy at a
controlled rate.

All this takes place within the reactor core and is the
first step in the electricity-generating process of a nuclear
power plant (**FIGURE 20.5**). The reactor core is housed within
a reactor vessel; and the vessel, steam generator, and associated plumbing loops are often protected within a containment building. Containment buildings, with their meter-thick
concrete and steel walls, are constructed to prevent leaks of
radioactivity due to accidents or natural catastrophes such as
earthquakes. Not all nations require containment buildings,
which points out the key role that government regulation
plays in protecting public safety.

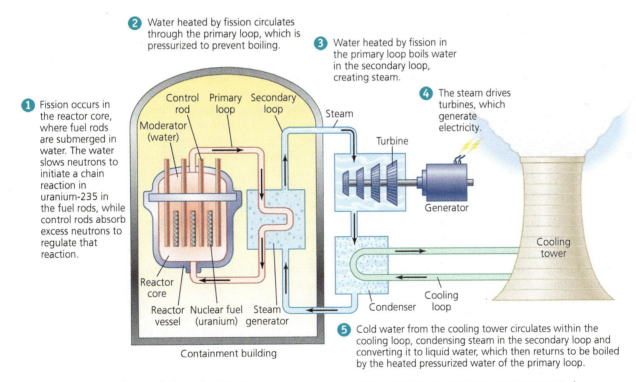

FIGURE 20.5 Nuclear reactors produce electricity. In a pressurized light water reactor (the most common type of
nuclear reactor), uranium fuel rods are placed in water, which slows neutrons so that fission can occur ❶. Control rods
are moved into and out of the reactor core, absorbing excess neutrons to regulate the chain reaction. Water heated by
fission circulates through the primary loop ❷ and warms water in the secondary loop, where water turns to steam ❸.
Steam drives turbines, which generate electricity ❹. The steam is cooled afterward in the cooling tower by water from
an adjacent river or lake and returns to the containment building ❺, to be heated again by heat from the primary loop.

Nuclear energy comes from processed and enriched uranium

We use the element uranium for nuclear power because it is radioactive. Radioactive isotopes, or radioisotopes (p. 24), emit subatomic particles and high-energy radiation as they decay into lighter radioisotopes until they ultimately become stable isotopes. The isotope uranium-235 decays into a series of daughter isotopes, eventually forming lead-207. Each radioisotope decays at a rate determined by that isotope's *half-life* (p. 24), the time it takes for half of the atoms to give off radiation and decay. The half-life of ^{235}U is about 700 million years.

We obtain uranium from various minerals in naturally occurring ore (rock that contains minerals of economic interest [p. 632]) that we extract by large-scale mining. Uranium-containing minerals are uncommon, and uranium ore is in finite supply, so nuclear power is generally considered a nonrenewable energy source. Currently, most uranium used in U.S. power plants is imported from abroad.

More than 99% of the uranium in nature occurs as the isotope uranium-238. Uranium-235 (with three fewer neutrons) makes up less than 1% of the total. Because ^{238}U does not emit enough neutrons to maintain a chain reaction when fissioned, we use ^{235}U for commercial nuclear power. Therefore, we must process the ore we mine to enrich the concentration of ^{235}U to at least 3%. The enriched uranium is formed into pellets of uranium dioxide (UO_2), which are incorporated into metallic tubes called *fuel rods* that are used in nuclear reactors.

After several years in a reactor, the decayed uranium fuel no longer generates adequate energy, so it must be replaced with new fuel. In some countries, the spent fuel is reprocessed to recover the remaining energy. However, this process is costly relative to the market price of uranium, so most spent fuel is disposed of as radioactive waste (p. 624).

Nuclear power delivers energy more cleanly than fossil fuels

Fission enables nuclear power plants to generate electricity without creating air pollution from stack emissions. In contrast, burning fossil fuels generally emits sulfur dioxide, which contributes to acid deposition; particulate matter, which threatens human health; and carbon dioxide and other greenhouse gases, which drive global climate change. Of course, the construction of nuclear plants has a large carbon footprint, but the actual power-generating process is essentially emission-free. Considering all the steps involved from construction through mining through power generation, researchers from the International Atomic Energy Agency (IAEA) have calculated that nuclear power releases from 4 to 150 times fewer emissions than fossil fuel combustion. Scientists estimate that nuclear power helps the United States avoid emitting 600 million metric tons of carbon dioxide each year, equivalent to the CO_2 emissions of almost all passenger cars in the nation

and 11% of total U.S. CO_2 emissions. Globally, using nuclear power in place of fossil fuels helps the world avoid emissions of 2.5 billion metric tons of carbon dioxide per year, about 7% of global CO_2 emissions.

Nuclear power has additional advantages over fossil fuels—coal in particular. For residents living downwind from power plants, scientists calculate that nuclear power poses far fewer chronic health risks from pollution. For instance, nuclear power prevents the emission of half a million tons of nitrogen oxide and 1.4 million tons of sulfur dioxide each year that might otherwise be generated by coal-fired power plants. And because uranium generates far more power than coal by weight or volume, less of it needs to be mined. As a result, uranium mining causes less damage to landscapes and generates less solid waste than coal mining. Moreover, in the course of normal operation, nuclear power plants have proved safer for workers than coal-fired plants.

Supporting the assessment that nuclear power is safer in many ways than fossil fuel power, a 2015 research paper published in the journal *Energy Policy* examined Sweden's plan to phase out nuclear power, and calculated the costs that this would have if the nation had to replace the lost power with coal-fired power. These scientists concluded that replacing Sweden's nuclear plants with coal would produce 1.8 billion tons of extra carbon dioxide emissions and cause more than 51,000 extra deaths due to the health impacts of coal pollution. Already Sweden's nuclear plants, by replacing coal, had prevented 2.1 billion tons of CO_2 emissions and saved 61,000 lives, the researchers calculated.

Nuclear power also has drawbacks, however. One main drawback is that the waste it produces is radioactive, and arranging for safe disposal of this waste has proved challenging. Another major drawback is that if an accident occurs at a power plant, or if a plant is sabotaged, the consequences can potentially be catastrophic.

Given this mix of advantages and disadvantages (TABLE 20.2), most governments (although not necessarily most citizens) have judged the good to outweigh the bad, and today the world has 439 operating nuclear plants in 31 nations.

weighing the ISSUES

Choose Your Risk

Examine Table 20.2. Given the choice of living next to a nuclear power plant or living next to a coal-fired power plant, which would you choose? What would concern you most about each option?

Fusion remains a dream

For as long as scientists and engineers have generated power from nuclear fission, they have tried to figure out how to harness nuclear fusion as well. **Nuclear fusion**—the process that drives our sun's vast output of energy, and the force behind hydrogen bombs (thermonuclear bombs)—involves forcing together the small nuclei of lightweight elements under extremely high temperature and pressure.

TABLE 20.2 Risks and Impacts of Coal-Fired versus Nuclear Power Plants

TYPE OF IMPACT	COAL	NUCLEAR
Land and ecosystem disturbance from mining	Extensive, on surface or underground	Less extensive
Greenhouse gas emissions	Considerable emissions	None from plant operation; much less than coal over the entire life-cycle
Other air pollutants	Sulfur dioxide, nitrogen oxides, particulate matter, and other pollutants	No pollutant emissions
Radioactive emissions	No appreciable emissions	No appreciable emissions during normal operation; possibility of emissions during severe accident
Occupational health among workers	More known health problems and fatalities	Fewer known health problems and fatalities
Health impacts on nearby residents	Air pollution impairs health	No appreciable known health impacts under normal operation
Effects of accident or sabotage	No widespread effects	Potentially catastrophic widespread effects
Solid waste	More generated	Less generated
Radioactive waste	None	Radioactive waste generated
Fuel supplies remaining	Should last several hundred more years	Uncertain; supplies could last longer or shorter than coal supplies

For each type of impact, the more severe impact is highlighted in red.

The hydrogen isotopes deuterium and tritium can be fused together to create helium, releasing a neutron and a tremendous amount of energy (**FIGURE 20.6**).

Overcoming the mutually repulsive forces of protons in a controlled manner is difficult, however, and fusion requires temperatures of many millions of degrees Celsius. As a result, researchers have not yet developed this process for commercial power generation. Despite billions of dollars of funding and decades of research, fusion experiments in the lab still require scientists to input more energy than they produce from the process. That is, they experience a loss in net energy (p. 519) and a ratio of energy returned on investment (EROI; p. 520) lower than 1.

If we could find a way to control fusion in a reactor, the potential payoffs would be immense. We could produce vast amounts of energy using water as a fuel, and the process would create only low-level radioactive wastes, without pollutant emissions or the risk of dangerous accidents, sabotage, or weapons proliferation. A consortium of industrialized nations is collaborating to build a prototype fusion reactor called the International Thermonuclear Experimental Reactor (ITER) in southern France. It aims to achieve an EROI of 10:1. Construction is scheduled for completion in 2019, and experiments would be conducted during the 2020s. Even if this multi-billion-dollar effort succeeds, however, commercial power from fusion seems unlikely in the near future.

Nuclear power poses small risks of large accidents

Although nuclear power delivers energy more cleanly than fossil fuels, the possibility of catastrophic accidents has spawned a great deal of public anxiety. Three events have been most influential in shaping public opinion about nuclear power: Three Mile Island, Chernobyl, and Fukushima.

Three Mile Island In Pennsylvania in 1979 a combination of mechanical failure and human error at the **Three Mile Island** plant caused coolant water to begin draining from the reactor vessel, temperatures to rise inside the reactor core, and metal surrounding the uranium fuel rods to start melting, releasing radiation. This process is termed a **meltdown,** and

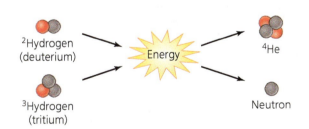

FIGURE 20.6 Can we harness nuclear fusion? In nuclear fusion, two small atoms, such as the hydrogen isotopes deuterium and tritium, are fused together, releasing energy along with a helium nucleus and a free neutron. So far, scientists have not been able to fuse atoms without supplying far more energy than the reaction produces, so this process is not used commercially.

These two cooling towers are linked to the damaged reactor that was decommissioned.

These two cooling towers are linked to the reactor that still operates today.

FIGURE 20.7 Pennsylvania's Three Mile Island nuclear power plant suffered a partial meltdown in 1979. This "near miss" alerted the world that a major accident could potentially occur.

at Three Mile Island (**FIGURE 20.7**) it proceeded through half of one reactor core. Residents of Harrisburg and nearby towns stood ready to be evacuated as the nation held its breath, but fortunately most radiation remained trapped inside the containment building.

The accident was brought under control within days, and the damaged reactor was shut down, but multi-billion-dollar cleanup efforts stretched on for years. Three Mile Island is best regarded as a near miss; the emergency could have been far worse had the meltdown proceeded through the entire stock of uranium fuel or had the containment building not contained the radiation. Although residents have shown no significant health impacts in the years since, the event raised safety concerns in the United States and abroad. In fact, it was Three Mile Island that inspired Sweden's referendum in 1980 that resulted in its national vote for a phaseout of nuclear power.

Chernobyl In 1986 an explosion at the **Chernobyl** power plant in Ukraine (part of the Soviet Union at the time) caused the most severe nuclear accident the world has yet seen (**FIGURE 20.8a**). Engineers had turned off safety systems to conduct tests, and human error, along with unsafe reactor design, led to explosions that destroyed the reactor and sent clouds of radioactive debris billowing into the atmosphere. For 10 days radiation escaped from the plant while emergency crews risked their lives putting out fires (some later died from radiation exposure). Most residents of the area remained at home for these 10 days, exposed to high levels of radiation, before the Soviet government belatedly began evacuating more than 100,000 people.

The accident killed 31 people directly and sickened or caused cancer in thousands more. Exact numbers are uncertain because of inadequate data and the difficulty of determining long-term radiation effects (see **THE SCIENCE BEHIND THE STORY,** pp. 560–561). Health authorities estimate that most of the 6000-plus cases of thyroid cancer diagnosed in

people who were children at the time resulted from radioactive iodine spread by the accident.

Atmospheric currents carried radioactive fallout from Chernobyl across much of the Northern Hemisphere, particularly Ukraine, Belarus, and parts of Russia and Europe (**FIGURE 20.8b**). Fallout was greatest where rainstorms brought radioisotopes down from the radioactive cloud. Parts of Sweden received high amounts of fallout, and the accident reinforced the Swedish public's anxieties about nuclear power. A poll taken after the event showed that nearly half of Swedish citizens had come to regret their own nation's investment in nuclear power.

In the months and years following the catastrophe at Chernobyl, workers erected a gigantic concrete sarcophagus around the demolished reactor and took measures to limit radioactive contamination of milk, meat, and produce from the region's farms. However, the landscape for at least 30 km (19 mi) around the plant remains contaminated today, the demolished reactor is still full of dangerous fuel and debris, and radiation is leaking from the hastily built, quickly deteriorating sarcophagus. Currently an international team is building a larger confinement structure (**FIGURE 20.8c**) and plans to slide it into place around the old sarcophagus to prevent a re-release of radiation.

Fukushima Daiichi On March 11, 2011, a magnitude-9.0 earthquake struck eastern Japan and sent an immense tsunami roaring onshore (pp. 21–22). More than 18,000 people were killed, and many thousands of buildings were destroyed. This natural disaster also affected the operation of several of Japan's nuclear power plants, most notably **Fukushima Daiichi.** At this plant, the earthquake shut down power and the tsunami flooded the plant's emergency power generators. The plant was protected by a 5.7-m (19-ft) seawall, but the tsunami reached 14 m (46 ft) high, and the generators were located in the basement of the plant (**FIGURE 20.9a**, p. 558). Without electricity, workers could not use moderators and control rods to cool the uranium fuel, and the fuel began to overheat as fission proceeded, uncontrolled.

Amid the damage and chaos across the region, help was slow to arrive, so workers began flooding the reactors with seawater in a desperate effort to prevent meltdowns. Several explosions and fires occurred over the next few days, and eventually three reactors experienced full meltdowns, while the plant's other three reactors were seriously damaged. Parts of the plant remained inaccessible for months because of radioactive water, and it will likely require decades to fully clean up the site.

Radioactivity was released during and after these events at levels about one-tenth of those from Chernobyl. Thousands of residents of areas near the plant were evacuated and screened for radiation effects (**FIGURE 20.9b**), and restrictions were placed on food and water from the region. Much of the radiation spread by air or water into the Pacific Ocean, and trace amounts were detected around the world (**FIGURE 20.9c**). In the years following the event, a slow flow of radioactive groundwater from beneath the plant has continued to leak into the ocean. Long-term health effects on the region's people remain uncertain (see **The Science Behind the Story,** pp. 560–561).

(a) The destroyed reactor at Chernobyl, 1986

(c) The containment dome under construction, 2015

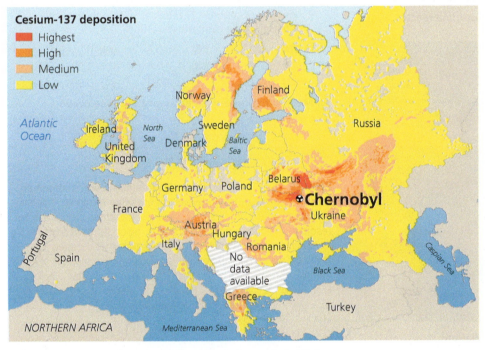

Cesium-137 deposition

- 🟥 Highest
- 🟧 High
- 🟧 Medium
- 🟨 Low

(b) The spread of radioactive fallout from Chernobyl, 1986

FIGURE 20.8 The world's worst nuclear accident unfolded in 1986 at Chernobyl. The destroyed reactor **(a)** was later encased in a massive concrete sarcophagus to contain further radiation leakage. Radioactive fallout **(b)** was deposited across Europe; complex patterns of cesium-137 deposition resulted from atmospheric currents and rainstorms in the days following the accident. Although Chernobyl produced 100 times more fallout than the U.S. bombs dropped on Hiroshima and Nagasaki in World War II, it was distributed over a much wider area. Thus, levels of contamination outside Ukraine, Belarus, and western Russia were relatively low. Today an international team is building a huge new confinement structure **(c)** that will be slid into place to encase the deteriorating sarcophagus around the plant. *Data from chernobyl.info, Swiss Agency for Development and Cooperation, Bern, Switzerland.*

In the aftermath of the disaster, Japan's government temporarily shut down all 50 of the nation's nuclear reactors and embarked on safety inspections. Afterward, efforts to restart them were met with public debate and street protests. Across the world, many nations responded by reassessing their own nuclear programs. In Sweden, where substantial majorities of citizens had long supported their nation's production of nuclear power, public opinion polls showed Swedes to be divided over whether to keep nuclear power or phase it out. Of all nations, Germany reacted most strongly, shutting down half of its nuclear power plants and deciding to phase out the rest by 2022. Sweden's environment minister at the time criticized Germany's decision, saying it would impede Germany's ability to move away from fossil fuels—and indeed, Germany has been burning more coal for power since that time. In

2015, Japan began restarting its nuclear plants, following four years of high electricity prices due to the import of replacement power fired by coal and natural gas. Yet despite the prospect of economic recovery, the majority of Japanese citizens remained opposed to restarting the nuclear plants.

We are managing risks from nuclear power and nuclear weapons with some success

It is fortunate that we have not experienced more accidents on the scale of Fukushima or Chernobyl. Yet smaller-scale incidents have occurred. A 1999 accident at a plant in Tokaimura, Japan, killed two workers and exposed more than 400 others

(a) The tsunami barrels toward the Fukushima reactors

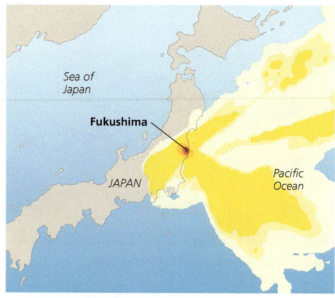

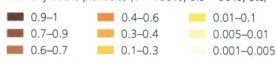

Intensity of cesium-137 radiation in fallout relative to intensity at the plant site (1 = 100%, 0.9 = 90%, etc)

■ 0.9–1	■ 0.4–0.6	■ 0.01–0.1
■ 0.7–0.9	■ 0.3–0.4	■ 0.005–0.01
■ 0.6–0.7	■ 0.1–0.3	■ 0.001–0.005

(c) Most radiation drifted eastward over the ocean

(b) A Japanese child is screened for radiation

FIGURE 20.9 The Fukushima Daiichi crisis was unleashed after an earthquake generated a massive tsunami. The tsunami **(a)** tore through a seawall and inundated the plant's nuclear reactors. Children evacuated from the region **(b)** were screened for radiation exposure. Most of the radiation that escaped from the plant drifted over the ocean, as shown in this map **(c)** of cesium-137 isotopes in the days following the accident. *Data (c) from Yasunari, T.J., et al. 2011. Cesium-137 deposition and contamination of Japanese soils due to the Fukushima nuclear accident. Proc. Natl. Acad. Sci. USA 108: 19530–19534.*

to leaked radiation. Sweden experienced a near miss in 2006, when the Forsmark power plant north of Stockholm narrowly avoided a meltdown after two generators failed to start up following a power outage. Each year several U.S. reactors undergo emergency shutdowns in response to tornadoes, hurricanes, earthquakes, and flooding.

We can be thankful that the designs of most modern reactors are safer than Chernobyl's. Designs for future plants promise even more safety features. And in most emergencies at reactors around the world, safety systems have functioned well. For instance, Japan's Onagawa power plant was closer to the epicenter of the 2011 quake, yet its safety systems protected it from serious damage.

However, as plants around the world age, they require more maintenance and become less safe. There is also the concern that radioactive material could be stolen from plants

and used in terrorist attacks. This possibility has been especially worrisome in the cash-strapped nations of the former Soviet Union, where hundreds of former nuclear sites have gone without adequate security for years. Finally, there is the ever-present concern that more nations may develop nuclear weapons.

To address concerns about stolen fuel and to reduce the world's nuclear weapons stockpiles, the United States and Russia embarked on a remarkably successful program nicknamed "Megatons to Megawatts." In this cooperative international agreement, the United States purchased weapons-grade uranium and plutonium from Russia, let Russia process them into lower-enriched fuel, and diverted the fuel to peaceful use in power generation. Between 1993 and 2013, 500 metric tons of highly enriched uranium were processed and transferred. As a result, in recent years fully 10% of America's electricity has been generated from fuel recycled from Russian warheads that used to be atop missiles pointed at American cities!

Waste disposal remains a challenge

Even if nuclear power generation could be made completely safe, and even if we could recycle all weapons-grade fuel into fuel for power plants, we still would be left with the conundrum of what to do with spent fuel rods and other radioactive waste. Recall that fission uses ^{235}U as fuel, leaving as waste the 97% of uranium that is ^{238}U. This ^{238}U, as well as all irradiated material and equipment that is no longer being used, must be disposed of safely. The half-lives of uranium, plutonium, and many other radioisotopes are inconceivably long—700 million years for ^{235}U and 4.5 *billion* years for ^{238}U. As a result, this waste will continue emitting radiation for as long as our civilization exists. Thus, radioactive waste must be placed in unusually stable and secure locations where radioactivity will not harm future generations.

Currently, nuclear waste from power generation is being held in temporary storage at nuclear power plants across the world. Spent fuel rods are sunk in pools of cooling water to minimize radiation leakage (**FIGURE 20.10a**). However, many plants have no room left for this type of storage, so they are now storing waste in thick casks of steel, lead, and concrete (**FIGURE 20.10b**). In total, U.S. power plants are storing nearly 70,000 metric tons of high-level radioactive waste—enough to fill a football field to the depth of 7 m (21 ft)—as well as much more low-level radioactive waste. This waste is held at more than 120 sites spread across 39 states (**FIGURE 20.11**). The majority of Americans live within 125 km (75 mi) of temporarily stored waste.

Because storing waste at many dispersed sites creates a large number of potential hazards, nuclear waste managers would prefer to send all waste to a single, central repository that can be heavily guarded. Sweden is far ahead of most nations in this regard; after 20 years of research looking for a suitable location, it selected the Forsmark power plant site

(a) Wet storage

(b) Dry storage

FIGURE 20.10 Nuclear waste is stored at nuclear power plants, because no central repository yet exists. Spent fuel rods are kept in wet storage **(a)** in pools of water, which keep them cool and reduce radiation release, or in dry storage **(b)** in thick-walled casks layered with lead, concrete, and steel.

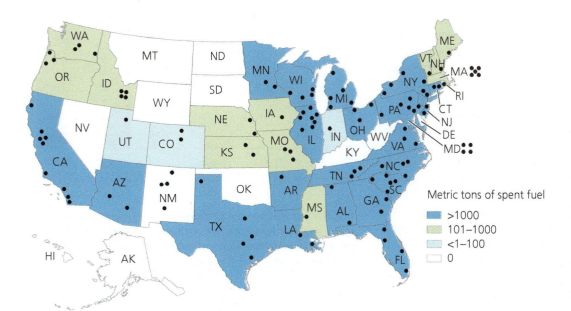

Metric tons of spent fuel

- >1000
- 101–1000
- <1–100
- 0

FIGURE 20.11 High-level radioactive waste from civilian reactors is currently stored at more than 120 sites in 39 states across the United States. In this map, dots indicate storage sites and colors indicate the amount of waste stored in each state. *Data from Nuclear Energy Institute.*

What Health Impacts Have Resulted from Chernobyl and Fukushima?

Chernobyl-area boy is tested for thyroid cancer in a Red Cross clinic.

In the wake of the meltdowns at Japan's Fukushima Daiichi nuclear power plant in 2011, Japanese authorities evacuated residents and tried to keep people safe, while medical scientists from around the world rushed to study how the release of radiation might affect human health. Both looked back to lessons learned from Chernobyl 25 years earlier.

Determining long-term health impacts of radiation exposure is enormously difficult, and initial attempts to predict Fukushima's health impacts have stirred vigorous debate. Most scientists expect that the reactor failures at Fukushima will have less severe health consequences than those at Chernobyl, because less radiation was emitted at Fukushima and because most of it drifted over the ocean away from populated areas (see Figure 20.9c and pp. 24–25). However, judging Fukushima's impacts will be challenging and will require long-term study.

Let's turn first to Chernobyl. The hundreds of researchers who have tried to pin down Chernobyl's health impacts have sometimes differed in their conclusions. In an effort to reach consensus, the World Health Organization (WHO) engaged 100 experts to review all studies through 2006 and issue a report summarizing what scientists had learned in the 20 years since the accident. In 2008, the United Nations Scientific Committee on the Effects of Atomic Radiation (UNSCEAR) issued a comprehensive report as well.

The most severe effects were documented in emergency workers who battled to contain the incident in its initial days. Medical staff treated and recorded the progress of 134 workers hospitalized with acute radiation sickness (ARS). Radiation destroys cells in the body, and if the destruction outpaces the body's abilities to repair the damage, the person will soon die. Symptoms of ARS include vomiting, fever, diarrhea, thermal burns, mucous membrane damage, and weakening of the immune system. In total, 28 people died from ARS soon after the accident. Those who died had the greatest estimated exposure to radiation.

The major health consequence of Chernobyl's radiation, however, has been thyroid cancer in children. The thyroid gland is where our bodies concentrate iodine, and one of the most common radioactive isotopes released early in the disaster was iodine-131 (^{131}I). Healthy children have large and active thyroid glands, so they are especially vulnerable to thyroid cancer induced by radioisotopes.

Predicting that thyroid cancer might be a problem, medical workers measured iodine activity in the thyroid glands of several hundred thousand people in Russia, Ukraine, and Belarus following the accident. They also measured food contamination and surveyed people on their food consumption. These data showed that drinking milk from cows that had grazed on contaminated grass was the main route of exposure to ^{131}I, although fresh vegetables also contributed.

As doctors had feared, rates of thyroid cancer rose among children in regions of highest exposure (**FIGURE 1**). Multiple studies found linear dose-response relationships (p. 371) in data from Ukraine and Belarus. By 2006, medical professionals estimated the number of cases of thyroid cancer in children at 6000 and rising. Fortunately, treatment of thyroid cancer has a high success rate, so as of that time, only 15 children had died from it.

Studies addressing other health aspects of the accident have found limited impact. Some research has shown an increase in cataracts due to radiation, especially among emergency workers. But neither the WHO nor the United Nations assessments found evidence that rates of leukemia or any other cancer (aside from thyroid cancer) had risen among people exposed to Chernobyl's radiation. Still, some cancers may appear decades after exposure, so it is possible that many illnesses have yet to

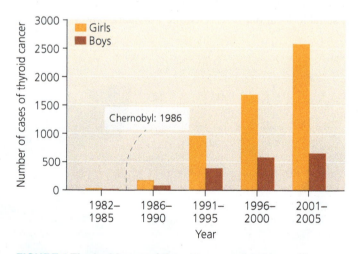

FIGURE 1 The incidence of thyroid cancer in girls and boys below age 18 rose in regions of Ukraine, Belarus, and Russia that experienced the heaviest fallout of radioactive iodine from Chernobyl. *Data from U.N. Scientific Committee on the Effects of Atomic Radiation, 2008. Sources and effects of ionizing radiation, Vol. 2. New York, 2011.*

arise. Moreover, conducting epidemiological studies (p. 370) with enough statistical power to detect minor increases in rare events requires observing huge numbers of people over many years.

When the Fukushima crisis struck, Japanese authorities applied lessons from Chernobyl. They provided iodine tablets to young people evacuated from the region (the tablets supply the thyroid gland with stable iodine, preventing it from taking up radioactive iodine). They also restricted agriculture near the plant, stopping contaminated food and milk from entering the market. These measures to prevent people's uptake of iodine-131 apparently were effective. In late March 2011, 1000 evacuated children were tested for ^{131}I exposure, and none showed evidence of a high dose affecting the thyroid.

As the crisis was gradually brought under control, some researchers began predicting what long-term health impacts might arise. Using data on cancer rates from studies on survivors of the atomic bombs dropped on Hiroshima and Nagasaki, researchers extrapolated to predict cancer rates at the lower radiation doses from Fukushima. Frank Von Hippel in a 2011 study calculated that about 1 million people were exposed to more than 1 curie per square kilometer of radiation from cesium-137, which should result in a 0.1% increase in cancer risk among them, or 1000 cancer cases.

In 2012, John Ten Hoeve and Mark Jacobson of Stanford University used a global atmospheric model together with data on radiation, population dispersion, and weather in the days following the event to estimate radiation levels people received (FIGURE 2). In the journal *Energy and Environmental Science*, they predicted that Fukushima's radiation would eventually produce 125 cancer-related deaths and 178 nonfatal cancer cases. Large uncertainties attended both estimates; the projected number of total cases ranged from 39 to 2900.

Meanwhile, their Stanford colleague Burton Richter wrote to the journal that—far from suggesting that nuclear power is dangerous—the number of fatalities from Fukushima is far less than Japan would have experienced from air pollution had its electricity been produced by fossil fuels instead!

As with Chernobyl, both the WHO and UNSCEAR issued status reports reviewing what was scientifically understood about health impacts from Fukushima in the years following the event. Both reports concluded that there was no evidence yet for any cancers or other health impacts, and that radiation levels were low enough that any increases in cancer risk were very small. Indeed, they cautioned that it will prove difficult to measure statistically rare instances of cancer in a large population exposed to very low doses.

Researchers will try, however. They will be assisted by Japan's government, which budgeted $1.2 billion for coordinated research into long-term health effects of radiation. Questionnaires were sent to all 2 million residents of Fukushima Prefecture, asking where they were in the days after the accident, and what they ate and drank. More than 500,000 of these surveys were returned. Thyroid exams are being given to all 360,000

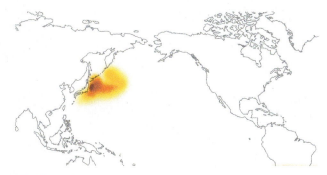

(a) After 36 hours

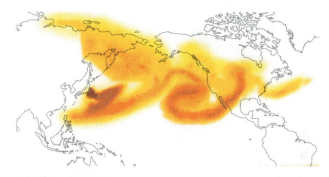

(b) After 180 hours

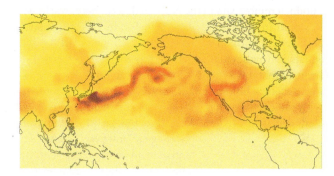

(c) After 324 hours

$10^0 \quad 10^1 \quad 10^2 \quad 10^3 \quad 10^4 \quad 10^5 \quad 10^6$
Cesium-137 concentration (micro-Becquerels per m³)

FIGURE 2 Radiation from cesium-137 spread around the world from Fukushima. Atmospheric modeling shows how this radioisotope spread for two weeks after the accident. *Data from Ten Hoeve, J., and M. Jacobson, 2012. Worldwide health effects of the Fukushima Daiichi nuclear accident. Energy & Environmental Science 5: 8743–8757.*

children and teens from the region, and 20,000 pregnant women and their babies are being closely monitored. All 200,000 people evacuated from the area are getting regular exams, and mental health support is being offered. If this research program can extend for 30 years, as the government intends, it could provide valuable information on the health effects of low-dose radiation.

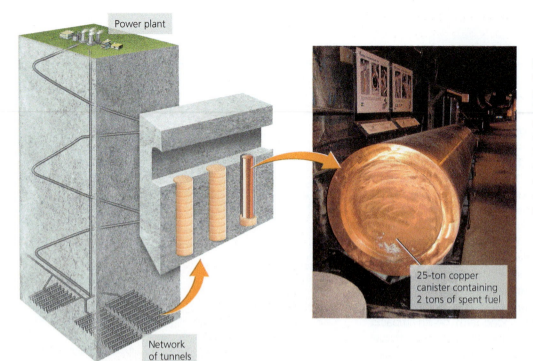

Power plant

Network
of tunnels

FIGURE 20.12 Sweden leads the world in attempts to establish a central repository for nuclear waste. Plans call for placing waste in copper canisters and burying them in a network of tunnels deep beneath the Forsmark nuclear power plant. Each canister will be surrounded by clay, and tunnels will be backfilled and sealed.

25-ton copper canister containing 2 tons of spent fuel

as its single location for final disposal (**FIGURE 20.12**). If the facility is approved and constructed as planned, spent fuel rods and other high-level waste will be buried in copper canisters 500 m (1650 ft) underground within stable bedrock.

In the United States, a lengthy search homed in on Yucca Mountain, a remote site in the Nevada desert. Here, waste was to be stored in a network of tunnels 300 m (1000 ft) underground, and $13 billion was spent on development of the site. Scientists and policymakers chose this location because they determined that it was unpopulated, has minimal risk of earthquakes, receives little rain that could contaminate groundwater with radioactivity, has a deep water table atop an isolated aquifer, and is on federal land that can be protected from sabotage. However, some scientists and antinuclear activists challenged these conclusions, and most Nevadans opposed the choice. In 2010 President Barack Obama's administration ended support for the project.

Without Yucca Mountain, the United States has no site designated to dispose of radioactive waste from commercial nuclear power plants. Until it does, this waste will remain spread among many locations.

Although storing nuclear waste at numerous widespread sites creates numerous risks, one concern with a centralized repository is that waste would need to be transported there from the many current storage sites and from each nuclear plant in the future. Because this would involve many thousands of shipments by rail and truck across hundreds of public highways through almost every state of the union, some people worry that the risk of an accident or of sabotage is unacceptably high.

Nuclear power's growth has slowed

Dogged by concerns over waste disposal, safety, and cost overruns, nuclear power's growth has slowed. Since the late 1980s, nuclear power has grown by only 2.5% per year worldwide, about the same rate as electricity generation overall. Public anxiety in the wake of Chernobyl made utilities less willing to invest in new plants, and reaction to Fukushima stalled a resurgence in the industry. The process of building, maintaining, operating, and ensuring the safety of nuclear facilities is enormously expensive, and almost every nuclear plant has overrun its budget. In addition, plants have aged more quickly than expected because of problems that were underestimated, such as corrosion in coolant pipes. The plants that have been shut down—well over 100 to date—have served on average less than half their expected lifetimes. Moreover, shutting down, or decommissioning, a plant is sometimes more expensive than the original construction.

As a result of these financial issues, electricity from nuclear power remains more expensive than electricity from fossil fuels. Governments are still subsidizing nuclear power to keep electricity prices down for ratepayers, but many private investors lost interest long ago. In the United States, the nuclear industry stopped building plants following Three Mile Island, and public opposition scuttled many that were under construction. Of the 259 U.S. nuclear reactors ordered since 1957, nearly half have been canceled. At its peak in 1990, the United States had 112 operating reactors; in 2016 it had 99.

weighing the
ISSUES

How to Store Waste?

Which do you think is a better option—to transport nuclear waste cross-country to a single repository or to store it permanently at numerous power plants scattered across the nation? Would your opinion be affected if you lived near the repository site? Near a power plant? On a highway route along which waste is transported?

Nonetheless, nuclear power remains one of the few currently viable alternatives to fossil fuels with which we can generate large amounts of electricity in short order. This is why a growing number of environmental advocates propose expanding nuclear capacity using a new generation of reactors designed to be safer and less expensive. For a nation wishing to cut its pollution and greenhouse gas emissions quickly and substantially, nuclear power is in many respects the leading option.

With future growth for nuclear power uncertain, and with climate change worsening, our society must determine where we will turn for clean and sustainable energy. Increasingly, people are looking to renewable sources of energy: energy sources that are replenished naturally and cannot be depleted by our use. Many promising renewable sources are still early in their development (Chapter 21), but two of them—bioenergy and hydroelectric power—are already well developed and widely used. We will examine these renewable sources of energy next.

weighing the ISSUES

More Nuclear Power?

Sweden and a number of other European nations have reduced their carbon emissions by expanding their use of nuclear power to replace fossil fuels. Do you think the United States should expand its nuclear power program? Why or why not?

Bioenergy

Bioenergy—also known as *biomass energy*—is energy obtained from biomass. **Biomass** (p. 80) consists of organic material derived from living or recently living organisms, and it contains chemical energy (p. 28) that originated with sunlight and photosynthesis. We harness bioenergy from many types of plant matter, including wood from trees, charcoal from wood charred in the absence of oxygen, and matter from agricultural crops, as well as from combustible animal waste products such as cattle manure.

The great attraction of bioenergy is that—in principle—it is renewable and releases no net carbon dioxide into the atmosphere. Although burning biomass emits plenty of carbon, this is balanced by the fact that photosynthesis had pulled this same amount of carbon from the atmosphere to create the biomass just years, months, or weeks before. Therefore, in theory, when we replace fossil fuels with bioenergy, we reduce net carbon flux to the atmosphere, helping to alleviate global climate change (Chapter 18). However, in practice it is not so simple, and judging the sustainability of any given bioenergy strategy requires careful consideration of what biomass source we are using and how we gain energy from it.

We gain bioenergy from many sources

To a poor farmer in Africa, bioenergy comes from cutting wood from trees or collecting livestock manure by hand and burning it to heat and cook for her family. To an industrialized farmer in Iowa, bioenergy means shipping his grain to a hi-tech refinery that converts it to liquid fuel to run automobiles.

TABLE 20.3 Sources and Uses of Bioenergy

DIRECT COMBUSTION FOR HEATING

- Wood cut from trees (fuelwood)
- Charcoal
- Manure from farm animals

BIOPOWER FOR GENERATING ELECTRICITY

- Crop residues (such as cornstalks) burned at power plants
- Forestry residues (wood waste from logging) burned at power plants
- Processing wastes (solid or liquid waste from sawmills, pulp mills, and paper mills) burned at power plants
- "Landfill gas" burned at power plants
- Livestock waste from feedlots for gas from anaerobic digesters

BIOFUELS FOR POWERING VEHICLES

- Corn grown for ethanol
- Bagasse (sugarcane residue) grown for ethanol
- Soybeans, rapeseed, and other crops grown for biodiesel
- Used cooking oil for biodiesel
- Plant matter treated with enzymes to produce cellulosic ethanol
- Algae grown for biofuels

The diversity of sources and approaches involved in bioenergy (**TABLE 20.3**) gives us many ways to address our energy challenges.

More than 1 billion people use wood from trees as their principal energy source. In rural regions of developing nations, people (generally women) gather fuelwood to burn in their homes for heating, cooking, and lighting (**FIGURE 20.13**;

FIGURE 20.13 Well over a billion people in developing countries rely on wood from trees for heating and cooking. In principle, biomass is renewable, but in practice it may not be if forests are overharvested.

also see Figure 17.32, p. 472). Although fossil fuels and electricity are replacing traditional energy sources as developing nations industrialize, fuelwood, charcoal, and manure still account for one-third of energy use in these nations—and up to 90% in the poorest nations.

Fuelwood and other traditional biomass sources constitute three-quarters of all renewable energy used worldwide. However, biomass is renewable only if it is not overharvested. Harvesting fuelwood at unsustainably rapid rates can lead to deforestation (p. 305), soil erosion (p. 221), and desertification (p. 222), thereby damaging landscapes, diminishing biodiversity, and impoverishing societies. Heavily populated arid regions that support meager woodlands are most vulnerable to overharvesting, and these include many regions of Africa and Asia. Another drawback of burning fuelwood and other biomass for cooking and heating is that it poses health hazards from indoor air pollution (p. 472).

Although much of the world still relies on fuelwood, charcoal, and manure, new and innovative bioenergy approaches are being developed using a variety of materials (see Table 20.3). Some of these biomass sources can be burned in power plants to produce **biopower,** generating heat and electricity. Other sources can be converted into **biofuels,** liquid fuels used primarily to power automobiles. Because many of these novel biofuels and biopower strategies depend on technologies resulting from extensive research and development, they are being developed primarily in wealthier industrialized nations, such as Sweden and the United States.

We use biomass to generate electricity

We harness biopower by combusting biomass to produce heat or generate electricity in the same way that we burn coal for power (see Figure 19.11, p. 527). This can be done using a variety of sources and techniques.

Waste products Some waste products can be used as sources for biopower. The forest products industry generates large amounts of woody debris in logging operations (**FIGURE 20.14**) and at sawmills, pulp mills, and paper mills. Sweden's efforts to promote bioenergy have focused largely on using forestry residues. Because so much of the nation is forested and the timber industry is a major part of the Swedish economy, plenty of forestry waste is available.

Other waste sources used for biopower include animal waste from feedlots and residue from agricultural crops, such as cornstalks and corn husks. In addition, the anaerobic bacterial breakdown of waste in landfills produces methane, and this "landfill gas" is being captured and sold as fuel (p. 617). Methane and other gases can also be produced in a controlled way in anaerobic digestion facilities. This *biogas* can then be burned in a power plant's boiler to generate electricity.

The Swedish city of Kristianstad showcases just how much can be done with biogas. Back in 1999, this city of

FIGURE 20.14 Forestry residues are a major source of material for biopower in some regions. Here a Swedish logging operation gathers woody residue.

81,000 people—the hub of an agricultural and food processing region and best known as the home of Absolut Vodka—aimed to free itself of fossil fuels. After building a power plant that used forestry waste, it constructed a facility to turn waste into biogas. The biogas plant takes in household garbage, crop waste, food industry waste, and manure and uses anaerobic digestion to turn all this into biogas (**FIGURE 20.15**). The city's landfill and wastewater treatment plant also collect methane, adding to the city's biogas supply. The biogas is burned to generate electricity and to heat homes by district heating. It is also refined and used to fuel hundreds of cars, trucks, and buses designed to run on biogas (see photo, p. 549). All in all, this waste-to-energy system replaces about 7% of the city's gasoline and diesel fuel per year, all of its district heating, and much of its electricity, while excess biogas is sold to neighboring communities. A by-product of biogas production is liquid bio-fertilizer, and nearly 100,000 tons per year are sold to area farms.

Bioenergy crops We are beginning to grow certain types of plants as crops to generate biopower. These include fast-growing grasses such as bamboo, fescue, and switchgrass, as well as trees such as specially bred willows and poplars. Many of these plants are also being grown to produce liquid biofuels.

Combustion strategies Power plants built to combust biomass operate like those fired by fossil fuels; combustion heats water, creating steam to turn turbines and generators,

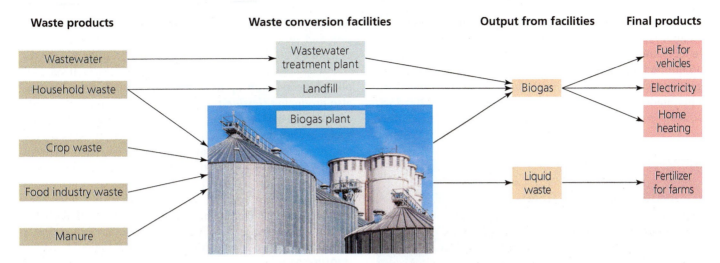

Waste products | **Waste conversion facilities** | **Output from facilities** | **Final products**

FIGURE 20.15 **In Kristianstad, Sweden, a diversity of waste products are converted into biogas, which provides vehicle fuel, electricity, and home heating.**

thereby generating electricity. Much of the biopower produced so far comes from power plants that use cogeneration (p. 542) to generate both electricity and heating. These plants are often located where they can take advantage of forestry waste.

In some coal-fired power plants, wood chips, wood pellets, or other biomass is introduced with coal into a high-efficiency boiler in a process called *co-firing*. We can substitute biomass for up to 15% of the coal with only minor equipment modification and no appreciable loss of efficiency. Co-firing is a relatively easy and inexpensive way for utilities to expand their use of renewable energy.

We also harness biopower by gasification (p. 538), in which biomass is vaporized at extremely high temperatures in the absence of oxygen, creating a gaseous mixture including hydrogen, carbon monoxide, carbon dioxide, and methane. This mixture can generate electricity when used to turn a gas turbine to propel a generator in a power plant. Another method of heating biomass in the absence of oxygen results in pyrolysis (p. 524), which produces a mix of solids, gases, and a liquid fuel called pyrolysis oil that can be burned to generate electricity.

Scales of production

At small scales, farmers, ranchers, or villages can operate modular biopower systems that use livestock manure to generate electricity. Small household bio-digesters provide portable and decentralized energy production for remote rural areas. At large scales, the forest products industry is using its waste to generate power, and industrialized farmers are growing bioenergy crops. In Sweden, nearly one-quarter of the nation's energy supply now comes from biomass, and biomass provides more than four times more fuel for electricity generation than coal, oil, and natural gas combined. Waste liquids from pulp mills, solid wood waste, municipal solid waste, biogas from digestion, and other sources are all used. In the United States, several dozen biomass-fueled power plants are now operating, and several dozen coal-fired plants are pursuing co-firing.

Benefits and drawbacks

By enhancing energy efficiency and putting waste products to use, biopower helps move utilities and industries in a sustainable direction. The U.S. forest products industry now obtains half its energy by combusting woody waste and liquid waste from pulp mill processing. Biopower helps alleviate climate change as well by reducing carbon dioxide emissions, and capturing landfill gas reduces emissions of methane, a potent greenhouse gas. Biopower also benefits human health as it cuts down on pollution; by replacing coal in co-firing and direct combustion, biopower reduces emissions of sulfur dioxide because plant matter, unlike coal, contains no appreciable sulfur content. In addition, biomass resources tend to be geographically dispersed, so using them can help support rural economies and reduce many nations' dependence on imported fuels.

A disadvantage is that when we burn plant matter for power, we deprive the soil of nutrients and organic matter it would have gained from the plant matter's decomposition. We essentially draw fertility from the soil and never return it, so that the soil becomes progressively depleted. This also is the case when we burn crops or plant matter as a liquid fuel, as discussed next. The depletion of soil fertility is a major long-term problem for bioenergy and is one reason that relying solely on bioenergy is not a sustainable option.

Ethanol can power automobiles

Liquid fuels from biomass sources are powering millions of vehicles on today's roads. The two primary biofuels developed so far are ethanol (for gasoline engines) and biodiesel (for diesel engines).

Ethanol is the alcohol in beer, wine, and liquor. It is produced as a biofuel by fermenting biomass, generally from carbohydrate-rich crops, in a process similar to brewing beer. In fermentation, carbohydrates are converted to sugars and then to ethanol. Spurred by the 1990 Clean

Air Act amendments and generous government subsidies, ethanol is widely added to gasoline in the United States to reduce automotive emissions. In 2015 in the United States, 56 billion L (14.8 billion gal) of ethanol were produced—46 gal for every American—mostly from corn (**FIGURE 20.16a**). This amount has grown rapidly, and more than 200 U.S. ethanol production facilities are now operating. Further growth seems assured, as Congress has mandated production and use of 136 billion L (36 billion gal) per year of ethanol by 2022.

Any vehicle with a gasoline engine runs well on gasoline blended with up to 10% ethanol, but automakers are also producing *flexible-fuel vehicles* that run on E-85, a mix of 85% ethanol and 15% gasoline. More than 17 million such cars are on U.S. roads today. Most gas stations do not yet offer E-85, so drivers often fill these cars with conventional gasoline, but this situation is changing as infrastructure for ethanol increases. In Brazil, almost all new cars are flexible-fuel vehicles, and ethanol from crushed sugarcane residue (called *bagasse*) accounts for half of all fuel that Brazil's drivers use. Together, Brazil and the United States currently produce about 85% of the world's ethanol supply (**FIGURE 20.16b**).

Ethanol is not our most sustainable energy choice

The enthusiasm for corn-based ethanol shown by U.S policymakers is not widely shared by scientists. Growing corn to produce ethanol intensifies pesticide use, fertilizer use, freshwater depletion, and other impacts of industrialized agriculture (p. 239). Fully 40% of the U.S. corn crop today is used to make ethanol. (Some by-products of ethanol production are used in livestock feed; with this accounted for, 28% of the U.S. corn crop goes solely toward ethanol.) Corn ethanol crops take up millions of acres of land. To produce all the automotive fuel used in the United States with ethanol from U.S. corn, the nation would need to expand its already immense corn acreage by more than four times (**FIGURE 20.17**). Even at our current level of production, ethanol already competes with food production and drives up food prices.

Growing corn for ethanol also requires substantial inputs of fossil fuel energy (for operating farm equipment, making petroleum-based pesticides and fertilizers, transporting corn to processing plants, and heating water in refineries to distill

(a) Corn grown for ethanol

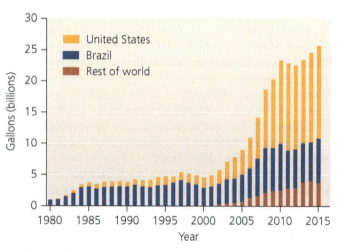

(b) Global ethanol production, 1980–2015

FIGURE 20.16 Ethanol is booming. About 40% of the U.S. corn crop **(a)** is used to produce ethanol to add to gasoline (28% after by-products are put to use). Next to the United States, Brazil produces most of the rest of the world's ethanol, from bagasse (sugarcane residue). Ethanol production **(b)** has grown rapidly in recent years. *Data (b) from EIA International Energy Statistics and Renewable Fuels Association.*

DATA Roughly what percentage of the world's ethanol is produced by the United States? Based on the data shown in the graph, give one explanation for why we might predict that U.S. and world ethanol production will be much higher in the future. Now explain one reason why we might predict that U.S. and world ethanol production will be about the same in the future as it is today.

Go to **Interpreting Graphs & Data** on **Mastering**EnvironmentalScience°

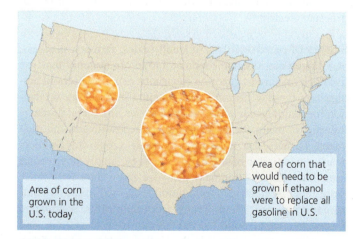

Area of corn grown in the U.S. today

Area of corn that would need to be grown if ethanol were to replace all gasoline in U.S.

FIGURE 20.17 Growing corn to produce ethanol uses up a great deal of land. To produce enough ethanol to replace all gasoline used by U.S. drivers, the United States would need to more than quadruple its area planted in corn.

FAQ

If we substitute ethanol for gasoline, won't that solve our problems with oil dependence?

Increasing the proportion of ethanol in gasoline does indeed help to conserve oil and reduce reliance on foreign imports. However, obtaining the amount of corn ethanol needed to replace gasoline entirely would require that impractically large amounts of land be converted to corn production. Moreover, so much corn would likely be diverted from food to fuel that food prices would rise sharply. This is why researchers are studying other plants as more efficient sources of ethanol and are trying to produce cellulosic ethanol from crop and forestry wastes.

ethanol). Thus, simply shifting from gasoline to corn ethanol for our transportation needs would not eliminate our reliance on fossil fuels.

Moreover, corn ethanol yields only a modest amount of energy relative to the energy that needs to be input. The EROI ratio (p. 520) for corn-based ethanol is variable and controversial, but recent estimates place it around 1.3:1 (see **THE SCIENCE BEHIND THE STORY**, pp. 568–569). This means that to gain 1.3 units of energy from ethanol, we need to expend 1 unit of energy. The EROI of Brazilian bagasse ethanol is considerably higher, but the extremely low ratio for corn ethanol makes this fuel inefficient. For this reason, many critics do not view corn ethanol as an effective path to sustainable energy use.

Biodiesel powers diesel engines

Drivers of diesel-fueled vehicles can use **biodiesel,** a fuel produced from vegetable oil, used cooking grease, or animal fat. The oil or fat is mixed with small amounts of ethanol or methanol (wood alcohol) in the presence of a chemical catalyst. In Europe, where most biodiesel is used, rapeseed oil is the oil of choice, whereas U.S. biodiesel producers use mostly soybean oil. Vehicles with diesel engines can run on 100% biodiesel, or biodiesel can be mixed with conventional diesel derived from petroleum; a 20% biodiesel mix (called B20) is common today.

Replacing diesel with biodiesel cuts down on emissions (**FIGURE 20.18**). Biodiesel's fuel economy is nearly as good, it costs just slightly more, and it is nontoxic and biodegradable. Growing numbers of people are fueling their cars with biodiesel from waste oils. Indeed, some college students are creating biodiesel from food waste from dining halls and fast-food restaurants (**FIGURE 20.19**). Some buses and recycling trucks now run on biodiesel, and many state and federal fleets use biodiesel blends.

Some enthusiasts have taken biofuel use further. Eliminating the processing step that biodiesel requires, they use straight vegetable oil in their diesel engines (which requires modifying the engine).

Using waste oil as a biofuel is sustainable, but most biodiesel today, like most ethanol, comes from crops grown specifically for the purpose—and this has environmental impacts. Growing soybeans in Brazil and oil palms in Southeast Asia hastens the loss of tropical rainforest (pp. 306–308). Growing soybeans in the United States and rapeseed in Europe takes up large areas of land as well. Because the major crops grown

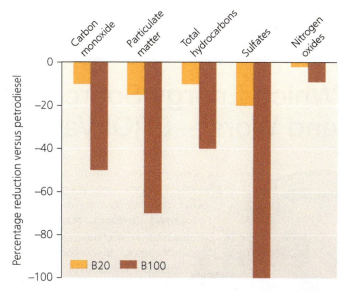

FIGURE 20.18 Burning biodiesel in a diesel engine emits less pollution than burning conventional petroleum-based diesel. Shown are percentage reductions in several major pollutants that one can attain by using B20 (a mix of 20% biodiesel and 80% petroleum-based diesel) and B100 (pure biodiesel). *Data from U.S. Environmental Protection Agency.*

 DATA Which type of pollutant does use of biodiesel reduce most effectively, relative to petroleum-based diesel?

Go to **Interpreting Graphs & Data** on **Mastering**EnvironmentalScience®

for biodiesel and for ethanol exert heavy impacts on the land, farmers and agricultural scientists are experimenting with a variety of other crops, from wheat, sorghum, cassava, and sugar beets to lesser known plants such as hemp, jatropha, and the grass miscanthus.

FIGURE 20.19 At Loyola University Chicago, students and staff produce biodiesel from waste vegetable oil from the dining halls and use it to fuel this van. They transport this mini-diesel reactor to local high schools to teach students about alternative fuels.

Which Energy Sources Have the Best—and Worst—EROI Values?

As our society develops energy sources as alternatives to fossil fuels, we need to be able to compare them so we can make informed decisions about which sources to prioritize. One important aspect to compare is the measure called EROI, or energy returned on investment (p. 520), a ratio that indicates how much energy is produced for each unit of energy that is invested in the energy-producing activity.

It's fairly straightforward to measure how much energy is produced when we simply burn a fuel or when we let water or wind turn a turbine. The difficult part is calculating how much energy is invested across all the many and diverse aspects of finding, extracting, transporting, and using a resource, and in manufacturing the equipment needed to do so. To fully and accurately measure the denominator in an EROI ratio, the researcher must conduct a thorough life-cycle analysis (pp. 262, 618), summing up all the energy invested at each stage in the life-cycle of the process. Now that more researchers are beginning to do this, we are gaining better data that help us compare energy sources.

Dr. Charles Hall, a pioneer of EROI research

Charles Hall of the State University of New York (SUNY) College of Environmental Science and Forestry in Syracuse, New York, has been one of the pioneers in calculating EROI values. In 2014, he and students Jessica Lambert and Stephen Balogh surveyed recent research and compiled EROI estimates for major energy sources (**FIGURE 1**). Their summary, published in

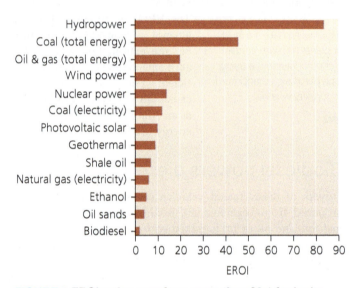

FIGURE 1 EROI ratios vary from more than 80:1 for hydropower down to 2:1 for biodiesel. *Adapted from Hall, C., et al. 2014. EROI of different fuels and the implications for society. Energy Policy 64: 141–152.*

Novel biofuels are being developed

One promising next-generation biofuel crop is algae (better known to most of us as green pond scum!). Several species of these photosynthetic organisms produce large amounts of lipids that can be converted to biodiesel. Alternatively, carbohydrates in algae can be fermented to create ethanol. In fact, a variety of fuels, including jet fuel (**FIGURE 20.20**), can be produced from algae.

FIGURE 20.20 Commercial aircraft are beginning to use biofuels. KLM and several other European airlines have been powering flights with various biofuels for several years. United Airlines has used algae-based biofuel in a 30% blend on flights between Houston and Chicago since 2011. In 2015 it began fueling flights between Los Angeles and San Francisco with fuel derived from waste oils, farm waste, and landfill trash.

the journal *Energy Policy*, revealed a wide variation in the efficiency of our major energy sources.

Hydroelectric power had the best EROI value, averaging 84:1. This means that for each unit of energy we invest in hydropower, we produce 84 units of energy. This is because once dams, reservoirs, and transmission lines are constructed, passing running water through turbines year after year is highly efficient.

Nuclear power places above the middle of the pack in the Hall team's study, with an EROI of 14:1. Nuclear power produces a large amount of electricity, but it also requires substantial investment in uranium mining and in the facilities, equipment, and maintenance of power plants.

Solar power technologies tend to have moderately low EROI values, but as these young renewable technologies are further developed, we can expect EROI values to rise. Indeed, wind power is already among the most efficient energy sources, with an EROI around 20:1.

At the low end of the spectrum in the Hall team's study were biodiesel, oil sands, and ethanol, with EROIs of only 2:1, 4:1, and 5:1, respectively. Mining oil sands and processing their bitumen into usable oil requires huge amounts of energy expenditure (Chapter 19). The biofuels ethanol and biodiesel have low values because most come from crops like corn and soybeans grown in monocultures with industrial agriculture. Large amounts of equipment, fertilizer, pesticides, and irrigation—all powered by or made from fossil fuels—are used to grow these crops and then transport the yields to refineries.

Ethanol has received more research attention than any other fuel in regard to EROI, and results have varied. Early attempts at making cellulosic ethanol resulted in EROI values below 1:1 (energy was lost), whereas, in contrast, one study showed an EROI of 48:1 for ethanol from molasses in India. Earlier work by Hall found EROI values of ethanol from Brazilian sugarcane to average 5:1 but those from U.S. corn ethanol to average only 1.3:1. In fact, researchers have invariably found the EROI of corn ethanol to be remarkably low, and this is a large part of the reason many environmental scientists oppose U.S. government subsidies for corn ethanol.

In one 2011 research paper, Hall, Bruce Dale, and David Pimentel sought to clarify why different studies had come up with different EROI estimates for corn ethanol. By carefully comparing the methods of each study, they determined that the differences were due to how the researchers measured energy investment in the many steps involved in producing ethanol. They recommended that future researchers be more transparent and explicit about their methods as they try to work out the EROI values of newer biofuels, such as those from switchgrass and cellulosic ethanol.

The 2014 review by Hall and his students made clear that EROI values for fossil fuels were historically very high but have been dropping through the years (see Figure 19.6, p. 520). This is because we have already extracted the easy deposits and now need to expend more and more energy to reach the less-accessible deposits that remain. Newer, unconventional, fossil fuel resources such as oil sands and shale oil have low EROI values. This means that if we switch to these sources as conventional oil begins to run out, we will deplete them rapidly (while producing a great deal of pollution), because a great deal of energy will need to be devoted to extracting and processing them.

As studies of EROI become more numerous and higher in quality, they will help scientists, engineers, and policymakers compare the relative advantages of different energy sources. This will allow us to make better decisions as we chart our energy future and shift from fossil fuels to alternative and renewable energy sources.

Algae can be grown at a large scale in outdoor ponds, or more controlled and intensive production can be achieved in closed tanks or in closed transparent tubes called photobioreactors (FIGURE 20.21). Algae grow faster and produce more oil than terrestrial biofuel crops, and they can grow in seawater, saline water, or nutrient-rich wastewater from sewage treatment plants. Because carbon dioxide speeds their growth, algae can even make use of smokestack emissions. Indeed, placing algae farms next to power plants as a source of carbon capture (pp. 504, 538) is a promising combination. Producing biofuels from algae is expensive, but private investment may bring costs down.

Relying on any monocultural crop for energy may not be sustainable. With this in mind, researchers are refining techniques to produce **cellulosic ethanol** by using enzymes to produce ethanol from cellulose, the substance that gives structure to all plant material. This would be a major advance because ethanol as currently made from corn or sugarcane

FIGURE 20.21 Algae are a leading candidate for next-generation biofuels. Here algae grow at a demonstration facility.

FIGURE 20.22 Switchgrass is being studied as a crop to produce cellulosic ethanol, and also provides fuel for biopower.

uses starch, which is nutritionally valuable to us, and proponents of food security voice ethical concerns about using crops for fuel when millions of people continue to die from hunger. Cellulose, in contrast, is of no food value to people yet is abundant in all plants. If we can produce cellulosic ethanol in commercially feasible ways, then ethanol could be made from low-value crop waste (residues such as corn stalks and husks) rather than from high-value crops.

Certain plants hold promise as crops specifically for cellulosic ethanol. Switchgrass (**FIGURE 20.22**) could potentially be a sustainable choice for the United States because it is a native grass of the North American prairies. It could provide wildlife habitat while serving as a crop, especially if planted in a polyculture mixed with other species. Cellulosic ethanol from switchgrass has an EROI ratio of 5:1 (much better than corn ethanol's), and growing this perennial grass could prevent the soil erosion and depletion of soil carbon that would result from removing too much crop waste for ethanol.

weighing the ISSUES

Biofuels

Do you think producing and using ethanol from corn is a good idea? Do the benefits outweigh the drawbacks? Explain why or why not. Do you think we should invest billions of dollars into developing next-generation biofuels such as algae and cellulosic ethanol? Can you suggest ways of using biofuels that would minimize environmental impacts?

Is bioenergy carbon-neutral?

In principle, energy from biomass is carbon-neutral, releasing no net carbon into the atmosphere. This is because burning biomass releases carbon dioxide that plants recently pulled from the air by photosynthesis. However, burning biomass for energy is not carbon-neutral if forests are destroyed to plant bioenergy crops. Forests sequester more carbon (in vegetation and in soil) than croplands do, so cutting forests to plant crops will increase carbon flux to the atmosphere. Bioenergy also is not carbon-neutral if we consume fossil fuel energy to produce the biomass (for instance, by using tractors, fertilizers, and pesticides to grow biofuel crops).

International climate change policy so far has failed to encourage sustainable bioenergy approaches. The Kyoto Protocol (p. 506) required nations to submit data on emissions from energy use and from changes in land use, but only the emissions from energy use were "counted" toward judging how well nations were controlling their emissions under the treaty. Negotiators trying to design a follow-up treaty to Kyoto have been trying to address this (p. 507). In the meantime, researchers are trying to develop means of using bioenergy that are truly renewable and carbon-neutral. With continued research and careful decision making, our many bioenergy options may provide promising avenues for sustainably replacing fossil fuels.

Hydroelectric Power

Next to biomass, we draw more renewable energy from the motion of water than from any other resource. In **hydroelectric power,** or **hydropower,** we use the kinetic energy of flowing river water to turn turbines and generate electricity. We examined hydroelectric power and its environmental impacts in our discussion of freshwater resources (pp. 400–402). Now we will take a closer look at hydropower as an energy source.

Hydropower uses three approaches

Most hydroelectric power today comes from impounding water in reservoirs behind concrete dams that block the flow of river water and then letting that water pass through the dam. Because water is stored behind dams, this is called the **storage** technique. As reservoir water passes through a dam, it turns the blades of turbines, which cause a generator to generate electricity (**FIGURE 20.23**). Electricity generated in the powerhouse of a dam is transmitted by transmission lines to the electrical grid that serves consumers, while the water flows to the riverbed below the dam and continues downriver. The amount of power generated depends on the distance the water falls and the volume of water released. By storing water in reservoirs, dam operators can ensure a steady and predictable supply of electricity, even during periods of naturally low river flow.

An alternative approach is the **run-of-river** technique, which generates electricity without greatly disrupting a river's flow. One method is to divert a portion of a river's flow through a pipe or channel, passing it through a powerhouse and then returning it to the river (**FIGURE 20.24**, p. 572). This can be done with or without a small reservoir that pools water temporarily, and the pipe or channel can be run along the surface or underground. Another method is to flow river water over a dam small enough not to impede fish passage, siphoning off water to turn turbines, and then returning the water to the river. Run-of-river systems are useful in areas remote from established electrical grids and in regions without the economic resources to build and maintain large dams. This approach cannot guarantee reliable water flow in all seasons, but it minimizes many impacts of the storage technique.

(a) Ice Harbor Dam, Snake River, Washington

(b) Generators inside McNary Dam, Columbia River

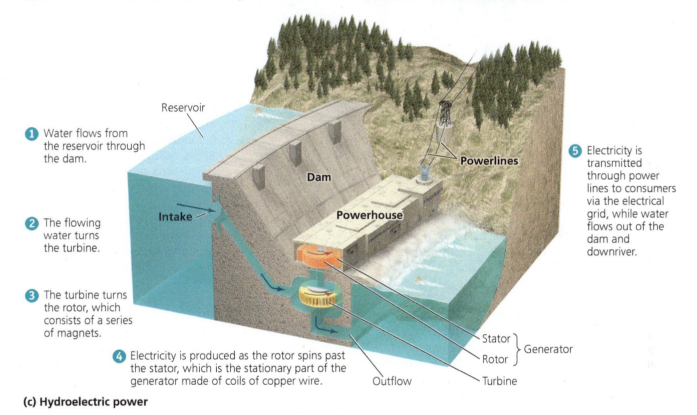

① Water flows from the reservoir through the dam.

② The flowing water turns the turbine.

③ The turbine turns the rotor, which consists of a series of magnets.

④ Electricity is produced as the rotor spins past the stator, which is the stationary part of the generator made of coils of copper wire.

⑤ Electricity is transmitted through power lines to consumers via the electrical grid, while water flows out of the dam and downriver.

Reservoir

Dam

Intake

Powerhouse

Powerlines

Stator ⎫
Rotor ⎭ Generator

Outflow

Turbine

(c) Hydroelectric power

FIGURE 20.23 We generate hydroelectric power with large dams. Inside these dams **(a)**, flowing water turns massive turbines, which rotate generators **(b)** to generate electricity. Water is funneled from the reservoir through the dam **(c)** to rotate turbines, which turn rotors containing magnets. The spinning rotors generate electricity as their magnets pass coils of copper wire. Electrical current is transmitted through power lines, and the river's water flows out through the base of the dam.

To better control the timing of flow, pumped-storage hydropower can be used. In **pumped storage,** water is pumped from a lower reservoir to a higher reservoir during times when demand for power is weak and prices are low. When demand is strong and prices are high, water is allowed to flow downhill through a turbine, generating electricity. Although energy must be input to pump the water, pumped storage can be profitable. When paired with intermittent sources such as solar and wind power, it can help balance a region's power supply by compensating for dips in power.

Hydropower is clean and renewable, yet has impacts

Hydroelectric power has three clear advantages over fossil fuels. First, it is renewable; as long as precipitation falls from the sky and fills rivers and reservoirs, we can use water to turn turbines. Second, hydropower is efficient. It is thought to have an EROI ratio of more than 80:1, higher than any other modern energy source. Third, no carbon compounds are burned in the production of hydropower, so no carbon dioxide

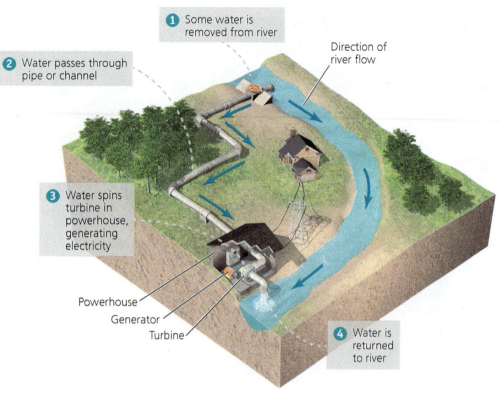

① Some water is removed from river

② Water passes through pipe or channel

Direction of river flow

③ Water spins turbine in powerhouse, generating electricity

Powerhouse
Generator
Turbine

④ Water is returned to river

FIGURE 20.24 Run-of-river systems divert a portion of a river's water. Water may be piped downhill through a powerhouse and released downriver, as shown.

Hydroelectric power is widely used, but it may not expand much more

Hydropower accounts for one-sixth of the world's electricity production (see Figure 20.2b). For nations with large amounts of river water and the economic resources to build dams, hydroelectric power has been a keystone of their development and wealth. Sweden receives 11% of its total energy and nearly half its electricity from hydropower. Canada, Brazil, Norway, Venezuela, and other nations obtain large amounts of their energy from hydropower (**TABLE 20.4**).

The great age of dam building for hydroelectric power (as well as for flood control and irrigation) began in the 1930s in the United States, when the federal government constructed dams as public projects, partly to employ people and help end the economic depression of the time. U.S. dam construction peaked in 1960, when 3123 dams were completed in a single year. American engineers subsequently exported their dam-building technologies, notably to nations of the developing world. In India, Prime Minister Jawaharlal Nehru commented

or other pollutants are emitted into the atmosphere, and this helps safeguard air quality, climate, and human health. Fossil fuels *are* used in constructing and maintaining dams—and large reservoirs may release the greenhouse gas methane as a result of anaerobic decay in deep water. But overall, hydropower accounts for only a small fraction of the greenhouse gas emissions typical of fossil fuel combustion.

Although it is renewable, efficient, and produces little air pollution, hydropower does exert negative impacts. Damming rivers (p. 400) destroys habitat for wildlife as riparian areas above dam sites are submerged and those below dam sites often are starved of water. Because water discharge is regulated to optimize electricity generation, the natural flooding cycles of rivers are disrupted. Suppression of flooding prevents river floodplains from receiving fresh nutrient-laden sediments. Instead, sediments become trapped behind dams, where they begin filling the reservoir. Dams also cause thermal pollution (p. 407), because water downstream may become unusually warm if water levels are kept unnaturally shallow. Moreover, periodic flushes of cold water occur from the release of reservoir water. Such thermal shocks, together with habitat alteration, have diminished or eliminated native fish populations in many dammed waterways. In addition, dams generally block the passage of fish and other aquatic creatures, effectively fragmenting the river and reducing biodiversity in each stretch.

All these ecological impacts generally translate into negative social and economic impacts on local communities. (We discussed the benefits, drawbacks, and impacts of dams more fully in Chapter 15; pp. 400–402.)

TABLE 20.4 Top Producers of Hydropower

NATION	HYDROPOWER PRODUCED (PERCENTAGE OF WORLD TOTAL)	PERCENTAGE OF NATION'S ELECTRICITY GENERATION FROM HYDROPOWER
China	23.8	16.9
Canada	10.1	60.1
Brazil	10.1	68.6
United States	7.5	6.7
Russia	4.7	17.3
India	3.7	11.9
Norway	3.3	96.1
Japan	2.2	8.1
Venezuela	2.2	67.8
France	2.0	13.2

Data from the International Energy Agency.

in 1963 that "dams are the temples of modern India," referring to their central importance in the development of energy capacity in that nation.

Today the world is witnessing some gargantuan hydroelectric projects. China's recently completed Three Gorges Dam (p. 400) is the world's largest. Its reservoir displaced 1.3 million people, and the dam is now generating as much electricity as dozens of coal-fired or nuclear plants.

However, hydropower has less potential for expansion than most other renewable energy sources. The main reason is that most of the world's large rivers that offer excellent opportunities for hydropower are already dammed. For instance, in the wake of Sweden's 1980 referendum to phase out nuclear power, many Swedes had hoped that hydropower could compensate for the electrical capacity that would be lost. However, Sweden had already dammed so many of its rivers that it could not gain much additional hydropower by erecting more dams. In addition, as people have grown more aware of the ecological impacts of dams, residents in some regions are resisting dam construction. This is the case in Sweden, where many citizens have made it clear that they want some rivers to remain undammed, preserved in their natural state. As a result, in Sweden, hydropower's contribution to the national energy budget has remained little changed for 40 years. In the United States, 98% of rivers appropriate for dam construction already are dammed, many of the remaining 2% are protected under the Wild and Scenic Rivers Act, and efforts are underway to dismantle some existing dams and restore river habitats (p. 401).

Overall, hydroelectric power will likely continue to increase in developing nations that have yet to dam their rivers, but in developed nations hydropower growth will likely slow down. The International Energy Agency forecasts that hydropower's share of electricity generation will decline very slightly between now and 2035, whereas the share of other renewable energy sources will nearly triple. Newer renewable energy sources have impressive potential for rapid growth (Chapter 21).

closing
THE LOOP

Given the considerable impacts of fossil fuel extraction and combustion on health, environmental quality, and the global climate, many nations have sought to diversify their energy portfolios with alternative energy sources. The three most developed and widely used alternatives to fossil fuels so far are nuclear power, bioenergy, and hydroelectric power.

Nuclear power showed promise to be a pollution-free and highly efficient form of energy. But escalating costs and public fears over safety in the wake of accidents at Three Mile Island, Chernobyl, and Fukushima stalled its growth, and a few nations are attempting to phase it out completely. Bioenergy sources are diverse and range from traditional fuelwood to newer biofuels and various means of generating biopower. These sources can be carbon-neutral but are not all strictly renewable. Hydropower is a renewable, pollution-free alternative, but it can expand only so much further and also involves substantial ecological impacts.

Some nations already use these three conventional alternatives extensively. Sweden has invested in them so greatly that it has managed to cut its fossil fuel use in half, and as we have learned, its government now aims to make Sweden the first nation in history to free itself of fossil fuels. Whether or not this goal is met, Sweden demonstrates that having a vibrant, healthy, and wealthy society does not in fact require a heavy reliance on fossil fuels. As Sweden navigates its ongoing societal debate over the safety of nuclear power, it can boast that it is already saving thousands of its citizens' lives each year by using far less coal, oil, or gas to generate its electricity. Regardless of the outcome of the debate over nuclear power, today Swedes—and growing numbers of people everywhere—are looking forward to a world powered by renewable energy.

REVIEWING Objectives

You should now be able to:

Discuss the reasons for seeking energy alternatives to fossil fuels

Fossil fuels are nonrenewable resources, and we are gradually depleting their easily accessible reserves. Burning fossil fuels causes air pollution that results in many environmental and health impacts and contributes to global climate change. (pp. 550–551)

Summarize the contributions to world energy supplies of conventional alternatives to fossil fuels

Alternatives to fossil fuels that are most widely used include nuclear power, bioenergy, and hydroelectric power. Biomass provides 10% of global primary energy use, nuclear power provides 5%, and hydropower provides 2%. Nuclear power generates 11% of the world's electricity, and hydropower generates 16%. (pp. 550–552)

Describe nuclear energy and explain how we harness it for electrical power

We gain nuclear power by converting the energy of subatomic bonds into thermal energy, using uranium isotopes. Uranium is mined, enriched, processed into pellets and fuel rods, and used in nuclear reactors. By controlling the reaction rate of nuclear fission, nuclear power plant engineers produce heat that powers electricity generation. (pp. 552–554)

Assess the benefits and drawbacks of nuclear power, and discuss the societal debate over this energy source

Nuclear power is a clean energy source because it does not emit the pollutants that fossil fuels do. It thereby reduces health impacts and helps address climate change. However, for many people the risk of a catastrophic power plant accident, like those at Chernobyl and Fukushima, outweighs these benefits. The disposal of nuclear waste also remains a dilemma: Both temporary storage and single-repository approaches involve risks. Adding to these concerns, economic factors and cost overruns have slowed the nuclear industry's growth. (pp. 554–563)

Describe established and emerging sources and techniques involved in harnessing bioenergy, and assess bioenergy's benefits and shortcomings

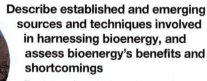

Fuelwood remains the major source of bioenergy, especially in developing nations. We use biomass to generate electrical power in several ways with a variety of source materials. Biofuels such as ethanol and biodiesel are used to power automobiles; some crops are grown for this purpose, and waste oils are also used. Bioenergy is renewable and in principle adds no net carbon to the atmosphere, but overharvesting of wood can lead to deforestation, and growing crops solely for fuel can be inefficient and ecologically damaging. New biofuels from algae and cellulosic ethanol need further research and development but hold promise for more sustainable biofuel production. (pp. 563–570)

Outline the scale, methods, and impacts of hydroelectric power

Hydroelectric power is generated when water from a river runs through a powerhouse and turns turbines. Three major approaches (storage, run-of-river, and pumped storage) each offer advantages. Hydropower produces little air pollution, but dams and reservoirs alter riverine ecology and local economies. Hydropower is the world's largest source of electricity outside of fossil fuels, but its long-term growth potential appears more limited than that of other renewable sources. (pp. 570–573)

TESTING Your Comprehension

1. How much of our global energy supply do nuclear power, bioenergy, and hydroelectric power contribute? How much of our global electricity do these three conventional energy alternatives generate?

2. Describe how nuclear fission works. How do nuclear plant engineers control fission and prevent a runaway chain reaction?

3. In terms of greenhouse gas emissions, how does nuclear power compare to coal, oil, and natural gas? How do hydropower and bioenergy compare?

4. In what ways did the events at Three Mile Island, Chernobyl, and Fukushima Daiichi differ? What consequences resulted from each incident?

5. List several concerns about the disposal of radioactive waste. What has been done so far about its disposal?

6. List five sources of bioenergy. What is the world's most used source of bioenergy? How does bioenergy use differ between developed and developing nations?

7. Describe two biofuels, identify where each comes from, and explain how each is used.

8. Evaluate two potential benefits and two potential drawbacks of bioenergy.

9. Compare and contrast the three major approaches to generating hydroelectric power.

10. Assess two benefits and two negative environmental impacts of hydroelectric power.

SEEKING Solutions

1. Nuclear power has by now been widely used for a half-century, and the world has experienced only two major accidents (Chernobyl and Fukushima) responsible for any significant number of health impacts. Would you call this a good safety record? Should we maintain, decrease, or increase our reliance on nuclear power? Why might safety at nuclear power plants be better in the future? Why might it be worse?

2. How serious a problem do you think the disposal of radioactive waste represents? Describe in detail how you would like to see this issue addressed.

3. There are many different sources of biomass and many ways of harnessing energy from biomass. Discuss one that seems particularly beneficial to you, and one with which you see problems. What bioenergy sources and strategies do you think our society should focus on investing in?

4. **CASE STUDY CONNECTION** Sweden's prime minister has appointed you as the nation's energy minister, and has asked you to recommend and direct energy policies for Sweden. Based on everything you have learned from this chapter, describe how you would approach the question of whether or not to phase out Sweden's use of nuclear power. What specific questions would you ask of scientists and energy experts before making a final policy recommendation? What would you like the nation's energy mix to look like 10 years in the future? And 25 years in the future? What actions would need to be taken to reach your goals?

5. **THINK IT THROUGH** You are the head of the national department of energy in a nation that has just experienced a minor accident at one of its nuclear plants. A partial meltdown released radiation, but the radiation was fully contained inside the containment building, and there were no health impacts on area residents. However, citizens are terrified, and the media are highlighting the dangers of nuclear power. Your nation relies on its five nuclear plants for 25% of its energy and 50% of its electricity needs. It has no fossil fuel deposits and recently began a promising but still-young program to develop renewable energy options. What will you tell the public at your next press conference, and what policy steps will you recommend taking to ensure a safe and reliable national energy supply?

6. **THINK IT THROUGH** You are an investor and would like to invest in alternative energy. You are considering buying stock in companies that (1) construct nuclear reactors, (2) build turbines for hydroelectric dams, (3) build corn ethanol refineries, (4) are ready to supply farm and forestry waste for cellulosic ethanol, and (5) are developing algae farms. For each of these companies, what questions would you research before deciding how to invest your money? How do you expect you might apportion your investments, and why?

CALCULATING Ecological Footprints

Our three conventional energy alternatives vary tremendously in their EROI (energy returned on investment; p. 520). Examine the data on EROI for each of the energy sources as provided in the figure from **The Science behind the Story** on p. 568, and enter the data in the table below.

ENERGY SOURCE	EROI
Coal (total energy)	46
Oil and gas (total energy)	
Oil sands	
Nuclear power	
Hydropower	
Ethanol	
Biodiesel	
Wind power	

1. How many units of energy would you generate by investing 1 unit of energy into producing hydropower? How many units of energy would you need to invest into producing nuclear power if you wanted to generate that same amount of energy? How many units of energy would you need to invest into producing ethanol if you wanted to generate that same amount of energy?

2. Based on EROI values, is it more efficient to get petroleum from conventional oil or from oil sands? Which source would you guess has a larger ecological footprint, based on EROI values?

3. Let's say you wanted to generate 130 units of energy from biodiesel. How many units of energy would you need to invest?

4. Based on EROI values alone, which energy sources would you advocate that society use? Which would you urge society to avoid? What other issues, besides EROI, are worth considering when comparing energy sources?

MasteringEnvironmentalScience®

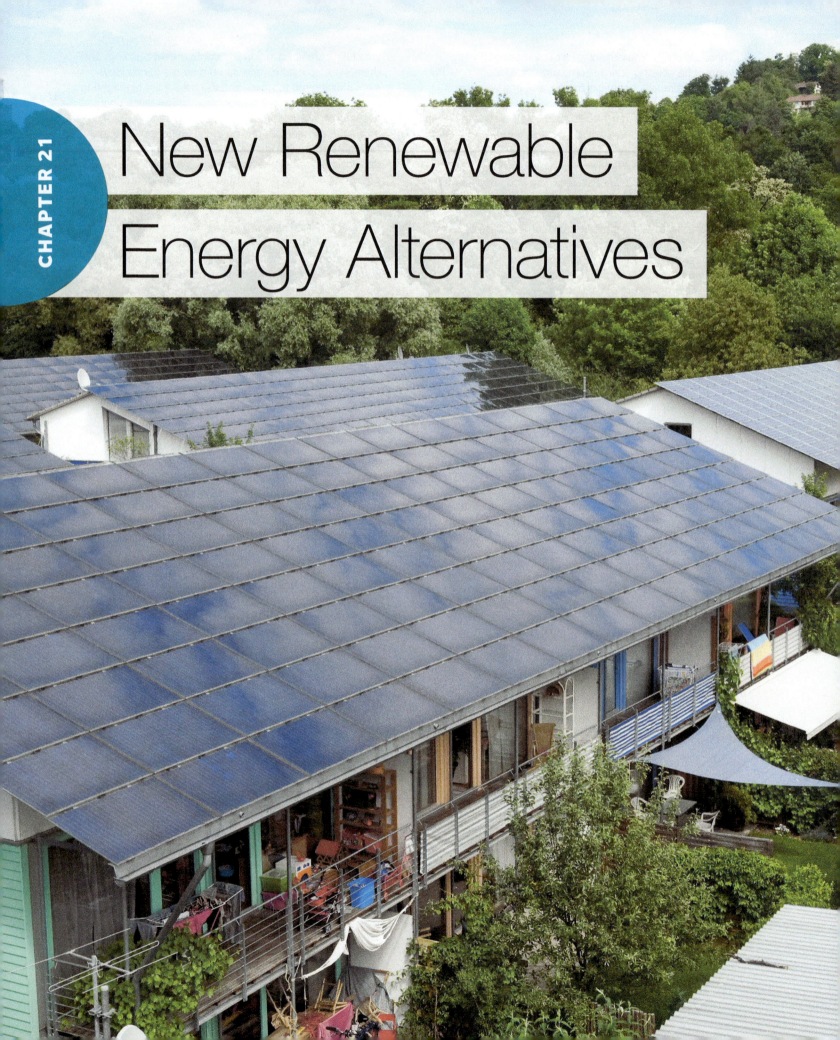

New Renewable Energy Alternatives

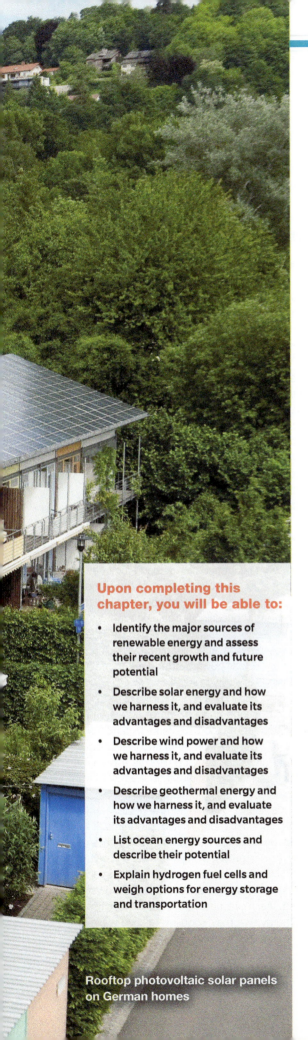

Germany Goes Solar

> [Renewable energy] will provide millions of new jobs. It will halt global warming. It will create a more fair and just world. It will clean our environment and make our lives healthier.
>
> Hermann Scheer, energy expert and member of the German parliament, 2009

> The nation that leads the clean energy economy will be the nation that leads the global economy.
>
> U.S. President Barack Obama, 2010

When we think of solar energy, most of us envision a warm, sunny place like Arizona or southern California. Yet the country that harnesses the most solar power per person in the world is Germany, a northern European nation that receives less sun than Alaska! In recent years Germany has been the world's top installer and user of photovoltaic (PV) solar power, a technology that produces electricity from sunshine. Germany now obtains more than 6% of its electricity from solar power—one of the highest rates in the world.

How is this possible in such a cool and cloudy country? A bold federal policy has used economic incentives to promote solar power and other forms of renewable energy. Germany has a **feed-in tariff** system whereby utilities are required to buy power from anyone who can generate power from renewable energy sources and feed it into the electrical grid. Under this system, utilities must pay guaranteed premium prices for this power under a long-term contract. In response, German homeowners and businesses have rushed to install PV panels and are selling their excess solar power to utilities at a profit.

The feed-in tariffs apply to all forms of renewable energy. As a result, Germany established itself as the world leader in wind power in the 1990s, until being overtaken by the United States and China. It ranks third in the world in using power from biomass, third in solar water heating, third in electrical power capacity from renewable sources, and second in renewable energy generated per person. In total, Germany now gets 28% of its energy from renewable sources.

Germany's push for renewable energy dates back to 1990. In the wake of the disaster at the Chernobyl nuclear power plant in the Soviet Union (p. 556), Germany decided to gradually phase out its own nuclear power plants. However, if these were shut down, the nation would lose virtually all its clean energy and would become utterly dependent on oil, gas, and coal imported from Russia and the Middle East.

Enter Hermann Scheer, a German parliament member and an expert on renewable energy. While everyone else assumed that technologies for harnessing solar, wind, and geothermal energy were costly, risky, and underdeveloped, Scheer saw them as a great economic opportunity—and as the only long-term answer. In 1990, Scheer helped push through a landmark law establishing feed-in tariffs. Ten years later, the law was revised and strengthened: The Renewable Energy Sources Act of 2000 aimed to promote renewable energy production and use, enhance the

Installing PV solar panels

security of the energy supply, reduce carbon emissions, and lessen the many external costs (pp. 141, 162) of fossil fuel use.

Under this landmark law, each renewable energy source is assigned its own payment rate according to market considerations. These rates have been gradually reduced to encourage increasingly efficient means of producing power. However, utilities pass along to consumers their costs of paying the feed-in-tariffs, and in recent years this had amounted to 15% of the average German citizen's electric bill. To reduce these costs, the German government in 2010 decided to slash PV solar tariff rates steeply. The new lower rates would ease the burden on consumers, while homeowners still had financial incentive to invest in solar panels because the feed-in tariffs had helped to cut PV market prices in half.

When German consumers heard that the tariff rate would be slashed after 2010, sales of PV modules skyrocketed as people rushed to lock in contracts at the existing rate. Announcements of additional tariff reductions the next two years also spurred rushes to install more PV systems before the rates dropped further. In 2010, 2011, and 2012, Germans installed more than 7 gigawatts of PV solar capacity each year—an amount surpassing the total cumulative capacity of solar power in the United States at the time. Solar installations in Germany then began slowing in 2013 as tariff rates became low. The German government plans to end subsidies completely once the nation reaches 52 gigawatts of cumulative capacity, which is roughly 35% beyond its total as of 2015 (FIGURE 21.1). Currently Germans are debating the reductions in tariff payments. Some people say they are being cut too much and too soon, endangering the 52-gigawatt goal; others say PV solar has grown too fast to be integrated well into the nation's power grid and that the cuts are wise.

By reducing the subsidies gradually, Germany's leaders aim to encourage technological innovation for efficiency within the solar industry, thereby creating a stronger industry that can sustain growth over the long term and can outcompete foreign companies for international business. Indeed, boosted by domestic demand, German industries have become global leaders in "green tech," designing and selling renewable energy technologies around the world. Germany is second in PV production behind China, leads the world in production of biodiesel, and has built several cellulosic ethanol (p. 569) facilities. Renewable energy industries in Germany today employ nearly 400,000 citizens.

By 2050, Germany aims to obtain 80% of its electricity from renewable sources. To achieve this historic energy transition, which Germans call the *Energiewende*, the government has been allotting more public money to renewable energy than any other nation—more than $25 billion annually in recent years.

As Germany replaces fossil fuels with renewable energy, it is improving its air quality and helping to fight climate change. Since 1990, carbon dioxide emissions from German energy sources have fallen by 25%, and emissions of seven other major pollutants (CH_4, N_2O, SO_2, NO_x, CO, volatile organic compounds, and dust) have been reduced by 12–95%. Much of this success was due to the shutdown of old polluting facilities in the former East Germany after reunification in 1990, but at least half of the emissions reductions are attributed to renewable energy paid for under the feed-in tariff system.

The planned transition from fossil fuels in Germany has not been completely smooth, however. Following the Fukushima nuclear disaster in Japan in 2011 (pp. 22, 556), the German government shut down 7 of its 15 nuclear power plants and promised to phase out the rest by 2022. The resulting dramatic reduction in electricity supply from nuclear power caused rates to rise and led to an increase in power generated from the combustion of coal, which had become cheaper after demand for it had fallen due to the growth of solar and wind power. At the same time, the rise of solar and wind, two intermittent power sources, has meant that the German grid occasionally gets flooded with excess electricity. The challenge for Germany's *Energiewende* will be to keep renewables growing at a steady rate that can compensate for the loss of nuclear power and reduce the reliance on coal without causing instability in supplies.

Germany's experience is serving as an international model. As of 2016, more than 70 nations and dozens of states and provinces had implemented some sort of feed-in tariff for renewable energy. Nations with high tariff rates such as Spain and Italy ignited their wind and solar development as a result. In North America, Vermont and Ontario established feed-in tariff systems similar to Germany's, while California, Hawai'i, Maine, New York, Oregon, Rhode Island, Washington, and utilities in several additional states conduct more limited programs. In 2010, Gainesville, Florida, became the first U.S. city to establish feed-in tariffs, and solar power grew quickly there as a result. Moreover, utilities in 46 U.S. states now offer net metering, in which utilities credit customers who produce renewable power and feed it into the grid. As more nations, states, and cities encourage renewable energy, we may soon experience a historic transition in the way we meet our energy demands.

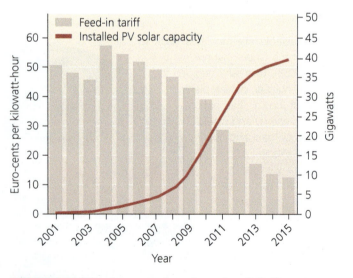

FIGURE 21.1 PV solar power has risen steeply in Germany, thanks to feed-in tariffs. The rapid expansion of installed capacity of PV solar is now slowing as Germany's feed-in tariffs have been reduced to lower levels. *Data from International Energy Agency, 2016.* Photovoltaic power systems programme annual report 2015.

"New" Renewable Energy Sources

Germany's bold federal policy is just one facet of a global shift toward renewable energy. Across the world, nations are searching for ways to move away from fossil fuels while ensuring a reliable, affordable, sustainable supply of energy for their economies. This is because the economic and social costs, national security risks, and health and environmental impacts of fossil fuel dependence (Chapter 19) are all intensifying.

The two renewable energy sources that are most widely used throughout the world are bioenergy, the energy from combustion of biomass (wood and other plant matter), and hydropower, the energy from running water (Chapter 20). These conventional alternatives to fossil fuels are renewable, but they can be depleted with overuse and they exert some undesirable environmental impacts.

In this chapter we explore a group of alternative energy sources that are often called "new renewables." These diverse sources include energy from the sun, from wind, from Earth's geothermal heat, and from ocean water. These energy sources are not truly new. In fact, they are as old as our planet, and people have used them for millennia. We commonly refer to them as "new" because (1) they are just beginning to be used on a wide scale in our modern industrial society, (2) they are harnessed using technologies still in a rapid phase of development, and (3) they will likely play much larger roles in the future.

New renewable sources are growing fast

As with other energy sources, the new renewables provide energy for three types of applications: (1) to generate electricity, (2) to heat air or water, and (3) to fuel vehicles. Their potential is enormous, yet so far their contribution to our society's overall energy budget remains small. Today we obtain just over 1% of our global energy from the new renewable energy sources, whereas fossil fuels still provide 81% of the world's energy (see Figure 20.2a, p. 551).

Likewise, the new renewable sources are making a similarly small contribution to our global generation of electricity thus far (see Figure 20.2b, p. 551). At this point, 22% of our electricity comes from renewable energy sources overall, but hydropower accounts for nearly three-quarters of this amount.

Nations and regions vary in the renewable sources they use. In the United States, most renewable energy comes from biomass and hydropower. As of 2015, wind power accounted for 19% of renewable energy, solar energy for 5.7%, and geothermal energy for 2.3% (**FIGURE 21.2a**). Of electricity generated in the United States from renewables, hydropower accounts for over 44%, while as of 2015 wind power contributed 34%, solar 6.9%, and geothermal 3.0% (**FIGURE 21.2b**).

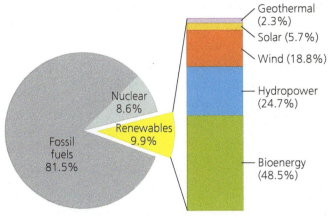

(a) U.S. consumption of renewable energy, by source

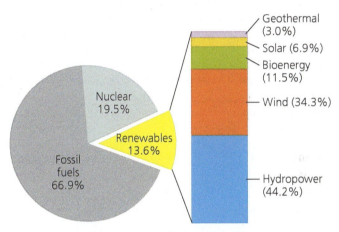

(b) U.S. electricity generation from renewable sources

FIGURE 21.2 About ten percent of the energy consumed in the United States each year comes from renewable sources. Of this amount **(a)**, most derives from bioenergy and hydropower. Wind power, solar energy, and geothermal energy together account for just over 25% of this amount. Similarly, just 13.6% of electricity generated in the United States **(b)** comes from renewable energy sources, predominantly hydropower. *Data are for 2015, from Energy Information Administration, U.S. Department of Energy, 2016.*

DATA What percentage of overall U.S. energy consumption does solar power contribute? What percentage of overall U.S. electricity generation does wind power contribute?

Go to **Interpreting Graphs & Data** on **Mastering**EnvironmentalScience®

Although they currently make up a small proportion of our energy budget, the new renewable energy sources are growing quickly. Over the past four decades, solar, wind, and geothermal energy sources have grown far faster than has the overall energy supply. The long-term leader in growth is wind power, which has expanded by nearly 50% *each year* since the 1970s. In recent years, solar power has grown faster than wind (**FIGURE 21.3**, p. 580). Because these sources started from such low levels of use, however, it will take them some time to catch up to conventional sources.

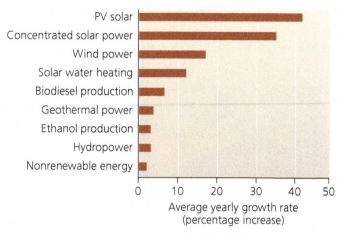

FIGURE 21.3 **The new renewable energy sources are growing far faster than conventional energy sources.** Shown are average rates of growth each year between 2010 and 2015. *Data from REN21, 2016. Renewables 2016: Global status report. Paris: REN21, UNEP.*

DATA At the rate shown, if you began with 10 units of PV solar capacity, how much would exist after 5 years?

Go to **Interpreting Graphs & Data** on **Mastering**EnvironmentalScience®

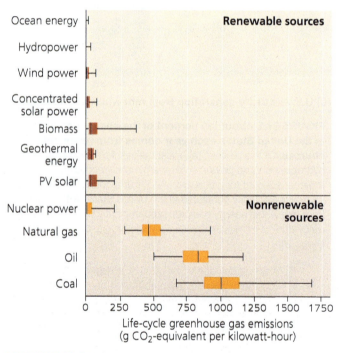

FIGURE 21.4 **Renewable energy sources release far fewer greenhouse gas emissions than fossil fuels.** Shown are ranges of estimates from scientific studies of each source when used to generate electricity. *Data from Intergovernmental Panel on Climate Change, 2012. Renewable energy sources and climate change mitigation. Special report. New York: Cambridge University Press.*

DATA For every unit of emissions released from 1 kilowatt-hour of electricity generated by PV solar, roughly how many units would be released from 1 kilowatt-hour of electricity generated by coal combustion?

Go to **Interpreting Graphs & Data** on **Mastering**EnvironmentalScience®

The new renewables offer advantages

Use of new renewables has been expanding because of growing concerns over diminishing supplies of conventional fossil fuels and because of the many environmental and health impacts of fossil fuel combustion (Chapter 19). The new renewables offer several benefits over fossil fuels. They help alleviate many types of air pollution (Chapter 17). In particular, they help reduce the greenhouse gas emissions that drive global climate change (Chapter 18) (**FIGURE 21.4**). Renewables also can diversify an economy's energy mix, helping to reduce price volatility and buffer us against restrictions in supply of imported fuels. Moreover, novel energy sources can generate income and property tax revenue for communities, especially in rural areas passed over by other economic development. And unlike fossil fuels, renewable sources are inexhaustible on timescales relevant to our society.

Shifting to renewable energy also creates new employment opportunities. The design, installation, maintenance, and management required to develop technologies and rebuild and operate our society's energy infrastructure are becoming major sources of employment today, through **green-collar jobs.** Already more than 8 million people work in renewable energy jobs around the world (**FIGURE 21.5**).

Whether we can—or even should—give up fossil fuels entirely and switch to 100% renewable energy remains a matter of some debate among energy experts. Many are doubtful we can make a complete transition soon, but some recent research makes a strong case that we can transition quickly and gain substantial benefits by doing so (see **THE SCIENCE BEHIND THE STORY**, pp. 582–583).

FAQ

Isn't renewable energy too expensive and untested to rely on?

Some approaches for harnessing renewable energy are not yet ready for prime time. For instance, many designs for ocean energy are in the planning stages and may never be used commercially. And researchers are experimenting with dozens of types of solar cells in the lab, yet only a few will ever be developed for widespread use. But already a number of highly reliable and well-established renewable energy sources and technologies have been supplying power to millions of people for decades. This goes for hydropower and bioenergy (Chapter 20), of course, but also for most of the "new renewable" sources. These range from geothermal power and ground-source heat pumps to solar water heating and PV solar cells to onshore and offshore wind power. All are becoming more affordable, and in many locations electricity from wind is already cheaper than electricity from fossil fuel combustion.

Policy and investment can accelerate our transition from fossil fuels

Rapid growth in renewable energy seems likely to continue as global population and consumption rise, easily accessible fossil fuel supplies decline, and people demand cleaner environments. Yet we cannot wholly convert to renewable energy

FIGURE 21.5 Renewable energy creates new green-collar jobs. More than 8 million people worldwide were employed in jobs directly and indirectly connected to renewable energy, as of 2015. *Data from REN21, 2016.* Renewables 2016: Global status report. *Paris, REN21, UNEP.*

JOBS CREATED

PV solar: 2,772,000
Biofuels: 1,678,000
Wind power: 1,081,000
Solar heating: 939,000
Biomass: 822,000
Other: 760,000

sources overnight, because we lack the infrastructure needed to transfer huge volumes of power from renewable sources inexpensively on a continent-wide scale. Moreover, although prices are falling and some sources have become cost-competitive, most renewable energy remains more expensive than fossil fuel energy (**FIGURE 21.6**).

However, renewables can overcome technological and economic barriers if they are lent political support. As more governments, utilities, corporations, and consumers promote and use renewable energy, market prices of renewables

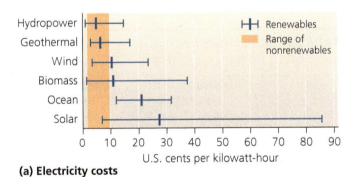

(a) Electricity costs

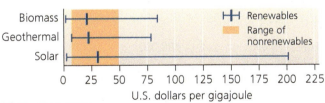

(b) Heating costs

FIGURE 21.6 Most renewable energy sources have market prices greater than those for nonrenewable sources, but some are competitive. Shown are price ranges for each source used for **(a)** electricity and **(b)** heating. *Data from Intergovernmental Panel on Climate Change, 2012.* Renewable energy sources and climate change mitigation. Special report. *New York: Cambridge University Press.*

should continue to fall, further hastening their adoption. Feed-in tariffs like Germany's are a prime example of an economic policy tool (p. 177) that can hasten the spread of renewable energy by creating financial incentives for businesses and individuals. Governments can also set goals or mandate that certain percentages of power come from renewable sources. As of 2016, more than 160 nations and 30 U.S. states had set official targets for renewable energy use. Governments also invest in research and development of technologies, lend money to renewable energy businesses as they start up, and offer tax credits and tax rebates to companies and individuals who produce or buy renewable energy.

When a government boosts an industry with such policies, the private sector often responds with investment of its own as private investors recognize an enhanced chance of success and profit. As a result, throughout the world in recent years, advances in renewable energy have been tightly linked to national, state, or local policies that encourage them and to the investment funding that flows more freely as a result. Global investment (public plus private) in renewable energy rose to $329 billion in 2015—six times the amount just a decade earlier. In a feedback process, investment may breed success, and success may breed further investment.

Yet the economics of renewable energy have been erratic. Technologies evolve quickly, and policies vary from place to place and can change unpredictably. As a result, renewable markets have been volatile, prices have fluctuated steeply, growth has often come in rapid bursts, and many promising companies have met with bankruptcy. Moreover, we are still learning how to apply policies wisely. In 2007 Spain adopted feed-in tariffs that were so generous (58 cents per kilowatt-hour) that thousands of people rushed to set up solar facilities. Within just a year, Spain increased its solar capacity fivefold. However, the generous payments had attracted a flood of unqualified contractors and speculators; as a result, many solar plants were hastily built, poorly designed, or badly located. The government realized that many of these new plants would never be economically self-sufficient, so it abruptly slashed the tariff rates.

Critics of public subsidies for renewable energy complain that funneling taxpayer money to particular energy sources is inefficient and skews the market. Instead, they propose, we should let energy sources compete freely. Proponents of renewable energy point out that governments have long

Can We Power the World with Renewable Energy?

Mark Jacobson of Stanford University

Despite the rapid growth of renewable energy in recent years, many experts remain skeptical that we will ever be able to replace fossil fuels entirely. Yet recent scientific research has outlined in detail how we could power our society completely with clean renewable energy—without fossil fuels, biofuels, or nuclear power.

Mark Jacobson, director of the Atmosphere/Energy Program at Stanford University, has been assessing energy sources for years with the goal of finding solutions to pollution, climate change, and energy insecurity. In 2009 he published a full life-cycle analysis (p. 618) of the social, health, and environmental impacts of all major energy sources and found that new renewable sources based on wind, water, and solar power were lower-impact than biofuels, nuclear power, and fossil fuels.

Jacobson then teamed up with Mark Delucchi of the University of California–Davis to examine whether and how the world could meet 100 percent of its energy needs with clean renewable energy from the sun, wind, and water. In 2011 these researchers published a pair of scientific papers in the journal *Energy Policy*.

In this research, Jacobson and Delucchi first calculated the likely global demand for energy in the years 2030 and 2050, using projections from the U.S. Energy Information Administration. They then examined the current outputs and limitations of renewable energy technologies and selected those that were technically and commercially proven and established. Because electrical power is more energy-efficient than fuel combustion, they chose to propose only electrical technologies. (For instance, battery-electric vehicles make use of power much more efficiently than gasoline-powered vehicles, so that far less energy is required.) The efficiency savings from electrification lower the projected global energy demand in 2030 by more than 30%.

The researchers calculated what it would take to manufacture these technologies at the needed scale and to build infrastructure for renewable energy storage and transmission throughout the world. Because sources such as solar power and wind power are intermittent (varying from hour to hour and day to day), Jacobson and Delucchi judged what balance of sources was needed to compensate for intermittency and ensure a consistent, reliable energy supply.

Once all the math was done, Jacobson and Delucchi concluded that the world *can*, in fact, fully replace fossil fuels, nuclear power, and biofuels and meet all its energy demands with clean renewable sources alone. They proposed a quantitative breakdown of the various wind, water, and solar technologies needed to power the world in 2030 (**TABLE 1**). The researchers judged that a full transition was technically feasible by 2030—but because this seemed politically unlikely, they suggested producing all energy newly added between now and 2030 with renewable sources and then gradually replacing fossil fuel and nuclear power facilities with new renewables by 2050.

To achieve this transition, society would need to greatly expand its transmission infrastructure, construct fleets of fuel-cell-powered vehicles and ships, and more. To deal with intermittency of sun and wind and prevent gaps in energy supplies, we would need to link complementary combinations of sources across large regions, and when there is oversupply, use the extra energy to produce hydrogen fuel.

With an eye toward impacts, Jacobson and Delucchi also added up the total area of land that all this new renewable energy infrastructure would require. For example, concentrated solar power plants sprawl across desert terrain, and wind turbines need to be widely spaced to avoid turbulence that hinders their operation. The researchers calculated that 0.74% of Earth's land surface would be occupied directly by energy

TABLE 1 Renewable Energy Infrastructure Needed to Power the World in 2030

TECHNOLOGY	NUMBER OF PLANTS OR DEVICES NEEDED	PERCENTAGE OF GLOBAL DEMAND SATISFIED
Wind turbines	3,800,000	50
Wave devices	720,000	1
Geothermal plants	5350	4
Hydroelectric plants	900	4
Tidal turbines	490,000	1
Rooftop PV systems	1,700,000,000	6
Solar PV plants	40,000	14
CSP plants	49,000	20

Data from Jacobson, M.Z., and M.A. Delucchi, 2011. Providing all global energy with wind, water, and solar power, Part I: Technologies, energy resources, quantities and areas of infrastructure, and materials. *Energy Policy* 39: 1154–1169.

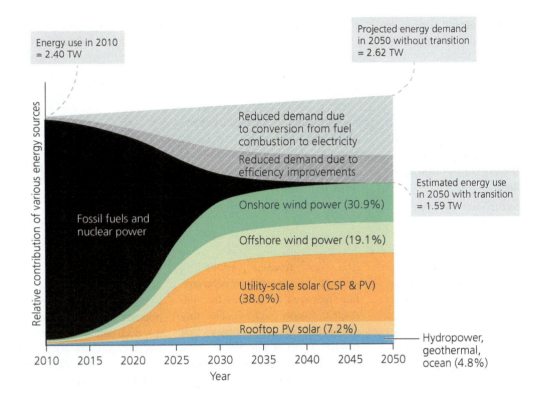

Energy use in 2010 = 2.40 TW

Projected energy demand in 2050 without transition = 2.62 TW

Reduced demand due to conversion from fuel combustion to electricity

Reduced demand due to efficiency improvements

Estimated energy use in 2050 with transition = 1.59 TW

Fossil fuels and nuclear power

Onshore wind power (30.9%)

Offshore wind power (19.1%)

Utility-scale solar (CSP & PV) (38.0%)

Rooftop PV solar (7.2%)

Hydropower, geothermal, ocean (4.8%)

Relative contribution of various energy sources

Year

FIGURE 1 Fossil fuels and nuclear power can be phased out and replaced entirely by renewable energy by 2050 in the United States, research indicates. This graph summarizes the transition to renewables proposed by Jacobson's team. Heights of each colored band at a given year show percentages of total U.S. energy demand to be met by each energy source in that year. *Data from Jacobson, M., et al., 2015. 100% clean and renewable wind, water, and sunlight (WWS) all-sector energy roadmaps for the 50 United States. Energy Environ. Sci. 8: 2093–2117.*

infrastructure, with 0.41% being newly required beyond what is already taken up. An additional 1.18% of area would be needed for spacing between structures (mostly wind turbines), but half of this area could be over water if half our wind power is offshore, and most land between turbines could continue to be used for farming, grazing, or other uses. The researchers calculated that the amount of available developable area for harnessing each resource far exceeds what would be needed.

In terms of economic costs, Jacobson and Delucchi estimated that the overall cost of energy in their proposed scenario would be roughly the same as the cost of energy today. One main challenge, however, is the limited availability of a handful of rare-earth metals (such as platinum, lithium, indium, tellurium, and neodymium) used in certain materials and equipment for wind, water, and solar technologies. While additional reserves of these metals may be discovered and mined, we will likely need to enhance efforts to recycle them.

Critics of Jacobson and Delucchi's work, such as Australian energy expert Ted Trainer, felt that they underestimated the costs of their proposal, overestimated efficiency gains from electrification, and failed to offer persuasive quantitative evidence that intermittency can be overcome. Jacobson and Delucchi staunchly defended their work against these critiques, while also agreeing that the scenario they propose may not be the optimal one and that research should continue.

More recently, Jacobson and Delucchi have worked with eight colleagues to design plans for how each of the 50 U.S.

states might power itself entirely with wind, water, and solar sources. This research team evaluated each state's resources and energy demands, and in 2015 published a study in the journal *Energy and Environmental Science* that summarized its state-specific "roadmaps." The researchers determined a mix of renewable sources and technologies that could feasibly meet the energy demand forecast for the United States in 2050. **FIGURE 1** shows their overall plan for transitioning to 100% renewable energy across the United States. Footprints of the infrastructure required would be 0.47% of land area (0.42% new), with spacing area of 2.4% (1.6% new).

Across all 50 states, the team estimated that a full conversion to renewable energy would raise the number of energy jobs from 3.9 million to 5.9 million and would eliminate up to 62,000 premature deaths per year due to pollution, saving $600 billion in costs. The proposal would also save an estimated $3.3 trillion in costs from predicted climate change impacts worldwide due to U.S emissions in 2050. As a result, they calculated, the average American citizen in 2050 would save $260 in energy costs, $1500 in health costs, and $8300 in climate change costs each year.

As with Jacobson and Delucchi's 2011 global analysis, the team concluded that most barriers to achieving a transition to renewables are social and political, not technological or economic. If governments implement policies to hasten the development of renewable alternatives, this body of research is telling us, we *can* soon switch fully to clean renewable energy.

subsidized fossil fuels and nuclear power far more than they are now subsidizing renewable energy. As a result, there has never been a level playing field, nor a truly free market.

One recent study by venture capitalist Nancy Pfund and Yale University graduate student Ben Healey dug into data on the U.S. government's many energy subsidies and tax breaks (p. 178) over the past century. The report revealed three main findings:

1. Together, oil and gas have received 75 times more subsidies—and nuclear power has received 31 times more—than new renewable energy sources (**FIGURE 21.7a**). This alone is not surprising, because oil, gas, and nuclear power have been major contributors to the U.S. energy supply for a longer time period than solar, wind, and geothermal sources.

2. Per year, oil and gas have received 13 times more subsidies than new renewable energy sources, and nuclear power has received more than 9 times more support than new renewable energy (**FIGURE 21.7b**). This is notable, because a per-year comparison controls for the amount of time each source has been used and thereby shows that oil, gas, and nuclear power have received far more support than solar, wind, and geothermal power.

3. Even in the earliest years of each energy source (when subsidies are most useful), oil, gas, and nuclear power received far more in subsidy support than the new renewables have received. In fact, only in one single year have solar, wind, and geothermal combined *ever* received as much support as the *lowest* amount ever offered to oil, gas, or nuclear power.

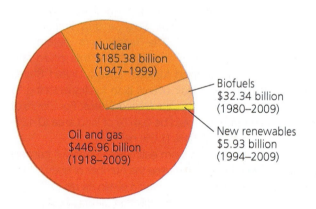

(a) Total subsidies

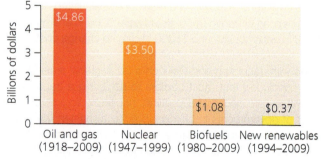

(b) Per-year subsidies

Many feel that the subsidies showered on nonrenewable energy sources have helped to enhance America's economy, national security, and international influence by establishing thriving global energy industries dominated by U.S. firms. Yet by this logic, if America wants to be a global leader to rival nations such as China and Germany in the transition to clean and renewable energy, then it will likely need to direct greater political and financial support toward these new energy sources.

Solar Energy

The sun releases astounding amounts of energy by converting hydrogen to helium through nuclear fusion (p. 554). The tiny proportion of this energy that reaches Earth is enough to drive most processes in the biosphere, helping to make life possible on our planet. Each day, Earth receives enough **solar energy,** or energy from the sun, to power human consumption for a quarter of a century. On average, each square meter of Earth's surface receives about 1 kilowatt of solar energy—17 times the energy of a lightbulb. As a result, a typical home has enough roof area to meet all its power needs with rooftop panels that harness solar energy. However, we are still in the process of developing solar technologies and learning the most effective and cost-efficient ways to put the sun's energy to use.

We can collect solar energy using passive or active methods

The simplest way to harness solar energy is through **passive solar energy collection.** In this approach, buildings are designed to maximize absorption of sunlight in winter yet to keep the interior cool in the heat of summer (**FIGURE 21.8**). South-facing windows maximize the capture of winter sunlight. Overhangs shade windows in summer, when the sun

FIGURE 21.7 Fossil fuels and nuclear power have received far more in U.S. government subsidies than have renewable energy sources. This is true both for **(a)** total amounts over the past century and for **(b)** average amounts per year. *Data are in 2010 dollars, from Pfund, N., and B. Healey, 2011. What would Jefferson do? The historical role of federal subsidies in shaping America's energy future. DBL Investors.*

Overhang shades summer sun from above.

Plants buffer house from temperature swings.

Thermal mass in floors and walls absorbs heat and then slowly releases it.

South-facing porch lets in low-angle sunlight in winter.

FIGURE 21.8 Passive solar design elements can be seen in this house built by college students. Stanford University students and staff designed and constructed this solar-powered house as part of the sixth biannual Solar Decathlon (p. 654). In this event, college and university teams compete to build the best houses fully powered by solar energy.

is high in the sky and when cooling, not heating, is desired. Planting vegetation around a building buffers it from temperature swings. Passive solar techniques also use materials that absorb heat, store it, and release it later. Such *thermal mass* (of straw, brick, concrete, or other materials) often makes up floors, roofs, and walls, or can be used in portable blocks. Passive solar approaches are an important component of green building design (p. 343). By heating buildings in cold weather and cooling them in warm weather, passive solar methods conserve energy and reduce energy costs.

Active solar energy collection makes use of devices to focus, move, or store solar energy. One common method involves installing *flat plate solar collectors* on rooftops to heat water and air for homes and businesses. These panels generally consist of dark-colored, heat-absorbing metal plates mounted in flat glass-covered boxes. Water, air, or antifreeze runs through tubes that pass through the collectors, transferring heat from the collectors to the building or its water tank (**FIGURE 21.9**). Heated water can be stored for later use and passed through pipes designed to release the heat into the building.

More than 300 million households and businesses worldwide heat water with solar collectors. China is the world's leader in this technology, having installed 70% of the world's solar collectors. The United States is a very distant second, and most American solar heating units are used for swimming pools. Germany is third; feed-in tariffs motivated Germans to install 200,000 new systems in one year (2008) alone. Solar heating systems are especially useful in isolated locations. In Gaviotas, a remote town on the high plains of Colombia in South America, residents use active solar technology for heating, cooling, and water purification. This is but one example illustrating that solar energy need not be confined to wealthy communities or to regions that are always sunny.

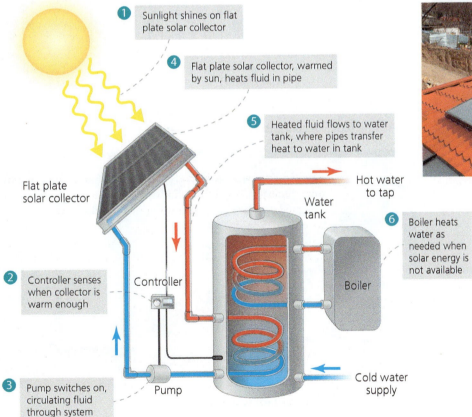

1 Sunlight shines on flat plate solar collector

4 Flat plate solar collector, warmed by sun, heats fluid in pipe

5 Heated fluid flows to water tank, where pipes transfer heat to water in tank

Flat plate solar collector

Hot water to tap

Water tank

6 Boiler heats water as needed when solar energy is not available

2 Controller senses when collector is warm enough

Controller

Boiler

3 Pump switches on, circulating fluid through system

Pump

Cold water supply

FIGURE 21.9 Solar collectors provide heating. Systems for heating water vary in design, but typically 1 sunlight is gathered on a flat plate solar collector, until a controller 2 switches on a pump 3 to circulate fluid through pipes to the collector. The sunlit collector heats the fluid 4, which flows through pipes 5 to a water tank. The hot fluid in the pipes transfers heat to the water in the tank, and this heated water is available for the taps of the home or business. Generally an external boiler 6 kicks in to heat water when solar energy is not available.

Concentrating sunlight focuses energy

We can intensify solar energy by gathering sunlight from a wide area and focusing it on a single point. This is the principle behind solar cookers, simple portable ovens that use reflectors to focus sunlight onto food and cook it (**FIGURE 21.10a**). Such cookers are proving extremely useful in the developing world.

At much larger scales, utilities are using the principle behind solar cookers to generate electricity. **Concentrated solar power (CSP)** is being harnessed by several methods (**FIGURE 21.10b**) in sunny regions in Spain, the U.S. Southwest, and elsewhere. The dominant technology used so far is the parabolic trough approach (leftmost diagram in Figure 21.10b), in which curved mirrors focus sunlight onto synthetic oil in pipes. The superheated oil is piped to an adjacent facility where it heats water, creating steam that drives turbines that in turn generate electricity. In another approach, numerous mirrors concentrate sunlight onto a receiver atop a tall "power tower" (**FIGURE 21.10c**). From this central receiver, heat is transported by air or fluids (often molten salts) through pipes to a steam-driven generator to create electricity. CSP facilities can harness light from lenses or mirrors spread across large areas of land, and the lenses or mirrors may move to track the sun's movement.

Although CSP facilities need to be located in sunny areas, they have great potential. The International Energy Agency has estimated that just 260 km² (100 mi²) of Nevada desert dedicated to CSP facilities could generate enough electricity to power the entire U.S. economy. Indeed, the world's deserts soak up enough sun in just six hours to power humanity's energy demands for an entire year!

To harness this potential, German scientists, industrialists, and investors a decade ago spearheaded an ambitious effort to create an immense CSP facility in the Sahara Desert. The $775 billion DESERTEC project called for thousands of mirrors to harness the Sahara's sunlight and transmit electricity to Europe, the Middle East, and North Africa. Critics of the project—including German parliament member Hermann Scheer—argued that it would be vulnerable to sandstorms and political disputes and would be less reliable and more expensive than decentralized production from rooftop panels. Such concerns led investors to back out of the project, and today its future is uncertain. Across the world, people are considering the environmental impacts that such large-scale developments may pose (see **THE SCIENCE BEHIND THE STORY**, pp. 588–589).

(a) Solar cooker in India

  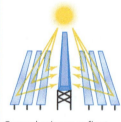 

Curved reflectors heat liquid in horizontal tubes

Each curved reflector focuses light on its own small receiver

Curved mirrors reflect light onto absorber tube

Field of mirrors focuses light on central power tower

(b) Four methods of concentrating solar power

FIGURE 21.10 By concentrating solar energy, we can provide heat and electricity. Solar cookers **(a)** focus solar radiation to cook food. Utilities concentrate solar power with several approaches **(b)** to generate electricity at large scales. At the Planta Solar 10 (PS10) facility in Spain **(c)**, 624 mirrors reflect sunlight onto a central receiver atop a 115-m (375-ft) power tower.

(c) The PS10 power tower facility in Spain

Photovoltaic cells generate electricity directly

The most direct way to produce electricity from sunlight involves photovoltaic (PV) systems. **Photovoltaic (PV) cells** convert sunlight to electrical energy when light strikes one of a pair of plates made primarily of silicon, a semiconductor that conducts electricity. The light causes one plate to release electrons, which are attracted by electrostatic forces to the opposing plate. Connecting the two plates with wires enables the electrons to flow back to the original plate, creating an electrical current (direct current, DC), which can be converted into alternating current (AC) and used for residential and commercial electrical power (**FIGURE 21.11**). Small PV cells may already power your watch or your calculator. Atop the roofs of homes and other buildings, PV cells are arranged in modules, which make up panels that can be gathered together in arrays. Arrays of PV panels can be seen on the roofs of the German homes in the photo that opens this chapter (p. 576).

Researchers are experimenting with variations on PV technology, including **thin-film solar cells,** photovoltaic materials compressed into ultra-thin sheets. Thin-film solar cells are lightweight and far less bulky than the standard crystalline silicon cells shown in Figure 21.11. Although less efficient at converting sunlight to electricity, they are cheaper to produce. Thin-film technologies can be incorporated into roofing shingles and potentially many other types of surfaces, even highways! For these reasons, some people view thin-film solar technologies as a promising direction for the future.

Photovoltaic cells of all types can be connected to batteries that store the accumulated charge until needed.

Alternatively, producers of PV electricity can sell their power to their local utility if they are connected to the regional electrical grid. In parts of 46 U.S. states, homeowners can sell power to their utility in a process called **net metering,** in which the value of the power the consumer provides is subtracted from the consumer's monthly utility bill. Feed-in tariff systems like Germany's go a step further by paying producers more than the market price of the power, thereby offering power producers the hope of turning a profit.

Solar energy offers many benefits

The fact that the sun will continue burning for another 4–5 billion years makes it inexhaustible as an energy source for human civilization. Moreover, the amount of solar energy reaching Earth should be enough to power our civilization once we develop technology adequate to harness it. These advantages of solar energy are clear, but the technologies themselves also provide benefits. PV cells and other solar technologies use no fuel, are quiet and safe, contain no moving parts, and require little maintenance. An average unit can produce energy for 20–30 years.

Solar systems also allow for local, decentralized control over power. In developed nations, most PV systems are connected to a regional electrical grid and homeowners sell excess solar energy to their utility through feed-in tariffs or net metering. In developing nations and in rural communities, many homes and businesses use solar power to produce electricity without being near a power plant or connected to a grid. Likewise, one can use flat plate solar collectors to heat water or air just about anywhere. And in many developing nations, inexpensive solar cookers (see Figure 21.10a) help

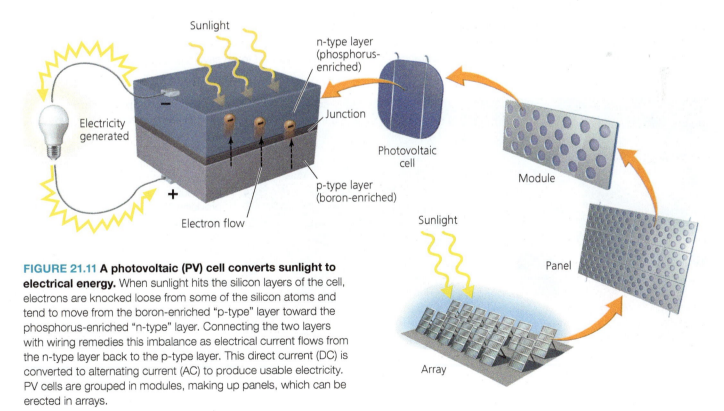

FIGURE 21.11 A photovoltaic (PV) cell converts sunlight to electrical energy. When sunlight hits the silicon layers of the cell, electrons are knocked loose from some of the silicon atoms and tend to move from the boron-enriched "p-type" layer toward the phosphorus-enriched "n-type" layer. Connecting the two layers with wiring remedies this imbalance as electrical current flows from the n-type layer back to the p-type layer. This direct current (DC) is converted to alternating current (AC) to produce usable electricity. PV cells are grouped in modules, making up panels, which can be erected in arrays.

What Are the Impacts of Solar and Wind Development?

Demonstrators for and against the proposed Cape Wind farm

Renewable energy eliminates many of the negative impacts of fossil fuel combustion. However, this does not mean that renewable energy is a cost-free panacea. The costs of renewable energy have become clearer in recent years as large solar and wind projects bring to the fore a host of issues that many clean energy proponents had not fully considered. As our society decides how to pursue renewable energy projects at large scales, we will need to conduct scientific study of their impacts and to consider these impacts alongside their benefits.

The several dozen concentrated solar power (CSP) installations currently in development or under review for the Mojave Desert and other arid regions of California and the U.S. Southwest would, if constructed, cover many square miles of land (**FIGURE 1**). Desert environments are particularly sensitive, so researchers say we should expect substantial impacts from such installations. Besides altering the pristine appearance of an undeveloped landscape, arrays of thousands of mirrors or panels affect communities of plants and animals by casting shade and altering microclimate. Altered conditions tend to hurt native desert-adapted species while helping invasive weeds. At existing solar facilities, the sites are graded and sprayed with herbicide, eliminating plants and damaging fragile soils. Human presence increases as workers travel to and fro to maintain the facilities. Solar power plants require water for cooling and cleaning, and water is scarce in the arid regions hosting most of these facilities. All these impacts have consequences for plants, animals, and ecosystems.

One dramatic impact of CSP plants is that birds may be scorched to death as they fly through the superheated airspace between mirrors and power towers. Observers have given the macabre nickname "streamers" to birds that burst into a puff of smoke and tumble through the air. Since this phenomenon came to light at California's Ivanpah solar plant, scientists have been working to assess how frequently this occurs and what birds, bats, and insects are being affected, while at the same time the industry is trying various methods to keep birds and insects away.

Large-scale projects need government approval and are subject to the environmental impact statement process (p. 171). As a result, teams of researchers study conditions at each site to determine what impacts energy development may have. If impacts are judged to be severe enough, government agencies

FIGURE 1 Solar power farms require large areas of land and exert substantial environmental impacts. This solar plant in Kramer Junction, California, is one of nine that spread across more than 650 hectares (ha; 1600 acres) of the Mojave Desert, providing power for more than 230,000 homes. Still, researchers estimate that impacts are less than those caused by fossil fuels; burning coal for energy uses at least as much land, once one includes the strip mining needed to obtain the coal.

can insist that plans be amended or that companies fund mitigation efforts to compensate for damage.

For instance, the California Energy Commission asked for limits on the proposed Calico Solar Project in southern California in 2010, after biologists concluded that the project would damage habitat of the desert tortoise and bighorn sheep. The company agreed to cut the size of its footprint in half, reducing estimated impacts to wildlife by 80%, and the Commission then approved the project.

In central California, a solar project underwent 18 months of environmental analysis and was approved after the Solargen company agreed to purchase and set aside 23,000 acres of preserved land as mitigation for the 3200 acres it was developing. A third California solar project, the Topaz Solar Farm, was scaled back in size after researchers found that this was needed to protect farmland; minimize aesthetic impacts; and lessen disturbance to tule elk, kit foxes, pronghorn antelope, burrowing owls, and seasonal freshwater pools.

Given the impacts of CSP facilities, researchers have determined that installing photovoltaic panels on rooftops is a low-impact solar alternative. Adding PV panels or roofing tiles to a rooftop has no effect on the landscape. One study led by Dutch, German, and American researchers compared impacts of various ground-based and rooftop PV systems in Germany and in Arizona. They found that besides avoiding land use impacts, the rooftop systems also emitted significantly fewer greenhouse gases over the systems' entire life-cycles (from production to installation to operation).

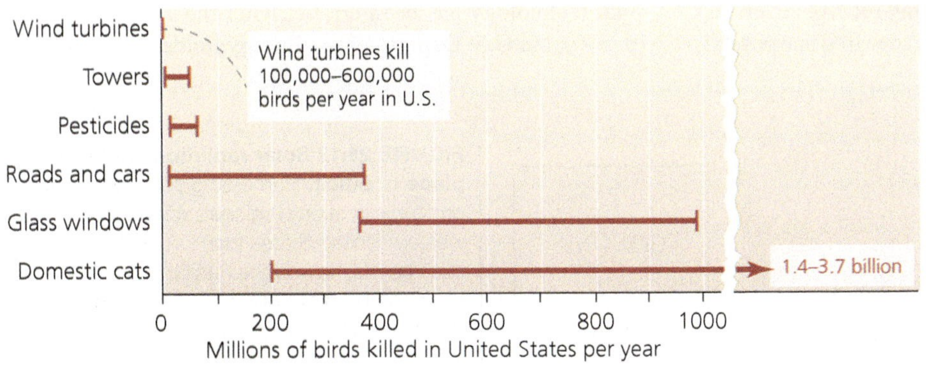

Scientists concerned about health impacts of manufacturing PV panels have found that PV cell production can pose risks to assembly-line workers because of exposure to toxic chemicals such as cadmium, arsenic, and fine silicon dust that can damage lung tissue. However, researchers have concluded that proper attention to worker safety can greatly reduce these risks and that the exposure risks are not much different from other exposure risks we all face in our modern industrialized society.

Overall, the messages from scientific studies so far are that (1) solar power, even with its impacts, is still cleaner and more sustainable than fossil fuel power; and that (2) we can minimize the impacts of solar power by using rooftop panels and developing better technologies.

Similar messages are emerging from the science on wind power. One major concern is that birds and bats are killed when they fly into the spinning blades of turbines. At California's Altamont Pass wind farm, turbines killed dozens of golden eagles and other raptors in the 1990s. Despite studies since then at many sites, uncertainty remains, because it is very challenging to document the full extent of bird and bat deaths. For instance, one European study indicated that migrating seabirds fly past offshore turbines without problem, but other data show that resident seabird densities have declined near turbines.

On land, the wind industry estimates that about two birds are killed per 1-megawatt-turbine per year, while peer-reviewed research and federal biologists peg the number about five times higher, at more than a half-million birds per year. Either way, this is far fewer than the hundreds of millions of birds being killed each year by television, radio, and cell phone towers; pesticides; automobiles; windows; and especially free-roaming domestic cats (FIGURE 2).

Further research is needed, but we can reduce wildlife impacts from wind turbines by avoiding wind development at sites along known migratory flyways or amid prime habitat for bat and bird species that are likely to fly into the blades. Continued scientific studies on the various impacts of wind and solar development should help us find ways to harness renewable energy and attain a sustainable energy future while minimizing the environmental and social impacts of this development.

FIGURE 2 Wind turbines kill hundreds of thousands of birds each year when they fly into the blades. Yet far more birds are killed by other human causes. Shown are ranges of recent estimates of yearly bird mortality in the United States from several main causes. Habitat alteration is responsible for still more deaths than any of the causes shown. *Data from American Bird Conservancy; Loss, S., et al., 2012. The impact of free-ranging domestic cats on wildlife of the United States. Nature Communications 4: Article 1396; and Loss, S., et al., 2014. Bird-building collisions in the United States: Estimates of annual mortality and species vulnerability. Condor 116: 8–23.*

families of any income level to cook food without gathering fuelwood. This lessens people's daily workload and helps reduce deforestation.

Developing, manufacturing, and deploying solar technology also create new green-collar jobs. Currently, among major energy sources, PV technology employs the most people per unit energy output, resulting in nearly 2.8 million jobs worldwide (see Figure 21.5).

Finally, a major advantage of solar power over fossil fuels is that it does not emit greenhouse gases and other air pollutants (see Figure 21.4). The manufacture of photovoltaic cells *does* currently require fossil fuel use, but once up and running, a PV system produces no emissions. Consumers can access online calculators offered by the U.S. National Renewable Energy Laboratory and other sources to estimate the economic and environmental results of installing a PV system. At the time of this writing, these calculators estimated that installing a standard 5-kilowatt PV system atop a home in Fort Worth, Texas, to provide most of its annual power needs would save the homeowners $834 each year on energy bills and would prevent more than 5 tons of carbon dioxide emissions per year—as much CO_2 as results from burning 525 gallons of gasoline. Even in overcast Seattle, Washington, a 5-kilowatt system producing half a home's energy needs will save $420 per year and prevent more than 3.6 tons of CO_2 emissions (equal to the emissions from burning 375 gallons of gas).

Location, timing, and cost of solar energy can be drawbacks

Solar energy currently has three main disadvantages. One is that not all regions are equally sunny (**FIGURE 21.12**). People in Seattle or Anchorage will find it more challenging to harness solar energy than people in Tucson or San Diego. (However, observe in Figure 21.12 that Germany receives even less sun than Alaska, and yet it is the world leader in solar power!)

A second drawback is that solar energy is an intermittent resource. Daily or seasonal variation in sunlight can limit stand-alone solar systems if storage capacity in batteries or fuel cells is not adequate or if backup power is not available from a municipal electrical grid. Pumped-storage hydropower (p. 571) can sometimes help by compensating for periods of low solar power production.

The third disadvantage of current solar technology is the up-front cost of the equipment. Because of the investment cost, solar power remains the most expensive way to produce electricity (see Figure 21.6). The high costs result from the fact that the technologies are still developing. Moreover, they are competing against energy sources (fossil fuels and nuclear power) that have remained relatively cheap as a result of decades of taxpayer support through subsidies and whose external costs (pp. 141, 162) are not included in market prices. As a result, market prices have given governments, businesses, and consumers little economic incentive to switch to solar energy thus far.

However, declines in price and improvements in efficiency of solar technologies are encouraging, even in the absence of significant funding from government and industry. At their advent in the 1950s, solar technologies had efficiencies of around 6% while costing $600 per watt. Today, PV cells are showing up to 20% efficiency commercially and 45% efficiency in lab research, suggesting that future solar cells could be more efficient than any solar energy technologies we have today. Solar systems are becoming less expensive; the latest installations pay for themselves in less than 10–20 years. After that time, they provide energy virtually for free as long as the equipment lasts.

Solar energy is expanding

Although active solar technology dates from the 18th century, it was pushed to the sidelines as fossil fuels came to dominate our energy economy. Funding for research and development of solar technology has been erratic. After the 1973 oil embargo (p. 540), the U.S. Department of Energy funded the installation

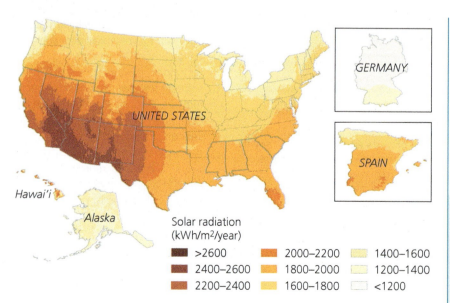

FIGURE 21.12 Solar radiation varies from place to place. Harnessing solar energy is more profitable in sunny regions such as the southwestern United States than in cloudier regions such as Alaska and the Pacific Northwest. However, compare solar power leaders Germany and Spain with the United States. Spain is similar to Kansas in the amount of sunlight it receives, and Germany is cloudier than Alaska. This suggests that solar power can be used with success just about anywhere. *Data from National Renewable Energy Laboratory, U.S. Department of Energy.*

GERMANY

UNITED STATES

SPAIN

Hawai'i

Alaska

Solar radiation (kWh/m²/year)

- >2600
- 2400–2600
- 2200–2400
- 2000–2200
- 1800–2000
- 1600–1800
- 1400–1600
- 1200–1400
- <1200

DATA Q Roughly how much more solar radiation does southern Arizona receive than Germany? How does your own state compare to Germany in its solar radiation?

Go to **Interpreting Graphs & Data** on **Mastering**EnvironmentalScience®

and testing of more than 3000 PV systems, providing a boost to fledgling companies in the solar industry. But later, as oil prices declined, so did government support for solar power.

Largely because of the lack of investment, solar energy contributes just 0.56%—56 parts in 10,000—of the U.S. energy supply and just 0.94% of U.S. electricity generation. Even in Germany, which gets more of its energy from solar than any other nation, solar energy is responsible for less than 7% of electricity generation. However, solar energy use has grown by 30% each year worldwide in the past four decades, a growth rate second only to that of wind power. Solar energy is proving especially attractive in developing countries rich in sun but poor in infrastructure, where hundreds of millions of people still live without electricity.

PV technology is the fastest-growing power source today, having recently doubled every two years (**FIGURE 21.13**). Germany led the world in installation of PV technology until China (with its much larger economy) passed it in 2015—and German rooftops host more than one-sixth of all PV cells in the world. Germany's investment began in 1998 when Hermann Scheer spearheaded a "100,000 Rooftops" program to install PV panels atop 100,000 German roofs. The popular program ended up easily surpassing this goal, and today more than *1.5 million* German rooftops have PV systems.

China leads the world in yearly production of PV cells, followed by Germany and Japan, while U.S. firms now account for only 2% of the industry. Recent federal tax credits and state-level initiatives may help the United States recover the leadership it has lost to other nations in this technology, but China is moving faster and is dominating the market. In fact, the Chinese government's support of its solar industry has led to so much production that supply has outstripped global demand in recent years. Highly subsidized Chinese firms have been selling solar products abroad at low prices (often at a loss), driving American and European solar manufacturers out of business. In response, the United States and European nations have slapped tariffs on Chinese imports, while both sides have filed complaints with the World Trade Organization in an escalating global trade dispute.

Despite volatility in the industry as firms try to deal with swings in national policies, global production of PV cells continues to rise sharply, while prices fall (see Figure 21.13). At the same time, efficiencies are increasing, making each unit more powerful. Use of solar technology should continue to expand as prices fall, technologies improve, and governments enact economic incentives to spur investment.

Wind Power

As the sun heats the atmosphere, it causes air to move, and the movement of differentially heated air masses produces wind. We can harness **wind power** from air's movement by using **wind turbines,** mechanical assemblies that convert wind's kinetic energy (p. 28), or energy of motion, into electrical energy.

Wind turbines convert kinetic energy to electrical energy

Today's wind turbines have their roots in Europe, where wooden windmills were used for 800 years to grind grain and pump water. The first wind turbine built to generate electricity was constructed in the late 1800s in Cleveland, Ohio. However, it was not until after the 1973 oil embargo that governments and industry in North America and Europe began funding research and development for wind power.

In a modern wind turbine, wind turns the blades of the rotor, which rotate machinery inside a compartment called a *nacelle*, which sits atop a tower (**FIGURE 21.14**, p. 592). Inside the nacelle are a gearbox, a generator, and equipment to monitor and control the turbine's activity. Today's towers average 80 m (260 ft) in height, and the largest are taller than a football field is long. Higher is generally better, to minimize turbulence (and potential damage) while maximizing wind speed.

Engineers design turbines to yaw, or rotate back and forth in response to changes in wind direction, ensuring that the motor faces into the wind at all times. Some turbines are designed to generate low levels of electricity by turning in light breezes. Others are programmed to rotate only in strong winds, generating large amounts of electricity in short time periods. Slight differences in wind speed yield substantial differences in power output, for two reasons. First, the energy content of wind increases as the square of its velocity; thus if wind velocity doubles, energy quadruples. Second, an increase in wind speed causes more air molecules to pass through the wind turbine per unit time, making power output equal to wind velocity cubed. Thus a doubled wind velocity results in an eightfold increase in power output.

Turbines are often erected in groups; such a development is called a **wind farm.** In the largest wind farms, hundreds of turbines spread across the landscape.

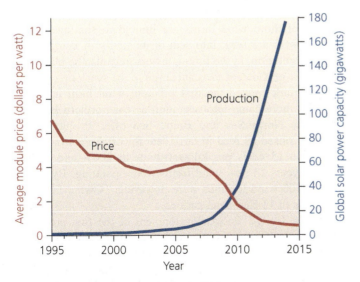

FIGURE 21.13 Global production of PV cells has been growing exponentially, and prices have fallen rapidly. *Data from REN21, 2015. Renewables 2015: Global status report. Paris: REN21, UNEP; U.S. Department of Energy, EERE, 2011. 2010 Solar technologies market report.*

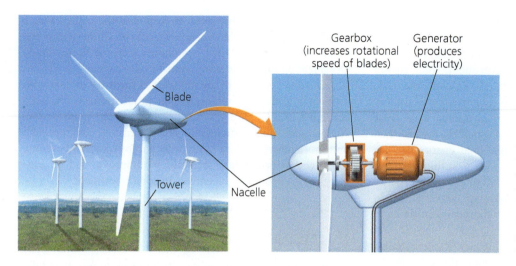

Gearbox (increases rotational speed of blades)

Generator (produces electricity)

Blade

Tower

Nacelle

FIGURE 21.14 A wind turbine converts wind's energy of motion into electrical energy. Wind spins a turbine's blades, turning a shaft that extends into the nacelle. Inside the nacelle, a gearbox converts the rotational speed of the blades, which can be up to 20 revolutions per minute (rpm) or more, into much higher speeds (over 1500 rpm). This provides adequate motion for the generator to produce electricity.

Wind power is growing fast

Like solar energy, wind provides just a small proportion of the world's power needs, but wind power is growing fast—doubling every three years (**FIGURE 21.15**). Five nations account for three-quarters of the world's wind power output (**FIGURE 21.16a**), but dozens of nations now produce wind power. Germany had long produced the most, but the United States overtook it in 2008, and China surpassed the United States two years later.

Denmark leads the world in obtaining the greatest percentage of its energy from wind power. In this small European nation, wind farms supply nearly 40% of Danish electricity needs (**FIGURE 21.16b**). Germany is sixth in this respect, and the United States is 15th. Texas generates the most wind power of all U.S. states, while Iowa and South Dakota each obtain more than 25% of their electricity from wind.

Wind power's growth in the United States has been haphazard because Congress has not committed to a long-term

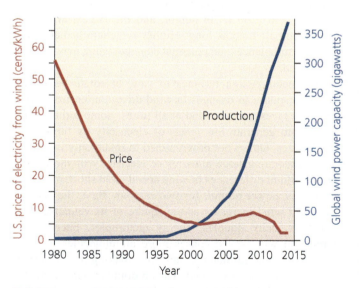

FIGURE 21.15 Global production of wind power has been doubling every three years in recent years, and prices have fallen slightly. *Data from Global Wind Energy Council; and U.S. Department of Energy, EERE, 2015. 2014 Wind technologies market report.*

federal tax credit for wind development, but instead has passed a series of short-term renewals, leaving the industry uncertain about how much to invest. However, experts agree that wind power's growth will continue, because only a small portion of this resource is currently being tapped and because wind power at favorable locations already generates electricity at prices nearly as low as fossil fuels (see Figure 21.6). Recent reports estimate that with adequate development, the United States could meet 20% of its electricity demands with wind power by 2030, and 35% by 2050.

Offshore sites for wind farms hold promise

Wind speeds on average are 20% greater over water than over land, and air is less turbulent (more steady) over water. For these reasons, offshore wind turbines are becoming popular (**FIGURE 21.17**). Costs to erect and maintain turbines in water are higher, but the stronger, steadier winds produce more power and make offshore wind potentially more profitable. Today's offshore wind farms are limited to shallow water, where towers are sunk into sediments singly or using a tripod configuration. In the future, towers may also be placed in deep water on floating pads anchored to the seafloor.

Denmark erected the first offshore wind farm in 1991, and soon more came into operation across northern Europe, where the North Sea and Baltic Sea offer strong winds. Germany raised its feed-in tariff rate for offshore wind from 9 cents to 15 cents per kilowatt-hour in 2009, and within just five years 14 new wind farms were operating. By 2015, nearly 2500 wind turbines were powering 74 wind farms in the waters of 11 European nations.

In the United States, development of the first proposed offshore wind farm was approved in 2010 after nine years of debate. The Cape Wind offshore wind farm, if constructed, would feature 130 turbines rising from Nantucket Sound 8 km (5 mi) off the coast of Cape Cod in Massachusetts. However, it ran into financial problems and local opposition, and a smaller project off nearby Block Island, in Rhode Island waters, is now set to become America's first commercial

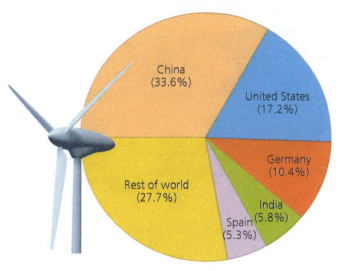

(a) Percentage of global wind power in each nation

China (33.6%)
United States (17.2%)
Germany (10.4%)
India (5.8%)
Spain (5.3%)
Rest of world (27.7%)

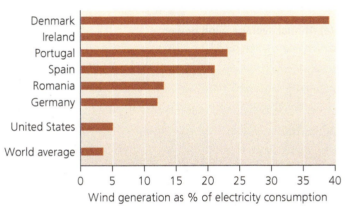

(b) Leading nations in proportion of electricity from wind power

Wind generation as % of electricity consumption

Denmark
Ireland
Portugal
Spain
Romania
Germany
United States
World average

FIGURE 21.16 Several nations are leaders in wind power. Most of the world's wind power capacity **(a)** is concentrated in a handful of nations led by China, the United States, and Germany. Yet tiny Denmark **(b)** obtains the highest percentage of its electricity needs from wind. *Data from (a) Global Wind Energy Council, 2016. Global wind report: Annual market update 2015. Brussels, Belgium: GWEC; and (b) U.S. Department of Energy, EERE, 2015. 2014 Wind technologies market report.*

FIGURE 21.17 More and more wind farms are being developed offshore. Offshore winds tend to be stronger and steadier.

offshore wind farm. As of 2016, roughly 20 more offshore wind developments were in the planning stages, mostly off the North Atlantic coast.

Wind power has many benefits

Like solar power, wind power produces no emissions once the equipment is manufactured and installed. As a replacement for fossil fuel combustion in the average U.S. power plant, running a 1-megawatt wind turbine for 1 year prevents the release of more than 1500 tons of carbon dioxide, 6.5 tons of sulfur dioxide, 3.2 tons of nitrogen oxides, and 60 lb of mercury, according to the U.S. Environmental Protection Agency. The amount of carbon pollution that all U.S. wind turbines together prevent from entering the atmosphere is equal to the emissions from 14 million cars, or from combusting the cargo of an 1100-car freight train of coal each and every day.

Under optimal conditions, wind power appears efficient in its energy returned on investment (EROI; pp. 520, 568). Most studies find that wind turbines produce roughly 20 times more energy than they consume—an EROI value better than that of most energy sources. Wind farms also use less water than do conventional power plants.

Wind turbine technology can be used on many scales, from a single tower for local use to farms of hundreds that supply large regions. Small-scale turbine development can help make local areas more self-sufficient, just as solar energy can. For instance, the Rosebud Sioux Tribe set up a single turbine on its reservation in South Dakota. The turbine has been producing electricity for 200 homes and brings the tribe $15,000 per year in revenue. The tribe now aims to develop a 30-megawatt wind farm nearby.

Another benefit of wind power is that farmers and ranchers can lease their land for wind development. A single large turbine can bring in $2000 to $4500 in annual royalties while occupying just a quarter-acre of land. Most of the land can still be used for agriculture. Royalties from the wind power company provide the farmer or rancher revenue while also increasing property tax income for his or her rural community.

Wind power involves up-front expenses to erect turbines and to expand infrastructure to transmit electricity, but over the lifetime of a project it requires only maintenance costs. Unlike fossil-fuel power plants, wind turbines incur no ongoing fuel costs. Startup costs of wind farms generally are higher than those of fossil fuel plants, but wind farms incur fewer expenses once they are up and running.

Finally, wind power creates job opportunities. Roughly 88,000 Americans and more than 1 million people globally are now employed in the wind industry. More than 100 colleges and universities now offer programs and degrees that train people in the skills needed for jobs in wind power and other renewable energy fields.

Wind power has limitations

Wind is an intermittent resource; we have no control over when it will occur. This is a major limitation in relying on wind as an electricity source, but it is lessened if wind is one of several sources contributing to a utility's power generation. Pumped-storage hydropower (p. 571) can help to compensate during windless times, and batteries or hydrogen fuel (p. 598) can store energy generated by wind and release it later when needed.

Just as wind varies from time to time, it varies from place to place. Global wind patterns combine with local topography—mountains, hills, water bodies, forests, cities—to make some places windier than others. Resource planners and wind power companies study wind patterns revealed by meteorological research when they judge prime areas for locating wind farms. A map of average wind speeds across the United States (FIGURE 21.18a) reveals that mountainous regions are best, along with areas of the Great Plains. Based on such data, the wind power industry has located much of its generating capacity in states with high wind speeds (FIGURE 21.18b), and is expanding in the Great Plains and mountain states. Provided that wind farms are strategically erected in optimal locations, an estimated 15% of U.S. energy demand could be met using only 43,000 km^2 (16,600 mi^2) of land (with less than 5% of this land area actually occupied by turbines, equipment, and access roads).

However, most of North America's people live near the coasts, far from the Great Plains and mountain regions that have the best wind resources. Thus, continent-wide transmission networks would need to be enhanced significantly to send wind-generated electricity to these population centers.

weighing the
ISSUES

Wind and NIMBY

If you could choose to get your electricity from a wind farm or a coal-fired power plant, which would you choose? How would you react if the electric utility proposed to build the wind farm that would generate your electricity atop a ridge running in back of your neighborhood, such that the turbines would be clearly visible from your living room window? Would you support or oppose the development? Why? If you would oppose it, where would you suggest the farm be located? Do you think anyone might oppose it in that location?

When wind farms *are* proposed near population centers, local residents often oppose them. Turbines are generally located in exposed, conspicuous sites, and some people object to wind farms for aesthetic reasons, feeling that the structures clutter the landscape. Although polls show wide public approval of existing wind projects and of the concept of wind power, newly proposed wind projects often elicit the **not-in-my-backyard (NIMBY)** syndrome among people living nearby. For instance, the previously mentioned Cape Wind project faced years of opposition from wealthy residents of Cape Cod, Nantucket, and Martha's Vineyard, even though many of these residents consider themselves progressive environmentalists.

Wind turbines also pose a threat to birds and bats, which are killed if they fly into the rotating blades. Large open-country raptors such as golden eagles are known to be at risk, and

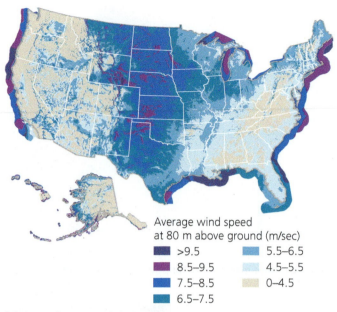

Average wind speed at 80 m above ground (m/sec)

- ■ >9.5
- ■ 8.5–9.5
- ■ 7.5–8.5
- ■ 6.5–7.5
- ■ 5.5–6.5
- ■ 4.5–5.5
- ■ 0–4.5

(a) Annual average wind power

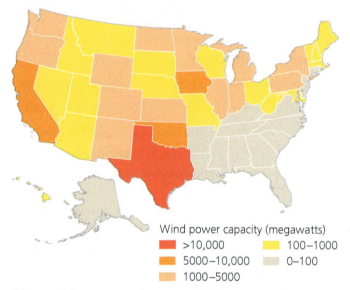

Wind power capacity (megawatts)

- ■ >10,000
- ■ 5000–10,000
- ■ 1000–5000
- ■ 100–1000
- ■ 0–100

(b) Generating capacity from wind, 2016

FIGURE 21.18 Wind speed varies from place to place.
Maps of average wind speeds **(a)** help guide the placement of wind farms. Another map **(b)** shows megawatts of wind-power-generating capacity developed in each U.S. state through early 2016. *Sources: (a) U.S. National Renewable Energy Laboratory; (b) American Wind Energy Association, 2016. 2nd Quarter 2016 Market Report.*

DATA Compare parts **(a)** and **(b)**. Which states or regions have high wind speeds but are not yet heavily developed with commercial wind power?

Go to **Interpreting Graphs & Data** on **Mastering**EnvironmentalScience®

turbines located on ridges along migratory flyways are likely most damaging. More research on wildlife impacts is urgently needed (see **The Science behind the Story**, pp. 588–589). One strategy for protecting birds and bats is to select sites that are not on migratory flyways or amid prime habitat for species likely to fly into the blades.

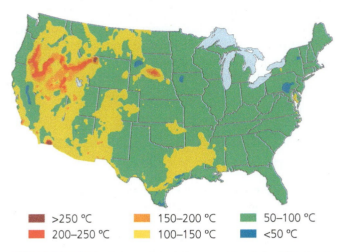

FIGURE 21.19 Geothermal resources in the United States are greatest in the western states. This map shows water temperatures 3 km (1.9 mi) belowground. *Data from Idaho National Laboratory.*

Geothermal Energy

Geothermal energy is thermal energy that arises from beneath Earth's surface. The radioactive decay of elements (p. 24) amid high pressures deep in the interior of our planet generates heat that rises to the surface through magma (molten rock, p. 31) and through cracks and fissures. Where this energy heats groundwater, natural spurts of heated water and steam are sent up from below and may erupt through the surface as terrestrial geysers or submarine hydrothermal vents (p. 31).

Geothermal energy manifests itself at the surface in these ways only in certain areas, and regions vary in their geothermal resources (**FIGURE 21.19**). One geothermally rich area is in California near Napa Valley's wine country. There, engineers have for years operated the world's largest geothermal power plants, The Geysers. The island nation of Iceland also has a wealth of geothermal energy resources, along with many volcanoes and geysers. This is because the island was formed from lava that extruded and cooled at the Mid-Atlantic Ridge (p. 33), along the spreading boundary of two tectonic plates.

We harness geothermal energy for heating and electricity

Geothermal energy can be harnessed directly from geysers at the surface, but most often wells must be drilled down hundreds or thousands of meters toward heated groundwater. Hot groundwater can be used directly for heating homes, offices, and greenhouses; and for driving industrial processes. Iceland heats nearly 90% of its homes through direct heating with piped hot water. Direct use of naturally heated water is efficient and inexpensive, but it is feasible only where geothermal energy is readily available and does not need to be transported far.

Geothermal power plants harness the energy of naturally heated underground water and steam to generate electricity (**FIGURE 21.20**). Generally, a power plant brings water at temperatures of 150–370°C (300–700°F) or more to the surface and converts it to steam by lowering the pressure in specialized compartments. The steam turns turbines to generate electricity. At The Geysers in California, electricity is generated for 725,000 homes.

At Iceland's Nesjavellir power station (**FIGURE 21.21**, p. 596), steam piped in from wells heats water from a lake. The heated water is sent through an insulated pipeline to the capital city, where residents use it for washing and space heating. Globally, one-third of the geothermal energy we use is for electricity, and two-thirds is used for direct heating.

Heat pumps make use of temperature differences

Heated groundwater is available only in certain areas, but we can take advantage of the mild temperature differences that exist naturally between the soil and the air just about anywhere. Soil varies in temperature from season to season less

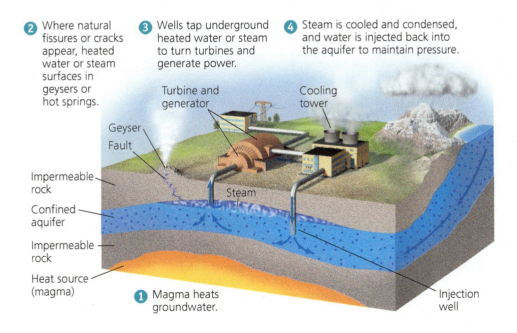

2 Where natural fissures or cracks appear, heated water or steam surfaces in geysers or hot springs.

3 Wells tap underground heated water or steam to turn turbines and generate power.

4 Steam is cooled and condensed, and water is injected back into the aquifer to maintain pressure.

Turbine and generator

Cooling tower

Geyser

Fault

Steam

Impermeable rock

Confined aquifer

Impermeable rock

Heat source (magma)

1 Magma heats groundwater.

Injection well

FIGURE 21.20 Geothermal power plants generate electricity using naturally heated water from underground. With geothermal energy **(a)**, magma heats groundwater **1**, some of which escapes through surface vents such as geysers **2**. A power plant may tap into heated water and channel steam through turbines to generate electricity **3**. Once used, steam may be condensed into water and pumped back into the aquifer to maintain pressure **4**.

FIGURE 21.21 **The Nesjavellir power station in Iceland uses geothermal energy to provide heated water and to generate electricity.**

than air does, because it absorbs and releases heat more slowly and because warmth and cold do not penetrate deeply belowground. Just several inches below the surface, temperatures are nearly constant year-round. Geothermal heat pumps, or **ground-source heat pumps,** make use of this phenomenon.

Ground-source heat pumps provide heating in the winter by transferring heat from the ground into buildings, and they provide cooling in the summer by transferring heat from buildings into the ground. This heat transfer is accomplished with a network of underground plastic pipes that circulate water and antifreeze (**FIGURE 21.22**). Because heat is simply moved from place to place rather than being produced using outside energy inputs, heat pumps can be highly energy-efficient.

More than 600,000 U.S. homes use ground-source heat pumps. Compared to conventional electric heating and cooling systems, ground-source heat pumps heat spaces 50–70% more efficiently, cool them 20–40% more efficiently, can reduce electricity use by 25–60%, and can reduce emissions by up to 70%.

Geothermal power has benefits and limitations

All forms of geothermal energy—direct heating, electrical power, and ground-source heat pumps—greatly reduce emissions relative to fossil fuel combustion. By one estimate, each megawatt of electricity produced at a geothermal power plant prevents the emission of 7.0 million kg (15.5 million lb) of carbon dioxide each year. Geothermally heated water can release dissolved gases, including carbon dioxide, methane, ammonia, and hydrogen sulfide, but they generally occur in small quantities, and many facilities use filters to cut down on their emission.

In principle, geothermal energy is renewable, because using it does not affect the amount of thermal energy produced underground. However, not every geothermal power plant will be able to operate indefinitely. If a plant uses heated water more quickly than groundwater is recharged, it will eventually run out of water. This was occurring at The Geysers in California, which began operating in 1960. In response, operators began injecting municipal wastewater into the ground to replenish the supply. Many geothermal plants worldwide are now injecting water back into aquifers to help maintain pressure and sustain the resource.

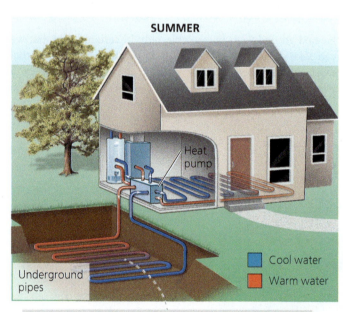

In summer, soil underground is **cooler** than surface air. Water flowing through the pipes transfers heat from the house to the ground, cooling air in ducts or radiant cooling system under floor.

In winter, soil underground is **warmer** than surface air. Water flowing through the pipes transfers heat from the ground to the house, warming air in ducts, water in tank, or radiant heating system under floor.

FIGURE 21.22 **Ground-source heat pumps provide an efficient way to heat and cool air and water in a home.** A network of pipes filled with water and antifreeze extends underground. Soil is cooler than air in the summer, and warmer than air in the winter, so by running fluid between the house and the ground, these systems adjust temperatures inside.

A second reason geothermal energy is not always renewable is that patterns of geothermal activity in Earth's crust shift naturally over time. As a result, an area that produces hot groundwater now may not always do so. In addition, some hot groundwater is laced with salts and minerals that corrode equipment and pollute the air. These factors may shorten the lifetime of plants, increase maintenance costs, and add to pollution.

The greatest limitation of geothermal power is that it is restricted to regions where we can tap energy from naturally heated groundwater. Places such as Iceland, northern California, and Yellowstone National Park are rich in naturally heated groundwater, but most areas of the world are not.

Enhanced geothermal systems might widen our reach

To broaden where we can harness geothermal energy, engineers are now developing **enhanced geothermal systems (EGS),** in which we drill deeply into dry rock, fracture the rock, and pump in cold water. The water becomes heated deep underground and is then drawn back up and used to generate power. EGS thereby uses natural thermal energy underground but supplies the water, which can be reused repeatedly. In theory we could use EGS in many locations. Germany, for instance, has little heated groundwater, but feed-in tariffs have enabled an EGS facility to operate profitably there.

EGS technology shows significant promise, and a 2006 report estimated that heat resources below the United States could power the world's energy demands for several millennia. However, EGS also appears to trigger minor earthquakes. Unless we can develop ways to use EGS safely and reliably, our use of geothermal power will remain localized.

Ocean Energy Sources

The oceans are home to several underexploited sources of energy resulting from continuous natural processes. Of the four approaches being developed, three involve motion and one involves temperature.

We can harness energy from tides, waves, and currents

Just as dams on rivers use flowing fresh water to generate hydroelectric power, we can use kinetic energy from the natural motion of ocean water to generate electrical power.

Scientists and engineers are working to harness the motion of ocean waves and convert their energy into electricity. Many designs for machinery to harness **wave energy** have been invented, but few have been thoroughly tested for commercial use. Some designs for offshore facilities involve floating devices that move up and down with the waves. An example is the snake-like wave energy converter shown in **FIGURE 21.23**. Built by the Scottish company Pelamis (a Latin word denoting a genus of sea snake), variations on this jointed, columnar design have been deployed in several parts of the world. Machinery and hydraulic fluids inside the

FIGURE 21.23 Various technologies harness energy from ocean waves. This wave energy converter stretches across the water off the coast of Scotland. As waves flex segments of the machinery, hydraulic equipment in the joints generates electricity and sends it by undersea cables to shore, where it powers 500 homes.

floating columns use wave motion to generate electricity, which is transmitted to shore via undersea cables.

Wave energy is greatest on the open ocean, but transmitting electricity to shore is expensive; thus, many efforts to harness wave energy are situated along coasts. Some designs for coastal onshore facilities funnel waves from large areas into narrow channels and elevated reservoirs, from which water then flows out, generating electricity as hydroelectric dams do. Other coastal designs use rising and falling waves to push air into and out of chambers, turning turbines (**FIGURE 21.24**). The first commercially operating wave energy facility began

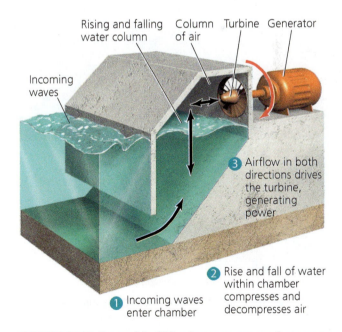

FIGURE 21.24 Coastal facilities harness energy from ocean waves. In one design, as waves enter and exit a chamber ❶, the air inside is alternately compressed and decompressed ❷, creating airflow that rotates turbines ❸ to generate electricity.

operating in 2011 in Spain, using technology tested in Scotland. Demonstration projects exist in Europe, Japan, and Oregon.

We are also developing ways of harnessing energy from tides. The rise and fall of ocean tides (p. 424) twice each day moves large amounts of water past any given point on the world's coastlines. Differences in height between low and high tides are greatest in long, narrow bays such as Alaska's Cook Inlet or the Bay of Fundy between New Brunswick and Nova Scotia. Such locations are best for harnessing **tidal energy,** generally accomplished by erecting dams across the outlets of tidal basins. In the most common design, the incoming tide flows through sluice gates and is trapped behind them. Then, as the outgoing tide passes through the gates, it turns turbines to generate electricity. The world's largest tidal generating station, South Korea's Sihwa Lake facility, generates electricity from the incoming tide (**FIGURE 21.25**). Some designs generate electricity from water moving in both directions.

The Sihwa Lake power station opened in 2011 and is just larger than the La Rance tidal facility in France, which has operated for 50 years. Smaller facilities operate in Canada, China, Russia, and the United Kingdom. The first U.S. tidal station began operating in 2012 in Maine, and one is now being installed in New York City's East River. Five more tidal stations are planned in South Korea, but some have been delayed amid concerns over environmental impacts. Tidal stations release few or no pollutant emissions, but they can affect the ecology of estuaries and tidal basins.

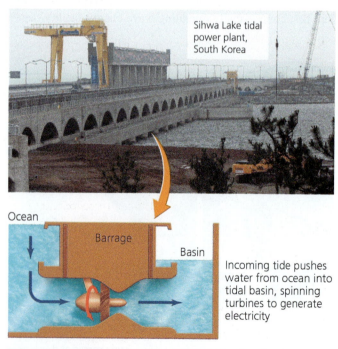

FIGURE 21.25 **We can harness tidal energy by allowing ocean water to spin turbines.** At the world's largest tidal power facility, water flows from the ocean into a huge enclosed basin as the tide rises, spinning turbines to generate electricity.

A third way to harness marine kinetic energy is to use the motion of ocean currents (p. 418), such as the Gulf Stream. Underwater turbines have been erected in European waters to test this approach.

The ocean stores thermal energy

Each day the tropical oceans absorb solar radiation with the heat content of 250 billion barrels of oil—enough to provide 20,000 times the electricity used daily in the United States. The ocean's sun-warmed surface is warmer than its deep water, and **ocean thermal energy conversion (OTEC)** relies on this gradient in temperature.

In the *closed cycle* approach, warm surface water is piped into a facility to evaporate chemicals, such as ammonia, that boil at low temperatures. These evaporated gases spin turbines to generate electricity. Cold water piped up from ocean depths then condenses the gases so they can be reused. In the *open cycle* approach, warm surface water is evaporated in a vacuum, and its steam turns turbines and then is condensed by cold water. Because ocean water loses salts as it evaporates, the water can be recovered, condensed, and sold as desalinized fresh water for drinking or agriculture. OTEC research has been conducted in Hawai'i and elsewhere, but costs remain high, and so far no facility operates commercially.

Hydrogen and Fuel Cells

Each renewable energy source we have discussed can be used to generate electricity more cleanly than can fossil fuels. However, electricity cannot be stored easily in large quantities for use when and where it is needed. This is why most vehicles rely on gasoline for power. The development of fuel cells and of fuel consisting of hydrogen—the simplest and most abundant element in the universe—offers a way to store considerable quantities of energy conveniently, cleanly, and efficiently. Like electricity and like batteries, hydrogen is an energy carrier, not a primary energy source. It holds energy that can be converted for use at later times and in different places.

Some yearn for a "hydrogen economy"

Some energy experts envision that hydrogen fuel, together with electricity, could serve as the basis for a clean, safe, and efficient energy system. In such a system, electricity generated from intermittent renewable sources, such as wind or solar energy, could be used to produce hydrogen. **Fuel cells**—essentially, hydrogen batteries (**FIGURE 21.26**)—could then use hydrogen to produce electricity to power vehicles, computers, cell phones, home heating, and more. NASA's space programs have used fuel-cell technology since the 1960s.

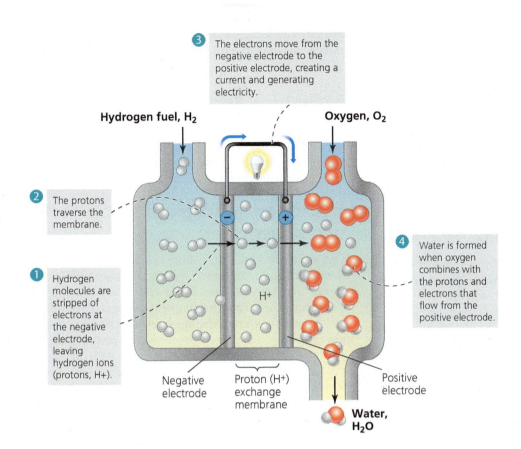

③ The electrons move from the negative electrode to the positive electrode, creating a current and generating electricity.

Hydrogen fuel, H₂

Oxygen, O₂

② The protons traverse the membrane.

① Hydrogen molecules are stripped of electrons at the negative electrode, leaving hydrogen ions (protons, H+).

H+

④ Water is formed when oxygen combines with the protons and electrons that flow from the positive electrode.

Negative electrode

Proton (H+) exchange membrane

Positive electrode

Water, H₂O

FIGURE 21.26 Hydrogen fuel drives electricity generation in a fuel cell, creating water as a waste product. Atoms of hydrogen are split ① into protons and electrons. The protons, or hydrogen ions ②, pass through a proton exchange membrane. The electrons, meanwhile, move from a negative electrode to a positive one via an external circuit ③ , creating a current and generating electricity. The protons and electrons then combine with oxygen ④ to form water molecules.

Basing an energy system on hydrogen could alleviate dependence on foreign fuels and help reduce greenhouse emissions that lead to climate change. For these reasons, governments have funded research into hydrogen and fuel-cell technology, and auto companies have developed vehicles that run on hydrogen. More than a decade ago, the nation of Iceland decided to set an example for the world by moving toward a "hydrogen economy." Iceland achieved several early steps in its 30- to 50-year plan to phase out fossil fuels,

such as converting buses in the capital city of Reykjavik to run on hydrogen fuel. But the global economic downturn in 2008–2009 and delays in manufacturing hydrogen cars stymied its progress, and today future efforts are uncertain. Meanwhile, Germany launched a number of hydrogen-fueled buses (**FIGURE 21.27**). Today certain cities in Germany, Japan, China, Brazil, Canada, the United Kingdom, California, and elsewhere feature modest fleets of hydrogen buses and cars dependent on a small number of hydrogen filling stations.

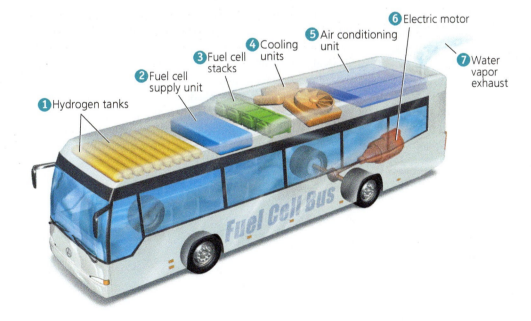

⑥ Electric motor

⑤ Air conditioning unit

④ Cooling units

③ Fuel cell stacks

② Fuel cell supply unit

① Hydrogen tanks

⑦ Water vapor exhaust

FIGURE 21.27 In one type of hydrogen-fueled bus, hydrogen is stored in nine fuel tanks ①. The fuel cell supply unit ② controls the flow of hydrogen, air, and cooling water into the fuel cell stacks ③. Cooling units ④ and the air conditioning unit ⑤ dissipate waste heat produced by the fuel cells. Electricity generated by the fuel cells is changed from direct current (DC) to alternating current (AC) by an inverter, and it is transmitted to the electric motor ⑥, which powers the operation of the bus. The vehicle's exhaust ⑦ consists simply of water vapor.

Hydrogen fuel may be produced from water or from other matter

Hydrogen gas (H_2) tends not to exist freely on Earth. Instead, hydrogen atoms bind to other molecules, becoming incorporated in everything from water to organic molecules. To obtain hydrogen gas for fuel, we must force these substances to release their hydrogen atoms, and this process requires an input of energy. Scientists are studying several ways of producing hydrogen. In the process of **electrolysis,** electricity is input to split hydrogen atoms from the oxygen atoms of water molecules:

$$2H_2O \longrightarrow 2H_2 + O_2$$

Electrolysis produces pure hydrogen, and it does so without emitting the carbon- or nitrogen-based pollutants of fossil fuel combustion. However, whether this strategy for producing hydrogen will cause pollution over its entire life-cycle depends on the source of the electricity used for the electrolysis. If coal is burned to generate the electricity, then the process will not reduce emissions. However, if the electricity is produced by some less-polluting renewable source, then hydrogen production by electrolysis would create much less pollution and greenhouse warming than relying on fossil fuels. The "cleanliness" of a future hydrogen economy would, therefore, depend largely on the source of electricity used in electrolysis.

The environmental impact of hydrogen production also depends on the source material for the hydrogen. Besides water, hydrogen can be obtained from biomass and from fossil fuels. This generally requires less energy input but results in pollution. For instance, extracting hydrogen from the methane (CH_4) in natural gas produces one molecule of the greenhouse gas carbon dioxide for every four molecules of hydrogen gas:

$$CH_4 + 2H_2O \longrightarrow 4H_2 + CO_2$$

Thus, whether a hydrogen-based energy system is cleaner than a fossil fuel system will depend on how the hydrogen is extracted.

Fuel cells produce electricity

Once isolated, hydrogen gas can be used as a fuel to produce electricity within fuel cells. The chemical reaction involved in a fuel cell is simply the reverse of that for electrolysis. Two hydrogen molecules and one oxygen molecule each split, and their atoms bind to form two water molecules:

$$2H_2 + O_2 \longrightarrow 2H_2O$$

Figure 21.25 shows how this occurs within one common type of fuel cell. As shown in the diagram, hydrogen gas (usually compressed and stored in an attached fuel tank) is allowed into one side of the cell, and the movement of the hydrogen's electrons from one electrode to the other electrode creates the output of electricity.

Hydrogen and fuel cells have costs and benefits

One major drawback of hydrogen is a lack of infrastructure to make use of it. To convert a nation such as Germany or the United States to hydrogen would require massive and costly development of facilities to transport, store, and provide the fuel.

Another concern is that some research suggests that leakage of hydrogen from its production, transport, and use could potentially deplete stratospheric ozone (p. 465) and lengthen the atmospheric lifetime of the greenhouse gas methane. Research into these questions is ongoing, because society needs to be clear about potential risks before switching from fossil fuels to hydrogen.

Hydrogen's benefits include the fact that we will never run out of it, because it is the most abundant element in the universe. Hydrogen can be clean and nontoxic to use, and—depending on its source and the source of electricity for its extraction—it may produce few greenhouse gases and other pollutants. Water and heat are the only waste products from a hydrogen fuel cell, along with negligible traces of other compounds. In terms of safety for transport and storage, hydrogen can catch fire and explode, but if kept under pressure, it may be no less safe than gasoline in tanks.

Hydrogen fuel cells are energy-efficient. Depending on the type of fuel cell, 35–70% of the energy released in the reaction can be used. If the system is designed to capture heat as well as electricity, then the energy efficiency of fuel cells can rise to 90%. These rates are comparable or superior to those of most nonrenewable alternatives. In addition, fuel cells are silent and nonpolluting. Unlike batteries (which also produce electricity through chemical reactions), fuel cells generate electricity whenever hydrogen fuel is supplied, without ever needing recharging. For all these reasons, hydrogen fuel cells remain promising for fueling vehicles—something that with more investment and infrastructure could one day happen on a large scale.

closing
THE LOOP

Rising concern over climate change, air pollution, health impacts, and security risks resulting from our dependence on fossil fuels has convinced many people that we need to shift to renewable energy sources that pollute far less and that will not run out. New renewable sources such as solar energy, wind power, geothermal energy, and ocean energy offer promise for sustaining our civilization far into the future without greatly degrading our environment. And by using electricity from renewable sources to produce hydrogen fuel, we may be able to use fuel cells to produce power when and where it is needed, helping to create nonpolluting vehicles.

Renewable energy sources have long been held back by limited funding for research and development and by competition with established and subsidized nonrenewable fuels whose market prices have not covered external costs. Despite these obstacles, renewable technologies are now progressing quickly enough to offer hope that we can shift from fossil fuels to renewable energy. Germany offers a prime example of how a government's economic policy can accelerate a transition to renewable energy. Its feed-in tariff programs have ignited the widespread adoption of PV solar, wind, and other renewable energy technologies throughout the country. In fact, Germans now produce so much electricity from renewable sources that their debates have shifted from questions of adequate supplies to ones of grid capacity, timing, and pricing. Another question that currently arises in Germany is what will happen once the feed-in tariffs expire.

As Germany blazes its path forward as a leader in new renewable energy sources, other nations are showing leadership as well. The United States has produced a great deal of research and technology with regard to renewable energy; Denmark is a leader in wind power; and many other European nations obtain high proportions of their energy from renewables. Developing nations are showing how to harness power off the grid, and China has thrown its economic might behind renewables, producing and deploying staggering amounts of technology. With these steps by many nations around the world, our global civilization is moving faster and faster toward a future fueled by renewable energy.

REVIEWING Objectives

You should now be able to:

Identify the major sources of renewable energy and assess their recent growth and future potential

The "new renewable" energy sources include solar, wind, geothermal, and ocean energy sources. These energy sources are not truly "new," but rather are in a stage of rapid development. They currently provide far less energy and electricity than we obtain from fossil fuels or other conventional sources, but they are growing quickly. Relative to fossil fuels, the new renewables alleviate air pollution, reduce greenhouse gas emissions, and help diversify a society's energy mix. Government subsidies have long favored nonrenewable energy, but investment and public policies such as feed-in tariffs can accelerate our transition to renewable sources. (pp. 579–584)

Describe solar energy and how we harness it, and evaluate its advantages and disadvantages

We can harness energy from the sun's radiation using passive methods or by active methods involving powered technology. Solar technologies include flat plate collectors for heating water and air, mirrors to concentrate solar rays, and photovoltaic (PV) cells to generate electricity. PV solar is today's fastest-growing energy source. Solar energy is perpetually renewable, creates no emissions, and enables decentralized power. However, solar radiation varies from place to place and time to time, and harnessing solar energy remains expensive. (pp. 584–591)

Describe wind power and how we harness it, and evaluate its advantages and disadvantages

Energy from wind is harnessed using wind turbines mounted on tall towers. Turbines are often erected in arrays at wind farms located on land or offshore, in locations with optimal wind conditions. Wind power is one of today's fastest-growing energy sources. Wind energy is renewable, turbine operation creates no emissions, wind farms can generate economic benefits, and the cost of wind power is competitive with that of electricity from fossil fuels. However, wind is an intermittent resource and is adequate only in some locations. Turbines kill some birds and bats, and wind farms often face opposition from local residents. (pp. 591–594)

Describe geothermal energy and how we harness it, and evaluate its advantages and disadvantages

Thermal energy from the naturally occurring radioactive decay in Earth's core rises toward the surface and heats groundwater. This energy may be harnessed by geothermal power plants and used to directly heat water and air or to generate electricity. Ground-source heat pumps heat and cool homes and businesses by circulating water whose temperature is moderated at shallow depths underground. Geothermal energy can be efficient, clean, and renewable, but naturally heated water occurs near the surface only in certain areas, and this water may be exhausted if overpumped. Enhanced geothermal systems allow us to gain geothermal energy from more regions, but the approach can trigger minor earthquakes. (pp. 595–597)

List ocean energy sources and describe their potential

Ocean energy sources include the motion of tides, waves, and currents, as well as the thermal heat of ocean water. Ocean energy is perpetually renewable and holds much promise, but technologies have seen limited commercial development thus far. (pp. 597–598)

Explain hydrogen fuel cells and weigh options for energy storage and transportation

Hydrogen can serve as a fuel to store and transport energy, so that electricity can be made portable and used to power vehicles. Hydrogen fuel cells generate electricity by controlling an interaction between hydrogen and oxygen, and they produce only water as a waste product. Hydrogen infrastructure that would facilitate storage and transportation requires much more development, but hydrogen can be clean, safe, and efficient. (pp. 598–600)

TESTING Your Comprehension

1. What proportion of U.S. energy now comes from renewable sources? What is the most prevalent form of renewable energy used in the United States? What form of renewable energy is most used to generate electricity?

2. What factors and concerns are causing renewable energy use to expand? Which two renewable sources are experiencing the most rapid growth?

3. Contrast passive and active solar heating. Describe how each works, and give examples of each.

4. Describe in your own words what a photovoltaic (PV) cell is, how it functions, and how it is used.

5. List several environmental and economic advantages of solar power. What are some disadvantages?

6. How do wind turbines generate electricity? How does wind speed affect the process? What conditions are most favorable for the placement of wind turbines?

7. Describe several environmental and economic benefits of wind power. What are some drawbacks?

8. Define *geothermal energy*, and explain three main ways in which it is obtained and used. Describe one sense in which it is renewable and one sense in which it is not renewable.

9. List and briefly describe four approaches for obtaining energy from ocean water.

10. How is hydrogen fuel produced? What factors determine the amount of pollutants hydrogen production will emit?

SEEKING Solutions

1. For each source of renewable energy discussed in this chapter, what factors stand in the way of an expedient transition to it from fossil fuel use? In each case, what could be done to ease a shift toward these renewable sources? Would market forces alone suffice to bring about this transition, or will we also need government? Do you think such a transition will be good for our economy? Why or why not?

2. Do some research online to find out what energy sources produce the most energy and electricity in your own state. Create a diagram like those in Figure 21.2 showing a quantitative breakdown of the energy your state's citizens use. Which renewable energy sources does your state use more than the United States as a whole, and which does it use less? What might be the reasons for these patterns?

3. Explain the circumstances under which using hydrogen fuel could be helpful in moving toward a low-emission future. If you could advise policymakers of Iceland, Germany, or the United States on hydrogen, what would you tell them?

4. **CASE STUDY CONNECTION** Explain how Germany accelerated its development of PV solar power and other renewable energy sources by establishing a system of feed-in tariffs. What steps did it take, and what have been the results so far? What future challenges does Germany face? Do you think the United States should adopt a similar system of feed-in tariffs to promote renewable energy across the nation? Explain in detail why, or why not.

5. THINK IT THROUGH You are the president of a nation the size of Germany, and your legislature is asking you to propose a national energy policy. Your country is located along a tropical coastline. Your geologists do not yet know whether there are fossil fuel deposits or geothermal resources under your land, but your country gets a lot of sunlight and a fair amount of wind, and broad, shallow shelf regions line its coasts. Your nation's population is moderately wealthy but is growing fast, and importing fossil fuels from other nations is becoming expensive.

What approaches would you propose in your energy policy? Name some specific steps you would urge your legislature to fund. Are there trade relationships you would seek to establish with other countries? What questions would you fund scientists to research?

6. THINK IT THROUGH You are the CEO of a company that develops wind farms. Your staff has presented you with three options, listed below, for sites for your next development. Describe at least one likely advantage and at least one likely disadvantage you would expect to encounter with each option. What further information would you like to know before deciding which to pursue?

- Option A: A remote rural site in North Dakota
- Option B: A ridge-top site in the suburbs of Philadelphia
- Option C: An offshore site off the Florida coast

CALCULATING Ecological Footprints

Assume that average per capita residential consumption of electricity is 12 kilowatt-hours per day, that photovoltaic cells have an electrical output of 15% incident solar radiation, and that PV panels cost $1000 per square meter. Now refer to Figure 21.12 on p. 590 and estimate the area and cost of the PV panels needed to provide all of the residential electricity used by each group in the table.

	AREA OF PHOTOVOLTAIC CELLS	COST OF PHOTOVOLTAIC CELLS
You		
A resident of Arizona		
A resident of Alaska		
Total for all U.S. residents		

1. What additional information would you need to increase the accuracy of your estimates for the areas in the table?

2. Considering the distribution of solar radiation in the United States, where do you think it will be most feasible to greatly increase the percentage of electricity generated from photovoltaic solar cells?

3. The purchase price of a photovoltaic system is considerable. What other costs and benefits should you consider, in addition to the purchase price, when contemplating "going solar"?

MasteringEnvironmentalScience®

CHAPTER 22

Managing Our Waste

A Mania for Recycling on Campus

> An extraterrestrial observer might conclude that conversion of raw materials to wastes is the real purpose of human economic activity.
>
> Gary Gardner and Payal Sampat, Worldwatch Institute

> Recycling is one of the best environmental success stories of the late 20th century.
>
> U.S. Environmental Protection Agency

OHIO

Miami • University

• Ohio University

At that time of year when NCAA basketball fever sweeps America's campuses, there's another kind of March Madness now taking hold: a mania for recycling.

It began in 2001, when waste managers at two Ohio campuses got the idea to use their schools' long-standing athletics rivalry to jump-start their recycling programs. Ed Newman of Ohio University, in Athens, and Stacy Edmonds Wheeler of Miami University, across the state in Oxford, challenged one another to see whose campus could recycle more in a 10-week competition. Come April, Miami University had taken the prize, recycling 41.2 pounds per student. Recyclemania was born.

Students at other colleges and universities heard about the event and wanted to get in on the action, and year by year more schools joined. Today Recyclemania pits several hundred institutions against one another, involving several million students and staff across North America. The event has grown to have a board of directors and major corporate sponsors.

Each year student leaders rouse their campuses to compete in 2 divisions and 11 different categories over 8 weeks in February and March. Every week during the competition, recycling bins are weighed and campuses report their data, which are compiled online at the Recyclemania website as the competition proceeds. The all-around winner gets a funky trophy made of recycled materials (a figure nicknamed "Recycle Dude," whose body is a rusty propane tank)—and, more important, global bragging rights for a year.

In spring 2016, 350 colleges and universities slugged it out. In the end, the battlefield was littered with stories of the victors and the vanquished (**FIGURE 22.1**, p. 606). Richland College in Dallas, Texas, took top honors as Grand Champion, recycling an impressive 82 percent of its waste, topping runners-up University of Missouri–Kansas City and New Mexico State University. Richland's cross-town neighbor, North Lake College in Irving, Texas, won the competition for least waste generated per capita, as its students limited their waste to just 4.1 pounds per person. Loyola Marymount University in Los Angeles took the prize for most recyclables per capita, with a hefty 67.4 pounds per student. Rutgers University claimed top honors in the Gorilla category, which measures total weight of items recycled, topping out at a staggering 2,057,581 pounds.

Campuses also compete to see which can collect the most of certain types of items per person. In 2016, Loyola Marymount collected the most paper, the most bottles and cans, and the most corrugated cardboard, while Union College saved the most food waste. University of San Diego recycled

Students at Pacific Lutheran University compete in the Recyclemania tournament.

The world's biggest collegiate waste management event ▲

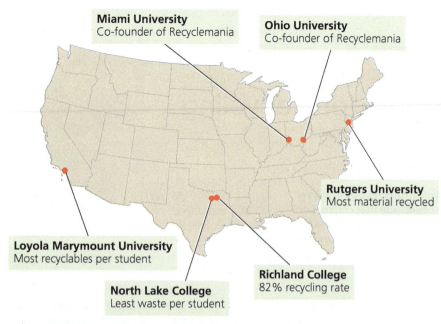

Miami University
Co-founder of Recyclemania

Ohio University
Co-founder of Recyclemania

Rutgers University
Most material recycled

Loyola Marymount University
Most recyclables per student

North Lake College
Least waste per student

Richland College
82% recycling rate

FIGURE 22.1 Four schools were top winners among 350 participating in Recyclemania 2016. The event began over 15 years ago at Ohio University and Miami University.

dioxide—equal to the emissions output of more than 24,000 cars or the electricity use of more than 15,000 homes. By focusing the attention of administrators on waste issues, Recyclemania facilitates the expansion of campus waste reduction programs. Most important, it gets a new generation of young people revved up about the benefits of recycling.

Recyclemania is the biggest of a growing number of campus competitions in the name of sustainability. In the 3-week springtime event called the Campus Conservation Nationals, schools have competed against one another for savings in water use and energy use. More than 345,000 students in 1374 buildings at 125 colleges and universities took part in the 2015 Campus Conservation Nationals, saving 394,000 gallons of water (equal to 2500 hours in the shower) and 1.9 million kilowatt-hours of electricity (equivalent to taking 182 homes off the power grid for a year).

Thanks in part to Recyclemania, recycling is the most widespread activity among campus sustainability efforts (pp. 16–17, 651–655). These efforts include water conservation, energy efficiency, green buildings, transportation options, sustainable food in dining halls, and campus gardens. Students restore native plants and habitats, promote renewable energy, and advocate for carbon neutrality on campus. A growing movement, campus sustainability is thriving because for students, faculty, staff, and administrators, it's satisfying to do the right thing and pitch in to help make campuses more sustainable. … And it's even more fun when you can compete and show that you can do it better than your rival school across the state!

the most electronic waste per capita and University of Oregon the most film plastic. Finally, Loyola University Chicago, Ohio State University, University of Virginia, and Cornell University were champions in the competition to see who could best reduce waste at a home basketball game.

By encouraging all this recycling, Recyclemania cuts down on pollution from the mining of new resources and the manufacture of new goods. Each year students in the event help to prevent the release of more than 120,000 tons of carbon

Approaches to Waste Management

As the world's population rises, and as we produce and consume more material goods, we generate more waste. **Waste** refers to any unwanted material or substance that results from a human activity or process. Waste can degrade water quality, soil quality, air quality, and human health. Waste also indicates inefficiency—so reducing waste can save money and resources. For these reasons, waste management has become a vital pursuit.

For management purposes, we divide waste into several categories. **Municipal solid waste** is nonliquid waste that comes from homes, institutions, and small businesses. **Industrial solid waste** includes waste from production of consumer goods, mining, agriculture, and petroleum extraction and refining. **Hazardous waste** refers to solid or liquid waste that is toxic, chemically reactive, flammable, or corrosive. Another type of waste is wastewater (p. 407), water we use in our households, businesses, industries, or public facilities and drain or flush down our pipes, as well as the polluted runoff from streets and storm drains. (We discuss wastewater in Chapter 15, pp. 409–411.)

There are three main components of **waste management:**

1. Minimizing the amount of waste we generate
2. Recovering discarded materials and finding ways to recycle them
3. Disposing of waste safely and effectively

We have several ways to reduce the amount of material in the **waste stream**, the flow of waste as it moves from its sources toward disposal destinations (**FIGURE 22.2**). Minimizing waste at its source—called **source reduction**—is the preferred approach. We can achieve source reduction when manufacturers use materials more efficiently, or when consumers buy fewer goods, buy goods with less packaging, or use those goods longer. Reusing goods you already own, purchasing used items, and donating your used items for others all help reduce the amount of material entering the waste stream.

The next-best strategy in waste management is **recovery,** which consists of recovering, or removing, waste from the waste stream. Recovery includes recycling and composting. **Recycling** is the process of collecting used goods and sending them to facilities that extract and reprocess raw

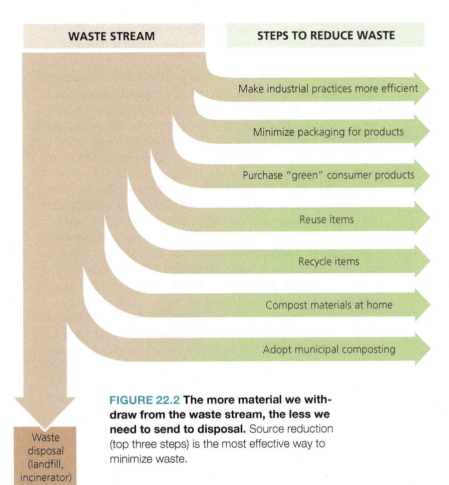

WASTE STREAM

STEPS TO REDUCE WASTE

Make industrial practices more efficient

Minimize packaging for products

Purchase "green" consumer products

Reuse items

Recycle items

Compost materials at home

Adopt municipal composting

Waste disposal (landfill, incinerator)

FIGURE 22.2 The more material we withdraw from the waste stream, the less we need to send to disposal. Source reduction (top three steps) is the most effective way to minimize waste.

are broken down at some point, and matter cycles through ecosystems (Chapter 5). People have taken these concepts from nature and applied them in our society to help cut down on waste and conserve resources.

Regardless of how well we reduce our waste stream through source reduction and recovery, there will likely always be some waste left to dispose of. Disposal methods include burying waste in landfills and burning waste in incinerators. The linear movement of products from their manufacture to their disposal is often described as "cradle-to-grave." As much as possible, however, the modern waste manager attempts to follow a **cradle-to-cradle** approach instead—one in which the materials from products are recovered and reused to create new products.

In our examination of these issues, we will first examine how waste managers use source reduction, recovery, and disposal to manage municipal solid waste, and then we will turn to industrial solid waste and hazardous waste.

Municipal Solid Waste

Municipal solid waste is what we commonly refer to as "trash" or "garbage." In the United States, paper, food scraps, yard trimmings, and plastics are the principal components of municipal solid waste, together accounting for two-thirds of what enters the waste stream (**FIGURE 22.3a**). Paper is recycled at a high rate and yard trimmings are composted at a high rate, so after recycling and composting reduce the waste stream, food scraps and plastics are left as the largest components of U.S. municipal solid waste (**FIGURE 22.3b**). In developing nations, food scraps are often the primary component, and paper makes up a smaller proportion.

materials that can then be used to manufacture new goods. **Composting** is the practice of recovering organic waste (such as food and yard waste) by converting it to mulch or humus (p. 214) through natural biological processes of decomposition. Recycling and composting are fundamental features of the way natural systems function; all materials in nature

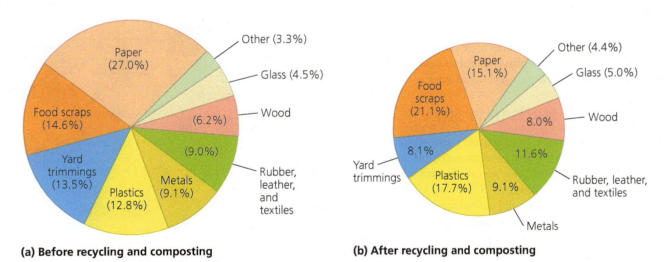

(a) Before recycling and composting

Paper (27.0%)
Other (3.3%)
Glass (4.5%)
Food scraps (14.6%)
Wood
(6.2%)
Yard trimmings (13.5%)
(9.0%)
Plastics (12.8%)
Metals (9.1%)
Rubber, leather, and textiles

(b) After recycling and composting

Paper (15.1%)
Other (4.4%)
Food scraps (21.1%)
Glass (5.0%)
8.0%
Wood
Yard trimmings 8.1%
11.6%
Plastics (17.7%)
9.1%
Rubber, leather, and textiles
Metals

FIGURE 22.3 Components of the municipal solid waste stream in the United States. Paper products make up the greatest portion by weight **(a)**, but after recycling and composting removes many items **(b)**, the waste stream becomes one-third smaller. Food scraps are now the largest contributor, because so much paper is recycled and yard waste is composted. *Data from U.S. Environmental Protection Agency, 2015.* Advancing sustainable materials management: Facts and figures 2013. *Washington, D.C.: EPA.*

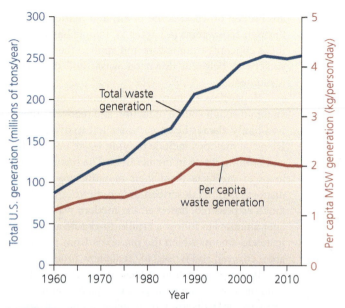

FIGURE 22.4 Rising U.S. waste generation has now leveled off. Total U.S. waste generation before recycling (**blue line**) nearly tripled after 1960, and U.S. per capita waste generation before recycling (**red line**) rose by 65%. In recent years, total and per-person waste generation have each leveled off, largely thanks to source-reduction efforts. *Data from U.S. Environmental Protection Agency, 2015.* Advancing sustainable materials management: Facts and figures 2013. *Washington, D.C.: EPA.*

DATA From these data, what would you infer has happened to the population size of the United States between 1990 and 2013? Explain how you know this.

Go to **Interpreting Graphs & Data** on **Mastering**EnvironmentalScience®

Most municipal solid waste comes from packaging and nondurable goods (products meant to be discarded after a short period of use). In addition, consumers throw away old durable goods and outdated equipment as they purchase new products. Plastics, which came into wide consumer use only after 1970, have accounted for the greatest relative increase in the waste stream during the past several decades.

Consumption leads to waste

As we acquire more goods, we generate more waste. In the United States since 1960, waste generation (before recovery) has nearly tripled (**FIGURE 22.4**), and per capita waste generation has risen by 64%. Today, U.S. citizens produce over 250 million tons of municipal solid waste (before recovery)—close to 1 ton per person. The average American generates 2.0 kg (4.4 lb) of trash per day—considerably more than people in most other industrialized nations. The relative wastefulness of the U.S. lifestyle, with its excess packaging and reliance on nondurable goods, has caused critics to label the United States "the throwaway society."

However, Americans are beginning to turn this around. Thanks to source reduction and reuse (especially by busi-

nesses looking to cut costs), Americans now generate slightly less waste per capita than they did in the period between 1990 and 2005.

In developing nations, people consume fewer resources and goods and, as a result, generate less waste. However, consumption is intensifying in developing nations as they become more affluent, and consequently these nations are generating increasing amounts of waste. This growth in waste reflects rising material standards of living, but it also results from an increase in packaging, manufacturing of nondurable goods, and production of inexpensive, poor-quality goods that wear out quickly. As a result, trash is piling up and littering the landscapes of countries from Mexico to Kenya to Indonesia.

Like U.S. consumers in the "throwaway society," wealthy consumers in developing nations often discard items that can still be used. In fact, at many dumps and landfills in the developing world, poor people support themselves by selling items that they scavenge (**FIGURE 22.5**).

In many industrialized nations in addition to the United States, per capita generation rates have begun to decline in recent years. Wealthier nations also can afford to invest more in waste collection and disposal, so they are often better able to manage their waste and minimize impacts on human health and the environment. Moreover, enhanced recycling and composting—fed by a conservation ethic growing among a new generation on today's campuses—have been removing more and more material from the waste stream (**FIGURE 22.6**). As of 2013, U.S. waste managers were landfilling 53% of municipal solid waste, incinerating 13%, and recovering 34% for composting and recycling.

FIGURE 22.5 Affluent consumers discard so much usable material that some people in developing nations support themselves by scavenging items from dumps. Tens of thousands of people used to scavenge each day from this dump outside Manila in the Philippines, selling material to junk dealers for 100–200 pesos (U.S. $2–$4) per day. The dump was closed in 2000 after an avalanche of trash killed hundreds of people.

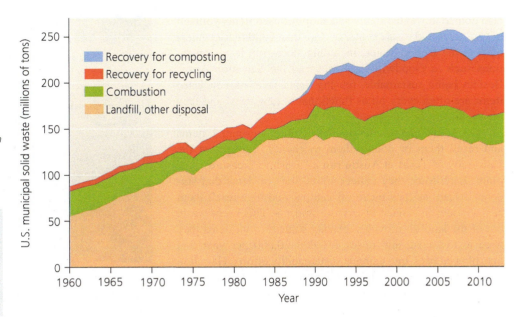

FIGURE 22.6 As recycling and composting have grown in the United States, the proportion of waste going to landfills has declined. As of 2013, 53% of U.S. municipal solid waste went to landfills and 13% to incinerators, whereas 34% was recovered for composting and recycling. *Data from U.S. Environmental Protection Agency, 2015. Advancing sustainable materials management: Facts and figures 2013. Washington, D.C.: EPA.*

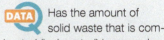 Has the amount of solid waste that is combusted (incinerated) increased or decreased since 1960?

Go to **Interpreting Graphs & Data** on **Mastering**EnvironmentalScience°

Reducing waste is our best option

Reducing the amount of material entering the waste stream is the preferred option for managing waste. Recall that preventing waste generation in this way is known as *source reduction*. This preventative approach avoids costs of disposal and recycling, helps conserve resources, minimizes pollution, and can save consumers and businesses money.

One means of source reduction is to lessen the materials used to package goods. Packaging helps to preserve freshness, prevent breakage, protect against tampering, and provide information—yet much packaging is extraneous. Consumers can give manufacturers incentive to reduce packaging by choosing minimally packaged goods, buying unwrapped fruit and vegetables, and buying food from the bulk sections of stores. Manufacturers can switch to packaging that is more recyclable. They can also reduce the size or weight of goods and materials, as they already have with aluminum cans, plastic soft drink bottles, personal computers, and much else.

Some governments have recently taken aim at a major source of waste and litter—plastic grocery bags. These lightweight polyethylene bags can persist for centuries in the environment, choking and entangling wildlife and littering the landscape—yet Americans discard 100 billion of them each year. A number of cities and more than 20 nations have now enacted bans or limits on their use. Financial incentives are also effective. When Ireland began taxing these bags, their use dropped 90%. IKEA stores began charging for them and saw similar drops in usage. Many businesses now give discounts if you bring your own reusable canvas bags.

Increasing the longevity of goods also helps reduce waste. Because companies seek to maximize sales, they often have a financial incentive to produce short-lived goods that need to be replaced frequently. As a result, increasing the longevity of goods is largely up to the consumer. If consumer demand for goods that last longer is great enough, manufacturers will respond.

Reuse is a main strategy to reduce waste

To reduce waste, you can save items to use again or substitute disposable goods with durable ones. **TABLE 22.1** presents a sampling of actions we all can take to reduce the waste we generate. Habits as simple as bringing your own coffee cup to coffee shops or bringing sturdy reusable cloth bags to the grocery store can, over time, have substantial impact. You can

---

TABLE 22.1 Some Everyday Things You Can Do to Reduce and Reuse

- Donate used items to charity
- Reuse boxes, paper, plastic wrap, plastic containers, aluminum foil, bags, wrapping paper, fabric, packing material, etc.
- Rent, borrow, or lend items instead of buying them
- Bring reusable cloth bags shopping
- Make double-sided photocopies
- Keep electronic documents rather than printing items out
- Bring your own coffee cup to coffee shops
- Pay a bit extra for durable, long-lasting, reusable goods rather than disposable ones
- Buy rechargeable batteries
- Select goods with less packaging
- Compost kitchen and yard wastes
- Buy clothing and other items at resale stores and garage sales
- Use cloth napkins and rags, not paper napkins and towels

Adapted from U.S. Environmental Protection Agency.

---

also donate unwanted items and shop for used items yourself at yard sales and resale centers. More than 6000 reuse centers exist in the United States, including stores run by organizations such as Goodwill Industries and the Salvation Army. Besides being sustainable, reusing items saves money. Used items are often every bit as functional as new ones, and they are usually much cheaper.

On some campuses, students collect unwanted items and resell them or donate them to charity. Students at the University of Texas at Austin run a "Trash to Treasure" program. Each May, they collect 40–50 tons of items that students discard as they move off campus and then resell them at low prices in August to arriving students. This keeps waste out of the landfill, provides arriving students with items they need at low cost, and raises $10,000–20,000 per year that gets plowed back into campus sustainability efforts. Hamilton College in New York runs a similar program, called "Cram & Scram." It reduces Hamilton's landfill waste by 28% (about 90 tons) each May.

FIGURE 22.7 Inside Edmonton's gigantic aeration building, a mix of solid waste and sewage sludge is exposed to oxygen and composted.

Composting recovers organic waste

Composting is the conversion of organic waste into mulch or humus (p. 214) through natural decomposition. We can place waste in compost piles, underground pits, or specially constructed containers. As waste is added, heat from microbial action builds in the interior and decomposition proceeds. Banana peels, coffee grounds, grass clippings, autumn leaves, and other organic items can be converted into rich, high-quality compost through the actions of earthworms, bacteria, soil mites, sow bugs, and other detritivores and decomposers (pp. 78, 212). The compost is then used to enrich soil. Home composting is a prime example of how we can live more sustainably by mimicking natural cycles and incorporating them into our daily lives.

On campus, composting is becoming popular. Ball State University in Indiana shreds surplus furniture and wood pallets and makes them into mulch to nourish campus plantings. Ithaca College in New York composts 44% of its food waste, saving $11,500 each year in landfill disposal fees. The compost is used on campus plantings, and student-run experiments showed that the plantings grew better with the compost mix than with chemical soil amendments.

Municipal composting programs—more than 3500 across the United States at last count—divert yard debris from the waste stream and send it to central composting facilities, where it decomposes into mulch that community residents can use for gardens and landscaping. Increasingly, these programs are also accepting food scraps for composting. About one-fifth of the U.S. waste stream is made up of materials that can easily be composted. Composting reduces landfill waste, enriches soil, enhances soil biodiversity, helps soil to resist erosion, makes for healthier plants and more pleasing gardens, and reduces the need for chemical fertilizers.

One of the world's top composting programs is in Edmonton, Alberta, where close to half the waste stream is composted. When Edmonton's residents put their trash out at the curb, city trucks take it to a facility the size of eight football fields. The waste is dumped on the floor, large items such as furniture are removed and landfilled, and the rest is mixed with dried sewage sludge for 1–2 days in five huge rotating drums, each the length of six buses. The resulting mix travels on a conveyor to a screen that removes nonbiodegradable items. It is aerated for several weeks (**FIGURE 22.7**), then passed through a finer screen and left outside for 4–6 months. The resulting compost—80,000 tons annually—is made available to farmers and gardeners.

Recycling consists of three steps

Recycling, too, offers many benefits. It involves collecting used items and breaking them down so that their materials can be reprocessed to manufacture new items. Recycling today in the United States diverts about 65 million tons of materials away from incinerators and landfills.

The recycling loop consists of three basic steps. The first step is to collect and process used goods and materials, as is being done on so many campuses. Some towns and cities designate locations where residents can drop off recyclables or receive money for them. Others offer curbside recycling, in which trucks pick up recyclable items in front of homes, usually in conjunction with municipal trash collection.

Items collected are taken to **materials recovery facilities (MRFs),** where workers and machines sort items using automated processes including magnetic pulleys, optical sensors, water currents, and air classifiers that separate items by weight and size. The facilities clean the materials, shred them, and prepare them for reprocessing.

Once readied, these materials are used to manufacture new goods—the second step in the recycling loop. Newspapers and many other paper products use recycled paper, many glass and metal containers are now made from recycled materials, and some plastic containers are of recycled origin. Benches, bridges, and walkways in city parks may now be made from recycled plastics, and glass can be mixed with asphalt (creating "glassphalt") to pave roads and paths.

If the recycling loop is to function, consumers and businesses must complete the third step in the cycle by purchasing ecolabeled products (p. 152) made from recycled materials. By buying recycled goods, consumers provide economic incentive for industries to recycle materials and for recycling facilities to open or expand.

Recycling has grown rapidly

Today nearly 10,000 curbside recycling programs across all 50 U.S. states serve 70% of all Americans. These programs, and the 800 MRFs in operation today, have sprung up only in the past few decades. Recycling in the United States rose from 6.4% of the waste stream in 1960 to 25.5% in 2013 (and 34.3% if composting is included) (**FIGURE 22.8**).

Many college and university campuses run active recycling programs, although attaining high recovery rates can be challenging in the campus environment. The most recent survey of campus sustainability efforts suggested that the average recycling rate was only 29%. Thus there appears to be much room for growth. Fortunately, waste management initiatives are relatively easy to conduct on campus because they offer many opportunities for small-scale improvements and because people generally enjoy recycling and reducing waste.

Besides participation in Recyclemania, there are many ways to promote recycling on campus. Louisiana State University students initiated recycling efforts at home football games, and over three seasons they recycled 68 tons of refuse

FIGURE 22.9 In a trash audit, students sort through rubbish and separate out recyclables. Events like this "Mt. Trashmore" exercise at Central New Mexico Community College in 2015 show passersby just how many recyclable items are needlessly thrown away.

that otherwise would have gone to the landfill. "Trash audits" or "landfill on the lawn" events involve emptying trashcans or dumpsters and sorting out recyclable items (**FIGURE 22.9**). When students at Ashland University in Ohio audited their waste, they found that 70% was recyclable, and they used this information to press their administration to support recycling programs.

The economics of recycling are complex

Across the United States, recycling rates vary greatly from one product or material type to another, ranging from nearly zero to almost 100% (**TABLE 22.2**, p. 612). Recycling rates among U.S. states also vary greatly (**FIGURE 22.10**, p. 612). This variation makes clear that opportunities remain for further growth in recycling.

Recycling's growth thus far has been propelled in part by economic forces as businesses see prospects to save money and as entrepreneurs see opportunities to start new businesses. It has also been driven by the desire of community and campus leaders to reduce waste and by the satisfaction people take in recycling. These latter two forces have driven the rise of recycling even when it has

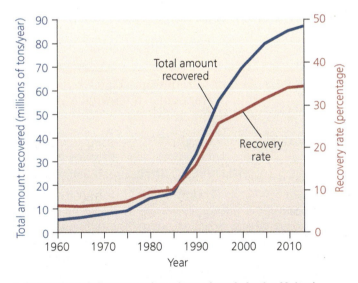

FIGURE 22.8 Recovery has risen sharply in the United States. Today more than 87 million tons of material are recovered (65 million tons by recycling and 22 million tons by municipal composting), making up one-third of the waste stream. *Data from U.S. Environmental Protection Agency.*

DATA From the data in this graph alone, what would you infer has happened to the total amount of municipal solid waste generated (before recovery) since 1960? Explain how you can determine this.

Go to **Interpreting Graphs & Data** on **Mastering**EnvironmentalScience®

weighing the
ISSUES

Managing Waste on Your Campus

Does your campus have a recycling program? Does it have any composting initiatives? Does it run programs to reduce or reuse materials? Think about the types and amounts of waste generated on your campus. Describe several examples of this waste that you feel could be prevented or recycled, and describe how this might be done in each case. If you could do one thing on campus to improve your school's waste management practices, what would it be?

TABLE 22.2
Recovery Rates for Various Materials in the United States

MATERIAL	PERCENTAGE RECYCLED OR COMPOSTED
Lead-acid batteries	99
Steel cans	71
Newspapers	67
Paper and paperboard	63
Yard trimmings	60
Aluminum cans	55
Tires	41
Glass containers	34
Total plastics	9

Data from U.S. Environmental Protection Agency.

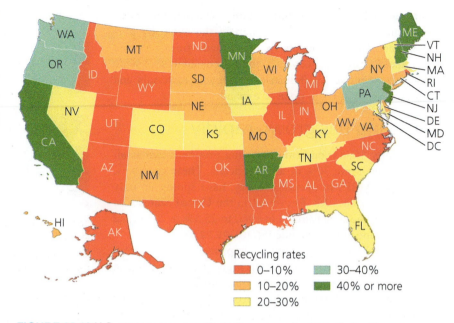

Recycling rates
- 0–10%
- 10–20%
- 20–30%
- 30–40%
- 40% or more

FIGURE 22.10 U.S. states vary greatly in the rates at which their citizens recycle. *Data from Shin, D., 2014.* Generation and disposition of municipal solid waste (MSW) in the United States—A national survey. *New York: Columbia University, Earth Engineering Center.*

not been financially profitable. In fact, many of our popular municipal recycling programs are run at an economic loss. The expense required to collect, sort, and process recycled goods is often more than recyclables are worth in the marketplace. In addition, the more people recycle, the more glass, paper, and plastic is available to manufacturers for purchase, which drives down prices. And transporting items to recycling facilities can sometimes involve surprisingly long distances (see **THE SCIENCE BEHIND THE STORY**, pp. 614–615).

The low commodity prices of recent years have also posed a challenge to recycling programs. When world oil prices are low, buying new plastic (made from petroleum) can be cheaper than buying recycled plastic; and when market prices of metals are low, buying newly mined metals can be cheaper than buying recycled metals. When recycling is no longer profitable for those in the recycling industry, MRFs may shut down, municipalities may cancel contracts, and recycling companies may go out of business.

Recycling advocates, however, point out that market prices do not take into account external costs (pp. 141, 162)—in particular, the environmental and health impacts of *not* recycling. For instance, it has been estimated that globally, recycling

saves enough energy to power more than 6 million households per year. Each year in the United States, recycling and composting together save energy equal to that of 230 million barrels of oil, and prevent carbon dioxide emissions equal to those of 39 million cars (**TABLE 22.3**). Recycling aluminum cans saves 95% of the energy required to make the same

weighing the ISSUES

Costs of Recycling and Not Recycling

Should governments subsidize recycling programs if they are run at an economic loss? What types of external costs—costs not reflected in market prices—do you think would be involved in not recycling, say, aluminum cans? Do you feel these costs justify sponsoring recycling programs even when they are not financially self-supporting? Why or why not?

TABLE 22.3 Annual Greenhouse Gas Reductions due to Recovery of Various Materials in the United States

MATERIAL	WEIGHT RECOVERED (MILLIONS OF TONS)	EQUAL TO NUMBER OF CARS TAKEN OFF THE ROAD
Paper and paperboard	43.0	31,000,000
Metals	7.9	4,500,000
Textiles	2.3	1,200,000
Wood	2.5	798,000
Plastics	3.0	760,000
Food	1.8	308,000
Yard trimmings	20.6	220,000
Glass	3.2	210,000
Rubber and leather	1.2	127,000

Data from U.S. Environmental Protection Agency.

amount of aluminum from mined virgin bauxite, its source material.

As more manufacturers use recycled products and as more technologies and methods are developed to use recycled materials in new ways, markets should continue to expand, and new business opportunities may arise. We are just beginning to shift from an economy that moves linearly, from raw materials to products to waste, to a more sustainable economy that moves circularly, taking a cradle-to-cradle approach and using waste products as raw materials for new manufacturing.

Financial incentives help address waste

To encourage recycling, composting, and source reduction, waste managers frequently offer consumers economic incentives to reduce the waste stream. In "pay-as-you-throw" garbage collection programs, municipalities charge residents for home trash pickup according to the amount of trash they put out. The less waste one generates, the less one has to pay.

Bottle bills are another approach hinging on financial incentives. In the 10 U.S. states and 23 nations that have these laws, consumers pay a deposit on bottles or cans upon purchase—often 5 cents per container—and then receive a refund when they return them to stores after use. The first bottle bills were passed in the 1970s to cut down on litter, but they have also served to decrease the waste stream. Beverage container recycling rates for states with bottle bills are 3.5 times higher than for states without them (**FIGURE 22.11**). U.S. states with bottle bills report that their beverage container litter has decreased by 69–84%, their total litter has decreased

by 30–65%, and their per capita container recycling rates have risen 2.6-fold.

As of 2016, bottle bill advocates in 13 states were seeking to establish or expand programs. It is a testament to the lobbying influence of the beverage industries and grocery retailers, which have traditionally opposed passage of bottle bills, that more states do not have such legislation.

States with bottle bills now face two challenges. One is to amend these laws to include new kinds of containers, particularly plastic containers. The second challenge is to adjust refunds for inflation. In the 45 years since Oregon passed the nation's first bottle bill, the value of a nickel has dropped such that today, the refund would need to be 28 cents to reflect the refund's original intended value. Proponents argue that increasing refund amounts will raise return rates, and available data support this view (see Figure 22.11).

Sanitary landfills are our main disposal method

Material that remains in the waste stream following source reduction and recovery needs to be disposed of, and landfills offer our main method of disposal. In modern **sanitary landfills,** waste is buried in the ground or piled up in large mounds engineered to prevent waste from contaminating the environment and threatening public health (**FIGURE 22.12**, p. 616). Most municipal landfills in the United States are regulated locally or by the states, but they must meet national standards set by the U.S. Environmental Protection Agency (EPA) under the **Resource Conservation and Recovery Act** (p. 172), a major federal law enacted in 1976 and amended in 1984.

In a sanitary landfill, waste is partially decomposed by bacteria and compresses under its own weight to take up less space. Soil is layered along with the waste to speed decomposition, reduce odor, and lessen infestation by pests. Some infiltration of rainwater into the landfill is good, because it encourages biodegradation by bacteria—yet too much is not good, because contaminants can escape if water carries them out.

To protect against environmental contamination, U.S. regulations require that landfills be located away from wetlands and earthquake-prone faults and be at least 6 m (20 ft) above the water table. The bottoms and sides of sanitary landfills must be lined with heavy-duty

How much does garbage decompose in a landfill?

You might assume that a banana peel you throw in the trash will soon decay away to nothing in a landfill. However, it just might survive longer than you do! This is because surprisingly little decomposition occurs in landfills. Researcher William Rathje, a recently retired archaeologist known as "the Indiana Jones of Solid Waste," made a career out of burrowing into landfills and examining their contents to learn about what we consume and what we throw away. His research teams would routinely come across whole hot dogs, intact pastries that were decades old, and grass clippings that were still green. Newspapers 40 years old were often still legible, and the researchers used them to date layers of trash.

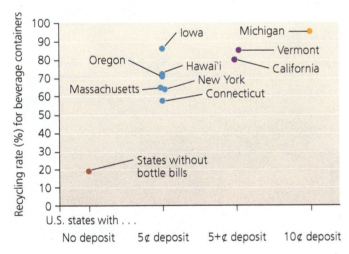

FIGURE 22.11 Bottle bills increase recycling rates, and higher redemption amounts boost these rates further, data suggest. States with bottle bills have much higher recycling rates for beverage containers than states without bottle bills. Maine also has a bottle bill but has not kept detailed data on recycling rates. *Data from Container Recycling Institute, Arlington, VA, 2015.*

DATA How many times greater is Michigan's bottle recycling rate than that of states without bottle bills?

Go to **Interpreting Graphs & Data** on **Mastering**EnvironmentalScience®

Where Does Trash Go?

Trash tags used in Seattle and New York City

Where does your trash go once you throw it away? What happens to your recycling? How far might it travel, and how much energy might it take to get rid of it?

With the help of the latest tracking technology, we can find out. Researchers from the SENSEable City Lab at the Massachusetts Institute of Technology (MIT) are affixing tiny sensors to everyday items in our trash and monitoring them to reveal their hidden travels. By documenting what actually happens to trash and to recyclables, they hope to help make the trash removal process more effective and to encourage better recycling.

The Trash Track project was launched in 2009 in New York City and in Seattle. Early results were unveiled at public exhibitions in both cities, and it is now expanding elsewhere.

Here's how trash tracking works: Project director Carlo Ratti and associate director Assaf Biderman, both architects at MIT, organize research teams and local volunteers in the target city to affix tiny electronic tags (see photo) to hundreds of items being thrown away. As each item journeys through the waste stream, its tag calculates its location every few hours and relays the information via the cell phone network to a central server at MIT. A computer plots the movements atop satellite maps, helping the researchers to visualize and interpret the migration of trash.

As an example, a plastic container of liquid soap was tagged on September 5, 2009, and placed in the trash at 457 Madison Avenue in Manhattan (**FIGURE 1**). Mapping reveals that the truck that picked it up looped through the city's streets a few times on its route, crossed the Hudson River via the

FIGURE 1 Follow that trash truck! A plastic soap container put out with the trash on Madison Avenue in New York City looped through midtown Manhattan, crossed the Hudson River, and was last detected traveling down the Bellevue Turnpike in New Jersey.

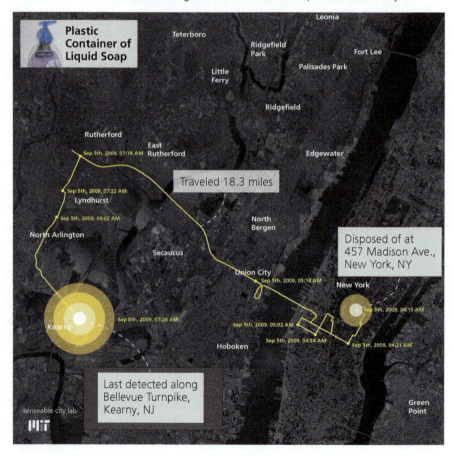

plastic and 60–120 cm (2–4 ft) of impermeable clay to help prevent contaminants from seeping into aquifers. Sanitary landfills also have systems of pipes, collection ponds, and treatment facilities to collect and treat **leachate,** liquid that results when substances from the trash dissolve in water as rainwater percolates downward.

Once a landfill is closed, it is capped with an engineered cover consisting of layers of plastic, gravel, and soil. Managers are required to maintain leachate collection systems for 30 years after a landfill has closed, and regulations require that groundwater be monitored regularly for contamination.

Lincoln Tunnel, and then headed to Rutherford, New Jersey. Here the truck turned south and was in transit along the Bellevue Turnpike in Kearny, New Jersey, three days later, when the tag's battery gave out.

Each tag calculates its position by measuring the signal strength from nearby cell phone towers and comparing this to a map of tower locations. Tags used in New York City and Seattle are not as accurate as global positioning systems (GPS), but their signals carry better through such barriers as building roofs, walls of garbage trucks, and deep piles of trash. Second-generation tags the project is now using are more accurate, combining GPS technology with better cell network triangulation.

As of 2016, the project had posted mapped data of all its Seattle items online for the public to view and had published several research papers. The results from Seattle reveal some expected patterns but also some odd surprises. Of the 760 items tagged, about 200 ended up at the city's Allied Waste Recycling Center and Transfer Station after brief journeys. Smaller numbers were transported to other city or regional landfills and recycling centers. But some items followed surprisingly circuitous routes, being transferred from one waste center to another, or back and forth between cities. Some ended up in seemingly random places, perhaps having fallen off a garbage truck along a roadside.

A few items made very long journeys. Two printer cartridges were driven down Interstate 5 to California's border with Mexico, perhaps to be disassembled at a factory. Two cell phones were transported halfway across the country to Dallas (one flown directly and one making apparent stops at three other cities). A compact fluorescent bulb went to St. Louis after traveling to Portland, Oregon, and back to Seattle. Chicago received a coffee cup that apparently made its way east on the nation's interstates. Shipments of batteries were flown 1500 miles to Minneapolis, 2500 miles to Pittsburgh, and 2600 miles to Atlanta. The longest-traveling piece of trash was a cell phone that was transported more than 3000 miles from Seattle to the other corner of the United States, ending up near Ocala, Florida.

In general, hazardous waste items and electronic waste items tended to travel farthest (**FIGURE 2**) because they were sent to special facilities for handling. This raises the question of whether special handling is worthwhile, given

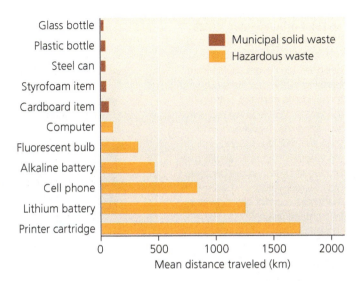

FIGURE 2 Items tagged in Seattle traveled vastly different distances. *Adapted from Offenhuber, D., et al. 2012. Putting matter in place: Tradeoffs between recycling and distance in planning for waste disposal. J. Amer. Planning Assoc. 78: 173–196.*

DATA Q Which type of trash tended to travel farther: municipal solid waste or hazardous waste? How far did each type of trash travel? What do you think accounts for this difference?

Go to **Interpreting Graphs & Data** on **Mastering**EnvironmentalScience®

the impacts of the extra gasoline use and greenhouse gas emissions it entails. In a 2012 research paper, the Trash Track team suggested that we seek to make processing systems for hazardous and electronic waste as efficient as those for curbside waste collection and recycling.

Why did each tracked item migrate so far? How much gasoline was used to transport them? When we move items great distances for disposal or recycling, does that reflect efficiency from an economy of scale—or does it indicate an excessive waste of resources? Do we need more recycling and disassembly facilities nearer to each major city? Researchers and waste managers hope to use data from the Trash Track project to address such questions and improve the way we handle waste.

Despite improvements in liner technology and landfill siting, however, liners can be punctured, and leachate collection systems eventually cease to be maintained. Moreover, landfills are kept dry to reduce leachate, but the bacteria that break down material thrive in wet conditions. Dryness, therefore, slows waste decomposition. In fact, the low-oxygen conditions of most landfills turn trash into a sort of time capsule. Researchers examining landfills often find some of their contents perfectly preserved, even after years or decades.

In 1988 the United States had nearly 8000 landfills, but today it has only 1900. Waste managers have consolidated the waste stream into fewer landfills of larger size.

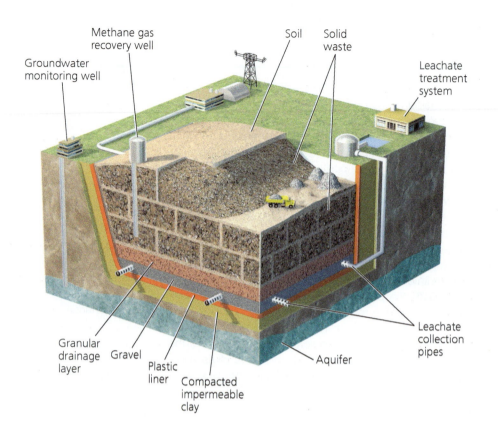

FIGURE 22.12 Sanitary landfills are engineered to prevent waste from contaminating soil and groundwater. Waste is laid in a large depression lined with plastic and impervious clay designed to prevent liquids from leaching out. Pipes draw out these liquids from the bottom of the landfill. Waste is layered with soil, filling the depression, and then is built into a mound until the landfill is capped. Landfill gas produced by anaerobic bacteria may be recovered, and waste managers monitor groundwater for contamination.

Methane gas recovery well
Groundwater monitoring well
Soil
Solid waste
Leachate treatment system
Granular drainage layer
Gravel
Plastic liner
Compacted impermeable clay
Aquifer
Leachate collection pipes

In many cities, landfills that were closed are now being converted into public parks or other uses (**FIGURE 22.13**). The world's largest landfill conversion project is at New York City's former Fresh Kills Landfill. This site, on Staten Island, was the primary repository of New York City's garbage for a half-century, and its mounds rose higher than the nearby Statue of Liberty! Today New York is transforming the site into a world-class public park—a verdant landscape of ball fields, playgrounds, jogging trails, rolling hills, and wetlands teeming with wildlife. Almost three times bigger than Central Park, the mounds offer panoramic views of the Manhattan skyline.

(a) THEN: Fresh Kills Landfill in use

(b) NOW: Fresh Kills Landfill site today

FIGURE 22.13 Old landfills, once capped, can serve other purposes. Visitors to Fresh Kills Park in New York City will enjoy this panoramic view of the Manhattan skyline from atop what used to be an immense mound of trash.

Incinerating trash reduces pressure on landfills

Just as sanitary landfills are an improvement over open dumping, incineration in specially constructed facilities is better than open-air burning of trash. **Incineration,** or combustion, is a controlled process in which garbage is burned at very high temperatures (**FIGURE 22.14**). At incineration facilities, waste is generally sorted and metals are removed. Metal-free waste is chopped into small pieces to aid combustion and then is burned in a furnace. Incinerating waste reduces its weight by up to 85% and its volume by up to 95%.

The ash remaining after trash is incinerated contains toxic components and therefore must be disposed of in hazardous waste landfills (p. 624). Moreover, when trash is burned, hazardous chemicals—including dioxins, heavy metals, and polychlorinated biphenyls (PCBs) (Chapter 14)—can be created and released into the atmosphere. Such emissions caused a backlash against incineration from citizens concerned about health hazards.

Most developed nations now regulate incinerator emissions, and some have banned incineration outright. Engineers have also developed technologies to reduce emissions. Scrubbers (see Figure 17.15, p. 456) chemically treat the gases produced in combustion to remove hazardous components and neutralize acidic gases, such as sulfur dioxide and hydrochloric acid, turning them into water and salt. Scrubbers generally do this either by spraying liquids formulated to neutralize the gases or by passing the gases through dry lime.

Particulate matter, called *fly ash*, often contains some of the worst dioxin and heavy metal pollutants in incinerator emissions. To physically remove these tiny particles, facilities may use a huge system of filters known as a *baghouse*. In addition, burning garbage at especially high temperatures can destroy certain pollutants, such as PCBs. Even all these measures, however, do not fully eliminate toxic emissions.

We can gain energy from trash

Incineration was initially practiced simply to reduce the volume of waste, but today it often serves to generate electricity as well. Most incinerators now are **waste-to-energy (WTE) facilities,** which use the heat produced by waste combustion to boil water, creating steam that drives electricity generation or that fuels heating systems. When burned, waste generates about 35% of the energy generated by burning coal. Roughly 80 WTE facilities are

weighing *the* ISSUES

Environmental Justice?

Do you know where your trash goes? Where is your landfill or incinerator located? Are the people who live closest to this facility wealthy, poor, or middle-class? What race or ethnicity are most people who live in this neighborhood? How might individuals or communities be compensated for the drawbacks of living near a waste disposal facility?

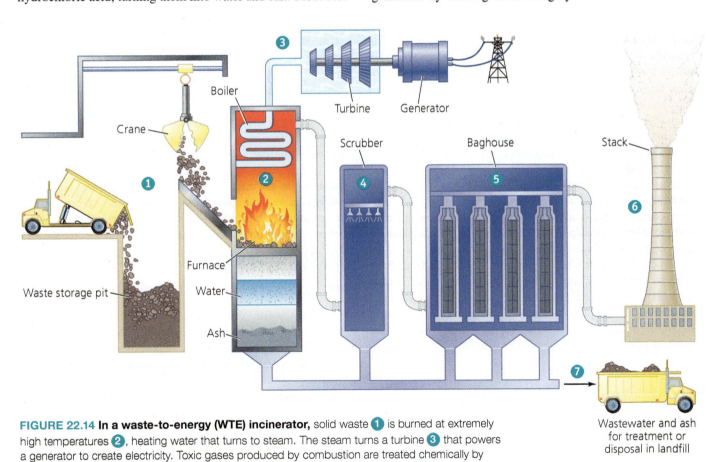

FIGURE 22.14 In a waste-to-energy (WTE) incinerator, solid waste ❶ is burned at extremely high temperatures ❷, heating water that turns to steam. The steam turns a turbine ❸ that powers a generator to create electricity. Toxic gases produced by combustion are treated chemically by a scrubber ❹, and airborne particulate matter is filtered physically in a baghouse ❺ before air is emitted from the stack ❻. Residual ash is disposed of ❼ in a landfill.

Boiler

Crane

Turbine   Generator

Scrubber   Baghouse   Stack

Furnace

Water

Ash

Waste storage pit

Wastewater and ash for treatment or disposal in landfill

operating across the United States (mostly in the Northeast and South), with a total capacity to process 95,000 tons of waste per day.

Revenues from power generation, however, are often not enough to offset the considerable financial cost of building and running incinerators. Because it can take many years for a WTE facility to become profitable, companies that build and operate these facilities sometimes require communities contracting with them to guarantee the facility a minimum amount of garbage. On occasion, such long-term commitments have interfered with communities' subsequent efforts to reduce their waste through recycling and source reduction.

Combustion in WTE plants is not the only way to gain energy from waste. Deep inside landfills, bacteria decompose waste in an oxygen-deficient environment. This anaerobic decomposition produces **landfill gas,** a mix of gases roughly half of which is methane (pp. 27, 482–483). Landfill gas can be collected, processed, and used in the same way as natural gas (pp. 522, 526). Today hundreds of landfills are collecting landfill gas and selling it for energy.

We can recycle material from landfills

Landfills can offer us useful by-products beyond landfill gas. With improved technology for sorting rubbish and recyclables, businesses and entrepreneurs are weighing the economic benefits and costs of rummaging through landfills to salvage materials of value that can be recycled. Steel, aluminum, copper, and other metals are abundant enough in some landfills to make salvage operations profitable when market prices for the metals are high enough. For instance, Americans throw out so many aluminum cans that at 2016 prices for aluminum, the nation buries $5 billion of this metal in landfills each year. If we could retrieve all the aluminum from U.S. landfills, it would exceed the amount the world produces from a year's worth of mining ore.

Landfills also offer organic waste that can be mined and sold as premium compost. Old landfill waste can also be incinerated in newer, cleaner-burning WTE facilities to produce energy. Some companies are even looking into gaining carbon offset credits (p. 510) by harvesting methane leaking from open dumps in developing nations.

Such approaches have been tried in places from New York to Israel to Sweden to Singapore. The costs of mining landfills and meeting regulatory requirements while commodity prices change unpredictably have meant that investing in landfill mining has been risky, but this could change in the future if prices rise and technologies improve.

Industrial Solid Waste

Industrial solid waste includes waste from factories, mining activities, agriculture, petroleum extraction, and more. Each year, U.S. industrial facilities generate about 7.6 billion tons of waste, according to the EPA, about 97%

of which is wastewater. Thus, very roughly, 230 million or so tons of solid waste are generated by 60,000 facilities each year—an amount approaching that of municipal solid waste.

Regulation and economics each influence industrial waste generation

Most methods and strategies of waste disposal, reduction, and recycling by industry are similar to those for municipal solid waste. Businesses that dispose of their own waste on site must design and manage their landfills in ways that meet state, local, or tribal guidelines. Other businesses pay to have their waste disposed of at municipal disposal sites. Whereas the federal government regulates municipal solid waste, state or local governments regulate industrial solid waste (with federal guidance). Regulation varies greatly from place to place, but in most cases, state and local regulation of industrial solid waste is less strict than federal regulation of municipal solid waste. In many areas, industries are not required to have permits, install landfill liners or leachate collection systems, or monitor groundwater for contamination.

The amount of waste generated by a manufacturing process is a good measure of its efficiency; the less waste produced per unit or volume of product, the more efficient that process is, from a physical standpoint. However, physical efficiency is not always reflected in economic efficiency. Often it is cheaper for industry to manufacture its products or perform its services quickly but messily. That is, it can be cheaper to generate waste than to avoid generating waste. In such cases, economic efficiency is maximized, but physical efficiency is not. Because our market system awards only economic efficiency, all too often industry has no financial incentive to achieve physical efficiency. The frequent mismatch between these two types of efficiency is a major reason why the output of industrial waste is so great.

Rising costs of waste disposal enhance the financial incentive to decrease waste. Once either government or the market makes the physically efficient use of raw materials economically efficient as well, businesses gain financial incentives to reduce their waste.

Industrial ecology seeks to make industry more sustainable

To reduce waste, growing numbers of industries today are experimenting with industrial ecology. A holistic approach that integrates principles from engineering, chemistry, ecology, and economics, **industrial ecology** seeks to redesign industrial systems to reduce resource inputs and to maximize both physical and economic efficiency. Industrial ecologists would reshape industry so that nearly everything produced in a manufacturing process is used, either within that process or in a different one.

The larger idea behind industrial ecology is that industrial systems should function more like ecological systems, in which organisms use almost everything that is produced. This principle brings industry closer to the ideal of ecological economists, in which economies function in a circular fashion rather than a linear one (p. 147). It means taking a cradle-to-cradle approach, in which products and manufacturing systems are designed to maximize reuse and recycling of materials into new products.

Industrial ecologists pursue their goals in several ways:

- They examine the entire life-cycle of a given product—from its origins in raw materials, through its manufacturing, to its use, and finally its disposal—and look for ways to make the process more efficient. This strategy is called **life-cycle analysis** (p. 262).

- They try to identify how waste products from one manufacturing process might be used as raw materials for another. For instance, used plastic beverage containers can be shredded and reprocessed to make other plastic items, such as benches, tables, and decks.

- They seek to eliminate environmentally harmful products and materials from industrial processes.

- They study the flow of materials through industrial systems to look for ways to create products that are more durable, recyclable, or reusable. For instance, they seek to design computers, automobiles, and appliances to be easily disassembled so more of their components can be reused or recycled.

Businesses are adopting industrial ecology

Attentive businesses are taking advantage of the insights of industrial ecology to save money while reducing waste. For example, American Airlines switched from hazardous to nonhazardous materials in its Chicago facility, decreasing its need to secure permits from the EPA. The company used more than 50,000 reusable plastic containers to ship goods, reducing packaging waste by 90%. Its Dallas–Fort Worth headquarters recycled enough aluminum cans and white paper in five years to save $205,000 and recycled 3000 broken baggage containers into lawn furniture. A program to gather suggestions from employees brought more than 700 ideas to reduce waste—and 15 of these ideas saved the company more than $8 million.

The Swiss Zero Emissions Research and Initiatives (ZERI) Foundation sponsors dozens of innovative projects worldwide that attempt to create goods and services without generating waste. One example involves breweries in Canada, Sweden, Japan, and Namibia. Brewers in these projects take waste from the beer-brewing process and use it to fuel other processes (**FIGURE 22.15**). As a result, the brewer can make money from bread, mushrooms, pigs, gas, and fish, as well as beer, all while producing little waste. By attempting to create closed-loop systems, ZERI projects cut down on waste while increasing output and income, often generating new jobs as well.

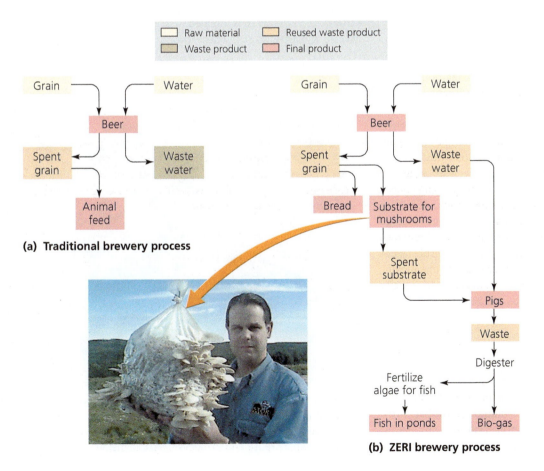

(a) Traditional brewery process

(b) ZERI brewery process

FIGURE 22.15 Creative use of waste products can help us approach zero-waste systems. Traditional breweries **(a)** produce only beer while generating much waste, some of which goes toward animal feed. Breweries sponsored by the Zero Emissions Research and Initiatives (ZERI) Foundation **(b)** use their waste grain to make bread and to farm mushrooms **(photo)**. Waste from the mushroom farming, along with brewery wastewater, goes to feed pigs. The pigs' waste is digested in containers that capture natural gas and collect nutrients used to nourish algae for growing fish in fish farms. The brewer derives income from bread, mushrooms, pigs, gas, and fish, as well as beer.

Few businesses have taken industrial ecology to heart as much as the carpet tile company Interface, which founder Ray Anderson set on the road to sustainability years ago. Interface asks customers to return used tiles for recycling and for reuse as backing for new carpet. It modified its tile design and production methods to reduce waste. It adapted its boilers to use landfill gas for energy. Through such steps, Anderson's company cut its waste generation by 80%, its fossil fuel use by 45%, and its water use by 70%—all while saving $30 million per year, holding prices steady for its customers, and raising profits by 49%.

Hazardous Waste

Hazardous wastes are diverse in their chemical composition and may be liquid, solid, or gaseous. By EPA definition, hazardous waste is waste that is one of the following:

- *Ignitable.* Likely to catch fire (for example, gasoline or alcohol).
- *Corrosive.* Apt to corrode metals in storage tanks or equipment (for example, strong acids or bases).
- *Reactive.* Chemically unstable and readily able to react with other compounds, often explosively or by producing noxious fumes (for example, ammonia reacting with chlorine bleach).
- *Toxic.* Harmful to human health when inhaled, ingested, or touched (for example, pesticides or heavy metals).

Hazardous wastes are diverse

Industry, mining, households, small businesses, agriculture, utilities, and building demolition all create hazardous waste. Industry produces the most, but in developed nations industrial waste disposal is often highly regulated. This regulation has reduced the amount of hazardous waste entering the environment from industrial activities. As a result, households are now the largest source of unregulated hazardous waste.

Household hazardous waste includes a wide range of items, including paints, batteries, oils, solvents, cleaning agents, lubricants, and pesticides. U.S. citizens generate 1.6 million tons of household hazardous waste annually, and the average home contains close to 45 kg (100 lb) of it in sheds, basements, closets, and garages. Although many hazardous substances become less hazardous over time as they degrade chemically, two types are particularly hazardous because their toxicity persists over time: organic compounds and heavy metals.

Organic compounds and heavy metals pose hazards

In our daily lives, we rely on synthetic organic compounds and petroleum-derived compounds to resist bacterial, fungal, and insect activity. Plastic containers, rubber tires, pesticides, solvents, and wood preservatives are useful to us precisely because they resist decomposition. We use these substances to protect buildings from decay, kill pests that attack crops, and keep stored goods intact. However, the capacity of these compounds to resist decay is a double-edged sword, for it also makes them persistent pollutants. Many synthetic organic compounds are toxic because they are readily absorbed through the skin and can act as mutagens, carcinogens, teratogens, and endocrine disruptors (p. 364).

Heavy metals such as lead, chromium, mercury, arsenic, cadmium, tin, and copper are used widely in industry for wiring, electronics, metal plating, metal fabrication, pigments, and dyes. Heavy metals enter the environment when paints, electronic devices, batteries, and other materials are disposed of improperly. Lead from fishing weights and from hunting ammunition accumulates in rivers, lakes, and forests. In older homes, lead from pipes contaminates drinking water, and lead paint remains a problem, especially for infants. Heavy metals that are fat soluble and break down slowly are prone to bioaccumulate and biomagnify (p. 368). In California's Coast Range, for instance, mercury that has washed downstream from abandoned mercury mines enters lakes and rivers, is consumed by bacteria and invertebrates, and accumulates in increasingly large quantities up the food chain, poisoning organisms at higher trophic levels and making fish unsafe to eat.

E-waste has grown

Today's proliferation of computers, printers, smartphones, tablets, TVs, DVD players, MP3 players, and other electronic technology has created a substantial new source of waste (**FIGURE 22.16**). These products have short life spans

FIGURE 22.16 Each day, Americans throw away about 350,000 cell phones. Phones that enter the waste stream can leach toxic heavy metals into the environment. Alternatively, we can recycle them for reuse and to recover valuable metals.

before people judge them obsolete, and most are discarded after just a few years. The amount of this **electronic waste**—often called **e-waste**—has grown rapidly, and now makes up more than 1% of the U.S. solid waste stream by weight. More than 7 billion electronic devices have been sold in the United States since 1980, and U.S. households discard more than 300 million per year—two-thirds of them still in working order.

Most electronic items we discard have ended up in conventional sanitary landfills and incinerators. However, electronic products contain heavy metals and toxic flame-retardants, and research suggests that e-waste should instead be treated as hazardous waste (see **THE SCIENCE BEHIND THE STORY**, pp. 622–623). The EPA and a number of states are now taking steps to keep e-waste out of conventional landfills and incinerators and instead treat it as hazardous waste.

Fortunately, the downsizing of many electronic items and the shift toward mobile devices and tablets mean that fewer raw materials by weight are now going into electronics being manufactured—and, as a result, U.S. e-waste generation appears to have recently peaked. In addition, more and more electronic waste today is being recycled (**FIGURE 22.17**), and Americans now recycle 40% by weight of their e-waste. Increasingly, used electronics are collected by businesses, nonprofit organizations, or municipal services and are processed for reuse or recycling. Campus e-waste recycling drives are proving especially effective. Devices collected are shipped to facilities and taken apart, and the parts and materials are refurbished and reused in new products. There are serious concerns, however, about health risks that recycling may pose to workers doing the disassembly. Wealthy nations ship much of their e-waste

FIGURE 22.18 Medals awarded to athletes at the 2010 Winter Olympic Games in Vancouver were made partly from precious metals recycled from discarded e-waste.

to developing countries, where low-income workers disassemble the devices and handle toxic materials with minimal safety regulations.

Another challenge is that the recent conversion of television and computer monitor technology from cathode ray tubes to LCD and plasma screens has meant that there is no longer much demand for recycled cathode ray tubes. As a result, old cathode ray tubes (rich in toxic lead) are piling up in recyclers' warehouses and are at risk of never being recycled.

Besides keeping toxic substances out of our waste stream, e-waste recycling helps us recover rare and lucrative trace metals used in electronics. A typical cell phone contains up to a dollar's worth of precious metals (p. 644). By one estimate, 1 ton of computer scrap contains more gold than 16 tons of mined ore from a gold mine, while 1 ton of iPhones contains more than 300 times more. Every ounce of metal we can recycle from a manufactured item is an ounce of metal we don't need to mine from the ground. Thus, "mining" e-waste for metals helps reduce the environmental impacts of mining the earth. In one example, the 2010 Winter Olympic Games in Vancouver produced its stylish gold, silver, and bronze medals (**FIGURE 22.18**) partly from metals recovered from recycled and processed e-waste!

Several steps precede the disposal of hazardous waste

For many years we discarded hazardous waste without special treatment. In many cases, people did not know that certain substances were harmful to human health. In other cases, it was assumed that the substances would disappear or be sufficiently diluted in the environment. The resurfacing of toxic chemicals in the 1970s in a residential area at Love Canal (p. 625) in upstate New York, years after their burial, convinced the public that hazardous waste deserves special attention and treatment.

Many communities now designate sites or special collection days to gather household hazardous waste, or designate facilities for the exchange and reuse of substances

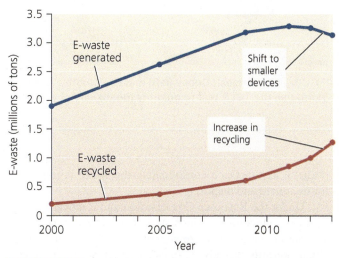

FIGURE 22.17 More and more electronic waste is being recycled. The total amount of electronic waste generated each year in the United States has risen, but the shift to mobile devices and tablets has helped decrease this amount since 2011. Meanwhile, we are recycling greater amounts of e-waste each year. *Data from U.S. Environmental Protection Agency, 2015. Advancing sustainable materials management: Facts and figures 2013. Washington, D.C.: EPA.*

How Hazardous Is E-Waste?

Most electronic waste, or "e-waste," is disposed of in conventional sanitary landfills. However, most electronics contain heavy metals, flame retardants, and other materials with the potential to cause environmental contamination and public health risks. For instance, more than 6% of a typical computer is composed of lead.

Researchers, engineers, and regulators have long debated how hazardous e-waste is and how best to dispose of it. All agree that encouraging reuse and recycling is vital, but what should we do with electronics and their components once they've truly reached their "end of life"?

Discarded electronic waste

One early set of studies was conducted when the U.S. EPA funded Dr. Timothy Townsend's lab at the University of Florida at Gainesville to determine whether e-waste is toxic enough to be classified as hazardous waste under the Resource Conservation and Recovery Act.

With students and colleagues, Townsend determined in 1999–2000 that cathode ray tubes (CRTs) from computer monitors and televisions leach an average of 18.5 mg/L of lead, far above the regulatory threshold of 5 mg/L. Following this research, the EPA proposed classifying CRTs as hazardous waste, and several U.S. states banned these items from conventional landfills.

In 2004, Townsend's lab group ran experiments on 12 other types of electronic devices. To measure toxicity, they used the EPA's standard test, the Toxicity Characteristic Leaching Procedure (TCLP), designed to mimic the process by which chemicals leach out of solid waste in landfills. In the TCLP, waste is ground up into fine pieces, and 100 g (3.5 oz) of it is put in a container with 2 L (0.53 gal) of an acidic fluid (**FIGURE 1**). The container is rotated for 18 hours, after which the leachate is chemically analyzed. Researchers look for eight heavy metals—arsenic, barium, cadmium, chromium, lead, mercury, selenium, and silver—and determine for each whether their concentration in the leachate exceeds that allowed by EPA regulations.

To conduct the standard TCLP, Townsend's team ground up the central processing units (CPUs) of personal computers, creating a jumbled mix made up by weight of 16% circuit board, 8% plastic, 68% ferrous metal, 5% nonferrous metal, and 3% wire and cable. However, grinding up a computer into small bits is no easy task, and it is hard to obtain a sample that accurately represents all components and materials. So the researchers

FIGURE 1 Dr. Brajesh Dubey (left) and Dr. Timothy Townsend (right) use the TCLP test to quantify the toxicity of e-waste.

(**FIGURE 22.19**). Once consolidated, the waste is transported for treatment and disposal.

Under the Resource Conservation and Recovery Act, the EPA sets standards by which states manage hazardous waste. The act also requires large generators of hazardous waste to obtain permits. Finally, it mandates that hazardous materials be tracked "from cradle to grave." As hazardous waste is

FIGURE 22.19 Many communities designate collection sites or collection days for household hazardous waste. Here, workers handle waste from a collection event in Brooklyn, New York.

also designed a modified TCLP test in which they placed whole CPUs—with the parts disassembled but not ground up—in a rotating 55-gallon drum full of acidic liquid. They tested their 12 types of devices using both standard and modified TCLP methods.

The team found lead to be the only heavy metal that exceeded the EPA's regulatory threshold, but this threshold (5 mg/L) was surpassed in the majority of trials. Desktop computer monitors leached the most lead (48 mg/L on average), because older-style monitors include cathode ray tubes. However, laptops, TVs, smoke detectors, cell phones, and computer mice also leached high levels of lead. Next came remote controls, VCRs, keyboards, and printers, all of which leached more lead on average than the EPA threshold—and in 50% or more of the trials.

The researchers found that items containing more ferrous metals (such as iron) tended to leach less lead. For instance, CPUs contain 68% ferrous metals (compared to just 7% in laptops), and laptops leached seven times as much lead as CPUs. Further experiments confirmed that ferrous metals chemically react with lead and stop it from leaching.

Since that time, researchers have conducted a number of other studies. Taken together, they have shown similar results: that most electronics in most cases leach lead beyond the EPA threshold (FIGURE 2).

Townsend and others say this research suggests that many electronic devices should be classified as hazardous waste. However, lab tests may or may not accurately reflect what actually happens in landfills. For this reason, Townsend's team has filled large tubular columns with e-waste and buried them in Florida landfills. The researchers are testing leachate from these tubes, and are reporting results for various materials as time passes. For example, a 2011 paper showed that the leaching of several heavy metals was influenced by whether the landfill was experiencing aerobic or anaerobic decomposition.

Scientists are also testing new materials that electronics manufacturers are using as they attempt to create safer

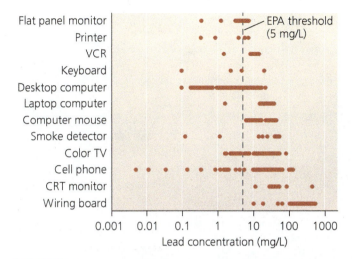

FIGURE 2 **Most devices tested have exceeded the EPA safety standard for lead leachate.** Data points for each type of device represent data from TCLP tests in different scientific studies between 2000 and 2008. *Adapted from Townsend, Timothy G., 2011. Environmental issues and management strategies for waste electronic and electrical equipment. J. Air & Waste Mgmt. Assoc. 61: 587–610.*

alternatives. For instance, Townsend and his colleagues found alternative solders to leach less than traditional tin/lead solders used on printed wire boards.

Scientific results from such projects are helping regulators and policymakers decide how best to deal with e-waste. Currently, federal regulations require special handling and disposal of the most hazardous types of e-waste, but exemptions are made to encourage reuse and recycling. Some states consider most e-waste to be hazardous and exercise strict controls, whereas others do not.

In our new age of smartphones and so much more, the number and variety of electronic devices continue to expand, and further research will be needed as our technologies evolve.

generated, transported, and disposed of, the producer, carrier, and disposal facility must each report to the EPA the type and amount of material generated; its location, origin, and destination; and the way it is handled. This process is intended to prevent illegal dumping and to encourage the use of reputable waste carriers and disposal facilities.

Because current U.S. law makes disposing of hazardous waste quite costly, irresponsible companies sometimes illegally dump waste, creating health risks for residents and financial headaches for local governments forced to deal with the mess (FIGURE 22.20, p. 624). Moreover, companies from industrialized nations often find it cheaper to pay cash-strapped developing nations to take the waste—or cheaper

still, to dump it illegally. In nations with lax environmental and health regulations, workers and residents are often uninformed of or unprotected from the health dangers of this waste. This global environmental justice issue (p. 137) continues despite the Basel Convention, an international treaty to limit such practices.

High costs of disposal, however, have also encouraged conscientious businesses to invest in reducing their hazardous waste. Many biologically hazardous materials can be broken down by incineration at high temperatures in cement kilns. Others can be treated by exposure to bacteria that break down harmful components and synthesize them into new compounds. In addition, various plants have been bred

FIGURE 22.20 Unscrupulous individuals or businesses sometimes dump hazardous waste illegally to avoid disposal costs.

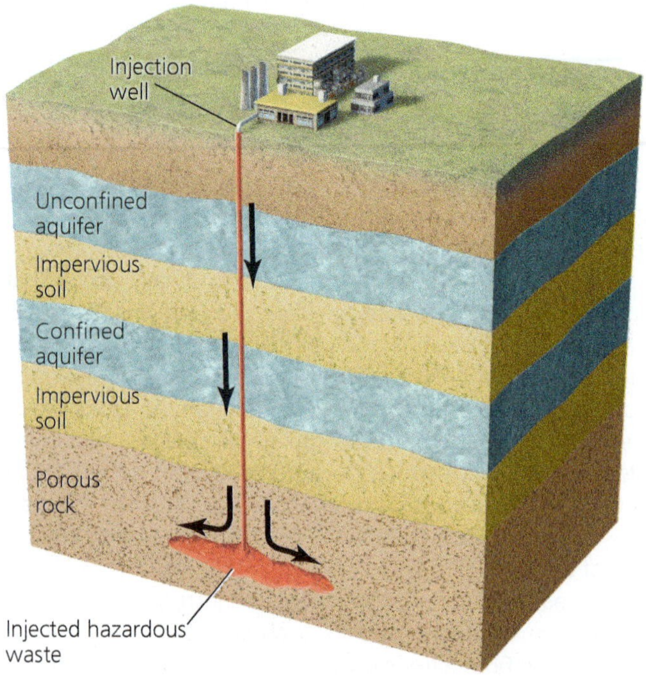

FIGURE 22.21 Liquid hazardous waste is pumped deep underground by deep-well injection. The well must be drilled below any aquifers, into porous rock isolated by impervious clay.

Labels in Figure 22.21: Injection well; Unconfined aquifer; Impervious soil; Confined aquifer; Impervious soil; Porous rock; Injected hazardous waste

or engineered to take up specific contaminants from soil and break down organic contaminants into safer compounds or concentrate heavy metals in their tissues. The plants are eventually harvested and disposed of.

We use three disposal methods for hazardous waste

We have developed three primary means of hazardous waste disposal: landfills, surface impoundments, and injection wells. These do nothing to diminish the hazards of the substances, but they help keep the waste isolated from people, wildlife, and ecosystems. Design and construction standards for landfills that receive hazardous waste are stricter than those for ordinary sanitary landfills. Hazardous waste landfills must have several impervious liners and leachate removal systems and must be located far from aquifers.

Liquid hazardous waste, or waste in dissolved form, may be stored in **surface impoundments,** shallow depressions lined with plastic and an impervious material, such as clay. The liquid or slurry is placed in the pond and water is allowed to evaporate, leaving a residue of solid hazardous waste on the bottom. This process is repeated and eventually the dry residue is removed and transported elsewhere for permanent disposal. Impoundments are not ideal. The underlying layer can crack and leak waste. Some material may evaporate or blow into surrounding areas. Rainstorms may cause waste to overflow and contaminate nearby areas.

For these reasons, surface impoundments are used only for temporary storage.

In **deep-well injection,** a well is drilled deep beneath the water table into porous rock, and wastes are injected into it (**FIGURE 22.21**). The process aims to keep waste deep underground, isolated from groundwater and human contact. However, wells can corrode and can leak wastes into soil, contaminating aquifers, and deep-well injection may very occasionally induce earthquakes. Roughly 34 billion L (9 billion gal) of hazardous waste are placed in U.S. injection wells each year.

Radioactive waste is especially hazardous

Radioactive waste is particularly dangerous to human health and is persistent in the environment. The dilemma of disposal has dogged the nuclear energy industry for decades. The United States has no designated site to dispose of its commercial nuclear waste if Yucca Mountain in Nevada is removed from consideration (p. 562). Instead, waste will continue to accumulate at the many nuclear power plants spread throughout the nation (see Figure 20.11, p. 559).

Currently, a site in the Chihuahuan Desert in New Mexico serves as a permanent disposal location for radioactive waste from military sources. The Waste Isolation Pilot Plant is the world's first underground repository for transuranic waste from nuclear weapons development. The mined caverns holding the military waste are located 655 m (2150 ft) belowground in a huge salt formation thought to be geologically stable. This site became operational in 1999 and receives shipments of waste from 23 other locations.

Contaminated sites are being cleaned up, slowly

Many thousands of former military and industrial sites remain contaminated with hazardous waste in the United States and virtually every other nation on Earth. For most nations, dealing with these messes is simply too difficult, time-consuming, and expensive. In 1980, however, the U.S. Congress passed the Comprehensive Environmental Response Compensation and Liability Act (CERCLA). This law established a federal program to clean up U.S. sites polluted with hazardous waste. The EPA administers this cleanup program, called the **Superfund.** Under EPA auspices, experts identify sites polluted with hazardous chemicals, take action to protect groundwater, and clean up the pollution. Later laws also charged the EPA with cleaning up **brownfields,** lands whose reuse or development is complicated by the presence of hazardous materials.

Two well-publicized events spurred creation of the Superfund legislation. In Love Canal, a residential neighborhood in Niagara Falls, New York, families were evacuated in 1978–1980 after toxic chemicals buried by a company and the city in past decades rose to the surface, contaminating homes and an elementary school and apparently leading to birth defects, miscarriages, and other health impacts. In Missouri, the entire town of Times Beach was evacuated and its buildings demolished after being contaminated in the 1970s by dioxin (p. 379, a chemical that causes cancer and other serious health problems) from waste oil sprayed on its roads.

Once a Superfund site is identified, EPA scientists evaluate how near the site is to homes, whether wastes are confined or likely to spread, and whether the pollution threatens drinking water supplies. Sites judged to be harmful are placed on the National Priorities List, ranked according to the risk to human health that they pose. Cleanup proceeds as funds are available. Throughout the process, the EPA is required to hold public hearings to inform area residents of its findings and to receive feedback.

The objective of CERCLA was to charge the polluting parties for the cleanup of their sites, according to the *polluter-pays principle* (p. 163). For many sites, however, the responsible parties cannot be found or held liable, and in such cases—roughly 30% so far—cleanups have been covered by taxpayers and from a trust fund established by a federal tax on industries producing petroleum and chemical raw materials. However, Congress let the tax expire and the trust fund went bankrupt in 2004, so taxpayers are now shouldering the entire burden. As the remaining cleanup jobs become more expensive, fewer are being completed.

As of 2016, 1328 Superfund sites remained on the National Priorities List, and only 391 had been cleaned up or otherwise deleted from the list. The average cleanup has cost more than $25 million and has taken nearly 15 years. Many sites are contaminated with hazardous chemicals we have no effective way to deal with. In such cases, cleanups simply aim to isolate waste from human contact, either by building trenches and barriers around a site or by excavating contaminated material and shipping it to a hazardous waste disposal facility. For all these reasons, the current emphasis in the United States and elsewhere is on preventing hazardous waste contamination in the first place.

closing
THE LOOP

We have made great strides in addressing our waste problems. Modern methods of waste management are far safer for people and gentler on the environment than past practices of open dumping and open burning. Recycling and composting efforts are advancing steadily, and Americans now divert one-third of all solid waste away from disposal.

Students on college and university campuses are making great contributions in accelerating these trends. The enthusiasm of students for recycling is apparent each year in the success of Recyclemania—and this competition is just the tip of the iceberg. Most campuses have their own recycling and waste reduction programs, and these continue to grow and evolve as students and staff find new and innovative ways of inspiring people to reduce waste. Across the greater society, continuing growth of recycling and composting, driven by market forces, public policy, and consumer behavior, shows potential for further advances.

Still, our prodigious consumption habits have created more waste than ever before. Our waste management efforts are marked by a number of difficult challenges, including the cleanup of Superfund sites and the safe disposal of hazardous and radioactive waste.

These dilemmas make clear that the best solution is to reduce our generation of waste and to pursue a cradle-to-cradle approach. Finding ways to reduce, reuse, and efficiently recycle the materials and goods that we use stands as a key ongoing challenge for our society.

REVIEWING Objectives

You should now be able to:

Summarize major approaches to managing waste, and compare and contrast the types of waste we generate

Source reduction, recovery, and disposal are the three main components of waste management. Source reduction is preferred, and recovery is next-best. Municipal solid waste comes from homes, institutions, and small businesses. Industrial solid waste comes from manufacturing, mining, agriculture, and petroleum extraction and refining. Hazardous waste is toxic, chemically reactive, flammable, or corrosive. (pp. 606–607)

Discuss the nature and scale of the waste dilemma

Developed nations generate far more waste than developing nations, but they are beginning to decrease their waste, whereas waste in developing nations is increasing as population and consumption grow. Open dumping and burning continue in developing nations. (pp. 607–609)

Evaluate source reduction, reuse, composting, and recycling as approaches for reducing waste

Reducing waste before it is generated is the best management approach. Reusing items helps to reduce waste. Composting creates organic matter for gardening and farming, while recycling now removes 26% of the U.S. waste stream. Bottle bills are one way in which financial incentives can motivate people to reduce waste. (pp. 609–613)

Describe landfills and incineration as conventional waste disposal methods

Sanitary landfills are our main disposal method. Properly built and maintained, they guard against contamination of groundwater, air, and soil. Incinerators reduce waste volume by burning it. Pollution control technology removes most pollutants from incinerator emissions, but some escape, and toxic ash needs to be disposed of in landfills. We are harnessing energy from landfill gas and generating electricity from incineration, and we are also starting to recycle materials from landfills. (pp. 613–618)

Discuss industrial solid waste and principles of industrial ecology

Regulations differ, but industrial solid waste management is similar to that for municipal solid waste. Industrial ecology provides ways for industry to enhance efficiency and studies how industrial systems can mimic ecological systems with a cradle-to-cradle approach. (pp. 618–620)

Assess issues in managing hazardous waste

Organic compounds, heavy metals, electronic waste, and radioactive waste are common types of hazardous waste. Hazardous waste is regulated and monitored, yet illegal dumping occurs, and no fully satisfactory method of disposing of hazardous waste has yet been devised. The Superfund program cleans up hazardous waste sites, but cleanup is a long and expensive process. (pp. 620–625)

TESTING Your Comprehension

1. Describe the three major components of managing waste. Why do we practice waste management?
2. Why have some people labeled the United States "the throwaway society"? How much solid waste do Americans generate, and how does this amount compare to that of people from other countries?
3. What is composting, and how does it help reduce the waste stream?
4. What are the three elements of the recycling process?
5. Name several guidelines by which sanitary landfills are regulated. Describe three problems with landfills.
6. Describe the process of incineration or combustion. What happens to the resulting ash? What is one drawback of incineration?
7. In your own words, describe the goals of industrial ecology.
8. What four criteria are used to define hazardous waste? What makes heavy metals and synthetic organic compounds particularly hazardous?
9. What are the largest sources of hazardous waste? Describe three ways we dispose of hazardous waste.
10. What is the Superfund program? How does it work?

SEEKING Solutions

1. How much waste do you generate? Look into your waste bin at the end of the day and categorize and measure the waste there. List all other waste you may have generated in other places throughout the day. How much of this waste could you have avoided generating? How much could have been reused or recycled?

2. Of the various waste management approaches covered in this chapter, which ones are your community or campus pursuing? Would you suggest pursuing any new approaches? If so, which ones, and why?

3. Can manufacturers and businesses benefit from source reduction if consumers were to buy fewer products as a result? How? Given what you know about industrial ecology, what do you think the future of sustainable manufacturing may look like?

4. **CASE STUDY CONNECTION** Does your college or university participate in Recyclemania? If so, describe how it has done so, what events it has staged, how successful these were, and how this success might be improved. If not, describe how you think your school could compete effectively in Recyclemania. What events, programs, or strategies do you think would be effective on your campus to give it a shot at winning one of the categories in Recyclemania? For more information, consult the Recyclemania Web page, http://recyclemaniacs.org.

5. **THINK IT THROUGH** You are the president of your college or university. Your students participate in Recyclemania each year, and want you to make the school a leader in waste reduction and industrial ecology. Consider the industries and businesses in your community and the ways they interact with facilities on your campus. Bearing in mind the principles of industrial ecology, can you think of any novel ways in which your school and local businesses might mutually benefit from one another's services, products, or waste materials? Are there waste products from one business, industry, or campus facility that another might put to good use? What steps would you propose to take as president?

6. **THINK IT THROUGH** You are the CEO of a major corporation that produces containers for soft drinks and a wide variety of other consumer products. Your company's shareholders are asking that you improve the company's image—while not cutting into profits—by taking steps to reduce waste. What steps would you consider taking?

CALCULATING Ecological Footprints

The biennial "State of Garbage in America" survey documents the ability of U.S. residents to generate prodigious amounts of municipal solid waste (MSW). According to the most recent survey, on a per capita basis, Missouri residents generate the least MSW (4.5 lb/day), and Hawai'i residents generate the least MSW (4.5 lb/day), and Hawai'i residents generate the most (15.5 lb/day). The average for the entire country is 6.8 lb MSW per person per day. Calculate the amount of MSW generated in 1 day and in 1 year by each of the groups listed, if they were to generate MSW at each of the rates shown in the table.

GROUPS GENERATING MUNICIPAL SOLID WASTE	PER CAPITA MSW GENERATION RATES					
	U.S. AVERAGE (6.8 LB/DAY)		MISSOURI (4.5 LB/DAY)		HAWAI'I (15.5 LB/DAY)	
	DAY	YEAR	DAY	YEAR	DAY	YEAR
You	6.8	2482				
Your class						
Your state						
United States						
World						

Data from Shin, D., 2014. *Generation and disposition of municipal solid waste (MSW) in the United States—A national survey.* New York: Columbia University, Earth Engineering Center.

1. Suppose your town of 50,000 people has just approved construction of a landfill nearby. Estimates are that it will accommodate 1 million tons of MSW. Assuming that the landfill is serving only your town, and that your town's residents generate waste at the U.S. average rate, for how many years will it accept waste before filling up? How much longer would a landfill of the same capacity serve a town of the same size in Missouri?

2. One study has estimated that the average world citizen generates 1.47 pounds of trash per day.

How many times more does the average U.S. citizen generate?

3. The same study showed that the average resident of a low-income nation generates 1.17 pounds of waste per day and that the average resident of a high-income nation generates 2.64 pounds per day. Why do you think U.S. residents generate so much more MSW than people in other "high-income" countries, when standards of living in those countries are comparable?

MasteringEnvironmentalScience®

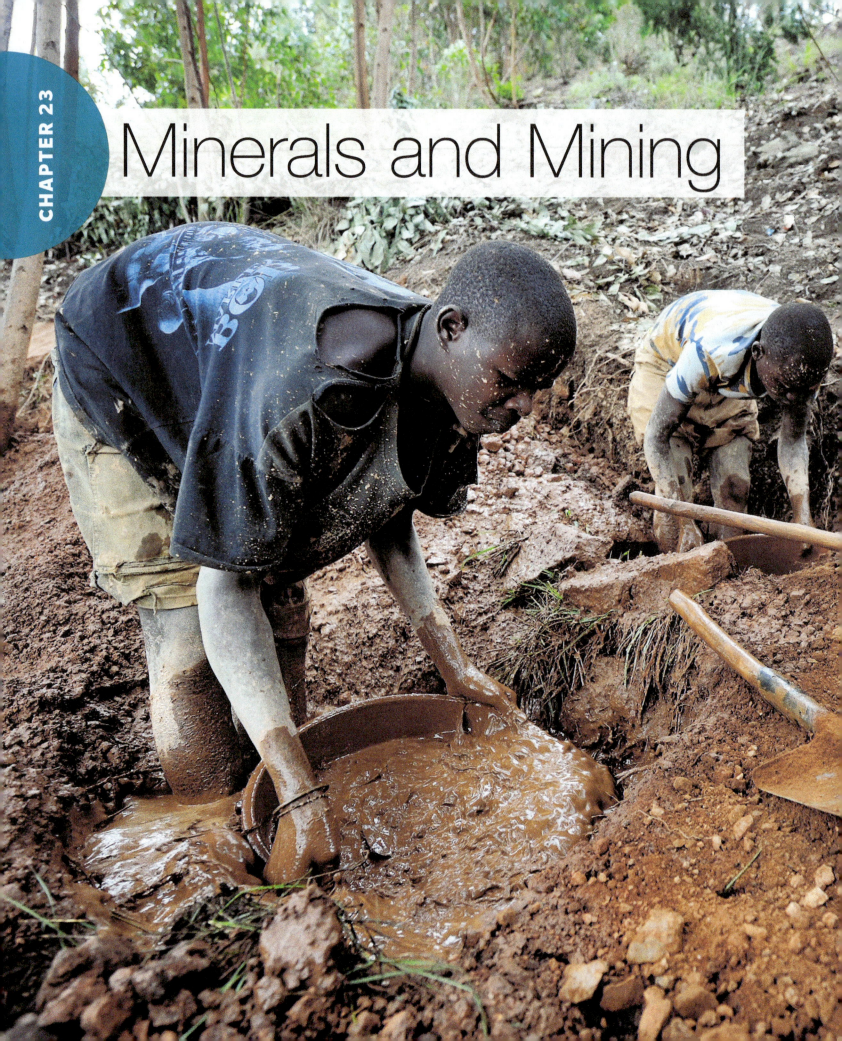

CHAPTER 23

Minerals and Mining

Children mining for coltan ore

Mining for ... Cell Phones?

> **The conflict in the Democratic Republic of the Congo has become mainly about access, control, and trade of five key mineral resources: coltan, diamonds, copper, cobalt, and gold.**
> Report to the United Nations Security Council, April 2001

> **Coltan . . . is not helping the local people. In fact, it is the curse of the Congo.**
> African journalist Kofi Akosah-Sarpong

D.R. CONGO

Region of coltan mining

You walk across your college campus to your next class. In the process, you pull out your cell phone and text a friend—and likely give very little thought to the technology that makes this text possible. What you probably don't know is that inside your phone is a little-known metal called tantalum—just a tiny amount—and without it, no cell phone can operate. Half a world away, a miner in the heart of Africa toils all day in a jungle streambed, sifting sediment for nuggets of coltan ore, which contain tantalum. At nightfall, rebel soldiers take most of his ore, leaving him to sell what little remains to buy food for his family at the squalid mining camp where they live.

In bedeviling ways, tantalum links our glossy global high-tech economy with one of the most abused regions on Earth. Democratic Republic of Congo (D.R. Congo) has long been embroiled in a sprawling regional conflict fueled by ethnic tensions and access to valuable natural resources. This conflict has involved six nations and various rebel militias and has claimed more than 5 million lives since 1998. The current conflict is the latest chapter in the sad history of a nation rich in natural resources—copper, cobalt, gold, diamonds, uranium, and timber—whose impoverished people have been repeatedly robbed of control of those resources.

Tantalum (Ta), element number 73 on the periodic table (**APPENDIX D**), is at the heart of the current battle over resources in D.R. Congo. As noted, we rely on this metal for our cell phones, but it is also vital for production of computer chips, DVD players, game consoles, and digital cameras. Tantalum powder is ideal for capacitors (the components that store energy and regulate current in miniature circuit boards) because it is highly heat resistant and readily conducts electricity.

Tantalum comes from a dull blackish mineral called tantalite, which often occurs with a mineral called columbite—so the ore is referred to as columbite–tantalite, or coltan for short. In eastern Congo, miners dig craters in rainforest streambeds, panning for coltan much as early California miners panned for gold.

As information technology boomed in the late 1990s, global demand for tantalum rose, and market prices for the metal shot up. High prices led some Congolese miners to mine coltan by choice, but many more were forced to work as miners.

Open war broke out in D.R. Congo in 1998 when rebel groups, supported by forces from neighboring

▲ **Mining is harming the endangered okapi in eastern Congo**

Upon completing this chapter, you will be able to:

- Describe the types of mineral resources and their uses

- Explain how minerals are processed prior to use

- Describe the major methods of mining

- Discuss the environmental and social impacts of mining

- Explain reclamation efforts and mining policy

- Evaluate ways to encourage sustainable use of mineral resources

Rwanda and Uganda, attempted to overthrow the government of President Laurent-Désiré Kabila. The country was fragmented by conflict between government forces and various rebel groups, with 11 other African nations becoming involved in the hostilities. In mineral-rich eastern D.R. Congo, fighting was particularly intense. Local farmers were chased off their land; villages were burned; and civilians were raped, tortured, and killed. Soldiers and rebels seized control of mining operations. They forced farmers, refugees, prisoners, and children to work, and skimmed profits from the coltan the people mined. Children and teachers abandoned school and worked in the mines. The conflict also caused ecological havoc as people streamed into national parks to avoid the fighting and locate new sources of minerals. This led to the clearing of rainforests for fuelwood and the killing of wildlife for food, including forest elephants, endangered gorillas, and okapi, a rare zebra-like relative of the giraffe. Miners disturbed streambeds in the search for coltan, increasing erosion rates and choking streams with sediments. Toxic metals used in the mining process, such as cyanide and mercury, were released into streams, endangering aquatic wildlife.

Most miners ended up with little, while rebels, soldiers, and bandits who took the lion's share of the valuable metal enriched themselves by selling it to traders, who in turn made profits by selling it to processing companies in the United States, Europe, and Asia. These companies refine and sell tantalum powder to capacitor manufacturers, which in turn sell capacitors to Nokia, Motorola, Sony, Intel, Compaq, Dell, and other high-tech corporations that use the capacitors in their products.

A grass-roots activist movement urged international action to stop the violence and exploitation in D.R. Congo and advanced the slogan, "No blood on my cell phone!" In 2001, an expert panel commissioned by the United Nations (UN) concluded that coltan riches were fueling, financing, and prolonging the war. The panel urged a UN embargo on coltan and other minerals from regions of D.R. Congo where conflict flourished.

Sony, Nokia, Ericsson, and other corporations rushed to assure consumers that they were not using tantalum from eastern D.R. Congo—noting that the region was producing less than 10% of the world's supply. Meanwhile, some observers felt an embargo could hurt the long-suffering Congolese people. The mining life may be miserable, they said, but it pays better than most jobs in a land where the average income is only 20 cents a day.

A 2002 peace treaty led to the withdrawal of foreign troops from D.R. Congo, but conflict with rebel groups within the country continues, bankrolled by the mineral riches of the region. Success by Congolese troops and an African-led UN intervention brigade against a major rebel group in eastern D.R. Congo has, however, helped to reduce conflict in the region. Unfortunately, D.R. Congo is not the only source of "conflict minerals" in the world today. A thriving black market in coltan is emerging in remote portions of Brazil, Colombia, and Venezuela in the northern Amazon jungle. Armed gangs and narcotics smugglers in the region are accused of using women, children, and indigenous people to mine and smuggle coltan ore. The recent discovery of vast mineral reserves in Afghanistan (p. 643), coupled with that nation's political unrest, suggests that it, too, could become a significant source of conflict minerals in the near future.

Steps are now being taken to help support legitimate Congolese mines while preventing the exploitation that has defined mining in D.R. Congo in the recent past. Industry groups, working with national governments and nongovernmental aid organizations, are creating a certification system for conflict-free coltan, and U.S. law now mandates that manufacturers of electronics report the origin of the tantalum in their products to aid consumers in avoiding products that contain conflict minerals.

These efforts and others provide an opportunity to significantly reduce trade in conflict minerals while promoting trade of minerals sourced from legitimate mines in poor nations such as D.R. Congo. It is hoped that ongoing regulatory efforts to certify minerals will provide a framework that not only satisfies the world's demand for mineral resources but also protects the people and ecosystems that provide them.

Earth's Mineral Resources

Coltan provides just one example of how we extract raw materials from beneath our planet's surface and turn them into products we use every day. We mine and process a wide array of mineral resources in the modern world. Indeed, without these resources—which we use to make everything from building materials to fertilizers—civilization as we know it could not exist. Just consider a typical scene from a student lounge at a college or university (FIGURE 23.1) and note how many items are made with elements from the minerals we take from the earth.

Both gradual geologic processes and catastrophic geologic hazards influence the distribution of rocks and minerals in the lithosphere and their availability to us. In the lithosphere (p. 32), the region that includes the uppermost layers of rock near Earth's surface, the rock cycle (p. 34) creates new rock and alters existing rock. Plate tectonics (pp. 32–34) builds mountains; shapes the geography of oceans, islands, and continents; and gives rise to earthquakes and volcanoes. Throughout the world, geologic processes are fundamental to shaping the world around us and affecting the distribution of the mineral resources our modern society requires.

Rocks provide the minerals we use

A **rock** is a solid aggregation of minerals; a **mineral** is a naturally occurring solid chemical element or inorganic compound with a crystal structure, a specific chemical composition, and distinct physical properties (p. 34). (See the periodic table in APPENDIX D for chemical elements.) For instance, the mineral tantalite consists of the elements tantalum, oxygen, iron, and manganese. Tantalite occurs most commonly in pegmatite, a type of igneous rock (p. 34) similar to granite. In addition to tantalite, pegmatite generally contains the

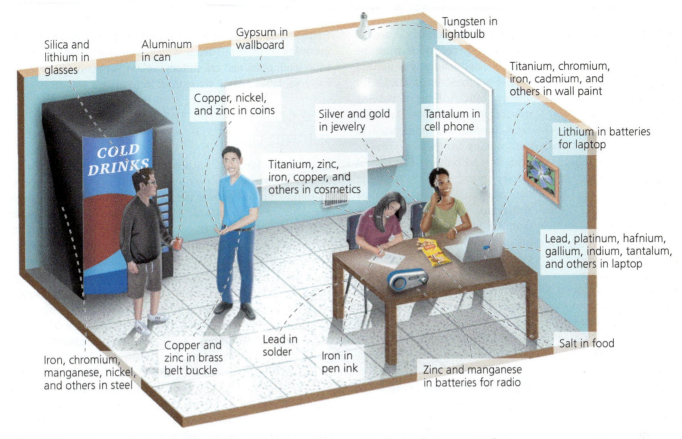

FIGURE 23.1 Elements from minerals that we mine are everywhere in the products we use in our everyday lives. This scene from a typical college student lounge points out just a few of the many elements from minerals that surround us.

minerals feldspar, quartz, and mica, and occasionally it even includes gemstones and other rare minerals.

In some cases, the rock itself is a resource, such as the gravel used to make concrete. In other cases, valuable mineral resources, such as iron or phosphate, are embedded within rocks. To secure these minerals, miners remove the mineral-containing rock, and then separate the desired mineral from the surrounding rock. In either case, the process begins by removing rocks, fossil fuels, or mineral-bearing rock from the lithosphere.

We obtain minerals by mining

We obtain the minerals we use through the process of mining. The term *mining* in the broad sense describes the extraction of any resource that is nonrenewable on the timescale of our society. In this sense, we mine fossil fuels and groundwater, as well as minerals. When used specifically in relation to minerals, **mining** refers to the systematic removal of rock, soil, or other material for the purpose of extracting minerals of economic interest. Because most minerals of interest are widely spread but in low concentrations, miners and mining geologists first try to locate concentrated sources of minerals before mining begins.

We use mined materials extensively

The average American consumes more than 17,900 kg (39,500 lb) of new minerals and fuels every year, according to 2015 estimates by the U.S. Minerals Education Coalition. At

current rates of use, a child born in 2015 will use more than 1.4 million kg (3.1 million lb) of minerals and fuel during his or her lifetime (**FIGURE 23.2**).

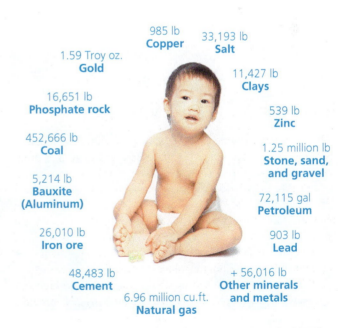

1.59 Troy oz. **Gold**

985 lb **Copper**

33,193 lb **Salt**

16,651 lb **Phosphate rock**

11,427 lb **Clays**

452,666 lb **Coal**

539 lb **Zinc**

5,214 lb **Bauxite (Aluminum)**

1.25 million lb **Stone, sand, and gravel**

26,010 lb **Iron ore**

72,115 gal **Petroleum**

48,483 lb **Cement**

903 lb **Lead**

6.96 million cu.ft. **Natural gas**

+ 56,016 lb **Other minerals and metals**

FIGURE 23.2 At current rates of use, a baby born in 2015 is predicted to use over 1.4 million kg (3.1 million lb) of minerals over his or her lifetime. *Data from Minerals Education Coalition, 2015.*

FAQ

How do geologists "see" large mineral deposits below the ground?

Searching for reserves of underground minerals, also called prospecting, can be pursued in a number of ways. The earliest prospectors explored promising areas on foot, looking for exposed seams of mineral-containing rocks or for minerals carried into streams by runoff. Today, geologists direct vibrations into underground rock strata and capture the vibrations with sensors as the vibrations bounce off underground rock layers and reflect back to the surface. This process enables scientists to visualize the underlying rock layers and identify likely locations for reserves, just as they do for fossil fuel deposits (p. 522). Geologists also measure the magnetic fields in rock layers to look for metal ores, and they conduct chemical analyses of stream water to detect minerals of interest. If promising sites are located, cores can be drilled deep into the ground and inspected for the desired mineral before actual mining begins.

More than half of the annual mineral and fuel use is from the coal, oil, and natural gas used to supply our intensive demands for energy. Much of the remaining mineral use is attributable to the sand, gravel, and stone used in constructing our buildings, roads, bridges, and parking lots. Metal use is dwarfed by these other two categories, but the average American will still use more than 2.5 tons of aluminum over his or her lifetime. This level of consumption clearly shows the potential of recycling and reuse (such as recycling stone and gravel from old highways into new construction) to make our modern, mineral-intensive lifestyle more sustainable.

Metals are extracted from ores

Some minerals can be mined for metals. As we have seen, the tantalum used in electronic components comes from the mineral tantalite (**FIGURE 23.3**). A **metal** is a type of chemical element, or a mass of such an element, that typically is lustrous, opaque, and malleable, and can conduct heat and electricity. Most metals are not found in a pure state in

Earth's crust but instead are present within **ore,** a mineral or grouping of minerals from which we extract metals. Copper, iron, lead, gold, and aluminum are among the many economically valuable metals we extract from mined ore.

We process metals after mining ore

Extracting minerals from the ground is the first step in putting them to use. However, most metals need to be processed in some way to become useful for our products. For example, after ores are mined, the metal-bearing rock is pulverized and washed, and the desired minerals are then isolated using chemical and/or physical means. For iron, this involves heating ore-bearing rocks to extremely high temperatures in a blast furnace and collecting the molten iron when it separates from the surrounding minerals, a process known as **smelting** (heating ore beyond its melting point and combining it with other metals or chemicals) (**FIGURE 23.4**). For aluminum, bauxite ore is first treated with chemicals to extract alumina (an aluminum oxide), and then an electrical current is used to generate pure aluminum from alumina. With coltan, processing facilities use acid solvents to separate tantalite from columbite. Other chemicals are then used to produce metallic tantalum powder. This powder can be consolidated by various melting techniques and can be shaped into wire, sheets, or other forms. Sometimes we mix, melt, and fuse a metal with another metal or a nonmetal substance to form an **alloy.** For example, steel is an alloy of the metal iron that has been fused with a small quantity of carbon.

Processing metals can exert substantial environmental impacts. First, most processing methods are water-intensive and energy-intensive, straining precious water resources and generating greenhouse gases that contribute to global climate change. Second, many chemical reactions and heating processes used for extracting metals from ore emit air pollution;

(a) Coltan ore

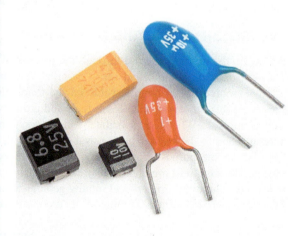

(b) Capacitors containing tantalum

FIGURE 23.3 Tantalum is used to manufacture electronics. Coltan ore **(a)** is mined from the ground and then processed to extract the pure metal tantalum. This metal is used in capacitors **(b)** and other electronic components.

FIGURE 23.4 **A worker guides molten iron out of a blast furnace.** The metal is separated from the surrounding rock by melting ore at high temperatures and collecting the heavier metal.

FIGURE 23.5 **This surface impoundment at the Upper Big Branch mine in West Virginia holds coal tailings from a surface mining operation.** The impoundment, shown in the bottom left of the image, is like a large in-ground swimming pool holding potentially toxic liquids produced by the mining or processing of metals, minerals, or fuels.

smelting plants in particular have long been hotspots of toxic air pollution (Chapter 17).

Pollution from mineral processing is also a concern. Soil and water commonly become polluted by **tailings,** portions of ore left over after metals have been extracted. Tailings contain small amounts of potentially toxic ore (such as metals), as well as chemicals applied in the extraction process. For instance, we use chemical cyanide to extract gold from ore, and sulfuric acid to extract copper, and both cyanide and sulfuric acid are highly toxic to organisms (Chapter 14).

Mining operations often pump a toxic slurry of tailings into large reservoirs called **surface impoundments** (**FIGURE 23.5**). Impoundment walls are designed to prevent leaks and collapse, but accidents can occur if the structural integrity of the impoundment is compromised. In 2000, a breach of an impoundment near Inez, Kentucky, released more than 1 billion liters (250–300 million gal) of coal slurry (water used to wash mined coal that contained numerous toxic compounds), blackening 120 km (75 mi) of streams, killing aquatic wildlife, and affecting drinking water supplies for many communities for weeks. The failure of two impoundments at an iron ore mine in Brazil in 2015 buried nearby villages in a toxic slurry of water and mining waste, claiming 19 lives.

Even absent catastrophic accidents, smaller-scale leaching of toxic materials also occurs, because it is often difficult to properly line and maintain such large impoundments.

We also mine nonmetallic minerals and fuels

We also mine and use many minerals that do not contain metals. **FIGURE 23.6** illustrates a selection of economically useful mineral resources, both metallic and nonmetallic. For each one, its major nation of origin and several main uses are shown.

Sand and gravel, the most commonly mined mineral resources, are used as fill and as construction materials for the manufacturing of products such as concrete. Each year more than $7 billion of sand and gravel are mined in the United States. Phosphate rock provides us phosphorus for fertilizers and industrial chemicals. We mine limestone, salt, potash, and other minerals for a number of diverse purposes.

Gemstones are treasured for their rarity and beauty. For instance, diamonds have long been prized—and like coltan, they have fueled resource wars. Besides the conflict in eastern Congo, the diamond trade has acted to fund, prolong, and intensify wars in Angola, Sierra Leone, Liberia, and elsewhere, as armies exploit local people for mine labor and then sell the diamonds for profit. This is why you may hear the phrase "blood diamonds," just as coltan has been called a "conflict mineral."

We also mine substances we use for fuel. Uranium ore is a mineral from which we extract the metal uranium, which we use in the production of nuclear power (pp. 552–563). One of the most common fuels we mine is coal. Coal (p. 521) is the modified remains of ancient swamp plants and comprises primarily the mineral carbon. Other fossil fuels— petroleum, natural gas, and alternative fossil fuels such as oil sands, oil shale, and methane hydrates—are also organic and are extracted from the earth (Chapter 19).

Mining Methods and Their Impacts

Mining for minerals is an important industry that provides jobs for people and revenue for communities in many regions. Mining supplies us raw materials for countless products we

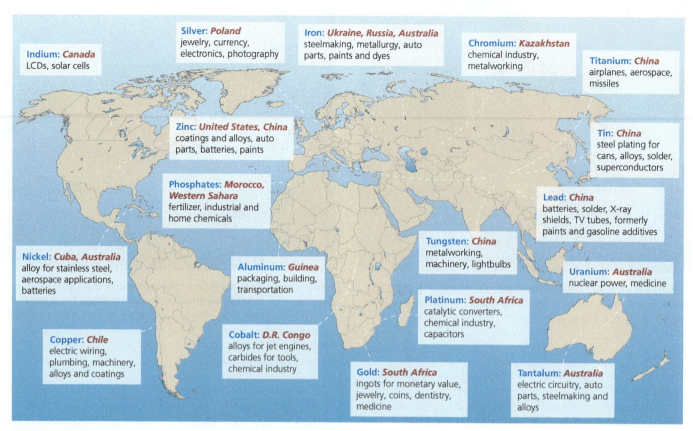

FIGURE 23.6 The minerals we use come from all over the world. Shown is a selection of economically important minerals (mostly metals, with several nonmetals), together with their major uses and their main nation of origin. Only a minority of minerals, uses, and origins is shown.

use daily, so it is necessary for the lives we lead. In 2015 raw materials from mining contributed $78 billion to the U.S. economy, and after processing, mineral materials contributed $630 billion. About 175,000 Americans were employed directly in mining for coal and metals in 2015, and the mining industry, together with processors and manufacturers of products from mined materials, employed more than 1.2 million people.

But as we have seen, mining also exacts a price in environmental and social impacts. Because minerals of interest often make up only a small portion of the rock in a given area, typically very large amounts of material must be removed to obtain the desired minerals. This frequently means that mining disturbs large areas of land, thereby exerting severe impacts on the environment and on people living nearby. The cleaning and processing of mined materials can add to these impacts, as described previously.

Depending on the nature of the mineral deposit, any of several mining methods may be used to extract the resource from the ground. If multiple methods are appropriate for a given resource, companies typically select a method based on its economic efficiency. We will examine major mining approaches commonly used throughout the world, and will also take note of the impacts of each approach as we proceed.

Strip mining removes surface layers of soil and rock

When a resource occurs in shallow horizontal deposits near the surface, the most effective mining method is often **strip mining,** whereby layers of surface soil and rock are removed from large areas to expose the resource. Heavy machinery removes the overlying soil and rock (termed *overburden*) from a strip of land, and the resource is extracted. This strip is then refilled with the overburden that had been removed, and miners proceed to an adjacent strip of land and repeat the process. Strip mining is commonly used for coal (**FIGURE 23.7a**) and oil sands (pp. 515–516), and sometimes for sand and gravel.

Strip mining can be economically efficient, but it obliterates natural communities over large areas, and the soil in refilled areas can easily erode away. Strip mining also pollutes waterways through the process of **acid mine drainage,** which occurs when sulfide minerals in newly exposed rock surfaces react with oxygen and rainwater to produce sulfuric acid (**FIGURE 23.8**). As the sulfuric acid runs off, it leaches metals from the rocks, and many of these metals are toxic to organisms. Acid drainage can affect fish and other aquatic organisms

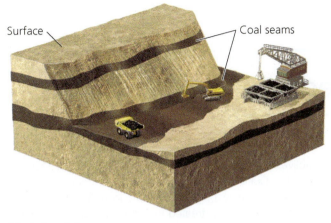

Surface

Coal seams

(a) Strip mining

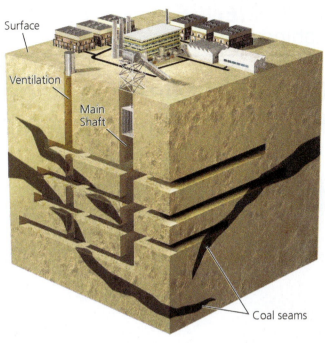

Surface

Ventilation

Main Shaft

Coal seams

(b) Subsurface mining

FIGURE 23.7 Coal mining illustrates two types of mining approaches. In strip mining **(a)**, soil is removed from the surface in strips, exposing seams from which coal is mined. In subsurface mining **(b)**, miners work belowground in shafts and tunnels blasted through the rock. These passageways provide access to underground seams of coal or minerals.

when it runs into streams and can pollute groundwater supplies people use for drinking water or irrigating crops. Although acid drainage is a natural phenomenon, mining greatly accelerates this process by exposing many new rock surfaces at once. New research has shown, however, that acidic mine drainage may be put to use to mitigate the effects of hydraulic fracturing (or "fracking") in areas where both occur (see **THE SCIENCE BEHIND THE STORY**, pp. 636–637).

FIGURE 23.8 Acidic drainage flows from a coal mine in Scotland. The orange color is due to iron from the drainage settling out on the soil surface and forming rust.

In subsurface mining, miners burrow into the Earth

When a resource occurs in concentrated pockets or seams deep underground, and the earth allows for safe tunneling, then mining companies pursue **subsurface mining.** In this approach, shafts are excavated deep into the ground, and networks of tunnels are dug or blasted out to follow deposits of the mineral (**FIGURE 23.7b**). Miners remove the resource systematically and ship it to the surface.

We use subsurface mining for metals such as zinc, lead, nickel, tin, gold, copper, and uranium, as well as for diamonds, phosphate, salt, and potash. In addition, a great deal of coal is mined using the subsurface technique. As with surface mining, the disturbance of rock in subsurface mining generates acid drainage that affects nearby waterways.

Subsurface mining is the most dangerous form of mining and indeed one of society's most dangerous occupations. Besides risking injury or death from dynamite blasts, natural gas explosions, and collapsing shafts and tunnels, miners inhale toxic fumes and coal dust, which can lead to respiratory diseases, including fatal black lung disease. Fatal accidents are not unusual. According to government figures, 931 miners lost their lives in Chinese coal mines in 2014 alone, and critics argue this number may be underestimated. As recently as 2000, more than 7600 miners were perishing annually in Chinese mines, but fatalities have dropped significantly in the past 15 years as government regulators strengthened safety rules and shut down many smaller mines where accidents were more common.

Can Acid Mine Drainage Reduce Fracking's Environmental Impact?

**Dr. Avner Vengosh,
Duke University**

In a strange twist of fate and chemistry, it appears that one environmental legacy of past mining for fossil fuels may help ameliorate the environmental impacts of mining today.

A study led by geologist Avner Vengosh of Duke University and published in 2014 in the scientific journal *Environmental Science & Technology* investigated whether the acid mine drainage leaking from old coal mines could be put to use to lessen the environmental impacts of the very modern process of hydraulic fracturing, or "fracking," used to extract natural gas (pp. 159–161, 530–531).

Hydraulic fracturing injects water laden with drilling chemicals into layers of shale rock deep underground, fracturing the rock and releasing natural gas trapped within. Along with natural gas, fracking wells pull up wastewater called "flowback fluid" made up of drilling chemicals mixed with water containing dissolved salts, toxic metals such as barium and strontium, and radioactive radium isotopes from deep underground. Previous work by the Duke team found that standard treatment of this wastewater was only partially effective in removing contaminants. Hence, when the treated wastewater was released into streams near the fracking site, its high salinity posed threats to freshwater organisms. The researchers also detected a buildup of radioactive radium in stream sediments near release sites, posing additional, long-term threats to wildlife. Clearly, a more effective method of treating wastewater was necessary to protect water quality near fracking sites.

To better treat wastewater, the team looked to something found in close proximity to many fracking wells—streams affected by acid mine drainage. Acidic drainage from coal mines fouls more than 20,000 km (12,400 mi) of streams in Pennsylvania alone (**FIGURE 1**). Vengosh's team assessed the chemical composition of acid mine drainage and hypothesized that it might chemically interact with dissolved substances in fracking wastewater and cause them to "drop out" of solution as a solid precipitate that could be collected and safely disposed of. To test this hypothesis, the team was provided with samples of fracking wastewater from three drilling sites and acid mine drainage from two sites in western Pennsylvania.

The researchers used two types of acid mine drainage: synthetic acid mine drainage (a solution mimicking natural acid mine drainage that researchers commonly use in laboratory

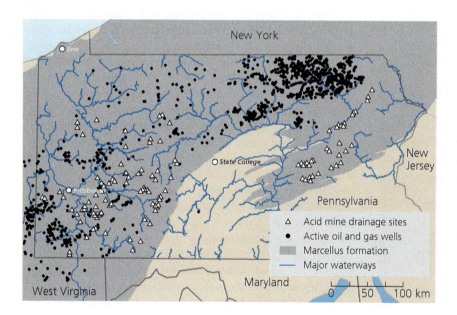

FIGURE 1 In parts of Pennsylvania, active hydraulic fracturing operations are very near streams affected by acid mine drainage. *Source: Kondash, A.J., et al., 2014. Radium and barium removal through blending hydraulic fracturing fluids with acid mine drainage. Environmental Science & Technology 48: 1334–1342.*

DATA In which part of Pennsylvania (eastern, central, or western) would the approach of using acid mine drainage in fracking operations be the most practical?

Go to **Interpreting Graphs & Data** on **Mastering**EnvironmentalScience®

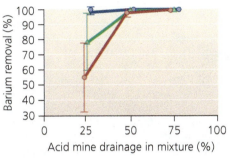

(a) Barium

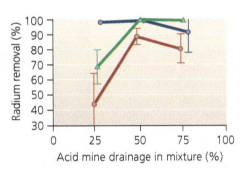

(b) Radioactive radium

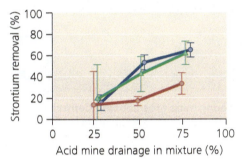

(c) Strontium

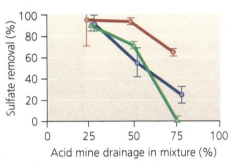

(d) Sulfate

FIGURE 2 **Acid mine drainage was highly effective in removing barium, radioactive radium, strontium, and sulfate from wastewater produced by hydraulic fracturing.** Synthetic acid drainage (**red lines**) was more effective at removing sulfate from wastewater than were field samples of lime-treated acid drainage from site #1 (**blue lines**) and site #2 (**green lines**), but generally less effective at removing barium, strontium, and radium. Increasing concentrations of acid mine drainage removed higher percentages of barium, strontium, and radium, but sulfate was more effectively removed when acid mine drainage concentrations were lower. *Source: Kondash, A.J., et al., 2014. Radium and barium removal through blending hydraulic fracturing fluids with acid mine drainage.* Environmental Science & Technology 48: *1334–1342.*

tests) and acid mine drainage from the two field sites (these samples were modified on-site in Pennsylvania by the addition of the mineral lime, which raises the pH and lowers the iron content of the acidic drainage, reducing its impact on waterways). In the laboratory at Duke University, each of these samples of acidic drainage was combined with fracking wastewater in mixtures containing 25%, 50%, or 75% acid mine drainage. The samples were shaken for 48 hours to promote mixing, and their concentrations of select chemical constituents subsequently analyzed.

The effectiveness of treatment varied both by type of acid mine drainage (synthetic or field-collected) and by acid mine drainage concentration (25%, 50%, or 75%), but the team found that within 10 hours, very high percentages of the barium, strontium, sulfate (which gives acidic drainage its low pH [p. 27]) and radioactive radium precipitated out of solution (**FIGURE 2**). The process was also effective in greatly reducing the salt content of the wastewater. After these reactions, the contaminants from wastewater were concentrated in precipitate, mostly in the form of strontium barite. Because they were not dissolved or suspended in solution, the contaminants were efficiently collected for disposal.

This study demonstrated that there is potential for treating fracking wastewater with mixtures of acid mine drainage,

removing some of the contaminants as precipitates, and producing water that could be less damaging to the environment or reused in drilling applications. Although laboratory experiments were successful, the next step is to test the process at actual fracking sites where conditions are less controlled.

Conducting this process at a large scale in the field is not without cost, because acid mine drainage would need to be pumped to drilling sites (or vice versa). Also, the precipitate produced by the reaction is radioactive and highly concentrated with contaminants, so it would require disposal in a specialized hazardous waste landfill, increasing disposal costs. The process does save money overall, though, because it produces water clean enough to be reused for additional drilling. This not only minimizes wastewater release into streams but also lowers drilling costs for companies because it reduces demand on precious freshwater resources for fracking operations. The water savings from using recycled water are significant: A single fracking well can use 7–49 million liters (2–13 million gal) of water while in production.

Although it would be far better for streams not to be polluted with mine drainage in the first place, this research shows that we may be able to "make lemonade out of lemons" by using a toxic legacy of past mining—acid mine drainage—to reduce the environmental impacts of present mining for fossil fuels.

While the United States has fewer subsurface mining operations than China and relatively stronger government safety inspections, accidents still occur. In 2010, an explosion at an underground coal mine killed 29 miners, and in 2015 mining accidents claimed 28 lives in the United States. This number seems small, but there are relatively few people employed in mining, and this makes mining one of the United States' statistically most dangerous ways to make a living.

Occasionally, subsurface mines can affect people for years after they are closed. Acidic drainage can contaminate surface and groundwater, sometimes for centuries, after mining operations cease. Natural disasters or accidents can lead to catastrophic releases of toxin-laden waters from abandoned mines.

In 2015, contractors hired by the U.S. Environmental Protection Agency (EPA) working at the abandoned Gold King mine in Colorado accidentally released millions of gallons of heavy-metal–laden mine drainage into the nearby Animas River. Concentrations of arsenic, mercury, and cadmium spiked in the river, threatening the people and wildlife that use the waterway. As of 2016, the EPA had dedicated $29 million to cleanup efforts and was facing lawsuits from New Mexico and the Navajo nation, who claimed the agency had failed to properly remediate the spill. Although the scale of this accident was unique, smaller-scale releases of toxic drainage from abandoned mines is common. The U.S. government estimates that as many as 500,000 abandoned mines exist in the United States, and that full cleanup of these sites could cost taxpayers up to $70 billion.

Abandoned mines can cause structural issues at the surface as well. The collapse of tunnels at the Retsof Salt Mine in Genessee Valley, New York, after a minor earthquake in 1994 created sinkholes at the surface that damaged roads, bridges, and homes and sucked groundwater from neighborhood wells. Coal veins in abandoned mines underneath Centralia, Pennsylvania, caught fire in 1961 and are still burning today. This once-thriving city is now a ghost town, because nearly all its residents accepted buyouts in the 1980s and relocated when it became clear the smoldering fires beneath them could not be contained (**FIGURE 23.9**).

FIGURE 23.9 Smoldering mine fires beneath Centralia, Pennsylvania, have led to the creation of a "ghost town." A nearly one-mile long section of State Route 61 was closed, as subsidence due to mine fires caused the road to buckle and crack. The roadway has become a tourist attraction for people visiting the area.

Open pit mining creates immense holes in the ground

When a mineral is spread widely and evenly throughout a rock formation, or when the earth is unsuitable for tunneling, the method of choice is **open pit mining.** This essentially involves digging a gigantic hole and removing the desired ore, along with waste rock that surrounds the ore. The ore-bearing rock is sent off for processing while waste rock is dumped in massive heaps outside the pit. Open pit mines are terraced so that workers and massive machinery can move about, and the pit is expanded until the resource runs out or becomes unprofitable to mine. Open pit mining is used to extract copper, iron, gold, diamonds, and coal, among other resources. We also use this technique to extract clay, gravel, sand, and stone such as limestone, granite, marble, and slate, but in these cases we generally call these pits quarries. Some open pit mines are inconceivably enormous. The world's largest, the Bingham Canyon Mine near Salt Lake City, Utah, is 4 km (2.5 mi) across and 1.2 km (0.75 mi) deep (**FIGURE 23.10**).

Open pit mines are so large because huge volumes of waste rock need to be removed to extract relatively small amounts of ore, which in turn contain still smaller traces of valuable minerals. The sheer size of these mines means that the degree of habitat loss and aesthetic degradation is considerable. Another impact is chemical contamination from acid drainage as water runs off the waste heaps or collects in the

FIGURE 23.10 The Bingham Canyon open pit mine outside Salt Lake City, Utah, is the world's largest human-made hole in the ground. This immense mine produces mostly copper.

pit. Once mining is complete, abandoned pits generally fill up with groundwater, which soon becomes toxic as water and oxygen react with sulfides from the ore. Acidic water from the pit can harm wildlife and can percolate into aquifers and spread through the region.

The Berkeley Pit, a former copper mine near Butte, Montana, is today one of the largest Superfund toxic waste cleanup sites (p. 625) in the United States. After its closure in 1982, the pit filled with groundwater and became so acidic (pH of 2.2) and concentrated with toxic metals that microbiologists discovered new species of microbes in the water—the harsh conditions were so rare in nature that scientists had never before encountered microbes adapted to them!

Mountaintop mining reshapes ridges and can fill valleys

When a resource occurs in underground seams near the tops of ridges or mountains, mining companies may practice **mountaintop removal mining,** in which several hundred vertical feet of mountaintop may be blasted off to allow recovery of entire seams of the resource (**FIGURE 23.11**). This method of mining is used primarily for coal in the Appalachian Mountains of the eastern United States. In mountaintop removal mining, a mountain's forests are clear-cut, the timber is sold, topsoil is removed, and then rock is repeatedly blasted away to expose the coal for extraction.

Afterward, overburden is placed back onto the mountaintop, but this waste rock is unstable and typically takes up more volume than the original rock, so generally a great deal of waste rock is dumped into adjacent valleys (a practice called "valley filling"). So far, mountaintop removal in the Appalachian Mountains has blasted away an area the size of Delaware and has buried nearly 3200 km (2000 mi) of streams.

Scientists are finding that dumping tons of debris into valleys degrades or destroys immense areas of habitat, clogs streams and rivers, and pollutes waterways with acid drainage. With slopes deforested and valleys filled with debris, erosion intensifies, mudslides become frequent, and flash floods ravage the lower valleys. Worsening the environmental impact of mountaintop removal mining is the fact that the Appalachian forests that are cleared in mountaintop mining are some of the richest forests for biodiversity in the nation.

People living in communities near the mining sites experience social and health impacts. Explosions that are a part of clearing the mountaintops crack house foundations and wells, loose rock tumbles down into yards and homes, and floods tear through properties when mining operations block or divert streams. Coal dust causes respiratory ailments, and contaminated water unleashes a variety of health problems. Studies have shown that people in mountaintop mining areas exhibit elevated levels of birth defects, lung cancer, heart disease, kidney disease, pulmonary disorders, hypertension, and early mortality.

Critics of mountaintop removal mining argue that valley filling violates the Clean Water Act (p. 172) because runoff flowing through waste rocks in valleys often contains high levels of salts and toxic metals that degrade water quality and affect aquatic organisms. Although hundreds of permits for mountaintop mining were issued during the Bill Clinton and George W. Bush administrations, in 2010 the U.S. Environmental Protection Agency (EPA) announced

FIGURE 23.11 Mountaintop mining removes entire mountaintops to obtain the coal underneath. The rocks removed during the process are dumped into adjacent valleys, burying streams, promoting flooding, and contaminating drinking water supplies.

new guidelines that prohibit valley fills unless strict measures of water quality can be attained. Critics of the policy argue that these new guidelines will essentially end the practice of mountaintop mining. In 2011, the EPA revoked the permit of an existing mountaintop mining operation in West Virginia, citing these new guidelines. A 2014 court ruling backed the EPA's decision, and the agency continued to reexamine other permits across Appalachia.

Placer mining uses running water to isolate minerals

Some metals and gems accumulate in riverbed deposits, after having been displaced from weathered rock and carried along by runoff to streams. To search for these metals and gems, miners sift through material in modern or ancient riverbed deposits, generally using running water to separate lightweight mud and gravel from heavier minerals of value (**FIGURE 23.12**). This technique is called **placer mining** (pronounced "plasser").

Placer mining is the method used by Congo's coltan miners, who wade through streambeds, sifting through sediments by hand with a pan or simple tools, searching for high-density tantalite that settles to the bottom while low-density material washes away. Today's African miners practice small-scale placer mining similar to the method used long ago by American miners who ventured to California in the Gold Rush of 1849, and later to Alaska in the Klondike Gold Rush of 1896–1899. Indeed, placer mining for gold is still practiced in areas of Alaska and Canada, although today miners in the regions use large dredges and heavy machinery.

FIGURE 23.12 Miners in eastern Congo find coltan by placer mining. Sediment is placed in plastic tubs, and water is run through them. A mixing motion allows the sediment to be poured off while the heavy coltan settles to the bottom.

Besides the many social and political impacts of placer mining in places such as D.R. Congo, placer mining is environmentally destructive because most methods wash large amounts of debris into streams, making them uninhabitable for fish and other life for many miles downstream. Gold mining in northern California's rivers in the decades following the Gold Rush washed so much debris all the way to San Francisco Bay that a U.S. district court ruling in 1884 finally halted this mining practice. Placer mining also disturbs stream banks, causing erosion and harming ecologically important riparian plant communities.

Solution mining dissolves and extracts resources in place

When a deposit is especially deep and the resource can be dissolved in a liquid, miners may use a technique called **solution mining.** In this technique, a narrow borehole is drilled deep into the ground to reach the deposit, and water, acid, or another liquid is injected down the borehole to leach the resource from the surrounding rock and dissolve it in the liquid. The resulting solution is sucked out, and the desired resource can then be isolated. Salts can be mined in this way; water is pumped into deep salt caverns, the salt dissolves in the water, and the salty solution is extracted. Sodium chloride (table salt), lithium, boron, bromine, magnesium, potash, copper, and uranium can be obtained with solution mining.

Solution mining generally exerts less environmental impact than other mining techniques, because less area at the surface is disturbed. The main potential impacts involve accidental leakage of acids into groundwater surrounding the borehole and the contamination of aquifers with acids, heavy metals, or uranium leached from the rock.

Some mining occurs in the ocean

The oceans hold many minerals useful to our society. We extract some minerals from seawater, such as magnesium from salts held in solution. We extract other minerals from the ocean floor, often using large vacuum-cleaner–like hydraulic dredges. Valuable minerals found on or beneath the seafloor include calcium carbonate (used in making cement), copper, zinc, silver, and gold ore. Many minerals are concentrated in manganese nodules, small ball-shaped accretions that are scattered across parts of the ocean floor. Over 1.5 trillion tons of manganese nodules may exist in the Pacific Ocean alone, and their reserves of metal may exceed all terrestrial reserves. The logistical difficulty of mining them, however, has kept their extraction too expensive thus far.

As land resources become scarcer and as undersea mining technology develops, mining companies may turn increasingly to the seas. Impacts of undersea mining are largely unknown, but increases in such mining would undoubtedly destroy marine habitats and organisms that have not yet been studied. It would also likely cause some metals to diffuse into the water column at toxic concentrations and enter the food chain.

Restoration helps to reclaim mine sites

Because of the environmental impacts of mining, governments of the United States and other developed nations now require that mining companies restore, or reclaim, surface-mined sites following mining. The aim of such restoration, or **reclamation,** is to restore the site to a condition similar to its condition before mining.

To restore a site, companies are required to remove buildings and other structures used for mining, replace overburden, fill in shafts, and replant the area with vegetation. Former mine sites have been reclaimed for wildlife habitat (**FIGURE 23.13**) or repurposed for other uses. Virginia Polytechnic Institute and State University, for example, is coordinating a project that is investigating the use of reclaimed coal mines in southwestern Virginia to grow fast-growing trees and other vegetation for bioenergy production (p. 563).

In the United States, the 1977 Surface Mining Control and Reclamation Act mandates restoration efforts, requiring companies to post bonds to cover reclamation costs before mining can be approved. This ensures that if the company fails to restore the land for any reason, the government will have the money to do so. Most other nations exercise less oversight regarding reclamation, and in nations such as D.R. Congo, there is essentially no regulation.

The mining industry has made great strides in reclaiming mined land. But even on sites that are restored, impacts from mining (such as soil and water damage from acid drainage) can be severe and long-lasting because the soil is often acidic and can contain elevated levels of metals that are toxic to native plant life. It is therefore often difficult to regain the same biotic communities that were naturally present before mining. Research is continuing on new strains of plants that can tolerate conditions on reclaimed sites and pave the way for the return of native vegetation. Establishing plant communities on reclaimed sites is key, because the plants stabilize the soil, prevent erosion, and can help establish conditions that favor the reestablishment of native vegetation and functioning ecosystems.

Water polluted by mining and acid drainage can also be reclaimed, if pH can be moderated and if toxic heavy metals can be removed. Like the reclamation of land, this is a challenging and imperfect process, but researchers and the mining industry are making progress in improving techniques. The need for treatment can be long-lasting. Mines in Spain from the era of the Roman Empire, some 2000 years ago, still leach acid drainage into waterways today.

FIGURE 23.13 More mine sites are now being restored. Here, bison graze on land reclaimed from a tar sands mining operation in Alberta, Canada.

weighing the
ISSUES

Restoring Mined Areas

Mining has severe environmental impacts, but restoring mined sites to their pre-mining condition can be costly and difficult. How extensively should mining companies be required to restore a site after a mine is shut down, and what criteria should we use to guide restoration? Should we require nearly complete restoration? What should our priorities be—to minimize water pollution, health impacts, biodiversity loss, soil damage, or other factors? What measures should we use to evaluate the results of restoration? Should the amount of restoration we require depend on how much money the company made from the mine? Explain your recommendations.

An 1872 law still guides U.S. mining policy

Government policy plays a role in the ways that mining companies stake claims and use land. In the United States, this has been controversial because policy is still guided by a law that is well over a century old. The **General Mining Act of 1872** encourages people and companies to prospect for minerals on federally owned land by allowing any U.S. citizen or any company with permission to do business in the United States to stake a claim on any plot of public land open to mining. The person or company owning the claim gains the sole right to take minerals from the area. The claim-holder can also patent the claim (buy the land) for only about $5 per acre. Regardless of the profits they might make on minerals they extract, the law requires no payments of any kind to the public and, until recently, no restoration of the land after mining.

The General Mining Act of 1872 was enacted partly in response to the chaos of the California Gold Rush and mining rushes, and it was designed to bring some order to mining activities. It also aimed to promote mining at a time when the government was trying to hasten settlement of the West in an orderly way. The law may have made good sense in 1872, but the United States has changed a great deal since then, and many question the law's suitability for today's nation.

Supporters of the policy say that it is appropriate and desirable to continue encouraging the domestic mining industry, which must undertake substantial financial risk and investment to locate resources that are vital to our economy. Critics counter that the policy gives away valuable public resources to private interests for next to nothing. They also point out that many claims made under this law have eventually led to lucrative land development schemes (such as condominium development) that have nothing to do with mining.

Critics and some legislators have tried to amend the law many times over the years, but the economic activity generated by the act has given it supporters in Congress, and reform efforts have mostly been without success. The latest effort, the Hardrock Mining and Reclamation Act of 2015, seeks to end the patenting process, consider environmental degradation in the permit-granting process, and charge a royalty of up to 8% on mining profits. The outlook for this bill, currently in committee, is not optimistic, as similar bills introduced in Congress over the past two decades have failed to become law.

The General Mining Act of 1872 covers a wide variety of metals, gemstones, uranium, and minerals used for building materials. In contrast, fossil fuels, phosphates, sodium, and sulfur are governed by the Mineral Leasing Act of 1920. This law sets terms for leasing public lands that vary according to the resource being mined, but in all cases the terms include the payment of rents for the use of the land and the payment of royalties on profits.

Toward Sustainable Mineral Use

Mining exerts plenty of environmental impacts, but we also have another concern to keep in mind: Minerals are nonrenewable resources (p. 4) in finite supply. Like fossil fuels, they form far more slowly than we use them, and if we continue to mine them, they will eventually be depleted. As a result, it will benefit us to find ways to conserve the supplies we have left and to make them last. Reducing waste and developing means of recovering and recycling used mineral resources are ways we can pursue the use of mineral resources more sustainably. We will likely never achieve 100% recovery, but we can do much better than we are doing today.

Minerals are nonrenewable resources in limited supply

Some minerals we use are abundant in their supply and will likely never run out, but others are rare enough that they could soon become unavailable. For instance, geologists in 2016 calculated that the world's known reserves of tantalum will last about 129 more years at today's rate of consumption. If demand for tantalum increases, it could run out faster, of course. And if everyone in the world began consuming tantalum at the rate of U.S. citizens, then it would last for only 31 years!

Most pressing may be dwindling supplies of indium. This obscure metal, which is used in LCD screens, might last only another 30 years. Because of these supply concerns and price volatility, industries now are working hard to develop ways of substituting other materials for indium. A lack of indium and the metal gallium would threaten the production of high-efficiency cells for solar power. Platinum is dwindling, too, and if it became unavailable, it would be harder to develop fuel cells for vehicles. However, platinum's high market price

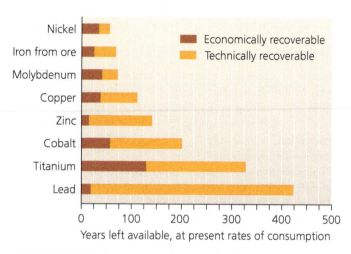

Years left available, at present rates of consumption

FIGURE 23.14 Minerals are nonrenewable resources, so supplies of metals are limited. Shown in red are the numbers of remaining years that certain metals are estimated to be economically recoverable at current prices. The entire lengths of the bars (red plus orange) show how long certain metals are estimated to be available using current technology on all known deposits, whether economically recoverable or not. *Data are for 2015, from U.S. Geological Survey, 2016.* Mineral commodity summaries 2016. *Reston, VA: USGS.*

 DATA
- Which metal has the highest proportion of its technically recoverable reserves that are currently economically recoverable?
- Approximately what percentage is economically recoverable?
- Which metal has the smallest proportion of its technically recoverable reserves that are currently economically recoverable and what is this value?

Go to **Interpreting Graphs & Data** on **Mastering**EnvironmentalScience®

encourages recycling, which may keep it available, albeit as an expensive metal.

FIGURE 23.14 shows estimated years remaining for several selected minerals at today's consumption rates. As minerals become scarcer, demand for them increases and price rises. Higher market prices make it more profitable for companies to mine the resource, so they become willing to spend more to reach further deposits that were not economically worthwhile originally. Thus the entire length of each bar in Figure 23.14 shows the number of years researchers calculate that we will have the mineral available in total, as prices rise. These two categories used by the U.S. Geological Survey (USGS) are similar to the way the USGS classifies fossil fuel deposits as being either "economically recoverable" or "technically recoverable" (pp. 524–525).

Several factors affect how long mineral deposits may last

Calculating how long a given mineral resource will be available to us, as shown by the data in Figure 23.14, is beset by a great deal of uncertainty. There are several major reasons why such estimates may increase or decrease over time, each of which we will discuss in turn.

Discovery of new reserves

As we discover new deposits of a mineral, the known reserves—and thus the years this mineral is available to us—increase. For this reason, some previously predicted shortages have not come to pass, and we may have access to these minerals for longer than currently estimated. For example, in 2010 geologists associated with the U.S. military discovered that Afghanistan holds immense mineral riches that were previously unknown. The newly discovered reserves of iron, copper, niobium, lithium, and many other metals are estimated to be worth more than $1 trillion—enough to realign the entire Afghan economy around mining. (Note, however, that such riches are not guaranteed to make Afghanistan a wealthy nation; history teaches us that regions rich in nonrenewable resources, such as D.R. Congo and Appalachia, have often not prospered from them.)

New extraction technologies

Just as rising prices of scarce minerals encourage companies to expend more effort to reach difficult deposits, rising prices also may favor the development of enhanced mining technologies that can reach more minerals at less expense. If more powerful technologies are developed, then these may increase the amounts of minerals that are technically feasible for us to mine.

Changing social and technological dynamics

New societal developments and new technologies in the marketplace can modify demand for minerals in unpredictable ways. Just as cell phones and computer chips boosted demand for tantalum, fiber-optic cables decreased demand for copper as they replaced copper wiring in communications applications. Today lithium–ion batteries are replacing nickel–cadmium batteries in many devices. Synthetically produced diamonds are driving down prices of natural diamonds and extending their availability. Additionally, health concerns sometimes motivate change: We have replaced toxic substances such as lead and mercury with safer materials in many applications, for example.

Changing consumption patterns

Changes in the rates and patterns of consumption also alter the speed with which we exploit mineral resources. For instance, economic recession depressed demand and led to a decrease in production and consumption of most minerals from 2007 to 2009 after rising for long stretches. However, over the long term, demand has been rising. This is especially true today as China, India, and other major industrializing nations rapidly increase their consumption.

Recycling

Advances in recycling technologies and the extent of recycling have been helping us to extend the lifetimes of many mineral resources. Further progress in recycling will likely continue to do so.

Despite these sources of uncertainty, we would be wise to be concerned about Earth's finite supplies of mineral resources and to encourage recycling to use them more sustainably. Sustainable use will benefit future generations by conserving resources for them to use. It will also benefit us today, because conserving mineral resources through reuse and recycling can prevent price hikes that result from reduced supply. And as we saw with fossil fuels, domestic conservation of resources helps make a national economy less vulnerable in instances when other nations decide to withhold resources. Currently, the United States is 100% dependent on imports from other nations for 17 of the 63 major minerals on which the USGS reports annually. For 24 more of these minerals, the United States relies on imports for 50% or more of its supply.

We can make our mineral use more sustainable

We can address both major challenges facing us regarding mineral resources—finite supply and environmental damage—by encouraging recycling of these resources. Metal-processing industries regularly save resources and money by reusing some of the waste products produced during their refining processes. In addition, municipal recycling programs help provide metals by handling used items that we as consumers place in recycling bins and divert from the waste stream. Currently, around 35% of metals in the U.S. municipal solid waste stream were diverted for recycling. For example, 80% of the lead we consume today comes from recycled materials, in particular, recycled car batteries. Similarly, 33% of our copper comes from recycled copper sources such as pipes and wires. We recycle steel, iron, platinum, and other metals from auto parts. Altogether, we have found ways to recycle much of our gold, lead, iron and steel scrap, chromium, zinc, aluminum, and nickel. **TABLE 23.1** (p. 644) shows minerals that currently boast high recycling rates in the United States.

In many cases, recycling can decrease energy use substantially. For instance, making steel by remelting recycled iron and steel scrap requires much less energy than producing steel from virgin iron ore. Because this practice saves money, the steel industry today is designed to make efficient use of iron and steel scrap. Over half its scrap comes from discarded consumer items such as cars, cans, and appliances; scrap produced within their plants and scrap produced by other types of plants in the industry each account for nearly one-fourth of the total. Similarly, more than 40% of the aluminum in the United States today is recycled. This is beneficial because it takes more than 20 times more energy to extract virgin aluminum from ore (bauxite) than it does to obtain it from recycled sources. Every ton of aluminum cans your community recycles saves the energy equivalent of more than 1600 gallons of gasoline.

Saving energy means reducing our emissions of heat-trapping greenhouse gases (p. 482). The 7 million tons of metals that U.S. consumers recycle each year reduce greenhouse gas emissions by 25 million metric tons of carbon dioxide, the Environmental Protection Agency estimates. This substantial reduction is like taking more than the yearly emissions of 4.5 million cars out of the atmosphere.

TABLE 23.1 Recycled minerals in the United States

MINERAL	U.S. RECYCLING RATE
Gold	Slightly less is recycled than is consumed
Iron and steel scrap	85% for autos, 82% for appliances, 72–98% for construction materials, 70% for cans
Lead	69% consumed comes from recycled post-consumer items
Tungsten	59% consumed is from recycled scrap
Nickel	45% consumed is from recycled nickel
Zinc	37% produced is recovered, mostly from recycled materials used in processing
Chromium	34% is recycled in stainless steel production
Copper	32% of U.S. supply comes from various recycled sources
Aluminum	30% produced comes from recycled post-consumer items
Tin	30% consumed is from recycled tin
Germanium	30% consumed worldwide is recycled. Optical device manufacturing recycles more than 60%
Molybdenum	About 30% gets recycled as part of steel scrap that is recycled
Cobalt	28% consumed comes from recycled scrap
Niobium (columbium)	Perhaps 20% gets recycled as part of steel scrap that is recycled
Silver	15% consumed is from recycled silver. U.S. recovers as much as it produces
Bismuth	All scrap metal containing bismuth is recycled, providing less than 10% of consumption
Diamond (industrial)	7% of production is from recycled diamond dust, grit, and stone

Data are for 2015, from U.S. Geological Survey, 2016. *Mineral commodity summaries 2016.* Reston, VA: USGS.

As a valuable mineral, tantalum is also recycled. It is recycled from scrap by-products generated during the manufacture of electronic components and also from scrap from tantalum-containing alloys and manufactured materials. Tantalum is also obtained from retired cell phones, after the mineral has completed its journey from mine to consumer.

Currently the industry estimates that recycling accounts for 20–25% of the tantalum available for use in products. This percentage has been growing quickly, but its future growth will depend on how quickly we expand recycling efforts for used cell phones and other electronic waste (pp. 620–621) and on how well we enable recycling facilities to recover metals from these products.

We can recycle metals from e-waste

Electronic waste, or e-waste, from discarded computers, printers, cell phones, handheld devices, and other electronic products is rising fast—and that e-waste contains hazardous substances (p. 622). Recycling old electronic devices helps keep them out of landfills and also helps us conserve valuable minerals such as tantalum.

In fact, each of the 1.7 billion cell phones sold each year contains about 200 chemical compounds and close to a dollar's worth of precious metals (**FIGURE 23.15**). Upgrades

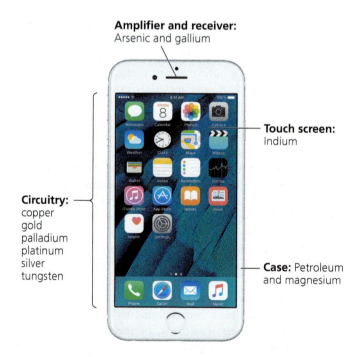

Amplifier and receiver: Arsenic and gallium

Touch screen: Indium

Circuitry: copper gold palladium platinum silver tungsten

Case: Petroleum and magnesium

FIGURE 23.15 Your cell phone contains a diversity of mined materials from around the world.

and improvements render more than 130 million cell phones obsolete each year in the United States alone, and an estimated 500 million old cell phones are currently lying inactive in people's homes and offices.

When you turn in your old phone to a recycling and reuse center rather than discarding it, the phone may be refurbished and resold in a developing country. People in African nations in particular readily buy used cell phones, because they are inexpensive and land-line phone service does not always exist in poor and rural areas. Alternatively, the phone may be dismantled in a developing country and the various parts refurbished and reused, or recycled for their metals. Either way, you are helping to extend the availability of resources through reuse and recycling and to decrease waste of valuable minerals.

Today only about 10% percent of old cell phones are recycled. That leaves a long way to go! As more of us recycle our phones, computers, and other electronic items, more tantalum and other metals may be recovered and reused. By recycling more, we can reduce demand for virgin ore and decrease pressure on people in Africa and South America and on the ecosystems where coltan is mined. Throughout the world, recycling to make better use of the mineral resources we have already mined will help minimize the impacts of mining and ensure us access to resources further into the future.

closing
THE LOOP

Along with tantalum, we depend on a diversity of minerals and metals to help manufacture products widely used in our society. We mine these nonrenewable resources by various methods, according to how the minerals are distributed. Economically efficient mining methods have greatly contributed to our material wealth, but they have also resulted in extensive environmental impacts, ranging from habitat loss to acid drainage. Restoration efforts and enhanced regulation help to minimize the environmental and social impacts of mining, although to some extent these impacts will always exist. As shown in the opening case study profiling the mining of tantalum in D.R. Congo, procuring these metals and minerals for products in our country can have profound impacts on people and ecosystems far away, a realization that prompted the current movement toward the certification of minerals and gemstones as "conflict free." These initiatives are vital, as conflict minerals are an emerging problem in regions other than central Africa. A thriving black market in coltan is emerging in remote portions of the northern Amazon jungle, and the recent discovery of vast mineral reserves in Afghanistan suggests that it, too, could become a significant source of conflict minerals.

Manufacturers, governments, and nongovernmental organizations are collaborating to make certification efforts practical and meaningful in the complex, multilayered global trade in metals and minerals. Through these efforts, and efforts to make our mineral use more sustainable by maximizing recovery and recycling, we are working toward a future in which we can simultaneously meet our demand for these vital resources while minimizing environmental impacts and adverse effects on the people who extract and produce them.

REVIEWING Objectives

You should now be able to:

Describe the types of mineral resources and their uses

We mine ore for metals, as well as nonmetallic minerals and fuels to provide raw materials for products we use every day. An average American uses over a million kg of mined materials over his or her lifetime. (pp. 630–632)

Explain how minerals are processed prior to use

Metal ore and valuable minerals must typically be separated from surrounding minerals through physical or chemical means. Alloys are produced when we combine a metal with another metal or a nonmetallic substance, such as carbon. (pp. 632–633)

Describe the major methods of mining

Many methods are available to miners to secure valuable minerals and fuels. Strip mining removes surface layers of soil and rock to mine resources from above, while subsurface mining tunnels underground to reach deeper resources. Open pit mining involves digging gigantic holes. Placer mining uses running water to isolate high-density minerals. Mountaintop removal coal mining removes immense amounts of rock from mountaintops and dumps them into valleys below. Solution mining uses liquids to dissolve minerals in place. (pp. 633–640)

Discuss the environmental and social impacts of mining

Many methods of mining completely remove vegetation, soil, and natural habitat. Acid drainage occurs when water leaches compounds from rocks exposed by mining, forming a low-pH solution that can be high in toxic metals. Mountaintop removal for coal destroys forests, mountaintops, and adjacent valleys and streams. People living near mines experience adverse health impacts and environmental impacts, but mining can offer employment in economically depressed regions. (pp. 633–640)

Explain reclamation efforts and mining policy

Mine reclamation efforts are challenging and sometimes fall short of effective ecological restoration due to the toxic soil and water conditions left behind by mining operations. Some mining effects, such as acid drainage, are difficult to contain and long-lasting. Mining policy in the United States is still being dictated by a law passed in 1872. (pp. 641–642)

Evaluate ways to encourage sustainable use of mineral resources

Minerals are nonrenewable resources. Reuse and recycling by industry and consumers are the keys to more sustainable practices of mineral use. (pp. 642–645)

TESTING Your Comprehension

1. Define each of the following: (1) mineral, (2) metal, (3) ore, (4) alloy. Compare and contrast the terms.

2. A mining geologist locates a horizontal seam of coal very near the surface of the land. What type of mining method will the mining company likely use to extract it? What is one common environmental impact of this type of mining?

3. How does strip mining differ from subsurface mining? How does each of these approaches differ from open pit mining?

4. What is acid drainage and where does it come from in a mining context? Why can such drainage be toxic to fish?

5. What type of mining is used for both coltan and gold? What does a miner do to conduct this type of mining?

6. Describe and then contrast how water is used in placer mining and in solution mining.

7. Describe three major environmental or social impacts of mountaintop removal mining.

8. Explain why reclamation efforts after mining frequently fail to effectively restore natural communities. Include reference to both soil and vegetation in your answer.

9. List five factors that can influence how long global supplies of a given mineral will last, and explain how each might increase or decrease the time span the mineral will be available to us.

10. Name three types of metal that we currently recycle, and identify the products or materials that are recycled to recover these metals.

SEEKING Solutions

1. List three impacts of mining on the natural environment, and describe how particular mining practices can lead to each of these impacts. How are these impacts being addressed? Can you think of additional solutions to prevent, reduce, or mitigate these impacts?

2. List three impacts of mining on people's health, lifestyles, or well-being, and describe how particular mining practices can lead to each of these impacts. How are these impacts being addressed? Can you think of additional solutions to prevent, reduce, or mitigate these impacts?

3. You have won a grant from the EPA to work with a mining company to develop a more effective way of restoring a mine site that is about to be closed. Describe a few preliminary ideas for carrying out restoration better than it is typically being done. Now describe a field experiment you would like to run to test one of your ideas.

4. **CASE STUDY CONNECTION** The story of coltan in Congo is just one example of how an abundance of exploitable resources can often worsen or prolong military conflicts in nations that are too poor or ineffectively

governed to protect these resources. In such resource wars, civilians often suffer the most as civil society breaks down. Suppose you are the head of an international aid agency that has earmarked $10 million to help address conflicts related to mining in Democratic Republic of Congo. You have access to government and rebel leaders in Congo and neighboring countries, to ambassadors of the world's nations in the United Nations, and to representatives of international mining corporations. Based on what you know from this chapter, what steps would you consider taking to help improve the situation in Congo?

5. **THINK IT THROUGH** In your public speaking class you have been assigned to (1) present one argument in favor of retaining the General Mining Act of 1872, (2) one argument in favor of repealing or reforming the law, and (3) a conclusion regarding your personal view of what to do regarding this legislation. Would you retain, repeal, or reform the 1872 law? Why? What arguments would you make in (1) or (2) that inform your ultimate decision?

6. **THINK IT THROUGH** As you finish your college degree, you learn that the mountains behind your childhood home in the hills of Kentucky are slated to be mined for coal using the mountaintop removal method. Your parents, who still live there, are worried for their health and safety and do not want to lose the beautiful forested creek and ravine behind their property. However, your brother is out of work and could use a steady, well-paying mining job. What would you attempt to do in this situation?

CALCULATING Ecological Footprints

As we saw in Figure 23.14, the supplies of some metals are limited enough that, at today's prices, these metals could be available to us for only a few more decades. After that, prices will rise as they become scarcer. The number of years of total availability (at all prices) depends on a number of factors: On the one hand, metals will be available for longer if new deposits are discovered, new mining technologies are developed, or recycling efforts are improved. On the other hand, if our consumption of metals increases, the number of years we have left to use them will decrease.

Currently the United States consumes metals at a much higher per-person rate than the world does as a whole. If one goal of humanity is to lift the rest of the world up to U.S. living standards, then this will sharply increase pressures on mineral supplies.

The table below shows currently known economically recoverable global reserves for several metals, together with the amount used per year (each figure in thousands of metric tons). For each metal, calculate and enter in the fourth column the years of supply left at current prices by dividing the reserves by the amount used annually.

The fifth column shows the amount that the world would use if everyone in the world consumed the metal at the rate that Americans do. Now calculate the years of supply left at current prices for each metal if the world were to consume the metals at the U.S. rate, and enter these values in the sixth column.

METAL	KNOWN ECONOMIC RESERVES	AMOUNT USED PER YEAR	YEARS OF ECONOMIC SUPPLY LEFT	AMOUNT USED PER YEAR IF EVERYONE CONSUMED AT U.S. RATE	YEARS OF ECONOMIC SUPPLY LEFT IF EVERYONE CONSUMED AT U.S. RATE
Titanium	790,000	6090	129.7	25,123	31.4
Copper	720,000	18,700		40,654	
Nickel	79,000	2530		5162	
Tin	4800	294		964	
Tungsten	3300	87		320	
Antimony	2000	150		560	
Silver	570	27.3		185	
Gold	56	3.0		3.43	

Data are for 2015, from U.S. Geological Survey, 2016. *Mineral commodity summaries 2016*. Reston, VA: USGS.

Notes: All numbers are in thousands of metric tons. World consumption data are assumed equal to world production data. "Known economic reserves" include extractable amounts under current economic conditions. Additional reserves exist that could be mined at greater cost.

1. Which of these eight metals will last the longest under current economic conditions and at current rates of global consumption? For which of these metals will economic reserves be depleted fastest?

2. If the average citizen of the world consumed metals at the rate that the average U.S. citizen does, the economic reserves of which of these eight metals would last the longest? Which would be depleted fastest if everyone consumed at the U.S. rate?

3. In this chart, our calculations of years of supply left do not factor in population growth. How do you think population growth will affect these numbers?

4. Describe two general ways that we could increase the years of supply left for these metals. What do you think it will take to accomplish this?

MasteringEnvironmentalScience®

CHAPTER 24

Sustainable Solutions

De Anza College Strives for a Sustainable Campus

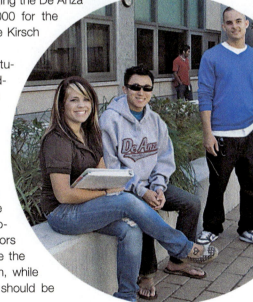

> "Sustainability considers our impact not just on air, land, and water, but includes our impact on community vibrancy, environmental stewardship, social equity, and financial responsibility.
>
> De Anza College's Sustainability Management Plan

> Sustainability isn't a distant, unattainable concept to be discussed in the abstract. It is a very real way of making decisions that can be integrated into every single person's daily actions.
>
> Yale University student
> Jacquelyn Maitram Truong

California's Silicon Valley has long been a hub for innovation. So perhaps it's no surprise that a college in the heart of this region is a leader in the burgeoning movement for campus sustainability.

De Anza College, located in Cupertino, California (home of Apple Computer, Inc.), serves 24,000 students, making it one of North America's largest community colleges. De Anza has now become one of the "greenest" community college campuses as well, thanks to the ongoing commitment of its students, faculty, staff, and administrators. Today more and more colleges and universities are serving as models for our larger society by taking steps to address their resource consumption, pollution, and ecological footprints.

At De Anza College, sustainability efforts reach back to 1990, when faculty member Julie Phillips incorporated concepts of sustainability into the curriculum and energized students with the idea of a green building project. Soon students, faculty, staff, administrators, and community members came together to create a College Environmental Advisory Group (CEAG) to encourage green building and other sustainable practices on campus. Eventually, those early students brought their classroom experiences, sustainable practices, and green building vision to reality by urging the De Anza Associated Student Body to allocate $180,000 for the conceptual design of what would become the Kirsch Center for Environmental Studies.

In 2005, the Kirsch Center opened to students, as the first community-college building in the nation to be certified LEED Platinum (p. 343). Showcasing hi-tech and innovative labs and classrooms that promote visual and hands-on learning, the Kirsch Center demonstrates to students and the public how to merge energy efficiency, pollution prevention, and biodiversity protection. Displays inside the entrance show real-time data on the energy generated by the 36.5-kilowatt photovoltaic energy system on the rooftop. Monitors show current temperature readings that guide the advanced radiant heating and cooling system, while red and green lights advise when windows should be

Upon completing this chapter, you will be able to:

- Describe approaches being taken on college and university campuses to promote sustainability

- Explain the concept of sustainable development and discuss how environmental protection can enhance economic well-being

- Assess key approaches to designing sustainable solutions

- Explain how time is limited yet human potential to solve problems is tremendous

LEED-certified Media and Learning Center, De Anza College

De Anza students in front of their LEED-certified Science Center

opened and closed. Students use labs, classrooms, and open study stations bathed in natural lighting from outdoors, which saves on electricity while creating a bright and pleasant environment that promotes learning. Everything from carpeting to furniture to structural steel to toilet seats is made with recycled materials, and features to conserve water and energy abound. The building's architects calculate that energy efficiency features save $65,000 per year (88% of energy costs) when compared to a typical building (**FIGURE 24.1**).

Sustainable architecture doesn't stop at a building's walls. The landscaping around the Kirsch Center features drought-tolerant native plants that require little watering. Adjacent to the Kirsch Center is a 1.5-acre arboretum called the Cheeseman Environmental Study Area. Here, 12 California native plant communities are represented, with more than 400 species of native

FIGURE 24.1 The Kirsch Center for Environmental Studies at De Anza College is a climate-responsive and energy-efficient building. It conserves water, generates renewable energy, uses recycled and nontoxic materials, and features outdoor learning spaces and labs. The spacious interior gathers natural daylight, providing a welcoming learning environment.

plants. Docents lead schoolchildren on field trips here, thousands of De Anza students gain knowledge of California native plants, and thousands more visitors learn about native landscaping.

Motivated by the Kirsch Center's success, De Anza's administrators in 2006 signed a sustainability policy, committing to green building renovation and construction and to the selection of vendors experienced with sustainability. Today a total of nine LEED-certified green buildings grace De Anza's campus, including the LEED-Platinum Media and Learning Center shown in the photo that kicks off this chapter.

The school took another major step in 2007 when it adopted the Sustainability Management Plan that CEAG had spearheaded. This plan identified environmental and health risks and helped to prioritize opportunities for addressing them. It specified six focus areas:

- Reducing solid and hazardous waste
- Conserving energy and reducing carbon emissions
- Conserving water
- Making sustainable purchasing decisions
- Pursuing ecologically responsible landscaping and maintenance
- Undertaking green building practices in construction and renovation

This plan continues to guide De Anza's greening efforts, while CEAG monitors progress and makes policy recommendations.

Challenges abound, and many today are financial as California and other states have suffered severe budget cutbacks in higher education. In tough financial times, more young people enroll in colleges, yet at the same time schools receive fewer resources. Sustainability initiatives often cost money up front, and their proponents generally need to argue their case to penny-conscious administrators time after time. But De Anza's administration, like growing numbers of others, has recognized that many of the short-term costs associated with sustainability efforts are actually investments that will save substantial amounts of money in the long term.

Despite these challenges, De Anza continues to make strides. Today, 60% of campus waste is diverted from landfills, and 90% of chemicals used by the custodial staff are environmentally friendly. The dining services use biodegradable containers and local and organic food. Students receive transit passes from the local bus service and can rent bicycles for use on and off campus. The college is gaining more and more renewable electricity from newly erected photovoltaic (PV) solar panels. De Anza's sustainability proponents have produced a policy handbook on energy efficiency and are leading workshops for other community colleges. And in 2013 students convinced directors of the school's foundation to divest from holdings in fossil fuel stocks. The Foothill–De Anza Foundation made De Anza and its sister institution, Foothill College, the first community colleges in the nation to divest from fossil fuels.

Today, thousands of students enroll in De Anza's environmental studies and environmental science courses each year. The learning extends far beyond the classroom. In one new certificate program, students are trained in wildlife tracking and help conduct research on wildlife populations and habitat corridors in the Coyote Valley and other areas in the region (**FIGURE 24.2**). The data they collect informs the work of regional planners and brings pragmatic benefits to their region as the students themselves gain hands-on experience in the science of conservation biology. In multiple ways, De Anza is providing an inspiring model for educational institutions nationwide as more and more colleges and universities strive for campus sustainability.

FIGURE 24.2 Sustainability-oriented curricula extend far beyond the classroom. In California's Coyote Valley, De Anza College instructor Ryan Phillips trains students in tracking and surveying wildlife as they help perform research into the importance of habitat corridors.

Sustainability on Campus

Whether on campus or around the world, sustainability means living in a way that can be lived far into the future. Sustainability involves conserving resources, reducing waste and pollution, and safeguarding ecological processes and ecosystem services to ensure that our society's practices can continue and our civilization can endure. Truly sustainable solutions will satisfy all three pillars of sustainability: environmental quality, economic well-being, and social justice (p. 153).

To attain a sustainable civilization, we will need to make efforts at every level, from the individual to the household to the community to the nation to the world. Governments, corporations, and organizations all must encourage and pursue sustainable practices. Among the institutions that can contribute to sustainability efforts are colleges and universities (pp. 16–17). In today's quest for sustainable solutions, students in higher education are playing a crucial role. Indeed, they are creating models for the wider world by leading sustainability initiatives on their campuses (**TABLE 24.1**, p. 652).

Student-led campus sustainability efforts are growing and thriving

All across North America and beyond, students are pioneering ways to enhance efficiency in energy use and water use on their campuses. They are running programs to recycle and cut down on waste. They are growing organic gardens and restoring natural areas on campus. They are advocating with administrators for renewable energy, sustainable green buildings, and the reduction of greenhouse gas emissions. Most efforts are home-grown and local, like the sustainable food and

dining initiatives at Kennesaw State University (pp. 209–210). Others are part of national or international programs such as Recyclemania (pp. 605–606). Altogether, today's campus sustainability efforts are diverse, numerous, and growing (**FIGURE 24.3**). These wide-ranging efforts by students, faculty, staff, and administrators are reducing the ecological footprints of college and university campuses.

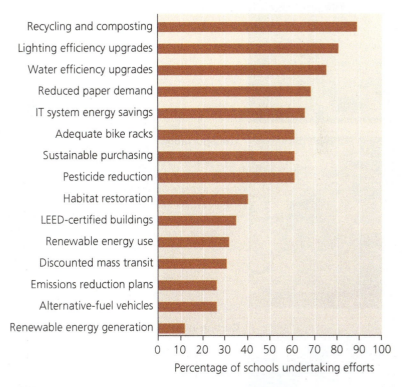

FIGURE 24.3 Campus sustainability efforts are diverse. Shown are the frequency of different pursuits at schools across North America that responded to a recent comprehensive survey. *Data from McIntosh, M., et al., 2008.* Campus environment 2008: A national report card on sustainability in higher education. *National Wildlife Federation Campus Ecology; survey conducted by Princeton Survey Research Associates International.*

TABLE 24.1 Major Approaches in Campus Sustainability

Waste reduction

Reusing, recycling, and composting offer abundant opportunities for tangible improvements, and people understand and enjoy these activities. Waste reduction events such as trash audits and recycling competitions can be fun and productive. Compost can be applied to plantings to beautify the campus. Some schools aim to become "zero-waste."

Wayne State University

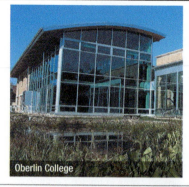

Green buildings

Constructed from sustainable materials, "green buildings" feature designs and technologies to reduce pollution, use renewable energy, and encourage efficiency in water and energy use. These sustainable buildings are certified according to LEED standards (pp. 342–344).

Oberlin College

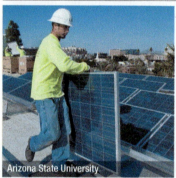

Water conservation

Indoors, schools are installing water-saving faucets, toilets, urinals, and showers in dorms and classroom buildings to reduce water waste. Outdoors, students are helping to landscape gardens that harvest rainwater and to build facilities to treat and reuse wastewater.

Loyola University Chicago

Energy efficiency

Campuses are installing energy-efficient lighting, motion detectors to shut off lights when rooms are empty, and sensors to record and display a building's energy consumption. Students are mounting campaigns to reduce thermostat settings, distribute efficient bulbs, and publicize energy-saving tips to their peers.

Xavier University of Louisiana

Renewable energy

Many schools are switching from fossil fuels to renewable heating and electricity. Some are installing solar panels; others use biomass in power plants; a few have built wind turbines on campus. Students have persuaded administrators—and have voted for student fees—to buy "green tags" or carbon offsets to fund renewable energy.

Arizona State University

Food and dining

More and more schools grow their own food on campus farms and in gardens, supplying students with local, healthy, organic food. In dining halls, trayless dining cuts down on waste (on average, 25% of food taken is wasted otherwise) because people take only what they really want. Food scraps are composted on many campuses.

Yale University

Transportation

Half the greenhouse gas emissions of the average college or university come from commuting to and from campuses in motor vehicles. To combat pollution, traffic congestion, delays, and parking shortages, schools are investing in bus and shuttle systems, hybrid and alternative-fuel fleet vehicles, and programs to promote carpooling, walking, and bicycling.

University of Virginia

Plants and landscaping

Students are helping to restore native plants and communities, remove invasive species, improve habitat for wildlife, and enhance soil and water quality. Some schools have green roofs, greenhouses, and botanical gardens. Enhancing a campus's natural environment creates healthier, more attractive surroundings, and makes for educational opportunities in ecology and natural resources.

University of Arizona

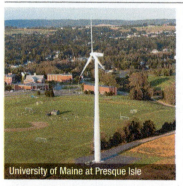

Carbon-neutrality

To combat climate change, many campuses aim to become carbon-neutral, emitting no net greenhouse gases. These schools seek to power themselves with clean renewable energy, although for most, buying carbon offsets is also necessary. Nearly 700 presidents have signed the American College and University Presidents' Climate Commitment, with carbon-neutrality as a goal.

University of Maine at Presque Isle

Fossil fuel divestment

Endowment money in U.S. higher education is estimated at more than $400 billion, and some of this is invested in stocks of coal, oil, and gas corporations. Since 2012, students have lobbied their administrators, boards of trustees, and fund managers to divest from, or sell off, stocks in fossil fuel companies.

Tufts University

An audit is a useful way to begin

Campus sustainability efforts often begin with a quantitative assessment of the institution's operations. An audit provides baseline information on what an institution is doing or how much it is consuming. Audits also help set priorities and goals. Students can conduct an audit themselves, as when University of Vermont graduate student Erika Swahn measured heating, transportation, electricity, waste, food, and water use to calculate her school's ecological footprint (it turned out to be 4.5 acres for each student, instructor, and staff member). If you set out to do an audit yourself, one useful tool is a portable device called a "Kill-A-Watt" meter (**FIGURE 24.4**), which can help measure energy use plug by plug and room by room.

Sometimes an audit can be done as part of a class, as when students at Stetson University in Florida conducted a greenhouse gas inventory of their campus. In contrast, Harford Community College in Maryland hired specialists to audit its energy use and pollutant emissions, which then served as the basis for setting reduction goals. Students at Georgian Court University in New Jersey worked with consultants from the Environmental Protection Agency to assess their campus's carbon footprint, then remeasured it after instituting changes. Their suite of actions prevented the emission of nearly 5500 tons of CO_2, as much as is emitted by 1100 cars in a year. At De Anza College, students helped conduct a comprehensive audit of various buildings on campus, describing the buildings' inefficiencies and identifying opportunities for improvement. The information from this audit was then included in De Anza's Sustainability Management Plan.

It is most useful in an audit to target items that can lead directly to specific recommendations. For instance, an audit should quantify the performance of individual appliances so that decision makers can identify particular ones to replace with higher-efficiency models. Once changes are implemented, the institution can monitor progress by comparing future measurements to the audit's baseline data.

In the absence of an audit, data from other institutions may prove helpful. For example, data show that half of collegiate greenhouse gas emissions come from commuting in

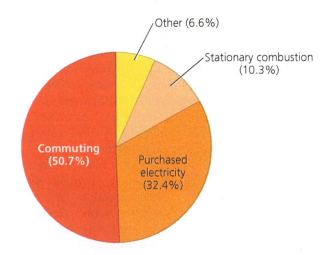

FIGURE 24.5 Commuting to and from campuses in automobiles accounts for half the greenhouse gas emissions of the average college or university. Data, measured in CO_2-equivalents, from College and University Presidents' Climate Commitment.

vehicles (**FIGURE 24.5**). This finding suggests that improving transportation options can be a promising way to lower a campus's carbon footprint.

Events and competitions make campus efforts fun

Pitching in to conserve resources is satisfying and most people enjoy doing so, but sometimes we all need a little incentive to embark on something new. This is why festive events and competitive contests have become common vehicles for many campus sustainability efforts.

Waste reduction is a popular focus for special events. Students at the University of Texas at Austin run an annual "Trash to Treasure" program. Each May, they collect 40–50 tons of items that students discard as they leave and then resell them at low prices in August to arriving students. This keeps waste out of the landfill, provides arriving students with items they need at low cost, and raises $10,000–20,000 per year that gets plowed back into campus sustainability efforts. Hamilton College in New York runs a similar program, called "Cram & Scram." It reduces Hamilton's landfill waste by 28% (about 90 tons) each May.

A friendly competition between two universities in Ohio to see which could recycle the most waste eventually led to the nationwide competition known as Recyclemania (pp. 605–606). This annual 10-week competition among schools has grown to involve several hundred institutions and millions of students each year competing across a number of categories.

A similar dynamic occurred with energy conservation and water conservation: As more schools began competitions on their own campuses, these efforts coalesced into a national program called the Campus Conservation Nationals. Students at Williams College in Massachusetts were some of the first to transform energy conservation into a kind of intramural sport. Their "Do It in the Dark" competition pitted residence halls against one another and soon produced a 13% cut in energy

FIGURE 24.4 A valuable tool for auditing energy use on campus is a Kill-a-Watt meter. This device measures the electrical current drawn by appliances and fixtures.

use as students got caught up in the fun. Connecticut College students ran a similar competition and then took 25% of the money saved by energy conservation and used it to fund a concert for the whole student body. At Denison University in Ohio, students staged a month-long "water wars" competition among dorms. The dorm that reduced its water use the most received a financial reward, which was donated to a community charity voted on by residents of the dorm.

As a natural outgrowth of the many programs being run independently on campuses across North America, the Campus Conservation Nationals grew tremendously fast, involving hundreds of schools competing to save water and energy over a three-week period. In its first five years, participants saved more than 6 million kilowatt-hours of electricity and 1.4 million gallons of water.

Some collegiate events reach out into the wider world. In a remarkable biennial event called the Solar Decathlon, run by the U.S. Department of Energy, teams of students travel to the National Mall in Washington, D.C., bringing material to build solar-powered homes they have spent months designing! The teams erect their homes on the Mall, where they are open to the public for 10 days (FIGURE 24.6). The homes are judged on various criteria, and prizes are awarded to winners in each category. In 2015, the seventh Solar Decathlon was held in Irvine, California, where 17 teams from 29 schools competed, erecting innovative homes that were admired by tens of thousands of visitors. The team from Stevens Institute of Technology took top honors. The event has inspired similar events in China and Europe.

Competitive events that boost enthusiasm for campus sustainability programs have helped these efforts proliferate among thousands of schools across North America (FIGURE 24.7). In all likelihood, your school is one!

Organizations assist campus efforts

If you choose to participate in—or initiate—a sustainability effort on your own campus, you won't be alone. Some schools now have sustainability programs and staff, and many campus efforts are supported by organizations such as the Association for the Advancement of Sustainability in Higher Education and the Campus Ecology program of the National Wildlife Federation. These organizations act as information clearinghouses for campus sustainability efforts. Each year the Campus Ecology program has recognized the most successful campus sustainability initiatives and has posted hundreds of case studies on its website. In addition, national and international conferences have grown, such as the biennial Greening of the Campus conferences at Ball State University. With the assistance of such organizations and events, it is easier than ever to kick-start sustainability efforts on your own campus. You can find links to resources for campus sustainability efforts at the end of the SELECTED SOURCES AND REFERENCES section at the back of this book.

FIGURE 24.6 In October 2015, college and university teams converged in Irvine, California, for the seventh Solar Decathlon. Each team designed and erected an entire house fully powered by solar energy.

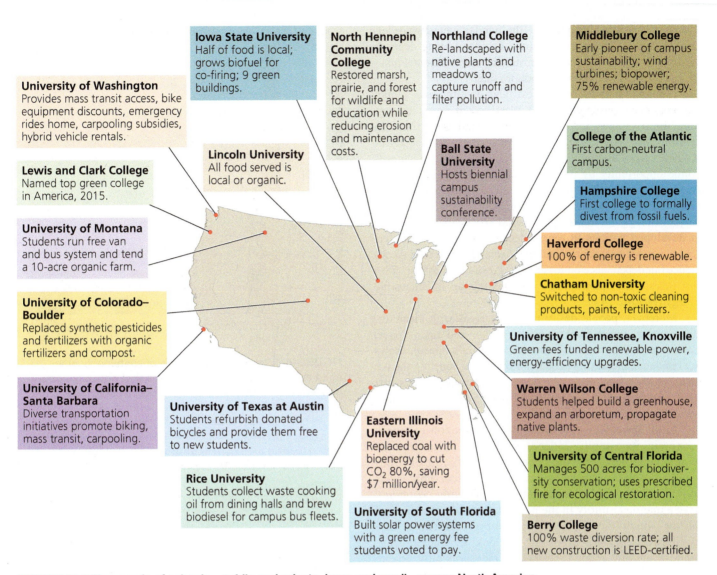

Iowa State University
Half of food is local; grows biofuel for co-firing; 9 green buildings.

North Hennepin Community College
Restored marsh, prairie, and forest for wildlife and education while reducing erosion and maintenance costs.

Northland College
Re-landscaped with native plants and meadows to capture runoff and filter pollution.

Middlebury College
Early pioneer of campus sustainability; wind turbines; biopower; 75% renewable energy.

University of Washington
Provides mass transit access, bike equipment discounts, emergency rides home, carpooling subsidies, hybrid vehicle rentals.

Lincoln University
All food served is local or organic.

Ball State University
Hosts biennial campus sustainability conference.

College of the Atlantic
First carbon-neutral campus.

Lewis and Clark College
Named top green college in America, 2015.

Hampshire College
First college to formally divest from fossil fuels.

University of Montana
Students run free van and bus system and tend a 10-acre organic farm.

Haverford College
100% of energy is renewable.

Chatham University
Switched to non-toxic cleaning products, paints, fertilizers.

University of Colorado–Boulder
Replaced synthetic pesticides and fertilizers with organic fertilizers and compost.

University of Tennessee, Knoxville
Green fees funded renewable power, energy-efficiency upgrades.

University of California–Santa Barbara
Diverse transportation initiatives promote biking, mass transit, carpooling.

University of Texas at Austin
Students refurbish donated bicycles and provide them free to new students.

Warren Wilson College
Students helped build a greenhouse, expand an arboretum, propagate native plants.

Eastern Illinois University
Replaced coal with bioenergy to cut CO_2 80%, saving $7 million/year.

University of Central Florida
Manages 500 acres for biodiversity conservation; uses prescribed fire for ecological restoration.

Rice University
Students collect waste cooking oil from dining halls and brew biodiesel for campus bus fleets.

University of South Florida
Built solar power systems with a green energy fee students voted to pay.

Berry College
100% waste diversion rate; all new construction is LEED-certified.

FIGURE 24.7 Thousands of schools—public and private, large and small—across North America have embarked on campus sustainability efforts. Shown are just a few of the leaders in recent years. Could your school be next?

Strategies for Sustainability

The pursuit of sustainable solutions on college and university campuses parallels efforts in the world at large. As more people come to appreciate Earth's limited capacity to accommodate our rising population and consumption, they are voicing concern that we will need to modify our behaviors, institutions, and technologies if we wish to sustain our civilization and the natural environment on which it depends. In the quest for sustainability, the strategies pursued on campuses can serve as models for our wider society.

Sustainable development aims to achieve a triple bottom line

As we explored sustainable development (p. 153), we considered the definition put forth by the United Nations: "development that meets the needs of the present without

compromising the ability of future generations to meet their own needs." Sustaining human institutions in a healthy and functional condition requires sustaining ecological systems in a healthy and functional condition. This is because the contributions of biodiversity (pp. 274–277) and ecosystem goods and services (pp. 4, 116, 149) are fundamental to human welfare. Yet sustainability means more than just environmental protection; it also means promoting economic well-being and social justice. Meeting this triple bottom line (p. 153) is the goal of modern sustainable development.

Environmental protection can enhance economic opportunity

Our society has long labored under the misconception that economic well-being and environmental protection are in conflict. This may have been true in frontier societies where natural resources were abundant yet populations were low and people struggled to survive. But in today's modern developed societies, the opposite is true: Our well-being depends on a healthy environment, and protecting environmental quality can improve our economic bottom line.

For individuals, businesses, and institutions, reducing resource consumption and waste often saves money. Sometimes savings accrue immediately, and other times an up-front investment brings long-term savings. For society as a whole, promoting environmental quality can enhance economic opportunity by providing new types of employment. As we transition to a more sustainable economy, some industries will decline while others spring up to take their place. As jobs in logging, mining, and manufacturing have dwindled in developed nations in recent decades, jobs have proliferated in service occupations and high-technology sectors. As we reduce our dependence on fossil fuels, green-collar jobs (p. 580) and investment opportunities are opening up in renewable energy (**FIGURE 24.8**).

Moreover, people desire to live in areas with clean air, clean water, intact forests, public parks, and open space. Environmental protection increases a region's appeal, drawing residents, increasing property values, and boosting the tax revenues that fund social services. As a result, regions that safeguard their environments tend to retain and enhance their wealth, health, and quality of life.

In all these ways, environmental protection enhances economic opportunity. Indeed, a recent U.S. government review concluded that the economic benefits of environmental regulations greatly exceed their economic costs (p. 182). Both the U.S. economy and the global economy have expanded rapidly in the past 50 years, the very period during which environmental protection measures have proliferated. What's more, if we look beyond conventional economic accounting (which measures only private

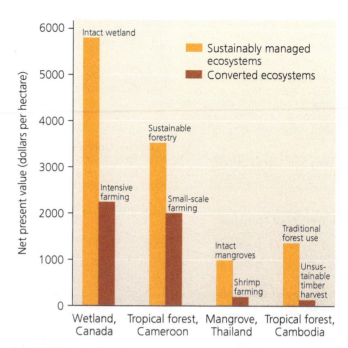

FIGURE 24.9 The economic value of sustainably managed ecosystems generally exceeds that of ecosystems converted for intensive private resource harvesting, once external costs and benefits are factored in. Shown are land values calculated by researchers in four such comparisons from sites around the world. *Adapted from Millennium Ecosystem Assessment, 2005. Ecosystems and human well-being: biodiversity synthesis. Washington, D.C.: World Resources Institute. By permission.*

DATA Q How does the value of 1 hectare of intact wetland in Canada compare to the value of 1 hectare of intensively farmed land there? Does it create more value in Thailand to convert mangrove forests for shrimp farming or to leave mangrove forests intact?

Go to **Interpreting Graphs & Data** on **Mastering**EnvironmentalScience®

economic gain and loss) and include external costs and benefits (pp. 141, 162) that affect the public at large, then environmental protection is recognized as still more valuable. Take several studies reviewed by the Millennium Ecosystem Assessment (pp. 15–16): They each show how overall economic value is maximized by conserving natural resources rather than exploiting them for short-term private gain (**FIGURE 24.9**).

We are part of our environment

On a day-to-day basis, it is easy to feel disconnected from our natural environment, particularly in industrialized nations and large cities. We live inside houses, work in shuttered buildings, travel in enclosed vehicles, and generally know little about the plants and animals around us. Millions of urban citizens have never set foot in a natural area. Just a few centuries or even decades ago, most of the world's people were able to name the plants and animals that lived nearby and describe their habits. They knew exactly where their food, water, and clothing came from. Today it seems that water comes from

FIGURE 24.8 Green-collar jobs, such as employment as a wind power technician, proliferate as we progress toward a sustainable economy.

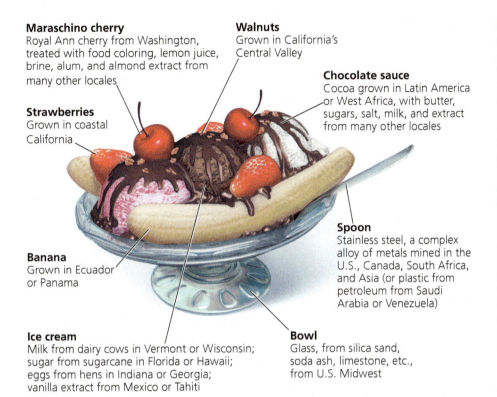

Maraschino cherry
Royal Ann cherry from Washington, treated with food coloring, lemon juice, brine, alum, and almond extract from many other locales

Strawberries
Grown in coastal California

Banana
Grown in Ecuador or Panama

Ice cream
Milk from dairy cows in Vermont or Wisconsin; sugar from sugarcane in Florida or Hawaii; eggs from hens in Indiana or Georgia; vanilla extract from Mexico or Tahiti

Walnuts
Grown in California's Central Valley

Chocolate sauce
Cocoa grown in Latin America or West Africa, with butter, sugars, salt, milk, and extract from many other locales

Spoon
Stainless steel, a complex alloy of metals mined in the U.S., Canada, South Africa, and Asia (or plastic from petroleum from Saudi Arabia or Venezuela)

Bowl
Glass, from silica sand, soda ash, limestone, etc., from U.S. Midwest

FIGURE 24.10 A banana split eaten at an ice cream shop in Tulsa, Denver, or Des Moines consists of ingredients from around the world, whose production has impacts on the environments of many locations. Ice cream requires milk from dairy cows that graze pastures or are raised in feedlots on grain grown in industrial monocultures. Ice cream is sweetened with sugar from sugar beet farms or sugarcane plantations. The banana was shipped thousands of miles by oil-fueled transport from a tropical country, where it grew on a plantation that displaced rainforest and where it likely was treated with fertilizers and fungicides. Fruits and nuts grown in California's Central Valley were irrigated generously with scarce water piped in from the Sierras. The spoon originated with metal ores mined along with thousands of tons of soil and processed into stainless steel using energy from fossil fuels.

the faucet, clothing from the mall, and food from the grocery store. It's little wonder we have lost track of the connections that tie us to our natural environment.

Yet this doesn't make those connections any less real. Consider a thoroughly un-"natural" (yet delicious!) invention of the human species: the banana split (**FIGURE 24.10**). Even in this triumph of human creation, seemingly concocted *de novo* at an ice cream shop, each and every element has ties to the resources of the natural environment, and our extraction or harvesting of each exerts environmental impacts.

Once we learn to consider where the things we use and value each day actually come from, it becomes easier to see how we are part of our environment. And once we reestablish this connection, it becomes readily apparent that our own interests are best served by preservation or responsible stewardship of the natural systems around us. What is good for our environment can also be good for people, so win-win solutions are very much within reach if we learn from what science can teach us, think creatively, and act on our ideas.

We can follow a number of strategies toward sustainable solutions

Truly lasting win-win solutions for humanity and our environment are numerous, and we have seen specific examples throughout this book. Let's now summarize 10 broad strategies or approaches that can help generate sustainable solutions (**TABLE 24.2**).

Political engagement Sustainable solutions often require policymakers to usher them through, and policymakers respond to whoever exerts influence. Corporations and

interest groups employ lobbyists to influence politicians all the time. Ordinary citizens have power as well, if they choose to exercise it. You can exercise your power at the ballot box; by attending public hearings; by donating to advocacy groups; and by writing letters, sending e-mails, and making phone calls to officeholders.

The environmental and consumer protection laws we all benefit from today came about because citizens pressured their representatives to act. The raft of legislation enacted in the 1960s through the 1980s in the United States

TABLE 24.2 Major Approaches to Sustainability

- Be politically active
- Vote with our wallets
- Pursue quality of life, not just economic growth
- Limit population growth
- Enhance local self-sufficiency, yet embrace some aspects of globalization
- Encourage and invest in green technologies
- Mimic natural systems by promoting closed-loop industrial processes
- Pursue systemic solutions
- Think in the long term
- Promote research and education

and other nations (p. 172) might never have come about had ordinary citizens not stepped up and demanded action. As we enjoy today's cleaner air, cleaner water, and greater prosperity, we owe it to future generations to engage ourselves so that they, in turn, have a better world in which to live. The words of anthropologist Margaret Mead are worth repeating: "Never doubt that a small group of thoughtful, committed people can change the world. Indeed, it's the only thing that ever has."

Consumer power Each of us also wields influence through the choices we make as consumers. When products produced sustainably are ecolabeled (p. 152), consumers can "vote with their wallets" by purchasing these products. Consumer choice has helped drive sales of everything from recycled paper to organic produce to sustainable seafood.

Quality of life It is conventional among economists and policymakers to speak of economic growth as an ultimate goal. Yet economic growth is merely a tool with which we try to attain the real goal of maximizing human happiness. Moreover, there are two ways to achieve economic growth. It can result from gains in efficiency—producing more with less—and this is a very good thing. But so far it has instead largely been driven by rising consumption of material goods and services, and thus the use of resources involved in their manufacture, transport, and sale (**FIGURE 24.11**). Advertisers are always seeking to sell us more goods more quickly, but accumulating possessions does not necessarily bring us contentment. Affluent people often fail to find happiness in their material wealth, and scientific research shows that money does not buy as much happiness as people typically believe (p. 150; and **FIGURE 24.12**).

We can enhance our quality of life by prioritizing friends, family, leisure time, and memorable experiences over material consumption. Economists and policymakers can help shift the current focus on economic growth toward

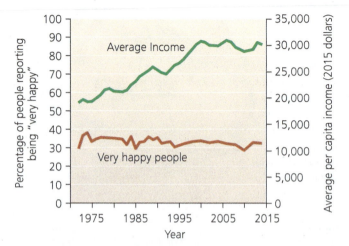

(a) Happiness versus income, through time

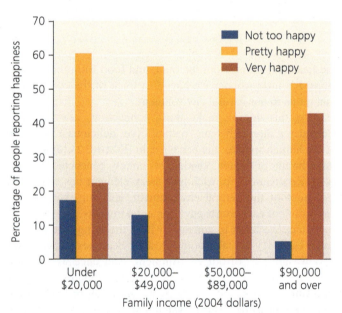

(b) Happiness versus family income

FIGURE 24.12 Money is not the only key to happiness. Although average U.S. income rose steadily in the past half-century **(a)**, the percentage of people reporting themselves as being "very happy" remained stable or declined slightly. A different study **(b)** found that Americans in higher income brackets were more likely to report themselves as "very happy" and less likely to report themselves as "not too happy." However, the difference was far less than people generally expect. For instance, when asked how much a fivefold increase in income improves a person's mood day to day, respondents guessed on average 32%, but the actual improvement, from respondents' self-reporting, was only 12%. *Data in (a) from NORC at the University of Chicago, 2015.* General Social Survey: Trends in Psychological Well-Being, 1972–2014, *Table 1; and U.S. Census Bureau, 2016.* Current Population Survey, Annual Social and Economic Supplements, *Table P-1. Data in (b) from Kahneman, D., et al., 2006. Would you be happier if you were richer? A focusing illusion.* Science 312: 1908–1910.

FIGURE 24.11 People worldwide have been consuming more and more material goods. However, consumption cannot continue rising unless we find ways to enhance the sustainability of our processes of manufacturing, packaging, transport, and waste management.

a focus on people's quality of life by incorporating external costs into market prices, introducing green taxes, eliminating harmful subsidies, and adopting full-cost accounting practices (Chapter 6). The market might then become a truly free market and a powerful tool for improving the quality of our lives.

Population stability

Just as continued growth in consumption is unsustainable, so is growth in the human population. We have used technology to increase Earth's carrying capacity for our species, but sooner or later our population, like all populations, will stop growing. The question is how: through war, plagues, and famine, or through voluntary means? Thanks to urbanization, wealth, education, and the empowerment of women, the demographic transition (p. 197) is already well advanced in many developed nations. If today's developing nations also pass through the demographic transition, then humanity may be able to rein in population growth while creating a more prosperous and equitable society.

Local and global approaches

Encouraging local self-sufficiency is one important element of building sustainable societies. When people feel closely tied to the area in which they live, they tend to value the area and seek to sustain its environment and its human community. Moreover, relying on locally made products cuts down on fossil fuel use from long-distance transportation. This argument is frequently made in relation to the cultivation and distribution of food (p. 262; and FIGURE 24.13a).

Many advocates of local self-sufficiency criticize globalization. They are troubled by the homogenization of the world's societies, in which traditional ways of life in many areas are being abandoned as people take up the material, commercial, and cultural trappings of a few dominant societies, particularly those of the United States.

The growing power of large multinational corporations is a key driver of these trends. In today's globalizing world, multinational corporations have attained greater and greater power over global trade while governments retain less and less. Critics of globalization consider corporations more likely than governments to promote high-consumption lifestyles and unlikely to support protection for workers and the environment, so they feel that corporate-driven globalization is hindering progress toward sustainability.

However, globalization is a broad and complicated phenomenon with diverse consequences. One very positive aspect is the way that people of the world's diverse cultures are increasingly communicating and learning about one another. Immigration, air travel, books, television, and the Internet have made us more aware of one another's cultures and more likely to respect and celebrate, rather than fear, differences among cultures (FIGURE 24.13b).

Moreover, globalization may foster sustainability because Western democracy, as imperfect as it is, serves as a model and a beacon for people living under repressive governments. Open societies allow for entrepreneurship and the flowering of creativity in business, art, science, and education. Millions of free minds thinking about issues are more likely to come up with sustainable solutions than the minds of a few holding authoritarian power.

weighing the ISSUES

Globalization

From your own experience, what advantages and disadvantages do you see in globalization? Have you personally benefited or been harmed by it in any way? In what ways might promoting local self-sufficiency be helpful for the pursuit of global sustainability? In what ways might it not?

(a) Local food at a farmers' market

(b) International cuisine at a Thai restaurant

FIGURE 24.13 Local and global approaches each offer advantages. Local food advocates celebrate the benefits of locally grown produce **(a)**. Farmers' markets and community-supported agriculture keep us aware of where our food is coming from, often reduce fossil fuel use, and allow us to invest in the economic health of our own region. At the same time, globalization enriches our lives in many ways. As a result of growing connections between Asia and America, Thai food has become wildly popular in the United States **(b)**, adding to the rich diversity of food here. Interestingly, many ingredients in Thai food—like those of the banana split in Figure 24.10—came originally from other parts of the world. In fact, the chili peppers so vital to modern Thai cuisine came from the Americas in an earlier wave of globalization centuries ago!

Green technologies

Technology—developed with the agricultural revolution, the industrial revolution, and advances in medicine and health—has facilitated our population growth, and it has magnified our impacts on Earth's environmental systems. Yet technology can also give us ways to reduce our impact. The *IPAT* equation (p. 189) summarizes human environmental impact (*I*) as the interaction of population (*P*), consumption or affluence (*A*), and technology (*T*). Technology can represent either a positive or a negative value in this equation. The shortsighted use of technology often gets us into a mess, but wiser use of environmentally friendly, or "green," technologies can help get us out.

In recent years, we have intensified environmental impacts in developing countries by exporting industrial technologies from the developed world to poorer nations eager to industrialize. Meanwhile, in developed nations, we have begun using green technologies to mitigate our impacts. Catalytic converters on cars have reduced emissions (see Figure 17.16, p. 456), as have scrubbers on smokestacks (see Figure 17.15, p. 456). Recycling technology and wastewater treatment are reducing our waste output. Solar, wind, and geothermal energy technologies are producing cleaner, renewable energy. Technological advances such as these help explain why people of the United States and western Europe today enjoy cleaner environments—although they consume far more—than people of eastern Europe or rapidly industrializing nations such as China. In turn, nations such as China today are racing to produce and install green technology themselves.

Mimicking nature

As industries seek to develop green technologies and sustainable practices, they have an excellent model: nature itself. Environmental systems tend to operate in cycles featuring feedback loops and the circular flow of materials. In natural systems, output is recycled into input. In contrast, our manufacturing processes have run on a linear model in which raw materials are input and processed to create products while by-products and waste are simply discarded. Some forward-thinking industrialists are now transforming linear pathways into circular ones, in which waste is recycled and reused (p. 619). For instance, several companies now produce carpets that can be retrieved from the consumer when they wear out, so the materials can be recycled to create new carpeting (p. 620). Some automobile manufacturers are planning cars that can be disassembled and recycled into new cars. Proponents of this industrial model see little reason why virtually all products cannot be recycled, given the right technology. Their ultimate vision is to create truly closed-loop industrial processes, generating no waste.

Systemic solutions

There are two general ways to respond to a problem. One is to take a symptomatic approach by addressing the symptoms of the problem. The other is to take a systemic approach by viewing the whole system and addressing the root cause of the problem. Addressing symptoms one by one as they appear is often easier in the short term, but generally it is not effective in resolving the problem. For instance, when pests evolve resistance to a pesticide (p. 247), we generally call on chemists to develop more potent pesticides. This addresses a symptom but does not resolve the overall problem, because pests will likely evolve resistance to the new chemical as well. A systemic solution would be to develop agricultural approaches that rely less heavily on chemical pesticides. Similarly, as we deplete easily accessible fossil fuel deposits, we have been choosing to reach further for new fossil fuel sources, even as environmental and health repercussions mount. A systemic solution to our energy demands would involve developing clean renewable energy sources instead. For many issues we face, it will prove worthwhile to pursue systemic solutions.

Long-term perspective

To be sustainable, a solution must work in the long term (**FIGURE 24.14**). Often the best long-term solution is not the best short-term solution, which explains why much of what we currently do is not sustainable. Corporate leaders often pursue short-term profits to please shareholders, and policymakers in democracies often act to produce near-term results that will help them be reelected. Yet many environmental dilemmas are cumulative, worsen gradually, and can be resolved only over long periods. Often the costs of addressing an environmental problem are short term, whereas the benefits are long term. In such a situation, citizen pressure on policymakers is especially vital.

Research and education

Finally, we each can magnify our own influence by educating others and by serving as role models through our actions. The campus sustainability efforts at De Anza College and so many other colleges and universities accomplish both approaches. The discipline of environmental science plays a key role in providing information that people can use to make wise decisions about a wide diversity of issues. By promoting scientific research and by educating the public about environmental science, we can all assist in the pursuit of sustainable solutions.

FIGURE 24.14 Sustainable solutions require long-term thinking. These students in India are caring for tree seedlings at their school's nursery and, by planting them on eroded hillsides, are helping to reforest the valley in which they live. In doing so, they are investing in their own future.

Precious Time

By shifting our conventional patterns of thinking and by following the approaches outlined above, we can bring sustainable solutions within reach. However, time is getting short; the natural systems we depend on are changing quickly, and human impacts continue to intensify. Deforestation, overfishing, wetland loss, and resource extraction are but a few examples. Indeed, today we are modifying the planet's climate itself. The many impacts of global climate change are beginning to influence virtually everything else in our environment and our society—and the impacts are accelerating. Even if we can visualize sustainable solutions to these many challenges, how can we find the time to implement them before we do irreparable damage to our environment and our future?

We need to reach again for the moon

In 1961, U.S. President John F. Kennedy announced that within a decade the United States would be "landing a man on the moon and returning him safely to the Earth." It was a bold and astonishing statement; the technology to achieve this unprecedented, almost unimaginable, feat did not yet exist. Kennedy's directive had powerful motivation behind it, however. The United States was dueling with the Soviet Union in the Cold War. In this competition for global hegemony, the two nations, held mutually at bay with nuclear weapons, tried to prove their mettle by other means, and the race for dominance in space became the centerpiece of the rivalry. The prospect of "losing" the space race prompted U.S. leaders to set a national goal on an ambitious timeline. Congress supplied funding, NASA performed the science and engineering, and in 1969—a short eight years after Kennedy's initial proclamation—astronauts walked on the moon.

America accomplished this historic milestone by harnessing public enthusiasm for a goal and giving its scientists and engineers the support they needed to meet that goal. Similarly great challenges were met when the United States confronted the Great Depression, when it threw itself into World War II, and when it conducted the Marshall Plan after that war to help rebuild Western Europe.

Today humanity faces a challenge more important than any previous one. Attaining sustainability is a larger and more complex process than traveling to the moon. However, it is one to which every person on Earth can contribute; in which government, industry, and citizens can cooperate; and toward which all nations can work together. If America was able to reach the moon in a mere eight years, then certainly humanity can begin down the road to sustainability with comparable speed. Human ingenuity can do it; we merely need to rally public resolve and engage our governments, institutions, and entrepreneurs in the race.

We must think of Earth as an island

We began this book with the vision of Earth as an island, and indeed it is (**FIGURE 24.15**). Islands can be paradise, as Easter Island (pp. 8–9) likely was when early Polynesians first reached it. Yet when Europeans arrived centuries later, they witnessed the aftermath of a civilization that had collapsed once its island's resources were depleted and its environment degraded.

It would be tragic folly to let such a fate befall our planet as a whole. By recognizing this fact, by shifting our behavior and our cultural institutions in ways that encourage sustainable practices, and by employing science to help us achieve these ends, we may yet be able to live happily and sustainably on our wondrous island, Earth.

FIGURE 24.15 This photo of Earth, taken by astronauts orbiting the moon, shows our planet as it truly is—an island in space. Everything we know, need, love, and value comes from and resides on this small sphere, so we had best treat it well.

The notion of sustainability has run throughout this book. In one case after another, we have seen how people are devising creative solutions to the dilemmas that arise when resources for future generations are being depleted. As a student, you can choose to play a key role in our transition to a sustainable society. Efforts at institutions of higher education—at De Anza College and at hundreds of campuses across North America—are advancing new ideas and are demonstrating how sound principles from environmental science can be transformed into effective sustainable practices. These efforts at so many educational institutions are already having positive impacts across our broader society.

Moreover, our world today benefits from many thousands of scientists who study Earth's processes and resources. As a result, we are amassing a detailed knowledge and an ever-developing understanding of our dynamic planet, what it offers us, and what impacts it can bear. It remains vital for our global society today, our one-world island of humanity, to support scientific research so that we can identify concerns and design ways to address them. In this way, environmental science, this study of Earth and of ourselves, offers us hope for our future. By applying what we learn from environmental science, our society can continue developing innovative and workable solutions that enhance our quality of life while protecting and restoring the natural environment that supports us.

REVIEWING Objectives

You should now be able to:

Describe approaches being taken on college and university campuses to promote sustainability

Students, faculty, staff, and administrators are pursuing all kinds of ways to reduce material waste; conserve energy and water; grow and serve sustainable food; construct green buildings; encourage sustainable transportation; restore plants, habitat, and landscapes; and address climate change concerns by promoting renewable energy, divesting from fossil fuels, and seeking carbon-neutrality. Audits offer a helpful first step by producing baseline data on how much a campus consumes and pollutes, while competitive events can boost enthusiasm and organizations can help provide guidance. (pp. 651–655)

Explain the concept of sustainable development and discuss how environmental protection can enhance economic well-being

Sustainable development entails environmental protection, economic development, and social justice. Proponents of sustainable development believe that economic opportunity and environmental quality go hand-in-hand. Environmental protection and green technologies and industries can create rich sources of new jobs, while safeguarding environmental quality enhances a community's desirability and economic well-being. (pp. 655–657)

Assess key approaches to designing sustainable solutions

Ten general strategies and actions that can be taken by individuals, businesses, governments, and communities can inspire specific sustainable solutions: (1) getting engaged politically, (2) harnessing consumer power, (3) prioritizing quality of life, (4) halting population growth, (5) deploying green technologies, (6) mimicking nature in our technology and practices, (7) weighing the pros and cons of local and global approaches, (8) striving for systemic solutions, (9) taking a long-term perspective, and (10) supporting research and education. (pp. 657–660)

Explain how time is limited yet human potential to solve problems is tremendous

Time for turning around our environmental impacts is running short. Yet the United States and other nations have met tremendous challenges before, so we have reason to hope that we will be able to achieve the ultimate challenge of attaining a sustainable society. (p. 661)

TESTING Your Comprehension

1. In what ways are campus sustainability efforts relevant to sustainability efforts in our broader society?

2. Describe one way in which campus sustainability proponents have addressed each of the following issues: (1) recycling and waste reduction, (2) "green" building, and (3) water conservation.

3. Describe one way in which campus sustainability proponents have addressed each of the following issues: (1) energy efficiency, (2) renewable energy, and (3) global climate change.

4. Describe one way in which campus sustainability proponents have addressed each of the following issues: (1) dining services, (2) transportation, and (3) habitat restoration.

5. What do environmental scientists mean by sustainable development?

6. Describe three ways in which environmental protection can enhance economic well-being.

7. Why are many people now living at the highest level of material prosperity in history? Is this level of consumption sustainable? How can we consume less while improving our quality of life?

8. In what ways can technology help us achieve sustainability? How do natural processes provide good models of sustainability for manufacturing? Provide examples.

9. Why do many people feel that local self-sufficiency is important? What consequences of globalization may threaten sustainability? How can open democratic societies help to promote sustainability?

10. How can thinking of Earth as an island help us in the quest for a sustainable future?

SEEKING Solutions

1. What sustainability initiatives are being pursued on your campus? What additional such initiatives would you like to see attempted? If you were to take the lead in promoting these activities on your own campus, how would you go about it? What obstacles would you expect to face, and how would you deal with them?

2. Choose one item or product that you enjoy, and consider how it came to be. Think of as many components of the item or product as you can, and determine how each of them was obtained or created. Now refer to Figure 24.10 as a guide. What steps were involved in creating your item's components, and where did the raw materials come from? How was your item manufactured? How was it delivered to you?

3. Do you think that we can increase our quality of life through development while also protecting the integrity of the environment? Discuss examples from your course or from other chapters of this book that illustrate possible win-win solutions. Are you familiar with any cases in your community or at your college that bear on this issue? Describe such a case, and state what lessons you would draw from it.

4. CASE STUDY CONNECTION You have been elected president of your college class, and your school's administrators promise to be responsive to student concerns. Many of your fellow students are asking you to promote sustainability initiatives on your campus. Consider the many approaches and activities pursued by De Anza College and by the many other colleges and universities mentioned in this chapter, and now think about your own school. Which of these approaches and activities are most needed at your school? Which might be most effective? What ideas would you prioritize and promote during your term as president of your class?

5. THINK IT THROUGH Think back across everything we have covered in this book and everything you have learned in your course. Of all the environmental challenges our planet and our society face, which one concerns you the most? Describe how people have tried to address this particular problem so far, and the degree to which they've succeeded. Now describe at least one way in which you think people should address this problem that has not yet been tried.

6. THINK IT THROUGH In this chapter and throughout this book, you have encountered a diversity of ideas for sustainable solutions to environmental problems. Many of these are approaches you can pursue in your own life. Name at least three ways in which you think you can make a difference—and would most like to make a difference—in helping to attain a more sustainable society. For each approach, describe one specific thing you could do today or tomorrow or next week to begin.

CALCULATING Ecological Footprints

Individuals can contribute to sustainable solutions for our society and our planet in many ways. Some of these involve advocating for change at high levels of government or business or academia. But plenty of others involve the countless small choices we make in how we live our lives day to day. Where we live, what we buy, how we travel—these types of choices we each make as citizens, consumers, and human beings determine how we affect the environment and the people around us. As you know, such personal choices are summarized (crudely, but usefully) in an ecological footprint.

Turn back to the first *Calculating Ecological Footprints* exercise (Chapter 1, p. 19), and recover the numerical value of your own personal ecological footprint that you calculated at the beginning of your course. Enter it in the table. Now return to the same online ecological footprint calculator that you used at that time. For many of you, this will have been http://www.footprintnetwork.org/en/index.php/GFN/page/personal_footprint. Take the footprint quiz again, and calculate your current footprint.

	FOOTPRINT VALUE (hectares per person)
World average	2.8
U.S. average	8.2
Your original footprint (from Chapter 1)	
Your footprint now	
Your footprint with three more changes	

1. Enter your current footprint, as determined by the online calculator, in the table. If you calculated your footprint at the beginning of the course, how does this value compare? By what percentage did your footprint decrease or increase? If it changed, why do you think it changed? What changes have you made in your lifestyle since beginning this course that influence your environmental impact?

2. How does your personal footprint compare to the average footprint of a U.S. resident? How does it compare to that of the average person in the world? What do you think would be an admirable yet realistic goal for you to set as a target value for your own footprint?

3. Now think of three changes in your lifestyle that would lower your footprint. These should be changes that you would like to make and that you believe you could reasonably make. Take the footprint quiz again, incorporating these three changes. Enter the resulting footprint in the table.

4. Now set as a goal reducing your footprint by 25% and experiment by changing various answers in your footprint quiz. What changes would allow you to attain a 25% reduction in your footprint? What changes would be needed to reduce your footprint to the hypothetical target value you set in Question 2?

MasteringEnvironmentalScience®

Answers to Data Analysis Questions

Chapter 1

Fig. 1.3 The graph shows that nearly 1 billion people were alive in 1800, whereas over 7 billion are alive today. Thus, for every person alive in 1800, there are more than 7 alive today.

Fig. 1.5 The global ecological footprint today is roughly 1.64 planet Earths. The global ecological footprint half a century ago (1961) was roughly 0.73 planet Earth. This makes for a difference of 0.91 planet Earth, and it means that today's footprint is more than twice the size of the footprint half a century ago.

Fig. 1.10 In part (**a**), time in weeks is shown on the *x* axis, the horizontal axis. In part (**b**), the dependent variable is the percentage of pond surface area covered by algae, because this is on the *y* axis and depends on whether ponds were fertilized. In part (**c**), the data show a positive correlation, with values of pond cover increasing along with increases in fertilizer application. In part (**d**), Species #1 is most numerous because it has the largest percentage and the largest pie slice, whereas Species #5 is least numerous. The thin black lines in part (b) are called *error bars*. They indicate the amount of variation the data show around the mean, or average, value (which is indicated by the height of each colored bar).

Fig. 1.12 Of the nations shown in the figure, the United States and Canada each have the largest per capita footprint, and Haiti has the smallest per capita footprint. The U.S. and Canadian footprints (8.2 ha) are 13.7 times larger than Haiti's footprint (0.6 ha), because 8.2 divided by 0.6 equals 13.7.

Chapter 2

Fig. 2.18 Comparison of the two figures reveals that this belt of intense earthquake and volcanic activity corresponds closely to the subduction zones at the boundaries of the tectonic plates that surround the Pacific Ocean. As shown in Figure 2.13, convergent plate boundaries dominate the length of the ring of fire. Note that other locations that experience earthquakes and volcanic activity—such as Indonesia, Iran, Turkey, and southern Italy—are similarly located along convergent plate boundaries.

2-SBS1 Fig. 2 Drawing a straight line through the clusters of points from 1960 to 2010 produces a downward-sloping line that shows a decreasing trend

in ocean radioactivity from cesium-137 over time. (This is because input of radioactive cesium into the ocean was reduced once nations stopped conducting nuclear weapons tests in the atmosphere.) Inputs of cesium-137 into the oceans from the Chernobyl accident did not significantly alter this downward trend, as radioactivity levels continued to decline following this "pulse" of increased radioactivity. Similar dilution of radioactivity over time of the radioactive material released from the Fukushima Daiichi accident would therefore be expected, leading to continued declines in radioactivity from cesium-137 in coming decades.

2-SBS2 Fig. 2 If you hold a ruler over the line on the graph that runs from 1973 to 2008 and follow it to 2015 on the graph, you see that the predicted number of earthquakes for 2015 would have been around 1000, were it not for the rapid increase seen starting in 2009. To determine the percentage difference between the predicted number of earthquakes in 2015 and the actual number (around 2750), you first subtract 1000 from 2750 to determine that there were about 1750 more earthquakes in 2015 than would have been predicted. Next, you divide the difference in predicted and actual earthquakes (1750) by the number of expected earthquakes (1000), and then multiply the answer by 100 to convert it to a percentage (1750/1000 × 100). Doing so reveals that there were 175% more earthquakes in 2015 than would have been predicted based on historical data.

Chapter 3

Fig. 3.7 All vertebrate groups shown, except for lampreys, are shown branching off "after" (to the right of) the hash mark for jaws. This indicates that lampreys diverged before jaws originated (and thus lack them), whereas all other vertebrate groups possess jaws. Birds are more closely related to crocodiles than to amphibians. We can tell this because birds and crocodiles share a recent common ancestor, having diverged from this ancestor much more recently than the lineages that lead to birds and to amphibians diverged in the tree.

Fig. 3.18 Because exponential growth cannot last forever, we would expect that growth of the western U.S. population of Eurasian collared doves will eventually slow down and that the population will reach carrying capacity. Thus the population growth graph for the western United States would come to have a shape more like the current graph for Florida, showing logistic growth.

Fig. 3.19 A human being has the highest rate of survival at a young age. As shown by the Type I survivorship curve, the vast majority of individuals survive during youth, and then mortality rates rise at older ages.

Chapter 3

3-SBS2 Fig. 1 Because the 21-year trend is upward and the trend of the past 9 years is downward, we can expect that the trend during the first 12 years will be upward at a steeper rate of increase than across the entire 21-year period. Running the data through regression analyses is the only way to plot a perfectly accurate trend line, but the line you drew just by eyeballing the data should show an upward trend and probably shows a slightly steeper rate of increase than the 21-year line.

Chapter 4

Fig. 4.10 In the generalized example shown, there are 100 grasshoppers for every 10 rodents; therefore, we can say there are 1/10 (or 1:10) as many rodents as grasshoppers. Thus, if there were 3000 grasshoppers, we would expect 300 rodents (300/3000, a 1:10 ratio). Likewise, in the example shown there is 1 hawk for every 10 rodents—again, a 1:10 ratio. Thus, if there were 300 rodents there would be 30 hawks.

Fig. 4.14 The first difference the map indicates is that zebra mussel occurrences are more numerous than quagga mussel occurrences. One hypothesis for this difference could be that zebra mussels are more successful invaders. Another hypothesis could be that because they arrived earlier, they have simply had more time to spread to more locations. The second difference shown by the map is that zebra mussels have spread to the lower Mississippi River Valley but not much to the West, whereas quagga mussels have spread to the West but not to the lower Mississippi River Valley. Many hypotheses could be offered. One might be that each type of mussel spread to a region where conditions happened to best suit it. Another might be that zebra mussels spread more effectively along river systems, while quagga mussels spread more effectively by cross-country transport on boat hulls.

Fig. 4.19 Note that the temperature curve is above the precipitation curve during July to September. High temperatures lead to increased evaporation. Thus, even though precipitation is roughly average at this time, the warm temperatures enhance evaporation and thereby cause dry conditions.

4-SBS1 Fig. 3 In 1991, nauplii and rotifers had densities of 300–500 per square meter, whereas zebra mussels were at nearly zero. In 2000, zebra mussels were at nearly 300/m², while nauplii and rotifers were close to zero. In 2008, nauplii and rotifers were again at 300–500/m², while zebra mussels had returned to near zero. These data indicate that nauplii and rotifer densities crashed as zebra mussel densities became high, but that eventually nauplii and rotifer populations recovered as zebra mussel densities declined. There appears to be a negative correlation between the densities of zebra mussels and the densities of nauplii and rotifers. The pattern can be explained by the likelihood that zebra mussels have harmful effects on these two types of zooplankton. Thus, when zebra mussels are abundant, the zooplankton are not abundant, and when zebra mussels become less numerous, zooplankton become more numerous.

4-SBS2 Fig. 2 The mudflow substrate gained the highest species richness, as seen by the fact that its curve is highest on the graph in part **(a)**. The maximum number of species it had in any one year was 22 or 23, about 17–18 years after the eruption. The pumice substrate showed the slowest increase in percentage plant cover, as seen by the fact that its curve is the lowest in the graph in part **(b)**.

Chapter 5

Fig. 5.6 The width of the arrows in each figure represents magnitude, so the wider the arrow, the larger the value. For chemical energy, the widest arrow goes from producers to detritus. This means the largest directional flux of chemical energy in the ecosystem is from producers to detritus. For nutrients, a comparison of arrow widths shows the flux from detritus to producers to be, by far, the largest.

Fig. 5.8 The nation of India stands out prominently on the map as a region with a high human footprint but comparatively few marine dead zones. As dead zones are heavily influenced by inputs of nitrogen and phosphorus from agriculture, one explanation is that India experiences fewer offshore dead zones because farmers there apply less synthetic fertilizer to their land than farmers in North America, Europe, and Australia. This is largely financial in nature, because synthetic fertilizers are relatively expensive for many farmers in industrializing nations. However, the large amount of land cleared for agriculture and development, coupled with extensive impacts from air pollution, give India a very high human footprint value.

Fig. 5.22 Reducing nitrogen inputs into the Chesapeake Bay through enhanced nutrient management programs costs $21.90 per pound, whereas a 1-lb reduction from forest buffers costs only $3.10. Dividing $21.90 per pound by $3.10 per pound shows us that for the same price, we could keep about 7 lb of nitrogen out of waterways by using forested buffers versus 1 lb of nitrogen by using nutrient management programs.

Chapter 6

Fig. 6.8 The arrows show the directions in which items are moving. When you work at a job, you give labor and you receive wages. When you buy a product, you pay money for that product and you receive the product in terms of goods or services. The environment provides to the economy both ecosystem goods (natural resources) and ecosystem services (such as waste acceptance). Some ecosystem services (such as climate regulation, nutrient cycling, air and water purification) act as natural recycling processes, helping to repurpose waste materials into the creation of new resources.

Fig. 6.12 Resources decline to near zero in **(a)** but stabilize at a moderate level in **(b)**. Population declines after a peak in **(a)** but stabilizes without falling in **(b)**. Food supplies drop sharply after a peak in **(a)** but stabilize at a high level in **(b)**. Industrial output crashes after a high peak in **(a)** but stabilizes at a moderately high level in **(b)**. Pollution increases exponentially in **(a)** but declines after a low peak in

(b). If a third graph were drawn showing faster, more intensive resource use, then we would expect that, in comparison to **(a)**, resources would decline more steeply; population, food, and industrial output would each peak earlier and drop more steeply; and pollution would rise sooner and more steeply.

Fig. 6.14 The three ecosystem services that provide the most benefits in dollar terms are shown by the three longest bars on the bar chart: treating waste ($22.6 trillion/yr), enabling recreation ($20.6 trillion/yr), and controlling erosion ($16.2 trillion/yr).

Fig. 6.16 In 1950, GDP was just under $12,000 per capita and GPI was just over $8000 per capita, making for a ratio of nearly 1.5. In 2004, GDP was nearly $37,000 per capita and GPI was about $15,000 per capita, making for a ratio of nearly 2.5. For most people, the ratio for the year they were born will be between these two values. Together these changes indicate that GDP has been growing faster than GPI.

Chapter 7

Fig. 7.6 After 2008, the percentage of wells documented as having integrity problems rose dramatically, then dropped. There are two potential explanations. One is that wells drilled after 2008 truly did have more structural problems, resulting in greater detection of problems by DEP inspectors. Another possibility, however, is that DEP inspectors became more attentive or more aggressive about noticing and citing violations. The latter possibility seems more likely, because the DEP felt pressure to be better watchdogs on the gas industry after scientific studies and media stories around this time began calling attention to possible water pollution as a result of fracking. The subsequent drop from 2011 to 2012 could be due either to industry improvements or to lax inspection by the DEP.

Fig. 7.14 Oil, coal, and natural gas have together received about $594 billion, while renewable energy (excluding hydropower) received about $81 billion in the United States during the period covered by the graph. If we divide 594 by 81, we get 7.33. Thus, about $7.33 has been spent on fossil fuel subsidies for every $1.00 that has gone to renewable energy.

7-SBS Fig. 1 Based on the data in the graph, we can predict that water in the well that is 250 m from the drilling site would be more likely to contain methane at high levels. All the existing data points in the vicinity of 5000 m are near zero, suggesting that well water 5000 m from a drilling site would be unlikely to contain methane. In contrast, wells around 250 m from drilling sites show values ranging from zero to 70 mg/L.

Chapter 8

Fig. 8.4 By examining the key that links colors on the map to growth rates, we see that red indicates the highest growth rates and that Africa has the highest overall growth rate of any region. Europe has the lowest growth rate of any region, as evidenced by its many nations with very low or negative growth rates.

Fig. 8.15 The transitional stage has the greatest growth of any stage in the demographic transition, because that is the period when birth and death rates are far apart and population increase is substantial. In the transitional stage, growth is the greatest at the end, when the difference between birth and death rates is far greater than at the stage's beginning.

Fig. 8.17 The best approach to answer a question such as this one is to draw a "best-fit" line through the points on the figure that minimizes the distance between each point and the line you draw. Doing this produces a line that slopes downward from left to right, suggesting a negative relationship between total fertility rate and the rate of enrollment of girls in secondary school. This relationship makes sense—one would expect that as more girls pursue education, they delay childbirth and reduce the nation's TFR.

Fig. 8.19 Africa will add about 1.3 billion people to the global population by 2050, more than the roughly 1 billion people added by Asia. Africa will also increase by the largest percentage, some 108%. This value is calculated by dividing the number of people added to Africa's population (1.3 billion) by its 2015 population (1.2 billion) and then multiplying by 100 to convert the resulting proportion to a percentage: (1.3/1.2) × 100 = 108%. Possible explanations for Africa growing the fastest of any world region would include relatively lower levels of women's rights than other regions, significantly less contraceptive use than other regions, and the lowest per capita income of any world region. Because all of these factors are correlated with high fertility, it follows that Africa's population growth will likely surpass that of other regions in coming decades.

Chapter 9

Fig. 9.5 By tracing the lines for 20% clay, 60% silt, and 20% sand inward, one arrives at a location within the region described as "Silt loam." Thus, a soil of this texture would be classified as silt loam.

Fig. 9.15 The erosion rate from conservation tillage is about 0.125 mm/yr. The erosion rate from conventional agriculture is about 3.9 mm/yr. Dividing 3.9 by 0.125 equals 31.2; therefore, the erosion rate from conservation tillage is about 31 times less than that from conventional agriculture. Likewise, dividing 0.125 by 3.9 equals 0.032; therefore, the erosion rate from conservation tillage is roughly 3.2% of that from conventional agriculture.

Fig. 9.22 In a figure like this, arrows point from causes to consequences. Thus, once you find the box "Exposes bare topsoil," you see that the box with the arrow pointing to it reads "Removes native grass." Thus, the removal of native grass is the immediate cause of exposing bare topsoil. Likewise, an arrow points from the "Exposing bare topsoil" box to the box marked "Wind and water erosion." Thus, wind and water erosion are the immediate consequences of exposure of bare topsoil. In the figure, four immediate consequences of wind and water erosion are shown with arrows: (1) compaction of soil and damage to soil structure, (2) invasive species gaining a foothold and outcompeting native species, (3) a decrease in grass growth and survival, and (4) removal of native grass.

Chapter 10

Fig. 10.2 Overall growth of the human population provides the answer. Between the two periods specified, our global population increased by several hundred million people, with the vast majority of this growth occurring in developing nations. Thus, although the absolute number of undernourished people in the developing world rose slightly, a great many people were added to the total population, so the percentage of people who were undernourished still fell.

Fig. 10.9 Beef requires 17.5 times more land to produce than chicken ($245\ m^2 / 14\ m^2 = 17.5$). Beef requires 15 times more water to produce than chicken ($750\ kg / 50\ kg = 15$). Beef releases 8.6 times more greenhouse gases than chicken ($342\ kg / 40\ kg = 8.6$).

Fig. 10.23 In the past 5 years, GM crops have been growing faster in developing nations, as indicated by the fact that the curve for developing nations rises upward more steeply than the curve for industrialized nations. In 2012, developing nations for the first time produced more GM crops than industrialized nations. If current trends continue, developing nations should be growing more GM crops than industrialized nations in 2020. To estimate how much more GM crops they might be growing in 2020, we can extend the two trend lines forward 5 years, keeping their slopes the same as in the preceding years. If we base our predictions for 2020 on the slopes in the data lines over the past 5 years (2010–2015), then this leads *approximately* to values of 120 million ha for developing nations and 85 million ha for industrialized nations. However, the way we project the trend lines forward into the future depends on how many years we choose to reach back through in the past. Note that the rate of adoption of GM crops has been slowing down in the most recent years. As a result, using data from only the most recent past years will lead to lower projections for 2020, whereas using a data from longer period of years into the past will lead to higher projections for 2020.

Fig. 10.29 About 5% of the scientific results fell in the interval 40–50%, as indicated by the height of the bar for that particular interval. To determine what percentage of results showed organic yields higher than conventional yields, we would sum up the heights of the bars for all intervals above 100%. Doing this adds up to a value of almost 12%.

Chapter 11

Fig. 11.4 The rightmost pie chart in the figure shows that there are 5900 species of mammals. The middle pie chart shows that there are about 65,000 species of vertebrates. Because 5900 is 0.091 of 65,000, this means that mammals make up about 9.1% of all vertebrates. The leftmost pie chart shows that there are about 1,552,000 known and described species of animals. Because 5900 is 0.0038 of 1,552,000, this means that mammals make up just 0.38% of all animals. One can add up the numbers in the leftmost pie chart to find that there are about 2,118,000 likely known species of organisms in total. Because 5900 is 0.28% of 2,118,000, this means that mammals make up just 0.28% of all organisms (or about 1 out of 350). In reality, the percentage is actually much lower, because virtually all mammal species have already been discovered, yet most species of other types of organisms have not yet been discovered. Finally, the center pie chart shows 1,014,000 insect species; therefore there are 1,014,000/5900 = 172 insect species for every mammal species.

Fig. 11.16 In the winter of 2015–2016, monarchs occupied just 4.01 hectares, whereas in 1994–1995 (the first year of data), they occupied 7.81 ha. Thus in 2015–2016 they occupied just 51.3% (4.01/7.81) of their original area. Compared with the year of greatest area occupied (18.19 ha in 1996–1997), monarchs in 2015–2016 occupied just 22.0% of that area (4.01/18.19).

Fig. 11.17 The bar for pollution stretches to a value of nearly 1200 species, second only to habitat loss; this indicates that pollution is the second greatest cause of amphibian declines overall. For threatened species alone, we need to look at the red portions of the bars. Comparing red portions of the bars, we can see that habitat loss is the primary cause of declines for threatened species of amphibians. For non-threatened species, we need to look at the yellow portions of the bars. Comparing these, we see that the portion for fires is shorter than that for pollution; thus, pollution is a greater cause of declines.

Fig. 11.22 In 2015, there were 268 condors in the wild—more than the 167 that were in captivity. In the 1980s and 1990s, there were far more birds in captivity than in the wild. The proportion and number of wild birds have generally increased since then, and wild birds have outnumbered captive birds since 2011. Today there are about 435 condors alive, which is about 20 times more than in the 1980s, when the number was down to about 20. The wild condor population in 1890 was about 500 birds, so today's wild population is about 54% of that (268/500), and today's total (wild plus captive) population is about 87% of the 1890 total (435/500).

Chapter 12

Fig. 12.11 The ratio of growth to removal is greatest for the land type for which the relative height of the two bars is most different. This is the case for the national forests, where annual growth exceeds 4 billion ft^3 while annual removal totals less than 0.4 billion ft^3.

12-SBS2 Fig. 2 According to the data in the graph, the forest plot held, on average, about 55 bird species in the 4 years it was censused before its fragmentation in 1984. After the plot became a fragment, the average number of bird species dropped to about 20 species.

12-SBS2 Fig. 3 According to the data in the graph, a tree 275 meters in from the edge of a forest fragment would be susceptible to elevated tree mortality (an edge effect that extends 300 m in) and increased wind disturbance (which extends 400 m in).

Chapter 13

Fig. 13.1 The dashed red line (which projects the urban population) for developing nations surpasses the dashed blue line (which projects the rural population) for developing nations between the year 2010 and the year 2020.

Fig. 13.10 Driving an SUV consumes about 4 MJ/passenger-km of energy, whereas riding commuter rail consumes only about 1.4 MJ/passenger-km—a difference of 2.6, and a ratio of 2.9 to 1. Driving a pickup truck emits about 380 g CO_2-eq./passenger-km of energy, whereas riding light rail consumes only about 120 g CO_2-eq./passenger-km—a difference of 260, and a ratio of 3.2 to 1. Roadway costs and parking costs are created by automobile traffic but not by rail traffic. Note the yellow and orange portions of the bars in the figure for part (c). These costs make automobile traffic more costly overall than rail traffic.

13-SBS Fig. 2 The blue curve represents the expected distribution of values based on the data obtained. The blue curve crosses the threshold for damage to land plants at about 14% of impervious surface, so that is the percentage at which one would expect to begin seeing this effect. The blue curve crosses the threshold for chronic toxicity to freshwater life at about 36% of impervious surface, so that is the percentage at which one would expect to begin seeing this effect.

Chapter 14

Fig. 14.3 In 2015, respiratory infections claimed approximately 3.2 million lives and diarrheal diseases about 1.8 million, for a total of 5 million lives. There were about 1.7 million deaths from AIDS. Dividing 5 million by 1.7 million reveals that about three times more lives were lost to respiratory infections and diarrheal diseases than were lost to AIDS. As this example shows, the diseases that garner the most attention are not always the ones that cause the greatest impacts on human health.

Fig. 14.4 Consulting the figure, note that in 2015 about 30% of Americans were obese and around 18% smoked. For the calculations in part (a), multiply 0.30 (30% written as a proportion) by 320 million people to find that around 96 million Americans were obese in 2015. Similarly, for part (b), multiply 0.18 (18%) by 320 million people to find that about 57.6 million Americans were smokers in 2015.

Fig. 14.18 The odds of perishing in a motor vehicle accident is 1 in 112, whereas the chance of dying in an air and space transport incident is 1 in 8015. Dividing 8015 by 112 shows that the odds of dying in a car accident is about 71 times that of dying in a plane crash—even though our instinctive risk assessment often makes us feel safer "behind the wheel."

14-SBS1 Fig. 2 Begin by connecting the data points in the figure to form a dose-response curve like that shown in Figure 14.17c. Mark the spot on the curve directly above 70 ng/g on the *x* axis. Then reference the *y* axis at the site of your mark, which indicates that an estimated 9% of the mice would likely suffer chromosomal effects at that dose.

Chapter 15

Fig. 15.3 Consulting the figure, note that 2.5% of the water on Earth is fresh water and that 1% of all fresh water is surface water. Within this surface water, 52% is found in lakes. To determine the percentage of water found in freshwater lakes, multiply 2.5% (0.025) by 1% (0.01) and by 52% (0.52). Multiplying these three values reveals that although freshwater lakes (such as the Great Lakes) seem massive, all of the world's freshwater lakes *combined* contain only 0.013% of Earth's water.

Fig. 15.10 Among the three regions, Latin America and the Caribbean have the greatest amount of water per capita. With its abundant river systems (including the mighty Amazon) and relatively small population compared to that of Africa and Asia, the per-person water quantities in the nations in this region are consistently large. Although Africa and Asia do contain abundant river systems, they also have larger populations, which means that less water is available per person. Africa and Asia are also home to large regions with arid climates, another factor that reduces the quantities of water available.

Fig. 15.14 The area between Fresno in the north and Bakersfield in the south experienced the most consistent decline in groundwater levels, as indicated by the swath of red dots (which indicate a reduction in the water table of 3 m [10 ft] or more) between these two cities. Given that these cities lie within California's agricultural valleys, the most likely reason for this area's precipitous decline in water table is over-extraction of water to irrigate crops.

Chapter 16

Fig. 16.3 Multiplying 1000 g by 3.5% (0.035) reveals that there are 35 g of salts in the 1000-g sample of seawater in the beaker. To determine the grams of negatively charged ions in the sample, sum the values for chloride (1.9%), sulfate (0.3%), and bicarbonate (0.01%) and find that 2.21% of the sample (or 22.1 g) is from such ions.

Fig. 16.4 Released off the southeastern coast of Japan, the buoy would be carried northeast by the Kuroshio Current and then eastward across the ocean on the North Pacific Current. Upon reaching North America, it could turn southward on the California Current and float by the western coast of the United States, passing Washington, Oregon, and California. Alternatively, the buoy might follow the Alaska Current northward upon reaching North America and pass Alaska, then return to Japan. So although Japan is closer to Australia, ocean currents would carry the buoy to the United States first.

Fig. 16.19 In 2012, the total global fisheries production was close to 160 million metric tons, with about 90 million metric tons coming from capture fisheries and about 70 million metric tons from aquaculture. To determine the percentage of the total derived from aquaculture, you divide 70 million metric tons by 160 million metric tons (and then multiply the answer by 100 to convert it to a percentage) and find that about 44% of the world's seafood originated from aquaculture operations in that year. Similar calculations for 1980 (65 million metric tons from capture, 5 million metric tons from aquaculture) reveals that only about 7% of total fisheries production came from aquaculture. One explanation for this increase is that as ocean stocks of wild fish dwindled due to overharvesting, it became more costly to locate schools of large fish, opening the door for aquaculture as an economically viable alternative to wild capture.

16-SBS1 Fig. 1 To interpret figures with two y-axis values, such as this one, carefully note which axis corresponds to which value on the graph. In this figure, the left y axis represents the area of the hypoxic zone (the bars on the figure), and the right y axis represents nitrogen flux (the line on the figure). To determine the year with the largest hypoxic zone, look for the highest bar (2002), and then consult the left y axis at its height. You'll see that the hypoxic zone that year was around 22,000 km^2. To determine the year with the largest nitrogen flux, find the highest point on the line on the figure (1993) and follow its value to the right y axis. You'll see that roughly 210,000 metric tons of nitrogen entered the northern Gulf that year.

Chapter 17

Fig. 17.14 As indicated by the numbers associated with the blue bars in part (b), population has increased by 56% since 1970, whereas emissions have decreased by 69% Thus, emissions per person have decreased by just over five times. (Imagine a population rising from 100 to 156, and emission dropping from 100 to 31. 31/156 = 0.199, or just less than one-fifth of the original 1-unit-per-person rate.)

Fig. 17.18 Answers will vary. For example, a person living in Los Angeles would be breathing dirtier air than people in most other cities, yet would find that L.A.'s air had improved over time, with only about 40% as many unhealthy days in 2015 as in 1985. Most cities have improved their air quality in recent years. Factors influencing air quality could include topography and climate, pollution sources such as power plants and the type of fuel they use, intensity of vehicle traffic, and more. Steps for reduction could include various policy measures and adoption of better pollution control technology.

Fig. 17.23 According to the data in the bars of the graph, in the late 1970s the L.A. basin suffered about 210 days per year of unhealthy air, and in recent years it has suffered about 100 such days—this is roughly a 52% reduction. According to the data in the line of the graph, in the late 1970s peak daily ozone levels averaged about 0.30 ppm, and in recent years they have averaged about 0.12–0.13 ppm, representing about a 57–60% reduction. Thus, both data sets show similar declines. One can tell this at a glance because the downward slopes of the two data sets appear more or less parallel on the graph.

Fig. 17.31 Answers will vary, but in virtually all locations, precipitation has become less acidic. For example, in many parts of the northeastern United States, pH increased from about 4.3 to about 5.0.

Chapter 18

Fig. 18.3 Since 1750 the atmospheric carbon dioxide concentration has increased from about 280 ppm to more than 400 ppm—a 43% increase.

Fig. 18.4 Changing land use accounts for 6 metric tons of carbon dioxide emissions per year, and industry emits 26 metric tons of carbon dioxide annually. Because 26/6 = 4.33, this means that for every 1 metric ton released by changing land use, 4.33 tons are released by industry.

Fig. 18.11 Answers will vary. In most regions temperature rose. In some areas of the Southeast it was stable or fell slightly.

Fig. 18.29 Many of the nations that reduced emissions are European. Three of the nations where emissions increased (Australia, Canada, and the United States) are large and less densely populated. Because they are geographically more spread out, long-distance transportation consumes more petroleum, giving rise to more emissions. In addition, these nations are more politically conservative than most European nations, and many conservatives tend to fear that emissions reductions will suppress economic activity.

18-SBS2 Fig. 3 The red-shaded area reflects modeling results of temperature change with both natural factors and human factors considered. Human impacts (in particular, greenhouse gas emissions) increased greatly during the course of the twentieth century due to steep growth in our population and resource consumption. Because the right-hand side of the graph shows data for the latter portion of the century, it is here that the influence of human impacts moves the red-shaded area upward, diverging from the blue-shaded area. The observed data track with the red-shaded area because human-caused emissions have raised global temperatures, so it is to be expected that the data actually observed on Earth would match what is simulated by a model that takes human, as well as natural, impacts into account.

Chapter 19

Fig. 19.2 Answers will vary. One should take the value at the far right end of the data line for oil (approximately 4.2 billion tons in 2014) and divide it by the value of that line in the year one was born. For example, for a person born in 1997, when oil consumption was about 3.4 billion tons per year, the percentage change by 2014 would be roughly 4.2/3.4 = 1.24, or about a 24% increase. For most people, coal has risen fastest since they were born; the same type of calculation can be performed. Note how the data line for coal rises more steeply than those for oil and gas.

Fig. 19.3 According to the information in the figure, the average U.S. citizen uses 7.24 tons per year, and the average person in the world uses 1.79 tons per year. Because 7.24/1.79 = 4.04, this tells us that the average U.S. citizen uses approximately four times more energy than the average person in the world. Likewise, the average U.S. citizen uses 7.24/0.49 = 14.78, or nearly 15 times more energy than the average citizen of India.

Fig. 19.20 Answers will vary. One should take the value at the far right end of the black ("Total") data line and divide it by the value of that line in the year one's mother or father was born. For example, if one's parent were born in 1970, when emissions were about 4.0 billion tons per year, then the percent change since then would be about 9.8/4.0 = 2.45, or roughly a 145% increase.

Fig. 19.22 The United States extracts 14.0 million barrels of oil per day and consumes 19.0 million barrels per day. Thus, for every barrel extracted, 19.0/14.0 = 1.36 barrels are consumed.

Chapter 20

Fig 20.1 The most reasonable hypothesis would be that nuclear power took the place of fossil-fuel-fired power and caused the decline in use of fossil fuels. This is because the decrease in fossil fuel use is tightly correlated in time with the increase in nuclear power; fossil fuels declined between 1970 and 1990—the same time period during which nuclear power was increasing. In contrast, both bioenergy and hydropower were showing slow growth at much lower rates of use at that time.

Fig 20.3 Since 1950, oil has shown the most growth in absolute terms, rising from about 13 quadrillion BTU to about 35 quadrillion BTU in 2015—an absolute gain of about 22 quadrillion BTU. Natural gas has risen by about the same amount. In percentage terms, nuclear power has seen the most growth since 1950; it started at a level of zero and rose to a little over 8 quadrillion BTU. Between 1950 and 1970, oil and natural gas grew fastest in both absolute and percentage terms. Between 1970 and 1990, coal and nuclear power grew fastest in both absolute and per-centage terms. Since 1990, oil has risen, then fallen; natural gas rose, then was flat, then rose again; coal rose and then fell; nuclear power rose and then plateaued off; hydropower remained about the same; and bioenergy increased slightly in the most recent years.

Fig. 20.16 In 2015, 25.7 billion gallons of ethanol were produced across the world, of which the United States produced 14.8 billion gallons. Thus U.S. production is 14.8/25.7 = 0.58. Therefore, U.S. production makes up 58% of the world total. From the data alone, we might reasonably predict ethanol production to rise sharply in the future because that has been the overall trend since 1980 and especially since 2000. However, we might just as reasonably predict that ethanol production will level off in the future, because production has stayed fairly constant for the most recent five years. This illustrates the risks of extrapolating data into the future; your predictions of trends can depend on how far back in time you go to assess past data. In actuality, you would want to know as much as you can about the factors driving ethanol production before making predictions.

Fig 20.18 Biodiesel reduces sulfates most effectively relative to petroleum-based diesel. For both B20 and B100, the percentage reduction for sulfates is greater than for any other pollutant shown.

Chapter 21

Fig 21.2 Note that in both part (**a**) and part (**b**), the bar on the right gives a breakdown of data from the pie slice for renewable energy. In part (**a**), the pie chart for energy consumption tells us that renewable energy as a whole provides 9.9% of U.S. energy consumption. The bar tells us that solar energy provides 5.7% of renewable energy. Therefore solar contributes 5.7% of 9.9%—or 0.6%—of total U.S. energy consumption. Similarly, in part (**b**), the pie chart for electricity generation tells us that renewable energy as a whole provides 13.6% of U.S. electricity generation. The bar tells us that wind power provides 34.3% of renewable energy. Therefore wind contributes 34.3% of 13.6%—or 4.7%—of total U.S. electricity generation.

Fig 21.3 The yearly growth rate of PV solar is 42%, so you would multiply 10 units by 1.42, then multiply that number by 1.50, and so on, for 5 years. The result is 57.7 units.

Fig 21.4 Using median values indicated by the thin black vertical lines within the colored bars, 1 kilowatt-hour of electricity from PV solar results in roughly 35 g CO_2-equivalent emissions, whereas 1 kilowatt-hour of electricity from coal results in roughly 1000 g CO_2-equivalent emissions. Thus for every unit of emissions from PV solar, we would expect roughly 1000/35 = 29 units from coal.

Fig 21.12 On average, southern Arizona receives roughly 2400–2600 kilowatt-hours per square meter per year, and most of Germany receives fewer than 1200 kilowatt-hours/m²/yr. Thus, southern Arizona receives more than twice as much sunlight as does Germany. For one's own state, answers will vary, but the comparison would be made in the same way.

Fig 21.18 Answers will vary, but regions that appear underutilized for wind power include (on land) South Dakota, Nebraska, Montana, and New Mexico; and (offshore) the entire Atlantic and Gulf coasts.

Chapter 22

Fig. 22.4 Between 1990 and 2013, the population size of the United States has grown at a rate that exceeds the rate of growth in total waste generation. We can infer this from the data in the graph alone because the per capita waste generation rate has decreased slightly, even though total generation has risen.

Fig. 22.6 The amount of solid waste that is combusted (incinerated) is shown in green. To determine increase or decrease, note whether the green band widens or narrows over time. (This is independent of the overall height of the graph data, which reflects cumulative, summed, totals.) Examining the green band, we see that the amount of solid waste that is combusted (incinerated) decreased from 1960 until about 1985, and then it increased until about 2002. As of 2013, the amount was roughly equal to the amount back in 1960.

Fig. 22.8 Between 1960 and 2013, the total amount of waste that was recovered increased by 14 times (from about 6 to about 87 million tons). However, in that same time period the recovery rate (percentage of waste generated that is recovered) increased by only about 6 times (from about 6% to about 34%). From this we can infer that the total amount of waste generated must also have risen. This is because had the total amount generated stayed the same, the amount recovered and percentage recovered would have changed by the same amount. Had the total amount generated fallen, the percentage recovered would have increased by more than the amount recovered.

Fig. 22.11 Michigan's recycling rate is nearly 100%, whereas states without bottle bills have a 20% rate. Thus, Michigan's rate is nearly 5 times higher.

22-SBS1 Fig. 2 Hazardous waste items (orange bars) tended to travel farther. Hazardous waste items traveled from about 100 km to about 1700 km. In contrast, the farthest-traveling standard municipal solid waste item shown traveled less than 100 km. Hazardous waste items tend to travel farther because they often need to be taken to specialized facilities for handling, and these can be few and far between.

Chapter 23

Fig. 23.15 At present rates of consumption, nickel has technically recoverable reserves that would last around 51 years, of which about 31 years of reserves are economically recoverable. Dividing 31 years of economically recoverable reserves by 51 years of technically recoverable reserves yields the value 0.61, indicating that 61% (multiply 0.61 by 100 to present the value as a percentage) of nickel reserves are economically recoverable. Molybdenum (56%) comes in a close second when similar calculations are performed. The metal with the lowest percentage of economically recoverable reserves is lead. Lead's 19 years of economically recoverable reserves divided by its 425 years of technically recoverable reserves finds that only 4% of the world's known lead reserves are currently economically recoverable.

23-SBS Fig. 1 The approach of using acid mine drainage in hydraulic fracturing operations is feasible only when mine sites are in very close proximity to fracking operations, as is the case in western Pennsylvania. There are fracking operations in central Pennsylvania, but few acid mine drainage sites, making the approach impractical in this part of the state. The eastern third of the state has both fracking operations and acid mine drainage sites, but they are too far apart to be paired with one another in this method.

Chapter 24

Fig 24.8 According to the bars at the left of the graph, one hectare of intact wetland in Canada is worth about $5800 (once external costs and benefits are considered), whereas one hectare (ha) of intensive farmland is worth about $2250. Thus, a hectare of wetland is worth about $3550 more in absolute terms, or about 58% more in percentage terms. Likewise, the pair of bars for mangroves in Thailand tell us that intact mangroves are more valuable (about $1000/ha) than shrimp farming (about $200/ha). Thus it creates five times more value to leave mangrove forests intact than to clear them for shrimp farming.

How to Interpret Graphs

Presenting data in ways that help make trends and patterns visually apparent is a vital element of science. For scientists, businesspeople, policymakers, and others, the primary tool for expressing patterns in data is the graph. Thus, the ability to interpret graphs is a skill that you will want to cultivate. This appendix guides you in how to read graphs, introduces a few vital conceptual points, and surveys the most common types of graphs, giving rationales for their use.

Navigating a Graph

A graph is a diagram that shows relationships among *variables*, which are factors that can change in value. The most common types of graphs relate values of a *dependent variable* to those of an *independent variable*. As explained in Chapter 1 (p. 11), a dependent variable is so named because its values "depend on" the values of an independent variable. In other words, as the values of an independent variable change, the values of the dependent variable change in response. In a manipulative experiment (p. 13), changes that a researcher specifies in the value of the independent variable *cause* changes in the value of the dependent variable. In observational studies, there may be no causal relationship, and scientists may plot a correlation (p. 13). In a positive correlation, values of one variable go up or down along with values of another. In a negative correlation, values of one variable go up when values of the other go down. Whether we are graphing a correlation or a causal relationship, the values of the independent variable are known or specified by the researcher, whereas the values of the dependent variable are unknown until the research has taken place. The values of the dependent variable are what we are interested in observing or measuring.

By convention, independent variables are generally represented on the horizontal axis, or *x axis*, of a graph, while dependent variables are represented on the vertical axis, or *y axis*. Numerical values of variables generally become larger as one proceeds rightward on the *x axis* or upward on the *y axis*. Note that the tick marks along the axes must be uniformly spaced so that when the data are plotted, the graph gives an accurate visual representation of the scale of quantitative change in the data.

As a simple example, **FIGURE B.1** shows data from the Breeding Bird Survey that reflect population growth of the Eurasian collared dove following its introduction to North America. The *x* axis shows values of the independent variable, which in this case is time, expressed in units of years. The dependent variable, presented on the *y* axis, is the average number of doves detected on each route. For each year, a data point is plotted on the graph to show the average number of doves detected. In this particular graph, a line (dark red curve) was then drawn through the actual data points (orange dots), showing how closely the empirical data match an exponential growth curve (p. 64), a theoretical phenomenon of importance in ecology.

Now that you're familiar with the basic building blocks of a graph, let's survey the most common types of graphs you'll see, and examine a few vital concepts in graphing.

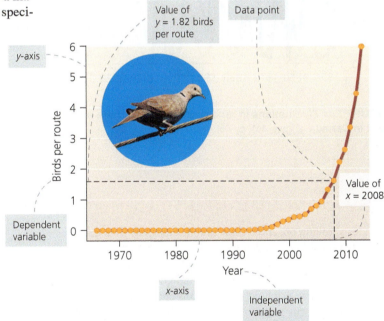

FIGURE B.1 Exponential population growth, demonstrated by the Eurasian collared dove in North America in recent years. (Figure 3.16, p. 64)

Mastering**Environmental**Science®

Graph Type: Line Graph

A line graph is used when a data set involves a sequence of some kind, such as a series of values that occur one by one and change through time or across distance. In a line graph, a line runs from one data point to the next. Line graphs are most appropriate when the *y* axis expresses a continuous numerical variable, and the *x* axis expresses either continuous numerical data or discrete sequential categories (such as years). **FIGURE B.2** shows values for the size of the ozone hole over Antarctica in recent years. Note how the data show that the size of the hole increases until 1987, when the Montreal Protocol (p. 468) came into force, and then begins to stabilize afterward.

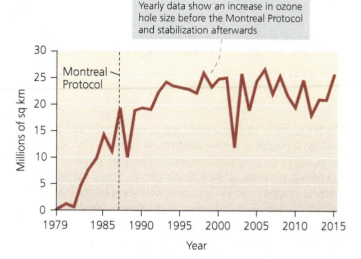

FIGURE B.2 Size of the Antarctic ozone hole before and after a treaty that was designed to address it. (Figure 17.27, p. 468)

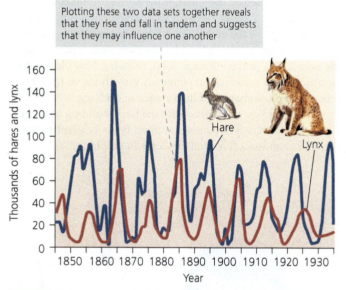

FIGURE B.3 Fluctuations in recorded numbers of hare and lynx in Canada. (Figure 4.4, p. 76)

One useful technique is to plot two or more data sets together on the same graph. This allows us to compare trends in the data sets to see whether they may be related and, if so, the nature of that relationship. In **FIGURE B.3**, recorded numbers of a predator species rise and fall immediately following those of its prey, suggesting a possible connection.

Key Concept: Projections

Besides showing observed data, graphs can show data that are predicted for the future. Such *projections* of data are based on models, simulations, or extrapolations from past data, but they are only as good as the information that goes into them—and future trends may not hold if conditions change in unforeseen ways. Thus, in this textbook, projected future data are shown with dashed lines, as in **FIGURE B.4**, to indicate that they are less certain than data that have already been observed. Be careful when interpreting graphs in the popular media and on the Internet, however; often newspapers, magazines, websites, and advertisements will show projected future data in the same way as known past data!

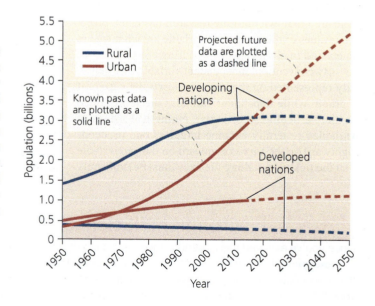

FIGURE B.4 Past population change and projected future population change for rural and urban areas in more developed and less developed regions. (Figure 13.1, p. 332)

Graph Type: Bar Chart

A bar chart is most often used when one variable is a category and the other is a number. In such a chart, the height (or length) of each bar represents the numerical value of a given category. Higher or longer bars mean larger values. In **FIGURE B.5**, the bar for the category "Respiratory infections" is higher than that for "Malaria," indicating that respiratory infections cause more deaths each year (the numerical variable on the *y* axis) than does malaria.

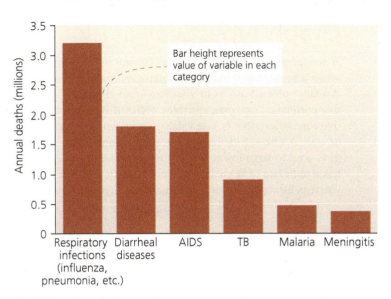

FIGURE B.5 Leading causes of death from infectious disease. (Figure 14.3b, p. 358)

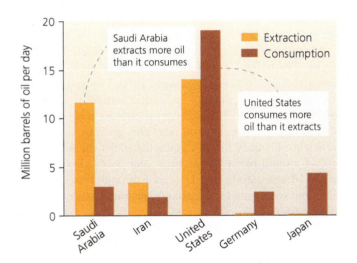

FIGURE B.6 Oil extraction and consumption by selected nations. (Figure 19.23, p. 540)

As we saw with line graphs, it is often instructive to graph two or more data sets together to reveal patterns and relationships. A bar chart such as **FIGURE B.6** lets us compare two data sets (oil extraction and oil consumption) both within and among nations. A graph that does double duty in this way allows for higher-level analysis (in this case, suggesting which nations depend on others for petroleum imports). Most bar charts in this book illustrate multiple types of information at once in this manner.

Graph Type: Pie Chart

A pie chart is used when we wish to compare the numerical proportions of some whole that are taken up by each of several categories. Each category is represented visually like a slice from a pie, with the size of the slice reflecting the percentage of the whole that is taken up by that category. For example, **FIGURE B.7** compares the percentages of genetically modified crops worldwide that are soybeans, corn, cotton, and canola.

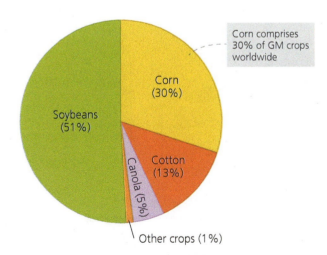

FIGURE B.7 Genetically modified crops grown worldwide, by type. (Figure 10.24a, p. 252)

Graph Type: Scatter Plot

A scatter plot is often used when data are not sequential and when a given *x*-axis value could have multiple *y*-axis values. A scatter plot allows us to visualize a broad positive or negative correlation between variables. **FIGURE B.8** shows a negative correlation (that is, one value goes up while the other goes down): Nations with higher rates of school enrollment for girls tend to have lower fertility rates. Jamaica, for example, has a high rate of school enrollment for girls and low fertility, whereas Ethiopia has low enrollment and high fertility.

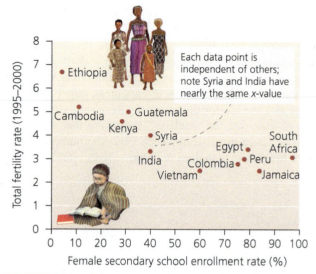

FIGURE B.8 **Fertility rate and education of females, by nation.** (Figure 8.17, p. 202)

Key Concept: Statistical Uncertainty

Most data sets involve some degree of uncertainty. When a graphed value represents the *mean* (average) of many measurements, the researcher may want to use mathematical techniques to show the degree to which the raw data vary around this mean. Results from such statistical analyses may be expressed in a number of ways, and the two graphs in this section show methods used in this book.

In a bar chart or scatter plot (**FIGURE B.9**), thin black lines called *error bars* may be shown extending above and below the tops of the reported data (or simply above them in the case of a bar chart). In this example of likelihood of death from air pollution, error bars show the most variation at the highest measured concentration of pollutants and no variation at the lowest measured concentration.

Sometimes shading is used to express variation around a mean. The black and red data lines in **FIGURE B.10** show mean global sea level readings since 1880. The data lines are surrounded by gray shading indicating statistical variation. Note how the amount of statistical uncertainty is exceeded by the sheer scale of the sea level rise. This gives us confidence that sea level is truly rising, despite the statistical uncertainty we find around mean values each year. Note also how the amount of uncertainty has decreased through time. This reflects improvements in technology, enabling more accurate measurements.

The statistical analysis of data is critically important in science. In this book, we provide a broad and streamlined introduction to many topics, so we often omit error bars from our graphs and details of statistical significance from our discussions. Bear in mind that this is for clarity of presentation only; the research we discuss analyzes its data in far more depth than any textbook could possibly cover.

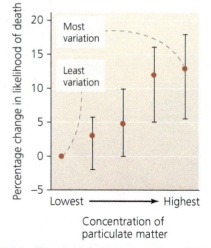

FIGURE B.9 Likelihood of death due to air pollution. (17-SBS1 Figure 3a, p. 461)

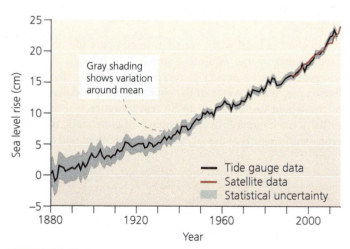

FIGURE B.10 Change in global sea level, measured since 1880. (Figure 18.16, p. 495)

Metric System

MEASUREMENT	UNIT AND ABBREVIATION	METRIC EQUIVALENT	METRIC TO ENGLISH CONVERSION FACTOR	ENGLISH TO METRIC CONVERSION FACTOR
Length	1 kilometer (km)	= 1000 (10^3) meters	1 km = 0.62 mile	1 mile = 1.61 km
	1 meter (m)	= 100 (10^2) centimeters	1 m = 1.09 yards	1 yard = 0.914 m
		= 1000 millimeters	1 m = 3.28 feet	1 foot = 0.305 m
			1 m = 39.37 inches	
	1 centimeter (cm)	= 0.01 (10^{-2}) meter	1 cm = 0.394 inch	1 foot = 30.5 cm
				1 inch = 2.54 cm
	1 millimeter (mm)	= 0.001 (10^{-3}) meter	1 mm = 0.039 inch	
Area	1 square meter (m^2)	= 10,000 square centimeters	1 m^2 = 1.1960 square yards	1 square yard = 0.8361 m^2
			1 m^2 = 10.764 square feet	1 square foot = 0.0929 m^2
	1 hectare (ha)	= 10,000 square meters	1 ha = 2.47 acres	1 acre = 0.405 hectares
	1 square kilometer (km^2)	= 1,000,000 square meters	1 km^2 = 0.386 square miles	1 square mile = 2.59 km^2
Mass	1 metric ton (t)	= 1000 kilograms	1 t = 1.103 tons	1 ton = 0.907 t
	1 kilogram (kg)	= 1000 grams	1 kg = 2.205 pounds	1 pound = 0.4536 kg
	1 gram (g)	= 1000 milligrams	1 g = 0.0353 ounce	1 ounce = 28.35 g
	1 milligram (mg)	= 0.001 gram		
Volume (solids)	1 cubic meter (m^3)	= 1,000,000 cubic centimeters	1 m^3 = 1.3080 cubic yards	1 cubic yard = 0.7646 m^3
			1 m^3 = 35.315 cubic feet	1 cubic foot = 0.0283 m^3
	1 cubic centimeter (cm^3 or cc)	= 0.000001 cubic meter	1 cm^3 = 0.0610 cubic inch	1 cubic inch = 16.387 cm^3
		= 1 milliliter		
	1 cubic millimeter (mm^3)	= 0.000000001 cubic meter		
Volume (liquids and gases)	1 kiloliter (kl or kL)	= 1000 liters	1 kL = 264.17 gallons	
	1 liter (l or L)	= 1000 milliliters	1 L = 0.264 gallons	1 gallon = 3.785 L
			1 L = 1.057 quarts	1 quart = 0.946 L
	1 milliliter (ml or mL)	= 0.001 liter	1 ml = 0.034 fluid ounce	1 quart = 946 ml
		= 1 cubic centimeter	1 ml = approx. 1/4 teaspoon	1 pint = 473 ml
				1 fluid ounce = 29.57 ml
				1 teaspoon = approx. 5 ml
Temperature	Degrees Celsius (°C)		$°C = \dfrac{5}{9} (°F - 32)$	$°F = \dfrac{9}{5} (°C) + 32$
Energy and Power	1 kilowatt-hour	= 34,113 BTUs = 860,421 calories		
	1 watt	= 3.413 BTUs/hr		
		= 14.34 calories/min		
	1 calorie	= the amount of heat necessary to raise the temperature of 1 gram (1 cm^3) of water 1 degree Celsius		
	1 horsepower	= 7.457 × 102 watts		
	1 joule	= 9.481 × 10^{-4} BTUs		
		= 0.239 cal		
		= 2.778 × 10^{-7} kilowatt-hours		
Pressure	1 pound per square inch (psi)	= 6894.757 pascals (Pa)		
		= 0.068045961 atmosphere (atm)		
		= 51.71493 millimeters of mercury (mm Hg = Torr)		
		= 68.94757 millibars (mbar)		
		= 6.894757 kilopascals (kPa)		
	1 atmosphere (atm)	= 101.325 kilopascals (kPa)		

Periodic Table of the Elements

Key (legend box):

6	— Atomic number
C	— Chemical symbol
12.011	— Atomic weight
Carbon	— Name

Representative (main group) elements — Transition metals — Rare earth elements

Period	IA	IIA	IIIB	IVB	VB	VIB	VIIB	VIIIB	VIIIB	VIIIB	IB	IIB	IIIA	IVA	VA	VIA	VIIA	VIIIA
1	1 **H** 1.0079 Hydrogen																	2 **He** 4.003 Helium
2	3 **Li** 6.941 Lithium	4 **Be** 9.012 Beryllium											5 **B** 10.811 Boron	6 **C** 12.011 Carbon	7 **N** 14.007 Nitrogen	8 **O** 15.999 Oxygen	9 **F** 18.998 Fluorine	10 **Ne** 20.180 Neon
3	11 **Na** 22.990 Sodium	12 **Mg** 24.305 Magnesium											13 **Al** 26.982 Aluminum	14 **Si** 28.086 Silicon	15 **P** 30.974 Phosphorus	16 **S** 32.066 Sulfur	17 **Cl** 35.453 Chlorine	18 **Ar** 39.948 Argon
4	19 **K** 39.098 Potassium	20 **Ca** 40.078 Calcium	21 **Sc** 44.956 Scandium	22 **Ti** 47.88 Titanium	23 **V** 50.942 Vanadium	24 **Cr** 51.996 Chromium	25 **Mn** 54.938 Manganese	26 **Fe** 55.845 Iron	27 **Co** 58.933 Cobalt	28 **Ni** 58.69 Nickel	29 **Cu** 63.546 Copper	30 **Zn** 65.39 Zinc	31 **Ga** 69.723 Gallium	32 **Ge** 72.61 Germanium	33 **As** 74.922 Arsenic	34 **Se** 78.96 Selenium	35 **Br** 79.904 Bromine	36 **Kr** 83.8 Krypton
5	37 **Rb** 85.468 Rubidium	38 **Sr** 87.62 Strontium	39 **Y** 88.906 Yttrium	40 **Zr** 91.224 Zirconium	41 **Nb** 92.906 Niobium	42 **Mo** 95.94 Molybdenum	43 **Tc** 98 Technetium	44 **Ru** 101.07 Ruthenium	45 **Rh** 102.906 Rhodium	46 **Pd** 106.42 Palladium	47 **Ag** 107.868 Silver	48 **Cd** 112.411 Cadmium	49 **In** 114.82 Indium	50 **Sn** 118.71 Tin	51 **Sb** 121.76 Antimony	52 **Te** 127.60 Tellurium	53 **I** 126.905 Iodine	54 **Xe** 131.29 Xenon
6	55 **Cs** 132.905 Cesium	56 **Ba** 137.327 Barium	57* **La** 138.906 Lanthanum	72 **Hf** 178.49 Hafnium	73 **Ta** 180.948 Tantalum	74 **W** 183.84 Tungsten	75 **Re** 186.207 Rhenium	76 **Os** 190.23 Osmium	77 **Ir** 192.22 Iridium	78 **Pt** 195.08 Platinum	79 **Au** 196.967 Gold	80 **Hg** 200.59 Mercury	81 **Tl** 204.383 Thallium	82 **Pb** 207.2 Lead	83 **Bi** 208.980 Bismuth	84 **Po** 209 Polonium	85 **At** 210 Astatine	86 **Rn** 222 Radon
7	87 **Fr** 223 Francium	88 **Ra** 226.025 Radium	89** **Ac** 227.028 Actinium	104 **Rf** 267 Rutherfordium	105 **Db** 268 Dubnium	106 **Sg** 269 Seaborgium	107 **Bh** 270 Bohrium	108 **Hs** 269 Hassium	109 **Mt** 278 Meitnerium	110 **Ds** 281 Darmstadtium	111 **Rg** 281 Roentgenium	112 **Cn** 285 Copernicium	113 **Nh** 286 Nihonium	114 **Fl** 289 Flerovium	115 **Mc** 289 Moscovium	116 **Lv** 293 Livermorium	117 **Ts** 293 Tennessine	118 **Og** 294 Oganesson

*Lanthanides

58 **Ce** 140.115 Cerium	59 **Pr** 140.908 Praseodymium	60 **Nd** 144.24 Neodymium	61 **Pm** 145 Promethium	62 **Sm** 150.36 Samarium	63 **Eu** 151.964 Europium	64 **Gd** 157.25 Gadolinium	65 **Tb** 158.925 Terbium	66 **Dy** 162.5 Dysprosium	67 **Ho** 164.93 Holmium	68 **Er** 167.26 Erbium	69 **Tm** 168.934 Thulium	70 **Yb** 173.04 Ytterbium	71 **Lu** 174.967 Lutetium

**Actinides

90 **Th** 232.038 Thorium	91 **Pa** 231.036 Protactinium	92 **U** 238.029 Uranium	93 **Np** 237.048 Neptunium	94 **Pu** 244 Plutonium	95 **Am** 243 Americium	96 **Cm** 247 Curium	97 **Bk** 247 Berkelium	98 **Cf** 251 Californium	99 **Es** 252 Einsteinium	100 **Fm** 257 Fermium	101 **Md** 258 Mendelevium	102 **No** 259 Nobelium	103 **Lr** 262 Lawrencium

The periodic table arranges elements by atomic number and atomic weight into horizontal rows called periods and vertical columns called groups.

Elements of each group in Class A have similar chemical and physical properties. This reflects the fact that members of a particular group have the same number of valence shell electrons, which is indicated by the group's number. For example, group IA elements have one valence shell electron, group IIA elements have two, and group VA elements have five. In contrast, as you progress across a period from left to right, properties of the elements change, varying from the very metallic properties of groups IA and IIA to the nonmetallic properties of group VIIA to the inert elements (noble gases) in group VIIIA. This reflects changes in the number of valence shell electrons.

Class B elements, or transition elements, are metals and generally have one or two valence shell electrons. In these elements, some electrons occupy more distant electron shells before the deeper shells are filled.

In this periodic table, elements with symbols printed in black exist as solids under standard conditions (25°C and 1 atmosphere of pressure); elements in red exist as gases; and those in dark blue as liquids. Elements with symbols in green do not exist in nature and must be created by some type of nuclear reaction.

Geologic Time Scale

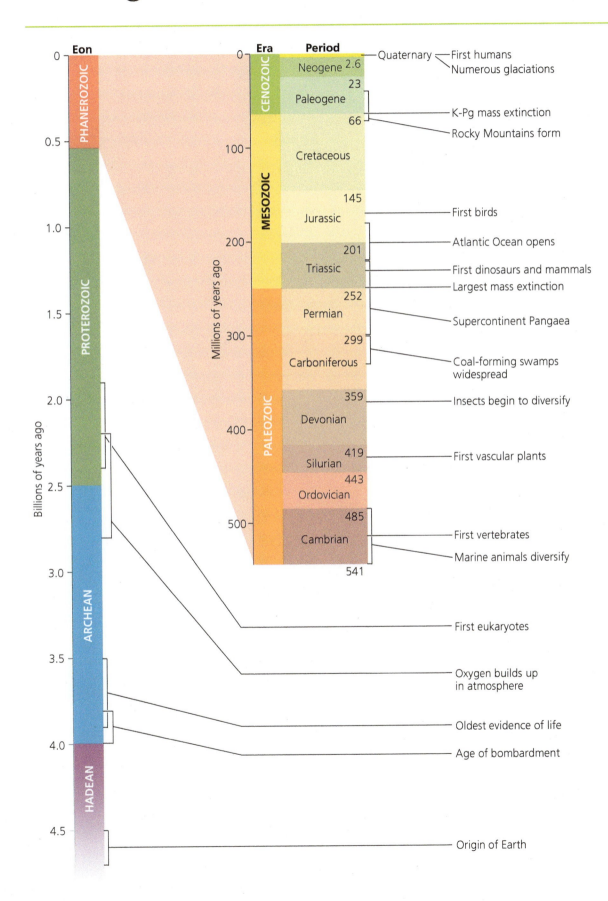

Glossary

acid deposition The settling of acidic or acid-forming pollutants from the *atmosphere* onto Earth's surface. This may take place by precipitation, fog, gases, or the deposition of dry particles. Compare *acid rain*.

acid mine drainage A process in which sulfide minerals in newly exposed rock surfaces react with *oxygen* and rainwater to produce sulfuric acid, which causes chemical *runoff* as it *leaches* metals from the rocks. Acid drainage is a natural phenomenon, but mining greatly accelerates it by exposing many new surfaces.

acid rain *Acid deposition* that takes place through rain.

acidic The property of a *solution* in which the concentration of *hydrogen* (H⁺) *ions* is greater than the concentration of hydroxide (OH⁻) ions. Compare *basic*.

active solar energy collection An approach in which technological devices are used to focus, move, or store *solar energy*. Compare *passive solar energy collection*.

acute exposure Exposure to a *toxicant* occurring in high amounts for short periods of time. Compare *chronic exposure*.

adaptation (re: *climate change*) The pursuit of strategies to protect ourselves from the impacts of climate change. Compare *mitigation*.

adaptation (re: *evolution*) (1) The process by which traits that lead to increased reproductive success in a given environment evolve in a *population* through *natural selection*. (2) A trait that confers greater likelihood that an individual will reproduce.

adaptive management The systematic testing of different management approaches to improve methods over time.

aerosols Very fine liquid droplets or solid particles aloft in the atmosphere.

age structure The relative numbers of individuals of different ages within a *population*. Age structure can have a strong effect on rates of population growth or decline and is often expressed as a ratio of age classes, consisting of organisms (1) not yet mature enough to reproduce, (2) capable of reproduction, and (3) beyond their reproductive years.

agricultural revolution The shift around 10,000 years ago from a hunter-gatherer lifestyle to an agricultural way of life in which people began to grow crops and raise domestic animals. Compare *industrial revolution*.

agriculture The practice of cultivating *soil*, producing crops, and raising livestock for human use and consumption.

air pollutant A gas or particulate material added to the atmosphere that can affect climate or harm people or other living things.

air pollution The release of *air pollutants*.

airshed The geographic area that produces air pollutants likely to end up in a waterway.

allergen A *toxicant* that overactivates the immune system, causing an immune response when one is not necessary.

alloy A substance created by fusing a *metal* with other metals or nonmetals. Bronze is an alloy of the metals copper and tin, and steel is an alloy of iron and the nonmetal *carbon*.

anthropocentrism A human-centered view of our relationship with the *environment*. Compare *biocentrism* and *ecocentrism*.

aquaculture The cultivation of aquatic organisms for food in controlled environments.

aquifer An underground water reservoir.

artificial selection *Natural selection* conducted under human direction. Examples include the selective breeding of crop plants, pets, and livestock.

asbestos Any of several types of *mineral* that form long, thin microscopic fibers—a structure that allows asbestos to insulate buildings for heat, muffle sound, and resist fire. When inhaled and lodged in lung tissue, asbestos scars the tissue and may eventually lead to lung cancer or *asbestosis*.

asbestosis A disorder resulting from lung tissue scarred by acid following prolonged inhalation of *asbestos*.

asthenosphere A layer of the upper *mantle*, just below the *lithosphere*, consisting of especially soft rock.

atmosphere The layer of gases surrounding planet Earth. Compare *biosphere*; *hydrosphere*; *lithosphere*.

atmospheric deposition The wet or dry deposition onto land of a wide variety of pollutants, including mercury, nitrates, organochlorines, and others. *Acid deposition* is one type of atmospheric deposition.

atmospheric pressure The weight (or gravitational force) per unit area produced by a column of air.

atom The smallest component of an *element* that maintains the chemical properties of that *element*.

autotroph (primary producer) An organism that can use the *energy* from sunlight to produce its own food. Includes green plants, algae, and cyanobacteria.

***Bacillus thuringiensis* (Bt)** A naturally occurring *soil* bacterium that produces a protein that kills many pests, including caterpillars and the larvae of some flies and beetles.

background extinction rate The average rate of *extinction* that occurred before the appearance of humans. For example, the *fossil record* indicates that for both birds and mammals, one *species* in the world typically became extinct every 500–1000 years. Compare *mass extinction event*.

basic The property of a *solution* in which the concentration of *hydroxide* (OH⁻) *ions* is greater than the concentration of hydrogen (H⁺) ions. Compare *acidic*.

bedrock The continuous mass of solid rock that makes up Earth's *crust*.

benthic Of, relating to, or living on the bottom of a water body. Compare *pelagic*.

bioaccumulation The buildup of *toxicants* in the tissues of an animal.

biocapacity A term in *ecological footprint* accounting meaning the amount of biologically productive land and sea available to us.

biocentrism A philosophy that ascribes relative values to actions, entities, or properties on the basis of their effects on all living things or on the integrity of the biotic realm in general. The biocentrist evaluates an action in terms of its overall impact on living things, including—but not exclusively focusing on—human beings. Compare *anthropocentrism* and *ecocentrism*.

biodiesel Diesel fuel produced by mixing vegetable oil, used cooking grease, or animal fat with small amounts of *ethanol* or methanol (wood alcohol) in the presence of a chemical catalyst.

biodiversity (biological diversity) The variety of life across all levels of biological organization, including the diversity of *species*, *genes*, *populations*, and *communities*.

biodiversity hotspot An area that supports an especially great diversity of *species*, particularly species that are *endemic* to the area.

bioenergy *Energy* harnessed from plant and animal matter, including wood from trees, charcoal from burned wood, and combustible animal waste products, such as cattle manure. *Fossil fuels* are not considered bioenergy sources because their organic matter has not been part of living organisms for millions of years and has undergone considerable chemical alteration since that time. Also called biomass energy.

biofuel Fuel produced from *biomass* sources and used primarily to power automobiles. Examples include *ethanol* and *biodiesel*.

biogeochemical cycle See *nutrient cycle*.

biological control Control of pests and weeds with organisms that prey on or parasitize them, rather than with chemical *pesticides*. Commonly called biocontrol.

biological diversity See *biodiversity*.

biological hazard Human health hazards that result from ecological interactions among organisms. These include *parasitism* by viruses, bacteria, or other *pathogens*. Compare *infectious disease; chemical hazard; cultural hazard; physical hazard*.

biomagnification The magnification of the concentration of *toxicants* in an organism caused by its consumption of other organisms in which toxicants have *bioaccumulated*.

biomass (1) In ecology, organic material that makes up living organisms; the collective mass of living matter in a given place and time. (2) In energy, organic material derived from living or recently living organisms, containing chemical *energy* that originated with *photosynthesis*.

biome A major regional complex of similar plant *communities*; a large *ecological* unit defined by its dominant plant type and vegetation structure.

biophilia An inherent love for and fascination with nature and an instinctive desire people have to affiliate with other living things. Defined by biologist E.O. Wilson as "the connections that human beings subconsciously seek with the rest of life."

biopower Power attained by combusting *biomass* sources to generate *electricity*.

bioprospecting Searching for organisms that might provide new drugs, medicines, foods, or other products of value or interest.

biosphere The sum total of all the planet's living organisms and the nonliving portions of the *environment* with which they interact.

biosphere reserve A tract of land with exceptional *biodiversity* that couples preservation with *sustainable development* to benefit local people. Biosphere reserves are designated by UNESCO (the *United Nations* Educational, Scientific, and Cultural Organization) following application by local stakeholders.

biotechnology The material application of biological *science* to create products derived from organisms. The creation of *transgenic* organisms is one type of biotechnology.

birth control The effort to control the number of children one bears, particularly by reducing the frequency of pregnancy. Compare *contraception; family planning*.

boreal forest A *biome* of northern coniferous forest that stretches in a broad band across much of Canada, Alaska, Russia, and Scandinavia. Also known as *taiga*, boreal forest consists of a limited number of *species* of evergreen trees, such as black spruce, that dominate large regions of forests interspersed with occasional bogs and lakes.

bottle bill A law establishing a program whereby consumers pay a deposit on bottles or cans upon purchase—often 5 or 10 cents per container—and then receive a refund when they return them to stores after use. Bottle bills reduce litter, raise *recycling* rates, and decrease the *waste stream*.

bottleneck In environmental science, a step in a process that limits the progress of the overall process.

breakdown product A compound that results from the degradation of a toxicant.

brownfield An area of land whose redevelopment or reuse is complicated by the presence or potential presence of hazardous material.

campus sustainability A term describing a wide array of efforts taking place on college and university campuses by which students, faculty, staff, and administrators are trying to reduce the environmental impacts of their institutions.

canopy The upper level of tree leaves and branches in a *forest*.

cap-and-trade An *emissions trading* system in which government determines an acceptable level of *pollution* and then issues polluting parties permits to pollute. A company receives credit for amounts it does not emit and can then sell this credit to other companies.

captive breeding The practice of keeping members of *threatened* and *endangered species* in captivity so that their young can be bred and raised in controlled *environments* and subsequently reintroduced into the wild.

carbohydrate An *organic compound* consisting of *atoms* of *carbon*, *hydrogen*, and *oxygen*.

carbon The chemical *element* with six protons and six neutrons. A key element in *organic compounds*.

carbon capture and storage Technologies or approaches to remove *carbon dioxide* from emissions of power plants or other facilities, and sequester, or store, it (generally in liquid form) underground under pressure in locations where it will not seep out, in an effort to mitigate *global climate change*.

carbon cycle A major *nutrient cycle* consisting of the routes that *carbon atoms* take through the nested networks of environmental systems.

carbon dioxide (CO₂) A colorless gas used by plants for *photosynthesis*, given off by *respiration*, and released by burning *fossil fuels*. A primary *greenhouse gas* whose buildup contributes to *global climate change*.

carbon footprint The cumulative amount of *carbon*, or *carbon dioxide*, that a person or institution emits, and is indirectly responsible for emitting, into the *atmosphere*, contributing to *global climate change*. Compare *ecological footprint*.

carbon monoxide (CO) A colorless, odorless gas produced primarily by the incomplete combustion of fuel. An EPA *criteria pollutant*.

carbon-neutrality The state in which an individual, business, or institution emits no net carbon to the atmosphere. This may be achieved by reducing carbon emissions and/or employing *carbon offsets* to offset emissions.

carbon offset A voluntary payment to another entity intended to enable that entity to reduce the *greenhouse gas* emissions that one is unable or unwilling to reduce oneself. The payment thus offsets one's own emissions.

carbon pricing The practice of putting a price on the emission of *carbon dioxide*, either through *carbon trading* or a *carbon tax*, as a means to address *global climate change*. Carbon pricing compensates the public for the external costs of fossil fuel use by shifting costs to emitters, and creates financial incentives to reduce emissions.

carbon tax A type of *green tax* charged to entities that pollute by emitting *carbon dioxide*. Carbon taxation is one approach to *carbon pricing*, and gives polluters a financial incentive to reduce emissions in order to address *global climate change*. Compare *carbon trading; fee-and-dividend; revenue-neutral carbon tax*.

carbon trading A form of *emissions trading* that focuses on the emission of *carbon dioxide*. In a carbon trading market, emitters buy and sell permits to emit CO₂. Carbon trading is one approach to *carbon pricing*, and gives polluters a financial incentive to reduce emissions in order to address *global climate change*. Compare *carbon tax*.

carcinogen A chemical or type of radiation that causes cancer.

carrying capacity The maximum *population size* of a given organism that a given *environment* can sustain.

case history Medical approach involving the observation and analysis of individual patients.

Cassandra A worldview (or a person holding the worldview) that predicts doom and disaster as a result of our environmental impacts. Compare *Cornucopian*.

catalytic converter Automotive technology that chemically treats engine exhaust to reduce *air pollution*. Reacts exhaust with metals that convert *hydrocarbons*, CO, and NOₓ into *carbon dioxide*, *water* vapor, and *nitrogen* gas.

cellular respiration The process by which a *cell* uses the chemical reactivity of *oxygen* to split glucose into its constituent parts, water and *carbon dioxide*, and thereby release chemical energy that can be used to form chemical bonds or to perform other tasks within the cell. Compare *photosynthesis*.

cellulosic ethanol *Ethanol* produced from the cellulose in plant tissues by treating it with enzymes. Techniques for producing cellulosic ethanol are under development because of the desire to make ethanol from low-value crop waste (residues such as corn stalks and husks), rather than from the sugars of high-value crops.

chaparral A *biome* consisting mostly of densely thicketed evergreen shrubs occurring in limited small patches. Its "Mediterranean" *climate* of mild, wet winters and warm, dry summers is induced by oceanic influences. In addition to ringing the Mediterranean Sea, chaparral occurs along the coasts of California, Chile, and southern Australia.

character displacement A phenomenon resulting from *competition* among *species* in which competing species evolve characteristics that better adapt them to specialize on the portion of the resource they use. The species essentially become more different from one another, reducing their competition.

chemical hazard Chemicals that pose human health hazards. These include *toxins* produced naturally, as well as many of the disinfectants, *pesticides*, and other synthetic chemicals that our society produces. Compare *biological hazard; cultural hazard; physical hazard*.

chemistry The study of the different types of *matter* and how they interact.

chemosynthesis The process by which bacteria in *hydrothermal vents* use the chemical energy of hydrogen sulfide (H_2S) to transform inorganic *carbon* into *organic compounds*. Compare *photosynthesis*.

Chernobyl Site of a nuclear power plant in Ukraine (then part of the Soviet Union), where in 1986 an explosion caused the most severe *nuclear reactor* accident the world has yet seen. The term is also often used to denote the accident itself. Compare *Fukushima Daiichi; Three Mile Island*.

chlorofluorocarbon (CFC) A type of *halocarbon* consisting of only chlorine, fluorine, carbon, and hydrogen. CFCs were used as refrigerants, as fire extinguishers, as propellants for aerosol spray cans, as cleaners for electronics, and for making polystyrene foam. They were phased out under the *Montreal Protocol* because they are *ozone-depleting substances* that destroy stratospheric *ozone*.

chronic exposure Exposure for long periods of time to a *toxicant* occurring in low amounts. Compare *acute exposure*.

city planning The professional pursuit that attempts to design cities in such a way as to maximize their efficiency, functionality, and beauty. Also known as *urban planning*.

classical economics Founded by Adam Smith, the study of the behavior of buyers and sellers in a capitalist market economy. Holds that individuals acting in their own self-interest may benefit society, provided that their behavior is constrained by the rule of law and by private property rights and operates within competitive markets. Compare *neoclassical economics*.

clay *Sediment* consisting of particles less than 0.002 mm in diameter. Compare *sand; silt*.

Clean Air Act U.S. *legislation* to control *air pollution*, first enacted in 1963 and amended multiple times since, most significantly in 1970 and 1990. Funds research into pollution control, sets standards for air quality, encourages emissions standards for automobiles and for stationary point sources such as industrial plants, imposes limits on emissions from new sources, funds a nationwide air quality monitoring system, enables citizens to sue parties violating the standards, and introduced an *emissions trading* program for *sulfur dioxide*.

clean coal technologies An array of techniques, equipment, and approaches to remove chemical contaminants (such as sulfur) during the process of generating *electricity* from *coal*.

clear-cutting The harnessing of timber by cutting all the trees in an area. Although it is the most cost-efficient method, clear-cutting is also the most ecologically damaging.

climate The pattern of atmospheric conditions that typifies a geographic region over long periods of time (typically years, decades, centuries, or millennia). Compare *weather*.

climate change See *global climate change*.

climate diagram A visual representation of a region's average monthly temperature and *precipitation*. Also known as a climatograph.

climate model A computer program that combines what is known about weather patterns, atmospheric circulation, atmosphere–ocean interactions, and feedback mechanisms, to simulate *climate* processes.

climax community In the traditional view of ecological *succession*, a *community* that remains in place with little modification until *disturbance* restarts the successional process. Today, ecologists recognize that community change is more variable and less predictable than originally thought and that assemblages of *species* may instead form complex mosaics in space and time.

coal A solid blackish *fossil fuel* formed from organic matter (generally woody plant material) that was compressed under very high pressure and with little decomposition, creating dense, solid carbon structures.

coevolution The process by which two or more species evolve in response to one another.

Parasites and hosts may coevolve, as may flowering plants and their pollinators.

cogeneration A practice in which the extra heat generated in the production of *electricity* is captured and put to use heating workplaces and homes, as well as producing other kinds of power.

cold front The boundary along which a mass of cold air displaces a mass of warmer air. Compare *warm front*.

colony collapse disorder A mysterious malady afflicting honeybees, which has destroyed roughly one-third of all honeybees in the United States annually over the past decade. Likely caused by chemical insecticides, pathogens and parasites, habitat and resource loss, or combinations of these factors.

command-and-control A top-down approach to policy, in which a legislative body or a regulating agency sets rules, standards, or limits and threatens punishment for violations of those limits.

community In *ecology*, an assemblage of *populations* of interacting organisms that live in the same area at the same time.

community-based conservation The practice of engaging local people to protect land and wildlife in their own region.

community ecology The scientific study of patterns of species diversity and interactions among *species*, ranging from one-to-one interactions to complex interrelationships involving entire *communities*.

community-supported agriculture (CSA) A system in which consumers pay farmers in advance for a share of their yield, usually in the form of weekly deliveries of produce.

competition A relationship in which multiple organisms seek the same limited resource. Competition can take place among members of the same species or among members of different species.

competitive exclusion An outcome of interspecific competition in which one *species* excludes another species from resource use entirely.

compost A mixture produced when *decomposers* break down organic matter, such as food and crop waste, in a controlled environment.

composting The conversion of organic *waste* into mulch or humus by encouraging, in a controlled manner, the natural biological processes of decomposition.

compound A *molecule* whose *atoms* are composed of two or more *elements*.

concentrated solar power (CSP) A means of generating *electricity* at a large scale by focusing sunlight from a large area onto a smaller area. Several approaches are used.

concession The right to extract a resource, granted by a government to a corporation. Compare *conservation concession*.

confined (artesian) aquifer A water-bearing, porous layer of rock, *sand*, or gravel that is trapped between an upper and lower layer of less permeable substrate, such as *clay*. The water in a confined aquifer is under pressure because it is trapped between two impermeable layers. Compare *unconfined aquifer*.

conservation biology A scientific discipline devoted to understanding the factors, forces, and processes that influence the loss, protection, and restoration of *biodiversity* within and among *ecosystems*.

conservation concession A type of *concession* in which a conservation organization purchases the right to prevent resource extraction in an area of land, generally to preserve habitat in developing nations.

conservation ethic An *ethic* holding that people should put *natural resources* to use but also have a responsibility to manage them wisely. Compare *preservation ethic*.

Conservation Reserve Program U.S. policy in farm bills since 1985 that pays farmers to stop cultivating highly erodible cropland and instead place it in conservation reserves planted with grasses and trees.

conservation tillage *Agriculture* that limits the amount of tilling (plowing, disking, harrowing, or chiseling) of *soil*. Compare *no-till*.

consumptive use Use of *fresh water* in which water is removed from a particular *aquifer* or surface water body and is not returned to it. *Irrigation* for *agriculture* is an example of consumptive use. Compare *nonconsumptive use*.

continental collision The meeting of two tectonic plates of continental *lithosphere* at a *convergent plate boundary*, wherein the continental *crust* on both sides resists *subduction* and instead crushes together, bending, buckling, and deforming layers of rock and forcing portions of the buckled crust upward, often creating mountain ranges.

continental shelf The gently sloping underwater edge of a continent, varying in width from 100 m (330 ft) to 1300 km (800 mi), with an average slope of 1.9 m/km (10 ft/mi).

contingent valuation A technique that uses surveys to determine how much people would be willing to pay to protect a resource or to restore it after damage has been done.

contour farming The practice of plowing furrows sideways across a hillside, perpendicular to its slope, to help prevent the formation of rills and gullies. The technique is so named because the furrows follow the natural contours of the land.

contraception The deliberate attempt to prevent pregnancy despite sexual intercourse. Compare *birth control*.

control The portion of an *experiment* in which a *variable* has been left unmanipulated, to serve as a point of comparison with the *treatment*.

controlled experiment An *experiment* in which a *treatment* is compared against a *control* in order to test the effect of a *variable*.

convective circulation A circular *current* (of air, water, magma, etc.) driven by temperature differences. In the atmosphere, warm air rises into regions of lower *atmospheric pressure*, where it expands and cools and then descends and becomes denser, replacing warm air that is rising. The air picks up heat and moisture near ground level and prepares to rise again, continuing the process.

Convention on Biological Diversity A 1992 treaty that aims to conserve *biodiversity,* use biodiversity in a *sustainable* manner, and ensure the fair distribution of biodiversity's benefits.

Convention on International Trade in Endangered Species of Wild Fauna and Flora (CITES) A 1973 treaty facilitated by the *United Nations* that protects *endangered species* by banning the international transport of their body parts.

conventional law International law that arises from *conventions*, or treaties, that nations agree to enter into. Compare *customary law*.

convergent evolution The evolutionary process by which very unrelated species acquire similar traits as they adapt to similar selective pressures from similar environments.

convergent plate boundary The area where tectonic plates converge or come together. Can result in *subduction* or *continental collision*. Compare *divergent plate boundary* and *transform plate boundary*.

coral Tiny marine animals that build *coral reefs*. Corals attach to rock or existing reef and capture passing food with stinging tentacles. They also derive nourishment from photosynthetic symbiotic algae known as *zooxanthellae*.

coral reef A mass of calcium carbonate composed of the skeletons of tiny colonial marine organisms called *corals*.

core The innermost part of Earth, made up mostly of iron, that lies beneath the *crust* and *mantle*.

Coriolis effect The apparent deflection of north–south air *currents* to a partly east–west direction, caused by the faster spin of regions near the equator than of regions near the poles as a result of Earth's rotation.

Cornucopian A worldview (or a person holding the worldview) that we will find ways to make Earth's natural resources meet all of our needs indefinitely and that human ingenuity will see us through any difficulty. Compare *Cassandra*.

correlation Statistical association (positive or negative) among variables. The association may be causal or may occur by chance.

corridor A passageway of protected land established to allow animals to travel between islands of protected *habitat*.

cost-benefit analysis A method commonly used in *neoclassical economics*, in which estimated costs for a proposed action are totaled and then compared to the sum of benefits estimated to result from the action.

covalent bond A type of chemical bonding where atoms share electrons in chemical bonds. An example is a water molecule, which forms when an oxygen atom shares electrons with two hydrogen atoms.

cradle-to-cradle An approach to *waste management* and industrial design in which the materials from products are recovered and reused to create new products.

criteria pollutant One of six *air pollutants*—*carbon monoxide, sulfur dioxide, nitrogen dioxide, tropospheric ozone, particulate matter,* and *lead*—for which the *Environmental Protection Agency* has established maximum allowable concentrations in ambient outdoor air because of the threats they pose to human health.

crop rotation The practice of alternating the kind of crop grown in a particular field from one season or year to the next.

cropland Land that people use to raise plants for food and fiber.

crude oil *Oil* in its natural state, as it occurs once extracted from the ground but before processing and refining.

crust The lightweight outer layer of the Earth, consisting of *rock* that floats atop the malleable *mantle*, which in turn surrounds a mostly iron *core*.

cultural hazard Human health hazards that result from the place we live, our socioeconomic status, our occupation, or our behavioral choices. These include choosing to smoke cigarettes, or living or working with people who do. Also known as *lifestyle hazard*. Compare *biological hazard; chemical hazard; physical hazard*.

culture The overall ensemble of knowledge, beliefs, values, and learned ways of life shared by a group of people.

current The flow of a liquid or gas in a certain direction.

customary law International law that arises from long-standing practices, or customs, held in common by most *cultures*. Compare *conventional law*.

dam Any obstruction placed in a river or stream to block the flow of water so that water can be stored in a *reservoir*. Dams are built to prevent floods, provide drinking water, facilitate *irrigation*, and generate *electricity*.

Darwin, Charles (1809–1882) English naturalist who proposed the concept of *natural selection* as a mechanism for *evolution* and as a way to explain the great variety of living things. Compare *Wallace, Alfred Russel*.

data Information, generally quantitative information.

debt-for-nature swap   A transaction in which a conservation organization pays off a portion of a developing nation's international debt in exchange for a promise by the nation to set aside reserves, fund environmental education, and better manage protected areas.

decomposer   An organism, such as a fungus or bacterium, that breaks down leaf litter and other nonliving matter into simple constituents that can be taken up and used by plants. Compare *detritivore*.

deep-well injection   A *hazardous waste* disposal method in which a well is drilled deep beneath an area's *water table* into porous rock below an impervious *soil* layer. Wastes are then injected into the well, so that they will be absorbed into the porous rock and remain deep underground, isolated from *groundwater* and human contact. Compare *surface impoundment*.

deforestation   The clearing and loss of *forests*.

demographer   A social scientist who studies the population size; density; distribution; age structure; sex ratio; and rates of birth, death, immigration, and emigration of human populations. See *demography*.

demographic fatigue   An inability on the part of governments to address overwhelming challenges related to population growth.

demographic transition   A theoretical *model* of economic and cultural change that explains the declining death rates and birth rates that occurred in Western nations as they became industrialized. The model holds that industrialization caused these rates to fall naturally by decreasing mortality and by lessening the need for large families. Parents would thereafter choose to invest in quality of life rather than quantity of children.

demography   A *social science* that applies the principles of *population ecology* to the study of statistical change in human *populations*.

denitrifying bacteria   Bacteria that convert the nitrates in *soil* or water to gaseous *nitrogen* and release it back into the *atmosphere*.

density-dependent   The condition of a *limiting factor* whose effects on a *population* increase or decrease depending on the *population density*. Compare *density-independent*.

density-independent   The condition of a *limiting factor* whose effects on a *population* are constant regardless of *population density*. Compare *density-dependent*.

deoxyribonucleic acid (DNA)   A double-stranded *nucleic acid* composed of four nucleotides, each of which contains a sugar (deoxyribose), a phosphate group, and a nitrogenous base. DNA carries the hereditary information for living organisms and is responsible for passing traits from parents to offspring. Compare *ribonucleic acid (RNA)*.

dependent variable   The *variable* that is affected by manipulation of the *independent variable* in an *experiment*.

deposition   The arrival of eroded *soil* at a new location. Compare *erosion*.

desalination (desalinization)   The removal of salt from seawater to generate fresh water for human use.

descriptive science   Research in which scientists gather basic information about organisms, materials, systems, or processes that are not yet well known. Compare *hypothesis-driven science*.

desert   The driest *biome* on Earth, with annual *precipitation* of less than 25 cm. Because deserts have relatively little vegetation to insulate them from temperature extremes, sunlight readily heats them in the daytime, but daytime heat is quickly lost at night, so temperatures vary widely from day to night and in different seasons.

desertification   A form of *land degradation* in which more than 10% of a land's productivity is lost due to *erosion,* soil compaction, forest removal, *overgrazing*, drought, *salinization, climate* change, water depletion, or other factors. Severe desertification can result in the expansion of desert areas or creation of new ones. Compare *land degradation; soil degradation*.

detritivore   An organism, such as a millipede or soil insect, that scavenges the waste products or dead bodies of other community members. Compare *decomposer*.

development   The use of natural resources for economic advancement (as opposed to simple subsistence, or survival).

directional drilling   A drilling technique (e.g., for oil or natural gas) in which a drill bores down vertically and then bends horizontally to follow layered deposits for long distances from the drilling site. This enables extracting more *fossil fuels* with less environmental impact on the surface.

discounting   A practice in *neoclassical economics* by which short-term costs and benefits are granted more importance than long-term costs and benefits. Future effects are thereby "discounted," because the idea is that an impact far in the future should count much less than one in the present.

disturbance   An event that affects environmental conditions rapidly and drastically, resulting in changes to the *community* and *ecosystem*. Disturbance can be natural or can be caused by people.

divergent plate boundary   The area where tectonic plates push apart from one another as *magma* rises upward to the surface, creating new *lithosphere* as it cools and spreads. A prime example is the Mid-Atlantic Ridge. Compare *convergent plate boundary* and *transform plate boundary*.

DNA   See *deoxyribonucleic acid*.

dose   The amount of *toxicant* a test animal receives in a dose-response test. Compare *response*.

dose-response analysis   A set of experiments that measure the *response* of test animals to different *doses* of a *toxicant*. The response is generally quantified by measuring the proportion of animals exhibiting negative effects.

dose-response curve   A curve that plots the *response* of test animals to different *doses* of a *toxicant*, as a result of *dose-response analysis*.

downwelling   In the ocean, the flow of warm surface water toward the ocean floor. Downwelling occurs where surface *currents* converge. Compare *upwelling*.

Dust Bowl   An area that loses huge amounts of *topsoil* to wind *erosion* as a result of drought and/or human impact. First used to name the region in the North American Great Plains severely affected by drought and topsoil loss in the 1930s. The term is now also used to describe that historical event and others like it.

dynamic equilibrium   The state reached when processes within a *system* are moving in opposing directions at equivalent rates so that their effects balance out.

e-waste   See *electronic waste*.

earthquake   A release of energy that occurs as Earth relieves accumulated pressure between masses of *lithosphere* and that results in shaking at the surface.

ecocentrism   A philosophy that considers actions in terms of their damage or benefit to the integrity of whole ecological systems, including both living and nonliving elements. For an ecocentrist, the well-being of an individual is less important than the long-term well-being of a larger integrated ecological system. Compare *anthropocentrism* and *biocentrism*.

ecolabeling   The practice of designating on a product's label how the product was grown, harvested, or manufactured, so that consumers can judge which brands use more sustainable processes.

ecological economics   A school of *economics* that applies the principles of *ecology* and *systems* thinking to the description and analysis of *economies*. Compare *environmental economics; neoclassical economics*.

ecological footprint   A concept that measures the cumulative area of biologically productive land and water required to provide the resources a person or population consumes and to dispose of or recycle the waste the person or population produces. The total area of Earth's biologically productive surface that a given person or population "uses" once all direct and indirect impacts are summed together.

ecological modeling   The practice of constructing and testing *models* that aim to explain and predict how ecological *systems* function.

ecological restoration Efforts to reverse the effects of human disruption of ecological systems and to restore *communities* to their condition before the disruption. The practice that applies principles of *restoration ecology*.

ecologist A scientist who studies *ecology*.

ecology The *science* that deals with the distribution and abundance of organisms, the interactions among them, and the interactions between organisms and their nonliving *environments*.

economic growth An increase in an economy's activity—that is, an increase in the production and consumption of goods and services.

economics The study of how we decide to use scarce resources to satisfy demand for *goods* and *services*.

economy A social *system* that converts resources into *goods* and *services*.

ecosystem In *ecology*, an assemblage of all organisms and nonliving entities that occur and interact in a particular area at the same time.

ecosystem-based management The attempt to manage the harvesting of resources in ways that minimize impact on the *ecosystems* and ecological processes that provide the resources.

ecosystem diversity The number and variety of ecosystems in a particular area. One way to express *biodiversity*. Related concepts consider the geographic arrangement of *habitats*, *communities*, or *ecosystems* at the landscape level, including the sizes, shapes, and interconnectedness of patches of these entities.

ecosystem ecology The scientific study of how the living and nonliving components of *ecosystems* interact.

ecosystem services Processes and the results of those processes that naturally result from the normal functioning of ecological systems and from which human beings draw benefits. Examples include nutrient cycling, air and water purification, climate regulation, *pollination*, waste recycling, and more.

ecotone A transitional zone where *ecosystems* meet.

ecotourism Visitation of natural areas for tourism and recreation. Most often involves tourism by more-affluent people, which may generate *economic* benefits for less-affluent communities near natural areas and thereby provide economic incentives for conservation of natural areas.

ED$_{50}$ (effective dose–50%) The amount of a *toxicant* it takes to affect 50% of a *population* of test animals. Compare LD_{50}; *threshold dose*.

edge effect An impact on organisms, populations, or communities that results because conditions along the edge of a habitat fragment differ from conditions in the interior.

El Niño An exceptionally strong warming of the eastern Pacific Ocean that occurs every 2–8 years and depresses local fish and bird *populations* by altering the marine *food web* in the area. Originally, the name that Spanish-speaking fishermen gave to an unusually warm surface *current* that sometimes arrived near the Pacific coast of South America around Christmastime. Compare *La Niña*.

El Niño–Southern Oscillation (ENSO) A systematic shift in atmospheric pressure, sea surface temperature, and ocean circulation in the tropical Pacific Ocean. ENSO cycles give rise to *El Niño* and *La Niña* conditions.

electricity A secondary form of energy that can be transferred over long distances and applied for a variety of uses.

electrolysis A process in which electrical current is passed through a *compound* to release *ions*. Electrolysis offers one way to produce *hydrogen* for use as fuel: Electrical current is passed through water, splitting the water *molecules* into hydrogen and *oxygen atoms*.

electron A negatively charged particle that moves about the nucleus of an *atom*.

electronic waste Discarded electronic products such as computers, monitors, printers, televisions, DVD players, cell phones, and other devices. *Heavy metals* in these products mean that this *waste* may be judged hazardous. Also known as *e-waste*.

element A fundamental type of *matter*; a chemical substance with a given set of properties, which cannot be broken down into substances with other properties. Chemists currently recognize 92 elements that occur in nature, as well as more than 20 others that have been artificially created.

emergent property A characteristic that is not evident in a *system*'s components.

eminent domain A policy in which a government pays landowners for their land at market rates and landowners have no recourse to refuse. In eminent domain, courts set aside private property rights to make way for projects judged to be for the public good.

emissions trading The practice of buying and selling government-issued marketable permits to emit pollutants. Under a *cap-and-trade* system, the government determines an acceptable level of *pollution* and then issues permits to pollute. A company receives credit for amounts it does not emit and can then sell this credit to other companies. Compare *cap-and-trade*.

endangered In danger of becoming extinct in the near future.

Endangered Species Act The primary *legislation*, enacted in 1973, for protecting *biodiversity* in the United States. It forbids the government and private citizens from taking actions (such as developing land) that would destroy *threatened* and *endangered species* or their *habitats*, and it prohibits trade in products made from *threatened* and *endangered* species.

endemic Native or restricted to a particular geographic region. An endemic *species* occurs in one area and nowhere else on Earth.

endocrine disruptor A *toxicant* that interferes with the *endocrine (hormone) system*.

energy The capacity to change the position, physical composition, or temperature of *matter*; a force that can accomplish work.

energy conservation The practice of reducing *energy* use as a way of extending the lifetime of our *fossil fuel* supplies, of being less wasteful, and of reducing our impact on the *environment*. Conservation can result from behavioral decisions or from technologies that demonstrate *energy efficiency*.

energy conversion efficiency The ratio of the useful output of *energy* to the amount that needs to be input. See also *EROI* and *net energy*.

energy efficiency The ability to obtain a given result or amount of output while using less energy input. Technologies permitting greater energy efficiency are one main route to *energy conservation*.

energy intensity A measure of energy use per dollar of *Gross Domestic Product (GDP)*. Lower energy intensity indicates greater efficiency.

enhanced geothermal systems (EGS) A new approach whereby engineers drill deeply into rock, fracture it, pump in water, and then pump it out once it is heated belowground. This approach would enable us to obtain *geothermal energy* in many locations.

environment The sum total of our surroundings, including all of the living things and non-living things with which we interact.

environmental economics A school of *economics* that modifies the principles of *neoclassical economics* to address environmental challenges. Most environmental economists believe that we can attain *sustainability* within our current economic systems. Whereas ecological economists call for revolution, environmental economists call for reform. Compare *ecological economics*; *neoclassical economics*.

environmental ethics The application of *ethical standards* to environmental questions.

environmental health The study of environmental factors that influence human health and quality of life and the health of *ecological* systems essential to environmental quality and long-term human well-being.

environmental impact statement (EIS) A report of results from detailed scientific studies that assess the potential effects on the *environment* that would likely result from development projects or other actions undertaken by the government.

environmental justice The fair and equitable treatment of all people with respect to environmental policy and practice, regardless of their income, race, or ethnicity. This principle is a response to the perception that minorities and the poor suffer more pollution than the majority and the more affluent.

environmental literacy A basic understanding of Earth's physical and living systems and how we interact with them. Some people take the term further and use it to refer to a deeper understanding of society and the environment and/or a commitment to advocate for *sustainability*.

environmental policy *Public policy* that pertains to human interactions with the *environment*. It generally aims to regulate resource use or reduce *pollution* to promote human welfare and/or protect natural systems.

Environmental Protection Agency (EPA) An administrative agency of the U.S. federal government charged with conducting and evaluating research, monitoring environmental quality, setting standards, enforcing those standards, assisting the states in meeting standards and goals for environmental protection, and educating the public.

environmental science The scientific study of how the natural world functions, how our *environment* affects us, and how we affect our environment.

environmental studies An academic *environmental science* program that emphasizes the social sciences as well as the natural sciences.

environmental toxicology The study of *toxicants* that come from or are discharged into the *environment*, including the study of health effects on humans, other animals, and *ecosystems*.

environmentalism A social movement dedicated to protecting the natural world and, by extension, people.

epidemiological study A study that involves large-scale comparisons among groups of people, usually contrasting a group known to have been exposed to some *toxicant* and a group that has not.

EROI (energy returned on investment) The ratio determined by dividing the quantity of *energy* returned from a process by the quantity of energy invested in the process. Higher EROI ratios mean that more energy is produced from each unit of energy invested. Compare *net energy*.

erosion The removal of material from one place and its transport to another by the action of wind or water. Compare *deposition*.

estuary An area where a river flows into the ocean, mixing *fresh water* with saltwater.

ethanol The alcohol in beer, wine, and liquor, produced as a *biofuel* by fermenting biomass, generally from *carbohydrate*-rich crops such as corn or sugarcane.

ethical standard A criterion that helps differentiate right from wrong.

ethics The academic study of good and bad, right and wrong. The term can also refer to a person's or group's set of moral principles or values.

European Union (EU) Political and economic organization formed after World War II to promote Europe's economic and social progress. As of 2016, the EU consisted of 27 member nations.

eutrophic Term describing a water body that has high-nutrient and low-oxygen conditions. Compare *oligotrophic*.

eutrophication The process of *nutrient* enrichment, increased production of organic matter, and subsequent *ecosystem* degradation in a water body.

evaporation The conversion of a substance from a liquid to a gaseous form.

even-aged Condition of timber plantations—generally *monocultures* of a single *species*—in which all trees are of the same age. Most *ecologists* view plantations of even-aged stands more as crop *agriculture* than as ecologically functional *forests*. Compare *uneven-aged*.

evolution Genetically based change in *populations* of organisms across generations. Changes in *genes* may lead to changes in the appearance, physiology, and/or behavior of organisms across generations, often by the process of *natural selection*.

executive order Specific legal instructions for a government agency issued directly by the president.

experiment An activity designed to test the validity of a *hypothesis* by manipulating *variables*. See *controlled experiment*.

exponential growth The increase of a *population* (or of anything) by a fixed percentage each year. Results in a J-shaped curve on a graph. Compare *logistic growth*.

external cost A cost borne by someone not involved in an economic transaction. Examples include harm to citizens from *water pollution* or *air pollution* discharged by nearby factories.

extinction The disappearance of an entire *species* from Earth. Compare *extirpation*.

extirpation The disappearance of a particular *population* from a given area, but not the entire *species* globally. Compare *extinction*.

exurb A region surrounding a city and beyond the suburbs, generally inhabited by affluent individuals seeking even more space than the suburbs provide.

family planning The effort to plan the number and spacing of one's children to offer children and parents the best quality of life possible.

farmers' market A market at which local farmers and food producers sell fresh, locally grown items.

fee-and-dividend A *carbon tax* program in which proceeds from the tax are paid to consumers as a tax refund or "dividend." This strategy seeks to prevent consumers from losing money if polluters pass their costs along to them.

feedback loop A circular process in which a *system*'s output serves as input to that same system. See *negative feedback loop; positive feedback loop*.

feed-in tariff A program of public policy intended to promote renewable energy investment, whereby utilities are mandated to purchase electricity from homeowners or businesses that generate power from renewable energy sources and feed it into the electrical grid. Under such a system, utilities must pay guaranteed premium prices for this power under long-term contract. Compare *net metering*.

feedlot A huge indoor or outdoor pen designed to deliver *energy*-rich food to animals living at extremely high densities. Also called a factory farm or concentrated animal feeding operation.

fertilizer A substance that promotes plant growth by supplying essential *nutrients* such as *nitrogen* or *phosphorus*.

first law of thermodynamics The physical law stating that *energy* can change from one form to another, but cannot be created or lost. The total energy in the universe remains constant and is said to be conserved.

flooding The spillage of water over a river's banks due to heavy rain or snowmelt.

floodplain The region of land over which a river has historically wandered and periodically floods.

flux The movement of nutrients among *reservoirs* in a *nutrient cycle*.

food chain A linear series of feeding relationships. As organisms feed on one another, energy is transferred from lower to higher *trophic levels*. Compare *food web*.

food security The guaranteed availability of an adequate, safe, nutritious, and reliable food supply to all people at all times.

food web A visual representation of feeding interactions within an *ecological community* that shows an array of relationships between organisms at different *trophic levels*. Compare *food chain*.

forensic science The scientific analysis of evidence to make an identification or answer a question relating to a crime or an accident. Often called forensics.

forest Any ecosystem characterized by a high density of trees.

forest type A category of *forest* defined by its predominant tree *species*.

forestry The professional management of *forests*.

fossil The remains, impression, or trace of an animal or plant of past geologic ages that has been preserved in rock or *sediments*.

fossil fuel A *nonrenewable natural resource*, such as *crude oil, natural gas*, or *coal*, produced by the decomposition and compression of organic matter from ancient life. Fossil fuels have provided most of society's *energy* since the *industrial revolution.*

fossil record The cumulative body of *fossils* worldwide, which paleontologists study to infer the history of past life on Earth.

fracking See *hydraulic fracturing.*

Frank R. Lautenberg Chemical Safety for the 21st Century Act U.S. legislation, enacted in 2016, that updates the *Toxic Substances Control Act* and directs the EPA to monitor and regulate industrial chemicals.

free rider A party that fails to invest in conserving resources, controlling *pollution*, or carrying out other responsible activities and instead relies on the efforts of other parties to do so. For example, a factory that fails to control its emissions gets a "free ride" on the efforts of other factories that do.

fresh water Water that is relatively pure, holding very few dissolved salts.

front The boundary between air masses that differ in temperature and moisture (and therefore density). See *warm front; cold front.*

fuel cell A device that can store and transport energy to produce electricity, much as a battery can. A hydrogen fuel cell generates electricity by the input of hydrogen fuel and oxygen, producing only water as a waste product.

Fukushima Daiichi Japanese nuclear power plant severely damaged by the tsunami associated with the March 2011 Tohoku earthquake that rocked Japan. Most radiation drifted over the ocean away from population centers, but the event was history's second most serious nuclear accident. Compare *Chernobyl; Three Mile Island.*

full cost accounting An accounting approach that attempts to summarize all costs and benefits by assigning monetary values to entities without market prices and then generally subtracting costs from benefits. Examples include the *Genuine Progress Indicator*, the Happy Planet Index, and others. Also called *true cost accounting.*

fundamental niche The full *niche* of a species. Compare *realized niche.*

gene A stretch of *DNA* that represents a unit of hereditary information.

General Mining Act of 1872 U.S. law that legalized and promoted *mining* by private individuals on public lands for just $5 per acre, subject to local customs, with no government oversight.

generalist A *species* that can survive across a wide array of *habitats* or that can use a wide array of resources. Compare *specialist.*

genetic diversity A measurement of the differences in *DNA* composition among individuals within a given *species.*

genetic engineering Any process scientists use to manipulate an organism's genetic material in the lab by adding, deleting, or changing segments of its *DNA.*

genetically modified food Food derived from a *genetically modified organism.*

genetically modified organism (GMO) An organism that has been *genetically engineered* using recombinant DNA technology.

gentrification The transformation of a neighborhood to conditions (such as expensive housing and high-end shops and restaurants) that cater to wealthier people. Often results in longtime lower-income residents being "priced out" of their homes or apartments.

Genuine Progress Indicator (GPI) An *economic* indicator that attempts to differentiate between desirable and undesirable economic activity. The GPI accounts for benefits such as volunteerism and for costs such as environmental degradation and social upheaval. Compare *Gross Domestic Product (GDP).*

geoengineering Any of a suite of proposed efforts to cool Earth's climate by removing carbon dioxide from the atmosphere or reflecting sunlight away from Earth's surface. Such ideas are controversial and are not nearly ready to implement.

geographic information system (GIS) Computer software that takes multiple types of data (for instance, on geology, hydrology, vegetation, animal species, and human development) and overlays them on a common set of geographic coordinates. GIS is used to create a complete picture of a landscape and to analyze how elements of the different datasets are arrayed spatially and how they may be correlated. A common tool of geographers, landscape ecologists, resource managers, and conservation biologists.

geology The scientific study of Earth's physical features, processes, and history.

geothermal energy Thermal energy that arises from beneath Earth's surface, ultimately from the radioactive decay of elements amid high pressures deep underground. Can be used to generate electrical power in power plants, for direct heating via piped water, or in *ground-source heat pumps.*

global climate change Systematic change in aspects of Earth's *climate*, such as temperature, *precipitation*, and storm intensity. Generally refers today to the current warming trend in global temperatures and the many associated climatic changes. Compare *global warming.*

global warming An increase in Earth's average surface temperature. The term is most frequently used in reference to the pronounced warming trend of recent decades. Global warming is one aspect of *global climate change* and in turn drives other components of climate change.

globalization The ongoing process by which the world's societies have become more interconnected, linked in many ways by diplomacy, commercial trade, and communication technologies.

Great Pacific Garbage Patch A portion of the North Pacific *gyre* where currents concentrate plastics and other floating debris that pose danger to marine organisms.

green building (1) A structure that minimizes the ecological footprint of its construction and operation by using sustainable materials, using minimal energy and water, reducing health impacts, limiting pollution, and recycling waste. (2) The pursuit of constructing or renovating such buildings.

green-collar job A job resulting from an employment opportunity in a more sustainably oriented *economy*, such as a job in *renewable energy.*

Green Revolution An intensification of the industrialization of *agriculture* in the developing world in the latter half of the 20th century that dramatically increased crop yields produced per unit area of farmland. Practices include devoting large areas to *monocultures* of crops specially bred for high yields and rapid growth; heavy use of *fertilizers, pesticides*, and *irrigation* water; and sowing and harvesting on the same parcel of land more than once per year or per season.

green tax A levy on environmentally harmful activities and products aimed at providing a market-based incentive to correct for *market failure.* Compare *subsidy.*

greenbelt A long and wide corridor of park land, often encircling an entire urban area.

greenhouse effect The warming of Earth's surface and *atmosphere* (especially the *troposphere*) caused by the *energy* emitted by *greenhouse gases.*

greenhouse gas A gas that absorbs infrared radiation released by Earth's surface and then warms the surface and *troposphere* by emitting *energy*, thus giving rise to the *greenhouse effect.* Greenhouse gases include *carbon dioxide* (CO_2), water vapor, *ozone* (O_3), nitrous oxide (N_2O), halocarbon gases, and *methane* (CH_4).

greenwashing A public relations effort by a corporation or institution to mislead customers or the public into thinking it is acting more sustainably than it actually is.

greenway A strip of park land that connects parks or neighborhoods; often located along a river, stream, or canal.

Gross Domestic Product (GDP) The total monetary value of final goods and services produced in a country each year. GDP sums all economic activity, whether good or bad, and does not account for benefits such as volunteerism or for *external costs* such as environmental degradation and social upheaval. Compare *Genuine Progress Indicator (GPI).*

gross primary production The *energy* that results when *autotrophs* convert solar energy (sunlight) to energy of chemical bonds in sugars through *photosynthesis.* Autotrophs use a portion of this production to power their own metabolism, which entails oxidizing *organic compounds* by *cellular respiration*. Compare *net primary production*.

ground-source heat pump   A pump that harnesses *geothermal energy* from near-surface sources of earth and water to heat and cool buildings. Operates on the principle that temperatures belowground are less variable than temperatures aboveground.

groundwater   Water held in *aquifers* underground. Compare *surface water*.

gyre   An area of the ocean where currents converge and floating debris accumulates.

Haber-Bosch process   A process to synthesize ammonia on an industrial scale. Developed by German chemists Fritz Haber and Carl Bosch, the process has enabled humans to double the natural rate of *nitrogen fixation* on Earth and thereby increase *agricultural* productivity, but it has also dramatically altered the *nitrogen cycle*.

habitat   The specific *environment* in which an organism lives, including both biotic (living) and abiotic (nonliving) elements.

habitat fragmentation   The process by which an expanse of natural *habitat* becomes broken up into discontinuous fragments, often as a result of farming, logging, road building, and other types of human development and land use.

habitat selection   The process by which organisms select *habitats* from among the range of options they encounter.

habitat use   The process by which organisms use *habitats* from among the range of options they encounter.

half-life   The amount of time it takes for one-half the atoms of a *radioisotope* to emit radiation and decay. Different radioisotopes have different half-lives, ranging from fractions of a second to billions of years.

halocarbon   A class of human-made chemical *compounds* derived from simple *hydrocarbons* in which *hydrogen* atoms are replaced by halogen atoms such as bromine, fluorine, or chlorine. Many halocarbons are *ozone-depleting substances* and/or *greenhouse gases*.

harmful algal bloom   A *population* explosion of toxic algae caused by excessive *nutrient* concentrations.

hazardous waste   Liquid or solid *waste* that is toxic, chemically reactive, flammable, or corrosive. Compare *industrial solid waste; municipal solid waste*.

heat capacity   A measure of the heat energy required to increase the temperature of a given substance by a given amount.

herbivory   The consumption of plants by animals.

heterotroph (consumer)   An organism that consumes other organisms. Includes most animals, as well as fungi and microbes that decompose organic matter.

high-pressure system   An air mass with elevated *atmospheric pressure*, containing air that descends, typically bringing fair *weather*. Compare *low-pressure system*.

homeostasis   The tendency of a *system* to maintain constant or stable internal conditions.

humus   A dark, spongy, crumbly mass of material made up of complex organic compounds, resulting from the partial decomposition of organic matter.

hurricane   A type of cyclonic storm that forms over the ocean but can do damage upon its arrival on land.

hydraulic fracturing   A process to extract *shale gas*, in which a drill is sent deep underground and angled horizontally into a shale formation; water, sand, and chemicals are pumped in under great pressure, fracturing the rock; and gas migrates up through the drilling pipe as sand holds the fractures open. Also called hydrofracking or simply *fracking*.

hydrocarbon   An *organic compound* consisting solely of *hydrogen* and *carbon atoms*.

hydroelectric power   The generation of *electricity* using the *kinetic energy* of moving water. Also called *hydropower*.

hydrogen   The chemical *element* with one proton. The most abundant element in the universe. Also a possible fuel for our future economy.

hydrogen bond   A weakly attractive interaction between *molecules* due to the attraction of partial positive and partial negative charges.

hydrologic cycle   The flow of water—in liquid, gaseous, and solid forms—through our biotic and abiotic *environment*.

hydropower   See *hydroelectric power*.

hydrosphere   All water—salt or fresh, liquid, ice, or vapor—in surface bodies, underground, and in the *atmosphere*. Compare *biosphere; lithosphere*.

hydrothermal vent   A location in the deep ocean where heated water spurts from the seafloor, carrying *minerals* that precipitate to form rocky structures. Unique and recently discovered *ecosystems* cluster around these vents; tubeworms, shrimp, and other creatures here use symbiotic bacteria to derive their energy from chemicals in the heated water rather than from sunlight.

hypothesis   A statement that attempts to explain a phenomenon or answer a *scientific* question. Compare *theory*.

hypothesis-driven science   Research in which scientists pose questions that seek to explain how and why things are the way they are. Generally proceeds in a somewhat structured manner, using *experiments* to test *hypotheses*. Compare *descriptive science*.

hypoxia   The condition of extremely low dissolved *oxygen* concentrations in a body of water.

igneous rock   One of the three main categories of rock. Formed from cooling *magma*. Granite and basalt are examples of igneous rock. Compare *metamorphic rock* and *sedimentary rock*.

incineration   A controlled process of burning solid *waste* for disposal in which mixed garbage is combusted at very high temperatures. Compare *sanitary landfill*.

independent variable   The *variable* that a scientist manipulates in an *experiment*.

indoor air pollution   *Air pollution* that occurs indoors.

industrial agriculture   *Agriculture* that uses large-scale mechanization and *fossil fuel* combustion, enabling farmers to replace horses and oxen with faster and more powerful means of cultivating, harvesting, transporting, and processing crops. Other aspects include large-scale *irrigation* and the use of *inorganic fertilizers*. Use of chemical herbicides and *pesticides* reduces *competition* from weeds and *herbivory* by insects. Compare *traditional agriculture*.

industrial ecology   A holistic approach to industry that integrates principles from engineering, chemistry, *ecology*, economics, and other disciplines and seeks to redesign industrial *systems* in order to reduce resource inputs and minimize inefficiency.

industrial revolution   The shift beginning in the mid-1700s from rural life, animal-powered agriculture, and manufacturing by craftsmen to an urban society powered by *fossil fuels*. Compare *agricultural revolution*.

industrial smog   "Gray-air" *smog* caused by the incomplete combustion of *coal* or oil when burned. Compare *photochemical smog*.

industrial solid waste   Nonliquid *waste* that is not especially hazardous and that comes from production of consumer goods, mining, *petroleum* extraction and *refining*, and *agriculture*. Compare *hazardous waste; municipal solid waste*.

industrial stage   The third stage of the *demographic transition* model, characterized by falling birth rates that close the gap with falling death rates and reduce the rate of *population* growth. Compare *pre-industrial stage, post-industrial stage, transitional stage*.

infant mortality rate   The number of deaths of infants under 1 year of age per 1000 live births in a population.

infectious disease   A disease in which a pathogen attacks a host. Compare *noninfectious disease*.

inorganic fertilizer   A *fertilizer* that consists of mined or synthetically manufactured mineral supplements. Inorganic fertilizers are generally more susceptible than *organic fertilizers* to *leaching* and *runoff* and may be more likely to cause unintended off-site impacts.

instrumental value   Value ascribed to something for the pragmatic benefits it brings us if we put it to use. Also called utilitarian value.

integrated pest management (IPM)   The use of multiple techniques in combination to achieve long-term suppression of pests, including *biological control*, use of *pesticides*, close monitoring of *populations*, *habitat* alteration, *crop rotation, transgenic* crops, alternative tillage methods, and mechanical pest removal.

intercropping   Planting different types of crops in alternating bands or other spatially mixed arrangements.

interdisciplinary   Involving or borrowing techniques from multiple traditional fields of study and bringing together research results from these fields into a broad synthesis.

Intergovernmental Panel on Climate Change (IPCC)   An international panel of *climate* scientists and government officials established in 1988 by the *United Nations Environment Programme* and the World Meteorological Organization. The IPCC's mission is to assess and synthesize scientific research on *global climate change* and to offer guidance to the world's policymakers, primarily through periodic published reports.

intertidal   Of, relating to, or living along shorelines between the highest reach of the highest *tide* and the lowest reach of the lowest tide.

intrinsic value   Value ascribed to something for its intrinsic worth; the notion that the thing has a right to exist and is valuable for its own sake. Also called inherent value.

introduced species   A species introduced by human beings from one place to another (whether intentionally or by accident). Some introduced species may become *invasive species*.

invasive species   A *species* that spreads widely and rapidly becomes dominant in a *community*, interfering with the community's normal functioning.

inversion layer   A band of air in which temperature rises with altitude (that is, in which the normal direction of temperature change is inverted). Cool air at the bottom of the inversion layer is denser than the warm air above, so it resists vertical mixing and remains stable. A key feature of a *temperature inversion*.

ion   An electrically charged *atom* or combination of atoms.

ionic bond   A type of chemical bonding where electrons are transferred between atoms, creating oppositely charged ions that bond due to their differing electrical charges. Table salt, sodium chloride, is formed by the bonding of positively charged sodium ions with negatively charged chloride ions.

ionizing radiation   A high-energy form of radiation that can damage the cells of living things. Sources of this radiation include the sun and radioactive particles from *nuclear energy* and natural sources.

IPAT model   A formula that represents how humans' total impact (I) on the *environment* results from the interaction among three factors: *population* (P), affluence (A), and technology (T).

irrigation   The artificial provision of water to support *agriculture*.

island biogeography theory   *Theory* initially applied to oceanic islands to explain how *species* come to be distributed among them. Researchers have increasingly applied the theory to islands of *habitat* (patches of one type of habitat isolated within "seas" of others). Aspects of the theory include *immigration* and *extinction* rates, the effect of island size, and the effect of distance from the mainland. Full name is the equilibrium theory of island biogeography.

isotope   One of several forms of an *element* having differing numbers of *neutrons* in the nucleus of its *atoms*. Chemically, isotopes of an element behave almost identically, but they have different physical properties because they differ in mass.

kelp   Large brown algae, or seaweed, that can form underwater "forests," providing habitat for marine organisms.

keystone species   A *species* that has an especially far-reaching effect on a *community*.

kinetic energy   *Energy* of motion. Compare *potential energy*.

Kyoto Protocol   An international agreement drafted in 1997 that called for reducing, by 2012, emissions of six *greenhouse gases* to levels lower than their levels in 1990. It has been extended to 2020 until a replacement treaty can be reached. An outgrowth of the *United Nations Framework Convention on Climate Change*.

La Niña   An exceptionally strong cooling of surface water in the equatorial Pacific Ocean that occurs every 2–8 years and has widespread climatic consequences. Compare *El Niño*.

land degradation   A general deterioration of land that diminishes its productivity and biodiversity, impairs the functioning of its ecosystems, and reduces the ecosystem services the land can offer us. Compare *soil degradation*; *desertification*.

land trust   A local or regional organization that preserves lands valued by its members. In most cases, land trusts purchase land outright with the aim of preserving it in its natural condition.

landfill gas   A mix of gases that consists of roughly half *methane* produced by anaerobic decomposition deep inside *landfills*.

landscape ecology   The study of how landscape structure affects the abundance, distribution, and interaction of organisms. This approach to the study of organisms and their *environments* at the landscape scale focuses on broad geographic areas that include multiple *ecosystems*.

landslide   The collapse and downhill flow of large amounts of rock or soil. A severe and sudden form of *mass wasting*.

lava   *Magma* that is released from the *lithosphere* and flows or spatters across Earth's surface.

law of conservation of matter   The physical law stating that *matter* may be transformed from one type of substance into others, but that it cannot be created or destroyed.

LD_{50} (lethal dose–50%)   The amount of a *toxicant* it takes to kill 50% of a *population* of test animals. Compare ED_{50}; *threshold dose*.

leachate   Liquid that results when substances from waste dissolve in water as rainwater percolates downward. Leachate may sometimes seep through liners of a *sanitary landfill* and leach into the *soil* underneath.

leaching   The process by which *minerals* dissolved in a liquid (usually water) are transported to another location (generally downward through *soil horizons*).

lead   A heavy metal that may be ingested through water or paint, or that may enter the *atmosphere* as a particulate pollutant through combustion of leaded gasoline or other processes. Atmospheric lead deposited on land and water can enter the *food chain*, accumulate within body tissues, and cause *lead poisoning* in animals and people. An EPA *criteria pollutant*.

lead poisoning   Poisoning by ingestion or inhalation of the heavy metal *lead*, causing an array of maladies including damage to the brain, liver, kidney, and stomach; learning problems and behavioral abnormalities; anemia; hearing loss; and even death. Lead poisoning can result from drinking water that passes through old lead pipes or ingesting dust or chips of old lead-based paint.

Leadership in Energy and Environmental Design (LEED)   The leading set of standards for certification of a *green building*.

legislation   Statutory law passed by a legislative body.

Leopold, Aldo (1887–1949)   American scientist, scholar, philosopher, and author. His book *The Land Ethic* argued that humans should view themselves and the land itself as members of the same *community* and that humans are obligated to treat the land *ethically*.

levee   A long raised mound of earth erected along a river bank to protect against floods by holding rising water in the main channel. Synonymous with dike.

life-cycle analysis   A quantitative analysis of inputs and outputs across the entire life cycle of a product—from its origins, through its production, transport, sale, and use, and finally its disposal—in an attempt to judge the sustainability of the process and make it more ecologically efficient.

life expectancy   The average number of years that individuals in particular age groups are likely to continue to live.

life history theory   Scientific study that seeks to explain how *natural selection* influences patterns in reproduction, survival, and life span, and how organisms allocate investment among reproduction, survival, and parental care.

light pollution   Pollution from urban or suburban lights that obscures the night sky, impairing people's visibility of stars.

limiting factor   A physical, chemical, or biological characteristic of the *environment* that restrains *population* growth.

lipids   A class of chemical compounds that do not dissolve in water and are used in organisms for energy storage, for structural support, and as key components of cellular membranes.

lithosphere   The outer layer of Earth, consisting of *crust* and uppermost *mantle* and located just above the *asthenosphere*. More generally, the solid part of Earth, including the rocks, *sediment*, and *soil* at the surface and extending down many miles underground. Compare *atmosphere*, *biosphere*, and *hydrosphere*.

loam   *Soil* with a relatively even mixture of *clay-*, *silt-*, and *sand-*sized particles.

logistic growth   The pattern of population growth that results as a *population* at first grows exponentially and then is slowed and finally brought to a standstill at *carrying capacity* by *limiting factors*. Results in an S-shaped curve on a graph. Compare *exponential growth*.

low-pressure system   An air mass in which air moves toward the low *atmospheric pressure* at the center of the system and spirals upward, typically bringing clouds and *precipitation*. Compare *high-pressure system*.

macromolecule   A very large molecule, such as a *protein*, *nucleic acid*, *carbohydrate*, or *lipid*.

macronutrient   Elements and compounds required in relatively large amounts by organisms. Examples include *nitrogen*, *carbon*, and *phosphorus*.

magma   Molten, liquid rock.

malnutrition   The condition of lacking *nutrients* the body needs, including a complete complement of vitamins and minerals.

Malthus, Thomas (1766–1834)   British economist who maintained that increasing human *population* would eventually deplete the available food supply until starvation, war, or disease arose and reduced the population.

mangrove   A tree with a unique type of roots that curve upward to obtain *oxygen*, which is lacking in the mud in which they grow, or that curve downward to serve as stilts to support the tree in changing water levels. Mangrove forests grow on the coastlines of the tropics and subtropics.

mantle   The malleable layer of rock that lies beneath Earth's *crust* and surrounds a mostly iron *core*.

marine protected area (MPA)   An area of the ocean set aside to protect marine life from fishing pressures. An MPA may be protected from some human activities but be open to others. Compare *marine reserve*.

marine reserve   An area of the ocean designated as a "no-fishing" zone, allowing no extractive activities. Compare *marine protected area (MPA)*.

market failure   The failure of markets to take into account the *environment*'s positive effects on *economies* (for example, *ecosystem services*) or to reflect the negative effects of economic activity on the environment and thereby on people (*external costs*).

mass extinction event   The *extinction* of a large proportion of the world's *species* in a very short time period due to some extreme and rapid change or catastrophic event. Earth has seen five mass extinction events in the past half-billion years.

mass transit   A public transportation system for a metropolitan area that moves large numbers of people at once. Buses, trains, subways, streetcars, trolleys, and light rail are types of mass transit.

mass wasting   The downslope movement of soil and rock due to gravity. Compare *landslide*.

materials recovery facility (MRF)   A *recycling* facility where items are sorted, cleaned, shredded, and prepared for reprocessing into new items.

matter   All material in the universe that has mass and occupies space. See *law of conservation of matter*.

maximum sustainable yield   The maximal harvest of a particular *renewable natural resource* that can be accomplished while still keeping the resource available for the future.

meltdown   The accidental melting of the uranium fuel rods inside the core of a *nuclear reactor*, causing the release of radiation.

meta-analysis   A scientific analysis that gathers together results from all scientific studies on a particular research question and statistically analyzes their data for significant patterns or trends that hold across all of them together. An analysis of analyses.

metal   A type of chemical *element*, or a mass of such an element, that typically is lustrous, opaque, and malleable and that can conduct heat and electricity.

metamorphic rock   One of the three main categories of rock. Formed by great heat and/or pressure that reshapes crystals within the rock and changes its appearance and physical properties. Common metamorphic rocks include marble and slate. Compare *igneous rock* and *sedimentary rock*.

metapopulation   A network of subpopulations, most of whose members stay within their respective landscape *patches*, but some of whom move among patches or mate with members of other patches.

methane hydrate   An ice-like solid consisting of molecules of *methane* embedded in a crystal lattice of water molecules. Most is found in sediments on the continental shelves and in the Arctic. Methane hydrate is an unconventional *fossil fuel*.

micronutrient   Elements and compounds required in relatively small amounts by organisms. Examples include zinc, copper, and iron.

Milankovitch cycle   One of three types of variations in Earth's rotation and orbit around the sun that result in slight changes in the relative amount of solar radiation reaching Earth's surface at different latitudes. As the cycles proceed, they change the way solar radiation is distributed over Earth's surface and contribute to changes in *atmospheric* heating and circulation that have triggered *glaciations* and other *climate* changes.

Millennium Development Goals   A program of targets for *sustainable development* set by the international community through the *United Nations* at the turn of this century.

mineral   A naturally occurring solid *element* or inorganic *compound* with a crystal structure, a specific chemical composition, and distinct physical properties. Compare *ore* and *rock*.

mining   (1) In the broad sense, the extraction of any resource that is nonrenewable on the timescale of our society (such as *fossil fuels* or *groundwater*). (2) In relation to *mineral* resources, the systematic removal of *rock*, *soil*, or other material for the purpose of extracting minerals of economic interest.

mitigation   The pursuit of strategies to lessen the severity of climate change, notably by reducing emissions of greenhouse gases. Compare *adaptation*.

model   A simplified representation of a complex natural process, designed by scientists to help understand how the process occurs and to make predictions.

molecule   A combination of two or more *atoms*.

monoculture   The uniform planting of a single crop over a large area. Characterizes *industrial agriculture*. Compare *polyculture*.

Montreal Protocol   International treaty ratified in 1987 in which 180 (now 196) signatory nations agreed to restrict production of *chlorofluorocarbons (CFCs)* in order to halt stratospheric *ozone* depletion. This was a protocol of the Vienna Convention for the Protection of the Ozone Layer. The Montreal Protocol is widely considered the most successful effort to date in addressing a global *environmental* problem.

mosaic In *landscape ecology*, a spatial configuration of *patches* arrayed across a landscape.

mountaintop removal mining A large-scale form of *coal* mining in which entire mountaintops are blasted away in order to extract the resource. While economically efficient, large volumes of rock and soil generally slide downhill, causing extensive impacts on surrounding *ecosystems* and human residents.

Muir, John (1838–1914) Scottish immigrant to the United States who eventually settled in California and made the Yosemite Valley his wilderness home. Today, he is most strongly associated with the *preservation ethic*. He argued that nature deserved protection for its own *intrinsic value* (an *ecocentrist* argument) but also claimed that nature facilitated human happiness and fulfillment (an *anthropocentrist* argument).

multiple use A principle guiding management policy for *national forests* specifying that forests be managed for recreation, wildlife habitat, mineral extraction, water quality, and other uses, as well as for timber extraction.

municipal solid waste Nonliquid *waste* that is not especially hazardous and that comes from homes, institutions, and small businesses. Compare *hazardous waste; industrial solid waste*.

mutagen A *toxicant* that causes *mutations* in the *DNA* of organisms.

mutation An accidental change in *DNA* that may range in magnitude from the deletion, substitution, or addition of a single nucleotide to a change affecting entire sets of chromosomes. Mutations provide the raw material for evolutionary change.

mutualism A relationship in which all participating organisms benefit from their interaction. Compare *parasitism*.

National Environmental Policy Act (NEPA) A U.S. law enacted on January 1, 1970, that created an agency called the Council on Environmental Quality and requires that an *environmental impact statement* be prepared for any major federal action.

national forest An area of forested public land managed by the U.S. Forest Service. The system consists of 191 million acres (more than 8% of the nation's land area) in many tracts spread across all but a few states.

National Forest Management Act *Legislation* passed by the U.S. Congress in 1976, mandating that plans for renewable resource management be drawn up for every *national forest*. These plans were to be explicitly based on the concepts of *multiple use* and *maximum sustainable yield* and be open to broad public participation.

national park A scenic area set aside for recreation and enjoyment by the public and managed by the National Park Service. The U.S. national park system today numbers more than 400 sites totaling 84 million acres and includes national historic sites, national recreation areas, national wild and scenic rivers, and other areas.

national wildlife refuge An area of public land set aside to serve as a haven for wildlife and also sometimes to encourage hunting, fishing, wildlife observation, photography, environmental education, and other uses. The system of more than 560 sites is managed by the U.S. Fish and Wildlife Service.

natural capital Earth's accumulated wealth of *natural resources* and *ecosystem services*.

natural gas A *fossil fuel* consisting primarily of *methane* (CH_4) and including varying amounts of other volatile hydrocarbons.

natural resource Any of the various substances and *energy* sources that we take from our environment and that we need in order to survive.

natural sciences Academic disciplines that study the natural world. Compare *social sciences*.

natural selection The process by which traits that enhance survival and reproduction are passed on more frequently to future generations of organisms than traits that do not, thereby altering the genetic makeup of populations through time. Natural selection acts on genetic variation and is a primary driver of *evolution*.

negative feedback loop A *feedback loop* in which output of one type acts as input that moves the *system* in the opposite direction. The input and output essentially neutralize each other's effects, stabilizing the system. Compare *positive feedback loop*.

neoclassical economics A mainstream economic school of thought that explains market prices in terms of consumer preferences for units of particular commodities and that uses *cost-benefit analysis*. Compare *classical economics; ecological economics; environmental economics*.

neonicotinoid A new class of chemical insecticide. Neonicotinoids are sprayed on plants or used to coat seeds. When seeds are treated, the poison becomes systemic throughout the plant, dispersing through its tissues as it grows and making the tissues toxic to insects.

net energy The quantitative difference between *energy* returned from a process and *energy* invested in the process. Positive net energy values mean that a process produces more energy than is invested. Compare *EROI*.

net metering Process by which homeowners or businesses with *photovoltaic* systems or *wind turbines* can sell their excess *solar energy* or *wind power* to their local utility.

Whereas *feed-in tariffs* award producers with prices above market rates, net metering offers market-rate prices.

net primary production The *energy* or biomass that remains in an ecosystem after *autotrophs* have metabolized enough for their own maintenance through *cellular respiration*. Net primary production is the energy or biomass available for consumption by *heterotrophs*. Compare *gross primary production; secondary production*.

net primary productivity The rate at which *net primary production* is produced. See *productivity; gross primary production; net primary production; secondary production*.

neurotoxin A *toxicant* that assaults the nervous system. Neurotoxins include heavy metals, *pesticides*, and some chemical weapons developed for use in war.

neutron An electrically neutral (uncharged) particle in the nucleus of an *atom*.

new forestry A set of *ecosystem-based management* approaches for harvesting timber that explicitly mimic natural disturbances. For instance, "sloppy clear-cuts" that leave a variety of trees standing mimic the changes a *forest* might experience if hit by a severe windstorm.

new urbanism An approach among architects, planners, and developers that seeks to design neighborhoods in which homes, businesses, schools, and other amenities are within walking distance of one another, so that families can meet most of their needs close to home without the use of a car.

niche The functional role of a *species* in a *community*.

nitrification The conversion by bacteria of ammonium ions (NH_4^+) first into nitrite ions (NO_2^-) and then into nitrate ions (NO_3^-).

nitrogen The chemical *element* with seven *protons* and seven *neutrons*. The most abundant element in the *atmosphere*, a key element in *macromolecules*, and a crucial plant *nutrient*.

nitrogen cycle A major *nutrient cycle* consisting of the routes that *nitrogen atoms* take through the nested networks of environmental *systems*.

nitrogen dioxide (NO_2) A foul-smelling reddish brown gas that contributes to *smog* and *acid deposition*. It results when atmospheric *nitrogen* and *oxygen* react at the high temperatures created by combustion engines. An EPA *criteria pollutant*.

nitrogen fixation The process by which inert *nitrogen* gas combines with *hydrogen* to form ammonium ions (NH_4^+), which are chemically and biologically active and can be taken up by plants.

nitrogen-fixing bacteria Bacteria that live independently in the soil or water, or those that form *mutualistic* relationships with many types of plants and provide *nutrients* to the plants by converting gaseous *nitrogen* to a usable form.

nitrogen oxide (NO_x) One of a family of compounds that includes nitric oxide (NO) and *nitrogen dioxide (NO_2)*.

no-analog community An ecological *community* composed of a novel mixture of organisms, with no current analog or historical precedent.

no-till *Agriculture* that does not involve tilling (plowing, disking, harrowing, or chiseling) the *soil*. The most intensive form of *conservation tillage*.

noise pollution Undesired ambient sound.

nonconsumptive use Use of *fresh water* in which the water from a particular *aquifer* or surface water body either is not removed or is removed only temporarily and then returned. The use of water to generate electricity in hydroelectric *dams* is an example. Compare *consumptive use*.

nongovernmental organization (NGO) An organization not affiliated with any national government, and frequently international in scope, that pursues a particular mission or advocates for a particular cause.

noninfectious disease A disease that develops as a result of the interaction of an individual organism's genes, lifestyle, and environmental exposures, rather than by pathogenic infection. Compare *infectious disease*.

nonmarket value A value that is not usually included in the price of a *good* or *service*.

non-point source A diffuse source of *pollutants*, often consisting of many small sources. Compare *point source*.

nonrenewable natural resources *Natural resources* that are in limited supply and are formed much more slowly than we use them. Compare *renewable natural resources*.

North Atlantic Deep Water (NADW) The deep portion of the *thermohaline circulation* in the northern Atlantic Ocean.

not-in-my-backyard (NIMBY) Syndrome in which people do not want something (e.g., a polluting facility) near where they live, even if they may want or need the thing to exist somewhere.

nuclear energy The *energy* that holds together *protons* and *neutrons* within the nucleus of an *atom*. Several processes, each of which involves transforming *isotopes* of one *element* into isotopes of other elements, can convert nuclear energy into thermal energy, which is then used to generate *electricity*. See also *nuclear fission; nuclear power; nuclear reactor*.

nuclear fission The conversion of the *energy* within an *atom's* nucleus to usable thermal energy by splitting apart atomic nuclei. Compare *nuclear fusion*.

nuclear fusion The conversion of the *energy* within an *atom's* nucleus to usable thermal energy by forcing together the small nuclei of lightweight *elements* under extremely high temperature and pressure. Developing a commercially viable method of nuclear fusion remains an elusive goal.

nuclear power The use of *nuclear energy* to generate *electricity*. This is accomplished using *nuclear fission* within *nuclear reactors* in power plants.

nuclear reactor A facility within a nuclear power plant that initiates and controls the process of *nuclear fission* to generate electricity.

nucleic acid A *macromolecule* that directs the production of *proteins*. Includes *DNA* and *RNA*.

nutrient An *element* or *compound* that organisms consume and require for survival.

nutrient cycle The comprehensive set of cyclical pathways by which a given *nutrient* moves through the *environment*.

ocean acidification The process by which today's oceans are becoming more *acidic* (attaining lower *pH*) as a result of increased *carbon dioxide* concentrations in the atmosphere. Ocean acidification occurs as ocean water absorbs carbon dioxide from the air and forms carbonic acid. This impairs the ability of corals and other organisms to build exoskeletons of calcium carbonate, imperiling coral reefs and the many organisms that depend on them.

ocean thermal energy conversion (OTEC) An *energy* source (not yet commercially used) that involves harnessing the solar radiation absorbed by tropical ocean water by strategically manipulating the movement of warm surface water and cold deep water.

oil A *fossil fuel* produced by the slow underground conversion of *organic compounds* by heat and pressure. Oil is a mixture of hundreds of different types of *hydrocarbon* molecules characterized by *carbon* chains of different lengths. Compare *crude oil; petroleum*.

oil sands *Fossil fuel* deposits that can be mined from the ground, consisting of moist sand and clay containing 1–20% bitumen. Oil sands represent *crude oil* deposits that have been degraded and chemically altered by water *erosion* and bacterial decomposition. Also called *tar sands*.

oil shale *Sedimentary rock* filled with kerogen that can be processed to produce liquid *petroleum*. Oil shale is formed by the same processes that form *crude oil* but occurs when kerogen was not buried deeply enough or subjected to enough heat and pressure to form oil.

oligotrophic Term describing a water body that has low-nutrient and high-oxygen conditions. Compare *eutrophic*.

open pit mining A *mining* technique that involves digging a gigantic hole and removing the desired *ore*, along with waste *rock* that surrounds the ore.

ore A *mineral* or grouping of minerals from which we extract *metals*.

organic agriculture *Agriculture* that uses no synthetic *fertilizers* or *pesticides* but instead relies on biological approaches such as *composting* and *biological control*.

organic compound A *compound* made up of *carbon atoms* (and, generally, *hydrogen* atoms) joined by covalent bonds and sometimes including other *elements*, such as *nitrogen*, *oxygen*, sulfur, or *phosphorus*. The unusual ability of carbon to build elaborate *molecules* has resulted in millions of different organic compounds showing various degrees of complexity.

organic fertilizer A *fertilizer* made up of natural materials (largely the remains or wastes of organisms), such as animal manure, crop residues, charcoal, fresh vegetation, and *compost*. Compare *inorganic fertilizer*.

Organization of Petroleum Exporting Countries (OPEC) A cartel of predominantly Middle Eastern nations that collaborate to determine rates and policies for the extraction and sale of their *oil* and *natural gas* on the world market.

outdoor air pollution *Air pollution* that occurs outdoors. Also called ambient air pollution.

overgrazing The consumption by too many animals of plant cover, impeding plant regrowth and the replacement of *biomass*. Overgrazing can worsen damage to *soils*, natural *communities*, and the land's productivity for further grazing.

overnutrition A condition of excessive food intake in which people receive more than their daily caloric needs.

overshoot The amount by which humanity's resource use, as measured by its *ecological footprint*, has surpassed Earth's long-term capacity to support us.

oxygen The chemical *element* with eight *protons* and eight *neutrons*. A key *element* in the *atmosphere* that is produced by *photosynthesis*.

ozone-depleting substance One of a number of airborne chemicals, such as *halocarbons*, that destroy *ozone* molecules and thin the *ozone layer* in the *stratosphere*.

ozone hole Term popularly used to describe the thinning of the stratospheric *ozone layer* that occurs over Antarctica each year, as a result of *chlorofluorocarbons (CFCs)* and other *ozone-depleting substances*.

ozone layer A portion of the *stratosphere*, roughly 17–30 km (10–19 mi) above sea level, that contains most of the *ozone* in the *atmosphere*.

paleoclimate Climate in the geologic past.

paradigm A dominant philosophical and theoretical framework within a scientific discipline.

parasitism A relationship in which one organism, the parasite, depends on another, the host, for nourishment or some other benefit while simultaneously doing the host harm. Compare *mutualism; predation.*

parent material   The base geologic material in a particular location.

particulate matter   Solid or liquid particles small enough to be suspended in the *atmosphere* and able to damage respiratory tissues when inhaled. Includes *primary pollutants,* such as dust and soot, as well as *secondary pollutants,* such as sulfates and nitrates. An EPA *criteria pollutant.*

passive solar energy collection   An approach in which buildings are designed and building materials are chosen to maximize direct absorption of sunlight in winter and to keep the interior cool in the summer. Compare *active solar energy collection.*

patch   In *landscape ecology,* spatial areas within a landscape. Depending on a researcher's perspective, patches may consist of habitat for a particular organism, or communities, or ecosystems. An array of patches forms a *mosaic.*

pathogen   A parasite that causes disease in its host.

pathway inhibitor   A *toxicant* that interrupts vital biochemical processes in organisms by blocking one or more steps in important biochemical pathways. Compounds in the herbicide atrazine kill plants by blocking key steps in the process of *photosynthesis.*

peak oil   Term used to describe the point of maximum production of *petroleum* in the world (or for a given nation), after which oil production declines. This is expected to be roughly the midway point of extraction of the world's oil supplies.

peer review   The process by which a scientific manuscript submitted for publication in an academic journal is examined by specialists in the field, who provide comments and criticism (generally anonymously) and judge whether the work merits publication in the journal.

pelagic   Of, relating to, or living between the surface and floor of the ocean. Compare *benthic.*

pest   A pejorative term for any organism that damages crops that we value. The term is subjective and defined by our own economic interests and is not biologically meaningful. Compare *weed.*

pesticide   An artificial chemical used to kill insects (called an insecticide), plants (called an herbicide), or fungi (called a fungicide).

petroleum   See *oil.* However, the term is also used to refer to both *oil* and *natural gas* together.

pH   A measure of the concentration of *hydrogen ions* in a *solution.* The pH scale ranges from 0 to 14: A solution with a pH of 7 is neutral; solutions with a pH below 7 are *acidic,* and those with a pH higher than 7 are *basic.* Because the pH scale is logarithmic, each step on the scale represents a 10-fold difference in hydrogen ion concentration.

phosphorus   The chemical element with 15 *protons* and 15 *neutrons.* An abundant element in the *lithosphere,* a key element in *macromolecules,* and a crucial plant *nutrient.*

phosphorus cycle   A major *nutrient cycle* consisting of the routes that *phosphorus atoms* take through the nested networks of environmental *systems.*

photic zone   In the ocean or a freshwater body, the well-lit top layer of water where *photosynthesis* occurs.

photochemical smog   "Brown-air" *smog* formed by light-driven reactions of *primary pollutants* with normal atmospheric *compounds* that produce a mix of over 100 different chemicals, *tropospheric ozone* often being the most abundant among them. Compare *industrial smog.*

photosynthesis   The process by which *autotrophs* produce their own food. Sunlight powers a series of chemical reactions that convert *carbon dioxide* and water into sugar (glucose), thus transforming low-quality *energy* from the sun into high-quality energy the organism can use. Compare *cellular respiration.*

photovoltaic (PV) cell   A device designed to collect sunlight and directly convert it to electrical *energy.* When light strikes one of a pair of metal plates in the cell, this causes the release of *electrons,* which are attracted by electrostatic forces to the opposing plate. The flow of electrons from one plate to the other creates an electrical current. The basis of PV solar technology.

phthalates   A class of endocrine-disrupting chemicals found in fire retardants and plasticizers.

phylogenetic tree   A treelike diagram that represents the history of divergence of *species* or other taxonomic groups of organisms.

physical hazard   Physical processes that occur naturally in our environment and pose human health hazards. These include discrete events such as *earthquakes, volcanic eruptions,* fires, floods, blizzards, *landslides,* hurricanes, and droughts, as well as ongoing natural phenomena such as ultraviolet radiation from sunlight. Compare *biological hazard; chemical hazard; cultural hazard.*

phytoplankton   Microscopic photosynthetic algae, protists, and cyanobacteria that drift near the surface of water bodies and generally form the first *trophic level* in an aquatic *food chain.* Compare *zooplankton.*

Pinchot, Gifford (1865–1946)   The first professionally trained American forester, Pinchot helped establish the U.S. Forest Service. Today, he is the person most closely associated with the *conservation ethic.*

pioneer species   A *species* that arrives earliest, beginning the ecological process of *succession* in a terrestrial or aquatic *community.*

placer mining   A *mining* technique that involves sifting through material in modern or ancient riverbed deposits, generally using running water to separate lightweight mud and gravel from heavier *minerals* of value.

plastics   Synthetic (human-made) *polymers* used in numerous manufactured products.

plate tectonics   The process by which Earth's surface is shaped by the extremely slow movement of tectonic plates, or sections of *crust.* Earth's surface includes about 15 major tectonic plates. Their interaction gives rise to processes that build mountains, cause *earthquakes,* and otherwise influence the landscape.

poaching   The illegal killing of wildlife, usually for meat or body parts.

point source   A specific spot—such as a factory—where large quantities of *air pollutants* or *water pollutants* are discharged. Compare *non-point source.*

policy   A rule or guideline that directs individual, organizational, or societal behavior.

pollination   A plant-animal interaction in which one organism (for example, a bee or a hummingbird) transfers pollen (containing male sex cells) from flower to flower, fertilizing ovaries (containing female sex cells) that grow into fruits with seeds.

polluter-pays principle   Principle specifying that the party responsible for producing *pollution* should pay the costs of cleaning up the pollution or mitigating its impacts.

polybrominated diphenyl ethers (PBDEs)   Synthetic compounds that provide fire-retardant properties and are used in a diverse array of consumer products, including computers, televisions, plastics, and furniture. Released during production, disposal, and use of products, these chemicals persist and accumulate in living tissue and appear to be *endocrine disruptors.*

polyculture   The planting of multiple crops in a mixed arrangement or in close proximity. An example is some traditional Native American farming that mixed maize, beans, squash, and peppers. Compare *monoculture.*

polymer   A chemical *compound* or mixture of compounds consisting of long chains of repeated *molecules.* Important biological molecules, such as *DNA* and *proteins,* are examples of polymers.

population   A group of organisms of the same *species* that live in the same area. Species are often composed of multiple populations.

population density   The number of individuals within a *population* per unit area. Compare *population size.*

population distribution   The spatial distribution of organisms in an area. Three common patterns are random, uniform, and clumped.

population ecology The scientific study of the quantitative dynamics of population change and the factors that affect the distribution and abundance of members of a *population*.

population growth rate The rate of change in a *population*'s size per unit time (generally expressed in percent per year), taking into accounts births, deaths, immigration, and emigration. Compare *rate of natural increase*.

population size The number of individual organisms present at a given time in a *population*.

positive feedback loop A *feedback loop* in which output of one type acts as input that moves the *system* in the same direction. The input and output drive the system further toward one extreme or another. Compare *negative feedback loop*.

post-industrial stage The fourth and final stage of the *demographic transition* model, in which both birth and death rates have fallen to a low level and remain stable there, and *populations* may even decline slightly. Compare *industrial stage, pre-industrial stage, transition stage*.

potential energy *Energy* of position. Compare *kinetic energy*.

precautionary principle The idea that one should not undertake a new action until the ramifications of that action are well understood.

precipitation Water that condenses out of the *atmosphere* and falls to Earth in droplets or crystals.

precision agriculture The use of technology to precisely monitor crop conditions, crop needs, and resource use to maximize production while minimizing waste of resources.

predation The process in which one *species* (the predator) searches for, captures, and ultimately kills its prey. Compare *parasitism*.

prediction A specific statement, generally arising from a *hypothesis*, that can be tested directly and unequivocally.

pre-industrial stage The first stage of the *demographic transition* model, characterized by conditions that defined most of human history. In pre-industrial societies, both death rates and birth rates are high. Compare *industrial stage, post-industrial stage, transitional stage*.

prescribed fire The practice of burning areas of *forest* or grassland under carefully controlled conditions to improve the health of *ecosystems*, return them to a more natural state, reduce fuel loads, and help prevent uncontrolled catastrophic fires.

preservation ethic An ethic holding that we should protect the natural *environment* in a pristine, unaltered state. Compare *conservation ethic*.

primary consumer An organism that consumes *producers* and feeds at the second *trophic level*.

primary forest Natural *forest* uncut by people. Compare *secondary forest*.

primary pollutant A hazardous substance, such as soot or *carbon monoxide*, that is emitted into the *troposphere* directly from a source. Compare *secondary pollutant*.

primary production The conversion of solar energy to the energy of chemical bonds in sugars during *photosynthesis*, performed by *autotrophs*. Compare *secondary production*.

primary succession A stereotypical series of changes as an ecological *community* develops over time, beginning with a lifeless substrate. In terrestrial *systems*, primary succession begins when a bare expanse of rock, *sand*, or *sediment* becomes newly exposed to the atmosphere and *pioneer species* arrive. Compare *secondary succession*.

primary treatment A stage of *wastewater* treatment in which contaminants are physically removed. Wastewater flows into tanks in which sewage solids, grit, and particulate matter settle to the bottom. Greases and oils float to the surface and can be skimmed off. Compare *secondary treatment*.

probability A quantitative description of the likelihood of a certain outcome.

producer An organism that uses energy from sunlight to produce its own food. Includes green plants, algae, and cyanobacteria.

productivity The rate at which plants convert solar *energy* (sunlight) to *biomass*. *Ecosystems* whose plants convert solar energy to biomass rapidly are said to have high productivity. See *net primary productivity; gross primary production; net primary production*.

protein A *macromolecule* made up of long chains of amino acids.

proton A positively charged particle in the nucleus of an *atom*.

proven recoverable reserve The amount of a given *fossil fuel* in a deposit that is technologically and economically feasible to remove under current conditions.

proxy indicator A source of indirect evidence that serves as a proxy, or substitute, for direct measurement and that sheds light on past climate. Examples include data from ice cores, sediment cores, tree rings, packrat middens, and coral reefs.

public policy *Policy* made by governments, including those at the local, state, federal, and international levels; it may consist of *legislation, regulations*, orders, incentives, and practices intended to advance societal welfare. See also *environmental policy*.

public trust doctrine A legal philosophy that holds that natural resources such as air, water, soil, and wildlife should be held in trust for the public and that government should protect them from exploitation by private interests.

This doctrine has its roots in ancient Roman law and in the Magna Carta.

pumped storage A technique used to generate *hydroelectric power*, in which water is pumped from a lower reservoir to a higher reservoir when power demand is weak and prices are low. When demand is strong and prices are high, water is allowed to flow downhill through a turbine, generating *electricity*. Compare *run-of-river, storage*.

pycnocline A zone of the ocean beneath the surface in which density increases rapidly with depth.

radiative forcing The amount of change in thermal energy that a factor (such as a *greenhouse gas* or an *aerosol*) causes in influencing Earth's temperature. Positive forcing warms Earth's surface, whereas negative forcing cools it.

radioactive The quality by which some *isotopes* "decay," changing their chemical identity as they shed atomic particles and emit high-energy radiation.

radioisotope A radioactive *isotope* that emits subatomic particles and high-*energy* radiation as it "decays" into progressively lighter isotopes until becoming a stable isotope.

radon A highly *toxic*, radioactive, colorless gas that seeps up from the ground in areas with certain types of bedrock and that can build up inside basements and homes with poor air circulation.

rainshadow A region on one side of a mountain or mountain range that experiences arid climate. This occurs because moisture-laden air rising over the terrain from the opposite direction releases precipitation on the windward slope as it cools, leaving the air's humidity low as it descends over the peak and into the rainshadow region.

rangeland Land used for grazing livestock.

rate of natural increase The rate of change in a *population*'s size resulting from birth and death rates alone, excluding migration. Compare *population growth rate*.

REACH Program of the European Union that shifts the burden of proof for testing chemical safety from national governments to industry and requires that chemical substances produced or imported in amounts of over 1 metric ton per year be registered with a new European Chemicals Agency. *REACH*, which stands for Registration, Evaluation, Authorisation, and Restriction of Chemicals, went into effect in 2007.

realized niche The portion of the *fundamental niche* that is fully realized (used) by a *species*.

rebound effect The phenomenon by which gains in efficiency from better technology are partly offset when people engage in more energy-consuming behavior as a result. This common psychological effect can sometimes reduce conservation and efficiency efforts substantially.

recharge zone   An area where water infiltrates Earth's surface and reaches an *aquifer* below.

reclamation   The act of restoring a *mining* site to an approximation of its pre-mining condition. To reclaim a site, companies are required to remove all mining structures, replace *overburden*, fill in mine shafts, and replant the area with vegetation.

recovery   Waste management strategy composed of *recycling* and *composting*.

recycling   The process by which materials are collected and then broken down and reprocessed to manufacture new items.

Red List   An updated list of *species* facing unusually high risks of *extinction*. The list is maintained by the World Conservation Union.

red tide   A *harmful algal bloom* consisting of algae that produce reddish pigments that discolor surface waters.

Reducing Emissions from Deforestation and Forest Degradation (REDD)   A proposed international program, still being developed, to encourage the conservation of *forests* globally for the purpose of reducing *greenhouse gas* emissions to control *climate change*. A key mechanism is the transfer of funds from wealthy nations to poorer forest-rich nations. Initially abbreviated as REDD, the program became known as REDD+ as it expanded in scope.

refining   The process of separating *molecules* of the various *hydrocarbons* in *crude oil* into different-sized classes and transforming them into various fuels and other petrochemical products.

regime shift   A fundamental shift in the overall character of an ecological community, generally occurring after some extreme *disturbance*, and after which the community may not return to its original state. Also known as a *phase shift*.

regional planning   Planning similar to *city planning* but conducted across broader geographic scales, generally involving multiple municipal governments.

regulation   A specific rule issued by an administrative agency, based on the more broadly written statutory law passed by Congress and enacted by the president.

regulatory taking   The deprivation of a property's owner, by means of a law or *regulation*, of most or all economic uses of that property.

relative humidity   The ratio of water vapor that air contains at a given temperature to the maximum amount it could contain at that temperature.

relativist   An ethicist who maintains that *ethics* do and should vary with social context. Compare *universalist*.

renewable natural resources   *Natural resources* that are virtually unlimited or that are replenished by the *environment* over relatively short periods (hours to weeks to years). Compare *nonrenewable natural resources*.

replacement fertility   The *total fertility rate (TFR)* that maintains a stable *population* size.

reproductive window   The portion of a woman's life between sexual maturity and menopause during which she may become pregnant.

reserves-to-production ratio   The total remaining reserves of a *fossil fuel* divided by the annual rate of production (extraction and processing). Abbreviated as R/P ratio.

reservoir   (1) An artificial water body behind a dam that stores water for human use. (2) A location in which nutrients in a *biogeochemical cycle* remain for a period of time before moving to another reservoir. Can be living or nonliving entities. Compare *flux; residence time*.

residence time   (1) In a *biogeochemical cycle*, the amount of time a nutrient typically remains in a given *reservoir* before moving to another. Compare *flux*. (2) In the *atmosphere*, the amount of time a gas molecule or a *pollutant* typically remains aloft.

resilience   The ability of an ecological *community* to change in response to disturbance but later return to its original state. Compare *resistance*.

resistance   The ability of an ecological *community* to remain stable in the presence of a disturbance. Compare *resilience*.

Resource Conservation and Recovery Act   U.S. law (enacted in 1976 and amended in 1984) that specifies, among other things, how to manage *sanitary landfills* to protect against environmental contamination. Often abbreviated as RCRA.

resource management   Strategic decision making about how to extract resources, so that resources are used wisely and conserved for the future.

resource partitioning   The process by which *species* adapt to *competition* by evolving to use slightly different resources, or to use their shared resources in different ways, thus minimizing interference with one another.

response   The type or magnitude of negative effects an animal exhibits in response to a *dose* of *toxicant* in a *dose-response analysis*. Compare *dose*.

restoration ecology   The study of the historical conditions of ecological *communities* as they existed before humans altered them. Principles of restoration ecology are applied in the practice of *ecological restoration*.

revenue-neutral carbon tax   A type of *fee-and-dividend* program in which funds from the *carbon tax* that a government collects are disbursed to citizens in the form of payments or tax refunds. It is "revenue-neutral" because the government neither gains nor loses revenue in the end.

revolving door   The movement of individuals working in the private sector to government agencies or vice versa. May create conflicts of interest.

ribonucleic acid (RNA)   A usually single-stranded *nucleic acid* composed of four nucleotides, each of which contains a sugar (ribose), a phosphate group, and a nitrogenous base. RNA carries the hereditary information for living organisms and is responsible for passing traits from parents to offspring. Compare *deoxyribonucleic acid (DNA)*.

risk   The mathematical probability that some harmful outcome (for instance, injury, death, *environmental* damage, or *economic* loss) will result from a given action, event, or substance.

risk assessment   The quantitative measurement of *risk*, together with the comparison of risks involved in different activities or substances.

risk management   The process of considering information from scientific *risk assessment* in light of economic, social, and political needs and values, to make decisions and design strategies to minimize *risk*.

RNA   See *ribonucleic acid*.

roadless rule   A 2001 Clinton administration executive order that put 31% of national forest land off-limits to road construction or maintenance.

rock   A solid aggregation of *minerals*.

rock cycle   The very slow process in which *rocks* and the *minerals* that make them up are heated, melted, cooled, broken, and reassembled, forming *igneous*, *sedimentary*, and *metamorphic* rocks.

run-of-river   Any of several methods used to generate *hydroelectric power* without greatly disrupting the flow of river water. Run-of-river approaches eliminate much of the *environmental* impact of large *dams*. Compare *pumped storage; storage*.

runoff   The water from *precipitation* that flows into streams, rivers, lakes, and ponds, and (in many cases) eventually to the ocean.

salinization   The buildup of salts in surface *soil* layers.

salt marsh   Flat land that is intermittently flooded by the ocean where the *tide* reaches inland. Salt marshes occur along temperate coastlines and are thickly vegetated with grasses, rushes, shrubs, and other herbaceous plants.

salvage logging   The removal of dead trees following a natural disturbance. Although it may be economically beneficial, salvage logging can be ecologically destructive, because *snags* provide food and shelter for wildlife and because removing timber from recently burned land can cause *erosion* and damage to *soil*.

sand   *Sediment* consisting of particles 0.005–2.0 mm in diameter. Compare *clay; silt*.

sanitary landfill A site at which solid waste is buried in the ground or piled up in large mounds for disposal, designed to prevent the waste from contaminating the *environment*. Compare *incineration*.

savanna A *biome* characterized by grassland interspersed with clusters of acacias and other trees. Savanna is found across parts of Africa (where it was the ancestral home of our *species*), South America, Australia, India, and other dry tropical regions.

science (1) A systematic process for learning about the world and testing our understanding of it. (2) The accumulated body of knowledge that arises from this dynamic process.

scientific method A formalized method for testing ideas with observations that involves a more-or-less consistent series of interrelated steps.

scrubber Technology to chemically treat gases produced in combustion in order to reduce smokestack emissions. These devices typically remove hazardous components and neutralize acidic gases, such as *sulfur dioxide* and hydrochloric acid, turning them into water and salt.

second law of thermodynamics The physical law stating that the nature of *energy* tends to change from a more-ordered state to a less-ordered state; that is, *entropy* increases.

secondary consumer An organism that consumes *primary consumers* and feeds at the third *trophic level*.

secondary forest *Forest* that has grown back after *primary forest* has been cut. Consists of second-growth trees.

secondary pollutant A hazardous substance produced through the reaction of primary pollutants with one another or with other constituents of the *atmosphere*. Compare *primary pollutant*.

secondary production The total *biomass* that *heterotrophs* generate by consuming *autotrophs*. Compare *primary production*.

secondary succession A stereotypical series of changes as an *ecological community* develops over time, beginning when some *disturbance* disrupts or dramatically alters an existing community. Compare *primary succession*.

secondary treatment A stage of *wastewater* treatment in which biological means are used to remove contaminants remaining after *primary treatment*. Wastewater is stirred up in the presence of *aerobic* bacteria, which degrade organic pollutants in the water. The wastewater then passes to another settling tank, where remaining solids drift to the bottom. Compare *primary treatment*.

sediment The eroded remains of *rocks*.

sedimentary rock One of the three main categories of rock. Formed when dissolved *minerals* seep through *sediment* layers and act as a kind of glue, crystallizing and binding sediment particles together. Sandstone and shale are examples of sedimentary rock. Compare *igneous rock* and *metamorphic rock*.

seed bank A storehouse for samples of seeds representing the world's crop diversity.

seed-tree Timber harvesting approach that leaves small numbers of mature and vigorous seed-producing trees standing so that they can reseed a logged area.

selection system Method of timber harvesting whereby single trees or groups of trees are selectively cut while others are left, creating an *uneven-aged* stand.

septic system A *wastewater* disposal method, common in rural areas, consisting of an underground tank and series of drainpipes. Wastewater runs from the house to the tank, where solids precipitate out. The water proceeds downhill to a drain field of perforated pipes laid horizontally in gravel-filled trenches, where microbes decompose the remaining waste.

sex ratio The proportion of males to females in a *population*.

shale gas Natural gas trapped deep underground in tiny bubbles dispersed throughout formations of shale, a type of sedimentary rock. Shale gas is often extracted by *hydraulic fracturing*.

shale oil A liquid form of petroleum extracted from deposits of *oil shale*.

shelterbelt A row of trees or other tall perennial plants that are planted along the edges of farm fields to break the wind and thereby minimize wind *erosion*.

shelterwood Timber harvesting approach that leaves small numbers of mature trees in place to provide shelter for seedlings as they grow.

sick building syndrome A building-related illness produced by indoor *pollution* in which the specific cause is not identifiable.

silicon The chemical *element* with 14 *protons* and 14 *neutrons*. An abundant element in rocks in Earth's crust.

silt *Sediment* consisting of particles 0.002–0.005 mm in diameter. Compare *clay; sand*.

sink In a *nutrient cycle*, a *reservoir* that accepts more *nutrients* than it releases.

sinkhole An area where the ground has given way with little warning as a result of subsidence caused by depletion of water from an *aquifer*.

slash-and-burn A mode of agriculture frequently used in the tropics in which natural vegetation is cut and then burned, adding nutrition to the soil, before farming begins. Generally farmers move on to another plot once the soil fertility is depleted.

SLOSS Abbreviation for "single large or several small." The debate over whether it is better to make reserves large in size and few in number or many in number but small in size.

smart growth A *city planning* concept in which a community's growth is managed in ways intended to limit *sprawl* and maintain or improve residents' quality of life.

smelting A process in which *ore* is heated beyond its melting point and combined with other *metals* or chemicals to form metal with desired characteristics. Steel is created by smelting iron ore with carbon, for example.

smog Term popularly used to describe unhealthy mixtures of air *pollutants* that often form over urban and industrial areas as a result of fossil fuel combustion. See *industrial smog; photochemical smog*.

snag A dead tree that is still standing. Snags are valuable for wildlife.

social cost of carbon An estimate of the total economic cost of damages resulting from the emission of *carbon dioxide* (from *fossil fuel* burning, *deforestation*, etc.) and resulting *global climate change*, on a per-ton basis. Estimates vary widely, but the U.S. government currently uses an estimate of roughly $40/ton CO_2.

social sciences Academic disciplines that study human interactions and institutions. Compare *natural sciences*.

socially responsible investing Investing in companies that have met criteria for environmental or social sustainability.

soil A complex plant-supporting *system* consisting of disintegrated rock, organic matter, air, water, *nutrients*, and microorganisms.

soil degradation A deterioration of soil quality and decline in soil productivity, resulting primarily from forest removal, cropland agriculture, and *overgrazing* of livestock. Compare *land degradation; desertification*.

soil horizon A distinct layer of *soil*.

soil profile The cross-section of a *soil* as a whole, including all *soil horizons* from the surface to the *bedrock*.

solar energy Energy from the sun. Solar energy is perpetually renewable and may be harnessed in several ways.

solution mining A *mining* technique in which a narrow borehole is drilled deep into the ground to reach a *mineral* deposit, and water, acid, or another liquid is injected down the borehole to *leach* the resource from the surrounding rock and dissolve it in the liquid. The resulting solution is then sucked out, and the desired resource is isolated.

source In a *nutrient cycle*, a *reservoir* that releases more *nutrients* than it accepts.

source reduction The reduction of the amount of material that enters the *waste stream* to avoid the costs of disposal and *recycling*, help conserve resources, minimize *pollution*, and save consumers and businesses money.

specialist   A *species* that can survive only in a narrow range of *habitats* or that depends on very specific resources. Compare *generalist*.

speciation   The process by which new *species* are generated. In one common mechanism, allopatric speciation, species form in the aftermath of the physical separation of populations over some geographic distance.

species   A *population* or group of populations of a particular type of organism whose members share certain characteristics and can breed freely with one another and produce fertile offspring. Different biologists may have different approaches to diagnosing species boundaries.

species-area curve   A graph showing how number of *species* varies with the geographic area of a landmass or water body. Number of species commonly doubles as area increases 10-fold.

species coexistence   An outcome of interspecific competition in which no competing *species* fully excludes others and the species continue to live side by side.

species diversity   The number and variety of *species* in the world or in a particular region.

sprawl   The unrestrained spread of urban or suburban development outward from a city center and across the landscape. Often specified as growth in which the area of development outpaces *population* growth.

steady-state economy   An *economy* that does not grow or shrink but remains stable.

storage   Technique used to generate *hydroelectric power*, in which large amounts of water are impounded in a reservoir behind a concrete *dam* and then passed through the dam to turn *turbines* that generate *electricity*. Compare *pumped storage; run-of-river*.

stratosphere   The layer of the *atmosphere* above the *troposphere;* it contains the *ozone layer* and extends 11–50 km (7–31 mi) above sea level.

strip mining   The use of heavy machinery to remove huge amounts of earth to expose *coal* or *minerals*, which are mined out directly. Compare *subsurface mining*.

subcanopy   The middle level of trees in a *forest*, beneath the *canopy*.

subduction   The *plate tectonic* process by which denser *crust* slides beneath lighter crust at a *convergent plate boundary*. Often results in *volcanism*.

subsidy   A government grant of money or resources to a private entity, intended to support and promote an industry or activity. A tax break is one type of subsidy. Compare *green tax*.

subsurface mining   Method of *mining* underground deposits of *coal, minerals*, or *fuels*, in

which shafts are dug deeply into the ground and networks of tunnels are dug or blasted out to follow coal seams. Compare *strip mining*.

suburb   A smaller community located at the outskirts of a city.

succession   A stereotypical series of changes in the composition and structure of an *ecological community* through time. See *primary succession; secondary succession*.

sulfur dioxide (SO$_2$)   A colorless gas that can result from the combustion of *coal*. In the *atmosphere*, it may react to form sulfur trioxide and sulfuric acid, which may return to Earth in *acid deposition*. An EPA *criteria pollutant*.

Superfund   A program administered by the *Environmental Protection Agency* in which experts identify sites polluted with hazardous chemicals, protect *groundwater* near these sites, and clean up the *pollution*. Established by the Comprehensive Environmental Response Compensation and Liability Act (CERCLA) in 1980.

surface impoundment   (1) A disposal method for *hazardous waste* or mining waste in which waste in liquid or slurry form is placed into a shallow depression lined with impervious material such as clay and allowed to evaporate, leaving a solid residue on the bottom. (2) The site of such disposal. Compare *deep-well injection*.

surface water   Water located atop Earth's surface. Compare *groundwater*.

sustainability   A guiding principle of *environmental science*, entailing conserving resources, maintaining functional ecological systems, and developing long-term solutions, such that Earth can sustain our civilization and all life for the future, allowing our descendants to live at least as well as we have lived.

sustainable agriculture   *Agriculture* that can be practiced in the same way and in the same place far into the future. Sustainable agriculture does not deplete *soils* faster than they form, nor reduce the clean water, genetic diversity, pollinators, and other resources essential to long-term crop and livestock production.

sustainable development   Development that satisfies our current needs without compromising the future availability of *natural capital* or our future quality of life.

sustainable forest certification   A form of *ecolabeling* that identifies timber products that have been produced using *sustainable* methods. The Forest Stewardship Council (FSC) and several other organizations issue such certification.

symbiosis   A relationship between different *species* of organisms that live in close physical proximity. People most often use the term "symbiosis" when referring to a *mutualism*, but symbiotic relationships can be either parasitic or mutualistic.

synergistic effect   An interactive effect (as of *toxicants*) that is more than or different from the simple sum of their constituent effects.

system   A network of relationships among a group of parts, elements, or components that interact with and influence one another through the exchange of *energy*, matter, and/or information.

tailings   Portions of *ore* left over after *metals* have been extracted in *mining*.

tar sands   See *oil sands*.

temperate deciduous forest   A *biome* consisting of midlatitude *forests* characterized by broad-leafed trees that lose their leaves each fall and remain dormant during winter. These forests occur in areas where *precipitation* is spread relatively evenly throughout the year: much of Europe, eastern China, and eastern North America.

temperate grassland   A *biome* whose vegetation is dominated by grasses and features more extreme temperature differences between winter and summer and less *precipitation* than *temperate deciduous forests*. Also known as steppe or prairie.

temperate rainforest   A *biome* consisting of tall coniferous trees; cooler and less *species*-rich than *tropical rainforest* and milder and wetter than *temperate deciduous forest*.

temperature inversion   A departure from the normal temperature distribution in the *atmosphere*, in which a pocket of relatively cold air occurs near the ground, with warmer air above it. The cold air, denser than the air above it, traps *pollutants* near the ground and can thereby cause a buildup of *smog*. Also called a thermal inversion.

teratogen   A *toxicant* that causes harm to the unborn, resulting in birth defects.

terracing   The cutting of level platforms, sometimes with raised edges, into steep hillsides to contain water from *irrigation* and *precipitation*. Terracing transforms slopes into series of steps like a staircase, enabling farmers to cultivate hilly land while minimizing their loss of *soil* to water *erosion*.

tertiary consumer   An organism that consumes *secondary consumers* and feeds at the fourth *trophic level*.

theory   A widely accepted, well-tested explanation of one or more cause-and-effect relationships that has been extensively validated by a great amount of research. Compare *hypothesis*.

thermohaline circulation   A worldwide system of ocean currents in which warmer, fresher water moves along the surface and colder, saltier water (which is denser) moves deep beneath the surface.

thin-film solar cell   A photovoltaic material compressed into an ultra-thin lightweight sheet that may be incorporated into various surfaces to produce photovoltaic solar power.

threatened Likely to become *endangered* soon.

Three Mile Island Nuclear power plant in Pennsylvania that in 1979 experienced a partial *meltdown.* The term is often used to denote the accident itself, the most serious *nuclear reactor* malfunction that the United States has thus far experienced. Compare *Chernobyl; Fukushima Daiichi.*

threshold dose The amount of a *toxicant* at which it begins to affect a *population* of test animals. Compare ED_{50}; LD_{50}.

tidal energy *Energy* harnessed by erecting a *dam* across the outlet of a tidal basin. Water flowing with the incoming or outgoing *tide* through sluices in the dam turns turbines to generate *electricity.*

tide The periodic rise and fall of the ocean's height at a given location, caused by the gravitational pull of the moon and sun.

topsoil That portion of the *soil* that is most nutritive for plants and is thus of the most direct importance to *ecosystems* and to *agriculture.* A *soil horizon* also known as the A horizon.

tornado A type of cyclonic storm in which warm air rises quickly in a funnel, potentially lifting up soil and objects and threatening life and great damage to property.

total fertility rate (TFR) The average number of children born per female member of a *population* during her lifetime.

toxic air pollutant *Air pollutant* that is known to cause cancer, reproductive defects, or neurological, developmental, immune system, or respiratory problems in humans, and/or to cause substantial ecological harm by affecting the health of nonhuman animals and plants. The *Clean Air Act* identifies 187 toxic air pollutants, ranging from the heavy metal mercury to *volatile organic compounds (VOCs)* such as benzene and methylene chloride.

Toxic Substances Control Act (TSCA) A 1976 U.S. law that directs the *Environmental Protection Agency* to monitor thousands of industrial chemicals and gives the EPA authority to regulate and ban substances found to pose excessive risk.

toxicant A substance that acts as a poison to humans or wildlife.

toxicity The degree of harm a chemical substance can inflict.

toxicology The scientific field that examines the effects of poisonous chemicals and other agents on humans and other organisms.

toxin A *toxic* chemical stored or manufactured in the tissues of living organisms. For example, a chemical that plants use to ward off *herbivores* or that insects use to deter predators.

traditional agriculture *Agriculture* in which human and animal muscle power, along with hand tools and simple machines, performs the work of cultivating, harvesting, storing, and distributing crops. Compare *industrial agriculture.*

tragedy of the commons The process by which publicly accessible resources open to unregulated use tend to become damaged and depleted through overuse. Term was coined by Garrett Hardin and is widely applicable to resource issues.

transform plate boundary The area where two tectonic plates meet and slip and grind alongside one another, creating *earthquakes.* For example, the Pacific Plate and the North American Plate rub against each other along California's San Andreas Fault. Compare *convergent plate boundary* and *divergent plate boundary.*

transgene A *gene* that has been extracted from the *DNA* of one organism and transferred into the DNA of an organism of another *species.*

transgenic Term describing an organism that contains *DNA* from another *species.*

transit-oriented development A development approach in which compact communities in the *new urbanism* style are arrayed around stops on a major rail transit line.

transitional stage The second stage of the *demographic transition* model, which occurs during the transition from the *pre-industrial stage* to the *industrial stage.* It is characterized by declining death rates but continued high birth rates. See also *post-industrial stage.* Compare *industrial stage, post-industrial stage, pre-industrial stage.*

transpiration The release of water vapor by plants through their leaves.

treatment The portion of an *experiment* in which a *variable* has been manipulated in order to test its effect. Compare *control.*

triple bottom line An approach to sustainability that attempts to meet environmental, economic, and social goals simultaneously.

trophic cascade A series of changes in the *population* sizes of organisms at different *trophic levels* in a *food chain,* occurring when predators at high trophic levels indirectly promote populations of organisms at low trophic levels by keeping species at intermediate trophic levels in check. Trophic cascades may become apparent when a top predator is eliminated from a system.

trophic level Rank in the feeding hierarchy of a *food chain.* Organisms at higher trophic levels consume those at lower trophic levels.

tropical dry forest A *biome* that consists of deciduous trees and occurs at tropical and subtropical latitudes where wet and dry seasons each span about half the year. Widespread in India, Africa, South America, and northern Australia. Also known as *tropical deciduous forest.*

tropical rainforest A *biome* characterized by year-round rain and uniformly warm temperatures. Found in Central America, South America, Southeast Asia, west Africa, and other tropical regions. Tropical rainforests have dark, damp interiors; lush vegetation; and highly diverse biotic communities.

troposphere The bottommost layer of the *atmosphere;* it extends to 11 km (7 mi) above sea level.

tropospheric ozone *Ozone* that occurs in the *troposphere,* where it is a *secondary pollutant* created by the interaction of sunlight, heat, *nitrogen oxides,* and volatile *carbon*-containing chemicals. A major component of *photochemical smog,* it can injure living tissues and cause respiratory problems. An EPA *criteria pollutant.* Also called ground-level ozone.

tsunami An immense swell, or wave, of ocean water triggered by an *earthquake, volcano,* or *landslide* that can travel long distances across oceans and inundate coasts.

tundra A *biome* that is nearly as dry as *desert* but is located at very high latitudes along the northern edges of Russia, Canada, and Scandinavia. Extremely cold winters with little daylight and moderately cool summers with lengthy days characterize this landscape of lichens and low, scrubby vegetation.

unconfined aquifer A water-bearing, porous layer of rock, *sand,* or gravel that lies atop a less permeable substrate. The water in an unconfined aquifer is not under pressure because there is no impermeable upper layer to confine it. Compare *confined aquifer.*

undernutrition A condition of insufficient nutrition in which people receive fewer calories than are needed on a daily basis for a healthy diet.

understory The layer of a *forest* consisting of small shrubs and trees above the forest floor and below the *subcanopy,* usually shaded by foliage above it.

uneven-aged Term describing stands consisting of trees of different ages. Uneven-aged stands more closely approximate a natural *forest* than do *even-aged* stands.

United Nations (UN) Organization founded in 1945 to promote international peace and to cooperate in solving international economic, social, cultural, and humanitarian problems.

United Nations Framework Convention on Climate Change (FCCC) An international treaty signed in 1992 outlining a plan to reduce emissions of *greenhouse gases.* Gave rise to the *Kyoto Protocol.*

universalist An ethicist who maintains that there exist objective notions of right and wrong that hold across cultures and situations. Compare *relativist*.

upwelling In the ocean, the flow of cold, deep water toward the surface. Upwelling occurs in areas where surface *currents* diverge. Compare *downwelling*.

uranium The chemical *element* with 92 *protons* and 92 *neutrons*. Uranium is used as a fuel source to produce electricity with *nuclear energy*.

urban ecology A scientific field of study that views cities explicitly as *ecosystems*. Researchers in this field apply the fundamentals of *ecosystem ecology* and *systems* science to urban areas.

urban growth boundary A line on a map established to separate areas zoned to be high-density and urban from areas intended to remain low-density and rural. The aim is to control *sprawl*, revitalize cities, and preserve the rural character of outlying areas.

urban heat island effect The phenomenon whereby a city becomes warmer than outlying areas because of the concentration of heat-generating buildings, vehicles, and people, and because buildings and dark paved surfaces absorb heat and release it at night.

urban planning See *city planning*.

urbanization A population's shift from rural living to city and suburban living.

variable In an *experiment*, a condition that can change. See *dependent variable* and *independent variable*.

vector An organism that transfers a *pathogen* to its host. An example is a mosquito that transfers the malaria pathogen to humans.

volatile organic compound (VOC) One of a large group of potentially harmful organic chemicals used in industrial processes. One of six major pollutants whose emissions are monitored by the *EPA* and state agencies.

volcano A site where molten rock, hot gas, or ash erupts through Earth's surface, often creating a mountain over time as cooled *lava* accumulates.

Wallace, Alfred Russel (1823–1913) English naturalist who proposed, independently of *Charles Darwin*, the concept of *natural selection* as a mechanism for *evolution* and as a way to explain the great variety of living things.

warm front The boundary along which a mass of warm air displaces a mass of colder air. Compare *cold front*.

waste Any unwanted material or substance that results from a human activity or process.

waste management Strategic decision making to minimize the amount of *waste* generated and to dispose of waste safely and effectively.

waste stream The flow of *waste* as it moves from its sources toward disposal destinations.

waste-to-energy (WTE) facility An incinerator that uses heat from its furnace to boil water to create steam that drives *electricity* generation or that fuels heating systems.

wastewater Any water that is used in households, businesses, industries, or public facilities and is drained or flushed down pipes, as well as the polluted *runoff* from streets and storm drains.

water A *compound* composed of two hydrogen atoms bonded to one oxygen atom, denoted by the chemical formula H_2O.

water pollution The release of matter or *energy* into waters that causes undesirable impacts on the health and well-being of humans or other organisms. Water pollution can be physical, chemical, or biological.

water table The upper limit of *groundwater* held in an *aquifer*.

waterlogging The saturation of *soil* by water, in which the *water table* is raised to the point that water bathes plant roots. Waterlogging deprives roots of access to gases, essentially suffocating them and damaging or killing the plants.

watershed The entire area of land from which water drains into a given river.

wave energy *Energy* harnessed from the motion of ocean waves. Many designs for machinery to harness wave energy have been invented, but few have been adequately tested.

weather The local physical properties of the *troposphere*, such as temperature, pressure, humidity, cloudiness, and wind, over relatively short time periods (typically minutes, hours, days, or weeks). Compare *climate*.

weathering The process by which rocks and *minerals* are broken down, turning large particles into smaller particles. Weathering may proceed by physical, chemical, or biological means.

weed A pejorative term for any plant that competes with our crops. The term is subjective and defined by our own economic interests, and is not biologically meaningful. Compare *pest*.

wetland A system in which the soil is saturated with water and which generally features shallow standing water with ample vegetation. These biologically productive systems include freshwater marshes, swamps, bogs, and seasonal wetlands such as vernal pools.

wilderness area Federal land that is designated off-limits to development of any kind but is open to public recreation, such as hiking, nature study, and other activities that have minimal impact on the land.

wildland-urban interface A region where urban or suburban development meets forested or undeveloped lands.

wind farm A development involving a group of *wind turbines*.

wind power A source of *renewable energy*, in which *kinetic energy* from the passage of wind through *wind turbines* is used to generate *electricity*.

wind turbine A mechanical assembly that converts the wind's *kinetic energy*, or energy of motion, into electrical energy.

work When a force acts on an object, causing it to be displaced (to move in space).

World Bank Institution founded in 1944 that serves as one of the globe's largest sources of funding for economic development, including such major projects as *dams*, *irrigation* infrastructure, and other undertakings.

world heritage site A location internationally designated by the *United Nations* for its cultural or natural value. There are more than 1000 such sites worldwide.

World Trade Organization (WTO) Organization based in Geneva, Switzerland, that represents multinational corporations and promotes free trade by reducing obstacles to international commerce and enforcing fairness among nations in trading practices.

worldview A way of looking at the world that reflects a person's (or a group's) beliefs about the meaning, purpose, operation, and essence of the world.

xeriscaping Landscaping using plants that are adapted to arid conditions.

zoning The practice of classifying areas for different types of development and land use.

zooplankton Tiny aquatic animals that feed on *phytoplankton* and generally constitute the second *trophic level* in an aquatic *food chain*. Compare *phytoplankton*.

zooxanthellae Symbiotic algae that inhabit the bodies of *corals* and produce food through *photosynthesis*.

Selected Sources and References for Further Reading

Chapter 1

Bahn, Paul, and John Flenley. 1992. *Easter Island, Earth island*. Thames and Hudson, London.

Blomqvist, L., et al. 2013. Does the shoe fit? Real versus imagined ecological footprints. *PLOS Biology* 11:1–6.

Brown, Lester R. 2009. *Plan B 4.0: Mobilizing to save civilization*. Earth Policy Institute and W.W. Norton, New York.

Brown, Lester R. 2011. *World on the edge: How to prevent environmental and economic collapse*. Earth Policy Institute and W.W. Norton, New York.

Campus Ecology. National Wildlife Federation. http://www.nwf.org/campus-ecology.

Diamond, Jared. 2005. *Collapse: How societies choose to fail or succeed*. Viking, New York.

Flenley, John, and Paul Bahn. 2003. *The enigmas of Easter Island*. Oxford University Press.

Global Footprint Network. http://www.footprintnetwork.org. Oakland, Calif.

Harrison, Paul, and Fred Pearce, eds. 2000. *AAAS atlas of population & environment*. University of California Press, Berkeley.

Hunt, Terry L., and Carl P. Lipo. 2006. Late colonization of Easter Island. *Science* 311: 1603–1606.

Hunt, Terry L., and Carl P. Lipo. 2011. *The statues that walked: Unraveling the mystery of Easter Island*. Simon & Schuster.

Kuhn, Thomas S. 1962. *The structure of scientific revolutions*, 2nd ed., 1970. University of Chicago Press, Chicago.

Lomborg, Bjorn. 2001. *The skeptical environmentalist: Measuring the real state of the world*. Cambridge University Press.

Millennium Ecosystem Assessment. 2005. *Ecosystems and human well-being: General synthesis*. Millennium Ecosystem Assessment and World Resources Institute.

Musser, George. 2005. The climax of humanity. *Scientific American* 293(3): 44–47.

Ponting, Clive. 1991. *A green history of the world: The environment and the collapse of great civilizations*. Penguin Books, New York.

Popper, Karl R. 1959. *The logic of scientific discovery*. Hutchinson, London.

Redman, Charles R. 1999. *Human impact on ancient environments*. University of Arizona Press, Tucson.

Sagan, Carl. 1997. *The demon-haunted world: Science as a candle in the dark*. Ballantine Books, New York.

Siever, Raymond. 1968. Science: Observational, experimental, historical. *American Scientist* 56: 70–77.

U.N. Environment Programme. 2011. *Keeping track of our changing environment: From Rio to Rio+20 (1992–2012)*. UNEP, Nairobi.

U.N. Environment Programme. 2012. *Global environment outlook 5 (GEO-5)*. UNEP, Nairobi.

Valiela, Ivan. 2001. *Doing science: Design, analysis, and communication of scientific research*. Oxford University Press.

Van Tilburg, Jo Anne. 1994. *Easter Island: Archaeology, ecology, and culture*. Smithsonian Institution Press, Washington, D.C.

Wackernagel, Mathis, and William Rees. 1996. *Our ecological footprint: Reducing human impact on the Earth*. New Society Publishers, Gabriola Island, British Columbia, Canada.

Wackernagel, Mathis, et al. 2002. Tracking the ecological overshoot of the human economy. *Proc. Natl. Acad. Sci. USA* 99: 9266–9271.

Worldwatch Institute. 2013. *State of the world 2013: Is sustainability still possible?* Worldwatch Institute, Washington, D.C.

Worldwatch Institute. 2015. *State of the world 2015: Confronting hidden threats to sustainability*. Worldwatch Institute, Washington, D.C.

Worldwatch Institute. 2015. *Vital signs, volume 22*. Worldwatch Institute, Washington, D.C. http://vitalsigns.worldwatch.org.

WWF–World Wide Fund for Nature. 2014. *Living planet report 2014*. WWF, Gland, Switzerland.

Chapter 2

Berardelli, Phil. 2008. Human-driven planet: Time to make it official? *Science NOW*. 24 Jan. 2008. http://www.sciencemag.org/news/2008/01/human-driven-planet-time-make-it-official.

Berry, R. Stephen. 1991. *Understanding energy: Energy, entropy and thermodynamics for every man*. World Scientific Publishing Co.

Buesseler, Ken O., et al. 2012. Fishing for answers off Fukushima. *Science* 26: 480–482.

Buesseler, Ken O., et al. 2012. Fukushima-derived radionuclides in the ocean and biota off Japan. *Proc. Natl. Acad. Sci. USA* 109: 5984–5988.

Christopherson, Robert W. 2011. *Geosystems: An introduction to physical geography*, 8th ed. Prentice Hall, Upper Saddle River, N.J.

Craig, James R., David J. Vaughan, and Brian J. Skinner. 2010. *Earth resources and the environment*, 4th ed. Benjamin Cummings, San Francisco.

Keller, Edward A. 2011. *Introduction to environmental geology*, 8th ed. Prentice Hall, Upper Saddle River, N.J.

Keller, Edward A., and Robert H. Blodgett. 2011. *Natural hazards: Earth's processes as hazards, disasters, and catastrophes*, 3rd ed. Prentice Hall, Upper Saddle River, N.J.

Keller, Edward A., and Nicholas Pinter. 2001. *Active tectonics: Earthquakes, uplift, and landscape*, 2nd ed. Prentice Hall, Upper Saddle River, N.J.

Keranen, Katie M., et al. 2013. Potentially induced earthquakes in Oklahoma, USA: Links between wastewater injection and the 2011 Mw 5.7 earthquake sequence. *Geology* 41: 699–702.

Lancaster, Mike, et al. 2010. *Green chemistry: An introductory text*. Royal Society of Chemistry, London.

Manahan, Stanley E. 2009. *Environmental chemistry*, 9th ed. Lewis Publishers, CRC Press, Boca Raton, Fla.

McMurry, John E. 2011. *Organic chemistry*, 8th ed. Brooks/Cole, San Francisco.

Montgomery, Carla. 2013. *Environmental geology*, 10th ed. McGraw-Hill, New York.

Rubinstein, J.L., and A.B. Mahani, 2015. Myths and facts on wastewater injection, hydraulic fracturing, enhanced oil recovery, and induced seismicity. *Seismol. Res. Lett.* 86: 1–8.

Skinner, Brian J., and Stephen C. Porter. 2003. *The dynamic earth: An introduction to physical geology*, 5th ed. John Wiley & Sons, Hoboken, N.J.

Tarbuck, Edward J., Frederick K. Lutgens, and Dennis Tasa. 2011. *Earth science*, 13th ed. Prentice Hall, Upper Saddle River, N.J.

Timberlake, Karen C. 2011. *Chemistry: An introduction to general, organic, and biological chemistry*, 11th ed. Pearson Benjamin Cummings, San Francisco.

Van Ness, H.C. 1983. *Understanding thermodynamics*. Dover Publications, Mineola, N.Y.

Zalasiewicz, Jan, et al. 2008. Are we now living in the Anthropocene? *GSA Today* 18(2) (Feb. 2008): 4–8.

Chapter 3

Alvarez, Luis W., et al. 1980. Extraterrestrial cause for the Cretaceous-Tertiary extinction. *Science* 208: 1095–1108.

Atkinson, Carter T., and Dennis A. LaPointe. 2009. Introduced avian diseases, climate change, and the future of Hawaiian honeycreepers. *J. Avian Med. Surg.* 23: 53–63.

Baldwin, Bruce, and Michael Sanderson. 1998. Age and rate of diversification of the Hawaiian silversword alliance (Compositae). *Proc. Natl. Acad. Sci. USA* 95: 9402–9406.

Begon, Michael, Colin R. Townsend, and John L. Harper. 2006. *Ecology: From individuals to ecosystems*, 4th ed. Wiley-Blackwell Publishing, U.K.

Camp, Richard J., et al. 2009. *Passerine bird trends at Hakalau Forest National Wildlife Refuge, Hawaii.* Hawaii Cooperative Studies Unit Tech. Rep. HCSU-011. University of Hawaii at Hilo.

Camp, Richard J., et al. 2010. Population trends of forest birds at Hakalau Forest National Wildlife Refuge, Hawaii. *Condor* 112: 196–212.

Darwin, Charles. 1859. *On the origin of species by means of natural selection.* John Murray, London.

Endler, John A. 1986. *Natural selection in the wild.* Monographs in Population Biology 21, Princeton University Press, Princeton, N.J.

Freed, Leonard A., and Rebecca L. Cann. 2009. Negative effects of an introduced bird species on growth and survival in a native bird community. *Current Biology* 19: 1736–1740.

Freed, Leonard A., and Rebecca L. Cann. 2010. Misleading trend analysis and decline of Hawaiian forest birds. *Condor* 112: 213–221.

Futuyma, Douglas J. 2013. *Evolution*, 3rd ed. Sinauer Associates, Sunderland, Mass.

Herron, Jon C., and Scott Freeman. 2013. *Evolutionary analysis.* 5th ed. Pearson Prentice Hall, Upper Saddle River, N.J.

Krebs, Charles J. 2016. *Why ecology matters.* University of Chicago Press, Chicago.

Lerner, Heather R.L., et al. 2011. Multilocus resolution of phylogeny and timescale in the extant adaptive radiation of Hawaiian honeycreepers. *Current Biology* 21: 1838–1844.

MacLeod, Norman. 2013. *The great extinctions: What causes them and how they shape life.* Firefly Books, Richmond Hill, Ontario.

Olson, Storrs L., and Helen F. James. 1982. Fossil birds from the Hawaiian islands: Evidence for wholesale extinction by man before Western contact. *Science* 217: 633–635.

Powell, James L. 1998. *Night comes to the Cretaceous: Dinosaur extinction and the transformation of modern geology.* W.H. Freeman, New York.

Price, Jonathan P., and David A. Clague. 2002. How old is the Hawaiian biota? Geology and phylogeny suggest recent divergence. *Proc. Roy. Soc. B* 269: 2429–2435.

Raup, David M. 1991. *Extinction: Bad genes or bad luck?* W.W. Norton, New York.

Reece, Jane B., et al. 2014. *Campbell Biology*, 10th ed. Pearson Education, San Francisco.

Ricklefs, Robert E., and Dolph Schluter, eds. 1993. *Species diversity in ecological communities.* University of Chicago Press, Chicago.

Scott, J. Michael, et al. 1988. Conservation of Hawaii's vanishing avifauna. *BioScience* 38: 238–252.

Smith, Thomas M., and Robert L. Smith. 2015. *Elements of ecology*, 9th ed. Pearson Education, San Francisco.

Steadman, David W. 1995. Prehistoric extinctions of Pacific island birds: Biodiversity meets zooarchaeology. *Science* 267: 1123–1131.

U.S. Fish and Wildlife Service. 2010. *Hakalau Forest National Wildlife Refuge Comprehensive Conservation Plan.* USFWS, Hilo, Hawaii.

Wagner, Warren L., and Vicki A. Funk. 1995. *Hawaiian biogeography: Evolution on a hot spot archipelago.* Smithsonian Institution Press, Washington, D.C.

Wilson, Edward O. 1992. *The diversity of life.* Harvard University Press, Cambridge, Mass.

Chapter 4

Baskin, Yvonne. 2002. *A plague of rats and rubbervines: The growing threat of species invasions.* Island Press, Washington, D.C.

Breckle, Siegmar-Walter. 2002. *Walter's vegetation of the Earth: The ecological systems of the geo-biosphere*, 4th ed. Springer-Verlag, Berlin.

Bright, Chris. 1998. *Life out of bounds: Bioinvasion in a borderless world.* Worldwatch Institute and W.W. Norton, Washington, D.C., and New York.

Bronstein, Judith L. 1994. Our current understanding of mutualism. *Quarterly Journal of Biology* 69: 31–51.

Clewell, Andre F., and James Aronson. 2013. *Ecological restoration: Principles, values, and structure of an emerging profession*, 2nd ed. Island Press, Washington, D.C.

Connell, Joseph H., and Ralph O. Slatyer, 1977. Mechanisms of succession in natural communities. *American Naturalist* 111: 1119–1144.

Dale, Virginia H., Frederick J. Swanson, and Charles M. Crisafulli, eds. 2005. *Ecological responses to the 1980 eruption of Mount St. Helens.* Springer, New York.

Drake, John M., and Jonathan M. Bossenbroek. 2004. The potential distribution of zebra mussels in the United States. *BioScience* 54: 931–941.

Estes, James A., et al. 2011. Trophic downgrading of planet Earth. *Science* 333: 301–306.

Hobbs, Richard J., Eric S. Higgs, and Carol M. Hall. 2013. *Novel ecosystems: Intervening in the new ecological world order.* Wiley-Blackwell, West Sussex, U.K.

Marris, Emma. 2013. *The rambunctious garden: Saving nature in a post-wild world.* Bloomsbury, USA, New York.

Menge, Bruce A., et al. 1994. The keystone species concept: Variation in interaction strength in a rocky intertidal habitat. *Ecological Monographs* 64: 249–286.

Molles, Manuel C. 2015. *Ecology: Concepts and applications*, 7th ed. McGraw-Hill, Boston.

Nijhuis, Michelle. 2007. Wish you weren't here. *High Country News*, 5 Mar. 2007.

Orion, Tao, and David Holmgren. 2015. *Beyond the war on invasive species: A permaculture approach to ecosystem restoration.* Chelsea Green Publishing Co., White River Junction, Vt.

Pace, M.L., et al. 2010. Recovery of native zooplankton associated with increased mortality of an invasive mussel. *Ecosphere* 1(1) Article 3.

Palmer, Margaret A., et al., eds. 2016. *Foundations of restoration ecology*, 2nd ed. Island Press, Washington, D.C.

Pearce, Fred. 2015. *The new wild: Why invasive species will be nature's salvation.* Beacon Press, Boston.

Pimentel, David, et al. 2005. Update on the environmental and economic costs associated with alien-invasive species in the United States. *Ecological Economics* 52: 273–288.

Power, Mary E., et al. 1996. Challenges in the quest for keystones. *BioScience* 46: 609–620.

Ricklefs, Robert, and Rick Relyea. 2013. *Ecology: The economy of nature*, 7th ed. W.H. Freeman, New York.

Rothlisberger, John D., et al. 2012. Ship-borne nonindigenous species diminish Great Lakes ecosystem services. *Ecosystems* 15: 462–476.

Sax, Dov F., John J. Stachowicz, and Steven D. Gaines, eds. 2012. *Species invasions: Insights into ecology, evolution, and biogeography.* Sinauer, Sunderland, Mass.

Shea, Katriona, and Peter Chesson. 2002. Community ecology theory as a framework for biological invasions. *Trends in Ecology and Evolutionary Biology* 17: 170–176.

Smith, Robert L., and Thomas M. Smith. 2001. *Ecology and field biology*, 6th ed. Benjamin Cummings, San Francisco.

Stokstad, Erik. 2007. Feared quagga mussel turns up in western United States. *Science* 315: 453.

Strayer, David L. 2009. Twenty years of zebra mussels: Lessons from the mollusk that made headlines. *Frontiers in Ecology and the Environment* 7: 135–141.

Strayer, David L. 2010. Alien species in fresh waters: Ecological effects, interactions with other stressors, and prospects for the future. *Freshwater Biology* 55(Suppl 1): 152–174.

Strayer, David L., et al. 1999. Transformation of freshwater ecosystems by bivalves: A case study of zebra mussels in the Hudson River. *BioScience* 49: 19–27.

Strayer, David L., et al. 2004. Effects of an invasive bivalve (*Dreissena polymorpha*) on fish in the Hudson River estuary. *Canadian Journal of Fisheries and Aquatic Sciences* 61: 924–941.

Strayer, David L., et al. 2014. Decadal-scale change in a large-river ecosystem. *BioScience* 64: 496–510.

Thompson, John N. 1999. The evolution of species interactions. *Science* 284: 2116–2118.

USDA Pacific Northwest Research Station. 2010. Mount St. Helens 30 years later: A landscape reconfigured. *Science Update*, Issue #19 (12 pp). U.S. Department of Agriculture, Washington, D.C.

U.S. Fish and Wildlife Service. 2012. The cost of invasive species. USFWS, Washington, D.C.

U.S. Geological Survey. Zebra and quagga mussel information resource page. http://nas.er.usgs.gov/taxgroup/mollusks/zebramussel.

Van Andel, Jelte, and James Aronson. 2012. *Restoration ecology: The new frontier*, 2nd ed. Wiley-Blackwell, West Sussex, U.K.

Weigel, Marlene, ed. 1999. *Encyclopedia of biomes*. UXL, Farmington Hills, Michigan.

Woodward, Susan L. 2003. *Biomes of Earth: Terrestrial, aquatic, and human-dominated*. Greenwood Publishing, Westport, Conn.

Chapter 5

Alexander, Richard B., et al. 2008. Differences in phosphorus and nitrogen delivery to the Gulf of Mexico from the Mississippi River Basin. *Environ. Sci. Technol.* 42: 822–830.

Carpenter, Edward J., and Douglas G. Capone, eds. 1983. *Nitrogen in the marine environment*. Academic Press, New York.

Chesapeake Bay Foundation. 2014. *2014 State of the Bay report*. Chesapeake Bay Foundation, Annapolis, MD.

Chesapeake Bay Program. 2011. *Bay barometer: A health and restoration assessment of the Chesapeake Bay and Watershed in 2010*. Chesapeake Bay Program, Annapolis, MD.

Committee on Environment and Natural Resources. 2000. *An integrated assessment: Hypoxia in the northern Gulf of Mexico*. CENR, National Science and Technology Council, Washington, D.C.

Committee on the Mississippi River and the Clean Water Act. 2009. *Nutrient control actions for improving water quality in the Mississippi River basin and northern Gulf of Mexico*. National Academies Press, Washington, D.C.

Diaz, Robert J., and Rutger Rosenberg. 2008. Spreading dead zones and consequences for marine ecosystems. *Science* 321: 926–929.

Field, Christopher B., et al. 1998. Primary production of the biosphere: Integrating terrestrial and oceanic components. *Science* 281: 237–240.

Gruber, Nicolas, and James N. Galloway. 2008. An Earth-system perspective of the global nitrogen cycle. *Nature* 451: 293–296.

Isebrands, J.G., et al. 2001. Growth responses of Populus tremuloides clones to interacting elevated carbon dioxide and tropospheric ozone. *Environmental Pollution* 115: 359–371.

Jacobson, Michael, et al. 2000. *Earth system science from biogeochemical cycles to global changes*. Academic Press.

Mississippi River/Gulf of Mexico Watershed Nutrient Task Force. 2008. *Gulf hypoxia action plan 2008*. U.S. EPA, Washington, D.C.

Mississippi River/Gulf of Mexico Watershed Nutrient Task Force. 2015. *Report to Congress*. U.S. EPA, Washington, D.C.

Mitsch, William J., et al. 2001. Reducing nitrogen loading to the Gulf of Mexico from the Mississippi River Basin: Strategies to counter a persistent ecological problem. *BioScience* 51: 373–388.

National Science and Technology Council, Committee on Environment and Natural Resources. 2003. *An assessment of coastal hypoxia and eutrophication in U.S. waters*. National Science and Technology Council, Washington, D.C.

Raloff, Janet. 2004. Dead waters: Massive oxygen-starved zones are developing along the world's coasts. *Science News* 165: 360–362.

Raloff, Janet. 2004. Limiting dead zones: How to curb river pollution and save the Gulf of Mexico. *Science News* 165: 378–380.

Ricklefs, Robert, and Rick Relyea. 2013. *Ecology: The economy of nature*, 7th ed. W.H. Freeman, New York.

Schlesinger, William H. 2013. *Biogeochemistry: An analysis of global change*, 3rd ed. Academic Press, London.

Schulte, D.M., et al. 2009. Unprecedented restoration of a native oyster metapopulation. *Science* 325: 1124–1128.

Smith, Thomas M., and Robert L. Smith. 2015. *Elements of ecology*, 9th ed. Pearson Education, San Francisco.

Takahashi, Taro. 2004. The fate of industrial carbon dioxide. *Science* 305: 352–353.

Turner, Monica, et al. 2003. *Landscape ecology in theory and practice: Pattern and process*. Springer.

U.S. Department of Energy. 2002. *An evaluation of the Department of Energy's free-air carbon dioxide enrichment (FACE) experiments as scientific user facilities*. U.S. DOE, Washington, D.C.

Vitousek, Peter M., et al. 1997. Human alteration of the global nitrogen cycle: Sources and consequences. *Ecological Applications* 7: 737–750.

Whittaker, Robert H. 1975. *Communities and ecosystems*, 2nd ed. Macmillan, New York.

Wu, Jianguo, and Richard J. Hobbs, eds. 2007. *Key topics in landscape ecology*. Cambridge University Press.

Chapter 6

Arriagada, Rodrigo A., et al. 2012. Do payments for environmental services affect forest cover? A farm-level evaluation from Costa Rica. *Land Economics* 88: 382–399.

Arriagada, Rodrigo A., et al. 2015. Do payments pay off? Evidence from participation in Costa Rica's PES program. *PLOS ONE* 10(7): e0131544.

Attfield, Robin. 2014. *Environmental ethics: An overview for the twenty-first century*. Polity, Cambridge, U.K.

Balmford, Andrew, et al. 2002. Economic reasons for conserving wild nature. *Science* 297: 950–953.

Barbour, Ian G. 1992. *Ethics in an age of technology*. Harper Collins, San Francisco.

Beddoe, Rachael, et al. 2009. Overcoming systemic roadblocks to sustainability: The evolutionary redesign of worldviews, institutions, and technologies. *Proc. Natl. Acad. Sci. USA* 106: 2483–2489.

Blewitt, John. 2014. *Understanding sustainable development*, 2nd ed. Routledge, Abingdon, U.K.

Brown, Lester. 2001. *Eco-economy: Building an economy for the Earth*. Earth Policy Institute and W.W. Norton, New York.

Carson, Richard T., et al. 1994. Valuing the preservation of Australia's Kakadu Conservation Zone. *Oxford Economic Papers* 46: 727–749.

Cole, Luke W., and Sheila R. Foster. 2001. *From the ground up: Environmental racism and the rise of the environmental justice movement*. New York University Press.

Costanza, Robert, et al. 1997. The value of the world's ecosystem services and natural capital. *Nature* 387: 253–260.

Costanza, Robert, et al. 2014. *An introduction to ecological economics*, 2nd ed. CRC Press, Boca Raton, Fla.

Daily, Gretchen, ed. 1997. *Nature's services: Societal dependence on natural ecosystems*. Island Press, Washington, D.C.

Daly, Herman E. 1996. *Beyond growth: The economics of sustainable development*. Beacon Press, Boston.

Daly, Herman E. 2005. Economics in a full world. *Scientific American* 293(3): 100–107.

Daniels, Amy E., et al. 2010. Understanding the impacts of Costa Rica's PES: Are we asking the right questions? *Ecological Economics* 69: 2116–2126.

De Graaf, John, et al. 2002. *Affluenza: The all-consuming epidemic*. Berrett-Koehler Publishers, San Francisco.

Esty, Daniel C., and Andrew S. Winston. 2009. *Green to gold: How smart companies use environmental strategy to innovate, create value, and build competitive advantage* (revised and updated ed.). John Wiley & Sons, Hoboken, N.J.

Fletcher, Robert, and Jan Breitling. 2012. Market mechanism or subsidy in disguise? Governing payment for environmental services in Costa Rica. *Geoforum* 43: 402–411.

Fox, Stephen. 1985. *The American conservation movement: John Muir and his legacy*. University of Wisconsin Press, Madison.

Francis, Pope. 2015. *Encyclical letter Laudato Si' of the Holy Father Francis on care for our common home*. Vatican Press, The Vatican.

Goodstein, Eban. 1999. *The tradeoff myth: Fact and fiction about jobs and the environment*. Island Press, Washington, D.C.

Goodstein, Eban, and Stephen Polasky. 2014. *Economics and the environment*, 7th ed. John Wiley & Sons, Hoboken, N.J.

Greiber, Thomas, and Simone Schiele, eds. 2011. *Governance of ecosystem services: Lessons learned from Cameroon, China, Costa Rica, and Ecuador*. IUCN, Gland, Switzerland.

Hawken, Paul, Amory Lovins, and L. Hunter Lovins. 1999. *Natural capitalism*. Little, Brown, and Co., Boston.

Her Majesty's (HM) Treasury. 2006. *Stern review on the economics of climate change.* Her Majesty's Treasury and Cambridge University Press.

Kinzig, Ann P., et al. 2011. Paying for ecosystem services—promise and peril. *Science* 334: 603–604.

Klein, Naomi. 2015. *This changes everything: Capitalism versus the climate.* Simon and Schuster, New York.

Leopold, Aldo. 1949. *A Sand County almanac, and sketches here and there.* Oxford University Press.

McCauley, Douglas J. 2006. Selling out on nature. *Nature* 443: 27–28.

Meadows, Donella, Jørgen Randers, and Dennis Meadows. 2004. *Limits to growth: The 30-year update.* Chelsea Green Publishing Co., White River Junction, Vt.

Millennium Ecosystem Assessment. 2005. *Ecosystems and human well-being: Opportunities and challenges for business and industry.* Millennium Ecosystem Assessment and World Resources Institute.

Nash, Roderick F. 1989. *The rights of nature.* University of Wisconsin Press, Madison.

Nash, Roderick F. 1990. *American environmentalism: Readings in conservation history,* 3rd ed. McGraw-Hill, New York.

National Research Council, Board on Sustainable Development. 1999. *Our common journey: A transition toward sustainability.* National Academies Press, Washington, D.C.

Nordhaus, William. 2007. Critical assumptions in the Stern Review on Climate Change. *Science* 317: 201–202.

Overseas Development Institute. 2011. *Costa Rica's sustainable resource management: Successfully tackling tropical deforestation.* ODI Publications, London.

Pagiola, Stefano. 2007. Payments for environmental services in Costa Rica. *Ecological Economics* 65: 712–724.

Pojman, Louis P., et al. 2016. *Environmental ethics: Readings in theory and application,* 7th ed. Wadsworth, Belmont, Calif.

Porras, Ina, et al. 2014. *Ecosystems for sale: Land prices and payments for ecosystem services in Costa Rica.* International Institute for Environment and Development, London.

Renner, Michael. 2008. *Green jobs: Working for people and the environment.* Worldwatch Report 177. Worldwatch Institute, Washington, D.C.

Ricketts, Taylor, et al. 2004. Economic value of tropical forest to coffee production. *Proc. Natl. Acad. Sci. USA* 101: 12579–12582.

Schmidtz, David, and Elizabeth Willott. 2011. *Environmental ethics: What really matters, what really works,* 2nd ed. Oxford University Press, Oxford, U.K.

Singer, Peter, ed. 2011. *Practical ethics.* Cambridge University Press, Cambridge, U.K.

Smith, Adam. 1776. *An inquiry into the nature and causes of the wealth of nations.* 1993 ed., Oxford University Press.

Smith, Stephen. 2011. *Environmental economics: A very short introduction.* Oxford University Press, Oxford, U.K.

Stone, Christopher D. 1972. Should trees have standing? Towards legal rights for natural objects. *Southern California Law Review* 1972: 450–501.

Talberth, John, et al. 2007. *The genuine progress indicator 2006: A tool for sustainable development.* Redefining Progress, Oakland, Calif.

TEEB. 2010. *The economics of ecosystems and biodiversity: Mainstreaming the economics of nature: A synthesis of the approach, conclusions, and recommendations of TEEB.* The Economics of Ecosystems and Biodiversity.

TEEB. 2012. *Nature and its role in the transition to a green economy.* The Economics of Ecosystems and Biodiversity.

Tietenberg, Tom, and Lynne Lewis. 2014. *Environmental and natural resource economics* (Pearson Series in Economics). Routledge, Abingdon, U.K.

United Nations. 2012. *Report of the United Nations Conference on Sustainable Development.* Rio de Janeiro, Brazil, 20–22 June 2012. http://www.un.org/ga/search/view_doc.asp?symbol=A/CONF.216/16&Lang=E

United Nations. 2015. *The Millennium Development Goals Report 2015.* U.N., New York.

Vucetich, John A., et al. 2015. Evaluating whether nature's intrinsic value is an axiom of or anathema to conservation. *Conservation Biology* 29: 321–332.

Walker, Gordon. 2012. *Environmental justice: Concepts, evidence, and politics.* Routledge, New York.

White, Lynn. 1967. The historic roots of our ecologic crisis. *Science* 155: 1203–1207.

Worldwatch Institute. 2008. *State of the world 2008: Innovations for a sustainable economy.* Worldwatch Institute, Washington, D.C.

Worldwatch Institute. 2010. *Transforming cultures.* Worldwatch Institute, Washington, D.C.

Worldwatch Institute. 2012. *State of the world 2012: Moving toward sustainable prosperity.* Worldwatch Institute, Washington, D.C.

Wunder, Sven, et al. 2008. A comparative analysis of payments for environmental services programs in developed and developing countries. *Ecological Economics* 65: 834–852.

Wunscher, Tobias, et al. 2008. Spatial targeting of payments for environmental services: A tool for boosting conservation benefits. *Ecological Economics* 65: 822–833.

Chapter 7

Bogojevic, Sanja. 2013. *Emissions trading schemes: Markets, states, and laws.* Hart Publishing, Oxford, U.K.

Boyer, Elizabeth W., et al. 2012. *The impact of Marcellus gas drilling on rural drinking water supplies.* The Center for Rural Pennsylvania, a legislative agency of the Pennsylvania General Assembly, Harrisburg, Penn.

Council on Environmental Quality. 2014. A citizen's guide to the NEPA: Having your voice heard. Executive Office of the President, Washington, D.C.

Darrah, Thomas H., et al. 2014. Noble gases identify the mechanisms of fugitive gas contamination in drinking-water wells overlying the Marcellus and Barnett Shales. *Proc. Natl. Acad. Sci. USA* 111: 14076–14081.

Dietz, Thomas, et al. 2003. The struggle to govern the global commons. *Science* 302: 1907–1912.

Energy Information Administration, U.S. Department of Energy. 2013. *Technically recoverable shale oil and shale gas resources: An assessment of 137 shale formations in 41 countries outside the United States.* Washington, D.C.

Environmental Law Institute. 2009. *Estimating U.S. government subsidies to energy sources: 2002–2008.* ELI, Washington, D.C.

Fogleman, Valerie M. 1990. *Guide to the National Environmental Policy Act.* Quorum Books, New York.

Fontenot, Brian E., et al. 2013. An evaluation of water quality in private drinking water wells near natural gas extraction sites in the Barnett Shale Formation. *Environ. Sci. Technol.* 47: 10032–10040.

Green Scissors*: Cutting wasteful and environmentally harmful spending.* http://www.greenscissors.com.

Hansjurgens, Bernd, ed. 2005. *Emissions trading for climate policy: US and European perspectives.* Cambridge University Press, Cambridge, U.K.

Hardin, Garrett. 1968. The tragedy of the commons. *Science* 162: 1243–1248.

Houck, Oliver. 2003. Tales from a troubled marriage: Science and law in environmental policy. *Science* 302: 1926–1928.

Ingraffea, Anthony R., et al. 2014. Assessment and risk analysis of casing and cement impairment in oil and gas wells in Pennsylvania, 2000–2012. *Proc. Natl. Acad. Sci. USA* 111: 10955–10960.

Jackson, Robert B., et al. 2013. Increased stray gas abundance in a subset of drinking water wells near Marcellus shale gas extraction. *Proc. Natl. Acad. Sci. USA* 110: 11250–11255.

Jackson, Robert B., et al. 2014. The environmental costs and benefits of fracking. *Annu. Rev. Environ. Resources* 39: 327–362.

Jacoby, Henry D., et al. 2012. The influence of shale gas on U.S. energy and environmental policy. *Economics of Energy and Environmental Policy* 1: 37–51.

Kubasek, Nancy K., and Gary S. Silverman. 2013. *Environmental law,* 8th ed. Pearson Prentice Hall, Upper Saddle River, N.J.

Llewellyn, Garth T., et al. 2015. Evaluating a groundwater supply contamination incident attributed to Marcellus Shale gas development. *Proc. Natl. Acad. Sci. USA* 112: 6325–6330.

Lustgarten, Abrahm. 2013. EPA's abandoned Wyoming fracking study one retreat of many. *Propublica.* http://www.propublica.org/article/epas-abandoned-wyoming-fracking-study-one-retreat-of-many.

Myers, Norman, and Jennifer Kent. 2001. *Perverse subsidies: How misused tax dollars harm the environment and the economy.* Island Press, Washington, D.C.

New York State Department of Health. 2014. *A public health review of high volume hydraulic fracturing for shale gas development.* New York State Department of Health, Albany, 186 pp.

Nordhaus, Ted, and Michael Shellenberger. 2007. *Break Through: From the death of environmentalism to the politics of possibility.* Houghton Mifflin, Boston.

OECD. 2012. *Inventory of estimated budgetary support and tax expenditures for fossil fuels 2013.* OECD Publishing.

Osborn, Stephen G., et al. 2011. Methane contamination of drinking water accompanying gas-well drilling and hydraulic fracturing. *Proc. Natl. Acad. Sci. USA* 108: 8172–8176.

Percival, Robert V., et al. 2013. *Environmental regulation: Law, science, and policy,* 7th ed. Aspen Publishers, New York.

Phillips, Susan. 2012. Dimock: A town divided. StateImpact Pennsylvania. 28 March 2012. http://stateimpact.npr.org/pennsylvania/2012/03/28/dimock-a-town-divided/.

Rosenbaum, Walter. 2013. *Environmental politics and policy,* 9th ed. CQ Press, Congressional Quarterly, Inc., Washington, D.C.

Schmidt, Charles W. 2011. Blind rush? Shale gas boom proceeds amid human health questions. *Environ. Health Perspect.* 119: A348–A353.

Shellenberger, Michael, and Ted Nordhaus. 2004. *The death of environmentalism: Global warming politics in a post-environmental world.* Presented at the Environmental Grantmakers Association meeting, October 2004.

Siegel, Donald I., et al. 2015. Methane concentrations in water wells unrelated to proximity to existing oil and gas wells in northeastern Pennsylvania. *Environ. Sci. Technol.* 49: 4106–4112.

StateImpact Pennsylvania. The Pennsylvania guide to hydraulic fracturing, or "fracking." https://stateimpact.npr.org/pennsylvania/tag/fracking/. National Public Radio.

Tietenberg, Tom H. 2006. *Emissions trading: Principles and practice.* 2nd ed. Resources for the Future, Washington, D.C.

Tietenberg, Tom, and Lynne Lewis. 2010. *Environmental economics and policy,* 6th ed. Prentice Hall, Upper Saddle River, N.J.

U.N. General Assembly. 2012. *The future we want.* Outcome document from Rio+20 Conference. Resolution 66/288. http://www.un.org/ga/search/view_doc.asp?symbol=A/RES/66/288&Lang=E

U.S. Environmental Protection Agency. Summary of the National Environmental Policy Act. https://www.epa.gov/laws-regulations/summary-national-environmental-policy-act.

U.S. Environmental Protection Agency. 2015. *Assessment of the potential impacts of hydraulic fracturing for oil and gas on drinking water resources.* External Review Draft EPA/600/R-15/047c. EPA Office of Research and Development, Washington, D.C.

U.S. Office of Management and Budget, 2016. *2015 report to Congress on the benefits and costs of federal regulations and agency compliance with the Unfunded Mandates Reform Act.* Executive Office of the President of the United States, Washington, D.C.

Urbina, Ian. 2011. Regulation lax as gas wells' tainted water hits rivers. *New York Times,* 26 Feb. 2011.

Vengosh, Avner, et al. 2014. A critical review of the risks to water resources from unconventional shale gas development and hydraulic fracturing in the United States. *Environ. Sci. Technol.* 48: 8334–8348.

Vig, Norman J., and Michael E. Kraft, eds. 2015. *Environmental policy: New directions for the twenty-first century,* 9th ed. CQ Press, Congressional Quarterly, Inc., Washington, D.C.

von Friedeburg, Christoph. 2015. Effects and sustainability of the U.S. shale gas boom. http://vitalsigns.worldwatch.org/vs-trend/effects-and-sustainability-us-shale-gas-boom. Worldwatch Institute, Washington D.C.

Warner, N.R., et al. 2014. New tracers identify hydraulic fracturing fluids and accidental releases from oil and gas operations. *Environ. Sci. Technol.* 48: 12552–12560.

Wilber, Tom. Shale Gas Review. http://tomwilber.blogspot.com/.

Wilber, Tom. 2012. *Under the surface: Fracking, fortunes, and the fate of the Marcellus Shale.* Cornell University Press, Ithaca, N.Y.

World Bank. 2016. *World development indicators 2016.* World Bank, Washington, D.C. http://data.worldbank.org/products/wdi.

Worldwatch Institute. 2014. *State of the world 2014: Governing for sustainability.* Worldwatch Institute, Washington, D.C.

Chapter 8

Ausubel, Jesse H. 1996. Can technology spare the earth? *American Scientist* 84: 166–178.

Ball, Philip. 2008. Where have all the flowers gone? [China's one-child legacy.] *Nature* 454: 374–375.

Cohen, Joel E. 1995. *How many people can the Earth support?* W.W. Norton, New York.

De Souza, Roger-Mark, et al. 2003. Critical links: Population, health, and the environment. *Population Bulletin* 58(3). Population Reference Bureau, Washington, D.C.

Ehrlich, Paul. 1968. *The population bomb.* 1997 reprint, Buccaneer Books, Cutchogue, N.Y.

Ehrlich, Paul R., and Anne H. Ehrlich. 1990. *The population explosion.* Touchstone, New York.

Ehrlich, Paul R., and John P. Holdren. 1971. Impact of population growth: Complacency concerning this component of man's predicament is unjustified and counterproductive. *Science* 171: 1212–1217.

Engelman, Robert. 2008. *More: Population, nature, and what women want.* Island Press, Washington D.C.

Engelman, Robert. 2010. *Population, climate change, and women's lives.* Worldwatch Report 183. Worldwatch Institute, Washington, D.C.

Gribble, James N., and J. Bremner. 2012. Achieving a demographic dividend. *Population Bulletin* 67 (2). Population Reference Bureau, Washington, D.C.

Haberl, Helmut, et al. 2007. Quantifying and mapping the human appropriation of net primary production in earth's terrestrial ecosystems. *Proc. Natl. Acad. Sci. USA* 104: 12942–12947.

Haub, Carl, and James N. Gribble. 2011. The world at 7 billion. *Population Bulletin* 66 (2). Population Reference Bureau, Washington, D.C.

Haub, Carl, and O.P. Sharma. 2006. India's population reality: Reconciling change and tradition. *Population Bulletin* 61(3). Population Reference Bureau, Washington, D.C.

Holdren, John P., and Paul R. Ehrlich. 1974. Human population and the global environment. *American Scientist* 62: 282–292.

Imhoff, Marc L., et al. 2004. Global patterns in human consumption of net primary production. *Nature* 429: 870–873.

Kane, Penny. 1987. *The second billion: Population and family planning in China.* Penguin Books, Australia, Ringwood, Victoria.

Kearney, Melissa S., and Phillip B. Levine. 2014. *Media influences on social outcomes: The impact of MTV's 16 and Pregnant on teen childbearing.* The National Bureau of Economic Research, Cambridge, MA.

Kent, Mary M., and Carl Haub. 2005. Global demographic divide. *Population Bulletin* 60(4). Population Reference Bureau, Washington, D.C.

La Ferrara, E., et al., 2012. Soap operas and fertility: Evidence from Brazil. *American Economic Journal: Applied Economics* 4: 1–31.

Lamptey, Peter R., et al. 2006. The global challenge of HIV and AIDS. *Population Bulletin* 61(1). Population Reference Bureau, Washington, D.C.

Malthus, Thomas R. *An essay on the principle of population.* 1983 ed., Penguin USA, New York.

McDonald, M., and D. Nierenberg, 2003. *Linking population, women, and biodiversity. State of the world 2003.* Worldwatch Institute, Washington, D.C.

Notestein, Frank. 1953. Economic problems of population change. pp. 13–31 in *Proceedings of the Eighth International Conference of Agricultural Economists.* Oxford University Press.

Population Reference Bureau. 2015. *2015 World population data sheet.* Population Reference Bureau, Washington, D.C.

Riley, Nancy E. 2004. *China's population: New trends and challenges. Population Bulletin* 59(2). Population Reference Bureau, Washington, D.C.

UNAIDS and World Health Organization. 2012. *UNAIDS report on the global AIDS epidemic.* UNAIDS and WHO, Geneva, Switzerland.

U.N. Population Division. 2009. *World population ageing 2009.* UNPD, New York.

U.N. Population Division. 2015. *World population prospects: The 2015 revision*. UNPD, New York.

U.N. Population Fund. 2010. *The millennium development goals report 2010*. UNFPA.

U.S. Census Bureau. http://www.census.gov.

Wackernagel, Mathis, and William Rees. 1996. *Our ecological footprint: Reducing human impact on the earth*. New Society Publishers, Gabriola Island, British Columbia, Canada.

WWF–World Wide Fund for Nature. 2018. *Living planet report 2018*. WWF, Gland, Switzerland.

Chapter 9

Buchmann, Stephen L., and Gary Paul Nabhan. 1996. *The forgotten pollinators*. Island Press/Shearwater Books, Washington, D.C./Covelo, California.

Curtin, Charles G. 2002. Integration of science and community-based conservation in the Mexico/U.S. borderlands. *Conservation Biology* 16: 880–886.

Diamond, Jared. 1999. *Guns, germs, and steel: The fates of human societies*. W.W. Norton, New York.

Diamond, Jared, and Peter Bellwood. 2003. Farmers and their languages: The first expansions. *Science* 300: 597–603.

Egan, Timothy. 2006. *The worst hard time: The untold story of those who survived the great American dust bowl*. Mariner Books, Houghton-Mifflin, Boston and New York.

Flack, Sarah 2016. *The art and science of grazing: How grass farmers can create sustainable systems for healthy animals and farm ecosystems*. Chelsea Green Publishing Co., White River Junction, Vt.

Garcia-Tejero, Iva, et al. 2011. *Water and sustainable agriculture*. Springer, New York.

Gemmill-Herren, Barbara. 2016. Pollination services to agriculture: Sustaining and enhancing a key ecosystem service. FAO and Routledge, Abingdon, U.K.

Glanz, James. 1995. *Saving our soil: Solutions for sustaining Earth's vital resource*. Johnson Books, Boulder, Colo.

Havlin, John L., et al. 2013. *Soil fertility and fertilizers*, 8th ed. Pearson Education, Upper Saddle River, N.J.

Hayes, Rhonda Fleming. 2016. *Pollinator friendly gardening: Gardening for bees, butterflies, and other pollinators*. Voyageur Press, New York.

Huggins, David R., and John P. Reganold. 2008. *No-till: How farmers are saving the soil by parking their plows. Scientific American* 299(1): 70–71.

Imeson, Anton. 2012. *Desertification, land degradation, and sustainability: Paradigms, processes, principles, and policies*. Wiley-Blackwell, Sussex, U.K.

Jenny, Hans. 1941. *Factors of soil formation: A system of quantitative pedology*. McGraw-Hill, New York.

Kaiser, Jocelyn. 2004. Wounding Earth's fragile skin. *Science* 304: 1616–1618.

Kennesaw State University Culinary and Hospitality Services. Sustainable Practices. http://dining.kennesawstateauxiliary.com/sustainability/sustainable-practices.

Lal, Rattan, ed. 1998. *Soil quality and agricultural sustainability*. Ann Arbor Press, Chelsea, Mich.

Lengnick, Laura. 2015. Resilient agriculture: Cultivating food systems for a changing climate. New Society Publishers, Vancouver, B.C.

Malpai Borderlands Group. Malpai Borderlands Group. http://www.malpaiborderlandsgroup.org.

Millennium Ecosystem Assessment. 2005. *Ecosystems and human wellbeing: Desertification synthesis*. Millennium Ecosystem Assessment and World Resources Institute.

Molden David. 2007. Water for food, water for life; A comprehensive assessment of water management in agriculture. Routledge, Abingdon, U.K.

Montgomery, David R. 2007. *Dirt: The erosion of civilizations*. University of California Press, Berkeley and Los Angeles.

Montgomery, David R. 2007. Soil erosion and agricultural sustainability. *Proc. Natl. Acad. Sci. USA* 104: 13268–13272.

Morgan, R.P.C. 2005. *Soil erosion and conservation*, 3rd ed. Blackwell, London.

Natural Resources Conservation Service. Soils. NRCS, USDA. http://www.soils.usda.gov.

Ohlson, Kristin. 2014. *The soil will save us: How scientists, farmers, and foodies are healing the soil to save the planet*. Rodale Inc., Emmaus, Penn.

Pierzynski, Gary M., et al. 2005. *Soils and environmental quality*, 3rd ed. CRC Press, Boca Raton, Fla.

Plaster, Edward. 2013. *Soil science and management*, 6th ed. Cengage Learning, Boston.

Pollinator Health Task Force. 2015. *National strategy to promote the health of honey bees and other pollinators*. The White House, Washington, D.C.

Soil Science Society of America. Soil basics. http://www.soils.org/discover-soils/soil-basics.

Trimble, Stanley W., and Pierre Crosson. 2000. U.S. soil erosion rates—myth and reality. *Science* 289: 248–250.

Troeh, Frederick R., and Louis M. Thompson. 2005. *Soil and soil fertility*, 6th ed. Blackwell Publishing, London.

U.N. Environment Programme. 2012. "Land." Chapter 3 in *Global environment outlook 5 (GEO-5)*. UNEP, Nairobi.

U.S. Department of Agriculture. 2014. *2012 Census of agriculture*. USDA, Washington, D.C.

U.S. Department of Agriculture. 2015. *Summary report: 2012 national resources inventory*. Natural Resources Conservation Service, Washington, D.C., and Center for Survey Statistics and Methodology, Iowa State University, Ames, Iowa.

U.S. Geological Survey. 2014. *Conservation Reserve Program (CRP) contributions to wildlife habitat, management issues, challenges, and policy choices: An annotated bibliography*. U.S. Department of the Interior, Washington, D.C.

Uri, Noel D. 2001. The environmental implications of soil erosion in the United States. *Environmental Monitoring and Assessment* 66: 293–312.

Vallentine, John F. 2000. *Grazing management*, 2nd ed. Academic Press, Cambridge, U.K.

Wilkinson, Bruce H. 2005. Humans as geologic agents: A deep-time perspective. *Geology* 33: 161–164.

Chapter 10

Australian Broadcasting Corporation. 2015. Organic farmer Steve Marsh loses GM appeal for compensation from neighbour Michael Baxter. *ABC News*, 3 Sept. 2015.

Australian Broadcasting Corporation. 2016. Organic farmer Steve Marsh loses bid for High Court review of genetic modification contamination case. *ABC News*, 12 Feb. 2016.

Benbrook, Charles M. 2012. Impacts of genetically engineered crops on pesticide use in the U.S.—the first sixteen years. *Environmental Sciences Europe* 24: 24, 13 pp.

Brar, Satinder Kaur. 2012. *Biocontrol: Management, processes, and challenges*. Nova Science Publishers, Hauppauge, N.Y.

Brown, Lester R. 2012. *Full planet, empty plates: The new geopolitics of food scarcity*. Earth Policy Institute, Washington, D.C.

Canadian Biotechnology Action Network. 2015. *Where in the world are GM crops and foods?* Report 1. CBAN, Ottawa, Ontario.

Center for Food Safety. 2007. *Monsanto vs. U.S. farmers: November 2007 update*. Center for Food Safety, Washington, D.C.

Cerdeira, Antonio L., and Stephen O. Duke. 2006. The current status and environmental impacts of glyphosate-resistant crops: A review. *J. Environ. Qual.* 35: 1633–1658.

Delate, Kathleen, et al. 2015. A review of long-term organic comparison trials in the U.S. *Sustainable Agriculture Research* 4: 5–14.

Farm Scale Evaluations of spring-sown genetically modified crops, The. 2003. A themed issue from *Philosophical Transactions of the Royal Society of London B: Biological Sciences* 358 (1439), 29 Nov. 2003.

Fedoroff, Nina, and Nancy Marie Brown. 2004. *Mendel in the kitchen: A scientist's view of genetically modified foods*. National Academies Press, Washington, D.C.

Flint, Mary Louise. 2010. *IPM in practice: Principles and methods of integrated pest management*. University of California Division of Agriculture and Natural Resources.

Food and Agriculture Organization of the United Nations. 2006. *Livestock's long shadow: Environmental issues and options*. FAO, Rome.

Food and Agriculture Organization of the United Nations. 2013. *Tackling climate change through livestock: A global assessment of emissions and mitigation opportunities.* FAO, Rome.

Food and Agriculture Organization of the United Nations. 2014. *The state of world fisheries and aquaculture 2014.* FAO Fisheries and Aquaculture Department, Rome.

Food and Agriculture Organization of the United Nations. 2015. *The state of food insecurity in the world 2015.* FAO, Rome.

Food and Water Watch and Organic Farmers' Agency for Relationship Marketing. 2014. *Organic farmers pay the price for GMO contamination.* Food and Water Watch, Portland, Ore., and OFARM, Aldrich, Minn.

Gardner, Gary, and Brian Halweil. 2000. *Underfed and overfed: The global epidemic of malnutrition.* Worldwatch Paper 150. Worldwatch Institute, Washington, D.C.

Goulson, Dave, et al. 2015. Bee declines driven by combined stress from parasites, pesticides, and lack of flowers. *Science* 347: 1435–1444.

Gurian-Sherman, Doug. 2009. *Failure to yield: Evaluating the performance of genetically engineered crops.* Union of Concerned Scientists, Cambridge, Mass.

Halweil, Brian. 2008. *Farming fish for the future.* Worldwatch Report 176. Worldwatch Institute, Washington, D.C.

James, Clive. 2016. *Global status of commercialized biotech/GM crops: 2015.* International Service for the Acquisition of Agri-biotech Applications.

Klümper, Wilhelm, and Matin Qaim. 2014. A meta-analysis of the impacts of genetically modified crops. *PLOS ONE* 9: e111629.

Kristiansen, P., et al., eds. 2006. *Organic agriculture: A global perspective.* CABI Publishing, Oxfordshire, U.K.

Lappé, Frances Moore, and Joseph Collins. 2015. *World hunger: 10 myths.* Grove Press, New York.

Liebig, Mark A., and John W. Doran. 1999. Impact of organic production practices on soil quality indicators. *J. Environ. Qual.* 28: 1601–1609.

Lim, XiaoZhi. 2014. In landmark case, Australian court rejects organic farmer's claim of GMO "contamination." *Genetic Literacy Project,* 28 May 2014.

Maeder, Paul, et al. 2002. Soil fertility and biodiversity in organic farming. *Science* 296: 1694–1697.

Mahgoub, Salah E.O. 2015. *Genetically modified foods: Basics, applications, and controversy.* CRC Press, Boca Raton, Fla.

Miller, Henry I., and Gregory Conko. 2004. *The frankenfood myth: How protest and politics threaten the biotech revolution.* Praeger Publishers, Westport, Conn.

Nierenberg, Danielle, and Brian Halweil. 2005. Cultivating food security. pp. 62–79 in *State of the world 2005.* Worldwatch Institute, Washington, D.C.

Nierenberg, Danielle, and Laura Reynolds. 2012. *Innovations in sustainable agriculture: Supporting climate-friendly food production.* Worldwatch Report 188. Worldwatch Institute, Washington, D.C.

Nordhaus, Hannah. 2011. *The beekeeper's lament: How one man and half a billion honey bees help feed America.* Harper Perennial, New York.

Norris, Robert F., et al. 2003. *Concepts in integrated pest management.* Prentice Hall, Upper Saddle River, N.J.

Paull, John. 2015. The threat of genetically modified organisms (GMOs) to organic agriculture: A case study update. *Agriculture & Food* 3: 56–63.

Pearce, Fred. 2002. The great Mexican maize scandal. *New Scientist* 174: 14.

Pedigo, Larry P., and Marlin Rice. 2014. *Entomology and pest management,* 6th ed. Waveland Press, Long Grove, Ill.

Polak, Paul. 2005. The big potential of small farms. *Scientific American* 293(3): 84–91.

Pollan, Michael. 2007. *The omnivore's dilemma.* Penguin Press, New York.

Pollan, Michael. 2009. *In defense of food: An eater's manifesto.* Penguin Press, New York.

Ponisio, Lauren C., et al. 2014. Diversification practices reduce organic to conventional yield gap. *Proc. Roy. Soc. B* 282: 20141396

de Ponti, Tomek, et al. 2012. The crop yield gap between organic and conventional agriculture. *Agricultural Systems* 108: 1–9.

Pretty, Jules. 2007. Agricultural sustainability: Concepts, principles, and evidence. *Phil. Trans. Roy. Soc. B* 363: 447–465.

Pretty, Jules. 2007. *Sustainable agriculture and food.* Earthscan, London.

Quist, David, and Ignacio H. Chapela. 2001. Transgenic DNA introgressed into traditional maize landraces in Oaxaca, Mexico. *Nature* 414: 541–543.

Randall, Rebecca. 2015. Zero tolerance policies on GMO "contamination" hurt organic and conventional farmers alike. *Genetic Literacy Project,* 12 Feb. 2015.

Roberts, Paul. 2008. *The end of food.* Houghton Mifflin, Boston.

Rodale Institute. 2011. *The farming systems trial: Celebrating 30 years.* Rodale Institute, Kutztown, Penn.

Sánchez–Bayo, Francisco. 2014. The trouble with neonicotinoids. *Science* 346: 806–807.

Science. 2013. Smarter pest control. (special section). pp. 730–759 in *Science* 341 (16 Aug. 2013).

Seufert, V., et al. 2012. Comparing the yields of organic and conventional agriculture. *Nature* 485: 229–232.

Shepard, Mark. 2013. *Restoration agriculture: Real-world permaculture for farmers.* Acres USA, Austin, Texas.

Shiva, Vandana. 2000. *Stolen harvest: The hijacking of the global food supply.* South End Press, Cambridge, Mass.

Smil, Vaclav. 2001. *Feeding the world: A challenge for the twenty-first century.* MIT Press, Cambridge, Mass.

Stewart, C. Neal. 2004. *Genetically modified planet: Environmental impacts of genetically engineered plants.* Oxford University Press, New York.

Sustainable Agriculture Network. 2010. *The new American farmer: Profiles of agricultural innovation,* 2nd ed. Sustainable Agriculture Network, Beltsville, MD.

Sustainable Agriculture Research and Education (SARE). 2010. *Exploring sustainability in agriculture.* SARE.

U.S. Department of Agriculture. 2015. *2012 Census of agriculture: Organic Survey (2014).* Vol. 3 Special Studies Part 4. USDA, Washington, D.C.

U.S. Department of Agriculture Advisory Committee on Biotechnology and 21st Century Agriculture (AC21). 2012. *Enhancing coexistence: A report of the AC21 to the Secretary of Agriculture.* USDA, Washington, D.C.

U.S. National Academy of Sciences. 2010. *The impact of genetically engineered crops on farm sustainability in the United States.* National Academies Press, Washington, D.C.

Wall Street Journal. 2015. Can organic food feed the world? Essays by Catherine Badgley and Steve Savage. *Wall Street Journal,* 12 July 2015.

Weasel, Lisa. 2008. *Food fray: Inside the controversy over genetically modified food.* AMACOM Books, New York.

Weber, Christopher L., and H. Scott Mathews. 2008. Food-miles and the relative climate impacts of food choices in the United States. *Environ. Sci. Technol.* 42: 3508–3513.

Willer, Helga, and Julia Lernoud, eds. 2016. *The world of organic agriculture: Statistics and emerging trends 2016.* Research Institute of Organic Agriculture (FiBL), Frick, and IFOAM—Organics International, Bonn.

Worldwatch Institute. 2011. *State of the world 2011: Innovations that nourish the planet.* Worldwatch Institute, Washington, D.C.

Xerces Society. 2011. *Attracting native pollinators.* Storey Publishing, North Adams, Mass.

Xie, Jian, et al. 2011. Ecological mechanisms underlying the sustainability of the agricultural heritage rice-fish coculture system. *Proc. Natl. Acad. Sci. USA* 108: E1381–E1387.

Zhou, Wanqing. 2015. Genetically modified crop industry continues to expand. http://vitalsigns.worldwatch.org/vs-trend/genetically-modified-crop-industry-continues-expand. Worldwatch Institute, Washington D.C.

Chapter 11

Baker, C. Scott, Frank Cipriano, and Stephen R. Palumbi. 1996. Molecular genetic identification of whale and dolphin products from commerical markets in Korea and Japan. *Molecular Ecology* 5: 671–685.

Balmford, Andrew, et al. 2002. Economic reasons for conserving wild nature. *Science* 297: 950–953.

Barnosky, Anthony D., et al. 2004. Assessing the causes of late Pleistocene extinctions on the continents. *Science* 306: 70–75.

Baskin, Yvonne. 1997. *The work of nature: How the diversity of life sustains us.* Island Press, Washington, D.C.

Brodie, Jedediah F., et al., eds. 2013. *Wildlife conservation in a changing climate.* University of Chicago Press, Chicago.

Chapman, Arthur D. 2009. *Numbers of living species in Australia and the world,* 2nd ed. Australian Biological Resources Study, Canberra.

Chivian, Eric, and Aaron Berstein, eds. 2008. *Sustaining life: How human health depends on biodiversity.* Oxford University Press, New York.

CITES Secretariat. Convention on International Trade in Endangered Species of Wild Fauna and Flora. https://cites.org/.

Collins, James P., and Martha L. Crump. 2009. *Extinction in our times: Global amphibian decline.* Oxford University Press, U.K.

Conniff, Richard, 2012. A bitter pill. *Conservation,* 9 Mar. 2012.

Convention on Biological Diversity. https://www.cbd.int.

Cooper, John E., and Margaret E. Cooper. 2013. *Wildlife forensic investigation: Principles and practice.* CRC Press, Boca Raton, Fla.

Daily, Gretchen C., ed. 1997. *Nature's services: Societal dependence on natural ecosystems.* Island Press, Washington, D.C.

Estes, R.D., et al. 2006. Downward trends in Ngorongoro Crater ungulate populations 1986–2005: Conservation concerns and the need for ecological research. *Biological Conservation* 131: 106-120.

Galatowitsch, Susan. 2012. *Ecological restoration.* Sinauer Associates, Sunderland, Mass.

Gaston, Kevin J., and John I. Spicer. 2004. *Biodiversity: An introduction,* 2nd ed. Blackwell, London.

Groom, Martha J., et al. 2005. *Principles of conservation biology,* 3rd ed. Sinauer Associates, Sunderland, Mass.

Groombridge, Brian, and Martin D. Jenkins. 2002. *Global biodiversity: Earth's living resources in the 21st century.* UNEP, World Conservation Monitoring Centre, and Aventis Foundation; World Conservation Press, Cambridge, U.K.

Groombridge, Brian, and Martin D. Jenkins. 2002. *World atlas of biodiversity: Earth's living resources in the 21st century.* University of California Press, Berkeley.

Hambler, Clive, and Susan M. Canney. 2013. *Conservation,* 2nd ed. Cambridge University Press., U.K.

Hanken, James. 1999. Why are there so many new amphibian species when amphibians are declining? *Trends in Ecology and Evolution* 14: 7–8.

Hoelzel, A. Rus. 2015. Can DNA foil the poachers? *Science* 349: 34–35.

Holdo, Ricardo, M. 2011. Predicted impact of barriers to migration on the Serengeti wildebeest population. *PLoS ONE* 6(1):e16370.

Hopcraft, J. Grant C. 2015. Conservation and economic benefits of a road around the Serengeti. *Conservation Biology* 29: 932–936.

Kaufman, Leslie. 2012. Zoos' bitter choice: To save some species, letting others die. *New York Times,* 27 May 2012.

Kolbert, Elizabeth. 2014. *The sixth extinction: An unnatural history.* Henry Holt and Co., New York.

Laurance, William F., et al. 2015. Estimating the environmental costs of Africa's massive "development corridors." *Current Biology* 25: 3202–3208.

Louv, Richard. 2005. *Last child in the woods: Saving our children from nature-deficit disorder.* Algonquin Books, Chapel Hill, N.C.

Lovejoy, Thomas E., and Lee Hannah, eds. 2006. *Climate change and biodiversity.* Yale University Press, New Haven, Conn.

MacArthur, Robert H., and Edward O. Wilson. 1967. *The theory of island biogeography.* Princeton University Press, Princeton, N.J.

Marris, Emma. 2013. *The rambunctious garden: Saving nature in a post-wild world.* Bloomsbury, USA, New York.

Mburu, John, and Regina Birner. 2007. Emergence, adoption, and implementation of collaborative wildlife management or wildlife partnerships in Kenya: A look at conditions for success. *Society and Natural Resources* 20: 379–395.

Millennium Ecosystem Assessment. 2005. *Ecosystems and human well-being: Biodiversity synthesis.* Millennium Ecosystem Assessment and World Resources Institute.

Mooney, Harold A., and Richard J. Hobbs, eds. 2000. *Invasive species in a changing world.* Island Press, Washington, D.C.

Mora, Camilo, et al. 2011. How many species are there on Earth and in the ocean? *PLoS Biology* 9(8): e1001127.

Ogutu, Joseph O., et al. 2011. Continuing wildlife population declines and range contraction in the Mara region of Kenya during 1977–2009. *Journal of Zoology* 285: 99–109.

Pearce, Fred. 2015. *The new wild: Why invasive species will be nature's salvation.* Beacon Press, Boston.

Primack, Richard B. 2014. *Essentials of conservation biology,* 6th ed. Sinauer Associates, Sunderland, Mass.

Quammen, David. 1996. *The song of the dodo: Island biogeography in an age of extinction.* Touchstone, New York.

Rohr, Jason R., et al. 2008. Evaluating the links between climate, disease spread, and amphibian declines. *Proc. Natl. Acad. Sci. USA* 105:17436–17441.

Rosenzweig, Michael L. 1995. *Species diversity in space and time.* Cambridge University Press.

Roskov, Y., et al. 2016. Species 2000 & ITIS Catalogue of Life. Digital resource at http://www.catalogueoflife.org/col. Species 2000: Naturalis, Leiden, the Netherlands.

Sepkoski, John J. 1984. A kinetic model of Phanerozoic taxonomic diversity. *Paleobiology* 10: 246–267.

Simberloff, Daniel. 1998. Flagships, umbrellas, and keystones: Is single-species management passé in the landscape era? *Biological Conservation* 83: 247–257.

Soulé, Michael E. 1986. *Conservation biology: The science of scarcity and diversity.* Sinauer Associates, Sunderland, Mass.

Stoner, Chantal, et al. 2007. Assessment of effectiveness of protection strategies in Tanzania based on a decade of survey data for large herbivores. *Conservation Biology* 21: 635–646.

Takacs, David. 1996. *The idea of biodiversity: Philosophies of paradise.* Johns Hopkins University Press, Baltimore.

TEEB. 2010. *The economics of ecosystems and biodiversity: Mainstreaming the economics of nature: A synthesis of the approach, conclusions, and recommendations of TEEB.* The Economics of Ecosystems and Biodiversity.

Thompson, Ken. 2014. *Where do camels belong? Why invasive species aren't all bad.* Greystone Books, Vancouver, B.C.

U.N. Environment Programme. 2012. "Biodiversity." Chapter 5 in *Global environment outlook 5 (GEO-5).* UNEP, Nairobi.

U.S. Environmental Protection Agency. Summary of the Endangered Species Act. https://www.epa.gov/laws-regulations/summary-endangered-species-act.

U.S. Fish and Wildlife Service. Endangered species program. http://www.fws.gov/endangered.

Ward, Peter D. 2007. *Under a green sky: Global warming, the mass extinctions of the past, and what they can tell us about our future.* Smithsonian Books and Collins, Washington, D.C., and New York.

Wasser, S.K., et al. 2015. Genetic assignment of large seizures of elephant ivory reveals Africa's major poaching hotspots. *Science* 349: 84–87.

Western, David. 2003. Conservation science in Africa and the role of international collaboration. *Conservation Biology* 17: 11–19.

Western, David, et al. 2009. The status of wildlife in protected areas compared to non-protected areas of Kenya. *PLoS One* 4(7): e6140.

Wilson, Edward O. 1984. *Biophilia.* Harvard University Press, Cambridge, Mass.

Wilson, Edward O. 1992. *The diversity of life.* Harvard University Press, Cambridge, Mass.

Wilson, Edward O. 2002. *The future of life.* Alfred A. Knopf, New York.

Wittemyer, George. 2014. Illegal killing for ivory drives global decline in African elephants. *Proc. Natl. Acad. Sci. USA* 111: 13117–13121.

World Conservation Union. IUCN Red List. http://www.iucnredlist.org.

Wren, Sally, et al., eds. 2015. Amphibian conservation action plan. IUCN, Gland, Switzerland. http://www.amphibians.org/publications/amphibian-conservation-action-plan.

WWF–World Wide Fund for Nature. 2014. *Living planet report 2014.* WWF, Gland, Switzerland.

Chapter 12

Agronne, Dianna M. 2013. *Deforestation and climate change.* Nova Science Publishers, Hauppauge, N.Y.

Aubertin, Catherine, and Estienne Rodary, eds. 2011. *Protected areas, sustainable land?* Ashgate Publishing, Surrey, U.K., and Burlington, Vermont.

Bettinger, Pete, et al. 2008. *Forest management and planning.* Academic Press, New York.

Bratkovich, Steve, et al. 2012. *Forests of the United States: Understanding trends and challenges.* Dovetail Partners Inc., Minneapolis, Minn.

Clary, David. 1986. *Timber and the Forest Service.* University Press of Kansas, Lawrence.

Donato, Daniel C., et al. 2006. Post-wildfire logging hinders regeneration and increases fire risk. *Science* 311: 352.

Duncan, Dayton, and Ken Burns. 2009. *The national parks: America's best idea.* Alfred A. Knopf, New York.

Ferraz, Goncalo N., et al. 2007. A large-scale deforestation experiment: Effects of patch area and isolation on Amazon birds. *Science* 315: 238–241.

Food and Agriculture Organization of the United Nations. 2014. *State of the world's forests 2014.* FAO, Rome.

Food and Agriculture Organization of the United Nations. 2015. *Global forest resources assessment.* FAO, Rome.

Forest Stewardship Council. http://www.fsc.org.

Gorte, Ross W., and Pervaze A. Sheikh. 2013. *Deforestation and climate change.* Congressional Research Service, Washington, D.C.

Haddad, Nick M., et al. 2015. Habitat fragmentation and its lasting impact on Earth's ecosystems. *Science Advances* 1(2): e1500052, 20 Mar. 2015.

Harris, Larry D. 1984. *The fragmented forest: Island biogeography theory and the preservation of biotic diversity.* University of Chicago Press, Chicago.

Jensen, Sara E., and Guy R. McPherson. 2008. *Living with fire: Fire ecology and policy for the twenty-first century.* University of California Press, Berkeley and Los Angeles.

Land Trust Alliance. 2011. *2010 National land trust census report: A look at voluntary land conservation in America.* Land Trust Alliance, Washington, D.C.

Laurance, William F., et al. 2011. The fate of Amazonian forest fragments: A 32-year investigation. *Biological Conservation* 144(1): 56–67.

Lindenmayer, David B., et al. 2008. *Salvage logging and its ecological consequences.* Island Press, Washington, D.C.

Lockwood, Michael, et al. 2006. *Managing protected areas: A global guide.* IUCN and Earthscan Publishing, London.

MacArthur, Robert H., and Edward O. Wilson. 1967. *The theory of island biogeography.* Princeton University Press, Princeton, N.J.

MacDicken, Kenneth, ed. 2015. Changes in global forest resources from 1990 to 2015. (Special issue). *Forest Ecology and Management* 352: 1–145.

McKenzie, Donald, Carol Miller, and Donald A. Falk, eds. 2011. The landscape ecology of fire. *Ecological Studies* 213. Springer, New York.

National Forest Management Act of 1976. http://www.fs.fed.us/emc/nfma/includes/NFMA1976.pdf.

National Interagency Fire Center. http://www.nifc.gov.

Newmark, William D. 1987. A land-bridge perspective on mammal extinctions in western North American parks. *Nature* 325: 430.

Noss, Reed F., et al. 2006. Managing fire-prone forests in the western United States. *Frontiers in Ecology and the Environment* 4: 481–487.

Oswalt, Sonja M., et al. 2014. *Forest resources of the United States, 2012: A technical document supporting the Forest Service update of the 2010 RPA assessment.* Gen. Tech. Rep. WO-91. USDA Forest Service, Washington, D.C. 218 pp.

Pimm, Stuart L. 1998. The forest fragment classic. *Nature* 393: 23–24.

Pye, Oliver, and Jayati Bhattacharya, eds. 2012. *The palm oil controversy in Southeast Asia: A transnational perspective.* ISEAS Publishing, Singapore.

Pyne, Stephen. 2001. *Fire: A brief history.* University of Washington Press, Seattle.

Runte, Alfred. 1979. *National parks and the American experience.* University of Nebraska Press, Lincoln.

Schwartz, John. 2015. As fires grow, a new landscape appears in the West. *New York Times,* 21 Sept. 2015.

Science. 2015. Forest health in a changing world. (Special issue). pp. 800–836 in *Science* 349 (21 Aug. 2015).

Sedjo, Robert A. 2000. *A vision for the U.S. Forest Service.* Resources for the Future, Washington, D.C.

Shatford, Jeffrey, et al. 2007. Conifer regeneration after forest fire in the Klamath-Siskiyous: How much, how soon? *J. Forestry* 105: 139–146.

Smith, David M., et al. 1996. *The practice of silviculture: Applied forest ecology,* 9th ed. Wiley, New York.

Smithsonian Tropical Research Institute. Biological Dynamics of Forest Fragments Project. http://www.stri.si.edu/english/research/programs/programs_information/biological_dynamics_forest_fragments.php.

Soulé, Michael E., and John Terborgh, eds. 1999. *Continental conservation.* Island Press, Washington, D.C.

Stegner, Wallace. 1954. *Beyond the hundredth meridian: John Wesley Powell and the second opening of the West.* Houghton Mifflin, Boston.

Thompson, Jonathan R., et al. 2007. Re-burn severity in managed and unmanaged vegetation in a large wildfire. *Proc. Natl. Acad. Sci. USA* 104: 10743–10748.

U.N. Environment Programme and World Conservation Monitoring Centre. 2014. *Protected planet report 2014: Tracking progress towards global targets for protected areas.* Cambridge, U.K.

USDA Forest Service. 2011. *National report on sustainable forests—2010.* FS-979. U.S. Department of Agriculture, Washington, D.C.

Westerling, A.L., et al. 2006. Warming and earlier spring increase western U.S. forest wildfire activity. *Science* 313: 940–943.

Williams, Michael. 2006. *Deforesting the Earth: From prehistory to global crisis: An abridgment.* University of Chicago Press, Chicago.

Wilson, Edward O. 2016. *Half Earth: Our planet's fight for life.* Liveright, New York.

Young, Raymond A., and Ronald L. Giese. 2002. *Introduction to forest ecosystem science and management,* 3rd ed. John Wiley & Sons, New York.

Chapter 13

Abbott, Carl. 2001. *Greater Portland: Urban life and landscape in the Pacific Northwest.* University of Pennsylvania Press.

Abbott, Carl. 2002. Planning a sustainable city. pp. 207–235 in Squires, Gregory D., ed. *Urban sprawl: Causes, consequences, and policy responses.* Urban Institute Press, Washington, D.C.

Adler, Frederick R., and Colby J. Tanner. 2013. *Urban ecosystems: Ecological principles for the built environment.* Cambridge University Press, Cambridge, U.K.

Bloomberg, Michael. 2015. City century: Why municipalities are the key to fighting climate change. *Foreign Affairs,* Sept./Oct. 2015. Council on Foreign Relations, Washington, D.C.

Brockerhoff, Martin P. 2000. An urbanizing world. *Population Bulletin* 55(3). Population Reference Bureau, Washington, D.C.

Calthorpe, Peter. 2010. *Urbanism in the age of climate change.* Island Press, Washington D.C.

Chester, Mikhail, et al. 2013. Infrastructure and automobile shifts: Positioning transit to reduce life-cycle environmental impacts for urban sustainability goals. *Environmental Research Letters* 8: 015041 (10 pp).

Cronon, William. 1991. *Nature's metropolis: Chicago and the great West.* W.W. Norton, New York.

Douglas, Ian and Philip James. 2015. *Urban ecology: An introduction.* Routledge, Abingdon, U.K.

Duany, Andres, et al. 2001. *Suburban nation: The rise of sprawl and the decline of the American dream.* North Point Press, New York.

Duany, Andres, et al. 2009. *The smart growth manual.* McGraw-Hill Professional, New York.

Ellin, Nan. 2012. *Good urbanism.* Island Press, Washington, D.C.

Ewing, Reid, et al. 2002. *Measuring sprawl and its impact.* Smart Growth America.

Forman, Richard T. T. 2014. *Urban ecology: Science of cities.* Cambridge University Press, Cambridge, U.K.

Girardet, Herbert. 2004. *Cities people planet: Livable cities for a sustainable world.* Academy Press.

Harnik, Peter. 2010. *Urban green: Innovative parks for resurgent cities.* The Trust for Public Land and Island Press, Washington, D.C.

Jacobs, Jane. 1992. *The death and life of great American cities.* Vintage.

Kriken, John Lund. 2010. *City building: Nine planning principles for the twenty-first century.* Princeton Architectural Press, New York.

Litman, Todd. 2012. *Rail transit in America: A comprehensive evaluation of benefits.* Victoria Transport Policy Institute and American Public Transportation Association.

Litman, Todd. 2015. *Evaluating public transit benefits and costs: Best practices guidebook.* Victoria Transport Policy Institute, Victoria, B.C.

Long Term Ecological Research Network: Baltimore Ecosystem Study. http://www.lternet.edu/sites/bes.

Long Term Ecological Research Network: Central Arizona—Phoenix LTER. http://www.lternet.edu/sites/cap.

Metro. http://www.metro-region.org.

Metro. 2010. *Urban growth report 2009–2030: Employment and residential.* Metro, Portland, Ore.

New Urbanism. http://www.newurbanism.org.

New York, City of. 2015. *One New York: The plan for a strong and just city.* City of New York, New York. http://www.nyc.gov/onenyc.

Northwest Environment Watch. 2004. *The Portland exception: A comparison of sprawl, smart growth, and rural land loss in 15 U.S. cities.* Northwest Environment Watch, Seattle.

Pearce, Fred. 2005. Cities lead the way to a greener world. *New Scientist,* 4 June 2005: 8–9.

Portland, City of. 2016. Adopted 2035 Comprehensive Plan. http://www.portlandoregon.gov/bps/70936.

Pugh, Cedric, ed. 1996. *Sustainability, the environment, and urbanization.* Earthscan Publications, London.

Speck, Jeff. 2012. *Walkable city: How downtown can save America, one step at a time.* Farrar, Straus, and Giroux, New York.

TEEB. 2011. *TEEB manual for cities: Ecosystem services in urban management.* The Economics of Ecosystems and Biodiversity.

U.N. Population Division. 2015. *World urbanization prospects: The 2014 revision.* UNPD, New York.

U.S. Green Building Council. http://www.usgbc.org.

Worldwatch Institute. 2007. *State of the world 2007: Our urban future.* Worldwatch Institute, Washington, D.C.

Worldwatch Institute. 2016. *State of the world 2016: Can a city be sustainable?* Worldwatch Institute, Washington, D.C.

Chapter 14

Ames, Bruce N., et al. 1990. Nature's chemicals and synthetic chemicals: Comparative toxicology. *Proc. Natl. Acad. Sci. USA* 87: 7782–7786.

Bloom, Barry. 2005. Public health in transition. *Scientific American* 293(3): 92–99.

Carson, Rachel. 1962. *Silent spring.* Houghton Mifflin, Boston.

Carwhile, Jenny R., et al. 2011. Canned soup consumption and urinary bisphenol A: A randomized crossover trial. *Journal of the American Medical Association* 306: 2218–2220.

Colburn, Theo, Dianne Dumanoski, and John P. Myers. 1996. *Our stolen future.* Penguin USA, New York.

Consumer Reports. 2009. Concern over canned foods: Our tests find wide range of bisphenol A in soups, juice, and more. *Consumer Reports* Dec. 2009.

Curtis, Kathleen, and Bobbi Chase Wilding. 2010. *Is it in us? Chemical contamination in our bodies.* Body Burden Working Group and Commonweal Biomonitoring Resource Center.

Environmental Defence. 2008. *Toxic baby bottles in Canada.* Environmental Defence, Toronto, Ontario.

European Commission, Environment Directorate General. 2007. *REACH in brief.* European Commission.

Gilliom, Robert J., et al. 2006. *Pesticides in the nation's streams and ground water, 1992–2001—A summary.* USGS National Water-Quality Assessment Program Circular 1291.

Gross, Liza. 2007. The toxic origins of disease. *PLoS Biology* 5(7): e193. doi:10.1371/journal.pbio.0050193.

Guillette, Elizabeth A., et al. 1998. An anthropological approach to the evaluation of preschool children exposed to pesticides in Mexico. *Environ. Health Perspect.* 106: 347–353.

Guillette, Louis J., Jr., et al. 2000. Alligators and endocrine disrupting contaminants: A current perspective. *American Zoologist* 40: 438–452.

Hayes, Tyrone, et al. 2003. Atrazine-induced hermaphroditism at 0.1 PPB in American leopard frogs (Rana pipiens): Laboratory and field evidence. *Environ. Health Perspect.* 111: 568–575.

Hunt, Patricia A., et al. 2003. Bisphenol A exposure causes meiotic aneuploidy in the female mouse. *Current Biology* 13: 546–553.

Kent, Mary M., and Sandra Yin, 2006. Controlling infectious diseases. *Population Bulletin* 61(2), 24 pp. Population Reference Bureau, Washington, D.C.

Kolpin, Dana W., et al. 2002. Pharmaceuticals, hormones, and other organic wastewater contaminants in U.S. streams, 1999–2000: A national reconnaissance. *Environ. Sci. Technol.* 36: 1202–1211.

Landis, Wayne G., et al. 2010. *Introduction to environmental toxicology,* 4th ed. CRC Press, Boca Raton, Fla.

Lang, Iain A., et al. 2008. Association of urinary bisphenol A concentration with medical disorders and laboratory abnormalities in adults. *Journal of the American Medical Association* 300: 1303–1310.

Loewenberg, Samuel. 2003. E.U. starts a chemical reaction. *Science* 300: 405.

Manahan, Stanley E. 2009. *Environmental chemistry,* 9th ed. Lewis Publishers, CRC Press, Boca Raton, Fla.

McGinn, Anne Platt. 2000. *Why poison ourselves? A precautionary approach to synthetic chemicals.* Worldwatch Paper 153. Worldwatch Institute, Washington, D.C.

Millennium Ecosystem Assessment. 2005. *Ecosystems and human well-being: Health synthesis.* World Health Organization.

Moeller, Dade. 2011. *Environmental health,* 4th ed. Harvard University Press, Cambridge, Mass.

Myers, Samuel S. 2009. *Global environmental change: The threat to human health.* Worldwatch Report 181. Worldwatch Institute, Washington, D.C.

National Center for Environmental Health; U.S. Centers for Disease Control and Prevention. 2005. *Third national report on human exposure to environmental chemicals.* NCEH Pub. No. 05-0570, Atlanta.

National Center for Health Statistics. 2009. *Health, United States, 2009, with special feature on medical technology.* Hyattsville, Md.

Our Stolen Future. http://www.ourstolenfuture.org/.

Pirages, Dennis. 2005. Containing infectious disease. pp. 42–61 in *State of the world 2005.* Worldwatch Institute, Washington, D.C.

President's Cancer Panel. 2010. *Reducing environmental cancer risk.* President's Cancer Panel 2008–2009 annual report. U.S. Department of Health and Human Services, Washington, D.C.

Renner, Rebecca. 2002. Conflict brewing over herbicide's link to frog deformities. *Science* 298: 938–939.

Stancel, George, et al. 2001. Report of the bisphenol A sub-panel. Chapter 1 in *National Toxicology Program's report of the endocrine disruptors low-dose peer review.* U.S. EPA and NIEHS, NIH.

Stockholm Convention on Persistent Organic Pollutants. http://www.pops.int.

United Health Foundation. 2015. *America's health rankings: 2015 annual report.* United Health Foundation, Minnetonka, Minn.

U.S. Environmental Protection Agency. Summary of the Toxic Substances Control Act. https://www.epa.gov/laws-regulations/summary-toxic-substances-control-act.

Vogel, Sarah. 2009. The politics of plastic: The making and unmaking of bisphenol A "safety." *Framing Health Matters, Am. J. Public Health* Suppl 3, vol. 99 #S3, S559–S566.

vom Saal, Frederick S., et al. 2012. The estrogenic endocrine disrupting chemical bisphenol A (BPA) and obesity. Molecular and Cellular Endocrinology 354: 74–84.

World Health Organization. 2008. *The global burden of disease: 2004 update.* WHO, Geneva, Switzerland.

World Health Organization. 2009. *Global health risks: Mortality and burden of disease attributable to selected major risks.* WHO, Geneva, Switzerland.

World Health Organization. 2016. *World health statistics 2016.* WHO, Geneva, Switzerland.

Zogorski, John S., et al. 2006. *Volatile organic compounds in the nation's ground water and drinking-water supply wells.* USGS National Water-Quality Assessment Program Circular 1292.

Chapter 15

American Rivers. 2002. *The ecology of dam removal: A summary of benefits and impacts.* American Rivers, Washington, D.C.

Coastal Protection and Restoration Authority. 2012. *2012 coastal master plan.* Coastal Protection and Restoration Authority of Louisiana. Baton Rouge, LA.

Cook, Benjamin I., et al., 2015, *Science Advances* 1(1): e1400082.

Gleick, Peter H. 2010. *Bottled and sold: The story behind our obsession with bottled water.* Island Press, Washington, D.C.

Gleick, Peter H., and H.S. Cooley. 2009. Energy implications of bottled water. *Environ. Res. Lett.* 4: 014009 (6 pp).

Gleick, Peter H., et al. 2009. *The world's water 2008–2009: The biennial report on freshwater resources.* Island Press, Washington, D.C.

Gulf Coast Ecosystem Restoration Council. 2013. Draft initial comprehensive plan: *Restoring the Gulf Coast's ecosystem and economy.* https://www.restorethegulf.gov.

Millennium Ecosystem Assessment. 2005. *Ecosystems and human well-being: Wetlands and water synthesis.* Millennium Ecosystem Assessment and World Resources Institute.

Naidenko, Olga, et al. 2008. *Bottled water quality investigation: 10 major brands, 38 pollutants.* Environmental Working Group. http://www.ewg.org.

New York Times. 2009–2010. *Toxic waters: A series about the worsening pollution in American waters and regulators' response.* http://projects.nytimes.com/toxic-waters.

Postel, Sandra. 2003. *Rivers for life: Managing water for people and nature.* Island Press, Washington, D.C.

Postel, Sandra. 2005. *Liquid assets: The critical need to safeguard freshwater ecosystems.* Worldwatch Paper 170. Worldwatch Institute, Washington, D.C.

Rabalais, Nancy N., et al. 2002. Beyond science into policy: Gulf of Mexico hypoxia and the Mississippi River. *BioScience* 52: 129–142.

Reisner, Marc. 1986. *Cadillac desert: The American West and its disappearing water.* Viking Penguin, New York.

TEEB. 2013. *The economics of ecosystems and biodiversity (TEEB) for water and wetlands.* The Economics of Ecosystems and Biodiversity. IEEP, London and Brussels, Ramsar Secretariat, Gland.

U.N. Environment Programme. 2008. *Water quality for ecosystem and human health,* 2nd ed. UNEP Global Environment Monitoring System (GEMS)/ Water Programme, Burlington, Ontario.

U.N. Environment Programme. 2012. "Water." Chapter 4 in *Global environment outlook 5 (GEO-5).* UNEP, Nairobi.

U.N. World Water Assessment Programme. 2012. *U.N. world water development report: Managing water under uncertainty and risk.* Paris, New York, and Oxford, UNESCO and Berghahn Books.

U.S. Environmental Protection Agency. Summary of the Clean Water Act. https://www.epa.gov/laws-regulations/summary-clean-water-act.

U.S. Environmental Protection Agency. 2009. *Water on tap: What you need to know.* EPA 816-K-09-002. Office of Water, Washington, D.C.

U.S. Government Accountability Office (GAO). 2009. *Bottled water: FDA safety and consumer protections are often less stringent than comparable EPA protections for tap water.* GAO Report to Congressional Requesters, GAO-09-610.

Wagner, Martin, and Jörg Oehlmann. 2009. Endocrine disruptors in bottled mineral water: Total estrogenic burden and migration from plastic bottles. *Environ. Sci. Pollut. Res. Int.* 16: 278–286.

Wagner, Martin, et al. 2013. Identification of putative steroid receptor antagonists in bottled water: Combining bioassays and high-resolution mass spectrometry. *PLoS One* 8(8): e72472.

Worldwatch Institute. 2011. *State of the world 2011: Innovations that nourish the planet.* Worldwatch Institute, Washington, D.C.

Chapter 16

Allsopp, Michelle, et al. 2007. *Oceans in peril: Protecting marine biodiversity.* Worldwatch Report 174. Worldwatch Institute, Washington, D.C.

Food and Agriculture Organization of the United Nations. 2014. *The state of world fisheries and aquaculture 2014.* FAO Fisheries and Aquaculture Department, Rome.

Frank, Kenneth T., et al. 2005. Trophic cascades in a formerly cod-dominated ecosystem. *Science* 308: 1621–1623.

Garrison, Tom. 2012. *Oceanography: An invitation to marine science,* 8th ed. Brooks/Cole, San Francisco.

Gell, Fiona R., and Callum M. Roberts. 2003. Benefits beyond boundaries: The fishery effects of marine reserves. *Trends in Ecology and Evolution* 18: 448–455.

Halweil, Brian. 2006. *Catch of the day: Choosing seafood for healthier oceans.* Worldwatch Paper 172. Worldwatch Institute, Washington, D.C.

Hoegh-Guldberg, O., et al. 2007. Coral reefs under rapid climate change and ocean acidification. *Science* 318: 1737–1742.

International Pacific Research Center. 2008. Tracking ocean debris. *IPRC Climate* 8: 14–16.

Jackson, Jeremy B.C., et al. 2001. Historical overfishing and the recent collapse of coastal ecosystems. *Science* 293: 629–638.

Kurlansky, Mark. 1998. *Cod: A biography of the fish that changed the world.* Penguin Books, New York.

Lotze, Heike L., et al. 2006. Depletion, degradation, and recovery potential of estuaries and coastal seas. *Science* 312: 1806–1809.

Maximenko, N.A., et al. 2012: Pathways of marine debris from trajectories of Lagrangian drifters. *Marine Pollution Bulletin,* 65: 51–62.

Moore, Charles James. 2008. Synthetic polymers in the marine environment: A rapidly increasing, long-term threat. *Environmental Research* 108: 131–139.

Morrissey, John F., and James L. Sumich. 2010. *Introduction to the biology of marine life,* 10th ed. Jones & Bartlett, Boston.

Myers, Ransom A., and Boris Worm. 2003. Rapid worldwide depletion of predatory fish communities. *Nature* 423: 280–283.

National Center for Ecological Analysis and Synthesis (NCEAS) and Communication Partnership for Science and the Sea (COMPASS), sponsors. 2001. *Scientific consensus statement on marine reserves and marine protected areas.* https://www.nceas.ucsb.edu/consensus.

National Research Council. 2003. *Oil in the sea III: Inputs, fates, and effects.* National Academies Press, Washington, D.C.

Norse, Elliott, and Larry B. Crowder, eds. 2005. *Marine conservation biology: The science of maintaining the sea's biodiversity.* Island Press, Washington, D.C.

Orr, James C. 2005. Anthropogenic ocean acidification over the twenty-first century and its impact on calcifying organisms. *Nature* 437: 681–686.

Pauly, Daniel, et al. 2002. Towards sustainability in world fisheries. *Nature* 418: 689–695.

Pauly, Daniel, et al. 2003. The future for fisheries. *Science* 302: 1359–1361.

Roberts, Callum M., et al. 2001. Effects of marine reserves on adjacent fisheries. *Science* 294: 1920–1923.

Rosenberg, A., et al. 2006. Rebuilding U.S. fisheries: Progress and problems. *Frontiers in Ecology and the Environment* 4(6).

TEEB. 2012. *Why value the oceans—A discussion paper.* The Economics of Ecosystems and Biodiversity.

Trujillo, Alan P., and Harold V. Thurman. 2013. *Essentials of oceanography,* 11th ed. Prentice Hall, Upper Saddle River, N.J.

Weber, Michael L. 2001. *From abundance to scarcity: A history of U.S. marine fisheries policy.* Island Press, Washington, D.C.

Weiss, Kenneth R., and Usha Lee McFarling. 2006. Altered oceans. (A special five-part series). *Los Angeles Times,* 30 July–3 Aug., 2006. Available at: http://www.latimes.com/world/la-fg-altered-oceans-sg-20060730-storygallery.html.

Worm, Boris, et al. 2006. Impacts of biodiversity loss on ocean ecosystem services. *Science* 314: 787–790.

Chapter 17

Ahrens, C. Donald, and Robert Henson. 2015. *Meteorology today,* 11th ed. Cengage Learning, Boston.

Akimoto, Hajime. 2003. Global air quality and pollution. *Science* 302: 1716–1719.

Bernard, Susan M., et al. 2001. The potential impacts of climate variability and change on air pollution-related health effects in the United States. *Environ. Health Perspect.* 109(Suppl 2): 199–209.

Borja-Aburto, Victor H., et al. 1997. Ozone, suspended particulates, and daily mortality in Mexico City. *Am. J. Epidemiology* 145: 258–268.

Bruce, Nigel, Rogelio Perez-Padilla, and Rachel Albalak. 2000. Indoor air pollution in developing countries: A major environmental and public health challenge. *Bull. World Health Organization* 78: 1078–1092.

Calderón-Garcidueñas, Lilian, et al. 2003. Respiratory damage in children exposed to urban pollution. *Pediatric Pulmonology* 36: 148–161.

Cooper, C. David, and F. C. Alley. 2010. *Air pollution control: A design approach,* 4th ed. Waveland Press, Long Grove, Ill.

Davis, Devra. 2002. *When smoke ran like water: Tales of environmental deception and the battle against pollution.* Basic Books, New York.

Davis, Devra L., et al. 2002. A look back at the London smog of 1952 and the half century since. *Environ. Health Perspect.* 110: A734.

Driscoll, Charles T., et al. 2001. *Acid rain revisited: Advances in scientific understanding since the passage of the 1970 and 1990 Clean Air Act Amendments.* Hubbard Brook Research Foundation.

Fabian, Peter, and Martin Dameris. 2014. *Ozone in the atmosphere: Basic principles, natural and human impacts.* Springer, Berlin.

Godish, Thad, et al. 2014. *Air quality,* 5th ed. CRC Press, Boca Raton, Fla.

Hall, Jane V., et al. 2008. *The benefits of meeting federal clean air standards in the South Coast and San Joaquin Valley air basins.* William and Flora Hewlett Foundation.

Hoffman, Matthew J. 2005. *Ozone depletion and climate change: Constructing a global response.* SUNY Press, New York.

Jacobs, Chip, and William J. Kelly. 2008. *Smogtown: The lung-burning history of pollution in Los Angeles.* Overlook Press, New York.

Jacobson, Mark Z. 2002. *Atmospheric pollution: History, science, and regulation.* Cambridge University Press, Cambridge, U.K.

Jenkins, Jerry C., et al. 2007. *Acid rain in the Adirondacks: An environmental history.* Comstock, Ithaca, N.Y.

Kaiman, Jonathan. 2013. Chinese struggle through 'airpocalypse' smog. *The Observer,* 16 Feb. 2013.

Kelly, William J., and Chip Jacobs. 2014. *The people's republic of chemicals.* Rare Bird Books, Los Angeles.

Likens, Gene E. 2004. Some perspectives on long-term biogeochemical research from the Hubbard Brook ecosystem study. *Ecology* 85: 2355–2362.

Loomis, Dana, et al. 1999. Air pollution and infant mortality in Mexico City. *Epidemiology* 10: 118–123.

Lutgens, Frederick K., Edward J. Tarbuck, and Dennis Tasa. 2013. *The atmosphere: An introduction to meteorology*, 12th ed. Pearson Education.

McDonnell, Patrick J. 2016. An old nemesis returns: Smoggy days in Mexico City. *Los Angeles Times,* 19 May 2016.

Molina, Mario J., and F. Sherwood Rowland. 1974. Stratospheric sink for chlorofluoromethanes: Chlorine atom catalyzed destruction of ozone. *Nature* 249: 810–812.

Parson, Edward A. 2003. *Protecting the ozone layer: Science and strategy.* Oxford University Press.

Solomon, Susan, et al. 2016. Emergence of healing in the Antarctic ozone layer. *Science* 353: 269–274.

U.N. Environment Programme. 2012. "Atmosphere." Chapter 2 in *Global environment outlook 5 (GEO-5).* UNEP, Nairobi.

U.S. Environmental Protection Agency. Air Quality Planning and Standards. https://www3.epa.gov/airquality/index.html.

U.S. Environmental Protection Agency. Summary of the Clean Air Act. https://www.epa.gov/laws-regulations/summary-clean-air-act.

U.S. Environmental Protection Agency. 2016. *2013 program progress: Clean Air Interstate Rule, Acid Rain Program, and former NO_x Budget Trading Program.* EPA, Washington, D.C.

U.S. Environmental Protection Agency. *Air quality trends.* EPA, Washington, D.C. https://www3.epa.gov/airtrends/aqtrends.html.

U.S. Environmental Protection Agency. 2015. *2011 National air toxics assessment.* EPA, Washington, D.C. https://www.epa.gov/national-air-toxics-assessment/2011-national-air-toxics-assessment.

Vallero, Daniel. 2014. *Fundamentals of air pollution*, 5th ed. Academic Press, Cambridge, U.K.

Wong, Edward. 2015. As Beijing shuts down over smog alert, worse-off neighbors carry on. *New York Times,* 9 Dec. 2015.

World Health Organization. Indoor air pollution. http://www.who.int/indoorair.

World Health Organization. 2016. *WHO global urban ambient air pollution database (update 2016).* http://www.who.int/phe/health_topics/outdoorair/databases/cities.

World Meteorological Organization. *Scientific assessment of ozone depletion: 2014.* WMO, Global Ozone Research and Monitoring Project, Report #55, Geneva, Switzerland, 416 pp.

Chapter 18

Alley, Richard B. 2000. *The two-mile time machine: Ice cores, abrupt climate change, and our future.* Princeton University Press, Princeton, N.J.

Baptiste, Nathalie. 2016. That sinking feeling: The politics of sea-level rise and Miami's building boom. *American Prospect,* 19 Feb. 2016.

Bianco, Nicholas M., and Franz T. Litz. 2010. *Reducing greenhouse gas emissions in the United States using existing federal authorities and state action.* WRI Report, World Resources Institute, Washington, D.C.

Bloom, Arnold J. 2009. *Global climate change: Convergence of disciplines.* Sinauer Associates, Sunderland, Mass.

Bogojevic, Sanja. 2013. *Emissions trading schemes: Markets, states, and laws.* Hart Publishing, Oxford, U.K.

Burney, Nelson E., ed. 2010. *Carbon tax and cap-and-trade tools: Market-based approaches for controlling greenhouse gases.* Nova Science Publishers, Hauppauge, N.Y.

Burroughs, William James. 2007. *Climate change: A multidisciplinary approach*, 2nd ed. Cambridge University Press.

Caldeira, Kenneth, and Michael E. Wickett. 2003. Anthropogenic carbon and ocean pH. *Nature* 425: 365.

Carbon Tax Center. Where carbon is taxed. http://www.carbontax.org/where-carbon-is-taxed.

Center for Climate and Energy Solutions. http://www.c2es.org.

Climate Central. http://www.climatecentral.org.

Davenport, Coral. 2015. A climate deal, 6 fateful years in the making. *New York Times,* 13 Dec. 2015.

Edenhofer, Ottmar, et al., eds. 2012. *Renewable energy sources and climate change mitigation.* Special Report of the Intergovernmental Panel on Climate Change. IPCC and Cambridge University Press.

EPICA community members. 2004. Eight glacial cycles from an Antarctic ice core. *Nature* 429: 623–628.

Field, Christopher, et al., eds. 2012. *Managing the risks of extreme events and disasters to advance climate change adaptation.* Special Report of the Intergovernmental Panel on Climate Change. IPCC and Cambridge University Press.

Flannery, Tim. 2005. *The weather makers: The history and future impact of climate change.* Text Publishing, Melbourne, Australia.

Francis, Jennifer A., and Stephen J. Vavrus. 2012. Evidence linking Arctic amplification to extreme weather in mid-latitudes. *Geophysical Research Letters* 39: L06801, 6 pp.

Gelbspan, Ross. 1997. *The heat is on: The climate crisis, the cover-up, the prescription.* Perseus Books, New York.

Gelbspan, Ross. 2004. *Boiling point: How politicians, big oil and coal, journalists, and activists are fueling the climate crisis—and what we can do to avert disaster.* Basic Books, New York.

Gillis, Justin, and Coral Davenport. 2016. Leaders roll up sleeves on climate, but experts say plans don't pack a wallop. *New York Times,* 21 Apr. 2016.

Goodell, Jeff. 2013. Goodbye, Miami. *Rolling Stone.* 20 June 2013.

Goodstein, Eban, 2007. *Fighting for love in the century of extinction: How passion and politics can stop global warming.* University Press of New England, Lebanon, N.H.

Gore, Al. 2006. *An inconvenient truth: The planetary emergency of global warming and what we can do about it.* Rodale Press and Melcher Media, New York.

Hansen, James, Makiko Sato, and Reto Ruedy. 2012. Perception of climate change. *Proc. Natl. Acad. Sci. USA* E2415-E2423, 6 Aug. 2012.

InsideClimate News. http://www.insideclimatenews.org.

Intergovernmental Panel on Climate Change. http://www.ipcc.ch.

Intergovernmental Panel on Climate Change. 2013. *Climate change 2013: The physical science basis. Working Group I contribution to the Fifth Assessment Report of the IPCC.* (Stocker, Thomas F., et al., eds.). IPCC, Geneva, and Cambridge University Press, Cambridge.

Intergovernmental Panel on Climate Change. 2014. *Climate change 2014: Impacts, adaptation, and vulnerability. Working Group II contribution to the Fifth Assessment Report of the IPCC.* (Field, Christopher B., and Vicente R. Barros, et al., eds.). IPCC, Geneva, and Cambridge University Press, Cambridge.

Intergovernmental Panel on Climate Change. 2014. *Climate change 2014: Mitigation of climate change. Working Group III contribution to the Fifth Assessment Report of the IPCC.* (Edenhofer, Ottmar, et al., eds.). IPCC, Geneva, and Cambridge University Press, Cambridge.

Intergovernmental Panel on Climate Change. 2014. *Climate change 2014: Synthesis report. Contribution of Working Groups I, II, and III to the Fifth Assessment Report of the IPCC* (Core writing team: Pachauri, R.K., and L.A. Meyer, eds.). IPCC, Geneva, Switzerland.

International Energy Agency. 2012. *CO_2 emissions from fuel combustion: Highlights: 2012* edition. IEA, Paris.

International Energy Agency. 2015. *Energy and climate change.* World Energy Outlook Special Report. IEA, Paris.

Jackson, Robert B., et al. 2016. Reaching peak emissions. *Nature Climate Change* 6: 7–10.

Jonzén, Niclas, et al. 2006. Rapid advance of spring arrival dates in long-distance migratory birds. *Science* 312: 1959–1961.

Kamp, David. 2015. Can Miami Beach survive global warming? *Vanity Fair,* 11 Nov. 2015.

Karl, Thomas R., and Kevin E. Trenberth. 2003. Modern global climate change. *Science* 302: 1719–1723.

Klein, Naomi. 2015. *This changes everything: Capitalism versus the climate.* Simon and Schuster, New York.

Kolbert, Elizabeth. 2015. The siege of Miami. *New Yorker*, 21 & 28 Dec. 2015.

Kraska, James, ed. 2013. *Arctic security in an age of climate change.* Cambridge University Press, Cambridge, U.K.

Mann, Michael, and Lee R. Kump. 2015. *Dire predictions: Understanding global warming*, 2nd ed. DK Publishing and Pearson Education, New York.

Marshall, George. 2015. *Don't even think about it: Why our brains are wired to ignore climate change.* Bloomsbury USA, New York.

Mayewski, Paul A., and Frank White. 2002. *The ice chronicles: The quest to understand global climate change.* University Press of New England, Hanover, N.H.

McKibben, Bill. 2012. Global warming's terrifying new math. *Rolling Stone*, 19 July 2012.

Melillo, Jerry M., et al. 2014. *Climate change impacts in the United States: The Third National Climate Assessment.* U.S. Global Change Research Program, Washington, D.C.

NOAA Climate.gov. http://www.climate.gov.

NOAA State of the Climate. http://www.ncdc.noaa.gov/sotc.

Oreskes, Naomi, and Erik Conway. 2010. *Merchants of doubt.* Bloomsbury, London.

Oreskes, Naomi, and Erik Conway. 2014. *The collapse of Western civilization: A view from the future.* Columbia University Press, New York.

Pacala, Stephen, and Robert Socolow. 2004. Stabilization wedges: Solving the climate problem for the next 50 years with current technologies. *Science* 305: 968–972.

Parmesan, Camille, and Gary Yohe. 2003. A globally coherent fingerprint of climate change impacts across natural systems. *Nature* 421: 37–42.

Real Climate: Climate science from climate scientists. http://www.realclimate.org.

Romm, Joseph. 2015. *Climate change: What everyone needs to know.* Oxford University Press, Oxford, U.K.

Root, Terry L., et al. 2003. Fingerprints of global warming on wild animals and plants. *Nature* 421: 57–60.

Royal Society, The. 2005. *Ocean acidification due to increasing atmospheric carbon dioxide.* The Royal Society, London, June 2005.

Schneider, Stephen H., and Terry L. Root, eds. 2002. *Wildlife responses to climate change: North American case studies.* Island Press, Washington, D.C.

Service, Robert. 2012. Rising acidity brings an ocean of trouble. *Science* 337: 146–148.

Solovitch, Sara. 2016. How Miami Beach is keeping the Florida dream alive—and dry. *Politico*, 14 Mar. 2016.

Southeast Florida Regional Climate Change Compact. http://www.southeastfloridaclimatecompact.org.

Surging Seas: Sea-level rise analysis by Climate Central. http://www.sealevel.climatecentral.org.

Taylor, David. 2003. Small islands threatened by sea level rise. pp. 84–85 in *Vital signs 2003.* Worldwatch Institute, Washington D.C.

Tietenberg, Tom H. 2006. *Emissions trading: Principles and practice,* 2nd ed. Resources for the Future, Washington, D.C.

Tompkins, Forbes, and Christina DeConcini. 2014. *Sea-level rise and its impact on Miami-Dade County.* World Resources Institute, Washington, D.C.

Toomey, Diane. 2015. How British Columbia gained by putting a price on carbon. *Yale Environment 360*, 30 Apr. 2015.

United Nations. U.N. Framework Convention on Climate Change. http://www.unfccc.int/2860.php.

United Nations. The Paris Agreement. http://www.unfccc.int/paris_agreement/items/9485.php.

U.N. Environment Programme. 2011. *Bridging the emissions gap: A UNEP Synthesis Report.* http://www.unep.org/publications/ebooks/bridgingemissionsgap/

U.S. Climate Change Science Program. 2008. *Climate models: An assessment of strengths and limitations.* Synthesis and Assessment Product 3.1. Washington, D.C.

U.S. Climate Change Science Program. 2009. *Coastal sensitivity to sea-level rise: A focus on the mid-Atlantic region.* Synthesis and Assessment Product 4.1, Washington, D.C.

U.S. Department of Defense. 2014. *2014 Climate change adaptation roadmap.* Office of the Deputy Under Secretary of Defense for Installations and Environment, Alexandria, Va.

U.S. Environmental Protection Agency. 2015. *Climate change in the United States: Benefits of global action.* EPA 430-R-15-001, June 2015. EPA, Washington, D.C.

Weiss, Jessica. 2016. Miami Beach's $400 million sea-level rise plan is unprecedented, but not everyone is sold. *Miami New Times*, 19 Apr. 2016.

Wilcox, Jennifer. 2012. *Carbon capture.* Springer, New York.

World Bank. 2012. *Turn down the heat: Why a 4° warmer world must be avoided.* A report for the World Bank by the Potsdam Institute for Climate Impact Research and Climate Analytics. World Bank, Washington D.C.

World Meteorological Organization. 2016. *WMO statement on the status of the global climate in 2015.* WMO No. 1167. Geneva, Switzerland.

Worldwatch Institute. 2009. *State of the world 2009: Into a warming world.* Worldwatch Institute, Washington, D.C.

Chapter 19

Al-Fattah, Saud M., et al. 2011. *Carbon capture and storage: Technologies, policies, economics, and implementation strategies.* CRC Press, Boca Raton, Fla.

American Petroleum Institute. 2009. *Offshore access to oil and natural gas reserves.* American Petroleum Institute, Washington, D.C.

Association for the Study of Peak Oil—USA. http://www.peak-oil.org.

Austen, Ian. 2015. Lower oil prices strike at heart of Canada's oil sands production. *New York Times*, 2 Feb. 2015.

Avery, Samuel. 2013. *The pipeline and the paradigm: Keystone XL, tar sands, and the battle to defuse the carbon bomb.* Ruka Press, Washington, D.C.

British Petroleum. 2016. *BP statistical review of world energy 2016.* BP, London.

Deffeyes, Kenneth S. 2005. *Beyond oil: The view from Hubbert's peak.* Farrar, Straus, and Giroux, New York.

Energy Information Administration, U.S. Department of Energy. http://www.eia.doe.gov.

Energy Information Administration, U.S. Department of Energy. Monthly energy review. Data updated monthly online at: http://www.eia.doe.gov/totalenergy/data/monthly.

Energy Information Administration, U.S. Department of Energy. 2013. *Technically recoverable shale oil and shale gas resources: An assessment of 137 shale formations in 41 countries outside the United States.* Washington, D.C.

Energy Information Administration, U.S. Department of Energy. 2016. *Annual energy outlook 2016.* DOE/EIA, Washington, D.C.

European Parliament, 2011. *Impacts of shale gas and shale oil extraction on the environment and on human health.* Directorate-General for Internal Policies.

Findley, David. 2010. *Do-it-yourself home energy audits: 140 simple solutions to lower energy costs, increase your home's efficiency, and save the environment.* McGraw-Hill, New York.

Haerens, Margaret. 2010. *Offshore drilling (opposing viewpoints).* Greenhaven Press, Farmington Hills, Mich.

Hall, Charles A.S., and Doug Hansen. 2011. New Studies in EROI (Energy Return on Investment). *Sustainability*, special issue. MDPI AG, Basel, Switzerland, 2011.

Hall, Charles A.S., et al. 2014. EROI of different fuels and the implications for society. *Energy Policy* 64: 141–152.

Herring, Horace, and Steve Sorrell, eds. 2009. *Energy efficiency and sustainable consumption: The rebound effect.* Palgrave Macmillan, London.

Hubbert, M. King. 1956. *Nuclear energy and the fossil fuels.* Publication No. 95, Shell Development Company, Houston, Tex.

International Energy Agency. 2015. *Key world energy statistics 2015.* IEA, Paris.

International Energy Agency. 2015. *World energy outlook 2015.* IEA, Paris.

Jackson, Robert B., et al. 2014. The environmental costs and benefits of fracking. *Annu. Rev. Environ. Resources* 39: 327–362.

Kitasei, Saya. 2010. *Powering the low-carbon economy: The once and future roles of renewable energy and natural gas.* Worldwatch Report 184. Worldwatch Institute, Washington, D.C.

Krauss, Clifford. 2015. New balance of power. *New York Times*, 22 Apr. 2015.

Kunstler, James H. 2005. *The long emergency*. Atlantic Monthly Press, New York.

Kunzig, Robert. 2009. Scraping bottom. *National Geographic*, Mar. 2009.

Leffler, William L., et al., 2011. *Deepwater petroleum exploration and production: A nontechnical guide*, 2nd ed. Pennwell Corp. Publishing, Tulsa, Okla.

Levant, Ezra. 2011. *Ethical oil: The case for Canada's oil sands*. McClelland & Stewart, Toronto, Ontario.

Lovins, Amory B. 2005. More profit with less carbon. *Scientific American* 293(3): 74–83.

Lovins, Amory B., et al. 2004. *Winning the oil endgame: Innovation for profits, jobs, and security*. Rocky Mountain Institute, Snowmass, Colo.

Ma, Haibing. 2014. Global energy and carbon intensity continue to decline. http://vitalsigns.worldwatch.org/vs-trend/global-energy-and-carbon-intensity-continue-decline. Worldwatch Institute, Washington D.C.

MacKay, David J.C. 2009. *Sustainable energy—without the hot air*. UIT Cambridge.

McGlade, Christophe, and Paul Ekins. 2015. The geographical distribution of fossil fuels unused when limiting global warming to 2°C. *Nature* 517: 187–190.

Miller, Bruce G. 2010. *Clean coal engineering technology*. Butterworth-Heinemann, Elsevier.

National Commission on the BP Deepwater Horizon Oil Spill and Offshore Drilling. 2011. *Deep water: The Gulf oil disaster and the future of offshore drilling*. Report to the President. Oil Spill Commission.

National Oceanic and Atmospheric Administration. 2012. *Natural resource damage assessment for the Deepwater Horizon oil spill: April 2012 status update*. NOAA, Washington, D.C.

OECD. 2012. *Inventory of estimated budgetary support and tax expenditures for fossil fuels 2013*. OECD Publishing.

Parfomak, Paul, et al., 2012. *Keystone XL pipeline project: Key issues*. Congressional Research Service CRS Report for Congress 7-5700 R41668.

Rao, Vikram. 2012. *Shale gas: The promise and the peril*. RTI International, RTI Press, Research Triangle Institute, N.C.

Renne, John L., and Billy Fields. 2013. *Transport beyond oil: Policy choices for a multimodal future*. Island Press, Washington, D.C.

Ristinen, Robert A., et al., 2016. *Energy and the environment*, 3rd ed. John Wiley & Sons, New York.

Sontag, Deborah, and Robert Gebeloff. 2014. The downside of the boom. *New York Times Magazine*, 22 Nov. 2014.

Spellman, Frank R. 2013. *Environmental impacts of hydraulic fracturing*. CRC Press, Boca Raton, Fla.

Sumper, Andreas, and Angelo Baggini. 2012. *Electrical energy efficiency: Technologies and applications*. John Wiley & Sons, Chichester, U.K.

U.S. Department of State. 2014. *Final supplemental environmental impact statement for the Keystone XL project*. Jan. 2014.

U.S. Environmental Protection Agency. 2015. *Light-duty automotive technology, carbon dioxide emissions, and fuel economy trends: 1975 through 2015*. EPA-420-S-15-001. EPA Office of Transportation and Air Quality, Washington, D.C.

U.S. Geological Survey. 2001. *Arctic National Wildlife Refuge, 1002 Area, petroleum assessment, 1998, including economic analysis*. USGS Fact Sheet FS-028-01.

U.S. Geological Survey. 2002. *Petroleum resource assessment of the National Petroleum Reserve Alaska (NPRA)*. USGS, Washington, D.C.

U.S. Government Accountability Office (GAO). 2007. *Crude oil: Uncertainty about future oil supply makes it important to develop a strategy for addressing a peak and decline in oil production*. Report to Congressional Requesters.

Vengosh, Avner, et al. 2014. A critical review of the risks to water resources from unconventional shale gas development and hydraulic fracturing in the United States. *Environ. Sci. Technol.* 48: 8334–8348.

White, Helen K., et al. 2012. Impact of the Deepwater Horizon oil spill on a deep-water coral community in the Gulf of Mexico. *Proc. Natl. Acad. Sci. USA* 109: 20303-20308.

White House, The, U.S. 2015. Statement by the President on the Keystone XL Pipeline. http://www.whitehouse.gov/the-press-office/2015/11/06/statement-president-keystone-xl-pipeline.

Wilcox, Jennifer. 2012. *Carbon capture*. Springer, New York.

Yergin, Daniel. 1991. *The prize: The epic quest for oil, money, and power*. Simon and Schuster, New York.

Yergin, Daniel. 2011. *The quest: Energy, security, and the remaking of the modern world*. Penguin Press, London.

Chapter 20

Ansar, Atif, et al. 2014. Should we build more large dams? The actual costs of hydropower megaproject development. *Energy Policy* 69: 43–56.

Biomass Research and Development Board. 2008. *National biofuels action plan*. Biomass Research and Development Initiative, U.S. DOE and USDA.

British Petroleum. 2016. *BP statistical review of world energy 2016*. BP, London.

Chernobyl Forum, The. 2006. *Chernobyl's legacy: Health, environmental and socio-economic impacts* and *Recommendations to the governments of Belarus, the Russian Federation and Ukraine*. The Chernobyl Forum: 2003–2005. Second revised version. World Health Organization and International Atomic Energy Agency, Vienna.

Earley, Jane, and Alice McKeown. 2009. *Red, white, and green: Transforming U.S. biofuels*. Worldwatch Report 180. Worldwatch Institute, Washington, D.C.

Energy Efficiency and Renewable Energy, U.S. Department of Energy. http://www.eere.energy.gov.

Energy Information Administration, U.S. Department of Energy. http://www.eia.doe.gov.

Energy Information Administration, U.S. Department of Energy. Monthly energy review. Data updated monthly online at: http://www.eia.doe.gov/totalenergy/data/monthly.

Energy Information Administration, U.S. Department of Energy. 2016. *Annual energy outlook 2016*. DOE/EIA, Washington, D.C.

European Commission/International Atomic Energy Agency/World Health Organization. 1996. One decade after Chernobyl: Summing up the consequences of the accident. Summary of the conference results. Vienna, Austria, 8–12 April 1996. EC/IAEA/WHO.

Flavin, Christopher. 2008. *Low-carbon energy: A roadmap*. Worldwatch Report 178. Worldwatch Institute, Washington, D.C.

Hall, Charles A.S., and Doug Hansen. 2011. New Studies in EROI (Energy Return on Investment). *Sustainability*, special issue. MDPI AG, Basel, Switzerland, 2011.

Hall, Charles A.S., et al. 2014. EROI of different fuels and the implications for society. *Energy Policy* 64: 141–152.

International Atomic Energy Agency. *Nuclear power and sustainable development*. IAEA Information Series 02-01574/FS Series 3/01/E/Rev.1. Vienna.

International Atomic Energy Agency. 2006. *Environmental consequences of the Chernobyl accident and their remediation: Twenty years of experience*. Report of the U.N. Chernobyl Forum Expert Group "Environment." IAEA, Vienna.

International Energy Agency. 2007. *Biomass for power generation and CHP*. IEA, Paris.

International Energy Agency. 2015. *Key world energy statistics 2015*. IEA, Paris.

International Energy Agency. 2015. *Renewables information 2015*. IEA, Paris.

International Energy Agency. 2015. *Technology roadmap: Nuclear energy*. IEA, Paris.

International Energy Agency. 2015. *World energy outlook 2015*. IEA, Paris.

Ishikawa, Tetsuo, et al., 2015. The Fukushima Health Management Survey: Estimation of external doses to residents in Fukushima Prefecture. *Nature Scientific Reports* 5: 12712.

Lynas, Mark. 2014. *Nuclear 2.0: Why a green future needs nuclear power*. UIT Cambridge Ltd.

Murphy, David J., and Charles A.S. Hall. 2010. Year in review—EROI or energy return on (energy) invested. *Annals of the New York Academy of Sciences* 1185: 102–118.

National Renewable Energy Lab, U.S. Department of Energy. http://www.nrel.gov.

Nature. 2006. Special report: Chernobyl and the future. *Nature* 440: 982–989.

Normile, Dennis. 2011. Fukushima revives the low-dose debate. *Science* 332: 908–910.

Nuclear Energy Agency. 2002. Chernobyl: Assessment of radiological and health impacts. 2002 update of *Chernobyl: Ten years on.* OECD, Paris.

Pearce, Fred. 2006. Fuels gold: Are biofuels really the greenhouse-busting answer to our energy woes? *New Scientist*, 23 Sept. 2006: 36–41.

Qvist, Staffan A., and Barry W. Brook. 2015. Environmental and health impacts of a policy to phase out nuclear power in Sweden. *Energy Policy* 84: 1–10.

REN21 Renewable Energy Policy Network for the 21st Century. 2016. *Renewables 2016 global status report.* REN21 Secretariat, Paris.

Renewable Fuels Association. 2010. *Fueling a high octane future: 2016 ethanol industry outlook.* RFA, Washington, D.C.

Rosenthal, Elisabeth. 2010. Using waste, Swedish city cuts its fossil fuel use. *New York Times*, 10 Dec. 2010.

Sawin, Janet L., and William R. Moomaw. 2009. *Renewable revolution: Low-carbon energy by 2030.* Worldwatch Report 182. Worldwatch Institute, Washington, D.C.

Schneider, Mycle, et al. 2015. *The world nuclear industry status report 2015.* Mycle Schneider Consulting, Paris and London.

Science. 2005. News Focus: Rethinking nuclear power. *Science* 309: 1168–1179.

Spadaro, Joseph V., et al. 2000. Greenhouse gas emissions of electricity generation chains: Assessing the difference. *IAEA Bulletin* 42(2).

Swedish Bioenergy Association (SVEBIO). 2003. *Focus: Bioenergy.* Nos. 1–10. SVEBIO, Stockholm.

Swedish Energy Agency. 2012. *Sustainable biofuels 2011.* Swedish Energy Agency, Eskilstuna, Sweden.

Swedish Energy Agency. 2015. *Energy in Sweden 2015.* Swedish Energy Agency, Eskilstuna, Sweden.

Ten Hoeve, John E., and Mark Z. Jacobson. 2012. Worldwide health effects of the Fukushima Daiichi nuclear accident. *Energy & Environmental Science* 5: 8743-8757.

U.N. Environment Programme. 2009. *Assessing biofuels.* UNEP, Nairobi, Kenya.

U.N. Scientific Committee on the Effects of Atomic Radiation. 2014. *Sources, effects, and risks of ionizing radiation.* UNSCEAR 2013 Report to the General Assembly with Scientific Annexes, Vol. I. United Nations, New York.

World Health Organization. 2006. *Health effects of the Chernobyl accident and special health care programmes.* Report of the U.N. Chernobyl Forum Expert Group "Health." WHO, Geneva.

World Health Organization. 2013. *Health risk assessment from the nuclear accident after the Great East Japan Earthquake and Tsunami.* WHO, Geneva.

World Nuclear Association. 2013. *Nuclear power in Sweden.* http://www.world-nuclear.org/information-library/country-profiles/countries-o-s/sweden.aspx#.UdLWiuCbIbA.

Worldwatch Institute. 2007. *Biofuels for transport: Global potential and implications for sustainable agriculture and energy in the 21st century.* Worldwatch Institute, Washington, D.C.

Worldwatch Institute and Center for American Progress. 2006. *American energy: The renewable path to energy security.* Washington, D.C.

Yasunari, Teppei J. 2011. Cesium-137 deposition and contamination of Japanese soils due to the Fukushima nuclear accident. *Proc Natl. Acad. Sci. USA* 108: 19530–19534.

Chapter 21

American Wind Energy Association. 2015. *U.S. wind industry annual market report 2015.* AWEA, Washington, D.C.

Ananthaswamy, Anil. 2003. Reality bites for the dream of a hydrogen economy. *New Scientist*, 15 Nov. 2003: 6–7.

Barbose, Galen, et al. 2011. *Tracking the sun: An historical summary of the installed costs of photovoltaics in the United States from 1998 to 2010.* Lawrence Berkeley National Laboratory, Berkeley, Calif.

Boyle, Godfrey. 2012. *Renewable energy: Power for a sustainable future.* 3rd ed. Oxford University Press USA, New York.

Brown, Lester R., and Emily Adams. 2015. *The great transition: Shifting from fossil fuels to solar and wind energy.* W.W. Norton, New York.

Davidson, Osha Gray. 2012. *Clean break: The story of Germany's energy transformation and what Americans can learn from it.* Inside Climate News.

Dunn, Seth. 2000. The hydrogen experiment. *World Watch* 13: 14–25.

Economist. 2012. Germany's energy transformation: Energiewende. *The Economist*, 28 July 2012.

Energy Efficiency and Renewable Energy, U.S. Department of Energy. http://www.eere.energy.gov.

Energy Efficiency and Renewable Energy, U.S. Department of Energy. 2011. *2010 solar technologies market report.* EERE, Washington, D.C.

Energy Efficiency and Renewable Energy, U.S. Department of Energy. 2015. *2014 wind technologies market report.* EERE, Washington, D.C.

Energy Information Administration, U.S. Department of Energy. http://www.eia.doe.gov.

Energy Information Administration, U.S. Department of Energy. Monthly energy review. Data updated monthly online at: http://www.eia.doe.gov/totalenergy/data/monthly.

Energy Information Administration, U.S. Department of Energy. 2016. *Annual energy outlook 2016.* DOE/EIA, Washington, D.C.

Environmental Law Institute. 2009. *Estimating U.S. government subsidies to energy sources: 2002–2008.* ELI, Washington, D.C.

Federal Ministry for the Environment, Nature Conservation, and Nuclear Safety [Germany]. http://www.erneuerbare-energien.de/en.

Federal Ministry for the Environment, Nature Conservation, and Nuclear Safety [Germany]. 2009. *Electricity from renewable energy sources: What does it cost?* Berlin.

Federal Ministry for the Environment, Nature Conservation, and Nuclear Safety [Germany]. 2011. *Renewable energy sources in figures: National and international development.* Berlin.

Federal Ministry for the Environment, Nature Conservation, and Nuclear Safety [Germany]. 2012. *Innovation through research: 2011 annual report on research funding in the renewable energies sector.* Berlin.

Federal Ministry of Economics and Technology [Germany]. http://www.renewables-made-in-germany.com/index.php?id=50&L=1.

Flavin, Christopher. 2008. *Low-carbon energy: A roadmap.* Worldwatch Report 178. Worldwatch Institute, Washington, D.C.

Global Wind Energy Council. 2016. *Global wind report 2015.* GWEC, Brussels, Belgium.

Grant, Paul M., et al. 2006. A power grid for the hydrogen economy. *Scientific American* 295(1): 76–83.

Hall, Charles A.S., and Doug Hansen. 2011. New Studies in EROI (Energy Return on Investment). *Sustainability*, special issue. MDPI AG, Basel, Switzerland, 2011.

International Energy Agency. 2009. *Technology roadmap: Wind energy.* IEA, Paris.

International Energy Agency. 2015. *Key world energy statistics 2015.* IEA, Paris.

International Energy Agency. 2015. *Renewables information 2015.* IEA, Paris.

International Energy Agency. 2015. *Technology roadmap: Hydrogen and fuel cells.* IEA, Paris.

International Energy Agency. 2015. *World energy outlook 2015.* IEA, Paris.

International Energy Agency Photovoltaic Power Systems Programme. 2015. *PVPS annual report 2014.* Imprimerie St. Paul, Fribourg, Switzerland.

International Renewable Energy Agency. 2015. *Renewable energy target setting.* IRENA, Masdar City, Abu Dhabi, UAE.

Jacobson, Mark Z. 2009. Review of solutions to global warming, air pollution, and energy security. *Energy & Environmental Science* 2: 148–173.

Jacobson, Mark Z., and Mark A. Delucchi. 2009. A path to sustainable energy by 2030. *Scientific American* 301(5): 58–65.

Jacobson, Mark Z., and Mark A. Delucchi. 2011. Providing all global energy with wind, water, and solar power, Part I: Technologies, energy resources, quantities and areas of infrastructure, and materials. *Energy Policy* 39: 1154–1169.

Jacobson, Mark Z., et al. 2005. Cleaning the air and improving health with hydrogen fuel-cell vehicles. *Science* 308: 1901–1905.

Jacobson, Mark Z., et al., 2015. 100% clean and renewable wind, water, and sunlight (WWS) all-sector energy roadmaps for the 50 United States. *Energy Environ. Sci.* 8: 2093–2117.

Kitasei, Saya. 2010. *Powering the low-carbon economy: The once and future roles of renewable energy and natural gas.* Worldwatch Report 184. Worldwatch Institute, Washington, D.C.

Knott, Michelle. 2003. Power from the waves. *New Scientist*, 20 Sept. 2003: 33–35.

McNamee, Gregory. 2008. *Careers in renewable energy: Get a green energy job*. Pixy Jack Press, Masonville, Colo.

Morris, Craig, and Martin Pehnt. 2015. *Energy transition: The German energiewende*. Heinrich Boll Foundation, Berlin.

National Renewable Energy Lab, U.S. Department of Energy. http://www.nrel.gov.

National Renewable Energy Lab. 2015. *2014–2015 Offshore wind technologies market report*. Tech Rep. NREL/TP-5000-64283, Sept. 2015. U.S. DOE, Washington D.C.

Ochs, Alexander, and Shakuntala Makhijan. 2012. *Sustainable energy roadmaps: Guiding the global shift to domestic renewables*. Worldwatch Report 187. Worldwatch Institute, Washington, D.C.

OECD. 2012. Inventory of estimated budgetary support and tax expenditures for fossil fuels 2013. OECD Publishing.

Pan, Jihua, et al. 2011. *Green economy and green jobs in China: Current status and potentials for 2020*. Worldwatch Report 185. Worldwatch Institute, Washington, D.C.

Pfund, Nancy, and Ben Healey. 2011. *What would Jefferson do? The historical role of federal subsidies in shaping America's energy future*. DBL Investors, San Francisco.

REN21 Renewable Energy Policy Network for the 21st Century. 2016. *Renewables 2016 global status report*. REN21 Secretariat, Paris.

Quashning, Volker. 2010. *Renewable energy and climate change*. Wiley-IEEE Press.

Ristinen, Robert A., et al., 2016. *Energy and the environment*, 3rd ed. John Wiley & Sons, New York.

Sawin, Janet L., and William R. Moomaw. 2009. *Renewable revolution: Low-carbon energy by 2030*. Worldwatch Report 182. Worldwatch Institute, Washington, D.C.

Weisman, Alan. 1998. *Gaviotas: A village to reinvent the world*. Chelsea Green Publishing Co., White River Junction, Vt.

Wirth, Harry. 2015. Recent facts about photovoltaics in Germany. Fraunhofer ISE, Freiburg, Germany.

World Future Council. 2010. *Feed-in tariffs—Boosting energy for our future*. Earthscan Publications, London.

Worldwatch Institute and Center for American Progress. 2006. *American energy: The renewable path to energy security*. Washington, D.C.

Chapter 22

Ayres, Robert U., and Leslie W. Ayres. 1996. *Industrial ecology: Towards closing the materials cycle*. Edward Elgar Press, Cheltenham, U.K.

Beede, David N., and David E. Bloom. 1995. The economics of municipal solid waste. *World Bank Research Observer* 10: 113–150.

Campbell, Stu. 1998. *Let it rot! The gardener's guide to composting*, 3rd ed. Storey Publishing, North Adams, Mass.

Container Recycling Institute. http://www.container-recycling.org.

Douglas, Ed. 2009. There's gold in them there landfills. *New Scientist*, 1 Oct. 2008.

Gourmelon, Gaelle. 2015. Global plastic production rises, recycling lags. http://vitalsigns.worldwatch.org/vs-trend/global-plastic-production-rises-recycling-lags. Worldwatch Institute, Washington D.C.

Graedel, Thomas E., and Braden R. Allenby. 2009. *Industrial ecology and sustainable engineering*. Prentice Hall, Upper Saddle River, N.J.

Hieronymi, Klaus, et al., eds. 2012. *E-waste management: From waste to resource*. Routledge, Abingdon, U.K.

Leonard, Annie. 2010. *The story of stuff*. Free Press, New York.

Lilienfeld, Robert, and William Rathje. 1998. *Use less stuff: Environmental solutions for who we really are*. Ballantine, New York.

Manahan, Stanley E. 1999. *Industrial ecology: Environmental chemistry and hazardous waste*. Lewis Publishers, CRC Press, Boca Raton, Fla.

McDonough, William, and Michael Braungart. 2002. *Cradle to cradle: Remaking the way we make things*. North Point Press, New York.

Navarro, Mireya. 2011. Lunch, landfills, and what I tossed. *New York Times*, 21 Oct. 2011.

New York City Department of Parks and Recreation. Fresh Kills Park. http://www.nycgovparks.org/sub_your_park/fresh_kills_park/html/fresh_kills_park.html.

Pichtel, John. 2014. *Waste management practices: Municipal, hazardous, and industrial*, 2nd ed. CRC Press, Boca Raton, Fla.

Rathje, William, and Colleen Murphy. 2001. *Rubbish! The archeology of garbage*. University of Arizona Press.

Recyclemania. http://www.recyclemaniacs.org.

Scott, Nicky. 2007. *Reduce, reuse, recycle: An easy household guide*. Chelsea Green Publishing Co., White River Junction, Vt.

Shin, Dolly. 2014. *Generation and disposition of municipal solid waste (MSW) in the United States—a national survey*. M.S. thesis, Columbia University Earth Engineering Center, New York.

Spalvins, E., B. Dubey, and T. Townsend. 2008. Impact of electronic waste disposal on lead concentrations in landfill leachate. *Environ. Sci. Technol.* 42: 7452–7458.

Townsend, Timothy. 2011. Environmental issues and management strategies for waste electronic and electrical equipment. *J. Air & Waste Manage. Assoc.* 61: 587–561.

Trash Track. http://senseable.mit.edu/trashtrack.

U.N. Environment Programme. 2012. "Chemicals and Waste." Chapter 6 in *Global environment outlook 5 (GEO-5)*. UNEP, Nairobi.

U.S. Environmental Protection Agency. Summary of the Comprehensive Environmental Response, Compensation, and Liability Act (Superfund). https://www.epa.gov/laws-regulations/summary-comprehensive-environmental-response-compensation-and-liability-act.

U.S. Environmental Protection Agency. Summary of the Resource Conservation and Recovery Act. https://www.epa.gov/laws-regulations/summary-resource-conservation-and-recovery-act.

U.S. Environmental Protection Agency. 2011. *Electronics waste management in the United States through 2009*. EPA530-R-11-002. EPA Office of Resource Conservation and Recovery, Washington, D.C.

U.S. Environmental Protection Agency. 2015. *Advancing sustainable materials management: Facts and figures 2013*. EPA530-R-15-002. EPA Office of Resource Conservation and Recovery, Washington, D.C.

van Haaren, R., et al. 2010. The state of garbage in America. *BioCycle* 51: 16–23.

Chapter 23

Brugge, Doug, and Rob Goble. 2002. The history of uranium mining and the Navajo people. *Am. J. Public Health* 92(9): 1410–1419.

Christopherson, Robert W. 2011. *Geosystems: An introduction to physical geography*, 8th ed. Prentice Hall, Upper Saddle River, N.J.

Cohen, David. 2007. Earth audit. *New Scientist*, 26 May 2007: pp. 34–41.

Craig, James R., David J. Vaughan, and Brian J. Skinner. 2010. *Earth resources and the environment*, 4th ed. Benjamin Cummings, San Francisco.

Essick, Kristi. 2001. Guns, money, and cell phones. *The Industry Standard Magazine*, 11 Jun. 2001.

Gordon, R.B., et al. 2006. Metal stocks and sustainability. *Proc. Natl. Acad. Sci. USA* 103(5): 1209–1214.

Hendryx, Michael, and Melissa M. Ahern. 2009. Mortality in Appalachian coal mining regions: The value of statistical life lost. *Public Health Reports* 124: 541–550.

Keller, Edward A. 2011. *Introduction to environmental geology*, 5th ed. Prentice Hall, Upper Saddle River, N.J.

Kondash, Avner J., et al., 2014. Radium and barium removal through blending hydraulic fracturing fluids with acid mine drainage. *Environ. Sci. Technol.* 48: 1334–1342.

Lovgren, Stefan. 2006. Can cell-phone recycling help African gorillas? *National Geographic News*, 20 Jan. 2006.

Mooallem, Jon. 2008. The afterlife of cellphones. *New York Times Magazine*, 13 Jan. 2008.

Palmer, Margaret, et al. 2010. Mountaintop mining consequences. *Science* 327: 148–149.

Perkins, Dexter. 2010. *Mineralogy*, 3rd ed. Prentice Hall, Upper Saddle River, N.J.

Rogich, D.G., and G.R. Matos. 2008. *The global flows of metals and minerals*. U.S. Geological Survey Open-File Report 2008-1355, 11 pp.

Sibley, Scott F., ed. 2004. *Flow studies for recycling metal commodities in the United States*. USGS Circular 1196-A-M. U.S. Geological Survey, Reston, Va.

Skinner, Brian J., and Stephen C. Porter. 2003. *The dynamic earth: An introduction to physical geology*, 5th ed. John Wiley & Sons, Hoboken, N.J.

Sullivan, Daniel E. 2006. *Recycled cell phones—A treasure trove of valuable metals*. USGS Fact Sheet 2006-3097. U.S. Geological Survey, Denver, Colo.

Tarbuck, Edward J., Frederick K. Lutgens, and Dennis Tasa. 2011. *Earth science*, 13th ed. Prentice Hall, Upper Saddle River, N.J.

U.N. Security Council. 2001. *Report of the panel of experts on the illegal exploitation of natural resources and other forms of wealth of the Democratic Republic of the Congo*. U.N. Security Council, 12 Apr. 2001.

U.N. Security Council. 2007. *Interim report of the group of experts on the Democratic Republic of the Congo, pursuant to Security Council resolution 1698 (2006)*. U.N. Security Council, 25 Jan. 2007.

U.S. Department of the Interior. 2003. *Surface coal mining reclamation: 25 years of progress, 1977–2002*. Office of Surface Mining, Washington D.C.

U.S. Geological Survey. 2016. *Mineral commodity summaries 2016*. Reston, VA: USGS U.S. Geological Survey. Minerals information. http://www.minerals.usgs.gov/minerals.

Chapter 24

Bartlett, Peggy, and Geoffrey W. Chase, eds. 2004. *Sustainability on campus: Stories and strategies for change*. MIT Press, Cambridge, Mass.

Blewitt, John. 2008. *Understanding sustainable development*. Earthscan, London.

Brower, Michael, and Warren Leon. 1999. *The consumer's guide to effective environmental choices: Practical advice from the Union of Concerned Scientists*. Three Rivers Press, New York.

Brown, Lester R. 2009. *Plan B 4.0: Mobilizing to save civilization*. Earth Policy Institute and W.W. Norton, New York.

Brown, Lester R. 2011. *World on the edge: How to prevent environmental and economic collapse*. Earth Policy Institute and W.W. Norton, New York.

Campus Ecology. National Wildlife Federation. http://www.nwf.org/campus-ecology.

Campus Ecology. 2008. *Campus environment 2008: A national report card on sustainability in higher education*. Campus Ecology program of the National Wildlife Federation.

Creighton, Sarah Hammond. 1998. *Greening the ivory tower: Improving the environmental track record of universities, colleges, and other institutions*. MIT Press, Cambridge, Mass.

Daly, Herman E. 1996. *Beyond growth: The economics of sustainable development*. Beacon Press, Boston.

Dasgupta, Partha, et al. 2000. Economic pathways to ecological sustainability. *BioScience* 50: 339–345.

De Anza College. De Anza students move Foundation to vote for fossil fuel divestiture. http://www.deanza.edu/news/2013fossildivest.html

De Anza College. Sustainability. http://www.deanza.edu/sustainability.

Durning, Alan. 1992. *How much is enough? The consumer society and the future of the Earth*. Worldwatch Institute, Washington, D.C.

French, Hilary. 2004. Linking globalization, consumption, and governance. pp. 144–163 in *State of the world 2004*. Worldwatch Institute, Washington, D.C.

Gardner, Gary. 2011. *Creating sustainable prosperity in the United States: The need for innovation and leadership*. Worldwatch Report 186. Worldwatch Institute, Washington, D.C.

Hawken, Paul. 1994. *The ecology of commerce: A declaration of sustainability*. Harper Business, New York.

Kahneman, Daniel, et al. 2006. Would you be happier if you were richer? A focusing illusion. *Science* 312: 1908–1910.

Keniry, Julian. 1995. *Ecodemia: Campus environmental stewardship at the turn of the 21st century*. National Wildlife Federation, Washington, D.C.

McMichael, A.J., et al. 2003. New visions for addressing sustainability. *Science* 302: 1919–1921.

Meadows, Donella, Jørgen Randers, and Dennis Meadows. 2004. *Limits to growth: The 30-year update*. Chelsea Green Publishing Co., White River Junction, Vt.

Millennium Ecosystem Assessment. 2005. *Ecosystems and human well-being: General synthesis*. Millennium Ecosystem Assessment and World Resources Institute.

National Research Council, Board on Sustainable Development. 1999. *Our common journey: A transition toward sustainability*. National Academies Press, Washington, D.C.

Renner, Michael. 2008. *Green jobs: Working for people and the environment*. Worldwatch Report 177. Worldwatch Institute, Washington, D.C.

Robertson, Margaret. 2014. *Sustainability principles and practice*. Routledge, Abingdon, U.K.

Sachs, Jeffrey D., and Ban Ki-moon. 2015. *The age of sustainable development*. Columbia University Press, New York.

Schor, Juliet B., and Betsy Taylor, eds. 2002. *Sustainable planet: Solutions for the twenty-first century*. The Center for a New American Dream. Beacon Press, Boston.

United Nations. 2002. *Report of the World Summit on Sustainable Development, Johannesburg, South Africa, 26 August–4 September 2002*. U.N., New York.

United Nations. 2012. *Report of the United Nations Conference on Sustainable Development*. Rio de Janeiro, Brazil, 20–22 June 2012. https://sustainabledevelopment.un.org/index.php?page=view&type=111&nr=1358&menu=35.

U.N. Development Programme. 2013. *Human development report 2013*. Oxford University Press.

U.N. Division for Sustainable Development. *Sustainable Development Goals*. http://www.sustainabledevelopment.un.org/sdgs.

U.N. Environment Programme. 2012. *Global environment outlook 5 (GEO-5)*. UNEP, Nairobi.

U.N. General Assembly. 2012. *The future we want*. Outcome document from Rio+20 Conference. Resolution 66/288. http://www.un.org/ga/search/view_doc.asp?symbol=A/RES/66/288&Lang=E

World Bank. *World development indicators*. World Bank, Washington, D.C. http://data.worldbank.org/products/wdi.

World Commission on Environment and Development. 1987. *Our common future*. Oxford University Press.

Worldwatch Institute. 2008. *State of the world 2008: Innovations for a sustainable economy*. Worldwatch Institute, Washington, D.C.

Worldwatch Institute. 2010. *State of the world 2010: Transforming cultures*. Worldwatch Institute, Washington, D.C.

Worldwatch Institute. 2012. *State of the world 2012: Moving toward sustainable prosperity*. Worldwatch Institute, Washington, D.C.

Worldwatch Institute. 2013. *State of the world 2013: Is sustainability still possible?* Worldwatch Institute, Washington, D.C.

Worldwatch Institute. 2015. *State of the world 2015: Confronting hidden threats to sustainability*. Worldwatch Institute, Washington, D.C.

Campus Sustainability Resources

350.org. *Fossil Free: A campus guide to fossil fuel divestment*. http://www.gofossilfree.org/wp-content/uploads/2014/05/350_FossilFreeBooklet_LO4.pdf.

American College and University Presidents' Climate Commitment. http://www.presidentsclimatecommitment.org.

Association for the Advancement of Sustainability in Higher Education. http://www.aashe.org.

Association for the Advancement of Sustainability in Higher Education. 2010. *Creating a campus sustainability revolving loan fund: A guide for students*. http://www.aashe.org/resources/pdf/CERF.pdf.

Association for the Advancement of Sustainability in Higher Education. 2010. *Sustainability curriculum in higher education: A call to action*. http://www.aashe.org/files/A_Call_to_Action_final(2).pdf.

Association for the Advancement of Sustainability in Higher Education. 2014. *Higher education sustainability review*. AASHE, Philadelphia.

Association for the Advancement of Sustainability in Higher Education. 2015. *Sustainable campus index 2015: Top performers, best practices, and trends*. AASHE, Philadelphia.

Ball State University: Greening of the Campus [conference series.] http://cms.bsu.edu/academics/centersandinstitutes/goc.

Barth, Matthias. 2013. *Implementing sustainability in higher education: Learning in an age of transformation*. Routledge, New York.

Bartlett, Peggy, and Geoffrey W. Chase, eds. 2004. *Sustainability on campus: Stories and strategies for change*. MIT Press, Cambridge, Mass.

Bartlett, Peggy, and Geoffrey W. Chase, eds. 2013. *Sustainability in higher education: Stories and strategies for transformation*. MIT Press, Cambridge, Mass.

Campus Conservation Nationals. http://www.competetoreduce.org/.

Campus Ecology. National Wildlife Federation. http://www.nwf.org/campus-ecology.

Campus Ecology. 2008. *Campus environment 2008: A national report card on sustainability in higher education*. Campus Ecology program of the National Wildlife Federation.

Campus Transport Management: Trip reduction programs on college, university, and research campuses. Online TDM Encyclopedia. Victoria Transport Policy Institute. http://www.vtpi.org/tdm/tdm5.htm.

Carlson, Scott. 2006. In search of the sustainable campus: With eyes on the future, universities try to clean up their acts. *Chronicle of Higher Education* 53: A10.

Creighton, Sarah Hammond. 1998. *Greening the ivory tower: Improving the environmental track record of universities, colleges, and other institutions*. MIT Press, Cambridge, Mass.

Erickson, Christina, and David J. Eagan. 2009. *Generation E: Students leading for a sustainable, clean energy future*. Campus Ecology program of the National Wildlife Federation.

Filho, Walter Leal, et al., eds. 2015. *Implementing campus greening initiatives: Approaches, methods, and perspectives*. World Sustainability Series. Springer, Berlin.

Fossil Free. http://www.gofossilfree.org.

Higher Education Associations Sustainability Consortium. http://heasc.aashe.org/.

Indvik, Joe, et al. 2013. *Green revolving funds: An introductory guide to implementation and management*. Sustainable Endowments Institute and Association for the Advancement of Sustainability in Higher Education, Cambridge, Mass.

Keniry, Julian. 1995. *Ecodemia: Campus environmental stewardship at the turn of the 21st century*. National Wildlife Federation, Washington, D.C.

Koester, Robert J., James Eflin, and John Vann. 2006. Greening of the campus: A whole-systems approach. *Journal of Cleaner Production* 14: 769–779.

Martin, James, and James E. Samels. 2014. *The sustainable university: Green goals and new challenges for higher education*. Johns Hopkins University Press, Baltimore.

Newman, Julie. 2009. *Reaching beyond compliance: The challenges of achieving campus sustainability*. VDM Verlag.

Princeton Review. 2015. *The Princeton Review's guide to 353 green colleges*. The Princeton Review, Princeton, N.J.

Recyclemania. http://www.recyclemaniacs.org.

Sierra Club. Sierra Magazine's 2016 "Cool Schools" rankings. http://www.sierraclub.org/sierra/coolschools-2016.

Simpson, Walter, ed. 2008. *The green campus: Meeting the challenge of environmental sustainability*. APPA Association of Higher Education.

Simpson, Walter. 2009. *Cool campus! How-to guide for college and university climate action planning*. AASHE, Lexington, Ky. http://www.aashe.org/files/resources/cool-campus-climate-planning-guide.pdf.

Sustainable Endowments Institute: College Sustainability Report Card. http://www.greenreportcard.org.

Sustainable Endowments Institute. 2012. *Greening the bottom line*. Sustainable Endowments Institute, Cambridge, Mass.

Thomashow, Michael. 2014. *The nine elements of a sustainable campus*. MIT Press, Cambridge. Mass.

Toor, Will, and Spenser W. Havlick. 2004. *Transportation and sustainable campus communities: Issues, examples, solutions*. Island Press, Washington, D.C.

University Leaders for a Sustainable Future. http://www.ulsf.org.

University of New Hampshire Sustainability Institute. CarbonMAP [carbon calculator for campuses]. http://www.campuscarbon.com.

U.S. Department of Energy: Solar Decathlon. http://www.solardecathlon.gov.

U.S. Green Building Council. 2014. LEED campus guidance v2009. http://www.usgbc.org/resources/leed-campus-guidance. U.S. Green Building Council, Washington, D.C.

Photo Credits

Index

sunflowers, 251*t*
Superfund, 172*t*, 173, **625**
Superfund sites, 624–625, 639
Superstorm Sandy, 493–494, 496, 497*f*, 498
supervolcano, 42
superweeds, 251*t*
Supplemental Generic EIS (SGEIS), 171
supply, 140–141, 141*f*
surface, land, 395–396, 396*f*
surface fresh water, 385*f*
surface impoundments, **624**, **633**, 633*f*
Surface Mining Control and Reclamation Act
　　(1977), 641
surface water, 108–109, 385*f*, **388**–390, 406
survivorship curve, **66**–67, 66*f*
sustainability, **15**. *See also* campus sustainability;
　　urban sustainability
　　of consumption, 658, 658*f*
　　corporations and, 152–153, 152*f*
　　of economic growth, 144–145, 144*f*, 145*f*
　　of ethanol, 566–567, 566*f*
　　food and, 209–210, 261–263
　　future of, 14–17
　　irrigation and, 393–394, 394*f*
　　science and, 3–17
sustainability goals, 348, 348*f*
sustainability index, 152–153
sustainability strategies
　　connectivity, 656–657, 657*f*
　　consumer power, 658
　　economic opportunities in, 655–656, 656*f*
　　green technologies, 660
　　local and global approaches, 659, 659*f*
　　long-term perspective, 660, 660*f*
　　mimicking nature, 660
　　political engagement, 657–658
　　population stability, 659
　　quality of life, 658–659, 658*f*
　　research and education, 660
　　systemic solutions, 660
　　triple bottom line, 153, 655
sustainable agriculture, **212**, **239**–240
　　in Japan, 263, 263*f*
　　negative feedback loop and, 262–263, 263*f*
　　transportation in, 262, 266, 266*t*
sustainable development, 139, **153**, 153*f*, 655
　　biosphere reserves in, 321–322, 321*f*
　　world, 154, 154*t*, 655
Sustainable Development Goals (U.N.), **154**,
　　154*f*
sustainable fertilizer, 219
sustainable food production, 261–263
sustainable forest certification, 301–302, **318**,
　　318*f*, 318*t*, 322
sustainable forest management, 301–302, 315,
　　318, 318*t*
sustainable growth, 153–154, 153*f*, 154*t*
sustainable yield, maximum, **309**, 439
Suzuki, Tatsujiro, 21
Svalbard Global Seed Vault, 245, 245*f*
swamps, 111*f*, 391
Sweden, 173, 361, 508, 573
　　biogas in, 564, 564*f*, 565*f*
　　ecological footprint of, 19*t*
　　fossil fuels in, 549–551, 550*f*
　　nuclear energy in, 549–550, 550*f*, 554, 556
　　nuclear waste in, 559, 562, 562*f*
　　renewable energy in, 549–550, 550*f*
swidden agriculture, 215, 216*f*
switchgrass, 570, 570*f*

Switzer, Victoria, 159–160
Switzerland, 43
　　organic agriculture in, 258–259, 258*f*, 259*f*
symbiosis, **78**
syncrude, 524
synergistic effects, **374**
syngas, 538
Syngenta, 236, 255
synthetic chemicals, 361, 363*f*, 472
　　air currents and, 366–367, 366*f*, 367*f*
　　as chemical hazards, 355–356, 355*f*
　　in organisms, 362–363, 362*t*
Syracuse University, 164
Syria, 500, 500*f*
systems, 105, 108. *See also specific systems*

T

table salt, 26
tailings, **633**
takings clause, 167
tantalite, 629–631
tantalum, 632*f*, 634*f*, 642, 644
　　in cell phones, 629–630
Tanzania, Africa, 277
　　biocapacity of, 299*t*
　　conservation in, 269–270, 270*f*
　　ecological footprint of, 19*t*
　　poaching in, 295, 295*f*
　　wildlife decline in, 280–281, 280*f*
tap water, 398–399, 399*f*
tar sands, **515**. *See also* oil sands
tariffs, 178, 577–578, 578*f*, 581
tax breaks, 178, 584, 584*f*
taxes, 584*f*
　　carbon, 178, 509–510, 584
　　gasoline, 544
　　green, 178, 509–510, 584
　　revenue-neutral carbon, 509–510, 584
taxonomists, 53, 56, 56*f*
Taxpayers for Common Sense, 178*f*
Taxus brevifolia (Pacific yew), 275*t*
TCLP. *See* Toxicity Characteristic Leaching
　　Procedure
technically recoverable, 524–525, 642, 642*f*
technologies, 243–244, 243*f*, 643
　　automobile, 543–544, 543*f*
　　biotechnology, 250, 251*t*
　　clean coal, 456, 538
　　e-waste from, 620–621, 620*f*, 621*f*
　　GM, 255–256
　　green, 660
　　in industrialized fishing, 438
telenovelas (soap operas), 200–201, 200*f*,
　　201*f*
temperate, 93*f*
temperate deciduous forest, 81, 92*f*, 93*f*, **94**,
　　94*f*, 111*f*
temperate grassland, 92*f*, 93*f*, **94**–96, 94*f*
temperate rainforest, 92*f*, 93*f*, **95**–96, 95*f*
temperature, 116*t*, 489*f*
　　rise of, 492–493, 492*f*, 493*f*
　　in U.S., 492, 492*f*, 502, 502*f*, 502*t*
temperature inversion, **450**–451, 450*f*
Ten Hoeve, John, 561, 561*f*
Tennessee, 319*f*, 339
teratogens, **364**
terracing, 225*f*, **226**
terrestrial index, 278, 278*f*
tertiary consumers, **78**, 79*f*, 80*f*

tessellated darter, 84, 84*f*
test, 10*f*, 13*f*
testing
　　animal, 370, 370*f*, 371, 374
　　of chemicals, 374, 376, 377*f*
Texas, 17*f*, 339, 496*f*
　　Barnett Shale in, 160, 165
　　Fort Worth, 333, 333*f*
　　fracking in, 168
　　population in, 70, 71*t*
TFR. *See* total fertility rate
Thailand
　　ecological footprint of, 19*t*
　　population growth rate of, 199
thalidomide, 364
theories, **14**
thermal expansion, 495*f*
thermal inversion, 450, 450*f*
thermal pollution, 407, 572
thermogram, 543*f*
thermohaline circulation, **422**–423, 422*f*
thermosphere, 447, 447*f*
thin-film solar cells, **587**
Thompson, Jonathan, 317
Thomson's gazelles, 281*f*
Thoreau, Henry David, 136
threatened, **289**
Three Gorges Dam, 400–401, 401*f*, 573
Three Mile Island, **555**–556, 556*f*
thyroid cancer, 556, 560–561, 560*f*
Tiananmen Square, 459*f*
tidal creeks, 425*f*
tidal energy, 517*t*, 582*t*, **598**, 598*f*
tides, **424**, 424*f*, 432, 432*f*, 598
tiger (*Panthera tigris*), 278*t*, 285
tight oil, 528
timber, 117*f*, 152, 169*f*, 308
　　clear-cutting of, 311–313, 311*f*, 312*f*,
　　322*f*, 355
　　even-aged, 311, 311*f*
　　FSC certification for, 301–302, 318, 318*f*,
　　318*t*, 322
　　locations of, 304–305, 310–311, 310*f*
　　subsidies for, 310–311
　　from temperate rainforests, 96
　　from U.S., 310–311, 310*f*
"timber famine," 309–310, 310*f*
tin, 634, 644*t*, 647*f*
titanium, 631*f*, 634*f*, 642*f*, 647*f*
Tohoku, Japan, earthquake, 21–22, 21*f*, 37*t*, 38,
　　42*f*, 43, 549
topography, 115*f*
　　seafloor, 416–417, 417*f*
　　soil and, 214
topsoil, **214**, 214*f*
tornadoes, 43, **452**, 452*f*
total fertility rate (TFR), **196**, 197*t*
　　in Bangladesh, 201–202
　　urbanization and, 202
tourism, 269–270, 276
　　ecotourism, 67, 67*f*, 132, 133*f*, 277, 297,
　　320–321
Townsend, Timothy, 622–623, 622*f*,
　　623*f*
toxic, 620
toxic air pollutants, **457**
toxic chemicals, 362–363, 405
toxic hazardous waste, 620
toxic substances, 361. *See also* synthetic
　　chemicals